Applied Polymer Science

21st Century

Applied Polymer Science

21st Century

Edited by

Clara D. Craver and Charles E. Carraher, Jr.

ELSEVIER

2000

Amsterdam - London - New York - Oxford - Paris - Shannon - Tokyo

ELSEVIER SCIENCE Ltd
The Boulevard, Langford Lane
Kidlington, Oxford OX5 1GB, UK

First edition 2000

British Library Cataloguing in Publication Data
A catalogue record from the British Library has been applied for.

Library of Congress Cataloging-in-Publication Data

Applied polymer science : 21st century / edited by Clara D. Craver and Charles E. Carraher, Jr.-- 1st ed.
     p. cm.
   Includes bibliographical references and indexes.
   ISBN 0-08-043417-7 (hardcover)
    1. Polymers. 2. Polymerization. I. Craver, Clara D. II. Carraher, Charles E.

QD381.7 .A66 2000
547'.7--dc21
                                        00-058697

ISBN: 0 08 0434177

♾ The paper used in this publication meets the requirements of ANSI/NISO Z39.48-1992 (Permanence of Paper).
Printed in The Netherlands.

PREFACE

The Division of Polymeric Materials: Science and Engineering of the American Chemical Society has literally grown up alongside the scientific and commercial development of polymeric materials. By the 1920's the ancient arts of coatings and adhesives from natural organic materials were being treated by methodologies of chemical science. The researchers were enthusiastic and collegial and shared together the vision of studying materials that would greatly aid society. The working groups petitioned to form a Division of the American Chemical Society.

The friendly spirit and enthusiastic cooperation which marks this group of researchers of complex macromolecular systems was celebrated on the event of the Division's Fiftieth Anniversary in 1975. One legacy of that celebration was the book called simply **Applied Polymer Science**, edited by J. Kenneth Craver and Roy W. Tess. Many members produced chapters in their areas of specialty, the editors handled the printing, and the book sales were managed by the Division officers. The book sold out at the great price of nine dollars a copy and provided the first investment funds of the Division.

Fifteen years later the need for another book with such a broad perspective on this field led the Books Department of the American Chemical Society to publish a second edition of **Applied Polymer Science** which was edited by Roy W. Tess and Gary W. Poehlein. It had very broad coverage and has long been sold out.

The 75th Anniversary Celebration of the Division in 1999 sparked this third edition with emphasis on the developments of the last few years and a serious look at the challenges and expectations of the 21st Century.

This book is divided into six sections, each with an Associate Editor responsible for the contents with the group of Associate Editors acting as a board to interweave and interconnect various topics and to insure complete coverage. These areas represent both traditional areas and emerging areas, but always with coverage that is timely. The areas and associated chapters represent vistas where PMSE and its members have made and are continuing to make vital contributions. The authors are leaders in their fields and have graciously donated their efforts to encourage the scientists of the next 75 years to further contribute to the well being of the society in which we all live.

Synthesis, characterization, and application are three of the legs that hold up a steady table. The fourth is creativity. Each of the three strong legs are present in this book with creativity present as the authors were asked to look forward in predicting areas in need of work and potential applications. The book begins with an introductory history chapter introducing readers to PMSE. The second chapter introduces the very basic science, terms and concepts critical to polymer science and technology. Sections two, three and four focus on application areas emphasizing emerging trends and applications. Section five emphasizes the essential areas of characterization. Section six contains chapters focusing of the synthesis of the materials.

The editors are pleased to acknowledge the help and support of the Division, its Board and members, and especially the Associate Editors David Lohse, Kenneth Edwards, Hermenegildo Mislang, Donald Schulz, Elsa Reichmanis, and Michael Jaffe.

CLARA D. CRAVER

Clara Craver was Chair-Elect of Polymeric Materials: Science and Engineering, at the time of its 50th anniversary in 1974, and Chair in 1975. Her services include twenty-three years as Councilor; symposium and FT-IR workshop organizer on polymer characterization, and editor of four books from these activities. Three were ACS Advances in Chemistry books, focussed on the relationship between polymer structures and properties. In 1999 she was awarded the PMSE Division Distinguished Service Award at its 75th Anniversary Celebration. She is a consultant and provides expert testimony in spectroscopy and polymer characterization, and teaches short courses at Universities and scientific meetings. She received her B. Sc. in chemistry from Ohio State University, *cum laude, phi beta kappa* in 1945. In 1974 Fisk University awarded her an honorary Doctor of Science degree based on internationally recognized contributions to the field of infrared spectroscopy. In 1976 she was Chair of the Vibrational Spectroscopy Gordon Research Conference and was an invited speaker to GRC conferences on Coatings in 1954, 1964, 1966. International courses on lectureships have included Sao Paulo, Brazil in 1965, Monaco for ISTA on Man and the Oceans in 1975, China in 1979, multiple lectures on spectroscopy, polymer science, and small business management, and FT-IR at U. of Glamorgan in Wales, 1999.

She is a Certified Professional Chemist and Fellow of the American Institute of Chemists; has been active on ASTM: Chair of E-13 on molecular spectroscopy, and received ASTM's highest award, the Award of Merit; she has managed the Coblentz Society's infrared data program from its inception, now serving as an investigator on a NIST CRADA. The Society honored her with the Williams-Wright Award in Industrial Spectroscopy in 1985 and she was named an honorary member in 1990. She served on the National Research Council Advisory Committee on the Strategic Highway Research Program 1985-89.

In her first research position, at Esso Research Laboratories in NJ, in 1945 she had an opportunity in a new field to develop applications for spectroscopic methods applied to petroleum, especially its use in oil exploration. This research led to patents in characterizing complex hydrocarbons and in instrumentation for highly absorbing materials such as crude oils, asphaltenes, gilsonite and shale oils. This research was followed on polymer related materials when she established the infrared laboratory at Battelle Memorial Institute, 1949-1958 and applied IR to other complex mixtures, tall oil, rosin acids, alkyds, paper chemicals *in situ* and drying oil chemistry. Research on the oxidative curing of varnishes and the role of driers earned her and collaborator Ernest Mueller the second Carbide and Carbon Award, in 1956.

She founded Chemir laboratories, 1959-1990, and Polytech Laboratories in the 1980's with her husband J. Kenneth Craver.

CHARLES E. CARRAHER Jr.

Charles E. Carraher, Jr. is Professor of Chemistry at Florida Atlantic University and Associate Director of the statewide Florida Center for Environmental Studies and Director of the Environmental Chemistry Secretariat. He previously was Dean of the College of Science at FAU, Chair of the Science Division at the University of South Dakota, and Chair of the Department of Chemistry at Wright State University.

He has been given many awards and recognitions including being named a Fellow in the American Institute of Chemists, received the Outstanding Scientists and Engineering Award from the Engineers and Scientist Affiliate Societies Council in 1984, was named as the outstanding Chemist in southeast USA by the American Chemical Society in 1992 and given a Distinguished Service Award for his work in science education in 1994 and the Saltarilli Sigma Xi Award for research in 1992. He has worked as

a science adviser for Sen. McGovern, served as a reader for the national science 2000 committee, served on a national testing committee, serves on the Governor's task-force committee for the Sustainability of South Florida, and headed up one of the national test committees. He is author or coauthor of some 40 books, and over 700 articles in a number of broad areas.

His research has led to the synthesis of over 70 new families of polymers as well as new methods for their characterization and synthesis. He has chaired numerous national and international committees. He is a founder and cochair of PolyEd which works with K-Post-graduate education through a number of groups. PolyEd has worked to develop the science education page of USA Today; recognize outstanding high school teachers of science on a national basis; publish PEN a science newsletter that goes to all the Universities and Colleges in the USA and Canada; and works with industry in developing short courses and other educational aids. He is also one of the founders of IPEC which is an inter-societal organization focusing on K-12 education. He serves on the Executive Board for Divisions of Polymer Chemistry and Polymeric Materials: Science and Engineering and the Editorial Board of numerous journals.

CONTENTS

NEW MATERIALS
Section Editors: Elsa Reichmanis and Michael Jaffe

SPECTROSCOPIC & PHYSICAL CHARACTERIZATION
Section Editor: Clara D. Craver

POLYMERIZATION AND POLYMERIZATION MECHANISMS
Section Editor: Donald Schulz

Section 1

INTRODUCTION
Section Editor: Charles E. Carraher, Jr.

1. History of the American Chemical Society Division of Polymeric Materials: Science and Engineering

2. Introduction to Polymer Science and Technology

3. Polymer Nomenclature

4. Polymer Education

HISTORY OF THE AMERICAN CHEMICAL SOCIETY DIVISION OF POLYMERIC MATERIALS: SCIENCE AND ENGINEERING

CLARA D. CRAVER
Craver Chemical Consultants, French Village, MO 63036-0265

CHARLES E. CARRAHER, Jr
Florida Atlantic University, Department of Chemistry and Biochemistry, Boca Raton, FL 33431 and Florida Center for Environmental Studies, Palm Beach Gardens, FL 33410

E.E. McSWEENEY
17 Nuevo Leon, Point St.Lucie, FL 334952

Importance of Polymers
Introduction and Early History
Acknowledgments of Assistance from Companies and Institutions
Missions, Values, and Principles
Changes in the Scope and Name of the Division
Programming
Preprint Book and Other Publications
Education
Short Courses
Society
Secretariats
Awards

Importance of Polymers

Life as we know it would not exist without the macromolecule. Polymers form the basis of the chemical replication system (nucleic acids, enzymes), food chain (polysaccharides, proteins), shelter (wood, concrete, polystyrene insulation, glass windows), furnishings (nylon, wool, acrylic rugs; coatings, polyethylene waste paper baskets, laminate counter tops), clothing (acrylic sweaters, cotton socks, leather shoes, segmented polyurethane underwear, wool suits, polyester shirts), recording and communication (written history on parchment, papyrus, paper; paintings with canvas and polymerizing oils; optical fibers, CDs, microelectronics), etc. Current employment figures cite over half of the US chemical industrial employment is in synthetic polymers. And the reach of the giant molecule is increasing as the demands for materials with specialized properties increase.

Unlike most areas of science, the majority of the research in polymers has been done in industrial and governmental laboratories. In recent years Universities have increasingly had polymer research programs, and are contributing strongly, sometimes through combined government and industry sponsorship. Leading philosophers and scientists have been attracted to polymeric materials because of their properties and importance to Society as we know it.

Introduction and Early History

The American Chemical Society Division of **Polymeric Materials: Science and Engineering, PMSE**, is one of the largest and most prestigious Divisions with a membership of over 6,000.

Its formation followed the path of the Division of Rubber Chemistry, which in 1919 was organized as the Society's ninth Division. Paint chemists and rubber chemists were particularly interactive at that time because of shared interest in pigments. Shortly, paint chemists began to think "in terms of forming their own organization".

"In the early 1920's the Scientific Section of the National Varnish Manufacturers Association and the Paint Manufacturers of the United States provided the impetus for bringing together paint chemists to discuss mutual problems and to provide a forum" for papers and discussion by scientists working in the field of coatings. Quotations are from a comprehensive summary of these events in the History chapter by Allen Alexander, *Applied Polymer Science*, Organic Coatings and Plastics Chemistry, 1975, J. K. Craver and R. W. Tess, Eds.

The initial meeting was organized by H. A. Gardner and occurred in June of 1922 in Washington, DC. "At this very first meeting... Mr. Gardner read a letter from Mr. Paul R. Croll of Pittsburgh Plate Glass... suggesting the organization of a Paint and Varnish Section of the ACS." At the second meeting in 1923 after thoughtful consideration by a group that included seven of the Division's chairmen-to-be it was decided to affiliate with the American Chemical Society. One of the persuading considerations expressed by Mr. John R. MacGregor of the Eagle-Picher Lead Co. was "that affiliation with the ACS would bring to bear on the problems of the Paint and Varnish Industry the knowledge of the very best of the country's scientists".

Details of publication of proceedings also received considerable attention with interest expressed in the world-wide prestige and acceptance of ACS publications, but with concern about the need to have manuscripts distributed in advance of meetings as a basis for discussion of the presented papers. The availability of Preprints to provide a basis for informed discussion of reported research data was recognized as essential for effective scientific meetings. This important emphasis on discussion is a concept that risks being overlooked in the tight scheduling requirements of papers today.

Organization of the section began at the ACS meeting in Milwaukee in September of 1923. In October of 1923, W. T. Pearce sent to the ACS a petition with about 100 signatures. In December 1923, Pearce was notified by the ACS that he had been appointed secretary and that H. A. Gardner was chair of the new ACS Section. The Section was part of the spring 1924 national ACS meeting held in Washington, DC where it presented its first "full" program. At this time, "sections" were required to be in operation three years before they were officially recognized as a Division. "At each of the national meetings during this three year period the Section scheduled substantial programs and in April, 1927, at its meeting in Richmond the Council approved the by-laws of the **Paint and Varnish Division** , thus giving it full status."

"Early the Division took a leading role in the study of macromolecules" in the form of resins, both natural and sythesized. By 1937 the participation of chemists from the plastics industry had grown so large that a petition by eighty-one members, including Leo H. Baekeland, led to the formation of a second Section of the Division , later called the Organic Plastics Group of the Division. Its first Chairman was Gordon M. Kline of the National Bureau of Standards.

"As more theoretical research in the polymer field began to emerge, papers dealing with more basic aspects of polymerization were finding their way into the programs of several other divisions". Beginning in 1944, C.S. Fuller and A.C. Elm , chairmen of the Division in 1944 and 1945-1946, conceived of the idea and enthusiastically worked to organize a High Polymer Forum, and "with the cooperation of the Divisions of Colloid, Cellulose, Physical and Inorganic, and Rubber Chemistry the plan was successfully launched in April 1946". "This proved to be a most successful experiment. It attracted papers of excellent quality from the leading authorities in all sections of the field with crowded sessions which in turn produced some of the more spirited discussions characteristic of an earlier era."

In 1949 an impressive trio, W.A. Noyes, Jr. and Robert Adams, past presidents of the ACS and Howard K. Nason of Monsanto supported forming a new division and a petition was submitted for forming a new Division of High Polymer Chemistry. This new Division has grown rapidly to become the Division of Polymer Chemistry and is now another of the largest ACS divisions, and a PMSE collaborator in joint programming, in forming the more recent Macromolecular Secretariat and in education activities described below.

An important aspect of the Division was that it provided mimeographed copies of the papers or of the data contained therein for the attendees at the beginning of the national meeting. A demand for all members to receive copies of an abbreviated paper soon developed. "At the 1941 Spring Meeting in St. Louis the officers directed the Secretary to investigate the feasibility of having all papers planographed and bound in booklet form for distribution to all members of the Division about two weeks in advance of each National Meeting. Through the individual efforts of A.C. (Mike) Elm this task was accomplished" by the Fall meeting in Atlantic City. The Preprints, further discussed below, were born.

From this beginning, and with changing research and development emphasis as research tools and polymer applications have developed rapidly, a highly collegial, dedicated membership has grown to over six thousand members.

Acknowledgements of Assistance from Companies and Institutions

Many industrial firms, research institutions and governmental bodies have been strong supporters of the Division, partly through direct contributions, but even more importantly through allowing and encouraging employees to be active in the division. Any listing will be flawed but following is an attempt to acknowledge and thank some of the ones that have been most active during these many decades emphasizing those most active for the past 25 years: Bell Labs/Bellcore, Shell, Ford Motor, General Electric, PPG, Glidden, Hoechst Celanese, Exxon, Hercules, Dow, Battelle Memorial Institute, which provided a secretarial and treasurer office through the terms of several Chairmen, E. McSweeney, E Mueller, L. Nowacki, and retained much of the membership mailing responsibilities from the 1940's into the 1970's, Akzo Nobel, ICI, Monsanto, Eastman/Eastman Kodak, Rohm & Haas, Bausch & Lomb, USDA, S. C. Johnson, IBM, Unilever, Solutia, Chiron, Polaroid, AT&T/Lucent, Coatings Research Institute, NBS/NIST, Naval Research Laboratory, and Sherwin Williams. The listing can continue and is a "Who's Who" in the polymer industry. The Division offers a strong and necessary interface with industry and the American Chemical Society through these major organizations which do research, development and engineering of polymers.

Missions, Values and Principles

The following is a summary of the "Mission Statement" of PMSE.

Mission-PMSE is a division of the American Chemical Society providing a forum for the exchange of technical information on the chemistry of polymeric materials including plastics, paints, adhesives, composites, and biomaterials. The mission is to meet the needs of divisional members and the community at large through technical programming, educational outreach, awards for excellence in scientific discovery and presentation, and publication.

Values-Fundamental to the success of PMSE are three basic values-
People-Members are the source of its strength with member involvement and volunteer efforts essential.
Products-The principal product is the sharing of technical information with the technical programs and publications, including the present book, being tangible results.
Finances-Financial stability is essential to the long-term viability of PMSE.

Principles
Quality- The quality of the Division's products and services is the number one priority. Excellence and continued improvement are sought.
Customer Focus- The Division's members are it's customers with the primary focus being to meet customer needs.
Value- Products and services must be offered to its members that are

affordable and sustainable.

       Membership Involvement- The success of the Division is dependent on membership involvement.

       Uncompromised Integrity- The conduct of the Division must command respect for its integrity and for its positive contributions to the technical community and the public good.

Changes in Scope and Name

To reflect the growth of the areas represented by the Division, it underwent several name changes during it's 75 years of existence. From its original name of **Paint and Varnish** in 1923, it became the Division of **Paint, Varnish and Plastics Chemistry** in 1940; Division of **Paint , Plastics and Printing Ink Chemistry** in 1952; **Organic Coatings and Plastics Chemistry** in 1960; and its current name **Polymeric Materials: Science and Engineering** in 1982/83. These name changes were intended to allow the Division to reflect research and development trends that were occurring within the polymer industry and within Division membership. For the present name, the inclusion of both "Science" and "Engineering" is intended to show the inter-relationship that often exists between these two segments of the polymer community.

       The present name change did not go unopposed. During the early 1980s the Division had requested the name "Applied Polymer Science" to better reflect its focus and programming and which for that reason had been the title of the 50th anniversary book, as chosen by Kenneth Craver. This was opposed in the Divisional Activities committee, where it lost by one vote. The present name was also offered by Craver and also received some opposition. Some of the Executive Committee wanted to simply accept "Polymeric Materials Science" or some reasonable combination of those words, but Ken and others argued that the new choice reflected what the Division was doing and where industry was "going" so that is was appropriate that both "Science and Engineering" be included in the name of the Division. However, an "applied" emphasis is reflected in the Divisional membership, Executive Committee composition, actions and activity, and in its programming. Throughout the name changes, the Division has retained its industrial and "applied" emphasis working with its sister Division, Polymer Chemistry, which has emphasized the more basic and fundamental aspects of polymer chemistry.

Programming

Historically, the source of PMSE programming has been coating's related topics, and along with adhesion, the investigation of solvent/polymer interactions, and modifiers, plasticizers, antioxidants and polymer curing and applications technology, and microscopic, spectroscopic and physical testing methods was the major programming emphasis in the first three decades of the Division. However, the last quarter century has seen new areas emerge such as biomedical materials (dental materials, artificial organs, biosensors, ophthalmic materials, membranes); biocatalysis, biotechnology, aerospace and harsh environment materials, high solids coatings, electronic/photonic (nanostructures, microelectronics, display technologies, microlithography), inorganic and organometallic polymers (Carraher, Sheats, Pittman, Zeldin), and computer modeling.   The major emerging areas were biomedical headed by Charles Gebelein and Charles Carraher, electronics headed by Larry Thompson, Elsa Reichmanis, Grant Willson, and Murrae Bowden, investigations of the relationship between chemical structure and properties and advanced molecular characterization methods such as spectroscopy, particle size and physical properties led by Clara Craver, Theodore Provder, John Gillham, Loren Hill and many others. These areas represent broadly "colored" programming and signaled the expansion in topic areas covered by the Division.

Continued areas were emphasized including corrosion resistant coatings, naturally occurring materials (L. Sperling, C.Carraher, E. Glass), polymerizations (anionic, emulsion), adhesion (L.H. Lee, D.R. Bauer and R. Dickie), membranes, and polymer additives. Environmental concerns and demands requiring dropping optimal organic solvents from filling film-forming applications made it necessary to obtain better performance from powder coatings (G. Brewer) , water soluble polymers (Glass), polymer recycling, toughened materials (K. Riew), chemical modification of polymeric materials (Carraher), polymer design (block and graft copolymers, dendrimers & hyperbranched materials (J.M.J. Frechet), IPNs (interpenetrating networks) and multicomponent systems (L. Sperling).

While recycling and environmental issues have become a relatively new emphasis of many, the Division has continued dealing with these issues with symposia following a tradition begun many years prior to this past quarter of a century and made more essential in the 1990's.

Co-sponsorship with Polymer Chemistry, has been important with the attempt by both divisions to offer the broad spectrum of their memberships, from industrial to governmental to academic, needed programming in areas of fundamental and applied polymeric materials. The two Divisions often share sponsorship in shared areas of interest and have a substantial overlap in membership.

Nomenclature has remained an area where the Division has contributed through on-going articles appearing in the PMSE Preprint (L. Sperling, W. V. Metanomski). Here the division cooperates with IUPAC and the Polymer Division.

Preprint Book and Other Publications

Copies of papers given at a meeting of the division were first distributed to all members in a bound preprint booklet in 1941. Previously most of the papers in mimeograph form were distributed to attendees. Demand for these developed to the point where then division Secretary A.C. Elm organized the papers, or somewhat condensed version of them, in a bound form to send to all members about two weeks prior to the meeting. An important objective of this was to stimulate informed discussion at the meeting. Also, distribution in condensed form did not obviate printing in an ACS journal at a later date.

Wartime restrictions in 1945 caused cancellation of the national ACS Meeting. The Division effectively held its own " Meeting-in Print" by mailing out its preprint book for the papers which would have been presented. This major publication twice each year is the major expense of the Division and is a major benefit that all members receive for nominal dues.

The most important PMSE-associated publication is the PMSE Preprints that contains short papers based on presentations that are to be given at the national ACS meetings. Second to Polymer Preprints, Polymeric Materials: Science and Engineering is the second most widely circulated polymer journal in the world. These two publications have evolved into a "journal" category that contains short papers. It is sent to both members and a number of public and private libraries around the world. Much of the work presented in the preprint journal is never published anywhere else and as such is the primary source for such work. It is often the first publication of late-breaking research. It is highly regarded and abstracted by Chemical Abstracts. Because of the high reputation of the publication and symposium organizers, the papers are generally of highest quality with the papers presenting experimental conditions and results as well as background information. The Division and membership share in the pride of offering such a high quality publication that has contributed much to the continued growth of polymer science and engineering.

In addition to the present volumes and its predecessors, the Division has encouraged the publication of books to serve both the immediate PMSE membership and the wider scientific community. Because of its outstanding symposia and the importance of the areas represented by these programs, a large number of books have been published based on these symposia. The major publishing companies of symposia based on PMSE programming have been the ACS and Plenum Press with others published by Technomic Publishers, Marcel Dekker, Inc., Academic Press, and Elsevier. One year, Divisional books were the two best selling ACS books.

While the Division receives royalties to help support Divisional activities, the major reason for the emphasis on books is to share with its membership and chemical community knowledge contained within the Division. The Division contributes to the growth and direction of science in the USA and within the world. It acts as a catalyst to bring together diversified efforts within a common theme; to allow exchange of ideas between various research teams, companies, and countries.

The Division also publishes a newsletter twice a year that goes to all of its members. The newsletter includes future programs and other information about Divisional activities.

Education

Because of shared interests and goals, many projects have been joint with the ACS Division of Polymer Chemistry. These include the Polymer Education Committee, PolyEd, and the Polymer Nomenclature Committee. A major achievement has been to bring about changes in University curricula to teach more polymer chemistry than had been traditional fifteen or twenty years ago.

The Division is one of two ongoing supporters of PolyEd. Through PolyEd, enhancement of science education has occurred within the K-12 sector, within Colleges and Universities, within private industry, and within society itself. Along with Polymer Chemistry, PMSE has supplied much of the financial support for PolyEd, but more importantly they have supplied PolyEd with the leadership and personnel needed to "make it work". More recently, the Division, again with the parallel support of the Division of Polymer Chemistry, has been a supporter of IPEC, an outgrowth of PolyEd focusing on the K-12 sector. These educational activities are described in greater detail in another chapter in this book.

Short Courses.

When there were explosive developments in several of PMSE's active programming areas, PMSE responded to the educational and training needs of its members by offering short courses. Topics were selected where the Division had expertise and the ability to rapidly assemble and deliver such courses to present the rapid development and technological changes that were occurring. Frequently a short course is associated with a symposium that is being offered during one of the ACS national meetings. The short course is offered just prior to the meeting allowing the participants a "double" exposure to the material and a greater opportunity to visit with fellow members that work in the particular area.

Major repeated short courses were organized during the 1980's directed toward bringing new and fast changing technology to the members. Prominent among these were microelectronics (L.Thompson, M. Bowden), water-soluble polymers (E.Glass), FT-IR (C. Craver) and GPC (T. Provder). For the latter two topics, operational laboratories were set up at the national ACS meetings, by the course organizers with cooperation by major instrument companies: Perkin-Elmer, Nicolet, Bio-Rad, and Waters. E.Glass served as short course administrator from the late 1980's into the early 1990's as the diversity of courses increased.

The Division has also supported and published results related to the naming of polymers. As polymeric materials become more varied and complex, the need for standardized naming of the materials has become increasingly important. The Nomenclature Committee works directly with IUPAC, the international committee responsible for scientific nomenclature.

Society

In the mid-1970's, leaders of the Division became activists within the ACS to gain four goals:

1) Financial contributions from National Meeting income to help defray some of the many expenses involved in organizing and presenting programs.

2) Recognition as the major technical programming arm within the ACS.

3) A greater voice in the governance of the ACS through additional councilor representation.

4) A voice equal to at least one-third of the membership in a new scientific body within the ACS dealing with broader scientific matters both nationally and internationally, and on both technical and broad social aspects of science: The Committee on Science.

The first of these has, among other things, added to divisional ability to partly fund expenses of eminent overseas speakers when needed for symposia, and to broaden the technical planning function, especially as it relates to selecting leading edge topics, by involving a planning group at National meetings. It is designated not to subsidize such optional programs as preprint books.

The second, enhanced by the third, enables Divisions to be assured of a reasonable number of technical sessions, with the subject matter of their choice, in satisfactory sites at National Meetings.

The achieved near doubling of councilors from Divisions is still believed not to be adequate representation. However, it has supplied the Divisions with some of the voice they badly needed and has become an important resource to the ACS as it tries to interface more with industry, to deal with National Meeting Planning, to attract councilors with business management experience into governance roles, to expand the exposition, and to serve as specific bases of technical information on subjects important in environmental and governmental matters which councilors may be called upon to supply.

The Science Committee has become a strong and valuable asset to the ACS, and it is the only place in the ACS where Divisions have appropriate representation.

On the strength of it's programming, business and management practices, and service to its membership and the ACS, PMSE received the "Best Section Award for a Large Section" three times during the period that the ACS Divisional Activities Committee gave the award.

The Division also contributed significantly to the ACS Books Department both in Advances in Chemistry Books and in the Symposium Series and in the membership of the editorial boards.

Secretariats

As polymers became more diversified in application, the need to cooperate with several divisions to provide broad coverage of related topics beyond the scope of any one or two Divisions resulted in the formal creation of the Macromolecular Secretariat. An echo of the High Polymer Forum! This broad programming has led to highly successful, often week-long, symposia, and   produced   major books. The rules and meeting format developed in forming the Macromolecular Secretariat have been the pattern for other Secretariats. PMSE assisted in the formation of, and participates in, the following secretariats: Macromolecular, Materials, Biotechnology, and Catalysis.

Awards

PMSE has an extensive awards program recognizing some of the many areas where its members make valuable contributions and encouraging continued growth of the scientific community. The longest standing of these awards is the Arthur K. Doolittle Award. In 1953 the Divisional Chair Arthur K. Doolittle, then Assistant Director of Research for the Carbide and Carbon Chemicals Company, suggested that the company donate to the Division royalties from his forthcoming book **The Technology of Solvents and Plasticizers** if a suitable use for the funds could be found. At the Chicago meeting in September it was proposed that the royalties be used to establish an annual award for the most outstanding paper presented within the Division's program each year, based on research merit and significance, and quality of presentation. The award was originally known as the Carbide Award, and is now known as the A.K. Dolittle Award. A list of awardees follows in this chapter.

The Award in Applied Polymer Science was first sponsored by the Borden Company in 1968 and latter sponsored by the Phillips Petroleum Company. This award encourages and recognizes outstanding achievements in the science and engineering of plastics, coatings, polymer composites, and related fields.

The Student Award in Applied Polymer Science was first awarded in 1985 and was originally sponsored by Sherwin-Williams and is currently sponsored by ICI. The award recognizes outstanding research and presentations made by graduate students that are currently in graduate school or are not more than one year beyond graduation.

The Roy W. Tess Award in Coatings was established with generous gifts by Roy Tess to reflect the continued importance of coatings to the Division. The award recognizes outstanding individual achievements and noteworthy contributions to coatings science, technology, and engineering. This award was first given in 1986.

The Cooperative Research Award in Polymer Science and Engineering was established in the fall of 1992 and is supported by the Eastman Kodak Company. The award recognizes, and thus encourages, sustained, intensive cooperative and collaborative research across the university/industry or national laboratory/industry interface.

The Division offers the Ford Travel Grant sponsored by the Ford Motor Company that provides travel support to graduate student women or under-represented minority men to attend and present their research at national ACS meetings. It also cooperates in offering the Unilever Award for outstanding graduate research in polymer science.

Finally, the Division offers, when merited, other awards to recognize special sustained substantial contributions to the Division.

Past award winners and a listing of Divisional Chairs are given in the following Tables.

Awardees and Chairs of ACS Division of Polymeric Materials Science and Engineering

Chairs of the PMSE Division

1924	Henry A. Gardner – *Inst. Paint & Varnish Research*
1925, 1926	John R. MacGregor – *Eagle-Pilcher Lead*
1927, 1928	W. T. Pearce
1929	Paul E. Marling – *Lowe Bros. Co.*
1930	John S. Long – *DeVoe Reynolds*
1931	Paul R. Croll - *PPG*
1932	H. A. Nelson – *N. J. Zinc*
1933	F. E. Bartell
1934	R. J. Moore - *Bakelite*
1935	W. R. Fuller-
1936	E. W. Boughton – *N. J. Zinc*
1937	R. H. Kienle – *Calco Chemical* Gordon M. Kline - *NBS*
1938	E. E. Ware – *Sherwin-Williams* Herman A. Bruson – *Rohm & Haas*
1939	W. H. Gardner - *NBS* H. R. Dittmar - *duPont*
1940	E. J. Probeck
1941	G. G. Sward - *NBS*
1942	Shailer L. Bass – *Dow Corning*
1943	W. W. Bauer
1944	C. S. Fuller – *Bell Labs*
1945, 1946	A. C. Elm – *N. J. Zinc*
1947	Ralph H. Ball - *Celanese*
1948	Paul O. Powers – *Batelle; Armstrong*
1949	Malcolm M. Renfrew - *duPont*
1950	C. R. Bragdon – *Interchemical Corp.*
1951	E. E. McSweeney - *Battelle*
1952	Francis Scofield – *Natl. Paint & Coatings Assoc.*
1953	Arthur K. Doolittle – *Union Carbide*
1954	John K. Wise – *US Gypsum*
1955	Albert C. Zettlemoyer - *Lehigh*
1956	Russell B. Akin - *duPont*
1957	J. Kenneth Craver - *Monsanto*
1958	L. Reed Brantley – *Occidental U*
1959	Allen L. Alexander - *NRL*
1960	W. A. Henson - *Dow*
1961	Edward G. Bobalek – *U Maine*
1962	Ernest R. Mueller - *Battelle*
1963	W. O. Bracken - *Hercules*
1964	George R. Somerville - *Shell*
1965	Raymond R. Myers - *Lehigh*
1966	Robert F. Helmreich - *Dow*
1967	Frank P. Greenspan – *Reeves Bros.*
1968	John C. Cowan - *NRRL*
1969	A. E. Rheineck – *DeVoe Reynolds; Hercules; ADM; ND State*
1970	K. N. Edwards – *Dunn-Edwards*
1971	Carleton W. Roberts - *Dow*
1972	Louis J. Nowacki - *Battelle*
1973	Lambertus H. Princen – *NRRL*
1974	George E. F. Brewer - *Ford*
1975	Clara D. Craver - *Battelle*
1976	Lieng-Huang Lee - *Xerox*
1977	Rufus F. Wint - *Hercules*
1978	Roy W. Tess - *Shell*
1979	Larry F. Thompson – *Bell Labs*
1980	John L. Gardon – *Akzo-Nobel*
1981	Sandy S. Labana - *Ford*
1982	Marco Wismer - *PPG*
1983	Richard H. Mumma - *Hercules*
1984	Murrae J. Bowden – *Bell Labs*
1985	Ronald S. Bauer - *Shell*
1986	John H. Lupinski - *GE*
1987	Ray A. Dickie - *Ford*
1988	Theodore Provder - *ICI*
1989	Theodore Davidson
1990	James F. Kinstle – *James River*
1991	Michael Jaffe - *Celanese*
1992	S. Richard Turner – *Kodak/Eastman*
1993	George R. Pilcher – *Akzo-Nobel*
1994	David R. Bauer - *Ford*
1995	Elsa Reichmanis – *Bell Labs*
1996	Donald N. Schulz - *Exxon*
1997	Frank N. Jones – *E Michigan U*
1998	David J. Lohse - *Exxon*
1999	David A. Cocuzzi – *Akzo-Nobel*
2000	Christopher Ober – *Cornell U*

PMSE Award Winners

Carbide and Carbon Award   /   A. K. Doolittle Award

1955	William A. Zisman, E. F. Hare, H. W. Fox - *NRL*
1956	Clara D. Craver, E. R. Mueller - *Battelle*
1958	Dean Taylor, Jr., John E. Rutzler, Jr. - *Case*
1959	Frank A. Bovey, George V. D. Tiers, G. N. Filipovich – *3M*
1961	Eugene R. Moore, Stanley S. Levy, Chi-Chiang Lee - *Case*
1963	Armand F. Lewis – *American Cyanamid*
1965	Edwin E. Bradford, John W. Vanderhoff - *Dow*
1966	R. H. Hansen, Harold Schonhorn – *Bell Labs*
1967	W. J. Jackson, Jr., John R. Caldwell – *Tennessee Eastman*
1968	Roger P. Kambour - *GE*
1969	Gilbert L. Burnside, George E. F. Brewer, Gordon G. Strosberg - *Ford*
1970	John Lynde Anderson - *AMTEK*
1971	Robert E. Baier, George I. Loeb – *Cornell/NRL*
1972	Saul M. Cohen, Raymond H. Young, Albert H. Markhart - *Monsanto*
1973	Donald R. Paul and D. Ray Kemp – *U Texas*
1975	Donald L. Schmidt, H. B. Smith, William E. Broxterman, Mary R. Thomas - *Dow*
1976	Bernhard Wunderlich, George Czornyj - *Rennselaer*
1977	James F. Kinstle, L. Eric Sepulveda – *U Tennessee*
1978	A. C. Ouano, J. A. Carothers - *IBM*
1979	Yehonathan Hazony – *Princeton U*
1980	Louis T. Manizone, John K. Gillham, C. Al McPherson – *Bell/Princeton*
1981	W. Harmon Ray, Francis J. Schork – *U Wisconsin*
1982	Alan G. MacDiarmid, Paul J. Nigrey, David F. Macinnes, Jr., David P. Nairns, Alan J. Heeger – *U Pennsylvania*
1983	William J. Hall, R. L. Kruse, Robert A. Mendelson, Q. A. Trementozzi - *Monsanto*
1984	Tamotsu Inabe, Joseph F. Lomax, Joseph W. Lyding, Carl R. Kannewurf, Tobin J. Marks – *Northwestern U*
1985	Jeffrey T. Koberstein, Thomas P. Russell – *Princeton U/IBM*
1986	Jean M. J. Fréchet, Francis M. Houlihan, C. Grant Willson – *U Ottawa/IBM*
1987	I. V. Yannas, E. Lee, A. Ferdman, D. P. Orgill, E. M. Skrabut, G. F. Murphy - *MIT*
1988	Mark G. Allen and Stephen D. Senturia - *MIT*
1989(a)	Hiroshi Ito, Mitsuru Ueda, Mayumi Ebina – *IBM/Yamagata U*
1989(b)	John L. West, J. William Doane, Zenaida Domingo, Paul Ukleja – *Kent State U*
1990(a)	Gary D. Friends, Jay F. Kunzler – *Bausch & Lomb*
1990(b)	Jae-Seung Kim, Victor C. Yang – *U Michigan*
1991(a)	Jack L. Koenig - *Case*
1991(b)	Mark R. Schure – *Rohm & Haas*
1992(a)	Willi Volksen, M. I. Sanchez, Jeffrey W. Labadié, T. Pascal – *IBM/CNRS*
1992(b)	Young C. Chung, Nicholas Leventis – *Molecular Displays*
1993	M. K. Georges, R. Veregin, P. M. Kazmaier, G. K. Hamer - *Xerox*

1994	Jane E. G. Lipson – *Dartmouth U*
1995	J. F. G. A. Jansen, E. M. M. de Brabander, E. W. Meijer – *Eindhoven/DSM*
1996	James J. Watkins, Thomas J. McCarthy – *U Massachusetts*
1997	Craig J. Hawker, Eva E. Malmstrom, Curtis W. Frank, J. Patrtick Kampf, Cristina Mio, John Prausnitz – *IBM/Stanford/ UC Berkeley*
1998(a)	John G. Curro, Jeffrey D. Weinhold – *Sandia*
1998(b)	Timothy P. Lodge, Ken J. Hanley, Ching-i Huang, Changyeol Ryu – *U Minnesota*
1999(a)	Geoffrey W. Coates, Ming Cheng, and Emil B. Lobkovsky – *Cornell U*
1999(b)	Joanna Aizenberg, Andrew J. Black, and George M. Whitesides – *Bell Lab*

Sherwin Williams / ICI   Student Award

1985	Stephen R. Holmes-Farley - *GelTex*
1986	Krishna Venkataswamy – *Advanced Elastomers Systems*
1987	Bruce M. Novak – *North Carolina State U*
1988	Peter J. Ludovice – *Georgia Tech*
1989	Rubing Cai - *Milliken*
1990	Lori P. Engle – *3M*
1991	Kathryn E. Uhrich – *Rutgers U*
1992	Nathan A. Mehl - *Milliken*
1993	Joan K. Vrtis
1994	Michael L. Greenfield - *Ford*
1995	Valerie V. Sheares – *Iowa State U*
1996	Hong Yee Low - *Case*
1997	Ellen C. Lee - *Ford*
1998	David B. Hall – *Northwestern U*
1999	Shu Yang – *Cornell U*

Roy W. Tess Award in Coatings

1986	William D. Emmons – *Rohm & Haas*
1987	Marco Wismer - *PPG*
1988	Zeno W. Wicks, Jr. – *North Dakota State U*
1989	Theodore Provder - *ICI*
1990	Walter K. Asbeck – *Union Carbide*
1991	Kenneth L. Hoy – *Union Carbide*
1992	Ray A. Dickie - *Ford*
1993	Larry F. Thompson – *Bell Labs*
1994	Werner Funke – *U Stutgart*
1995	John L. Gardon – *Akzo-Nobel*
1996	John K. Gillham – *Princeton U*
1997	Werner J. Blank – *King Industries*
1998	Loren W. Hill – *Solutia*
1999	Mitchell A. Winnik – *U Toronto*

Unilever Award for Outstanding Research

1991	Christopher B. Gorman – *North Carolina State U*
1992	Richard A. Register – *Princeton U*
1993	Christopher N. Bowman – *U Colorado*
1994	Timothy J. Deming – *UC Santa Barbara*
1995	Rangaramanujam M. Kannan – *Wayne State U*
1996	Kristi Anseth – *U Colorado*
1997	Dong Yu Kim – *U Massachusetts Lowell*
1998	James J. Watkins – *U Massachusetts Amherst*
1999	Scott G. Gaynor – *Carnegie Mellon U*

Cooperative Research Award

1994	Jean M. J. Fréchet, Hiroshi Ito, C. Grant Willson – *Cornell/IBM*
1995	Leo Mandelkern, C. Stanley Speed, Ferinand C. Stehling – *Florida State U/Exxon*
1996	Ray H. Baughman – *Allied-Signal*
1997	Henry K. Hall, Jr. – *U Arizona*
1998	Lynda K. Johnson, Maurice S. Brookhart – *duPont/U North Carolina*
1999	Jose E. Valentini, Yee C. Chiew, Leslie J. Fina, John Q. Jiang – *Sterling/Rutgers*
2000	James A. Schwindenman, Roderic P. Quirk – *FMC/U Akron*

ACS Award in Applied Polymer Science sponsored by Phillips Petroleum Company

1968	Harry Burrell - *Inmont*
1969	Sylvan O. Greenlee
1970	Raymond F. Boyer
1971	Raymond R. Myers – *Lehigh U*
1972	Richard S. Stein – *U Massachusetts*
1973	Carl S. Marvel – *U Illinois/U Arizona*
1974	Vivian T. Stannett – *North Carolina State U*
1975	Maurice L. Huggins
1976	Herman F. Mark – *Polytechnic U*
1977	William A. Zisman - *NRL*
1978	John K. Gillham – *Princeton U*
1979	Roger S. Porter – *U Massachusetts*
1980	John W. Vanderhoff – *Lehigh U*
1981	Eric Baer - *Case*
1983	Frank A. Bovey – *Bell Labs*
1984	Donald R. Paul – *U Texas*
1985	James Economy - *IBM*
1986	William J. Bailey – *U Maryland*

1987	Orlando A. Battista - *FMC*
1988	David S. Breslow - *Hercules*
1989	Leo Mandelkern – *Florida State U*
1990	Otto Vogl – *U Massachusetts*
1991	Edwin J. Vandenberg - *Hercules*
1992	Robert S. Langer - *MIT*
1993	Owen W. Webster - *duPont*
1994	James E. Mark – *U Cincinnatti*
1995	Joseph P. Kennedy – *U Akron*
1996	Jean M. J. Fréchet – *Cornell U*
1997	Jack L. Koenig - *Case*
1998	Alan D. English – *duPont*
1999	Elsa Reichmanis – *Bell Labs*
2000	Lewis J. Fetters - *ExxonMobil*

Distinguished Service Award

1993	Roy W. Tess - *Shell*
1994	E. E. McSweeney – *Battelle;Union Camp*
1995	Louis J. Nowacki - *Battelle*
1996	Kenneth N. Edwards – *Battelle; Dunn Edwards*
1999	Clara D. Craver - *Battelle; Chemir Laboratories*
2000	Santokh S. Labana - *Ford*

Excellence in Service to POLYED

1995	John Droske and Charles Carraher

ACS Division of Polymeric Materials: Science and Engineering
75th *Anniversary Celebration*
March 21st to 25th, 1999   Anaheim, California

In 1924, the newly organized *Paint and Varnish Section* of the ACS presented its first program at a national meeting. This became the *Paint and Varnish Division* in 1927 and has continued to grow for 75 years. We have changed our name four times, reflecting the growth of our division and the maturation of polymer science. As the *Division of Polymeric Materials: Science and Engineering* we are now one of the largest divisions of the ACS with well over 6000 members.

Our division was born just as the macromolecular hypothesis of Staudinger was beginning to be accepted, and we have grown as the understanding and use of polymers has developed. Polymeric materials now pervade the life of every citizen, from coatings to biomaterials, from electronic goods to plastics. PMSE is proud of the role the division has played in promoting the science behind these developments and in applying such understanding to improvements in our society. So we took the opportunity to celebrate our 75th anniversary at the 217th National Meeting in Anaheim in a number of ways. These included:

- **Symposium on *Celebrating 75 Years of Progress in Applied Polymer Science***: This was organized by Prof. Ed Glass of North Dakota State University to recognize those who have contributed to the growth and commercial development of the polymer industry and to the many contributions that polymers have made to the layman's quality of life over the past 75 years. The symposium ran for two days and reviewed technical aspects associated with the successful development of the industry. The speakers included many of those who have contributed to the amazing advances in polymer science and its application over this time.

- **Symposium on the *Future of Applied Polymer Science***: Dr. Michael Jaffe, now at Rutgers University and formerly with Hoechst-Celanese, arranged this one-day symposium. It was intended to open a dialogue into the agenda for applied polymer science in the twenty-first century. The participants were all recognized experts in their fields. They broadly outlined the potential commercial implications of the latest insights of polymer science, from the synthesis of complex molecular architectures and the controlled assembly of these molecules, to the application of these new materials in structural and functional parts and devices.

- **A new edition of *Applied Polymer Science***: This book was first published at our 50th anniversary in 1974, and a second edition appeared several years later. Many of the speakers in the two special symposia listed above have contributed chapters to this volume, and it covers the whole range of polymer science as applied in our industry.

• **The** *PMSE 75ᵗʰ Anniversary Banquet*: On the evening of Monday, March 22nd, we held a banquet to recognize all of those who have served the division as chair or who have won one of the PMSE awards. This combined list appears in this book, and the list of more than 240 people reads like a *Who's Who* of American polymer science. About 120 of them were able to attend the banquet. There we also presented the PMSE awards that are customarily given out at the Spring meeting. These included the 1998 Doolittle Awards (John G. Curro, Jeffrey D. Weinhold, Timothy P. Lodge, Ken J. Hanley, Ching-i Huang, and Changyeol Ryu), the 1998 ICI Student Award (David B. Hall), the 1999 Cooperative Research Award (Jose E. Valentini, Yee C. Chiew, Leslie J. Fina, and John Q. Jiang), the 1999 Distinguished Service Award (Clara D. Craver), the 1999 Ford Travel Grant recipients (Elizabeth Juang and Yanina Goddard), the 1999 ACS Award in Applied Polymer Science (Elsa Reichmanis), and the 1999 Chemistry of Materials Award (Alan G. MacDiarmid).

After dinner, Prof. David Tirrell of Cal Tech presented a discussion of one of the most exciting future areas of polymer science, the use of biomaterials and the ways by which we can mimic nature to create new polymers. This was entitled *"Polymers and Biology: An Old Relationship with New Life"*. This banquet and the reception before it were supported by Union Carbide Corp., Exxon Research & Engineering, and Lucent Technologies, and we thank them for their generosity.

All of these events helped to celebrate this milestone in the history of our division. The most lasting legacy will probably be this book, which captures the state of applied polymer science as we begin the new millennium. We hope that you will find it useful and that it will help you remember the part that PMSE has played in the first 75 years of polymer science.

Chairs Photo:

PMSE Chairs Attending the 75th Anniversary Banquet.   Seated (left to right): David A. Cocuzzi (1999); Lieng-Huang Lee (1976); Elsa Reichmanis (1995); E. E. McSweeney (1951); Kenneth N. Edwards (1970); George R. Pilcher (1993); David R. Bauer (1994). Standing (left to right): David J. Lohse (1998), Frank N. Jones (1997); S. Richard Turner (1992); Murrae J. Bowden (1984); James F. Kinstle (1990); John H. Lupinski (1986); Michael Jaffe (1991).

INTRODUCTION TO POLYMER SCIENCE AND TECHNOLOGY

CHARLES E. CARRAHER, JR.

Department of Chemistry and Biochemistry, Florida Atlantic University, Boca Raton, FL 33431 and Florida Center for Environmental Studies, Palm Beach Gardens, FL 33410

Introduction
Polymer Synthesis
Synthetic Routes
Polymer Testing
Polymer Companies
Polymer Structure-Property Relationships
Solubility and Flexibility
Size
Polymer Shape
Polymer Structures

Introduction

The term **polymer** is derived from the Greek's **poly** and **mers** meaning many parts. Some prefer the term **macromolecule** or large molecule.

There are many ways to measure the importance of a specific discipline. One is to consider its pervasiveness. Polymers serve as the basis of life in the form of nucleic acids, proteins and polysaccharides. They permit replication, energy transformation, transmission of foods within plants and animals, act as essential natural building materials, ... Polymers have served as the very building blocks of society-clays for jars, wood for fuel, hides for clothing, vegetation for food and shelter,... They are present in a variety of forms-as fibers and cloths, paper, lumber, elastomers, plastics, coatings, adhesives, ceramics, enzymes, DNA, concretes, and are major ingredients in soils and plant life.

The basic concepts of polymer science apply equally to natural and synthetic polymers and to inorganic and organic polymers, and as such are important in medicine, nutrition, engineering, biology, physics, mathematics, computers, environment, space, ecology, health, ...

Today, synthetic polymers are produced at the annual rate of over 200 pounds for every person in the USA. Paper products account for another over 200 pounds per person. This does not include the inorganic polymers such as graphite, glass and diamonds; natural polymers such as polysaccharides, lignin, and proteins; and regenerated materials such as rayon and rayon acetate.

Polymers are involved in all of the major new technologies including synthetic blood and skin; computer chips, CDs, liquid crystals, and circuit boards-information visualization, storage and retrieval; energy creation, storage, and transmission(portable electrical power (batteries), efficient, light, and low-emission transportation; high temperature superconductors; medicines, targeting and control of drug delivery, and synthetic limbs and other replacement parts; transportation; space craft; solar and nuclear energy; and photonics (optical fibers).

Science, in its broadest sense is our search to understand what is about us. This quest is marked by observation, testing, inquiring, gathering data, explaining, questioning, predicting, ... **Technology** had been described as simply applied science. The line dividing technology and science is often non-existent, though technology almost always follows science. The term **"scientific method"** is actually one that both scientists and each of us use everyday. As we bake cookies, we may vary the ratios and actual ingredients-testing (eating) the cookies and then drawing conclusion-and the next time using those changes we felt made a "better" cookie and laying aside those changes that were not positive. This is the process of inquiry, observation, testing, etc. noted above that allows us to understand, test, develop, and produce materials for the 21st century.

While the "technology" of polymers is ancient, the science of polymers is relatively new. This polymer science waited until an understanding of basic scientific principles and concepts such as atoms, electrons, periodic table, balanced equations and chemical structures were in place. Even then, the concept of very large molecules was at times debated to 1950s, even after the birth of the polymer industry producing plastic products that replaced metal products and other materials. The importance of polymeric materials is today accepted as basic chemistry within the chemistry curriculum of most chemistry courses.

Polymer Synthesis

Carothers brought together many of the polymer terms and concepts we use today. He noted a correlation between so-called vinyl polymers, polymers generally derived from the reaction with vinyl monomers such as vinyl chloride and styrene and the chain type of kinetics. He also noted a correlation between condensation polymers such as nylons and polyesters and step-wise kinetics. While these associations, condensation polymers/step-wise kinetics and vinyl polymers/chain kinetics, are generally accurate, there are exceptions.

Chain Type Kinetics. Chain type kinetics are probably the most studied and best understood kinetic system. With vinyl reactants, the reaction can be understood in terms of three main reactions. Using the free radical process employing an initiator, I, monomer M and free radical with a '♥' we can describe these three groupings.

Initiation

$I \longrightarrow 2R^\bullet$ (1)

$R^\bullet + M \longrightarrow R\text{-}M^\bullet$ or simply $M^\bullet$ (2)

Since the activation energy for decomposition is higher than that for addition of R$\cdot$ to M, the first step is the rate determining step with an associated specific rate constant k_d so that the rate of formation of initiator radical chains is

$$\text{Rate}_d = k_d[I] \tag{3}$$

This expression is generally modified to reflect that two active free radicals are created for each decomposition of I by inclusion of a "2". Further, all R$\cdot$'s do not create growing polymer-producing free radical chains so that some factor, f, is added that is the fraction of chains that do successfully begin chain growth. Thus the rate of radical formation is generally given as

$$\text{Rate}_d = 2k_df[I] \tag{4}$$

Propagation

$$\text{M}\cdot + \text{M} \text{ ---> M-M}\cdot \tag{5}$$
$$\text{M-M}\cdot + \text{M} \text{ ---> M-M-M}\cdot$$
$$\text{M-M-M}\cdot + \text{M} \text{ ---> M-M-M-M}\cdot \text{ Etc.}$$

Since it is experimentally found that the rates of addition of the various monomers to the growing free radical chain are similar, it is customary to treat the entire propagation sequence as follows.

$$\text{M}\cdot + \text{M} \text{ ---> M}\cdot \tag{6}$$

with an associated rate expression for chain propagation (p)

$$\text{Rate}_p = k_p\,[\text{M}\cdot][\text{M}] \tag{7}$$

Termination

Termination is generally via two mechanisms

combination $\text{M}\cdot + \text{M}\cdot \text{ ----> M-M} \tag{8}$

and

disproportionation $\text{M}\cdot + \text{M}\cdot \text{ ---> M + M} \tag{9}$

The associated rate expression for termination (tr) is

$$\text{Rate}_{tr} = 2k_{tr}\,[\text{M}\cdot][\text{M}\cdot] \tag{10}$$

The "2" is added to recognize the fact that two growing chains are involved and are destroyed each time termination occurs.

The rate of monomer use can be described as

$$\text{Rate} = k_p [M\bullet][M] + k_i [R\bullet][M] \tag{11}$$

where k_i is the specific rate constant for the monomer-consuming initiation step.

Since most of the monomer consumption occurs in the growth or propagation stage the second term can be neglected giving

$$\text{Rate}_p = k_p [M\bullet][M] \tag{12}$$

While this is a simple rate expression, the concentration of free radical monomer is not easily measured. Thus, the typical kinetic treatment seeks a way to describe the reaction rate in terms of more easily measurable values.

Experimentally it is found that the total number of "live" or growing free radical chains is generally about constant during the polymerization process so that the rate of initiation and propagation are the same.

$$k_i [R\bullet][M] = 2k_{tr}[M\bullet][M\bullet] \tag{13}$$

Further, there is no change in R• with time so that the two rate expressions for the initiation reactions involving the creation and demise of R• are equal to one another.

$$2k_d f[I] = k_i [R\bullet][M] \tag{14}$$

Through manipulation of equations 13 and 14, it is possible to get an expression for [M•] that contains easily measurable quantities that can be substituted in for [M•].

$$\text{Rate}_p = k_p [M\bullet][M] = k_p [M] (k_d f[I]/k_t)^{\frac{1}{2}} \tag{15}$$

The kinetic expressions for condensation reactions are similar to those of most other Lewis acid base reactions. For the formation of polyesters from the reaction of a diacid, A, and a diol, B, in the absence of added catalysts it is found that the rate expression is

Rate = k [A][B] and where [A] = [B] the expression becomes
$$\text{Rate} = k [A]^2 \tag{16}$$

Given that "p" is the extent of reaction so that "1-p" is the fraction of functional groups that have not reacted, we have on integration and subsequent substitution

$$kt = (1/[A_t]) - (1/[A_o]) \tag{17}$$

and that

$$[A_t] = [A_o](1-p) \tag{18}$$

giving after rearrangement

$$[A_o]kt = (1/1-p) - 1 \tag{19}$$

Thus, a linear relationship of 1/1-p with time exists.

Polymer molecular weight is quite sensitive to the presence of impurities in step-wise processes. This is demonstrated in the following calculation. The degree of polymerization, DP, or number of repeat units is defined as

$$[A_o]/[A_t] \tag{20}$$

Multiplying Equation 19 by $[A_o]$ and rearranging gives

$$DP = [A_o]kt + 1 \tag{21}$$

Combining this with Equation 18 gives

$$DP = 1/1-p \tag{22}$$

Thus, high purity and high extent of reaction is necessary to achieve high molecular weight polymers.

Free Radical Copolymerization. While the mechanism of copolymerization is similar to that described for homo-polymerizations, the reactivities of monomers may differ when more than one monomer is present in the reaction giving polymer chains with amounts varying with the co-monomer concentrations. The difference in the reactivity of monomers can be expressed with reactivity ratios, r.

The **copolymer equation**, which expresses the composition of growing chains at any reaction time t, was developed in the 1930s by a group of investigators including Wall, Mayo, Simha, Alfrey, Dorstal, and Lewis.

Four chain extension reactions are possible when monomers A and B are present in a polymerization reaction mixture. Two of these steps are self-propagation steps and two are cross-propagation steps. The difference in the reactivity of the monomers can be expressed in terms of reactivity ratios that are the ratios of the propagating steps. The difference in the reactivity of the monomers can be expressed in terms of reactivity ratios that are the ratios of the propagating rate constants where $r1 = k_{aa}/K_{bb}$ and $r2 = k_{bb}/k_{ba}$.

$$A\bullet + A \longrightarrow AA\bullet \qquad\qquad Rate = k_{aa}[A\bullet][A] \tag{23}$$

$$A\bullet + B \longrightarrow AB\bullet \qquad\qquad Rate = k_{ab}[A\bullet][B] \tag{24}$$

$$B\bullet + B \longrightarrow BB\bullet \qquad\qquad Rate = k_{bb}[B\bullet][B] \tag{25}$$

B•+ A ----> BA•   Rate = k_{ba}[B•][A]   (26)

Experimentally, as in the case of chain polymerization, the specific rate constants are found to be approximately independent of chain length, with monomer addition primarily dependent only on the adding monomer unit and the growing end. Thus, the four equations above are sufficient to describe the monomer consumption. The rate of monomer consumption can be described by the following:

Rate of A consumption = -d[A]/dt = k_{aa} [A•][A] + k_{ba}[B•][A] and   (27)

Rate of B consumption = -d[B]/dt = k_{bb}[B•][B] + k_{ab}[A•][B]   (28)

The relative rate of monomer consumption is found by simply dividing the two expressions describing the rate of monomer consumption.
Again, experimentally it is found that the number of growing chains   remains essentially constant giving a steady state concentration of monomer radical. Further, the concentrations of A• and B• are found to be the same and the rate of conversion of A• to B• is equal to the rate of conversion of B• to A•. Thus,

k_{ba}[B•][A] = k_{ab}[A•][B]   (29)

Solving for [A•] gives [A•] = k_{ba}[B•][A]/k_{ab}[B]   (30)

Substitution of this expression for [A•] into the expression describing the relative rate of monomer consumption and rearrangement gives

d[A]/d[B] = ([A]/[B])(r1[A] + [B]/[A] + r2[B])   (31)

which gives the copolymer composition at any monomer concentration without the need to know any free radical concentrations.
From knowledge of the r values it is possible to predict the type of copolymer composition that is being formed. For instance, when the r values are both less than 1, meaning growing chain ends tend to add monomer that are unlike the growing end, gives alternating polymer structures; when both r values are larger than one meaning growing chains will add monomer that is like that of the growing chain end, block copolymers are formed; when both r values are near one, a polymer chain formed from random addition of the two monomers is formed; etc.

Step-Wise Kinetics. Step-wise reactions generally require heating whereas chain processes are generally run below room temperature and normally require cooling. Chain polymerizations are generally exothermic and more rapid than the step-wise reactions. The difference in the reaction speed is a result of the differences in activation energy for the controlling step. For chain polymerizations, the rate determining step is the initial initiation step, **but** the polymer forming step has a low activation energy, generally between 2 to 10 kcal/mole. This allows ready polymer formation anytime free radicals are present. Thus, during most of the polymerization, high polymer is formed and the most abundant species are polymers and monomers.

The activation energy for the step-wise polymerization is generally much higher being on the order of 30 kcal/mole. During the polymerization sequence, high polymer is formed only near the completion of the reaction with molecular weight steadily increasing as the reaction progresses.

To illustrate typical step-wise polymerization kinetics, we will look at the formation of polyesters from reaction of a diol and a dicarboxylic acid. For uncatalyzed reactions where the dicarboxylic acid and diol are present in equal molar amounts it is found that one diacid molecule is used as a catalyst. This leads to the following kinetic expression.

$$\text{Rate of polycondensation} = -d[A]/dt = k[A][A][D] = k[A]^3 \qquad (32)$$

where [A] is the diacid concentration, [D] is the diol concentration and [A] = [D].

$$\text{Rearrangement gives } -d[A]/[A]^3 = kdt \qquad (33)$$

Integration gives

$$2kt = 1/[At]^2 - 1/[Ao]^2 = 1/[At]^2 + \text{Constant} \qquad (34)$$

It is convenient to express this equation in terms of extent of reaction, p, where p is the fraction of functional groups that have reacted at a time "t". Thus, 1-p is the fraction of unreacted groups and

$$At = Ao(1-p) \qquad (35)$$

Substitution of the expression for At and rearrangement gives

$$2Ao2kt = 1/(1-p)^2 + \text{Constant} \qquad (36)$$

A plot of $1/(1-p)^2$ as a function of time should be linear with a slope of 2Ao2k from which k is determined. Determination of k as a function of temperature allows the calculation of the activation energy.

The number average degree of polymerization, DP, can be expressed as

$$DP = \text{number of original molecules/number of molecules at time "t"} = No/N = Ao/At = Ao/Ao(1-p) = 1/(1-p) \qquad (37)$$

This relationship is called the Carothers equation. Because the value of k at any temperature can be determined from the slope of the line (2Ao2k) when $1/(1-p)^2$ is plotted against t, DP can be determined at any time from the expression

$$DP^2 = 2kt[Ao]^2 + \text{Constant} \qquad (38)$$

Synthetic Routes

Step-Wise Condensation Polymerizations. Following is a brief discussion of the three major techniques utilized to synthesize condensation polymers.
      The melt synthetic technique is also referred to by other names including high melt, bulk melt, bulk or neat. The **melt** process is an equilibrium-controlled process in which polymer is formed by driving the reaction toward completion, usually through removal of the by-product or condensate. Thus, in the reaction of a diacid and a diol to form a polyester, the water is removed causing the reaction to proceed towards polymer formation.
      The reactants are added to the reaction vessel along with any other needed material such as catalysts. Heat is applied to melt the reactants, permitting them to come into contact with one another. Additional heat can be added and the pressure reduced. These reactions typically take several hours to days before the desired polymer is formed. The product yield is necessarily high.
      Solution polymerizations are also equilibrium processes, with the reaction also often driven by removal of the small by-product. Because the reaction is often run at a lower temperature, more reactive reactants are generally required. Solvent entrapment, recovery and reuse are problems. Counter, the use of lower temperature provides an energy savings and minimizes thermally induced side reactions and rearrangements.
      The **interfacial** polycondensation reactions are heterophasic, with two fast-reacting reactants dissolved in a pair of immiscible liquids, one of which is usually water. Reaction occurs near the interface under somewhat non-equilibrium conditions and unlike typically step-wise processes, high polymer is formed throughout the reaction.   These latter conditions are due to the removal of the forming polymer from at least one phase and the use of reactive reactants where the energy of reaction is lower that for reactant usually employed in the solution and melt processes. Because reaction can occur at or below room temperature, thermally induced side-reactions are minimized. The reactions are often completed within a matter of seconds and minutes also minimizing side-reactions.

Chain-Reaction Polymerizations. Most free radical, and ionic polymerizations employing vinyl reactants can be run at or below room temperature. Heating is normally employed only when melting is required or to decompose initiators. Most are rapid occurring within a matter of minutes and hours.   The principle methods employed in free radical processes are bulk, solution, suspension, and emulsion polymerizations.   The **bulk** process can be carried out in a batch or continuous process. In the bulk process, the reactants are employed "neat" (without solvent). Heat control is important since most of these reactions are exothermic. In the **solution** processes, the reactant(s) is dissolved and the product may be recovered from the reaction system through addition of the reaction liquid to a non-solvent, removal of the solvent or through direct precipitation of the polymer from the reaction system.
      Water-insoluble monomers can be polymerized as suspended droplets in a process called **suspension** polymerization. Coalescing of the droplets is prevented by use of small amounts of water-soluble polymers such as poly(vinyl alcohol). It allows good heat control and easy removal of discrete polymer particles.

The **emulsion** process differs from the suspension polymerization in the size of the suspended particles and in the mechanism. While the particles in the suspension systems vary from about 10 to 1000 nm, those in the emulsion process range from about 0.05 to 5 nm. The small beads produced in the suspension process may be separated by filtering, but the latex produced in the emulsion systems is normally stable where the charged particles cannot be removed by ordinary separation procedures. A typical "recipe" for emulsion polymerizations includes monomer, water, a surfactant (normally a "soap"), and an initiator such as potassium persulfate. When the concentration of soap exceeds some critical micelle concentration, the molecules are present in micelles where the hydrophilic ends of the molecules are oriented toward the water-micelle interface, and the lyophilic portions of the molecules are oriented towards the center of the micelle. Since the initiation occurs in the aqueous phase, little polymerization occurs in the globules. Thus, the globules generally serve as reservoirs for the monomer that is supplied to the micelles as monomer is converted to polymer.

Polymer Testing

As with non-polymeric materials, the use and acceptance of polymer-containing products is generally based on testing. Many of the tests are standardized and included as standard tests in the American Society for Testing and Materials (ASTM) procedures. The ASTM cooperates on a world-wide basis with other similar organizations including the International Standards Organization (ISO), American National Standards Institute (ANSI), British Standards Institute (BSI), and the Deutsche Normenausschuss (DNA).

Here we will concentrate on ASTM standardized tests because they are the ones used in America and they are generally accepted internationally. The ASTM, through a group of committees that emphasize specific materials (for instance Committee D-1 is concerned with paints and Committee D-20 with plastics), manage "accepted" tests. Tests are continually being developed and submitted to the appropriate ASTM committee. After adequate verification through "round robin" testing, tests are accepted by consensus as "standard tests". Some tests are developed within a company to measure a certain property peculiar to that particular company and product. These tests may or may not be submitted to the ASTM. In order to insure, as well as possible, that different companies are using similar test conditions, the ASTM tests specify as many possible variables as believed to influence the test results. These variables may include temperature, atmosphere, rate of addition of stress, humidity, etc. It must be remembered that for many tests, it is not always clear what particular property is being measured since a number of properties are related and dependent on one another. Even so, the results from such tests form the basis for product reliability and reproducibility.

Rheology is the branch of science related to the study of deformation and flow of materials. Rheology includes two quite varied branches called fluid and solid mechanics. Many polymers are viscoelastic and thus can act as both solids and fluids.

The elastic component is dominant in solids with the basic mechanical properties described by Hooke's law which states that the applied stress, S, is proportional to the resultant strain but is independent of the rate of this strain, d(strain)/dt. Crystalline solids generally obey Hooke's law and amorphous materials, below the glass transition temperature exhibit some Hooke-like behavior.

S = G x Strain   (39)
where G is the Shear Modulus

   The fluid or viscous component is dominant in liquids and are thus described using Newton's law that states that the applied stress S is proportional to the rate of strain, d(strain)/dt but is independent of the strain or applied velocity gradient.

S = Viscosity x d(Strain)/dt   (40)

   Since many polymers have both amorphous and crystalline regions, they can be described using combinations of both Hooke and Newton laws.
   Hookean behavior is believed to describe bond bending while Newtonian behavior is believed to describe chain movement and slippage. Thus, effort is made in rheology to describe polymer stress-strain behavior in terms of modes employing components that "mimic" Hookean and components that "mimic" Newtonian behavior in an attempt to understand the importance and/or effect of the various polymer components on particular properties.
   Temperature is important when discussing polymer behavior. Thus, plastics that are measured at temperatures below their glass transition temperature will have a larger component of their stress-strain behavior be Hookean-like while plastics that are measured at temperatures above their glass transition temperature will have more of their behavior be Newtonian-like. Since one desired plastic property is flexibility, the use-temperature of plastics must be above the glass phase transition temperature of that particular plastic material.
   Polymer characterization generally involves both property characterization and structural characterization. Property characterization may include thermal and electrical behavior, chemical behavior, stress-strain-related properties (flexural strength, tensile strength, compression strength, impact resistance, shear strength), hardness (Rockwell, scratch), fatigue or endurance, etc. Structural characterization typically involves instrumentation that is also applied to smaller molecules including infrared spectroscopy, molecular weight (and molecular weight distribution), NMR, UV-VIS-NIR, scanning electron microscopy, X-ray spectroscopy, electron paramagnetic resonance, Auger electron spectroscopy, X-ray photoelectron spectroscopy, etc. Generally, with specific information about how a polymer might behave differently from a small molecule, scientists can directly apply knowledge gained from small molecule experiments to polymers.

Polymer Companies

About 10,000 American companies are active in the general area of synthetic polymers. These companies can be divided into three groupings as follows-
Manufacturers-Over 200 companies produce the "bulk" polymers that are used by the other two groupings of companies. While most of these produce the bulk polymers in large quantities, some produce what are called "specialty polymers", those polymers that are

used in special applications on a small scale volume wise.

Processors-While some companies produce their own polymers, most purchase the raw polymer material from one of the 200 manufacturing companies. Processors may specialize in the use of selected polymers, such a polypropylenes, polyethylenes, nylons; or on a particular mode of processing; or on the production of particular markets such as films, sheets, laminates, adhesives and coatings.

Fabricators and finishers-The large majority of companies are involved in the fabrication and finishing of polymer-containing products. Fabrication can be divided into three broad areas-

        *machining
        *forming, and
        *fashioning.

Polymer Structure-Property Relationships

The properties of polymers are dependent on many factors including inter and intrachain bonding, nature of the backbone, processing events, presence/absence of additives including other polymers, chain size and geometry, and molecular weight distribution.

Interchain and Intrachain Forces. The forces present in molecules can be divided into primary (generally greater than 50 kcal/mole of interactions) and secondary forces (generally less than 10 kcal/mole of interactions). Most synthetic polymers are connected by covalent bonds. These bonds are directional with bond lengths on the order of 9 to 20 nm.

      Secondary forces, frequently called van der Waals forces, interact over longer distances-generally on the order of 25 to 50 nm. The force of these interactions is inversely proportional to some power of r, generally 2 or greater, and thus is dependent on the distance between the interacting molecules.

      Atoms of individual polymer molecules are joined to each other through primary covalent bonds. Polymer molecules, as with small molecules, are also attracted to one other through a variety of secondary forces. These intermolecular forces include London dispersion forces, induced permanent forces, and dipolar forces including hydrogen bonding. Non-polar molecules, such as ethane and polyethylene, are attracted to one another by weak London or dispersion forces. The transient dipoles are due to instantaneous fluctuations in the electron cloud density and amount to about 2 kcal for each mole of methylene units. Thus, a polyethylene chain of about 1000 units long with have a London attractive force of about 2000 kcal/mole. This London force is sufficient to render the polyethylene non-volatile so that chain degradation occurs prior to volatilization of the polyethylene chain. Polar molecules such as ethyl chloride and poly(vinyl chloride), PVC, are attracted to each other by both the London forces and through dipole-dipole interaction. The combined attraction forces are of the order of about 5 kcal per (mole) repeat unit. Thus, for a PVC of 1000 units, the attractive forces are of the order of 5 kcal/(mole)chain. Again, this attractive force is sufficient to render PVA non-volatile.

These secondary forces, combined with structural considerations, are major factors that determine the shape of polymer chains both in the solid and mobile (melt or solution) states.

Crystalline-Amorphous Structures. A three-dimensional crystalline common synthetic polymer such as polyethylene is often described as a fringed micelle (where the chains are packed as a sheaf of grain) or as a folded chain. Regions where the polymer chains exist in an ordered array are called crystalline domains. Imperfections in polymer crystalline domains are more frequent than found for smaller molecules. Regions where the polymer chains exist in a more non-ordered fashion are referred to as amorphous regions.

For the same polymer composition, generally the more crystalline the polymer the higher is the density, higher the melting point, greater is the resistance to swelling and dissolution, higher is the moduli of rigidity (more stiff), and greater is the resistance to gas and solvent (including a decreased flow of materials through it).

The amount and kind of crystallinity depends on both the polymer structure and on its treatment (pre-history). This dependence on treatment is much greater than for small molecules. For instance, the proportion of crystallinity can be controlled by controlling the rate of formation of crystalline areas. Thus, polypropylene can be heated above its melting range and cooled quickly (quenched) to produce a product that has only a moderate amount of crystalline areas because the chains were not given enough time to rearrange themselves in crystalline regions. Cooling at a slower rate allows the chains to fold, etc. giving a product with a higher degree of crystallinity.

Following are a number of factors that influence the shape a polymer can take. Chain flexibility is related to the activation energies required to initiate rotational and vibrational segmental motions. For some polymers, as flexibility increases, the tendency towards crystalline formation increases. Polymers containing regularly spaced single C-C, C-N, and C-O bonds allow rapid conformational changes that contribute to the flexibility of the polymer chain and to the tendency towards crystalline formation. Yet, chain stiffness can also enhance crystalline formation by permitting or encouraging only certain "well-ordered" conformations to occur within the polymer chain. Thus p-polyphenylene is a rigid linear polymer that is highly crystalline.

Crystallization is favored by the presence of regularly spaced units that permit strong inter and intra-chain associations. This is reenforced by the presence of regularly spaced polar groups that can form secondary dipole-dipole interactions such as is present in polycarbonates, polyesters and polyamides (nylons).

Structural regularity and the absence of large substituents enhance the tendency for crystallization. The precise effect of substituents depends on a number of factors including location, size, shape, and mutual interactions. For instance, the presence of ethyl to hexyl substituents tends to lower the tendency for crystallization because their major contribution is to increase the average distance between chains and thus decrease the secondary bonding forces. When the substituents become longer (from 12 to 18 carbons) and remain linear, a new phenomenon may occur-the tendency of the side chains to form crystalline domains of their own.

<u>End Uses As Related to Structure.</u> The usefulness of polymers depends not only on their properties, but also on their abundance and availability and on cost including the cost of manufacture, shipping, machining, fabrication and finishing. Even so, polymer properties is an essential element in their usefulness. Polymer properties are related to a number of factors including molecular weight, distribution of chain sizes, previous treatment (history), nature of the polymer, additives, etc. These properties, in turn, are reflected in polymer properties-flex life, strength, biological response, weatherability, resistance to chemical attack, degradation (synthetic and biological), electrical resistance, flammability, dyeability, machinability, comfort, hardness, ...

Today, polymers are used in may ways that defy easy classification. Even so, we will look at a few of these major polymer divisions.

Elastomers (or rubbers)-are polymeric material that can be (relatively) shaped through application of force (stress), and when the force is released, the material will return to its original shape. This return to original shape is called "memory" and it is a result of the presence of crosslinks (either physical or more normally chemical). Further, the driving force for the return to the original shape is entropy. Products with low amounts of crosslinking will be easily deformable and "stretchable". As the amount of crosslinking increases, the material becomes harder. In order for elastomers to easily stretch, the attractions between chains should be low and the material should be above the temperature where local segmental mobility occurs. These properties are found in many hydrocarbon-intense polymers such as those listed in Table 1. In the normal, unstretched state the polymer is amorphous with a relatively high entropy or level of disorder. When stretched, the material should be more crystalline with a greater degree of order.

Table 1. Synthetic elastomers.

Polychloroprene	Epichlorhydrin Copolymers
Styrene-Butadiene, SBR	Polybutadiene
Nitrile	Ethylene-Propylene
Neoprene	Polyfluorocarbon
Silicon	Polyurethane (Segmented)
Polyisoprene	Butadiene-Acrylonitrile
Styrene-Isoprene	

Fibers-are polymeric materials that have high tensile strength and high modulus (high stress for small strains). Fibers are usually drawn (oriented) in one direction producing high mechanical properties in that direction. They are typically symmetrical allowing for the chains to more closely approach one another, increasing the bonding strength between the various chain segments. The attractive forces between the chains are, relative to elastomers, high. These materials should have no local mobility and only small amounts of crosslinking is introduced after fabrication to lock in a preferred structure. Table 2 contains a listing of some of the more popular fiber families.

Table 2. Industrially important synthetic fibers

Acrylic	Modacrylic
Polyester	Polyurethane
Nylon	Rayon
(Rayon) Acetate	(Rayon) Triacetate
Fibrous Glass	Olefins

Plastics-are materials that have "mid-way" properties between fibers and elastomers. Thus, it is expected that many of the polymers that are either fibers or elastomers are also plastics. Thus, crystalline nylon is a good fiber, whereas less crystalline nylon material is a plastic. It is unusual for a polymer to be a plastic, fiber and an elastomer. Table 3 contains a listing of plastics.

Table 3. Industrially important plastics

Epoxies	Polyesters
Urea-Formaldehydes	Melamine-Formaldehydes
Phenolics (Phenol-Formaldehydes)	Polyethylenes
Polypropylene	Styrene-Acrylonitriles
Polystyrene	Polyamides
Poly(vinyl chloride) and Co-polymers	
Polytetrafluoroethylene	Poly(methyl methacrylate)
Polycarbonates	Silicons
Polysulphone	Poly(phenylene oxide)
Polyimides	

Coatings and adhesives are generally derived from polymers that "fit" into the plastics grouping though there are major groups that do not. For instance, silicone rubbers are elastomers but also can be used as adhesives. **Coatings**, or coverings, are generally highly viscous (low flowing) materials. The major purposes of coatings are to protect and to decorate. Coatings protect much that is about us from wear and tear and degradation from the "elements" including oils, oxidative chemical agents, effects of temperature and temperature change, rain, snow, and ionizing radiation. They protect our stoves, chairs, cabinets, cars, planes, bridges, homes (inside and outside), etc. Coatings must adhere to the surface they are applied to. Coatings are typically a mixture of a liquid (vehicle or binder (adhesive) and one or more colorants (pigments). Coatings often also contain a number of so-called additives that can furnish added protection against ionizing radiation, increased rate of drying and/or curing (crosslinking), microorganisms, etc. Coatings are specially formulated for specific purposes and locations.

Coatings of today and tomorrow may fulfill purposes beyond those of beauty and protection. They may also serve as energy collective devices and burglar alarm systems.

In contrast to coatings that must adhere to only one surface, **adhesives** are used to join two surfaces together (Table 4). Adhesion for both adhesives and coatings can occur through a number of mechanisms including physical interlocking, chemical adhesion where primary bonding occurs between the adhesive and the surfaces being joined, secondary bonding where hydrogen bonding or polar bonding occurs and viscosity adhesion where movement is restricted because of the viscous nature of the adhesive material.

The combination of an adhesive and adherent is a **laminate**. Commercial laminates are produced, on a large scale, with wood as the adherent and phenolic, urea, epoxy, resorcinol, or polyester resins as the adhesives. Plywood is one such laminate. Laminates of paper or textile include Formica (TM) and Micarta (TM). Laminates of phenolic, nylon, or silicone resins with cotton, asbestos, paper, or glass textiles are used as mechanical, electrical, and general purpose structural materials.

Table 4. Synthetic polymeric adhesives.

Aromatic Polyamides	Acrylic Acid & Acrylic Ester Polymers
Acrylonitrile-Butadiene Copolymers	
Butyl Rubber	Cellulose Derivatives
Epoxy Resins	Phenol-Formaldehyde
Polychloroprene	Polyisobutylene
Polyurethane Resins	Poly(vinyl Alcohol)
Poly(vinyl Acetate)	Polyamides
Polyethylene	Poly(vinyl Butyral)
Poly(alkyl Cyanacrylates)	Resorcinol-Formaldehyde
Silicone Polymers	Styrene-Butadiene Copolymers
Unsaturated Polyester Resins	Vinyl Acetate-Ethylene Copolymers

Sealants and **caulks** provide a barrier to the passage of gases, liquids and solids, maintain pressure differences, moderate mechanical and thermal shock, etc. While adhesives are used for "load transfer", requiring high tensile and shear strengths, sealants act as insulators and shock attenuators and do not require high tensile and shear strengths.

Films are two dimensional forms of plastic, thick enough to be coherent, but thin enough to be flexed, creased or folded without cracking. **Sheeting** is two dimensional forms of plastic that are thicker (generally >250 um) than films and generally they are not easily flexed, creased or folded without cracking.

Composites are materials that contain strong fibers, reinforcement, embedded in a continuous phase called the matrix. Today's composites are often called "space-age" or "advanced materials" composites. They are found in the new jet fighters such as the stealth fighters and bombers, in the "reusable" space shuttle, graphite (a composite material) golf clubs, as synthetic human body parts, and for many years in marine craft (fibrous glass).

Polyblends are made by mixing components together in extruders or mixers, on mill rolls, etc. Most are heterogeneous systems consisting of a polymeric matrix in which another polymer is imbedded. The components of polyblends adhere through secondary bonding forces.

Liquid crystals, LCs, are materials that undergo physical reorganization where at least one of the rearranged structures involve molecular alignment along a preferred direction causing the material to exhibit non-isotropic behavior and associated molecular birefringent properties, ie. molecular asymmetry.

The term **ceramic** comes from the Greek word keramos which means "potter's clay" or "burnt stuff". Most ceramics contain large amounts of inorganic polymers. While traditional ceramics were largely based on natural clays, the ceramic of today generally contains synthetic materials. Ceramics are generally brittle, strong; resistent to chemicals such as acids, bases, salts, and reducing agents; and they are high melting.

The term **cement** covers a large grouping of materials. Some of these have bulk as their commonality, such as Portland cement (concrete), while others are termed cements and are actually performing as an adhesive, such as dental cements.

Solubility and Flexibility

Most linear and some two-dimensional polymers are soluble. Polymers with moderate to high degrees of crosslinking are insoluble.

The large size of polymers makes their solubility, in comparison to smaller molecules, poorer. This is due to both kinetic (motion) and thermodynamic factors. With respect to kinetic factors, solvent molecules must be able to come into contact with the materials that they are to dissolve. With polymers, this means that solution occurs "one layer" at a time with the solvent molecules having to penetrate outer layers before contact to inner layers is possible. Thus, solubilization of polymers may take hours to days and even weeks with the approach to solubilization often passing through what is referred to as a gel state where the solvent molecules have become entrenched in the polymer network, but the concentration of solvent molecules is not great enough as to dissolve the polymer.

In thermodynamic considerations, solubility can be described in terms of free energy relationships such as

$$\Delta G = \Delta H - \Delta(TS) \tag{41}$$

where G = Gibbs Free Energy, H = Enthalpy or Energy term, T = (Absolute) Temperature and S = Entropy or Order term.

At constant T, the free energy relationship becomes

$$\Delta G = \Delta H - T\Delta S \tag{42}$$

Since "like-likes-like" (or a material will have its greatest solubility in itself) better than anything else, the energy (ΔH) term will be against solution occurring. Thus, it is the entropy, ΔS, term that generally is the "driving force" for solubility to occur. (Nature tends to go from ordered systems to disordered systems so that systems go from greater order to less order "naturally".)

The number of geometric arrangements of connected polymer segments in a chain are much less than if the segments were free to act as individual units. Thus, the increase in randomness, the Δ S term, is much lower for polymers in comparison to small molecules making them less soluble and soluble in a lower number of liquids, in general, in comparison to small molecules.

As noted above, the energy term acts against solution occurring. Thus, there is an effort to "match" the solvent and polymer such that their "energy patterns" (shape, size and polarity, solubility parameter and cohesive energy density) are similar so that while the energy term will be against solution occurring, the entropy term is more likely to overcome the energy term allowing solubility to be achieved.

Many polymers are themselves brittle at room temperature. For these polymers to become more pliable, additives called plasticizers, that allow segmental solubility, and consequently segmental flexibility, are added. The proteins and nucleic acids in our bodies are actually inflexible in themselves, but flexibility is essential for them to carry out their functions and for replication to occurs. In this case, water acts as the plasticizer allowing the natural macromolecules to remain "solid" yet be flexible.

As noted above, flexibility is critical for many applications. Thus, elastomers and plastics must be flexible to allow them to be bent and reshaped. Polymer morphology or shape can be divided into two general groupings-amorphous or disordered and crystalline or ordered. Thus, elastomers in their "rest" state have a high degree of amorphous character (that is the entropy or disorder is high), yet in the stretched state, crystallization occurs. The amorphous areas of a polymer contribute to the flexibility of the polymer while the crystalline regions contribute to the strength of the polymer. Polymers, unlike small molecules, undergo two primary solid state transitions. As a polymer is heated, energy is added. At a given temperature, segmental motion begins. The temperature where segmental motion begins is called the glass transition temperature, T_g. Because polymer chains are present in many different orientations and because the addition of temperature is generally done in a relatively rapid manner (even increases of 0.1 C/ minute will contribute to T_g ranges), the T_g is generally reported as a range rather than as a specific temperature. As heating continues, there is a temperature, again actually a temperature range, where wholesale entire chain motion occurs. This temperature is called the melting point or melting range, T_m. Amorphous regions of a polymer exhibit T_g values while crystalline regions exhibit T_m values. Since most polymers have both amorphous and crystalline regions, they have both T_g and T_m values.

Polymers, such as elastomers, that have a moderate to high degree of crosslinking, exhibit only T_g since whole-sale movement of the polymer chain is not possible because of the presence of the crosslinks. Further, in order for plastics and elastomers to exhibit the essential property of flexibility, the "use" temperature must be above the T_g.

The inflexible regions of a polymer are often referred to as the "hard" regions while the flexible regions of a polymer are referred to as "soft" regions. This combination of hard and soft can be illustrated with so-called segmented polyurethanes. Here, the urethane portion of the polymer is involved in hydrogen bonding and is considered as "hard" while the polyether portion is considered "soft". These segmented polyurethane are sold under a number of trade names including Spandex.

$$\overset{\displaystyle O}{\underset{}{\|}} \qquad \overset{\displaystyle O}{\underset{}{\|}}$$

-(-O-CH$_2$-CH$_2$-)$_x$-(-C-NH-R-NH-C-)$_y$-
    Soft   Hard

Segmented Polyurethane

Size

Unlike small molecules where there is a single molecular weight, polymers often are produced where there is a range of chain lengths. Because there is a range of molecular weights, the particular molecular weight average is dependent on the type of measurement used to determine the molecular weight, which in turn is dependent on the particular mathematical relationship that relates the measured polymer property and molecular weight. Polymer chemists mainly use two types of molecular weights referred to as number average molecular weight and weight average molecular weight.

The **number average molecular weight**, Mn, is measured by any technique that "counts" the molecules, These techniques include vapor phase and membrane osmometry, boiling point elevation, end-group analysis, and freezing point lowering. Mn can be described using a jar filled with plastic capsules such as those in the circuses that contain tiny prizes. Each capsule is the same size and contains one polymer chain, regardless to the size of the polymer chain. Capsules are then withdrawn, opened, and the individual chain length determined and recorded. The probability of drawing a capsule containing a chain with a specific length is dependent on the fraction of capsules containing such a chain and independent of the length of the chain. The most probable value is the number average molecular weight.

The **weight average molecular weight**, Mw, is measured by any technique that is dependent on the size of the polymer chain such as light scattering photometry. Using the capsule scenario described above except where the size of the capsule is directly proportional to the chain size, the capsules can be withdrawn, opened, and the individual chain length determined and recorded. The probability of drawing a capsule containing a chain with a specific length is dependent on both the fraction of capsules containing such a chain and on the size of the capsule. The most probable value is the weight average molecular weight.

The ratio between Mw and Mn is referred to as the **polydispersity index**. The closer to one this ratio is, the less disperse or heterogeneous are the polymer chain lengths. Natural polymers that perform specific functions such as DNA and enzymes have a dispersity index of 1 so that all the chains are of the same size.

The major polymer properties are related to size. This large size allows polymers to behave in a more or less coherent manner. Such factors that influence one part of a polymer chain can be "felt" further along the polymer chain.

One prominent polymer property is the increase in viscosity of polymer melts and dilute solutions containing polymers. Polymers that are heated so that they can flow under applied pressure or simply through gravity effects are referred to as polymer melts.

Since many polymers must have some mobility (often under high pressure and when heated) to be processed, the energy that must be applied to process the polymer is dependent on the polymer molecular weight. There is often a polymer molecular weight where the desired polymeric property (such as strength) and molecular weight is such that further molecular weight increases result in minimal, and unnecessary property increases, but does result in increased effort needed to process the material. This "favorable" molecular weight range is called the **commercial molecular weight**.

Since polymer chains are long, they can reside in several "flow planes" acting to retard the flow of polymer-containing solutions. This resistance to polymer flow is called **viscosity**.

Polymer Shape

Two terms, configuration and conformation, are often confused. **Configuration** refers to arrangements fixed by chemical bonding that cannot be altered except by breakage of primary bonds. Terms such as heat-to-tail, *d-* and *l-*, *cis*, and *trans* refer to the configuration of a chemical species. **Conformation** refers to arrangements around primary bonds. Polymers in solution or in melts continuously undergo conformational changes.

Natural polymers utilize a combination of primary and secondary forces to form polymer structures with both long-range (supra or multi-macromolecular) and short-range structures with both structures critical for the "proper" functioning of the macromolecule. While most synthetic chemists focus on what is referred to as primary and secondary (short-order structural control) structures, work has just begun on developing the appropriate structural control to allow tertiary and quaternary structural (long-range) control.

Here we will focus on primary and secondary polymer shapes. Overall, flexible linear polymer chains will have some tendency to minimize size constraints while keeping "like" polymer moieties together and they will utilize strong (polar and hydrogen bonding) secondary bonding. Most polymers exist as some form of helix with the amount, extent, of helical nature varied. The second most common secondary polymer form is similar to that of a pleated skirt where the hydrogen or dipolar bonding is taken advantage of.

Unsymmetrical reactants, such as substituted vinyl monomers, react almost exclusively to give what is called "head-to-tail" products where the substitutants occur on alternative carbon atoms.

$-CH_2-CHX-CH_2-CHX-CH_2-CHX-CH_2-CHX-$

Occasionally a heat-to-head, tail-to-tail configuration occurs. For most vinyl polymers this structure occurs less than 1% of the time in a random manner throughout the chain.

$-CH_2-CHX-CHX-CH_2-CH_2-$

Even with the head-to-tail configuration, a variety of structures are possible. These include simply a linear polymer structure

-M-M-M-M-M-M-M-M-M-M-M-M-M-M-M-M-M-

and branched structures with varying amounts and lengths of branching.

-M-M-M-M-M-M-M-M-M-M-M-M-M-M-M-M-M-
     M   M   M
     M   M   M
     M   M   M
     M   M
     M
     M

Copolymers, polymers derived from two different monomers, M and N, are also important polymer groups. some of the more important ones are as follows.

-M-N-M-N-M-N-M-N-M-N-M-N-M-N-   -N-N-M-N-M-M-N-M-N-N-M-N-M-M-N-
  Alternating   Random

-M-N-M-M-M-M-M-M-M-M-M-M-M-N-   -N-N-N-N-N-N-M-M-M-M-M-M-M-N-N-
  Block in "M"   Block in both "M" & "N"

-M-M-M-M-M-M-M-M-M-M-M-M-M-M-
   N   N
   N   N Graft
   N   N
   N   N
   N   N

Crosslinked or network (three-dimensional) polymers offer a wide variety of structures dependent on extent and type of crosslinking, type of polymer, type, functionality, and amount of reactants, etc.

Polymers and associated polymer properties also vary with the structural regularity with respect to the substituted carbon atoms in the polymer chain. Each substituted carbon atom is a chiral site with different geometries possible. There are three main geometries or configurations. When the chiral carbons contained within a polymer chain are present in a random fashion the geometry is referred to as **atactic**. When the pendant group is attached so that the geometry on the carbon atom alternates, the polymer geometry is said to be **syndiotactic**. When the geometry about the various chiral carbons is all alike, the polymer chain is said to be **isotactic**.

Linear crystalline polymers often form spherulites. For linear polyethylene the initial structure formed is a single crystal with folded-chain lamellae. These then form sheaf-like structures called axialites or hedrites. As growth continues, the lamellae develop on either side of a central plane. The lamellae continue to fan out, occupying increasing volume sections through the formation of additional lamellae at appropriate branch points. The end result is the formation of three-dimensional spherulites.

C.E. Carraher, Jr.

Polymer Structures

Following are structures of a number of the more important synthetic polymers.

Acrylonitrile-Butadiene-Styrene   **Butyl Rubber**
Terpolymer, ABS

$$\left[CH_2CHCH_2CH=CHCH_2CH_2CH \right]_n \qquad \left[CH_2-\overset{\overset{\displaystyle CH_3}{|}}{\underset{\underset{\displaystyle CH_3}{|}}{C}}-CH_2CH=CCH_2 \right]_n$$

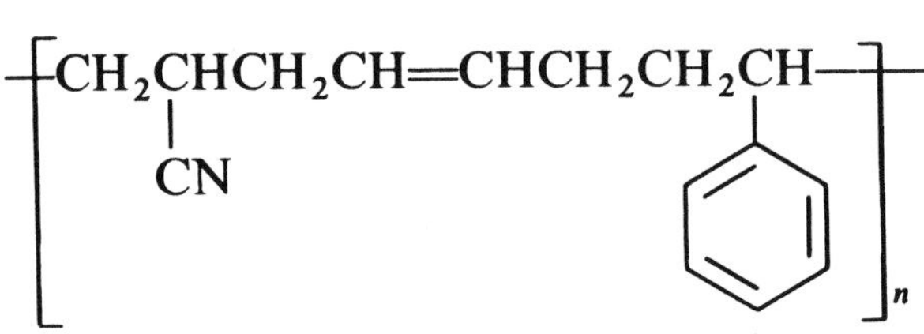

with CN and phenyl substituents

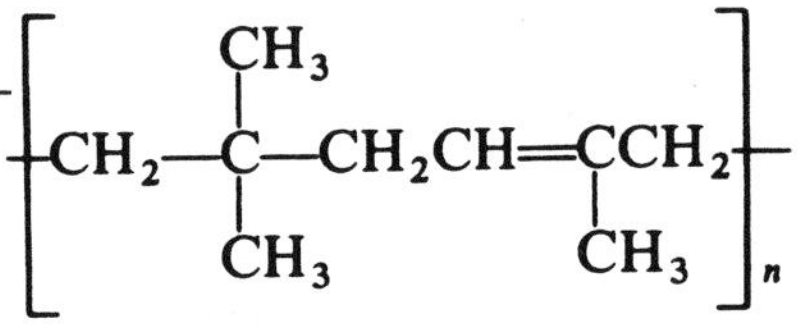

Ethylene-Methacrylic Acid
Ionomer

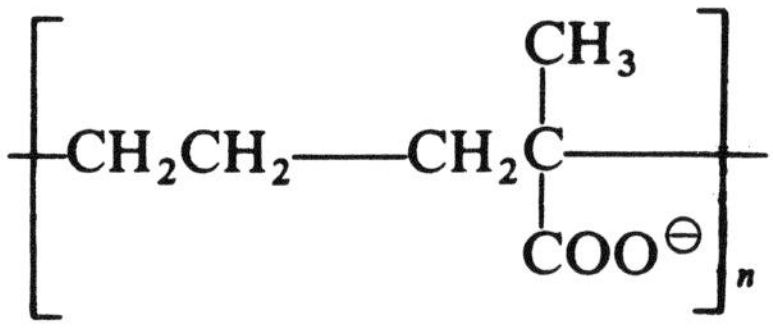

$$\left[CH_2CH_2 - CH_2\overset{\overset{\displaystyle CH_3}{|}}{\underset{\underset{\displaystyle COO^{\ominus}}{|}}{C}} \right]_n$$

Melamine-Formaldehyde Resin Structure

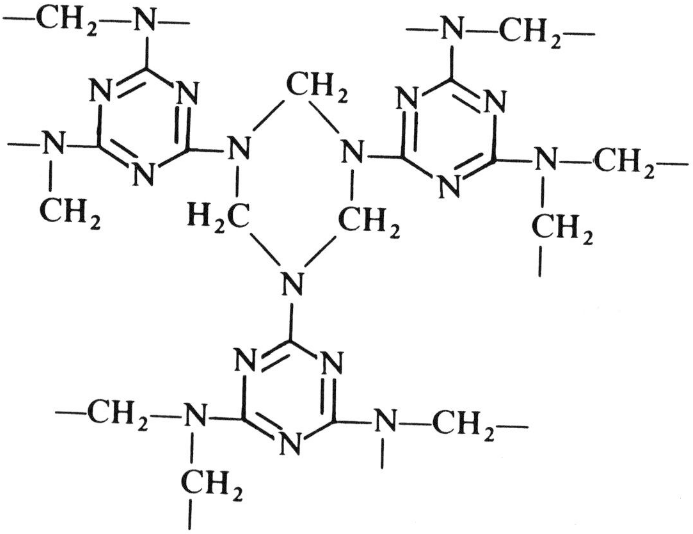

Nitrile Rubber, NBR

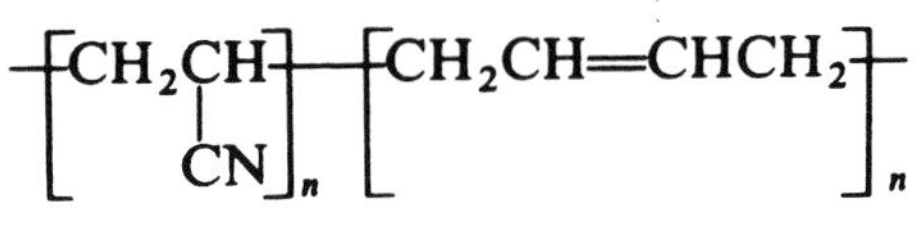

$$\left[CH_2CH\atop CN\right]_n \left[CH_2CH=CHCH_2\right]_n$$

Nylon 6

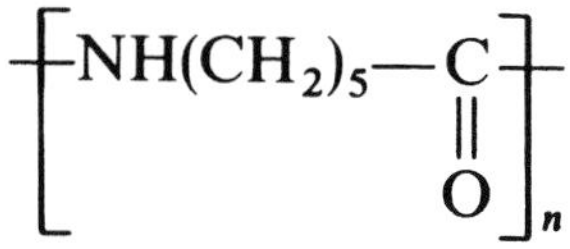

$$\left[NH(CH_2)_5-C\atop \|\atop O\right]_n$$

Nylon 6,6

$$\left[NH-(CH_2)_6-NH-\overset{O}{\overset{\|}{C}}-(CH_2)_4-\overset{O}{\overset{\|}{C}}\right]_n$$

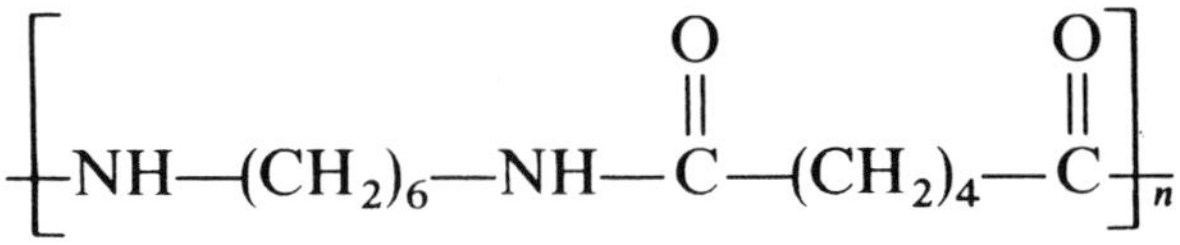

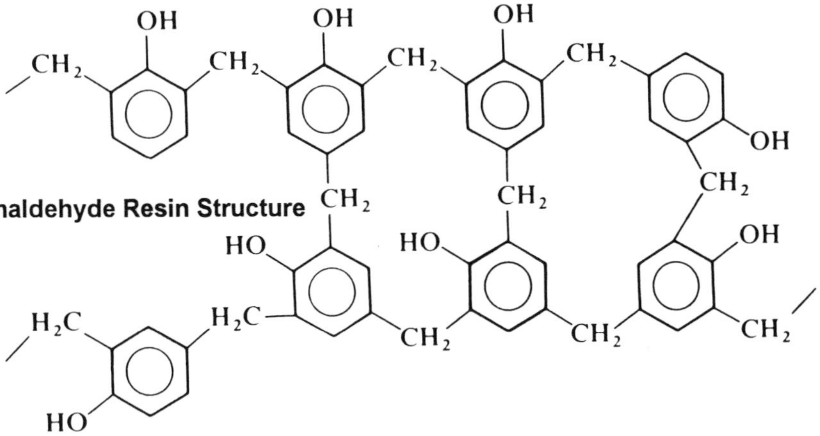

Phenol-Formaldehyde Resin Structure

Polyacrylonitrile

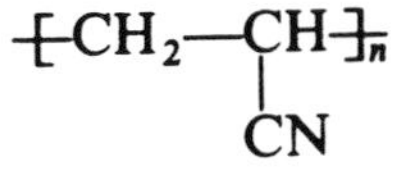

$$\left[CH_2-CH\atop CN\right]_n$$

Polybenzimidazole

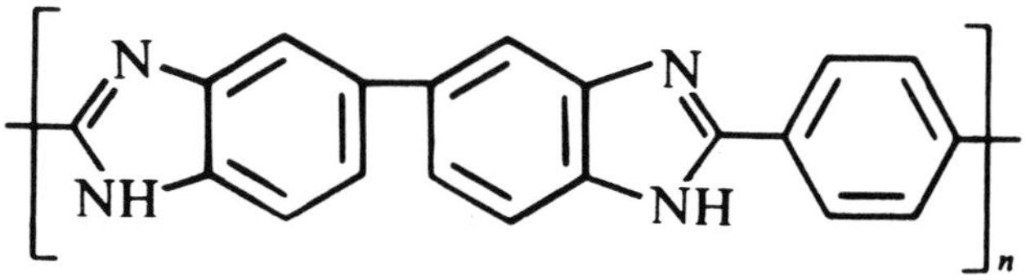

Poly(butylene Terephthalate), PBT

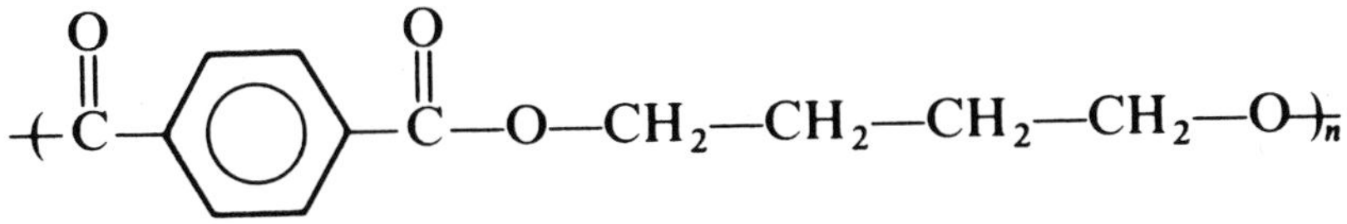

Polycarbonate, PC

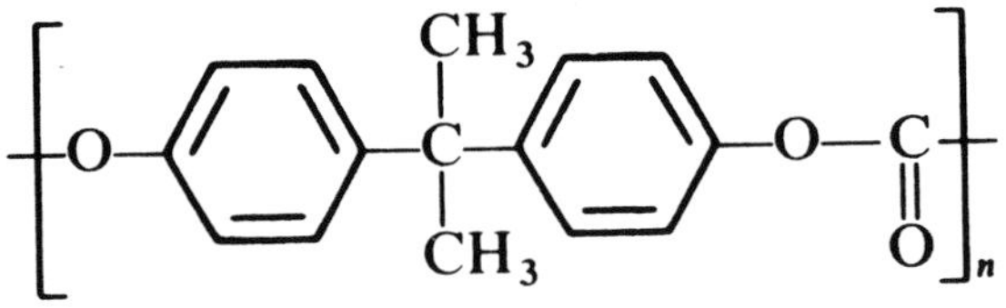

$$\left[\!\!\left[O\!-\!\!\!\bigcirc\!\!\!-\!\!\underset{\underset{CH_3}{|}}{\overset{\overset{CH_3}{|}}{C}}\!\!\!-\!\!\!\bigcirc\!\!\!-\!O\!-\!\underset{\overset{\|}{O}}{C}\!\!\right]\!\!\right]_n$$

Polychloroprene

$$\left[\!CH_2\!-\!\underset{\underset{Cl}{|}}{C}\!\!=\!\!CH\!-\!CH_2\!\right]_n$$

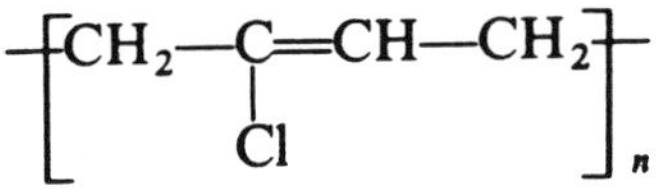

Poly(dimethyl Siloxane)

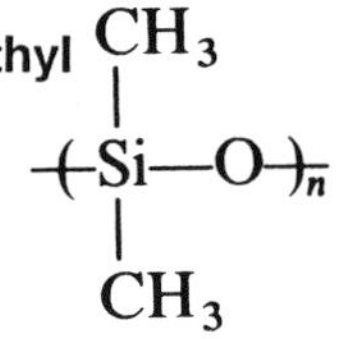

$$\left(\!\!\overset{\overset{CH_3}{|}}{\underset{\underset{CH_3}{|}}{Si}}\!\!-\!O\!\right)_n$$

Polyethylene, PE

$$\left(\!CH_2CH_2\!\right)_n$$

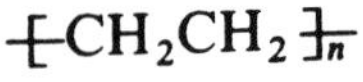

Poly(ethylene Glycol), PEG; Poly(ethylene Oxide), PEO

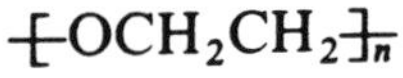

$$\left(\!OCH_2CH_2\!\right)_n$$

Poly(ethylene Terephthalate), PET

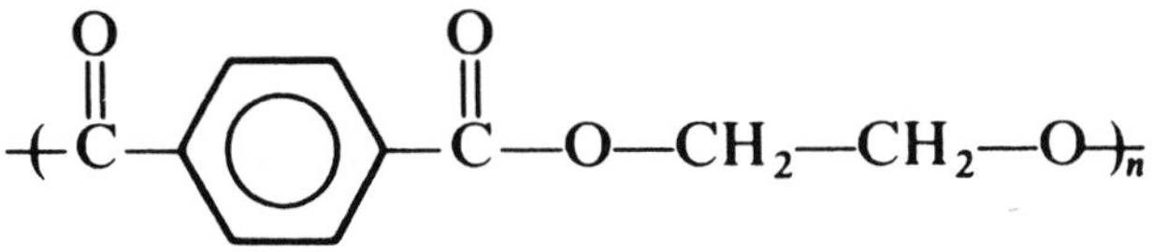

$$\left(\!\overset{\overset{O}{\|}}{C}\!-\!\!\bigcirc\!\!-\!\overset{\overset{O}{\|}}{C}\!-\!O\!-\!CH_2\!-\!CH_2\!-\!O\!\right)_n$$

Polyimide

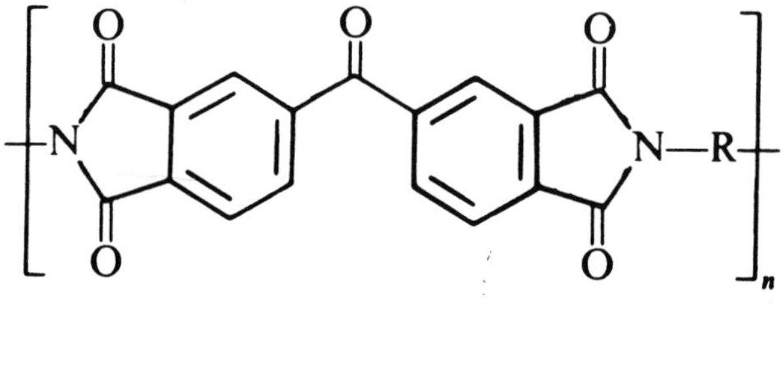

Polyisobutylene

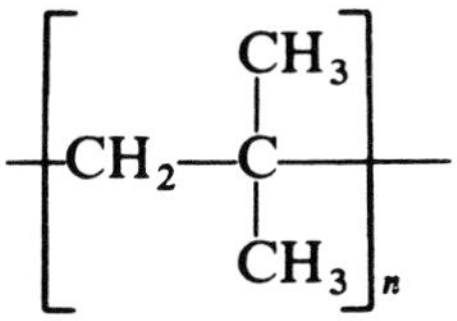

$$\left[CH_2-\underset{\underset{CH_3}{|}}{\overset{\overset{CH_3}{|}}{C}} \right]_n$$

Polyisoprenes

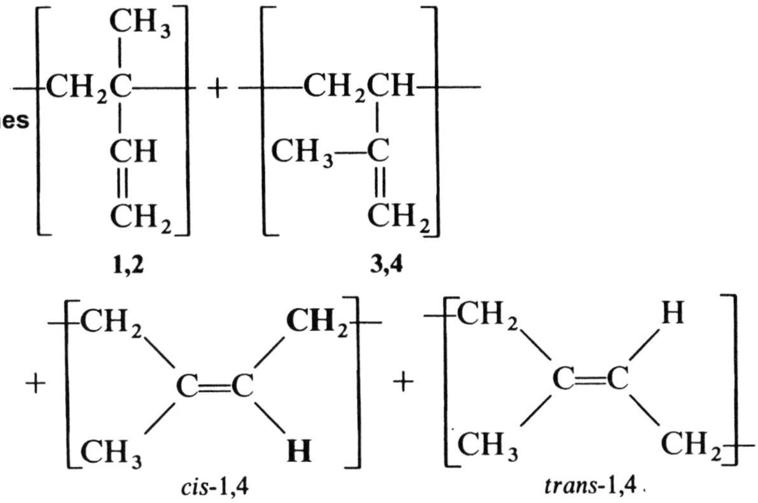

$$\left[CH_2C \underset{\underset{\overset{\|}{CH_2}}{CH}}{\overset{\overset{CH_3}{|}}{}} \right] + \left[CH_2CH \underset{\underset{\overset{\|}{CH_2}}{CH_3-C}}{} \right]$$

1,2   3,4

$$+ \left[\begin{array}{c} CH_2 \quad\quad CH_2 \\ C=C \\ CH_3 \quad\quad H \end{array} \right] + \left[\begin{array}{c} CH_2 \quad\quad H \\ C=C \\ CH_3 \quad\quad CH_2 \end{array} \right]$$

cis-1,4   *trans*-1,4.

Poly(methyl Acrylate)

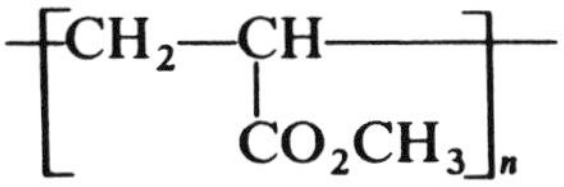

$$\left[CH_2-\underset{\underset{CO_2CH_3}{|}}{CH} \right]_n$$

Poly(methyl Methacrylate), PMMA

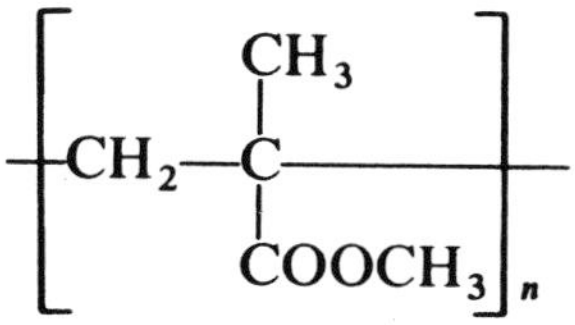

$$\left[CH_2-\underset{\underset{COOCH_3}{|}}{\overset{\overset{CH_3}{|}}{C}} \right]_n$$

Polyoxymethylene

$$[OCH_2]_n$$

Poly(phenylene Oxide), PPO

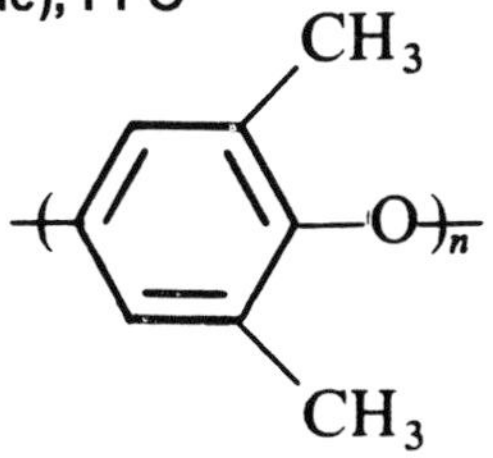

Poly(phenylene Sulfide), PPS

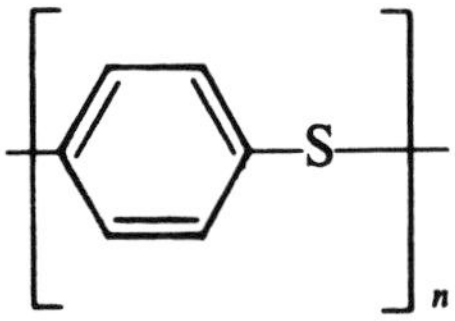

$$\left[\begin{array}{c} \end{array} \right]_n$$ structure with benzene ring — S

Polyphosphazene

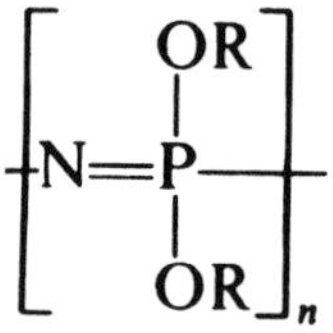

$$\left[\begin{array}{c} OR \\ -N=P- \\ OR \end{array} \right]_n$$

Polypropylene, PP

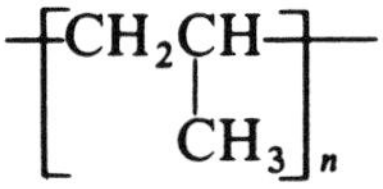

$$\left[\begin{array}{c} CH_2CH- \\ CH_3 \end{array} \right]_n$$

Polystyrene, PS

$$\left[CH_2-CH \right]_n$$ with benzene ring

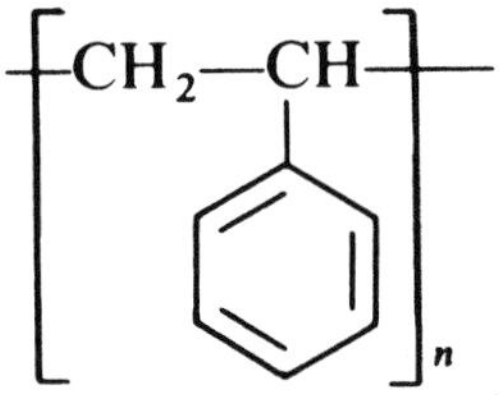

Polytetrafluoroethylene

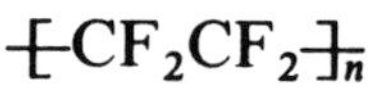

$$\left[CF_2CF_2 \right]_n$$

Polyurethane, PU

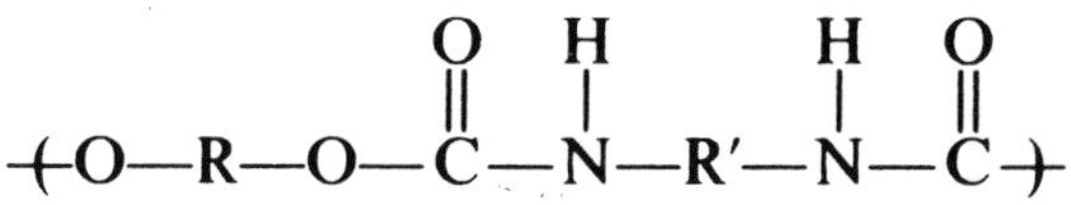

$$\left[O-R-O-\overset{O}{\overset{\parallel}{C}}-\overset{H}{\overset{\mid}{N}}-R'-\overset{H}{\overset{\mid}{N}}-\overset{O}{\overset{\parallel}{C}} \right]$$

Poly(vinyl Acetate)

$$\left[\begin{array}{c} CH_2-CH- \\ OCOCH_3 \end{array} \right]_n$$

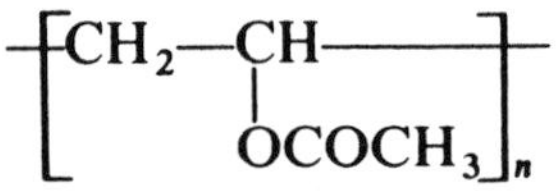

Poly(vinyl Alcohol), PVA or PVAI

$$\left[\begin{array}{c} CH_2CH- \\ OH \end{array} \right]_n$$

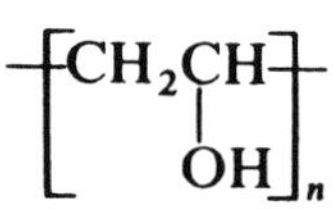

Poly(vinyl Butyral)

$$\left[\begin{array}{c} CH_2 \\ CH_2-CH \quad CH- \\ O \qquad O \\ CH \\ (CH_2)_2CH_3 \end{array} \right]_n$$

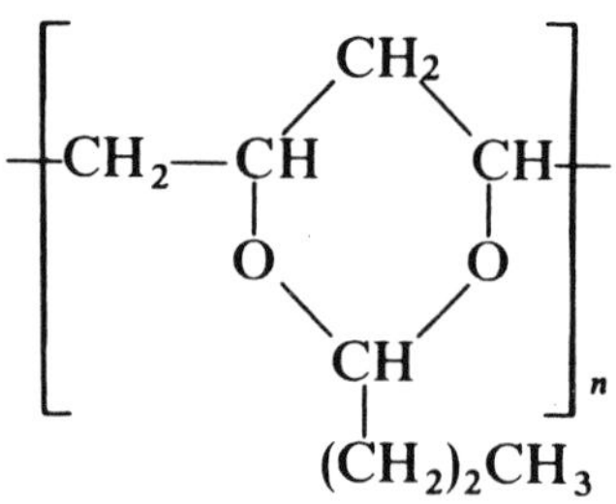

Poly(vinyl Chloride), PVC

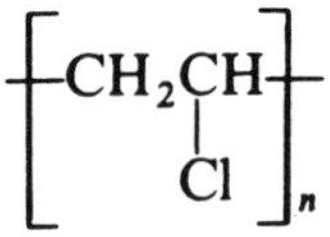

$$\left[\begin{array}{c} CH_2CH- \\ Cl \end{array} \right]_n$$

Poly(vinylidene Chloride)

$\left[CH_2CCl_2 \right]_n$

Poly(vinyl Isobutylether)

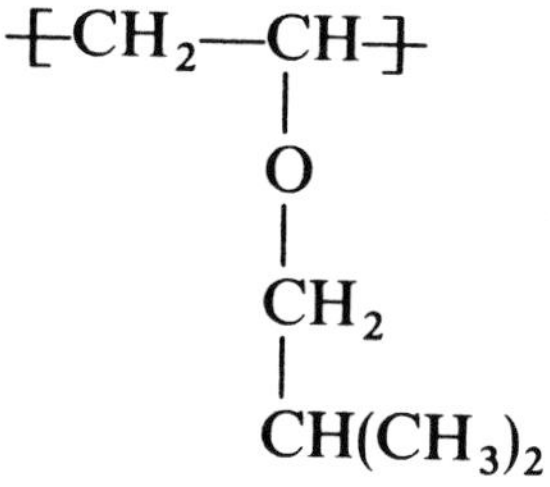

$$\left[CH_2-CH \right]$$
$$|$$
$$O$$
$$|$$
$$CH_2$$
$$|$$
$$CH(CH_3)_2$$

Poly(vinyl Pyridine)

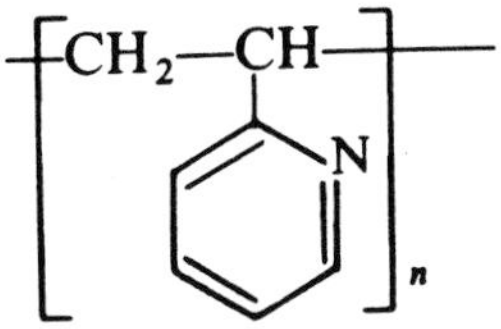

Poly(vinyl Pyrrolidone

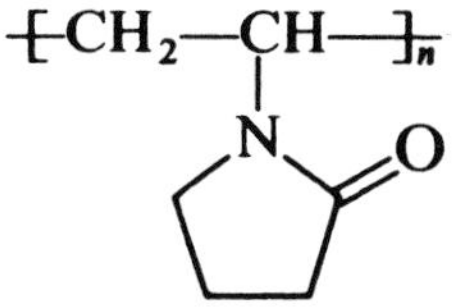

Styrene-Acrylonitrile Copolymer, SAN

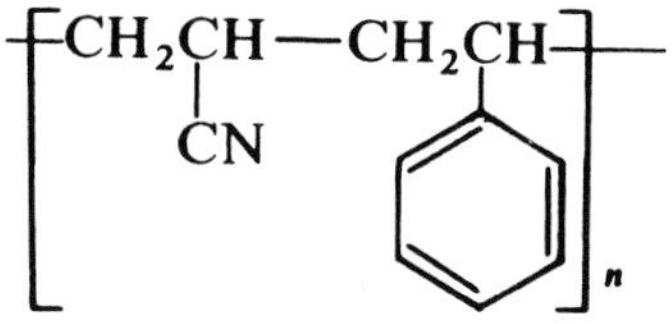

Styrene-Butadiene rubber, SBR
(1,4-addition of butadiene)

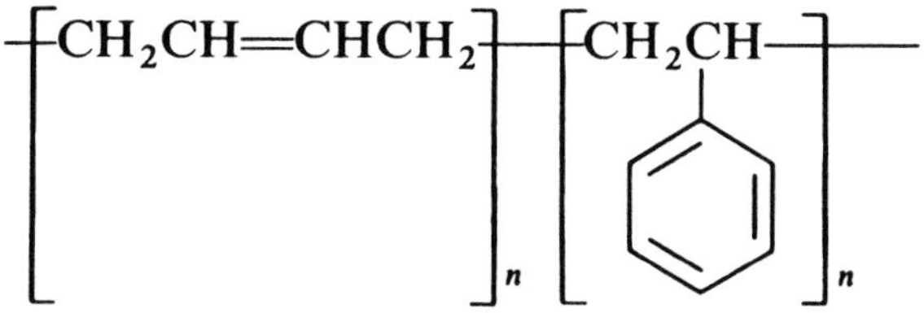

POLYMER NOMENCLATURE

L. H. SPERLING[a], W. V. METANOMSKI[b], AND CHARLES E. CARRAHER, JR.[c]

a. Center for Polymer Science and Engineering, Materials Research Center and Departments of Chemical Engineering and Materials Science and Engineering, Lehigh University, Bethlehem, PA 18015-3194

b. Chemical Abstracts Service, PO Box 3012, Columbus, OH 43210-0012

c. Department of Chemistry and Biochemistry, Florida Atlantic University, Boca Raton, FL 33431 and , Florida Center for Environmental Studies, NorthCorp Center, Palm Beach Gardens, FL 33410

Introduction
Homopolymers
Trade Names, Brand Names and Abbreviations
Copolymers
Readings

Introduction

Polymers come in a wide variety of shapes and composition. Here we will focus only on the naming of synthetic polymers. Because of the diversity and universality of polymeric materials there existed few guiding principles in the naming of polymeric materials as polymers were emerging.

Homopolymers

Common Names. Little reason is associated with the common names of many important polymeric materials. Some are based on the place of origin such as *Hevea brasiliensis*, which literally means "rubber from Brazil". Others are named after the discoverer. Bakelite, which is a three-dimensional polymer synthesized from the reaction of phenol and formaldehyde, was discovered by Leo Baekeland. Some are named after an element found within the polymer such as thiokols which are polymers that contain sulfur.

Special systems and names were developed to name some polymers and polymer groupings (see Table 1). Thus, the products of the reaction between diacids and diamines forming polyamides are called "nylons" after DuPont's tradename product, Nylon. . In turn, nylons are named according to the number of carbons in the diamine and dicarboxylic acid or acid chloride used for their synthesis. The nylon produced from reaction of 1,6-hexanediamine (six carbons) and sebacic acid (ten carbons) is given the common name nylon-6,10, nylon 610 or nylon 6,10. The product derived from reaction of 1,6-hexanediamine and adipic acid (six carbons) is called nylon-6,6. The product derived from reaction of the single reactant caprolactam (six carbons) is called nylon-6. The recommended name today is polyamide. Thus, the materials above become polyamide 6,10, etc.

Table 1. Common polymer names and groupings.

Common name/grouping	Description
Nylons	Polymeric products from reactions of amines with acids or acid chlorides
Acrylics	Polymers made from acrylic acid or a derivative of acrylic acid
Alkyds	Polyester containing a fatty acid
Ionomers	Polymer generally containing large amounts of ethylene units but containing units containing ionic groups-such as acrylic acid units.
Phenolics	Polymer produced from reaction of an aromatic alcohol, generally phenol, and an aldehyde, generally formaldehyde.
Polyolefins	Polymers derived from hydrocarbon monomers including polyethylene and polypropylene; also known as hydrocarbon polymers..
Polyvinyls	Polymers produced from vinyl monomers.

Source-Based Names. The majority of names used in the common literature for "simple" polymers are source-based. They are based on the common name of the reactant monomer preceded by the prefix **poly**. Thus, the polymer formed from reaction of vinyl chloride is called poly(vinyl chloride). The polymer formed from the monomer 1-phenylethene (or simply phenylethene), which has the common name of styrene, is called polystyrene. This practice holds for many of the polymers formed from reaction of the vinyl group.

$$HCH{=}CH \longrightarrow -(-HCH{-}CH{-})_n -$$
$$| \phantom{{=}CH \longrightarrow -(-HCH-}|$$
$$X \phantom{{=}CH \longrightarrow -(-HCH-}X$$

The appearance of this practice holds true for vinyl-derived polymers that are not directly derived from the monomer. Thus, poly(vinyl alcohol), which is not derived from from the monomer vinyl alcohol but rather from hydrolysis of poly(vinyl acetate), is named as though it were derived from vinyl alcohol.

Many condensation polymers are similarly named. In the case of the polyester poly(ethylene terephthalate) the glycol portion of the name of the monomer, ethylene glycol, is employed in constructing the polymer name so that the name is really a hybrid of a source-based named and a structure-based name.

Structure-Based Names. While the majority of common polymers are known by common or source-based names, the "correct" or so-called scientific names are based on rules described by the international body responsible for systematic nomenclature of chemicals-compounds. IUPAC and others have published a series of reports for naming polymers that are used for more complex polymers (1-6). Along with these formal reports, Polymer Chemistry, PC, and Polymeric Materials: Science and Engineering, PMSE periodically publishes updates derived from several study groups (7,8). For linear polymers, the IUPAC system names the components of the repeating unit, arranged in a prescribed order. The rules for selecting the order of the components to be used as the repeating unit are found elsewhere (1-8). Once the order is selected, the naming is straightforward. Following are several illustrations. For polystyrene (source-based name), the IUPAC name is poly(1-phenylethylene); for poly(methyl methacrylate) the IUPAC name is poly[1-(methoxycarbonyl)-1-methylethylene]; and for nylon-6,6 (common name) or polyamide 6,6 (source-based) the IUPAC name is poly(iminohexamethyleneiminoadipoyl).

Linkage-Based Names. Many polymer "families" are referred to by the name of the particular linkage that connects the polymers (Table 2). The family name is "poly" followed by the linkage-name. Thus, those polymers that contain the amide linkage are known as polyamides; those containing the ester linkage are called polyesters, etc.

Tradenames, Brandnames, and Abbreviations

Trade (and or brand) names and abbreviations are often used to describe materials. They may be used to identify the product of a manufacturer, processor or fabricator and may be associated with a particular product or with a material or modified material. Trade names are used to describe specific groups of materials that are produced by a specific company or under licence of that company. Thus, a rug whose contents are described as containing Fortrel (TM) polyester fibers contain polyester fibers that are "protected" under the Fortrel trademark and produced or licenced to be produced by the holder of the Fortrel (TM) trademark. Also, Lexan, Merlon, Baylon and Lupilon (TM) are all trade names for polycarbonates manufactured by different companies. Some polymers are better known by their trade name than their generic name. For instance polytetrafluoroethylene is better known as Teflon (TM), the trade name held by DuPont.

Table 2. Linkage-based names.

Family name	Linkage	Family name	Linkage
Polyamide	O‖ -N-C-	Polyvinyl	-C-C-
Polyester	O‖ -O-C-	Polyanhydride	O O‖ ‖ -C-O-C-
Polyurethane	O‖ -O-C-N-	Polyurea	O‖ -N-C-N-
Polyether	-O-	Polycarbonate	O‖ -O-C-O-
Polysiloxane	-O-Si-	Polyphosphate ester	O‖ -O-P-O-R- O R
Polysulfide	-S-R-		

Abbreviations are also employed to describe materials. Table 3 contains a listing of some of the more widely employed abbreviations and the polymer associated with the abbreviation.

Table 3. Abbreviations for selected polymeric materials.

Abbreviation	Polymer	Abbreviation	Polymer
ABS	Acrylonitrile-butadiene-styrene copolymer	CA	Cellulose acetate
EP	Epoxy	HIPS	High-impact polystyrene
MF	Melamine-formaldehyde polymer	PAA	Poly(acrylic acid)
PAN	Polyacrylonitrile	SBR, PBS	Butadiene-styrene copolymer
PBT	Poly(butylene terephthalate)	PC	Polycarbonate
PE	Polyethylene	PET	Poly(ethylene terephthalate)
PF	Phenol-formaldehyde polymer	PMMA	Poly(methyl methacrylate)
PP	Polypropylene	PPO	Poly(phenylene oxide)
PS	Polystyrene	PTFE	Polytetrafluoroethylene
PU	Polyurethane	PVA, PVAc	Poly(vinyl acetate)
PVA, PVAl	Poly(vinyl alcohol)	PVB	Poly(vinyl butyral)
PVC	Poly(vinyl chloride)	SAN	Styrene-acrylonitrile copolymer
UF	Urea-formaldehyde polymer		

Copolymers

Generally, copolymers are defined as polymeric materials containing two or more kinds of mers. It is important to distinguish between two kinds of copolymers-those with statistical distributions of mers, or at most, short sequences of mers (Table 4) and those containing long sequences of mers connected in some fashion (Table 5).

Table 4. Short sequence copolymer nomenclature.

Type	Connective	Example
Homopolymer	none	PolyA
Unspecified	*-co-*	*Poly*(A-co-B)
Statistical	*-stat-*	Poly(A-*stat*-B)
Random	*-ran-*	Poly(A-*ran*-B)
Periodic	*-per-*	Poly(A-*per*-B-per...)
Alternating	*-alt-*	Poly(A-*alt*-B)
Network	*net-*	*net*-PolyA

Table 5. Long sequence copolymer nomenclature.

Type	Connective	Example
Block copolymer	*-block-*	PolyA-*block*-polyB
Graft copolymer	*-graft-*	PolyA-*graft*-polyB
AB-Crosslinked	*-net-*	PolyA-*net*-polyB
Polymer blend	*-blend-*	PolyA-*blend*-polyB
Interpenetrating network		
polymer	*-ipn-* or *-inter-*	*net*-PolyA-*ipn*-*net*-polyB
Starblock	*star-*	*Star*-(PolyA-*block*-polyB)

Literature Cited

I. IUPAC Report on Nomenclature in the Field of Macromolecules, <u>J. Polym. Sci.</u>,1952, <u>8</u> 257.

2. A Structure-Based Nomenclature for Linear Polymers, <u>Macromolecules</u>, 1968, <u>1</u>, 193.

3. IUPAC, <u>Pure Appl. Chem.</u>, 1976, <u>48</u>, 373; 1985, <u>57</u>, 149, 1427; 1993, <u>65</u>, 1561; 1994, <u>66</u>, 873; 1997, <u>69</u>, 2511, 1998, <u>70</u>, 701.

4. IUPAC, "Compendium of Macromolecular Nomenclature", W. V. Metanomski, Ed., Blackwell, Oxford, 1991.

5. Carraher, C. E., Hess, G., Sperling, L. H., <u>J. Chem. Ed.</u>, 1987, <u>64</u>, 36.

6. Bikales, N. M., "Encyclopedia of Polymer Science and Engineering", 2nd Ed., Vol. 10, 191, Wiley, NY, 1987.

7. Polymer Prepr., 1991, <u>32(1)</u>, 655; 1992, <u>33(2)</u>, 6; 1993, <u>34(1)</u>, 6; 1993, <u>34(2)</u>,6; 1994, <u>35(1)</u>6; 1995, <u>36(1)</u>, 6; 1995, <u>36(2)</u>, 6; 1996, <u>37(1)</u>, 6; 1998, <u>39(1)</u>, 9; 1998, <u>39(2)</u>, 6; 1999, <u>40(1)</u>, in press.

8. Polym. Mater. Sci. Eng., 1993, <u>68</u>, 341; 1993, <u>69</u>, 575; 1995, <u>72</u>, 612; 1996, <u>74</u>, 445; 1998, <u>78</u>, Back Page; 1998, <u>79</u>, Back Page; 1999, <u>80</u>, Back Page.

POLYMER EDUCATION

John Droske[a] , Charles E. Carraher, Jr.[b], and Ann Salamone[c]

a. University of Wisconsin-Stevens Point, Stevens Point, WI 54481

b. Florida Atlantic University, Boca Raton, FL 33431 and
Florida Center for Environmental Studies, Palm Beach Gardens, FL 33410

c. Enterprise Development Corporation, Palm Beach Gardens, FL 33410

Introduction
Current

Introduction

William Bailey's American Chemical Society presidential address challenged academic centers to come to grips with the lack of proper recognition of polymer science. This call had previously been given by other ACS Presidents, but with little success.   Dr. Bailey's call had two additional elements beyond prior calls. First, key appointments within the ACS-related academic structure were made. Second, the divisions of Polymer Chemistry and Organic Coatings and Plastics Chemistry supported the call. The initial meeting where the formulation of an overall plan occurred at the Atlantic City 1974 national ACS meeting. It was agreed that the emphasis should be on undergraduate education and that the elements of polymer science be integrated within the entire undergraduate education of chemistry students. Further, that polymers offered a much needed natural bridge between "the real world" and chemistry.

While the initial meeting was initiated by committees from the Division of Polymer Chemistry, the Division of Organic Coatings and Plastics Chemistry became the second major player allowing the formation of what is now called the Joint Polymer Education Committee or PolyEd. The Board of Organic Coatings and Plastics Chemistry embraced the idea that something had to be done to assist in the integration of polymers into the undergraduate education of chemistry majors. Lieng-Huang (Sam) Lee, Clara Craver, and George Brewer were early "champions" of PolyEd.

Significant events that occurred due to the formation of PolyEd include:

*Inclusion of the statement "In view of the current importance of inorganic chemistry, biochemistry, and polymer chemistry, advanced courses in these areas are especially recommended and students should be strongly encouraged to take one or more of them. Furthermore, the basic aspects of these three important areas should be included at some place in the core material." This statement appeared in the 1978 edition of **Undergraduate Professional Education in Chemistry: Criteria and Evaluation Procedures** , published by the ACS Committee on Professional Training, CPT-the committee that certifies the ACS undergraduate programs. Eli Pearce was a member of CPT and played an important role in this event.

*Committees were formed, under the direction of Charles Carraher, to develop topics (and associated material) that would be useful, appropriate, and applicable for introduction of polymer concepts and examples in each of the core courses-General Chemistry, Inorganic Chemistry, Biochemistry, Physical Chemistry, Analytical Chemistry and Chemical Engineering. These committees also developed guidelines as to the level and depth of coverage of these topics; created specific illustrations, and developed broad guidelines as to the proportion of time to be spent on polymer related topics and examples. These reports were published in the Journal of Chemical Education.

*Development of the Polymer Education Newsletter, PEN, under the direction of Charles Carraher and Guy Donaruma. PEN is published twice yearly and goes to every College within the USA and Canada.

*Establishment of the National Information Center for Polymer Education at the University of Wisconsin-Stevens Point under the direction of John Droske.

*Establishment of the first standardized ACS exam in 1978 under the direction of Charles Carraher. This exam has been translated into a number of languages and has been used extensively in the USA. It assisted in the standardization of introductory polymer courses.

*Formation of the Intersociety Polymer Education Council, IPEC. IPEC is a multisociety education effort emphasizing use of polymers in teaching pre-college science. This grouping was initiated by a number of individuals including Ann Salamone, Charles Carraher, John Droske and Vivian Malpass (SPE President, 1988)

*NSF funding for K-12 teacher science training under the direction of John Droske. This funding has allowed the creation of the IPEC's MaTR Institute that is housed at the UWSP.

Current

PolyEd and IPEC are dedicated to service and education and are supported by many volunteers with direct financial support by the ACS divisions of Polymeric Materials: Science and Engineering , PMSE, and Polymer Chemistry, PC. They are dedicated to the education of the general public, including students, with regard to the basic nature of polymer science as it underpins our daily living and helps our understanding of the world about us. In addition, polymer science is an important vehicle for enhancing the appreciation of and understanding of the role of science and technology in today's society.

PolyEd is involved in education at all levels (pre-school through post graduate) involving giant molecules-macromolecules-polymers. It has working relationships and cooperates with many education related groups including the Division of Chemical Education (ACS), Society Committee on Education (ACS; SOCED), Committee on Professional Training (ACS; CPT), American Institute of Chemical Engineering, Society of Plastics Engineers (SPE), National Science Teachers Association (NSTA) and with polymer education groups throughout the world.

PolyEd programs include :
*Education symposia at regional and national meetings
*Media-both production and location; "Polymers in Introductory Chemistry Courses: A Sourcebook" is one such project
*Web Site-http://chemdept.uwsp.edu/polyed; This Web site acts as a holder for material developed by PolyEd and IPEC as well as a locator for other material appropriate for use in teaching
*Award for Excellence in Polymer Education-supported by the Dow Company foundation, Inc. This program encourages and gives special recognition to those teachers who are pioneering the teaching of polymers at the pre-college level. Areas included in selecting those receiving the awards include use of polymers in the classroom, developing novel approaches to the teaching of polymers, influencing other teachers, and educating the general public.
*Unilever Award for Outstanding Graduate Research-recognizes contributions to thesis research considering the innovation and impact of the research on the science of polymers and biopolymers.
*Catalogue of short courses
*Undergraduate Summer Scholarship Program
*Textbook Author Committee-encourages, alerts and supplies information to authors of textbooks including potential authors, editors and publishers of chemistry and chemical engineering textbooks encouraging them to integrate polymer topics in their text. Also develops and works with authors, editors and publishers in developing polymer-related materials.
*Outstanding Organic Chemistry Award-recognizes the outstanding organic chemistry students in over 300 schools.
*Visitation Program-members of PolyEd visit college and university campuses helping them to develop courses in polymers and assisting them in the integration of polymer studies into their curriculum.
*Polymer Curriculum Development Award-allows PolyEd to develop materials in a number of areas including computerized polymer simulations and special laboratory programs. One such supported program is Macrogalleria which features a "shopping mall" where students of all ages can learn about polymers by visiting different "stores". Macrogalleria was developed by the University of Southern Mississippi.
*Special minority programs
*Industrial teachers-locates industrial scientists that are willing to teach polymer courses or present polymer topics at the college level. This effort also seeks to "connect" the teachers to schools requesting the service. A related effort is underway to identify industrial sites that are willing to give tours to local K-12 and college level groups.
*Assist in developing, with other organizations, polymer demonstration kits
*Survey of polymer-intense courses and programs
*National Chemistry Week assistance

The Intersociety Polymer Education Council, IPEC, is a non-profit corporation funded by the Society of Plastics Engineers, the Society of the Plastics Industry, American Plastics Council, the Federation of Societies for Coatings Technology, and the three polymer-related ACS Divisions: PMSE, POLY, and Rubber. IPEC's mission is to significantly increase student interest and participation in science and technology subjects by incorporating the teaching of polymers and polymeric materials into K-12 grade curricula by utilizing the combined resources and infrastructures of the participating scientific societies. IPEC supports three programs-

> *Polymer Ambassadors,
> *MaTR Institute, and the
> *Plastics Processing Workshop

The Polymer Ambassador program is "teachers teaching teachers." Each of the current Polymer Ambassadors is an award winning pre-college teacher recognized nationally for their educational contributions in designing, developing, and implementing polymers and activities for K-12 teachers and their students. The Polymer Ambassadors provide other teachers with accurate, easily understood information and materials that can be directly incorporated into existing courses. The purpose of the Polymer Ambassador program is to assist K-12 teachers in linking science and technology subjects with the "real" world by helping them become more familiar with polymer topics that can be used in their classrooms. Students encounter polymers in all facets of their daily lives. K-12 science education can be improved by providing strong visual and tactile examples of science utilizing polymer and plastic examples.

The MaTR Institute and the Plastics Processing workshop are both intensive training courses for current and future Polymer Ambassadors.

More than 10,000 teachers have been directly affected by IPEC's programs since IPEC's inception in 1990. This statistic translates to more than 50,000 teachers being indirectly "in-serviced" and more than 3,000,000 students benefiting from the program.

Section 2

POLYMER SCIENCE AND TECHNOLOGY
Section Editor: David J. Lohse

POLYMER CHAIN CONFIGURATIONS: MEASUREMENT AND APPLICATIONS

LEWIS J. FETTERS

Corporate Research Labs, Exxon Research & Engineering Co., Rte. 22 East, Annandale, NJ 08801-0998

Outline
Introduction
Unperturbed Chain Dimensions
The Packing Length Influence in Linear Polymer Melts
Applications

Outline

A primary goal of polymer science has been to relate macromolecular structure to macroscopic properties. In particular, it has been hoped that the sizes of polymer coils could be related to the degree with which they entangle, and hence, to their viscoelasticity in the melt. This notion has been realized via the use of the concept of the packing length (p) which can be defined as the volume of a chain divided by its root mean square end to end distance. This has led to the development of simple correlations between such properties as the chain dimension ($<R^2>_0$), density (ρ) and plateau modulus (G_N^0). The interplay of these observable parameters leads to $G_N^0 \propto Tp^{-3}$ and $M_e \propto \rho p^3$ where M_e denotes the entanglement molecular weight. These relations seem to be universal for Gaussian chains in the melt-state and can be extended to include M_c, which marks the onset of entanglement effects, and M_r, the crossover to the reptation form. These expressions are useful for their predictive powers. A practical example in the design of adhesives is given.

Introduction

In 1952 Bueche (1) proposed that the melt viscosity of short polymer chains (pre-entanglement regime) could be expressed as follows:

$$\eta = \left[\rho N_a \left\langle R^2 \right\rangle_0 / 36M \right] N f_o \qquad (1)$$

Here, the observable parameters are the chain density (ρ), the Avogadro number (N_a), the chain molecular weight M, the unperturbed chain dimension ($<R^2>_0$) expressed in terms of the root-mean-square end-to-end distance, N, the number of backbone atoms and f_0 the molecular friction factor. Fox and Allen (2) demonstrated the validity of Eq. 1 in 1964.

Bueche (1) also showed that the combination of the melt viscosity and the diffusion constant led to the now well-known relation:

$$D\eta = \left[\rho N_a \langle R^2 \rangle_0 / 36M\right]kT \qquad (2)$$

Thus, Bueche's work was the first to show that density and the unperturbed chain dimension were directly relatable to polymer melt rheological properties.

In 1969 Flory (3) offered the following observation: "Comprehension of the configurational statistics of chain molecules is indispensable for a rational interpretation and understanding of their properties". The verity of his statement has become more apparent, in the intervening years, with the realization that these chain dimensions were related to the plateau modulus, G_N^0, of a polymer. Empirically, it was recognized that G_N^0 values decrease as chain 'thickness' increased. For example (4), polyethylene shows G_N^0 of ~2.5 MPa, polyisobutylene the value of 0.32 MPa, polystyrene the value of 0.20 MPa and poly(cyclohexyl ethylene) the value of ~0.07 MPa over the temperature range of 413 to 433 K. This decrease in G_N^0 is accompanied by a corresponding decrease (ca. four-fold) in the respective unperturbed chain dimensions (4).

The first rigorous attempt to relate Gaussian chain dimensions to G_N^0 was that of Graessely and Edwards (5). Shortly thereafter Ronca (6) presented the basics of what now is referred to as the packing model. Later contributions to the packing model were made by Lin (7) and Noolandi and Kavassalis (8,9). At the same time Witten, Milner, and Wang (10) introduced the concept of the packing length (p), which is defined as the chain volume divided by the unperturbed mean square chain end-to-end distance. Thus:

$$p = \frac{M}{<R^2>_0 \, \rho N_a} = \frac{V}{<R^2>_0} \quad (\text{Å}) \qquad (3)$$

where $M/\rho N_a$ is the inverse of the number of molecules per unit volume. It is apparent that the Bueche equations, (1) and (2), can be recast in terms of the packing length.

The pivotal role in melt rheological behavior of the packing length is discussed below. The importance of these findings is to illustrate the overriding importance of the unperturbed chain dimension in the physical characteristics of melt polymer chains. This article provides information upon the acquisition of this parameter and recent developments (4,11,12) in its use in predicting the viscoelastic parameters of G_N^0, M_e (the entanglement molecular weight), M_c (the critical molecular weight), M_r (the reptation molecular weight) and the entanglement length. M_c and M_r are the molecular weights observed, respectively, at the point where the melt Newtonian viscosity enters the regime where viscosity commences to scale with $M^{3.4}$ and at which the crossover to the reptation gradient of 3 occurs.

Unperturbed Chain Dimensions

Historically (13), the measure of an unperturbed chain dimension has been carried out in dilute solution where the technique of assay has been via dilute solution viscosity measurements or light scattering. The primary proviso for these evaluations is that the solvent used provides an environment where excluded volume effects are totally screened out at a certain temperature. This is the state that is referred to as the theta condition and at which the second virial coefficient is zero. For a given solvent-polymer pair this state exists at a single temperature. A parallel parameter of interest has been the chain dimension temperature coefficient (14): $\kappa = d(\ln <R^2>_0)/dT$. Experimentally, the values found embrace the range of negative to positive. Again, the basic experimental protocols are the same as those involved in the chain dimension evaluation with the added need of finding a series of theta solvents over as broad a temperature range as possible. Such an approach requires the needed assumption that the theta solvents used do not in their own right influence (via specific solvent effects) the necessary unperturbed chain posture. As will be seen later this assumption can be invalid. A second approach to the measurement of κ involves the thermoelastic approach (15,16), a concept that allows the evaluation of κ for chains in the crosslinked melt-state. This involves the thermodynamic quantity, f_e, the energetic component of the total elastic force, f, and temperature, T. This in turn leads to the relation (15,16)

$$\kappa = T f_e / f \qquad (4)$$

The evaluation of κ in this fashion has the advantage of being done in the bulk state. That is the natural environment of the chain and thus dispenses with the possibility of distortions due to specific solvent effects and the necessity of finding a suitable series of theta solvents useable over a convenient temperature range. However, the absolute value of $<R^2>_0/M$ is not obtainable by this technique.

The measurement of $<R^2>_0/M$ in dilute solution has for the most part used intrinsic viscosity measurements as the analytical tool of choice. Relative to light scattering the viscosity method is simpler and requires relatively inexpensive equipment. However, the method is not an absolute one (3). This is shown in the following:

$$[\eta]_\theta = K_\theta M^{1/2} = \Phi\left[<R^2>_0/M\right]^{3/2} M^{1/2} \qquad (5)$$

$$\left[<R^2>_0/M\right] = (K_\theta/\Phi)^{2/3} \qquad (6)$$

The observable chain dimension parameter, $<R^2>_0/M$, is obtainable from light scattering which directly yields the mean square radius of gyration, $<R_g^2>_0$, $(<R^2>_0 = 6<R_g^2>_0)$ via Zimm's classic evaluation (17). In this fashion, and coupled with intrinsic viscosity measurements, the Flory hydrodynamic constant (13), Φ, has been evaluated to be $2.5(\pm 0.1) \cdot 10^{-3}$ for $<R_g^2>_0$ in Å and $[\eta]$ in dL g^{-1} units.

The introduction of small angle neutron scattering as a practical tool for the measurement of polymer chain dimensions occurred (18) in 1972. This provided the route through which single chain properties could be accessed in the bulk state. This is due to a gift from nature in that the scattering lengths of hydrogen and deuterium are different thus causing the needed contrast. The generic scattering equation is as follows:

$$\frac{K_c}{I} = \frac{1}{M}\left[1 + \frac{1}{3}R_g^2 Q^2\right] + 2A_2 c \qquad (7)$$

Here, c is the polymer concentration, I is the measured intensity and Q is the wave vector which is defined as:

$$Q = \frac{4\pi}{\lambda}\sin\left(\frac{\theta}{2}\right) \qquad (8)$$

The measurement angle is denoted by θ and λ represents the wavelength. The second virial coefficient, A_2, is zero both in the theta state and in the melt

The advent of SANS as an analytical tool for unperturbed chain dimensions has revealed that differences can exist between the dilute solution theta-state value and its melt-state counterpart. A flavor of this is given in Table I where the chain dimension is expressed in the dimensionless terms of Flory's characteristic ratio (3):

$$C_\infty = \left[m_b \langle R^2 \rangle_0 / (M l_0^2)\right] \qquad (9)$$

where m_b is the average molecular weight per backbone bond and l_0^2 is the mean square bond length.

In some cases the data (19-29) in Table I show good agreement between the theta condition environment and that of the melt-state. Such agreement is exemplified by PIB. Contrary behavior, though, is evident for a-PP, i-PP. a-PEE, a-PMMA and a-PS. This is highlighted for a-PEE where the theta condition based values of κ are markedly negative. That behavior is seen to hold over the rather extensive temperature range of 227 to 414 K. In contrast the thermoelastic and SANS melt data yield positive values of κ as does SANS scattering on three different theta solvents (283 to 353 K). This contradictory behavior demonstrates that, relative to melt state-based data, the dilute solution viscosity theta-state approach can yield highly misleading results. Bianchi (30) reached this conclusion in 1963 when the database was smaller than currently exists. Time's passage has fortified his assessment. This state of play regarding κ is, in varying degrees, also seen for a-PP, i-PP, a-PMMA, and a-PS. To date the only available rationalization for this behavior involves specific solvent effects, which are generally viewed as influencing the relative populations of a chain's available conformer content (31,32). When this occurs the measured chain dimension is no longer unperturbed insofar as comparison with melt-state is concerned. In contrast, the agreement between the SANS and thermoelastic based κ values has been, without exception, excellent.

Table I. A Comparison of Theta and Melt Condition Single Chain Properties

POLYMER	C_∞		TEMP[a]	κ (10^{-3} DEG^{-1})		REFS.
	THETA	SANS	(K)	THETA	SANS	
PE	6.9-7.7	7.7	413	-1.1	-1.2	3,4,11,19
alt-PEP	6.2	6.9	298	-1.0	-1.1	4,20,21
a-PP	5.9	6.0	311	-1.4 to -2.9	0	4,20,21
i-PP	4.7 to 6.4	6.2	397 - 456	-1 to -4	0	11,20
a-PEE	5.3 to 7.1	5.2	298	-0.1 to -2.3	0.4 $(0.5)^{[b]}$	4,11,22-24
a-PMMA	7.3 to 11.1	9.1	413	-4 t0 2	0.1$(0.1)^{[b]}$	4,11,26
i-PMMA	11.5	10.9	413	---	---	27,28
a-PS	10.8	9.6	413	0.5 to -2.2	0	11,29
i-PS	17.3	9.3	523	---	~0	11,29
PIB	6.9	6.8	298	-0.2	$(-0.2)^{[b]}$	3,4,23
1,4-PBd	5.1	5.5	298	---	$(\sim0)^{[c]}$	4,30
1,4-PI	5.0	4.6	298	---	0.4$(\sim0.4)^{[b]}$	4,31

a Temp. of characteristic ratio measurements. b Thermoelastic results.
c Athermal conditions

Another facet of interest in Table I is the behavioral commonality of atactic and isotactic structures as seen for polystyrene and polypropylene. Identical behavior has been observed for poly(hexadecene) (33). The melt viscosity behavior (34) and the SANS-based (11) $<R^2>_0/M$ for atactic and isotactic polypropylene demonstrate that no significant differences exist between the atactic and isotactic formats of polyolefins. This behavior is in stark contrast to the behavior predicted by rotational isomeric state (RIS) calculations (13,27). The RIS approach consistently advocates notable differences in C_∞ and κ for these two tactic formats. Conversely, the RIS approach of Vacatello and Flory (35) successfully handles the atactic and isotactic forms of PMMA insofar as C_∞ is concerned.

The Packing Length Influence in Linear Polymer Melts

It has been shown (4,11) that the entanglement molecular weight for a polymer melt, M_e, is related by a power law to p, the packing length of the polymer species. Thus, the species dependence of entanglement molecular weight can be expressed by what appears to be a universal power law (for flexible linear Gaussian chains) in terms of the species packing length:

$$M_e = \rho RT / G_N^0 = n_t^2 N_a \rho \, p^3 \tag{10}$$

The coefficient n_t of eqn 10 is taken (7,11) as insensitive to temperature and equal to 21.3(±7.5%). It is dimensionless and denotes the number of entanglement strands present per cube of the tube diameter. Figure 1 displays a large data set where $M_e \rho^{-1}$ is plotted vs. packing length. The polymer acronyms are identified in the Appendices of refs.4, 11 and 12. Within experimental uncertainties the gradient for this data set is three which agrees with the prediction of the packing model (6).

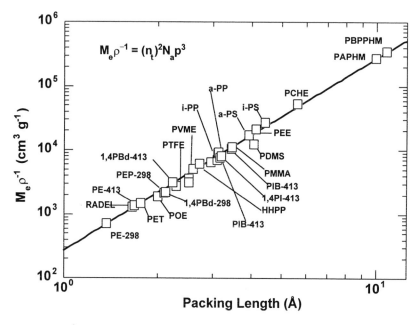

Figure 1. $M_e \rho^{-1}$ vs. p for assorted polymers.

As a supplement to Eq (10) it was found (12) that power laws also describe the molecular weights characterizing the melt viscosity M_c and M_r. We were able to identify 13 polymer species with well established values of M_c and:

$$M_c = n_t^2 N_a p^3 \left[\frac{p^*}{p} \right]^{0.65} = n_t^2 N_a p^{2.35} [p^*]^{0.65} \tag{11}$$

where $p^* = 9.2$ Å. Note that $M_e = M_c$ at $p = p^*$. The M_c/M_e ratio can thus be expressed as:

$$\frac{M_c}{M_e} = 4.24p^{-0.65} = \left[\frac{p^*}{p}\right]^{0.65} \tag{12}$$

Equations (8) and (9) demonstrate the packing length exponents for M_e and M_c, differ significantly. Thus, the notion that $M_c/M_e \approx 2$ for all species is incorrect. This is shown in Figure 2.

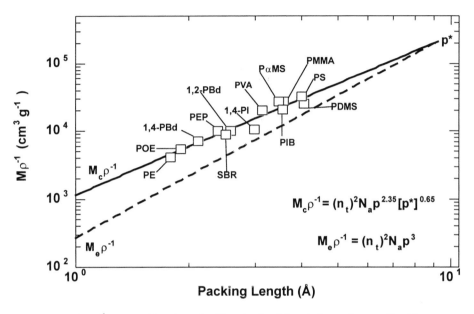

Figure 2. Plot of $M\rho^{-1}$ vs. packing length. The dashed line is based upon Eq 10.

The remaining molecular weight is the reptation value, which is that deemed to represent the crossover from the 3.4 power to that of 3, the event which signals the onset of pure reptation. A formula for estimating M_r can be obtained as follows. The observed viscosity behavior above M_c can be written as:

$$\eta_0(M) = \left[\frac{M}{M_c}\right]^{3.4} \eta_0(M_c) \tag{13}$$

while the prediction for pure reptation is:

$$\eta_0(M) = Q \frac{M^3}{M_e^2 M_c} \eta_0(M_c) \tag{14}$$

where M_e is defined in Eq (10), and Q is equal to 15/4. By one estimate (37), constraint release reduces Q to a value of about $0.3(15/4) = 1.13$. The viscosities denoted by Eq (12) and (13) become equal at $M = M_r$. By setting the viscosities from Eqs (12) and (13) equal we find that for $Q = 1.13$:

$$\frac{M_r}{M_e} = 1.36 \left[\frac{M_c}{M_e} \right]^{6.0}$$

(15)

or, on making use of Eq (11):

$$M_r = n_t^2 N_a \rho p^{3.0} \left[\frac{1.08p^*}{p} \right]^{4.0} = M_e \left[\frac{1.08p^*}{p} \right]^{4.0}$$

(16)

Thus, Eqs (10), (11) and (16) are governed by the same dimensionless coefficient, n_t^2.

An additional point to make is that for η_0, in the molecular weight regime of M^3, the constraint release correction becomes trivial. From Eqs (10), (11) and (16), within the uncertainties of the data, it seems that M_e, M_c and M_r converge toward the same value when the packing length for the species approaches 9-10 Å (Figure 3). Thus:

$$M_r \cong M_e \left[\frac{p^*}{p} \right]^{4.0}$$

(17)

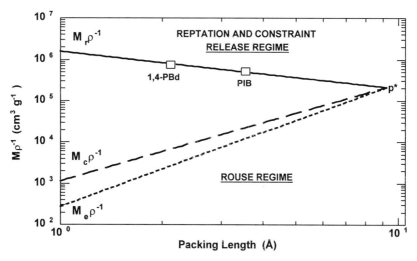

Figure 3. Plots of $M\rho^{-1}$ as a function of packing length. The lines are calculated via Eq 10, 11 and 17. The symbols denote the experimental $M_r\rho^{-1}$ values (12) for 1,4-PBd and PIB.

The M_r data of Table II seem to demonstrate that pure reptation behavior in melt polymer systems is obtainable between $M\rho^{-1}$ of ~5 $\cdot 10^6$ to ~10^6. This is in contrast to earlier notions (38) that estimated M_r to be about 800 times larger than the corresponding M_c for all polymer systems. However, the M_r/M_c ratio seemingly ranges from ~250 (PE) to ~14 (PS) for polymers over the p range of 1.5 to 4.5 Å, a range which covers virtually all common polymers (see Figure 1). This ratio for poly(cyclohexyl ethylene) (11), where p is ~5.6 Å, is projected to be about 8.

Table II. Molecular Characteristics and Melt Based Rheological Molecular Weights Refs (4,11,12)

POLYMER	T (K)	ρ (g cm^{-3})	$<R^2>_o/M$ (Å^2)	p (Å)	$M_e \cdot 10^{-3}$ (g/mol)	$M_c \cdot 10^{-3}$ (g/mol)	$M_r \cdot 10^{-3}$ (g/mol)
PE	443	0.768	1.21	1.79	1.15	3.48	*680* [a]
POE	353	1.081	0.805	1.91	2.00	5.87	*920*
1.4-PBd	298	0.895	0.876	2.12	2.00	6.38	*~630* [b]*/610*
PEP	373	0.812	0.851	2.40	3.10	8.10	*590*
SBR	298	0.930	0.708	2.52	3.00	8.21	*460*
1,2-PBd	300	0.889	0.720	2.59	3.82	8.20	*540*
1,4-PI	243	0.950	0.583	3.00	6.40	10.0	*510*
PVA	428	1.08	0.490	3.14	9.10	24.5	*600*
PαMS	459	1.04	0.460	3.47	13.3	28.0	*600*
PIB	490	0.817	0.570	3.55	10.5	17.0	*~430/420*
PMMA	490	1.09	0.425	3.58	13.6	29.5	*540*
PS	490	0.959	0.434	3.99	18.1	31.2	*470*
PDMS	298	0.970	0.422	4.06	12.0	24.5	*290*

a *Italicized values* are calculated via Eq (17). b Measured, see ref. 12.

Table III summarizes the basic equations involving packing length and the rheological equations for G_N^0, M_e, M_c, M_r and the tube diameter (entanglement length), d_t. These expressions are simple scaling relations with the seeming power to predict basic melt viscoelastic properties. Furthermore, a new route to the melt chain dimension emerges when M_e and ρ are known.

Applications

The expressions in Table III constitute a rheological 'design' kit that permits the calculation of basic viscoelastic parameters when ρ and $<R^2>_o/M$ are in hand. This can clearly be useful in understanding and controlling the flow of polymer melts during various kinds of processing. Another area where this capacity can be exploited is adhesion. One of the recognized adhesive criteria is that of Dahlquist, which requires that G_N^0 be below 0.33 MPa at room temperature. This permits sufficient flow so that good contact can be achieved between surface and adhesive. The relations in Table III can be applied to find polymers that will pass the Dahlquist criterion for an adhesive. The tendency of polymers to fail by crazing has also been related to the entanglement density. The ideas briefly outlined here have an impact on a wide range of polymer properties.

Table III. Packing Length Based Rheological Equations

$G_N^o = k_b T / n_t^2 p^3$	
$M_e = n_t^2 N_a \rho p^3$	
$M_c = M_e [p^* / p]^{0.65}$	$M_r = M_e [p^* / p]^{3.9}$
$n_t = [M_e < R^2 >_o / M]^{0.5} p^{-1}$	
Entanglement length = $p\, n_t$	
$[< R^2 >_o / M] = [n_t / \rho N_a]^{2/3} M_e^{-1/3}$	

Conclusions

The reported behavior demonstrates the critical importance of the packing length in melt rheological behavior. It is now apparent that M_c scales with packing length in a different fashion than M_e such that M_c/M_e is not a constant but depends on p as given in Eq 11. M_r/M_c depends on the packing length and decreases with increasing p. The results further seemingly show that $M_e = M_c = M_r$ for packing lengths in the 10 Å range.

Literature Cited

1. Bueche, F. J. Chem. Phys., 1952, 20, 1959.
2. Fox, T. G.; Allen, V. R. J. Chem. Phys., 1964, 41, 344.
3. Flory, P. J. "Statistical Mechanics of Chain Polymers" Interscience: New York, 1969.
4. Fetters, L. J.; Lohse, D. J.; Richter, D.; Witten, T. A.; Zirkel, A. Macromolecules, 1994, 27, 4639.
5. Graessley, W. W.; Edwards, S. F. Polymer, 1981, 22, 1329.
6. Ronca, G. J. Chem. Phys., 1983, 79, 1031.
7. Lin, T. Macromolecules, 1987, 20, 3080.
8. Kavassalis, T. A.; Noolandi, J. Macromolecules, 1988, 21, 2869.
9. Kavassalis, T. A.: Noolandi, J. Phys. Rev. Lett., 1988, 59, 2674.
10. Witten, T. A.; Milner, S. T. and Wang, Z.-G. in "Multiphase Macromolecular Systems" Culbertson, B. M. Ed.; Plenum: New York, 1989.
11. Fetters, L. J.; Lohse, D. J.; Graessley, W. W. J. Polym. Sci., Polym. Phys. Ed., 1999, 3, 1023.
12. Fetters, L. J.; Lohse, D. J.; Graessley, W. W. Macromolecules, 1999, 32, in press.

13. Flory, P. J. "Principles of Polymer Chemistry"; Cornell University Press: Ithaca, NY, 1953.
14. Flory, P. J.; Ciferri, A.; Chiang, R. J. Am. Chem. Soc., 1961, 83, 1023.
15. Flory, P. J. Trans. Faraday Soc., 1961, 57, 829.
16. Ciferri, A. J. Polym. Sci. Part A, 1961, 2, 3089.
17. Zimm, B. H. J. Chem. Phys., 1948, 16, 157.
18. Kirste, R. G.; Kruse, W. A.; Schelten, J. Makromol. Chem., 1972, 162, 299.
19. Boothroyd, A. T.; Rennie, A. R.; Boothroyd, C. B. Eur. Phys. Lett., 1991, 15, 715.
20. Mays, J. W.; Fetters, L. J. Macromolecules, 1989, 22, 921.
21. Zirkel, A.; Urban, V.; Richter, D.; Fetters, L. J.; Huang, J. S.; Kampmann, R.; Hadjichristidis, N. Macromolecules, 1992, 25, 6148.
22. Zirkel, A.; Richter, D.; Fetters, L. J.; Schneider, D.; Graciano, V.; Hadjichristidis, N. Macromolecules, 1995, 28, 5262.
23. Mark, J. E. J. Polym. Sci., Macromolecular Rev., 1976, 11, 135.
24. Fetters, L. F.; Graessley, W. W.; Krishnamoorti, R.; Lohse, D. J. Macromolecules, 1997, 30, 4973.
25. Boothroyd, A. T.; Rennie, A. R.; Wignall, G. D. J. Chem. Phys., 1993, 99, 9135.
26. Mays, J. W.; Hadjichristidis, N. Rev. Macromol. Chem.-Phys., 1998, C28, 371.
27. O'Reilly, J. M.; Teegarden, D. M.; Wignall. G. Macromolecules, 1985, 18, 2747.
28. Hayward, R. C.; Graessley, W. W. Macromolecules, 1999, 32, 3502.
29. Mattice, W. L.; Sutter, U. W. "Conformational Theory of Large Molecules" John Wiley and Sons: New York, 1994.
30. Mays, J. W.; Hadjichristidis, N.; Graessley, W. W.; Fetters, L. J. J. Polym. Sci. Part B, Polym. Phys., 1986, 24, 2553.
31. Fetters, L. J.; Krishnamoorti, R.; Zirkel, A.; Richter, D. in preparation.
32. Bianchi, U. J. Polym. Sci., Part A, 1964, 2, 3083.
33. Lifson, S.; Oppenheim, I. J. Chem. Phys., 1960, 33, 109.
34. Bahar, I.; Baysal, B. M. Macromolecules, 1986, 19, 1703.
35. Pena, B.; Aroca, M.; Perez, E.; Bello, A.; Riande, E.; Benavente, R. Macromol. Chem., 1997, 198, 1691.
36. Younghouse, L. B.; Fetters, L. J.; Pearson, D. S.; Mays, J. W. Macromolecules, 1988, 21, 478.
37. Graessley, W. W. J. Polym. Sci., Polym. Phys. Ed., 1980, 18, 227.
38. Doi, M.; Edwards, S. F. "The Theory of Polymer Dynamics", Clarendon Press: Oxford 1986, p. 234.

POLYOLEFINS

DAVID J. LOHSE

Corporate Research Labs, Exxon Research & Engineering Co., Rte. 22 East, Annandale, NJ 08801-0998

Introduction
General Description of Polyolefins
Polyethylene
Polypropylene
Ethylene/Propylene Elastomers
Isobutylene-based Polymers
Other Polyolefins
Polyolefin Blends

Introduction

Polyolefins are by far the largest class of synthetic polymer made and used today. There are several reasons for this, such as low cost of production, light weight, and high chemical resistance. A wide range of mechanical properties is possible through the use of copolymerization, blending, and additives to make products from elastomers to thermoplastics to high strength fibers. Although they were first produced in the 1930's, important advances are still being made in improving the process and performance of these materials, and their use is growing at a rate well above the GDP. Several factors have been principally responsible for the great success that polyolefins have enjoyed: an abundant supply of cheap and simple monomers; advances in reactor engineering and catalysis; and the ability to compound these polymers with fillers and other polymers. The combination of all of these factors has led to the enormous number of ways in which polyolefins are now being used to improve our lives. The field of polyolefins is now so vast that there is no way to do it justice in the space we have here. In this chapter the properties, history, and future of these remarkable materials are briefly summarized and a set of references for further information is given.

General Description of Polyolefins

In general polyolefins are defined as polymers made from olefins, which are principally ethylene and propylene, but also 1-butene, 1-hexene, 1-octene, isobutylene, and other monomers. (By this definition, one could also include polymers made from styrene as polyolefins, but for the purposes of this book the large family of styrenic polymers is discussed in the next chapter.) These are thus fully saturated hydrocarbon molecules (much less than one double bond per molecule), which is the basis of their properties (Table 1). The density of all amorphous polyolefins (at 25°C) is about 0.855 g/cc (the exception being polyisobutylene at 0.917 g/cc), and the density of a crystalline phase is never more than 1.00 g/cc. This means that a polyolefin will nearly always be lighter in weight for a given application than any alternate material. Because they are fully saturated, their interactions are completely by dispersive, van der Waals forces. As a result, they have a high degree of chemical resistance to many of the solvents and liquids encountered in use. The lack of double bonds means that they are low in reactivity, and so are highly stable to oxidation. These general properties explain much of the usefulness of polyolefins.

Chemical Structure	Property	Advantages	Disadvantages
Fully saturated	Low chemical reactivity	Oxidative stability	Difficult to functionalize
van der Waals interactions	Low crystallization temperature	Easy processing; Easy to recycle	Low use temperatures
Hydrocarbon	Low density	Light weight parts	Flammable
Made from olefins	Wide availability; Polymerizes by several means	Low cost; High volumes	Fluctuations in price

Table 1. Basic properties of polyolefins.

The weak van der Waals forces between polyolefin molecules result in lower melting and crystallization temperatures than for polymers with stronger interactions such as the hydrogen bonding in polyamides. This makes them inappropriate for high temperature applications, but also means that they are easier to process because of the lower melting point.

Other features of polyolefins are related to the means of synthesizing them. The main monomers used, ethylene and propylene, are produced by cracking of petroleum feeds or the dehydrogenation of alkanes. For this reason they are available in large quantities at low cost. There has been a continual evolution in methods to polymerize olefins over the past sixty years, which is expanding the range of properties offered by these versatile materials. One of the main parameters that differentiates the various olefins is crystallinity, which is based on the regularity of the chemical structure of the polyolefin chains. Figure 1 shows how density, a reflection of the crystallinity, varies for ethylene-propylene copolymers between the polyethylene and polypropylene homopolymers. There are many different products in this range as can be seen, from high strength fibers to soft elastomers. As catalyst and process technology advance, the boundaries between these product categories are beginning to fade. In this chapter I hope to give the reader at least a glimpse of the remarkable breadth of the properties and utility of these materials.

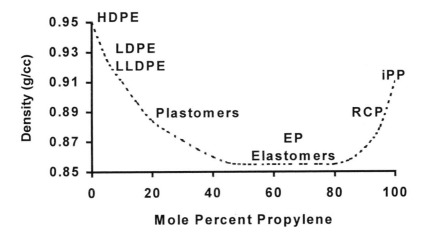

Figure 1. Dependence of density (crystallinity) on propylene content of ethylene-propylene copolymers. The labels on the graph show the major polyolefin product categories.

Polyethylene (PE)

This section might also be labeled 'polyethylenes', as there is now a broad range of ethylene homo- and copolymers that fit under this category. By convention these include any polymer for which ethylene is the main comonomer and that has some degree of crystallinity at room temperature. Here I will discuss the various types of PE in the order in which they became commercially available.

Low Density Polyethylene (LDPE). This first form of PE to become commercially viable was discovered by Fawcett *et al.* at ICI in 1933 (1). Although this was an unintentional consequence of research on general chemical reactions at high pressure, the value of the new material was quickly realized and exploited as, for example, cable insulation for radar in World War II. In the sixty years of its history, its use has grown until its production is now one of the largest of any material in the world.

The basic process is the free radical polymerization of supercritical ethylene (2). Typically this means that the reactions are done at high pressure (over 150 MPa) and temperatures of 150 to 350 °C. Some sort of free radical initiator is needed for these reactions; oxygen or peroxides are commonly employed. Two main types of reactors are used – autoclaves and tubular. The high pressure process is used to make not only ethylene homopolymers, but also copolymers, especially with 'polar' comonomers such as vinyl acetate and methacrylic acid. These higher value materials represent an increasingly larger fraction of the production of these plants. The total capacity worldwide is now more than 30 billion pounds (see Table 2). Even though the basic technology is sixty years old and there have been many developments and even 'revolutions' in PE production that have led to competitors for LDPE (see below), these products continue to grow, and new plants continue to be built.

The polymers made in this process are characterized as having a highly branched structure. This is believed to come about from two main processes (3). The free radical on the growing end of a chain can loop back to some other portion of the chain. The loop breaks at the point of reaction, to which the radical is transferred. The loop then becomes a branch off of the chain, which continues to grow from the reaction point.

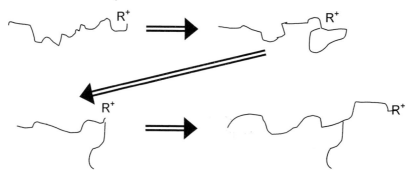

Figure 2. Backbiting Reaction

This backbiting reaction mainly leads to short branches a few repeat units long. Much longer branches are produced from the second mechanism, in which the growing end of a chain terminates on another molecule:

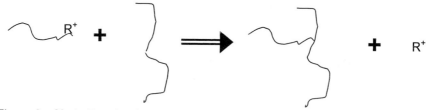

Figure 3. Chain Termination Reaction

Several other mechanisms have also been proposed, and the complete suite of reactions that occur has not been firmly elucidated. It is clear though, that these result in a very complicated, tree-like architecture with both short and long chain branches.

These are called 'low-density' polyethylenes in contrast to the higher density, linear versions that were first produced somewhat later (below). The reason that the products of the high pressure process have lower density is that they are lower in crystallinity. Only long methylene sequences can participate in the paraffin-like crystals of PE, so the side branches serve to lower the crystallizability of LDPE. This effect is mainly due to the short branches (*e. g.,* ethyl, butyl, hexyl) that arise from the backbiting mechanism, simply because there are many more of these. Since the frequency of such branching can be controlled by various process variables (*T, P,* initiators), so can the density or crystallinity of the polymers.

LDPE finds application in a wide variety of markets (see Fig. 4), but its main application is in film. The mechanical properties of the resulting films can be varied by controlling the density and molecular weight of the polymer, but in general they are not as tough as those made from linear polyethylenes without long side branches. On the other hand, these long branches are responsible for the superior ease of processing of LDPE in comparison to its linear counterparts, and this factor has kept it as an important product to this day.

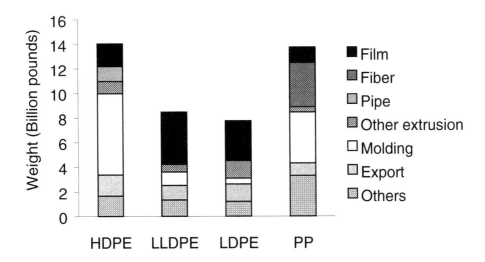

Figure 4. US markets for polyolefin plastics in 1998. (From *Modern Plastics*, January 1999, pp. 72-73.)

Among the ethylene copolymers made by the high pressure process, poly(ethylene – *co* - vinyl acetate) (EVA) is the one most commonly produced (4).

These are made with ~ 3 to 25 % VA at a variety of molecular weights and used in a wide range of applications. Film is one of the largest uses, particularly as a sealing layer. EVA is also important in adhesives, as a diesel fuel additive, and as a modifier for asphalt, among many others. Ethyl acrylate and methyl acrylate copolymers find their uses in such items as disposable gloves, hoses, and gaskets. Acrylic acid and methacrylic acid copolymers exhibit stronger interactions and are used in extrusion coating of metals, wire and cable applications, and various film uses. These copolymers can be modified by reactions with metal salts to produce ionomers, which have even stronger interactions and which are mainly used in heat seal layers for film and as protective coatings for golf balls and the like.

Table 2 shows the major producers of HP-LDPE in the world today. Although this is a mature technology, new plants continue to be built, especially in Asia where most film manufacturers have low-powered equipment that requires the processability of LDPE. It is likely to be an important material for many years to come.

Producer	HDPE	LLDPE	LDPE	Total
Exxon	1270	870	953	3093
Dow	260	1765	937	2962
Equistar	1500	500	823	2823
Borealis	815	150	975	1940
Union Carbide	820	800	227	1847
Polimeri Europa	300	380	655	1335
Solvay	1270	-	-	1270
Chevron	650	200	357	1207
Nova	325	760	120	1205
Elenac	400	-	720	1120
Others	14,675	8,810	11,935	35,420
Total	22,285	14,235	17,702	54,222

Table 2. Worldwide capacity for polyethylene production in thousand metric tons per year. (From *Chemical Week*, 23 September 1998, pg. 78 and 26 May 1999, pg. 44). Ten largest producers listed.

High Density Polyethylene (HDPE). The next major advance in polyethylene production was the discovery of catalysts that permit the synthesis of linear versions of polymethylene by Ziegler in 1953 (5,6). The new linear polymers were more crystalline than LDPE (hence the name 'high density') and thus had a very different set of mechanical properties. It has become one of the largest volume materials around the world.

Although HDPE can be made in a homogeneous catalytic process such as bulk or solution, the great majority is made with supported, heterogeneous catalysts. This is done mainly to be able to separate the catalyst from the polymer more easily. The catalysts are typically titanium chlorides supported on silica, and these are used in slurry or gas-phase reactors. The activity of these catalysts has increased greatly over the years, so that currently rates of 1,000,000 g of PE per g of catalyst or more can be achieved. Many compounds that catalyze the polymerization of ethylene have been found (7). A list of the largest current producers of HDPE can be found in Table 2. Note that the ten largest producers account for only about one third of the capacity worldwide.

The molecular weight of the HDPE is controlled by the addition of chain transfer agents, principally hydrogen. As a result, a wide range of molecular weights is produced. The density of these products varies slightly with molecular weight (being somewhat less for very high molecular weights), but the main way that density is controlled is by adding small amounts of a comonomer such as butene. This breaks up the methylene sequences and so reduces crystallinity. So modern HDPE plants have a great deal of flexibility in the kinds of products they can make.

Because of this, HDPE finds use in a great variety of applications as well. These are outlined in Fig. 4. Among the most commonly encountered in everyday life are molded containers such as milk bottles, pails, detergent bottles, stadium cups, and lubricating oil bottles. A large fraction of the piping used for such applications as water distribution and sewage is now made from HDPE. It is increasingly used for gasoline tanks on motor vehicles. There are many more exotic applications as well, such as medical prostheses and high strength fibers. The latter are made from HDPEs of extremely high molecular weight by several processes, and are used for sails and bulletproof vests. It is indeed one of the most versatile materials available.

Linear Low Density Polyethylene (LLDPE). As mentioned above, α-olefin comonomers can be used to reduce the density of PE. It was clear quite early in the history of Ziegler catalyzed PE that tougher (although less stiff) polymers could be produced by copolymerizing ethylene with sufficient amounts of butene or octene to lower density. However, since such monomers polymerize more slowly than ethylene, the commercial production of large amounts of linear versions of PE with low density (LLDPE) had to await the invention of high activity catalysts.

The first methods to produce LLDPE used a solution process. These were fairly expensive due to the need to remove the polymer from the solution. Karol and coworkers at Union Carbide (8) developed a cheaper process in the 1970's. This led to the Unipol process for LLDPE that is the most widely used today. This is a gas-phase process using a fluidized bed reactor to make LLDPE. Because of the high activity of the catalysts used, there is no need to remove it from the polymer, which reduces the process costs. Several other companies have also developed successful gas-phase technology to make LLDPE. One of the main limitations of these processes is that only propylene and butene can be used as comonomers when using Ziegler-Natta catalysts. This is because the low reactivity of such monomers as 1-octene means that a large amount of it must be kept volatile in the gas-phase, which makes energy costs too high to be feasible. The solution processes for LLDPE have the advantage of being able to use octene or other large α-olefins. Although the ethylene-octene copolymers are more expensive to produce than ethylene-butene ones, they also have superior mechanical properties, such as tear strength, when made into films (9). With either the solution or gas-phase process a wide range of molecular weights can be made through the use of transfer agents, and density can be controlled by the amount of comonomer incorporated. The main producers of LLDPE are shown in Table 2.

Just as for LDPE, the largest use of LLDPE is for film applications; see Fig. 4. The fact that LLDPE has no long branches gives it mechanical properties superior to LDPE (at the same density), principally in terms of toughness attributes such as puncture or tear resistance. However, the long branches of LDPE also give it superior processing characteristics. In many regions, especially in Asia, film processors have very low-powered extruders and need to use LDPE or perhaps blends of LDPE with LLDPE. As a result, LLDPE has not completely replaced LDPE as was expected upon its introduction on the 1970's.

Plastomers. This is a class of ethylene copolymers that stand between LLDPE and the totally amorphous elastomers, so they have been dubbed 'plastomers'. They have also been called 'very low density PE' or VLDPE. They have a large enough amount of comonomer so that there is only a very small amount of crystallinity left. The crystalline phase in these plastomers is so small that the domains cannot order into periodic lamellae, but more likely are arranged in something of a fringed micelle fashion. As would be expected, they are more elastic and less stiff than LLDPE. They have become much more important with the advent of metallocene catalysts, and find use as, among other uses, a sealing layer in film applications and controlled permeation packaging for fruits and vegetables. They are a good example of how the various categories of polyolefins are merging into one another, as well as the remarkable diversity of uses for polyolefins.

Impact of New Catalysts. Probably the most important developments in polyolefin technology in the last two decades have been improvements in catalysis (10). A huge number of new catalysts have been found that polymerize olefins. The most prominent of these are the metallocenes. These have been examined for many years, and even Ziegler tested their potential for polymerizing ethylene. The key advance came in 1978 when Sinn and Kaminsky discovered the efficacy of methyl alumoxanes as cocatalysts for zirconocenes (11). These had very high polymerization activities, but required excessively high levels of alumoxane to be commercially feasible. However, this led to renewed efforts in many industrial labs, leading to commercialization of metallocene-produced PE, first by Exxon in 1991, then by Dow in 1993, and now by many polyolefin producers (see Table 3). The main advantage that metallocenes bring is that they are single-site catalysts that allow much greater control of parameters such as molecular weight distribution than is afforded by traditional Ziegler catalysts. For polyethylenes, other benefits include greater ability to incorporate comonomer. This leads to new products such as ethylene-hexene copolymers made in gas-phase reactors and to more highly efficient processes. These catalysts also provide the ability to introduce long branches in otherwise linear polymers (12). Metallocenes have also been commercialized for the synthesis of polypropylene and ethylene-propylene elastomers (see below). Moreover, they have allowed the production of polymers that were not commercially feasible before, such as syndiotactic polystyrene and the so-called cyclic olefin copolymers. These are copolymers of ethylene and comonomers such as norbornene, which are under development as engineering plastics.

The success of metallocenes has led organometallic chemists to examine the catalytic activity of other compounds. Chemists at the University of North Carolina and duPont have found that nickel and palladium complexes with diimine ligands polymerize ethylene and other olefins (13). Interestingly, under certain conditions the homopolymers of ethylene can contain many short branches. On the other hand, homopolymers of an olefin such as 1-hexene can be made to contain fewer branches than the number of monomers, meaning that some of the monomers have been incorporated end-to-end. There have also been reports of olefin catalysis with other late-transition-metals such as aluminum (14). The success of such catalysts is significant as they are more tolerant of polar impurities that can poison polymerization, and in fact such compounds can copolymerize monomers such as acrylates. As more systems are examined it is likely that many more advances in polyolefins will be found.

Company	LLDPE	HDPE	PP
Borealis	√	√	
Dow	√	√	
Elenac	√		
Evolue	√		
Exxon	√	√	√
Fina		√	√
Japan Polychem	√		
Japan Polyolefin	√		
Phillips	√		
Targor			√
Ube	√		

Table 3. Producers of metallocene-catalyzed polyolefins in 1998. A check mark (√) indicates that the company has a commercial plant in operation for this polymer. (From *European Polymer News*, March 1999, p. 27.)

Polypropylene (PP)

The hallmarks of variety and versatility that I have just described for the forms of polyethylene apply equally to polypropylene. Although it has not been around for quite as long, having only become available in the late 1950's, PP is continuing to evolve and new forms and processes continue to be developed. PP products can be found as part of nearly every aspect of daily life, from fibers and fabrics for upholstery and carpeting to automotive parts, from diapers to recipe boxes. Here I can only give a brief overview of these materials; for a more complete story, see refs. 15 and 16.

Forms of PP. The main reason for the great versatility of propylene polymers is that their properties can be controlled in several different ways. Unlike PE, the crystallinity of PP is determined by the regularity of the placement of the monomers along the chain, that is, by the tacticity of the polypropylene. The first highly tactic forms of PP that were made by the group of Natta and at Phillips Petroleum in the 1950's were isotactic (6, 17), with the methyl groups predominantly to one side of the chain:

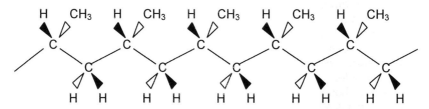

Fig. 5. Structure of isotactic polypropylene.

The first versions of isotactic polypropylene (iPP) were not highly isotactic, and contained many chain defects. This reduces the crystallinity because the average lengths of the isotactic sequences are reduced. By the late 1950's catalysts had been discovered that increased the degree of isotacticity to the point where commercially successful products could be made (6). These increases in crystallinity meant the corresponding enhancement of physical properties, such as increased stiffness, higher heat distortion temperature, and better clarity. Through the years the capability of catalysts has continued to increase until now better than 99% isotacticity can be achieved.

There are three important crystalline textures of iPP, the α, β, and γ forms. The α form is the most common, and the one thermodynamically stable at ambient conditions. Many nucleating agents are available to control the nature of α crystallinity. In order to get a large degree of β crystallinity, it is necessary to use nucleators for β. The γ form is only dominant under high pressure, but some fraction of it can occur along with α and β. The properties of the polymer can be controlled to some extent by manipulating the amount of each of these phases.

The other crystalline form of propylene homopolymers is the syndiotactic form, where the methyl groups are on alternate sides of the chain. While this was also synthesized early by Natta, it is only with the advent of metallocene catalysts that it has become commercial (18). Syndiotactic polypropylene (sPP) has a similar melting temperature to iPP, but a very different crystalline structure and a different balance of stiffness and toughness. A good review of the crystalline structures of both iPP and sPP has been written by Lotz, Wittmann, and Lovinger (19).

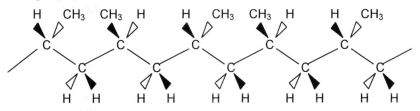

Figure 6. Structure of syndiotactic polypropylene.

Atactic or amorphous forms of PP (aPP) have also been useful products. At first these were simply the fraction of an iPP product that was extracted by a solvent such as heptane. As an inexpensive byproduct, these found use mostly due to their low cost in applications such as modification of asphalt for roofing and in adhesives. However, as iPP catalysts were perfected, less and less of aPP was available. There has been a small amount of deliberate synthesis of aPP to satisfy those markets where the heptane soluble aPP had become common, but this is a very small production.

Just as much work has gone into raising the crystallinity of PP, and so make it stiffer, there have been similar efforts to make it tougher. This is especially important for the low temperature uses of PP, since its T_g is 0°C. One way by which this is done is to copolymerize the iPP with a small amount of ethylene to reduce crystallinity. These are the so-called reactor copolymers (RCP). Another way that this was done is to blend it with rubber. As they became available, the elastomers of choice were ethylene-propylene copolymers. The last section discusses these in more detail. However, here I will point out that there has been a great deal of technology developed to make PP and EP in the same reactor train so that an in-situ blend can be made. A common form of these are the impact copolymer PP (ICP), which are blends with about 15 – 25% of EP copolymer.

Other forms of PP are becoming possible with the application of metallocene catalysts (10). I have already mentioned that these have made sPP commercially feasible. It is also now possible to produce copolymers of propylene with higher α-olefins that may increase the variety of propylene polymers available. Another intriguing development are the so-called elastomeric PPs (20), which are made from solely propylene monomer and have atactic blocks that alternate with isotactic ones, presumably due to fluctuations in the nature of the catalyst during the polymerization process. Such block copolymers also hold the promise to give thermoplastic elastomers from propylene homopolymers. There is clearly still much potential for polyolefins made with propylene.

Catalysts and Processes. The original catalysts that Natta used to make iPP had too low an activity and too low a degree of stereospecificity to be commercially viable. Within a few years, however, useful systems based on $TiCl_3$ and aluminum alkyl activators had been developed by several companies. Further advances in both the activity of the catalyst and the tacticity of the PP produced were made over the years by the addition of electron donors, and by supporting the titanium compounds on magnesium chloride. Activities as high as 100,000 g PP per g of catalyst can now be achieved, eliminating the need to remove the catalyst residues from the polymer. Moreover the degree of tacticity is now generally high enough that very little heptane soluble amorphous PP is made (16). Metallocenes have not yet made as large an impact on PP as they have on PE, although there is commercial production of PP by Exxon, Fina, and Targor. Perhaps the biggest impact of the new catalysts will be the novel forms of PP, such as sPP and the elastomeric PPs mentioned above (10).

Producer	PP
Montell	2938
Targor	1550
Fina	1375
Borealis	1255
BP Amoco	1235
Exxon	890
Solvay	751
Huntsman	745
Grand Polymer	731
DSM	730
Others	17,786
Total	29,986

Table 4. Worldwide capacity for polypropylene production in thousand metric tons per year. (From *Chemical Week*, 15 July 1998, pg. 38). Ten largest producers listed.

Essentially all commercial PP is made using supported catalysts. Much of this has been in a slurry process using an inert diluent like hexane, but most of the plants being constructed today use bulk propylene as the medium. This can be either liquid propylene, as in Montell's Spheripol process, or in the gas phase, such as BASF's Novolen process. In fact, in many cases several reactors are linked in series, for example to make the PP/EP blends known as ICP (see above). Typically the homopolymer iPP is made first in liquid propylene loop reactors, and then the EP is made in subsequent gas phase reactors. The process requirements called for to deal with the complexities of such multiple-reactor trains is justified by the ability to make a wide range of useful products from such a plant. A key factor in process improvements for PP has been to increase the fraction of rubber made in the reactors (while still having a product that can be made into easily transported pellets) so that softer and tougher products can be made directly in the polymerization plant. Such products are increasingly replacing melt-mixed blends due to their lower cost of production. Innovations are still being made in the PP process, aimed at both cost reductions and new product development. One intriguing suggestion is to run the polymerization in supercritical propylene (16), which would combine the advantages of the bulk and gas phase processes in terms of removal of the heat of polymerization, monomer recycle, and productivity. This will not be feasible, however, until PP catalysts can be found that operate well above the critical point of propylene.

The annual production of PP is shown in Table 4. The demand for PP is still growing quite rapidly, at 6 to 8 % per year worldwide. As in PE, it is interesting to note that the size and market share of the largest producers continues to grow, both by expansion and also by merger, but that the ten largest still only account for a little over 40% of the capacity.

Applications. The range of uses for propylene polymers is now so large that some have said that PP usage correlates with the GDP of a country. Products made from PP are used in essentially every industry and are ubiquitous in everyday use. This breadth can be seen in Fig. 4. No one application dominates. A large fraction is now used in fibers for applications such as carpets, twine, and furniture fabrics. Another significant slice of PP production is molded into parts for the automotive and appliance industries. Large sheets of PP fabric (both woven and nonwoven) are used as so-called 'geotextiles' or 'geomembranes' for various civil engineering applications such as soil stabilization, pond liners, and reinforcement of asphalt pavements. Nonwoven PP fabrics are also used for disposable diapers and hospital gowns. Filled PP, especially using glass fibers, is used in applications where stiffness is more important than toughness, such as some automotive components. PP is also increasingly being used for disposable food containers that are made by thermoforming. The wide variety of applications of PP indicated here shows why its use is still growing at a high rate.

Ethylene/Propylene Elastomers

Copolymers of ethylene and propylene with intermediate levels of each comonomer do not display either sort of crystallinity and so are completely amorphous (Fig. 1). As such they are completely saturated elastomers, and fill a different set of applications than highly unsaturated elastomers such as polydienes. When first introduced in the early 1960's, it was thought that ethylene-propylene elastomers might become an important component of tires, due to the low cost of their monomers. While this has not occurred, they have become extremely useful in a number of uses and the production of EP is growing faster than other rubbers (21, 22).

The copolymers of ethylene and propylene have found many uses, but suffer from the inability to be crosslinked by the means used for general purpose rubbers such as polyisoprene. In order to make them curable, a diene is introduced. These are called EPDM, for ethylene-propylene diene monomer, or EPR, for ethylene-propylene rubber. The most commonly used diene is ethylidene norbornene. The strained ring unsaturation reacts much more readily than that on the short branch. As a result, the former polymerize quickly (about as fast as ethylene) and the latter much more slowly. This allows one to run the polymerization without the production of much gel and keeps the pendant unsaturation available for subsequent crosslinking. Related compounds such as vinyl norbornene have similar reactivity. Other commonly used dienes are dicyclopentadiene and 1,4-hexadiene. While there is greater danger of gelation during polymerization with these dienes, the EPDM made from them has faster curing rates.

EPDM is currently made exclusively in solution processes, which makes the separation of the amorphous rubber and the removal of catalyst residues easier. The homogeneous catalysts used are vanadium compounds, with aluminum alkyl cocatalysts. Hydrogen can be used for molecular weight control. This often has characteristics of a single catalyst site (e. g., narrow molecular weight distribution), but often it is clear that several catalyst sites are active. After recovery of the polymer from solution and removal of the catalyst by steam stripping, the elastomer is formed into bales or pellets, depending on the cold flow properties of the polymer. Worldwide there is more than one million metric tons of annual capacity in EPDM, with the major producers including DSM, Exxon, Uniroyal, DuPont-Dow Elastomers, and Bayer.

Metallocene catalysts will also have an impact on EPDM production. DuPont-Dow Elastomers has started production of a metallocene-based EPDM plant, and Exxon Chemical is considering such a plant (23). The advantages lie not in the single-sited nature of the metallocenes, as this is possible with the vanadium systems, but in a greater ability to incorporate the comonomers, including perhaps hexene or octene in place of propylene. Union Carbide is currently developing a gas-phase process for EPDM, which has the promise of lowering the production costs (23). In order to make pellets from the amorphous rubber, a small amount of carbon black is added to prevent agglomeration.

The main uses of EP copolymers are for the toughening of PP and as a viscosity modifier for lubricating oils. The former application is shrinking in volume as PP reactor blends advance (see above), but the latter continues to increase. The terpolymers are used where the nearly complete saturation gives EPDM an advantage over highly unsaturated elastomers, such as in oxidative stability. A good example of this is cable insulation, where the saturation of these rubbers helps them avoid electrical failure. Other important applications for EPDM are found in the roof sheeting and hose markets. Finally, one of the fastest growing uses of EPDM is in thermoplastic vulcanizate (TPV's) blends with PP (24). These materials are mixed while the EPDM is being cured, and end up with a continuous PP matrix and dispersed, crosslinked EPDM domains that have a volume fraction of as much as 80%. As such they can be processed as a thermoplastic but show true elastomeric properties in terms of recovery. The major applications of these TPV's are in the automotive field.

Isobutylene-based Polymers

Polyisobutylene (PIB) and the various isobutylene copolymers that are called butyl rubber were among the first synthetic rubbers made, but they still hold a unique place among elastomers (25). All of these isobutylene-based polymers have a much lower permeability and higher damping than other elastomers, which have made them highly prized in applications such as tire innerliners and engine mounts. Moreover, since they were available so early in the history of polymer science they have been well studied and have served as models for polymers in general.

Polyisobutylene was first synthesized at I. G. Farben in the 1920's (26). It is one of the few examples of a commercial polymerization by cationic catalysis. The initiator most commonly used in AlCl$_3$, with a 'coinitiator' such as water or HCl. Low molecular weights (~ 5,000 g/mol) can be made at ambient temperatures, and very high molecular weights (well over 1,000,000 g/mol) are made at low temperatures when transfer reactions are suppressed. In recent years quasi-living cationic polymerization methods have been developed by the group of Kennedy that allow for nearly monodisperse versions of PIB to be made (27).

Due to the symmetry of the isobutylene monomer, one might expect that PIB would be highly crystalline like PE. In fact, it will crystallize under stress, but not under quiescent conditions. Because of the geminal dimethyl groups on every other backbone carbon, the bond angles are distorted to around 123° (as opposed to the tetrahedral angle of 109.5°), straightening out the chain:

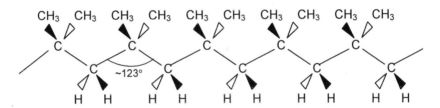

Figure 7. Structure of polyisobutylene.

Because of this, the chains can pack quite efficiently, which makes even the amorphous density of PIB high (0.917 g/cc *vs.* 0.855 g/cc for most amorphous polyolefins), and so reduces the incentive for crystallization (28). This high density is also the reason for the very low permeability of PIB. For most gases the permeability in PIB is about an order of magnitude lower than in other elastomers. Despite the high density, the T_g of PIB is well below ambient temperature, at -60°C, making it useful as an elastomer.

As a fully saturated polymer, PIB cannot be cured as a thermoset elastomer. In order to form a curable rubber from an isobutylene polymer, Sparks and Thomas developed copolymers of isobutylene and isoprene in 1936, which are now known as butyl rubber (29). Only around 1 to 2% of isoprene is needed to make the polymer well curable. However, isoprene is a powerful transfer agent in this cationic polymerization, so high molecular weight butyl rubber can only be made at very low temperatures around -90°C. The most common polymerization technology is a continuous slurry process with methyl chloride as the diluent. A critical part of the process is the system to remove the heat of polymerization and maintain the temperature at the low values needed.

At 1 to 2 % isoprene butyl rubber cures well, but much more slowly than fully saturated rubbers such as polyisoprene or polybutadiene. This is not a problem in 100% butyl rubber formulations, but in blends with a polydiene the difference in cure rate causes difficulties. In 1955 Baldwin discovered that chlorination of some or all of the isoprene groups would greatly increase the cure rate of butyl rubber (30). The halogenation is generally done in a hexane solution after polymerization. Both chlorobutyl and bromobutyl rubber are now important parts of the isobutylene polymer family.

More recently, another member of this family has been introduced by Exxon (31). These are copolymers of isobutylene and brominated paramethyl styrene (PMS). By replacing isoprene with PMS, the copolymer has fewer double bonds and so is more stable oxidatively, but the curing reactivity is still very good.

The majority of applications take advantage of the low permeability of butyl polymers. Chief among these are inner tubes from butyl rubber (now mostly restricted to bicycles and large vehicles) and the innerliners of tubeless tires from halobutyl rubber. Other examples that take advantage of the low permeability include medical closures and sealants. Butyl rubber also has a huge damping peak, so it is often used for applications that need energy absorption, such as engine mounts and dock barriers. The low molecular weight PIB homopolymer is principally used in sealing applications, while the high molecular weight ones are used to toughen plastics.

Total worldwide capacity of isobutylene-based elastomers is a little more than 750,000 metric tons per year. Nearly all of this belongs to the two principal producers, Exxon and Bayer. A very small amount is produced in Russia.

Other Polyolefins

Polymers made mainly from higher α-olefins are produced in much lower volumes than the main types described above. As the molecular weight of the olefin increases, its cost rises while polymerization rates drop. Both of these factors raise the cost of the polymers made from them, so they are only used when they possess some special property that cannot be obtained from PE or PP. A summary of the chemistry of these polymers can be found in ref. 15.

The production capacity of poly(1-butene) (PB), the most common of these higher α-olefins polymers, is around 30,000 metric ton per year, which is three orders of magnitude less than PE or PP. These are isotactic polymers that show better creep resistance than PP or PE. For this reason it has been used to make pipes for water transport, which is the major use of this polymer. However, the rate of failure of these pipes has been higher than anticipated, which is hurting the growth of this market. The production capacity of poly(4-methyl pentene-1) is another order of magnitude still lower. This polymer is characterized by a high melting point and high crystallinity, which has led to its use in sterilizible medical articles. Some polyolefins with very long alkyl side chains, such as poly(1-dodecene) have been used as drag reducers, most notably for flow in oil pipelines. The lack of success in the commercialization of polymers made from higher α-olefins relative to the high-production polyolefins indicates how successful people have been in modifying PE and PP with copolymerization and the use of additives to meet a wide range of market needs.

Polyolefin Blends

It is not widely appreciated that the majority of blends used around the world are made from polyolefins. Many of them are made in the reactor, either because of multiple catalyst sites, as in many versions of LLDPE, or because two or more reactors are used to make different polyolefins, as in ICP polypropylene. A large amount of melt mixing of polyolefins also is still performed, to make products such as the PP/EP blends that are called 'thermoplastic olefins' TPO (32). Here I will briefly outline how these are used.

PERCENT EPR		Product Class	Polymer Type	Typical Parts	Relevant Properties
	0	PP homopolymer	plastic	battery case	stiffness
		ICP high impact PP	toughened plastic	battery case	stiffness
				dashboard	impact strength
	25	'thermoplastic olefinic elastomer" (TPO)	stiff TPE	bumper	impact strength
					processability
	50	thermoplastic vulcanizate (TPV, DVA)	true TPE	weatherstrip	softness
				hose	tensile strength
	85				
	100	EPDM	thermoset elastomer	engine mount hose, belt	fatigue elongation

Table 5. Examples of typical uses of PP/EPR blends in an automobile.

A great deal has been learned about the thermodynamics of these blends, which has been summarized in a recent review article (33). Because these are all saturated hydrocarbon polymers, there are no 'specific interactions' to increase their miscibility; but because the van der Waals interactions between them are so weak, a large fraction of polyolefin pairs are at or near a phase boundary in common use. Thus the effects of shear and pressure, for instance, can be quite important for such systems.

The wide variety and utility of polyolefin blends is illustrated in Table 5. This shows the range of PP/EPDM blends that are used in automotive applications, going all the way from stiff PP homopolymer through to elastomeric EPDM. In between, nearly every blend composition finds use somewhere in a car, as the properties vary with EPDM content. Another advantage of the ability of polyolefin blends to satisfy many applications in a car is that this makes the recycling of such parts much easier, as there is less need to separate the various components of the car. Because of this kind of versatility of performance, a large fraction, perhaps as much as 30%, of polyolefins are used in blends with another polyolefin. Moreover, other parts of a typical automobile will be made from polyolefins, from glass fiber reinforced PP for child seats to HDPE gasoline tanks. Polyolefins are clearly an extremely useful class of materials, and the advances being made in controlling their structure through catalysis and in understanding how this structure controls their performance promise to enhance this utility.

Literature Cited

1.   British Patent No. 471,590 (1937)
2.   Doak, K. W. in "Encyclopedia of Polymer Science and Engineering", vol. 6: Wiley; NY, 1986, p. 386.
3.   Axelson, D. E.; Levy, G. C.; Mandelkern, L. Macromolecules, 1979, 12, 41; Usami, T.; Takayama, S. Macromolecules, 1984, 17, 1756.
4.   Ehrlich, P.; Mortimer, G. A. Adv. Polym. Sci., 1970, 7, 386.
5.   German Patent No. 878,560 (1953)
6.   McMillan, F. M. "The Chain Straighteners", MacMillan: London, 1979.
7.   Choi, K. Y.; Ray, W. H. J. Macromol. Sci. – Rev. Macromol. Chem. Phys., 1985, C25, 1.
8.   Karol, F. J. Catal. Rev. – Sci. Eng., 1984, 26, 557; Levine, I. J.; Karol, F. J., US Patent No. 4,011,382, 8 March 1977 (to Union Carbide).
9.   Plumley, T. A.; Sehanobish, K.; Patel, R. M.; Lai, S. Y.; Chum, S. P.; Knight, G. W. ANTEC Proceedings, 1995, p. 2263.
10.   Benedikt, G. M.; Goodall, B. L. "Metallocene-Catalyzed Polymers"; Plastics Design Library, Norwich, NY, 1998.
11.   Sinn, H.; Kaminsky, W. Adv. Organomet. Chem., 1980, 18, 99.
12.   Malmberg, A.; Kokko, E.; Lehmus, P.; Löfgren, B.; Seppälä Macromolecules, 1998, 31, 8448.
13.   Mecking, S.; Johnson, L.K.; Wang, L.; Brookhart, M. J. Am. Chem. Soc. 1998, 120, 888.
14.   Gibson, V. Angew. Chem. Int. Ed., 1999, 38, 428

15. van der Ven, S. "Polypropylene and Other Polyolefins"; Elsevier, Amsterdam, 1990.
16. Karger-Kocsis, J. "Polypropylene: An A-Z Reference"; Kluwer, Dordrecht, 1999.
17. Pino, P.; Moretti, G. Polymer, 1987, 28, 683.
18. Ewen, J. A.; Jones, R. L.; Razavi, A.; Ferrara, J. D. J. Amer. Chem. Soc., 1988, 110, 339.
19. Lotz, B.; Wittmann, J. C.; Lovinger, A. J. Polymer, 1996, 37, 4979.
20. Hu, Y.; Krejchi, M. T.; Shah, C. D.; Myers, C. L.; Waymouth, R. M. Macromolecules, 1998, 31, 6908.
21. Baldwin, F. P.; Ver Strate, G. Rubber Chemistry and Technology, 1972, 45, 709; Ver Strate, G. in "Encyclopedia of Polymer Science and Engineering", vol. 6: Wiley; NY, 1986.
22. Agoos, A. Chemical Week, 19 April 1989, pp. 44-50; Chemical & Engineering News, 28 June 1999, p. 38.
23. Wood, A. Chemical Week, 3 April 1996, pp. 41-42.
24. Coran, A. Y. in "Thermoplastic Elastomers", Legge, N. R.; Holden, G.; Schroeder, H. E., eds.; Hanser: Munich, 1987; chapter 7, pp. 133-161.
25. Kresge, E. N.; Schatz, R. H.; Wang, H.-C. in "Encyclopedia of Polymer Science and Engineering", vol. 8: Wiley; NY, 1986.
26. Tornqvist, E. G. M. in "Polymer Chemistry of Synthetic Elastomers: Kennedy, J. P.; Tornqvist, E. G. D., Eds.; Wiley: New York, 1968; p.76.
27. Faust, R.; Kennedy, J. P. J. Polym. Sci. – Polym. Chem., 1987, A18, 1397.
28. Boyd, R. H.; Breitling, S. M. Macromolecules, 1972, 5, 1; Ver Strate, G. W.; Lohse, D. J. in "Polymer Data Handbook", J. E. Mark, Ed.; Oxford University Press: New York, 1999. pp. 600-606
29. Thomas, R. M.; Lightbown, I. E.; Sparks, W. J.; Frolich, P. K.; Murphree, E. V. Ind. Eng. Chem., 1940, 32, 1283.
30. Baldwin, F. P. Rubber Chemistry and Technology, 1979, 52, 677.
31. Wang, H. C.; Powers, K. W. Elastomerics, 1992, 124 (1), 14.
32. Kresge, E. N. in "Polymer Blends", Paul, D. R.; Newman, A., Eds.; Academic Press: New York, 1978; Vol. 2, p. 293.
33. Lohse, D. J.; Graessley, W. W.; in "Polymer Blends: Formulation and Performance", ed. by D. R. Paul and C. B. Bucknall, Wiley, New York, 1999, p. 219.

STYRENE POLYMERS AND COPOLYMERS

MEHMET DEMIRORS
Polystyrene R&D, The Dow Chemical Co., Midland, MI 48667

Introduction
Styrene Family of Plastics
Chemistry of Styrene Polymers
Fabrication Processes
Future Trends

Introduction

History of styrene and polystyrene spans a period of over two hundred years starting with the discovery of the styrene monomer by Newman (1) in 1786, followed by the first polymerization by Simon in 1839 (1). Transformation of styrene to a resinous gum on standing, and its formation by ethyl benzene pyrolysis were all discovered in the nineteenth century. The mechanism of polymerization was not discovered until early twentieth century. Staudinger, using styrene as the principal model, had identified the polymerization process as a chain reaction, involving the build up of high molecular weight products from low molecular weight units in 1920.

Low yields of styrene, due to thermal cracking rather than the dehydrogenation of the ethyl benzene had discouraged attempts to manufacture styrene by the ethyl benzene dehydrogenation route. Developments in catalysis coupled with the use of steam have eventually increased the yields to an extent where styrene monomer had become commercially viable.

The commercial utilization of styrene polymers was not established until 1938. The Dow Chemical Company became the first company to successfully commercialize polystyrene (PS). At around the same time BASF also started to develop small-scale polymerization processes in Germany (2).

Developments in styrene manufacturing processes, coupled with readily availability of polybutadiene rubbers and acrylonitrile monomer lead to commercialization of high impact polystyrene (HIPS), copolymers of styrene-acrylonitrile (SAN), and copolymers of acrylonitrile-butadiene-styrene (ABS).

Initial production process that was developed by The Dow Chemical Company involved filling styrene monomer to 10 gallon (38 L) cans and heating it in a liquid bath at progressively higher temperatures until the styrene was completely converted to polystyrene (3). The resin was then removed from the cans, ground and packaged. Today bulk polymerization is used for PS almost exclusively. For the manufacture of

ABS resins mostly emulsion polymerization is employed as well as some mass polymerization processes also being operational.

Styrene Family of Plastics

Polystyrene (PS) also known as general purpose polystyrene is unmodified polystyrene. It is a transparent, rigid and a brittle polymer which is easy to process. It is almost exclusively made by bulk polymerization. It's primary uses include insulation foam for construction, packaging foam such as meat trays and egg cartons, and bi-axially oriented film for food packaging. Injection molded articles include picnic cutlery, compact disc jewel boxes, audiocassette housings and toys.

High Impact Polystyrene (HIPS) is a rubber modified version of PS in which higher toughness is achieved by incorporation of micron sized polybutadiene rubber particles. Improvement in toughness is achieved by in-situ grafting of polybutadiene rubber with polystyrene leading to increase adhesion between the rubber particles and polystyrene matrix. A major portion of HIPS is used in food packaging such as yogurt containers, drinking cups, toys, refrigerator linings, audio and video equipment. A special version of HIPS, modified by incorporation of additives to make it ignition resistant is used for TV housings as well as other electronic equipment such as printers, copiers and computers.

Due to its properties, low cost and ease of fabrication PS finds a diversity of applications in a wide range of markets. In table 1 gives the global PS and HIPS use for the main markets.

Table 1: Global PS Usage in 1996*

Market	North America	Western Europe	Asia
Packaging	1115	907	341
Appliances	134	282	645
Electrical/ Electronics	264	68	349
Furniture/ House-ware	104	230	110
Other	862	450	1625

* Data taken from Polystyrene Report January 1998 SRI Consulting; all numbers are in metric tons.

Styrene-Acrylonitrile Copolymers (SAN) Incorporation of acrylonitrile (AN) improves the chemical resistance, scratch resistance and rigidity of PS. Due to the copolymerization characteristics of acrylonitrile and styrene monomer, most SAN resins contain 15 to 30% by weight of AN monomer. SAN is used in application where it's additional attributes such as better chemical resistance, toughness, and heat distortion temperature are most beneficial. Those include shower doors, cosmetic bottles, and toys. Most of SAN resins are consumed in the production of emulsion polymerized ABS resins through compounding.

Acrylonitrile-Butadiene-Styrene (ABS) Copolymers : Most of ABS used is manufactured by utilizing the emulsion process. In this process, butadiene is first polymerized to form sub micron rubber particles and styrene-acrylonitrile copolymer is then grafted on the outside of rubber particles to form grafted rubber concentrate (GRC). Based on the needs of a given application, GRC is mixed with SAN through compounding to provide the final product with appropriate rubber level for required toughness. ABS is also manufactured by mass process in a manner similar to HIPS above. ABS polymers generally are tougher than HIPS, and have higher chemical resistance, but are somewhat more difficult to process. ABS polymers are used in a large variety of applications. These include household piping, luggage, power tool housings, vacuum cleaner housings, toys and automotive components such as interior trim.

Other Styrene Polymers: Several other important classes of styrene polymers are also used. Those include copolymers of styrene and maleic anhydride for high heat applications, transparent impact polystyrene (TIPS), manufactured by anionic block copolymerization of styrene and butadiene, and expandable polystyrene, usually used as molded foam for packaging of electronic equipment, such as TV's, audio equipment and computers.

Chemistry of Styrene Polymers

Free Radical Polymerization of Styrene: Styrene monomers can be free radically polymerized by either heating it up or by introducing free radical generalizing species, usually peroxide initiators.

When heated to a temperature above 100 °C styrene monomer will form a Diels-Alder adduct (4),(5) as shown in figure 1.

The Diels-Alder reaction produces two stereo isomers of 1-phenyltetralin. One of the isomers, axial 1-phenyltetralin is capable of initiating styrene polymerization.

Typically the rate of polymerization is governed by the monomer concentration, temperature of polymerization, and if used the concentration of peroxide initiators and their decomposition temperatures.

Molecular weight control is provided by diluents such as ethyl benzene as well as use of initiators or chain transfer agents. Most solvents such as ethyl benzene and toluene have a very mild chain transfer activity with a chain transfer coefficient of around 10^{-5}. Most commonly used initiators are peroxide initiators such as 1,1-di(tertiary butyl peroxy) cyclohexane (6, 7), 1-1di(tertiary butyl peroxy) - 3,3,5 -

trimethylcyclohexane (8),tert - butyl peracetate (9) (10) and tertiary butyl perbenzoate (10).

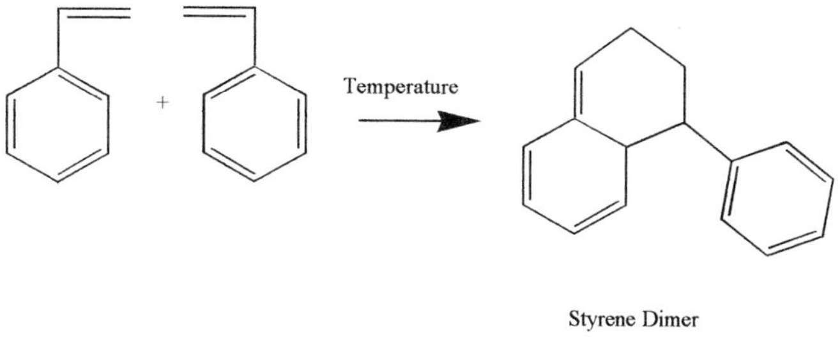

Styrene Dimer

Figure 1: Formation of Diels-Alder adduct in thermal polymerization of styrene monomer.

The factors that make those initiators suitable for polystyrene production is low cost, decomposition temperatures well suited for styrene polymerization, high efficiency in polymerization as well as capability to graft the polybutadiene rubber. The temperature regime in which styrene is polymerized is usually around 100 to 180 °C. Initiators that are active in the lower half of this range can be used for styrene polymerization. Initiators that have a one-hour half-life temperature in the range of 110 to 130 °C are ideal. In order to improve the temperature range of peroxide initiated polymerization a mixture of two initiators with two different half lives are used. This way a continuous supply of peroxy radicals is maintained over a broader temperature range.

As the rate of polymerization has a negative impact on the molecular weight of the polymer (see figure 2), several initiators with higher functionalities have been developed and their utility demonstrated in improving the rate - MW relationship for styrene polymerization (7),(11),(12).

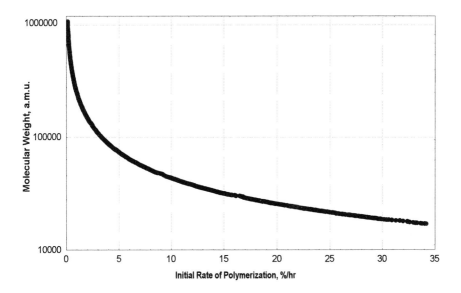

Figure 2: Relationship between the initial rate of polymerization and the weight average molecular weight of the polymer formed (13).

Chain transfer agents are extensively used for molecular weight control in styrene polymerization. The key attributes of a chain transfer agent can be summarized as low cost, no influence on rate of polymerization, high chain transfer coefficient and efficiency. The class of chemicals that fit the above criteria best is mercaptans. Dodecyl mercaptans are probably the most widely used chain transfer agents (14).

The mass polymerization process for polystyrene utilizes a series of reactors either stirred tube reactors of Continuously Stirred Tube Reactor (CSTR) types or a combination of the two. First a polymerization feed is prepared by mixing monomers, diluents and other additives. This feed is then fed into the first of a series of reactors (see figure 3) where the polymerization is carried out by heating the polymerization feed to the required temperature. Polymerization of styrene is a highly exothermic reaction and generates a lot of heat. As a consequence the reactors are designed in such a way that they can remove large amounts of heat very quickly. As the styrene monomer gets converted to polystyrene, the reaction rate is reduced due to decreasing monomer concentration. In order to maintain the reaction rate, the temperature is gradually increased.

Once the desired level of conversion of styrene monomer to polystyrene is achieved, then the unreacted monomer and the diluents are removed form the polymer by exposing the polymerization mixture to high vacuum at high temperatures. The styrene monomer and diluents are flashed off and condensed to be cycled back to the start of the process. The polymer at this stage contains less than 1000 parts per million of unreacted styrene monomer and the polymer melt then is extruded, cooled in a water bath and granulated for storage.

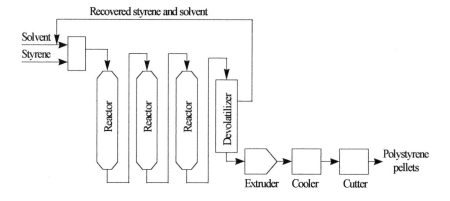

Figure 3: Mass process for the production of polystyrene

While free radical polymerization is exclusively used for the commercial production of polystyrene, styrene can be polymerized by all the major polymerization mechanisms. Anionic polymerization of styrene can be carried out very conveniently using butyl lithium initiation (15) in conventional CSTR reactors (16) or continuous flow reactors (17). Anionic polymerization process for styrene has the unique capability to produce narrow molecular weight distribution if operated on a batch mode. As the rate of polymerization is extremely high, high molecular weight polymers can be made at much faster rates than in free radical polymerization process. High cost of butyl lithium initiator coupled with the stringent purity requirements for the polymerization feed required by anionic polymerization mechanism make this process more costly than free radical polymerization process.

Cationic polymerization of styrene can be achieved by a variety of acidic initiators. Due to strong chain transfer to the monomer, cationic process produces only low molecular weight resin even at fairly low temperatures (18). Those polymers are not used as conventional plastics due to their low molecular weight but find uses as tackifiers for adhesives and printing inks.

Styrene can also be polymerized by coordination mechanisms. (19), (20). Both isotactic and syndiotactic polystyrene can be produced by this way. Syndiotactic polystyrene is by far the most important one of the two, as its crystallization rate is much faster than that of isotactic polystyrene (21). Currently The Dow Chemical Company and Idemitsui Petrochemical Company of Japan are in the process of commercializing syndiotactic polystyrene. Syndiotactic polystyrene is truly an engineering thermoplastic with very high use temperature and resistance to a range of chemicals including a broad group of organic solvents.

The physical properties such as flexural strength and flexural modulus of polystyrene depends on the molecular weight of the resin and the additives present (22), (23). Higher molecular weight products have high flexural strength but suffer from low processability. While lower molecular weight products have very good processability, they have somewhat lower flexural strength and toughness. Most of the time a balance is provided by intermediate molecular weight products, depending on the requirements of the final application. Commercially useful range of molecular weights (weight average molecular weight, Mw) is typically from about 150,000 a.m.u. to 350,000 a.m.u. If the Mw is lower than this range, then the polymer is very brittle while for Mw values higher than this range, processing of the resin becomes very difficult.

While all of the commercially available products have a standard molecular weight distribution typically 2.2 to 3.5 higher and lower distributions can be made by free radical processes. There is no clear evidence that suggest any significant change in physical properties with variations of polydispersity in this range (24),(25). At high polydispersities the physical properties such as flexural strength and toughness tend to suffer.

Plasticizers, mostly white mineral oil, are used to improve processability. Typically amounts are in 0 to 4.5% by weight ranges. While plasticizers improve the processability they tend to reduce the heat distortion temperature and rigidity of the resin (26).

Polymerization of HIPS: To improve the toughness of general purpose polystyrene, a two phase system of polybutadiene rubber and polystyrene was developed early on (27, 28).

HIPS polymerization is usually carried out by dissolving polybutadiene rubber in styrene monomer and the diluent. Polymerization process of HIPS is unique in that two processes take place simultaneously, i.e. formation of matrix polystyrene by polymerization of styrene and formation of graft polymerization between polybutadiene and styrene monomer. The grafting of polybutadiene rubber provides the necessary compatibility between the matrix phase and the rubber phase as well as defining the structure and the size of the rubber particles formed.

The grafting reaction takes place primarily by H-abstraction from polybutadiene backbone either by growing polystyrene radicals or alkoxy radicals if peroxide initiators are utilized (29), (30), (31). At the early stages of polymerization, polybutadiene rubber forms a continuous phase in styrene monomer while newly formed polystyrene becomes phase separated in the rubber solution in styrene, stabilized by the grafted copolymer of styrene and butadiene (27),(28),(32). As the reaction proceeds further and more polystyrene is formed, the polybutadiene phase becomes minor component and phase inversion takes place (33-36). Phase inversion usually takes place when the total polymer content of the system reached 2.5 times the initial polybutadiene rubber content.

At the point of phase inversion discrete rubber particles are formed with the help of agitation (27),(28),(34). If there is no agitation, then a three-dimensional network of polybutadiene in polystyrene is formed that is not processible.

Typically the most desirable particle size range for HIPS is 0.5 to 10 microns (37),(38). The rate of agitation has a strong influence on the rubber particle size. If the agitation rate is increased the rubber particle size becomes smaller eventually reaching a plateau depending on the solution viscosity of the rubber used and the level of grafting (34),(39). The structure of rubber particles is determined by the degree of grafting (40), (41). The structure of rubber particles is determined by the degree of grafting (40). At low to moderate levels of grafting salami type particles are obtained. These particles contain occlusions of polystyrene inside rubber particles. As the level of grafting increases, particle size continues to decrease, eventually reaching a single occlusion particle, usually referred to as core-shell particles. At very high levels of grafting, extremely small rubber particles with no polystyrene occlusions are observed. A transmission electron micrograph of a typical HIPS product is given in figure 4 showing salami type rubber particles (dark areas) with polystyrene occlusions (white areas) in a matrix of polystyrene.

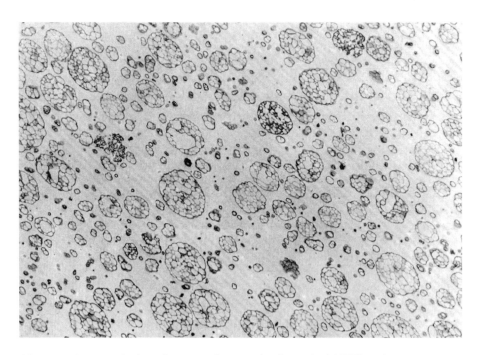

Figure 4: A transmission electron micrograph of a typical HIPS resin.

Most commercial HIPS products have 4 to 12 % (by weight) polybutadiene rubber dispersed as discrete particles of about 0.5 to 10 microns. Even though the percentage of rubber is fairly low, rubber phase volume (the total volume enclosed by the rubber particles including polystyrene occlusions) can be as high as 40% of the total for a HIPS resin containing 8% polybutadiene rubber by weight. Incorporation of high levels of occlusions increases the utility of the rubber. Depending on the intended use, rubber particle size is controlled within a certain range. For high rigidity and good surface aesthetics smaller rubber particle sizes are found to be most desirable, while larger particles provide better toughness at the expense of rigidity and surface aesthetics (41),(42). For toughness the most desirable particle size range is usually 2 to 3 microns (43), (44). For increased environmental stress crack resistance (ESCR) a rubber particle size of 5 - 8 micron is most desirable (45),(46). Larger particles are more capable of coping with the stresses introduced by the stress cracking agents. ESCR characteristics are required in applications where the resin comes in contact either with fatty foods as in food packaging or with blowing agents that are typically used in manufacture of refrigerators.

While most HIPS products have normal rubber particle size distribution, bimodal or broader particle size distribution HIPS products can combine the desirable properties of both small and large rubber particle size based HIPS resins. Such bimodal products have the high rigidity, good surface aesthetics of small particles with improved toughness of larger particles (47-51),(52, 53). These products are most useful in applications requiring a combination of rigidity, improved surface aesthetics such as TV cabinets, small appliance housings and kitchen utensils.

The optimum particle size for toughness in HIPS is a consequence of the energy absorption mechanism. The primary modes of energy absorption in HIPS are crazing (54-57) and rubber cavitation (58-60). Due to the Poisson's ratio difference between rubber and polystyrene phases, under load, a concentration of stresses takes place in the equatorial plane of the rubber particles, which can be two to three times higher then the bulk of the material (61). Once a critical stress level is achieved in the vicinity of the rubber particles, then as a consequence localized deformation takes place. This way the material can absorb energy without breaking. The crazes that are formed consist of highly oriented polystyrene fibrils, perpendicular to the plane of the stress. The typical craze thickness for polystyrene is usually around 1 micron. As the rubber particles act both as initiators of crazes and also as terminators, preventing formation of cracks, their optimum size is around 2 to 3 microns (43). The rubber particles themselves are also involved in energy absorption through cavitation process (59, 60).

Typical properties of Styrene Polymers are given in the table below.

Table 2: Physical properties of styrene polymers

Property	GPPS	HIPS	SAN**	ABS
Tensile yield Stress MPa*	45	24	70	40
Elongation to Break %	2	30	2	20
Tensile Modulus MPA	3200	3800	2000	2100
Notched Izod Impact J/M***	20	120	20	250
Vicat Heat Distortion Temperature °C	96	96	105	102

* To convert MPa to PSI multiply by 145
** AN content 24% by weight
*** To convert J/M to ft.lb./ln. Divide by 53.4

SAN Copolymers: Copolymers of acrylonitrile are the most important group of styrene copolymers. Due to the reactivity ratios of styrene and acrylonitrile, ideal SAN copolymer composition is one in which 24% of the copolymer is acrylonitrile by weight. This corresponds to the azeotropic point, where the composition of the polymer formed is the same as that of the polymerization feed. On either side of this point composition drift takes place. Outside the azeotropic range a reactor in which a high degree of back mixing is present needs to be used such as a coil or a CSTR type reactor. Commercially SAN resins with AN levels between 15 to 30% are available.

Copolymerization of SAN can be carried out thermally or by using peroxide or azo initiators. Due to color concerns mostly azo initiators are used (62).

For the modification of properties of SAN polymers, similar techniques as in polystyrene are used. For higher toughness and chemical resistance higher levels of AN is employed.

ABS Copolymers: ABS Copolymers are manufactured by two distinct processes. Those are the mass process and the emulsion process.

Mass Process for ABS is almost identical to that of the HIPS process. Polybutadiene rubber is dissolved in the polymerization feed consisting of styrene, diluents and acrylonitrile. This feed is then fed into a series of reactors where grafting of polybutadiene rubber by SAN copolymer takes place prior to formation of particles. Modification of the final product is achieved by changing process conditions, feed and other components in a manner similar to the HIPS process (63) with the exception of the devolatilization part. In the manufacture of HIPS resins a single stage devolatilizer can be used (even though two stage devolatilization is more common). In the case of mass process ABS resins, two devolatilizers are used due to a large separation of boiling points between acrylonitrile and styrene monomer.

The second process that is used for the manufacture of ABS is the emulsion process (64). Into an emulsion of butadiene in water, a water soluble initiator system is introduced. As the initiator radicals enter the micelles, they polymerize the butadiene to form a seed. Later on styrene acrylonitrile monomers are introduced into the system. As they form a thin layer on the seed substrate, the polymerization of SAN phase takes place generating grafted SAN as well as free SAN polymers. Once the reaction is complete, the emulsion is coagulated, washed and dried to obtain grafted rubber concentrate (GRC) which can be blended with SAN polymers to make ABS.

Typical stress strain curves of styrene polymers are given in figure 5.

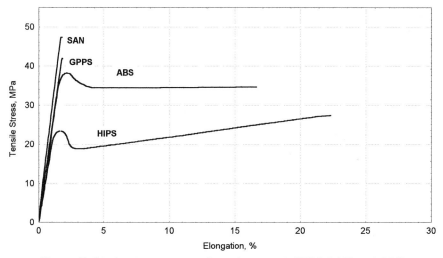

Figure 5: Strain-stress curves for polystyrene, HIPS SAN and ABS.

Fabrication Processes

Extensive utilization of styrene polymers is in part due to their ease of processing with many fabrication techniques. Since styrene family of polymers are all amorphous (with the exception of syndiotactic polystyrene) they can be injection molded with little shrinkage. This enables molding of articles with close dimensional tolerances and high speed injection molding.

Styrene polymers show one of the best extrusion thermoforming behavior. The resins usually have sufficient melt strength, enabling deep draws on thermoforming.

Such characteristics are important in food packaging area for yogurt containers and drinking cups. The refrigerator liners are thermoformed from a single sheet of resin with extensive deep draw.

Polystyrene can be foamed into a cellular structure with a suitable blowing agent. Both the extruded sheet for insulation materials (such as Styrofoam® brand insulation foam from The Dow Chemical Co.) for construction as well as for packaging applications are made by blending a molten polystyrene stream with a blowing agent under pressure and extruding the mixture. Under pressure the gas expands, generating a cellular foam structure. Historically flourocarbon blowing agents have been used. In recent years due to environmental concerns hydrocarbon blowing agents such as pentane and carbon dioxide are more commonly used.

Brittle polystyrene can also be converted to a tough, flexible film by bi-axial orientation.

Future Trends

Several significant developments are taking place in styrene polymerization area that could potentially have a major impact in styrene polymers. Developments in the area of living free radical polymerization opens up the possibility of formation of block copolymers of styrene and other monomers (65, 66). One such area is the formation of styrene butadiene block copolymers by functionalization of polybutadiene with a living free radical group such as a nitroxy unit. Such materials could potentially lead to development of newer grades of polystyrene with unique property balances.

Incorporation of fillers and additives such as nano composites offer unique property enhancements to all plastics. These materials while being available in large quantities can be incorporated into the plastics to give enhanced rigidity and other desirable properties.

Recent developments in anionic polymerization (16),(17) also offer significant opportunities for enhancing utility of styrene polymers. Ability to form random and block copolymers and high reaction rates even for high molecular weight resins opens up possibilities for future. One such example is copolymers of styrene and alpha methyl styrene for high heat applications.

The future of styrene resins continues to be dominated by the need to develop higher performance and lower cost products. In industries where these plastics are used, there is a strong desire to reduce the cost as TV cabinets, computer equipment and disposable food packaging. These overall trends will continue to direct strong research emphasis at increasing the rate of polymerization, improving utility of the additives and raw materials used as well as enhancing performance attributes of the resins.

Literature Cited

1.   Bounty, R.H. and Boyer, R.F., "Styrene , Its Polymers, Copolymers and Derivatives", Reinhold Publishing Corp. New York, 1952, p.3 and p.1.
2.   Wullf, C. and E. Dorer, DRP Patent 643 278, 1930, to BASF.

3.   Amos, J.L., Polym. Eng. Sci., (1974) 14, 1.
4.   Pryor, W.A. and Lasswell, L.D., in G. H. Williams, ed. "Advances in Free Radical Chemistry" Vol. 5, Elek. Science, London, 1975, p.32.
5.   Mayo, F.R., J. Am. Chem. Soc., 1968, 75, 1289.
6.   Villalobos, M.A., H.A. E., and P.E. Wood, J. Appl. Poly. Sci., 1991, 42, 629.
7.   Kamath, V.R., Mod. Plast., 1981, 58, 106.
8.   In Technical Bulletin 85-12, Noury Chemicals, Chicago, 1985.
9.   Kamath, V.R., Rubber Plast. News, 1980, 46, 48.
10.   Sheppard, C.S., Polym. Eng. Sci., 1979, 19, 597.
11.   Priddy, D.B., in Kirck-Othmar ed. in "Encyclopedia of Chemical Technology", 4th ed. John Wiley & Sons Inc., New York, 1997, p. 1015.
12.   Demirors, M., Polym. Prepr. (Am. Chem. Soc., Div. Polym. Chem.), 1996, 37, 2, 531.
13.   Boundy, R.H., Boyer, R. F., and Stoesser, S., "Styrene: Its Polymers, Copolymers and Derivatives", American Chemical Society; Monograph No: 115, Reinhold Publishing Corp. New York, 1952.
14.   Young, L.J., in J. Brandrup and E.H. Immergut, eds. Polymer Handbook 2nd ed. John Wiley & Sons Inc. New York, 1975, pp 2-57
15.   Schwarc, M., "Living Polymers and Electron Transfer Processes", Wiley-Interscience, New York, 1968.
16.   Priddy, D.B. and Pirc, M., J., Polym. React. Eng., 1993, 1, 343.
17.   Thiele, R., Chem. Eng. Technol., 1994, 17, 127.
18.   Roe, J.M. and Priddy, D. B., US Patent 4,087,599, 1978, to The Dow Chemical Co.
19.   Kern, R.J. and Hurst, H.G., J. Polym. Sci, 1960, 45, 195.
20.   Zambelli, A., et al., Macromolecules, 1987, 20, 2035.
21.   Moore, G., Edt. Chapter on "Styrene Polymers", Encyclopedia of Polymer Science and Engineering, Vol. 16, John Wiley & Sons, New York, 1989.
22.   Hauss, A., Angew. Makromol. Chem., 1969, 8, 73, 79.
23.   Szwarc, M., Levy, M., and Milkovich, R., J. Am. Chem. Soc., 1956, 78, 2656.
24.   Martin, J.R., Johnson, J.F., and Cooper, A.R., J. Macromol. Sci. Rev. Macromol. Chem., 1972, 8, (57), 59.
25.   Nunes, R.W., Martin, J.R., and Johnson, J.F., Polym. Eng. Sci, 1982, 22, 205.
26.   Kruse, R.L. and Southern, J.H., Polym. Eng. Sci., 1979, 19, 11, 815.
27.   Molau, G.E., Jour. Poly. Sci. Part A, 1965, 3, 4235.
28.   Molau, G.E. and Keskkula, H., J. Poly. Sci. Part A-1, (1966) 4, 1595.
29.   Brydon, A., Burnett, G.M., and Cameron, G.G., J. Polym. Sci. Poly. Chem. Ed., 1973, 11, 3255.
30.   Brydon, A., Burnett, G.M., and Cameron, G.G., J. Polym. Sci. Poly. Chem. Ed, 1974, 12, 1011.
31.   Reiss, C. and Locatelli, J.L., Adv. Chem. Ser., 1975, 142, 186.
32.   Riess, G., et al. "Grafting kinetics in the case of rubber modified polymers. ABS and HIPS systems" in Polym. Sci. Technol. (Polym. Alloys: Blends, Blocks, Grafts, and Interpenetrating Networks), 1977.
33.   Freeguard, G.F., Br. Polym. J., 1974, 6, 4, 205.

34.   Freeguard, G.F. and Karmarkar, M., <u>Jour. Applied Poly. Sci.</u>, 1971, <u>15</u>, 1657.
35.   Freeguard, G.F. and Karmarkar, M., <u>J. Appl. Polym. Sci.</u>, 1972, <u>16</u>, 1, 69.
36.   Freeguard, G.F. "Simulating and interpreting the HIPS process" <u>in Toughening Plast., Int. Conf.</u>, [Prepr.], 1978, London, Plast. Rubber Inst.
37.   Bucknall, C.B., <u>Jounal of Material Science</u>, 1987, <u>22</u>, 1341.
38.   Keskkula, H., Adv. Chem. Ser., 1989, 222 "Rubber-Toughened Plast.", 289.
39.   Turley, S.G. and H. Keskkula, <u>Polymer</u>, 1980, <u>21</u>, 466.
40.   Echte, A., <u>Angew. Makromol. Chem.</u>, 1977, <u>58/59</u>, 175.
41.   Van Dyke, J.D. "<u>The effect of rubber phase structure on physical properties in high impact polystyrene</u>" in Technol. Plast. Rubber Interface, Eur. Conf. Plast. Rubber Inst., [*Prepr.*], 2nd. Ed. Plast. & Rub. Inst., 1976, London, England.
42.   Hall, R.A. and Burnstein, I., <u>J. Mater. Sci.</u>, 1994, <u>29</u>, 24, 6523.
43.   Michler, G.H., <u>Acta Polymer.</u>, 1993, <u>44</u>, 113.
44.   Cigna, G., Matarrese, S., and Biglione, G.F., <u>J. Appl. Polym. Sci.</u>, 1976, <u>20</u>, 8, 2285.
45.   Bubeck, R.A., *et al.*, <u>Polym. Eng. Sci.</u> 1981, <u>21</u>, 10, 624.
46.   Corbett, P.J. and Bown, D.C.,, <u>Polymer</u>, 1970, <u>11</u>, 8, 438.
47.   Dupre, C.R., in <u>US Patent</u>: 4,334,039, 1982, to Monsanto Chemical Co.
48.   Burk, R.D., <u>Eur. Pat. Appl.</u> 15 752, 1980, to Monsanto Chemical Co.
49.   Kim, S., *et al.*, <u>Eur. Pat. Appl.</u> 620 236, 1994, to Fina Technology Inc.
50.   Mark, J.E., <u>Macromol. Chem.</u>, 1978, <u>3</u>, 2, 128.
51.   Okamoto, Y., *et al.*, <u>Macromolecules</u>, 1991, <u>24</u>, 20, 5639.
52.   Demirors, M., <u>Polym. Mater. Sci. Eng.</u>, 1998, <u>79</u>, 162.
53.   Demirors, M., <u>Eur. Pat. Appl.</u> 418 042, 1991, to The Dow Chemical Co.
54.   Bucknall, C.B., <u>Br. Plast.</u>, 1967, <u>40</u>, 12, 84.
55.   Bucknall, C.B. and Clayton, D., Nature (London), Phys. Sci., 1971, <u>231</u>, 22, 107.
56.   Bucknall, C.B., <u>J. Mater.</u>, 1969, <u>4</u>, 1, 214.
57.   Bucknall, C.B. "Role of crazing in the deformation and fracture of rubber-toughened plastics" in <u>Conf. Eur. Plast. Caoutch.</u>, [C. R.], 5th. 1978, Dep. Mater., Cranfield Inst. Technol., Cranfield/Bedford, MK43 OAL, England.
58.   Starke, J.U., *et al.*, <u>J. Mater. Sci.</u>, 1997, <u>32</u>, 7, 1855.
59.   Bucknall, C.B., Karpodinis, A., and Zhang, X.C., <u>J. Mater. Sci.</u>, 1994, <u>29</u>, 13, 3377.
60.   Bucknall, C.B., *et al.*, <u>Macromol. Symp.</u>, 1996, <u>101</u>, (5th International Polymer Conference "Challenges in Polymer Science and Technology", 1994, 265.
61.   Matsuo, M., Wang, T.T., and Kwei, T.K., <u>J. Polym. Sci.</u>, Part A-2, 1972, <u>10</u>, 6, 1085.
62.   Kent, R.W.Jr. <u>US Patent</u>: 4,268,652, 1981, to The Dow Chemical Co.
63.   Bredeweg, C.J., <u>US Patent</u>: 4,239,863, 1980, to The Dow Chemical Co.
64.   Calvert, W.C., <u>US Patent</u>: 3,238,275, 1966, to Borg-Warner Corp.
65.   Li, I.Q., *et al.*, <u>Macromolecules</u>, 1997, <u>30</u>, 18, 5195.
66.   Li, I.Q., *et al.* "<u>Block copolymer preparation using normal/living tandem polymerization</u>" in <u>Book of Abstracts</u>, 213th ACS National Meeting, San Francisco, April 13-17. 1997: American Chemical Society, Washington, D. C.

POLY(VINYL CHLORIDE)

A. WILLIAM COAKER

A. W. Coaker and Associates, Inc., 6726 Chadbourne Drive, North Olmsted, Ohio 44070

Introduction
Historical Background
Manufacture of Vinyl Chloride Monomer
PVC Resin Manufacture and Properties
PVC Usage, Compound Design and Additives
Processes for Compounding PVC
Processes for the Fabrication of PVC-based Items
Specifications for, and Quality Control Testing of PVC Products
Environmental and Occupational Safety/Health Considerations and Green Politics
Future Projections
Literature Cited

Introduction

The composition of poly(vinyl chloride) consists of 56.8% chlorine, 38.4% carbon and 4.8% hydrogen. The relative abundance by mass of these elements in the earth's crust, hydrosphere and atmosphere have been estimated to be 0.205% chlorine, 0.19% carbon and 0.97% hydrogen (1). Assuming that electrical energy is available to electrolyze brine, chlorine will be an element mankind can tap inexpensively for use in chemical reactions, structural materials, medicines, etc. A high and growing proportion, approximately 50%, of the chlorine produced in the United States finds a safe and useful home in PVC.

In this chapter an overview of the current status of the PVC industry in the United States is presented, along with some relatively short-term projections. Long term projections have been made by the Vinyl Institute. For comparative interest, a few statistics on the world-wide PVC industry are also shown.

Historical Background

In 1835 Justus von Liebig discovered that the monomer, vinyl chloride, was produced by the reaction of ethylene dichloride with alcoholic potash. Von Liebig assigned confirmation of the reaction to his student, Victor Regnault, who did so and was then allowed to publish in his name alone and be credited with the discovery. In 1838 Regnault reported what he thought to be polymerized vinyl chloride, but later study showed it to be poly (vinylidene chloride). In 1872 E. Baumann exposed vinyl chloride in sealed tubes to sunlight. He obtained a solid product with the properties of poly(vinyl chloride) (PVC) (2, 3, 4, 5).

Over the period 1912 to 1915, Fritz Klasse in Germany studied potential uses for calcium carbide available in excess as the use of acetylene for lighting declined. Klasse found that vinyl chloride can be made by a process suitable for commercial scale-up by addition of hydrogen chloride to acetylene.

Various companies in Germany continued to build on Klasse's work. When World War II started a considerable number of both flexible and rigid products were being fabricated from PVC in Germany. PVC pipe was important as a replacement for corrosion-resistant metals urgently needed for military purposes.

Over the years 1926 to 1933 Waldo Semon at BF Goodrich found that some organic high-boiling liquids such as dibutyl phthalate and tricresyl phosphate acted as flexibilizers for PVC and developed products using flexible plasticized PVC. In 1934 Frazier Groff at Union Carbide discovered that alkaline earth and lead soaps work as heat stabilizers for PVC. Meanwhile, Waldo Semon was using basic carbonate of white lead as a heat stabilizer and tricresyl phosphate or dibutyl phthalate as plasticizers in PVC products. By this time, Union Carbide was able to produce vinyl chloride by dehydrochlorinating ethylene dichloride, of which they had an excess. Monsanto sold their DOP patent to Union Carbide since Carbide was concentrating on PVC, and Monsanto on polyvinyl butyral.

In 1936 Union Carbide initiated commercial production of both vinyl chloride homo- and co-polymers. Waldo Semon preferred the homopolymers for his flexible, plasticized products, an example of which was the flexible vinyl sheeting discovered and trademarked Koroseal in 1931. General Electric developed plasticized PVC wire insulation trademarked Flamenol. Gresham at BF Goodrich selected di-2-ethylhexyl phthalate (DOP) as the best plasticizer for PVC. Both Union Carbide and BF Goodrich then became producers of PVC and DOP. Yngve and Quattlebaum at Union Carbide discovered organo tin soap stabilizers for PVC.

In both the United States and Europe wartime shortages of numerous conventional materials such as rubber, leather, copper and brass stimulated development and manufacture of flexible and unplasticized PVC as replacements for many war-short items.

After the war, the Allies investigated German industry and published extensive information about both the polymerization and processing of PVC in the wartime Reich. In 1945 the United States Armed Forces were specifying flexible PVC for use in a number of flame-retarded and mildew-resistant items. By 1950 five United States' companies were manufacturing PVC. In 1952 the U. S. Navy started buying impact-modified rigid PVC pipe developed by Parks and Jennings at BF Goodrich. They also developed a process for molding the rigid PVC fittings needed for tough, corrosion-resistant piping installations (6, 7).

Due to its versatility, some unique performance characteristics, relatively low cost and adequate or superior processing behavior in particular types of forming equipment, PVC has become the polymer of choice in enough applications to be the world's second largest selling thermoplastic, behind polyethylene. In 1997, global sales of PVC were approximately 52 billion pounds. This translates into considerably more pounds of PVC-based products after adding in the many other formulating ingredients such as fillers, impact modifiers, plasticizers, pigments, stabilizers, lubricants and specialty additives used along with the PVC resin in commercial products.

In studying published statistics relating to products using significant amounts of PVC in their composition, it is always necessary for correct interpretation of the numbers to determine whether the weights quoted refer to the whole product or only to the amount of PVC used in them. Floor tile is an example of a popular product in which the total product weight may be greater than the weight of PVC used by a factor of six or seven to one.

Manufacture of Vinyl Chloride Monomer.

Vinyl chloride monomer (VCM), sometimes abbreviated VCl, is the precursor of PVC homopolymers and the main ingredient of copolymers and terpolymers of vinyl chloride. It is usually depicted as:

$$\begin{array}{ccc} H & & Cl \\ | & & | \\ C & = & C, \text{ or } CH_2 = CHCl \\ | & & | \\ H & & H \end{array}$$

Among the earlier processes for making VCM, pyrolysis of 1, 2-dichoroethane (EDC) was preferred because ethylene was cheaper than acetylene and this process yielded a purer VCM than did the acetylene-based process. But it yielded large quantities of by-product HCl which were not always easy to use or dispose of. Therefore some companies built so-called balanced plants in which ethylene was chlorinated to EDC and then pyrolyzed:

$$CH_2 = CH_2 + Cl_2 \rightarrow CH_2Cl - CH_2Cl \tag{1}$$

$$CH_2Cl - CH_2Cl \xrightarrow[\text{Heat}]{} CH_2 = CHCl + HCl \tag{2}$$

The HCl produced in equation (2) was then reacted with acetylene to produce an equal amount of VCM:

$$HC \equiv CH + HCl \rightarrow CH_2 = CHCl \tag{3}$$

The acetylene used in equation [3] was usually less pure than the ethylene used in reaction [1]. The net effect was that the VCM from equation [3] was less pure than that from equation [2]. Quality of PVC made from EDC-derived VCM was better than that from the less pure acetylene-derived VCM.

Economic pressure to devise a different kind of balanced VCM plant led to development of the oxychlorination reaction represented as:

$$H_2C = CH_2 + 2 HCl + \tfrac{1}{2}O_2 \rightarrow CH_2Cl - CH_2Cl + H_2O \tag{4}$$

Combining equation [4] with equation [1] and equation [2] taken twice gives the overall oxychlorination process:

$$2 H_2C = CH_2 + Cl_2 + \tfrac{1}{2}O_2 \rightarrow 2 CH_2 = CHCl + H_2O \tag{5}$$

It is reported that this type of ethylene-based process today is used for about 94% of vinyl chloride production capacity (8).

The Transcat process has been known for some time. In this, VCM is produced in a series of steps after reacting chlorine with ethane. It has not been commercially favored because some of the steps required temperatures in excess of 500°C. EVC International has announced a one thousand metric tons per day semi-works plant using their new Transcat catalyst system whereby they achieve 90% conversion at temperatures below 500°C. This plant is in Wilhelmshaven in Germany.

World production of VCM for 1996 is summarized in Table 1 (8):

Table 1---World Production of VCM—1996

Millions of Pounds

	North America	Western Europe	Japan	Other Asia	Other Regions	Total
Production	16,277	12,577	6,150	8,523	6,885	50,412
Operating Rate	92%	89%	88%	91%	70%	87%

Future consumption of VCM will depend on the volume of PVC and related resins produced. About 99% of VCM used goes into PVC.

Industrially, VCM is handled as a liquid under pressure. Its boiling point is -13.4°C at 760 mm pressure. The vapor pressure of vinyl chloride is 115 psia at 50°C and 180 psia at 70°C. At typical ambient temperatures and pressures, VCM is a colorless, sweet-smelling vapor, whose explosive limits in air fall between 4 and 22% by volume. Both the liquid and the vapor are flammable. It is conveyed by pipeline where this is possible, or shipped in railroad tank cars or ocean tankers. Depending on ambient temperatures and time in transit, VCM may need to be inhibited against polymerization. Low levels of phenol are effective. These can be removed in a caustic wash. During handling, exposure to atmospheric oxygen is normally denied by use of nitrogen blankets. This limits possibilities of accidental explosion or adventitious initiation of polymerization. The heat of polymerization for VCM is 660 Btu/lb.

In 1974 the occurrence of angiosarcoma of the liver was reported among polymerization reactor cleaners who had a history of exposure to high concentrations of VCM. The United States Occupational Safety and Health Administration (OSHA) proposed and later established maximum exposure limits for workers in the VCM and PVC industries as: a time-weighted average (TWA) of one part per million over any eight-hour period, with a ceiling of five parts per million for any period not exceeding fifteen minutes. Most countries with PVC industries have adopted these limits or established analogous ones.

The international environmental organization, Greenpeace, is conducting a world-wide campaign against actually or allegedly polluting industries, with particular emphasis on chlorine-containing chemicals. In the United States, the Vinyl Institute, the Manufacturing Chemists Association, the Chlorine Institute and individual VCM and PVC manufacturers dispute Greenpeace's allegations relating to issues such as the relative amounts of dioxins produced in the course of making VCM or incinerating PVC, on the one hand, and by forest fires and volcanoes on the other (i.e., anthropogenic versus "natural" sources). Questionable attacks by environmentalists on the PVC industry have forced members of the industry to become more aware of public relations and political issues. However, the VCM and PVC industries need to become far more pro-active in presenting the demonstrable environmental benefits of their products to the public. A recent example of the procedures followed by Greenpeace in an attack on the use of flexible vinyls in toys and some industry responses was discussed by W. Storck (9)

In 1998, the eight largest producers of VCM in the United States were(10):

Company	Annual Capacity Millions of Pounds
Dow Chemical Company	3,600.
The Geon Company	2,400.
Formosa Plastics	2,200.
Oxy Mar	2,100.
Georgia Gulf	1,600.
Westlake Chemical	1,100.
PHH Monomers	1,100.
CONDEA Vista	1,000.

PVC Resins Manufacture and Properties.

Molecular Structure and Crystallinity.   The definition of PVC resins for the purpose of this chapter is polymers containing 50 weight percent or more of vinyl chloride units:

$$\begin{matrix} H & Cl \\ | & | \\ -C ----- C-, & or - CH_2 - CHCl- \\ | & | \\ H & H \end{matrix}$$

The molecular weight of vinyl chloride monomer, and thus of the repeating unit in vinyl chloride homopolymers, is roughly 62.5, comprising 56.8 weight percent chlorine, 38.4 per cent carbon and 4.8 per cent hydrogen.

In typical PVC homopolymers, an average of 400 to 1000 vinyl chloride units are joined to each other. In vinyl chloride copolymers and terpolymers, other monomers such as vinyl acetate, vinylidene chloride, ethylene or propylene constitute less than 50 weight percent of the polymer molecules.

Manufacture of vinyl chloride-vinyl acetate copolymers peaked around 1974. Since then the volume of these copolymers has steadily declined in the United States. Ways were found to replace them with relatively low molecular weight PVC homopolymers which were more cost-effective. Other copolymers have mostly been used in relatively low volume specialty applications involving plastisols and powder-coating resins, for instance. The discussion of molecular structure and crystallinity will concentrate, therefore, on PVC homopolymers.

Commercial PVC resins are made by addition polymerization of vinyl chloride. Chlorination of PE results in products referred to as chlorinated polyethylene (CPE) and these all differ markedly from PVC.

If the group in vinyl chloride containing the chlorine is referred to as the head (–CH Cl–) and the methylene group is referred to as the tail (–CH$_2$–), possibilities for addition polymerization include:

– orderly head-to-head followed by tail-to-tail;   – random head-to-head or tail;
– orderly head-to-tail

The third one predominates in the manufacture of commercial PVC. In PVC homopolymers, the chlorine atoms, typically, are attached to every other carbon atom. Growth of the polymer molecules during polymerization may generally be represented as:

$$
\begin{array}{cccc}
\text{H} & \text{Cl} & \text{H} & \text{Cl} \\
| & | & | & | \\
\text{R}-\text{C}-\text{C}\cdot + \text{C} = \text{C} \\
| & | & | & | \\
\text{H} & \text{H} & \text{H} & \text{H}
\end{array}
\rightarrow
\begin{array}{cccc}
\text{H} & \text{Cl} & \text{H} & \text{Cl} \\
| & | & | & | \\
\text{R}-\text{C}-\text{C}-\text{C}-\text{C}\cdot \\
| & | & | & | \\
\text{H} & \text{H} & \text{H} & \text{H}
\end{array}
\tag{6}
$$

Evidence exists that sometimes head-to-head addition does occur, and that when it does a re-arrangement may then produce a "defect structure" in the PVC molecule:

$$
\begin{array}{cccc}
\text{H} & \text{Cl} & \text{H} & \text{H} \\
| & | & | & | \\
\text{R}-\text{C}-\text{C}\cdot + \text{C} = \text{C} \\
| & | & | & | \\
\text{H} & \text{H} & \text{Cl} & \text{H}
\end{array}
\rightarrow
\begin{array}{cccc}
\text{H} & \text{Cl} & \text{H} & \text{H} \\
| & | & | & | \\
\text{R}-\text{C}-\text{C}-\text{C}-\text{C}\cdot \\
| & | & | & | \\
\text{H} & \text{H} & \text{Cl} & \text{H}
\end{array}
\rightarrow
\begin{array}{ccccc}
\text{H} & \text{Cl} & & \text{H} & \\
| & | & & | & \\
\text{R}-\text{C}-\text{C}-\text{CH}-\text{C}-\text{Cl} \\
| & | & \cdot & | & \\
\text{H} & \text{H} & & \text{H} &
\end{array}
$$

$$
\begin{array}{cc}
\text{H} & \text{Cl} \\
| & | \\
+ \text{C} = \text{C} \\
| & | \\
\text{H} & \text{H}
\end{array}
\rightarrow
\begin{array}{cccccc}
\text{H} & \text{Cl} & \text{H} & & \text{H} & \text{Cl} \\
| & | & | & & | & | \\
\text{R}-\text{C}-\text{C}-\text{C}- & & \text{C}-\text{C}\cdot \\
| & | & | & & | & | \\
\text{H} & \text{H} & \text{CH}_2\text{Cl} & & \text{H} & \text{H}
\end{array}
\tag{7}
$$

which goes on growing beyond the chloromethyl branch.

Another outcome may be the occurrence of chain transfer to monomer by the following mechanism:

$$
\begin{array}{ccccc}
\text{H} & \text{Cl} & & \text{H} & \\
| & | & & | & \\
\text{R}-\text{C}-\text{C}-\text{CH}-\text{C}-\text{Cl} \\
| & | & \cdot & | & \\
\text{H} & \text{H} & & \text{H} &
\end{array}
\rightarrow
\begin{array}{ccccc}
\text{H} & & \text{H} & & \text{H} \\
| & & | & & | \\
\text{R}-\text{C}-\text{C}=\text{C}-\text{C}-\text{Cl} + \text{Cl}\cdot \\
| & & | & & | \\
\text{H} & & \text{H} & \text{H} & \text{H}
\end{array}
\tag{8}
$$

followed by:

$$
\text{Cl}\cdot +
\begin{array}{cc}
\text{H} & \text{Cl} \\
| & | \\
\text{C} = \text{C} \\
| & | \\
\text{H} & \text{H}
\end{array}
\rightarrow
\begin{array}{cc}
\text{H} & \text{Cl} \\
| & | \\
\text{Cl}-\text{C}-\text{C}\cdot \\
| & | \\
\text{H} & \text{H}
\end{array}
\tag{9}
$$

These types of reactions are credited with both controlling the molecular weight of PVC and reducing its molecular weight as polymerization temperature is increased.

Another important feature of the polymerization of vinyl chloride is that after the addition of only about eight monomer units, the telomer precipitates out of solution in the monomer and continues growing as a solid swollen by its own monomer, provided polymerization is carried out below the glass-transition temperature of PVC. This behavior leads to the formation of so-called primary particles, which are much smaller than the 3 to 16.5 mil (74 to 420 micron) "grains" produced in suspension polymerization.

If short sequences of vinyl chloride units are visualized as planar zig-zags with the carbon atoms on the average in the plane of the paper on which they are depicted, the asymmetric carbon atoms (those having both a hydrogen and a chlorine) can have different spatial relationships to each other. The carbons having two hydrogens may be considered to be below the plane of the paper, and the asymmetric carbons marked with asterisks may be

```
 (1)      (2)      (3)      (4)      (5)
 Cl    H   Cl   H   Cl   H    H    H   Cl    H
 |     |   |    |   |    |    |    |   |     |
 C  —  C — C  — C — C  — C — C —  C — C  —  C
 |     |   |    |   |    |    |    |   |     |
 H     H   H    H   H    H    Cl   H   H     H
 *         *        *         *         *
```

visualized as being above the plane of the paper. Chlorines (1), (2) and (3) which are on the same side of the polymer chain are said to be isotactic or meso (m) with respect to each other. Chlorines (3), (4) and (5) which are on opposite sides of the polymer chain to each other are said to be syndiotactic or racemic (r) to each other. A pair of adjacent asymmetric carbon atoms is called a diad; three pairs in a row are called a triad, and so forth. If three in a row are syndiotactic, this may be called an rr triad. If four in a row are isotactic, this is referred to as an mmm tetrad. If three in a row are heterotactic they are described as rm or mr. If this type of arrangement predominates, the polymer is described as atactic or amorphous, since only syndiotactic or isotactic conformations tend to crystallise.

Pham, et al., (11) used ^{13}C NMR to measure the relationship between syndiotactic fraction in PVC and polymerization temperature. They reported:

Syndiotactic Fraction	Polymerization Temperature °C
0.55	55.
0.57	25.
0.60	0.
0.64	-30.
0.66	-50.
0.68	-76.

Measurements of syndiotacticity of PVC, based on infrared methods, have indicated higher syndiotacticity of PVC polymerized at temperatures below 0°C. (12, 13)

Even though commercial PVC is generally reported to be only slightly crystalline (5 to 10%) the effects of this low degree of crystallization on performance properties and behavior are considered profound, by comparison with amorphous, atactic polymers. It is documented that PVC can form at least three types of crystallites and that these melt over wide ranges of temperature. Witenhafer (14) and Illers (15) have shown by DTA that PVC crystallites can melt over the temperature range 105 to 210 °C. Wenig (16) has proposed a lamellar crystal structure for PVC. Blundell (17) found evidence of PVC crystallites having a more spherical structure.

It is currently accepted that highly syndiotactic PVC produced by heroic polymerizaiton methods is very difficult to characterize quantitatively. It is reluctant to dissolve except in the form of multi-molecular agglomerates and intractable in processing even into very simple test specimens. Methods for obtaining very precise control of the syndiotactic sequences in PVC will be necessary before it will be possible to take commercial advantage of the superior properties potentially available by decoupling and independently controlling crystallinity, molecular weight and defect structures in PVC. Isotactic PVC, if feasible to make inexpensively, may turn out to have interesting properties.

Molecular Weight of PVC. Commercial PVCs manufactured by suspension, emulsion, mass or solution polymerization today are all initiated by free radicals. The Edison Polymer Innovation Corporation (EPIC) is putting together a consortium to study innovative ways to polymerize VCM and process PVC (18). Until such studies bear fruit, the vinyl industry is locked into use of free radical addition processes with their associated limitations regarding the molecular weights and crystalline fractions of the resultant polymers.

In the processes, the molecular weight of the PVC produced is regulated by the net effects of chain propagation, chain transfer to monomer and chain termination by disproportionation or by coupling. Molecular weight modifiers, such as chain transfer agents and cross-linking agents may be added to lower or raise the averages corresponding to particular polymerization temperatures.

In the early days of the PVC industry it was common practice in suspension polymerization to use monomer-soluble initiators with relatively long half-lives at polymerizaiton temperatures. To get polymerization started at a reasonable rate, these initiators would be charged at relatively high concentrations to batch processes, and their concentration in the residual monomer would increase as monomer became depleted. Simultaneously, batches would become more viscous as the concentration of polymer particles increased, resulting in decreased heat transfer capability. The net effect was considerable acceleration of polymerization towards the end of a batch, faster heat evolution and a so-called "heat kick", i.e., non-isothermal final stage of the batch. During this end stage conversion rapidly increased from about 80 to 96% and much of the chain termination was by disproportionation or coupling. When polymerization virtually stopped due to monomer starvation, the last 4% of unreacted monomer was vented off and a fair amount of unreacted initiator, such as lauroyl peroxide, was left entrained in the PVC particles. Some degradation of the PVC was then common during heated drying of the PVC particles after the loose suspending water had been centrifuged off. In these early days it was also usual for the premium quality resins to be polymerized to only about 70% conversion. These premium batches were run isothermally to the target conversion at which unreacted monomer was vented off before occurrence of a heat kick.

For the last 25 years there has been much better availability of initiators with relatively short half-lives at polymerization temperatures. Examples are di-isopropylperoxydicarbonate or acetyl cyclohexylsulfonylperoxide. Using these initiators, batches can be brought up to the full desired polymerization rate quickly and maintained at full rate at the desired temperature until the target conversion is reached. Batches may then be short-stopped and transferred to a blow-down vessel for removal of most of the unreacted monomer followed by stripping to a low residual monomer content prior to final drying. In today's plants, once a satisfactory polymerization protocol has been established for a given product the process is controlled by computers. An operator only intervenes when something unexpected happens (e.g., failure of a so-called uninterruptible power supply).

The active part of the PVC polymerization cycle in most current suspension processes does not deviate from a simplified model nearly as much as the early processes did. However, the kinetics are still complex. M. Langsam has provided a summary of suspension, bulk and emulsion processes with extensive references (19). The recipes, temperatures and statistics of polymerization control the number average and weight average molecular weights of the PVCs obtained.

Number-average molecular weight,

$$M_n = \frac{\sum_i N_i M_i}{\sum_i N_i}$$

and, weight-average molecular weight,

$$M_w = \frac{\sum_i N_i M_i^2}{\sum_i N_i M_i}$$

where N_i is the number of PVC molecules of molecular weight M_i.

The primary standards for measuring number average molecular weight are based on osmotic pressure. Light-scattering techniques are the primary standard for measuring weight-average molecular weight.

Due to cost and efficiency requirements secondary standards based on gel permeation chromatography (GPC) are most often used for resolving complex issues relating to the average molecular weights of particular PVC resins. A referee method based on high-performance size-exclusion chromatography (HP-SEC) has been described by Perkins and Haehn. (20).

For routine commercial quality control and description of most PVC resins, dilute solution viscosity methods are used for measuring their average molecular weights. Despite years of effort to standardize, discrepancies are still encountered between data published by different manufacturers on the molecular weights of PVC resins. Coaker and Wypart presented a description of popular dilute solution viscosity methods and a table of comparative results for inherent viscosity by ASTM D 1243-66, intrinsic viscosity in cyclohexanone at 30°C, Fikentscher K Value, three commonly used specific viscosity tests and a relative viscosity test compared with number average molecular weights (21).

PVC molecules tend to aggregate in solution. Heating PVC solutions and then cooling them prior to using them for determining molecular weights overcomes this problem, if handled correctly. Daniels and Collins (22) and Abdel-Alim and Hamielic (23) have addressed and shown ways to solve some of these problems.

Care is necessary in comparing molecular weight data from Europe, Japan and the United States due to regional methodological preferences which may lead to misunderstandings between technicians from different regions.

<u>Particulate Architecture of PVC.</u> PVC resins made for varying uses are tailored for those applications with different particle sizes, shapes and internal structures.

Average Size and Size Distribution. PVC suspension and mass resins made for general-purpose uses typically have particle size specifications such as: > 99.8% through a U. S. 40-mesh sieve, whose nominal openings are 16.5 mils (420 microns) square; 10% maximum by weight retained on a U. S. 60-mesh sieve whose openings are 9.8 mils (250 microns) square; and 2% maximum through a U. S. 200-mesh sieve whose openings are 2.9 mils (74 microns) square. The aim point for average particle size (APS) is normally between 100 and 80-mesh for PVC general purpose resins; that is between 5.9 mils (149 microns) and 7.0 mils (177 microns). A relatively narrow particle size distribution is desirable and is mandated by many users' and manufacturers' specifications.

Typical PVC suspension and mass resins made for special purposes may have their APSs skewed to the fine or coarse ends of the range 140 to 80 mesh (4.1 to 7.0 mils or 105 to 177 microns).

Plastisol extender resins made by suspension polymerization need their APSs skewed considerably further to the fine end. Extender resins in a plastisol, if too coarse, tend to settle in the plastisol and also give a rough surface to the plastisol after fusion in an open oven, such as those typically used for plastisol-derived flooring sheet goods. These resins usually require >99.6% through 140 mesh, 5% maximum on 200 mesh and >65% through 325 mesh. Their APS thus is equal to or less than 1.7 mils (44 microns), which is the size of the 325 mesh screen openings.

Plastisol resins made by emulsion polymerization have controlled particle size distributions which are proprietary to individual manufacturers. In some cases the desired distribution of particle sizes is achieved by using a 0.3 micron monodisperse seed latex whose particles are grown to about 1 micron in the plastisol resin batch. Other particles are initiated during the second polymerization. On completion of polymerization the final latex contains a controlled range of particle sizes from 1 micron in diameter down to about 0.1 micron in some cases.

During concentration and dewatering of plastisol resin latexes, some particle-to-particle agglomeration occurs, but most of these agglomerates are broken up in a grinding step which follows dewatering and reduction of residual monomer content of the resins. A few of the primary emulsion-polymerized particles shatter during grinding. The resin particle size distribution in a typical plastisol based solely on emulsion-polymerized resin consists primarily of emulsion resin particles together with some agglomerates and some fragments of particles shattered during grinding. Solids such as fillers, pigments, fire retardants and smoke suppressants, along with any plastisol extender resins used, contribute to the overall particle size and shape distribution in a commercial plastisol, which often is very complex.

Microsuspension resins used in plastisols differ from emulsion-polymerized plastisol resins both in surfactant content and particle size distribution. Microsuspension plastisol resins generally have lower surfactant content than their emulsion-polymerized counterparts. Microsuspension resins are targeted for applications requiring high clarity, low moisture sensitivity and special properties such as little or no contribution to fogging in automobile interiors.

The PVC resins made for use in specialty applications such as rotational molding from powders and powder-coating of preheated parts are made to exacting particle size distribution specifications. Some of them are copolymers for blending with homopolymers to get special performance results.

Shape, Internal Structure and Porosity. The comments in this section apply to the particles of PVC resins polymerized in suspension, mass and emulsion processes. Solution vinyl resins are a specialty dealt with in dedicated texts (24). Solution vinyl resins in most cases are used for surface coatings applied to substrates from solution. These resins are normally studied in solution or after coating onto substrates. Their particle size, composition and morphology in the dry state are tailored to make it easy for them to go into solution quickly and completely. Their molecular weight distributions and compositions provide good clarity and excellent film-forming properties. Most of them are copolymers. Some of the comonomers used are selected to enhance adhesion to substrates.

General purpose PVC homopolymers typically are made into grains between 2.9 mils (74 microns) and 16.5 mils (420 microns) in average diameter. When studied internally at progressively smaller size levels, these grains are found to have an elaborate microstructure which profoundly affects their performance properties.

In the PVC industry low particle-porosity is targeted for applications benefiting from high resin bulk density and in which only small amounts of or no liquid formulating ingredients are used. High porosity historically was built into the resin particles when they were designed for dry blending with large amounts of high molecular weight monomeric plasticizers, or with viscous polymeric plasticizers. High particle porosity is obtained by running batches to low final conversions, such as 70%, and venting off the remaining vinyl chloride. This property can also be obtained by use of proprietary porosity enhancers as formulating ingredients in suspension polymerizations run to conversions as high as 95%.

The size, size-distribution and internal porosity of PVC grains formed in suspension polymerization of vinyl chloride is a function of many factors, including the types and amounts of suspending agents used, agitation conditions and porosity modifiers, if any, used. Good suspending agents stabilize monomer droplets over the desired range of sizes early (i.e., at low conversions) in the polymerization cycle. They also prevent batches from "setting up" later in the cycle when the particles have become predominantly solid and somewhat sticky (PVC swollen with unreacted VCM). The ratio of water to VCM plus PVC in suspension batches has to be high enough to maintain potential fluidity given the packing efficiency of the PVC particles in the liquid water and free monomer. Reproducible control of particle size and porosity distributions from one batch to the next is essential for meeting the demands of today's markets.

Authorities are not agreed on all of the mechanisms contributing to formation of porosity in PVC particles during suspension and mass polymerizations. One obvious mechanism is the shrinkage of vinyl chloride as it forms PVC. Monomer density at polymerization temperatures is about 0.85 g/ml. PVC slightly swollen by vinyl chloride has a density about 1.35 g/ml., and particle size is established fairly early in the cycle. By the end of the batch there must be open space within each particle unless the particle shrinks as polymer is formed. The amount of porosity in each particle is enhanced if polymerization is interrupted and unreacted monomer is vented off. Absorption or entrainment of water into the grains under some circumstances may be a contributing mechanism for maintaining high particle porosity (large pore volume and pore diameter). Another factor which depends on polymerization rate and heat transfer factors is that monomer inside particles may boil during polymerization, particularly if some of the heat of polymerization is removed from the reactors by reflux condensers.

Within the range of general-and-special-purpose PVC resins made by suspension and mass polymerizations, measurable particle porosities may be controlled at levels between 0.18 and 0.57 ml/g (25).

Suspension and mass PVC grain shape aims generally, but not exactly, at the spherical. Sphericity provides good bulk flow properties of the resins, along with efficient particle packing. Irregular, knobby particle shapes, along with high porosity and low bulk density give poor bulk flow.

Emulsion and microsuspension resins for use in plastisols are deliberately made to have essentially solid particles. This is done to provide low viscosity and to enhance the viscosity stability of the plastisols made from them by inhibiting plasticizer absorption by the resin particles until the plastisols are purposely heated to bring about gelation. Similarly, plastisol extender resins are also made with relatively solid particles.

Geil proposed terminology relating to the morphology of PVC suspension- and mass-polymerized resin particles (26). The lower end of Geil's size hierarchy also applies to emulsion polymerized resins. The following is adapted from Geil:

Table 2. Size Hierarchy of PVC Particulate Phenomena

Term	Size	Description
Grain	70-420 μm Diameter	Free-flowing powder produced in suspension or mass polymerizations
Agglomerates of Primary Particles	3-10 μm Diameter	Formed during polymerization by coalescing of primary particles.
Primary Particles	1 μm Diameter	Formed at single polymerization sites by precipitation of newly-formed polymer into discrete molecular aggregates.
Domains	0.1 μm Diameter	Observed after certain types of mechanical working of the PVC.
Microdomains	0.01 μm Spacing	Crystallites capable of holding portions of tie molecules which maintain integrity of primary particles, and span between primary particles in well fused PVC.
Secondary Crystallinity	0.01 μm Spacing	Crystallites re-formed during cooling of a fused PVC melt.

Summers has rationalized many aspects of the relationship between PVC particle morphology and fusion phenomena (27). Many authors, including Berens and Folt (28), Faulkner (29), Singleton, et al. (30), Summers (31), Pezzin (32), Rosenthal (33), Collins and Krier (34), Collins and Daniels (35), and Lyngaae-Jorgensen (36) have contributed to understanding relationships between PVC particle morphology, fusion and the processing rheology of PVC.

Practical processing technologists running extrusion, calendering, injection molding and thermoforming operations use rheological theory where it is helpful, and follow the bottom-line rule, which is, "Run your line as fast as you can while producing acceptable product quality."

An important feature in the processing of both rigid and flexible PVC formulations is that so much frictional heat is generated during processing PVC at high shear rates that the stock temperatures reached are primarily shear-rate dependent for formulations of a given melt viscosity in a standardized test.

Visible defects due to trying to run a process too fast include "shark skin" finishes, "melt fracture," failure to hold dimensional tolerances, air occlusion in and visible discoloration of products.

Hidden defects are often more damaging because a large quantity of off-grade may have been produced before the operator realizes anything is wrong. These include loss of needed residual heat stability, loss of needed light stability, subsequent warpage due to built-in strains, low sag temperature and low heat distortion temperature.

Incipient degradation causing loss of residual stability may often be detected by checking parts for fluorescence under UV light, which reveals the presence of short sequences of double bonds in incipiently degrading PVC molecules.

PVC Usage, Compound Design and Additives

Handling and Processing PVC Resins. All PVC resins as produced in polymerization facilities need to be mixed with formulating ingredients and fluxed in the course of processing them into useful articles. Rules imposed by OSHA and EPA in the United States require manufacturers of suspension and mass PVC to strip their resins to low levels of residual vinyl chloride monomer, typically 1 to 5 ppm (37). Exporters and importers of PVC resins need to pay attention to the rules, if any, imposed by the country of destination regarding VCM residuals in PVC resins. In the United States, PVC prime, off-grade and scrap suspension resins are required by OSHA to carry special cautionary labels whenever their residual VCM content exceeds 8.5 ppm (37). In handling PVC resins, VCM vapor concentrations within explosive limits (4 to 22% in air) must not be allowed to exist anywhere even momentarily. For workers in the resin user's location, the U. S. OSHA action level for triggering attention to worker exposure to VCM vapors is 0.5 ppm averaged over any 8-hour work day. A good feature of PVC is that it does not de-polymerize and generate free VCM.

In dust-explosion studies in air, for large particle size PVC resins (130 microns APS or higher) and medium particle size resins (75 microns APS) no hazard was found (37). However, a low order of dust explosion risk was determined for fine particle size PVC resins (APS 2 microns) (37).

Most suspension and mass PVC resins have apparent bulk densities of 0.47 g/ml (29.3 lb/ft^3) or higher and flow well in and out of hopper cars, bulk trucks and silos in normal circumstances. Bulk flow may be stopped by generation of static electric charges due to unusual atmospheric dryness combined with other aggravating conditions, such as unfavorable system design. Excessive moisture in the resin combined with cold weather can cause freeze-ups in conveyor lines and discharge valves of bulk-handling vehicles. To minimize bulk flow problems, recommended types of resin hoppers, air-conveying systems, weighing systems, etc., have been described (38).

Schwaegerle (39) found that the densities and internal and external voids for a range of PVC powders are related to one another by the expression:

$$\frac{1}{ABD} = \frac{1}{d} + P + V$$

where ABD is the apparent density, g/ml;
    d is the fused density, g/ml;
    P is the porosity, ml/g;
and   V is the external void space between grains of the PVC, ml/g.
He studied PVC resins over the ABD range of 0.432 to 0.620 g/ml (27.0 to 38.7 lb/ft^3).

PVC resins stored in silos fitted with suitable air pads can be fluidized by back-flushing with air when their flow behavior gets impeded by compaction. Back-flushing with air decompacts the resin. Hammering the hopper usually exacerbates poor flow.

Dry blending, also called powder mixing is the first step in compounding for many processing operations such as extrusion into profiles, calendering into sheets, injection molding into parts, and so forth. Dry blending consists of mixing liquid and solid formulating ingredients with PVC resin in such a way that the resulting mixture is a free-flowing powder

For pipe compounds in which PVC resin was the main ingredient, Schwaegerle (39) found that the properties of the resin from which they were made correlated very highly with the flow properties of the dry blends. He also found that the output rates of twin-screw extruders operating under flood-fed conditions were directly proportional to the bulk densities of the powder compounds introduced into their feed throats. For pressure pipe, it is essential that there be no adventitious contamination in the resin or any of the other formulating ingredients which would cause weak spots in the pipe from which cracks may form and propagate during cycling of pressures in or on the pipe.

Large volume dry blending operations for which pipe grade PVC resins are unsuitable are the manufacture of flexible PVC compounds for use in wire insulation or electrical cable sheathing. For these, higher porosity resins capable of absorbing the flexibilizing plasticizers used in these types of compounds are selected. The electrical properties of the PVC resin are also important for resins used in electrical insulation.

Some dry blends are made for processing as such into finished items after fluxing in extrusion systems, calendering lines, injection molders and so forth. However, considerable quantities of dry blend are made by so-called "compounders" who sell pelletized or cubed compounds for subsequent forming into finished goods. These dry blends must perform well in pelletizing equipment and the pellets must satisfy the end user.

Compounding of liquid plastisols is entirely different. It requires use of plastisol grade resins and suitable plasticizers in liquids-mixing equipment. These may consist of inverted conical Nauta mixers, Pony mixers, high speed Cowles dissolvers, medium speed Ross power mixers and three-roll mills. A. C. Shah has described typical procedures for preparation of vinyl plastisols and organosols (40).

On a global basis, the five largest PVC producers, representing one-third of global capacity, are (10):

Company	Capacity, Millions of lb.
Shin Etsu	6,100.
Formosa Plastics	5,400.
Oxy Vinyls	4,200.
Solvay/BASF	3,500
EVC	3,200

PVC Compound Design. There are two general approaches to formulating vinyl compounds for particular applications. If the physical, stability, decorative, electrical (if any), toxicological (if any), allowable cost, density, odor, etc., requirements for the application are known, formulations which meet or come close to meeting these can be developed and trial items submitted for field testing and test marketing. If the stated requirements are such as to rule out use of vinyl compounds, industry practice is not to submit a vinyl trial part for evaluation unless there is doubt about the validity of the stated needs for the application.

The second general approach to formulating and submitting vinyl parts for a new application is used when the requirements have not been established because the application is a new, untried concept. For this case a set of tentative requirements is guestimated, and parts or items bracketing these tentative needs are submitted for field testing.

In the early days of designing PVC compounds, a great deal of effort was devoted to developing compounds which replaced existing natural materials and had some clear cut advantages. For vinyl electrical wire insulation versus a particular rubber insulation the advantages might be better electrical properties, better flammability, no need to go through a curing operation, better overall economics, superior colorability and better weatherability. Today there are still markets in undeveloped countries which have similar opportunities. In developed countries, most markets are well defined and very competitive. There may be several plastics and plastics' alloys competing and whose economics are close together.

Malpass, Petrich and Lutz (41) published typical formulations for the following types of well established PVC pipe, which constitute the single largest market area for PVC: pressure, drain-waste-and-vent (DWV), sewer, drain, telephone ducts, electrical conduit, small diameter and irrigation pipe. For most of these applications, the formulations are different depending on whether the pipe is extruded in a twin-screw or single-screw extruder. The twin-screw formulations are more economical by virtue of needing less stabilizer. For irrigation pipe, Malpass, *et al.*, give only a twin-screw formulation. Illustrative twin-screw formulations are:

INGREDIENT	PRESSURE PIPE PHR	TELEPHONE DUCT PHR
PVC Resin (K 67-68, IV 0.92)	100	100
Tin stabilizer	0.4	0.4
Calcium stearate	0.8	0.8
Paraffin wax (mp165°F)	1.2	1.2
Titanium dioxide	1.0	1.0 -2.0
Acrylic processing aid	0.75	—
Acrylic lubricant-processing aid	—	0.75
Acrylic impact modifier	—	2.0
Calcium carbonate	0. to 5.	20. to 30

Tin stabilizers are used in most United States pipe compounds, whereas in Europe and Japan lead stabilizers are fairly common. Pressure pipe formulations for most installations are designed for maximum tensile strength and creep and fatigue resistance. PVC resins are selected for optimal purity (absence of adventitious particulate contamination) and low residual VCM content. For special applications such as supplying high pressure water to mines, the pipes are made thicker for extra strength and often contain 8 to 12 phr of a high efficiency impact modifier to give them greater resistance to brittle failure on impact.

Telephone duct formulations normally contain some weather-resistant acrylic impact modifier for extra toughness, often have additional titanium dioxide (TiO_2) for enhanced weatherability and fairly high levels of calcium carbonate for added stiffness and lower cost per pound.

Prior to 1974 which was the year in which it became common knowledge that VCM is a potential carcinogen for human reactor cleaners exposed to VCM at high levels as well as for laboratory animals, the levels of residual VCM in pipe grade PVC were not held below any prescribed target level. To the best of its ability, the PVC industry has replaced old PVC pipe known to or suspected of having been made with relatively high VCM content resins. The VCM has been found to diffuse out of these pipes slowly over many years.

Specific requirements for PVC resins used for pressure pipe include that they must flow well in bulk handling systems with minimal dusting and must not contribute porosity to pipe cross-sections. Bone dry and significantly wet resin must be avoided. Bulk density greater than 0.6 g/ml is desirable. Antistatic properties are built into some pipe grade resins. Consistent and optimized particle size and porosity distributions are maintained. Optimum molecular weight is a compromise. Higher molecular weight provides greater strength, better holding of dimensional tolerances and creep resistance. Lower molecular weight contributes fast fusion, ease of processing and low melt viscosity. The practical limits are between I. V. 0.88 and 0.97 (K65 and K 68) with the optimum at I. V. 0.92 (K 66). Resins for use in potable water pipe must contribute low extractables which are acceptable to the National Sanitation Foundation (NSF) and must have low residual VCM content. NSF's 1998 Standard 61 requires that rigid PVC water pipe also comply with hydrodynamics stress resistance requirements.

Roman (42) reports that PVC currently has a 40% share of the North American pipe market on a footage basis. Its advantages include toughness, durability, corrosion resistance, improved flow characteristics and lower installation cost than competitive materials. PVC use is growing particularly in large-diameter pipe, displacing concrete, iron and clay.

Malpass, et al., (41) describes typical formulations used for injection molded PVC fittings for use with PVC pipe. These are based on much lower molecular weight PVC resins to provide the lower melt viscosity needed for good mold filling as opposed to what performs best in extrusion of pipe. PVC resins used for fittings have I. V. 0.68 to 0.79 (K 57 to K 61).

In 1997, pipe and fittings together were reported to have consumed 43% of the PVC marketed in the United States (42). Siding was next in market size, taking 13% of the PVC sold. After siding came exports, which accounted for 10% of United States PVC sales, but the end uses for exports cannot be tracked.

Rapacki (43) has recently reviewed the materials used for rigid PVC construction products and developments such as introduction of rigid PVC foams. The generic formulations he cites for vinyl siding are for capstock and siding substrate produced by twin-screw extrusion:

INGREDIENT	SIDING CAPSTOCK PHR	SIDING SUBSTRATE PHR
PVC (K 66)	100.0	100.0
Hi-tin stabilizer	1.2	0.9
Calcium stearate	1.3	1.3
Paraffin wax	1.0	1.0
Polyethylene wax	0.1	0.15
Titanium dioxide	10.0	1.0
Calcium carbonate	3.0	10.0
Acrylic impact modifier	5.0	5.0
Processing aid	—	1.0
Lubricating processing aid	0.5	0.5

It is normal to laminate the capstock onto the substrate and to sell the siding as one piece of goods with two layers well adhered together. The capstock and substrate need to have essentially the same coefficient of thermal expansion to prevent bowing or delamination with temperature changes. The capstock has a high concentration (10 phr) of TiO_2 for light stability and to reflect solar heat. The substrate is cheapened by use of 10 phr of calcium carbonate.

Several composite sidings are being offered today made with PVC substrates and highly weather-resistant capstocks made from other polymers such as polyvinyl fluoride (PVF).

Rapacki (43) also lists a generic formulation for other large volume uses for rigid PVC such as window profiles. This formulation is fairly similar to siding capstock.

The majority of items made from calendered PVC film and sheet are flexible or semi-rigid. Exceptions are blister-packaging film, (mostly clear, which is generally calendered and then thermoformed into specific package shapes), credit card stock and floppy disc jacketing. Additional significant calendered products include various flooring constructions, other than those made from dispersion-coating vinyls, pond and pool liners, furniture upholstery, films used in automobiles such as skins for covering crash pads, doors, arm rests, ceilings and artificial leather seating. Vinyl shower curtains and some vinyl wall-coverings are also made by calendering.

In many instances, vinyl items made by dispersion-coating with plastisols compete with and complement items made from calendered flexible vinyl sheet. Most flooring sheet goods are made from vinyl plastisols along with many vinyl-coated fabrics. Vinyl backings for carpets are mostly made from plastisols. Many furniture coverings made with a foam layer overlaying a fabric and covered by a vinyl skin are based on plastisol technology. Vinyl adhesives, sealants and automobile undercoatings are made from plastisols or modified plastisols (sometimes called organosols because they contain significant amounts of organic solvents and/or thinners). Vinyl bottles are mostly blow-molded from rigid vinyl compositions. Vinyl bottle-closures are mostly made from plastisols. Numerous other flexible items, such as vinyl toys are made by rotational molding vinyl plastisols. Traffic cones are often made by molding vinyl plastisols.

A specialised application area for flexible vinyls is wire and cable. In the United States, most insulated wires and jacketed cables run outdoors use low cost polyethylene (PE) wire insulations and cable jacketings. These have very high calorific values and rapid heat release once they are ignited. Inside buildings the United States National Electrical Code (NEC) (44) calls for use of insulations and jacketings which are more difficult to ignite, more reluctant to burn and give lower heat release. As far back as 1946, The Bell Telephone Company started using PVC in central office switchboard wiring. Additional uses for PVC have been phased in every few years since then in the fields of communications and electronic wiring. Most of these are now covered in the NEC. Today there are very cost-effective and satisfactory flexible vinyl formulations for use in most classes of indoor power cables operated at 600 volts or less, as well as indoor communications and electronic cables. These include coaxial and specialty constructions such as riser and plenum cables, many of which use more than one polymer type in their insulation, shielding and jackets if flame resistant. Vinyl formulations are also used in a number of fiber optic cables, particularly when good FR (Flame Resistant) behavior is a requirement. Fluoropolymers are used where the optimum combination of electrical properties, low smoke evolution and low heat release is needed and the application can bear the cost. A couple of illustrative generic formulations for commonly encountered flexible PVC applications give a feel for this type of vinyl compound.

This is an extraction-resistant upholstery fabric covering like those used in automobile seating, commercial and residential chairs and upholstered sofas.

INGREDIENT	PHR
PVC (GP-4, K 69, IV 0.98)	100.
Medium molecular weight polymeric plasticizer	60.
Linear phthalate ester	20.
Epoxidized Soy Bean Oil	3.
Barium-Zinc-Phosphite Stabilizer	3.
Stearic acid lubricant	0.25
Calcium carbonate	15.
Pigment	5.—10
Antimony Oxide	0.----10

5 phr of antimony oxide render this type of upholstery resistant to ignition by cigarettes and small open flames. The interest in FR behavior is a growing trend due to the desire to reduce deaths, injuries and property loss in building fires involving furniture.

Insulation for 70-90°C-rated building wire (Exxon Chemical Company Formula 310).

INGREDIENT	PHR
PVC (GP-5, IV 1.04, K 70)	100.
TINTM	23.
DIDP-E	19.5
Calcined clay	8.
Calcium carbonate	7.
Oncor®	1.5

```
Tribase® EX  . . . . . . . . . . . . . . . . . . . . .  4.
Dythal® XL  . . . . . . . . . . . . . . . . . . . . . .  4.
Stearic Acid  . . . . . . . . . . . . . . . . . . . . .  0.25
```
Oncor® is a registered trademark of Great Lakes Chemical. Tribase® and Dythal® are registered trademarks of the Halstab Division of the Hammond Group.

Low cost shower curtain:

INGREDIENT	PHR
PVC (G P-5, IV 1.04, K 70)	100.
DIDP	55.
Epoxidized soy bean oil	3.
Barium-zinc-phosphite stabilizer	2.5
Fungicide	1.
Lubricant	0.25

<u>Additives Used in PVC Compounds</u>. Without stabilizers, process aids and lubricants, PVC is intractable. Much of the progress in adapting PVC to its many valuable uses is due to contributions from the additives industry. For instance, experimental sampling of an early tin-mercaptide (RS-31) to BF Goodrich by M & T Corporation in 1951 made it possible to produce rigid PVC pipe and fittings economically. Some additives serve dual functions, such as stabilizers which also have strong lubricating or plasticizing action.

Stabilizers. The primary function of PVC stabilizers is subject to the rules of thermodynamics, kinetics and mass action which control chemical reactions. The sites at which the "unzippering effect" of dehydrohalogenation of degrading PVC start are generally agreed to be labile chlorine atoms which are allylic, tertiary or terminal in a PVC molecule. The stabilization reaction can be depicted as:

$$---- \; C \; - \; Cl^T \; + \; M\text{-}S \; \rightarrow \; --- \; C \; - \; S \; + \; MCl \tag{10}$$

tertiary chlorine   stabilizer   stabilized PVC   spent stabilizer
on PVC

It is highly desirable that the spent stabilizer be unreactive, not a species which catalyzes dehydrohalogenation of stable PVC. Secondary stabilizers are often needed, because the metallic chlorides formed by primary stabilizers may be Lewis acids which promote degradation of PVC. The function of the secondary stabilizer is to chelate the Lewis acid or otherwise prevent or mitigate the degradating influence of the Lewis acid on PVC.

The desirability parameters for stabilizers include a large number of effects they should not have: stabilizers preferably should be colorless, nontoxic, odorless, tasteless, non-staining, non-volatile, nonconductive, nonextractible, nonmigrating, nonplating, nonplasticizing, resistant to moisture and oxidation, either nonlubricating or only weakly lubricating, non-exuding and nonchalking.

Positive desirability parameters for stabilizers include that they should be readily available, low in cost, shelf stable, readily dispersible, compatible with PVC and other additives, environmentally acceptable, homogeneous, heat stable, light stable, easy processing and themselves chemically stable in addition to being efficient in stabilizing action.

Organotins. The most popular class of stabilizers for rigid PVC applications in the United States is the organotins. They are used less in Europe and the Far East, where lead-based stabilizers became established in rigid applications in the early days of the PVC industry.

Most commercial tin stabilizers sold today are proprietary even though a great deal of information has been published relating to their basic chemistry (45, 46). In approximate time sequence, the broad acceptance of tin stabilizers for use in rigid PVC in the United States was dibutyl tin dilaurate, dibutyltin bis (iso-octyl mercaptoacetate), dimethyltin bis (isooctyl mercaptoate) and dibutyltin bis (2-mercaptoethyl oleate). In the1980s, a stabilizer structure used as part of proprietary systems was thiobis [mono-methyltin bis (2-mercaptoethyl oleate)]. The first tin stabilizer to gain FDA sanction for many uses was di-n-octyltin bis (isooctyl mercaptoacetate). Some di-n-octyltin maleate derivatives and some so-called ester tins which use acrylic esters to alkylate tin have since been FDA sanctioned (45).

During the last ten years economic competition in the stabilizer business has been intense. Stabilizer marketers have been promoting so-called "one pack" tin systems in which the tin content is relatively low, there is a secondary lubricating stabilizer such as calcium stearate, and a blend of lubricants fortified with an antioxidant or other specifically functioning stabilizing chemical species, such as synergistic secondary stabilizers.

Most of the tin mercaptide stabilizers have a discernable odor, which is a disadvantage.

Lead Stabilizers. In the early days of the PVC industry, litharge (PbO) was found to be a fairly effective stabilizer for PVC, but was abandoned because of its yellow color. Waldo Semon soon switched to basic carbonate of white lead (BCWL). It has been largely replaced by tribasic lead sulfate (TBLS), usually manufactured as a fine white powder of high purity, which is an outstanding heat stabilizer for both rigid and flexible PVC. Drawbacks to TBLS are that it is basic enough to hydrolyze some polyester plasticizers during high temperature processing and it does not enhance the light stability of PVC products. Dibasic lead phthalate was developed to overcome the plasticizer hydrolysis problem. Dibasic lead phosphite, also a fine white powder, was found to be an effective heat and light stabilizer with intermediate and satisfactory basicity for a wide range of applications. All these lead stabilizers sulfur-stain to some extent, on contact with mercaptides or hydrogen sulfide.

For several years there has been increasingly persistent pressure from environmentalists for the PVC industry to stop using lead-containing stabilizers, pigments and lubricants. Lead compounds get bad ratings in terms of scientific, political and emotional toxicologies. TBLS, dibasic lead phthalate and dibasic lead phosphite are fine white powders with refractive indices between 2.0 and 2.25. These are high enough to make them act as white pigments and be unusable in transparent or translucent PVC items. They are among the most cost-effective stabilizers for PVC.

By virtue of handling the powdered lead stabilizers in closed bulk air pallet systems, in preweighed batch charges each in its own PVC bag or in prilled stabilizer-lubricant one-packs, worker exposure problems have been effectively overcome. The permissible exposure limit (PEL) for airborne lead is 0.05 mg/m^3 in the United States.

In the early stages of their degradation, lead-stabilized PVC compounds form lead mono-and dichlorides which are not Lewis acids, nor are they readily soluble in water, hygroscopic or strongly ionizing. These characteristics make the lead stabilizers good performers in electrical insulations, particularly those rated for use in wet locations which, in the United States need to pass long-term insulation-resistance tests (immersion in hot water at 75 or 90°C for 26 or more weeks). In 1985 Mitchell proposed a free-radical mechanism of PVC stabilization rather than simple scavenging of HCl or atomic chlorine by basic lead stabilizer-stearic acid or lead stearate combinations (47, 48).

After several years of intense effort, more cost-effective and technically superior lead-replacement stabilizer systems have not yet been fully qualified. Nevertheless, wire and cable companies and other users of lead stabilizers have evaluated or are evaluating low-lead and non-lead stabilizer systems. They plan to be ready in case the use of lead stabilizers in PVC is further restricted or banned.

TBLS and dibasic lead phthalate have low solubilities in neutral water (pH 6-8). Authorities wishing to show that lead is extractible from lead-stabilised PVC compounds specify that the compound be ground to a fairly fine particle size and that the extractant be buffered sufficiently to the alkaline or acidic side of neutral (considering the amphoteric nature of lead) to make the lead relatively soluble. Marginal to failing results are often experienced on lead-stabilized vinyl insulation subjected to EPA's Toxic Characteristic Leaching Procedure (TCLP) which allows a maximum lead concentration in the leachate of 5 mg/l. Calcium and zinc are not regulated in the TCLP test. A maximum of 100 mg/l of barium is allowed in TCLP leachate. Lead-stabilized wire and cable PVC scrap usually is sent to expensive secure land-fills.

Mixed Metal Stabilizers. The successful mixed metal stabilizer systems mostly use zinc compounds to exchange their anions for labile chlorine atoms on PVC molecules. Since zinc chloride is a potent Lewis acid capable of promoting catastrophic dehydrochlorination of PVC, a more basic metal such as barium or calcium has to be included in the stabilizer at a higher level than the zinc. The barium and calcium stabilizers do not react directly with the labile chlorines on PVC as actively as the zinc compounds do. But the barium and calcium compounds steal the chlorine ions from zinc by anion exchange. Barium and calcium chlorides are much weaker Lewis acids than zinc chloride.

From about 1950 to 1990, cadmium was used along with zinc or along with barium as the primary metal in most mixed metal stabilizers for PVC. Since 1990 use of cadmium compounds has steadily been phased out on the grounds that cadmium is dangerously toxic. However, cadmium may be encountered in PVC rework over eight years old, derived from now discontinued pigment and/or stabilizer systems.

Today's barium-zinc and calcium-zinc stabilizers may be either solids or liquids. Barium stearate plus zinc stearate together with various synergists constitute the workhorse solid stabilizers in this category. Mixed fatty acid salts including palmitates and laurates of barium and calcium also work well. In liquid systems barium alkyl phenates and zinc octoate may be used along with solvents suitable for use with PVC. Other synergistic ingredients in liquid mixed metal stabilizers include epoxides and phosphite antioxidants whose solubility parameters are close to those of the other formulation ingredients such as plasticizers or impact modifiers (if any) in the PVC formulation.

Alkyl aryl phosphites used in mixed metal stabilizers for many years were referred to as "chelators" because they improved clarity under ideal use conditions and helped maintain "good, early color." Pentaerythritol was added to many mixed metal stabilizers in the belief that it facilitated the transfer of chloride from the primary to the secondary metal.   Pentaerythritol was empirically shown to be beneficial. Many phenolic antioxidants are also used, including butylated hydroxytoluene (BHT) and Bisphenol A.

The liquid mixed metal stabilizers are nearly all hygroscopic and have to be protected from exposure to atmospheric moisture. This is done by closed bulk or semi-bulk handling systems. More than trace quantities of moisture in liquid mixed metal stabilizers cause phase separation and catastrophic loss of properties by hydrolyzing the phosphites and adding to the epoxides.

A number of calcium zinc stabilizers are sanctioned by FDA for use in food contact applications and for medical packaging of pills. Some are also approved by The National Sanitation Foundation (NSF) and other regulatory authorities. Regulated phosphites and polyols are used as synergists in these formulations, usually as part of a one-pack stabilizer.

Miscellaneous Stabilizer Applications. In the manufacture of PVC foams, presence of a particular metal as a kicker for the blowing agent may be required. This is usually achieved by using a stabilizer recommended by the blowing agent manufacturer to work both as a kicker and to provide adequate stability for the end-use for which the foam is intended. With azodicarbonamide blowing agents, lead stabilizers are effective kickers for blowing in the range 160-180°C (320-356°F). Zinc stabilizers are more effective kickers when blowing is to be carried out above 180°C.

Antimony tris (isooctyl mercaptoacetate) has been known to be a good stabilizer for PVC since 1954 (49). It functions similarly to the dialkyltin mercaptides, except that the tri-valent antimony is easily oxidized unless used along with an efficient antioxidant and enough calcium stearate to provide adequate lubricity. The calcium stearate acts as a secondary stabilizer, converting scavenged HCl into calcium chloride. This prevents the formation of antimony trichloride which is a strong enough Lewis acid to promote degradation of PVC.

Lead-Replacement Stabilizers for Electrical Applications. The most promising lead-replacement stabilizers are proprietary and shrouded in secrecy due to unresolved patent issues. Proprietary lead-replacement stabilizers typically contain combinations of primary and secondary metals, metallic chloride deactivators, inorganic acid acceptors, metal co-ordinators and antioxidants.

For applications in which electrical properties are not important, replacement of lead stabilizers is relatively straight forward except that it is very difficult to match lead stabilizer economics.

An approach adopted by several companies involves testing of various hydrotalcites resembling the medically bland antacid called Maalox®. The hydrotalcites have aluminum, magnesium, hydroxyl and carbonate functionality and are claimed to be efficient absorbers of HCl. However, when used alone with PVC they produce an orange or purple discoloration, as do most grades of magnesium oxide or magnesium hydroxide. This discoloration may be greatly reduced by use of β-diketones such as Rhodiastab 83® or Rhodiastab 50® which were first patented in 1975 by Rhone-Poulenc as supplemental PVC stabilizers. (Maalox® is copyrighted by Rhone-Poulenc Rorer.)

Plasticizers. See the chapter on Plasticizers for a more general and detailed treatment covering plasticization of PVC and other polymers.

The definition of plasticizers adopted by IUPAC in 1951 is still accepted: a substance or material incorporated in a material (usually a plastic or elastomer) to increase its flexibility, workability, or distensibility. A plasticizer may reduce the melt viscosity, lower the temperature of a second-order transition, or lower the elastic modulus of a product.

Sometimes it is useful to distinguish between liquid plasticizers for PVC and solid flexibilizers by referring to the latter as plastifiers, whose blends with PVC may be categorised as thermoplastic elastomers (TPEs) or alloys.

The literature on plasticizers for PVC is extensive and should be consulted for information beyond the scope of this brief review (50, 51, 52).

Principal requirements for a successful commercial plasticizer for PVC are that it should be cost-effective, stable, low in color, compatible with PVC, readily dispersible in PVC, low in volatility, low in odor, have good permanence and should not interact unfavorably with other needed formulating ingredients. More arcane requirements such as having solubility parameters, dielectric constants, dipole moments, hydrogen parameters and Flory-Huggins interaction parameters within certain ranges are sometimes prescribed. Going from one chemical plasticizer type to another, the correlations between the pragmatic performance parameters and the scientific measurements on simple systems become too loose to maintain the latter as absolute standards for predicting performance in PVC in the marketplace, especially when mixtures of plasticizers are used.

Plasticizers marketed for use in compounded PVC fall into the following chemical families: dialkyl ortho-phthalates, dialkyl tere-phthalates, epoxides, aliphatic carboxylic diesters, polyester-type polymerics, phosphate esters, trimellitate esters, benzoate and dibenzoate esters, and miscellaneous types.

Secondary plasticizers only achieve good compatibility with PVC when used together with significant amounts of primary plasticizers. The secondaries include chlorinated paraffins and hydrocarbon extenders.

Diluents and some solvents are used primarily for viscosity reduction in modified plastisols and so-called organosols.

The current "Modern Plastics Encyclopedia" lists about 500 plasticizers offered by 48 manufacturers in the United States (53). Many of these are incompatible with or only partially compatible with PVC and are used in other polymer systems.

General theories as to the mechanisms whereby plasticizers work in PVC have been called the lubricity, gel, mechanistic and free-volume theories (50, 51) They provide insights into why certain phenomena occur, but cannot be applied quantitatively to complex commercial formulations. In these, some degradation products appear, particularly as the service life of a plasticized product is extended. Wear-and-tear on the PVC, plasticizers, pigments, etc., may show up as fading of swimming pool liners, mold growing on a shower curtain, embrittlement of a frequently overloaded electric wire insulation, etc. Plastics technologists invoke plasticization theory when it is helpful, but mostly deal with observed performance phenomena.

Dialky ortho-phthalates and terephthalates. These comprise about 70% of all plasticizers sold in the United States. They are generally designated by acronym, such as DOP for di-octyl phthalate, aka di-2-ethylhexyl phthalate, (Toxicologists prefer the more precise DEHP for DOP), and DINP for di-isononyl phthalate. Various alcohols are reacted with phthalic anhydride to produce the range of ortho-phthalates. The number of carbons in the alcohols range from butyl (C_4) to tridecyl (C_{13}) and the alcohols may be linear or branched.

Increasing the degree of branching gives: higher volatility, greater susceptibility to oxidation, higher volume resistivity in formulated PVC and poorer low temperature brittleness.

Several plasticizer suppliers offer services whereby they calculate the concentrations of their products required to produce a desired set of physical properties, if attainable in plasticized PVC.

DOP is considered by most to be the PVC industry standard general pupose (GP) plasticizer due to its long use history and excellent balance of properties in non-demanding applications. DOP is sanctioned by FDA and corresponding organizations in many other countries for use in numerous food-packaging applications. DOP has a history of safe usage in medical devices such as blood bags and tubing used in kidney dialysis machines.

DIDP is well established in $60°C$-rated insulation for power cables used at voltages up to 600 and in CMG, CM and CMX communications cables and many other cable constructions.

DHP (dihexyl phthalate), BOP (butyl octyl phthalate) and BBP (butyl benzyl phthalate) are fast fusing plasticizers for PVC. BBP is used in many vinyl flooring constructions to impart superior stain resistance.

610 P and 711 P are mixed alcohol phthalates made respectively from linear 6-8-10 and predominantly linear 7-9-11 alcohols. DUP is di-undecyl phthalate. L9P is a linear C 9 phthalate. DOTP is di-2-ethyhexyl terephthalate, which has somewhat better low temperature brittleness and volatility than DOP, but is less compatible with PVC.

Most phthalates lower the melting temperatures of PVC crystallites to varying degrees, facilitating fusion of the total formulation. The selection of the best phthalate for an application is guided by economics, ease of processing and performance in the end use.

Epoxides, aka. Epoxy Plasticizers. Epoxides have oxirane oxygen groups in their molecules, formed by the epoxidation of olefinic double bonds in their starting raw materials:

$$R — CH = CH — + CH_3COOOH \xrightarrow[cat]{} R — CH \overset{\overset{\displaystyle O}{/ \ \backslash}}{—} CH — + CH_3COOH \qquad (11)$$

The virtue of the epoxides is that the oxirane oxygen ring acts synergistically with metallic stabilizers based on Ba, Ca and Zn and shows promise with some of the newer stabilizers based on Mg. They impart superior heat and ultraviolet (UV) light stability to vinyl compositions containing a suitable co-stabilizer.

The most widely used epoxides are epoxidised soy bean oil (ESO) and epoxidised linseed oil (ELO).

Epoxy plasticizers have good initial compatibility with PVC, but can lose their compatiblizing oxirane oxygen through photo-oxidation or hydrolysis. Sound formulators use epoxides conservatively at low levels. A tacky exudation or "spew" may develop when they are used at higher levels and exposed to UV light and/or warm, moist air for long periods. Most epoxides made from natural products have the disadvantage of being good nutrients for molds and fungi.

Aliphatic Carboxylic Diesters. These plasticizers are also generally identified by acronyms. They are based on aliphatic dibasic acids esterified with alcohols ranging from C_8 to C_{10}. The carbon numbers of the dibasic acids used range from C_5 (glutaric) to C_{10} (sebasic).

DOA represents di-2-ethylhexyl adipate. Dilinear C_7 C_9 adipate is abbreviated as L 79A. Di-isononyl adipate is called DINA and n-ocytl n-decyl adipate is NODA. DIDA is di-isodecyl adipate. DOZ is di-2-ethylhexyl azelate. DOS is di-2-ethyl hexyl sebacate.

The aliphatic dicarboxylic acid esters are used mainly for their excellent efficiency and low temperature properties and in FDA-sanctioned applications. The adipates and azelates do not lower the melting points of PVC crystallites as much as the phthalates do. These plasticizers are less compatible with PVC than the corresponding phthalates, and they are more volatile: DOA, e.g., is considerably more volatile than DOP. Increasing the carbon number of the alcohol to reduce volatility makes the already poor compatibility worse in aliphatic dicarboxylic diesters.

DOA is used as a primary plasticizer in some clear meat-wrap and produce-wrap films. In most other applications the adipates are used along with more compatible and lower volatility phthalates to reduce brittleness temperatures.

Polyester-type Polymeric Plasticizers. Most polyester plasticizers are condensation products of glycols and dibasic organic acids.

The range of molecular weights in use is from about 1,000 to about 8,000 average, with some lower and some higher molecular weights present according to the statistics of the polymerization reaction and any stripping which may be part of the manufacturing process.

1, 3 butylene glycol and adipic acid are the most frequently used starting materials. C_8 to C_{10} alcohols, or mixtures of these are commonly used for termination.

The principal advantages for the polymerics over general purpose plasticizers are:
— Lower volatility during compounding and in end use;
— Lower migration into lacquers, protective coatings, adhesives, fabrics and other plastics in contact with the plasticized item during end use;
— Reduced extraction by oils, fats, hydrocarbon solvents, oxygenated solvents and most aqueous media.
The chief disadvantages are:
— Higher cost;

— Lower plasticizing efficiency;
— Poorer low temperature brittleness;
— Poorer environmental stability of end products exposed to combinations of warmth, humidity, UV light and active microbial cultures.

In many applications the most practical formulations contain mixtures of polymeric and monomeric plasticizers.

Examples of applications in which polymeric plasticizers are used include refrigerator gaskets, electrical tape, migration-resistant shoe soles, high temperature-rated oil-resistant electrical insulation and jacketing, gasoline-resistant hose and extraction-resistant babywear and hospital sheeting.

Phosphate Esters. The general structure of these, made from phosphorus oxychloride, is:
$$(R_1O)\,(R_2O)\,(R_3O)\,P = 0$$
For many years TCP (tricresyl phosphate) was the industry standard phosphate plasticizer. This was mostly discontinued as a plasticizer due to medical concerns about contact with even trace amounts of the ortho isomer in derivatives of cresylic acid.

Most of the triaryl phosphate plasticizers sold today in the United States are made from blends of phenol and alkylated phenols such as isopropylphenol or t-butylphenol.

The other main class of phosphate plasticizers consists of alkylaryl phosphates, such as 2-ethylhexyl diphenyl phosphate or isodecyl diphenyl phosphate.

Tri-2-ethylhexyl phosphate, known as TOF, is an outstanding low temperature plasticizer, but is too expensive for most applications.

Monoalkyl diaryl phosphate esters have outstanding solvating properties for PVC with which they are very compatible. If they were less expensive, they would be used much more than they are for improving processability of balky formulations.

The principal reason for using phosphate ester plasticizers is to enhance the flame retardancy of flexible PVC. By use with proper choice of solid flame retardants and smoke suppressants, phosphate esters are valuable components of FR formulations. The alkyl aryl phosphates give lower smoke evolution than the triaryls. Phosphates are used in clear, flexible FR applications in which a pigmenting solid cannot be used.

A new low volatility alkyl aryl phosphate being offered by Solutia, Inc., has been welcomed by the cable industry for use in highly flame retarded, low smoke jacketing designed for use in plenum cables of types such as CMP, CL3P, CL2P, FPLP and MPP.

Trimellitate Esters. These are made by reacting plasticizer-grade alcohols with trimellitic anhydride. They represent the state of the art in low-volatility monomeric plasticizers. TOTM is tri-2-ethylhexyl trimellitate. TINTM is tri-isononyl trimellitate. NODTM is tri-n-C_8C_{10} trimellitate.

They are used in 90°C-rated and 105°C-rated wire insulations and cable jacketings and other uses requiring low plasticizer volatility. Their use is limited primarily by their costs being higher than the phthalates. For some applications they compete with polymeric plasticizers, where low volatility is the main concern.

Benzoate and Dibenzoate Esters. Dipropylene glycol dibenzoate (DPGDB) is used in stain resistant flooring, alone or in blends with other benzoate esters. Texanol® benzoate is another stain resistant plasticizer, made from a specialty high molecular weight alcohol from Eastman. This family of plasticizers is less versatile than the phthalate esters and is mainly used in plastisol-derived flooring sheet goods.

Miscellaneous Plasticizers. Some citrate esters such as acetyl tri-n-hexyl citrate and butyryl tri-n-hexyl citrate are promoted for specialty uses, particularly in medical applications such as blood bags and in food wraps.

Several polymerizable plasticizers are available for specialty applications.

Alkyl sulfonate esters of phenol are promoted in Europe as plasticizers for PVC under the trade name Mesamoll®. (BASF trademark)

Texanol® isobutyrate (TXIB) is used as a viscosity-reducing plasticizer in plastisol-derived flooring sheet goods. (Registered Eastman trademark)

Tetrabromophthalate esters, such as Great Lakes DP-45®, are excellent flame retardant plasticizers whose flexibilizing action is conditioned by their high molecular weight.

Chlorinated paraffins are used as secondary plasticizers for cost reduction and for improving flame retardance.

Naphthenic hydrocarbons, alkylated aromatics and some linear paraffins are used as extender plasticizers and/or as diluents in plastisols.

Toxicology and Environmental Effects of Plasticizers. When materials used (or sometimes used) as plasticizers for PVC and other polymers have been proven to have adverse toxicological effects on some animal species, industry has usually responded by withdrawing the material (e.g., PCBs and TCP). When erroneous allegations about toxicity are made extensive testing to understand the scientifically provable relevant facts, is normally undertaken.

DEHP aka. DOP is probably the most studied plasticizer. DINP has also undergone extensive biological testing.

For many plasticizers, LD_{50} values are in excess of 20,000 mg/kg body weight for oral, dermal and intraperitoneal routes of exposure in animal tests. Thus they have very low acute toxicity.

Repeated oral exposures of several animal species have revealed that:
— phthalates tested are not genotoxic;
— oral administration of phthalates to rodents causes proliferation of peroxisomes in the livers and are linked to liver tumors.
— oral administration of phthalates to non-rodent species and primates does not cause formation of peroxisomes in their livers and no resulting liver damage has been observed.

Based on differences between primate and rodent responses to phthalates, the European Union Commission decided in 1990 that DEHP should not be classified as a carcinogenic or irritant substance to humans (54). Similar conclusions have been reached in recent studies in the United States (55). Nevertheless, there are people who do not understand or acknowledge the observed differences in metabolism between rodents and humans and keep repeating that phthalates are carcinogenic without the qualification "only to rodents."

Greenpeace has recently publicized allegations that phthalate plasticizers are endocrine disruptors without providing scientific documentation to this effect. CMA and various industry committees have published a rebuttal (55). Some *in vitro* tests have been inconclusive, but none of the *in vivo* tests have shown evidence that dialkyl phthalate esters are endocrine modulators in any animal species. The results of early studies of plasticizers in the environment were flawed by contamination of field samples when they were brought into laboratories which used Tygon® tubing and similar equipment items containing phthalate ester plasticizers. Also, many early aquatic toxicity tests on phthalates failed to take their low solubility in water into account. Early testers using protocols designed for more soluble substances would charge enough plasticizer to a test pond to form a plasticizer film over its entire surface. Then deaths of species trapped in the plasticizer films were erroneously ascribed to toxicity of the plasticizer, very little of which had dissolved in the water. More recent and realistic tests often exonerate the plasticizer.

Processing Aids, Impact Modifiers and Heat Distortion Temperature Improvers.

Processing Aids. Rigid and semi-rigid PVC compounds fuse more easily and process more readily if they contain low levels of processing aids.

The most effective of the early processing aids were high molecular weight methyl methacrylate (MMA)-alkyl acrylate copolymers. It has since been found that these can be optimized by preparing them as core-shell particles with a lower molecular weight core and a higher molecular weight shell. This gives faster fusion and dispersion in PVC. They are usually tailored to have a refractive index close to that of rigid PVC, which allows them to be used in clear compounds without causing unwanted haze.

Agreement as to how processing aids work on a microparticulate and molecular level has not been reached. Compounders and processors like to use processing aids, in spite of their high cost, because they fuse faster than PVC, stick to the PVC like a high viscosity putty and rapidly generate shear heating in fluxing equipment such as Banbury mixers, Buss Ko-Kneaders, twin screw extruders, planetary gear extruders and so forth and in laboratory equipment such as Brabender Plasticorders in which the progress of fusion can be closely monitored. Processing aids seem to expedite the break up of PVC as-polymerized grains into primary particle flow units rapidly and uniformly. They greatly decrease the "ropiness" of rolling banks of stock in the nips of two-roll mills and calenders, and increase the smoothness of extrudates from dies. Processing aids also improve the hot strength of extrudates from dies, of webs on calenders, of webs in vacuum formers and of molded parts. Remarkably, acrylic processing aids have very few, if any, unfavorable interactions with other vinyl formulating ingredients. Typical use levels are 1-10 phr.

Acrylic processing aids and lubricating processing aids, if handled carelessly, are capable of causing high grade dust explosions. Air suspensions of processing aids can be ignited by static electric sparks or other sources of ignition. Explosion suppressing means are recommended for air conveyors handling acrylic processing aids.

Recently, as the popularity of rigid PVC foams has increased, it has turned out that processing aids valuably decrease the tendency for gases to blow through expanding cell-wall membranes and the surface skin on foamed products.

One of the theories as to how processing aids work in rigid PVC was developed by Krzewki and Collins (56, 57). Practical information on their use is also available from manufacturers and texts (58, 59).

Impact Modifiers. Rigid PVC, in the medium molecular weight range used for most extrusions, is a relatively tough polymer when properly fused. Low molecular weight rigid PVC used for some high shear rate injection molding jobs is more brittle. However, both medium and low molecular weight rigid PVC parts are notch-sensitive (i.e., embrittled by stress-concentrators) and become brittle at low temperatures. Impact modifiers are very widely used to impart some or a lot of extra toughness, depending on the demands of the application.

An impact is a stress concentrated in time and place. If the stress exceeds the strength of the rigid PVC, one or more cracks start. In PVC, impact modifiers do not increase its strength, but they inhibit initiation of a crack by promoting yield, and they inhibit crack propagation.

The two general types of organic impact modifiers are referred to as predefined elastomers (PDEs) and non-predefined elastomers (NPDEs). The difference is that PDEs are usually emulsion-polymerized with a cross-linked rubbery core and a harder shell which adheres well to PVC. The shell is grafted onto a 0.1 to 0.2 μm rubber latex particle. NPDEs are rubbery polymers having limited compatibility with PVC.

As with plasticizers, impact modifiers for use in PVC are usually described by acronyms, some of which are decoded here.

Various grades of precipitated calcium carbonate and aluminum trihydrate are promoted as inorganic impact modifiers. They are mostly used synergistically with organic impact modifiers. Used alone, the effectiveness of inorganic impact modifiers is limited. Accepted acronyms for PDE impact modifiers include the following:

- ACR— all acrylic
- MACR— modified acrylic, usually containing some polymerized butadiene mono mer
- MBS — methacrylate-butadiene-styrene copolymers
- MABS — methacrylate-acrylonitirile-butadiene-styrene copolymers
- ASA — acrylate-styrene-acrylonitrile copolymers
- ABS — acrylonitrile-butadiene-styrene copolymers. Depending on how the ABS impact modifiers are made, they are sometimes referred to as "transi tional modifiers" because they are more subject to loss of properties on continued processing or exposure to sunlight than the first five above types.

Polymers included in the NPDE group are:

- CPE— chlorinated polyethylene
- EVA — ethylene vinyl acetate copolymers
- ACR grafts — PVC grafted onto an acrylic rubber.
- PU elastomers — used in niche markets
- PO elastomers — used in niche markets

In selecting impact modifiers for use in particular PVC market areas it is necessary to consult with the manufacturer. Within each category there are many differences in composition, molecular weight, particle size, etc. The material safety data sheets on each product should always be reviewed for precautions related to handling in case of dust explosion hazards.

Detailed information on impact modifiers is available in manufacturers' product bulletins and texts (60, 61).

Heat Distortion Temperature Improvers. For many applications the heat distortion temperature of rigid PVC is marginal or inadequate. Polymers which can be blended with PVC to improve its heat distortion temperature include:
• AMS-S-AN — alpha methyl styrene-styrene-acrylonitrile.
• AMS-SAN-BD — alpha methyl styrene-styrene acrylonitrile-butadiene.
• ACR-GI — acrylic-glutarimide
• S-MA — styrene-maleic anhydride
• A-S-AN — acrylic-styrene-acrylonitrile.
They should be handled carefully to avoid dust explosion hazards.

The glass transition temperature (Tg) of medium molecular weight PVC is about 87°C. All the above polymers have substantially higher Tgs. In some cases the HDT of a PVC-HDT improver blend may be as much as 30 to 40°C higher than that of PVC, but some of the other properties (e.g., toughness, flame resistance) may be impaired.

Fillers. Generically, fillers may be any low cost solid, liquid or gas which occupies volume in a part or product and reduces its volume-cost. In the PVC industry, the term "fillers" unqualified refers to inert particulate solids added to formulations for various reasons, including volume-cost reduction and stiffening. Functional fillers are products added to improve particular properties, such as calcined clays to increase electrical volume resistivity, fumed silica or bentonite clay added to a plastisol to increase its yield value, or hollow microspheres to lower specific gravity while achieving other desired filler effects. To be a filler, a substance must not dissolve in the PVC matrix.

The ratio of the average lengths of the major to minor axes of filler particles is called the "aspect ratio". Filler particles usually have aspect ratios less than 4 to 1. Reinforcements, such as glass fibers, usually have aspect ratios in excess of 10 to 1.

The most widely used fillers in PVC are various grades of dry ground, wet ground or precipitated calcium carbonate derived from limestone, which is predominantly calcite, the stable crystal structure at ordinary temperature and pressure. Marble consists of small, interlocking calcite crystals.

Important considerations in selecting as a filler a particular grade of calcium carbonate include the purity of the original ore, if it is dry- or wet-ground, the particle size and size distribution, and whether the filler has had a surface treatment. The efficiency with which fine filler particles fill voids between the coarser particles is measured by the "packing fraction." Presence of free silica or hard silicates makes the filler much more abrasive to processing equipment, because calcite is soft (Mohs hardness 3). Presence of iron oxide (Fe_2O_3) in the filler compromises the heat stability of vinyl compounds and pigments them a yellow-brown.

Size of filler particles is usually defined in terms of equivalent spherical diameter (esd). This is the diameter of a sphere having the same volume as the filler particle.

A major use for relatively coarse, dry-ground calcium carbonate is floor tile made by calendering, which tolerates particle sizes up to 99% through U. S. Standard 50 mesh

(11.7 mils, 297 microns openings). Typical flexible extrusions, such as electrical insulation and cable jacketing require fillers with an average esd of 3 microns and coarsest particles of 12 microns diameter. Surface-treated precipitated calcium carbonates used as impact modifiers in rigid PVC have 0.07 micron esd. Precipitated calcium carbonates used at relatively high levels in cable jacketing designed to meet low HCl emission specifications when burned have 0.6 micron esd. Experience determines the best particle size filler to use to optimize properties in each application.

Of the non-carbonate fillers used in PVC, talc (represented as 3 MgO 4SiO_2, H_2O) is the softest. Zero or very low content of asbestos-related minerals is specified for talcs used in or with PVC. Talc may be added to calendering formulations to reduce plate-out on the rolls or to extrusion formulations to reduce plate-out on screws or dies. Talc may also be dusted onto PVC pellets at 0.1 to 0.25% to improve bulk flow of the pellets in handling systems and hopper cars.

Mica is added to PVC compounds to impart a non-blocking surface and to stiffen them.

Diatomite is added to PVC plastisols to increase viscosity and/or reduce surface gloss after fusion. Fumed silica is added as a scrubbing agent to hot-processed compounds and to plastisols for the same reasons as diatomite.

Many miscellaneous low cost, locally available fillers are used for special purposes and cost reduction.

The refractive index (RI) of PVC is 1.55; that of phthalate plasticizers about 1.48 to 1.50. In PVC the matrix RI is usually between 1.55 and 1.53. Fillers with refractive indices in the 1.55 to 1.60 range contribute little to haze and may be used in "clear" formulations. Calcium carbonate with RI 1.65 is a weak pigment as well as a filler. TiO_2 with an RI of about 2.76 contributes a high degree of opacity to PVC.

Modern Plastics annually publishes a list of manufacturers of non-fibrous fillers and their product lines (62). These parties should be consulted for up-to-date supply and technical service information on fillers.

The negative side of using fillers in PVC compositions, particularly at high levels, is that they generally reduce tensile and tear strength, elongation at failure, toughness at low temperatures, abrasion resistance, resistance to attack by moisture and chemicals and they compromise processability by raising melt viscosity.

The principal advantages of fillers are cost reduction, stiffening, reducing coefficients of thermal expansion and improving FR behavior.

Flame Retardants and Smoke Suppressants. Due to its 56.8% chlorine content, PVC is relatively difficult to ignite and reluctant to burn unless subjected to a strong external and continuing heat flux. However, when diluted with flammable process aids, impact modifiers or plasticizers, the inherent flame retardant properties of PVC are reduced.

To meet various specifications calling for low flammability, low heat release and low smoke evolution on being subjected to particular test conditions, it is often necessary to include various flame retardants and/or smoke suppressants in PVC formulations. Particular circumstances are too specialized to cover in this chapter. Space allows only a partial listing of ingredients frequently used to these ends.

Organic flame retardants and smoke suppressants used in plasticized PVC include members of the families of phosphate esters and brominated phthalate plasticizers.

Inorganic flame retardants and smoke suppressants include alumina trihydrate, antimony trioxide, barium metaborate, huntite (hydrous magnesium-calcium carbonate), magnesium hydroxide, ammonium octamolybdate, various mixed metal complexes, zinc borate and zinc hydroxystannate.

Light Stabilizers. PVC and many of the compounding ingredients used with it experience degradation on long exposure to ultraviolet (UV) light of wavelengths between 295 and 400 nm. Since PVC is the most economical polymer having excellent resistance to creep for use in house siding, window and door profiles, outdoor furniture and many flexible applications such as swimming pool liners, it is desirable to protect it from these aggressive radiations.

The means used include selection of suitable pigment systems. Since PVC is degraded by heat as well as by UV light, in geographical areas where the sun is strong it is necessary to include a good heat stabilizer and a reflecting pigment to reflect as much of the sun's heat as possible. One or more of the pigments used should be capable of absorbing UV radiation and emitting the energy as heat, much of which is then radiated. Titanium dioxide (rutile) is widely used in outdoor applications for PVC because it both absorbs UV and reflects heat radiation. Carbon black is a very effective UV absorber, but it also absorbs heat radiations. The life of exposed black PVC may be shortened by its heat history in climates where the sun is strong.

Organic UV absorbers widely used in PVC are of the types classified as:
• 2 — (2' — hydroxyphenyl) benzotriazoles, and,
• 2 — hydroxybenzophenones.
Hindered amine light stabilizers (HALS) are effective in PVC stabilized against thermal degradation with a mixed metal or tin carboxylate system. HALS do not work well with tin mercaptide heat stabilizers. HALS are believed to function as radical scavengers by virtue of generating nitroxyl radicals which react with peroxy and alkyl radicals produced by photo-oxidation. HALS may be used in combination with benzotriazoles.

A caveat is that the outdoor performance of UV absorbers and HALS is hard to predict from accelerated laboratory tests. Proving the suitability of light stabilizers requires the patience to run actual outdoor exposures. UV-induced degradation of PVC manifests itself by discoloration which may be accompanied by chalking and pitting, and by loss of physical properties.

Significant improvement in the UV-stability of PVC formulations is fairly expensive because TiO_2, benzotriazoles, benzophenones and HALS are all expensive ingredients.

Lubricants. Successful hot melt processing of PVC depends on selection of a suitable lubricant system which works well in the formulation as a whole. After pinning down the rest of the formulation, it usually takes some Edisonian experimentation to optimise the lubricant system in the full scale manufacturing process.

The primary functions of the lubricant systems in PVC compounds subject to hot melt processing are to prevent the hot compound from sticking to processing equipment, to moderate generation of internal frictional heat in the compound being sheared and to maintain the desired degree of adhesion between solid and rubbery particulate ingredients and the PVC matrix. The lubricant should also discourage or prevent formulation ingredients or their degradation products from plating out on hot metal processing equipment surfaces such as extruder screws and dies, mill rolls or calender rolls and then inducing sticking or a marred surface appearance of the product.

Unless desired, the lubricant(s) should not bloom on the surface of the finished product.

The types of lubricants commonly used in PVC include:
Paraffin waxes, polyethylene waxes, carboxylic acids, amide waxes such as ethylene-bis-stearamide (EBS), lubricious esters such as stearyl stearate, ester waxes, metal carboxylates, *e.g.*, of Ba, Ca, Mg or Pb and proprietary mixtures.

Other Additives. These include: colorants, biocides, blowing agents, antistats, antioxidants, antifog agents, rheology control agents for plastisols, coupling agents, reinforcing fibers, gloss control agents, odor-masking agents and fragrances.

Some leading-edge formulating research relates to members of the miscellaneous group, such as treated glass reinforcing fibers.

Processes for Compounding PVC

Dry Blending (Powder Mixing). More than 85% of suspension and bulk PVC resins are initially processed by dry blending. This is a mixing procedure whereby the solids and liquids (if any) in a PVC formulation are intermixed with the solids mutually dispersed down to the particulate level and the liquids uniformly absorbed into the PVC resin. The resulting product is a free flowing powder.

Batch high speed (high intensity) dry blending mixers generally consist of a vertical drum with a dish-bottom through the center of which a drive shaft penetrates. One or more sets of mixing blades are attached to the drive shaft. A baffle penetrates the top of the mixer which can be tightly sealed to the drum. By running the drive shaft at an appropriate speed rapid mixing occurs with the generation of sufficient frictional heat to raise the batch temperature to a desired drop temperature such as 105°C in 4 to 8 minutes.

Detailed mixer designs vary from manufacturer to manufacturer. Some can sustain a high vacuum during mixing for removing traces of unwanted volatiles such as moisture or vinyl chloride monomer residuals. These intensive mixers all are characterized by having Froude numbers much greater than 1. Typical shaft speeds are 500 to 1200 rpm with blade tip speeds between 10 and 50 m/sec. Typical batch capacities are 80 to 200 kg (176 to 440 lb.).

These high speed mixers usually dump their batches into lower speed rotary coolers having Froude numbers less than 1. The coolers introduce little frictional heat and cool by conduction into their cold water jackets. To avoid adventitious condensation of moisture from the air, freezing brines are not used in the cooler jackets.

Note: Froude number = $\dfrac{R\omega^2}{g}$

where R = blade radius

ω = angular velocity, radians/sec.

g = gravitational acceleration (consistent units).

Vinyl formulators refer to mixers of this type by their trade names, some of which are Henschel, Papenmeier, Welex, and Littleford.

Intermediate speed mixers with Froude numbers somewhat greater than 1 and low speed mixers (generally ribbon blenders) with Froude numbers considerably less than 1 are also used. In large ribbon blenders, heating and cooling of batches is slow and the time to achieve uniformity in a batch is long (1 to 3 hours in many cases). A European manufacturer (Buss) offers a continuous mixing turbine for preparation of dry blends.

In production lines where many different formulations are scheduled, batch mixers are generally preferred for their versatility. Many other types of mixers such as Mix Mullers have been adapted to dry blending PVC, but are not commonly used today.

To maximize throughputs without sacrificing quality, high bulk density, low porosity PVC resins are normally used for rigid formulations. Higher porosity resins are used for flexible formulations in which the resin has to absorb large amounts of plasticizer to provide a free-flowing powder. To shorten mixing cycles, high viscosity plasticizers are pre-heated before addition to the mixer.

Compounding of PVC. In the PVC industry, the term compounding has different meanings when applied to liquid and solid systems. In the case of plastisols, it means mixing the solid resin and other solid ingredients uniformly into the liquid plasticizer along with the other liquid ingredients to achieve a targeted rheology suitable for the follow-up forming operation (spread-coating, rotational molding, dip-coating, spray-coating, etc.). By one set of definitions, a plastisol contains no liquid diluent or solvent, a modified plastisol may contain up to 5% of solvent or diluent and an organosol contains more than 5% of solvent by total weight.

In the case of compounding PVC latexes, formulating ingredients are normally added in waterborne solution, emulsion or dispersion and mixed into the base latex by gentle agitation, such that it does not coagulate the latex. In some cases, two or more latexes are mixed together to achieve a desired set of properties. The mixed latex is then applied to the substrate to be treated by dipping, impregnation, spraying, reverse roll coating, coagulation or any of the other common latex application methods.

When vinyl formulations are to be shaped by hot-melt processing, compounding involves bringing a dry blend (powder) or wet mix to a fluxed condition which will flow properly in the forming equipment and develop optimal properties in the end product. Alternatively, the fluxed melt can be formed into pellets or cubes such that they can be fed into the final forming equipment (extruder, injection molder, etc.)

Batch-type compounding equipment for PVC includes Banbury mixers, which were developed for processing rubber. Modern Banbury models adapted for versatile use in compounding PVC have variable speed rotors cored for circulation of a heat transfer fluid (oil, water or steam) operating in an enclosed mixing chamber sealed by an air-operated ram which can apply controlled pressure to the batch being fluxed. The mixing chamber walls are also cored for a heat transfer fluid. The non-intermeshing rotors are designed to apply kneading, shearing and mixing actions to the stock in the Banbury. Typical compounding cycles are 2 to 4 minutes in the Banbury.

Somewhat differently designed batch compounders having the same intent as the Banbury mixer include the Stewart-Bolling mixer, the Adamson-Shaw Intermix, and Gelimat K-Mixer, remarkable for its short cycles.

These mixers are usually set up to discharge their batches onto a sheet-forming 2-roll mill, having cored rolls designed to be held at a pre-assigned temperature by a heat transfer fluid (oil, steam or high-pressure hot water). The mills are usually provided with mixing plows and adjustable strip take-off units so that one operator can supervise several mills. Banbury or other mixer-mill combinations are substantially automated in modern plants.

Continuous compounding mixers used on both rigid and flexible PVC include Buss Ko-Kneaders, Farrel Continuous Mixers (FCMs), Kombiplasts and Plastificators from Werner & Pfleiderer. Single screw and various types of twin-screw extruders are also used for compounding prepared PVC dry blends. Davis-Standard, Berstorff, Krupp and Leistritz are among the manufacturers of these extruder-compounders.

The fluxed PVC compounds may be immediately fed into forming equipment, such as extruder dies, calenders, blow molders, injection molders and so forth. Alternatively, they may be extruded through strand-dies and hot cut or air- or water-cooled as they are cut into pellets. Or, they may be sheeted out on two-roll mills and diced. Cumberland and Bolton dicers are types that are used on cooled strips of PVC compounds in the United States, to produce cubes or octagons for subsequent processing into finished goods.

Processes for the Fabrication of PVC-Based Items.

Table 3---Major PVC Markets in North America Served by Particular Processes

Process and Market	Million Pounds by Year			
Calendering	1995	1996	1997	1998
Flooring	217	230	240.	248.
Textiles	65	80	115	102
Other	841	920	822	824
Total Calendering	1123	1230	1177	1192
Coating				
Flooring	207	225	220	223
Textile/Paper	109	110	100	90
Protective	53	60	65	66
Other	25	25	25	24
Total Coating	394	420	410	403
Extrusion				
Pipe and Conduit	4545	5360	5780	5904
Wire and Cable	416	450	460	482
Siding	1440	1740	1860	2148
Windows and Doors	323	355	510	538
Packaging	364	342	275	352
Non-packaging	60	77	110	80
Other	494	562	640	630
Total Extrusion	7642	8886	9635	10,134
Molding				
Bottles	171	167	150	145
Fittings	253	275	290	308

Other Injection	105	120	130	149
Total Molding	529	562	570	602
Compounders & Reselllers	514	582	672	622
Export	1480	1335	1275	1410
Paste Processes NOC	216	212	210	205
All Other	134	161	100	130
Grand Total	12,032	13,388	14,049	14,698

The versatility of compounded PVC is so great that it works well in a very wide range of fabrication processes, provided the processes are adapted to the particular behavior patterns of PVC.

Statistics on the markets into which PVC resins went in the United States and Canada during the last four years have been presented in Modern Plastics (63, 64). Defining markets partly by the manufacturing process serving them is not very satisfactory. For instance, some vinyl packaging films are made by calendering, and others by extrusion. From the statistics presented by Modern Plastics, it is not possible to determine the amount of PVC used in packaging applications as a whole. However, Modern Plastics is presenting an abbreviated form of the statistics formerly kept by the Society of the Plastics Industry, Inc., (SPI) and now by the American Plastics Council (APC). These are the most accurate available for the North American market. Modern Plastics adds Canadian consumption to the figures for the United States. The figures for the last four years are shown in Table 3, and are probably among the most accurate available.

Calendering of PVC. Calendering of rigid and flexible PVC formulations for various markets has been reviewed in several publications (65, 66, 67, 68). Hybrid extrusion-calendering is finding favor for some film and sheet manufacturing operations (69). The three-roll calenders normally used in these latter lines are often referred to as calenderettes.

Modern calender lines include equipment for measuring web thickness by attenuation of β-ray emissions passing through the web. For repetitive or long production runs closed-loop computerized web gauge control may be used.

Successful calendering is a capital-intensive operation. Coordinated materials' handling, weighing, feeding, fluxing, calendering and take-off systems are required. Effective raw materials and end-product quality control (QC) testing and procedures are essential in today's competitive markets.

Vinyl Coating Processes. A widely used process for coating vinyl finishes onto fabric or primed flexible metal substrates is to laminate calendered or extruded vinyl films onto the substrate, itself in the form of a web of about the same width as the vinyl film. Typical laminators pass the substrate and the vinyl film over synchronously run heated drums and then bring the heated feeds together under a laminating nip having controlled pressure applied between a steel roll contacting the vinyl film and a rubber roll against the backing. If the steel laminating roll has an appropriate finish, embossing of the vinyl layer and laminating can be achieved simultaneously. The laminated goods are then passed over a series of cold rolls (cooling cans) before wind-up.

In instances such as the manufacture of premium quality credit cards, two layers of thin, clear vinyl film are press-laminated over both sides of previously printed core stock to which a magnetic memory stripe and a hologram have already been affixed.

Vinyl dispersion coatings may be effected using unsupported knife coaters, knife-over-roll coaters, or blanket coaters. However, in larger volume applications, dispersion coating is nearly always done using reverse roll coaters or rotary screen coaters. Vinyl films with special embossments and finishes are made from plastisols by coating the plastisol onto a release paper or belt with the finish and embossment on it, passing the coated paper or belt through a fusion oven, followed by cooling means and then stripping the fused plastisol layer from the paper or belt.

Vinyl sheet flooring is the largest application of dispersion coating. This was originally developed using asbestos felt backings which could pass through the gelation and fusion ovens without deteriorating. Thermally stable replacements for asbestos felt had to be found when most industrial uses for asbestos were banned to avoid compromising the health of workers who might inhale asbestos fibers while making the felt or using it. Today, using dispersion coating processes plastisol-derived vinyl sheet flooring is made as rolled goods 6, 9, 12 or 15 feet wide and sold in rolls commonly between 100 and 200 yards long.

Typically, the manufacturing process involves coating a foamable plastisol onto the felt backing, and gelling it in a warm oven set at about 300°F (149°C). The felt with the gelled plastisol on it is then passed through a multistage gravure printing press. Several different inks are applied to the gelled surface in a pattern maintained in strict register. One or more of the inks normally contains a foaming inhibitor for the azodicarbonamide in the gelled foamable plastisol. Maleic anhydride or trimellitic anhydride are effective foaming inhibitors. Next a clear vinyl wear layer plastisol is coated over the gelled, printed, foamable layer. The composite is then passed into a multi-stage fusion oven whose hottest zone is set at about 400°F (204.4°C). The foamable layer then expands in exact register with the printed pattern and both the foamable layer and the wear layer become fully fused, i.e., they develop their ultimate physical properties. As a finishing step, a clear, very thin (0.0005 inch, 0.0127 mm) UV-curable urethane top coating may be placed over the clear vinyl wear layer to enhance gloss.

These patterned and chemically embossed plastisol-derived vinyl sheet goods are very popular in kitchens and bathrooms in the United States.

Extrusion of Rigid and Flexible PVC. About 80% of PVC products made in the United States involve the use of extrusion somewhere in their manufacturing process. In some cases the final shaping is done by a mold, as in extrusion-blow molding and reciprocating-screw-injection-molding.

Extrusion of rigid PVC developed differently in the United States and Europe. Highly formulated rigid vinyl compositions run through single-screw extruders were often preferred in the United States. Simpler formulations fluxed in twin-screw extruders were more common in Europe. Knowledge of the morphological changes affecting PVC rheology developed slowly. Much of the commercial development of extruders and their related tooling was done empirically. Early extruders for flexible PVC were actual or modified rubber extruders. Extruders for rigid PVC were then further modified to handle the high viscosity PVC melts without generating so much frictional heat that the PVC would degrade.

Best results were obtained with two fairly different flow regimens, either of which could be applied to pre-compounded pellets/cubes or to dry blends (powder mixes) fluxed directly in the extruder.

In so-called low-temperature rigid profile extrusion, rigid PVC melt temperatures are held to the range 177-185°C (350-365°F) and medium molecular weight PVC is generally used (IV 0.88-0.95, K 66-68). For this type of extrusion, the primary PVC resin particles formed during polymerization are the principal flow units. Molecular entanglements are kept low, the extrudate is non-tacky, fairly stiff and strong, making extrudate handling easy, and die swell is low, making sizing of dies and extrudates simple. If the stock temperature is allowed to drop too low, proper knitting of the primary PVC particles in the melt does not occur. This allows parts to be weaker, cracks to propagate and weathering-related crazing to develop. Low-temperature extrusion is preferred by many for vinyl window profiles, door lineals and other complex shapes, whose definition is hard to maintain in high temperature extrudates.

Twin-screw extrusion of pipes through suitable dies is mostly done at stock temperatures around 180 to 185°C (356°-365°F) when sizing is critical. This allows minimizing stabilizer content in the formulations.

Extrusion of pipes from powder using single-screw extruders is generally done at 200-206°C (392-403°F) because faster extrusion is attainable in this equipment at these stock temperatures. Under these conditions, a high proportion of the crystallites in the PVC resin melt. This obliterates most of the primary resin particles formed during polymerization and brings about molecular rather than particulate flow in the melt. Die swell increases substantially and surface gloss is increased.

Other rigid vinyl extrusions generally run at the high temperature regimen are house siding, extrusion blow molding, film extrusion and reciprocating screw injection molding.

At stock temperatures between 185-200°C (365-392°F), in many rigid vinyl formulations some of the primary resin particles melt and others do not, producing a mixed flow behavior, partly particulate and partly molecular. This usually produces rough and variable extrudates in which die-swell is impossible to control closely. Exceptions to this are that very low molecular weight PVC resins and copolymers can be processed in this temperature range.

For all extrusion formulations, selection of the lubricant system is extremely critical to attainment of good quality at maximum output. It not only prevents sticking of the vinyl melt to metal parts (extruder screw and dies), controls plate-out on the same metal parts, it conditions the way primary resin particles flow and generate frictional heat while flowing.

Theoretical and practical information on the extrusion of rigid PVC has been well summarized by Gomez (70) and by Batiuk, Korney and Cocco (71). Complex crystallite melting phenomena, gradual obliteration of primary particle flow units and the multi-component nature of many PVC extrusion formulations together result in non-ideal behavior of many PVC melts, in terms of theory, particularly if they include rework.

Table 4   Data on Typical Extruder Outputs (71)

Twin-Screw Extruder Capabilities

Extruder Type	Pipe	Siding	Window
55 mm, conical	≤ 600 lb/hr	------	≤ 400 lb/hr
92 mm, conical	≤ 2300 lb/hr	≤2300 lb/hr	≤1500 lb/hr
110 mm, cylindrical	≤1500 lb/hr	≤1500 lb/hr	≤1200 lb/hr
130 mm, parallel	≤3000 lb/hr	≤3000 lb/hr	---------
160 mm, parallel	≤4500 lb/hr	--------	---------

Single-Screw Extruder Capabilities Using Sophisticated Screw Designs

Barrel Diameter	L/D Ratio	Capacity
3.5 in.	24/1	≤ 550 lb/hr., 250 kg/hr.
3.5 in.	32/1	≤ 650 lb/hr., 295 kg/hr.
4.5 in.	24/1	≤750 lb/hr., 340 kg/hr.
4.5 in.	32/1	≤900 lb./hr., 410 kg/hr.

Larger diameter single screw extruders are generally not successful on rigid PVC because they generate unacceptably high and non-uniform stock temperatures at the high outputs which could justify their use.

Co-extrusion is commonly used where a product benefits from having two or more layers. Much siding has 80% of relatively non-weatherable, lower cost PVC core stock and 20% of a highly weather-resistant cap-stock of a weather resistant polymer, or PVC with high TiO_2 content.

Wire and cable primary insulation and jacketing for indoor cables needing good to excellent fire-resistant properties are the largest volume flexible and semi-rigid vinyl extrudates in the United States. These are usually run through single-screw extruders using cross-head dies. Thin (0.005 inch, 0.127 mm) semi-rigid insulation is routinely coated onto 18 gage (0.0403 in., 1.024 mm diameter) communications wire at linear speeds of 3500 ft/min (1067 m/min.). Typical building wire insulations coated onto 12 gage (0.0808 in., 2.052 mm. diameter) copper wire are run at linear speeds in the range of 600 to 1,000 ft./min. Depending on the cable construction, jackets over multiconductor cables are generally applied at somewhat lower speeds in order to meet and maintain the required dimensional tolerances.

Automobile side moldings, gasketing, window splines, shoe welting, floor mats, car mats and flat films over 0.004 in. (0.102 mm.) thick are other extruded flexible PVC products. Tubular film dies in a vertical conformation are used for thin gauge blown film used for meat and produce wrap and shrink films.

A specialty dual hardness co-extrusion is a window profile which carries a flexible PVC sealing lip permanently bonded to it. For this purpose, the soft PVC must be made with relatively high molecular weight plasticizers which do not readily migrate into rigid PVC, because if they do, the rigid PVC is apt to be embrittled.

Injection Molding of PVC. The largest volume category of products injection-molded from rigid PVC is pipe fittings. Most of these are heavy items with relatively thick cross-sections, not considered a challenge to mold. The first rigid vinyl injection-molding compounds were designed for these services and were formulated to meet specifications for fittings for use in potable water distribution, DWV pipe, electrical conduit or various grades of chemically resistant piping at the lowest attainable materials' cost. Since the physical requirements for fittings for pressure piping are extensive and rigorous, the highest molecular weight PVC resins which would allow the compounds to fill these molds well were used.

Rigid PVC then got a bad reputation from trial and error injection molders who tried to use these compounds for thin-wall moldings, which the compounds would not fill at their recommended processing temperatures. Raising stock temperatures to improve mold fill caused degradation of the PVC, evolution of HCl and corrosion of molds and injection molding equipment. Rigid PVC has not overcome the resulting corrosion stigma in the United States' injection-molding industry. Materials' suppliers have responded to the challenge by introducing readily moldable rigid PVC compounds based on lower molecular weight resins and better (more expensive) lubricant and stabilizer systems.

Old-fashioned plunger-type injection molding machines and sprues and molds having sharp corners are unsuitable for injection-molding rigid PVC. They were designed for sharp-melting polymers such as PE, PS and PP. The melt viscosity of rigid PVC drops gradually with increasing stock temperature and PVC stocks are shear-sensitive in the sense that rapid shearing of rigid vinyl stocks generates a lot of frictional heat which can only be removed relatively slowly by conduction due to the low thermal conductivity of PVC. The high shear rates in injection-molding systems designed for other polymers often degrade rigid PVC when it is substituted without reconforming the system.

Today's general purpose rigid PVC injection molding compounds are not much different in thermal stability and melt viscosity from molding grades of ABS. They give efficient molding cycles in reciprocating-screw machines fitted with PVC screws having low compression ratios, a "smear tip" and, no or specially designed back-flow check valves. There must be no places in the system where polymer hangs up. Long residence times in hot runners must be avoided. Molds must be designed to accommodate the low shrink characteristics of PVC. Undercuts which must strip out on part ejection have to be small. Molded-in inserts work very well with PVC due to its low shrink factor.

For low volume, short production runs aluminum molds are satisfactory. For other applications, more corrosion-resistant molds are recommended for service with PVC as insurance against machine malfunction or operator malpractice. Good quality general purpose PVC injection molding compounds are designed to withstand up to 45 minutes' exposure to 410°F (210°C), but discolor and emit HCl if their thermal history exceeds this.

Due to its combination of good properties including excellent flammability compared to PE, PP, PS and ABS, excellent chemical resistance to cleaning agents and staining materials, toughness, tensile strength ($>$6000 psi or 41.3 MPa) and flexural modulus ($>$350,000 psi or 2411 Mpa), and weatherability, rigid PVC is being both considered and used for many applications, including electrical fittings, electrical junction boxes, business machine housings, typewriter housings, various parts of personal computers, office furniture, pump components, mobile home components, small appliance housings and many other injection-molded parts. (72)

Where conventional brominated FR agents for ABS are being regulated out in Europe, an opportunity exists for enhanced heat distortion grades of FR rigid PVC.

The combination of the current economic slow-down in Asia and low prices of oil-based feedstocks and their polymeric derivatives has made the injection-molding industry very competitive.

Flexible items molded from plasticized PVC are often rotationally molded from plastisols rather than injection-molded, due to relative equipment cost factors. However, the relatively low melt viscosities of flexible PVC compounds make it technically easy to injection-mold them, and this is done when economics are favorable.

Miscellaneous Other Fabrication Processes Used for PVC Products. These processes include thermoforming of rigid PVC sheets, injection-blow molding and extrusion-blow molding of rigid PVC containers, compression molding, molding and extrusion of PVC foams, powder molding, powder coating, joining, sealing and decoration of parts. All of these processes work well on PVC where economics are favorable.

Specifications for, and Quality Control Testing on PVC Products.

There exist a formidable array of specifications, quality control tests, regulations and health and safety standards affecting PVC products in the United States. Many are listed in general references (73, 74). The caveat is that all of these are periodically updated. It is necessary to check with the issuing authority before assuming that a file copy is current.

Classifications for PVC resins include ASTM D 1755 which covers general purpose (GP) and dispersion (D) homopolymers. Specifications for particular resins for stated uses can be based on these classifications. ASTM D 2474 provides a classification protocol for vinyl chloride (VCl) copolymers of the following types: VCl—vinyl acetate; VCl—vinylidene chloride; VCl—maleic ester and VCl—acrylonitrile. The first three of the above co-polymers are available commercially.

Tests for particular PVC resin properties include ASTM D 1243 for dilute solution viscosity as a measure of molecular weight, with inherent viscosity (IV) preferred for every day use over limiting viscosity number (intrinsic viscosity). In Europe PVC resin molecular weights are reported as K-values, but there are minor differences in test method between some laboratories. The official methods are ISO 174-1974 and DIN 53 726 based on the relative viscosity of 0.5 g of PVC in 100 ml of cyclohexanone at 25°C. In precise work it is necessary to know how K-values were determined to avoid misunderstandings.

In Japan and much of the Far East PVC molecular weights are reported in terms of the JIS K 6721 method, which measures specific viscosities in nitrobenzene at 30°C, and provides transformations for calculating K-value and mean polymerization degree, P.

Other frequently used and referenced tests for PVC resins include ASTM D 1705 for wet sieve analysis, D 3596 for gel count, D 2873 for pore diameter and pore volume (measures of particle porosity), D 2396 for powder-mix time, D 3367 for plasticizer sorption by centrifuge, D 2538 for fusion characteristics by torque rheometer, D 1895 for bulk density. ASTM D 4202 and ISO/R 182-1970 provide virtually meaningless results on the heat stability of unstabilized PVC resins (which is always bad, but nobody uses unstabilized PVC). Taubinger, et al., (75), modified the ISO/R 182 Method B to measure the HCl evolution conductimetrically, The author in unpublished work has obtained useful results on the induction period for HCl evolution from formulated compounds by an adaptation of this method.

ASTM D3030 measures volatile content of PVC resins, D 2222 measures residual emulsifier on emulsion resins. D 4368 summarizes tests used on resins, including emulsion resins for plastisol use.

Residual VCM in PVC homopolymers may be measured by ASTM D 3680, D 3749 or D4443, which is finicky but accurate in the parts per billion range when run properly.

ASTM E 442 is useful for measuring the chlorine contents of PVC copolymers.

The range of products made from PVC is so wide that tests on the products are too numerous to cover here. See references (73 and 74) for details.

Many standards and/or regulations covering PVC products have been published by ASTM, ISO in Europe, the American National Standards Institute (ANSI), FDA, NSF, Underwriters' Laboratories (UL) and the National Fire Protection Association (NFPA). The 1999 National Electrical Code (NEC) is an example of an NFPA publication dealing with electrical safety and covering many PVC products including wire and cable constructions.

Safe manufacture of PVC and in-plant handling of formulating ingredients in the United States is the realm of the Occupational Safety and Health Administration (OSHA). Many OSHA workplace standards affect the PVC industry. Permissible exposure levels (PELs) and time-weighted averages (TWAs) are established as standards for many regulated materials including VCM and lead-containing stabilizers as examples. Material safety data sheets on all products used must be kept on file, and customers must be provided with material safety data sheets on all products sold or sampled to them.

Environmental concerns are the province of the Environmental Protection Agency (EPA) which handles regulations concerning substances which leave a manufacturing plant covered by TSCA, RCRA, CERCLA and other environmental acts such as SARA Title III, Sections 331-313 requiring disclosure of inventories and accidental releases of listed substances.

Because they are such large volume products, some pipe standards are very important to the PVC industry. These include ASTM D2241 for pressure-rated PVC pipe, D 2665 for PVC DWV pipe and fittings and D 3034 for PVC sewer pipe and fittings. Chlorinated PVC (CPVC) plastic hot and cold water distribution systems are covered in D 2864 and F442. These specifications are recognised by HUD and many local authorities having jurisdiction as code enforcers.

ASTM D 4099 covers PVC prime windows; D 3679 covers PVC siding, and D 3678 covers PVC interior-profile extrusions. Electrical non-metallic tubing (ENT) made from rigid PVC is accepted for many uses in the 1999 NEC.

Wire and cable specifications are covered by UL, the National Electrical Manufacturers Association (NEMA) and the Insulated Cable Engineers Association (ICEA) in addition to the NEC.

Standards for United States-made automobiles are established by the Society of Automotive Engineers (SAE) and individual automobile manufacturers.

Some years ago, the UPITT smoke toxicity test was run on a large number of wire insulation and jacketing compounds in the United States. PVC compounds in general gave average results and continued to be accepted in all U. S. jurisdictions. ISO committees are currently deliberating on new specifications which may classify halogen-containing insulations and jackets as more hazardous than those based on polyolefins in terms of smoke toxicity in fire situations. This is a matter of current controversy between the United States and Europe at the committee discussion level.

Environmental and Occupational Safety/Health Considerations and Green Politics.

Gottesman described the state of knowledge in 1974 relating to carcinogenicity of vinyl chloride monomer (VCM) and the proposed permanent standard for employee exposure to VCM to be imposed on VCM/PVC plants. The vinyl industry and the Society of the Plastics Industry (SPI) opposed it on the grounds that it was not technologically feasible to comply with, and if enforced would shut down the entire vinyl industry (76).

Lehman and Flury in 1943 had stated of the toxicity of VCM that "vinyl chloride is one of the least dangerous of the chlorinated hydrocarbons" (77) which may have given the vinyl industry a false sense of security. Barr has described the early and the more recent history of VCM toxicity and implementation of the current OSHA worker exposure regulation (78) which became effective in April 1975. This allows an 8-hour time-weighted average (TWA) of 1 ppm maximum and a 15-minute exposure ceiling of 5 ppm for workers. The regulation also sets an action level of 0.5 ppm below which no action is required. At exposure levels above 0.5 ppm in the workplace a number of rules are invoked and have to be followed, involving restriction of access to facilities, labeling, monitoring, medical examination protocols and training of employees. Several smaller and older plants shut down because they could not afford to comply. The larger plants which modified their processes lost 5 to 10% of production capacity and spent about $200 million over the next few years in added capital to comply with the OSHA regulation and related EPA rules regarding VCM emissions and VCM residuals in PVC resins.

EPA requires control of potential fugitive emissions, special work practices for vessel openings and sampling of VCM, stripping of VCM residuals from PVC resins and waste-water, no point-source emissions of VCM above 10 ppm, no relief valve discharges except for non-preventible emergencies (Acts of God), use of particular analytical procedures and detailed monitoring, reporting and record-keeping related to use of VCM. The vinyl industry estimates that it spent almost $1 billion during the period 1981-1990 complying with the EPA rules on handling and using VCM, and controlling VCM residuals in PVC which were issued in 1976.

Regarding acute toxicity, 7 to 10% concentrations of VCM in air have a narcotic (anesthetic) effect on humans and animals. Concentrations above 12% may be fatal. It is reported that VCM is metabolized by cytochrome P-450, but that this mechanism is saturable. When liver capacity is overwhelmed, metabolism produces carcinogenic intermediates. Therefore, repetitious, intermittent high exposures to VCM are dangerous. This was the condition to which PVC reactor cleaners were exposed prior to reports by Viola and Maltoni that exposure to high levels of VCM vapor produced tumors in rodents (1972-3) and in January 1974, BF Goodrich's report that three workers had died of angiosarcoma of the liver. Events moved swiftly. In April 1974, OSHA reduced the permissible worker exposure level from 500 to 50 ppm of VCM. Then in May 1974, OSHA announced a proposed permanent standard of "no detectable" VCM by a test accurate to 1 ± 0.5 ppm. After a number of hearings, the more reasonable April 1975 standard became final. Data in reference (78) show that among PVC reactor cleaners cases of angiosarcoma prior to 1985 were 111 world-wide. Of these, all but two had died. Alternate means of reactor cleaning were implemented throughout the PVC industry shortly after April 1975. Use of high pressure water sprays to dislodge PVC scale from reactor interiors was preferred by most companies. Clean reactor technology was then developed whereby periodic treatment of reactor insides with special coatings discouraged PVC scale formation during polymerization. The two most widely used clean reactor systems are those offered by BF Goodrich (now Geon) and Shinetsu of Japan.

Over the years the PVC industry has made great efforts to minimize the likelihood of disasters in VCM and PVC plants. Reference (78) lists sixteen major accidents worldwide between 1955 and 1981 in VCM/PVC plants, half of which involved operator error and all of which included flash fires or explosion of VCM/air mixtures.

VCM and PVC plants are generally designed with emergency back-up power sources. Loss of cooling system flow, flow control failure in a VCM plant, loss of agitation in a PVC plant or inability to charge short-stop and disperse it into a polymerization batch threatening to run away all have disaster potential. Relief valves protected by rupture discs are provided to vent runaway reactors before they explode. Emergency venting of VCM is usually arranged to go to a flare system provided with a steam sparge automatically activated by venting. Presence of the steam keeps the flame burning cleanly without massive production of soot. Large releases of free VCM (more than 1 lb. must be reported) are dangerous and illegal in the United States.

For safety, large reactor plants are run under computer control with redundant back-up computer systems and an additional last resort system which provides for orderly manual shut-down in case of a double computer failure.

In the early days of the industry, allowing VCM to contact copper or brass in pipes, instruments or pump impellers was strictly prohibited due to the presence in VCM of low levels of acetylenic impurities. Historically, formation of copper acety-lides, which are shock-sensitive and explosive was experienced.

EPA requires that waste water which has been in contact with VCM be stripped to below 10 ppm residual VCM before it is discharged or mixed with other wastes.

Uncompounded PVC is regarded as biologically inert. It does not depolyme-rize and may be disposed of in landfills provided it has been stripped of residual VCM to shipping levels. However, most off-grade PVC, including "BB" batches, is sold to recyclers rather than landfilled. PVC wastes mixed with combustible materials can be safely incinerated, provided the temperature of the incinerators is high enough to destroy any dioxins which may be formed in the initial cooler regions of the incinerator (79).

Safety information relating to the handling of initiators used in production of PVC, trichlorethylene (TCE) chain transfer agent, vinyl acetate monomer and literature for follow-up reading on safety matters are provided in reference (78).

Gottesman provided a description of the status of recycling post-consumer PVC (80). In general, the PVC industry does an outstanding job of re-using in-process scrap and re-work whose compositions are clearly known in terms of formulation ingredients. With regard to recycling of post-consumer vinyl packaging: SPI has assigned the number "3" to PVC packaging materials. This number is embossed into the bases of rigid PVC containers. Nevertheless, manual sorting of PVC containers from a mixed plastics stream has not proved accurate enough. Various automated sorting systems based on analytical machines have been proposed, but these are fairly expensive, and do not appear to be as widely used as recycling advocates had hoped.

An additional factor negative to post-consumer recycling has been the low prices at which virgin PVC is periodically available. Markets have not been willing to pay enough for post-consumer PVC recyclates on a sustainable basis to make PVC recycling a reliably profitable business. Theoretically, PVC producers could build feedstock recycling plants to recover HCl and hydrocarbons from post-consumer PVC. More work is being done in Germany and Japan along these lines than in the United States. The overall logistics for large-scale economically satisfactory recycling of PVC in terms of collecting, sorting, shipping, and de-contaminating PVC post-consumer wastes have not been worked out yet on a commercially viable basis in the United States.

The PVC industry funds the Vinyl Institute, recently moved to APC (American Plastics Council) headquarters in Arlington, Virginia to collect, analyze and report on safety and environmental information related to the safe manufacture and use of PVC. Tim Burns is the key contact (81). The Vinyl Institute has to be on the alert to counter negative publicity from periodic press releases by Greenpeace, which for many years, has been trying to harm the public image of PVC. A typical example was the April 1996 blaze at Düsseldorf Airport in Germany. The fire, which was started by a careless welder, killed 16 travelers. Greenpeace immediately alleged that burning PVC wire insulation was the prime culprit. The final report by fire expert Heinz Schiffers showed that other plastics (PS and PU foamed roofing insulations) played a much more prominent part in releasing heat and smoke than the small amount of vinyl wire insulation did. Traces of hydrochloric acid from burning PVC were minimal. The small amounts of PCBs found in the ashes of the burned terminal almost certainly came from the PCB-containing condensers in the neon lights placed throughout the terminal (82). The problem the PVC industry has not resolved, however, is that most of the public who read or heard reports of Greenpeace's allegations in the media have not realized that later scientific evaluation of the Düsseldorf fire exonerated PVC. The PVC industry would welcome creative ideas on how to approach and resolve this type of public relations' problem.

People concerned with preserving forests acknowledge the outstanding environmental role played by PVC in replacing wood cut from virgin forests for use as a structural material. Leaversuch reviewed uses of PVC in window frames, vinyl siding, vinyl cladding, and various wood-vinyl composites (83). Some of these have a wood veneer over a solid or foamed vinyl substrate. Others comprise 55 to 65% virgin PVC combined with 35 to 45% reclaimed milled-wood fiber and vinyl scrap and are used for door jambs, sash, stops, rails, casings, sills, stiles, grills, sliding window and door tracks and various trim parts. As top quality wood becomes scarcer and more expensive, wood-vinyl composites and other vinyl products become more attractive. Several European companies are developing wood-like vinyl products along lines similar to United States companies such as Andersen Window, Crane Plastics and Vinyl Building Products.

Gribble has refuted the constantly repeated erroneous claims by some environmentalist zealots that all, or nearly all chlorine-containing organic compounds are anthropogenic (man-made) (84). He stated in 1994 that the number of identified naturally-occuring organohalogens was about 2000, and that PCBs were found in ash from the 1980 eruption of Mt. St. Helens, Washington. He also references the conclusion of two research groups that forest and brush fires are the major sources of polychlorinated dibenzodioxins (PCDDs) and polychlorinated dibenzofurans (PCDFs) in the environment. Gribble states that horse-radish peroxidase enzyme (HRP) can convert natural chlorphenols into PCDDs and PCDFs. Some of these compounds have been found to be highly toxic to rodents, but Vietnam veterans and human inhabitants of Seveso, Italy (the site of a major dioxin release in 1976) have not experienced the drastic increases in incidence of cancer which pessimists confidently extrapolated from rodent-susceptibility data. Gribble's conclusion is that it is illogical to shut down chlorine-using industries because naturally occurring PCDDs and PCDFs are found to be ubiquitous at very low levels. Industrial releases can be controlled, as necessary, but naturally occurring PCDD/Fs cannot be eliminated. Greenpeace should acknowledge natural products.

Alcock, et al., found significant levels of PCDD/Fs in a sample of English soil taken in 1881 and kept in a sealed jar (85, 86). This establishes that these compounds have been around for more than 100 years and are not exclusively produced by the PVC industry as stated or implied by some environmental zealots.

Aylward, et al., have shown that humans are as much as 90 times less sensitive to dioxins than is assumed on the basis of rodent tests (87, 88, 89). The reason is that on repeated dosing of rodents with low doses of dioxin, in 4 months they reach a fairly high steady-state concentration of dioxin in their blood serum which does not decrease when dosing is stopped. The concentration of dioxin in human blood serum starts dropping as soon as dosing is stopped. Reference (89) points out that some workers in the field disagree with Aylward.

Yosie has discussed the "chlorine debate" from the United States EPA's point of view (90). He predicts that environmentalists will be successful in shaping public policy by "personalizing" the risks they allege are due to man-made chlorine-based organic chemicals, including vinyl products. The PVC industry and their allies should equally "personalize" the risks to the public due to taking technically wrong decisions in incorrectly restricting or banning valuable uses of vinyl products.

Dolan and McCarty showed that VCM under suitable conditions is readily biodegradable (91). This makes it relatively certain that accidental releases of VCM into the environment and VCM which has been found at low levels in some landfills have no long-term environmental consequences.

An organization called the "Chlorophiles" was formed in Europe in 1994 to counter inaccurate allegations from so-called environmentalists about organo-chlorine compounds (92).   They have a United States contact (93).

Future Projections

The Vinyl Institute has four predictions for the future of the vinyl industry in the United States.   For the "continuation" scenario, (50% probability) they projected 3.4% average growth per year from 1995 to 2010; and 3.2% average growth per year from 2010 to 2020.   For the "maturation" scenario (20% probability) they forecast 2.0% and 0.6% growth for the respective periods. For the "acceleration" scenario (20% of probability) they forecast 4.5% and 3.9% growth for the periods, respectively. With a 10% probability they forecast a "restriction" scenario having 1.0% and 0.0% growth for the respective periods.   They pointed out that PVC is "greener" than aluminum because of its lower energy requirements for manufacture.   They stated that vinyl is as green as a tree because its use as a wood replacement slows the world-wide deforestation, of which genuine environmentalists are complaining.

A few shorter-term projections from the 1997 Chemical Economics Handbook are:
— United States' demand for PVC will continue to grown at an average of 4.5% per year until 2001 (94).
— Siding and windows and doors are the fastest growing markets for PVC and are expected to continue to grow faster than 5% per year until 2001.
— Use of PVC in consumer goods, packaging, transportation, home furnishings and wire and cable is projected to grow more slowly than 5% per year until 2001. United States' exports are favored by the good economics of the United States' Gulf Coast plants, but will depend on the health of the Asia/Pacific economies.

Technical projections can be inferred from a recent review of advances in PVC chemistry by Starnes (95).

Literature Cited.

1. Partington, J. R., "General and Inorganic Chemistry"; Macmillan, London, 1946, p. 3.
2. Kaufman, M., "The Chemistry and Industrial Production of Polyvinyl Choride: The History of PVC"; Gordon and Breach, New York, 1969.
3. Regnault, V., <u>Annalen</u>, 1835, <u>15</u>, 63.
4. Regnault, V., <u>Ann. Chem. Phys.</u>,1838, <u>69</u>, 151.
5. Baumann, E., <u>Ann. Chem. Pharm.</u>, 1872, <u>163</u>, 308.
6. Society of Plastics Engineers, Inc. "Chronology of Significant Developments in the History of Polyvinyl Chloride", 1992, Brookfield Center, CT,.
7. Vinyl Institute, "Note: Vinyl— An Enlightening Look at the 50-year History of the World's Leading Electrical Material", <u>Journal of Vinyl Technology</u>, 1991, <u>13</u> (4), 223.
8. Vinyl Chloride Monomer (VCM) in "1997 Chemical Economics Handbook"; SRI International, Menlo Park, California.
9. Storck, W., <u>Chemical and Engineering News</u>, 1998, <u>76</u>, (49), 33.
10. Private communication.
11. Pham, Q. T.; Millan, J.; Madruga, E. L., <u>Macromolecular Chemistry</u>, 1974, <u>175</u>, 945.
12. Nakajima, A.; Hamada H.; and Hayashi, S., <u>Macromolecular Chemistry</u>, 1966, <u>95</u>, 40.

13. Natta,G.; Corradini, P., Journal of Polymer Science, 1956, 20, 251.
14. Witenhafer, D. E., Journal of Macromolecular Science-Physics, B4, 1970, 915.
15. Illers, K. H., Macromolecular Chemistry, 1969, 127, 1.
16. Wenig, W. J., Journal of Polymer Science and Physics, 1978, 16 1635.
17. Blundell, D. J., Polymer, 1979, 20, 934.
18. Wilkes,C. E., Proceedings of the Vinyl Technical Conference, 1998, East Brunswick, New Jersey, p. 21.
19. Langsam, M., "PVC Processes and Manufacture",in, "Encyclopedia of PVC",2nd Edition, Nass, L.I; Heiberger, C., Eds, Marcel Dekker, Inc., New York, 1986, Vol.1, pp. 49-103.
20. Perkins, G.; Haehn, J., Journal of Vinyl Technology, 1990, 12, (1), 2.
21. Coaker, A. W.; Wypart, R. W., "Handbook of PVC Formulating", E. J. Wickson, Ed.; John Wiley and Sons, New York, 1993, pp. 94-96; 98-100.
22. Daniels, C. A.; Collins, E. A., Journal of Macromolecular Science and Physics,1974, Vol. B4, 287.
23. Abdel-Alim, A. H.; Hamielec, A. E., Journal of Applied Polymer Science, 1973, 17, 3033.
24. Ibid., 21, Szoc, K. R.; Burns, R. J., "Solution Vinyl Resins", pp. 115 -135.
25. For instance, see Data Sheet G-081W, BF Goodrich, April 1987.
26. Geil, P. H., Journal of Macromolecular Science—Physics,. 1977, B 14, (1), 171.
27. Summers, J. W., "A Review of PVC Morphology and Fusion", SPE Vinyl RETEC Proceedings, 1986, Chicago, p. 229.
28. Berens, A. R.; Folt, V. L., "Polymer Engineering Science", 1969, 9 (1), 27.
29. Faulkner, P. G., Journal of Macromolecular Science and Physics, Vol. B, 11, (2), 251, (1975).
30. Singleton,C.; Isner, J., et al, PolymerEngineering Science, 1974, 14 (5), 371.
31. Summers, J. W., Journal of Vinyl Technology, 1981, 3, (2), 107.
32. Pezzin, G., Pure Applied Chemistry, 1971, 26 (2), 241.
33. Rosenthal, J., The Journal of Vinyl Technology,1983, 5 (3), 104.
34. Collins, E. A.; Krier, C. A., "Trans. Soc. Rheology", 11 (2), 225 (1967).
35. Collins, E. A.; Daniels, C. A., Polymer Engineering Science, 14 (5), 357 (1974).
36. Lyngaae-Jorgensen, J., Journal of Macromolecular Science and Physics, Vol.B 14, 1977 (2) 213.
37. Bulletin G-62 RES, "Geon Vinyl Resin Material Safety Data", BF Goodrich Chemical Group (Predecessor to The Geon Company), 1985.
38. "Bulk Handling of PVC", Technical Service Bulletin No. 10, BF Goodrich Chemical Group (Predecessor of The Geon Company),1983.
39. Schwaegerle, P. R., Journal of Vinyl Technology, 1985, 7, (1), 16; 1986, 8, (1) 32.
40. Shah, A. C., Plastisol and Organosol Preparation, in ibid (21), pp. 783-803.
41. Malpass, V. E.; Petrich, R. P.; and Lutz, J. T., Jr., "Encyclopedia of PVC",Vol. 2, 2nd Ed., Nass,L. I; Heiberger, C. A., Eds., Marcel Dekker, New York., 1986, 480-3.
42. Roman, R. H., "Modern Plastics Encyclopedia, '99", 75 (12), B11-13.
43. Rapacki, S. R., Journal of Vinyl and Additive Technology, 1998, 4 (1), 12.
44. "National Electrical Code", NFPA 70, 1999 Edition, National Fire Protection Association, Quincy, Massachusetts, 1998.
45. Ibid., 41, Jennings, T. C.; Fletcher, C. W., pp. 60-67.

46. Ibid. 21, Baker, P.; Grossman, R. F., pp. 307-311.
47. Mitchell, E. W. J., Journal of Material Science and Technology, 1985, 20, 3816-3830.
48. Mitchell, E. W. J., Journal of Vinyl Technology, 1986, 8 (2), 55.
49. Weinberg, E. L., U. S. Patent 2,680,726 (1954).
50. Sears, J. K.; Darby, J. R., "The Technology of Plasticizers", John Wiley and Sons, New York, 1982.
51. Ibid. 19, Sears, J. D; Darby, J. R. "Solvation and Plasticization", pp. 432-554.
52. Ibid., 21, Krauskopf, L. G., Chapter 5; Lutz, J. T., Chapters 6 & 7; and Touchette, N. W. , Chapter 8.
53. "Modern Plastics Encyclopedia '99," 75 (12), C105-115.
54. Official Journal of the European Communities, 17 August 1990, L 222/49.
55. CMA Phthalate Esters Panel, "Repsonse to Greenpeace Allegations about Phthalates in Children's Toys," CMA, May 1998.
56. Krzewki, R. J.; Collins, E. A., Journal of Vinyl Technology, 1981, 3 (2), 116-119.
57. Krzewki, R. J.; Collins, E. A., Journal of Macromolecular Science and Physics, 1981, B20 (4), 465-478.
58. Dunkelberger, D. L, *et al,* Journal of Macromolecular Science and Physics, 1981, 21, 639-667.
59. Malpass, V. E., *et al.*, Journal of Macromolecular Science and Physics, 1981, 41 392-409.
60. Lutz, J. T., Jr.; Szamborski, E. C., Journal of Macromolecular Science and Physics, 21, 579-600.
61. Ibid. 41, 409-470.
62. Ibid. 53, C34-35.
63. Modern Plastics, 1999, 76 (1), 72-3.
64. Modern Plastics, 1997, 74 (1), 76-7.
65. Coaker, A. William, "Engineering with Rigid PVC—Processability and Applications, Gomez, I. Louis, Ed., Marcel Dekker, Inc., New York, 1984, pp. 385-438.
66. Coaker, A. William, "Encyclopedia of Polymer Science and Engineering", 2nd Edition, John Wiley and Sons, Inc., New York, 1985, Vol. 2, pp. 606-622.
67. Arena, Arthur A., "Encyclopedia of PVC", 2nd Edition, Nass, L.I.;Grossman, R. I., Eds. Marcel Dekker, Inc., New York, 1998, Vol.4, pp. 103-148.
68. Blum, John B., "Proceedings of the Society of Plastics Engineers Vinyl RETEC '93", Atlanta, GA, pp. 254-261.
69. Hanson, Dana, "Modern Plastics Encyclopedia '99", p. D-29.
70. Ibid. 65, Gomez, I. Louis, pp. 99-244.
71. Ibid. 67, Batiuk, M.; Korney, A.; and Cocco, D., pp. 1-62.
72. Ibid. 67, Harris, B. C., pp. 63-101.
73. Ibid. 21, by Marx, M. F.; Whitley, M., pp. 817-867.
74. "Encyclopedia of PVC", Vol. 3, Nass, L. I., Ed., Marcel Dekker, New York, 1992, Gomez, I. L.; Lloyd, P., pp. 231-292; Ronezka, R., pp. 293-319; Weisfeld, L. B., pp. 321-353.
75. Taubinger, R. P.; Allsop, M. W.; Van der Loo; H. J. M.; Mooij, J., Polymer Testing, 1986 6, p. 337.
76. Gottesman, R. T., Applied Polymer Science, ACS Books, 1975, pp. 379-384.
77. Lehman, L. B.; Flury, F., "Toxicology and Hygiene of Industrial Solvents", Translated by

King, E.; Smyth, H. F., Jr., William and Wilkins, Baltimore, 1943.

78. Ibid. 19, Barr, J. T., pp. 239-308.
79. Randall, D.; Shoraka-Blair, S., "An Evaluation of the Cost of Incinerating Wastes Containing PVC", CRTD-Vol. 31, American Society of Mechanical Engineers, New York, 1994.
80. Ibid. 67, Gottesman, R. T., "Waste Reclamation of PVC Materials and Products", pp. 279-297.
81. The Vinyl Institute, A Member of the American Plastics Council, 1300 Wilson Blvd., suite 800, Arlington, VA 22209.
82. <u>Modern Plastics</u>, 1997, <u>74</u> (4) p. 16.
83. Leaversuch, R. D., <u>Modern Plastics</u>,1996, <u>73</u> (3) , pp. 52-57.
84. Gribble, G. W., <u>Environmental Science and Technology</u>, 1994, <u>28</u> (7), pp. 310A-319A.
85. Alcock, R. E.; McLachlan, M. S.; Johnston, A. E.; Jones, K. C., <u>Environmental Science and Technology</u>, 1998, <u>32</u> (11), pp. 1580-1587.
86. Ibid.,1999, <u>33</u> (1), pp. 206-207.
87. Aylward, L. L.; Hays, S. M.; Karch, N. J.; Paustenbach, D. J., <u>Environmental Science and Technology</u>, 1996, <u>30</u> (12), pp. 3534-3543.
88. Ibid., 1998, <u>32</u> (4), pp. 551-552.
89. Ibid., Phibbs, P., 1997, <u>31</u> (3), pp. 130A-131A.
90. Ibid., Yosie, T. P., 1996, <u>30</u> (11), pp. 498A-501A.
91. Ibid., Dolan, M. E.; McCarty, P. L., 1995, <u>29</u> (8), pp. 1892-1897.
92. Chlorophiles c/o Ferdinand Engebeen, Oude Ertbrandstraat 12, B-2940 Stabroek, Belgium.
93. Elliot Weinberg, P. O. Box 822, 8 Clovis Road, East Brunswick, NJ 08816.
94. "Polyvinyl Chloride (PVC) Resins", 1997 Chemical Economics Handbook, SRI International, Menlo Park, California.
95. Starnes, W. H., <u>Journal of Vinyl and Additive Technology</u>, 1996, <u>2</u>, p. 277.

PLASTICIZERS

ALLEN D. GODWIN

Exxon Chemical Company, Basic Chemicals and Intermediates Technology,
5200 Bayway Drive, Baytown Texas 77586

Introduction
Mechanism of Plasticization
Types of Plasticizers
Plasticizer Characteristics and Performance of Flexible PVC
Plasticizer Selection for Specific Applications
Plasticizers for Other Polymers
Health Aspects of Plasticizers
Future of Plasticizers

Introduction

A plasticizer is a material that when added to another substance, makes that material softer or more flexible. Although this definition can include many products, from water added to clay for production of pottery to lime mixed with concrete to make it more workable, the word **"plasticizer"** generally refers to materials used in plastics. The Council of the International Union of Pure and Applied Chemistry (IUPAC) adopted the following definition of a plasticizer: A plasticizer is a substance or material incorporated in a material (usually a plastic or elastomer) to increase its flexibility, workability, or distensibility. In addition, a plasticizer may reduce the melt viscosity, lower the temperature of a second-order transition, or lower the elastic modulus of the product.

Early plasticizer technology and development focused on nitrocellulose. One of the first commercially important plasticized products was Celluloid, which was cellulose nitrate plasticized with camphor. John Wesley Hyatt patented this technology in 1870 (1). Later developments included the use of triphenyl phosphate as a low flammability plasticizer for cellulose nitrate. In the 1920's, the increasing demand for plasticized nitrocellulose paints lead to the development of two more important plasticizers, tricresyl phosphate and dibutyl phthalate.

Although polyvinyl chloride (PVC) was first produced in the middle of the nineteenth century, it was Waldo Semon's vision of the usefulness of plasticized PVC that initiated the growth of the flexible PVC industry. In 1933, Semon patented the use of solvents such as o-nitrodiphenyl ether, dibutyl phthalate, and tricresyl phosphate to produce a "rubber-like composition" with many different uses (2).   Soon afterwards, the use of di-2-ethylhexyl phthalate or DOP (patented by Monsanto in 1929 for plasticization of nitrocellulose (3)) was applied to PVC. In the period that followed, thousands of products were tested and evaluated as plasticizers. DOP became commercially available in the 1940's and since then has remained one of the most important plasticizers for polyvinyl chloride.

There are presently about 450 different plasticizers produced worldwide, although only about 50 of these are classified as commercially important. Approximately 90% of these products are used in the production of plasticized or flexible polyvinyl chloride (PVC) products.   Other polymer systems that use small amounts of plasticizers include polyvinyl butyral, acrylic polymers, poly(vinyldiene chloride), nylon, polyolefins, and certain fluoroplastics. Plasticizers can be used with rubber, although more often these materials fall under the definition of extenders, rather than plasticizers. The 1998 estimated worldwide production of plasticizers approached 9 billion pounds. Over 85% of this volume were phthalic acid esters.   A listing of the major plasticizers and their compatibilities with different polymer systems is shown in Table 1.

Mechanism of Plasticization

For a plasticizer to be effective, it must be thoroughly mixed and incorporated into the polymer matrix. With organic polymers, this is typically performed by heating and mixing until either the resin dissolves in the plasticizer or the plasticizer dissolves in the resin.   The material is then formed into the useful product and cooled.   Different plasticizers will exhibit different characteristics in both the ease with which they form the plasticized material and in the resulting physical properties of the flexible product.

Several theories have been developed to account for the observed characteristics of the plasticization process; Sears and Darby (4) have prepared a significant review of the theoretical treatment of plasticization.   Although most mechanistic studies have focused on PVC, much of this information can be adapted to other polymer systems. According to the lubricating theory of plasticization, as the system is heated, the plasticizer molecules diffuse into the polymer and weaken the polymer-polymer interactions.   The plasticizer molecules act as shields to reduce polymer-polymer interactive forces and prevent the formation of the rigid network. This reduction in intermolecular or van der Waals forces along the polymer chains increases the flexibility, softness, and elongation of the polymer.

The gel theory considers the plasticized polymer to be neither solid nor liquid but an intermediate state loosely held together by a three-dimensional network of weak secondary bonding forces.   These bonding forces acting between plasticizer and polymer are easily overcome by applied external stresses allowing the plasticized polymer to flex, elongate, or compress.

Free volume is a measure of the internal space available within a polymer. As free volume is increased, more space or free volume is provided for molecule or polymer chain movement. This has the effect of making the polymer system more

Table 1 - Plasticizers and Plasticizer Compatibility [a]

Plasticizers	Compatibility with plastics [b]								
	CA	CAB	CN	EC	PM	PS	VA	VB	VC
Adipic acid derivatives									
di-n-hexyl adipate (DHA)	P	C	C		C		P	C	C
heptyl nonyl adipate (79A)	I	C	C		C		P	C	C
di-2-ethylhexyl adipate (DOA)	P	C	C	C	P	C	I	P	C
diisodecyl adipate (DIDA)	P	C	C	C	P	C	P	P	C
diisononyl adipate (DINA)	P	C	C	C	P	P	P	P	C
Azelaic acid derivative									
Di-2-ethylhexyl azelate (DOZ)	C	C	C	C			P	I	C
Benzoic acid derivatives									
diethylene glycol dibenzoate	P	C	C	C	C	C	C	C	C
dipropylene glycol dibenzoate	P	C	C	C	C	C	C	C	C
2,2,4-trimethyl-1,3-pentaneodiol-isobutyrate benzoate	I	C	C	P		C	C		C
Citric acid derivatives									
tri-n-butyl citrate	C	C	C	C		C	C	C	C
tri-n-butyl acetylcitrate	P	P	C	C		C	C	C	C
Epoxy derivatives									
epoxidized soybean oil (ESO)	I	P	C	C	I	I	I	I	C
epoxidized linseed oil (ELO)	P	C	C	P	P	I	I	I	C
2-ethylhexyl epoxy tallate	I	C	C	C	I	I	I	C	C
Glycol derivatives									
diethylene glycol dipelargonate	I	C	C	C			P	C	C
triethylene glycol di-2-ethylbutyrate	I	P	C	C	C		C	C	C
Hydrocarbons									
hydrogenated terphenyls	I	C	C	C		C	C	C	C
chlorinated paraffin (52 wt% Cl)					C	C	C		C
Isophthalic acid derivatives									
di-2-ethylhexyl isophthalate	I	C	C	C	P	C	I	C	C
Oleic acid derivatives									
butyl oleate	I	C	C	C		C	C	C	P
Phosphoric acid derivatives									
tri-2-ethylhexyl phosphate (TOF)	P	P	C	C	I		I	C	C
triphenyl phosphate (TPP)	C	C	C	C	C	C	C	C	P
tricresyl phosphate (TCP)	C	C	C	C	P	C	C	C	C
2-ethylhexyl diphenyl phosphate	P	C	C	C	C	C	C	C	C
isodecyl diphenyl phosphate	P	C	C	C	C	C	C	C	C
Phthalic acid derivatives									
dimethyl phthalate (DMP)	C	C	C	C	C	C	C	C	C
diethyl phthalate (DEP)	C	C	C	C	C	C	C	C	C
dibutyl phthalate (DBP)	C	C	C	C	C	C	C	C	C
butyl octyl phthalate (BOP)	I	C	C	C	P	C	P	C	C
diisohexyl phthalate (DHP)	P	C	C	C	C	C		C	C
diisoheptyl phthalate (DIHP)	I	C	C					C	C
heptyl nonyl phthalate (79P)	I	C	C	C		C	I	C	C
heptyl nonyl undecyl phthalate (711P)	I	C	C	C	C	C	I	P	C
diisooctyl phthalate (DIOP)	I	C	C	C	C	C	C	P	C
di-2-ethylhexyl phthalate (DOP)	I	C	C	C	C	C	I	P	C

Table 1 (Continued), Plasticizers and Plasticizer Compatibility [a]

Plasticizers	Compatibility with plastics [b]							VB	VC
	CA	CAB	CN	EC	PM	PS	VA		
Phthalic acid derivatives, continued	P	C	C	C	P	C	P	C	C
(n-hexyl, octyl, decy) phthalate (610)									
(n-octyl, decyl) phthalate (810P)									
diisodecyl phthalate (DIDP)	P	C	C	C	C	C	C	C	C
diundecyl phthalate (DUP)									
ditridecyl phthalate (DTDP)	P	C	C	C	P	C	P	C	C
butyl benzyl phthalate (BBP)	P	C	C	C	C	C	C	C	C
alkyl benzyl phthalate, Santicizer ® 261	I	C	C	C	C	C	C	C	C
Polyesters									
adipic acid polyester (mol wt 6000), Paraplex ® G-40	I	C	C	P				C	I	C
adipic acid polyester (mol wt 2000), Santicizer ® 334F	I	C	C	P	I		C	C	P	C
azelaic acid polyester (mol wt 2200), Platolein ® 9720	P	C	C	I	I	I	P	P	C
sebacic acid polyester (mol wt 8000), Paraplex ® G-25	I	I	C	I			P	P	C
Ricinoleic acid derivatives									
methyl ricinoleate	P	C	C	C	P		C	C	I
n-butyl acetylricinoleate	I	C	C	C	P	P	C	C	C
castor oil		P	P	C	P			C	C
Sebacic acid derivatives									
di-2-ethylhexyl sebacate (DOS)	I	P	C	C	C	C	P	P	C
Stearic acid derivatives									
n-butyl stearate	I	C	P	C	I	P	I	I	P
Sucrose derivatives									
sucrose acetate-isobutyrate (SAIB)	C	C	C	C	C	C	C	P	I
Sulfonic acid derivatives									
(o,p)-toluenesulfonamide	C	C	C	C	C	P	C	C	P
N-ethyl-(o,p)-toluenesulfonamide	C	C	C	C	C	C	C	C	P
alkylsulfonic acid ester of phenol and creosol, (Mesamoll ®)	I	C	C	C	I	C	I		C
Terephthalic acid derivatives									
bis(2-ethylhexyl)terephthalate (DOTP)	I	C	C		C	C		C	C
Trimellitic acid derivatives									
tris(2-ethylhexyl) trimellitate (TOTM)	C	C	C	C	C	P	P	P	C
heptyl nonyl trimellitate (79TM)	I	C	C	C	C	C	I	C	C
triisononyl trimellitate (TINTM)							I		C
Terpenes and derivatives									
camphor	C	C	C		C				C
hydrogenated methyl ester of rosin	I	C	C	C		C	C	C	C

[a] data from Sears and Touchette (5)
[b] Resins used: CA, cellulose acetate; CAB, cellulose acetate-butyrate; CN, cellulose nitrate; EC, ethylcellulose; PM, polymethyl methacrylate; PS, polystyrene;   VA, polyvinyl acetate; VB, polyvinyl butyral; VC, polyvinyl chloride. Code for compatibility: C, compatible; P, partially compatible; and I, incompatible.

flexible. Free volume can be increased through modifying the polymer backbone, such as by adding more side chains or end groups or by adding small molecules with flexible end groups. The free volume theory builds on both the lubricity and gel theories of plasticization.

The mechanistic explanation of plasticization looks at the interactions of the plasticizer with the resin macromolecules. It assumes that the plasticizer molecules are not permanently bound to the resin molecule but are free to self-associate and to associate with the polymer molecule at certain sites. As these interactions are weak, there is a dynamic exchange process whereby as one plasticizer molecule becomes attached at a site or center it is rapidly dislodged and replaced by another. Different plasticizers will yield different plasticization effects because of the differences in the strengths of the plasticizer-polymer and plasticizer-plasticizer interactions. At low plasticizer levels, the plasticizer-PVC interactions are the dominant interactions while at high plasticizer concentration ranges, plasticizer-plasticizer interactions can become more important.

For a plasticizer to be effective and useful in PVC, it must contain two types of structural components. The polar part of the molecule must be able to bind reversibly with the polymer while the non-polar portion of the molecule adds free volume and contributes shielding effects at other polar sites on the polymer chain. The balance between the polar and non-polar portions of the molecule is critical; if a plasticizer is either too polar or too non-polar, compatibility problems can arise. Useful tools in estimating plasticizer compatibility are the Ap/Po Ratio method developed by Van Veersen and Merelaenberg (6) and the solubility parameter methods (7,8).

Types of Plasticizers

Flexible polymeric systems can be obtained through internal and external plasticization techniques. Internally plasticized PVC is produced by the copolymerization of vinyl chloride with another monomer to obtain a more flexible polymer backbone or by grafting another polymer onto the PVC backbone. External plasticization is achieved by incorporating a plasticizer into the PVC matrix through mixing and heat. These external plasticizers are classified as either monomeric or polymeric plasticizers depending upon their molecular weight. Plasticizers can also be characterized on the basis of their chemical structure. The three most important chemical classes of plasticizers are the phthalic acid esters, adipic acid esters, and trimellitic acid esters; these structures are shown in Figure 1. Di- and Tri-esters with a molecular weight range from 300 to 600 typically offer a balance of solvency and compatibility with the resin, yielding plasticized or flexible materials with useful properties and good aging abilities. Polymeric plasticizers are generally polyesters of molecular weights less than 6000.

Plasticizers can be divided into two classes based on their solvating power and compatibility with the polymer. **Primary** plasticizers are used as either the sole plasticizer or the major plasticizer component of the flexible polymer system. These materials can effectively solvate the polymer and will retain good compatibility upon aging. **Secondary** plasticizers are products typically blended with a primary plasticizer to improve certain performance attributes such as flame resistance, enhanced low temperature flexibility, or to reduce formulation costs. Generally secondary plasticizers are not used as primary plasticizers because of high costs or performance disadvantages such as limited compatibility with the polymer system. The distinction between the primary and secondary plasticizer classifications can be arbitrary. It is possible that a plasticizer used in one formulation as a primary plasticizer could be used in a second formulation as a secondary plasticizer.

Figure 1. Primary Chemical Classes of Plasticizers

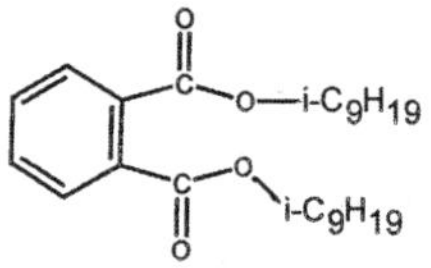

Phthalate Ester, DINP

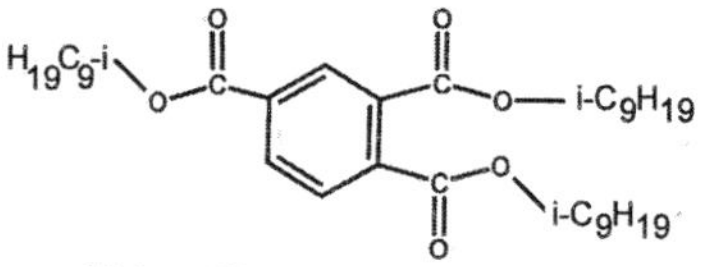

Trimellitate Ester, TINTM

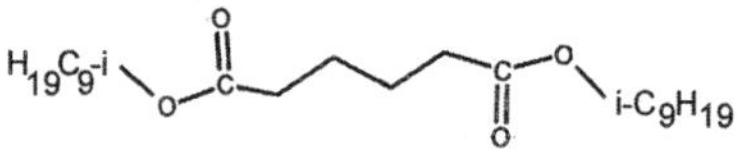

Adipate Ester, DINA

Most commercial plasticizers for PVC can be classified as belonging to one of three grouping (9); **general-purpose plasticizers**, **performance plasticizers**, and **specialty plasticizers**. General-purpose plasticizers are those products that offer optimum performance at the best economics and represent the largest class of plasticizers. This group includes diisononyl phthalate (DINP), di-2-ethylhexyl phthalate (DOP), and diisoheptyl phthalate (DIHP).

Performance plasticizers offer added performance benefits over general-purpose plasticizers, usually with additional formulation costs. This group of plasticizers can be subdivided into three groupings; fast solvating plasticizers, low temperature plasticizers, and low volatility plasticizers. Examples of performance plasticizers include the fast fusing plasticizers butyl benzyl phthalate (BBP) and dihexyl phthalate (DHP), the low temperature plasticizers di-2-ethylhexyl adipate (DOA) and di-n-undecyl phthalate (DUP), and the permanent plasticizers diisodecyl phthalate (DIDP), ditridecyl phthalate (DTDP), tri-2-ethylhexyl trimellitate (TOTM), and triisononyl trimellitate (TINTM).

The specialty plasticizer class includes several types of primary and secondary plasticizers. Generally these are used for unusually rigorous performance requirements and provide important attributes such as increased stabilization, reduced migration, improved permanence, flame resistance, and reduced plastisol viscosity. Polymeric plasticizers, esters of citric acid, benzoic acid esters, brominated phthalates, and phosphate esters are a few examples of specialty plasticizers.

Phthalate Esters. Phthalate esters are prepared by the esterification of two moles of a monohydric alcohol with one mole of phthalic anhydride as shown in Figure 2. Although phthalate esters can be prepared from many different alcohols, the range of alcohols used to make plasticizers for PVC applications is generally limited from C4 to C13 alcohols. Phthalate esters prepared from alcohols below C4 are too high in volatility while phthalate esters prepared from alcohols greater than C13 have limited compatibility. Di-2-ethylhexyl phthalate (DOP), which is prepared from 2-ethyl hexanol, is still regarded as the industry standard. Diisononyl phthalate and diisodecyl phthalate have significant market presence in many parts of the world and are used in applications where improved performance over DOP is required. Together these three plasticizers account for over 80% of all PVC plasticizers usage.

Figure 2. Synthesis of Di-2-ethylhexyl Phthalate

phthalic anhydride   2-ethyl hexanol

Di-2-ethylhexyl phthalate (DOP)

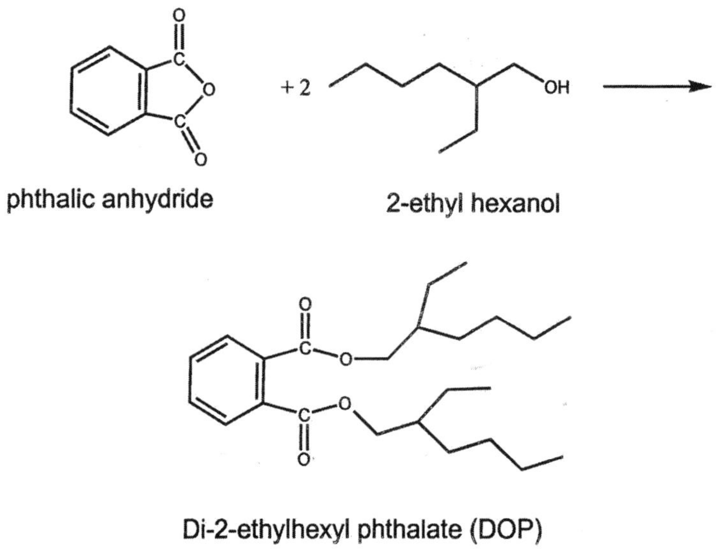

Improvements in plasticizer performance can be obtained through choice of the alcohol used to make the phthalate esters. Krauskopf has examined (10) the alcohol structural relationships with plasticizer performance and has developed a number of correlations that can be used to predict plasticizer performance in flexible PVC as the molecular weight and the linearity of the alcohol is varied. The use of linear alcohols favors improved performance in reduced volatility, lower plastisol viscosities, and low temperature flexibility over branched alcohols of the same carbon number. The phthalate esters prepared from C6 or C7 alcohols, for example dihexyl phthalate and diisoheptyl phthalate, fuse with PVC faster than the general-purpose plasticizers. The higher molecular weight C10 to C13 alcohols are used to prepare plasticizers for low volatility applications, such as those required for electrical cable insulation and sheathing.

Adipate Esters. Aliphatic dicarboxylic acid esters are prepared by the esterification of diacids such as adipic or azelaic acid with C6 to C10 monohydric alcohols. This class of plasticizer is used to extend the useful temperature range of plasticized PVC products, by providing increased flexibility at lower temperatures. Di-2-ethylhexyl adipate (DOA), which is prepared by the esterification of one mole of adipic acid with two moles of 2-ethyl hexanol, is the most important plasticizer in this class. Another important adipate is diisononyl adipate which offers greater permanence over DOA. Di-2-ethylhexyl azelate (DOZ), di-2-ethylhexyl sebacate (DOS), and diisodecyl adipate are used for extremely demanding low temperature applications or low temperature applications requiring lower plastisol volatility over that of DOA. The adipate and azelate esters may be used as primary or as secondary plasticizers.

Trimellitate Esters.   Trimellitate esters are used as primary plasticizers in those applications where greater permanence is required.   These esters are similar in structure to the phthalic acid esters, except for having a third ester functionality on the aromatic ring. Trimellitate esters offer advantages with greater permanence, either by reduced volatility losses or losses attributed to lower migration rates into other materials. Plasticized PVC electrical wire insulation prepared from either tri-2-ethylhexyl trimellitate (TOTM) or the more permanent plasticizer, triisononyl trimellitate (TINTM) will survive longer periods of high temperature service versus those products prepared from more volatile phthalate plasticizers.

Other Plasticizers. Polymeric plasticizers have found specialty uses due to their low volatility and high resistance to migration. These products are typically polyesters, with a molecular weight range from 1000 to 6000, and often prepared by the esterification of propylene glycol or butylene glycol with adipic acid. The higher viscosity of these products can cause processing problems in dry blending and in plastisol applications.

In order to improve the fire resistance characteristics of flexible PVC, phosphate esters can be used. PVC is inherently fire resistant compared to other polymer systems, but the presence of plasticizers reduces this property. Phosphate esters prepared from mixtures of either C8 or C10 alcohols with aromatic alcohols offer a compromise of flame reduction, volatility, and efficiency. Brominated plasticizers such as di-2-ethylhexyl tetrabromophthalic acid ester are also gaining importance as low-flammability plasticizers.

Although used only to a small extent in the US, the majority of the secondary plasticizers in use worldwide are chlorinated paraffins.   Chlorinated paraffins are produced by chlorination of hydrocarbons and have a chlorine content in the range of 30 -70%. These secondary plasticizers are used to reduce cost and to improve fire resistance.   Naphthenic and aliphatic mineral oils are also often used as secondary plasticizers to help reduce costs. Secondary plasticizers have limited compatibility with plasticized PVC and as the molecular weight of the primary plasticizer increases, compatibility problems increase.

Plasticizer Characteristics and Performance of Flexible PVC

A change in the type and level of plasticizer will affect the properties of the finished flexible article. The choice of plasticizer is almost always a compromise between performance requirements and product cost. Some of the more important properties that should be considered are listed in Table 2.

Plasticizer Efficiency. Plasticizers are added to materials to make them softer and more flexible; some plasticizers are more efficient at this than others. **Plasticizer efficiency** is used to describe the ability of a plasticizer to make the product softer and is reported as a ratio of the slope of the hardness versus plasticized concentration to the slope of that found for DOP. The relationship of hardness and plasticizer concentration expressed in phr (parts per hundred resin) is shown in Figure 3. Within a given series of esters having a common acid group, plasticizer efficiency increases as the molecular weight of the plasticizer decreases. Plasticizer efficiency improves as the linearity of the alcohol chain also increases.

Figure 3. Hardness versus Plasticizer Concentration

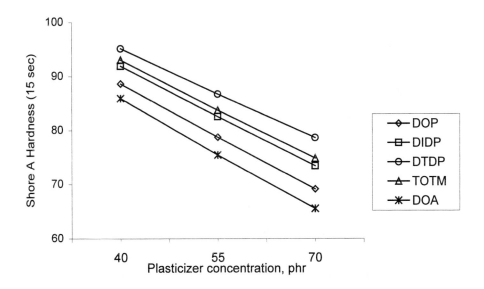

Plasticizer Solvency. To achieve a useful plasticized PVC product, the PVC polymer and the plasticizer must be fully solvated or fused. The fusion or gelation temperatures

A.D. Godwin

are related to the solvating strength of the plasticizer and to the size of the plasticizer molecule. Plasticizers with solubility parameters close to that of PVC require less energy to solvate or fuse the PVC polymer while larger plasticizer molecules require more thermal energy to plasticize the polymer. Esters of aromatic acids, being of greater polarity, tend to have lower fusion temperatures than the less polar esters of dicarboxylic acids. For phthalate esters, the ease of fusion decreases in the order

BBP > DHP > DIHP > DOP > DINP > DIDP > DTDP.

Table 2. Performance Properties of Selected Plasticizers[a]

	Viscosity cP, 20°C	Specific gravity 20/20°	Plasticizer Efficiency		Initial Gel Temp °C[d]	Low Temp Flex,T_f, °C	Loss after aging, wt%[e]
			Shore A[b]	Ratio[c]			
General-purpose plasticizers							
Diisoheptyl phthalate (DIHP)	51	0.994	80.6	0.97	70	-23.6	19.1
Di-2-ethylhexyl phthalate (DOP)	80	0.986	81.9	1.00	71	-24.9	10.6
Diisooctyl phthalate (DIOP)	85	0.985			71		
Diisononyl phthalate (DINP)	102	0.973	83.8	1.06	80	-23.6	5.4
Di-2-ethylhexyl terephthalate (DOTP)	79	0.984	82.9	1.03		-27.7	20.4
Fast solvating Plasticizers							
Butyl benzyl phthalate (BBP)	49[f]	1.119[f]	79.8	0.93	58	-11.0	15.0
Diisohexyl phthalate (DHP)	37	1.008	80.4	0.97	65	-25.1	20.4
Low Temperature Plasticizers							
n-heptyl, nonyl, undecyl phthalate (711P)	49	0.972	81.7	0.98	76	-33.5	4.5
n-hexyl, octyl, decyl phthalate (610P)	34[f]	0.971[f]	81.7	0.98		-33.5	4.5
Di-2-ethylhexyl adipate (DOA)	15	0.927	78.5	0.93	70	-53.3	27.5
Diisononyl adipate (DINA)	22	0.922	80.6	1.16	89	-50.5	9.2
Low Volatility Plasticizers							
Diisodecyl phthalate (DIDP)	129	0.968	85.6	1.11	80	-23.6	3.0
Ditridecyl phthalate (DTDP)	322	0.957	89.5	1.26	108	-24.2	1.9
Di-n-undecyl phthalate (DUP, L11P)	77	0.954	86.3	1.14	88	-34.3	1.5
Tri-2-ethylhexyl trimellitate (TOTM)	310	0.992	86.5	1.11	85	-19.2	<1
Triisononyl trimellitate (TINTM)	430	0.978	89.7	1.24	99	-19.2	<1

[a] 50 phr plasticizer except as noted
[b] Shore A Hardness, 15 seconds
[c] Efficiency ratio or substitution factor versus DOP
[d] Determined by dynamic mechanical analysis, onset of 70 phr plastisol gelation (11)
[e] weight loss after heating for 7 days at 100 °C
[f] 25 °C

Plasticizer Compatibility. An important performance requirement with any plasticizer is that it must be compatible with the polymer. **Compatibility** is the ability of two or more substances to mix with each other and form a useful product. Plasticizers with solubility parameters and polarity and hydrogen bonding characteristics similar to that of the polymer, would be expected to have good compatibility. Plasticizers with greatly differing solubility parameters would have poor compatibility. When the attractive forces of either the plasticizer or the polymer are greater for themselves than for each other, the result could be exudation or spewing. Lower molecular weight phthalates such as dimethyl phthalate and diethyl phthalate are too polar to be compatible with PVC, while phthalate esters prepared from C14 and higher alcohols are too non-polar to be considered useful.

Plasticizer Permanence. Plasticizers are not permanently bound to the PVC polymer, but are free to self-associate and to associate with the polymer at differing sites. Thus under certain conditions, plasticizers can leave the flexible PVC product. The three most common mechanisms for plasticizer loss are volatilization, extraction, and migration.

Plasticizer volatilization is directly related to the vapor pressure of the plasticizer. Volatilization losses will occur during processing and during use at elevated temperatures. Changes of as little as one carbon number of the alcohol group in a common series of esters can lead to significant reductions in losses. Trimellitate esters or polymeric plasticizers are often used to produce products designed for applications at elevated temperatures.

When plasticized PVC comes in contact with other materials, plasticizers can migrate. The rate of this migration depends both upon the structure of the plasticizer and the nature of the contact material. Plasticizers can also be extracted from PVC by a range of solvents including water. The rate of this extraction is related to the solvating strength of the solvent for the plasticizer. Water extracts plasticizers from PVC very slowly; oils are slightly more aggressive, while lower molecular weight organic solvents are very aggressive. Plasticizer molecular size is the most important factor in providing resistance to plasticizer migration or extraction. As the plasticizer molecular size increases, the tendency for plasticizer migration or extraction is reduced; polymeric plasticizers are very effective in providing migration or extraction resistance. Within a given series of esters, branched plasticizers offer better extraction and migration resistance than linear plasticizers.

Plasticizers can also migrate into other substances. Lacquer marring is controlled by the proper choice of plasticizer. Partially hydrogenated terphenyls added to DOP reduce the tendency to migrate into nitrocellulose lacquers. Terephthalate and isophthalate esters migrate to nitrocellulose slower than the corresponding ortho-phthalate esters. Migration to adhesives can destroy the cohesive nature of the adhesive. More polar plasticizers such as BBP or DHP migrate less than general-purpose plasticizers such as DOP. PVC gaskets used in either ABS or polystyrene refrigerator doors are typically prepared with polymeric plasticizers to minimize migration problems.

Low Temperature Flexibility. The addition of plasticizer to a PVC product extends the lower useful temperature limit of the finished product. Generally, the lower temperature limit is enhanced as the plasticizer concentration is increased.   However some plasticizers are more effective at this than others. The low temperature flexibility of the finished product results from depression of the polymer's Tg by the plasticizer. Low polarity plasticizers that lack aromatic moieties have more degrees of rotational freedom than their higher polarity counterparts of similar molecular weight.   Thus, aliphatic diesters of adipic, azelaic, and sebacic acids are very effective in improving the low temperature performance of flexible PVC. A major contribution to low temperature performance is also realized from the alcohol portion of the esters; the greater the linearity of the plasticizer, the greater the low temperature performance. For flexible PVC, the torsion modulus method of Clash-Berg and the Bell brittleness temperatures are commonly used. The low temperature flexibility for PVC plasticized with several esters is shown in Figure 4.

Figure 4. Low Temperature Effects in Plasticized PVC

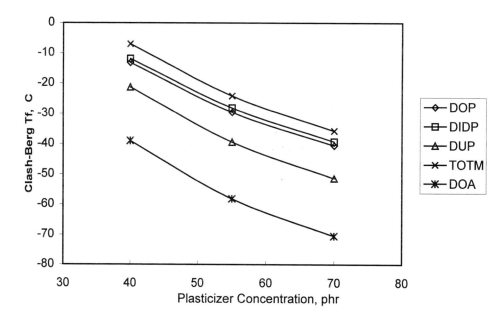

Flame Resistance.   Rigid PVC is inherently fire resistant and self-extinguishing; however the addition of plasticizers such as phthalates, adipates, and trimellitates contributes to flammability.   The triaryl and alkylaryl phosphates and brominated phthalate plasticizers inhibit burning of plasticized PVC. These products are often blended with other plasticizers to achieve a balance of flame resistance, physical properties, and cost. Chlorinated paraffins can also be used as secondary plasticizers, to help reduce flammability and smoke; however high volatility and potential compatibility problems restricts their use to but a few end uses. Flame resistance or smoke suppression may also be gained through addition of various additives into the flexible PVC formulation. These additives include antimony trioxide, alumina trihydrate, molybdenum ammonium octanoate, and zinc borate.

Plastisol Viscosity. In the processing of plastisols, plastisol viscosities at shear rates associated with the manufacturing process technology and plastisol viscosity stability are important properties. Using lower molecular weight or linear plasticizers will yield lower plastisol viscosity when compared at equal concentrations. However once the formulations are adjusted for equal hardness by taking into account the efficiency differences of the plasticizers, less-efficient plasticizers will often yield equal or lower plastisol viscosities, as they require higher plasticizer levels to achieve the same hardness. For example, even though DOP gives a lower plastisol viscosity than DINP at the same concentration, at equal hardness formulations, the DINP based plastisol would have the lower viscosity.

Plastisols are often mixed and stored before use. In this case, it is of great importance for the plasticizer to show little or no viscosity increase upon storage at room temperature. Plastisol viscosity for BBP based plastisols will increase significantly upon storage, while plastisols prepared with less polar plasticizers such as DINP and DIDP will remain relatively stable for some time. For phthalate esters, plastisol viscosity stability generally follows the order

DIDP > DINP > DOP > DHP > BBP.

Volume Costs. Most finished plastic products or articles are sold by the unit or by volume; however the raw materials used to prepare the product are purchased and formulated by weight. Thus when the costs to prepare the product are calculated, it should be determined on a cost per unit or cost per volume basis. Ingredients with lower specific gravities will give an increase in the number of units and thus plasticizers with lower specific gravities will yield lower costs per unit or lower volume costs.

Plasticizer efficiency is also important in determining volume costs. The volume costs for PVC polymer are higher than the volume cost of the plasticizers because of the high specific gravity of the polymer. Thus as more of the lower-volume-cost plasticizers are added to the polymer, the volume costs of the finished product decrease. Less efficient plasticizers require higher levels of addition to achieve the same degree of hardness or tensile properties; thus less efficient plasticizers lead to lower volume costs. For example when comparing the volume costs of a DOP plasticized product to the volume costs of the same article prepared with DINP at the same Shore A hardness, the DINP product will have lower volume costs because of the lower specific gravity of the DINP and the reduced efficiency.

Plasticizer Selection for Specific Applications

Plasticizer selection involves trying to meet a combination of end use performance requirements, processing needs, cost, and in some cases, specific regulatory requirements. Products produced from flexible PVC are used in a wider range of applications and industry market segments, than any other polymeric system. Table 2 lists a breakdown of the usage of plasticizers by different market segments.

Flexible PVC products are produced by a wide variety of differing manufacturing technologies. There are two main process routes to producing plasticized PVC products, melt processing and plastisol processing. In melt processing, suspension-grade PVC polymer, plasticizers, stabilizers, fillers, and other additives are compounded at elevated temperatures. The compound can be further shaped into final form using calendering, extrusion, or injection molding techniques.

The melt processing process often begins with formation of a **dry blend**. As the plasticizer and the polymer and other additives are mixed under shear, the plasticizer is slowly absorbed by the porous polymer particles. With continued mixing, the temperature of the mix will increase and eventually yield a dry power. This powder can be stored, compounded into pellets, directly extruded into a finished product, or fed to a Banbury mixer or mill and then calendered. Plasticizer selection does impact the dry blending process. Plasticizers with lower viscosities will dry blend faster than higher viscosity plasticizers (12). Higher viscosity plasticizers such as trimellitate esters and heavier phthalate esters are often preheated prior to the dry blending step to reduce the time required to complete the dry blending step.

Plastisols are liquid dispersions formed by mixing emulsion-grade PVC polymer into the liquid plasticizer system. Fillers, stabilizers, and other additives are also added, as needed. Plastisols are mixed and shaped at relatively low temperatures. After shaping, the temperature is raised to form the fused plasticized products. Plastisol shaping techniques include rotomolding, spread coating, dipping, pouring, and extrusion. Often plasticizers are chosen more for the benefits they bring to the plastisol manufacturing process, for example in reduced fusion temperature, reduced volatility, or improved viscosity stability, than for product end use performance.

The largest market sector for plasticizers is the film, sheet, and coated fabric market sector. These products are manufactured by calendering, extrusion, cast films, and spread coating processes and are used to produce a wide variety of products including table cloths, swimming pool liners, tarpaulins, agriculture films, wall coverings, shower curtains, upholstery fabrics, and shoe fabrics. The majority of these products will use the general-purpose plasticizers DOP and DINP. Specialty plasticizers are often used to improve performance attributes. For example, DIDP is used to make a product last longer and to improve the extraction resistance. Linear phthalate esters such as linear dinonyl phthalate (L9P) or linear diundecyl phthalate (DUP) are used for improved UV resistance, and fast fusing plasticizers (BBP, DHP) to help improve the processability in spread coating. Adipate esters (DOA, DINA) can be added to the phthalate esters in other film and sheet products to improve the low temperature performance.

Table 3. Plasticized PVC Major Market Segments[a]

Film, Sheet, and Coated Fabrics	31 %
Wire and Cable	24 %
Molding and Extruded products	17 %
Flooring	14 %
Adhesives, sealants, coatings, inks	5 %
Food, medical, other regulated	3 %
Other miscellaneous	6 %

Another large end use for flexible PVC is wire and cable. With electrical cables, plasticizer selection is dependent upon the performance specifications of the insulation material and the jacketing. Higher temperature rated products require more permanent plasticizers. In the US, building wire is typically formulated with trimellitate plasticizers such as TOTM or TINTM. Often to reduce costs, heavier phthalates such as DTDP or DUP are blended with the trimellitates. In other parts of the world where the performance requirements for building wire are not as severe, general purpose plasticizers can be used, but the trend remains to uses higher molecular weight plasticizers such as DIDP to improve product performance.

Cushion vinyl flooring is prepared by coating a substrate with various layers of plastisols. These products generally use a mixture of general-purpose plasticizers such as DIHP, DOP, and DINP often in blends with DHP or BBP. These faster fusing plasticizers improve processability and stain resistance of the finished flooring product. Vinyl floor tiles and other sheet flooring are produced by calendering; again general-purpose plasticizers, often used in blends with fast fusing plasticizers, are commonly used. Plasticized PVC is also a major component of some types of carpets and carpet tiles, either as an adhesive or as a backing material. Faster fusing plasticizers such as DIHP offer advantages over DOP because of the lower fusion temperature, which is desirable to protect the carpet fibers.

Most molded and extruded products are prepared from general purpose plasticizers such as DOP and DINP, as these products offer the best compromise between price and performance. Products designed for outdoor applications will often use more permanent plasticizers such as DIDP. Often choice of regulatory requirements will dictate plasticizer choice. Molded medical devices will use DOP while DINP will be the plasticizer of choice for toys. Molded products used for automotive interiors will use heavier phthalates such as DIDP or DUP to meet automotive manufacturer fog performance requirements.

In addition to these major market segments, plasticizers are used in a variety of smaller segments. Adhesives and sealants generally use lower molecular weight phthalate esters because of improved solvency and lower fusion temperatures. One major use of plasticized PVC is automotive underbody coating; these products are highly filled plastisols that are mostly commonly based upon DIHP, DOP, or DINP.

Another important use of plasticizers is to produce vinyl Medical devices. Most of these products use DOP although DINP is used for vinyl examination gloves.

Plasticizers for Other Polymers

Polymers other than PVC account for less than 10% of the current world wide plasticizer usage. While the combination of PVC polymer, plasticizers, and other additives can yield an extensive range of flexible products with varying performance characteristics, plasticizer usage with other polymers does not produce the same range of versatility. More often plasticizers are used with other polymer systems more for processing improvements than for producing flexible products.

Plasticizers are used with acrylic polymers to produce flexible coatings, caulks, and sealants. Most of the common phthalates and adipates have been described as having some compatibility with acrylic systems, although the number of plasticizers having acceptable performance is smaller than those listed for PVC. The most common plasticizers used are BBP, DBP, DHP, DIHP, and certain benzoate esters.

Polyvinyl acetate (PVAc) emulsion polymers are used for adhesives, sealants, and paints. PVAc homopolymer is too brittle for most end uses, so much of the PVAc used in the paint industry is internally plasticized through the incorporation of comonomers into the polymer backbone. Externally plasticized PVAc is used in adhesives. The most common plasticizers used with adhesives are the dibutyl phthalates and benzoate esters.

Nylon is a highly crystalline material and, as such, plasticization can only occur at very low levels. Plasticizers used with nylon are typically sulfonamides, as these products are significantly more compatible with nylon than are phthalates. Sulfonamides can be used as nylon flow aids, to retard degradation and to speed up processing.

Many rubbers and elastomers are comprised of long hydrocarbon segments, and thus can accept petroleum oils and other predominately hydrocarbon products for use as plasticizer or extenders. These products are available at a significantly lower cost than the synthetic ester plasticizers. Phthalates and adipates of linear alcohols are used to enhance low temperature properties of certain rubber applications, which cannot be met using the hydrocarbon extenders. Polar elastomers such as nitrile rubber and polychloroprene have low compatibility with hydrocarbons and require more polar products such as phthalates or adipates.

Plasticized polyvinyl butyral (PVB) is used as a laminating film between layers of glass to provide strength and shatter resistance. This forms the "safety" glass used in automobile windshields and architectural glass. PVB can be plasticized with a variety of esters, but in practice only a few products are used as the plasticizer selection is a complicated process. The most common plasticizers used with PBV in safety glass are di-n-hexyl adipate and the di-n-heptanoic acid esters of tri- or tetra-ethylene glycol. These esters give excellent compatibility, clarity, and resistance to sunlight.

Many cellulosic materials, including cellulose nitrate are compatible with relatively high levels of plasticizers. Plasticizers are used to reduce processing temperatures, improve impact resistance, and to increase flexibility and resistance to cracking in these materials. Cellulose acetate can be plasticized with more polar esters such as dimethyl and diethyl phthalate. Cellulose nitrate, on the other hand, shows better compatibility with BBP and dibutyl or other heavier phthalates.

Plasticizers can be used to reduce both the viscosity and the cost of polyurethane and polysulfide automotive and construction sealants. Higher molecular weight phthalate esters such as DINP, DIDP, and DUP are generally suitable for most polyurethane sealant applications. Polysulfide sealants require more polar plasticizers such as the alkyl benzyl phthalates. For automotive interior applications, the low fog requirements generally excludes the use of the more volatile polar plasticizers such as BBP. In those cases, higher molecular weight polar products such as Texanol® benzyl phthalate can be used.

Diluents or plasticizers for epoxy resins lower the viscosity of the uncross-linked resin for greater ease in application of surface coatings and adhesives. Non-reactive plasticizers such as phthalates and phosphates will reduce the viscosity but lead to poor impact resistance and lower the extraction resistance. Reactive plasticizers are low molecular weight epoxy compounds, typically having one reactive epoxy group per molecule. These contribute to lengthening the polymer segments between cross-links and produce a slight softening and flexibility effect with impaired impact strength. The products used include butyl glycidyl ether and the glydicyl esters of neodecanoic acid.

Health Aspects of Plasticizers (13)

Toxicity. The toxicity of a substance is a measure of its effects on living organisms. It is not possible to simply state whether or not plasticizers or any other substances are toxic, because this is determined by the dose. The acute toxicity of a substance is an assessment of the effect that a single dose will have upon a living organism. The relevant parameter is the dose that is lethal for 50% of the test animals, expressed relative to the body weight (LD_{50}). The acute toxicity of phthalate and adipate esters is extremely low; in fact lower than that found for common everyday substances such as ethyl alcohol and common table salt.

Chronic toxicity describes the effects that take place on a living organism as a consequence of long term exposure. Some phthalate esters were shown to cause increased liver tumors in rats and mice when these animals were fed phthalates at very high levels (comparable to human consumption of 300 g DOP per day), More recent studies have shown that phthalates are not genotoxic; plasticizers do not interact with genetic material. Other studies have shown that while oral administration of plasticizers to rodents causes increases in microbodies in the liver called peroxisomes, administration of the same plasticizers to non-rodent species such as marmosets does not lead to peroxisome proliferation and liver damage.

Phthalates are also alleged to be endocrine disrupters, with the capability to produce changes in the hormone system, which can lead to reproductive problems. Definitive studies have shown that the major commercial phthalate esters are not estrogenic (14) while other studies suggest that DINP, and DIDP are not endocrine disrupters. Recently the health risks associated with flexible PVC products was examined by an independent panel of scientists led by Dr. C. E. Koop, former US

Surgeon General. This panel concluded that products plasticized with DOP and DINP are not harmful(15).

<u>The level of human exposure to plasticizers.</u>   Experimental investigations and assessments in the USA and in Europe have shown that the average human intake of plasticizer amounts to only 2 g per person per year. This is mostly due to traces of DOA migrating from food packaging. The so-called No Observed Effect Level (NOEL) for DOA and DOP in rodents is around 40 mg per kg body weight per day or higher. Extrapolating this to the average human, this would equate to at least 1000 g plasticizer per person per year. This is a 500 fold safety factor between estimated plasticizer and the No Observed Effect Levels. If the differences in responses between rodents and primates are considered, the safety factor is even greater.

<u>Plasticizers in the Environment.</u> The release of plasticizers into the environment can occur during manufacture and distribution of the plasticizer, during fabrication of the flexible PVC product, from plasticizer loss during the use of the product or after its disposal. Plasticizer losses during production and distribution are tightly controlled and are very small. Plasticizers are very insoluble in water and thus plasticizer releases during product use or disposal are minimal.   Most plasticizer releases into the environment occur by evaporation during the processes required to incorporate the plasticizer into the PVC polymer. Plasticizer losses by this mechanism are continually being reduced by the installation of incinerators, scrubbing, and filtration equipment and by switching to less volatile plasticizers. For example, fusion oven studies have shown that by replacement of DOP with DINP, the plasticizer emission can be reduced by almost 50%. There are no indications of an accumulation of plasticizers in water, soil, or air because they biologically and photochemically degrade.   Degradation is particularly rapid under aerobic conditions to produce carbon dioxide and water.

<u>Future of Plasticizers</u>

The future of plasticizers is tied closely to the future of PVC. In recent years, several environmental activist groups have criticized PVC because it contains chlorine and a mistaken belief that the production, use, and disposal of PVC products caused serious damage to the environment. Although the initial attacks were relatively unsuccessful, it has raised awareness of alternative materials that could replace flexible PVC in some applications.

Recently, the environmental focus has shifted towards additives used in PVC; metal stabilizers and plasticizers. Plasticizers are criticized because under certain conditions, small amounts can migrate from the flexible PVC product. The future of plasticizers is strongly dependent on the continued availability of cost-effective plasticizers that do not present unacceptable health and environmental risks.

Up to now this demand has been met with the three primary plasticizers, DOP, DINP, and DIDP. As discussed, higher molecular weight phthalates are more permanent, have lower water solubility, and migrate slower than lower molecular weight products. It is expected that the trend of replacing DOP with either DINP or DIDP will continue, perhaps at accelerated rates. In the absence of any specific environmental or health drivers or new regulations, it is not expected that any new plasticizers or new plasticizer technology will be developed to replace sizable volumes of these plasticizers in the near future.

One consequence of the emotive attacks on flexible PVC, is that product designers will express interest in non-PVC products while producers of competitive flexible polymer systems will promote their products using the perceived negative PVC attributes. Flexible PVC products still offer cost, processing, and performance advantages over potential replacement systems; however, it is expected that future developments will bring forth new polymers which will begin to slowly replace flexible PVC in selected applications.

Literature Cited

1. J. W. Hyatt and I. S. Hyatt, U. S. Patent 105,338, 1870.
2. W. L. Semon, U. S. Patent 1,929,453, 1933.
3. L. P. Kyrides, U. S. Patent 1,923,838, 1933.
4. J. K. Sears and J. R. Darby, "The Technology of Plasticizers", John Wiley & Sons, New York, 1982, pp 35-74.
5. J. K. Sears and N. W. Touchette in "Kirk-Othmer Encyclopedia of Chemical Technology", Vol 18, John Wiley & Sons, USA, 1982, pp. 113.
6. G. J. Veersen and A. J. Meulenberg, Kunststoffe, 1966, 56, 23.
7. J. K. Sears and J. R. Darby, ibid., pp. 92-108.
8. L. G. Krauskopf, Journal of Vinyl & Additive Technology, 1999, 5(2), 101.
9. L. G. Krauskopf, Journal of Vinyl Technology, 1993, 15(3), 140.
10. L. G. Krauskopf in "Encyclopedia of PVC, Second Edition, Revised and Expanded", edited by L. I. Nass and C. A. Heiberger, Marcel Dekker, Inc., 1993, p. 214.
11. P. H. Daniels, C. M. Brofman, and G. D. Harvey, Journal of Vinyl Technology, 1986, 8, 160.
12. L. G. Krauskopf and A. D. Godwin, Journal of Vinyl & Additive Technology, 1999, 5(2), 107.
13. A review of the health aspects of plasticizers is available by A. S. Wilson in "Plasticisers Principles and Practice", Chapter 9, University Press, Cambridge, 1995
14. T. R. Zacharewski, et. al., Toxicological Science, 1998, 46, pp 282-293.
15. American Council on Science and Health technical report, June 22, 1999.

ENGINEERING THERMOPLASTICS

EDWARD N. PETERS

GE Plastics
One Noryl Avenue
Selkirk, NY 12158

R. K. ARISMAN

GE Silicones
260 Hudson River Road
Waterford, NY 12188

Introduction

Engineering polymers comprise a special, high-performance segment of synthetic plastic materials that offer premium properties. When properly formulated, they may be shaped into mechanically functional, semi-precision parts or structural components.

The term "mechanically functional" implies that the parts will continue to function even if they are subjected to factors such as mechanical stress, impact, flexure, vibration, sliding friction, temperature extremes, and hostile environments.[1]

As substitutes for metal in the construction of mechanical apparatus, engineering plastics offer advantages such as corrosion resistance, transparency, lightness, self-lubrication, and economy in fabrication and decorating. Replacement of metals by plastics is favored as the physical properties and operating temperature ranges of plastics improve and the cost of metals and their fabrication increases.[2] Plastic applications in transportation, a major growth opportunity, have been greatly accelerated by the current awareness of the interplay of vehicle weight and fuel requirements.

The ability to replace metals in many areas has resulted in tremendous growth in engineering thermoplastics.

A significant driving force behind the growth in engineering thermoplastics is the continuing expansion of electrical/electronic markets, which demands smaller, lighter components that operate at higher speeds. In addition the same requirements are driving the automotive market segment. Original Equipment Manufacturers strive towards lower production cost, style flexibility, lower maintenance and more efficient, lower polluting vehicles that utilize better performing materials under the hood and in exterior components. The consumption of engineering plastics increased from 10 million to more than 15 billion pounds from 1953 to 1999. Engineering polymers are the fastest growing segment of the plastics industry with an anticipated growth rate from 12 to 15%. This chapter focuses on the development of engineering thermoplastics during the past 45 years.

Polyamides

Nylon, the first commercial thermoplastic engineering polymer, is the prototype for the whole family of polyamides. Nylon 6,6 began at duPont with the polymer experiments of Wallace Carothers in 1928, and made its commercial debut as a fiber in 1938 and as a molding compound in 1941. By 1953, 10 million pounds of nylon 6,6 molding compound represented the entire annual engineering plastic sales.[3, 4]

Nylon was a new concept in plastics for several reasons. Because it was semi-crystalline, nylon underwent a sharp transition from solid to melt; thus it had a relatively high service temperature. A combination of toughness, rigidity, and "lubrication-free" performance peculiarly suited its mechanical bearing and gear applications. Nylon acquired the reputation of a quality material by showing that a thermoplastic could be tough, as well as stiff, and perform better than metals in some cases. This performance gave nylon the label "an engineering thermoplastic."

Nylon 6,6 derived from the condensation of a six-carbon diamine and a six-carbon dibasic acid and normally chain terminated with mono-functional reactants presented some unusual characteristics (Figure 1).

Figure 1.

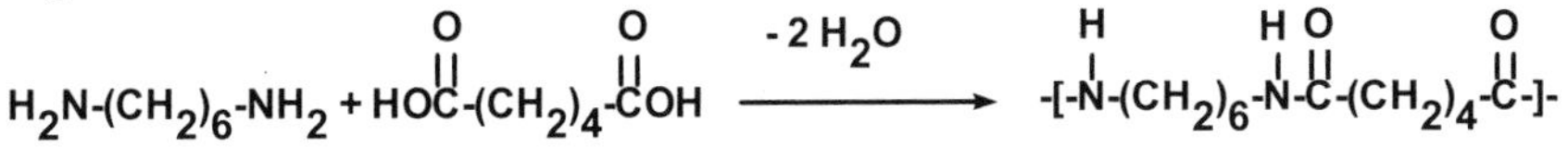

$$\underset{H_2N\text{-}(CH_2)_6\text{-}NH_2}{} + \underset{HO\overset{O}{\overset{\|}{C}}\text{-}(CH_2)_4\text{-}\overset{O}{\overset{\|}{C}}OH}{} \xrightarrow{-2\,H_2O} \underset{\text{-[-}\overset{H}{N}\text{-}(CH_2)_6\text{-}\overset{H}{N}\text{-}\overset{O}{\overset{\|}{C}}\text{-}(CH_2)_4\text{-}\overset{O}{\overset{\|}{C}}\text{-]-}}{}$$

The crystallinity and polarity of the molecule permitted dipole association that conveyed to relatively low molecular weight polymers the properties normally associated with much higher molecular weight amorphous polymers. At its crystalline melting point (T_m of 269°C) the polymer collapsed into a rather low-viscosity fluid in a manner resembling the melting of paraffin wax. It lacked the typical familiar broad thermal plastic range that is normally encountered in going from a glassy solid to a softer solid to a very viscous taffy stage. This factor led to some complications in molding because very close tolerances were required in mold construction, and very precise temperature and pressure monitoring was necessary to prevent flash or inadvertent leaking of the mobile melt. Early molders of nylon were highly skilled – they had to be because the industry was young.

Nylon based on ω-aminocarboxylic acids, although briefly investigated by Carothers, was commercialized first in Germany around 1939 (Figure 2).[4] Of particular interest to the plastic industry is nylon 6 (based on caprolactam), which became available in 1946 in Europe. Allied Chemical Company initially introduced it to the United States for fiber purposes in 1954. Polycaprolactam is crystalline, has a lower melting point (T_m of 228°C) than nylon 6,6 and has been successfully applied as a molding compound.

Figure 2.

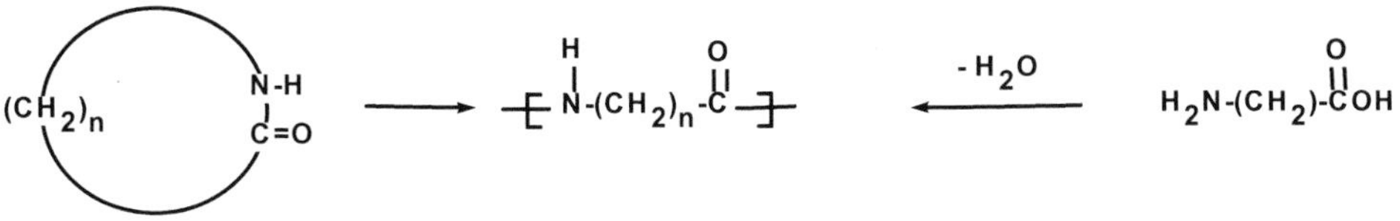

Nylon 4,6 was developed and commercialized in 1990 by DSM to address the need for a polyamide with higher heat and chemical resistance for use in automotive and electrical/electronic applications. The resin was prepared by the condensation of a four-carbon diamine and adipic acid. Nylon 4,6 has a T_m of 295°C and has higher crystallinity than nylon 6 or 6,6.[3]

The key features of nylon are fast crystallization, which means fast molding cycling; high degree of solvent/chemical resistance; toughness; lubricity; fatigue resistance; and excellent flexural-mechanical properties that vary with degree of water plasticization. Deficiencies include a tendency to creep under applied load.

Varying the monomer composition has produced many different varieties of polyamides. Variations include nylon 6,9; nylon 6,10; and nylon 6,12 (made from the 9-, 10-, and 12-carbon di-carboxylic acids, respectively); and nylon 11 and nylon 12 (via the self-condensation of 11-aminoundecanoic acid and lauryl lactam, respectively). These specialty nylons exhibit lower moisture absorption – only one-third or one-fourth that of nylon 6 or nylon 6,6.

When unsymmetrical monomers are used, the normal ability of the polymer to crystallize can be disrupted; amorphous (transparent) nylons can then be formed. For example, the polyamide prepared from the condensation of terephthalic acid with a mixture of 2,2,4- and 2,4,4-trimethylhexamethylenediamines was developed at W. R. Grace and Company, later produced under license by Dynamit Nobel AG, and currently available from Huls America, Inc. This polyamide is sold under the trade name Trogamid T.

Another amorphous nylon was developed at Emser Werke AG and is based on aliphatic, as well as cycloaliphatic, amines and terephthalic acid. It is marketed under the Grilamid trade name by Ems-America Grilon, Inc. These amorphous nylons are not as tough as nylon 6 or 6,6 but they do offer transparency and good chemical resistance in some environments.

Aromatic Polyamides, Aramids

Nylons prepared from aromatic di-amines and aromatic di-acids (aramids) can lead to very high-heat aromatic nylons such as DuPont's [poly (m-phenyleneisophthalamide)] (Figure 3; trade name Nomex), or the highly oriented crystalline fibers [poly (p-phenyleneterephthalamide)] (Figure 4; trade name Kevlar) derived from liquid-crystalline technology. Both of these aramids are sold as fibers. They have excellent inherent flame-retardant properties, and Kevlar fibers exhibit a very high modulus.

Figure 3.   Figure 4.

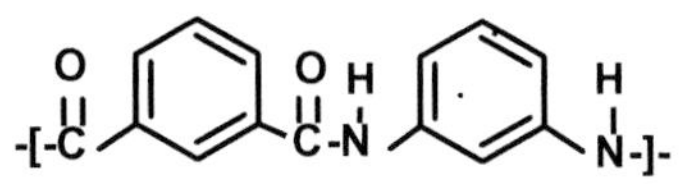

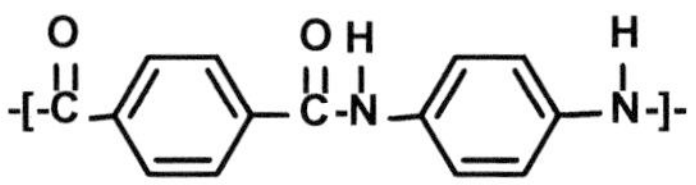

Semi-Aromatic Polyamides

Modified co-polymers based on poly (hexamethylene terephthalate), nylon 6,T (Figure 5), have been examined extensively because of use of inexpensive monomers and enhanced properties over aliphatic polyamides. Pure nylon 6,T exhibits a very high T_m of 370°C and a T_g of 180°C. The high T_m results in expensive polymerization processes and difficulty in molding. Hence terpolymers using an inexpensive, third monomer such as isophthalic acid, adipic acid, caprolactam, or 1,5-hexyl diamine have lead to the commercialization of semi-aromatic polyamides by Amoco, BASF, and duPont under the trademarks Amodel R, Ultramid T and Zytel HTN, respectively. These terpolymers exhibit T_m's from 290-320°C and T_g's from 100 to 125°C and offer enhanced performance – i.e., stiffer, stronger, greater thermal and dimensional stability – over nylon 6,6 and 6.

Figure 5.

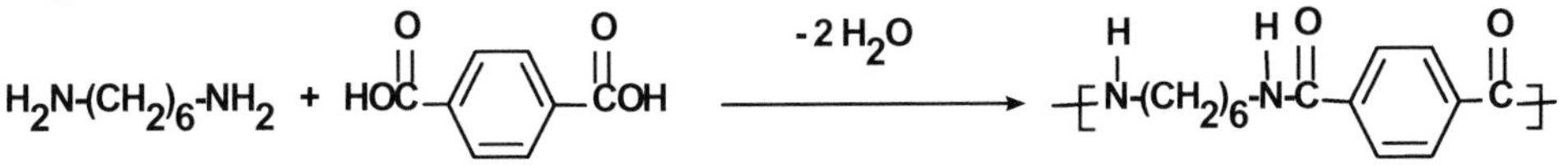

Polyacetals

After nylon, the next engineering polymers to be commercially introduced were polyacetals (3). Polyacetals are polymerized from formaldehyde and are called polyoxymethylenes (POM). Staudinger explored the basic polyformaldehyde structure rather thoroughly in the late 1920's and early 1930's, but he failed to produce a sufficiently high molecular weight polymer with requisite thermal stability to permit processing (4). Pure formaldehyde could be readily polymerized, but the polymer equally, readily, spontaneously de-polymerized – i.e., unzipped.

In 1947, duPont began a development program on the polymerization and stabilization of formaldehyde and its polymer. Twelve years later, duPont brought the unzipping tendency under control with proprietary stabilizers and commercially announced polyacetal polymer under the Delrin trade name (Figure 6). The key to the stabilization of polyformaldehyde resins appears to be a blocking of the terminal hydroxyl (OH) groups that participate in, or trigger, an unzipping action. Post etherification or esterification apparently may block/cap the hydroxyl groups. This material has a T_g of -75°C and a T_m of 181°C.

Figure 6.

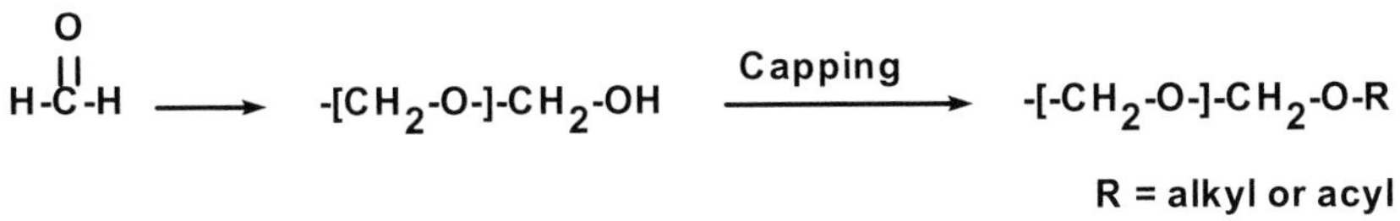

R = alkyl or acyl

Celanese joined duPont in the market with their proprietary polyacetal polymer under the Celcon trademark within a year (Figure 7). Celanese managed to obtain basic patent coverage, despite DuPont's prior filing, on the basis of a copolymer variation that led to an enhanced stabilization against thermal depolymerization. This copolymer has a T_m of 170°C.

Both Celanese and duPont aimed their products directly at metal replacement. Items such as plumbing hardware, pumps, gears, and bearings were immediate targets. In many respects, the acetals resemble nylons. They are highly crystalline, rigid, cold-flow resistant, solvent resistant, fatigue resistant, mechanically tough and strong, and self-lubricating. They also tend to absorb less water and are not plasticized by water to the same degree as the polyamides.

Figure 7.

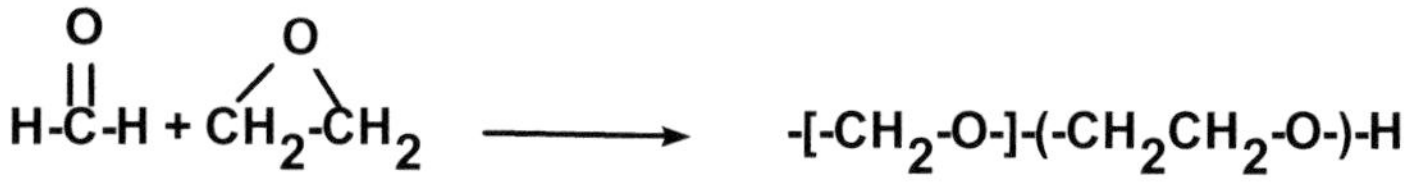

$$\underset{\text{H-C-H}}{\overset{O}{\overset{\|}{}}} + \underset{CH_2-CH_2}{\overset{O}{\overset{/\backslash}{}}} \longrightarrow -[-CH_2-O-]-(-CH_2CH_2-O-)-H$$

Rapid crystallization of acetals from the melt contributes to fast mold cycles. Crystallization also causes a significant amount of mold shrinkage. Thus, it is necessary to compensate in mold design for dimensional changes that occur during the transformation from a hot, low-density, amorphous melt to a denser, crystalline solid.

Key areas of use for POMs are industrial and mechanical products that include molded or machined rollers, bearing, gear, conveyor chains, and housings. POMs are widely used in plumbing and irrigation because they resist scale build up, and have excellent thread strength, creep resistance, and torque retention. A deficiency of polyacetals is a tendency to thermally unzip and an essentially un-modifiable flammability.

Polycarbonates

The aromatic polycarbonates were the next engineering polymers to be introduced. D. W. Fox at General Electric Company and H. Schnell at Bayer AG (Germany) independently discovered the same unique super tough, heat-resistant, transparent, and amorphous polymer in 1953.[7, 8, 9] When the companies became aware of each other's activities, agreements were reached that enabled both parties to continue independent commercialization activities without concern for possible subsequent adverse patent findings. The General Electric Company's polycarbonate was introduced United States under the Lexan trademark in 1959 at about the same time as the polyacetals, and a commercial plant was brought on stream in 1960.

Polycarbonates of numerous bisphenols have been extensively studied. However, most commercial polycarbonates are derived from bisphenol A (BPA). Both solution and solvent free, melt-transesterification processes are used to manufacturer polycarbonates.

In the solvent processes, phosgene reacts with bisphenol A to produce a polymer in a solution. Polycarbonate is produced by an interfacial adaptation of the reaction in Figure 8.[10] The bisphenol plus 1-3 m% mono-functional phenol, which controls molecular weight, is dissolved or slurried in aqueous sodium hydroxide; methylene chloride is added as a polymer solvent; a tertiary amine is added as a catalyst, and phosgene gas is dispersed in the rapidly stirred mixture. Additional caustic solution is added as needed to maintain basicity. The growing polymer dissolves in the methylene chloride, and the phenolic content of the aqueous phase diminishes.

Figure 8.

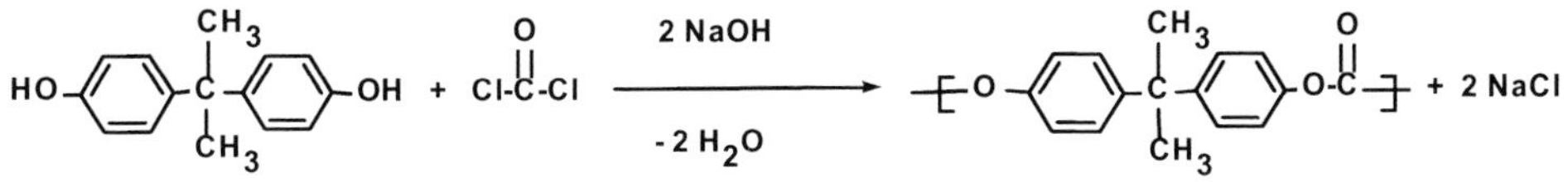

In the solvent free, melt-transesterification, diphenyl carbonate reacts with bisphenol A to regenerate phenol for recycle and molten, solvent-free polymer (Figure 9). Transesterification is reported to be the least expensive route.

Figure 9.

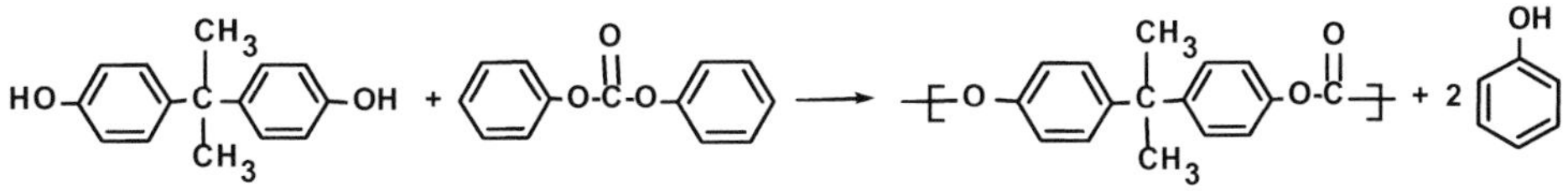

BPA-PC is an amorphous polymer with a T_g of 150°C. It is characterized by outstanding impact strength, glass-like transparency, heat resistance, excellent electrical properties, intrinsic flame retardancy, and high dimensional stability up to just below its Tg. Its outstanding combination of properties and processing versatility have made PCs the ideal resin for many applications.

The polycarbonates, like the nylons and acetals, were directed toward metal replacement applications. Glass fiber filled versions of polycarbonates are available, and this combination is particularly well suited to compete with metal parts. As in the case of other semi amorphous polymers, glass acts as stiffening and strengthening agent but does not raise operating temperatures significantly. In crystalline polymers, fillers tend to act as a pseudo cross-link or crutch to bridge the soft, amorphous regions that have T_g-dependent properties, thereby permitting the plastic to maintain structural integrity up to its crystalline melting point. Without filler crystalline polymers tend to creep under static load at relatively low temperatures because their T_g values are generally comparatively low.

In addition transparency gave polycarbonates another dimension and have lead to applications in safety glazing, light covers, automotive headlamp lenses, water bottles, compact disc, and ophthalmic applications.

Polycarbonates are readily modified via copolymerization with long chain aliphatic dicarboxylic acids to give a resin with improved flow; other bisphenols or phthalates to increase T_g to give a PC with higher heat resistance; and trisphenols to give a branched PC with improved melt strength for blow molding/extrusion. In addition PC is amenable for the development of many different commercial polyblends.[7, 11] For example, blending polycarbonates with thermoplastics polyesters, and ABS resins further widens the performance window.

Poly (Phenylene Ether)

In 1956, A. Hay of General Electric discovered a convenient catalytic oxidative coupling route to high molecular weight aromatic ethers.[12, 13] These polymers could be made by bubbling oxygen through a copper-amine-catalyzed solution of phenolic monomer at room temperature (Figure 10). A wide variety of phenolic compounds were explored, but the cleanest reactions resulted from those that contained small, electron-donor substituents in the two ortho positions. Work quickly focused on 2,6-dimethylphenol because it was the most readily synthesized. The poly (phenylene ether) (PPE) derived from 2,6-dimethylphenol had excellent hydrolytic resistance, an extremely high T_g (215°C), a high melting range (260-275°C), a very high melt viscosity, and a pronounced tendency to oxidize and gel at process temperatures. The polymer was called PPE resin in anticipation of developing the coupling process to produce unsubstituted poly (phenylene oxide) from phenol. With this objective the program was redirected to 2,6-dimethylphenol and a parallel effort was initiated to develop a synthetic route to this monomer. Thus, both monomer and polymer commercialization proceeded simultaneously. They went on stream in 1964. The very high melt viscosity made PPE very difficult to process. In contrast however PPE shows total compatibility with polystyrene, particularly high impact polystyrene, which results in a wide range of high temperature, easy to process, tough, dimensionally stable plastics.[14] The phenylene ether-based polymers might never have achieved commercial success if their unique compatibility with styrenic polymers had not been discovered at an early stage of development. This fortuitous and rather rare compatibility provided the basis for a family of polymers under the Noryl trade name.

Figure 10.

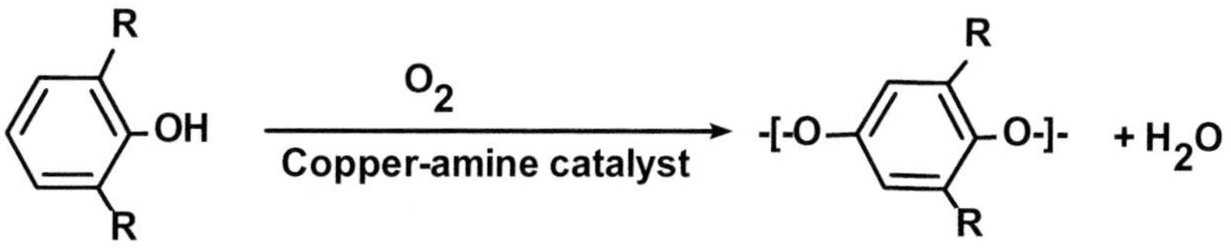

Noryl resins became the world's most successful and best known polymer blends or alloys because combinations of PPE resins and styrene polymers tend to assume the best features of each:

• PPE resins with very high heat distortion temperatures (HDTs) can readily raise the HDT of styrenics to over 100°C, which is a significant temperature because this qualifies the product for all boiling water applications.

• Styrene polymers, with ease of processing and well-established impact modification, balance the refractory nature of PPE resins.

• PPO resins bring fire retardance to the system.

• Both PPE resins and styrene polymers have excellent water resistance and outstanding electrical properties.

• In addition PPE/PS blends exhibit lower specific gravity that other engineering thermoplastics.

The first applications were those requiring autoclaving (medical equipment) and outstanding electrical properties at elevated temperatures. As compounding, stabilization, and processing skills improved, markets for Noryl expanded to include office equipment, electronic components, automotive parts, water distribution systems, and general metal replacement.

Noryl phenylene ether-based resins are relatively resistant to burning, and judicious compounding can increase their burn resistance without the use of halogenated flame-retardants. They may be modified with glass and other mineral fillers. Because of low moisture absorption, dimensional stability, and ability to be used over a wide temperature range, Noryl phenylene ether-based resins are especially adaptable to metallizing.

Polysulfones

Polyarylsulfones are a class of high-use temperature thermoplastics that characteristically exhibit excellent thermal-oxidative resistance, good solvent resistance, hydrolytic stability, and creep resistance.[16] In 1965 A. G. Farnham and R. N. Johnson of Union Carbide (this business was sold by Union Carbide to Amoco Polymers in 1986) announced the preparation of thermoplastic polysulfones.[17] The first commercially available polysulfone was prepared polymer prepared by made by the nucleophilic displacement of the chloride on bis (p-chlorophenyl) sulfone by the anhydrous di sodium salt of bisphenol A (Figure 11). The reaction is conducted in a dipolar aprotic solvent reported to be dimethyl sulfoxide. This polysulfone became commercially available in 1966 under the Udel trademark. This amorphous polymer exhibits a T_g of 190°C. Polysulfones are somewhat polar aromatic ethers with outstanding oxidation resistance, hydrolytic stability, and high heat-distortion temperature.

Figure 11.

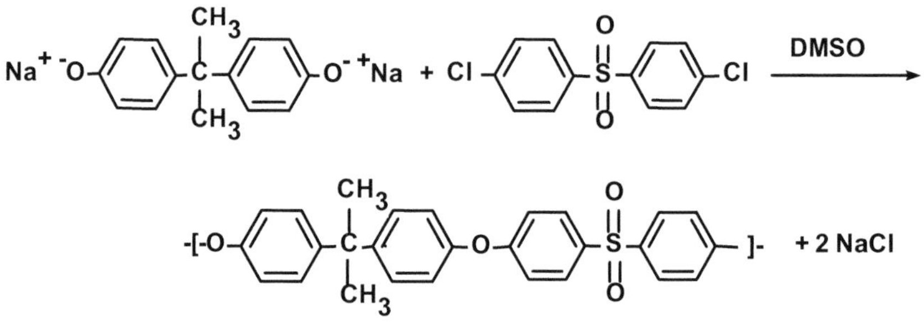

In 1976 Union Carbide introduced a second-generation polysulfone resin under the Radel polyphenylsulfone trade name. This higher performing resin is prepared from biphenol and bis (p-chlorophenyl) sulfone (Figure 12).[18] The biphenyl moiety imparts enhanced chemical/solvent resistance, outstanding toughness, greater resistance to combustion, a T_g of 220°C, and greater thermo-oxidative stability.

In addition Union Carbide and ICI offered polyethersulfone (PES) which is prepared from hydroquinone and bis (p-chlorophenyl) sulfone (Figure 13).[19] (In 1992 ICI withdrew from the business). PES offers high heat (T_g of 230°C) and thermal stability, better chemical/solvent resistance, and improved toughness over PSF.

Figure 12.   **Figure 13.**

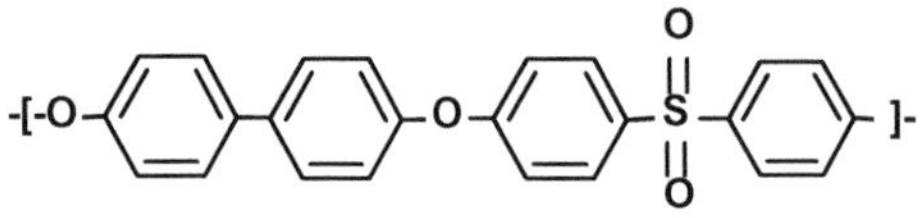

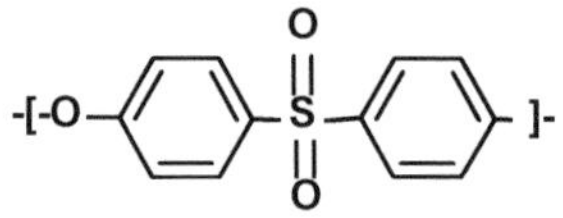

The hydrolytic stability and very high thermal endurance of this plastic in conjunction with a good balance of mechanical properties suit it for hot water and food handling equipment, range components, TV applications, alkaline battery cases, and film for hot transparencies. The unmodified product is transparent with a slightly yellow tint. Low flammability and low smoke suit it for aircraft and transportation applications. In addition it can withstand rigorous handling and repeated steam sterilization cycles and is used in medical application. Thus polyarylsulfones are used in surgical equipment, laboratory equipment, life support parts, and autoclavable tray systems. Blow molding polyarylsulfones produces suction bottles, tissue culture bottles and surgical hollow shapes.

Thermoplastic Polyesters

Thermoplastic polyesters had their beginning in 1941 when J. R. Whinfield and J. T. Dickson discovered terephthalate-based polyesters.[5, 20] Earlier, J. W. Hill and W. R. Carothers had examined aliphatic polyesters and found them inadequate as fiber precursors because of their low melting points. The aliphatic polyesters were bypassed by polyamides with much higher crystalline melting points. Whinfield and Dickson, in a subsequent investigation of polyesters as fiber precursors, substituted terephthalic acid for the previously investigated aliphatic dibasic acids and discovered high melting crystalline polymers. ICI, duPont, and others developed these polymers into the familiar polyester fibers and films.

Whinfield and Dickson quickly realized that the polymer based on ethylene glycol and terephthalic acid was the best suited for fibers (Figure 14). They did, however make and describe several other polyesters including poly (butanediol terephthalate) (PBT). Many years later a number of polyester fiber producers became interested in PBT. One producer explained that they were interested in PBT because it resembled nylon. Because nylon was becoming popular as a carpet yarn, and because they were not in the nylon business, they considered PBT a means of competing in carpet yarns.

Figure 14.

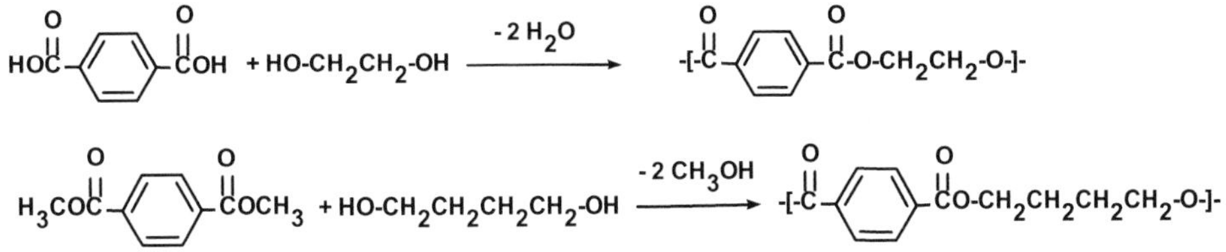

While the fiber producers were busily expanding their fiber activities, a number of companies were simultaneously trying to adapt poly (ethylene terephthalate) (PET) to molding in much the same way as the nylons had been made to double in brass as molding compound and fiber. This objective has been particularly attractive because the manufacturing capacity of PET the United States has surpassed 5 billion lbs. per year, and all the economy of scale has been obtained to yield high-performance polymers at commodity prices.

A number of companies have tried to promote PET as a molding compound. In 1966, the first injection molding grades of PET were introduced; however, these early materials were not very successful. The primary problem was that PET does not crystallize very rapidly; a molded object composed of a crystallizable polymer caught in an amorphous or partially crystallized state would be rather useless. In service such a part could crystallize, shrink, distort, crack, or fail. The obvious solution was to use hot molds and hold the parts in the mold until the crystallization process was completed. Post annealing also permits continued crystallization. These approaches, especially with glass fiber incorporation, led to acceptable parts at economically unacceptable molding cycles. Alternately, some developers tried to use very low molecular weight PET-glass products that crystallized more rapidly; however, because of their low molecular weights, these products lacked essential properties. A very broad search has been conducted for such things as nucleating agents and crystallization accelerators. An improved PET injection-molding compound was introduced by duPont in 1978 under the trade name Rynite. A number of other companies have followed duPont into the market. The PET-based molding compounds are gaining acceptance at a substantial rate, but actual volume is relatively small because of the recent beginning. While other companies sought means of increasing the rate of crystallization of PET, Celanese chemists turned their attention to PBT and found that it met all the requirements for a molding compound.

The basic composition of matter patents had long since expired when Celanese sampled the market in 1970 with a glass fiber reinforced PBT product designated X-917. This PBT molding compound was subsequently available under the Celanex trade name. Eastman Kodak followed Celanese early in 1971, and General Electric followed Eastman Kodak later in the same year with a PBT polyester resin under the Valox trade name. Since that time a dozen or more additional companies around the world have entered (and some have subsequently exited) the business.

Basically, PBT seems to have a unique and favorable balance of properties between nylons and acetal resins. It has relatively low moisture absorption, extremely good self-lubrication, fatigue resistance, solvent resistance, and good maintenance of mechanical properties at elevated temperatures. Maintenance of properties up to its crystal melting point is excellent if it is reinforced with glass fiber. Very fast molding cycles with cold to moderately heated molds complete the picture. Key markets include "under the hood" automotive applications such as ignition systems and carburetion, which require thermal and solvent resistance; electrical and electronic applications; power tools, small and large appliance components; and athletic goods.

Liquid Crystalline Polymers (Polyesters)

Most commercially important polyester liquid crystalline polymers (LCP) are based on p-hydroxy benzoic acid. In the late 1970's J. Economy of Carborundum developed poly (1,4 benzoate) (Figure 15) which could be compression-sintered and was marketed under the Ekanol trade name.[21, 22] This material did not melt and hence did not form liquid crystalline melts. By copolymerizing with biphenol and terephthalic acid (Figure 16), melt processible liquid crystalline polyesters were obtained and commercialized under the Ekkel trade name. These polymers were difficult to manufacture and process and were subsequently withdrawn from the marketplace. The technology was sold to Dartco.

Figure 15. **Figure 16.**

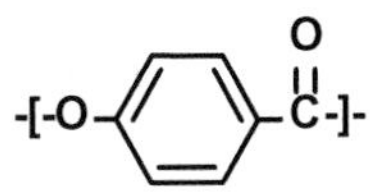

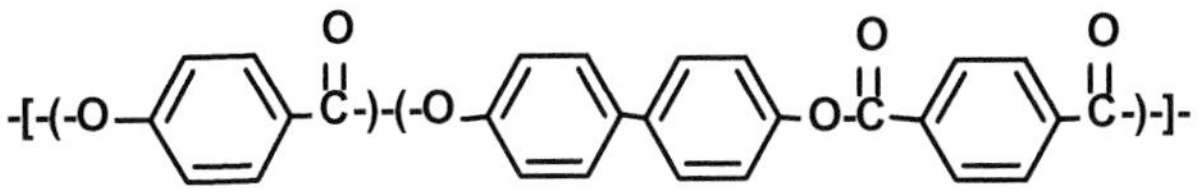

Dartco introduced their Xydar product line in the fall of 1984 for use in dual ovenable cookware. Their LCP was based on an improved p-hydroxy benzoic acid, biphenol, and terephthalic acid based resin. This technology was later sold to Amoco Polymers.

In the 1970s G. W. Calundann of Celanese (now Ticona) was developing LCPs based on copolymers of p-hydroxy benzoic acid with 6-hydroxy-2-naphthoic acid or with 2,6-dihydroxynaphthalene and 2, 6-naphthylene dicarboxylic acid (Figure 17). In the fall of 1985 Celanese commercialized their LCP under the Vectra trade name.

Principal U. S. and European suppliers of polyester LCPs are Ticona, Amoco, and duPont with products under the Vectra, Xydar, and Zenite trademarks, respectively.

LCPs have a rod like aromatic structure. The rod-like molecules arrange themselves in parallel domains in both the melt and solid states. In the molten state the molecules readily slide over one another giving the resin very high flow under shear. Thus melt viscosities decrease significantly as shear rates increase. Thus with LCPs molders lower viscosity by increasing injection velocity rather than temperature.

Figure 17.

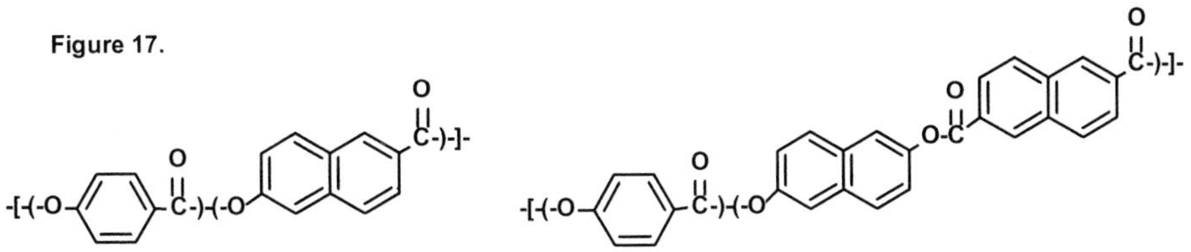

Highly crystalline LCPs offer high strength and rigidity in the direction of flow, dimensional stability, excellent solvent/chemical resistance, high heat resistance, and inherent flame retardancy. LCPs have anisotropic properties – i.e., properties that differ in the direction of flow and perpendicular to the flow direction.

Polyester LCPs are used in electronic connectors, surgical devices, and other parts where thin walls are essential.

Poly (Phenylene Sulfide)

The first poly (phenylene sulfides), PPS, was made in 1897 by the Friedel-Crafts reaction of sulfur and benzene.[23] Various other early attempts have been reported, all of which resulted in amorphous resinous materials that decomposed between 200-300°C. These materials were probably highly branched and partially cross-linked.

In 1973, Phillips Petroleum introduced linear and branched products under the Ryton trade name.[24] PPS were prepared by reacting 1,4-dichlorobenzene with sodium sulfide in a dipolar aprotic solvent (Figure 18). The polymer precipitates out of solution as a crystalline white powder. The polymer exhibits a T_g of 85°C and melts at 285°C.

Figure 18.

$$Cl-\langle\!\langle\bigcirc\rangle\!\rangle-Cl + Na_2S \longrightarrow -[\langle\!\langle\bigcirc\rangle\!\rangle-S-]- + 2\ NaCl$$

PPS exhibits high heat resistance, excellent chemical resistance, low friction coefficient, good abrasion resistance, and good electrical properties. Physical characteristics include high flexural modulus, very low elongation, and generally poor impact strength. Glass, glass-mineral, and carbon fiber reinforced grades that have high strength and rigidity are available. The un-reinforced resin is used only in coatings. The reinforced materials are finding applications in aerospace technology, pump systems, electrical and electronic equipment, appliances, and automotive vehicles and machines.

Poly (phenylene sulfides) are reported to be somewhat difficult to process because of their very high melting temperatures and relatively poor flow characteristics, and because some chemistry appears to continue during the fabrication step. Molded pieces have limited regrindability. Annealing of molded parts enhances mechanical properties but leads to almost total loss of thermoplastic character.

Polyetherimide

Polyetherimde (PEI) was formally announced by General Electric Company in 1982.[25] This amorphous polymer with the Ultem trade name resulted from the research work of a team headed by J. G. Wirth in the early 1970's (9). The early laboratory process involved a costly and difficult synthesis. Further development resulted in a number of breakthroughs that led to a simplified, cost-effective production process. The final step of the process involves the imidization of a di acid anhydride with m-phenylene diamine (Figure 19). PEI has a T_g of 217°C.

Figure 19.

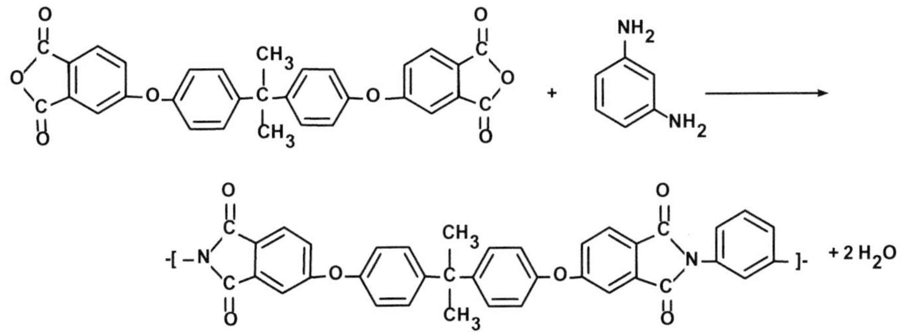

Polyetherimide offers an impressive collection of attributes such as high heat resistance, stiffness, impact strength, transparency, high mechanical strength, good electrical properties, high flame resistance, low smoke generation, and broad chemical resistance. In addition to its unique combination of properties matching those of high-priced specialty plastics, polyetherimide exhibits the processability of traditional engineering thermoplastics, although higher melt temperatures are required. The excellent thermal stability is demonstrated by the maintenance of stable melt viscosity after multiple regrinds and remolding. The processing window is nearly 100°C, and polyetherimide can be processed on most existing equipment. Furthermore, this excellent flow resin can be used for the molding of complicated parts and thin sections (as thin as 5 mil). Polyetherimide is suitable for use in internal components of microwave ovens, electrical and electronic products, and automotive, appliance, and aerospace, and transportation applications.

Aromatic Polyketones

The aromatic polyketones are a family of semi-crystalline engineering thermoplastics in which the combination of ketone and aromatic moieties imparts outstanding high temperature properties combined with exceptional thermal stability.[26] In addition, aromatic polyketones offer excellent environmental resistance, high mechanical properties, resistance to chemical environments at high temperatures, inherent flame retardancy, excellent friction and wear resistance, and impact resistance. A further property of aromatic polyketones is chemical purity that has led to applications in silicon chip manufacture. Aromatic polyketones were first produced inn the 1960s and 1970s but did not become commercially available until the early 1980s.

Polyetherketone (PEK) is a semi-crystalline polyether that was first developed by K. Dahl of Raychem in the 1970s for captive use. The Friedel-Crafts process required strong solvents or an excess of aluminum chloride to keep the polymer in solution. PEK has a T_m of 364°C and a T_g of 166°C. In the 1980s Raychem has licensed their patents to BASF and discontinued production. The first commercial quantities of PEK were made available in 1978 by ICI (now Victrex Ltd.).

Figure 20.

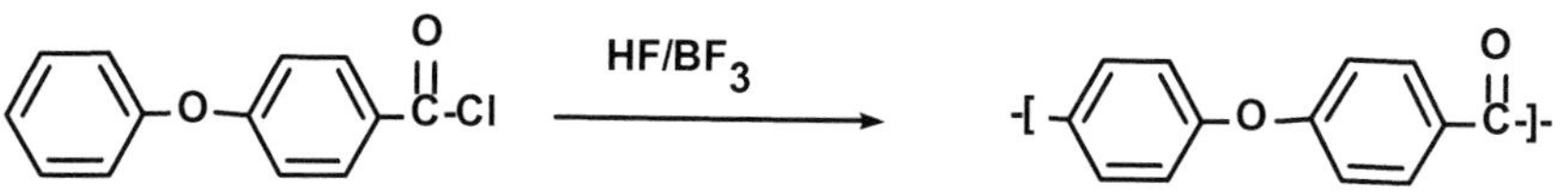

Polyetheretherketone (PEEK) is a semi-crystalline polyether prepared by nucleophilic aromatic substitution reaction. Several companies conducted early work on PEEK but they encountered various problems in its production. Victrex (neé ICI) solved these difficulties and commercialized PEEK in 1980 under the Victrex PEEK trade name. PEEK has a T_m of 335°C and a T_g of 145°C. PEEK is the mostly widely sold member of aromatic polyketones.

Figure 21.

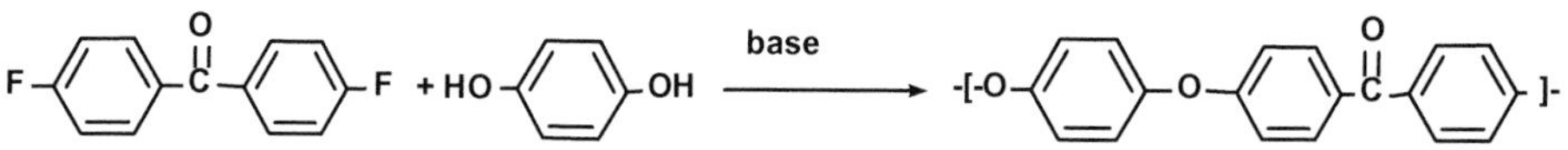

Applications are in the chemical process industry (compressor plates, valve seats, pump impellers, thrust washers, bearing cages), aerospace (aircraft fairings, radomes, fuel valves, ducting), and electrical (wire coating, semiconductor wafer carriers) industries.

Aliphatic Polyketones

Aliphatic polyketones are produced from olefin monomers and carbon monoxide. Basic patents on catalyst and composition appeared in the early 1970's.[27] However these early resins were not processible due to residual catalyst. In 1982 J. Dent of Royal Dutch/Shell discovered a new class of catalyst systems that were capable of co-polymerizing carbon monoxide and ethylene into linear, perfectly alternating polyketone polymers of high molecular weight. Shell commercialized a terpolymer of carbon monoxide, ethylene, and a small amount of propylene in 1996 under the Carilon trade name.

Figure 22.

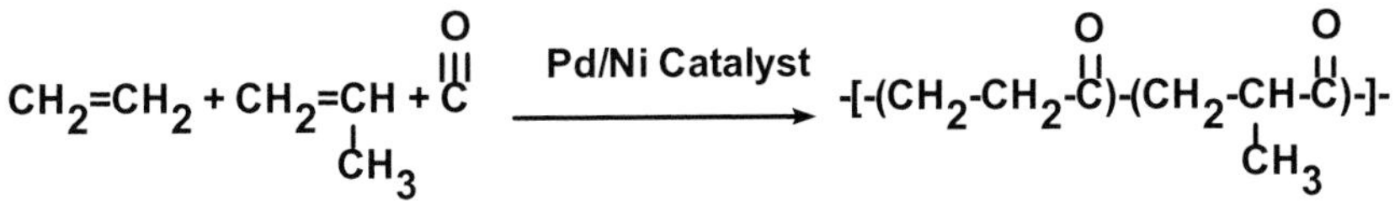

$$CH_2=CH_2 + CH_2=CH + \overset{\overset{O}{\|}}{C} \xrightarrow{\text{Pd/Ni Catalyst}} -[-(CH_2-CH_2-\overset{\overset{O}{\|}}{C})-(CH_2-CH-\overset{\overset{O}{\|}}{C})-]-$$

(with CH_3 groups on the propylene units)

The semi-crystalline, ethylene/propylene/CO terpolymer has a T_m of 200°C and a T_g of about 15°C. Aliphatic polyketones offer toughness, chemical resistance, and wear resistance and will compete with polyamides, thermoplastic polyesters and syndiotactic polystyrene in the electrical connector market and polyoxymethylenes in gear and barring applications.

Syndiotactic Polystyrene (SPS)

New developments in metallocene catalyst have resulted in the production of syndiotactic polystyrene, a semi-crystalline polymer. Amorphous polystyrene has no steroregularity and is a relatively brittle, amorphous polymer with poor chemical resistance. In contrast, SPS has high steroregularity, a T_m of 270°C, a T_g of 100°C, and good chemical/solvent resistance.[28]

Figure 23.

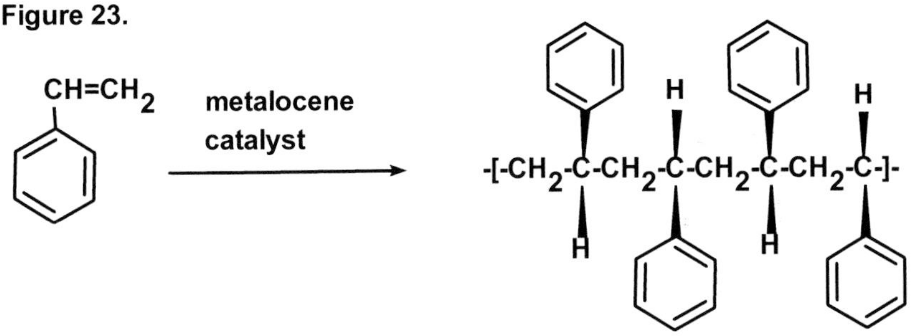

Dow Chemical commercialized SPS in 1997 under the Questra trade name.
SPS is targeted at automotive under-the-hood and specialty electronic applications. It will complete with polyamides, thermoplastic polyesters, aliphatic polyketones, and PPS.

Blends and Alloys

An interesting trend that appears to presage the wave of the future in engineering thermoplastics is the current focus on polymer blending and alloying. Metals and their alloys have been coeval with the spread of civilization. Early man used available metals in their naturally occurring state. The progress of civilization was literally determined by man's ability to modify natural metals, allowing mankind to induce the properties necessary for increasingly sophisticated tools. Indeed, societies that learned to exploit blends of metals developed distinct advantages over their monometallic neighbors; this was exemplified by the advent of bronze and later steels.

Metal alloying probably made the transition from art to science in the past century. The basic ingredients are essentially known and fixed in number. The plastic age began in 1909 with the discovery by Leo Baekeland of synthetic phenol-formaldehyde resin. As pointed out in the beginning of this chapter, the engineering plastic age began 45 years ago. While metal alloy components are essentially fixed, polymer alloy components are unlimited from a technical standpoint, but somewhat fixed from an economic point of view. It is still possible to make totally new and useful polymers if their value will support the cost of synthesizing new monomers and polymers. However, it is much more economically attractive to try to combine available polymers to produce desirable and novel alloys. The available degrees of freedom make the opportunity challenging and provide almost infinite possibilities. Variables include base polymers (20 or more), impact modifiers, additives, and fillers.

It is hard to say where or when the concept of polymer alloys was born, but within General Electric Company, Robert Finhold was writing and talking about alloys of PPO resin with styrene polymers in the very early 1960's. Since then other blends have been introduced. Blends of polycarbonate and ABS improve the flow and low temperature ductility. In addition PC/ABS blends can be flame retarded without the use of halogenated compounds.[2, 7] General Electric (neé Borg Warner) introduced PC/ABS blends under the Cycoloy trade name in 1971. In addition Bayer and Dow offer PC/ABS resins under the Bayblend and Pulse trade names, respectively.

In 1982 General Electric Company introduced a polycarbonate/poly (butylene terephthalate) impact modifier (PC/PBT/IM) polymer blend under the Xenoy trade name.[29] This polyblend was used on front-end, rear-end exterior bumper system for automobiles. It provides an outstanding combination of mechanical strength, impact strength, dimensional stability, and high modulus with chemical resistance. More recently the first in-colored interchangeable automotive body panel systems were made from PC/PBT/IM polyblend.

General Electric commercialized a unique polyamide/polyphenylene ether (PA/PPE) alloy in 1984 under the Noryl GTX trade name.[30, 31] This advanced-engineered alloy has a rigid-rigid matrix-particle system with interparticle rubber morphology. It is designed for use in exterior automotive parts. It is the first injection moldable thermoplastic with both the strength and rigidity needed for large panels, along with the high heat resistance for in-line or on-line painting. PA/PPE alloys provide 50% weight savings over traditional materials, and hence improves fuel efficiency.

Clearly, the trend toward alloys and blends is gaining momentum.

Conclusion

The future of engineering plastics looks bright. Those industries served by these plastics, and many others who use traditional materials such as metals, glass, and ceramics, will look to the benefits of engineering polymers to provide them with cost-effective materials to help overcome the pressures of spiraling costs. The world economy will continue to influence technical trends. The commercialization of any new engineering polymer based on a new monomer, although not impossible, is unlikely. Rather, the major thrust will take place in molecular shuffling with existing monomers, blend activity, and processing improvements. Key examples are the commercialization of aliphatic polyketones and syndiotactic polystyrene, which use commodity monomers.

Literature Cited

1. Fox, D. W.: Peters, E. N. in "Applied Polymer Science"; 2nd ed., Tess, R. W.; Poehlein, G. W., Eds.; ACS: Washington, DC, 1985; p, 495.
2. Peters, E. N. in "Mechanical Engineer's Handbook"; 2nd ed., Kutz, M., Ed.; Wiley-Interscience: NY, 1998; p. xxx.
3. Kohan, M. I., Ed. "Nylon Plastics Handbook"; SPE Monograph, Hanser: NY 1995.
4. Ahorani, S. M. "n-Nylons: Their Synthesis, Structure and Properties"; John Wiley & Sons: NY, 1997.
5. Bottenbruch, L., Ed. "Engineering Thermoplastics: Polycarbonates - Polyacetals - Polyesters - Cellulose Esters"; Hanser: NY 1996.
6. Staudinger, H. "Die Hochmolekularen Organischen Verbindungen Kautschuk und Cellulose"; Springer-Verlag: Berlin, 1932.
7. LeGrand, D.G.; Bendler, J. T., Eds. "Polycarbonates: Science and Technology"; M. Dekkers, Inc.: NY 1999.
8. Fox, D. W. "Kirk-Othmer Encyclopedia of Chemical Technology", 3rd ed.: Interscience: NY, 1982; Vol 18, p 479.
9. Schnell, H. "Chemistry and Physics of Polycarbonates"; Wiley-Interscience: NY, (1964).
10. Vernaleken, H. in "Interfacial Synthesis", Millich, F.; Carraher, C., Eds. Dekker: NY, 1997; Vol II.
11. Fox, D. W.: Gallucci, R. R.: Peters, E. N.: Smith, G. F. SPE, ANTEC, 1985, 85, 951.
12. Hay, A. S. J. Polym. Sci., 1962, 58, 581.
13. White, D. M.: Cooper, G. D. in "Kirk-Othmer Encyclopedia of Chemical Technology", 3rd ed. Interscience: NY, 1982; Vol 18, p. 595.
14. Cizek, E. P., U. S. Patent 3,338,435 (1968).
15. Kramer, M, Appl. Polym. Symp. 1970, 13, 227.
16. Harris, J. E. "Handbook of Plastic Materials and Technology", Rubin, I. I., Ed.; Wiley-Interscience: NY, 1990; p. 487.
17. Johnson, R. N. : Farnham, A. G.: Clendinning, R. A.: Hale, W. F.: Merriam, C. N. J. Polym. Sci., Part A-1, 1967, 5, 2375.

18. Robeson, L. M. in "Handbook of Plastic Materials and Technology", Rubin, I. I., Ed.; Wiley-Interscience: NY, 1990; p. 385.
19. Haas, T. W. in "Handbook of Plastic Materials and Technology", Rubin, I. I., Ed.; Wiley-Interscience: NY, 1990; p. 295.
20. Jaquiss, D. B. G.: Borman, W. F. H.: Campbell, R. W. in "Kirk-Othmer Encyclopedia of Chemical Technology", 3rd ed. Interscience: NY, 1982; Vol. 18, p. 549.
21. Economy, J.: Storm, R. S.: Matkovich, V. I.: Cottis, S. G.: Nowak, B. E. J. Polym. Sci., Polym. Chem. Ed., 1976, 14, 2207.
22. Kwolek, S. L.: Morgan, P. W.: Schaefgen, J. R. in "Encyclopedia of Polymer Science and Engineering", Kroschwitz, J. L., Ed. Wiley-Interscience: NY 1987; Vol. 9, p. 1.
23. Genvresse, P., Bull. Soc. Chim. Fr., 1897, 17, 599.
24. Short, J. M.: Hill, H. W. Chemtech, 1972, 2, 481.
25. Floryan, D. E.: Serfaty, I. W. Modern Plastics, 1982, p. 146.
26. Haas, T. W. in "Handbook of Plastic Materials and Technology" Rubin, I. I., Ed.; Wiley-Interscience: NY, 1990; p. 277.
27. Danforth, R. L.: Machado, J. M.: Jordaan, J. C. M. SPE ANTEC, 1995, p. 2316.
28. Brentin, R.: Bank, D.: Hus, M. SPE ANTEC, 1997, p. 3245.
29. Heuschen, J. M. in "High Perform. Polym. , Proc. Symp.", Seymour, R. B.; Kirshenbaum, G. S., Eds.; Elsevier: NY, 1986: p. 215.
30. Gallucci, R. R. SPE ANTEC, 1986, p. 48.
31. Peters, E. N. SPE, ANTEC, 1997, 97, 2322.

THERMOSETTING PLASTICS

RAYMOND A. PEARSON

Materials Science and Engineering
Lehigh University
Bethlehem, PA 18015-3195

Introduction

Thermosetting plastics have a large number of uses including uses as adhesives, coatings, encapsulants, and molding compounds. For example, one can go to the hardware store and purchase a "2-ton" epoxy adhesive. Paints often contain a thermosetting binder. Expensive integrated circuits are protected with thermosetting epoxy encapsulants. Thermosetting polyesters can be used to mold parts for automobiles. The traditional term "thermosetting resin" is now generally used for all reactive systems that form a crosslinked network.

   Thermosetting resins are sold in their unpolymerized form and must be reacted during processing. The reaction consists of a complex combination of both polymerization and crosslinking that involves conversion of low molecular weight molecules into a tight three dimensional network. The reactions can be separated into heat activated and catalyst activated systems. Typical temperature activated systems include phenol-formaldehyde, epoxies, polyimides, etc. The most common catalyst-activated resins are liquid unsaturated esters. In general, the properties of both types of thermosetting resins are often improved by the addition of additives.

   A thermosetting plastic consists of a thermosetting polymer resin along with various additives. There are a number of thermosetting polymer resins to choose from and a discussion of these resins can be found in the following section. Each thermosetting polymer resin possesses a unique combination of properties such as strength, stiffness, dimensional stability, thermal stability, processability, dielectric strength, refractive index, *etc.* However, many thermosetting resins would be commercial failures if they did not contain certain additives. The third section contains a discussion of the types of additives and the properties that they impart/enhance.

The purpose of this chapter is to familiarize the reader with the major types of thermosetting plastics, list possible additives, discuss processing methods and review properties.

Types of Resin Systems

Formaldehyde-based systems are based on reacting formaldehyde with active hydrogen containing compounds such as phenol, urea, or melamine. This family of thermosets has low flammability, high rigidity, good dimensional stability, and low cost. Formaldehyde-based systems are heat-activated and are processed using compression or injection molding.

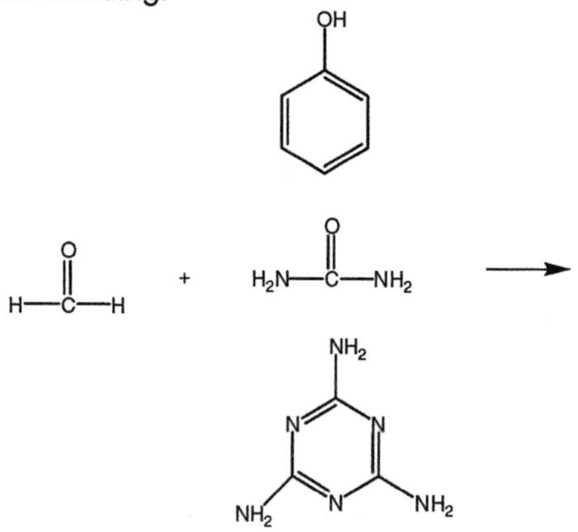

Figure 1. Formaldehyde can react with phenol, urea, or melamine to produce a cured thermosetting resin.

Allyl-based thermosets involve the reaction of a monofunctional unsaturated alcohol with a difunctional acid that forms a monomeric ester containing two vinyl groups. The two vinyl groups per monomer allows the formation of a three-dimensional network by the use of suitable catalysts such as benzoyl peroxide and elevated temperatures. Allyl resins systems are often used in high performance molding compounds and heat-cured laminates.

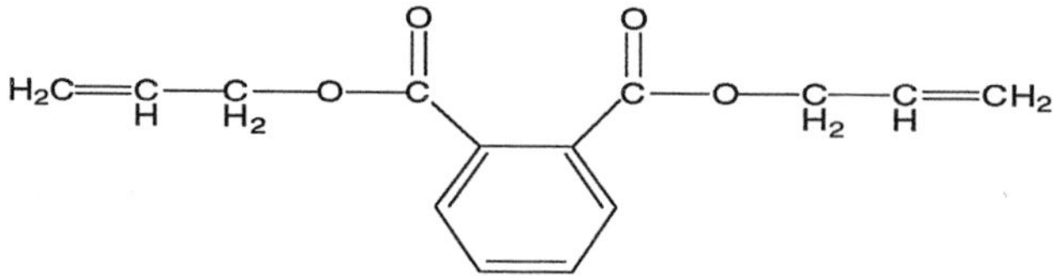

Figure 2: The diallylphthalate (DAP) molecule has two vinyl groups that enable the formation of a three dimensional network.

Unsaturated polyester systems consist of short oligomers (1000-3000 g/mole) derived from the reaction of difunctional acids (or anhydrides) with glycols. These oligomers contain ester linkages and are combined with vinyl monomers capable of polymerizing as well as reacting with the unsaturation in the oligomeric polyester molecules. Alpha-methylstyrene is the least expensive and most common monomer used. Heat can trigger the crosslinking reaction but catalysts and accelerators are often used. In general, unsaturated polyesters offer good weathering, low flammability, and good toughness. Typically, these resins are used to produce fiber-reinforced plastics.

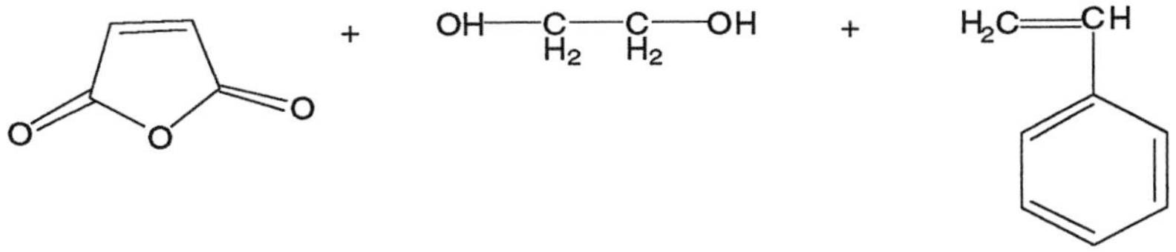

Figure 3: Unsaturated polyester resins consist of an anhydride (or difunctional acid) and a difunctional glycol. Some of the anhydride must have double bonds to react with the styrene.

Vinyl ester thermosets are based on small linear molecules that result from the reaction of epoxy terminated molecules with saturated acids. Like unsaturated polyester resins, a monomer such as styrene is used to complete the three-dimensional network. Vinylester resins are also primarily used as the matrix for fiber-reinforced plastics since they adhere well to glass and impart corrosion resistance.

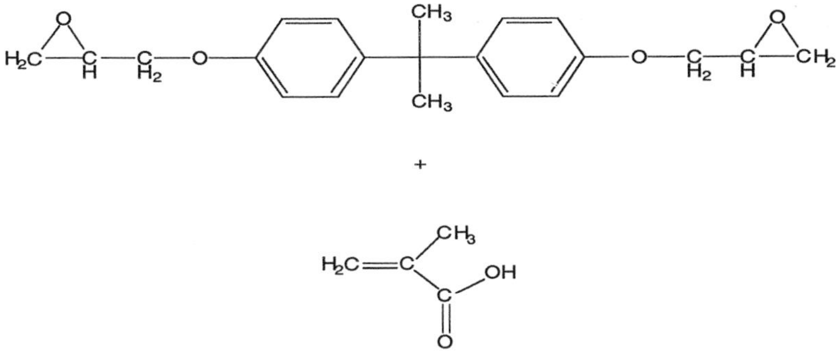

Figure 4: Vinyl esters can be prepared by reacting epoxide monomers with methacrylic acid and must be with a vinyl monomer such as styrene.

Epoxy resins consist of organic molecules featuring "epoxide groups," which are also called oxirane rings. These groups will react with active hydrogens on amine or anhydride molecules. The most common resin is based on bisphenol A and is referred to as the "diglycycidyl ether of bisphenol A" or DGEBA. The formation of the three-dimensional network can take place at room temperature or elevated temperatures depending upon the nature of the amine or anhydride. The properties of epoxies resins depends upon the chemical structure of the monomer used and, in general, are superior to less expensive thermosetting systems.

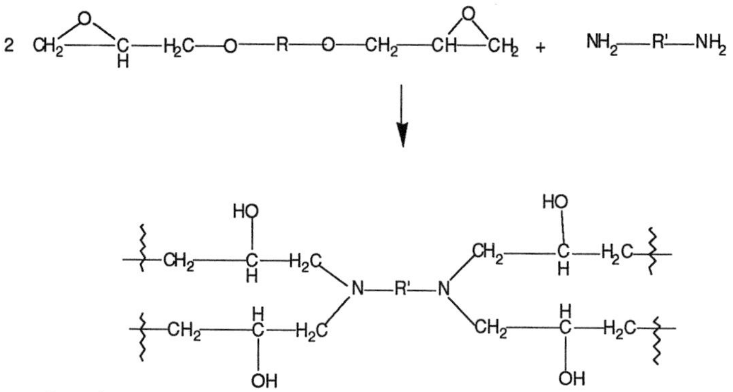

Figure 5: Cure chemistry of epoxy resins (amine addition reaction).

Polyurethanes are based on isocyanate chemistry. The basic reaction involves isocyanate and hydroxyl end-groups to form urethane linkages, although isocyanates can also be reacted with amines to form urea linkages. Multifunctional monomers must be used to produce three-dimensional networks. The properties of polyurethanes can very widely, depending upon the nature of the monomers used. Rigid and flexible polyurethanes can be found in either solid or foamed forms. In fact, reaction injection molding was initially developed for polyurethane resins.

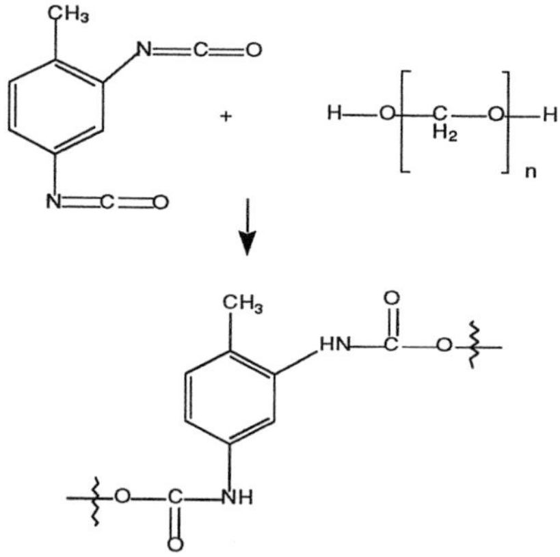

Figure 6: Cure chemistry of polyurethanes.

Bismaleimides (BMI) are synthesized from a nucleophilic reaction of a diamine with maleic anhydride then dehydration of the amic acid intermediate. See Figure 7. Polymerization and curing occurs via a radical-type addition reaction that occurs during the melting of the BMI monomer. BMIs are easy to process and are very heat resistant. They are often used as matrices for advanced composites.

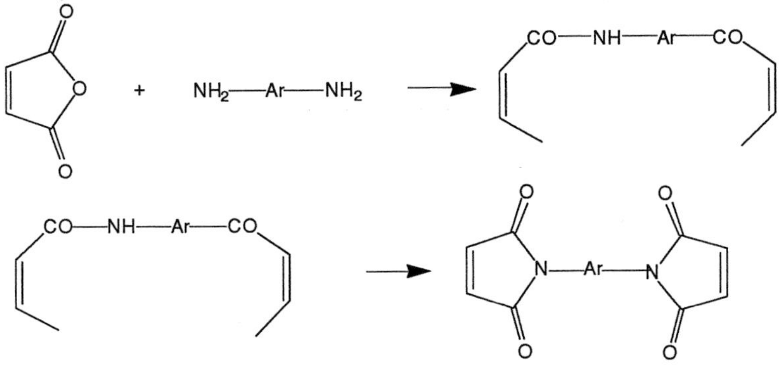

Figure 7: Synthesis of bismaleimide monomer.

Polyimide thermosets are a class of high temperature thermosets that have gained considerable importance in the electronic and aerospace industries. Typically, the cure involves the imidization of the polyamic acid precursor. Such reactions involve an elaborate thermal curing cycle with the last cycle being as high as 400 C. Polyimides possess excellent mechanical strength, high glass transition temperatures, low dielectric constant, and low thermal expansion.

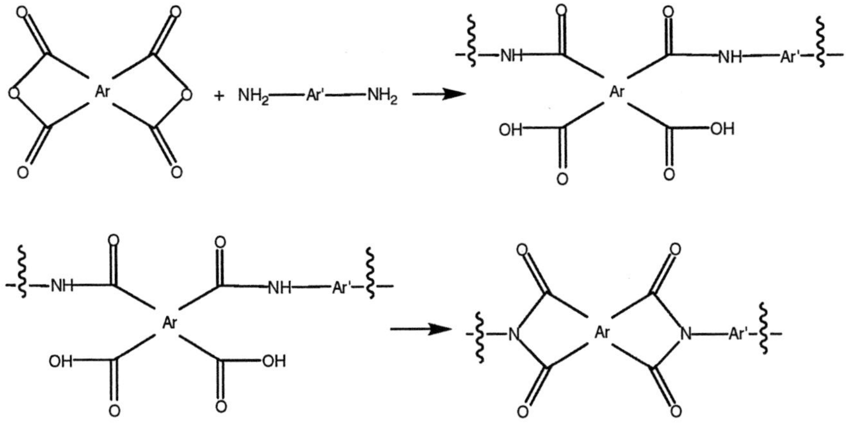

Figure 8: Polymerization of polyimides.

Cyanate ester resins contain cyanate monomers or oligomers and are usually cured by a transition metal carboxylate or chelate in the presence of an active hydrogen co-catalyst. An advantage of cyanate esters is that they process like an epoxy resin yet they have superior mechanical strength. Low moisture absorption, excellent electrical properties and good flammability characteristics make cyanate esters direct competitors with epoxies and bismaleimides in high performance applications.

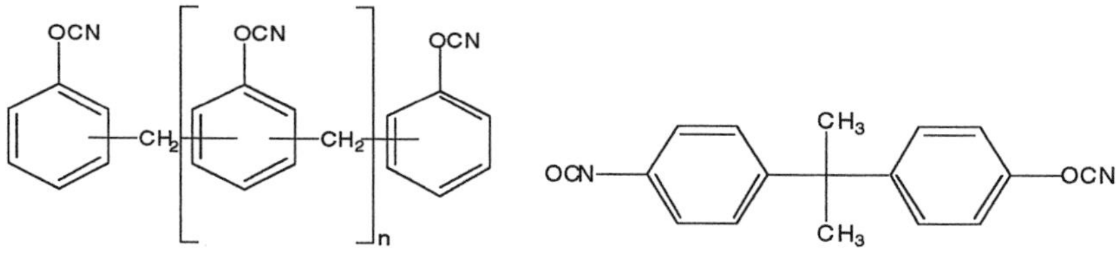

Figure 9: Novolac and bishphenol A cyanate ester resins.

Benzocyclobutene resins are cured in a two-step process. The first step involves a thermally activated opening of the benzocyclobutenes to form an o-quinodimethane intermediate. The second step is a Diels-Alder reaction where the o-quinodimethane groups react with residual alkene groups forming a tri-substituted tettrahydronaphthalene. In general benzocyclobutene resins have low dielectric constants, high planarization, and low moisture absorption. These attributes make them ideal candidates for organic passivation and dielectric layers in microelectronics.

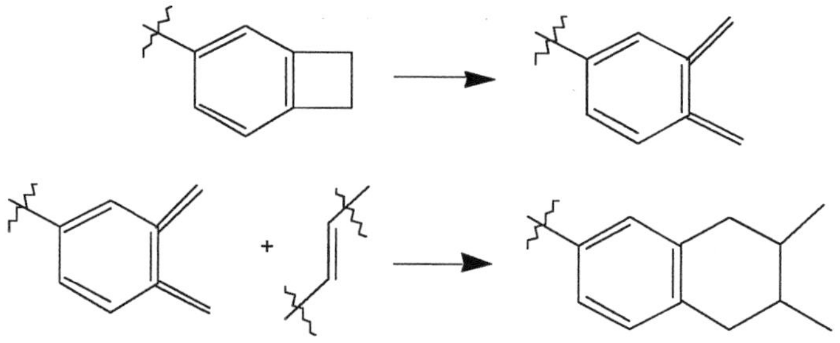

Figure 10: The two step cure reaction of benzocyclobutenes.

Additives

Thermosetting plastics are not viable without the use of additives. Additives are compounds that are included in the formulation of the thermosetting resin to improve properties or processing. The description below discusses some of the more common additives.

Antioxidants are a wide class of compounds that interfere with the oxidative cycles to inhibit or retard the oxidative degradation of polymers. Chain breaking antioxidants interrupt the primary oxidation cycles by removing the propagating radicals. Preventative oxidants interrupt the generation of free radicals. Antioxidants are often added to prevent degradation during processing, exposure to high temperatures, and exposure to UV light. They are also added to deactivate or retard the metal catalyzed oxidation of polymers.

Biocides are added to protect plastics against attack by bacteria, fungi, mildew, *etc.* Most plastics are relatively resistant to microbial attack, however, certain cellulosic polymers are susceptible to attack. Polyester based polyurethanes are also susceptible because most ester groups can undergo an enzyme-induced chain scission. A very widely used metal organic biocide is 10,10-oxybisphenoxyarsine. Heterocyclic compounds are used in coatings rather than bulk molding compounds. Biocides must be chosen carefully since they are toxic, strongly colored, can be detrimental to thermal stability, and can interfere with other formulation constituents.

Carbon black is one of the most versatile additives. It can provide color, UV protection, thermal conductivity, reinforcement and changes in electrical properties. Carbon black is a particulate form of industrial carbon produced by the thermal cracking of a raw hydrocarbon material. Carbon black is said to have a semi-graphitic structure consisting of three to four hexagonal layers that combine as crystallites or bundles. Most of today's carbon black is produced using an oil furnace process that results in spherical particles with diameters ranging from 10 to 100 nanometers. These particles fuse together to form aggregates. The smaller the primary particles the larger the surface area and the more 'jetness,' i.e. intensity of the black color. All carbon blacks have some level of chemisorbed oxygen complexes such as carboxylic, phonemic, quinine or laconic groups. These oxygen complexes are acidic and control the pH value of the carbon black. While carbon blacks are often classified on the basis of surface area / particle size and surface chemistry, one must also consider the amount of ash and grit that results from the furnace process.

Metallic flakes and fibers are added to plastics to impart electrical conductivity. A wide range of metal particles have been examined including aluminum, copper, nickel, silver and brass. Metals have electrical resistivities in the range of 10^6 ohm cm whereas polymers are typically on the order of 10^{15} ohm cm. A critical volume fraction of metal particles is need to impart electrical conductivity. This critical concentration is called the percolation threshold and is the concentration of particles that form a three-dimensional network of particles in intimate contact. Although copper particles are very conductive, they tend to react unfavorably with some polymers. Small aluminum flakes have a significant oxide layer and can behave as insulators. Silver is an excellent conductor and is often used when its high cost can be justified.

Coupling agents are chemicals that improve the adhesion between two phases and are often used in composites and adhesives. Organosilanes are perhaps the most important coupling agents, although zirconates and titantes can be used to treat fillers. There is a wide selection of organosilanes to choose from. Typically, the head contains a trimethoxy silane that can react with surface hydroxyls. The tail consists of a short alkyl chain with the appropriate functional group at its end (to react with the thermosetting resin). γ-Aminopropyl silanes are common coupling agents for epoxy-glass interfaces. Coupling agents are often used to improve the strength of interfaces in hot-humid environments.

can be used to lower the viscosity of the resin, increase potlife, reduce surface tension and improve wetting. They are often used in epoxy resins and we will limit our discussion to this group of diluents. Epoxy resins can contain reactive diluents that are based on derivatives of glycidyl ether. Butyl glycidyl ether is often used because a maximum viscosity reduction is achieved with a minimum amount of diluent. Reactive diluents are incorporated into the matrix and can decrease mechanical properties. Non-reactive diluents contain no reactive groups and function primarily as viscosity modifiers and plasticizers. The first requirement of such modifiers is that they must be miscible in the resin. Pine oil, dibutyl phthalate and glycol ethers have been used sparingly because of their adverse effects on the cured resin properties.

Fillers can be added to thermosetting polymers to increase toughness, strength, stiffness, dimensional stability, viscosity, thermal conductivity and electrical conductivity. Fillers can also be added to reduce mold shrinkage, coefficient of thermal expansion (CTE), vapor transport, and in some cases, cost. A wide range of fillers exist and each filler has a unique set of attributes. Particular fillers are discussed below.

Short, inorganic fibers are often added to thermosetting polymer in order to increase stiffness, strength and toughness. The fibers are either made out of boron, carbon or glass. Glass is the least expensive fiber but also the weakest and the least stiff. A potential disadvantage of using short fibers as fillers is that processing can align the fibers and cause strength and stiffness to be directionally dependent.

Inorganic platelets can also be used as fillers for thermosetting resins. The surface area to volume ratio is as large as the short fibers so an increase in mechanical properties can be obtained. Mica and talc are often found in platelet form and are inexpensive.

Flakes are formed using a grinding process. Silver flakes are commonly used to impart electrical conductivity. A number of inorganic fillers can be used to impart thermal conductivity.

Glass spheres are commonly used to improve CTE and fracture toughness. The advantage of glass spheres are that they flow more easily than platelets and fibers.

Toughening Agents_are used in two forms. The first type is that of the traditional rubber tougheners, which consist of soft rubber phases that can trigger plastic deformation in the thermosetting matrix. There are two kinds of rubber-tougheners: pre-formed particles and those that form particles *in situ*. Core-shell latex particles and reground rubber particles are examples of the pre-formed particle type. Reactive oligomers such as CTBN rubber (carboxyl terminated, butadiene-acrylonitrile copolymers) and hyperbranched copolymers form particles when they phase separate during the cure process. The second type of organic toughener are the rigid type. For example, polyetherimide and polyethersulphone can be added to epoxies or cyanate ester resins to improve toughness.

Internal release agents are often used in molding compounds. Mold release agents are substances that help separate the cured part from the tool. Examples of internal release agents include zinc and aluminum stearate, which are used in low temperature molding.

Flame retardants are additives that interfere with the chemistry and physics of the combustion process. The vapor phase combustion of polymers is a free radical process and organochlorine and organobromines are added to scavenge the free radicals. Antimony trioxide is often added to increase the efficiency of the organohalide additives. Alumina trihydrate can be added to dilute the volatile products with water molecules (a degradation product) and remove some of the heat of combustion. Phosphorous additives can sometimes be used to enhance char formation.

Processing Methods

The processing methods for thermosetting plastics are quite diverse. For the sake of brevity, we will focus on those techniques that utilize bulk molding compounds and fiber-reinforced thermosets.

Closed mold techniques are used to make complex parts from bulk molding compounds. The simplest closed mold method is casting which involves pouring the liquid into a mold. For more viscous systems, compression molding is used to squeeze the material into two halves of a heated mold. Resin transfer molding is similar to compression molding except the a plunger pushes the reactive melt through a runner system into the two-piece mold. Reaction injection molding is similar to resin transfer molding except that a mixing nozzle is used to mix the reactants before injecting them into a closed cavity.

Open tool processing is often employed to process fiber-reinforced parts. The simplest open mold process is called hand lay-up, which involves manual placement of mats of fibers impregnated with thermosetting resin on an open mold. Room temperature and elevated cure temperature resins can be used. A spray-up method is used for faster production rates but it is limited to chopped fibers. The combination of viscous resins and high fiber loads requires the used of the vacuum bag technique to eliminate voiding and increase compaction. In extreme circumstances, laminates enclosed in a vacuum bag can be place in an autoclave to cure under high pressure and temperature.

Typical Properties

In general, thermosetting resins have greater strength and solvent resistance than their thermoplastic counterparts. When bulky monomers are used or when the crosslink density is high, the glass transition temperatures can exceed 400° C! Of course, high temperature resistant polymers are from the more modern family of thermosets: the polyimides, bismaleimides and the cyanate ester resins. Although the basic chemistry of the conventional thermosets have been around for a long time, the synthesis of new monomers creates new thermosetting resins with new properties. However, one drawback to thermosetting plastics is their poor resistance to fracture.

Although thermosets are often inherently brittle, organic additives can be used to improve the fracture toughness of these materials. For thermosets that are lightly crosslinked, reactive oligomers that phase separate during cure can be used. Pre-formed core-shell rubber particles can also be used. For highly crosslinked thermosets a series of thermoplastic toughening agents have been developed. These toughening agents are discussed in more detail below.

Reactive elastomeric oligomers have been used to toughen lightly crosslinked thermosets for over 30 years. The original modifiers were based upon reactive oligomers of carboxyl terminated polybutadiene-acrylonitrile. The acrylonitrile content can be adjusted to enhance miscibility in the particular resin of interest. An emerging technology is the use of hyperbranched molecules as toughening agents for thermosets. As with the CTBN elastomers, the hyperbranched molecules phase separate upon cure.

Recently, core-shell latex particles have been studied as toughening agents for thermosets. The advantage of pre-formed particles is that the particle size is no longer controlled by the cure chemistry of the thermosetting resin. These particles are effective in promoting plasticity in the thermosetting matrix and thus improve fracture toughness. In general, the toughness can increase from 100 J/m^2 to 2,000 J/m^2 upon the addition of rubber particles.

Unfortunately, the addition of rubber particles to highly crosslinked thermosets does not result in a significant amount of fracture toughness. This is due to the fact that these materials are very resistant to plastic deformation. Therefore, an alternative type of organic toughening agent must be used. The alternative tougheners are thermoplastic resins.

The use of thermoplastic resins to toughen thermosets is a relatively new technology. Polyethersulfones, polyimides, polyetherimides, and nylons are often used. These ductile particles are able to create microcracks and bridge the crack surface. The amount of toughness increase is much more modest than their counterparts for lightly crosslinked thermosets but such increases allow these highly crosslinked thermosets to be successfully applied to high performance composites.

<u>Summary</u>

Thermosetting resins have been around for nearly seventy years. Phenol-novolac resins are the earliest examples of a commercial thermosetting resin. Subsequent chemistries have been developed to produce thermosets that process more easily and are more heat resistant. The discovery of the various cure chemistries do not represent the end of the development cycle but mark the beginning of families of materials that grow with each new monomer discovered. However, the success of these new materials depends in part on the additives used to enhance properties and fortunately, additive research is still thriving.

Additional Reading

Pritchard, G. "Plastic Additives: An A-Z Reference"; Chapman and Hall: New York, 1998.

Charrier, J-M "Polymeric Materials and Processing"; Hanson Publishing: New York, 1990.

McCrum, N. G.; Bucknall, C. P. "Principles of Polymer Engineering"; Oxford University Press: New York, 1997.

Riew, C. K.; Kinloch, A. J., eds. "Toughened Plastics I: Science and Engineering", American Chemical Society: Washington, D. C., 1993.

THERMOSET ELASTOMERS

J. E. MARK

Department of Chemistry and the Polymer Research Center
University of Cincinnati, Cincinnati, Ohio 45221-0172

Introduction
      Basic Concepts
      Some Historical High Points
      Some Rubberlike Materials
      Preparation of Networks
      Some Typical Applications
Experimental Details
      Mechanical Properties
      Swelling
      Optical and Spectroscopic Properties
      Scattering
Stress-Strain Behavior
Control of Network Structure
Networks at Very High Deformations
      Non-Gaussian Effects
      Ultimate Properties
Multimodal Chain-Length Distributions
Other Types of Deformation
      Biaxial Extension
      Shear
      Torsion
      Swelling
Filler-Reinforced Elastomers and Elastomer-Modified Ceramics
Current Problems and Future Trends

Introduction

Basic Concepts. Elastomers are defined by their very large deformability with essentially complete recoverability. In order for a material to exhibit this type of elasticity, three molecular requirements must be met: (i) the material must consist of polymeric chains, (ii) the chains must have a high degree of flexibility and mobility, and (iii) the chains must be joined into a network structure (1-3).

The first requirement arises from the fact that the molecules in a rubber or elastomeric material must be able to alter their arrangements and extensions in space dramatically in response to an imposed stress, and only a long-chain molecule has the required very large number of spatial arrangements of very different extensions. The second characteristic required for rubberlike elasticity specifies that the different spatial arrangements be *accessible*, i. e., changes in these arrangements should not be hindered by constraints as might result from inherent rigidity of the chains, extensive chain crystallization, or the very high viscosity characteristic of the glassy state (1,2,4 ,5-8). The last characteristic cited is required in order to obtain the elastomeric recoverability. It is obtained by joining together or "cross linking" pairs of segments, approximately one out of a hundred, thereby preventing stretched polymer chains from irreversibly sliding by one another. The network structure thus obtained is illustrated in Figure 1, in which the cross links are generally chemical bonds (as would occur in

Figure 1.   Sketch of a typical elastomeric network, with an interchain entanglement depicted in the lower right-hand corner.

sulfur-vulcanized natural rubber). These elastomers are frequently included in the category of "thermosets", which are polymers having a network structure which is generated or "set" by thermally-induced chemical cross-linking reactions. The term has now frequently taken on the more specific meaning of networks that are very heavily cross linked and below their glass transition temperatures. Such materials, exemplified by the phenol-formaldehyde and the epoxy resins, are very hard materials with none of the high extensibility associated with typical elastomers.

The cross links in an elastomeric network can also be temporary or physical aggregates, for example the small crystallites in a partially crystalline polymer or the glassy domains in a multi-phase triblock copolymer (3,6). The latter materials are considered separately in the chapter on "Thermoplastic Elastomers". Additional information on the cross linking of chains is given below.

<u>Some Historical High Points</u>. The earliest elasticity experiments involved stress-strain-temperature relationships, or network "thermoelasticity". They were first carried out many years ago, by J. Gough, back in 1805 (<u>1,2,4,5,9,10</u>). The discovery of vulcanization or curing of rubber into network structures by C. Goodyear and N. Hayward in 1839 was important in this regard since it permitted the preparation of samples which could be investigated in this regard with much greater reliabiity. Such more quantitative experiments were carried out by J. P. Joule, in 1859. This was, in fact, only a few years after the entropy was introduced as a concept in thermodynamics in general! Another important experimental fact relevant to the development of these molecular ideas was the fact that mechanical deformations of rubberlike materials generally occurred essentially at constant volume, as long as crystallization was not induced (<u>1</u>). (In this sense, the deformation of an elastomer and a gas are very different).

A molecular interpretation of the fact that rubberlike elasticity is primarily entropic in origin had to await H. Staudinger's much more recent demonstration, in the 1920's, that polymers were covalently-bonded molecules, and not some type of association complex best studied by the colloid chemists (<u>1</u>). In 1932, W. Kuhn used this observed constancy in volume to point out that the changes in entropy must therefore involve changes in orientation or configuration of the network chains (<u>4,6</u>).

Later in the 1930's, W. Kuhn, E. Guth, and H. Mark first began to develop quantitative theories based on this idea that the network chains undergo configurational changes, by skeletal bond rotations, in response to an imposed stress (<u>1,2</u>). More rigorous theories began with the development of the "Phantom Network" theory by H. M. James and E. Guth in 1941, and the "Affine Model" theory by F. T. Wall, and by P. J. Flory and J. Rehner, Jr. in 1942 and 1943. Modern theories generally begin with the phantom model and extend it, for example by taking into account interchain interactions (<u>6</u>).

<u>Some Rubberlike Materials</u>. Since high flexibility and mobility are required for rubberlike elasticity, elastomers generally do not contain stiffening groups such as ring structures and bulky side chains (<u>2,4</u>). These characteristics are evidenced by the low glass transition temperatures T_g exhibited by these materials. Such polymers also tend to have low melting points, if any, but some do undergo crystallization upon sufficiently large deformations. Examples of typical elastomers include natural rubber and butyl rubber (which do undergo strain-induced crystallization), and poly(dimethylsiloxane), poly(ethyl acrylate), styrene-butadiene copolymer, and ethylene-propylene copolymer (which generally don't).

The most widely used elastomers are natural rubber, synthetic polyisoprene and butadiene rubbers, styrene-butadiene copolymers, ethylene-propylene rubber (specifically EPDM), butyl and halobutyl elastomers, polyurethanes, polysiloxanes, polychloroprenes, nitrile rubber, polyacrylic rubbers, fluorocarbon elastomers, and thermoplastic elastomers (<u>11</u>). The examples which have unsaturation present in the repeat units (such as the diene elastomers) have the advantage of easy cross linkability, but the disadvantage of increased vulnerability to attack by reactants such as oxygen and ozone.

Some polymers are not elastomeric under normal conditions but can be made so by raising the temperature or adding a diluent ("plasticizer"). Polyethylene is in this category because of its high degree of crystallinity. Polystyrene, poly(vinyl chloride), and the biopolymer elastin are also of this type, but because of their relatively high glass transition temperatures require elevated temperatures or addition of diluent to make them elastomeric (4,5).

A final class of polymers is inherently nonelastomeric. Examples are polymeric sulfur, because of its chains are too unstable, poly(*p* - phenylene) because its chains are too rigid, and thermosetting resins because their chains are too short (4).

Preparation of Networks. One of the simplest ways to introduce the cross links required for rubberlike elasticity is to carry out a copolymerization in which one of the comonomers has a functionality ϕ of three or higher (4,12). This method, however, has been used primarily to prepare materials so heavily cross linked that they are in the category of hard thermosets rather than elastomeric networks (13). The more common techniques include vulcanization (addition of sulfur atoms to unsaturated sites), peroxide thermolysis (covalent bonding through free-radical generation), end linking of functionally-terminated chains (isocyantes to hydroxyl-terminated polyethers, organosilicates to hydroxyl-terminated polysiloxanes, and silanes to vinyl-terminated polysiloxanes).

For commercial materials, the compounding recipe generally contains numerous ingredients in addition to the polymer and cross-linking agent (for example, sulfur, a peroxide, or a isocyanate) (14). Examples are activators (to increase cross-linking efficiency), retarders (to prevent premature cross linking or "scorch"), accelerators, peptizing agents, antioxidants and antiozonants, softeners, plasticizing aids, extenders, reinforcing fillers (typically carbon black or silica), and processing aids. Specific applications can require still additional additives, for example blowing agents in the case of elastomeric foams, thermally conducting particles in the case of heated rollers, fiber meshes in the case of high pressure tubing, etc.

A sufficiently stable network structure can also be obtained by physical aggregation of some of the chain segments onto filler particles, by formation of microcrystallites, by condensation of ionic side chains onto metal ions, by chelation of ligand side chains to metal ions, and by microphase separation of glassy or crystalline end blocks in a triblock copolymer (4,5). The main advantage of these materials is the fact that the cross links are generally only temporary, which means that such materials frequently exhibit reprocessability. This temporary nature of the cross linking can, of course, also be a disadvantage since the materials are rubberlike only so long as the aggregates are not broken up by high temperatures, presence of diluents or plasticizers, etc.

Some Typical Applications. Typical non-biological applications are tires, gaskets, conveyor belts, drive belts, rubber bands, stretch clothing, hoses, balloons and other inflatable devices, membranes, insulators, and encapsulants. Biological applications include parts of living organisms (skin, arteries, veins, heart and lung tissue, etc.), and various biomedical devices (contact lens, prostheses, catheters, drug-deliver systems, etc.). It is interesting to note that most of these applications require only small deformations; relatively few take advantage of the very high extensibility that is characteristic of most elastomeric materials!

Frequently, specific applications require a particular type of elastomer (15). For example, a hose should have as large a mismatch of solubility parameters with the fluid it will be transporting. Thus, a polar elastomers such as polychloroprene would be best for hoses used with hydrocarbon fluids such as gasoline, jet fuel, greases, oils, lubricants, etc.

Some Experimental Details

Mechanical Properties. The great majority of studies of mechanical properties of elastomers have been carried out in elongation, because of the simplicity of this type of deformation (3,4). Results are typically expressed in terms of the nominal stress $f^* \equiv f/A^*$ which, in the simplest molecular theories, is given by

$$f^* = (\nu kT/V)\,(\alpha - \alpha^{-2}) \tag{1}$$

where ν/V is the density of network chains, i. e., their number per unit volume V, k is the Boltzmann constant, T is the absolute temperature, and α is the elongation or relative length of the stretched elastomer. Also frequently employed is the modulus, defined by

$$[f^*] \equiv f^* v_2^{1/3}/\,(\alpha - \alpha^{-2}) = \nu kT/V \tag{2}$$

where v_2 is the volume fraction of polymer in the (possibly swollen) elastomer. There are a smaller number of studies using types of deformation other than elongation, for example, biaxial extension or compression, shear, and torsion. Some typical studies of this type are mentioned below.

Swelling. This non-mechanical property is also much used to characterize elastomeric materials (1,2,4,12). It is an unusual deformation in that volume changes are of central importance, rather than being negligible. It is a three-dimensional dilation in which the network absorbs solvent, reaching an equilibrium degree of swelling at which the free energy decrease due to the mixing of the solvent with the network chains is balanced by the free energy increase accompanying the stretching of the chains. In this type of experiment, the network is typically placed into an excess of solvent, which it imbibes until the dilational stretching of the chains prevents further absorption. This equilibrium extent of swelling can be interpreted to yield the degree of cross linking of the network, provided the polymer-solvent interaction parameter χ_1 is known. Conversely, if the degree of cross linking is known from an independent experiment, then the interaction parameter can be determined. The equilibrium degree of swelling and its dependence on various parameters and conditions provide, of course, important tests of theory.

Optical and Spectroscopic Properties. An example of a relevant optical property is the birefringence of deformed polymer network (12). This strain-induced birefringence can be used to characterize segmental orientation, both Gaussian and non-Gaussian elasticity, crystallization and other types of chain ordering, and short-range correlations (2,4). Other optical and spectroscopic techniques are also important, particularly with regard to segmental orientation. Some examples are fluorescence polarization, deuterium NMR, and polarized infrared spectroscopy (4,12,16).

Scattering. The technique of this type of greatest utility in the study of elastomers is small-angle neutron scattering, for example, from deuterated chains in a non-deuterated host (4,12,17). One application has been the determination of the degree of randomness of the chain configurations in the undeformed state, an issue of great importance with regard to the basic postulates of elasticity theory. Of even greater importance is determination of the manner in which the dimensions of the chains follow the macroscopic dimensions of the sample, i. e., the degree of "affineness" of the deformation. This relationship between the microscopic and macroscopic levels in an elastomer is one of the central problems in rubberlike elasticity.

Some small-angle X-ray scattering techniques have also been applied to elastomers. Examples are the characterization of fillers precipitated into elastomers, and the corresponding incorporation of elastomers into ceramic matrices, in both cases to improve mechanical properties (4,18-21).

Typical Stress-Strain Behavior

A typical stress-strain isotherm obtained on a strip of cross-linked elastomer such as natural rubber is shown schematically in Figure 2 (1-3). The units for the force are gen-

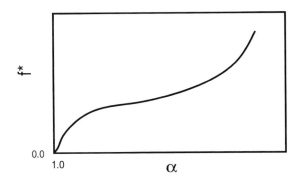

Figure 2. Stress-elongation curve for an elastomer showing an upturn in modulus at high elongations.

erally Newtons, and the curves obtained are usually checked for reversibility. In this type of representation, the area under the curve is frequently of considerable interest since it is proportional to the work of deformation $w = \int f dL$. Its value up to the rupture point is thus a measure of the toughness of the material.

The upturn in modulus at high elongation is of particular interest since it corresponds to an increase in toughness. It is generally due to strain-induced crystallization, resulting from increase in melting point of the network chains. This is, in turn, due to the decreased entropy of the stretched chains and the fact that the melting point is inversely proportional to the entropy of melting. In some cases, however, the upturns can be due to the limited extensibility of the chains. These instances are easy to identify, since these upturns will not be diminished by decreasing the amount of crystallilzation by increase in temperature or by addition of a diluent. It is in this sense that the stretching "induces" the crystallization of some of the network chains (5).

The initial part of the stress-strain isotherm shown in Figure 2 is of the expected form in that f^* approaches linearity with α as α becomes sufficiently large to make the subtractive α^{-2} term in Equation (1) negligibly small. The large increase in f^* at high deformation in the particular case of natural rubber is due largely, if not entirely, to strain-induced crystallization.

Additional deviations from theory are found in the region of moderate deformation upon examination of the usual plots of modulus against reciprocal elongation (2,22). Although Equation (2) predicts the modulus to be independent of elongation, it generally decreases significantly upon increase in α (22). The intercepts and slopes of such linear plots are generally called the Mooney-Rivlin constants $2C_1$ and $2C_2$, respectively, in the semi-empirical relationship $[f^*] = 2C_1 + 2C_2\alpha^{-1}$. As described above, the more refined molecular theories of rubberlike elasticity (6,23-26) explain this decrease by the gradual increase in the non-affineness of the deformation as the elongation increases toward the phantom limit.

Control of Network Structure

Until recently, there was relatively little reliable quantitative information on the relationship of stress to structure, primarily because of the uncontrolled manner in which elastomeric networks were generally prepared (1-4). Segments close together in space were linked irrespective of their locations along the chain trajectories, thus resulting in a highly random network structure in which the number and locations of the cross links were essentially unknown. Such a structure is shown above in Figure 1. New synthetic techniques are now available, however, for the preparation of "model" polymer networks of known structure (3,4,6,27-44). An example is the reaction shown in Figure 3, in which hydroxyl-terminated chains of poly(dimethylsiloxane) (PDMS) are end-linked using tetraethyl orthosilicate. Characterizing the uncross-linked chains with respect to molecular weight M_n and molecular weight distribution, and then running the specified reaction to completion gives elastomers in which the network chains have these characteristics, in particular a molecular weight M_c between cross links equal to M_n, and cross links having the functionality of the end-linking agent.

The end-linking reactions described above can also be used to make networks having unusual chain-length distributions (45-48). Those having a bimodal distribution are of particular interest with regard to their ultimate properties, as will be described below.

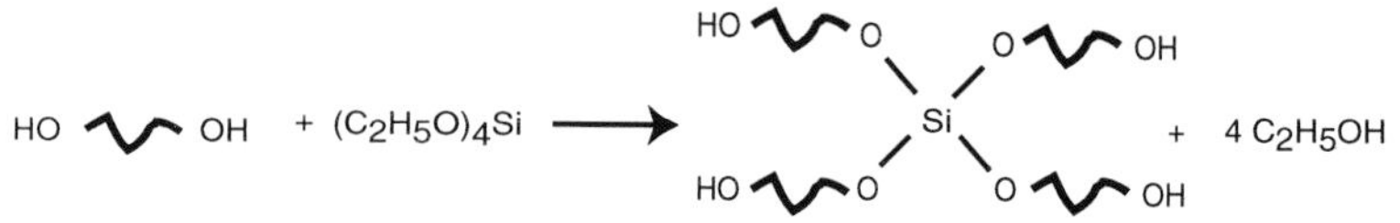

In this endlinking reaction, HO ∿ OH represents a hydroxyl-terminated poly(dimethylsiloxane) chain. The average molecular weight and its distribution for the precursor chains becomes the average molecular weight and its distribution for the network chains.

Figure 3. A typical end-linking scheme for preparing an elastomeric network of known structure.

Networks at Very High Deformations

Non-Gaussian Effects. As already described in Figure 2 (1-3), some (unfilled) networks show a large and rather abrupt increase in modulus at high elongations. This increase (49,50) is very important since it corresponds to a significant toughening of the elastomer. Its molecular origin, however, has been the source of considerable controversy (2,4,49,51-57). It had been widely attributed to the "limited extensibility" of the network chains (55), i. e., to an inadequacy in the Gaussian distribution function.

The issue has now been resolved (6,55,58-60), however, by the use of end-linked, non-crystallizable model PDMS networks. These networks have high extensibilities, presumably because of their very low incidence of dangling-chain network irregularities. They have particularly high extensibilities when they are prepared from a mixture of very short chains (molecular weights around a few hundred g mol^{-1}) with relatively long chains (around 18,000 g mol^{-1}), as further discussed below. Apparently the very short chains are important because of their limited extensibilities, and the relatively long chains because of their ability to retard the rupture process.

Comparisons of stress-strain measurements on such bimodal PDMS networks with those in crystallizable polymer networks such as natural rubber and *cis*-1,4-polybutadiene were carried out, particularly as a function of temperature and presence of a plasticizing diluent (55,61). The results showed that the anomalous upturn in modulus observed for crystallizable polymers such as natural rubber is largely if not entirely due to strain-induced crystallization.

Ultimate Properties. The ultimate properties of interest are the tensile strength, maximum extensibility, and toughess (energy to rupture), and all are affected by strain-induced crystallization (58). The higher the temperature, the lower the extent of crystallization and, correspondingly, the lower the ultimate properties. The effects of increase in swelling parallel those for increase in temperature, since diluent also suppresses network crystallization. For non-crystallizable networks, however, neither change is very important, as is illustrated by the results reported for PDMS networks (62).

In the case of such non-crystallizable, unfilled elastomers, the mechanism for network rupture has been elucidated to a great extent by studies of model networks similar to those already described. For example, values of the modulus of bimodal networks formed by end-linking mixtures of very short and relatively long chains as illustrated in Figure 4, were used to test the "weakest-link" theory (6) in which rupture

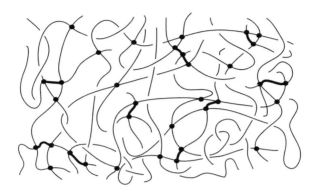

Figure 4. Sketch of a network having a bimodal distribution of network chain lengths. The very short and relatively long chains are arbitrarily shown by the thick and thin lines, respectively.

was thought to be initiated by the shortest chains (because of their very limited extensibility). It was observed that increasing the number of very short chains did *not* significantly decrease the ultimate properties. The reason (55) is the very non-affine nature of the deformation at such high elongations. The network simply reapportions the increasing strain among the polymer chains until no further reapportioning is possible. It is generally only at this point that chain scission begins, leading to rupture of the elastomer. The weakest-link theory implicitly assumes an affine deformation, which leads to the prediction that the elongation at which the modulus increases should be independent of the number of short chains in the network. This assumption is contradicted by relevant experimental results, which show very different behavior (55); the smaller the number of short chains, the easier the reapportioning and the higher the elongation required to bring about the upturn in modulus.

Multimodal Chain-Length Distributions

As already mentioned, there turns out to be an exciting bonus if one forms a multimodal distribution of network chain lengths by end linking a very large number of short chains into a long-chain network. The ultimate properties are then actually improved! Bimodal networks prepared by these end-linking techniques have very good ultimate properties, and there is currently much interest in preparing and characterizing such networks

(4,35,37,44,63-71), and developing theoretical interpretations for their properties (72-77). The types of improvements obtained are shown schematically in Figure 5. The

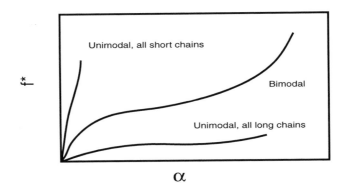

Figure 5. Typical plots of nominal stress against elongation for unimodal and bimodal networks obtained by end linking relatively long chains and very short chains. The area under each curve represents the rupture energy (a measure of the "toughness" of the elastomer).

results are represented in such a way that the area under a stress-strain isotherm corresponds to the energy required to rupture the network. If the network is all short chains it is brittle, which means that the maximum extensibility is very small. If the network is all long chains, the ultimate strength is very low. In neither case is the material a tough elastomer because the areas under the curves are relatively small. As can readily be seen from the figure, the bimodal networks are much improved elastomers in that they can have a high ultimate strength without the usual decrease in maximum extensibility.

A series of experiments were carried out in an attempt to determine if this reinforcing effect in bimodal PDMS networks could possibly be due to some intermolecular effect such as strain-induced crystallization. In the first such experiment, temperature was found to have little effect on the isotherms (5,47). This strongly argues against the presence of any crystallization or other type of intermolecular ordering. So also do the results of stress-temperature and birefringence-temperature measurements (47). In a final experiment, the short chains were pre-reacted in a two-step preparative technique so as possibly to segregate them in the network structure (45,61) as might occur in a network cross linked by an incompletely soluble peroxide. This had very little effect on elastomeric properties, again arguing against any type of intermolecular organization as the origin for the reinforcing effects. Apparently, the observed increases in modulus are due to the limited chain extensibility of the short chains, with the long chains serving to retard the rupture process. This can be thought of in terms of what executives like to call a "delegation of responsibilities".

There is an another advantage to such bimodality when the network can undergo strain-induced crystallization, the occurrence of which can provide an additional toughening effect (78). Decrease in temperature was found to increase the extent to which the values of the ultimate strength of at least some bimodal networks exceed those of the corresponding unimodal ones. This suggests that bimodality facilitates strain-induced crystallization.

In practical terms, the above results demonstrate that short chains of limited extensibility may be bonded into a long-chain network to improve its toughness. It is also possible to achieve the converse effect. Thus, bonding a small number of relatively long elastomeric chains into a relatively hard short-chain PDMS thermoset greatly improves its impact resistance (79).

Since dangling chains represent imperfections in a network structure, one would expect their presence to have a detrimental effect on the ultimate properties $(f/A^*)_r$ and α_r, of an elastomer. This expectation is confirmed by an extensive series of results obtained on PDMS networks which had been tetrafunctionally cross linked using a variety of techniques (80). The largest values of the ultimate strength $(f/A^*)_r$ were obtained for the networks prepared by selectively joining functional groups occurring either as chain ends or as side groups along the chains. This is to be expected, because of the relatively low incidence of dangling ends in such networks. Also as expected, the lowest values of the ultimate properties generally occurred for networks cured by radiation (UV light, high-energy electrons, and γ radiation) (80). The peroxide-cured networks were generally intermediate to these two extremes, with the ultimate properties presumably depending on whether or not the free-radicals generated by the peroxide are sufficiently reactive to cause some chain scission. Similar results were obtained for the maximum extensibility α_r (80). These results were supported by more definitive results obtained by investigation of a series of model networks prepared by end-linking vinyl-terminated PDMS chains (80).

Other Types of Deformation

Biaxial Extension. There are numerous other deformations of interest, including compression, biaxial extension, shear, and torsion (1,2). Some of these deformations are considerably more difficult to study experimentally than simple elongation and, unfortunately, have therefore not been as extensively investigated.

Measurements in biaxial extension are of particular importance since they are important in packaging applications. This deformation can be imposed by the direct stretching of a sample sheet in two perpendicular directions within its plane, by two independently-variable amounts. In the equi-biaxial case, the deformation is equivalent to compression. Such experimental results (81) have been successfully interpreted in terms of molecular theories (4,6).

Biaxial extension studies can also be carried out by the inflation of sheets of the elastomer (2). Upturns in the modulus (82) were seen to occur at high biaxial extensions, as expected.

<u>Shear</u>. Experimental results on natural rubber networks in shear (<u>83</u>) are not well accounted for by the simple molecular theory of rubberlike elasticity. The constrained-junction theory, however, was found to give excellent agreement with experiment (<u>6</u>). The upturns in modulus in shear (<u>84</u>) were found to be very similar to those obtained in elongation.

<u>Torsion</u>. Very little work has been done on elastomers in torsion. There are, however, some results on stress-strain behavior and network thermoelasticity (<u>2,85</u>). More results are presumably forthcoming, particularly on the unusual bimodal networks and on networks containing some of the unusual in-situ generated fillers described below.

<u>Swelling</u>. Most studies of networks in swelling equilibrium give values for the cross-link density or related quantities that are in satisfactory agreement with those obtained from mechanical property measurements (<u>1,2</u>).

A more interesting area involving some swollen networks or "gels" is their abrupt collapse (decrease in volume) upon relatively minor changes in temperature, pH, solvent composition, etc. (<u>4,6,86,87</u>). Although the collapse is quite slow in large, monolithic pieces of gel, it is rapid enough in fibers and films to make the phenomenon interesting with regard to the construction of switches and related devices.

Gels are also formed, of course, when elastomers are used to absorb liquids, for example in diapers and in attempts to control oil spills over bodies of water.

Filler-Reinforced Elastomers and Elastomer-Modified Ceramics

One class of multi-phase elastomers are those capable of undergoing strain-induced crystallization, as was mentioned above. In this case, the second phase is made up of the crystallites thus generated, which provide considerable reinforcement. Such reinforcement is only temporary, however, in that it may disappear upon removal of the strain, addition of a plasticizer, or increase in temperature. For this reason, many elastomers (particularly those which cannot undergo strain-induced crystallization) are generally compounded with a permanent reinforcing filler (<u>4,6,8,88-98</u>). The two most important examples are the addition of carbon black to natural rubber and to some synthetic elastomers (<u>90,92,99</u>), and the addition of silica to siloxane rubbers (<u>91</u>). In fact, the reinforcement of natural rubber and related materials is one of the most important processes in elastomer technology. It leads to increases in modulus at a given strain, and improvements of various technologically-important properties such as tear and abrasion resistance, resilience, extensibility, and tensile strength (<u>90,92,94,97,100-103</u>). There are also disadvantages, however, including increases in hysteresis (and thus of heat build-up), and compression set (permanent deformation).

There is an incredible amount of relevant experimental data available, with most of these data relating to reinforcement of natural rubber by carbon black (<u>92,94,100</u>). Recently, however, other polymers such as PDMS, and other fillers, such as precipitated silica, metallic particles, and even glassy polymers, have become of interest (<u>18,104-139</u>).

The most important unsolved problem in this area is the nature of the bonding between the filler particles and the polymer chains (<u>103</u>). The network chains may adsorb strongly onto the particle surfaces, which would increase the effective degree of cross linking. This effect will be especially strong if particles contain some reactive surface groups which may cross link (or end link) the polymer chains. Chemisorption, with permanent chemical bonding between filler particles and polymer chains, can be dominant, particularly if the filler is precipitated into the elastomer in-situ during curing (18,107,108,126). Another type of adsorption which can occur at a filler surface is physisorption, arising from long-range van der Waals forces between the surface and the polymer. Contrary to chemisorption, this physical adsorption does not severely restrict the movement of polymer chains relative to the filler surface when high stresses are applied. The available experimental data suggest that both chemisorption and physisorption contribute to reinforcement phenomena, and that the optimal degree of chemical bonding is quite low (of the order 0.2 bonding sites per nm^2 of filler surface) (97). Excessive covalent bonding, leading to immobilization of the polymer at the filler surface, is highly undesirable. A filler particle may thus be considered a cross link of very high functionality, but transient in that it can participate in molecular rearrangements under strain.

There are probably numerous other ways in which a filler changes the mechanical properties of an elastomer, some of admittedly minor consequence (8,103). For example, another factor involves changes in the distribution of end-to-end vectors of the chains due to the volume taken up by the filler (102,103,140). This effect is obviously closely related to the adsorption of polymer chains onto filler surfaces, but the surface also effectively segregates the molecules in its vicinity and reduces entanglements. Another important aspect of filler reinforcement arises from the fact that the particles influence not only an elastomer's static properties (such as the distribution of its end-to-end vectors), but also its dynamic properties (such as network chain mobility). More specifically, the presence of fillers reduces the segmental mobility of the adsorbed polymer chains to the extent that layers of elastomer close to the filler particles are frequently referred to as "bound rubber" (141-144).

As is obvious from the above comments, the mechanism of the reinforcement is only poorly understood. Some elucidation might be obtained by precipitating reinforcing fillers into network structures rather than blending badly agglomerated fillers into the polymers prior to their cross linking. This has, in fact, been done for a variety of fillers, for example silica by hydrolysis of organosilicates, titania from titanates, alumina from aluminates, etc. (4,6,106,108,138,139). A typical, and important, reaction is the acid- or base-catalyzed hydrolysis of tetraethylorthosilicate:

$$Si(OC_2H_5)_4 + 2H_2O \longrightarrow SiO_2 + 4C_2H_5OH \tag{3}$$

Reactions of this type are much used by the ceramists in the new sol-gel chemical route to high-performance ceramics (145-152). In the ceramics area, the advantages are the possibility of using low temperatures, the purity of the products, the control of ultrastructure (at the nanometer level), and the relative ease of forming ceramic alloys. In the elastomer reinforcement area, the advantages include the avoidance of the difficult, time-consuming, and energy-intensive process of blending agglomerated filler into high molecular weight and high-viscosity polymers, and the ease of obtaining extremely good dispersions.

In the simplest approach to obtaining elastomer reinforcement, some of the organosilicate material is absorbed into the cross-linked network, and the swollen sample placed into water containing the catalyst, typically a volatile base such as ammonia or ethylamine. Hydrolysis to form the desired silica-like particles proceeds rapidly at room temperature to yield the order of 50 wt % filler in less than an hour (4,6,106,108,152).

Impressive levels of reinforcement can be obtained by this in-situ technique (6,19,20). The modulus [f*] generally increases substantially, and some stress-strain isotherms show the upturns at high elongation that are the signature of good reinforcement. As generally occurs in filled elastomers, there can be considerable irreversibility in the isotherms, which is thought to be due to irrecoverable sliding of the chains over the surfaces of the filler particles.

If the hydrolyses in organosilicate-polymer systems are carried out with increased amounts of the silicate, bicontinuous phases can be obtained (with the silica and polymer phases interpenetrating one another) (18). At still-higher concentrations of the silicate, the silica generated becomes the continuous phase, with the polymer dispersed in it (6,153-167). The result is a polymer-modified ceramic, variously called an "ORMOCER" (153-155), "CERAMER" (156-158), or "POLYCERAM" (162-164). It is obviously of considerable importance to determine how the polymeric phase, often elastomeric, improves the mechanical properties of the ceramic in which it is dispersed.

Current Problems and Future Trends

There is a real need for more high-performance elastomers, which are materials that remain elastomeric to very low temperatures and are relatively stable at very high temperatures. Some phosphazene polymers, [–PRR'N–] (168-170), are in this category. These polymers have rather low glass transition temperatures in spite of the fact that the skeletal bonds of the chains are thought to have some double-bond character. There are thus a number of interesting problems related to the elastomeric behavior of these unusual semi-inorganic polymers. There is also increasing interest in the study of elastomers that also exhibit mesomorphic behavior (6).

A particularly challenging problem is the development of a more quantitative molecular understanding (171-174) of the effects of filler particles, in particular carbon black in natural rubber and silica in siloxane polymers (5,90,92,175,176). Such fillers provide tremendous reinforcement in elastomers in general, and how they do this is still only poorly comprehended. A related but even more complex problem involves much the same components, namely one that is organic and one that is inorganic. When one or both components are generated in-situ, however, there is an almost unlimited variety of structures and morphologies that can be generated (6). How physical properties such as elastomeric behavior depend on these variables is obviously a challenging but very important problem.

An example of an important future trend is the study of single polymer chains, particularly with regard to their stress-strain isotherms (177-187). Although such studies are obviously not relevant to the many unresolved issues that involve the interactions of chains within an elastomeric network, they are certainly of interest in their own right.

Literature Cited

(1)   Flory, P. J. "Principles of Polymer Chemistry"; Cornell University Press: Ithaca NY, 1953.
(2)   Treloar, L. R. G. " The Physics of Rubber Elasticity"; 3rd ed.; Clarendon Press: Oxford, 1975.
(3)   Mark, J. E. J. Chem. Educ. 1981, 58, 898.
(4)   Mark, J. E.; Erman, B. "Rubberlike Elasticity. A Molecular Primer"; Wiley-Interscience: New York, 1988.
(5)   Mark, J. E. In "Physical Properties of Polymers"; 2nd ed.; J. E. Mark, A. Eisenberg, W. W. Graessley, L. Mandelkern, E. T. Samulski, J. L. Koenig and G. D. Wignall, Ed.; American Chemical Society: Washington, DC, 1993; p 3.
(6)   Erman, B.; Mark, J. E. "Structures and Properties of Rubberlike Networks"; Oxford University Press: New York, 1997.
(7)   Mark, J. E.; Erman, B. In "Polymer Networks"; R. F. T. Stepto, Ed.; Blackie Academic, Chapman & Hall: Glasgow, 1998.
(8)   Mark, J. E.; Erman, B. In "Performance of Plastics"; W. Brostow, Ed.; Hanser: Cincinnati, 1999.
(9)   Mason, P. "Cauchu. The Weeping Wood"; Australian Broadcasting Commission: Sydney, 1979.
(10)   Morawetz, H. "Polymers: The Origins and Growth of a Science"; Wiley-Interscience: New York, 1985.
(11)   "Rubber Technology"; Third ed.; Morton, M., Ed.; Van Nostrand Reinhold: New York, 1987.
(12)   Erman, B.; Mark, J. E. Ann. Rev. Phys. Chem. 1989, 40, 351.
(13)   "Characterization of Highly Cross-Linked Polymers"; Labana, S. S.; Dickie, R. A., Ed.; American Chemical Society: Washington, DC, 1984.
(14)   "Elastomers and Rubber Compounding Materials"; Franta, I., Ed.; Elsevier: Amsterdam, 1989.
(15)   "Engineering with Rubber. How to Design Rubber Components"; Gent, A. N., Ed.; Hanser Publishers: New York, 1992.
(16)   Noda, I.; Dowrey, A. E.; Marcott, C. In "Fourier Transform Infrared Characterization of Polymers"; H. Ishida, Ed.; Plenum Press: New York, 1987.
(17)   Wignall, G. D. In "Physical Properties of Polymers Handbook"; J. E. Mark, Ed.; American Institute of Physics Press: Woodbury, NY, 1996; p 299.
(18)   Schaefer, D. W.; Mark, J. E.; McCarthy, D. W.; Jian, L.; Sun, C.-C.; Farago, B. In "Polymer-Based Molecular Composites"; D. W. Schaefer and J. E. Mark, Ed.; Materials Research Society: Pittsburgh, 1990; Vol. 171; p 57.
(19)   McCarthy, D. W.; Mark, J. E.; Schaefer, D. W. J. Polym. Sci., Polym. Phys. Ed. 1998, 36, 1167.
(20)   McCarthy, D. W.; Mark, J. E.; Clarson, S. J.; Schaefer, D. W. J. Polym. Sci., Polym. Phys. Ed. 1998, 36, 1191.
(21)   Breiner, J. M.; Mark, J. E. Polymer 1998, 39, 5483.
(22)   Mark, J. E. Rubber Chem. Technol 1975, 48, 495.
(23)   Ronca, G.; Allegra, G. J. Chem. Phys. 1975, 63, 4990.
(24)   Flory, P. J. Proc. R. Soc. London, A 1976, 351, 351.

(25)   Flory, P. J. Polymer 1979, 20, 1317.
(26)   Flory, P. J.; Erman, B. Macromolecules 1982, 15, 800.
(27)   Gottlieb, M.; Macosko, C. W.; Benjamin, G. S.; Meyers, K. O.; Merrill, E. W.
       Macromolecules 1981, 14, 1039.
(28)   Mark, J. E. Adv. Polym. Sci. 1982, 44, 1.
(29)   "Elastomers and Rubber Elasticity"; Mark, J. E.; Lal, J., Ed.; American Chemical
       Society: Washington, 1982; Vol. 193.
(30)   Queslel, J. P.; Mark, J. E. Adv. Polym. Sci. 1984, 65, 135.
(31)   Mark, J. E. Acc. Chem. Res. 1985, 18, 202.
(32)   Mark, J. E. Polym. J. 1985, 17, 265.
(33)   Mark, J. E. Brit. Polym. J. 1985, 17, 144.
(34)   Miller, D. R.; Macosko, C. W. J. Polym. Sci., Polym. Phys. Ed. 1987, 25, 2441.
(35)   Lanyo, L. C.; Kelley, F. N. Rubber Chem. Technol. 1987, 60, 78.
(36)   Mark, J. E. In "Frontiers of Macromolecular Science"; T. Saegusa, T.
       Higashimura and A. Abe, Ed.; Blackwell Scientific Publishers: Oxford, 1989; p
       289.
(37)   Smith, T. L.; Haidar, B.; Hedrick, J. L. Rubber Chem. Technol. 1990, 63, 256.
(38)   Mark, J. E. J. Inorg. Organomet. Polym. 1991, 1, 431.
(39)   Mark, J. E. Angew. Makromol. Chemie 1992, 202/203, 1.
(40)   Mark, J. E.; Eisenberg, A.; Graessley, W. W.; Mandelkern, L.; Samulski, E. T.;
       Koenig, J. L.; Wignall, G. D. "Physical Properties of Polymers"; 2nd ed.;
       American Chemical Society: Washington, DC, 1993.
(41)   Sharaf, M. A.; Mark, J. E.; Hosani, Z. Y. A. Eur. Polym. J. 1993, 29, 809.
(42)   Sharaf, M. A.; Mark, J. E. Makromol. Chemie 1994, 76, 13.
(43)   Mark, J. E. J. Inorg. Organomet. Polym. 1994, 4, 31.
(44)   Mark, J. E. Acc. Chem. Res. 1994, 27, 271.
(45)   Mark, J. E.; Andrady, A. L. Rubber Chem. Technol. 1981, 54, 366.
(46)   Llorente, M. A.; Andrady, A. L.; Mark, J. E. Coll. Polym. Sci. 1981, 259, 1056.
(47)   Zhang, Z.-M.; Mark, J. E. J. Polym. Sci., Polym. Phys. Ed. 1982, 20, 473.
(48)   Mark, J. E. In "Elastomers and Rubber Elasticity"; J. E. Mark and J. Lal, Ed.;
       American Chemical Society: Washington, DC, 1982.
(49)   Mullins, L. J. Appl. Polym. Sci. 1959, 2, 257.
(50)   Mark, J. E.; Kato, M.; Ko, J. H. J. Polym. Sci., Part C 1976, 54, 217.
(51)   Smith, K. J., Jr.; Greene, A.; Ciferri, A. Kolloid-Z. Z. Polym. 1964, 194, 49.
(52)   Morris, M. C. J. Appl. Polym. Sci. 1964, 8, 545.
(53)   Treloar, L. R. G. Rep. Prog. Phys. 1973, 36, 755.
(54)   Chan, B. L.; Elliott, D. J.; Holley, M.; Smith, J. F. J. Polym. Sci., Part C 1974, 48,
       61.
(55)   Andrady, A. L.; Llorente, M. A.; Mark, J. E. J. Chem. Phys. 1980, 72, 2282.
(56)   Doherty, W. O. S.; Lee, K. L.; Treloar, L. R. G. Br. Polym. J. 1980, 15, 19.
(57)   Furukawa, J.; Onouchi, Y.; Inagaki, S.; Okamoto, H. Polym. Bulletin 1981, 6, 381.
(58)   Su, T.-K.; Mark, J. E. Macromolecules 1977, 10, 120.
(59)   Chiu, D. S.; Su, T.-K.; Mark, J. E. Macromolecules 1977, 10, 1110.
(60)   Mark, J. E. Polym. Eng. Sci. 1979, 19, 409.

(61) Mark, J. E.; Eisenberg, A.; Graessley, W. W.; Mandelkern, L.; Koenig, J. L. "Physical Properties of Polymers"; 1 st ed.; American Chemical Society: Washington, DC, 1984.
(62) Chiu, D. S.; Mark, J. E. Coll. Polym. Sci. 1977, 225, 644.
(63) Mark, J. E. Makromol. Chemie, Suppl. 1979, 2, 87.
(64) Silva, L. K.; Mark, J. E.; Boerio, F. J. Makromol. Chemie 1991, 192, 499.
(65) Hanyu, A.; Stein, R. S. Macromol. Symp. 1991, 45, 189.
(66) Roland, C. M.; Buckley, G. S. Rubber Chem. Technol. 1991, 64, 74.
(67) Oikawa, H. Polymer 1992, 33, 1116.
(68) Hamurcu, E. E.; Baysal, B. M. Polymer 1993, 34, 5163.
(69) Subramanian, P. R.; Galiatsatos, V. Macromol. Symp. 1993, 76, 233.
(70) Sharaf, M. A.; Mark, J. E.; Al-Ghazal, A. A.-R. J. Appl. Polym. Sci. Symp. 1994, 55, 139.
(71) Besbes, S.; Bokobza, L.; Monnerie, L.; Bahar, I.; Erman, B. Macromolecules 1995, 28, 231.
(72) Erman, B.; Mark, J. E. J. Chem. Phys. 1988, 89, 3314.
(73) Termonia, Y. Macromolecules 1990, 23, 1481.
(74) Kloczkowski, A.; Mark, J. E.; Erman, B. Macromolecules 1991, 24, 3266.
(75) Sakrak, G.; Bahar, I.; Erman, B. Macromol. Theory Simul. 1994, 3, 151.
(76) Bahar, I.; Erman, B.; Bokobza, L.; Monnerie, L. Macromolecules 1995, 28, 225.
(77) Erman, B.; Mark, J. E. Macromolecules 1998, 31, 3099.
(78) Sun, C.-C.; Mark, J. E. J. Polym. Sci., Polym. Phys. Ed. 1987, 25, 2073.
(79) Tang, M.-Y.; Letton, A.; Mark, J. E. Coll. Polym. Sci. 1984, 262, 990.
(80) Andrady, A. L.; Llorente, M. A.; Sharaf, M. A.; Rahalkar, R. R.; Mark, J. E.; Sullivan, J. L.; Yu, C. U.; Falender, J. R. J. Appl. Polym. Sci. 1981, 26, 1829.
(81) Obata, Y.; Kawabata, S.; Kawai, H. J. Polym. Sci., Part A-2 1970, 8, 903.
(82) Xu, P.; Mark, J. E. J. Polym. Sci., Polym. Phys. Ed. 1991, 29, 355.
(83) Rivlin, R. S.; Saunders, D. W. Philos. Trans. R. Soc. London, A 1951, 243, 251.
(84) Wang, S.; Mark, J. E. J. Polym. Sci., Polym. Phys. Ed. 1992, 30, 801.
(85) Wen, J.; Mark, J. E. Polym. J. 1994, 26, 151.
(86) Tanaka, T. Phys. Rev. Lett. 1978, 40, 820.
(87) Tanaka, T. Sci. Am. 1981, 244, 124.
(88) Oberth, A. E. Rubber Chem. Technol. 1967, 40, 1337.
(89) Boonstra, B. B. In "Rubber Technology"; M. Morton, Ed.; Van Nostrand Reinhold: New York, 1973; p 51.
(90) Boonstra, B. B. Polymer 1979, 20, 691.
(91) Warrick, E. L.; Pierce, O. R.; Polmanteer, K. E.; Saam, J. C. Rubber Chem. Technol. 1979, 52, 437.
(92) Rigbi, Z. Adv. Polym. Sci. 1980, 36, 21.
(93) Queslel, J. P.; Mark, J. E. In "Encyclopedia of Polymer Science and Engineering, Second Edition"; R. A. Meyers, Ed.; Wiley-Interscience: New York, 1986; p 365.
(94) Donnet, J.-B.; Vidal, A. Adv. Polym. Sci. 1986, 76, 103.
(95) Ahmed, S.; Jones, F. R. J. Mater. Sci. 1990, 25, 4933.
(96) Enikolopyan, N. S.; Fridman, M. L.; Stalnova, I. O.; Popov, V. L. Adv. Polym. Sci. 1990, 96, 1.
(97) Edwards, D. C. J. Mater. Sci. 1990, 25, 4175.

(98)   Medalia, A. I.; Kraus, G. In "Science and Technology of Rubber"; 2nd ed.; J. E. Mark, B. Erman and F. R. Eirich, Ed.; Academic: New York, 1994; p 387.
(99)   Karasek, L.; Sumita, M. J. Mats. Sci. 1996, 31, 281.
(100)   Kraus, G. Adv. Polym. Sci. 1971, 8, 155.
(101)   "Reinforcement of Elastomers"; Kraus, G., Ed.; Interscience: New York, 1965.
(102)   Kloczkowski, A.; Sharaf, M. A.; Mark, J. E. Comput. Polym. Sci. 1993, 3, 39.
(103)   Kloczkowski, A.; Sharaf, M. A.; Mark, J. E. Chem. Eng. Sci. 1994, 49, 2889.
(104)   Matijevic, E.; Scheiner, P. J. Coll. Interfacial Sci. 1978, 63, 509.
(105)   Mark, J. E. Kautschuk + Gummi Kunstoffe 1989, 42, 191.
(106)   Mark, J. E. CHEMTECH 1989, 19, 230.
(107)   "Polymer-Based Molecular Composites"; Schaefer, D. W.; Mark, J. E., Ed.; Materials Research Society: Pittsburgh, 1990; Vol. 171.
(108)   Mark, J. E.; Schaefer, D. W. In "Polymer-Based Molecular Composites"; D. W. Schaefer and J. E. Mark, Ed.; Materials Research Society: Pittsburgh, 1990; Vol. 171; p 51.
(109)   Yasrebi, M.; Kim, G. H.; Gunnison, K. E.; Milius, D. L.; Sarikaya, M.; Aksay, I. A. In "Better Ceramics Through Chemistry IV"; B. J. J. Zelinski, C. J. Brinker, D. E. Clark and D. R. Ulrich, Ed.; Materials Research Society: Pittsburgh, 1990; Vol. 180; p 625.
(110)   Chung, Y. J.; Ting, S.-J.; Mackenzie, J. D. In "Better Ceramics Through Chemistry IV"; B. J. J. Zelinski, C. J. Brinker, D. E. Clark and D. R. Ulrich, Ed.; Materials Research Society: Pittsburgh, 1990; Vol. 180; p 981.
(111)   Saegusa, T.; Chujo, Y. J. Macromol. Sci. - Chem. 1990, A27, 1603.
(112)   Mauritz, K. A.; Jones, C. K. J. Appl. Polym. Sci. 1990, 40, 1401.
(113)   Mauritz, K. A.; Scheetz, R. W.; Pope, R. K.; Stefanithis, I. D.; Wilkes, G. L.; Huang, H.-H. Preprints, Div. Polym. Chem., Inc., Am. Chem. Soc. 1991, 32(3), 528.
(114)   Bianconi, P. A.; Lin, J.; Strzelecki, A. R. Nature 1991, 349, 315.
(115)   Okada, A.; Fukumori, K.; Usuki, A.; Kojima, Y.; Sato, N.; Kurauchi, T.; Kamigaito, O. Preprints, Div. Polym. Chem., Inc., Am. Chem. Soc. 1991, 32(3), 540.
(116)   Yano, K.; Usuki, A.; Okada, A.; Kurauchi, T.; Kamigaito, O. Preprints, Div. Polym. Chem., Inc., Am. Chem. Soc. 1991, 32(1), 65.
(117)   Calvert, P. In "U.S.-Japan Workshop on Smart/Intelligent Materials and Systems"; I. Ahmad, A. Crowson, C. A. Rogers and M. Aizawa, Ed.; Technomic Pub. Co.: Lancaster, 1991; p 162.
(118)   Calvert, P. In "Ultrastructure Processing of Advanced Materials"; D. R. Uhlmann and D. R. Ulrich, Ed.; Wiley: New York, 1992; p 149.
(119)   Mackenzie, J. D.; Chung, Y. J.; Hu, Y. J. Non.-Cryst. Solids 1992, 147&148, 271.
(120)   Hu, Y.; Mackenzie, J. D. J. Mat. Sci. 1992, 27, 4415.
(121)   Brennan, A. B.; Rodrigues, D. E.; Wang, B.; Wilkes, G. L. In "Chemical Processing of Advanced Materials"; L. L. Hench and J. K. West, Ed.; Wiley: New York, 1992; p 807.
(122)   Huang, H.; Glaser, R. H.; Brennan, A. B.; Rodigues, D.; Wilkes, G. L. In "Ultrastructure Processing of Advanced Materials"; D. R. Uhlmann and D. R. Ulrich, Ed.; Wiley & Sons: New York, 1992; p 425.

(123) Schmidt, H. In "Ultrastructure Processing of Advanced Materials"; D. R. Uhlmann and D. R. Ulrich, Ed.; Wiley: New York, 1992; p 409.

(124) Schmidt, H. K. In "Submicron Multiphase Materials"; R. H. Baney, L. R. Gilliom, S.-I. Hirano and H. K. Schmidt, Ed.; Materials Research Society: Pittsburgh, PA, 1992; Vol. 274; p 121.

(125) Ellsworth, M. W.; Novak, B. M. In "Submicron Multiphase Materials"; R. H. Baney, L. R. Gilliom, S.-I. Hirano and H. K. Schmidt, Ed.; Materials Research Society: Pittsburgh, 1992; Vol. 274; p 67.

(126) Mark, J. E.; Wang, S.; Xu, P.; Wen, J. In "Submicron Multiphase Materials"; R. H. Baney, L. R. Gilliom, S.-I. Hirano and H. K. Schmidt, Ed.; Materials Research Society: Pittsburgh, PA, 1992; Vol. 274; p 77.

(127) Sun, L.; Aklonis, J. J.; Salovey, R. Polym. Eng. Sci. 1993, 33, 1308.

(128) Matijevic, E. Chem. Mater. 1993, 5, 412.

(129) Novak, B. M. Adv. Mats. 1993, 5, 422.

(130) Mark, J. E.; Calvert, P. D. J. Mats. Sci., Part C 1994, 1, 159.

(131) Mark, J. E. In "Frontiers of Polymers and Advanced Materials"; P. N. Prasad, Ed.; Plenum: New York, 1994; p 403.

(132) "Hybrid Organic-Inorganic Composites"; Mark, J. E.; Lee, C. Y.-C.; Bianconi, P. A., Ed.; American Chemical Society: Washington, 1995; Vol. 585.

(133) Mark, J. E. In "Diversity into the Next Century"; R. J. Martinez, H. Arris, J. A. Emerson and G. Pike, Ed.; SAMPE: Covina, CA, 1995; Vol. 27.

(134) Schmidt, H. K. Macromol. Symp. 1996, 101, 333.

(135) Calvert, P. In "Biomimetic Materials Chemistry"; S. Mann, Ed.; VCH Publishers: New York, 1996; p 315.

(136) Giannelis, E. P. In "Biomimetic Materials Chemistry"; S. Mann, Ed.; VCH Publishers: New York, 1996; p 337.

(137) Wen, J.; Wilkes, G. L. In "Polymeric Materials Encyclopedia: Synthesis, Properties, and Applications"; J. C. Salamone, Ed.; CRC Press: Boca Raton, 1996.

(138) Mark, J. E. Hetero. Chem. Rev. 1996, 3, 307.

(139) Mark, J. E. Polym. Eng. Sci. 1996, 36, 2905.

(140) Yuan, Q. W.; Kloczkowski, A.; Mark, J. E.; Sharaf, M. A. J. Polym. Sci., Polym. Phys. Ed. 1996, 34, 1674.

(141) Litvinov, V. M.; Spiess, H. W. Macromol. Chem. 1992, 193, 1181.

(142) Meissner, B. J. Appl. Polym. Sci. 1993, 50, 285.

(143) Karasek, L.; Meissner, B. J. Appl. Polym. Sci. 1994, 52, 1925.

(144) Leblanc, J. J. Appl. Polym. Sci. 1997, 66, 2257.

(145) "Ultrastructure Processing of Ceramics, Glasses, and Composites"; Hench, L. L.; Ulrich, D. R., Ed.; Wiley: New York, 1984.

(146) "Ultrastructure Processing of Advanced Ceramics"; Mackenzie, J. D.; Ulrich, D. R., Ed.; Wiley: New York, 1988.

(147) Ulrich, D. R. J. Non-Cryst. Solids 1988, 100, 174.

(148) Ulrich, D. R. CHEMTECH 1988, 18, 242.

(149) Mackenzie, J. D. J. Non-Cryst. Solids 1988, 100, 162.

(150) Ulrich, D. R. J. Non-Cryst. Solids 1990, 121, 465.

(151) Brinker, C. J.; Scherer, G. W. "Sol-Gel Science"; Academic Press: New York, 1990.

(152) "Better Ceramics Through Chemistry VIII: Hybrid Materials"; Brinker, C. J.; Giannelis, E. P.; Laine, R. M.; Sanchez, C., Ed.; Materials Research Society: Warrendale, PA, 1998; Vol. 519.

(153) Schmidt, H. In "Inorganic and Organometallic Polymers"; M. Zeldin, K. J. Wynne and H. R. Allcock, Ed.; American Chemical Society: Washington, DC, 1988; p 333.

(154) Schmidt, H.; Wolter, H. J. Non-Cryst. Solids 1990, 121, 428.

(155) Nass, R.; Arpac, E.; Glaubitt, W.; Schmidt, H. J. Non-Cryst. Solids 1990, 121, 370.

(156) Wang, B.; Wilkes, G. L. J. Polym. Sci., Polym. Chem. Ed. 1991, 29, 905.

(157) Wilkes, G. L.; Huang, H.-H.; Glaser, R. H. In "Silicon-Based Polymer Science"; J. M. Zeigler and F. W. G. Fearon, Ed.; American Chemical Society: Washington, DC, 1990; Vol. 224; p 207.

(158) Brennan, A. B.; Wang, B.; Rodrigues, D. E.; Wilkes, G. L. J. Inorg. Organomet. Polym. 1991, 1, 167.

(159) Sobon, C. A.; Bowen, H. K.; Broad, A.; Calvert, P. D. J. Mat. Sci. Lett. 1987, 6, 901.

(160) Calvert, P.; Mann, S. J. Mat. Sci. 1988, 23, 3801.

(161) Azoz, A.; Calvert, P. D.; Kadim, M.; McCaffery, A. J.; Seddon, K. R. Nature 1990, 344, 49.

(162) Doyle, W. F.; Uhlmann, D. R. In "Ultrastructure Processing of Advanced Ceramics"; J. D. Mackenzie and D. R. Ulrich, Ed.; Wiley-Interscience: New York, 1988; p 795.

(163) Doyle, W. F.; Fabes, B. D.; Root, J. C.; Simmons, K. D.; Chiang, Y. M.; Uhlmann, D. R. In "Ultrastructure Processing of Advanced Ceramics"; J. D. Mackenzie and D. R. Ulrich, Ed.; Wiley-Interscience: New York, 1988; p 953.

(164) Boulton, J. M.; Fox, H. H.; Neilson, G. F.; Uhlmann, D. R. In "Better Ceramics Through Chemistry IV"; B. J. J. Zelinski, C. J. Brinker, D. E. Clark and D. R. Ulrich, Ed.; Materials Research Society: Pittsburgh, 1990; Vol. 180; p 773.

(165) Mark, J. E.; Sun, C.-C. Polym. Bulletin 1987, 18, 259.

(166) Ning, Y. P.; Zhao, M. X.; Mark, J. E. In "Frontiers of Polymer Research"; P. N. Prasad and J. K. Nigam, Ed.; Plenum: New York, 1991; p 479.

(167) Zhao, M. X.; Ning, Y. P.; Mark, J. E. In "Advanced Composite Materials"; M. D. Sacks, Ed.; American Ceramics Society: Westerville, OH, 1993; p 891.

(168) Mark, J. E.; Yu, C. U. J. Polym. Sci., Polym. Phys. Ed. 1977, 15, 371.

(169) Andrady, A. L.; Mark, J. E. Eur. Polym. J. 1981, 17, 323.

(170) Mark, J. E.; Allcock, H. R.; West, R. "Inorganic Polymers"; Prentice Hall: Englewood Cliffs, NJ, 1992.

(171) Heinrich, G.; Vilgis, T. A. Macromolecules 1993, 26, 1109.

(172) Witten, T. A.; Rubinstein, M.; Colby, R. H. J. Phys. II France 1993, 3, 367.

(173) Kluppel, M.; Heinrich, G. Rubber Chem. Technol. 1995, 68, 623.

(174) Kluppel, M.; Schuster, R. H.; Heinrich, G. Rubber Chem. Technol. 1997, 70, 243.

(175) Polmanteer, K. E.; Lentz, C. W. Rubber Chem. Technol. 1975, 48, 795.

(176) Kraus, G. Rubber Chem. Technol. 1978, 51, 297.

(177) Chu, S. Science 1991, 253, 861.
(178) Perkins, T. T.; Smith, D. E.; Chu, S. Science 1994, 264, 819.
(179) Perkins, T. T.; Quake, S. R.; Smith, D. E.; Chu, S. Science 1994, 264, 822.
(180) Perkins, T. T.; Smith, D. E.; Larson, R. G.; Chu, S. Science 1995, 268, 83.
(181) Smith, S. B.; Cui, Y.; Bustamante, C. Science 1996, 271, 795.
(182) Kellermayer, M. S. Z.; Smith, S. B.; Granzier, H. L.; Bustamante, C. Science 1997, 276, 1112.
(183) Rief, M.; Oesterhelt, F.; Heymann, B.; Gaub, H. E., Science 1997, 275, 1295.
(184) Oberhauser, A. F.; Marszalek, P. E.; Erickson, H. P.; Fernandez, J. M. Nature 1998, 393, 181.
(185) Li, H.; Rief, M.; Oesterhelt, F.; Gaub, H. E., Adv. Mater. 1998, 3, 316.
(186) Marszalek, P. E.; Oberhauser, A. F.; Pang, Y-P.; Fernandez, J. M. Nature 1998, 396, 661.
(187) Ortiz, C.; Hadziioannou, G. Macromolecules 1999, 32, 780.

THERMOPLASTIC ELASTOMERS AND THEIR APPLICATIONS

GEOFFREY HOLDEN

Holden Polymer Consulting, Incorporated, 1042 Willow Creek Road, A111 - 273, Prescott, AZ 86305. (520)-771-9938, Fax (520)-771-8389, gbholden@northlink.com

Introduction
Classification and Structure
Production
Structure / Property Relationships
Applications
Economic Aspects and Tradenames
Literature Cited

Introduction

The use of thermoplastic elastomers has significantly increased since they were first produced about thirty-five years ago. A recent article(1) estimates their worldwide annual consumption at about 1,000,000 metric tons/year in 1995 and this is expected to rise to about 1,400,000 metric tons/year in 2000. Several books(2-4) and articles (5,6) have covered this subject in detail. The first two books (2,3)concentrate mostly on the scientific aspects of these polymers while the other(4) concentrates on their end uses.

The properties of thermoplastic elastomers in relation to other polymers are summarized in Table I. This table classifies all polymers by two characteristics - how they are processed (as thermosets or as thermoplastics) and the physical properties (rigid, flexible or rubbery) of the final product. All commercial polymers used for molding, extrusion, etc., fit into one of the six resulting classifications - the thermoplastic elastomers are the newest. Their outstanding advantage can be summarized in a single phrase - they allow rubberlike articles to be produced using the rapid processing techniques developed by the thermoplastics industry. They have many physical properties of rubbers, e.g., softness, flexibility, and resilience. However they achieve their properties by a physical process (solidification) compared to the chemical process (cross-linking) in vulcanized rubbers.

In the terminology of the plastics industry, vulcanization is a thermosetting process. Like other thermosetting processes, it is slow, irreversible and takes place upon heating. With thermoplastic elastomers, on the other hand, the transition from a processable melt to a solid, rubberlike object is rapid, reversible and takes place upon cooling (Figure 1).

Table I. Comparison of thermoplastic elastomers with conventional plastics and rubbers

	Thermosetting	Thermoplastic
Rigid	Epoxies Phenol-Formaldehyde Urea-Formaldehyde	Polystyrene Polypropylene Poly(vinyl chloride) High Density Polyethylene
Flexible	Highly filled and/or highly vulcanized rubbers	Low Density Polyethylene EVA Plasticized PVC
Rubbery	Vulcanized Rubbers (NR, SBR, IR etc.)	**Thermoplastic Elastomers**

Thus thermoplastic elastomers can be processed using conventional plastics techniques, such as injection molding and extrusion; scrap can be recycled. Additionally, some thermoplastic elastomers are soluble in common solvents and so can be processed as solutions.

Figure 1. Polymer transitions

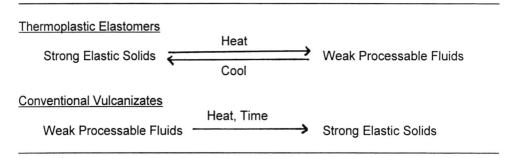

At higher temperatures, the properties of thermoplastic elastomers are usually not as good as those of the conventional vulcanized rubbers. Applications of thermoplastic elastomers are, therefore, in areas where these properties are less important, e.g., footwear, wire insulation, adhesives, polymer blending, and not in areas such as automobile tires.

Classification and structure

Thermoplastic elastomers can be divided into four basic types:

1. Styrenic Thermoplastic Elastomers
2. Multiblock Copolymers
3. Hard Polymer / Elastomer Combinations
4. Graft Copolymers

Almost all thermoplastic elastomers contain two or more distinct polymeric phases and their properties depend on these phases being finely and intimately mixed. In some cases the phases are not chemically bonded but in others they are linked together by block or graft copolymerization (Table II).

Table II. Thermoplastic elastomers based on block copolymers

Hard Segment, A	Soft or Elastomeric Segment, B	Structure[a]	Refs.
Polystyrene	Polybutadiene and polyisoprene	T, B	7-9,11
Polystyrene	Poly(ethylene-co-butylene) and Poly(ethylene-co-propylene)	T	8,9
Polystyrene and Substituted Polystyrenes	Polyisobutylene	T, B	10
Poly(α-methylstyrene)	Polybutadiene, polyisoprene	T	11
Poly(α-methylstyrene)	Poly(propylene sulfide)	T	11
Polystyrene	Polydimethylsiloxane	T, M	12
Poly(α-methylstyrene)	Polydimethylsiloxane	T	11,12
Polysulfone	Polydimethylsiloxane	M	13
Poly(silphenylene siloxane)	Polydimethylsiloxane	M	14
Polyurethane	Polyester and Polyether	M	15-17
Polyester	Polyether	M	18-19
Poly(β-hydroxyalkanoates)	Poly(β-hydroxyalkanoates)	M	20
Polyamide	Polyester and Polyether	M	21,22
Polycarbonate	Polydimethylsiloxane	M	23-25
Polycarbonate	Polyether	M	26,27
Polyetherimide	Polydimethylsiloxane	M	28
Polymethyl methacrylate	Poly(alkyl acrylates)	T, B	29
Polyurethane	Poly(diacetylenes)	M	30
Polyethylene	Poly(α-olefins)	M	31,32
Polyethylene	Poly(ethylene-co-butylene) and Poly(ethylene-co-propylene)	T	11,31
Polypropylene(isotactic)	Poly(α-olefins)	M*	31
Polypropylene(isotactic)	Polypropylene(atactic)	M*	31,32

a. T = triblock, A-B-A,   B = Branched, $(A-B)_n x$,   M = Multiblock, A-B-A-B-A-.........
M* = Mixed Structures, including multiblock,

At least one elastomeric phase and one hard phase must be present, and the hard phase (or phases) must become soft and fluid at higher temperatures so that the material as a whole can flow as a thermoplastic.

Styrenic thermoplastic elastomers. These are based on simple molecules such as an A-B-A block copolymer, where A is a polystyrene and B an elastomer segment. If the elastomer is the main constituent, the polymers should have a morphology similar to that shown in Figure 2. Here, the polystyrene end segments form separate spherical regions, i.e., domains, dispersed in a continuous elastomer phase. Most of the polymer molecules have their polystyrene end segments in different domains. At room temperature, these polystyrene domains are hard and act as physical cross-links, tying the elastomeric mid-segments together in a three-dimensional network. In some ways, this is similar to the network formed by vulcanizing conventional rubbers using sulfur cross-links. The difference is that in thermoplastic elastomers, the domains lose their strength when the material is heated or dissolved in solvents. This allows the polymer or its solution to flow. When the material is cooled down or the solvent is evaporated, the domains harden and the network regains its original integrity.

Figure 2. Morphology of styrenic block copolymers

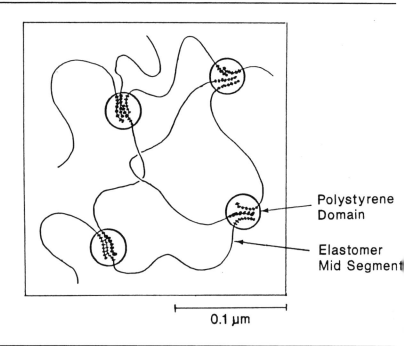

Polystyrene
Domain

Elastomer
Mid Segment

├───────────────┤
0.1 μm

Analogous block copolymers with only one hard segment (e.g., A-B or B-A-B) have quite different properties. The elastomer phase cannot form a continuous interlinked network since only one end of each elastomer segment is attached to the hard domains. These polymers are not thermoplastic elastomers, but are weaker materials similar to unvulcanized synthetic rubbers(7).

In commercial applications, three elastomeric mid-segments have been used for many years - polybutadiene, polyisoprene and poly(ethylene-butylene). The corresponding block copolymers are referred to as S-B-S, S-I-S and S-EB-S. Later, polymers with poly(ethylene-propylene) mid-segments (S-EP-S) were introduced. A more recent development, although not yet commercialized, is styrenic block copolymers with an isobutylene mid-segment (S-iB-S) (10). These can also be produced with substituted polystyrene end segments

Multiblock copolymers. The multiblock copolymers have structures that can be written as A-B-A-B-A-B-A-B-.... or $(A-B)_n$. For those of commercial importance, the hard (A) segments are crystalline thermoplastics while the softer, elastomeric (B) segments are amorphous. In best known types, the hard segments are thermoplastic polyurethanes, thermoplastic polyesters or thermoplastic polyamides and the soft segments are either polyesters or polyethers. Similar materials have been recently introduced in which the hard segments are polyethylene and the soft segments are either homopolymers or copolymers of α-olefins such as 1-butene, 1-hexene and 1-octene . The morphology of these $(A-B)_n$ multiblock copolymers is shown diagrammatically in Figure 3.

Figure 3. Morphology of multiblock copolymers with crystalline hard segments

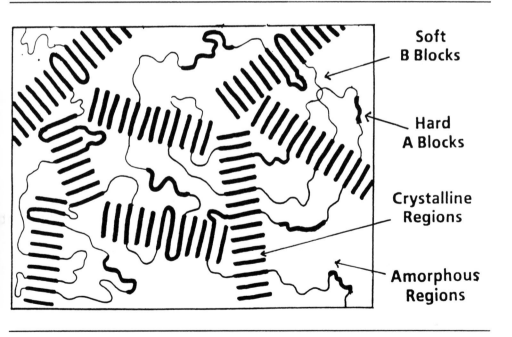

Soft
B Blocks

Hard
A Blocks

Crystalline
Regions

Amorphous
Regions

This structure has some similarities to that of a poly(styrene-b-elastomer-b-styrene) equivalents (Figure 2) and also some important differences. First, the hard domains are much more interconnected; Secondly, they are crystalline. Thirdly each long $(A-B)_n$ molecule may run through several hard and soft regions.

Hard polymer / elastomer combinations. Some thermoplastic elastomers are not block copolymers, but instead are fine dispersions of a hard thermoplastic polymer and an elastomer. Some are simple blends while others are produced by dynamic vulcanization (see later). A list of the various polymers used to produce thermoplastic elastomers based hard polymer / elastomer combinations of all types is given in Table III.

Table III. Thermoplastic elastomers based on hard polymer / elastomer combinations.

Hard Polymer	Soft or Elastomeric Polymer	Structure[a]	Refs.
Polypropylene	EPR or EPDM	B	31-34
Polypropylene	EPDM	DV	33,35,36
Polypropylene	Poly(propylene / 1-hexene)	B	33
Polypropylene	Poly(ethylene / vinyl acetate)	B	33
Polypropylene	Butyl Rubber	DV	35,37
Polypropylene	Natural Rubber	DV	35,38
Nylon	Nitrile Rubber	DV	35
Polypropylene	Nitrile Rubber	DV	35,36
PVC	Nitrile Rubber + DOP[b]	B, DV	39-41
Halogenated Polyolefin	Ethylene Interpolymer	B	41,42
Polyester	EPDM	B, DV	33
Polystyrene	S-B-S + Oil	B	43
Polypropylene	S-EB-S + Oil	B	43

a. B = Simple Blend,   DV = Dynamic Vulcanizate
b. DOP = Dioctyl phthalate. Other plasticizers can also be used.

The two materials usually form interdispersed co-continuous phases with a final morphology similar to that shown in Figure 4.

Polypropylene is often chosen as the hard thermoplastic because it is low priced, solvent resistant and has a high crystal melting point (165°C). Combinations with ethylene-propylene-diene monomer (EPDM) or ethylene-propylene copolymer (EPR) are the most important commercial products based on polypropylene(31-36); other elastomers that can be used include nitrile(35,36), butyl(37) and natural(38) rubbers. Softer, more impact resistant materials can be produced by using propylene copolymers as the hard phase(31,33). Halogen-containing polyolefins (41) are another option. Two examples are blends of PVC with nitrile rubber(39-41) and blends of halogenated polyolefins with ethylene inter-polymers. Mixtures of the last two polymers are claimed to give a single phase system.

In these blends dispersion of the two phases is most often achieved by intensive mechanical mixing but in the polypropylene / EPR combinations, polymerizing the finely dispensed elastomer phase simultaneously with the hard polypropylene is possible(31-33).

Figure 4. Morphology of simple blends based on hard polymer / elastomer combinations.

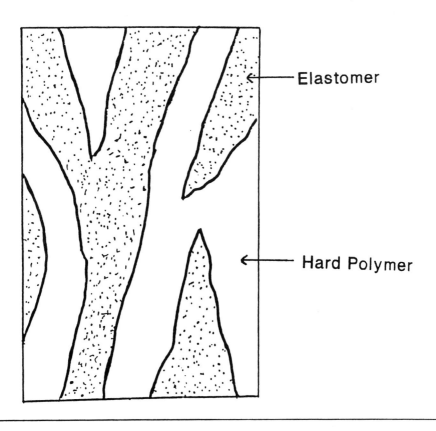

Elastomer

Hard Polymer

Sometimes, the elastomer phase is deliberately cross-linked during the intensive mechanical mixing. This is described as "dynamic vulcanization"(35,36). It produces a finely dispersed, <u>discontinuous</u>, cross-linked elastomer phase (see Figure 5). The products are called dynamic vulcanizates or thermoplastic vulcanizates.

This process is more complex than simple mixing, but the products have two important advantages. First, the cross-linked elastomer phase is insoluble and so oil and solvent resistance is improved. Secondly, cross-linking reduces or eliminates the flow of this phase at high temperatures and/or under stress. This improves resistance to compression set.

Figure 5. Morphology of dynamic vulcanizates

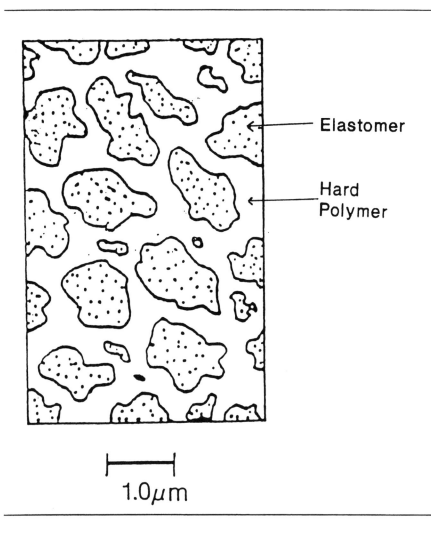

Elastomer

Hard
Polymer

|←——→|
1.0µm

<u>Graft copolymers.</u> Thermoplastic elastomers have also been produced from graft copolymers. A list of the various polymers used to produce thermoplastic elastomers based graft copolymers of all types is given in Table IV. Graft copolymers may be represented as

B-B-B-B-B-B ... B
|
$(A)_n$

This represents a polymer where each elastomeric B chain has (on average) n random grafts of hard A blocks. B chains that do not have at least two A blocks grafted onto them will not be elastically effective, because they cannot form a continuous interlinked network similar to that shown in Figure 2. To ensure that almost all the B chains have at least two A blocks grafted onto them, n should be greater than two, perhaps as high as ten (10).

Although much effort has been expended in research on thermoplastic elastomers based on graft copolymers, they have not become commercially important.

Table IV. Thermoplastic elastomers based on graft copolymers[a]

Hard Pendant Segment, A	Soft or Elastomeric Backbone Segment, B	Refs.
Polypivalolactone	Poly(ethylene-co-propylene)	31
Polystyrene & Poly (α-methyl styrene)	Polybutadiene,	44,45
Polyindene	Polybutadiene,	46
Polystyrene & Poly (α-methyl styrene)	Poly(ethylene-co-propylene)	47-49
Polyindene	Poly(ethylene-co-propylene)	46
Polystyrene & Poly (α-methyl styrene)	Polyisobutylene	50-52
Polyindene	Polyisobutylene	46,51
Polyacenapthylene	Polyisobutylene	51
Poly (para chlorostyrene)	Polyisobutylene	53
Polystyrene	Chlorosulfonated polyethylene	54
Poly(α-methyl styrene)	Polychloroprene	52
Polystyrene	Poly(butyl or ethyl-co-butyl) acrylates	55
Polymethylmethacrylate	Polybutylacrylate	56

a) For more detailed information, see Chapters 5, 13 and 14 of Reference 3

Production

As noted above, many copolymers and polymer combinations can give thermoplastic elastomers. This section covers the production of only the most significant.

Styrenic thermoplastic elastomers. The block copolymers on which these materials are based are made by anionic polymerization(57-60). In principle, this is a very simple system in which the polymer segments are produced sequentially from the monomers. The first step in the polymerization is the reaction of an alkyl-lithium initiator (R^-Li^+) with styrene monomer:

$$R^-Li^+ + nS \rightarrow R\text{-}(S)_n^-\, Li^+$$

For simplicity, we denote the product as S^-Li^+. It is called a "living polymer" because it can initiate further polymerization. If a second monomer, such as butadiene, is added:

$$S^-Li^+ + mB \quad \rightarrow \quad S\text{-}(B)_m^- \ Li^+$$

We denote this product as S-B $^-$Li$^+$. It is also a "living polymer" and by repeating these steps, block copolymers with multiple alternating blocks (S-B-S-B-S....) can be produced. In practice there are no apparent advantages in going beyond triblocks (i.e., S-B-S). Another variation is to use a coupling reaction to make linear or branched structures such as (S-B)$_n$x (where x represents an n-functional junction point). A typical example is:

$$2 \ S\text{-}B^- \ Li^+ + X\text{-}R\text{-}X \quad \rightarrow \quad S\text{-}B\text{-}R\text{-}B\text{-}S \ + 2LiX$$

Many coupling agents have been described, including esters, organo-halogens and silicon halides(59,63). The example above shows the reaction of a difunctional coupling agent, but those of higher functionality (for example SiCl$_4$) can also be used. These give branc/hed or star-shaped molecules such as (S-B)$_n$x.

The third method of producing these block copolymers uses multifunctional initiation(60,62,63). In this method a multifunctional initiator ($^+$Li $^-$R$^-$ Li $^+$) is first reacted with the diene (in this case, butadiene).

$$2nB + {}^+Li \ ^-R^- \ Li^+ \quad \rightarrow \quad {}^+Li \ ^-(B)_n\text{-}R\text{-}(B)_n^- \ Li^+$$

We denote this product as $^+$Li $^-$B$^-$ Li$^+$. The final two steps are similar to the corresponding steps in the sequential polymerization described above. When the reaction to produce the $^+$Li $^-$B$^-$ Li$^+$ is completed, styrene monomer is added and it in turn initiates its polymerization onto the "living" chain ends to give $^+$Li $^-$S-B-S$^-$Li$^+$. A protonating species is then added to stop the reaction and give final product, S-B-S. This example shows the use of a difunctional initiator. There is no reason in principle why initiators of higher functionality could not be used but none appears to have been reported in the literature.

All these reactions take place only in the absence of terminating agents such as water, oxygen or CO_2; thus they are usually carried out under nitrogen and in an inert hydrocarbon solvent. These conditions produce polymers with narrow molecular weight distributions and precise molecular weights.

Only three common monomers - styrene, butadiene and isoprene - are easily polymerized anionically and so only S-B-S and S-I-S block copolymers are directly produced on a commercial scale. In both cases polymerization of the elastomer segments in a non-polar solvent predominantly gives the 1,4 polymeric structures:

$$\{CH_2\text{-}CH=CH\text{-}CH_2\}_n\text{-} \qquad\qquad \{CH_2\text{-}C=CH\text{-}CH_2\}_n\text{-}$$
$$\qquad\qquad\qquad\qquad\qquad\qquad\qquad\qquad |$$
$$\qquad\qquad\qquad\qquad\qquad\qquad\qquad\qquad CH_3$$

      Polybutadiene   Polyisoprene

Both these polymers contain one double bond per molecule of the original monomer. These bonds are quite reactive and limit the stability of the product. More stable analogues can be produced from S-B-S polymers in which the polybutadiene mid-segments are polymerized in relatively polar solvents. These conditions produce a random copolymer of the 1,4 and 1,2 isomers. After hydrogenation this gives a saturated elastomer that can be considered a copolymer of ethylene and butylene (EB).

$$
\begin{array}{ccccc}
1,4 & 1,2 & & E & B \\
| & | & H_2 & | & \\
--CH_2\text{-}CH{=}CH\text{-}CH_2\text{-}CH_2\text{-}CH-- & & \rightarrow & --CH_2\text{-}CH_2\text{-}CH_2\text{-}CH_2\text{-}CH_2\text{-}CH-- \\
& |\quad | & & |\quad | \\
& CH & & |\quad C_2H_5 \\
& \| & & | \\
& CH_2 & & |
\end{array}
$$

Polybutadiene   Poly(ethylene-butylene)

S-EP-S block copolymers can be produced by hydrogenating S-I-S precursors.

Similar block copolymers with polyisobutylene mid-segments (e. g., S-iB-S) are made by carbocationic polymerization(10,66). This is a more complex system than the anionic system described above. The initiators have functionalities of two or more. They have the general formula $(X\text{-}R)_n x$ (where X-R represents a hydrocarbon moiety with a functional group X and x represents an n-functional junction point). X can be a chlorine, hydroxyl or methoxy group. Polymerization is carried out at low temperatures (about -80°C) in a moderately polar solvent and in the presence of a co-initiator ($TiCl_4$ or BCl_3). As in anionic polymerization, the polymer segments are produced sequentially from the monomers. Thus, an S-iB-S block copolymer would be produced in two stages:

$$X\text{-}R\text{-}X + 2n(iB) \rightarrow {}^+(iB)_n\text{-}R\text{-}(iB)_n{}^+$$

The product , which we can denote as ${}^+iB{}^+$ is a difunctional living polymer. It can initiate further polymerization, so if a second monomer, such as styrene, is added.

$$ {}^+iB{}^+ + 2m(S) \rightarrow {}^+(S)_m\text{-}iB\text{-}(S)_m{}^+ $$

After termination, this gives the block copolymer S-iB-S. Polyisobutylene is the only elastomeric mid-segment than can be produced by carbocationic polymerization. There are many aromatic polymers (mostly substituted polystyrenes) that can form the end-segments(10).

Multiblock cpolymers. The thermoplastic elastomers based on polyurethanes, polyesters and polyamides are produced from pre-polymers by condensation reactions. For those based on polyurethanes, three starting materials are used

1. A long chain diol, also called a polyglycol ($HO\text{-}R_L\text{-}OH$)
2. A short chain diol, also called a chain extender ($HO\text{-}R_S\text{-}OH$)
3. A diisocyanate ($OCN\text{-}R^*\text{-}NCO$) that can react with the hydroxyl groups in the diols to give a polyurethane,

The basic reaction can be written:

$$\wedge\!\!\vee\text{-NCO} + \text{HO-}\wedge\!\!\vee \quad \rightarrow \quad \wedge\!\!\vee\text{-NHCOO-}\wedge\!\!\vee$$

In the first stage of polymerization, an excess of the diisocyanate is reacted with the long chain diol. This gives a prepolymer terminated with the reactive isocyanate groups,

$$\text{OCN-R*-NCO} + \text{HO-R}_L\text{-OH} \rightarrow \text{OCN-(R*-}\underline{\text{U}}\text{-R}_L\text{-}\underline{\text{U})}_n\text{-R*-NCO}$$

Prepolymer

where $\underline{\text{U}}$ represents the urethane linking group, -NHCOO-. We can denote the prepolymer as OCN-Prepoly-NCO. It will further react with the short chain diol and more diisocyanate:

$$\text{OCN-Prepoly-NCO} + \text{HO-R}_S\text{-OH} + \text{OCN-R*-NCO} \rightarrow$$

$$\text{OCN-[Prepoly-}\underline{\text{U}}\text{-(R*-}\underline{\text{U}}\text{-R}_S\text{-}\underline{\text{U})}_n\text{-]}_m\text{-R*-NCO}$$

The final product is an alternating block copolymer with two types of segments:

1. Those formed in the first stage. These are based on the prepolymer. They are alternating copolymers of the long chain diols and the diisocyanate.

2. Those formed in the second stage. These are alternating copolymers of the short chain diols and the diisocyanate.

The long chain diols have a broad molecular weight distribution. Thus the prepolymers formed from them and the diisocyanate monomers do not have a regular repeating structure and are amorphous. Typical glass transition temperatures of the long chain diols are in the range of -45°C to -100°C (16), so at room temperatures these prepolymers are elastomeric. They form the soft elastomeric phase in the final polymer. In contrast, the short chain diols are single molecular species (e. g., 1,4-butanediol or ethylene glycol). Thus the copolymers formed from them and the diisocyanate or diacid monomers do have a regular repeating structure and so are crystalline. Typical crystallization temperatures of these segments are above 150°C (16), and so at room temperatures they are hard. They form the hard phase in the final polymer.
MDI (Diphenylmethane 4,4'-diisocyanate) and TDI (2,4 Toluene-diisocyanate) are the most common diisocyanates used to produce polyurethane thermoplastic elastomers. The long chain diols are usually polyesters (e. g., poly(ethylene adipate) glycol) or polyethers (e. g., poly(oxytetramethylene) glycol). Polycaprolactone glycol is used in premium products.
The various possible combinations of all three starting materials (diisocyanates, long chain diols and short chain diols) give a very wide variety of commercial thermoplastic

polyurethanes (16). In contrast, although thermoplastic polyesters are produced in a similar way (with diacids or diesters replacing diisocyanates) only three starting materials are used commercially(18). These are:

Poly(oxytetramethylene) glycol (the long chain diol)
1,4-butanediol (the short chain diol)
Terephthalic acid (the diacid) or its methyl diester

There are two ways to produce polyamide thermoplastic elastomers (21). The first is based on the reaction of a carboxylic acid with an isocyanate to give an amide:

$$\wedge\!\!\vee\!\!-NCO + HOOC\!\!-\!\!\wedge\!\!\vee \quad \rightarrow \quad \wedge\!\!\vee\!\!-NHOOC\!\!-\!\!\wedge\!\!\vee \quad + CO_2$$

The reaction scheme is similar to that shown above for the production of thermoplastic polyurethane and polyester elastomers. Again, the product is a copolymer with alternating segments.

In the second method of producing polyamide thermoplastic elastomers, a polyamide terminated by carboxylic acid groups, HOOC-PA-COOH (or the corresponding ester, ROOC-PA-COOR) reacts with a long chain diol:

$$HOOC\text{-}PA\text{-}COOH + HO\text{-}R_L\text{-}OH \quad \rightarrow \quad HO\text{-}(R_L\text{-}\underline{E}\text{-}PA\text{-}\underline{E})_n\text{-}R_L\text{-}OH + 2nH_2O$$

$$ROOC\text{-}PA\text{-}COOR + HO\text{-}R_L\text{-}OH \quad \rightarrow \quad HO\text{-}(R_L\text{-}\underline{E}\text{-}PA\text{-}\underline{E})_n\text{-}R_L\text{-}OH + 2nROH$$

where $\underline{E}$ represents an ester link.

Essentially, this amounts to preparing an alternating block copolymer from two prepolymers, one (the polyamide) crystalline, the other (the long chain diol) amorphous.

The block copolymers of ethylene with α-olefins are produced using metallocene catalysts (31-33). The α-olefins are typically 1-butene, 1-hexene or 1-octene. These copolymerize with ethylene to give segments with pendant groups, usually arranged atactically. Because of their random and atactic structures, these segments cannot crystallize. Instead, they are amorphous materials with low glass transition temperatures and so are soft and rubberlike at room temperature. They form the soft phase. The remainder of the polymer is polyethylene. Except for a very few side groups, this has a linear, symmetrical structure and therefore does not exhibit tacticity. Thus, the long polyethylene segments in the polymer chain cannot have significant irregularities and so can crystallize. They form the hard phase.

In all these multiblock $(A\text{-}B)_n$ polymers, both the number of segments and their individual molecular weights have a very broad distribution, in contrast to the simple A-B-A triblocks in the styrenic thermoplastic elastomers.

Hard polymer / elastomer combinations. There are two types of these materials - simple blends of the hard polymer and the elastomer and the dynamically vulcanized products in which the elastomer is cross-linked during the mixing process. Both the hard polymers and the elastomers used to make these products can be obtained "off the shelf". Thus an almost unlimited range of combinations can be investigated quickly and easily. Similarly, commercial products can be made without the very high capital investment required to produce novel polymers.

To produce simple unvulcanized blends, the two polymers are mixed on high shear compounding equipment. For the dynamically vulcanized versions, vulcanizing agents must be added and the temperature controlled so as to cross link the rubber particles during mixing. In both cases, only fine dispersions will produce optimum properties. A good match of the viscosities of the two polymers will aid the production of a fine dispersion, as will a match in solubility parameters. If the two polymers have very different solubility parameters (e.g., one is polar while the other is not), a coarse dispersion with poor adhesion between the phases can result. This can often be avoided by using block or graft copolymers as compatibilizing agents(33,35).

Graft Copolymers. Graft Copolymers are typically produced from elastomers with active sites (e. g., EPDM or halobutyl rubbers). These sites can be coupled to small blocks of the hard phase polymer or used to initiate further polymerization of the monomer that will form the hard phase polymer.

Structure / property relationships

With such a variety of materials, it is to be expected that the properties of thermoplastic elastomers cover an exceptionally wide range. Some are very soft and rubbery where others are hard and tough, and in fact approach the ill-defined interface between elastomers and flexible thermoplastics.

Since most thermoplastic elastomers are phase separated systems, they show many of the characteristics of the individual polymers that constitute the phases. For example, each phase has its own glass transition temperature (T_g), (or crystal melting point (T_m), if it is crystalline). These, in turn, determine the temperatures at which a particular thermoplastic elastomer goes through transitions in its physical properties. Thus, when the modulus of a thermoplastic elastomer is measured over a range of temperatures, there are three distinct regions (see Figure 6).

At very low temperatures, both phases are hard and so the material is stiff and brittle. At a somewhat higher temperature the elastomer phase becomes soft and the thermoplastic elastomer now resembles a conventional vulcanizate. As the temperature is further increased, the modulus stays relatively constant (a region often described as the "rubbery plateau") until finally the hard phase softens. At this point, the thermoplastic elastomer becomes fluid.

Figure 6. Stiffness of typical thermoplastic elastomers at various temperatures

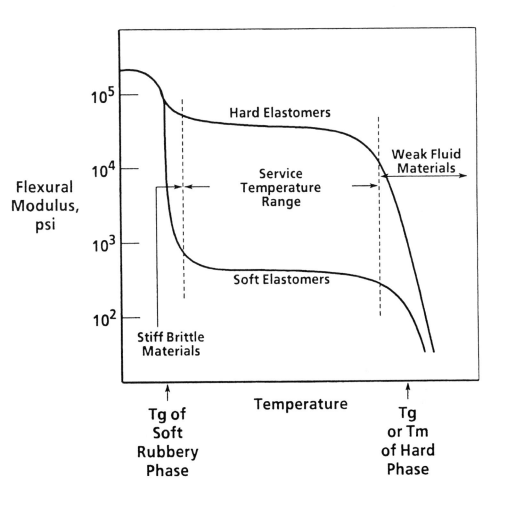

Thus, thermoplastic elastomers have two service temperatures. The lower service temperature depends on the T_g of the elastomer phase while the upper service temperature depends on the T_g or T_m of the hard phase. Values of T_g and T_m for the various phases in some commercially important thermoplastic elastomers are given in Table V.

Table V - Glass Transition and Crystal Melting Temperatures [a]

Thermoplastic Elastomer Type	Soft, Rubbery Phase T_g (^oC)	Hard Phase T_g or T_m (^oC)
Polystyrene / Elastomer Block Copolymers		
S-B-S	-90	95(T_g)
S-I-S	-60	95(T_g)
S-EB-S	-60	95(T_g) &165(T_m)[b]
S-iB-S	-60	95 - 240(T_g)[c]
Multi-block copolymers		
Polyurethane / Elastomer Block Copolymers	-40 to -60[d]	190(T_m)
Polyester / Elastomer Block Copolymers	-40	185 to 220(T_m)
Polyamide/Elastomer Block Copolymers	-40 to -60[d]	220 to 275(T_m)
Polyethylene / Poly(α-olefin) Block Copolymers	-50	70(T_m) [e]
Hard Polymer / Elastomer Combinations	-	
Polypropylene/EPDM or EPR combinations	-50	165(T_m)
Polypropylene/Nitrile Rubber combinations	-40	165(T_m)
PVC/Nitrile Rubber/DOP combinations	-30	80(T_g)

----

a.   Measured by Differential Scanning Calorimetry
b.   In compounds containing polypropylene
c.   The higher values are for substituted polystyrenes and polyaromatics (see Table 3.3)
d.   The values are for polyethers and polyesters respectively
e.   This low value for T_g is the result of the short length of the polyethylene segments.

As noted above, many different polymers are used to make the hard and soft phases in all these types of thermoplastic elastomers. Their influence on some properties of the products can be summarized as follows:

Hard Phase   The choice of polymer in the hard phase strongly influences the oil and solvent resistance of the thermoplastic elastomers. Even if the elastomer phase is resistant to a particular oil or solvent, if this oil or solvent swells the hard phase, all the useful physical properties of the thermoplastic elastomer will be lost. In most commercial thermoplastic elastomers, this hard phase is crystalline and so resistant to oils and solvents. Styrenic thermoplastic elastomers are an exception. As pure polymers, they have poor oil and solvent resistance (although this can be improved by compounding - see later). However, this gives them the advantage that they can be applied from solution.

<u>Soft Elastomer Phase</u>   In the styrenic thermoplastic elastomers, analogous S-B-S, S-I-S and S-EB-S polymers have somewhat different properties. S-B-S polymers are lowest in cost, S-I-S equivalents are the softest while the S-EB-S polymers are the most stable but also the highest priced. In the thermoplastic elastomers with crystalline hard segments, those with polyester-based elastomer segments are tougher and have better resistance to oils and solvents. The polyether-based materials are more flexible at low temperatures and show better hydrolytic stability.   In the hard polymer/ elastomer combinations, resistance to oil and solvents and to compression set are dramatically improved if the elastomer phase is dynamically vulcanized. Oil and solvent resistance can be still further improved if the elastomer is a polar material such as nitrile rubber.

<u>Hard/Soft Phase Ratio</u>   The hardness of these materials depends on the ratio of the volume of the hard phase to that of the softer elastomer phase. In the styrenic thermo-plastic elastomers, this ratio can be varied within quite wide limits. Thus, in an S-B-S block copolymer, as the ratio of the S to B segments is increased, the phase morphology changes from a dispersion of spheres of S in a continuous phase of B to a dispersion of rods of S in a continuous phase of B and then to a lamellar or "sandwich" structure in which both S and B are continuous. If the proportion of S is increased still further, the effect is reversed in that S now becomes disperse and B continuous. As the polystyrene phase predominates, the block copolymer gets harder and stiffer until eventually it becomes a clear flexible thermoplastic, such as K-Resin(Phillips) and Versaclear (Shell).

In hard polymer / elastomer combinations, there are limits on the proportions of the elastomer phase in both the simple blends and in the thermoplastic vulcanizates. In the simple blends, if too much elastomer is used, the morphology changes from an interdispersed structure (in which both phases are continuous) to a dispersion of the hard polymer in the elastomer. Since the elastomer is not vulcanized, it has little strength. Thus when it becomes the only continuous phase, the properties of the blend are unsatisfactory.

In the thermoplastic vulcanizates, the dispersed elastomer phase is cross-linked and so cannot flow. It can thus be considered as an elastomeric filler, and when too much is present, processability suffers. Because of these limits on the amount of the elastomer phase, producing very soft products from hard polymer / elastomer combinations is difficult.

The multiblock polymers with crystalline hard segments also have limits on softness. The hard phase segments must have high enough molecular weights so that they can crystallize.   Softer versions of these polymers require the molecular weights of the elastomer segments to be higher still, so as to increase the soft/hard phase ratio. Thus for very soft products, the total molecular weight of the block polymer is increased to the point where processing can be difficult.

<u>Applications</u>

Applications of thermoplastic elastomers of all types have been extensively described (4,43). Some highlights are:

<u>Styrenic Thermoplastic Elastomers.</u>   Only the anionically polymerized versions of these block copolymers (i.e., S-B-S, S-I-S, S-EB-S and S-EP-S) are produced commercially. Thus all the information in this section is based on experience with these materials.   If

S-iB-S and similar block copolymers are ever produced commercially, they should have similar applications.

Styrenic block copolymers differ from the other thermoplastic elastomers in at least two significant ways. First, both the hard and soft phases are amorphous, and thus the pure polymers are soluble in common solvents such as toluene. Secondly, in their various end uses, these polymers are always compounded with large amounts of ingredients such as other polymers, oils, resins and fillers. In the majority of their applications, the styrenic block copolymer comprises less than 50% of the final product

Compounding significantly changes many properties - for example, solubility. Thus although the pure styrenic thermoplastic elastomers are completely soluble in solvents such as toluene, compounded products containing insoluble polymers (e.g., polypropylene) are not. The properties of compounded products produced from styrenic thermoplastic elastomers cover an exceptionally wide range and so their applications are more varied than those of the other thermoplastic elastomers.

For injection molding, extrusion, etc. (i. e. processing on conventional thermoplastics equipment) end users prefer to buy pre-compounded products, and grades have been developed for the various specialized end uses. Products based on S-B-S are typically compounded with polystyrene, hydrocarbon oils and fillers. In those based on S-EB-S, polypropylene often replaces polystyrene. This polymer gives better solvent resistance and increases the upper service temperature. Typical applications include footwear, wire and cable insulation, automotive and pharmaceutical items. Processing of these compounded products is simple. Usually, compounds based on S-B-S block copolymers are processed under conditions suitable for polystyrene while those based on S-EB-S block copolymers are processed under conditions suitable for polypropylene.

Another major application of styrenic thermoplastic rubbers is in adhesives, sealants and coatings. Tackifying and reinforcing resins are used to achieve a desirable balance of properties. Oils and fillers can also be added. These adhesives and sealants can be applied either from solvents or as hot melts. One very important application is in pressure sensitive hot melt adhesives. S-I-S block copolymers are softer and stickier and so they are often used to formulate adhesives of this type - in fact it is probably their largest single end-use.

A final application is in blends with thermoplastics or other polymeric materials. Styrenic block copolymers are technologically compatible with a surprisingly wide range of other polymers. Blends with many other thermoplastics have improved impact resistance. These block copolymers can also be used as compatibilizers - that is, they can produce useful products from blends of thermoplastics that otherwise have poor properties(65).

Multiblock Polymers with Crystalline Hard Segments. The very tough and relatively hard materials based on polyurethane, polyester or polyamide hard segments are generally regarded as premium products(16-19,21,22). Articles made from them are produced by the typical techniques used to process thermoplastics (e. g., injection molding, blow molding, extrusion). Because of their crystalline hard segments and polar elastomer segments, they have excellent oil resistance. Thus they are used in demanding applications as blow molded boots for automobile steering gear assemblies, grease seals, drive belts and hydraulic hose. They can also be blended with polar polymers such as PVC or used as the hard phase in hard polymer / elastomer combinations (33).

The polymers with polyethylene hard segments are lower in cost. Their suggested applications include wire and cable insulation, PVC replacement (66) and blends with polypropylene, either to improve impact resistance or as the soft phase in a hard polymer / elastomer combination

Hard Polymer / Elastomer Combinations. Polypropylene / EPDM or EPR combinations are the most important (31,33) and are used to make products such as injection molded bumpers for automobiles, where a combination of toughness, low temperature flexibility and low cost makes them very attractive. However, their use is limited because they can only be used to produce fairly hard products (typically above 60 Shore A hardness). Almost all applications for the polypropylene / EPDM thermoplastic vulcanizates(35,36) are as replacements for vulcanized rubber. They are used in automotive and appliance parts and also in the construction industry for window seals, etc.. Generally, they have better compression set and can give softer products (as low as 35 Shore A hardness). Similar thermoplastic vulcanizates based on polypropylene and nitrile rubber blends have improved solvent resistance.

Blends based on halogen containing polymers are also significant (41). Those based on halogenated polyolefin / ethylene interpolymer blends are claimed to be single phase systems (41,42). They are often used where solvent and fire resistance is important. PVC / nitrile rubber / Dioctyl phthalate blends are used in similar applications and in footwear (39-41).

Finally, the S-B-S / polystyrene / oil and S-EB-S / polypropylene / oil compounds described above can also be considered as blends of a hard polymer (polystyrene or polypropylene) with a soft elastomer phase (S-B-S / oil or S-EB-S / oil).

Economic aspects and trade names

Worldwide, about 1,000,000 metric tons of thermoplastic rubbers of all types were estimated to be used in 1995(1), with a value of about $3 billion. Consumption should increase to at least 1,400,000 metric tons by 2000. North America consumed about 43% of this amount, Western Europe about 36% and Japan accounted for most of the rest. The styrenic block copolymers represent about 50% of the total market and polypropylene/ EPDM or EPR combinations (including thermoplastic vulcanizates) about another 30%. The thermoplastic polyurethanes and the thermoplastic polyesters together made up another 15% Major end uses were transportation, footwear, industrial goods, wire insulation, medical (growing very rapidly), adhesives, coatings, etc.

Table VI gives values of three important properties (price, specific gravity and hardness) for typical commercially available thermoplastic elastomers. Trade names and suppliers of commercial thermoplastic elastomers of all types are given in Tables VII - IX.

Table VI - Approximate Price and Property Ranges for Thermoplastic Elastomers [a]

	Price Range (cents/lb.)	Specific Gravity	Hardness
Polystyrene / Elastomer Block Copolymers			
S-B-S (Pure)	85-130	0.94	65A-75A
S-I-S (Pure)	100-130	0.92	32A-37A
S-EB-S (Pure)	185-280	0.91	65A-75A
S-B-S (Compounds)	90-150	0.9-1.1	40A-45D
S-EB-S (Compounds)	125-225	0.9-1.2	5A-60D
Polyurethane / Elastomer Block Copolymers	225-375	1.05-1.25	70A[b]-75D
Polyester / Elastomer Block Copolymers	275-375	1.15-1.40	35D-80D
Polyamide / Elastomer Block Copolymers	450-550	1.0-1.15	60A-65D
Polyethylene / Poly(α-olefin) Block Copolymers	80-110	0.85-0.90	65A-85A
Polypropylene / EPDM or EPR Blends	80-120	0.9-1.0	60A-65D
Polypropylene / EPDM Dynamic Vulcanizates	165-300	0.95-1.0	35A-50D
Polypropylene / Butyl Rubber Dynamic Vulcanizates	210-360	0.95-1.0	50A-80D
Polypropylene / Natural Rubber Dynamic Vulcanizates	140-160	1.0-1.05	60A-45D
Polypropylene / Nitrile Rubber Dynamic Vulcanizates	200-250	1.0-1.1	70A-50D
PVC / Nitrile Rubber Blends	130-150	1.20-1.33	50A-90A
Chlorinated Polyolefin / Ethylene Interpolymer Blends	225-275	1.10-1.25	50A-80A

a) These price and property ranges do not include fire retardant grades or highly filled materials for sound deadening.
b) As low as 60A when plasticized

Table VII - Some Trade Names of Thermoplastic Elastomers Based on Styrenic Block Copolymers

Trade Name (Mfr.)	Type	Elastomer Segment	Notes
KRATON[R] D and	Linear and	B or I	General purpose, soluble.
CARIFLEX TR (Shell)	branched		Also compounded products
VECTOR (Dexco)[a]	Linear	B or I	
SOLPRENE[b] (Phillips)	Branched	B	
TAIPOL (Taiwan Synthetic	Linear and	B or I	
Rubber Company	Branched		
QUINTAC (Nippon Zeon)	Linear	I	General purpose, soluble.
FINAPRENE (Fina)	Linear	B	Not available as compounded
COPERBO (Petroflex)	Linear	B	products.
TUFPRENE & ASAPRENE (Asahi)	Linear	B	
CALPRENE (Repsol)	Linear and branched	B	
EUROPRENE SOL T (Enichem)	Linear and branched	B or I	
STEARON (Firestone)	Linear	B	High polystyrene content
K-RESIN (Phillips)	Branched	B	Very high polystyrene content. Hard and rigid.
KRATON G (Shell)	Linear	EB or EP	Improved stability. Soluble
SEPTON (Kuraray)	Linear	EB or EP	when uncompounded
DYNAFLEX (GLS)	Linear	B or EB	
MULTI-FLEX (Multibase)	Linear	EB	
HERCUPRENE[c] (J-VON)	Linear	B or EB	Only compounded products
FLEXPRENE (Teknor Apex)	Linear	B	
TEKRON (Teknor Apex)	Linear	EB	
ELEXAR[Rd] (Teknor Apex)	Linear	EB	Wire and Cable compounds
C-FLEX (Concept)[e]	Linear	EB	Medical applications. Contains silicone oil

a) Joint venture of Dow and Exxon.
b) No longer made in U.S.A. Similar products are produced by Taiwan Synthetic Rubber Company.
c) Formerly J-PLAST.   d) Formerly produced by Shell.
e) Now Consolidated Polymer Technologies Inc.

Table VIII - Some Trade Names of Thermoplastic Elastomers Based on Multiblock Copolymers with Crystalline Hard Segments

Trade Name (Mfr.)	Hard Segment	Elastomer Segment	Notes
ESTANE (B.F. Goodrich) MORTHANE[a] (Morton International) PELLETHANE[a] (Dow) ELASTOLLAN (BASF) DESMOPAN and TEXIN (Bayer)[b]	Polyurethane	Polyether or amorphous Polyester	Hard and Tough. Abrasion and oil resistant. Good tear strength. Fairly high priced
HYTREL (DuPont) LOMOD (GE) URAFIL (Akzo) ECDEL (Eastman) RITEFLEX (Hoechst) ARNITEL (DSM)	Polyester	Polyether	Similar to polyurethanes but more expensive. Better low temperature flexibility. Low hysteresis.
PEBAX (Elf Atochem) VESTAMIDE (Huls) GRILAMID and GRILON (EMS America) MONTAC (Monsanto)[c] OREVAC (Atochem)[c]	Polyamide	Polyether or amorphous polyester	Similar to polyurethanes but can be softer. Expensive. Good low temperature flexibility.
ENGAGE & AFFINITY (Dow) EXACT (Exxon) FLEXOMER (Union Carbide)	Polyethylene	Poly(α-olefins)	Flexible and low cost. Good low temperature flexibility but limited at higher temperatures.

a) Including some with polycaprolactone segments.
b) Formerly marketed by Mobay and Miles.
c) For hot melt adhesives

Table IX - Some Trade Names of Thermoplastic Elastomers Based on Hard Polymer / Elastomer Combinations

Trade Name (Mfr.)	Type	Hard Polymer	Elastomer	Notes
REN-FLEX (D&S)[a] HIFAX (Himont) POLYTROPE (Schulmam) TELCAR (Teknor Apex) FERROFLEX (Ferro) FLEXOTHENE (Equistar)[b]	Blend	Polypropylene	EPDM or EPR	Relatively hard, low density, not highly filled
SANTOPRENE (AES)[c] SARLINK 3000 & 4000 (Novacor)[d] UNIPRENE (Teknor Apex) HIFAX MXL (Himont)	DV[f]	Polypropylene	EPDM	Better oil resistance, low compression set, softer
TREFSIN (AES) and SARLINK 2000 (Novacor)[d]	DV	Polypropylene	Butyl Rubber	Low permeability, high damping
VYRAM (AES)	DV	Polypropylene	Natural Rubber	Low Cost
GEOLAST (AES)	DV	Polypropylene	Nitrile Rubber	Oil resistant
ALCRYN (Advanced Polymer Alloys)[e]	Blend	Chlorinated Polyolefin	Ethylene Interpolymer	Single phase, oil resistant
SARLINK 1000 (Novacor)[d]	DV	PVC	Nitrile Rubber	Oil Resistant
CHEMIGUM (Goodyear)	Blend			
APEX N(Teknor Apex)	Blend			
RIMPLAST (Petrarch Systems)		Blends of TPEs with Silicone Rubbers		Medical applications

a) A joint venture between Dexter and Solvay
b) Formerly Quantum. Product is a blend of PP and EPR produced in the polymerization reactor
c) Advanced Elastomer Systems - a joint venture between Solutia (formerly Monsanto) and Exxon Chemical   d) Now a part of DSM   e) Formerly DuPont
f) Dynamic Vulcanizate - a composition in which the soft phase has been dynamically vulcanized, i.e., cross-linked during mixing

Literature Cited

1.   Chemical and Engineering News, August 5, 1996, p 10-14.
2.   Thermoplastic Elastomers - A Comprehensive Review (N.R.Legge, G.Holden and
     H.E. Schroeder, Eds), Hanser & Oxford Univ. Press - Munich/ New York (1987).
3.   Thermoplastic Elastomers - A Comprehensive Review, 2nd Ed. (G. Holden, N. R.
     Legge, R. P. Quirk and H. E. Schroeder, Eds), Hanser & Hanser/Gardner - Munich /
     Vienna / New York / Cincinnati, (1996).
4.   Handbook of Thermoplastic Elastomers, 2nd Ed. (B.M. Walker & C.P.Rader, Eds),
     Van Nostrand Reinhold, New York, 1988.
5.   G. Holden, "Elastomers, Thermoplastic" in Kirk-Othmer Encyclopedia of Chemical
     Technology, 4th ed, J. I. Kroschwitz, Ed. John Wiley & Sons, New York,
     NY, 1994.
6.   G. Holden, "Thermoplastic Elastomers (Overview)" in Polymeric Materials
     Encyclopedia, J. C. Salamone, Ed. CRC Press, Boca Raton, FL, 1996.
7.   G. Holden, E. T. Bishop and N. R. Legge, J. Poly. Sci., C26, 37 (1969).
8.   G. Holden and N. R. Legge in Ref. 3, Chapter. 3.
9.   W. M. Halper and G. Holden in Ref. 4, Chapter. 2.
10. J. P. Kennedy in Ref. 3, Chapter. 13.
11   R. P .Quirk and M. Morton in Ref. 3, Chapter. 4.
12. J. C. Saam, A. Howard and F. W. G. Fearon, J. Inst. Rubber Ind. 7, 69 (1973)
13. A. Noshay, M. Matzner and C. N. Merriam, J. Poly. Sci. A- 1, 9: 3147 (1971).
14. R. L. Merker, M. J. Scott and G. G. Haberland, J. Poly. Sci., A, 2, 31 (1964).
15. S. L. Cooper and A. V. Tobolsky, Textile Research Journal 36, 800 (1966).
16. W. Mekel, W. Goyert and W. Wieder in Ref. 3, Chapter. 2.
17. E. C. Ma in Ref. 4, Chapter. 7.
18. R. K. Adams, G. K. Hoeschele and W. K. Wisiepe in Ref. 3, Chapter. 8.
19. T. W. Sheridan in Ref. 4, Chapter. 6.
20. K. D. Gagnon in Ref. 3, Chapter 15B
21. R. G. Nelb and A. T. Chen in Ref. 3, Chapter. 9.
22. W. J. Farrissey and T. M. Shah in Ref. 4, Chapter. 8.
23. H. A. Vaughn, J. Poly. Sci. B, 7, 569 (1969).
24. R. P. Kambour, J. Poly. Sci. B, 7, 573 (1969).
25. D. G. LeGrand, J. Poly. Sci. B, 7, 579 (1969).
26. E. P. Goldberg, J. Poly. Sci. C, 4, 707 (1963).

27. K. P. Perry, W. J. Jackson, Jr. and J. R. Caldwell, J. Appl. Poly. Sci. **9**, 3451 (1965).
28. J. Mihalich, paper presented at the 2nd International Conference on Thermoplastic Elastomer Markets and Products sponsored by Schotland Business Research, Orlando, FL, March 15-17, 1989.
29. R. Jerome et al. in Ref. 3, Chapter 15D
30. P. T. Hammond and M. F. Rubner in Ref. 3, Chapter 15E
31. E. N. Kresge in Ref. 3, Chapter. 5.
32. J. L. Laird, Rubber World, **217(1)** 42 (1997)
33. E. N. Kresge, Rubber World, **217(1)** 30 (1997)
34. C. D. Shedd in Ref. 4, Chapter. 3.
35. A. Y. Coran and R. P. Patel in Ref. 3, Chapter. 7.
36. C. P. Rader in Ref. 4, Chapter. 4.
37. R. C. Puydak, paper presented at the 2nd International Conference on Thermoplastic Elastomer Markets and Products sponsored by Schotland Business Research, Orlando, FL, March 15-17, 1989.
38. A. J. Tinker, paper presented at the Symposium on Thermoplastic Elastomers sponsored by the ACS Rubber Division, Cincinnati, OH, October 18-21, 1988.
39. M. Stockdale, paper presented at the Symposium on Thermoplastic Elastomers sponsored by the ACS Rubber Division, Cincinnati, OH, October 18-21, 1988.
40. P. Tandon and M. Stockdale, paper presented at the 4th International Conference on Thermoplastic Elastomer Markets and Products sponsored by Schotland Business Research, Orlando, FL, February 13-15, 1991.
41. G. H. Hoffman in Ref. 3, Chapter. 6
42. J. G. Wallace in Ref. 4, Chapter. 5.
43. G. Holden in Ref. 3, Chapter. 16
44. J. P. Kennedy and J. M. Delvaux, Adv. Polym. Sci. **38**, 141 (1981)
45. R. Ambrose and J. J. Newell, J. Polym. Sci. Polym. Chem. Ed. **17**, 2129 (1979)
46. P. Sigwalt, A. Polton and M. Miskovic, J. Polym. Sci. Symp. No. 56 , 13 (1976)
47. J. P. Kennedy and R. R. Smith, In Recent Advances in Polymer Blends, Grafts and Blocks (L. H. Sperling Ed.) Plenum Press, New York /London (1974)
48. R. R. Smith, Ph. D. Thesis, The University of Akron, 1984
49. A. Gadkari and M. Farona, Polym. Bull **17**, 229 (1987)
50. J. P. Kennedy and J. J. Charles, J. Appl. Polym. Sci. Appl. Polym. Symp. 30 , 119 (1977)
51. J. J. Charles, Ph. D. Thesis, The University of Akron, 1983

52. J. P. Kennedy and S. C. Guhaniyogi, J. Macromol. Sci. Chem. A18 , 103 (1982)

53. J. P. Kennedy and F P. Baldwin, Belgian Patent 701,850 (1968)

54. J. P. Kennedy and D. M. Metzler, J. Appl. Polym. Sci. Appl. Polym. Symp. 30,105 (1977)

55. G. O. Schultz and R. Milkovich, J. Appl. Polym. Sci. 27, 4473 (1982)

56. H. Xie and S. Zhoui, J. Macromol. Sci. Chem. A27, 491 (1990)

57. P. Dreyfuss, L. J. Fetters and D. R. Hansen, Rubber Chem. Technol. 53 738 (1980).

58. M. Morton, Anionic Polymerization: Principles and Practice, Academic Press, New York, NY (1983).

59. H. L. Hsieh and R. P .Quirk, Anionic Polymerization: Principles and Practical Applications, Marcel Dekker, Inc., New York, NY (1993).

60. M. Szwarc, M. Levy and R. Milkovich, J . Am. Chem. Soc. 78, 2656 (1956).

61. N. R. Legge, S. Davison, H. E. DeLaMare, G. Holden and M. K. Martin in "Applied Polymer Science, 2nd Ed." R. W. Tess and G. W. Poehlein, Eds. ACS Symposium Series No. 285, American Chemical Society, Washington, D.C., 1985, ch.9.

62. a) L. H. Tung and G. Y-S. Lo, Macromolecules 27, 2219 (1994).

    b) C. J. Bredeweg, A. L. Gatzke, G. Y-S. Lo and L. H. Tung, Macromolecules 27, 2225 (1994).

    c) G. Y-S. Lo, E. W. Otterbacher, A. L. Gatzke and L. H. Tung,Macromolecules 27, 2233 (1994).

    d) G. Y-S. Lo, E. W. Otterbacher, R. G. Pews and L. H. Tung, Macromolecules 27, 2241 (1994).

    e) A. L. Gatzke and D. P. Green, Macromolecules 27, 2249 (1994).

63. a) L. H. Tung, G. Y-S. Lo and D. E. Beyer, (to Dow Chemical Co.), U. S. Patent 4,196,154, (1980).

    b) L. H. Tung, G. Y-S. Lo, J. W. Rakshys and B. D. Beyer, (to Dow Chemical Co.), U. S. Patent 4,201,729, (1980).

64. K. Matyjaszewski, Cationic Polymerizations: Mechanisms, Synthesis and Applications, Marcel Dekker, Inc., New York, NY (1996)

65. D. R. Paul in Ref. 3, Chapter 15C.

FIBERS

LUDWIG REBENFELD

TRI and Department of Chemical Engineering, Princeton University
P. O. Box 625, Princeton, NJ 08542

Introduction

Synthetic organic polymers have become widely accepted as materials of choice in many consumer and industrial applications on the basis of their desirable combination of physical properties, favorable economics, and broad versatility. As we celebrate the 75th Anniversary of the ACS Division of Polymeric Materials Science and Engineering, we also must celebrate the 75 years of pioneering research on the synthesis, structure, and properties of this class of materials. Of course, research on naturally occurring polymers had been underway for many decades prior to the revolutionary concept of linear macromolecules introduced by Staudinger and others in the 1920s. The earliest research on natural cellulosic fibers was conducted primarily by microscopists and botanists who attempted to explain in structural terms the reversible aqueous swelling of the these fibers. The first proposition of a micellar structure was put forward by Nageli in 1858, who postulated that cellulosic fibers were composed of discrete submicroscopic particles that he referred to as micelles. It was proposed that water could enter into a cellulosic fiber by penetrating the spaces between micelles without affecting the micelles themselves, thereby allowing the observed reversibility. The Nageli micelles introduced the concept of a two-phase structure for cellulosic fibers, a concept that survives to this day in our recognition of the co-existence of crystalline regions and disordered domains in many synthetic polymers in the solid state.

In this paper we shall examine some of the important technical developments in the field of fibers that have taken place since the 50th Anniversary of the Division. While it is the 1950s and 1960s that are generally regarded as the golden years of man-made fibers, the more recent decades have also seen important far-reaching developments in both the production and utilization of man-made fibers. Among these are high-speed spinning of melt spun fibers, the introduction of microfibers, the development of high-performance fibers, the increasing use of fiber-reinforced composites, and others. Before discussing some of these topics, it may be instructive to provide perspective by looking at current fiber production data.

Fiber Production

Man-made fibers are conveniently divided into those based on synthetic polymers and those based on cellulose, the most common naturally occurring polymer. Worldwide production of man-made fibers in 1997 amounted to about 28.5 million metric tons (1). As shown in Table I, the cellulosic fibers are a relatively small component compared to the fibers based on synthetic polymers, and it is noteworthy that man-made fiber production is a truly worldwide enterprise, with Asia being the dominant producing region. This is rather recent, reflecting the rapid economic and industrial development of the Asiatic continent during the past two decades. More than half of the fiber producing capacity in Asia is accounted for by China, Taiwan, and South Korea.

Table I. 1997 Worldwide Man-Made Fiber Production
(Million Metric Tons) (1).

	Synthetic Polymer	Cellulosic
Europe	5.08	0.69
Americas	6.20	0.29
Asia	14.47	1.31
M.E., Africa, Oceania	0.53	0.02
Total	26.28	2.31

Among the 26.3 million metric tons of synthetic polymer fibers produced in 1997 worldwide, polyester type fibers were by far the most dominant, amounting to about 14.7 million metric tons. The enormous versatility of this type polymer, principally poly(ethyleneterephthalate), allows a wide range of fiber characteristics and properties to be achieved by variations in processing conditions and thermomechanical treatment. Polyamide fibers, including both the aliphatic nylons and the aromatic high performance aramid fibers amounted to 4.0 million metric tons, while olefin fibers, principally polypropylene, accounted for 4.6 million metric tons. The use of polypropylene fibers in a wide range of consumer products has grown dramatically in recent years, reflecting not only favorable economics but also new catalyst technologies in polymer synthesis leading to improved fiber properties. Acrylic polymers, containing mainly polyacrylonitrile, accounted for about 2.7 million metric tons, while several specialty polymer fibers amounted to a 1997 production of about 0.3 million metric tons. The fiber production data in Table I do not include about 2.6 million metric tons of glass fiber, and about 0.6 million metric tons of cellulose acetate fiber in the form of tow used in cigarette filters.

Cellulosic Fibers

The cellulosic fibers, rayon and cellulose acetate, with a worldwide production of 2.3 million metric tons, continue to be an important category of man-made fibers. The relatively simple carbohydrate polymer, cellulose, is just about the perfect fiber-forming polymer. It has a high degree of inherent chain stiffness, arising from its cyclic anhydroglucose monomer units and its anomeric configuration, and it is highly hydrophilic due to its three free hydroxyl groups. These groups are involved in extensive inter- and intramolecular hydrogen bonding which serves to stabilize further the highly ordered three dimensional structure. At the same time, the free hydroxyl groups are active reaction sites for the formation of various cellulose derivatives. The major obstacle in the production of regenerated man-made cellulosic fibers from naturally occurring cellulose is the lack of solubility of this polymer in common non-degrading solvents. It is necessary to use complexing solvents, such as aqueous cuprammonium hydroxide or cupriethylene diamine, or to form partial derivatives, such as the acetates or nitrates, which are then soluble in organic solvents. Several processes for the manufacture of regenerated (and derivative) cellulosic fibers are based on these solubilization methods, but none has achieved the success of the well known viscose process. This process for the production of rayon from wood pulp and/or cotton linters was the first commercially successful process for the manufacture of man-made fibers.

The viscose process involves the formation of cellulose xanthate by reaction of cellulose impregnated with sodium hydroxide with carbon disulfide, with subsequent extrusion of the solubilized cellulose xanthate in a wet spinning process to form the regenerated cellulosic fiber (2). It is an enormously versatile process that allows the production of a wide range of diverse fibers with applications not only in traditional textile products but also in industrial markets, notably tire cord. As environmental issues became of increasing concern, and despite many important technological improvements in process control, the viscose process did not meet the air and water pollution standards established in the United States and in many other industrialized nations. The viscose process continues to be the major process for the production of rayon, but in view of environmental regulations it is practiced only in South America, Eastern Europe, Asia, and in the developing regions of the world.

With the obvious environmental limitations of the viscose process, and in view of the desirable properties of man-made cellulosic fibers, a major research effort was undertaken for alternate processes that would be environmentally acceptable. The search focused on means of dissolving cellulose without derivitization directly in a solvent from which the fiber could be formed by solvent or dry spinning. The intensive research efforts culminated in a new type of cellulosic fiber, generically designated as lyocell, which is produced by direct solvent or dry spinning of cellulose dissolved in aqueous N-methyl-morpholine oxide (3). With complete solvent recovery, the lyocell process is an environmentally acceptable technology for the production of man-made cellulosic fibers, and represents an important innovation in fiber technology. Several major lyocell production plants are now operative in the United States, Canada, and Europe.

Natural Fibers

The production and consumption of man-made fibers now accounts for about 55% of the worldwide production of textile fibers. The natural fibers, cotton with an annual worldwide production of about 20.0 million metric tons and wool with a production of about 1.5 million metric tons, add to the fiber mix that allows the manufacture of a nearly infinite number of products for the highly competitive textile markets. Natural and man-made fibers are extensively used in two and three component blends adding versatility in terms of properties and performance of apparel, household and industrial textiles. While cotton and wool are the dominant natural fibers used in textile applications, it is noteworthy that there are an additional 5 million metric tons of other natural cellulosic fibers produced and used annually. Among these are jute, flax (linen), ramie, hemp, sisal, and kapok (4). The properties of these fibers are quite interesting and useful, but their more widespread use is currently limited by the rather complex and expensive methods that are necessary for their isolation and purification. Extensive research in the agricultural community on enzyme based retting processes promises important future developments in the production of flax and other natural cellulosic fibers (5).

The recent advent of genetic engineering can be expected to play an important role in the development of improved natural fibers. Transgenic cotton that is resistant to non-selective herbicides has already been developed and is used widely (6). A recent development has been the synthesis of a thermoplastic aliphatic polyester, poly(3-hydroxybutyrate), within the internal structure of cotton fibers by transgenesis (7). The potential of modifying the properties of cotton and other natural fibers by polymer blending is an exciting one that may lead to new classes of fibers for the textile industry.

Figure 1.   Internal structure of enzyme retted flax fiber (5).

High Speed Spinning

In melt spinning of thermoplastic fibers, molten polymer is extruded under pressure through a spineret, each orifice producing a jet of molten polymer which passes through a cooling zone where it solidifies into a nascent filament. The as-spun filaments, wound-up at speeds in the range of 1000 m/min, do not have physical properties characteristic of fibers - they are weak and highly extensible.   These properties reflect an undeveloped internal fiber structure, i.e., little or no crystallinity and poor molecular orientation. The development of fiber structure takes place during a subsequent drawing step in which the fibers are extended at temperatures above their glass transition by factors, referred to as draw ratios, of anywhere from 3 to 10 and even higher.   Upon cooling from the draw temperature, the fibers are set in the extended state and have a well developed fiber structure with high crystallinity and orientation.   The highly birefringent fibers now yield characteristic x-ray diffraction patterns, and are mechanically strong with high elastic moduli and low extensibilities. Fiber mechanical properties are largely controlled by the extent of drawing.

In recent years the two-step extrusion and drawing process has been widely replaced by a one-step high-speed spinning process in which the filaments are wound-up after extrusion at take-up speeds up to 8,000 m/min (8).   Speeds up to 12,000 m/min have been explored experimentally. At these high take-up speeds, stress induced orientation and crystallization can take place directly on the spinline. Crystallinity as indicated by fiber density, and orientation as reflected in fiber birefringence are directly related to the take-up speed, as shown in Figure 2 for polyester fibers (9). High-speed spun polyester, nylon and polypropylene fibers have excellent fiber properties, and the one-step process is economically advantageous to the older two-step extrusion and drawing process.

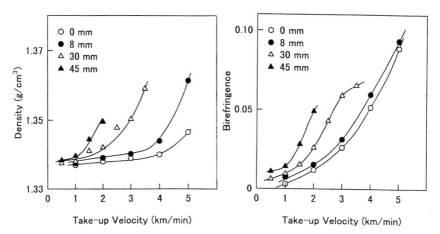

Figure 2. Birefringence and density as functions of take-up velocity (the parameter refers to the length of wire inserted in the extrusion nozzle) (9).

Microfibers

It has always been well known that fiber diameter, frequently referred to as fiber fineness, has an important influence on the properties of textile structures such as yarns and fabrics. Prior to the advent of man-made fibers, the approximately 10-20 μm diameter of a cotton fiber and the 10-15 μm diameter of silk filaments were considered as the finest fibers achievable. When the man-made fibers were introduced, it was necessary to match the geometric properties of these fibers (diameter, length, crimp curvature) to that of either cotton or wool so that these new fibers could be processed on textile equipment that evolved over centuries for the processing of natural fibers. Thus, for many years the lowest linear densities of the man-made fibers were in the 2 to 3 denier range. With increased understanding of how fiber linear density or diameter influenced the properties of yarns and fabrics, particularly softness and other tactile characteristics, it became of interest to develop small diameter fibers, referred to as microfibers, for certain applications.

Pioneering work in this area of fiber technology was done with polyester fibers in Japan in the 1980s. The first approach taken was to reduce the fiber diameter by controlled surface dissolution of the polyester by treatment in aqueous NaOH solutions at elevated temperatures. Typically weight losses achieved were in the 20 to 30% range producing fibers with linear densities as low as 1 denier. Even finer polyester fibers may be achieved by this approach, but the economics of controlled fiber dissolution are not particularly attractive. Depending on treatment conditions, the reduction in fiber linear density and diameter may be accompanied by changes in fiber surface texture, such as the fiber pitting shown in Figure 3.

More innovative approaches have been taken to the development of microfibers involving the extrusion of bicomponent fibers of the island-in-the-sea and citrus structure type. After extrusion, the lesser component can be removed by dissolution or by splitting. These methods have produced polyester, nylon, and polypropylene microfibers not only with linear densities down to 0.1 denier, but also with controlled cross-sectional shapes. The subject has recently been reviewed by Cooke (10). The advent of microfibers has enormously broadened the range of fibrous products for textile consumers. Polyester fabrics with the softness and drapeability previously associated only with the finest silks are now routinely available. Silklike polyesters and related innovations of this type were recently discussed by Fukuhara (11).

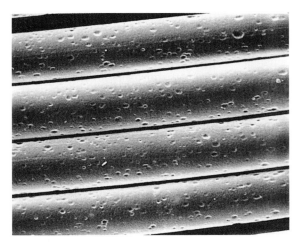

Figure 3. NaOH treated polyester fibers (courtesy of S. B. Ruetsch, TRI/Princeton)

Aramid Fibers

The last two decades have seen the development of several specialty fibers with exceptional physical properties and performance characteristics that makes each particularly useful in certain applications. Among these properties are thermal stability and retention of mechanical strength at high temperatures, flame resistance, high strength and stiffness, and low extensibility. It has long been recognized that fully aromatic structures would be able to provide these high-performance properties. Many polymers were synthesized and spun into fiber, frequently with a great deal of difficulty, to confirm the expectation of exceptional properties. The aromatic polymer backbone would confer a high degree of chain stiffness that would allow high orientation and crystallinity to be achieved in the final fiber. In most cases, however, the difficulties of spinning the fiber by acceptable extrusion technologies prevented these fibers from commercialization, except in special military and aerospace end uses. The aromatic polyamide (aramid) fiber Kevlar® with a nominal structure poly(p-phenyleneterephthalamide) was the first fully aromatic fiber to overcome the processing difficulties and gain commercial acceptance. It has a tenacity at break in excess of 20 gpd and a melting point, actually a decomposition temperature, of about 460°C. Due to the rigidity of this linear aromatic polymer, the chains are rod-like and in nematic liquid crystalline form in sulfuric acid solution, from which the fibers are spun in a modified solvent spinning process (12). Kevlar and related aramid fibers have found widespread use in fabric constructions that require high impact strength, such as in military and police protective vests. These fibers are also extensively used as reinforcing elements in composites, and in many other industrial applications.

The isomeric structure poly(m-phenyleneisophthalamide) is the polymer for the aramid fiber Nomex® which is flame resistant and retains physical properties even after long exposures to temperatures close to its melting point of 380°C (13). Nomex® aramid fibers have significantly broadened the range of available fiber properties for applications where flame resistance and thermal stability are required. Another nonflammable fiber with excellent thermal stability is produced from a polybenzimidazole polymer and is referred to as PBI fiber (14). It has good mechanical properties and a high equilibrium moisture regain which is advantageous in fire protective clothing applications. This fiber is also used in industrial high-temperature filtration systems.

Carbon Fibers

Carbon fibers have become an important category of high performance fibers with principal use in fiber-reinforced composites (15). These fibers combine low density with exceptional mechanical properties of high stiffness and high strength. Carbon fibers consist of small crystallites of turbostratic graphite with layer planes of hexagonally arranged carbon atoms held together by strong covalent bonds. Weaker van der Waal interactions are operative between the layer planes. The high performance characteristics of carbon fibers depend on the preferential orientation of the graphite layers parallel to the fiber axis. Carbon fibers are typically produced from continuous precursor filaments by a three stage procedure, involving stabilization of the precursor at relatively low temperatures, carbonization of the precursor to remove noncarbon elements at temperatures in the range of 500 to 1200°C, and graphitization at temperatures up to 2500°C. Increasing orientation is achieved at all three stages. Carbon fibers have been produced from many precursor polymers, but commercial production is currently limited to those produced from polyacrylonitrile, cellulose, and pitch. Physical properties of carbon fibers depend on internal structure, which in turn is determined by the nature of the precursor and processing conditions. Typically a PAN based carbon fiber may have a diameter of 8 μm, a specific gravity of 1.95, an elastic modulus of 390 GPa and a tensile strength of 2.2 GPa. The elastic modulus increases with increasing graphitization temperature, but the tensile strength generally passes through a maximum at temperatures of about 1300°C. Mechanical properties of carbon fibers and several organic and inorganic fibrous materials are summarized in Table II.

Table II. Mechanical Properties of Some Fibrous Materials (15).

Fibrous Material	Specific Strength (GPa)	Specific Modulus (GPa)
Carbon fiber (UHM)	1.135	282
Carbon fiber (HM)	1.405	192
Carbon fiber (HT)	1.805	118
Boron	2.000	180
Aramid	2.496	86
S-glass	1.807	34
Steel	0.054	26
Aluminum alloy	0.101	27
Titanium alloy	0.220	25
Magnesium alloy	1.556	23

Due to their use as reinforcing elements in composites, the surface properties of carbon fibers are of particular importance. Carbon fibers are relatively smooth with some longitudinal striations. To maximize adhesion between fibers and matrix, carbon fibers are normally subjected to controlled surface treatments, involving oxidative etching, chemical vapor deposition, and reactive plasma exposure.

Fiber Reinforced Composites

While we consider fiber-reinforced polymer composites to be a keystone of the materials revolution that has taken place during the past two decades or so, it is appropriate that we recognize that two-phase materials are abundant in nature. For example, the outstanding mechanical properties of wood are due to the fact that wood fibers are themselves fiber reinforced composites, with cellulosic fibrils embedded in a lignin matrix. Also the fine structure of wool fibers reveals them to have a composite structure of crystalline alpha-helix fibrils, now referred to as intermediate filaments, embedded in a matrix of globular protein. Even in the realm of man-made materials, we find fiber-reinforced composites preceding the current materials revolution in the form of pneumatic tires, which are excellent examples fiber-reinforced elastomers. The desirable mechanical properties of tires reflect not only the inherent properties of the tire cord and the rubber, but also the geometric arrangement of the cord.   However, the modern fiber-reinforced polymer composites that have found widespread application in the construction, automotive and aerospace industries, as well as in a variety of consumer products, are highly engineered structures where fibers in various geometric form reinforce a wide range of thermosetting and thermoplastic polymers to yield relatively light weight yet strong and durable materials that rival and even exceed the properties of metals.

Nearly all fibers have found some use as reinforcing elements in composites; however, aramid, glass and carbon fibers are the principal fibers used in high performance composites. Also of importance are gelspun high-density polyethylene, liquid crystalline polyesters, and several specialty polyimide fibers. To this ever growing list of organic polymer fibers, we must add inorganic fibers such as those based on aluminosilicates, alumina, silica, boron, boron nitride, and silicon carbide (16).   A wide range of thermosetting and thermoplastic polymers have found extensive use as matrices in composite materials. Epoxies, aliphatic polyesters, and polyimides are among the most important thermosets, while polycarbonates, aromatic and liquid crystalline polyesters, polyphenylene sulfides, polyetheretherketones, and certain polyimides are among the important thermoplastics. The performance characteristics of fiber composites depend on many factors, including the physical properties of the fibers, the geometric arrangement of the fibers in the reinforcing fiber network, the properties of the plastic matrix, and the interaction between the two major components of a composite - the fibers and the matrix. The interfacial effects are critically important, and in most cases the quality of the interface between fibers and matrix controls ultimate performance and durability of the composite. To improve the adhesion between fibers and matrix, the fibers are normally subjected to a variety of chemical treatments and surface finishes.

Crystalline thermoplastic matrices, in contrast to those that are completely amorphous, are being used increasingly in fiber reinforced composites because of the added strength, thermal stability, and durability that can be associated with crystalline matrix structure. Crystallinity develops in these polymers during the cooling period after composite formation and the properties of the composite depend on the degree of crystallinity and the crystalline texture. Many studies have shown how reinforcing fibers influence the crystallization process in terms of such characteristics as degree of crystallinity, the crystalline morphology, and thermal transition temperatures (17).

In certain cases the morphology of the matrix near the fiber surface may be quite different from that which develops in the bulk of the matrix, giving rise to a so-called transcrystalline region or interphase. A particularly striking example of this transcrystalline morphology for a model thin film composite of polyphenylene sulfide (PPS) reinforced with a highly graphitized carbon fiber is shown in Figure 4. The fiber surface clearly nucleates crystallization which is followed by highly constrained columnar crystal growth. The high nucleation density on the fiber surface prevents the normal development of a spherulite structure. Away from the fiber, in the bulk of the matrix, the bulk-nucleated crystals grow to form well developed spherulites. Not all carbon fibers will induce this transcrystalline structure, others will appear quite inert. Aramid fibers generally induce PPS transcrystallinity, while glass fibers do not. In the model composite shown in Figure 4 one can clearly see the interface between the PPS bulk spherulites and the transcrystalline region. In a real three-dimensional composite with a typical fiber loading of about 50% by weight, the fibers may be so close together that no bulk spherulites will form and the entire matrix morphology will be that of the transcrystalline region.

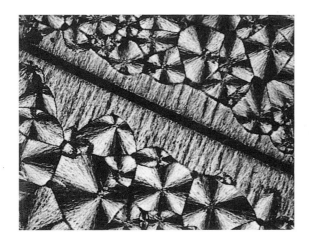

Figure 4. Model thin film composite of a carbon fiber in a PPS matrix (17)

There does not appear to be an agreed upon understanding of the factors that control the formation of a transcrystalline region. It is clear that transcrystallinity is a systems characteristic, and a given fiber may induce a transcrystalline morphology with one matrix but not with another, and conversely a given matrix may exhibit transcrystallinity with one fiber but not with another. In recent studies with a polypropylene matrix and aramid and carbon fibers, Dean et al. (18) have shown by means of x-ray diffraction measurements that epitaxial register between the fiber surface and the matrix crystal structure is necessary for transcrystallinity. Another not fully resolved issue in connection with transcrystallinity is whether it is desirable or not with regard to physical properties and performance characteristics of the composite. For the case of aramid fibers in PPS, Chen et al. (19) have rather convincing data to indicate that a system capable of transcrystallinity formation shows higher interfacial shear strength. However, this is not necessarily generally true. Among other factors, the apparent strength of a two-phase composite will depend on the mode of deformation. Furthermore, a dual morphology such as that shown for the model composite in Figure 4 would probably not be advantageous in terms of interfacial shear strength because the system has two interfaces, one between fiber and transcrystalline region, and the other between transcrystalline region and bulk matrix. However, as indicated previously, an extensive dual morphology probably does not exist in real highly fiber loaded composites.

Crystallization Rates in Composites

Fibers have an important influence not only on the developing morphology but also on the crystallization rate of the matrix. Such effects must be known in order to optimize processing conditions for the composite. Again using PPS as an example of a crystallizable thermoplastic matrix, we have used differential scanning calorimetry to quantify the influence of fibers on the kinetics of the crystallization (17, 20). The rate is analyzed by evaluating the crystalline content, which is the area under the crystallization exotherm, as a function of time. The kinetics of the crystallization process may be reported in terms of the crystallization half-times (inverse of the crystallization rate) under isothermal conditions. The crystallization half-time is a processing related parameter in that it denotes the time required for completion of 50% of the crystallization process.

In Figure 5 are shown the crystallization half-times as a function of crystallization temperature for unreinforced PPS, a commercial prepreg, and several experimental carbon fiber reinforced composites. The fiber loadings are quite similar (about 60% by weight) so that comparison among the systems can be made. In all cases crystallization half-time increases with increasing temperature, that is, the crystallization rate decreases as we approach the PPS melting temperature of approximately 280°C. This is because the thermodynamic driving force for crystallization is the undercooling from the melt temperature. However, the important point is that at any given crystallization temperature, the presence of the carbon fibers enhances the crystallization rate (lower half-time), particularly the graphitized Thornel® carbon fiber. Similar effects are observed with Kevlar® aramid fibers, but glass fibers generally do not enhance the crystallization rate of PPS. In fact, there are some data that show that glass fibers can depress the crystallization rate of a given polymeric matrix, for example, poly(ethylene terephthalate) (21). The explanation for these effects becomes clear when we realize that the presence

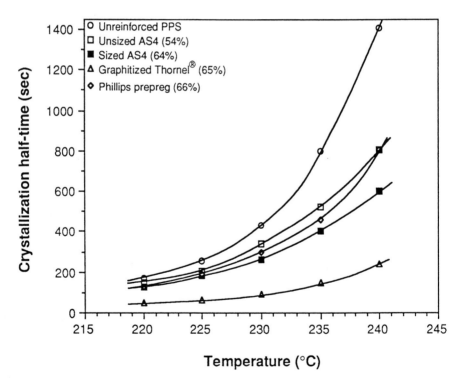

Figure 5. Crystallization half-times of PPS composites (17).

of fibers or other fillers in a crystallizing polymer melt can serve two functions. The fibers can serve as a source of nucleation sites which would tend to increase the overall crystallization rate, but they can also serve as impingement points or surfaces to impede the growth of bulk polymer nucleated crystals. How a given fiber affects the crystallization of a given matrix depends on the relative extent of these two competing effects.

Mehl and Rebenfeld (22) have shown by means of a computer simulation of the crystallization process that it is the relative rates of nucleation and crystal growth of fiber surface and bulk spherulites that controls the manner in which fibers affect the overall crystallization process of a composite, both in terms of rates and the developing morphology. These effects are illustrated in Figure 6.

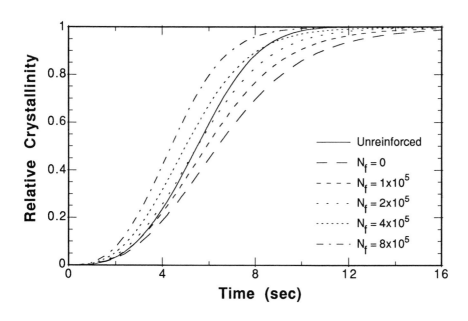

Figure 6. Simulation of polymer crystallization rate (22).

These and many other studies of fiber reinforced composites have served to optimize the properties of these materials and have allowed them to become the materials of choice in many cases. The combination of high performance and low weight makes them particularly attractive in automotive, aerospace, and transportation applications, and also in sports equipment. Active research continues on these important materials, and one can confidently expect further improvements in their performance and processing as new fibers and new matrices are developed and as we gain more complete understanding of the behavior of these two-phase systems. We can probably anticipate major innovations in the development and utilization of ceramic microfibers as reinforcements in polymer composites as well as in ceramic matrices. Mixed organic-inorganic systems may be quite promising because in many cases they mimic naturally occurring biomaterials which have exceptional properties.

One of the major obstacles to the more widespread use of all types of two-phase materials lies in the area of processability. At this time, fabrication methods for composites are largely batch processes, with many still relying on hand lay-up procedures. Emphasis must be placed on the development of automated continuous production technologies to allow not only improved efficiencies and economics but also more consistent supply of composite materials for end-product fabrication.

Conclusion

While many have claimed that fibers represent a mature technology and industry, with few further opportunities for major innovations, I think it is more likely that we shall see continuing evolution of new fiber based products and processes. Fiber production will continue to increase, simply because of steadily increasing world population, even if we assume no increase in the current average per capita fiber consumption of about 8 kg. On the other hand if we assume even a modest increase in worldwide standards of living, the per capita fiber consumption could rise, with major impact on fiber demand and production. Worldwide R&D in fibers and textiles is a healthy and dynamic enterprise, and there is vigorous competition in the marketplace. These factors alone promise new technologies, products and processes.

Today we utilize natural and man-made fibers in about equal amounts in the production of a nearly infinite number of fibrous products. It is interesting to note that just five basic polymer types (polyesters, polyamides, acrylics, polypropylene, and cellulose) provide the basis of the wide range of man-made fibers that are produced worldwide. It is a testament to the enormous versatility of these polymers that allows structure modification and optimization to achieve diverse material properties. At the same time, vigorous research on the structure and properties of natural fibers has resulted in improved fiber characteristics and product performance. However, in the next millenium we may see the slow disappearance of the distinction between natural and man-made fibers, as we manipulate the properties of cotton, wool and other "natural" fibers by genetic engineering. Of course, we are today just at the beginning of this genetic revolution, but it is not farfetched to speculate that its impact will dwarf that of the materials revolution that has dominated our field during the past decades. Under any circumstance, there will continue to be active and dynamic scientific research and technological development in fibers in the years ahead.

Literature Cited

1. Fiber Organon, Volume 69, June 1998, Fiber Economics Bureau, Washington, D.C.

2. Sisson, W. A. Textile Res. J. 1960, 60, 153.

3. Chanzy, H.; Paillet, M.; Hagege, R.; Polymer, 1990, 31, 400.

4. Batra, S. K., in "Handbook of Fiber Chemistry", 2nd edition, Lewin, M.; Pearce, E.M., Eds.; Marcel Dekker: NY, 1998.

5. Akin, D. E.; Morrison, W.H.; Gamble, G. R.; Rigsby, L. L.; Henriksson, G.; Eriksson, K.-E. L. Textile Res. J. 1997, 67, 279.

6. Keller, G.; Spatola, L.; McCabe, D.; Martinell, B.; Swain, W.; John, M. E. Transgenic Research 1997, 6, 385.

7. John, M. E.; Keller, G. Proc. Natl. Acad. Sci USA, 1996, 93, 12768.

8. Ziabicki, A.; Kawai, H.. Eds.; "High Speed Fiber Spinning"; John Wiley & Sons: NY, 1985.

9. Jeon, H. J.; Ito, H.; Kikutani, T.; Okui, N.; Okamoto, M. J. Appl. Polymer Sci. 1998, 70, 665.

10. Cooke, T. F. "Bicomponent Fibers - A Review of the Literature", Technical Information Center Report No. 44, TRI/Princeton, Princeton, N.J. 08542.

11. Fukuhara, M. Textile Res. J. 1993, 63, 387.

12. Yang, H. H. "Kevlar® Aramid Fiber"; John Wiley & Sons: NY, 1993.

13. Yang, H. H. in "High Technology Fibers, Part C", Lewin, M.; Preston, J., Eds.; Marcel Dekker: NY, 1993.

14. Conciatori, A. B.; Buckley, A.; Stuetz, D.E. in "High Technology Fibers, Part A", Lewin, M.; Preston, J., Eds.; Marcel Dekker: NY, 1993.

15. Donnet, J. B.; Bausal, R. C. "Carbon Fibers" 2nd Edition, Marcel Dekker: NY, 1990.

16. "High Performance Synthetic Fibers for Composites", Publication NMAB-458, National Academy Press, Washington, D.C., 1992.

17. Desio, G. P.; Rebenfeld, L. J. Appl. Polymer Sci. 1992, 44, 1989.

18. Dean, D. M.; Rebenfeld, L.; Register, R. A.; Hsiao, B. S. <u>J. Materials Sci.</u> 1998, <u>33</u>, 4797.

19. Chen, E. J. H.; Hsiao, B. S. <u>Polymer Eng. Sci.</u> 1992, <u>32</u>, 280.

20. Mehl, N. A.; Rebenfeld, L. <u>J. Appl. Polymer Sci.</u> 1995, <u>57</u>, 187.

21. Reinsch, V. E.; Rebenfeld, L. <u>J. Appl. Polymer Sci.</u> 1994, <u>52</u>, 649.

22. Mehl, N. A.; Rebenfeld, L. <u>J. Polymer Sci., Polymer Physics</u> 1993, <u>31</u>, 1677 and 1687.

ADHESIVE AND SEALANT CHEMISTRY

LIENG-HUANG LEE*

Consultant for Adhesion Science, 796 John Glenn Blvd., Webster, NY 14580

*Honorary professor of the Chinese Academy of Sciences

Introduction

A previous review (1) on adhesive and sealant chemistry was published in 1987. In the past fifteen years, the technology has moved forward under the external pressure on the reduction of volatile organic compounds (VOC) used in the adhesive and sealant industry, the drive in energy saving and the need for recyclability. We shall demonstrate that most polymers adopted before 1984 are essentially the same. However, the fine tuning of each polymer system and the application of new technologies have propelled this industry to a steady level of growth. In this paper, we attempt to highlight the chemistry and recent advances. For a detailed description, readers should consult two comprehensive reviews (2-3) published in recent years.

Historical Background

Natural adhesives, such as tar and starch, were used by ancient people for various purposes of gluing things together (4). Even up to this day, natural adhesives still occupy about one-third of the total consumption. The first synthetic polymer, nitrocellulose, was used as adhesive in 1869. Then came phenolics, urea-formaldehyde, polyurethane, polyvinyl acetate, chloroprene rubber, carboxylic-terminated nitrile rubber. Since the beginning of World War II, styrene-butadiene rubber (SBR) replaced a good portion of natural rubber (NR) as adhesives. In 1941, de Bruyne (5) discovered the Redux system using the synergistic effect of resorcinol-formaldehyde and polyvinyl butyral to form a structural adhesive that can bond metal to metal, especially for aircraft.

Between 1940 and 1950, several important classes (4,6) of adhesives were introduced. Epoxy resin, butyl rubber, silicone sealant, polysulfide sealant, and cyanoacrylate expanded the spectrum of applications. Between 1950 and 1960, tackifiers and block copolymer thermoplastic elastomers (TPE) laid the foundation for today's pressure sensitive adhesives (PSAs). In 1965, ethylene-vinyl acetate (EVA) emulsion was introduced. This may be one of the earlier water based synthetic adhesives. During the period of 1965 and 1980, the second generation epoxy resin was introduced and several high temperature adhesives were synthesized for aerospace applications. Table 1 is a brief summary of the early innovation on adhesives and sealants. This list, by no means complete, does serve to provide our readers with a background about the early contributions of chemistry to this industry.

Present Status and Future Trends of Adhesive and Sealant Industry

In 1996, about one-third of adhesives in volume (7), or nearly 4 billion pounds was based on natural products. The remaining 8.5 billion pounds were various synthetic polymers supplied by about 500 companies in the U.S. The distribution of these polymers are illustrated in Fig. I. Phenolics (37%) and urea and melamine (19%) are the leading families of adhesives. Figure 2 describes the end-use of various materials. Among them, construction (40%) and packaging (39%) nearly make up 80% of the total consumption. Figure 3 shows the projected revenues (in %) of adhesive and sealant related products to be produced in the year 2001.

Table 1.
Early Innovation of Adhesives and Sealants

Year	Innovation	Innovator	Patent or Reference
1690	First glue factory in Holland		
1808	First glue factory in Boston, U.S.A.		
1845	Natural rubber adhesive plaster (tape)	W.H.Sheout & N.H.Day	US Pat. 3,965
1869	Cellulosics, nitrocellulose (pyroxalin)	Hyatts Bros (UK); Parks (US)	
1905	Phenolics	L.H. Bakeland	US Pat. 1,019,406 (1912) (to General Bakelite Co.)
1911	"Hexa"-hexamethylene diamine, curing agent for Novolac	Aylsworth	
1920-30	Urea-formaldehyde	H. John	US Pat. 1,355,834 (1920)
1930	Isocyanates and polyurethane	O. Bayer	Angew. Chem. 59 (9), 257 (1947)
1912-32	Polyvinyl acetate		Ger.Pat. 281,687 (1913) & US Pat. 1,241,738 (1918)
1930-39	Adhesive Tape	R. G. Drew	US Pat. 1,760,820 (to 3M Co.)
1928	Neoprene - first synthetic rubber	du Pont de Nemours Co.	US Pat. 2,038,539 (1931)
1936	Stein-Hall process to gelatinize starch	J. V.Bauer	US Pat.2,051,025 (to Stein-Hall Mfg. Co.)
1939	Carboxylic containing nitrite rubber	H.P.Brown & J.F.Anderson	US Pat 2,671,074 (1957) to B.F. Goodrich Co.
1940	Styrene-butadiene rubber, styrene-isoprene rubber	L.E.Puddefoot & W.H.Swire	BP.567,096 (1945) (to B.B. Chemical Co.Ltd.)
1941	Redux (resorcinol-formaldehyde and polyvinyl butyral), first metal adhesive	N.A. de Bruyne	U.S.Pat.2,499,134 (1952)
1943	Epoxy resin curing	P.Castan	US Pat.2,324,483 & 2,444,333 (to de Trey Freres Co.)
1944	Butyl rubber	R.M.Thomas& W.J.Sparks	US Pat.2,356,128 & 2,356,130 (to Standard Oil Co.)
1945	Silicone Sealant	E.G.Roochow	US Pat.2,380,995 (to G.E.Co.)
1946	Polysulfide rubber	J.C.Patrick & H.R.Ferguson	US Pat.2,466,963 (to Thiokol Chemical Co.)
1941-45	Nitrile rubber-phenolics, first metal-to-metal bonding epoxy adhesive	Davis, et al.	US Pat.2,571,217 (to Shell Development Co.)
1947	Epoxy resin	Marples, et al.	US Pat.2,428,235 (to Shell Development Co.)
1949	Cyanoacrylate	A.E.Ardis	US Pat.2,467,926 (to B.F.Goodrich Co.)
1950	Epoxy resin	Bradley	US Pat.2,500,449 & 2,500,600 (to Shell Development Co.)
1957	Cyanoacrylate adhesive	H.W.Coover, Jr.	US Pat.2,794,788 (to Tennessee Eastman Co.)
1957	Tackifiers for PSA	Thompson	US Pat.2,610,910 & 2,918,442 (to 3M Co.)
1945-55	Nitrile rubber-epoxy film	Gerrard	US Pat.2,918,442 (to 3M Co.)
1959	Anaerobic	V.K.Krieble	US Pat.2,895,950 (to G.E.Co.)
1955-65	Block copolymer thermoplastic elastomers		US Pat.3,239,478 (1966) (to Shell Development Co.)
1965	Ethylene-vinyl acetate emulsion	Air Reduction Co.	
1965-80	High-temperature adhesives		
1970-80	2nd generation toughened epoxy resin		

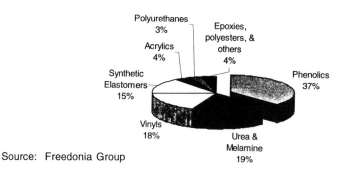

Source: Freedonia Group

Figure 1. Leading Synthetic Polymers for Adhesives and Sealants

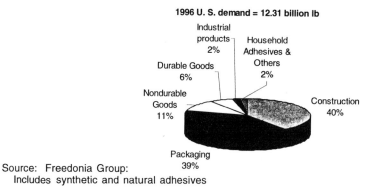

Source: Freedonia Group:
    Includes synthetic and natural adhesives

Figure 2. End-use of Various Adhesives and Sealants

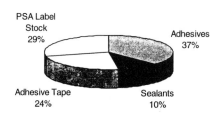

Source: Frost & Sullivan report 3186

Figure 3. Projected revenues (in %) of various adhesive and sealant related products in the year 2001

Table 2
U.S. Adhesives Demand (in Millions of Pounds)

Item	1987	1996	2001	96/87 %/year	01/96 %/year
Construction	4,074	4,975	5,619	2.2	2.5
Packaging	3,911	4,843	5,551	2.4	2.8
Nondurable goods	1,150	1,325	1,515	1.6	2.7
Durable goods and other markets	895	1,165	1,365	3.0	3.2
Total	10,030	12,307	14,050	2.3	2.7

Source: The Fredonia Group Inc., Cleveland, OH
Ref.: *Adhesives Age*, 40 (12), 38, Sept. 1997.

By the year of 2001, the US demand for adhesives (8,9) is expected to reach 14 billion pounds, an estimated value of $9 billion (Table 2). The growth of PSAs (10) alone has been estimated to be between 5% to 6% per year (Table 3). Worldwide growth of the industry has been estimated to be 3% to 4% per year depending on the area of the globe (Table 4). Prior to the recent crisis, the Asia/Pacific area has been projected to be the leader with an annual growth of 4.7%. In that region, the rapid growth of the adhesives industry in China has been phenomenal in the past several years and far above the average in comparison to other countries in the Pacific Rim (11).

Table 3
U.S. Pressure Sensitive Tape Shipments (in Millions of Dollars)

Item	1987	1997	2002	87/97 %/year	97/2002 %/year
Pressure sensitive tape shipments	2,254	4,410	5,885	6.9	5.9
Plastic	1,163	2,603	3,591	8.4	6.6
Paper	606	842	988	3.3	3.2
Other	485	964	1,306	7.1	6.3

Source: The Fredonia Group Inc., Cleveland, OH
Ref.: *Adhesives Age*, 42,33, January, 1999.

Table 4
1996-2001 Incremental New Demand and Growth (in Millions of Pounds)

Region	1996-2001 Incremental New Demand	Old (1996-2001) % Rate of Growth	Adjusted* % Rate of Growth
North America	640	2.2	2.0
Western Europe	660	2.3	2.2
Asia Pacific	575	4.8	3.5
Latin America	70	2.8	2.8
Rest of World	225	3.0	2.7
Total	2,110	2.8	2.4

Source: CHEM Research/DPNA International   *Adjusted in 1998 for the period of 1996-2001.
Ref.: *Adhesives Age*, 42,18,Jan.1999.

Types of Adhesives

For the convenience of our discussion, we can arbitrarily divide adhesives into five types (2,3) according to their functions: structural, hot melt, pressure sensitive, water based and UV/EB cured. However, the division of adhesives is not always clear-cut, and there are unavoidable overlaps. For example, pressure sensitive adhesives can be made from solvent based, water based, hot melt or UV/EB curable systems. Characteristics of various types of adhesives are summarized in Table 5. In addition, we also list advantages and limitations for each type of adhesives in Table 6. Now, let us briefly discuss each type in the following section.

Table 5
Characteristics of Various Types of Adhesives

Type of Adhesive	Characteristics	Polymer
Structural	Bonds can be stressed to a high proportion of maximum failure load under service environments Most are thermosets, one or two-component systems	Epoxies, polyurethane, modified acrylics, cyanoacrylates, anaerobics, silicones, phenolics, high-temperature adhesives
Hot Melt	100% solid thermoplastics Melt sharply to a low-viscosity liquid Rapid setting, no cure Nonpressure sensitive and pressure sensitive Compounded with additives for tack and wettability	Ethylene-vinyl acetate copolymer, polyethylene, polyvinyl acetates, polybutene, thermoplastic elastomers, polyamides (nylons), and polyesters
Pressure Sensitive	Hold substrate together upon brief application of pressure at room temperature Available in solvent-based, water-based and hot melt Some require compounding (rubber base) to achieve tackiness, where others (acrylate base) don't Available supported (most) or unsupported on substrate	Natural rubber, reclaimed rubber, styrene-butadiene rubber, butadiene-acrylonitrile rubber, TPE, polyacrylates, polyvinyl alkylethers, and silicones
Water Based	Including adhesives dissolved or dispersed (latex) in water On porous substrates, water is absorbed or evaporated to achieve bonding Some are bonded following reativation of dried adhesive film under heat and pressure Nonpressure sensitive (most) or pressure sensitive applications	Before 1960, polyvinyl alcohol, polyvinyl acetate, urea-formaldehyde, phenolics, SBR, neoprene, nitrile rubber and polyvinyl methyl ether After 1960, acrylics, multi-component systems are acrylated silicone, acrylated urethanes, and acrylated silicone urethanes
UV/EB Cured	100% reactive liquids cured to solids One substrate must be transparent for UV cure, except when dual-curing adhesives are used For the dual-curing system, the second cure is achieved with heat or moisture or the elimination of oxygen In EB curing, density of material affects penetration UV/EB-curable formulations are for laminating or PSA applications UV-curable formulations are for laminating, PSA, and structural adhesive applications	Acrylates, epoxies, urethanes, polyesters or polyethers Cationic-cured systems are based on epoxies with reactive diluents and cyclic monomers UV/EB PSA are acrylics, synthetic rubbers, and/or silicones

Ref.: *Engineered Materials Handbook, Desk Edition*, 638, ASM International, OH, 1995.

Table 6
Advantages and Limitations of Various Types of Adhesives

Type of Adhesive	Advantages	Limitations
Structural	High strength Load resisting Good elevated-temperature resistance Good solvent resistance Some available in film form	Two-component system requires careful, proper mixing Some have poor peel strength Some are difficult to remove and repair Some produce unwanted by-products upon curing
Hot Melt	100% solid, and solventless Able to bond to impervious surfaces Rapid bond formation Good gap-filling capability Forming rigid and flexible bonds Good barrier properties	Limited elevated-temperature resistance Poor creep resistance Little penetration due to fast viscosity increase upon cooling Limited toughness at usable viscosities
Pressure Sensitive	Uniform thickness for both tapes and labels Permanent tack at room temperature No activation required by heat, water, solvents Crosslinking of some formulations possible Soft or firm tapes as labels Easy to apply	Many are rubber-based, requiring compounding Poor gap fillers Limited heat resistance
Water Based	Low-cost, nonflammable, nonhazardous solvent Long shelf life Easy to apply Good solvent resistance Crosslinking for some formulations possible High mw dispersions at high solids content with low viscosity	Poor water resistance Slow drying Tendency to freeze Low strength under loads Poor creep resistance Limited heat resistance Shrinkage of certain substrates in supported film and tapes
UV/EB Cured	Fast cure (some in 2 to 60 seconds) One-component liquid; no mixing and no solvents Heat-sensitive substrates can be bonded; cure is "cool" Many are optically clear High production rates Good tensile strength	Equipment, e.g., EB is expensive Relatively high material cost UV-cures only through transparent materials Difficult curing on parts with complex shapes Many UV cures have poor weatherability due to continuous absorbing of UV rays

Ref. : *Engineered Materials Handbook, Desk Edition*, 638, ASM International, OH, 1995.

Structural Adhesives

Structural adhesives generally bond the load-bearing parts of a product. In practice, epoxies, polyurethanes, modified acrylics, cyanoacrylates, and anaerobics are considered to be chemically reactive structural adhesives. Typical properties (3) of these adhesives are shown in Table 7. Advantages and limitations of each type of adhesive are described in Table 8.

Epoxies are one of the old work horses for structural adhesives. However, epoxies lack toughness and durability. Fortunately, both of these are somewhat remedied by the incorporation of a carboxylic-terminated nitrile rubber, CTBN, developed by B. F. Goodrich (12). The rubber phase is dispersed in the epoxy matrix to form the second generation adhesive, which has been successfully employed in structure bonding.

In the middle 1980s, 3M Company introduced formulations allowing the rubber phase to disperse in a controlled fashion. That technology produced the third generation epoxies (13). Toughness of the adhesives was superior to previous generations.

Another problem with epoxies is flexibility. Several forms of flexible epoxies have been offered by Dow Chemical, Shell Chemical and Cardolite Corp. (14). The last is based on Novolac resin with long alkyl side chains (Fig. 4). In addition to flexible epoxies, there are also flexible curing agents, as shown in Fig. 5.

Polyurethanes (PUs) (15) are unique in their wettability to most substrates. They can readily form hydrogen bonds with polar surfaces. The oldest type of polyurethane was the one-component PU. In that system moisture reacted with isocyanate groups to form urea and biuret. Through crosslinking, both strength and molecular weight increased. The isocyanate groups can be blocked from polymerizing until the heat activation takes place.

Applications for water based PUs have grown in recent years because they are environmentally friendly, nonflammable, and non VOC (vapor of organic compound) emitting. The water-dispersed PU prepolymers can be made by reacting a carboxy-functional PU with a diamine to form a water-soluble anionic polyurea-polyurethane.

The two-component PUs are adaptable to fast-cure and lamination systems (16). Typically, they consist of a low equivalent-weight isocyanate or prepolymer, which is then cured with a low equivalent-weight polyol or polyamine. They can be either solvent based or 100% solid.

Modified acrylics, or the second generation acrylics, contain a modifier and an activator. A typical example is a crosslinkable poly(methyl methacrylate) grafted to a vinyl-terminated or a carboxylic-terminated nitrile rubber (17). Since it is cured by free radical addition, there is no need to apply the accurate proportion of a two-component system. Modified acrylates are good in flexibility and good in peel and shear strengths. However, the curing process is slower than those for cyanoacrylates and anaerobics. They also have a limited open time.

Cyanoacrylates, commonly known as superglue, are not gap-fillers (18). They are especially suited for large-area bonding, but there are problems associated with odor, fast cure and runniness. Thus, one approach is to thicken the monomer to high Newtonian viscosities (3,000 to 4,000 cps). Another approach is to apply a non-flowable gel that has been made by incorporating hydrophobic silicas.

Cyanoacrylates are brittle. Thus, toughening is a necessary step to insure its load-bearing. A toughened cyanoacrylate (18) was produced by blending the monomer with a rubbery polymer of ethylene and methyl methacrylate. An interesting application in the space industry is the use of cyanoacrylates in space shuttles. Prism 410, a high-viscosity toughened cyanoacrylate, was chosen to bond the rubber seal to the front skirt on the the solid rockets for the space shuttle (19). The bonded seal can prevent seawater corrosion of electrical components during splashdown.

Quick curing can be achieved by removing the inhibitor-containing crown ethers, silacrown compounds, certain polyglycoethers (20), or calixarens (21). The problem of odor has also been addressed. The substitution of n-butyl group with a methoxyethyl group produced an odorless cyanoacrylate. Loctite developed a thermal-resistant family of cyanoacrylates with an extended use for greater than 1,000 hours at 121°C.

Cyanoacrylates were found to be one of the important medical adhesives (22). N-butyl cyanoacrylate has been used routinely in certain veterinary procedures and in a minor wound closure kit, introduced by Loctite. In China, there have been attempts to use this polymer as a contraceptive to control the birthrate (23).

A. DER 732 and DER 736 (Dow Chemical)

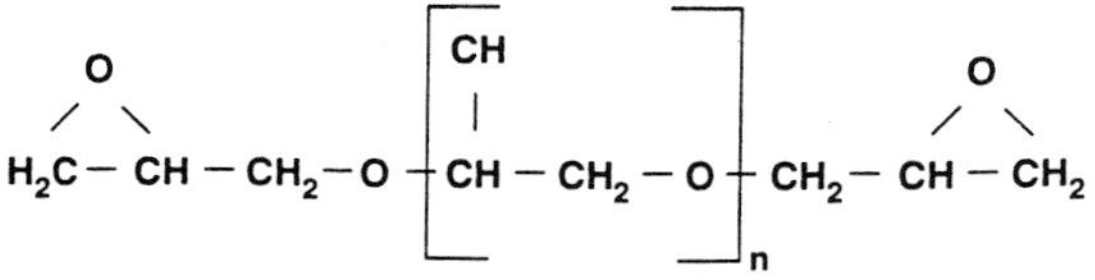

$$H_2C-CH-CH_2-O\left[CH-CH_2-O\right]_n CH_2-CH-CH_2$$
with CH substituent

Note: for DER 732, n=9 and for DER 736, n=4

B. Epon 871 (Shell Chemical)

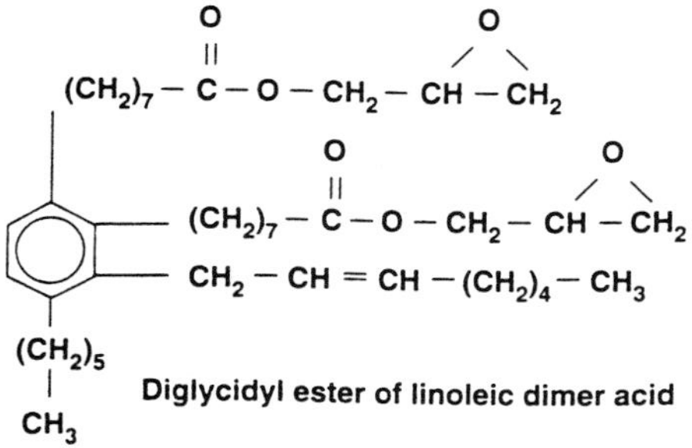

$$(CH_2)_7-C-O-CH_2-CH-CH_2$$
$$CH_2-CH=CH-(CH_2)_4-CH_3$$

$(CH_2)_5$
CH_3

Diglycidyl ester of linoleic dimer acid

C. Cardolite NC 547 (Cardolite Corp.)
Epoxy novolac resin

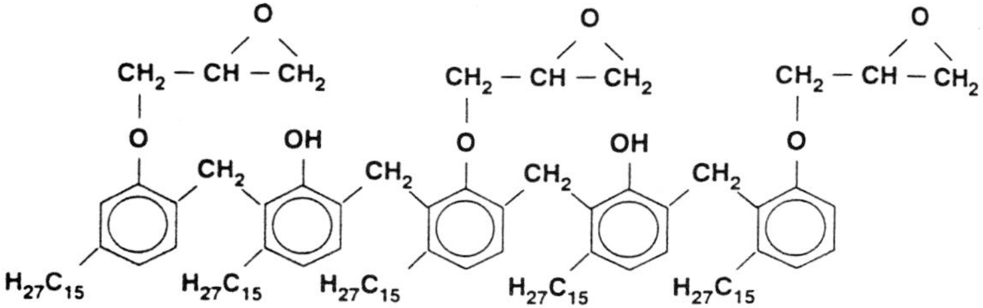

$CH_2-CH-CH_2$ (repeated)
O
OH CH_2
$H_{27}C_{15}$

Figure 4. Commercially available flexible epoxy resins

A. Dytek A (Du Pont)

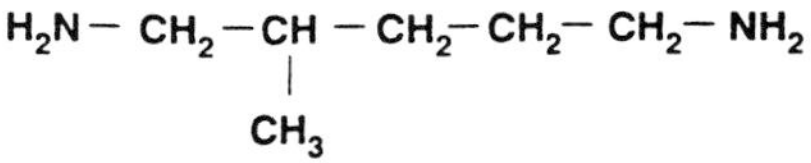

$$H_2N-CH_2-CH-CH_2-CH_2-CH_2-CH_2-NH_2$$
CH_3

2-Methyl-pentamethylene diamine

B. Kenamine DP-3680 (Witco Chemical)

$(CH_2)_7-CH_2-NH_2$
$(CH_2)_7-CH_2-NH_2$
$CH_2-CH=CH-(CH_2)_4-CH_3$
$(CH_2)_5$
CH_3

Di-primary amine of linoleic dimer acid

Figure 5. Flexible curatives

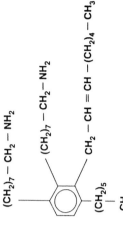

Table 7
Typical Properties of Chemically Reactive Structural Adhesives

Property	Epoxy	Polyurethane	Modified Acrylic	Cyano-acrylate	Anaerobic
Substrates bonded	Most	Most smooth, nonporous	Most smooth, nonporous	Most nonporous metals or plastics	Metals, glass, thermosets
Service temp. °C. (°F)	-55 to 121 (-67 to 250)	-157 to 79 (-250 to 175)	-73 to 121 (-100 to 250)	-55 to 79 (-67 to 175)	-55 to 149 (-67 to 300)
Impact resistance	Poor	Excellent	Good	Poor	Fair
Tensile shear strength MPa (ksi)	15.4 (2.2)	15.4 (2.2)	25.9 (3.7)	18.9 (2.7)	17.5 (2.5)
T-peel strength, N/M (lbf/in)	<525 (3)	14,000 (80)	5,250 (30)	<525 (3)	1,700 (10)
Heat cured or mixing	Yes	Yes	No	No	No
Solvent resistance	Excellent	Good	Good	Good	Excellent
Moisture resistance	Excellent	Fair	Good	Poor	Good
Gap limitation,nm (in)	None	None	0.762 (0.03)	0.254 (0.01)	0.635 (0.025)
Odor	Mild	Mild	Strong	Moderate	Mild
Toxicity	Moderate	Moderate	Moderate	Low	Low
Flammability	Low	Low	High	Low	Low

Ref.: *Engineered Materials Handbook, Desk Edition*, 641, ASM International, OH, 1995.

Table 8
Advantages and Limitations of Chemically Reactive Structural Adhesives

Adhesive	Advantages	Limitations
Epoxy	High strength Good solvent resistance Good gap-filling capabilities Good elevated-temperature resistance Wide range of formulations Relatively low cost	Exothermic reaction Exact proportions needed for optimum properties Two-component formulations require exact measuring and mixing One-component systems often require cold storage and an elevated-temperature cure Short pot life (more waste)
Poly-urethane	Varying cure times Generally tough Good flexibility at low temperature One- or two-component, room- or elevated-temperature cure Moderate cost	Moisture-sensitive, before or after curing Poor elevated-temperature resistance Depolymerizing with heat and moisture Short pot life Special mixing/dispensing equipment required
Modified acrylic	Good flexibility Moderate cost Good peel and shear strengths No mixing require Tolerates dirty surfaces Room-temperature cure	Low elevated-temperature strength Slower cure than anaerobics or cyanoacrylates Toxic, objectionable odor Flammable Limited open time Dispensing equipment required
Cyano-acrylate	Rapid room-temperaure cure Single component High tensile strength Long pot life Good adhesion to metal Dispenses easily from package	High cost Poor durability on some surfaces Limited solvent resistance Limited elevated-temperature resistance Sticks to skin
Anaerobic	Rapid room-temperature cure Good solvent resistance Nontoxic Good elevated-temperature resistance No mixing Indefinite pot life High strength on some substrates	Not good for permeable surfaces Unable to cure in air as a wet fillet Limited gap cure

Ref. : *Engineered Materials Handbook, Desk Edition*, 640, ASM International, OH, 1995.

Anaerobics (24) are known to cure in the absence of air. The curing mechanism has been postulated to involve a redox initiation of methacrylate monomers. The initiation is catalyzed by a transition metal species on the substrate surface and inhibited by air to form an inactive species. Initiators include an amine, sulfimide and a peroxide. Metal species must be carefully controlled during production and packaging in order to obtain a complete polymerization. Generally, anaerobics are good in solvent resistance and elevated-temperature resistance, but not suitable for porous surfaces. They have indefinite pot life, but possess limited gap cure.

Hot Melt Adhesives

In general, hot melt (HM) adhesives (25) are 100% thermoplastic solids. Bond formation is achieved by heating the adhesive well above its melting point (250-350°F). The molten adhesive wets both surfaces. Upon cooling, a bond is finally formed. A typical hot melt formulation consists of 33% polymer, 33% resins, 32.5% waxes and 0.5% antioxidant. Suitable polymers are polyvinyl acetate (PVAc), ethylene-vinyl acetate (EVA), polyurethane (PU), polyamides, polyamide copolymers, and aromatic polyamides. Table 9 shows typical properties of hot melt adhesives.

In recent years hot melt adhesives have gained popularity because of the following advantages: speed of bond formation, 100% solids, versatility, gap filling capability, and lower energy consumption. One limitation of hot melt adhesives is temperature sensitivity. The polymer tends to undergo oxidative or thermal degradation at elevated temperatures.

To alleviate the high temperature exposure, a new trend is to develop warm melt adhesives to be used at somewhat lower temperatures. Dahmane (26) used a new ethylene-vinyl acetate polymer as the basic component to prepare warm melt adhesive. Applications for these adhesives are for packaging in the forms of case sealing, carton sealing, tray erecting, container labeling, bookbinding and bag and sack manufacturing.

Most hot melt adhesives do not require curing. However, a new technique (27) is based on the curing of a block copolymer by peroxide or sulfur or UV/EB radiation. A reactive urethane can also be slowly cured by reacting with water.

Water Based Adhesives

The industry trend is to produce zero-emission adhesives such as the water based (WB) adhesives. During the last decade, the volume of WB adhesives has increased to over 50% of total production. Consequently, the VOC for the adhesive industry has dropped approximately to 10%. The largest application of WB adhesives is in the packaging industry, and the next is in construction.

The differences between the WB adhesives and the solvent based (SB) adhesives are not only the medium of dispersion, but also the following characteristics (28):
- WB adhesives are heterophase, e.g., emulsion, while SB adhesives are in a continuous phase.
- WB adhesives can have high molecular weight (or viscosity), while SB adhesives are mostly low molecular weight to maintain coatable viscosity.
- WB adhesives contain surfactants, defoamers, fillers, etc., while SB adhesives are generally formulated with no additives harmful to adhesion. For example, surfactants have always been a problem for WB systems.
- WB adhesives involve coalescence, while SB adhesives form films through diffusion, etc.

Table 9
Typical Properties of Hot Melt Adhesives

Property	Polyvinyl acetate	Ethylene-vinyl acetate and polyolefins	Polyure-thane	Polyamide	Polyamide copolymer	Aromatic polyamide
Brookfield viscosity, Pa.s	1.6-10	1-30	2	0.5-7.5	11	2.2
Viscosity test temperature °C (°F)	121 (250)	204 (400)	104 (220)	204 (400)	230 (446)	204 (400)
Softening temperature °C (°F)	–	99-139 (211-282)	–	93-154 (200-310)	–	129-140 (265-285)
Application temperature °C (°F)	121-177 (250-350)	–	–	–	–	–
Service temperature °C range (°F)	-1 to 120 (30 to 248)	-34 to 80 (-30 to 176)	–	-40 to 185 (140 to 365)	–	–
Bonding substrates	Paper, wood leather, glass, selected plastics selected metals	Paper, wood, selected plastics, selected metals, selected glass	Plastics	Wood, leather, selected plas-tics, selected metals	Selected metals, selected plastics	Selected metals, selected plastics
Applications	Tray forming, packaging, binding, sealing cases and cartons, bottle labels	Bookbinding, packaging, toys, automotive, furniture, electronics	Laminates	Packaging, electronics, furniture, footwear	Packaging, electronics, binding	Electronics, packaging, binding
Relative cost (to other hot-melt adhesives	Low to medium	Lowest	Medium to high	High	High	High

Ref.: *Engineered Materials Handbook, Desk Edition,* 644, ASM International, OH 1995.

Generally, WB adhesives have some problems in wetting low surface energy polymeric films. Overall properties usually inferior to SB adhesives: low moisture resistance, lower shear strength at elevated temperatures, lower peel strength at room temperature, and less flexibility in adhesion to a broad range of backings.

Water dispersible polyesters for HM adhesives have been marketed by Eastman Chemical Co. (29). The polymer is made dispersible by attaching an ionic group to the polyester chain, *e.g.*, 5-iodosulfoisophthalate. Structurally, the branched polyester has a broad distribution of molecular weight (M_w /M_N, 3.5 to 4 versus 2.3 for the linear counterpart).

Archer (30) described several new developments for the WB contact adhesives: the use of polymer blending, the emulsification of an SB adhesive, and the use of a foam adhesive. The foam adhesive provides an improvement to coverage, a decrease in drying time and an improvement in bond strength. Most WB contact adhesives are based on chloroprene.

The idea of hybridization has also been applied to WB adhesives. An ultrahigh solids VAE emulsion (31) has been blended with PU emulsions. In general, the blends possess enhanced properties, *e.g.*, cohesive strength, vinyl adhesion, heat sealability and crosslinking performance.

Pressure Sensitive Adhesives

There are at least three types (32) of pressure sensitive adhesives (PSAs): acrylates, rubber, and silicone. Generally, the polymer has to pass the Dahlquist criterion (33) in order to be qualified as a PSA. In other words, the polymer should have a modulus of 10^5 Pa or a T_g of 40-70°C below room temperature. The advantages and limitations of PSAs are summarized in Table 10. Historically, natural rubber was first used to make adhesive tape or medical plaster in 1845 by Sheout and Day (34). Between 1930 and 1939, Drew of 3M Company (35) patented the making of adhesive tapes. Nowadays, the acrylates and block copolymers have become the core polymers for PSAs. In recent years, silicone PSAs have also achieved significant applications, *e.g.*, medical adhesives.

Hoffman and Miles (36) compared various technologies in manufacturing PSAs (Table 11): 1. solvent based, 2. water based, 3. hot melt, 4. transfer, and 5. UV-curable. Each of these processes will be discussed separately.

There are two interesting works pertaining to WB PSAs. BASF researchers (37) developed a formulation for the WB PSAs. A tetramer, vinyl acetate-2-ethylhexyl acrylate/methyl methacrylate/acrylic acid and other copolymers can be made to a solids content of about 70%. Frezee (38) introduced a polymeric surfactant, that is a low molecular weight copolymer of butyl acrylate/acrylic acid (70:30). With this surfactant, the moisture sensitivity of the WB PSA could be reduced.

For hot melt application, the acrylates generally need to be high in molecular weight in obtaining acceptable properties. Bateman (39) discovered a method by complexing an acid group of the acrylate copolymers with zinc octoate in the presence of o-methoxybenzoic acid. Interestingly, at low temperature, Zn-acrylate complex is the predominant product, but at high temperature, Zn-methoxybenzoate is favored, so the low molecular weight polymer is set free to be coated on the substrate.

Since the 1970s, block copolymers have replaced natural rubber in many PSA formulations. Basically, there are three kinds of block copolymers (40): linear, hydrogenated, and multi-arm architectures. Linear polymers are based on SIS (styrene-isoprene-styrene), and SBS (styrene-butadiene-styrene) developed by Shell Chemical. Upon hydrogenation, the SBS and SIS yield, respectively, styrene-ethylene-butylene-styrene (SEBS) and styrene-ethylene-propylene-styrene (SEPS) polymers. SEPS is suitable for hot melt applications; the hydrogenated versions are more oxidative resistant.

Another innovation is to produce radial or star achitectures which allow higher molecular weight in a given hydrodynamic volume relative to linear polymer. Phillips Petroleum Co. (41) first introduced a radial polymer (SB)4 by reacting the (SB) anion with SnCl4 agent. Shell Chemical (42) then introduced Kraton 1320 with a structure of $(SI)_n$ prepared with the use of divinylbenzene. It should be noted that these star polymers are more sensitive to UV or electron beam than linear polymers. Thus, radiation curing enhances high temperature resistance of the $(SI)_n$ adhesives.

Recently, the styrene content of the SIS copolymer was raised from 15 to 40%, but at a lower molecular weight (43). This modification enables the resultant hot melt to be suitable for spray and fine line applications. On the other hand, silicone PSAs differ from the acrylate and the rubber based PSAs, especially in flexibility and thermal stability. Silicone PSAs are suitable for continuous use between -70° to 250°C and for intermittent high temperature exposure. In addition they can adhere to relatively lower surface energy polymers, *e.g.* polytetrafluoroethylene (PTFE). Other characteristics are shown in Table 10.

Table 10
Advantages and Limitations of Various Polymers for Pressure-Sensitive Adhesives

Type	Advantages	Limitations
Acrylates	Good UV resistance Good hydrolysis resistance Excellent adhesion buildup Good solvent resistance Good temperature use range (-45 to 121°C, or -50 to 250°F) Easier to apply (than rubbers) Good shear strength Good service life	Poor creep resistance (worse than rubbers) Fair initial adhesion Moderate cost (compared to rubbers, silicones) Poor adhesion to low polar surfces Poor adhesion at low temperatures
Rubbers	Good flexibility High initial adhesion Ease of tackification (with additives) Lower cost Good shear strength Good adherence to low- and high- energy surfaces Suitable for temporary and permanent holding	Low tack and adhesion (without additives) Poor aging, subject to yellowing Limited upper service temperature use Moderate service life Poor solvent resistance
Silicones	Excellent chemical and solvent resistance Wide temperature use range (-73 to 260°C or -100 to 500°F) Good oxidation resistance Good adherence to low- and high- energy surfaces	Highest cost (compared to rubbers, acrylates) Less aggressive adhesion than acrylates at ambient temperatures Sensitive to strong acids and bases

Ref.:   1. *Engineered Materials Handbook, Desk Edition*, 644, ASM International, OH, 1995.
         2. S. R. Kerr III, *Adhesives Age*, 35 (13), 32, Dec., 1992.

Table 11
Advantages and Limitations of Various Pressure-Sensitive Adhesive Technologies

Technology	Advantages	Limitations
Solvent-Based	High strength available High temperature resistance	Premature drying may affect application Drying oven needed Solvent disposal and clean-up VOC emissions Solvent attack on substrate
Water-Based	Low-medium cost High solids with low viscosity possible Nonflammable Low VOC	May freeze Difficult cleanup once dry Premature drying during application Dry-off oven required
Hot Melts	100% solids (no VOC) Rapid set-up; no cure Very thick film possible	Apply at elevated temperatures, us.>300°F Specialized application requirement Temperature resistance limited
Transfer	No liquids to work with Performance simillar to solvent-based	Must be die cut Significant wastes Very high cost
UV-Cured	Fast cure (seconds) Minimal floor space required Optically clear versions Applicable to heat-sensitive substrates Applicable by spray, roller, screen pad	Deep tints and cure through non-transparent films difficult Perceived high cost High-temperature resistance limited

Ref.: W.E. Hoffman and D.E. Miles, *Adhesives Age,* 35 (4), 20 , April, 1992.

Traditional Silicone PSA Technology

$$\text{HO-}\underset{\underset{CH_3}{|}}{\overset{\overset{CH_3}{|}}{Si}}\text{-(-O-}\underset{\underset{CH_3}{|}}{\overset{\overset{CH_3}{|}}{Si}}\text{-)}_n\text{-OH} \quad + \quad \text{MQ Resin} \xrightarrow[\geq 150°C]{peroxide} \quad \text{Cured PSA} \; + \; Solvent$$

(40-60 % VOC)

High M. W. "gum"

High Solids Silicone PSA Technology

$$\text{~O-Si-CH=CH}_2 \; + \; \text{HSi-O~} \; \text{+MQ Resin} \xrightarrow[\geq 110°C]{Platinum} \quad \text{Cured PSA} \; + \; solvent$$

(5-15 % VOC)

"Functionalized" silicones

Figure 6. Traditional versus high solids silicone technology

Regarding silicone PSAs, in 1953 Dexter (44) discovered that the addition of a silicone resin to a silicone rubber could produce a silicone PSA. A phenyl silicone PSA (39) has been optimized to yield better high temperature performance. Lin of GE Silicone (45-47) reported that a high solid (80-90%) silicone PSA could be produced by using platinum instead of peroxide as a catalyst for the polymerization (Fig. 6). One of the resins used in the formulation is the MQ resin, which is a highly branched siloxane copolymer. This glassy resin is made up of trimethylsiloxy (M or $(CH)_3Si\text{-}O$) units and quadri-functional siloxane (Q or SiO_4^{2-}). Later, GE Silicone (48) introduced the two-component silicone PSA having about 95% solids with a viscosity of 50,000 cps at room temperature.

Radiation-curable Adhesives

Both electron beam (EB) and ultraviolet light (UV) have been successfully applied (49) to cure adhesives and coatings. The EB accelerators are equipped to generate an electron beam capable of curing thicker, pigmented films. For UV-curing, the energy sources are typically medium-pressure mercury lamps, pulsed xenon lamps or lasers. However, for these the film should be thin and transparent.

The composition for a UV-curing system consists of an unsaturated monomer, an oligomer and a photoinitiator. The photon energy induce the photoinitiator to polymerize the monomer. On the other hand, the EB-curing is caused by the penetration of high energy electron beam to interact directly with the atoms generating reactive species that initiate polymerization.

The advantages of the EB/UV technology are improved productivity, adaptability to sensitive substrates, and environment- and user-friendly. Specifically, the EB technology benefits PSA product manufacture in the following ways (50):

1. New products can be developed by EB-curing, by using a variety of substrates, coatings and adhesives with various combinations of desirable properties of the product.
2. Higher productivity is provided because modern EB units operate at higher speed and process wider webs with a minimum waste.
3. Dependable product uniformity can be achieved. The specific treatment level can be automatically controlled.

It is widely realized that EB equipment is generally expensive, and the capital investment in purchasing takes a lot of consideration. However, the number of EB machines worldwide has doubled since 1981 presumably because of their versatility.

Recently Hül Troisdorf (formerly Dynamite Nobel) AG in West Germany used EB-curing (51) with a full-line of saturated polyester-based resins. The new polyesters were designed to provide low levels of toxicity, low odor and high resistance to light, oxygen, water, most chemicals, and temperature. Moreover, these adhesives have a shelf-life of more than six months. Recently, National Starch (52) has developed an ultra low voltage (78 kV) EB-curing system used to cure acrylic hot melt adhesives.

BASF AG (53) revealed an UV-crosslinkable acrylic hot melt PSA system. Specifically, they incorporate copolymerizable photoinitiators which are acrylic esters with a substituted benzophenone terminal group (Fig. 7). Upon activation, the photoinitiator starts the crosslinking reaction with a neighboring polymer chain. These scientists speculate the mechanism to be "insertion" of a neighboring C-H group. In this case, only one link per initiator molecule is formed. However, a different photo-initiator is required for a different polymer.

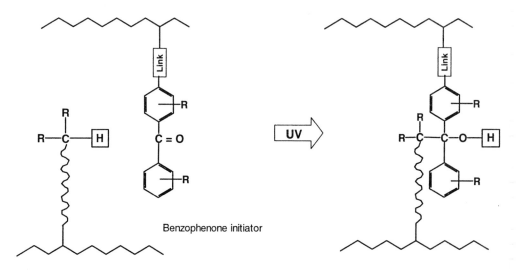

Figure 7. Incorporation of a photoinitiator in the polymer chain

A conjugated diene butyl rubber (54) has been cured with UV or EB to form PSAs. The introduction of a conjugated diene structure converts a polymer that normally undergoes scission into one that yields gel to as high as 80% solids on exposure to UV light. Gel formed *in situ* tackifies the rubbery phase to form the PSA.

Several warm melt (WM) adhesives (55) were made with SolarCure acrylated polyester developed by H. F. Fuller Company. These adhesives are compatible with a variety of commercially available monomers and polar tackifying resins. They have proven useful as unique polymeric bases for the development of UV curable RT PSAs and laminating adhesives.

Bachmann (56) introduced several aerobic acrylates. These acrylates consist of urethane oligomers dispersed in a methacrylic monomer and a special adhesion promoter. He compared their UV-curable aerobic acrylates with anaerobics and epoxy resin cured thermally (Table 12). In general, UV-cured aerobic acrylates yield excellent impact resistance and low shrinkage. This class of adhesives can achieve full curing in seconds at ambient temperature and possesses a long shelf life (>6 months).

Kerr of Rhone-Poulenc (57) prepared the next generation UV silicone release coating by introducing their new ionic photoinitiator, diaryl iodonium, tetrakis-(pentafluorophenyl) borate (Fig. 8). The borate anion excels over other initiators by shortening the induction time, speeding up the kinetics, producing crisp cure response with little reaction tail, and achieving a higher degree of epoxy reaction.

Table 12
Properties of UV-Cured Aerobic and Anaerobic Adhesives

Property	Acrylated Epoxides	Epoxy	Anaerobic	Aerobic (Rigid)	Aerobic (Elastomeric)	Aerobic (Flexible)
Type of Curing	Thermal	UV	UV	UV	UV	UV
Commercial availability	mid-1970s	late 1970s	mid-1970s	1981	1982	1980
Apparent gap filling, in.	0.15	0.02	0.05	0.15	0.25	0.30
Tensile on cold rolled steel using optional activator or primer, psi	NA	NA	2,500	3,800	3,000	1,000
Shore D hardness	85	80	80	80	50	10
Thermal shock[1] time to failure	15 sec.	15 sec.	5 min.	2 hr.	4+ hr.	4+ hr.
Impact resistance on glass	poor	poor	good	excellent	excellent	excellent
Adhesion to cold rolled steel	poor	poor	poor	excellent	good	fair
Relative cost index[2] (epoxy = 100)	2,000	100	1,000	500	400	400
Shrinkage on cure, %	10-25	10-25	8-15	1-5	1-5	1-5
Temperature resistance, °F	NA	250	250	400	300	300

Notes: 1. 1/4" square soft glass was bonded over a 1/2" area. The glass was not treated with adhesion promoter or cleaned. Each specimen was then cured for 5 min. under a 275-watt mercury arc lamp and cooled to 0°C. After 30 minutes, specimens were dropped into boiling water and tested to failure with tongs. The test was terminated after 4 hours.

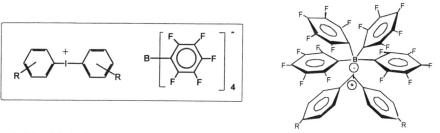

Figure 8. Diaryl iodonium tetrakis (pentafluorophenyl) borate

This photoinitiator can cure epoxy-silicone systems, presumably by the following mechanism.

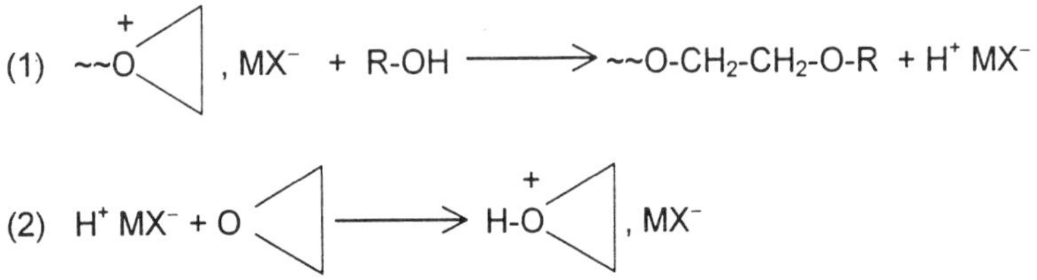

(1) $\sim\sim O{-}\triangleleft^{+}$, MX$^-$ + R-OH $\longrightarrow$ $\sim\sim$O-CH$_2$-CH$_2$-O-R + H$^+$ MX$^-$

(2) H$^+$ MX$^-$ + O$\triangleleft$ $\longrightarrow$ H-O$\triangleleft^{+}$, MX$^-$

    Another innovation in the radiation-curable adhesive is to use an H-Nu fluorone photoinitiator (Fig. 9) (58) at a very low concentration, 0.01 to 0.1 weight percent. This series of photoinitiators has been claimed to cure acrylate monomers with commonly available visible light or with commercial UV/visible sources. The initiator bleaches, resulting in a water-clear or lightly colored adhesive. The bleaching can also increase the depth of cure to several centimeters. Moreover, these initiators are able to cure through clear, colored or translucent substrates.

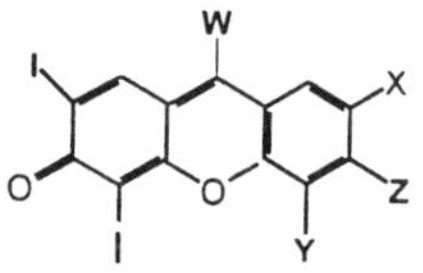

W	X	Y	Z	λmax(nm)	ε
H	H	H	OMe	470	23500
H	H	H	OEt	470	21100
H	H	H	OBu	470	30200
H	H	H	OOc	472	31600
H	I	I	O-	536	91200
H	COC$_7$H$_{15}$	I	O-	538	84800
H	C$_8$H$_{17}$	I	O-	534	92600
CN	I	I	O-	638,586	80000,35000
CN	COC$_7$H$_{15}$	I	O-	636,590	17100,9900
CN	C$_8$H$_{17}$	I	O-	630,580	40100,23300

Figure 9. Structures and absorption of H-Nu fluorone photoinitiators

Classification of Sealants

Construction sealants can be classified by their ability in handling movement of the joint width (3). Generally, there are three classes defined according to the movement capacity of the sealant:
 • Low-performance sealants - Typically, the low-performance sealants possess movement capacities between 0 to 5% of the joint width, and they are highly filled. These sealants are low-cost crack-fillers having a limited useful service life. The base material can be oil, resin, bituminous or polyvinyl acetate.
 • Medium-performance sealants - The medium-performance sealants demonstrate movement capacities of 5 to 12% for the joint width, and they have a longer service life than the low-performance sealants. This group includes butyl rubber, latex acrylics, and solvent based acrylics. Their characteristics are shown in Table 13. Generally, the shrinkage of these sealants is between 10 to 30% after application.
 • High-performance sealants - The high-performance sealants are mostly chemically curable elastomers with a movement capacity greater than 12%. This class includes polyurethane, the best of acrylics, polyether modified silicones, polysulfides and silicones. Their properties are compared in Table 13.

Table 13
Advantages and Limitations of Various Polymers for Sealants

Polymer	Type of Curing	Advantages	Limitations
Silicones	One-component (moisture-induced condensation) Two components (condensation or addition)	Best weathering Highest flexibility Good adhesion Heat resistance	High MVTR[2] Low depth of cure Slow curing[3]
Polyurethanes	One-component (moisture-induced condensation) Two components (condensation)	Good weathering Best adhesion High flexibility	Weak UV resistance Weak heat resistance
Polysulfides	One-component (moisture-induced condensation) Two components (condensation)	Low MVTR Fuel resistance Good flexibility	Slow curing[3] Low depth of cure
Acrylic latex	Water evaporation	Easy to use	High shrinkage Poor weathering Fair flexibility
Butyls	Sulfur vulcanization	Lowest MVTR Good flexibility	Fair weathering
Anaerobics	Metal/peroxide initiated free radical curing	Fast curing Chemical resistance Heat resistance	Brittleness Poor gap filling
Vinyl plastisols	Heat fusion	Good adhesion Low cost	Fair flexibility Fair weathering
Asphalts/coal tar resins	Cooling oxidation	Low cost Fuel resistance	Poor weathering
Polypropylene hot melts	Cooling	Low cost Expandable	Limited adhesion Fair flexibility

Notes: 1. Ref. *Engineered Materials Handbook, Desk Edition*, 673, ASM International, OH 1995.
2. MVTR-moisture vapor transmission rate
3. Two-component cured faster

New Applications of Adhesives and Sealants

Potential Electronic Encapsulants

Electronic encapsulants may be considered as one kind of sealants: however, most of them are extremely moisture resistant for use as hermetic encapsulants (59). In Table 14, we list the properties of several potential candidates for encapsulants: epoxies, polyimides, polyxylylene, siloxane polyimide, silicones and bisbenzocyclo-butene. Among them, siloxane polyimides and bisbenzocyclobutene are relatively new materials. One of the siloxane polyimides was available from Microsil (60), and bisbenzocyclobutene has been commercialized by Dow Chemical as Cyclotene (61). Cyclotene 5021 integrated circuit system is ultrahigh purity, low-k thin film coating material derived from B-stage bisbenzocyclobutene (BCB) monomers.

Besides the polymers in Table 14, there are cyanate esters for hermetic packaging, produced by Quantum Materials (62). The cyanate esters can be cured at 120°C for 60 minutes. The end product has very low stress and excellent dispensability.

Table 14
Characteristics of Potential Electronic Encapsulants

Polymer	Dispensing	Advantages	Limitations	Modulus (psi)	TCE (ppm/°C)
Epoxides	Normal	Good solvent resistance Excellent mechanical strength	Non-repairable Marginal electrical	$1\text{-}5 \times 10^6$	40-80
Polyimides	Normal	Good solvent resistance Good thermal stability	High-temperature Non-repairable High stress	1×10^6	3-80
Polyxylylenes (Parylene®)	Thermal	Good solvent resistance Conformable coating	Non-repairable Thin film only	0.4×10^6	35-40
Siloxane polyimide	Normal	Less stress than polyimide Better solvent resistance than polyimide	High stress Thin film only	0.4×10^6	5-100
Silicones (RTV, gel, etc.)	Normal	Good temperature cycling Good electrical Very low modulus	Weak solvent resistance Low mechanical strength	0-400	200-1000
Benzocyclobutene (Dow developmental polymer)	Normal	Good solvent resistance Low moisture absorption Low dielectric constant	High temperature cure	0.4×10^6	67

Ref.: C. P. Wong, *IEEE Trans Components, Hybrids and Manuf. Technol.* 12 (4) 421, Dec., 1989.

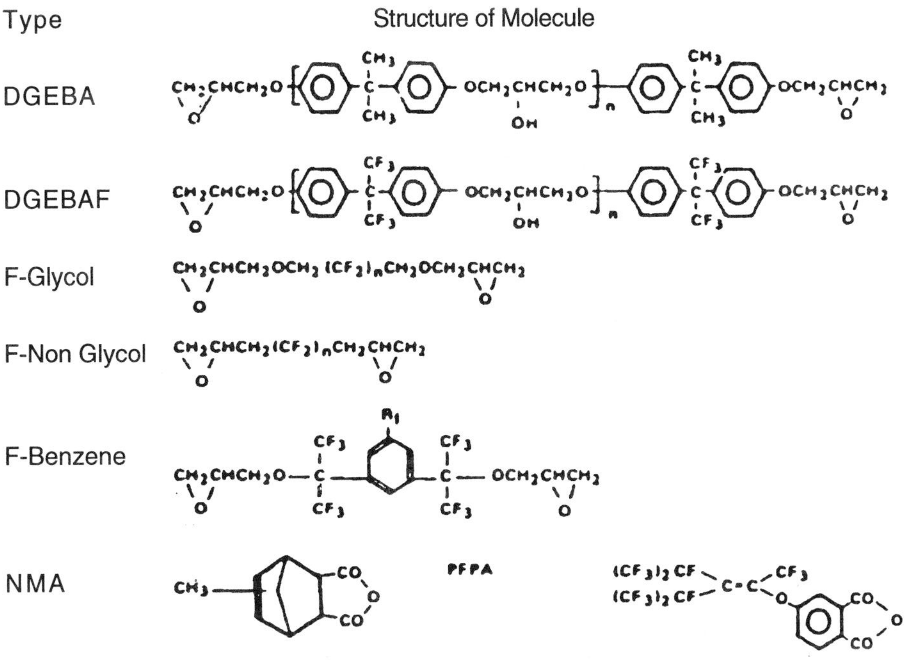

Type   Structure of Molecule

DGEBA CH_2CHCH_2O—⬡—$C(CH_3)_2$—⬡—OCH_2CHCH_2O—[⬡—$C(CH_3)_2$—⬡]$_n$—OCH_2CHCH_2

DGEBAF CH_2CHCH_2O—⬡—$C(CF_3)_2$—⬡—OCH_2CHCH_2O—[⬡—$C(CF_3)_2$—⬡]$_n$—OCH_2CHCH_2

F-Glycol $CH_2CHCH_2OCH_2(CF_2)_nCH_2OCH_2CHCH_2$

F-Non Glycol $CH_2CHCH_2(CF_2)_nCH_2CHCH_2$

F-Benzene CH_2CHCH_2O—$C(CF_3)_2$—⬡(R_1)—$C(CF_3)_2$—OCH_2CHCH_2

NMA

PFPA $(CF_3)_2CF$—C=C—CF_3 ...

Figure 10. Fluorinated epoxy resins as optical adhesives

UV-Curable Fluorinated Epoxy Resin for Optical Adhesives

With the advance of optical fiber technology, there is a need for optical adhesives. One of the requirements for optical adhesives is to match the refractive index of the fiber, which is about 1.46. In the U.S. marketplace, there has been no polymer which can meet this qualification. However, there are a series of optical adhesives developed by NET in Japan (63). These resins are based on the fluorinated epoxy resins (or acrylates) (64) as shown in Fig. 10. The properties of four fluorinated resins are shown in Table 15. It is noteworthy that the refractive indices of these resins are between 1.45 and 1.51, which can match that of the fiber.

Medical Adhesives

One of the fast growing segments of the adhesives and sealants industry is medical adhesives and sealants. We do not intend to discuss this subject in detail. We merely cite one example, skin adhesives. Cyanoacrylates have been used extensively as medical adhesives, especially as skin adhesives. In China, cyanoacrylate has been used to plug up reproductive ducts inside male and female bodies (23).

Table 15
Characteristics of UV-Curable Optical Adhesives

Property	Low Tg Type (refractive index control)	High Tg Type (refractive index control)	AT 6001 (high water-resistant)	AT 6117 (high water-resistant)
Refractive index, n_D (23°C)	1.45-1.57	2.47-1.57	1.51	1.51
Light transmittance, % (1.3 µm, 1 mm thick)	83-90	85-91	91	90
Adhesive strength, kgf/cm^2 (with BK-7glass)	110->200	80>180	>150	>185
Glass transition temperature, °C	57-52	132-147	-3	100
Thermal expansion coefficient (10^{-5} °C^{-1})	8-11	6-7	-	-
Curing time, minutes	About 1	About 1	About 1	About 1
Shrinkage during curing, %	4-8	3-5	6	10
Viscosity, cP	300-500	-	440	1470
Main components	Epoxy	Epoxy	Acrylate	Acrylate

Ref.: *Technical Bulletins,* NTT International, Tokyo, Japan, Oct., 1997.

In Table 16, Pfister (65) compared three types of skin adhesives: polyisobutylene, polyacrylate and polysiloxane. In general, polyisobutylene is inferior to both polyacrylate and polysiloxane. Polysiloxane excels in thermal stability, water vapor transmission and good adhesion to low surface energy substrates.

Recently, a new trauma-free-removal skin adhesive was disclosed by Webster (66) of Smith and Nephew Group Research Center in England. The light-switchable adhesive is based on an acrylate copolymer. They used methacrylate as the cross-linking functional group. Itaconic anhydride is copolymerized into the acrylic adhesive copolymer.

Table 16
Typical Properties of Skin Adhesives

Property	Polyisobutylene	Polyacrylate	Polysiloxane
Tack	Low to high	Moderate to high	Low to moderate
Skin adhesion	Moderate to high	Moderate to high	Low to moderate
Adhesion to nonpolar substrates	Good	Fair to good	Excellent
Peel force	Low	Low	Low
Creep resistance	Low to moderate	Moderate to high	Low to moderate
Cohesive strength	Low to moderate	Moderate to high	Low to moderate
Thermal stability	Poor to fair	Fair to good	Excellent
Oxygen permeability	Low	Medium	High
Water vapor transmission	Low	Medium	High
Polarity	Low	Medium	Low
Solvent resistance	Poor	Fair to good	Good to excellent
Commercial availability	Compound in-house	Yes	Yes
Release liners	Many	Many	Few

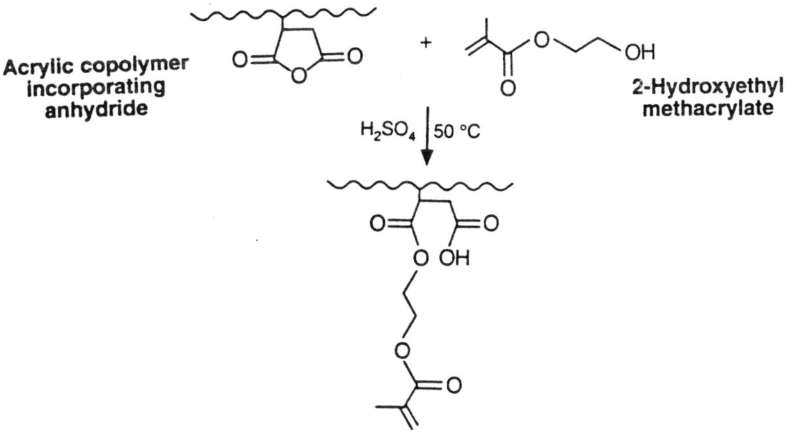

Figure 11. Light-switchable pressure-sensitive adhesive

The methacrylate PSA is produced by opening the pendant anhydride ring with 2-hydroxylethyl methacrylate. That PSA is blended with a commercially available titanium-containing visible light photo-initiator. The cured adhesive is then easier to remove. The reaction scheme is shown in Fig. 11.

Aerospace Adhesives and Sealants

Research on high temperature adhesives has been carried out mostly in government laboratories. In recent years, this field of research has slowed considerably. In the 1980s, the three most commonly used high temperature adhesives, polyimides (3), polybenzimidazole (3), and polyphenylquinoxaline (3), were reviewed and are commercially available.

There are several major requirements for high temperature / high performance structural resins (67):
1. Acceptable handlability (non-toxic, tack with tack and drape, etc.).
2. Processable under mild conditions—this requirement is very important because several early high temperature polymers were unprocessable.
3. Volatileless (or low volatile systems).
4. Capable of forming large structure.
5. Mechanically strong (durable, damage tolerant, etc.) over the temperature range -65° to 370°C for at least 6,000 hours.
6. Environmentally stable (moisture, fuel, etc.).
7. Long shelf life.
8. Acceptable in manufacturing cost.
9. Amenable to different fabrication methods (automated tape laying, RTM, filament winding, advanced tow placement, etc.)
10. Reproducible and reliable.
11. Repairable.
12. Compatible with various adherents and surface treatments.

Many of the above requirements are also applicable to adhesives. These requirements may have inhibited a fast development of high temperature adhesives.

In this paper we shall briefly mention two polymers developed in the late 1980s. The first example is a series of novel imide/arylene copolymers, investigated by Jensen and Working (68). The polymers were synthesized from the reaction between an amorphous polyarylene ether and a semi-crystalline polyimide (Fig.12). One block copolymer was selected for end-capping. The processability of the end-capped copolymer was better than that of polyimide, and adhesive properties were also good.

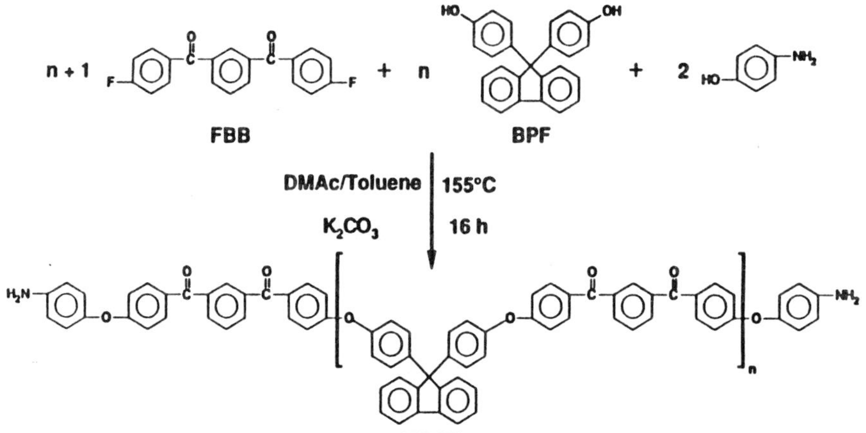

ATPAE

$\overline{M}_n$ = 3110 g/mole and $\overline{M}_n$ = 6545 g/mole
r = 0.816 r = 0.908
r = ratio of BFP to FBB

Figure 12. Amine-terminated Poly(arylene ether) (ATPAE)

The second example is LaRC^{TM}-SI, which is a versatile semicrystalline copolyimide with a unique combination of properties (69). This polymer was made from commercially available aromatic monomers (Fig. 13). It can also be used as a self-bonding and hot melt adhesive.

Future Research in Adhesive and Sealant Chemistry

High temperature adhesives remain an important field for the aerospace industry. Recently, we have noted several publications on some exotic materials used for adhesives and sealants. These materials may hold some promise for the future.

Dendrimers as Adhesives and Adhesion Promoters

Dendrimers (70), or the tree-like polymers, have been predicted to be potential candidates for adhesives. For example, on the surface of DSM dendrimer (71), there are 64 functional groups of amine (Fig. 14). What will happen to these amine groups, if this polymer contacts with another one containing an equal number of carboxylic groups? If a portion of both functional groups are interacting, a super strong adhesive may be formed.

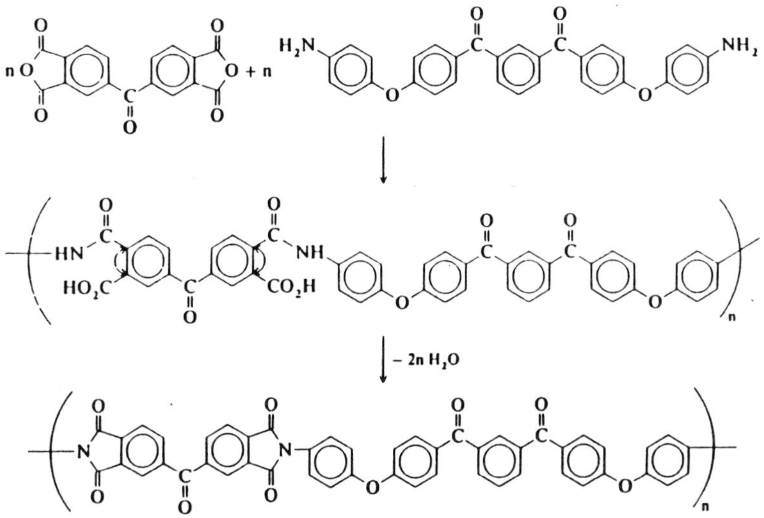

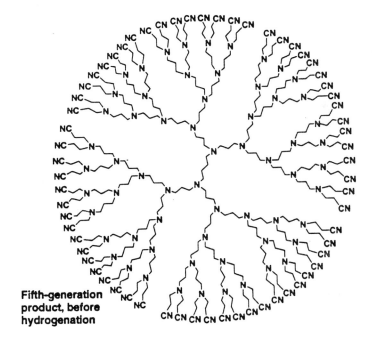

Figure 13. Reaction route to LARC-CPI, a semicrystalline polyimide

Fifth-generation product, before hydrogenation

Figure 14. DSM dendrimer

McKnight, Palmese and Phillips (72) reported that the PAMAM dendrimers can serve as adhesion promoters between aluminum and epoxy resin. Each G2 (second generation) PAMAM molecule possesses 16 terminal primary amine groups available for further reactions.

Supramolecular Polymers as Adhesives

In recent years, supramolecular polymers have proven a fertile field of research. There have been several reports on the use of supramolecular polymers for adhesives. Rowan, Feiters and Nolte (73) reported their findings of molecular clips and molecular bowls. The clip-shape molecules were synthesized from glycoluril with 12 long aliphatic side chains. When the supramolecules are mixed with copolymers of styrene and dihydroxystyrene, these clips resemble lobster claws cramping or adhering to the polymer side chains to reveal liquid-crystalline like properties. Presumably, the attachment is caused by hydrogen bonding with urea groups of the clips and π-π stacking interactions with the clip side-walls.

Interestingly, the crown ether of the molecular clips is bowl-shaped with dangling side chains. After it is dispersed in water, these molecular bowls transform again into nanospheres with dimples on the surface serving as anchoring sites for the organic molecules. Upon the addition of an acid, nanotubes are formed with the binding sites on their inner and outer surfaces. Moreover, the supramolecular structure can be stabilized by polymerization.

Adhesion of Fullerene C_{60} Films

Since the discovery of fullerene, much work has been done in the synthesis and elucidation of the structure of derivatives of fullerene. Recently, Manika *et al.* (74) reported the adhesion and properties of C_{60} films on silicon wafer. The cohesive fracture inside the fullerite film due to strong bonding at the film/substrate interphase was predominantly observed. The important results indicate that the adhesion of C_{60} film to a silicone substrate is high enough for different technological applications.

Silica Sol As a Nanoglue

A newcomer to nanotechnology may be the nanoglue recently disclosed by Morris, *et al.* (75). Low density nanoscale mesoporous composites may be readily prepared by a colloidal or dispersed solid to an about-to-gel silica sol. The silica gel can adhere or glue a range of chemically or physically diverse particles into a 3-D silica network resulted from the gelation. If the composite gel is dried supercritically, a composite aerogel is formed to retain all properties of the solid composite. The composite aerogel can be designed for chemical, electronic, and optical applications.

Summary

In this brief overview, we examined the early innovation of adhesives and sealants made from synthetic polymers. Then, from some published projections, we can visualize some future trends of various segments of the industry.

We then followed the classifications of various adhesives, and compared characteristics of each type of adhesives. We also tried to mention some distinctive developments of each kind of adhesives in the past decade. We discussed in detail the water based adhesives, hot melt and EB/UV-curable adhesives because these three are important current choices in saving energy and preventing pollution. We also briefly mentioned various kinds of sealants.

We demonstrated the importance of adhesives and sealants by citing four examples: electronic encapsulants, UV-curable optical adhesives, skin adhesives, and aerospace high-temperature adhesives.

Finally, we speculate on future materials, some of which are still in the research stage. We can see the possibility of dendrimers, supramolecular materials, fullerene and silica sol being used as adhesives and sealants for specific functions in the future.

Literature Cited

1. Lee, L. H., In "Adhesive Chemistry-Developments and Trends," Lee, L. H. , Ed., Plenum Press, New York, 1987, pp. 3-52.
2. Brinson, H. F., Ed. , "Engineering Materials Handbook," Vol. 3, "Adhesives and Sealants," ASM International, Ohio, 1991.
3. Gauthier, M. M., Ed., "Engineering Materials Handbook, Desk Edition," ASM International, Ohio, 1995.
4. Skeist, I. ; Myron, J., J. Macromol. Sci.-Chem. Part A15 1981, 1151.
5. De Bruyne, N. A., Laboratory Notebooks, 1941-42, U. S. Pat. 2,499,134, (1952).
6. Delquist, C. A., Kautschuk and Gummi, Kunstoffe, 1985, 38, 617.
7. Morse, P. M., Chem. & Eng. News, 21, April 20, 1998.
8. Chem. Weeks, 20, April 7, 1999.
9. Adhesives Age, 43 (1), 15, January, 1999.
10. Adhesives Age, 42 (1), 33, January, 1998.
11. Lee, L. H.; Lu. S.-D.; Wang, Z.-L., Adhesives Age, 38 (4), 30 , April, 1996.
12. Drake, R., "Proceedings of 20th Anniversary Meeting of Adhesion Society," Drazal, L. T.; Schreiber, H. P., Eds., February, 1997, p.187.
13. Hartshorn, S. R., Matl. Eng. 10, Oct., 1992.
14. Hermansen, R. D.; Lau, S. E., Adhesives Age, 36 (8), 38, July, 1993.
15. Lay, D. G.; Cranley, P., Adhesives Age, 37 (6), 6, May, 1994.
16. Elliot, C. ; Millard, T.; Wing, P., Ahesives Age, 33 (8), 22, July, 1990.
17. Shields, J., "Adhesives Handbook," 3rd. edn., Butterworth, 1984, pp. 32-79.
18. Connor, J. T., CHEMTECH, 51, Sept., 1994.
19. Novak, H. L. ; Comer, D. A., Adhesives Age, 33 (2), 28, Feb, 1990.
20. Motegi, H. I. ; Isowa, E.; Kimura, K., U. S. Pat. 4,171,416 (1979).
21. Liu, J. C., U. S. Pat. 4,906,317 (1990).
22. Harris, S. J.; McKervey, M. A.; Melody, D. P.; Woods, J.; Rooney, J. M., U.S. Pat. 4,556,700 (1985).
23. Ruan, C.-L.; Liu, J.-H.; Qin, Z.-Y.; in "Adhesives, Sealants, and Coatings for Space and Harsh Environments," Lee, L. H., Ed.; Plenum Press, New York ,1987, p. 81.
24. Catena, W.; Edelman, R., Adhesives Age, 36 (7), 25, June, 1993.
25. Nelson, K., Adhesives Age, 36 (6), 6, May, 1993.
26. Dahmane, H., Intl. J. Adhesion and Adhesives, 1996, 16 (1), 43.
27. Aucher, G.; Barwich, J.; Rehmer, G.; Jager, H., Adhesives Age, 37 (8), 20, July, 1994.
28. McMinn, B. W.; Snow, W. S.; Bowman, D.T, Adhesives Age, 38 (12), 34, Nov., 1995.
29. Miller, R. A.; G. M. Althen, Adhesives Age, 39 (12), 21, Nov., 1996.
30. Archer, B., Intl. J. Adhesives and Adhesion, 1998, 18, 15.
31. Lorenz, R. T., Ashesives Age, 40 (11), 42, Oct., 1997.
32. Clemens., L. M., "Proceedings of 20th Anniversary Meeting of Adhesion Society," Drazal, L. T.; Schreiber, H. P., Eds., February, 1997, p.351.
33. Dalquist, C. A., in "Adhesion, Fundamentals and Practice," Part 3, Chapt. 5, Maclauren, London (1970).
34. Sheout, W. H.; Day, N. H., U. S. Pat. 3,965 (1845).
35. Drew, R. G., U. S. Pat. 1,760,820 (1928), to 3 M Co.
36. Hoffmann, W. E.; Miles, D. E., Adhesives Age, 35 (4), 20, April, 1992.
37. Druschke, W.; Kerckow, A.; Stange, B., U. S. Pat. 4,371,659 (1983), to BASF.
38. Frezee, G. R., U. S. Pat. 4, 845,149 (1989), and 4,879,333 (1989), to S. C. Johnson & Sons.
39. Bateman, B., U. S. Pat. 4, 360,638 (1982), to Rohm & Haas Co.
40. Dreyfuss, P.; Fetters, L. J.; Hansen, D. R., Rubber Chem. Technol. ,1980, 53, 728.
41. Hsieh, H. L.; Naylor, F. E., U. S. Pat. 4,136,137 (1979), to Phillips Petroleum.

42. Erickson, J. R., Adhesives Age, 29 (4), 22, April, 1986.
43. Chin, S. S., Adhesives Age, 34 (8), 26, July, 1991.
44. Dexter, J. F., U. S. Pat. 2,736,721 (1956).
45. Lin, S. W.; Cooper, W., U. S. Pat. 5,441, 811 (1995).
46. Lin, S. W., Adhesives Age, 39 (8), 14, July, 1996.
47. Kerr, III, S. R.; Lin, S. W., Adhesives Age, 37 (10), 36, Sept., 1996.
48. Mazurek, M.; Kantner, S. S.; Galkiewicz, "Proceedings of Adhesion Society Annual Meeting," H. M. Clearfield, Ed., Feb, 1991, p. 39.
49. Drukenbrod, M., "Adhesives and Sealants Online, "Jan. 11, 1999.
50. S. R. Kerr, III, Adhesives Age, 41 (10), 27, Nov., 1998.
51. Adhesives Age, 41 (4), 24, April, 1998.
52. Ramharack, R.; Chandran, R.; Shah, S; Hariharan, D.; Orloff, J.; Foreman, P., Adhesives Age, 39 (13), 40, 1996.
53. Duis, J. J., Adhesives Age, 41 (11), 34, Oct., 1998.
54. Merrill, N. A.; Gardner, I. J.; Hughes, V. L., Adhesives Age, 35 (8), 24, July, 1992.
55. Kaufmann, T.; Chappell, J; Avereds, M.; Mitry, M., Adhesives Age, 41(8), 32,29, July, 1998.
56. Bachman, T., A. G., J. Radiation Curing/Radiation Curing,1992, 19 (4), 10.
57. Kerr, III, S. R., Adhesives Age, 39 (8), 26, July, 1996.
58. Marino, T. L.; Martin, D; Necers, D. C., Adhesives Age, 38 (9), 26, August, 1996.
59. Wong, C. P., IEEE Trans. Components, Hybrids and Manufacturing Technology, 12 (4), 421, Dec., 1989.
60. Technical Bulletin, Microsil Corp.
61. Technical Bulletin, "Cyclotene", Dow Chemical Co.
62. Technical Bulletin, "Die Attached Products," Quantum Materials, San Diego, CA.
63. Technical Bulletins, NTTT International, Tokyo, Japan, Oct., 1997.
64. Nakamura, K.; Marumo, T., Techno Japan, 1987, 20 (1), 8.
65. Pfister, W. R., Adhesives Age, 33 (13), 20, Dec., 1990.
66. Webster, I., Intl. J. Adhesion and Adhesives, 1999, 19, 29.
67. Hergenrother, P. M., Angew. Chem., Intl. Ed., 1990, 29, 1262.
68. Jensen, B., J.; Working, D. C., High Performance Polymers, 1992, 4 (1), 55.
69. Bryant, R. G., "Proceedings of Adhesion Society Annual Meeting," Ward, T.C. Ed., Feb., 1996, p. 36.
70. Tomalia, D. A., Scientific American. 1995, 272 (5), 62.
71. O'Sullivan, D. A., Chem. Eng. News, 20, Aug. 16, 1993.
72. Mcknight, S. H.; Palmer, G. R.; Phillips, C. B., "Proceedings of Adhesion Society 20th Anniversary Meeting, " Drazal, L. T.; Schreiber, H P.; Eds., Feb., 1997, p. 449.
73. Rowan, A. E.; Feiter, M. C.; Nolte, R. J. M., "Proceedings of Adhesion Society 20th Anniversary Meeting, " Drazal, L. T.; Schreiber, H P., Eds., Feb., 1997, p.271.
74. Manika, I.; Maniks, J.; Simanovskis, A.; Kalnacs, J. Fullerene Sci. Technol. 1998, 6, 981.
75. Morris, C. A.; Anderson, M. L.; Stroud, R. M.; Merzbacher, C. I.; Rolison, S. R., Science, 284, 622, April 23, 1999.

INDUSTRIALLY IMPORTANT POLYSACCHARIDES

EMMETT M. PARTAIN, III

Union Carbide Corporation
Bound Brook Technical Center
P. O. Box 670
Bound Brook, New Jersey 08805

Introduction
Cellulose
Cellulose solvents
Cellulose fiber formation
Starch
Free-radical grafting
Chitin and chitosan
Miscellaneous polysaccharides

Introduction

Polysaccharides are high molecular weight polymers or copolymers of monosaccharide repeat units joined through glycosidic bonds. In living systems, polysaccharides can be found as a structural material and as an energy storage system, and they represent a critically important class of polymers to man for their nutritional value as well as industrial utility. Virtually all animals on earth are dependent on the photosynthetic capabilities of green plants to produce cellulose and starch, which are important animal foods. Today, starch and modified food starches are used extensively in commercial food products, and other polysaccharide gums such as alginates, agar, carrageenan, guar, pectin, and xanthan are also used as thickeners and texture modifiers in food products. For the most part, these polysaccharide gums used to modify food products are not chemically modified but simply used in a purified form. The technology of many of these vegetable gums has been reviewed elsewhere, and they will not be discussed further here [1].

Many polysaccharides are non-toxic and benign to mammalian tissues, and for this reason there is extensive commercial interest in the use of polysaccharides in cosmetic and pharmaceutical applications. This overview on industrially important polysaccharides focuses primarily on chemical derivatives of the largest volume (or potential volume) polysaccharides -- cellulose, starch, and chitin/chitosan -- with emphasis on current and future industrial utility. Because polysaccharides are obtained from renewable plant and animal sources, there is continuing interest in them as non-petroleum based polymer feedstocks, especially regarding their biodegradability. The biodegradation of cellulose and other polysaccharides has recently been reviewed [2-3].

Cellulose

Cellulose, discovered in 1838 by Payen, is the principle structural constituent of the cell walls of many plants (hence the name). Cellulose is found in most plants (albeit not in pure form) in the leaves and stalks, with cotton fiber (95% cellulose) and wood (about 50%) being the principle industrial sources of cellulose. Because of its ubiquity in the plant kingdom, cellulose is arguably the most abundant polymer on earth, with billions of tons produced annually through photosynthesis. For thousands of years, cellulose in various forms has been used by man as a writing material (paper), as a building material (lumber), and as a textile (cotton, flax, and ramie). As chemists began to elucidate the structure of cellulose, they also began to chemically modify cellulose to alter or improve properties for a variety of commercial purposes.

Structurally, cellulose is a linear homopolymer consisting of D-anhydro-glucopyranose units linked by β-(1→4)-glycosidic bonds (Figure 1). Despite being a homopolymer of glucose, cellulose itself is not water-soluble because of extensive intra- and inter-molecular hydrogen bonding, and cellulose itself is a highly crystalline polymer. However, disruption of this hydrogen bonding either by chemical modification to the cellulosic backbone or by the use of suitable solvents can render cellulose or the cellulose derivative soluble. In the absence of chemical derivatization of the cellulosic backbone, cellulose is insoluble in practically all common organic solvents, and the development of cellulose solvents has been a continuing challenge to polymer chemists.

The earliest commercial work on cellulose involved chemical and physical modification of cotton fibers to improve tensile properties and to improve dyeability. If native cotton fibers are treated with dilute sodium hydroxide, stretched, and washed, the cellulose chains in the cotton fiber are aligned and additional tensile strength is conferred to the fiber. This process, known as mercerization after its discoverer (John Mercer, died 1866), is still used today to improve the tensile strength and luster of cotton thread. The dyeing of textiles, especially cotton, was in large part a driving force in the expansion of the German chemical industry in the latter part of the 19th century.

As the chemical structure of cellulose was elucidated, the potential reactivity of the primary and secondary hydroxyl groups on the polymer backbone to esterification and etherification was appreciated. The earliest derivative of cellulose was the nitric acid ester, cellulose nitrate, also known as gun cotton or nitrocellulose. Combined with camphor, cellulose nitrate was marketed as Celluloid, the first wholly synthetic material and regarded as the first commercial "plastic". Cellulose nitrate is still manufactured today for use in explosives and propellants (smokeless powder) and as a component in clear wood coatings and other lacquers.

Figure 1. Chemical structure of cellulose, chitin, chitosan, and amylose

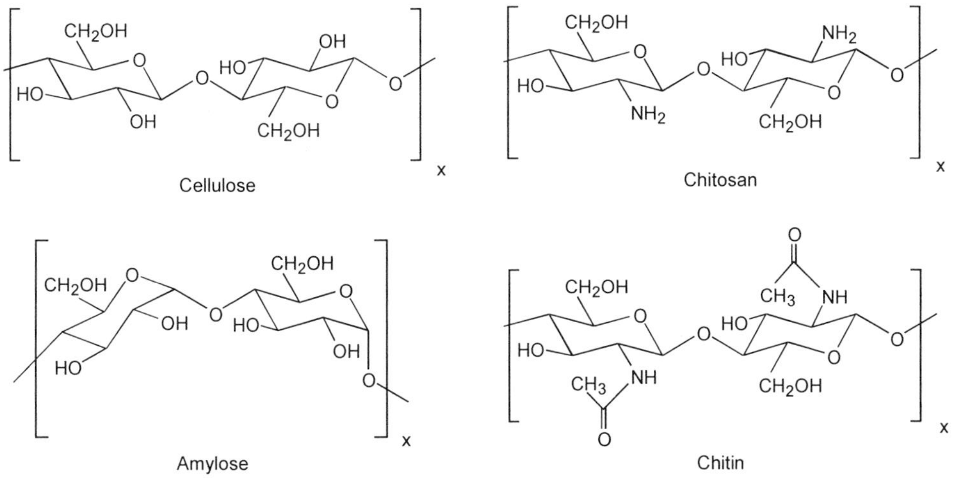

Polymers like cellulose may be chemically reacted like any other organic material, but the degree of reactivity is usually limited by the accessibility of the reactive sites in the polymer to the reactant of choice. Cellulose may be reacted either homogeneously or heterogeneously. In the homogeneous case, cellulose is physically dissolved in a solvent that insures essentially complete accessibility of the reactive sites on the polymer backbone, but such polymer solutions are highly viscous making mass transfer and heat exchange problematic. Further, the recovery of a solid polymer product from solution involves either spray drying or non-solvent precipitation, and in both cases, large volumes of solvent and non-solvent must be handled and recovered. Therefore, most commercial chemical reactions on cellulose are carried out heterogeneously, in which the polymer is slurried and swollen in a diluent system, but not physically dissolved. In some cases such as the manufacture of certain cellulose esters, the reaction begins heterogeneously and finishes homogeneously, but this polymer dissolution occurs with a reduction in the polymer molecular weight so that the final solution viscosity is substantially lower and the polymer solution can still be processed.

Several cellulose ethers and cellulose esters are commercially manufactured today in large volume. Cellulose nitrate is the only inorganic cellulose ester of commercial importance, although several organic cellulose esters are manufactured including cellulose acetate, cellulose propionate, cellulose butyrate, and some mixed cellulose esters. The general process used for the manufacture of these organic esters of cellulose is generally the same and has been reviewed in the literature [4-5]. To produce cellulose acetate, cellulose is slurried in acetic acid and reacted with a mixture of acetic acid, acetic anhydride, and sulfuric acid. As the reaction progresses, the cellulose acetate dissolves so that the reaction becomes homogeneous. Invariably there is significant degradation in the molecular weight of the cellulose backbone, and the typical molecular weights of cellulose esters obtained is on the order of 50,000. The cellulose acetate (generally obtained as the triacetate) can then be hydrolyzed to reduce the degree of substitution (DS) to the diacetate, and the final polymer can be recovered by precipitation into water. Esterification of cellulose in other acids such as methanesulfonic acid and trifluoroacetic acid have been described.

The organic esters of cellulose are thermoplastic and can be injection-molded to fabricate a variety of objects. In addition, these organic esters of cellulose find a variety of specialized commercial uses. Cellulose acetate is used in cigarette filters, photographic film, dialysis membranes, and textile fibers (see below), while cellulose acetate butyrate (CAB) and cellulose acetate propionate (CAP) are used in printing inks, and automotive & furniture lacquers. A specialized cellulose ester, cellulose acetate phthalate, is used in photography and in pharmaceutical compounding as an enteric coating on tablets. When dissolved in suitable solvents, many of these cellulose esters form liquid crystals, and spinning of these liquid crystals can yield fibers with high tensile strength. While there has been on-going interest in cellulose sulfate (heparin-like anticoagulant) and cellulose phosphate (flame retardant), both of these inorganic esters are still laboratory curiosities. Cellulose formate is known, but the ester is hydrolytically unstable. Attempts to exploit this instability to fabricate fibers have not been successful. The esterification of cellulose ethers with alkenylsuccinic anhydrides yields hydrophobic anionic polymers useful in construction applications, and the use of hydrophobic esters of cellulose in enzymatic immobilization has been reported [6-7].

Large scale commercial cellulose ethers include carboxymethyl cellulose (CMC), methyl cellulose (MC), hydroxyethyl cellulose (HEC), hydroxypropyl methyl cellulose (HPMC), hydroxypropyl cellulose (HPC), ethyl hydroxyethyl cellulose (EHEC), and methyl hydroxyethyl cellulose (MHEC). In contrast to the manufacture of organic cellulose esters which is partially a homogeneous process, cellulose ethers are usually manufactured by a wholly heterogeneous process in which wood pulp or cotton linters are slurried in some organic diluent, the cellulose is causticized by the addition of aqueous sodium hydroxide, and etherified by the reaction of an appropriate electrophile with the alkoxide groups on the cellulose backbone. Except for CMC which is an anionic polymer (pK_a = 4.4), these cellulose ethers are all non-ionic polymers. All of these cellulose ethers are soluble in water to varying degrees, and in most applications, the high efficiency of these polymers to viscosify water is one of the principle properties exploited commercially. Typical molecular weights vary between about 50,000 to over a million, and in water these cellulose ethers uncoil and yield extended polymer chains that have a large hydrodynamic volume. These extended polymer chains entangle to a high degree yielding high solution viscosity. The chemistry to produce these cellulose ethers has also been well documented [4,8].

These cellulose ethers find utility in a variety of commercial applications, and some of the more common uses for these cellulose ethers are summarized here. These examples of the uses of common cellulose ethers are not exhaustive, and the reader is referred to the published literature for additional information and examples [1,9]. The largest single use of carboxymethyl cellulose is as an antideposition aid in laundry detergents, although large volumes of CMC are also used as a food additive, petroleum drilling fluid thickener, sizing in paper manufacture, toothpaste thickener, and in pharmaceutical excipients. The alkyl and alkylhydroxyalkyl cellulose ethers (MC, HPMC, EHEC, MHEC) are non-ionic cellulose ethers that are soluble in cold water but insoluble in hot water. This temperature at which the cellulose ether becomes insoluble is referred to as a "cloud point", and manufacturers exploit this behavior by washing by-products out of these cellulose ethers after synthesis by using hot water. These cellulose ethers are used in cementitious materials (tape-joint compound, ceramic adhesives, stuccos, etc.) to improve workability and water retention, and also find uses in thickening cosmetics, food additives, and as an excipient in oral pharmaceuticals. HPMC is also used as a protective colloid in the suspension polymerization of vinyl chloride. Hydroxypropyl cellulose (HPC), which is soluble in both water and a wide range of organic solvents, exhibits liquid crystalline behavior in solution [10-12]. HPC is a thermoplastic that can be injection-molded. In contrast to the alkylhydroxyalkyl cellulose ethers, hydroxyethyl cellulose is a non-ionic cellulose ether that is soluble in both cold and hot water. It is used extensively throughout the world as a thickener in latex paints, and its use in latex paints has been reviewed [13-14]. HEC is also used as a protective colloid in emulsion polymerization, and HEC also finds uses in household products, oil-well cementing, and cosmetics. Cellulose and most cellulose derivatives are subject to enzymatic degradation by cellulase enzymes, but by adjusting the conditions of ethoxylation, it is possible to obtain an HEC product with a more uniform distribution of hydroxyethyl substituents resulting in substantially improved enzyme resistance. Unlike HPC, HEC is not thermoplastic and cannot be thermally fabricated.

In addition to these cellulose ethers manufactured on a large scale, a number of new cellulose ethers have been developed in the last several years for specialized applications (Figure 2). HEC can be modified with either glycidyl trimethylammonium chloride or the corresponding chlorohydrin to give cationic cellulose ethers [15]. Originally developed as flocculants, these cationic cellulose ethers show a strong affinity for anionically charged surfaces such as keratin, which is the principle protein of hair and skin [16-17]. This binding of cationic polymers to these substrates is also known in the cosmetics industry as substantivity. The affinity of these cationic cellulosic polymers for these proteins makes these polymers useful in cosmetics to modify the texture, strength, and appearance of skin and hair, and in pharmaceutical applications for the topical delivery of actives through skin and mucosa. HEC and other cellulose ethers can also be modified with hydrophobic quaternary ammonium halide electrophiles such as glycidyl dodecyl dimethylammonium chloride to obtain cellulose ethers with both hydrophobic and cationic functionality [18-19]. Besides uses in cosmetic applications, these hydrophobically-modified cationic cellulose ethers have been found to improve contraceptive compositions by reducing sperm motility [20-21].

Another active area of research is the hydrophobe-modification (HM) of various polysaccharides, specifically cellulose ethers. Much of this research on hydrophobe-modified polysaccharides has been driven by the needs of the water-borne coatings industry to market paints with improved application properties. For many years, latex paints have been thickened almost exclusively with cellulose ethers such as HEC or HPMC, which viscosify water by polymer chain entanglement. The efficiency of viscosification is proportional to the polymer chain length (degree of polymerization) and therefore the molecular weight. Increasing polymer molecular weight increased the thickener efficiency, but also impaired leveling and roller spatter behavior. In the 1970's, it was appreciated that an alternative thickening mechanism to polymer chain entanglement might be exploited in water-borne coatings. If a water-soluble polymer backbone were modified with sufficient numbers of hydrophobic groups, these hydrophobic groups would aggregate with each other as well as with other hydrophobic components in the water-borne coating (such as the latex), building a pseudo three dimensional network which would raise the apparent viscosity of the system. However, under high shear, these hydrophobe aggregates would be broken and reformed rapidly, allowing the formulation to flow. The aggregation of hydrophobes is similar to micelle formation, and this general mechanism is often referred to in the coatings industry as "associative" thickening [22]. The adsorption of hydrophobe-modified hydroxyethyl cellulose (HMHEC) polymers onto various latexes has been examined, and HMHEC adsorb more readily than do non-hydrophobic HEC polymers [23-27].

Figure 2. Structures of selected cellulose ethers

Selected cellulose ethers
(idealized structures)

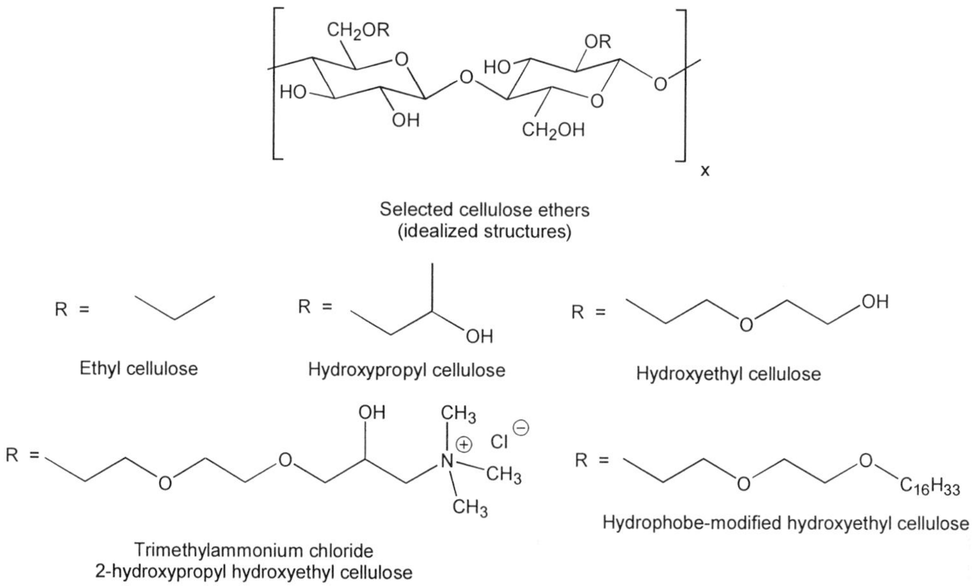

The modification of HEC or EHEC with alkyl (linear hexadecyl) or alkyl aryl (nonyl phenoxy) groups is practiced commercially, and like all cellulose ethers, these products are made by a heterogeneous process [28-33]. To produce these hydrophobe-modified cellulose ethers, causticized HEC or EHEC is suspended in an inert diluent and reacted with a hydrophobic electrophile (usually either a halide or glycidyl ether). One inherent problem in these industrial processes is the poor alkylation efficiency of the hydrophobes, which are usually relatively expensive materials. To improve the alkylation efficiency of these hydrophobes with cellulose ethers, work has been undertaken to use quaternary ammonium hydroxides to catalyze these heterogeneous reactions by a phase-transfer mechanism [34-35]. The uses of hydrophobe-modified cellulose ethers have been claimed in the rheology control of latex paints, personal care products, paper coatings, laundry care, and emulsion polymerization. Current work is exploring the modification of cellulose ethers with fluorinated hydrophobes [36-37] and the use of cross-linkable unsaturated hydrophobes [38-39].

The modification of cellulose ethers with pendent vicinal diol functionality has been developed to give gelling agents for hydraulic fracturing in petroleum recovery. Aqueous solutions of cellulose ethers with pendent vicinal diol functionality can be cross-linked with a variety of polyvalent metal ions such as borate or titanate to give strong, cohesive hydrogels [40]. Because of the presence of *cisoid* diol functionality on the polysaccharide backbone, guar (Figure 3) and hydroxypropyl guar also exhibit the same cross-linking behavior with polyvalent metal ions. There has also been recent work on the silylation of water-soluble cellulose ethers, but no significant commercial uses have yet been found for these polymers [41-42].

Figure 3. Structural comparison of guar and dihydroxypropyl hydroxyethyl cellulose

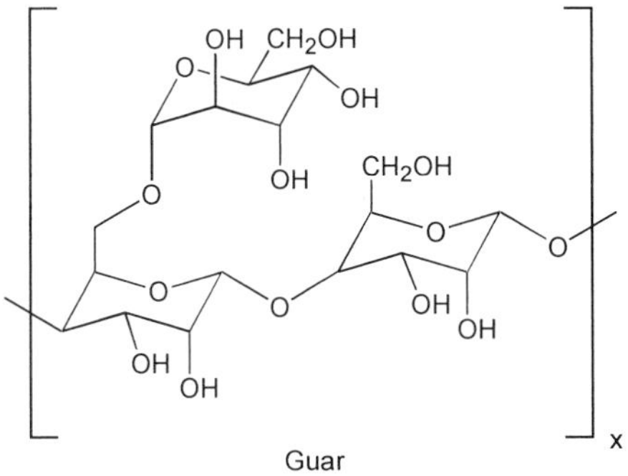

Guar 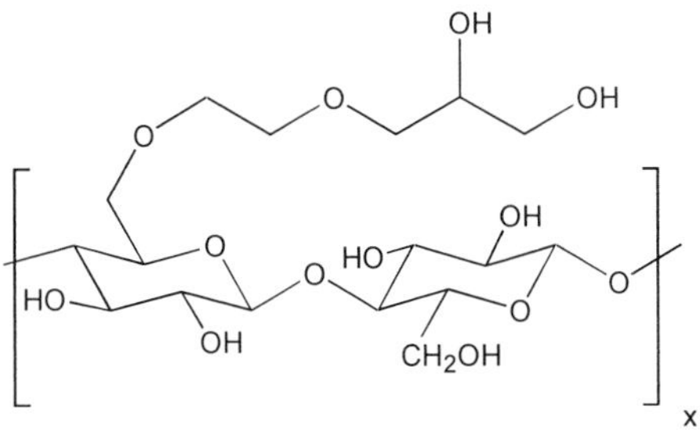   Dihydroxypropyl hydroxyethyl cellulose

Beside the ongoing work to better understand cellulose and its chemistry, the principle thrust of recent cellulosic research has focused on improved process technology, novel cellulose derivatives for specific, niche commercial applications, new fiber/membrane formation methods (liquid crystals), novel solvent systems, and applications of cellulose derivatives to new biotechnology fields such as enzyme immobilization and protein separation technologies. Future work on cellulose chemistry is likely to include the utilization of non-traditional sources of cellulose such as bagasse, bamboo, and cereal straws, as well as exocellular bacterial celluloses such as that produced by *Acetobacter xylinum*. Work is underway now in many laboratories to prepare a variety of poly(hydroxyalkanoates) from organisms such as *Pseudomonas oleovorans* by merely using different hydroxyalkylcarboxylic acid carbon sources. A potentially rich area for research is the analogous synthesis of exocellular cellulose derivatives using suitable bacteria and growth media containing judiciously chosen glucose derivatives as carbon sources.

<u>Cellulose solvents</u>
Most large-scale commercial cellulose derivatives are manufactured by a heterogeneous process. Unless cellulose dissolution is part of the fabrication process, the homogeneous derivatization of cellulose is usually impractical commercially because the viscosity of cellulose solutions is usually very high. The high solution viscosity necessitates the production and handling of dilute solutions which adversely affects throughput, and the cost of recovering expensive and toxic solvents is usually prohibitive. However, there is continued interest in the development of cellulose solvents as a way to study cellulose molecules, prepare cellulose derivatives with controlled microstructure, and produce specialized cellulose derivatives for high value-added use such as in pharmaceuticals in which the cost of manufacture is not critical.

The first cellulose solvents were transition metal complexes with either ammonia or various amines, and these solvents are still used in the physical characterization of cellulose and cellulose derivatives. Cuen, cadoxen, zincoxen, nioxen, and pd-en are the copper, cadmium, zinc, nickel, and palladium complexes of ethylenediamine [43]. Most of these solvents are deeply colored, which limits their utility in light scattering studies, but cadoxen is colorless and stable at ambient temperatures. Cadoxen has been used in the hydrodynamic volume, viscometry, and light scattering molecular weight determinations of cellulose and cellulose derivatives. Another transition metal cellulose solvent not based on ammonia or a primary amine is the iron-sodium tartrate system, which has also been used to study the solution rheology and light scattering molecular weight of cellulose [44].

Besides these transition metal complexes of cellulose, other solvent systems have been explored. Cellulose was found to be soluble in mixtures of ammonia/ ammonium thiocyanate, various tertiary amine oxides (see below), nitrogen dioxide/dimethyl formamide, and paraformaldehyde/dimethyl sulfoxide [45-48]. Work was undertaken to develop methods to spin regenerated cellulose fibers using these technologies, but only the tertiary amine oxide route has been conducted on a commercial scale. However, one of the most broadly applicable cellulose solvent systems in use today to prepare covalent derivatives is the dimethylacetamide/lithium chloride system. The DMA/LiCl system is also reported to be a solvent for chitin and chitosan. Solutions of cellulose in DMA/LiCl are reported to be stable, and a wide variety of covalent derivatives of cellulose have been prepared by this technique [49-51]. A wide variety of covalent derivatives of cellulose have been prepared homogeneously using the DMA/LiCl method including the cellulose carbamate derivative of the pesticide metribuzin, cellulose esters, cellulose ethers including the trityl derivative, and various silyl ethers [52-53]. The use of the DMA/LiCl system to fabricate cellulose fibers and membranes has also been proposed [54].

Cellulose fiber formation

Cellulose in the form of cotton staple is widely used in textiles throughout the world. However, beside using natural vegetative cellulose fibers such as cotton or ramie, a separate class of artificial textile fibers can be manufactured from regenerated cellulose. In this type of process for fiber manufacture, cellulose or some cellulose derivative is dissolved in a suitable solvent, then the polymer dope is spun through a spinneret, and the dissolved cellulose is precipitated (regenerated) as a fiber. The fiber can then be fabricated into cloth, or alternatively, the dissolved cellulose can be regenerated as sheets or films for non-woven applications such as hemodialysis or ultrafiltration membranes.

The two established methods for manufacturing artificial textile fibers from cellulose are the viscose and acetate processes [55]. In the acetate process, cellulose (wood pulp) is converted to cellulose acetate or triacetate, which is then dissolved in acetone or methylene chloride, and the polymer dope is spun into fibers by evaporation of the acetone. Extensive solvent recovery equipment is needed in this process. Cellulose acetate fibers are often referred to as "acetate" or "acetate rayon", and such fibers are lustrous and silk-like. Strictly speaking however, cellulose acetate fibers are not regenerated cellulose.

The viscose process takes cellulose (wood pulp) and dilute sodium hydroxide and forms alkali cellulose that is allowed to "age" to give an oxidatively degraded alkali cellulose. The addition of carbon disulfide forms cellulose xanthate, which is soluble in aqueous caustic. The solution of cellulose xanthate in aqueous caustic is then spun into fibers by passing the cellulosic solution into a dilute sulfuric acid bath. The regenerated cellulose produced by this process is usually referred to as "rayon". The carbon disulfide is recovered from the acid bath for reuse, although a by-product from the spinning of cellulose xanthate is hydrogen sulfide. Although reliant on toxic materials and generating toxic by-products, the viscose process is still extensively used today to manufacture rayon, cellophane, dialysis tubing, and cellulose sponges.

Although the viscose process is still used worldwide to manufacture rayon and other regenerated cellulose products in existing plants, it is unlikely that any industrialized country would permit the construction of a "new" viscose plant, primarily because of environmental concerns. For this reason, alternative routes to spinning regenerated cellulose into textile fibers are being considered. One is the so-called carbamate method being explored by Neste Oy in Finland. Cellulose can be reacted with urea around its melting point to yield cellulose carbamate. Around 140°C, urea begins decomposing into isocyanic acid, which reacts with the hydroxyl groups in cellulose to yield cellulose carbamate. Certain metal salts, particularly zinc sulfate, are often used as catalysts. Unreacted urea is then removed from the cellulose carbamate by washing with either water or ammonia [56-57]. Cellulose carbamate is relatively stable in aqueous acid, but gradually hydrolyzes in aqueous base to yield cellulose, ammonia, and carbonate. When dissolved in aqueous base, the carbamate groups will gradually hydrolyze, but cellulose carbamate dissolved in caustic is stable enough so that it can be extruded into an acid bath to form stable cellulose carbamate fibers.

These cellulose carbamate fibers can then be treated with dilute aqueous caustic to hydrolyze the carbamate groups and generate a cellulose fiber of quality similar to viscose rayon.

The carbamate process uses urea, a nontoxic, inexpensive raw material, with ammonia as the principle by-product. However, during the formation of cellulose carbamate, the degree of substitution (DS) of carbamate groups must be controlled to be between 0.2 and 0.3 and be uniformly distributed on the cellulosic backbone to insure adequate caustic solubility. Carbamate DS values above 0.3 tend to favor cross-linking and yield cellulosic polymers that are insoluble in aqueous caustic and therefore not extrudable.   Further, complete hydrolysis of the cellulose carbamate fiber to cellulose is necessary as cellulose carbamate fibers have poor wet strength.

Another environmentally benign approach to spinning regenerated cellulose is a process based on N-methylmorpholine-N-oxide. It has been known for many years that certain tertiary amine oxides were good solvents for cellulose, and that cyclic tertiary amine oxides such as N-methylpiperidine-N-oxide, N-methylmorpholine-N-oxide, and N-methylpyrrolidine-N-oxide seem to give the best results [58-59].   In this approach developed by Akzo-Nobel [60-64] and being commercialized by Courtaulds [65], cellulose is dissolved in a solution of 50% aqueous on N-methylmorpholine-N-oxide at elevated temperature. The dissolved cellulose dope is extruded into water through a spinneret, which precipitates the cellulose. The tertiary amine N-oxide is recovered from the water and recycled, so the system is closed and there are no toxic effluents. Courtaulds has demonstrated this technology on a commercial scale in South Africa and the United Kingdom. The principal problem in this approach to the manufacture of regenerated cellulose fibers is that existing viscose plants cannot be easily retrofitted to this technology, and the cost to construct this type of cellulose fiber plant is relatively high.

Starch

Starch is a polymer of glucose and commonly produced in plants as an energy and carbon storage medium, and large volumes of starch are consumed by man for food. Starch is actually a mixture of two different, but related polysaccharides, amylose and amylopectin.   Structurally, amylose is a linear homopolymer consisting of D-anhydroglucopyranose units linked by α-(1$\rightarrow$4)-glycosidic bonds (Figure 1), while amylopectin is a branched polymer of glucose with α-(1$\rightarrow$6) branch points. In contrast to cellulose, whose β-(1$\rightarrow$4)-glycosidic bonds yield a polymer with strong crystalline regions, the α-(1$\rightarrow$4) bonds of amylose yield a polymer chain that is less extended and with weaker crystalline regions, so that amylose is dispersible in cold water and soluble in hot water. The ratio of amylose to amylopectin varies depending on the botanical source of the starch, but typical starches are about 25% amylose. In aqueous solution, amylose behaves either as a random coil or as a helix, and the familiar blue color seen in starch treated with iodine is caused by the formation of an inclusion complex of iodine and amylose. A variety of inclusion complexes of lipids, polar organic solvents, and other hydrophobic materials can be prepared with amylose [66], and a recent paper describes the cross-linking of amylose and cationic hydrophobic HEC through a helical clathrate of amylose and the hydrophobic group [67].

Like cellulose, the hydroxyl groups in starch can be etherified or esterified to give the corresponding starch ethers or starch esters. Starch ethers are prepared by the base-catalyzed reaction of electrophiles such as ethylene oxide and propylene oxide with starch, and these reactions can be conducted heterogeneously either in a suspension of an organic liquid such as an alcohol, or in an aqueous slurry process. Hydroxyethyl starch is used extensively in papermaking and textile sizing [68], while hydroxypropyl starch is used as a food modifier to improve storage and freeze-thaw stability [69]. The degree of substitution (DS) of hydroxyalkylated starches is generally much lower than their cellulosic counterparts: derivatized starches with higher DS values are hard to process, and starch is inherently easier to disperse and dissolve in water, obviating the need for high degrees of substitution to achieve water solubility. For example, hydroxyethyl DS values for starch are typically between 0.05 and 0.1, while the DS for hydroxyalkyl celluloses is usually greater than 0.7. Cationic starches are prepared from either glycidyl quaternary ammonium halides or tertiary aminoalkyl halides, and have found extensive use as textile sizing agents, papermaking (fines retention, sizing, and coatings), and flocculation [70].

The principle inorganic ester of starch is starch phosphate, usually manufactured as the monoester. These esters are usually produced by the solid-state reaction of starch with polyphosphate salts such as sodium tripolyphosphate at elevated temperatures [71]. The pH of the reaction is generally controlled below about 8 to prevent starch cross-linking through phosphate diester formation. The degree of substitution (DS) for starch phosphates is usually in the range of 0.02 to 0.2. Starch phosphates are useful commercially as a wet strength additive in paper production, as a sizing agent in textile manufacture, as an antideposition agent in detergents, and in food as a freeze-thaw and emulsion stabilizer. Other inorganic esters such as starch nitrate and sulfate are known, but are not produced commercially.

Like cellulose, starch can also be esterified by reaction with suitable carboxylic anhydrides, although unlike cellulose, these reactions with starch are usually not acid-catalyzed to avoid excessive polymer degradation. Starch acetate is usually made industrially by the reaction of acetic anhydride in aqueous alkali, although starch acetate has been made by the reaction of starch with ketene or transesterification with vinyl acetate. Starch acetate is useful in papermaking, textile finishing, low temperature food stabilizers, and extrudable films. Like cellulose, starch can be converted to the corresponding xanthate by reaction with caustic and carbon disulfide, and starch xanthate has been used to encapsulate pesticides [72]. Starch esters of alkenylsuccinic anhydrides yield hydrophobic anionic derivatives that are useful in the stabilization of food emulsions, encapsulation, and in slow release of herbicides and pesticides [66].

Free-radical grafting

There is continuing interest in the free-radical grafting of various vinylic, styrenic, and acrylic monomers onto cellulose, starch, and chitosan to modify the native properties of these polysaccharides, and several comprehensive reviews have been published [73-79]. Free-radical grafting onto cellulose, starch, dextran, or chitin/chitosan has been initiated by ionizing radiation, ultraviolet/visible light, charge-transfer complexes, and various redox and oxidative reagents. One popular technique for free-radical grafting onto these polysaccharides is by using cerium (IV) salts to initiate the graft polymerization because the free-radical forms on the polysaccharide backbone and thus minimizes homopolymerization of the desired monomer [80-81].

A large portion of the work to date on the free-radical grafting of various monomers onto cellulose was conducted with the intent of modifying the tensile properties (strength and abrasion resistance), care properties (crease and wrinkle resistance, soil release), water absorption behavior, and combustibility of cellulosic fabrics for use in textile applications. Grafting of acrylamide and/or cationic acrylate monomers onto starch yields polymers useful as retention aids in paper production. Hydroxyethyl cellulose can be modified through the free-radical grafting of diallyldimethylammonium chloride to yield a cellulose ether with cationic grafts that exhibits keratin substantivity and thus has utility in cosmetic applications [82-84].

One area of extensive work on the modification of starch and cellulose by graft polymerization is the production of superabsorbent materials. Besides their use in diapers, sanitary undergarments, and feminine hygiene products, these superabsorbent materials are also of interest in wound dressings. The prototypical superabsorbent material developed in the 1970's was the hydrolyzed graft copolymer of acrylonitrile onto starch, but widespread commercialization of this technology was not achieved because of toxicity concerns with acrylonitrile and better performance with all-synthetic cross-linked acrylate systems [85-86]. However, research in this area is continuing.

Chitin and chitosan

Chitin is a naturally occurring polysaccharide that is found widely in nature as a structural constituent in fungi and yeast as well as in the exoskeletons of insects and crustaceans. The exoskeletons of crabs and lobsters, which are attractive sources of the raw material for chitin production, contain between 20% and 60% chitin, the balance being calcium carbonate and proteins. By comparison, the fiber in trees is about 50% cellulose, with the remainder being hemicelluloses and lignin. Next to cellulose, chitin is arguably the most abundant polysaccharide on earth, and for this reason there is continuing interest in finding new commercial uses for what is now essentially a waste product from the shrimp and crab industry. Not surprisingly, areas with extensive commercial shellfish industries (Japan, Massachusetts, Maryland, Washington State) have been at the forefront of chitin funding and research.

Chitin is present as part of the arthropod exoskeleton in combination with protein and calcium carbonate. Before the chitin in shellfish can be utilized, the calcium carbonate and protein must be removed. In a typical isolation procedure, crab or shrimp shells are ground, and the protein is extracted with dilute (2%) sodium hydroxide solution. The shells are then washed, and the calcium carbonate is removed with hydrochloric acid. The remaining chitin can be deacetylated by further treatment with caustic (40% to 50% sodium hydroxide) and steam to yield chitosan. The removal of protein and calcium carbonate from shellfish cuticle to isolate chitin is analogous to wood pulping, in which lignin and hemicellulose are removed from wood to yield cellulose. The process of chitin extraction from shellfish wastes has been reviewed [87].

Structurally, chitin is a polysaccharide consisting of β-(1→4) 2-acetamido-2-deoxy-D-glucose repeat units, some of which are deacetylated (Figure 1). The degree of deacetylation usually varies between 8% and 15%, but depends on the species from which the chitin is obtained and the method used for isolation and purification. Chitin is not one polymer with a fixed stoichiometry, but a class of polymers of N-acetyl-glucosamine with different crystal structures and degrees of deacetylation, with fairly large variability from species to species. The polysaccharide obtained by more extensive deacetylation of chitin is chitosan. Like chitin, chitosan is a generic term for a group of polymers of N-acetylglucosamine, but with a degree of deacetylation of between 50% an 90%. Chitosan is the β-(1→4) polysaccharide of D-glucosamine, and is structurally similar to cellulose except that the C-2 hydroxyl group in cellulose has been replaced with a primary amine group. The large number of primary amine groups (pK_a = 6.3) makes chitosan a weak base. Both chitin and chitosan are insoluble in water, dilute aqueous bases, and most organic solvents. However, unlike chitin, chitosan is soluble in dilute aqueous acids (usually carboxylic acids) as the chitosonium salt. Typical commercial chitosan has a degree of deacetylation of about 0.85 (85%).

When free of its naturally associated proteins, chitin is not antigenic to human tissue and may be inserted on or under the skin, or in contact with body fluids without harm. Chitin in the body is slowly hydrolyzed by lysozyme and is absorbed. In addition, chitin and chitosan may be safety ingested by humans. For example common foods such as bread, beer, mushrooms, shrimp, and crabs all contain some chitin. Much of the commercial interest in chitin and chitosan is based on this low order of toxicity in mammalian systems, which suggests a variety of dietary, cosmetic, and biomedical applications for chitin, chitosan, and their derivatives.

There are a number of reports that chitin and chitosan accelerate wound healing [88-89]. Glycosaminoglycans are a class of polysaccharides that occur in connective tissue of mammals and include hyaluronic acid, chondroitin sulfate, and heparin. Some of these polysaccharides (hyaluronic acid in particular) have been used successfully to accelerate wound healing and tissue regeneration in both laboratory animals and humans. The exact mechanism is not clearly understood, but oligomeric metabolites of N-acetylglucosamine and glucosamine may be responsible. The same N-acetyl-glucosamine and glucosamine functionality present in many glycosaminoglycans are also present in chitin and chitosan, and similar accelerated wound healing has been reported for chitin and chitosan (Figure 4). Work has also been undertaken to use chitosan and its derivatives to aid in the regeneration of both periodontal tissue and bone [90], and there is continuing interest in fabricating chitin/chitosan into sutures and other implantable biomedical devices. The fabrication of chitosan and chitin fibers and films has been reviewed [91].

In addition to its reported ability to accelerate wound healing, chitosan is also reported to have fungicidal and bacteriacidal activity. Grow inhibition of *Pseudomonas aeruginosa*, *Staphylococcus aureus*, and *Staphylococcus epidermis* was reported on agar plates with 1% chitosan in dilute acetic acid, and parallel experiments with *Candida tropicalis* also showed fungal growth inhibition. Chitosan solutions were reported to be effective against topical fungal infections such as athlete's foot. The active agent in these cases may be chitosan oligomers [92-94].

A recurring problem for the biomedical engineer is the incompatibility of most materials with blood, and the tendency for blood to form clots (thrombosis) on the surfaces of these materials. Heparin (Figure 4) is an anticoagulant, non-toxic polysaccharide that prevents the formation of these clots when coated on vascular implants. While chitosan itself is a hemostatic material (e.g., stops bleeding by enhancing clotting) [95], chitosan sulfate exhibits the same anticoagulant behavior as heparin, and there is on-going interest in chitosan sulfate as a coating on vascular grafts and an injectable substitute for heparin.

Cardiovascular disease is a leading cause of death in the industrialized world. A contributing factor to cardiovascular disease is serum cholesterol, and it is also known that chitosan ingested orally possesses hypocholesterolemic activity. It is believed that chitosan dissolves in the low pH gastric juices in the stomach and reprecipitates in alkaline intestinal fluid, entrapping cholic acid as an ionic salt and thus prevents its absorption by the liver. The cholic acid is then digested by bacteria in the large intestine [96-97]. Ingested chitosan may also increase the ratio of high density lipoprotein to total cholesterol.

Figure 4. Structural similarities of chitin/chitosan to selected glycosaminoglycans

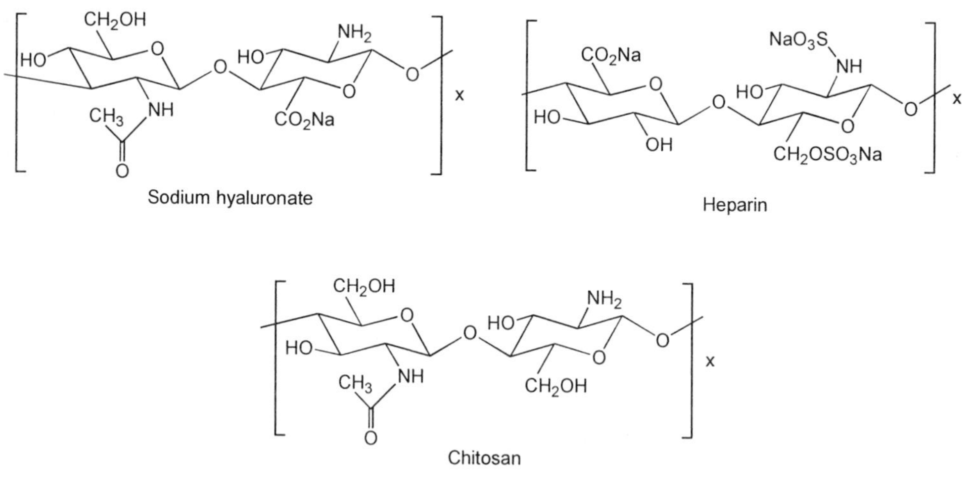

Sodium hyaluronate

Heparin

Chitosan

Because of the unique solubility and biocompatibility characteristics of chitosan, there is also interest in using chitosan in various biotechnology applications [98]. Chitosan can be dissolved in many dilute aqueous carboxylic acids such as acetic or lactic acid and films can be easily cast and insolubilized by treatment with aqueous bases. These films can also be cross-linked with a variety of dialdehydes such as glutaraldehyde. Alternatively, chitosan dissolved in dilute acids (as a polycation chitosonium salt) can be fabricated into membranes or gels by reaction with suitable polyanions such as sodium alginate or carrageenan. Because these various membrane/gel formation conditions are relatively mild, these approaches should have utility in enzyme as well as whole cell immobilization, and these approaches have been used to immobilize enzymes, pharmaceuticals, and mammalian cells [99-103].

One of the earliest uses of chitosan was as a flocculant in wastewater treatment. The presence of protonated primary amine groups on the chitosan polymer backbone can lead to coacervation with negatively charged materials such proteins, and for this reason chitosan can be used to flocculate proteinaeous wastes. The high density of primary amine groups on chitosan can also chelate various heavy metals, so chitosan has been used to remove heavy metals such as copper, lead, mercury, and uranium from wastewater. Chelating efficacy can be improved by chemical modification or cross-linking of chitosan [104-106]. A process has also been described using chitosan to remove polychlorinated biphenyls (PCB) from water [107-109].

Conceptually, both chitin and chitosan can be chemically modified by the same general processes that are used to prepare cellulose ethers and cellulose esters, and indeed many of these analogous chitin/chitosan esters and ethers are known [87,104]. Because of the presence of a primary amine as well as primary and secondary hydroxyl groups in chitosan, both N- and O-alkylation and acylation are possible, which leads to a wide range of functionalized polymers. Fundamental to the chemical modification of any polymer is the requirement that the substrate polymer be accessible to the reactant, either by dissolving the polymer in a suitable solvent or otherwise swelling the polymer to permit the permeation of the reactant throughout the solid polymer matrix. In the case of cellulose ethers, aqueous sodium hydroxide works well to swell cellulose and render it reactive to O-alkylation. However, chitin and chitosan are considerably less swollen and reactive in aqueous sodium hydroxide, and prolonged steeping and/or cyclical freezing and thawing are often needed to activate chitin and chitosan to base-catalyzed reactions. Like cellulose, chitin and chitosan are reported to be soluble in DMA/LiCl solvent, and other solvents for chitin such as trichloroacetic and formic acid have been reported. The xanthanation of alkali chitin by a process analogous to cellulose xanthate has been reported, and chitin xanthate has been used to prepare chitin fibers and sponges [110].

One novel approach to the preparation of covalent chitosan derivatives involves the suspension of ground chitosan in an aqueous organic diluent, followed by the addition of a suitable carboxylic acid to swell and decrystallize the chitosan, thus rendering it reactive [111-112]. This entirely heterogeneous process is referred to as "acid decrystallization", and has been used to prepare a number of covalent chitosan derivatives including carboxymethyl chitosan, hydroxyethyl chitosan, trimethyl-ammonium hydroxypropyl chitosan as well as a number of chitosonium salts of carboxylic acids. Many of these chitosonium salts of carboxylic acids exhibit substantivity to skin and hair have utility in either cosmetic or pharmaceutical applications [112].

Miscellaneous polysaccharides

A number of unique polysaccharides are produced extracellularly by various bacteria and fungi, and in a few cases, the yields are high enough to warrant commercial isolation. Xanthan, dextran, curdlan, pullulan, and scleroglucan are some of these extracellular polysaccharides of commercial interest [113]. Part of the interest in these polysaccharides is in their unique structures, and also the possibility to alter these polysaccharide structures by genetically altering the microorganisms that elaborate these polymers. Xanthan consists of a polymer backbone of β-(1→4)-glucopyranosyl units like cellulose, but in addition, xanthan has a trisaccharide side chain in the 3 position on alternating D-glucopyranose repeat units consisting of two mannopyranosyl and one glucopyranosyluronic acid units. Xanthan is a polyanionic polymer, but unlike most polyelectrolytes, its aqueous viscosity is not significantly affected by high electrolyte concentrations, so the aqueous viscosity of xanthan is not significantly affected by dissolved salts. For this reason, xanthan is used in the oil-field in drilling fluids, workover and completion, fracturing, and enhanced oil recovery. Xanthan is also used extensively in foods as a thickener with good food compatibility.

Dextran is a polysaccharide of repeating α-(1→6)-glucopyranosyl units with some branching. Dextran is produced enzymatically from sucrose and is used primarily in medical and pharmaceutical applications. Hydrolyzed dextran is used as a plasma extender, cross-linked dextran is used in the separation and purification of proteins, and there is interest in dextran sulfate as a substitute for heparin.

Literature cited

1.   Whistler, R. L.; BeMiller, J. N., "Industrial Gums: Polysaccharides and Their Derivatives", 3rd edition, Academic Press, New York, 1993.
2.   Chandra, R.; Rustgi R., Prog. Polym. Sci., 1998, 23, 1273.
3.   Hocking, P. J., J. Macromol. Sci., Rev. Macromol. Chem. Phys., 1992, C32(1), 35.
4.   Klemm, D.; Philipp, B.; Heinze, T.; Heinze, U.; "Comprehensive Cellulose Chemistry", Volume 2, Wiley-VCH, Weinheim, 1998.
5.   Wadsworth, L. C.; Daponte, D.; in "Cellulose Chemistry and Its Applications", Nevell, T. P.; Zeronian, S. H.; eds.; Ellis Horwood, Chinchester, 1985; Chap. 14.
6.   Peuscher, M.; Engelskirchen, K.; Grunberger, E.; U. S. Patent 4,906,744, March 6, 1990.
7.   Dixon, J.; Andrews, P.; Butler, L. G.; Biotech. Bioeng., 1979, 21, 2113.
8.   Nicholson, M. D.; and Merritt, F. M.; in "Cellulose Chemistry and Its Applications", Nevell, T. P.; Zeronian, S. H., Eds.; Ellis Horwood, Chinchester, 1985; Chap. 15.
9.   Davidson, R. L.; "Handbook of Water-Soluble Gums and Resins", McGraw-Hill, New York, 1980.

10. Robitaille, L.; Turcotte, N.; Fortin, S.; Chalet, G.; <u>Macromolecules</u>, 1990, <u>24</u>, 2413.
11. Ritcey, A. M.; Gray, D. G.; <u>Cellulose</u>, 1990, 367.
12. Wignall, G. D.; Annis, B. K.; Triolo, R.; <u>J. Polym. Sci., Polym. Phys.</u>, 1991, <u>29</u>, 349.
13. Detemmerman, A.; <u>Polymer Paint Colour J.</u>, 1990, <u>180</u>, 385.
14. Partain, E. M.; <u>Paint & Coatings Industry</u>, November 1998, 40.
15. Stone, F. W.; Rutherford, J. M.; U. S. Patent 3,472,840, October 14, 1969.
16. Goddard, E. D.; Harris, W. C.; <u>J. Soc. Cosmet. Chem.</u>, 1987, <u>38</u>, 233.
17. Faucher, J. A.; Goddard, E. D.; Hannan, R. B.; Kligman, A. M.; <u>Cosmetic Toiletries</u>, 1977, <u>92</u>, 39.
18. Brode, G. L.; Goddard, E. D.; Harris, W. C.; Salensky, G. A.; in "Cosmetic and Pharmaceutical Applications of Polymers", Gebelein, C. G., Ed.; Plenum, New York, 1991; pp. 117-128.
19. Marlin L.; Yamamoto, R. K.; U. S. Patent 5,358,706, October 25, 1994.
20. Brode, G. L.; Kreeger, R. L.; Salensky, G. A.; U. S. Patent 5,407,919, April 18, 1995.
21. Brode, G. L.; Doncel, G. F.; Gabelnick, H. L.; Kreeger, R. L.; Salensky, G. A.; U. S. Patent 5,595,980, January 21, 1997.
22. Shay, G. D., "Paint and Coating Testing Manual", Koleske, J. V., Ed.; ASTM, Philadelphia, 1995; Chap. 30.
23. Partain, E. M.; <u>Polym. Mater. Sci. Eng.</u>, 1992, <u>66</u>, 19.
24. Goodwin, J. W.; Hughes, R. W.; Lam, C. K.; Miles, J. A.; Warren, B. C. H.; in "Water-Soluble Polymers: Performance Through Association", Glass, J. E., Ed.; American Chemical Society, Washington, 1989, Chap 19.
25. Svanholm, T.; Molenaar, F.; Toussaint, A.; <u>Prog. Org. Coat.</u>, 1997, <u>30</u>, 159.
26. Fu, E.; Young, T. S.; <u>Polym. Mater. Sci. Eng.</u>, 1989, <u>61</u>, 614.
27. Liang, S. J.; Fitch, R. M., <u>J. Colloid Interface Sci.</u>, 1982, <u>90</u>, 51.
28. Landoll, L. M.; <u>J. Polym. Sci., Polym. Chem. Ed.</u>, 1982, <u>20</u>, 443.
29. Landon. L. M.; U. S. Patent 4,228.277, October 14, 1980.
30. Partain, E. M.; Brode, G. L.; Hoy, R. C.; European Patent Specification 0 384 167 B1, July 6, 1994.
31. Just, E. K.; Majewicz, T. G.; Sau, A. C.; U. S. Patent 5,124,445, June 23, 1992.
32. Bostrom, P.; Ingvarsson, I.; Sundberg, K.; U. S. Patent 5,140,099, August 18, 1992.
33. Jenkins, R. D.; Bassett, D. R.; U. S. Patent 5,426,182, June 20, 1995.
34. Partain, E. M.; in "Polymer Modification", Swift, G.; Carraher, C.E.; Bowman, C. N.; Eds., Plenum Press, New York, 1997.
35. Partain, E. M.; <u>Polym. Mater. Sci. Eng.</u>, 1996, <u>75</u>, 381.
36. Angerer, J. D.; Majewicz, T. G; Meshreki, M. H.; U. S. Patent 5,290,829, March 1, 1994.
37. Hwang, F. S.; Hogen-Esch, T. E.; <u>Macromolecules</u>, 1993, <u>26</u>, 3156.
38. Partain, E. M.; <u>Polymer Preprints</u>, 1998, <u>39</u>, 82.
39. Partain, E. M.; U. S. Patent 5,583,214, December 10, 1996.

40. Brode, G. L.; Stanley, J. P.; Partain, E. M.; Kreeger, R. L. in "Biotechnology 3: Industrial Polysaccharides"; Yalpani, M.; Ed, Elsevier, New York, 1987.
41. Billmers, R. L.; U. S. Patent 4,973,680, November 27, 1990.
42. Sau, A. C.; U. S. Patent 4,992,538, February 12, 1991.
43. Burger, J.; Kettenbach, G.; Klüfers, P.; Macromol. Symp., 1995, 99, 113.
44. Valtasaari, L.; Makromol. Chem., 1971, 150, 117.
45. Turbak, A. F.; Hammer, R. B.; Davies, R. E.; Hergert, H. L.; Chemtech., 1980, 10, 51.
46. Philipp, B.; Schleicher, H.; Wagenknecht, W.; Chemtech, 1977, 7, 702.
47. Hudson, S. M.; Cuculo, J. A.; Wadsworth, L. C.; J. Polym. Sci., Polymer Chem. Ed., 1983, 21, 651.
48. Johnson, D. C.; Nicholson, M. D.; Haigh, F. C.; Appl. Polym. Symp., 1976, 28, 931.
49. Turbak, A. F.; El-Kafrawy, A.; Snyder, F. W.; Auerbach, A. B.; U. S. Patent 4,302,252, November 24, 1981.
50. McCormick, C. L.; U. S. Patent 4,278,790, July 14, 1981.
51. Dawsey, T. R.; McCormick, C. L.; J. Macromol. Sci. Rev., Macromol. Chem. Phys., 1990, C30, 405.
52. McCormick C. L.; Lichatowich, D. K.; J. Polym. Sci., Polym. Lett. Ed., 1979, 17, 479.
53. Camacho Gomez, J. A.; Erler, U. W.; Klemm, D. O.; Macromol. Chem. Phys., 1996, 197, 953.
54. Turbak, A. F.; "Newer Cellulose Solvent Systems", pages 105-110 in Proceedings of the 1983 International Dissolving and Specialty Pulps Conference, TAPPI, Atlanta, USA, 1983.
55. Treiber, E. E. in "Cellulose Chemistry and Its Applications"; Nevell T. P.; Zeronian, S. H., Eds.; Ellis Horwood, Chinchester, 1985; Chap. 18.
56. Huttunen, J.; Turunen, O.; Mandell, L.; Eklund, V.; Ekman, K.; U. S. Patent 4,404,369, September 13, 1983.
57. Ekman, K.; Eklund, V.; Fors, J.; Huttunen, J. I.; Selin, J.-F.; Turunen, O. T. in "Cellulose: Structure, Modification, and Hydrolysis"; Young R. A.; Rowell, R. M., Eds., Wiley, New York, 1986; Chap. 7.
58. Johnson, D. L.; U. S. Patent 3,447,939, June 3, 1969.
59. Johnson, D. L.; U. S. Patent, 3,508,941, April 28, 1970.
60. McCorsley, C. C.; Varga, J. K.; U. S. Patent 4,142,913, March 6, 1979.
61. Franks, N. E.; Varga, J. K.; U. S. Patent 4,145,532, March 20, 1979.
62. McCorsley, C. C.; Varga, J. K.; U. S. Patent 4,211,574, July 8, 1980.
63. McCorsley, C. C; U. S. Patent 4,246,221, January 20, 1981.
64. Peguy, A.; in "Cellulosics Utilization", Inagaki H.; Phillips, G. O.; Eds, Elsevier Applied Science, London, 1989, p. 19.
65. Stinson, S. C.; Chem. & Eng. News, July 17, 1995, page 11.
66. Whistler, R. L.; BeMiller, J. N.; Paschall, E. F.; "Starch: Chemistry and Technology", Academic Press, New York, 1984.
67. Gruber, J. V.; Konish, P. N.; Macromolecules, 1997, 30, 5361.

68. Moser, K. B.; in "Modified Starches: Properties and Uses", Wurzburg, O. B.; Ed, CRC Press, Boca Raton, 1987; pp. 80-88.
69. Tuschhoff, J. V.; in "Modified Starches: Properties and Uses", Wurzburg, O. B.; Ed, CRC Press, Boca Raton, 1987; pp. 90-96.
70. Solarek, D. B.; in "Modified Starches: Properties and Uses", Wurzburg, O. B.; Ed, CRC Press, Boca Raton, 1987; pp. 113-129.
71. Solarek, D. B.; in "Modified Starches: Properties and Uses", Wurzburg, O. B.; Ed, CRC Press, Boca Raton, 1987; pp. 97-112.
72. Shasha, B. S.; Doane, W. M.; Russell, C. R.; J. Polym. Sci., Polym. Lett. Ed., 1976, 14, 417.
73. Hebeish, A.; Guthrie, J. T.; "The Chemistry and Technology of Cellulosic Copolymers", Springer-Verlag, Berlin, 1981.
74. McDowall, D. J.; Gupta, B. S.; Stannett, V. T.; Prog. Polym. Sci., 1984, 10, 1.
75. Bains, M. S.; J. Polym. Sci. C, 1972, 37, 125.
76. Samal, R. K.; Sahoo, P. K.; Samantaray, H. S.; J. Macromol. Sci., Rev. Macromol. Chem. Phys., 1986, C26, 81.
77. Mohanty, A. K.; J. Macromol. Sci., Rev. Macromol. Chem. Phys., 1987, C27, 593.
78. Fanta G. F.; Doane, W. M.; in "Modified Starches: Properties and Uses", Wurzburg, O. B.; Ed, CRC Press, Boca Raton, 1987; pp. 149-178.
79. Kurita, K.; in "Applications of Chitin and Chitosan", Goosen, M. F. A.; Ed, Technomic, Lancaster, 1997; Chap. 17.
80. Mino, G.; Kaizerman, S.; J. Polym. Sci., 1958, 31, 242.
81. Nagarajan S.; Srinivasan, K. S. V.; J. Macromol. Sci., Rev. Macromol. Chem. Phys., 1998, C38(1), 53.
82. Iovine C. P.; Ray-Chaudhuri, D. K.; U. S. Patent 4,131,576, December 26, 1978.
83. Neigel D.; Kancylarz, J.; U. S. Patent 4,464,523, August 7, 1984.
84. Iovine, C. P.; Nowak, R. A.; U. S. Patent 4,803,071, February 7, 1989.
85. Gross, J. R.; in "Absorbent Polymer Technology", Brannon-Peppas L.; Harland, R. S.; Eds.; Elsevier, Amsterdam, 1990, Chap. 1.
86. Pó, R.; J. Macromol. Sci., Rev. Macromol. Chem. Phys., 1994, C34, 607.
87. Roberts, G. A. F.; "Chitin Chemistry", Macmillan, London, 1992.
88. Muzzarelli, R. A. A.; Mattioli-Belmonte, M.; Muzzarelli, B.; Mattei, G.; Fini, M.; Biagini, G.; in "Advances in Chitin Science, Volume 2, Proceedings of the 7th International Conference on Chitin & Chitosan", Domard, A.; Roberts, G. A. F.; Vårum, K. M.; Eds., Jacques Andre, Lyon (France), 1997; pp. 580-589.
89. Minami, S.; Okamoto, Y.; Mori, T.; Fujinaga, T.; Shigemasa, Y.; in "Advances in Chitin Science, Volume 2, Proceedings of the 7th International Conference on Chitin & Chitosan", Domard, A.; Roberts, G. A. F.; Vårum, K. M.; Eds., Jacques Andre, Lyon (France), 1997; pp. 633-639.
90. Klokkevold, P.; Redd, M.; Salamati, A.; Kim, J.; Nishimura, R.; in "Advances in Chitin Science, Volume 2, Proceedings of the 7th International Conference on Chitin & Chitosan", Domard, A.; Roberts, G. A. F.; Vårum, K. M.; Eds., Jacques Andre, Lyon (France), 1997; pp. 656-663.
91. Rathke T. D.; Hudson, S. M.; J. Macromol. Sci., Rev. Macromol. Chem. Phys., 1994, C34(3), 375.

92. Allan, G. G.; Altman, L. C.; Bensinger, R. E.; Ghosh, D. K.; Hirabayashi, Y.; Neogi, A. N.; Neogi, S.; in "Chitin, Chitosan, and Related Enzymes", Zikakis, J. P.; Ed., Academic Press, New York, 1984.
93. Tanigawa, T.; Tanaka, Y.; Sashiwa, H.; Saimoto, H.; Shigemasa, Y.; in "Advances in Chitin and Chitosan", Brine, C. J.; Sandford, P. A.; Zikakis, J. P.; Eds., Elsevier Applied Science, London, 1992; pp. 206-215.
94. Ueno, K.; Yamaguchi, T.; Sakairi, N.; Nishi, N.; Tokura, S.; in "Advances in Chitin Science, Volume 2, Proceedings of the 7th International Conference on Chitin & Chitosan", Domard, A.; Roberts, G. A. F.; Vårum, K. M.; Eds., Jacques Andre, Lyon (France), 1997; pp. 156-161.
95. Klokkevold, P.; Fukayama, H.; Sung, E.; in "Advances in Chitin Science, Volume 2, Proceedings of the 7th International Conference on Chitin & Chitosan", Domard, A.; Roberts, G. A. F.; Vårum, K. M.; Eds., Jacques Andre, Lyon (France), 1997; pp.698-704.
96. Maezaki, Y.; Tsuji, K.; Nakagawa, Y.; Biosc. Biotech. Biochem., 1993, 57, 1439.
97. Muzzarelli R. A. A.; DeVincenzi, M.; in "Applications of Chitin and Chitosan", Goosen, M. F. A.; Ed, Technomic, Lancaster, 1997; Chap.7.
98. Yao, K. D.; Peng, T.; Yin, Y. J.; Xu, M. X.; Goosen, M. F. A.; J. Macromol. Sci., Rev. Macromol. Chem. Phys., 1995, C35(1) 155.
99. Ohtakara, A.; Mukerjee, G.; Mitsutomi, M.; in "Chitin and Chitosan", Skjåk-bræk, G.; Anthonsen, T.; Sandford, P.; Eds., Elsevier Applied Science, London, 1989; pp. 643-651.
100. Kise, H.; Hayakawa, A.; Noritomi, H.; Biotech. Lett., 1987, 9, 543.
101. Muzzarelli, R. A. A.; Barontini, A.; Rocchetti, R.; Biotech. Bioeng., 1976, 18, 1445.
102. Remuñán-López, C.; Lorenzo, M. L.; Portero, A.; Vila Jato, J. L.; Alonso, M. J.; in "Advances in Chitin Science, Volume 2, Proceedings of the 7th International Conference on Chitin & Chitosan", Domard, A.; Roberts, G. A. F.; Vårum, K. M.; Eds., Jacques Andre, Lyon (France), 1997; pp. 600-607.
103. Kim S.-K.; Rha, C.; in "Chitin and Chitosan", Skjåk-bræk, G.; Anthonsen, T.; Sandford, P.; Eds., Elsevier Applied Science, London, 1989; pp. 617-626.
104. Muzzarelli, R. A. A.; "Chitin", Pergamon Press, Oxford, 1977.
105. Kurita, K.; Chikaoka, S.; Koyama, Y.; Chem Letters, 1988, 9.
106. Kim, Y. B.; in "Advances in Chitin Science, Volume 2, Proceedings of the 7th International Conference on Chitin & Chitosan", Domard, A.; Roberts, G. A. F.; Vårum, K. M.; Eds., Jacques Andre, Lyon (France), 1997; pp. 837-844.
107. Van Daele, Y.; Thomé, J. P.; Bull. Environ. Contam. Toxicol., 1986, 37, 858.
108. Thomé, J. P.; Hugla, J. L.; Weltrowski, M.; in "Advances in Chitin and Chitosan", Brine, C. J.; Sandford, P. A.; Zikakis, J. P.; Eds., Elsevier Applied Science, London, 1992; pp. 639-647.
109. Thomé, J. P.; Patry, J.; Thys, I.; Weltrowski, M.; in "Advances in Chitin Science, Volume 1, Proceedings of the 1st International Conference on Chitin & Chitosan", Domard, A.; Jeuniaux, C.; Muzzarelli, R.; Roberts, G.; Eds., Jacques Andre, Lyon (France), 1996; pp. 470-475.

110. Balassa L. L.; Prudden J. F.; in "Proceedings of the 1st International Conference on Chitin/Chitosan", Muzzarelli, R. A. A.; Pariser, E. R.; Eds., MIT Sea Grant Program, Cambridge (Massachusetts), 1977.
111. Partain, E. M.; Brode, G. L.; U. S. Patent 4,929,722, May 29, 1990.
112. Partain, E. M.; Brode, G. L.; U. S. Patent 4,946,870, August 7, 1990.
113. Sandford, P. A.; Baird, J.; in "The Polysaccharides", Volume 2, Aspinall, G. O.; Ed., Academic Press, New York, 1983; Chap. 7.

ELECTRICALLY CONDUCTING POLYMERS

ABHIMANYU O. PATIL

Corporate Research Laboratory, Exxon Research & Engineering Company, Route 22 East, Clinton Township, Annandale, NJ 08801, USA

Introduction
Synthesis of Electrically Conducting Polymers
Polyacetylene
Polythiophenes and Polypyrroles
Poly(Arylene Vinylenes)
Polyaniline
Applications of Conducting Polymers
Summary
Literature Cited

Introduction

Most polymers are insulators, with desirable properties such as light weight, processability, durability, and low cost. By designing the molecular structures of polymers, chemists have developed new materials that exhibit electrical conductivities comparable to metals while retaining the advantages of polymers. There are three approaches to making conducting materials: pyrolysis to produce a conducting residue (mostly carbon); producing a composite structure from a conducting material and an insulating organic polymer; and making organic conjugated polymers. This review addresses the third approach. Generally, these electrically conducting polymers are composed of conjugated polymer chains with π-electrons delocalized along the backbone. In the neutral, or undoped, form the polymers are either insulating or semiconducting. The polymers are converted to the electrically conductive, or doped, form via oxidation or reduction reactions that create delocalized charge carriers. Charge balance is accomplished by incorporating an oppositely charged counterion into the polymer matrix. Over the past two decades, there has been a number of books[1-5] and reviews[6-15] on conducting polymers. A journal called *Synthetic Metals* reports on various aspects of conducting-polymer research. Since 1976, the biennial "International Conference of Synthetic Metals" (ICSM) has been the most important international platform in conducting polymers.

Initially, the drive for research in conducting polymers was the possibility of combining electronic properties with the attractive mechanical properties and processing advantages of polymers. The first generation of conjugated polymers exhibited insolubility, infusibility, and instability: they could not be processed and were generally, unstable in air. It became evident that the initial expectation of quickly replacing metals by plastics could not be met. In fact, the journey from the invention of conducting polymers to practical applications may be long. To be potentially useful in electronic applications, a material must be environmentally stable and have excellent electronic and mechanical properties, and it should be solution or melt-processable. The delocalized electronic structures of π-conjugated polymers that are responsible for their unusual electronic properties tend to yield relatively stiff chains with little flexibility but with relatively strong inter-chain attractive interaction that makes them insoluble and non-processable.

To place conducting polymers in context, copper has a conductivity of about 5 x 10^5 S/cm (equivalent to per ohm per cm), and polystyrene has a value of 1 x 10^{-18} S/cm. Nylon has a value of 10^{-14} S/cm; Hg, 10^4 S/cm. Typically, undoped conducting polymers have values comparable to those of other insulating polymers (10^{-12} S/cm), which on doping are increased to 10^2 S/cm.

Synthesis of Electrically Conducting Polymers

Conducting polymers are prepared either directly, by electro- or oxidative-polymerization, or are polymerized and then oxidized chemically or electrochemically. The different preparations have been driven by the desire to examine many different types of polymers and to create polymers that are soluble in water and common organic solvents and are processable in various forms. Other goals include novel structures, easier synthesis, and stability both in conducting and non-conducting states.

Among well-known conducting polymers, polyacetylene and poly(phenylene vinylene) are synthesized almost exclusively by chemical polymerization or from precursor route. Polyheterocycles such as polythiophene or polypyrrole are synthesized by both chemical and electrochemical polymerization.

Polyacetylene

Polyacetylene, a prototype conducting polymer, was first synthesized over 24 years ago, with flexible, free-standing films prepared by Ziegler-Natta polymerization. These films have become known as Shirakawa polyacetylene.[16-17] Replacing the toluene solvent by a silicone oil and modifying other preparative conditions led to improved synthesis of polyacetylene polymers containing fewer defects in their structure and with very high (100,000 S/cm) conductivities after doping. The specific conductivity of this material approaches that of copper or silver.[18-19] For p-type doping, iodine is the most popular dopant because of its ease of handling, availability, and effectiveness. The polymers obtained by these procedures were, however, nonprocessable.

The metathesis polymerization of 7,8-bis(trifluoromethyl)tricyclo[4.2.2.0^{2,5}]deca-3,7,9-triene by WCl_6 and Me_4Sn leads to a high molecular weight, soluble precursor polymer that can be thermally converted to polyacetylene.[20-23] The precursor polymer is readily soluble in common organic solvents and can therefore be easily purified by reprecipitation. In this process, the bis(trifluoromethyl)benzene group is eliminated. This soluble precursor route, the Durham route,[20-21] produces a prepolymer that can be oriented to obtain highly ordered structures with high conductivity. The main difference between the Shirakawa and Durham routes is the morphology of the final product. Instead of fibrillar Shirakawa polyacetylene with a low bulk density, the Durham route gives compact samples with higher bulk density (1.05-1.1 g/cm^3).

Using ring-opening metathesis polymerization (ROMP), Grubbs et al.[23] polymerized benzvalene by a tungsten alkylidene initiator system. Because no molecule is eliminated during conversion (unlike Durham polyacetylene), this method can produce thick films. Later, ROMP has been used to polymerize cyclooctatetraene and substituted cyclooctatetraenes to obtain one more form of polyacetylene. Substituted polyacetylenes obtained by this route are soluble in several solvents.[4]

Polythiophenes and Polypyrroles

Among the conducting polyheterocycles, the most intensively studied are polypyrrole, polythiophene, and their derivatives. Polypyrrole was shown to be a conducting polymer in 1968.[24] This work was extended by workers at IBM, who showed that polypyrrole films could be obtained by electrochemical polymerization. The films could be cycled electrochemically between conducting (doped) and insulating states with conductivities varying from 10^2 S/cm to 10^{-10} S/cm.[25] Unlike Shirakawa polyacetylene, which is fibrillar, polypyrrole films are dense.

Table 1. Representative Polythiophenes and Their Maximum conductivities

Polymer	Maximum Conductivity S/cm	Reference
Polythiophene	14^a	26
	190^b	27
Poly (3-methylthiophene)	3^a	28,29
	500^b	27
Poly (3-hexylthiophene)	10^a	28,29
	95^b	30
Poly 3-(2-ethanesulfonate)thiophene, Na-salt	10^a	31
Poly (3-alkylether)thiophene	1100^b	36

a: chemically prepared, b: electrochemically prepared

Research activity has been catalyzed by the discovery that heterocyclic derivatives such as substituted polythiophenes and polypyrroles could be prepared as relatively stable films with high conductivity. Several organic-solvent-soluble 3-substituted polythiophenes with high conductivities have been prepared. Poly(3-

hexylthiophene) has a room temperature conductivity of 95 S/cm. Poly(3-methylthiophene) has much higher conductivity (500 S/cm). Substituted polythiophenes have been reported with conductivities above 1100 S/cm. The parent polythiophene has a room temperature conductivity of 50-100 S/cm. These values suggest that alkyl substituents in polythiophene have little effect on conductivity and that there may be little twisting of the backbone conjugation. Substituted polythiophenes with some of the highest conductivities are given in Table 1.

Both electrochemical methods and chemical methods using Grignard reagents were employed to synthesize these polymers. The use of $FeCl_3$ to prepare the poly(3-alkylthiophenes) is still the simplest and most effective procedure.[26] Although polythiophene itself cannot be stretched, poly(3-hexylthiophene) film was stretched to a draw ratio of 5 with an increase in conductivity of an iodine-doped samples from 5 S/cm to 200 S/cm. It has been reported that poly(3-dodecylthiophene) can be melt-spun to give a tough, flexible fiber which, after doping with iodine, has a conductivity of 55 S/cm.[37] Since the alkylthiophenes are soluble, gel permeation chromatography of these polymers is feasible for characterization. The majority of these studies have used polystyrene standards. As with soluble poly(alkoxythiophene vinylenes), one should be cautious in interpreting the results.[63] It should be noted that these are comparisons of hydrodynamic volume, not molecular weight; nevertheless, relative values allow comparison of various polymers.

The optical properties of the conducting polymers are important to understanding their basic electronic structure. π-Conjugation in the polymers is implied by their color and electronic spectra; thus spectroscopy is a powerful probe for characterizing the electronic processes in the undoped and doped states and during doping. The changes of the optical spectra accompanying doping are significant and have played a key role in elucidating the mechanism of doping and the nature of the charge storage species in the polymer chain. Since these polythiophene derivatives are soluble in common organic solvents (e.g., chloroform, THF) in both their neutral and conductive (doped) forms, they have opened the way for studying optical and magnetic properties in solution and charge storage configurations of the doped isolated macromolecules in dilute solution. The electronic absorption spectra of the polyalkylthiophenes indicate that the band edge occurs at about 2 eV, a value typical for the entire polyalkylthiophene series, and that absorbance is essentially independent of the alkyl substituent. The optical properties of conducting polymers have been reviewed.[7]

Similar to organic-solvent-soluble 3-substituted polythiophenes, water-soluble polythiophenes have been prepared. In these polymers, the counterions are covalently bound to the polymer backbone, leading to "self-doping".[31] The sodium salts and the proton "salts" (acids) of poly-3-(2-ethane sulfonate)thiophene (P3-ETSNa and P3-ETSH, respectively) and poly-3-(4-butanesulfonate)thiophene (P3-BTSNa) and P3-BTSH, respectively) have been prepared. In these self-doped, conjugated polymers, charge injected into the π-electron system is compensated by protons (or Li^+, Na^+, etc.) ejection, leaving behind the oppositely charged counterion.[31,32]

The scheme takes advantage of the addition of a flexible side chain to a thiophene unit that make long-chain polyalkylthiophenes soluble. More importantly, these polymers are water soluble; the class of self-doped polymers therefore contains the first known water-soluble conducting polymers. Self-doped polymers also have been

M = H, Na

prepared by direct chemical polymerization of the monomer with FeCl₃.[33] Very strong evidence for self-doping has been provided without any ambiguity.[34,35] The self-doped polymer concept was later extended to polypyrrole.[38-39] Numerous N-substituted polypyrroles have been studied; they usually exhibit much lower conductivity of the doped material, presumably resulting from sterically twisting the aromatic rings out of planarity. In contrast, polyalkylpyrroles behave similarly to polyalkylthiophenes. Several poly (3-alkylpyrroles) have been prepared by electrochemical and chemical polymerization. These polymers were soluble in organic solvents.[40-41] Substituted polypyrroles with some of the highest conductivities are given in Table 2. Polyalkylthiophenes with chiral or redox-active substituents result in optical and electrocatalytically active electrode coatings, respectively.[42-43]

In addition to making processable polymers, another important synthetic challenge for electroactive conducting polymers is to prepare polymers with small bandgaps. The goal is to produce polymers that do not need doping to exhibit high conductivities, since at room temperature thermal energy would be enough to excite electrons from the valence to the conduction band (π- π^*). These polymers could also yield transparent conducting polymers. The relatively large bandgaps in common conducting polymers are generally attributed to bond alternation. By benzannelation of polythiophene, a new polymer, polyisothianaphthene (PITN), was synthesized with the smallest bandgap among all known conjugated conducting polymers: 1 eV compared with 2 eV for the parent polythiophene.[44,45] Doping-undoping cycles are electrochemically reversible and are accompanied by a high contrast electrochromic color change. In the undoped state, thin films of PITN are blue-black; upon doping they become transparent yellow. Doped PITN is thus the first example of a transparent conducting organic polymer. Unlike other polythiophenes, this polymer is very unstable in the atmosphere. The reduction in bandgap compared to the parent thiophene is presumably due to the aromatic benzene ring built onto the thiophene ring stabilizing the quinonoid contribution in the ground state.

Based on a similar concept, bromine-treated poly[α-(5,5'-bithiophenediyl)benzylidene] was claimed to be a small bandgap semiconductor polymer with a gap as low as 0.75 eV.[52,53] Unfortunately, synthesis and characterization of the polymer was poor. Later it was learned[54] that the reaction of the polymer with bromine vapor is complex, depending on the phase (solid-gas vs. solution-gas), not a simple dehydrogenation as reported earlier. In fact, the product is an overbrominated and not easily characterized solid of unknown bandgap. The synthesis of well-defined alternating aromatic and quinonoid polymers may lead to a low bandgap polymer.[55]

Lateron polymers, with alternating donor and acceptor units, lead to a remarkably low bandgap of 0.5 eV.[56]

Table 2. Representative Polypyrroles and Their Maximum Conductivities

Polymer	Maximum Conductivity S/cm	Reference
Polypyrrole	90^a	46
	10^{3},^{b}	47,48
Poly (N-methylpyrrole)	10^{-3},^{a}	49
	10^{-3},^{b}	50
Poly (3-methylpyrrole)	4^b	51
Poly (3-alkanesulfonatepyrrole)	0.1^b	38,39
Poly (3-alkylketonepyrrole)	360^b	36

a: chemically prepared, b: electrochemically prepared

Poly(arylene vinylenes)

One important way to process otherwise intractable materials is via a precursor route. Poly(p-phenylene vinylene) (PPV) prepared using Wittig condensation or dehydrohalogenation gave only oligomers.[57]

PPV

     The (indirect) precursor polymer route has been used to synthesize soluble, polymeric precursors that can be cast into films and drawn into fibers. Subsequent conversion to the desired conjugated polymers, such a poly(p-phenylene vinylene) (PPV), can be achieved by thermal or chemical treatment of the films and fibers. Heating the Cl-precursor polymer causes the loss of dimethylsulfide and HCl by an E1cB mechanism to give PPV. This route produces particularly attractive PPV, not only because it is very stable in the undoped state but also because its precursor can be obtained in very high yield and with very high molecular weight (on the order of 10^5). Films of the precursor polymer can be obtained free standing or supported on suitable substrates, such as polyethylene or glass, by casting them from viscous aqueous solutions of the polyelectrolyte.

     The conventional procedure for obtaining oriented PPV consists of stretching the precursor film while it is being heated to convert it to a conjugated polymer. Later, an even higher degree of orientation can be achieved by orienting the precursor polymer before thermal treatment, followed by further stretch-alignment during heating to give the highly oriented PPV.[65] Using an ion exchange technique, several high-molecular-weight organic-solvent-soluble polyelectrolytes were prepared from a water-soluble polyelectrolyte.

     The films and fibers of these organic-soluble precursor polymers can be treated in the same manner as the water-soluble polyelectrolyte to give highly oriented PPV. However, the precursor pclymer, an AsF_6 salt, gave directly doped PPV on heating. This process is called Incipient Doping.[67] Alkoxy substituted PPV has been prepared.[58,68] Incorporating these electron-donating substituents stabilizes the doped cationic form of the polymer and thus lowers the ionization potential of the species. Although iodine is a rather poor dopant for PPV, it is a good dopant for dimethoxy-PPV, reflecting the greater ease of oxidation of this polymer.[58,68] When the alkoxy groups are substituted in PPV, the polyelectrolyte precursor polymers from which they are made are soluble and processable, as are the conjugated polymers because of the long alkyl substituent.[69-71] These polymers both exhibit solvatochromism (color change depending on solvent) and thermochromism (color change depending on temperature).

Table 3. Poly(arylene vinylenes) With Their Maximum Conductivities

Polymer	Maximum Conductivity S/cm	Reference
Poly (p-phenylene vinylene)	5×10^3	58
Poly (2,5-dimethoxy-p-phenylene vinylene)	4.3×10^2	59
Poly (2,5-dibutoxy-p-phenylene vinylene)	7×10^3	59
Poly (2,5-thienylene vinylene)	62	60,61
Poly (3-methoxy-2,5-thienylene vinylene)	0.8	62,63
Poly (2,5-furylene vinylene)	36	64

Poly(thienylene vinylene) and poly(furylene vinylene) have also been prepared via the processable precursor route. The chemistry of poly(3-alkoxythienylene vinylene) is interesting because, on doping, the absorption band in the visible region of the spectrum (band gap of 1.50 eV) decreases significantly while the absorption in the infrared region increases, making the film semitransparent.[72-74]

Polyaniline

Although one of the oldest polymers, polyaniline has only recently been looked upon as a promising conducting polymer. Polyaniline can be made conductive by partial oxidation of the fully reduced (leucoemeraldine base) form or by partial protonation of the half oxidized (emeraldine base) form.

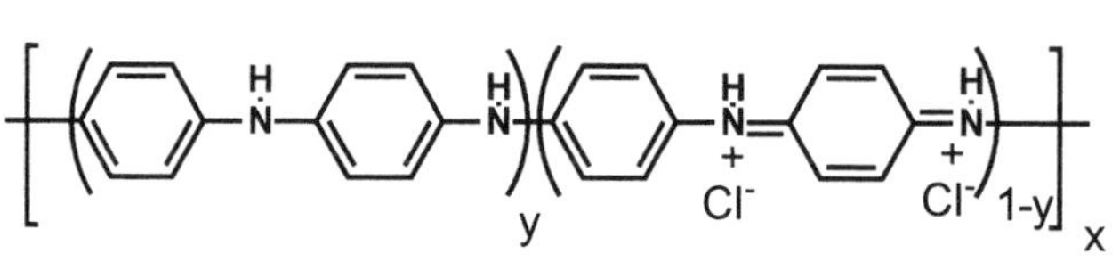

Emeraldine salt form

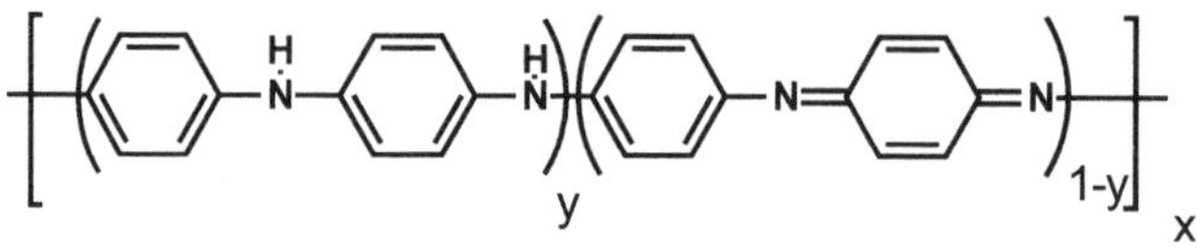

Emeraldine base form

Polyaniline, in which a simple protonation process can lead to an insulator-to-conductor transition, is a new concept. To elucidate the structure of polyaniline, an octamer of aniline was prepared and its properties were compared with those of polyaniline. The data suggest that polyaniline has *para*-linked poly(phenylnene amineimine) structure.[75] The most common procedure for polyaniline preparation is polymerization of aniline using ammonium persulfate in HCl, yielding the "emeraldine hydrochloride."[76] Using solution viscosity as a guide, attempts to vary the polymerization conditions to produce a higher molecular weight polymer have resulted in significant improvement.[77] The emeraldine hydrochloride form of polyaniline has been reported to be soluble in 97% sulfuric acid with polymer concentrations up to 20 wt%. A major benefit of solubility in the doped form is the ability to spin fibers of the polymer as conductors. The sulfuric acid solution was used to spin fibers into water to obtain shiny metallic fibers with conductivities of 20 to 60 S/cm.[78]

Conducting polymer blends based upon polyaniline form a self-assembled network morphology. The threshold for the onset of electrical conductivity can be reduced to volume fractions of polyaniline below 1% in a variety of insulating host polymers [such as polyolefins, poly(methyl methacrylate), polyesters, ABS, poly(vinyl butyral)]. Since the conducting polyblends are stable and retain the mechanical properties of the host polymer, film, fibers, and coatings can be fabricated from solution or by melt-processing for use in antistatic applications, for electromagnetic shielding, for transparent conducting films, etc. [80]

Applications of Conducting Polymers

Several potential applications of conducting polymers have been envisaged since 1969.[81] For some applications, conducting polymers are suitable replacements for metals. One of the greatest advantages of organic polymers compared with inorganic materials is their flexibility: they can be chemically modified and easily shaped according to the requirements of a particular device. Their versatility and compatibility, coupled with ease of fabrication and light weight, make them useful materials for electronic devices. The potential applications of these conjugated polymers derived from both their conducting or neutral (non-conducting) forms.

There is a huge commercial potential for conducting polymers as corrosion inhibiting coatings.[6] Some estimates indicate that corrosion costs U.S. industries tens of billions of dollars per year. Current methods of corrosion protection (especially marine coatings) do not last very long and are coming under increased scrutiny by the Environmental Protection Agency. Much of the work on corrosion protection has focused on polyaniline.[84,85] The corrosion protection ability of polyaniline is pH dependent. In lower pH, polyaniline-coated mild steel coupons corrode 100 times more slowly than their counterparts, whereas in pH 7 media, the polyaniline-coated material corrodes twice as slowly.[86,87] The pH of seawater is approximately 8 to 9.4 depending upon the season and location. It is unclear whether polyaniline can provide any additional practical corrosion protection for ocean-going vessels.

Other potential applications of conducting polymers are in sensors, photocells, solar batteries, xerography, electrostatic shielding, electronic devices (diodes, transistors, capacitors), memory devices, optical display devices, and recording.[88-90] The unusual apparent instability of polymers has been used in chemical sensor technology. Small-scale indicator devices have been designed from conducting polymers to detect moisture, radiation, chemicals, and mechanical abuse.[82]

The most popular application seems to be solid-state rechargeable polymer batteries. One Japanese company, Bridgestone, has developed a rechargeable, coin-type plastic battery mainly for use in watches and personal computers. Despite the low densities, the charge capacity per unit weight for conducting polymers is marginally better than metals. Many other issues, not related to conducting polymers, affect battery performance.[91]

Conducting polymers are candidates for application in electrochromic cells. In particular, polyheterocycles have been cycled thousands of time. Recently, Reynolds and coworkers demonstrated over 10,000 deep cycles, using derivatives of poly(3,4-ethylenedioxythiophenes) (PEDOT). Retention of 60% of electrochromic activity after 16,000 deep cycles using EDOT devices, and retention of approximately 60% of electrochromic activity after 10,000 deep cycles using devices with complementary EDOT polymer derivatives, has been observed.[92-94] Infrared polarizers based upon polyaniline have been demonstrated to be about as good as commercially available metal wire polarizers.[95]

Conducting polymers may find application in nonlinear optical devices, particularly in optoelectronics, e.g., signal processing and optical communication. These applications benefit from polymer flexibility, mechanical strength, high damage threshold, and ultra-fast response, in the subpicosecond range.[96,97] The third-order, or χ^3, properties of polypyrrole, polythiophene, and poly(phenylene vinylene) have been studied.[98-101] Friend and coworkers published electroluminescence studies on the neutral form of poly(phenylene vinylene) that could open a potential market for the material.[102,103] A transparent speaker was developed by coating polypyrrole on the surface of piezoelectric poly(vinylidenecyanide-vinylacetate). The speaker generates sound of very high quality, especially in the high frequency region.[83] Polyheterocyclic conducting polymers show wide variation in color when the applied voltage switches them between their oxidized and reduced states.[104,105]

In addition to pure materials, blends or composites of other non-conducting polymers may have wide applications. For practical applications, the new products should have much better cost performance or unique features not achievable by current technology. The ability to process polymers with reproducible properties may be a important factor.

Blends of conducting polymers for electromagnetic interference (EMI) shielding have been reported.[106] Coprocessing of conducting polymers such as polyalkylthiophenes with ultra-high molecular weight polyethylene (UHMW-PE) to form highly oriented fibers might be interesting due to different orientation in the conjugated polymer phase.[107] A route was developed for producing fibers with significantly enhanced mechanical properties by blend processing polyaniline with the rigid chain polymer poly (p-phenylene terephthalamide).[108]

Summary

This review discusses progress in the field of conjugated conducting polymers in the last two decades, as well as the potential of these materials in future technologies. The conducting polymer field is an interdisciplinary one, with unique challenges, and opportunities. From a materials science point of view, the design, synthesis, characterization and processing of materials are important factors. Although the ability to process most of the polymers from solution represents genuine progress, the mechanical properties of fibers and films made from these solutions are not adequate for many applications, mainly because the low molecular weight of the polymers used. One aim of synthetic studies should be to identify target molecules with enhanced properties. Novel properties, substantial increases in conductivity, and third order nonlinearities perhaps can be achieved through desired synthesis. Although both of the latter properties are based on extended π conjugation, conductivity is found only in doped polymers, while large third-order susceptibilities values are, to date, found only in neutral polymers. Emphasis should be given to the synthesis of well-defined conjugated polymers. A special challenge is attaining perfect conjugation as well as controlling the alignment of the conjugated chains, their morphology, and crystallinity. Naarmann's polyacetylene apparently contain no defects (such as sp^3 hybridized carbon) leading to an almost perfect chemical structure with extremely high conductivity and optical nonlinearity. Polyheterocyclic polymers, on the contrary, lack such sophisticated synthetic methodology. Several routes are now available to increase the processability of extended conjugated polymers. Excellent progress has been made in physical processing of precursor polymers. Other more sophisticated techniques can be applied as materials with better solubility and stability (thermal, hydrolytic, and oxidative) are prepared.

The last two decades saw laboratory efforts progress from producing intractable, insoluble materials to soluble and processable polymers. The study of these materials has generated new scientific concepts as well as the potential for new technology. In this period new improved synthetic routes were developed that gave much better characterized materials with controlled molecular weights and procedures for processing polymeric systems. Conducting polymers have been prepared in a variety of forms in which electrical conductivity can be systematically controlled over 10 orders of magnitude. The main purpose of these investigations was to synthesize and study new materials combining the electrical properties of metals or semiconductors and the mechanical properties of plastics. These conjugated materials also give highly nonlinear optical response, either for second-order or third-order effects. Although it is difficult to evaluate the technogical significance of such a new field, the research has established a foundation of materials and fundamental scientific principles ready to be cleverly applied.

Literature Cited

1.   "Polyacetylene - Chemistry, Physics and Material Science", Chien, J. W. Academic Press, Ireland, FL 1984.
2.   "Polyacetylene and Polyarylenes" Krivoshei, I. V.; Skorobogatov, V. M., Gordon and Breach, New York, 1991.
3.   "Handbook of Conducting Polymers" Skotheim, T. J. Ed. Marcel Dekker: New York, 1986.
4.   "Conjugated Polymeric Materials: Opportunities in Electronics, Optoelectronics and Molecular Electronics" Bredas, J. L.; Chance, R. R. eds., Kluwer Academic Publishers, Dordrecht Netherland and Boston USA, 1990.
5.   "Handbook of Organic Conductive Molecules and Polymers" Volume 1-4. Editor, Nalwa, H. S. Wiley Ney York, 1997.
6.   Stenger-Smith, J. D. Prog. Polym. Sci. 1998, 23, 57.
7.   Patil, A. O. ; Heeger, A. J.; Wudl, F. Chem Rev. 1988, 88, 183.
8.   Reynolds, J. R.; Child, A. D.; Gieselman, M. B. Kirk-Othmer Encyclopedia of Chemical Technology, 4th edn, Vol. 9. John Wilay, New York, 1994, p. 61.
9.   Green, R. L.; Street, G. B. Science 1984, 226, 651.
10.   Patil, A. O. Polymer News 1989, 234.
11.   Reynolds, J. R. Chemtech 1988, 447.
12.   Roncali, J. Chem Rev. 1992, 92, 711.
13.   Pethrick, R. A. Ref. Funct. Polym. 1997, 463. Editor: Arshady, R. ACS, Washington, D. C.
14.   Patil, A. O. Bull. Electrochem. 1992, 8, 509.

15.   Kanatzidis, M. Chem. Eng. News, 1990. December 3, 36.
16.   Ito, T.; Shirakawa, H.; Ikida, S. J. Polym. Sci. Polym. Chem. Ed., 1974, 12, 11.
17.   Chiang, C. K.; Druy, M. A.; Gau, S. C.; Heeger, A. J.; Louis, E. J.; MacDiarmid,
      A. G.; Park, Y. W.; Shirakawa, H. J. Am. Chem. Soc. 1978, 100, 1013.
18.   Naarmann, H.; Theophilou, N. Synth. Met. 1987, 22, 1.
19.   Basescu, N.; Liu, Z. X.; Moses, D.; Heeger, A. J.; Naarmann, H.; Theophilou, N.
      Nature 1987, 327, 403.
20.   Edwards, J. H.; Feast, W. J. ; Bott, D. C. Polymer, 1984, 25, 395.
21.   Feast, W. J.; Winter, J. N. J. Chem. Soc., Chem. Commun. 1985, 202.
22.   Leising, G. Polym. Commun. 1984, 25, 201.
23.   Swager, T. M.; Grubbs, R. H. J. Am. Chem. Soc. 1989, 111, 4413.
24.   Dall'Olio, A.; Dascola, Y.; Varacco, V.; Bocci, C. R. C. R. Seances Acad. Sci.,
      Ser. C 1968, 267, 433.
25.   Kanazawa, K. K.; Diaz, A. F.; Geiss, R. H.; Gill, W. D.; Kwak, J. F.; Logan, J. A.;
      Rabolt, J. F.; Street, G. B. J. Chem. Soc., Chem. Commun. 1979, 854.
26.   Yoshino, K.; Hayashi, S.; Sugimoto, R. Jpn. Appl. Phys. 1984, 23, 899.
27.   Sato, M.; Tanaka, S.; Kaeriyama, K. J. Chem. Soc., Chem. Commun. 1985, 713.
28.   Elsenbaumer, R. L; Jen, K. Y.; Oboodi, R. Synth. Met. 1986, 15, 169.
29.   Jen, K. Y.; Miller, G. G.; Elsenbaumer, R. L. J. Chem. Soc., Chem. Commun.
      1986, 1346.
30.   Sato, M.; Tanaka, S.; Kaeriyama, K. J. Chem. Soc., Chem. Commun. 1986, 873.
31.   Patil, A. O.; Ikenoue, Y.; Wudl, F.; Heeger, A. J. J. Am. Chem. Soc. 1987, 109,
      1858.
32.   Patil, A. O.; Ikenoue, Y.; Basescu, N.; Coleneri, N.; Chen, J.; Wudl, F.; Heeger,
      A. J. Synth. Met. 1987, 20, 151.
33.   Ikenoue, Y.; N. Saisa, Y.; Kira, M.; Tomozawa, Y.; Yashima, H.; Kobayashi, M.
      J. Chem. Soc., Chem. Commun. 1990, 1694.
34.   Ikenoue, Y.; Chiang, J. Patil, A. O.; Wudl, F.; Heeger, A. J. J. Am. Chem. Soc.
      1988, 110, 2983.
35.   Havinga, E. E.; Van Horssen, L. W. Makromol. Chem. Makromol. Symp. 1989,
      24, 67.
36.   Bryce, M. R.; Chissel, A.; Kathirgamanathan, P.; Parker, D.; Smith, N. R. M.
      J. Chem. Soc., Chem. Commun. 1987, 466.
37.   Yoshino, K.; Nakajima, S.; Fujii, M.; Sugimoto, S. Polym. Commun., 1987, 28,
      309.
38.   Sundaresan, N. S.; Basak, S.; Pomerantz, M.; Reynolds, J. R. J. Chem. Soc.,
      Chem. Commun. 1987, 621.
39.   Havinga, E. E.; Hoeve, W.; Meijer, E. M.; Wynberg, H. Chem. Mater. 1989, 1,
      650.
40.   Ruhe, J.; Ezquerra, T.; Wegner, G. Makromol. Chem., Rapid Commun., 1989,
      10, 131.
41.   Ruhe, J.; Ezquerra, T.; Wegner, G. Synth. Met. 1989, 28, 177 .
42.   Lemaire, M.; Delabouglise, D.; Garreau, R.; Guy, A.; Roncali, J. J. Chem. Soc.
      Chem. Commun., 1988, 658.
43.   Bauerle, P.; Gaudl, K. U. Gotz,G. Springer Ser. Solid State Sci., Vol. 107,
      Springer, Berlin 1992, 384.

44. Wudl, F.; Kobayashi, M.; Heeger, A. J. J. Org. Chem. 1984, 49, 3382.
45. Yashima, H.; Kobayashi, M.; Lee, K. B.; Chung, D.; Heeger, A. J. ; Wudl, F. J. Electrochem. Soc., Electrochem. Sci. Tech. 1987.
46. Myers, R. E. J. Electron. Mater. 1986, 15, 61.
47. Ogasawara, M.; Funahashi, K.; Demura, T.; Hagiwara, T.; Iwata, K. Synth. Met. 1986, 14, 61.
48. Yamaura, M.; Hagiwara, T.; Iwata, K. Synth. Met. 1988, 26, 209.
49. Kovacic, P.; Khoury, I.; Elsenbaumer, R. L. Synth. Met. 1983, 6, 31.
50. Diaz, A. F.; Castillo, J.; Kanazawa, K. K.; Logan, J. A.; Salmon, M.; Fajardo, O. J. Electroanal. Chem. 1982, 133, 233.
51. Waltman, R. J.; Bargon, J.; Diaz, A. F. J. Phys. Chem. 1983, 87, 1549.
52. Jenekhe, S. A. Nature 1986, 322, 345.
53. Jenekhe, S. A. Macromolecules 1986, 19, 2663.
54 Patil, A. O.; Wudl, F. Macromolecules 1988, 21, 540.
55. Hanack, M.; Hieber, G.; Dewald, G.; Rohrig, V. Polym. Mater. Sci. Eng. 1991, 64, 330.
56. Havinga, E. E.; tenHoeve, W.; Wynberg H. Synth. Met. 1993, 55 1251.
57. a) Wessling R.A.; Zimmerman, R.G. U.S. Patent 1968, 3,401,152. b) U.S. Patent 1972 3,706,677.
58. Murase, I.; Ohnishi, T.; Noguchi, T.; Hirooka, M. Synth. Met. 1987, 17, 639.
59. Han, C. C.; Lenz, R. W.; Karasz, F.E. Polym. Comm. 1987, 28, 261.
60. Jen, K. Y.; Maxfield, M.; Shacklette, L. W.; Elsenbaumer, R. L. J. Chem. Soc., Chem. Commun. 1987, 309.
61. Yamada, S.; Tokito, S.; Tsutsui, T.; Saito, S. J. Chem. Soc., Chem. Commun. 1987, 1448.
62. Jen, K. Y.; Eckhardt, H.; Jow, T. R.; Shacklette, L. W.; Elsenbaumer, R. L. J. Chem. Soc., Chem. Commun. 1988, 215.
63. Blohm, M. L.; Dort, C. V.; Pickett, J. E. Polym. Mater. Sci. Eng. 1991, 64, 210.
64. Jen, K. Y.; Jow, T. R.; Elsenbaumer, R. L. J. Chem. Soc., Chem. Commun. 1987, 1113.
65. Gagnon, D.R.; Karasz, F.E.; Thomas, E.L.; Lenz, R.W. Synth. Met. 1987, 20, 85.
66. Murase, I.; Ohnishi, T.; Noguchi, T.; Hirooka, M. Polym. Comm. 1984, 25, 328.
67. Patil, A. O.; Rughooputh, S. D. D. V.; Patil, A.; Heeger, A. J.; Wudl, F. Polym. Mater. Sci. Eng. 1988, 59, 1071.
68. Antoun, S.; Gagnon, D. R.; Karasz, F. E.; Lenz, R. W. Polym. Bull.1986, 15, 181.
69. Hans C. C.; Elsenbaumer, R. L. Synth. Met. 1989, 30, 123.
70. Askary, S. H.; Rughooputh, S. D. D. V.; Wudl, F. ACS meeting, Los Angeles 1988, Polym. Mater. Sci. Eng. 1988, 59, 1068.
71. Synth. Met. 1989, 29, 129.
72. Jen, K. Y.; Maxfield, M.; Shacklette, L. W.; Elsenbaumer, R. L. J. Chem. Soc., Chem. Commun. 1987, 309.
73. Jen, K. Y.; Eckhardt, H.; Jow, T. R.; Shacklette, L. W.; Elsenbaumer, R. L. J. Chem. Soc., Chem. Commun. 1988,
74. Jen, K. Y.; Jow, T. R.; Shacklette, L. W.; Maxfield, M.; Eckhardt, H.; Elsenbaumer, R. L. Mol. Cryst. Liq. Cryst. 1988, 160, 69.

75. Wudl, F.; Angus, R. O.; Lu, F. L.; Allemand, P. M.; Vachon, D. J.; Novak, M.; Liu, Z. X.; Heeger, A. J. J. Am. Chem. Soc. 1987, 109, 3677.
76. "Conducting Polymers" Alcacer, L. ed., Reidel, Dordrecht, Holland, 1987, MacDirmid, A. G.; Chiang, J. C.; Richter, A. E.; Somasiri, N. L. D.; Epstein, A. J. p. 105.
77. Cao, Y.; Andreatta, A.; Heeger, A. J.; Smith, P. Polymer 1989, 30, 2305.
78. Andreatta, A.; Cao, Y.; Chiang, J. C.; Heeger, A. J.; Smith, P. Synth. Met. 1988, 26, 383.
79. Polym. Prepr. (Am. Chem. Soc. Div. Polym. Chem.) 1989, 30, 149.
80. Heeger, A. J. Trends Polym. Sci. 1995, 3(2), 39.
81. Naarmann, H. Naturwissenschaften 1969, 56, 308. US Patent 4,560,593.
82. b) Baughman, R. H.; Elsenbaumer, R. L.; Iqbal, Z.; Miller, G. G.; Eckhardt, H. U. S. patent 1987, 4,646,066.
83. Ojio, T.; Miyata, S. Polym. J. 1986, 18, 95.
84. Wessling, B. German Patent DE 3834526A1, 1990.
85. May, P. Phys. World, 1995,8(3), 52.]
86. Lu, W. K.; Elsenbaumer, R. L.; Wessling, B. Synth. Met. 1995, 71, 2163.
87. Wei, Y.; Wang, J.; Jia, X.; Yeh, J. M.; Spellane, P. Polymer 1995, 36, 4535.
88. "Electroresponsive Molecular and Polymeric Systems", Skotheim, T. J. Ed. Marcel Dekker: New York, 1991. see Chapter 5. by Techagumpuch, A.; Nalwa, H. S.; Miyata, S.
89. Potember, R. S.; Hoffman, R. C.; Hu, H. S.; Cocchiaro, J. E.; Viands, C. A.; Murphy, R. A.; Poehler, T. O. Polymer, 1987, 28, 574.
90. Wnek, G. Polym. Mater. Sci. Eng. 1991, 64, 338.
91. Osaka, T.; Momma, T.; Nishimura, K.; Kakuda, S.; Ishii, T. J. Electrochem. Soc. 1994, 141, 1994.
92. Sankaran, B.; Reynolds, J. R. Macromolecules 1997, 30(9), 2582-2588.
93. Sapp, S. A.; Sotzing, G. A.; Reddinger, J. L.; Reynolds, J. R. Adv. Mater. (Weinheim, Ger.) (1996), 8(10), 808-811.
94. Sotzing, G. A.; Reddinger, J. L.; Reynolds, J. R.; Steel, P. J. Synth. Met. 1997, 84(1-3), 199-201.
95. Cao, Y.; Colaneri, N.; Smith, P.; Heeger, A. J. Polym. Prepr. 1994, 35, 253.
96. "Nonlinear optical properties of polymers", MRS symposium. proc. Heeger, A. J.; Orenstein, J.; Ulrich, D. R. eds., 1988, 109.
97. Yoshino, K. Synth. Met. 1989, 28, 669.
98. Chandrasekhar, P.; Thorne, J. R. G.; Hochstrasser, R. M. Appl. Phys. Lett. 1991, 59, 1661.
99. Jenekhe, S. A.; Lo, S. K.; Flom, S. R. Appl. Phys. Lett. 1989, 54, 2524.
100. Kang, I. N.; Lee, G. J.; Kim, D. H.; Shim, H. K. Polym. Bull. 1994, 33, 89.
101. Shim, H. D.; Yoon, C. B.; Lee, J. I.; Hwang, D. H. Polym. Bull. 1995, 34, 161.
102. Burroughes, J. H.; Bradley, D. D. C.; Friend, R. H.; Brown, A. R.; Marks, R. N.; Mackay, K.; Burn, P. L.; Holmes, A. B. Nature 1990, 347, 539.
103. Holmes, A.; Bradley, D. D. C.; Friend, R. H.; Kraft, A.; Burn, P.; Brown, A. U. S. Patent 5,402,827 (1995).
104. Diaz, A. F.; Costello, J. J.; Logan, J. A.; Lee, Y. W. J. Electroanal. Chem. 1981, 129, 115.

105. Garnier, F.; Tourillon, G.; Gazard, M.; Dubois, J. C. <u>J. Electroanal. Chem.</u> 1983, <u>148</u>, 299.

106. The thermoplastic polyaniline/PVC blend has been developed by Zipperling Kessler & Co. (FRG) in cooperation with Allied Signal Inc. and Americhem Inc. and will be sold under the trade name INCOBLEND. The polyaniline dispersed therein is offered by Allied Signal Inc. under the trade name VERSICON. also see Wessling, B. <u>Adv. Mater.</u> 1991, <u>3</u>, 507.

107. Moulton, J.; Ihn, K. J.; Smith, P. <u>Polym. Mater. Sci. Eng.</u> 1991, <u>64</u>, 256.

108. Andreatta, A.; Heeger, A. J.; Smith, P. <u>Polymer Comm</u> 1990, <u>31</u>, 275.

HISTORY AND DEVELOPMENT OF POLYMER BLENDS AND IPNS

L. H. SPERLING

Dept. Chemical Engineering, Dept. Materials Science and Engineering, Center for Polymer Science and Engineering, Materials Research Center, and the Polymer Interfaces Center, Lehigh University, Bethlehem, PA 18015-3194

Introduction
Early IPN History
How Does One Know Its An IPN?
Polymer Blend Development
The Current Status of Multicomponent Polymer Materials
Polymer Blend Interface Characteristics
Applications of Polymer Blends and IPNs
Conclusions

Introduction

Polymer blends and composites became a central part of polymer science and engineering because people could make compositions that had properties substantially unattainable with homopolymers and statistical copolymers. Such properties include greater toughness and impact resistance, higher modulus, higher use temperature, broader temperature range of sound and vibration damping, etc.

The art and science of polymer blends and interpenetrating polymer networks, IPNs, dates back to nearly the beginning of the century, even before Staudinger enunciated the *Macromolecular Hypothesis* in 1920(1). Long ago, of course, people had made polymer blends and composites from natural materials. One of the oldest examples is given in the Bible, where mud bricks were strengthened with the addition of straw *(Exodus, 5:7-19)*. Paper sized with starch constitutes an ancient polymer blend. Natural rubber was compounded with zinc oxides and many other components for rubber tires around the turn of the century. The early history of polymers, which does not emphasize polymer blends and IPNs is given by Furakawa(2) and Hounshell and Smith(3). Today, polymer blends and composites are often designated multicomponent polymer materials(4).

While this article will emphasize synthetic polymers, significant attention has been paid to blends, grafts, and IPNs of synthetic polymers with natural polymers such as cellulose(5,6) and triglyceride oils(7,8). In the 21st Century, greater attention will have to be paid to our renewable resources, to keep the planet green.

Early IPN History

Almost unrecognized by the literature, the first commercial polymer composition containing two distinct polymers was invented by Jonas Aylsworth in 1914(9). Actually, the material was a simultaneous interpenetrating network composed of phenol-formaldehyde resins (phenolics) and natural rubber crosslinked with sulfur(10). Table I(11) summarizes the historical development of polymer blends, blocks, and IPNs, putting them in context with the emergence of important polymer concepts and materials.

Of course, phenol and formaldehyde react to form a crosslinked polymer. This material, known as Bakelite, had been invented and commercialized by Leo Baekeland in Yonkers, NY around 1907 as Bakelite(12). Some years before, Thomas Edison had invented the phonograph. The earliest form of the phonograph was as a spool. When Edison switched to the platter style, the material he choose was a phenolic. However, the platters were very brittle, and needed to be very thick to prevent breakage.

At this time, Aylsworth was Edison's chief chemist. Aylsworth was working in Edison's East Orange, New Jersey, across the river from Yonkers. Aylsworth added almost every imaginable material to the phenol-formaldehyde resins in an effort to toughen them, and the patent literature of that day is replete with his efforts. The successful material was natural rubber and sulfur(9). In Aylsworth's own words, "...the rubber adheres with great tenacity to a hard phenolic condensation product when vulcanized therewith, making an improved rubber product very superior to rubber filled or loaded with fillers used in the art." The rubber-toughened composition remained commercial until at least 1929, when Edison closed the plant.

Since Aylsworth's patent predates the concept of a polymer chain, no less polymer blend and IPN terminology, such terminology is never mentioned. The idea was invented and reinvented several times through the 20th century.

J. J. P. Staudinger, H. Staudinger's son, began his efforts in the early 1940's, resulting in a 1951 patent(13). The product was an improved, smooth-surfaced transparent plastic made out of polystyrene or poly(methyl methacrylate). The idea was to take a relatively rough surfaced crosslinked plastic, and swell it with its own monomer mix, and repolymerize. The swelling action tended to even out the surfaces.

The next independent invention of IPNs was by Solt in 1955(14). Solt invented a cationic-anionic IPN ion exchange resin using suspension-sized particles. The two networks contained in each particle were oppositely charged. The idea of both charges on the same particle (rather than on different particles) was that the exchange efficiency should be improved if both charges were in juxtaposition, but still separated in space. Thus, microphase separation, although still not well understood, was important for the invention to work. The

Table I. Development of Multicomponent Polymeric Materials

Event	Investigator	Year	Ref.
Vulcanization of rubber	Goodyear	1844	a
First synthetic polymer (Bakelite)	Baekeland	1912	b
IPN Structure	Aylsworth	1914	c
Macromolecular Hypothesis	Staudinger	1920	d
Graft Copolymers	Ostromislensky	1927	e
Model of the random coil quantified	Guth and Mark	1934	f
Nylon polyamide synthesized	Carothers	1937	g
Block Copolymers	Dunn and Melville	1952	h
High-impact polystyrene, HIPS, and acrylonitrile-butadiene-styrene, ABS	Amos, McCurdy, and McIntire	1953	I
Homo-IPNs	Millar	1960	j
Thermoplastic Elastomers	Holden and Milkovich	1966	k
Liquid Crystalline Polymers	Kwolek	1966	l
AB Crosslinked Copolymers	Bamford, Dyson, and Eastmond	1967	m
Sequential IPNs	Sperling and Friedman	1969	n
Latex interpenetrating elastomer networks, IENs	Frisch, Klempner, and Frisch	1969	o
Reptation Theory	de Gennes	1971	p
Simultaneous Interpenetrating Networks	Sperling and Arnts	1971	q
Thermoplastic IPNs	Davison and Gergen	1977	r

a. C. Goodyear, U.S. 3,633, 1844.

b. L. H. Baekeland, U.S. 1,019, 406 and U.S. 1,019, 407, 1912.

c. J. W. Aylsworth, U.S. 1,111,284, 1914.

d. H. Staudinger, *Ber.*, 1920, 53, 1073.

e. I. Ostromislensky, U.S. 1,613,673, 1927.

f. E. Guth and H. Mark, Monatschefte Chemie, 65, 93 1934.

g. W. H. Carothers, U.S. 2,071,250 and U.S. 2,071,251, 1937.

h. A. S. Dunn and H. W. Melville, Nature, 1952, 169, 699.

i. J. L. Amos, J. L. McCurdy, and O. R. McIntire, U.S. 2,694,692, 1954.

j. J. R. Millar, J. Chem. Soc., 1960, 1311.

k. G. Holden and R. Milkovich, U.S. 3,265,765, 1966.

l. S. L. Kwolek, BP 1,198,081, 1966.

m. C. H. Bamford, R. W. Dyson, and G. C. Eastmond, J. Polym. Sci., 1967, 16C, 2425.

n. L. H. Sperling and D. W. Friedman, *J. Polym. Sci.*, 1969, A-2 7, 425.

o. H. L. Frisch, D. Klempner, and K. C. Frisch, Polym. Lett., 1969, 7, 775.

p. P. G. de Gennes, J. Chem. Phys., 1971, 55, 572.

q. L. H. Sperling and R. R. Arnts, J. Appl. Polym. Sci., 1971, 15, 2371.

r. S. Davison and W. P. Gergen, U.S. 4,041,103, 1977.

charges must be separated, even by a fraction of a nanometer, to prevent coacervation.

The first use of the term *interpenetrating polymer network* was by Millar, to make larger suspension-polymerized polystyrene ion-exchange particles(15). Millar also did the first scientific research on the topic.

The field of interpenetrating polymer networks developed later, though the efforts of Frisch, *et al.*(16), Lipatov and Sergeeva(17), and Sperling and Friedman(18).

How Does One Know Its An IPN?

The question of how to identify multicomponent polymer materials, and IPNs in particular, has come up repeatedly. There are very few papers which address this question in general, especially on the need to identify unknown materials.

Surely, simple polymer blends, blocks and grafts are soluble in the appropriate solvents. IPNs and AB-crosslinked polymers and related materials in general are not soluble. Further identification may require chemical or instrumental attention to identify the chemical nature of the grafts or crosslinks. Electron microscopy provides a powerful tool, since sequential IPN domains are usually smaller than blend domains. However, the blends where monomer II is polymerized in the presence of polymer I may have small domains and insoluble regions due to AB-crosslinked polymer formation. While block copolymer domains are as small or smaller than IPN domains, they tend to be more regular in spacing and appearance.

One related aspect of the problem has been given some attention: There is experimental evidence for dual phase continuity in sequential IPNs. Widmaier and Sperling(19) examined the system *net*-poly(*n*-butyl acrylate)-*inter-net*-polystyrene sequential IPNs. Assuming phase separation *via* spinodal kinetic mechanisms, theory suggests the formation of interconnected cylinders for polymer II.

First a network of poly(*n*-butyl acrylate) was synthesized, using acrylic acid anhydride, AAA, as the crosslinker. (The value of the AAA is that its crosslinks easily hydrolyze in dilute, warm ammonia water.) Then, styrene and divinyl benzene were swollen in and polymerized to synthesize the polymer network II.

Samples of the IPN with the labile AAA crosslinker were soaked in a 10% aqueous ammonium hydroxide solution for about 12 hours. Decrosslinking was effective, as indicated by the complete dissolution of the poly(*n*-butyl acrylate) homopolymer network in organic solvents while the original network only swelled in the same solvents. After decrosslinking, the resulting linear polymer was extracted from the former IPN in a Soxhlet extractor. Both the soluble and insoluble fractions were dried and characterized.

The sol fraction of decrosslinked poly(*n*-butyl acrylate) substantially followed the fraction inserted in the IPN, plus a small amount of polystyrene solubles. The molecular weights of the soluble polymer approximated that of the homopolymer prepared without the crosslinker.

Above about 20% of polymer network II, its phase domain structure was continuous after the extraction. A brittle porous solid of polystyrene was obtained, with densities in the 0.7-0.8 g/cm^3 range. This indicated a probable partial collapse of the morphology during the hot extraction.

In summary, polymer network I initially formed a continuous film. Polymer network II was formed inside of network I, with the normal volume increase. Polymer network I was decrosslinked and extracted, leaving a continuous, albeit porous, brittle polymer network II behind. This evidence suggests dual-phase continuity for the original IPN, since both phases were shown to be continuous. More recently, Gankema, *et al.*(20) used thermoreversible gelation to achieve a microporous membrane with a similar concept.

Polymer Blend Development

The idea of rubber-toughened polystyrene was developed first by Ostromislensky in 1927(21). He dissolved rubber into styrene monomer, and polymerized *without stirring*. The result was a continuous rubber phase and a discontinuous plastic phase with little toughening action, see Figure 1(22), upper right portion.

A most important advance was made by Amos, *et al.*(23), who added agitation with shearing to the reaction. In part:

> "It is important that the solution of the rubber and the polymerizable monovinyl aromatic compound, e.g., styrene, be agitated, preferably with a shearing action throughout its mass, during the early stages, or first part, of the polymerization reaction in order to obtain homogeneous linear interpolymerization products which are free, or substantially free, of cross-linked or highly branched-chain interpolymer molecules...
>
> The tendency toward the formation of cross-linked interpolymers appears to be greatest during the early part of the polymerization, e.g., when 10 per cent by weight or less of the starting materials have been polymerized. During this stage of polymerization, the interpolymer molecules appear to be attached to each other, or held together by relatively weak forces, i.e., the cross-linked interpolymers or highly branched-chain polymers are apparently agglomerates of polymer molecules which are held together by only a few or by relatively weak bonds...
>
> The tendency toward the formation of such cross-linked interpolymer molecules, which cause inhomogeneities in the polymeric product, can be prevented or substantially reduced by application of a shearing action to the polymerizing mass, i.e., by agitating the polymerizing mass, particularly during the early or first stages of the polymerization."

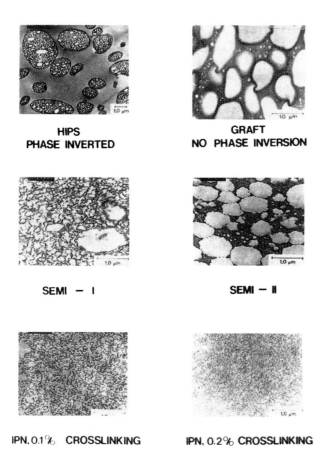

Figure 1. Transmission electron microscopy of osmium tetroxide PB-stained thin sections of HIPS and IPN compositions based on polybutadiene and polystyrene. Upper left, HIPS, upper right, Ostromislensky's material, middle left, polybutadiene crosslinked only, middle right, polystyrene crosslinked only, lower left, IPN, lower right, IPN with higher crosslinking.

If only Ostromislensky had used a stirrer! Since the invention preceded the use of osmium tetroxide stained transmission electron microscopy studies, the concepts of phase separation and inversion are not mentioned. In fact, however, Amos, et al. had invented high-impact polystyrene, known widely as HIPS. While Amos et al. received most of the credit for discovering the field of polymer blending, this invention was a generation later than the Aylsworth material. There immediately followed the development of ABS resins, and a host of other multicomponent polymer materials(4,24).

The Current Status of Multicomponent Polymer Materials

Today, research and engineering programs involving polymer blends and IPNs are growing at a rapid rate. People understand much better the influence of phase separation, phase inversion, and interfacial bonding on material properties. Concepts of fracture and healing of interfaces were developed and discussed by Wool(25). In fact, the field of polymer surfaces and interfaces only dates from 1989, when several theoretical and experimental papers were published. Instruments are now available to examine the individual polymer chains at surfaces and interfaces, as described by Adamson and Gast(26), Lohse, *et al.*(27), and Garbassi, *et al*(28), and others. Utracki(29), Arends(30), Riew and Kinloch(31), Sperling(4), and many others delineated the current status of polymer blend materials. While this chapter does not discuss block copolymers in detail, major reviews of phase separation, rubbery properties, location of block junction points, etc. were published by Hamley(32) and Holden, *et al.*(33).

Polymer Blend Interface Characteristics

Some of the findings of interest include a fractionation of polymers near surfaces and interfaces, with lower molecular weight materials tending to be at the interface. Unattended to, this causes the material to be weaker than expected. Chain ends tend to be at free surfaces(34) as well as polymer blend(35,36) and composite(37) interfaces; hence the ends of the chains tend to be placed vertical to the interface, while the center of the chain tends to lie parallel to the interface.

Chain end interdiffusion and entanglement are thought to contribute to interfacial strength and adhesion, although most polymer blend interfaces tend to be weak. The existence of the interface is usually stable, since most polymer blends are immiscible.

The polymer blend interface is sometimes called an interphase, because it has finite thickness. Helfand the Tagami(36) derived a thermodynamically based equation for the interphase thickness,

$$S_{th} = 2b/(6\chi)^{1/2} \qquad (1)$$

where χ represents the Flory-Huggins interaction parameter, and b is the statistical segment step length, equal to 6.5 for a number of polymers such as polystyrene and poly(methyl methacrylate).

Russell, *et al.*(38) determined the interphase thickness for the polystyrene-*blend*-poly(methyl methacrylate) interphase thickness *via* neutron reflection, finding it to be 50±2 Å. Fernandez, *et al.*(39), who also carried out neutron reflection found 20±5 Å. Equation (1) yields 27 Å. Assuming an M_c value of 31,200 g/mol for polystyrene(40), the radius of gyration, R_g, of a 31,200 g/mol segment equals 48 Å. The value for poly(methyl methacrylate) is somewhat smaller. Thus, the interphase thickness has approximately one entanglement per chain portion present.   According to Wool(25), approximately eight entanglements are necessary for optimum mechanical strength, total. This could be interpreted as a minimum of four being in the interphase, the remaining four being anchored in the corresponding pure phase. According to this analysis, one would expect such interfaces to be mechanically weak.   This brief section provides only a taste of what is now known about polymer blends and their interfaces; extensive research is now in progress in many laboratories.

Applications of Polymer Blends and IPNs

        Although space does not allow for the development of the current research status of these fields, a brief description of some of the more important application areas will be described.
        Perhaps the most important application of polymer blends is that of impact resistant plastics. At first, it was thought that the rubber particles bridged growing cracks, holding them from opening *via* rubber elasticity effects. Then, it was thought that since the modulus is lower in the rubber than in the plastic, if the craze were propagating at maximum velocity it would exceed same in the rubber (like exceeding the sound barrier), causing the craze to divide. This would divide the crack energy into ever smaller portions. Many polymers, especially rubber-toughened polymers, develop shear bands which are oriented portions of polymer, another mechanism to absorb fracture energy and prevent failure. The latest theories and experiments both show that such cavitation relieves triaxial stresses, preventing plastic rupture(41-44). In reality, it may be that all of these mechanisms play a role. Which one dominates may depend on the system.
        The major properties required of rubber-toughened plastics include(4):
1. The toughening elastomer must have a T_g at least 60°C below ambient, or about -40°C for room temperature applications.
2. The rubber domain size must be of the order of a fraction of a micrometer. Optimum diameter frequently is about 0.3 μm.
3. The spacing between the rubber particles should be in the range of 1-5 μm.
4. The modulus of the elastomer must be low enough so that it can cavitate easily.
5. The elastomer particle must adhere well to the plastic. Frequently, chemical bonding or grafting is employed, or through the use of core-shell particles.
6.   Multiple sizes of rubber particles allow for alleviating different kinds of stresses.
7. Larger rubber volumes produce greater toughness. For compositions where monomer II is polymerized in the presence of polymer I, such as HIPS and ABS, this means greater incorporation of the plastic in the cellular structure, as illustrated in Figure 1.
        Table II provides a summary of selected applications of polymer blends, blocks, and IPNs. A simple example is that of the ubiquitous ice tray. Many of the standard white ice trays are made out of ABS resins. (Most ABS resins are latex based; they contain a polybutadiene core and a styrene-acrylonitrile shell.

Table II. Selected Multicomponent Polymer Applications

Trade Name	Manufacturer	Composition	Application
Styron	Dow Chemical	PS/PB	Automotive, packaging
Cycolac	G.E.	PB/SAN	Appliances, telephones
Zytel	Du Pont	PA-6/elastomer	Outdoor applications
Hostalen GC	Hoechst	HDPE/LDPE	Photographic paper
Lupoy	LG Chemical	PC/ABS	Automotive, business equipment
Kraton D	Shell Chemical	SBS or SIS	Shoe soles
Kraton IPN	Shell Chemical	SEBS-polyester	Automotive wire insulation
Kelburon	DSM N.V	PP/EP	Automotive parts
Vistalon	Exxon	PP/EPDM	Paintable automotive parts
Trubyte	Dentsply	Acrylic-based	Artificial teeth
Silon-TSR	BioMed Sciences	PDMS/PTFE	Burn dressing
-	Hitachi Chemical	Vinyl/phenolics	Damping compounds

On forming a plastic, the rubber domains remain dispersed.) In average service, the trays are emptied daily either by twisting or banging them against the sink. On being refilled with water, the water undergoes freezing with an expansion force able to crack rocks. After 20 years of such service, the refrigerator usually needs replacing before the ice trays fail! Impact resistant plastics form the basis of many toys, surviving even generations of children. They are widely used in automotive, appliance, and business machine housings.

The shoe sole industry was transformed with the introduction of block copolymer elastomers. Sliding friction heat changes the elastomer into an adhesive, increasing the coefficient of friction with increasing sliding velocity. These were first introduced into sporting goods sneakers, but are used on many of today's shoes. Polyurethane segmented elastomer block copolymers are widely used as elastomeric fibers, suitable for undergarments and swimming attire.

Because of their broad glass transition temperatures, IPNs make useful sound and vibration damping materials, especially for outdoor or motor housing applications where the temperature varies. Another interesting application is for artificial teeth. Densely crosslinked, they resist swelling by salad oils, and exhibit better grinding characteristics in the dentist's office than the corresponding single network. This latter arises because of the suspension particle synthesis of polymer network I(45,46).

A proposed IPN application involves gradient-refractive index soft contact lenses(47). Polymer I has a different refractive index than polymer II. Monomer II is swollen into a proto-lens of polymer I, and polymerized with a computer driven laser, to fit individual astigmatisms(48).

Conclusions

The advent of polymer blending began almost as soon as man had polymers. In fact, one of the most active fields within polymer science and engineering is the practice of mixing every conceivable polymer with every other one, in an attempt to prepare materials with synergistic behavior. Sometimes, the intent is to produce lower price materials with the same property range. The overall result has been the development of an astonishing array of new materials.

One of the major applications of polymer blend technology has been to rubber toughened plastics. Here, one may obtain significant toughening just by simple mixing. However, the toughest materials available today has required great theoretical understanding and very sophisticated synthetic approaches.

Many of the new or proposed applications for IPNs involve biomedical uses; these materials can often be easily prepared in the elastomeric to leathery modulus range, rather than as harder materials. Other applications depend on their broad glass transition ranges, such as sound and vibration damping.

Literature Cited

1.   Staudinger, H. Ber. 1920, 53, 1073.
2.   Furukawa, Y. "Inventing Polymer Science"; University of Pennsylvania Press: Philadelphia, 1998.
3.   Hounshell, D. A.; Smith, J. K. "Science and Corporate Strategy: DuPont R&D"; 1902-1980; Cambridge University Press: Cambridge and New York, 1988.
4.   Sperling, L. H. "Polymeric Multicomponent Materials: An Introduction"; Wiley: New York, 1997.
5.   Gilbert, R. D., Ed.: "Cellulosic Polymers, Blends and Composites"; Hanser/Gardner: Cincinnati, 1994.
6.   Vigo, T. L. Polym. Adv. Technol., 1998, 9, 539.
7.   Parida, D.; P. Nayak; Mishra, D. K.; Lenka, S.; Nayak, P. L.; Mohanty, S.; Rao, K. K. J. Appl. Polym. Sci., 1995, 56, 1731.
8.   Barrett, L. W.; Sperling, L. H.; Gilmer, E.; Mylonakis, S. G. J. Appl. Polym. Sci., 1993, 48, 1035.
9.   Aylsworth, J. W. In U.S. Patent U.S. 1,111,284, 1914.
10.   Sperling, L. H. Polymer News, 1987, 132, 332.
11.   Sperling, L. H. "Interpenetrating Polymer Networks and Related Materials"; Plenum Press: New York, 1981.
12.   Baekeland, L. H. US 1,019,406 and 1,019,407, 1912.
13.   Staudinger, J. J. P.; Hutchinson, H. M. U.S. 2,539,377, 1951.
14.   Solt, G. S. U.K. 728,508, 1955.
15.   Millar, J. R. J. Chem. Soc. 1960, 1311.
16.   Frisch, H. L.; Klempner, D.; Frisch, K. C. Polym. Lett. 1969, 7, 775.
17.   Lipatov, Y. S.; Sergeeva, L. M. Russ. Chem. Rev. 1967, 45(1), 63.
18.   Sperling, L. H.; Friedman, D. W. J. Polym. Sci. 1969, A-2, 7, 425.

19. Widmaier, J. M.; Sperling, L. H. <u>Macromolecules</u>, 1982, <u>15</u>, 625.
20. Gankema, H.; Hempenius, M. A.; M. Moller; Johansson, G.; Percec, V. <u>Macromol. Symp.</u> 1996, <u>102</u>, 381.
21. Ostromislensky, I. 1,613,673, 1927.
22. Donatelli, A. A.; Sperling, L. H.; Thomas, D. A. <u>Macromolecules,</u> 1976, <u>9</u>, 671, 676.
23. Amos, J. L.; McCurdy, J. L.; McIntire, O. R. U.S. 2,694,692, 1954.
24. Klempner, D.; Sperling, L. H.; Utracki, L. A., Eds.; "Interpenetrating Polymer Networks"; American Chemical Society: Washington, DC, Vol. 239, 1994.
25. Wool, R. P. <u>Polymer Interfaces: Structure and Strength</u>; Hanser: Munich, 1995.
26. Adamson, A. W.; Gast, A. P. <u>Physical Chemistry of Surfaces</u>; 6th ed.; Wiley: New York, 1997.
27. Lohse, D. J.; Russell, T. P.; Sperling, L. H., Eds. <u>Interfacial Aspects of Multicomponent Polymer Materials</u>; Plenum: New York, 1997.
28. Garbassi, F.; Morra, M.; Occhiello, E. <u>Polymer Surfaces: From Physics to Technology</u>; 2nd ed.; Wiley: Chichester, England, 1998.
29. Utracki, L. A. <u>Commercial Polymer Blends</u>; Chapman and Hall: London, 1998.
30. Arends, C. B., Ed; <u>Polymer Toughening</u>; Dekker: New York, 1996.
31. Riew, C. K.; Kinloch, A. J. <u>Toughened Plastics II: Novel Approaches in Science and Engineering</u>, 1996.
32. Hamley, I. W. <u>Block Copolymers</u>; Oxford University Press: Oxford, England, 1998.
33. Holden, G.; Legge, N. R.; Quirk, R.; Schroeder, H. E., Eds.; <u>Thermoplastic Elastomers</u>; 2nd ed.; Holden, G.; Legge, N. R.; Quirk, R.; Schroeder, H. E., Eds.; Hanser: Munich, 1996.
34. Kajiyama, T.; Tanaka, K.; Takahara, A. <u>Macromolecules</u>, 1997, <u>30</u>, 280.
35. Helfand, E.; Tagami, Y. <u>J. Polym. Sci.</u>, 1971, <u>B, 9</u>, 741.
36. Helfand, E.; Tagami, Y. <u>J. Chem. Phys.</u>, 1972, <u>56</u>, 3592.
37. Cifra, P.; Nies, E.; Karasz, F. E. <u>Macromolecules,</u> 1994, <u>27</u>, 1166.
38. Russell, T. P.; Menelle, A.; Hamilton, W. A.; Smith, G. S.; Satija, S. K.; Majkrzak, C. A. <u>Macromolecules</u>, 1991, <u>24</u>, 5721.
39. Fernandez, M. L.; Higgins, J. S.; Penfold, J.; Ward, R. C.; Shackleton, C.; Walsh, D. <u>Polymer</u>, 1988, <u>29</u>, 1923.
40. Doi, M.; Edwards, S. F. <u>J. Chem. Soc., Faraday Trans. II</u>, 1978, <u>74</u>, 1789, 1802, 1818.
41. Azimi, H. R.; Pearson, R. A.; Hertzberg, R. W. <u>J. Mater. Sci. Lett.</u>, 1994, <u>13</u>, 1460.
42. Azimi, H. R.; Pearson, R. A.; Hertzberg, R. W. <u>J. Mat. Sci.</u>, 1996, <u>31</u>, 3777.
43. Cheng, C.; Hiltner, A.; Baer, E.; Soskey, P. R.; Mylonakis, S. G. <u>J. Appl. Polym. Sci.</u> 1995, <u>55</u>, 1691.
44. Cheng, C.; Hiltner, A.; Baer, E.; Soskey, P. R.; Mylonakis, S. G. <u>J. Mater. Sci.</u>, 1995, <u>30</u>, 587.

45.   Roemer, F. D.; Tateosian, L. H. US 4,396,476, 1983.
46.   Roemer, F. D.; Tateosian, L. H. US 4,396,377, 1984.
47.   Calderara, I.; Baude, D.; Joyeux, D.; Lougnot, D. <u>J. Polym. Mat. Sci. Eng.</u> <u>(Prepr.)</u>, 1996, <u>75</u>, 244.
48.   Lougnot, D.; Baude, D. EP 486638B1, 1995.

POLYMER PROCESSING

MARINO XANTHOS

Department of Chemical Engineering, Chemistry and Environmental Science
NJ Institute of Technology, Newark, NJ 07102, USA

and

Polymer Processing Institute, GITC Building, Suite 3901, NJ Institute of Technology, Newark, NJ 07102, USA

Introduction

"Polymer Processing" may be defined as the manufacturing activity of converting raw polymeric materials into finished products of desirable shape, microstructure and properties. Thermoplastic resins, usually supplied as pellets, when heated above their glass transition, T_g, and/or melting temperatures, T_m, soften and flow as viscous fluids; following shaping, rapid solidification by cooling results in the development of specific microstructures having different degrees/types of crystallinity and/or molecular orientation. Thermosetting resins, usually supplied as low viscosity liquids or low MW solids are formulated with suitable cross-linking agents and additives; they liquefy when heated and solidify with continued heating. By contrast to thermoplastics which may be reprocessed by heating, thermosets undergo permanent crosslinking to infusible, insoluble products that retain their shape during subsequent cooling/heating cycles.

Traditionally, "polymer processing" has been analyzed in terms of specific processing methods, such as extrusion, injection molding, compression molding, calendering, etc. Table 1 summarizes processing methods applicable to thermoplastics and thermosets, as well as methods common to both types of materials.

A more general analysis of converting the raw material, (solid or liquid), to a finished product would include, in addition to the shaping operation which is the essence of polymer processing, a series of pre-shaping and post-shaping operations. Such operations are included in the processes specified in Table 1. Thus, following the methodology of Tadmor and Gogos (1), "polymer processing" consists of:

1. Operations preceding shaping. Similar experiences undergone by a polymeric material may be common to all processing machinery (single and twin screw extruders, batch mixers, roll mills, calenders); they can be described by a set of elementary steps that prepare the polymer for shaping. These steps may involve all or some of the following unit operations:
- Handling of particulate solids (particle packing, agglomeration, gravitational flow, compaction and others)
- Melting or heat softening; this is the rate determining step in polymer processing and depends on the thermal and physical properties of the polymer (T_g, T_m, degradation temperature, viscosity, and others)
- Pressurization and pumping; moving and transporting the melt to the shaping operation largely depends on the polymer rheological characteristics
- Mixing for melt homogenization or dispersion of additives
- Devolatilization and stripping of residual monomers, solvents, contaminants

The common goal of the above operations is to deliver thermoplastics or cross-linkable thermosets in a deformable fluid state that will allow them to be shaped by the die or mold; solidification by cooling below T_g or T_m (thermoplastics) or by chemical reaction (thermosets) would then follow.

Table 1. Principal Processing Methods

Thermoplastics	Thermosets / Cross-linkable thermoplastics
Extrusion *Pipe, tubing, sheet, cast film, profile* *Blown film* *Coextrusion, Extrusion coating* *Wire &Cable coating* *Foam extrusion*	
	Compression Molding *Transfer molding*
Injection Molding *Structural foam Molding*	Injection Molding *Resin Injection Molding*
Expandable bead molding	PUR Foam
Thermoforming *Vacuum, pressure forming*	
Rotational Molding	
Calendering	
	Reinforced Plastics Molding *Open mold, Pultrusion, RIM, RTM,* *Filament Winding*
Blow Molding *Extrusion, Injection, Stretch*	

2. Shaping operations during which "structuring" occurs (morphology development, molecular orientation to modify and improve physical and mechanical properties). Shaping includes:

- Die forming, (fiber spinning, sheet and flat film, tube, pipe and tubular blown film, wire and cable coating, profile extrusion)
- Molding and casting, (injection molding, compression molding, transfer molding, casting of monomers or low MW materials, in situ polymerization)
- Secondary shaping, (thermoforming, blow molding, film blowing , cold forming)
- Calendering and coating, (knife, roll)
- Mold coating for hollow articles, (dip coating, slush molding, powder coating, rotational molding)

3. Post-shaping operations, (decorating, fastening, bonding, sealing, welding, dyeing, printing, metallizing).

Understanding and developing engineering models of the above operations is essential for the polymer process engineer who aims at:
  • improving process time and part uniformity,
  • substituting materials in a given application,
  • designing or modifying molds or dies,
  • saving R&D time in the selection and fabrication of processing equipment.

The future polymer technologist will increasingly be asked to define or predict the complex interrelationships between process equipment/process conditions, micro- or macrostructure developed after the shaping operation and the effects of these factors on the final properties of the fabricated part.   An understanding of engineering principles, (transport, mixing, solid mechanics, rheology), polymer physics and polymer chemistry are a necessity to accomplish this task.   The availability of improved computer-aided design software for mold or screw design (Fig. 1), cavity filling simulation, flow calculations, process simulation, and the increasing availability of multicomponent polymer systems such as blends, alloys and composites are expected to continue expanding the horizons of product concepts and novel applications during the next decade.

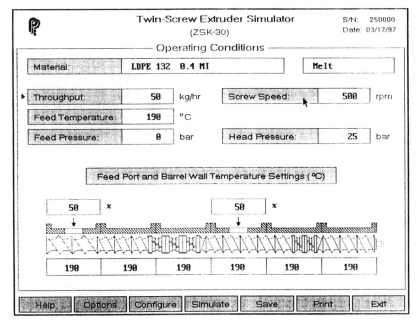

Fig. 1. Axial pressure profile simulation in a co-rotating twin screw extruder (courtesy of Polymer Processing Institute).

<u>Historical/ Current Status</u>

Machinery and processing methods adopted by the polymer industry evolved from the rubber industry (rubber masticator, two-roll mill, ram extruders in the first half of the 19th Century). In the second half of the 19th Century, the ram extruder and then the single most important development in the processing field - the single screw extruder- made their appearance. In the late 19th Century, the plunger injection molding and in the ensuing years the "torpedo" and then the reciprocating screw injection molding machines were developed (1).

Childs, (2), in his 1975 review of the status of the polymer industry, (the year celebrating the 50th Anniversary of the ACS Organic Coatings and Plastics Chemistry Division) compared developments in materials and processes during this 50 year period. Data for 1925, when the US polymer industry was in its infancy, were as follows:

- Production of polymers was about 10 million pounds
- The types of polymers were mostly thermosets such as alkyd resins, bitumen, casein, cellulosics, phenol/formaldehyde, phenol/furfural, shellac, urea based resins.
- Conversion processes were obviously those for thermosets and included casting, compression molding, laminating and stuffing/extrusion.

Following the explosive development of thermoplastics after World War II many improvements and new developments led to today's diversity of polymer processing machines and technologies. Machinery to convert polymers into finished products was developing side-by side with polymer capacity. Major markets in 1975, (but also today), were building and construction, packaging, transportation, electrical/electronics, furniture, housewares, toys/appliances. Childs' data for 1975 showed that:

- Production of polymers was about 30.7 billion pounds (with over 40 different families)
- More than 80% of all polymers were now thermoplastics with only about 15% thermosets,
- Principal conversion processes were listed as molding, film and sheet conversion, extrusion, adhesives conversion, foaming, laminating, surface coatings and other.

Today's data indicate a more than double growth in U.S. polymer production vs. 1975 (about 70 billion pounds) with still almost 85% of the resins produced being thermoplastics (3). Numerous new materials but also mixtures of existing materials with other polymers (blends, alloys) or reinforcing elements (fiber, ribbon composites) have been introduced. Over 70% of the total amount of thermoplastics is accounted for by the large volume, low cost commodity resins: polyethylenes of different densities, isotactic polypropylene, polystyrene and polyvinyl chloride. Next in performance and in cost are acrylics, cellulosics, and acrylonitrile-butadiene-styrene terpolymers. Engineering plastics such as acetals, polyamides, polycarbonate, polyesters, polyphenylene oxide and their blends are increasingly used in high performance applications; high-performance advanced materials such as liquid crystal polymers, polysulfones, polyimides, polyphenylene sulfide, polyetherketones and fluoropolymers often present processing challenges due to their high T_g or T_m (290-350°C).

Among the thermoplastics processes listed in Table 1 and to be discussed in this Chapter, extrusion is the most popular. Approximately 50% of all commodity thermoplastics are used in extrusion process equipment to produce profiles, pipe and tubing, film, sheet, wire and cable. Injection molding follows as a preferred processing method accounting for about 15% of all commodity thermoplastics. (4) Increased polymer consumption over the past twenty years did not only stimulate machinery sales but also a parallel growth in usage of a large variety of additives, fillers and reinforcements. This has lead to today's estimated sale of 20 kg of additives per 100 kg sales of plastics (5) and the significant advances that have been made to improve the efficiency of polymer melt mixing equipment.

Principal Thermoplastics Processes

Given the popularity, higher production rates than thermosets and significant growth of thermoplastics, only the principal thermoplastics processes, (i.e., extrusion, injection molding, blow molding, thermoforming and rotational molding), will be discussed in this overview. Principal sources of information are, [in addition to Ref. (4)], a series of 1998 dated articles on opportunities for the next decade in plastics processing technology (6-11) and author's notes.

Extrusion. In extrusion, a molten material usually supplied in the form of pellets is forced through a shaping device. Because the viscosity of most plastic melts is high, extrusion requires the development of pressure in order to force the melt through a die. Melts are extruded as specific profiles, (rods that are then cut into cylindrical pellets, film, sheet, tubing, etc), as a molten tube of resin (parison) for blow molding or into molds, as in injection molding.

Single screw extruders. The extruder shown in Figure 2, accepts dry solid feed (F,E,J) and melts the plastic by a combination of heat transfer through the barrel (B,C) and dissipation of work energy from the extruder drive motor (I). During melting, and in subsequent sections along the barrel, mixing of additives and homogenization is usually achieved. Venting may also be accomplished to remove undesirable volatile components, usually under vacuum through an additional deep channel section and side vent port. The final portion of the extruder (L) is used to develop the pressure (up to 50 MPa) for pumping the homogenized melt through an optional filtering screen and then through a shaping die attached to the end of the extruder.

Extruders are defined by their screw diameter and length; length is expressed in terms of the length-to-diameter ratio (L/D). Single screw extruders range from small laboratory size (6-mm D) to large commercial units (450-mm D) capable of processing up to 20 t/h. Melt-fed extruders run about 8 L/D; solids fed extruders run from 20 to 40 L/D depending upon whether intermediate venting is provided.

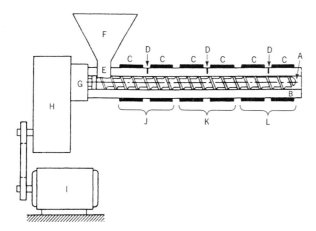

Fig. 2. Parts of an extruder: A, screw; B, barrel; C, heater; D, thermocouple; E, feed throat; F, hopper; G, thrust bearing; H, gear reducer; I, motor; J, deep channel feed section; K, tapered channel transition section; L, shallow channel metering section (4)

An extruder employs drag flow to perform a conveying action that depends on the relative motion between the screw and the barrel. With higher friction on the barrel than on the screw, the solids are conveyed almost as solid plug in the deep feed channel section. As the channel depth becomes shallower, the compressive action causes more frictional heat, which, combined with the conduction supplied by the barrel heaters, causes the plastic to melt. The molten plastic then enters a constant shallow depth section of the screw called the metering section, where the pumping pressure necessary for extrusion through the final shaping die is developed.

In addition to the conveying and melting steps, extruders perform the vital task of homogenization of additives, such as fillers, stabilizers, and pigments, into the base resin. If the cross-channel mixing in single screw extruders is insufficient, various mixing enhancers are available. These generally assure dispersion by forced passage over a higher shear restricting slot or an improved distribution by multiple reorientation of flow and sequences of extensional flow.

Twin-screw extruders. Varieties of twin-screw extruders are gaining in popularity, particularly when the ingredient mixing requirements are very difficult or require multiple staging, such as in *reactive extrusion* or powder feeding is required. Much of the recent polymer developmental work has been, and much of the future work will be based on mixing two or more polymers to obtain blends or alloys with unique properties. These opportunities have been developed and commercialized by taking advantage of the flexibility of the individual screw and barrel segments of the twin-screw extruders, but at a significantly higher capital cost compared with single-screw extruders. Mixing in twin screw extruders benefits from the additional interaction of the two screws with each other, as well as with the barrel. Twin-screw extruders are classified as being tangential or intermeshing, and the latter as being counter- or co-rotating. These extruders are generally supplied with slip-on conveying and kneading screw elements and segmented barrels. These elements, shown in Figure 3, give the processor improved mixing and pumping versatility by causing extensive melt reorientation, back mixing, and elongational flow patterns.

Reactive extrusion is the term used to describe the use of an extruder as a continuous reactor for polymerization or polymer modification by chemical reaction (12). Extruders are uniquely suitable for carrying out such reactions because of their ability to pump and mix highly viscous materials. Twin-screw extruders readily permit multiple process steps in a single machine, including melting, metering, mixing, reacting, side stream addition and venting.

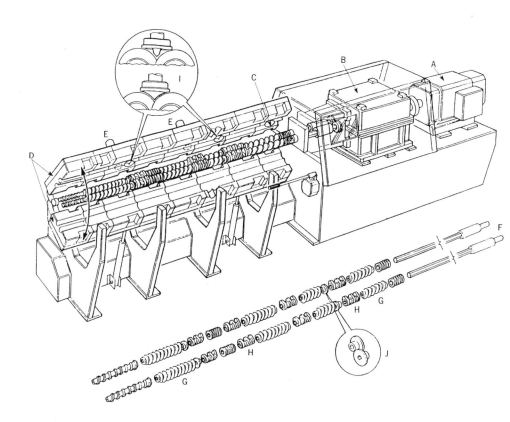

Fig. 3. Intermeshing co-rotating twin screw extruder: A, motor; B, gear box; C, feed port; D, clam shell barrel; E, vent port; F, screw shafts; G, conveying screws; H, kneading paddles; I, barrel valve; J, blister rings. (4)

Die Forming. Shaping operations associated with extrusion involve a variety of dies and take-off equipment. Additional information and schematics can be found in Reference 4. *Rod* dies in combination with cooling troughs and pelletizers are used to produce pellets for injection molding. In *profile* extrusion special dies are used to produce a variety of asymmetrical products such as housing siding, window sash molding, decorative trim, plastic lumber, gaskets, and channels. Dies for profile extrusion are designed to allow for shape changes that occur during quenching. Usually, a water trough is used for cooling, but air or cooled metal surfaces are also used. Profile extrusion typically relies on amorphous thermoplastics since the rapid shrinkage accompanying crystallization can result in severe distortion of the profile.

Tubular dies are used for pipe and tubing extrusion in combination with vacuum calibrators to control cooling and maintaining a constant extrudate diameter and thickness. Tubing extrusion usually involves drawdown of the resin from the die, i.e., the diameter and wall thickness of the tubing are less than that of the die opening. The amount of drawdown during tubing extrusion usually range from 2:1 to 8:1. Drawdown in pipe extrusion is often less than 1.1:1. Pipe and tubing may be cut in straight lengths or may be coiled, depending upon use and stiffness. The blown film process also uses a tubular die from which the extrudate expands in diameter while travelling upward to a film tower. The top of the tower has a collapsing frame followed by guide and pull rolls to transport the collapsed film to subsequent slitting and windup rolls. The film may be treated for subsequent printing. The tubular bubble from the die is inflated to the desired diameter by air passing through the center of the die. Primary cooling to solidify the melt is supplied by an external air ring. The tube is characterized by its blow-up ratio, which is expressed as the ratio of bubble diameter to the die diameter. Typical blow-up ratios range from 2:1 to 4:1 with typical film thickness from 0.007 to 0.125 mm. The process requires a resin with high melt viscosity and melt strength so that the molten extrudate can be pulled from the die in an upward direction. Polyethylene is the primary plastic used in most films, especially for packaging and trash bags. Coaxial dies can be used for manufacture of coextruded multilayer films.

Films or sheets may be produced in *slit* or *sheet* dies with adjustable gap openings. In the cast film process the die is positioned very carefully with respect to a highly polished and plated casting or chill roll which is cooled by rapid circulation of water. The melt needs to maintain good contact with the chill roll, i.e., air must not pass between the film and the roll to ensure uniform cooling and impart a smooth and virtually flawless surface to the film. The process used to make an extruded plastic sheet is illustrated in Figure 4. Sheeting thickness is 0.25-5 mm and widths are as great as 3 m. Cooling is controlled by a three-roll stack with individually cooled rolls 25-50 cm in diameter, highly polished, and chrome plated. Pressure between the rolls may be adjusted to produce sheet of the proper thickness and surface appearance. For a given width, thickness depends upon the balance between extruder output rate and the take-off rate of the pull rolls. Draw down from the die to the nip is typically ca 10%. Sheet extrusion requires that the resin be of relatively high viscosity to prevent excessive sag of the melt between the die and the nip between the top and middle rolls. Through the proper use of several extruders and a die feeding manifold sheets containing several layers of similar or dissimilar materials including regrinds and tie layers may be coextruded. Most sheeting is used for thermoforming.

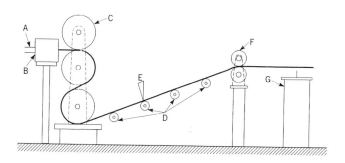

Fig. 4. Sheet extrusion: A, die inlet; B. die; C, three-roll finisher; D, support rollers; E, edge-trim cutter; F, pull rolls; G, saw or shear (4)

Flat or circular dies are used in a variety of *extrusion coating* operations. A coating of an appropriate thermoplastic, such as polyethylene, may be applied to a substrate of paper, thin cardboard, or foil, to provide a surface property which enables heat sealing or better barrier performance. In this process a molten web of resin (narrower than the die) is extruded downward, and the web and substrate make contact at the nip between a pressure roll and a water-cooled chill roll. The pressure roll pushes the substrate and the molten resin against the chill roll. Pressure and high melt temperatures are needed for adhesion of resin and substrate. Typical coating thickness are 0.005-0.25 mm; the die opening is ca 0.5 mm. Protective and insulating coatings can be applied continuously to wire as it is drawn through a cross-head die. A typical wire coating line consists of a wire payoff, wire preheater, extruder, die, cooling trough, capstan, and wire take-up. The die exit usually is the same diameter as that of the coated wire and there is little drawdown. Die openings are small and pressures inside the die are high ca 35 MPa. Wire take-up systems now operate as high as 2000 m/min.

Foamed thermoplastics provide excellent insulating properties because of their very low thermal conductivity, good shape retention, and good resistance to moisture pickup. As such, cylindrical shapes are extruded for pipe insulation, and flat sheets for building panel insulation. In *foam extrusion* with physical blowing agents, the gas or low boiling point liquid is dissolved and held in solution by the pressure developed in the extruder. As the molten thermoplastic exits the extruder die, the pressure release causes instantaneous foaming. The chlorinated fluorocarbons formerly used as blowing agents have been replaced with more environmentally friendly substitutes, such as HCFC's or low molecular weight hydrocarbons, such as butane or iso-pentane. For some thermoplastics inert atmospheric gases such carbon dioxide, nitrogen, or argon are increasingly used. Critical to the success of most foaming extrusion operations is cooling of the melt just prior to entry to the die. Cooling is most effectively accomplished with a tandem arrangement of two extruders, as shown in Figure 5, wherein the first extruder assures complete dissolution of the blowing agent, and the second extruder is operated at slow speed for optimum cooling.

Developments/Trends . Some of the following recent developments and future trends in extrusion technology have been discussed in Ref. (6):

Twin Screw Extruders (TSE)
- Increasing popularity of TSE for compounding, reactive extrusion, powder feeding.
- Understanding of dispersive and distributive mixing and design of mixing elements; need for better understanding of basic melting mechanisms, devolatilization and mixing role of each section of a TSE.

Single Screw Extruders (SSE)
- Longer barrel for foams, improved abrasion resistance barrel linings.
- Screw design: multistage, barrier, segmented, computer designed targeted for specific end uses to improve productivity, mixing, output, lower energies.

Overall
- More use of microprocessor control systems that provide precise and efficient control for the overall extrusion and downstream.
- Monitoring systems that measure viscosity or monitor extrudate can be combined with feedback control to automatically adjust the process.
- Materials: New, more easily processable resins (metallocenes, blends, long fiber composites).

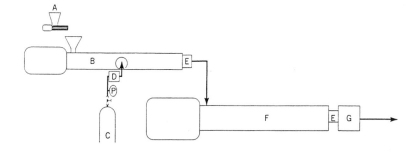

Fig. 5. Foam extrusion line: A, resin feed; B, extruder for melting and gas incorporation; C, gas supply; D, gas metering and compression; E, optical windows; F, cooling extruder; G, die. (4)

Injection Molding

Equipment/Process. The machine consists of an injection unit and a clamp unit. The injection unit is usually a reciprocating single screw extruder which melts the plastic and injects it into the mold. The clamp unit opens, closes, and holds the mold closed against the pressure of injection. An injection molding machine is operated by hydraulic power and equipped with an electric motor and hydraulic pump. A hydraulic cylinder opens and closes the mold and holds the mold closed during injection; another cylinder forces the screw forward, thereby injecting the melt into the mold; a separate hydraulic motor turns the screw to plasticate, homogenize and pressurize the melt. Control of these movements is a combined function of the hydraulic and electrical systems. Additional information on equipment, molding cycle and molds can be found in Reference 4.

In the injection molding machine shown in Figure 6, the clamp unit is on the left and the injection unit on the right. A mold is shown in position between the platens of the clamp unit with one-half of the mold fastened to the fixed platen and the other to the movable platen. When the mold is opened, the movable platen moves away from the fixed platen and the molded part can be removed. After part ejection, the mold is closed in preparation for the next injection cycle. In contrast to extrusion, the screw in the injection unit rotates only during part of molding cycle. When the screw turns, it pumps melt forward. The rearward movement of the screw is controlled by the placement of a limit switch, which stops the hydraulic motor. The front of the screw is usually equipped with a check valve. While the screw rotates, the melt can move freely forward through the valve. However, the valve closes to prevent any reverse flow. The position of the screw is adjusted in such a way that there is always some melt in front of the screw when the mold is full. This pad or cushion of melt transfers pressure from the screw to the plastic in the mold. After initial filling, some additional melt flows into the mold; this is called packing.

A common characteristic of all injection molded parts is an inherent skin-core structure with the relative skin to core thickness depending on materials characteristics and process conditions, essentially flow rate and melt-mold temperature difference. A typical shot consists of sprue, runners, gates and parts, with shot size being directly proportional to the amount of screw reciprocation. Gate design is influenced by part geometry, resin type and processing conditions. Sprue and runners channel the melt into the cavities. After ejection they are separated from the parts, ground, and fed back into the injection unit for reprocessing. Modern mold design tends to reduce or eliminate sprue and runner scrap through a variety of techniques such as hot runners, insulated runners or by designs that place the nozzle directly against the mold cavity. Cycle time and quality of moldings are optimized through microprocessor-based controllers that receive inputs from temperature, pressure and ram position sensors.

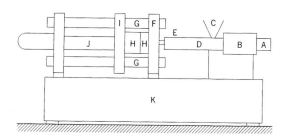

Fig. 6. Injection molding machine: A, hydraulic motor for turning the screw; B, hydraulic cylinder and piston allowing the screw to reciprocate about three diameters; C, hopper; D, injection cylinder (a single screw extruder); E, nozzle; F, fixed platen; G, tie rods; H, mold; I, movable platen; J, hydraulic cylinder and piston used to move the movable platen and to supply the force needed to keep the mold closed; K, machine base. (4)

Structural Foam Molding. Structural foam is a modified injection molding process for large articles having a cellular core and an integral solid skin with an overall 20-50% reduction in density, compared to their solid counterparts. The most common blowing agent is azodicarbonamide which by thermal decomposition releases nitrogen, carbon monoxide, and carbon dioxide. Instead of using a solid chemical blowing agent nitrogen may also be mixed with the melt and the mixture is maintained under pressure until it is injected into the mold. The injection molding machine must be equipped with a shutoff nozzle, which maintains the melt under pressure while the mold is opened. The screw is retracted only part of the way needed for a full shot and a short shot is injected into the mold. The empty space allows the blowing agent to expand the melt, forming the foam structure. Structural-foam molding is limited to parts with wall thickness of at least 6 mm which, typically, have a dense skin and a foamed interior with various pore sizes. Compared to injection molded surfaces, the surfaces of structural-foam moldings are poor, characterized by a rough, swirly finish. Maximum pressure in the mold during foaming is much lower than in injection molding and as a result some cost savings are possible in mold and press construction. Molding cycles are somewhat longer than for injection molding.

Developments/Trends. A variety of process modifications aimed at improving surface finish, controlling part weight, reducing cycle time, improving weld line integrity, and modifying part macrostructure and properties have been described and/or are under development (7). They include:

- *co-injection* or *sandwich* molding where the outer skin layer is molded from a different material than the center;
- *gas-assisted* molding where gas comprising the core of the part is injected after the cavity is partially filled with the skin material;
- *low-pressure* molding which minimizes overpressurizing the melt at the start of injection and better controls the flow behavior of the viscoelastic melt;
- *injection-compression* molding where cavity walls move perpendicular to the parting line during filling or packing and allow large area parts to be molded with more uniform distribution of built-in stresses;
- *lost-core (fusible core)* molding where easy removal of low M.P metallic or water soluble plastic cores offer manufacturing flexibility for parts with intricate internal channels;
- *multiple-live feed* and *push-pull* injection molding for mitigating the effects of weldlines particularly in blends and composites.

An important development of the past few years, which is also expected to continue providing insight and guidance to future process and part designers, is the development of CAD methods. Simulation of a proposed mold design and analysis of melt flow has helped to shorten cycles, dimension channels, and balance flow, determine gate location and number of gates, and estimate shrinkage and warpage.

Blow Molding

Equipment/Process. Blow molding is the most common process for making hollow thermoplastic objects including disposable containers and industrial or structural products. High density polyethylene is the most common blow molding resin to produce containers ranging in size from 30 cm^3 to 200 liters. It is followed by PET and to a lesser extent by other resins. Among the advantages of the process is the ability to produce complex, irregularly shaped hollow objects weighing from few grams to 500 kg with excellent part detailing. Inexpensive cast aluminum molds may be used, given the relatively low inflation pressures involved. Process limitations are residual stresses in parts, non-uniform wall thickness, relatively high levels of trim for industrial products and need for resins with specific rheological characteristics. More information and schematics of the process maybe found in (4,8).

General categories of blow molding include continuous or intermittent extrusion, injection blow and injection stretch molding. In extrusion blow molding a molten tube of resin called a parison is extruded from a die into an open mold. The mold is closed around the parison, and the bottom of the parison is pinched together by the mold. Air under pressure is fed through the die into the parison, which expands to fill the mold. The part is cooled as it is held under internal air pressure. As the parison is extruded, the melt is free to swell and sag. The process requires a viscous resin with consistent swell and sag melt properties. For a large container the machine is usually equipped with a cylinder and a piston called an accumulator. The accumulator is filled with melt from the extruder and emptied at a much faster rate to form a large parison; this minimizes the sag of the molten tube. With a simple parison the large diameter sections of the bottle have a thin wall and the small diameter sections have a thick wall. Certain modifications of the die can control the thickness of the parison wall along its length, which results in a bottle with improved wall thickness distribution and better mechanical properties.

In injection blow molding, a parison is injection molded onto a core pin; the parison is then rapidly transferred via the core pin to a blow mold, where it is inflated by air against a split mold wall into an article. This process is usually applied to small and intricate bottles.

Development of injection stretch blow molding followed the desire in the mid-1970s to produce rigid barrier plastic containers to compete with glass bottles. A two-step process is mostly applicable to soft drink bottles made from PET. First, a test-tube-shaped preform is molded, which is then reheated to just above its glass transition temperature, and mechanically stretched prior to and during inflation. Stretching the PET produces biaxial orientation which improves transparency, strength and toughness of the bottle.

Developments/Trends Growth of blow molding is connected to continuing developments in machinery for container and industrial products, and materials. Improved recyclability of PET and HDPE containers through improved design and elimination of metal caps, paper labels or a second incompatible polymer has been undoubtedly a significant factor affecting growth. With respect to materials, efforts will continue to optimize viscoelastic properties and improve thermal stability in attempts to develop new extrusion/injection blow molding grades from polymers other than HDPE. Multilayer blow molding for small containers with improved oxygen barrier and moisture-barrier properties or hydrocarbon–barrier shaped fuel tanks is an area of continuing activity; economic viability of such structures, however, is affected by difficulties in recycling the mixed regrind. Continuing efforts are under way to predict wall thickness for parisons and the final part in order to minimize materials usage, although software packages are in the early stages of development (8).

Thermoforming.

Equipment/Process. Thermoforming is a process for converting an extruded plastic sheet into small items such as thin wall packaging containers, cups, plates and trays, or deep drawn thicker items such as sailboats, bath tubs, freezer liners and cabinetry. "Skin" packaging, which employs a flexible plastic skin drawn tightly over an article on a card backing, is made by thermoforming. The steps involved are heating the sheet few degrees above its T_g or T_m, stretching it against a rigid mold, allowing the formed part to cool and trimming the formed part from the non-part web. Thermoforming is suitable for forming parts that have high surface-to-thickness ratio and uses inexpensive molds made of wood, aluminum, steel, epoxy. Limitations are residual stresses, high levels of trim, non-uniform wall thickness and high energy costs. The latter are typically the first or second part of the total manufacturing cost. and hence the continuing efforts for the development of more energy efficient heaters. Under vacuum the process is called vacuum forming and requires differential pressures of less than one atmosphere. The sheet is clamped in a frame and exposed to radiant heaters. The sheet softens to a formable condition, is moved over a mold and sucked against the mold by vacuum. Excess plastic is trimmed and recycled. In some plants the vacuum forming line is run in-line with the sheeting extruder. Timers control the length of the heating and cooling periods, which depend on composition and sheet thickness. Frequently, the process is improved by prestretching using mechanical aids (plug assisted, air-cushioned plug assisted, drape assisted). More information and schematics of the process maybe found in (4,9).

Amorphous resins such as styrenics including foamed polystyrene, acrylics, and PVC represent 80% of polymers used in thermoforming. LLDPE and HDPE represent the majority of the semicrystalline thermoformable resins. For typical amorphous polymers heated near its T_g the resulting sag develops over a relatively wide temperature range and is controllable. By contrast, for some semicrystalline resins such as PP, certain polyesters and fluoropolymers sag occurs over a narrow temperature range making thermoforming very difficult. Broader forming window may be achieved by resin modification for improved melt elasticity through branching, copolymerization, crystallization modifiers or other additives.

Developments/Trends. As for other processes the most rapidly developing segment in thermoforming is the development of software for comparative cost analysis, sheet temperature and wall thickness prediction. Much of the continuing growth of thermoforming is attributed to the development of technologies that allow it to compete with injection blow molding or rotomolding. This includes, for example, the refinement of the twin sheet process to produce hollow parts by simultaneous or sequential forming of both halves and then compressing at the periphery to effect closure; twin-sheet forming is expected to provide additional design freedom since the two halves could be from different materials. The increasing use of rapid prototyping for mold making and the ability of the thermoforming process to respond to recent trends in industrial design requiring fewer parts at shorter marketing times are expected to be contributing to the growth of the thermoforming industry (9).

Rotational Molding

Equipment/Process. Hollow articles and large, complex shapes are made by rotational molding, primarily from polyethylene fine powder (about 35 mesh) of relatively low viscosity, PVC plastisols and to a lesser extent resins such as nylon 6. The process is based on the heating and cooling of a biaxially rotating mold that defines the shape of the required part. A rotomolding machine has three long arms extending from a central driving mechanism; each arm rotates several molds in two planes. The arms are moved from one process station to the next, i.e., from unloading and loading to heating and cooling. More information and schematics of the process maybe found in (4,10,11).

A pre-measured amount of resin placed inside a mold cavity is heated in an oven by rotating the mold in two directions at low speed to uniformly distribute the material over the inside cavity wall. The resin begins to adhere to the wall and upon further heating it sinters and fuses, coating and duplicating the inside of the mold. The mold is then moved to the cooling chamber and the part solidifies. The melt is forced without pressure against the mold surface during heating or cooling, resulting in uniform wall thickness, zero orientation and high physical properties. Cycle times are long because of the heating and cooling required and depend heavily on wall thickness; initial financial nvestment in molds (usually cast aluminum and sheet metal) and machines, however, s relatively low by comparing with other processes for producing hollow parts. Common rotomolded products include large tanks and boxes, drums, furniture, toys, playballs and automotive arm rests.

Developments/Trends. Recent improvements in rotomolding technology include techniques to optimize and analyze the process such as the use of microprocessors, real–time temperature sensors controllers inside the rotating mold and the direct observation of the sintering stage through a video camera and light source. Also, refinement of methods to produce multiwalled solid or foamed structures, new machine designs that may incorporate heating and cooling in the walll of the mold cavity, use of micropellets instead of powders (10,11). Future developments will be depending on the availability of resins with suitable viscosities and capable of retaining their physical properties at high oven temperatures, (250-400°C), and prolonged periods of time. The availability of PP, nylons and metallocene-catalyzed polyolefins with improved thermooxidative stability and the development of suitable grades of engineering resins such as ABS, PPE, PEEK. PPS, glass reinforced nylons could expand the use of rotomolding to a variety of applications for the construction and transportation industries.

Literature Cited

1. Tadmor Z. and. Gogos C.G., "Principles of Polymer Processing", John Wiley & Sons, Inc., New York, 1979
2. Childs E.S., in "Applied Polymer Science", J.K. Craver and R.W. Tess, Eds. Organic Coatings and Plastics Chemistry, Div. of ACS, Washington D.C. p. 447, (1975)
3. Modern Plastics, 71, 1, 73 (1994).
4. Xanthos M. and Todd D.B., "Plastics Processing", Kirk-Othmer 'Encyclopedia of Chemical Technology', 4th Ed., Vol. 19, p. 290, John Wiley & Sons, New York (1996).
5. Xanthos M., "The Physical and Chemical Nature of Plastics Additives", Chapter 14 of "Mixing and Compounding of Polymers - Theory and Practice", p. 471, I. Manas-Zloczower and Z. Tadmor, Eds., Carl Hanser Verlag, Munich, New York (1994).
6. Gould R.J., Plast. Eng., 54, 10, .33, (1998)
7. Schmidt L.R., Plast. Eng., 54, 10, 27, (1998)
8. Throne J.L., Plast. Eng., 54, 10, 41, (1998)
9. Throne J.L., Plast. Eng., 54, 10, 37, (1998)
10. Beall G.L., Plast. Eng., 54, 2, 33, (1998)
11. Wigotsky . V, Plast. Eng., 54, 2, 18 , (1998)
12. Xanthos M, Ed., "Reactive Extrusion: Principles and Practice", Carl Hanser Verlag, Munich, New York (1992).

FIRE AND POLYMERS

GORDON L. NELSON

College of Science and Liberal Arts, Florida Institute of Technology, 150 West University Boulevard, Melbourne, Florida 32901-6975

Introduction
What It Means to Be Fire Retardant
Fire Retardant Chemistry
Recent Activity in Fire Retardants
Commercial and Regulatory Issues at Work for Change of Fire Retardants
Future Trends

Introduction

In the book "Applied Polymer Science," commemorating the fiftieth anniversary of the Division of Polymeric Materials Science and Engineering, there was not a chapter on the fire behavior of polymers (1). Twenty-five years later it is appropriate that such a chapter be included. It is indeed unfortunate that most polymer science texts include only a few pages, if that, on the subject of polymer flammability, given that the chemistry is an extension of that of thermal and oxidative degradation, and that a number of major markets for polymers include fire performance as part of the key selection criteria: markets including transportation, building/construction, electrical/electronics, and furniture/furnishings (2).

Fire is an ever present hazard. American homes suffer an unwanted fire every ten seconds. Every 60 seconds there is a fire serious enough to call the fire department. In our lifetime we each have a 40 percent chance of having a fire large enough to cause a fire department to arrive at our door. Every two hours someone in the United States is killed in a home fire, about 4,000 people per year. Some 19,300 people are injured in home fires in a typical year. Over the period 1985-1994 the United States had an average of 2.2 million fires reported each year, causing an average 5,300 civilian deaths, 29,000 injuries, and a $9.4 billion property loss. While the fire problem in the United States is less severe now than it was a decade ago, because the population has grown faster than fire casualties, the U. S. fire death rate is still higher than that in most developed nations. Of western industrialized nations the United States ranked behind only Hungary as having the highest per capita fire death rate, with an average of 26.5 deaths per million population. The rate in the United States is more than five times that of Switzerland (5.2 deaths per million population), which had the lowest rate of countries considered. The higher rate of fire in the United States has its origins in several factors including the commitment of fewer resources to fire prevention activities than other nations, a greater tolerance for "accidental" fires, the practice of riskier and more careless behavior, less use of compartmentation and fire resistance in home construction, and the presence of more "stuff" (higher fire load) and a higher use of energy (more ignition sources) than in other countries (3).

What It Means to Be Fire Retardant

The terms "fire retardant" or "flame retardant" are not absolute. Combustibility can depend as much on fire conditions as on polymer composition. Whether a material is fire retardant is in the context of the application. And whether a material is fire retardant is very much in the context of tests, tests specifically relevant to the intended application. Fire retardancy is not a single property but a sequence of properties, some more important to a given application than others. The behavior of materials in a fire can be described as follows (4-5):
  (1) Ease of Ignition - how readily a material ignites.
  (2) Flame Spread - how rapidly fire spreads across a surface
  (3) Fire Endurance - how rapidly fire penetrates a wall or barrier.
  (4) Rate of Heat Release - how much heat is released and how quickly.
  (5) Ease of Extinction - how rapidly or how easily the flame chemistry leads to extinction.
  (6) Smoke Evolution - amount, evolution rate, and composition of smoke released during stages of a fire.
  (7) Toxic Gas Evolution - amount, evolution rate, and composition of gases released during stages of a fire.
  A multitude of flammability tests measuring one or more of these properties have been developed over the years. There are more than 100 tests used in the United States. Other countries use different tests and rating systems, often making comparisons difficult. A material acceptable in one country for an application may be rejected in another. A polymer may be more ignition resistant than the parent polymer or another polymer and thus be "ignition resistant" or "flame retardant" but still have insufficient fire performance in the context of tests appropriate for the application. One unifying approach which is increasingly successful is the use of bench scale tests which measure fundamental properties, then using those properties in mathematical models to calculate the impact of candidate materials on the fire performance of the system.
  Figure 1 (6) shows fire as a series of stages (7). Fire starts with an ignition source and a first item to be ignited. For an electrical appliance one wants a material which will resist a small electrical arc or a small, short duration flame. One does not want the appliance to be the source of fire. One does not expect an appliance, however, to survive a house fire. Thus ignition resistance is what is required.

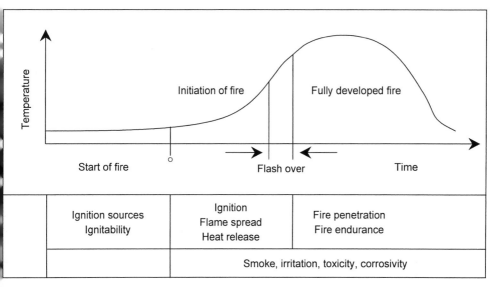

Figure 1. Stages of a fire. Following ignition there is a period of reaction of materials to fire. Once fire fills a compartment, the ability of the compartment to contain fire is at issue. Throughout, smoke and toxic and corrosive gases are emitted.

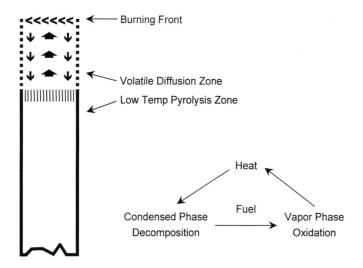

Figure 2. The fire cycle. Fire involves the thermal decomposition of materials to give volatile fuel, which oxidizes in the vapor phase to give heat, which continues the cycle.

Given ignition of an initial item, fire begins to spread and heat is released. For a wall covering, for example, flame spread should not be rapid. Indeed, wall coverings and other interior finish are regulated by rate of flame spread.

As fire spreads from item to item, more heat is released, and combustible gases rise. A point is reached when fire engulfs the upper part of the room or compartment (flashover). The upper part of the room exceeds 600°C. If a door is open, fire moves down from the ceiling and begins to exit the room. The fire is fully developed.

Given a fully developed fire the issue then is the ability of barriers to contain the fire. Can the assembly of materials resist penetration (fire endurance), to allow evacuation and to protect adjacent property? Walls and floors/ceilings are rated for their fire endurance against standard time temperature curves.

The answer as to whether a material is fire retardant clearly depends upon its use. A material suitable as an appliance enclosure because of its ignition resistance, would be totally unsuitable as a seal in a wall or floor where the expectation might be resistance to a room burnout on the other side of the wall or floor for several hours duration.

Fire Retardant Chemistry

Polymer combustion occurs in a continuous cycle (Figure 2) (4-5,8). Heat generated in the flame is transferred back to the polymer surface producing volatile polymer fragments, or fuel. These fragments diffuse into the flame zone where they react with oxygen by free radical processes. This in turn produces more heat and continues the cycle. Flame retardancy is achieved by interrupting the cycle by one of two methods.

In the first method, vapor phase inhibition, one introduces changes in the flame chemistry. Reactive species are built into the polymer which are transformed into volatile free-radical inhibitors during burning. These species diffuse into the flame and inhibit radical reactions, particularly the branching radical reaction. As a result, increased energy is required to maintain the flame and the cycle is interrupted.

In the second method, solid phase inhibition, one introduces changes in the polymer substrate. Systems that promote extensive polymer crosslinking at the surface, form a carbonaceous char upon heating. Char insulates the underlying polymer from the heat of the flame, preventing production of new fuel and further burning. Other systems evolve water during heating, cooling the surface and increasing the amount of energy needed to maintain the flame.

Many polymers are intrinsically fire retardant; examples include high temperature aromatic polymers and polyvinyl chloride. Some high temperature materials are stable for a few minutes at extreme temperatures, 600-1000°C, while others can perform at 200-300°C for long periods of time. There are three general types of aromatic structures: linear single-strand polymers such as aromatic polyamides and polyamides based on benzenoid systems, ladder polymers consisting of an uninterrupted sequence of cyclic aromatic or heterocyclic structures, and spiro polymers in which one carbon is common to two rings. Behind each type of structure is the thought that polymers with high aromatic character and very strong bonds between rings produce more char on heating, retaining most of the potential fuel of the original polymer as residue. A good, though impractical example, is polyphenylene, a crystalline, high-melting substance, with thermal degradation beginning at 500 to 550°C and continuing to 900°C with only 20-30 percent weight loss. Such intrinsically fire retardant materials unfortunately tend to be relatively expensive.

An inexpensive intrinsically fire retardant polymer is polyvinyl chloride. The first step in PVC degradation is the release of hydrogen chloride and the formation of double bonds. The hydrogen chloride provides vapor phase inhibition and the double bonds crosslink to provide extensive char (rigid PVC).

At the other end of the scale from intrinsically fire retardant polymers are those polymers modified with monomeric fire retardant additives. Fire retardant additives used with synthetic polymers include organic phosphorus compounds, organic halogen compounds, and combinations of organic halogen compounds with antimony oxide. Inorganic fire retardants include hydrated alumina, magnesium hydroxide, and borates. Not all fire retardant additives function well in all polymers. Additives may interact with other resin formulation components such as fillers, stabilizers, and processing aids. To be effective, the fire retardant must decompose near the decomposition temperature of the polymer in order to do appropriate chemistry as the polymer decomposes, yet be stable at polymer processing temperatures.

An intermediate approach between intrinsically fire retardant polymers and adding simple additives to a resin formulation is the incorporation of fire retardant chemical units directly into the polymer backbone. In this way one achieves permanent fire retardancy and is better able to maintain the desirable physical properties of the polymer. Monomeric additives may not be permanent, may come to the surface, and may change the polymer's physical characteristics. Fire retardant monomers used as coreactants in polymer synthesis contain appropriate functional groups for incorporation into the chain. Many are halogen compounds, the source of fire retardancy. Generally 10-25 percent halogen is necessary to impart suitable fire retardancy in polymers, either by additive or backbone incorporation. Phosphorus comonomers have also been used, as have phosphorus-halogen comonomers.

The specific functionality and modes of retardancy for fire retardant additives and comonomers are as follows (4,5 8):

Phosphorus additives. The thermolysis of phosphorus compounds yields phosphorous acids and anhydrides which promote dehydration leading to water vapor and promotion of carbonaceous char. Reaction cooling is accomplished by the endothermic reduction of phosphorus species by carbon. Formation of a stable, glassy, protective mask of polymetaphosphoric acid is considered a mode of action of phosphorus fire retardants. Phosphorus compounds can also be volatile and show effectiveness in the vapor phase.

Halogen additives. Halogens show activity in quenching the chain-carrying free radicals present in the flame (O•, H•, and OH•). In the solid state decomposition zone, halogen acids catalyze char formation OH• (particularly for polyolefins).

Halogen-antimony systems. A chain of reactions leading to volatile antimony halides and oxyhalides provides free-radical quenching in the vapor phase. The end product of the reactions occurring when antimony and halogen are present in the combustion zone is thought to be solid antimony oxide in a finely divided form in the flame. Injecting fine particles into the reaction zone is known to reduce flame propagation rates. Solid state catalysis directs formation of highly stable, intumescent, carbonaceous chars.

Phosphorus-halogen systems. Phosphorus halides and oxyhalides are flame free-radical quenchers and may be more effective than the halogen acids alone. When a single chemical moiety contains both halogen and phosphorus, the relative ease of breakdown of the phosphorus compound enhances the availability of the halogen species.

Phosphorus-nitrogen systems. Phosphorus-nitrogen interaction can alter the course of thermal decomposition to enhance char development. Not all polymers display responses to phosphorus-nitrogen interaction, but composites of plastics with cellulosic materials often show strong positive responses.

Inorganic materials. A number of filler materials of mineral origin contribute volume, structure, and thermal stability to carbonaceous chars once created by other routes. Vaporization of water of hydration is involved in cases like hydrated alumina. The detailed mechanism of the release of water during thermal decomposition of magnesium hydroxide has recently been reported (9). In other additives catalysis of dehydrochlorination reactions may occur. Nanocomposites, for example involving silica, have shown recent promise.

Metals and organometallic additives. A variety of metals and organometallic additives have been used. These range from platinum at the parts per million level in silicones, which involves changing decomposition chemistry to yield less energetic volatiles, to incorporation of ferrocene moieties into the backbone of polyurethanes and other polymers. In the ferrocene case char formation is enhanced without enhancement of smoke (10).

As noted above, many fire retardants have both solid phase and vapor phase activity. Identification of the precise mechanism is difficult given that one must uncover complex flame reactions, or reactions in the solid which are transient in nature or which are in an insoluble solid, degraded substrate or char.

Recent Activity in Fire Retardants

The use of fire retardant chemicals and fire retardant polymers continues to increase. The total consumption of fire retardant chemicals in the United States is increasing at 4-5 percent per year and is expected to reach 837 million pounds in the year 2000 (11). Fire retardant chemicals include alumina trihydrate, antimony oxide, bromine-based, chlorine-based, phosphorus-based additives and comonomers, magnesium hydroxide, and others including boron-based, molybdenum-based, and nitrogen-based additives. Patent activity in the United States shows a broad base of activity from a chemical prospective as follows (12):

	1/1/95-3/31/96	4/1/96-3/31/97	% 2yrs
Alumina Trihydrate	25	14	8
Antimony	12	28	8
Bromine	42	41	16
Chlorine	24	21	9
Phosphorus	70	60	25
Boron	13	3	3
Magnesium Hydroxide	20	10	6
Fluorine	17	16	7
Nitrogen	10	20	6
Silicone	0	11	2
Other	43	10	10
Total Patents	276	234	100%

While there were only 510 patents in the two-year period, it is interesting that halogen and phosphorus each account for 25 percent of patents and that no other single class accounts for 10 percent of patent activity. Work continues on advances in each area. This is as it should be. While fire retardancy is an important property, materials are used for many purposes. The presence of fire retardants can degrade mechanical properties, molding characteristics, electrical properties, color, etc. Even classical halogen or phosphorus fire retardants can thus be enhanced to provide a better, more effective formulation. The remainder of this section will be devoted to recent examples of work in each category.

Bromine compounds. One continues to look for reactive monomers which have high bromine content, good thermal stability, long shelf life, and which are readily usable at normal residence times to achieve high polymerization conversions in commercial plastics processing equipment. For example, pentabromobenzyl acrylate has been reacted in homo- and copolymerization and via reactive processing (13). Pentabromobenzyl acrylate has been reacted in an ethylene-propylene-diene terpolymer (EPDM) to produce a fire retardant concentrate rich in bromine (36 percent). The concentrate was then used in polypropylene. Other applications include glass-reinforced PBT and Nylon 6,6, where it is said to serve as a processing aid as well (14).

Markezich and coworkers have demonstrated that mixed halogens (bromine and chlorine) may be used at lower levels than single halogen additives to achieve needed fire retardancy and thus offer improved physical properties and lower costs (15-16). Effectiveness in polypropylene, polyethylene, HIPS, and ABS has been demonstrated. For example, polypropylene formulations which are UL-94 V-0 (a measure of ignitability) at 0.8 mm thickness require 30 percent of a chlorine compound (19.5% chlorine), 30 percent of a bromine compound (25% bromine), or 25 percent of a mixed halogen additive combination (18.5% halogen). For an ABS formulation at constant total halogen (11%) the oxygen index (a measure of ease of extinction) peaked at 31.3 for equal chlorine and bromine additives versus 25.8 for a chlorine additive alone or 27.3 for a bromine additive alone.

Camino and coworkers have studied the effects of small amounts of polytetrafluoroethylene (PTFE) in ABS fire retarded with a bromine/antimony oxide system (17). PTFE makes ignition more difficult and decreases the rate of flame propagation in the early stages of combustion. The action of PTFE is shown to occur in the condensed phase by accumulation of fluorine in the residue and by reaction of PTFE with antimony oxide to give volatile moieties. Catalytic action of fluorinated species formed in the presence of PTFE is thought to assist reactions of the bromine/antimony oxide system.

Phosphorus Compounds. Ebdon and coworkers have been studying new reactive phosphorus containing monomers for free radical copolymerization with styrene and methyl methacrylate (18). These include various substituted allyldiphenylphosphine oxides.

One issue with monomeric fire retardants can be volatility. For example resorcinol bis(diphenyl phosphate) exhibits lower volatility and higher thermal stability than triaryl phosphates (19-20). Little phosphorus volatilizes in PPO/HIPS blends during thermal decomposition or in combustion with resorcinol bis(diphenyl phosphate). In contrast triphenyl phosphate can be "distilled" unchanged in large percentages from PPO/HIPS blends (21).

McGrath and coworkers have synthesized bis(4-carboxyphenyl)phenylphosphine oxide and bis(4-hydroxyphenyl)phenylphosphine oxide (22-23). These compounds when reacted in Nylon 6,6 and polycarbonate, respectively, show the expected reduction in peak heat release rates as measured in the Cone Calorimeter, for example, from 1200 to 450 kW/m2 for the former at 40kW/m2 external heat flux.

Phosphorus-halogen Compounds. Howell and coworkers have reported on conversion of halogenated triaryl phosphines to the phosphonium salt of 1,3,5-tri(bromomethyl)benzene (24). These salts are well behaved and decompose at higher temperatures than many traditional phosphorus flame retardants (greater than 350 $^\circ$C).

Inorganics, metals and organometallic compounds. Starnes and coworkers have focussed on low-valent metals as reductive cross-linking agents for PVC (25-26). They have shown the preferential reductive coupling of allylic chloride structures when the coupling agent is Cu(0). In PVC the evidence for reductive coupling consists of rapid gel formation accompanied by reductions in the rates of total mass loss, carbon-carbon double bond formation, and HCl evolution. Additives that promote the coupling process are sources of a zero or low-valent metal on pyrolysis. These additives include a number of transition-metal carbonyls, divalent formates or oxalates of the late transition metals, simple Cu(I) halides, and various complexes of Cu(I) containing phosphites or other ligands. Since the reductive coupling agents have low acidities, they are not expected to promote the cationic cracking of chars and are expected to be effective smoke suppressants in PVC.

Shen and coworkers have discussed advances in the use of zinc borates of various compositions and other boron compounds as fire retardants with halogen-containing and halogen-free systems (27-28). Borates not only function as fire retardants, but as afterglow suppressants, smoke suppressants, corrosion inhibitors, and anti-tracking agents.

Antimony trioxide, antimony pentoxide, and sodium antimonate are used with halogen fire retardants. Colloidal antimony pentoxide, in sub-micron particle size, is found to be less detrimental to physical properties in ABS (tensile strength, elongation to break, izod impact), and its non-tinting properties allow lower loading of color concentrates (29).

With antimony oxide price levels showing marked escalation in recent years, efforts to reduce or eliminate antimony oxide in favor of other "synergists" with halogen have been studied. Among the compounds studied have been zinc oxide, zinc borate, zinc stannate, zinc phosphate, zinc sulfide, and iron oxide. Improved thermal stability on processing and improved tracking have been seen (16).

Molybdates have also found some interest (30). A fine-particle ammonium octamolybdate has been introduced as a smoke suppressant for PVC (31).

Chandrasiri and Wilkie have studied the effects of tetrachlorotin, phenyltin trichloride, diphenyltin dichloride, triphenyltin chloride, and tetraphenyltin, on the thermal decomposition of poly(methyl methacrylate) (32). Good Lewis acids, such as tetrachlorotin, may facilitate the formation of non-volatile, cross-linked ionomers that provide stabilization under thermal conditions.

Silicones and silica have shown fire retardant activity in a variety of polymers at low levels (33). Benrashid and Nelson showed that silicones as blocks in polyurethanes provided significant fire retardancy (34-36). The silicone soft block microphase segregated and was preferentially at the surface, providing silicone-like fire performance for a 70 percent polyurethane polymer.

Fire retardant polymer systems do not necessarily give identical performance. For a study of state-of-the-art fire retardants in polyurethanes using the Cone Calorimeter, see ref. 37.

Commercial and Regulatory Issues at Work for Change of Fire Retardants

Fire is a safety issue. Fire is a toxicity issue. Fire is an environmental issue. Fire is a world-wide issue. It should thus not be surprising that external forces play a significant role in the continuing evolution of the use of fire retardant polymers. The level of fire retardancy required may not be static. Concerns get raised over the toxicity of particular materials or compounds in a product, or in a fire. Increasingly one is looking to recycle materials in applications at the end of life. What is the impact of particular fire retardants on the recyclability of polymers? Such issues generate lots of opinions. Whether in response to good science, or the lack of it, changes are initiated. A lot of the change in fire retardancy and in the use of fire retardants is generated by these commercial "pressures" (38-40).

Clearly there are some applications where concern for fire safety is a prime concern (example: a spacecraft interior). In others there is minimal concern (example: consumer good packaging). One application where there is increased concern for fire retardancy is television sets manufactured for sale in Europe (41). There are 217 million TVs in 140 million households in Europe, with sales of 22-23 million new TVs per year. In 1996 there were an estimated 2200 TV fires in Europe or 12.2 TV fires per million TV sets. There were 16 deaths and 197 injuries per year. With the voluntary

382 G.L. Nelson

use of fire retardant plastics TV fires had fallen about 50 percent since the mid-1980s, but recent cost reduction efforts and environmental issues have caused a substantial reduction in the use of fire retardant enclosures. A TV fire is defined as a home consumer incident fire where the first point of ignition is within the structure of a TV or ancillary equipment. The first ignition may be smoldering, a flame, explosion (or implosion), or damage that causes the TV to cease to function. The resultant fire must breach the enclosure of the TV, or emit noxious fumes, or produce molten droplets that breach the enclosure to be included in the data.

In a specific country, the United Kingdom, the number of television fires increased from 436 to 693 from 1992 to 1997. On the whole, deaths resulting from domestic fires in the UK, which fell sharply after legislation controlling the flammability of upholstered furniture was introduced in 1988, have started to rise again. In 1994 there were 490 deaths caused by domestic fires in Britain; in 1997, 560 deaths (42).

Changes have been proposed to IEC 60065, the international standard for TVs, to increase the fire performance of TV sets through incorporation of good engineering practice in design and production to avoid potential ignition sources, through the use of materials of low flammability for internal parts in the vicinity of potential ignition sources, and through the use of a fire enclosure to limit the spread of flame. A French company has indicated its preference in design for identifying high risk areas, then designing component layout so as to contain any potential fire that might occur. Components are set apart from each other to prevent fire spread. The use of fire retardant materials is thought by them to be environmentally unfriendly and thus a last option. A fire started by an outside source would, however, not be stopped in the French approach. Likewise, in Germany, some believe that halogen fire retardants give off small amounts of toxic gases when heated and that their presence in plastics renders them unsuitable for conventional recycling. For these reasons the use of fire retardants is increasingly resisted in Germany. Past voluntary use of fire retardant resins has given way to non-fire retardant UL-94 HB resins for "environmental" and cost reduction reasons. However, in Germany 20-30 percent of TV fires are due to candles or accelerants, or caused by an external source that then spreads to the TV. And interestingly, in the Netherlands the average TV fire causes 50 percent more damage than the average house fire.

The origin of the environmental concern mentioned above has its beginning in 1986 when a Swiss scientist reported the formation of polybrominated di-benzo-p-dioxins and dibenzofurans when polybrominated diphenyloxides were burned, generating a public outcry in Germany after articles appeared in a major news magazine (43-46). Decabromodiphenyl oxide was the major fire retardant used in polystyrene TV enclosures. More than a decade and a lot of work later shows that the two largest fire retardants, decabromodiphenyl oxide and tetrabromobisphenol-A, do not present significant environmental threats (47). Now, given 2200 TV fires in Europe, the pressure for increased safety may necessitate the use of fire retardant TV enclosures and the increased use of fire retardant components. This would follow current practice in the United States. When adopted in the United States in the late 1970s the use of ignition resistant (UL-94 V-0) enclosures (among other changes) led to a 10 to 100 fold reduction in TV fire deaths and injuries.

Similarly, in the United States the US Consumer Product Safety Commission is currently concerned that there are thousands of portable appliance fires reported each year (48). Many are unattended products. Internal components fail and ignite non-fire retardant plastic enclosures. Examples include electric heaters, fans, monitors, and coffeemakers. The kinds of applications are those listed by Underwriters Laboratories. UL has been provided examples of incidents by CPSC. From those examples it would appear that design flaws and misuse of materials also play a role. A similar concern expressed by CPSC to UL in the mid-1980s resulted in design upgrades but not the use of ignition resistant plastics (UL-94 V-2, V-1, V-0, or 5V). The approach now offered by CPSC is the increased use of fire retardant materials in enclosures via voluntary standards. CPSC has had discussions with UL, with the Society of the Plastics Industry, and appliance manufacturers. In response to the issue UL formed an adhoc group under the UL 746 industrial advisory group on "Risk of Fire Hazard in Electrical/Electronic Enclosures." Much of the industry discussion was on "hazard based analysis." There is agreement that the concern is the appliance being the source of fire rather than the appliance being the second item of ignition. The CPSC staff have proposed that UL-94 V-0 be a starting place for enclosures and that most appliances be considered as unattended. After much discussion UL has proposed for household electrical appliances that for plastics within 3mm of connections not welded in place that UL-94 V-0 or 94 V-1 plastics be used or plastics that pass the glow wire test at 750°C (30 seconds). For plastics bearing components a ball pressure test at 125°C would be used. Such changes when approved will take several years to implement.

Personal experience shows that large amounts of non-fire retardant plastics, given ignition, can present a significant hazard. An amount under ten pounds is sufficient, if ignited, to cause flashover of a home-sized room. In the United States televisions and computers are key examples of the use of fire retardant enclosures. The concern of CPSC related to household appliances is an example of regulatory analysis which led in the late 1980s to better electronic design, but which in the late 1990s finds that the use of fire retardant materials is also required to achieve an acceptable level of risk.

Materials are used in applications for well-defined reasons. For example, plastics used in electrical and electronic applications for enclosures must resist impact and heat as well as serve their intended engineering function. In the world marketplace there has recently come about the need for "environmentally preferable" products, products which have less impact upon the environment in manufacture and life-cycle than competitive products (38-40). Environmentally preferable products generally include a requirement for recyclability. The elements of design that must be considered include ease of product disassembly, labeling of each part for material identification, minimization of the number of materials to maximize the single material content in the resulting recycle stream, ease of reuse or refurbishment of the equipment or components, maintenance of properties of materials on recycling, and compliance with "eco-label" requirements as well as applicable safety and regulatory requirements. In fact eco-labels have become an important element in the reuse-recycle equation. Eco-labels are issued by private, semi-regulatory, or regulatory bodies. They are found around the world, but particularly in Europe. General features of eco-labels for electrical equipment, as they affect the use of plastics, are design for disassembly, consolidation of resins, absence of contaminants (for example, molded-in inserts), absence of certain heavy metals (lead, and cadmium), absence of certain fire retardants, and labeling of plastics materials used for parts in the manufacture of components. In September of 1991, the German environmental agency launched an eco-label for copiers called "Blue Angel." By mid-1992, almost every copier manufacturer had at least one copier with this eco-label. Requirements for personal computers and printers have also been issued. Largely based on concerns related to incineration in inefficient incinerators, provisions have specified that there be no use of polybrominated biphenyls (PBBs) or polybrominated diphenyl ethers (PBDEs) and "flame retardant materials used must have no carcinogen- or dioxin- or furane-forming effect." This latter has its origins in the concerns over dioxins and furans in Germany mentioned earlier.

Requirements for environmental labels for computers have also been issued in Sweden by TCO, the Swedish Confederation of Professional Employees, in cooperation with the Swedish Society for Nature Conservation and NUTEK, the Swedish National Board for Industrial and Technical Development. TCO-95 requirements included keeping the variety of plastics used to a minimum and labeling plastic components for identification. Plastic parts that weigh more than 25 grams "may not contain organically bound chlorine or bromide." This latter is much broader than Blue Angel. For the Nordic environmental label, "White Swan," the requirements for personal computers also specify that large plastic parts (over 25g) may not be painted. Requirements for eco-labels continue to evolve.   Requirements For TCO-99 and the Nordic Ecolabel-98, White Swan, exclude PVC and brominated fire retardants in housings for computer applications. Blue Angel has repeated the halogen exclusion for plastic housings for printers and copiers. The draft of a European Ecolabel, EU-99, for computers excludes brominated fire retardants in housings. Such voluntary labels have found marketplace importance. Over 300 computer related products have been listed to TCO-95 (49).

The exclusions of PVC and of brominated fire retardants have little technical justification. The European brominated chemicals industry has filed a formal complaint with the European Union that such restrictions, despite being voluntary labels, are a barrier to trade and improper under EU rules.   Indeed, in product tenders labeled products have been specified. Fire incident analyses and government evaluations increasingly show no justification for restrictions on the use of PVC in buildings and products. Despite the lack of technical justification, if materials become part of someone's "green" agenda, such requirements lead to changes in the materials used. Given product safety requirements, that may mean using different fire retardant systems (replacing bromine by phosphorus fire retardants, for example).   If fire performance requirements permit non-fire retardant plastics, redesign may occur to use non-fire retardant materials.

Eco-labels are voluntary, but particularly in Europe, increasingly there are "product take-back" requirements. The advent of such requirements already serves as a driving force for design for disassembly and for the development of recycling schemes. On 11 July 1991, the German Ministry for the Environment published a "Draft Ordinance on the avoidance, reduction, and utilization of waste from used electrical and electronic appliances" (Electronic Scrap Ordinance), which brought forth comments on practical collection and recycling schemes (50). A second draft was issued on 15 October 1992 (51). A similar proposal was developed in the Netherlands. The proposed take-back requirement in Germany was broadly based and included everything from small household appliances and office equipment to large industrial items. After the date of effectiveness of the regulation the seller would have the duty to take back used equipment from the end user without charge. The manufacturer would have the responsibility to reuse and recycle the returned equipment.

Because such requirements have broad implications, the European Union accepted a proposal from Orgalime, a liaison group of European engineering industries, to develop a guideline for the "electronic waste" stream (52). A working group under the auspices of the European Union and the Italian government began meeting in 1993. The finalization of take-back requirements awaited completion of acceptable guidelines. In late 1995 a voluntary take-back and recycling program was advanced in Germany by German industry. In the Netherlands it was proposed that such recycling efforts be supported financially by charging a tariff on new electronic equipment or introducing a return premium for old equipment. In June 1997 the Swedish government proposed a producer responsibility law covering electrical and electronic products encompassing every stage of the distribution chain: everyone who manufactures, imports, or sells electrical/electronic products (53). The producers must see to it that there is a functional collection system including transportation. They are obliged to accept, free of charge, used products that consumers submit for collection.   This proposal was submitted (the process called official notification) to the EU Commission in Brussels.

In early 1998, the EU (DG XI) itself issued a draft directive on "Waste from Electrical and Electronic Equipment (WEEE)." This take-back directive included numerous prescriptive requirements on design for recycling and exclusion of specific materials including halogenated fire retardants, requirements similar to those found in the eco-labels discussed above. A second draft was issued in July 1998. As of this writing a final draft may be submitted to the Council of Ministers for implementation in 2000. The directive would seriously challenge industry, particularly as member states add their own interpretations.   Despite scientific/technical concerns about the justification of requirements in eco-labels, the Directive may codify the same or more severe requirements.

Take-back programs are of course not just limited to electrical/electronic products but encompass the general notion of "producer responsibility." Producers who put a product (including packaging) into the marketplace must be responsible for taking back as much as sold. Sweden, for example, has a general product responsibility law covering packaging, waste paper, tires, automobiles, and furniture as well as electrical/electronic products (53). The effect on all of this on plastics and on fire retardants is that take-back requirements will accelerate the design of products for recyclability and materials consolidation (use of fewer materials) as well as encourage the use of materials that do not have environmental issues of their own or complicate or are perceived to complicate the recycle stream. All of this is not to say that fire retardant plastics are not recyclable. IBM has investigated the mechanical, electrical, thermal and flammability performance of plastics in closed-loop recycling when computers are returned from the field (54-55). Results indicated that the extent of degradation of materials is negligible. Researchers at General Electric Plastics and Dow have also shown that fire retardant plastics can go through multiple molding cycles without serious property degradation (56-58). For such materials as polycarbonate, modified poly(phenylene oxide), and ABS it has been found that fire retardancy is also little affected by recycling.

In an allied issue, does the presence of different fire retardants cause resin compatibility problems in recycle streams? Does the presence of fire retardants complicate or hinder non-fire retardant applications for recycled plastics? In regions of the world with insufficient local volume (example, Scandinavia) mixed resin blends may be needed (59). Fire retardant materials could potentially complicate mixed resin recycle programs. While we have seen above that mixed halogen fire retardants may actually improve fire retardancy, some fire retardants may not be compatible.

Despite considerable regulatory activity during the decade of the 90's, particularly in Europe, as of this writing there are no regulations in effect which directly ban the use of brominated fire retardants (other than PBBs) anywhere in the world (47). The EU Waste Electrical/Electronic Equipment Directive would be a first, if adopted. Indirect regulations, however, do exist. In 1993 Germany proposed amending its existing Hazardous Substance Ordinance to include the regulation of brominated dibenzodioxins and furans. The previous regulations specified maximum levels of chlorinated dibenzodioxins and furans that can be present in materials marketed in Germany. The amendment added eight 2,3,7,8-substituted brominated dioxins and furans to the regulation. On 15 July 1994 a combined ordinance was published in the official German journal as an amendment to the Chemicals Banning Ordinance. This regulation affects brominated fire retardants only to the extent that greater-than-allowed limits of brominated dioxins or furans are present as impurities or degradation products. Decabromodiphenyl oxide (DBDPO or DBDPE) and tetrabromobisphenol-A (TBBPA), the two largest volume brominated fire retardants, meet the requirements of the regulation as a material in commerce. Other brominated fire retardants have been tested and meet the requirements of the ordinance. One fire retardant supplier has supplied compliance data on fire retardants in generic resins such as high-impact polystyrene (HIPS), poly(butylene terephthalate) (PBT), acrylonitrile-butadiene-styrene (ABS), and poly(ethylene terephthalate) (PET). On the negative side, there is one published study on the formation of polybrominated dibenzodioxins (PBDDs) and polybrominated dibenzofurans (PBDFs) that notes that when polybrominated diphenyl ether (PBDE) containing material are processed, the dioxin and furan levels exceed the final limits of the ordinance of 1 and 5 parts per billion (pbb) for PBDDs and PBDFs (effective 15 July 1999) (60). Currently the ordinance does not require testing, does not specify the analytical methods to be used, establish sampling protocols, or give a reporting mechanism for compliance. Are there fire retardants in use which do exceed the limits in processed resins? What happens on recycling? What happens on abusive molding. Likewise, in the European Union directive on hazardous waste of 12 December 1991 (91/689/EWC), antimony compounds were mentioned in the annex describing substances that classify a waste as being hazardous. How does this apply to fire retardant plastics?

Polyvinyl chloride (PVC) is a widely used material, particularly in the building and construction industry for pipes and fittings, rigid profiles, film and sheet, and wire and cable. There have been allegations about its environmental performance during production, use, and disposal. In Europe alone PVC finished goods are estimated to have a value of $46.1 billion (61). Little technical support, however, has emerged related to the issues raised by certain environmental groups. Such lack of support has not deterred those groups. PVC is low cost. Its high chlorine content as discussed above makes it intrinsically fire retardant. Its replacement by other commodity resins would likely be with materials of significantly less fire retardancy.

Cost reduction, recycling, and environmental preference lists have put pressure on specific fire retardants and on the use of fire retardant resins. Fire retardant resins cost 20 to 40 cents per pound more than corresponding non-fire retardant resins. The absence of fire retardants may make recycling easier as well as make compliance with environmental lists easier. The lack of use of fire retardants will likely ignore, however, the fire hazard of products from fire sources external to the product and expose consumers to potentially life threatening fires from sources as small as a match or candle.

Future Trends

As one looks to the 21st century, fewer resins may be used in products to accommodate recycling, and those used may be broader in their property profiles. Limited-volume tailored resins may no longer be used because of the need to increase volumes of recycle streams.

Fire retardant resins will need to be more robust and processible under more stringent molding conditions. Thin wall molding, for example, requires better flow at higher pressures and shorter machine cycle times to yield equivalent or higher mechanical properties in thinner part cross sections (62). Fire retardancy will need to be met in thinner wall sections. Changes in resin physical properties by use of a fire retardant will need to be less. More efficient fire retardants will be needed.

Fire retardants and fire retardant resins which are free of environmental issues will be in demand. Issues are currently with brominated fire retardants. Issues two decades ago were with phosphorus fire retardants (63). The presence of fire retardants does not necessarily increase toxic hazards in a fire, does not increase toxics on incineration, does not present hazards in use. But substantive data will be needed, otherwise facts may be overshadowed by environmental fears fueled by groups with a specific agenda.

Fire retardant resins will need to be recyclable with minimal degradation of physical properties and fire performance after multiple molding cycles. There will be the need for more facile identification of resins as well as the specific fire retardants they contain. Perhaps tagging will be needed. Research on the effects of fire retardants on mixed resin recycling will be needed.

As chemical analytical techniques improve more work is needed to elucidate the details of fire retardant mechanisms. One can expect this work to lead to more efficient fire retardants.

One promising research area is the effect of metals (metal compounds) on the degradation pathways of specific polymer structures. This approach offers enhanced char, diversion of high energy volatiles, and thus less smoke evolution and lower heat release rates for the target polymers.

Finally, more versatile "commodity" fire retardants will be needed for a market dominated by changing environmental standards and knowledge. And of course customers always want higher performance at lower cost.

<u>Literature Cited</u>

1. Craver, J. K.; Tess, R. W. "Applied Polymer Science"; Organic Coatings and Plastics Division, American Chemical Society: Washington, D.C., 1975.
2. For a discussion of polymers and markets see: Alper, J.; Nelson, G. L. "Polymeric Materials - Chemistry for the Future"; American Chemical Society: Washington, D.C.,1989, 110.
3. For the latest data see: National Fire Programs, United States Fire Administration, http://www.fema.gov.
4. Nelson, G. L. <u>Chemistry</u>, 1978, <u>51</u>, 22-27.
5. Nelson, G. L. Intern. <u>J. Polymeric Mater</u>, 1979, <u>7</u>, 127-145.
6. Troitzsh, J. "International Plastics Flammability Handbook"; Hansen: Munchen, Germany, 1983, 11.
7. Nelson, G. L. in "Fire and Polymers II, Materials and Tests for Hazard Prevention"; Nelson, G. L., Ed.; American Chemical Society: Washington, D. C., 1995, Chap. 1.
8. Nelson, G. L.; Kinson, P. L.; Quinn, C. B. <u>Annual Review of Materials Science</u>, 1974, <u>4</u>, 391-414.
9. L'vov, B. V.; Novichikhin, A. V.; Dyakov, A. O. <u>Thermochimica Acta</u>, 1998, <u>315</u>, 135-143.
10. Najafi-Mohajeri, N; Nelson, G. <u>Polymer Preprints</u>, 1998(2), <u>39</u>, 380-381.
11. "Flame Retardant Chemicals: How Much? Where? Why?"; Business Communications Company, Inc.: Norwalk, CT, 1996.
12. Hilton, R. D. "Proceedings The Eighth Annual BCC Conference on Flame Retardancy"; Business Communications Company, Inc.: Norwalk, CT, 1997.
13. Utevskii, L; Scheinker, M; Georlette, P; Reyes, J., D. "Proceedings The Eighth Annual BCC Conference on Flame Retardancy"; Business Communications Company, Inc.: Norwalk, CT, 1997.
14. "Additives- Non-halogens Hike HIPS Processability," <u>Modern Plastics</u>, 1998, <u>75(9)</u>, 92.
15. Markezich, R. L.; Aschbacher, D. G. in "Fire and Polymers II: Materials and Tests for Hazard Prevention"; Nelson, G., Ed.; American Chemical Society: Washington, D.C., 1995, Chap. 5.
16. Markezich, R. L.; Mundhenke, R. F. "Proceedings of the International Conference on Fire Safety"; Product Safety Corporation: Sissionville, WV, 1998, <u>25</u>, 344-354.

17. Roma, P; Camino, G.; Luda, M. P. <u>Fire and Materials</u>, 1997, 21(5).
18. Ebdon, J. R.; Joseph, P.; Hunt, B. J.; Price, D.; Milnes, G. J.; Gao, F. "Proceedings The Eighth Annual BCC Conference on Fire Retardancy"; Business Communications Company, Inc: Norwalk, CT, 1997.
19. Murashko, E. A.; Levchik, G. F.; Levchik, S. V.; Bright, D. A.; Dashevsky, S. "Fire Retardant Action of Resorcinol Bis(Diphenyl Phosphate in a PPO/HIPS Blend"; <u>J. Fire Sciences</u>, 1998, <u>16</u>, 233-249.
20. Murashko, E. A.; Levchik, G. F. ; Levchik, S. V; Bright, D. A.; Dashevsky, S. <u>J. Fire Sciences</u>, 1998, <u>16</u>, 278-296.
21. Carnahan, J.; Haaf W.; Nelson, G.; Lee, G.; Abolins, V.; Shank, O. "Proceedings of the 4th International Conference on Fire Safety"; (University of San Francisco, Jan. 15-19, 1979) 1979, <u>4</u>, 312-323.
22. Wan, I; McGrath, J. E.; Kashiwagi, T. in "Fire and Polymers II: Materials and Tests for Hazard Prevention"; Nelson, G., Ed.; American Chemical Society: Washington, D.C., 1995, Chap. 2.
23. Knauss, D. M.; McGrath, J. E.; Kashiwagi, T. in "Fire and Polymers II: Materials and Tests for Hazard Prevention"; Nelson, G., Ed.; American Chemical Society: Washington, D.C., 1995, Chap. 3.
24. Howell, B. A.; Uhl, F. M; Liu, C.; Johnston, K. F. "Proceedings The Eighth Annual BCC Conference on Flame Retardancy"; Business Communications Company Inc.: Norwalk, CT, 1997.
25. Jeng, J. P.; Terranova, S. A.; Bonaplata, E.; Goldsmith, K.; Williams, D. M.; Wojciechowski, B. J.; Starnes, Jr., W. H. in "Fire and Polymers II: Materials and Tests for Hazard Prevention"; Nelson G., Ed.; American Chemical Society: Washington, D.C., 1995, Chap. 8.
26. Pike, R. D.; Starnes, Jr., W. H.; Adams, C. W.; Bunge, S. D.; Kang, Y M.; Kim, A. S.; Kim, J. H.; Macko, J. A.; O'Brien, C. P. "Proceedings The Eighth Annual BCC Conference on Flame Retardancy"; Business Communications Company, Inc.: Norwalk, CT, 1997.
27. Shen, K. K.; Ferm, D. "Proceedings The Eighth Annual BCC Conference on Flame Retardancy"; Business Communications Company, Inc.: Norwalk, CT, 1997.
28. Shen, K. K.; Ferm, D. J. "Proceedings of the International Conference on Fire Safety"; Product Safety Corporation: Sissionville, WV, 1998, <u>25</u>, 288-302.
29. Bartlett, J. "Proceedings The Eighth Annual BCC Conference on Flame Retardancy"; Business Communications Company, Inc.: Norwalk, CT, 1997.
30. Cook, P. M.; Musselman, L. L. "Proceedings The Eighth Annual BCC Conference on Flame Retardancy"; Business Communications Company, Inc.; Norwalk, CT, 1997.
31. "Additives-Smoke-suppressant Tames PVC"; <u>Modern Plastics</u>, 1998, <u>75(9)</u>, 92.
32. Chandrasiri, J. A.; Wilkie, C. A. in "Fire and Polymers II: Materials and Tests for Hazard Prevention"; Nelson, G., Ed.; American Chemical Society: Washington, D.C., 1995, Chap. 9.
33. Pape, P. G.; Romenesko. "Proceedings The Seventh Annual BCC Conference on Flame Retardancy"; Business Communications Company, Inc.: Norwalk, CT., 1996.

34. Benrashid, R.; Nelson, G. L.; Linn, J. H.; Hanley, K. H.; Wade, W. R. J. Appl. Polym. Sci., 1993, 49, 523-537.
35. Benrashid, R.; Nelson, G. L. in "Fire and Polymers II: Materials and Tests for Hazard Prevention"; Nelson G., Ed.; American Chemical Society: Washington, D.C., 1995, Chap. 14.
36. Benrashid, R.; Nelson, G. J. Polym. Sci., 1994, 32, 1847-1865.
37. Nelson, G. L.; Jayakody, C. "Proceedings Fire Retardant Chemicals Association"; 1998 Fall Conference 1-28.
38. Nelson, G. L. In "Fire and Polymers II: Materials and Tests for Hazard Prevention", Nelson, G., Ed.; American Chemical Society: Washington, D. C., 1995; Chap. 35.
39. Nelson, G. L. CHEMTECH 1995, 25(12), 50-55.
40. Nelson, G. L. in "Fire Retardancy of Polymeric Materials"; Grand, A. F.; Wilkie, C. A Eds.; Marcel Dekker, Inc.: New York, 1999. Chap. 1.
41. "TV Fires (Europe)"; Department of Trade and Industry, Sambrook Research International: Newport, Shropshire, 14 March 1996, p. 98.
42. Nuki, P.; Winhett, R. "Cheap TVs Blamed in Spate of Fires"; The Sunday Times, 6 December 1998.
43. Buser, H. R. Environ. Sci. Technol., 1986, 20(4), 404-408.
44. "Dioxin-Dangerous Protection Against Fire"; Der Stern, 13 November 1986, 250-253.
45. "Is Watching Television Toxic?"; Der Stern, 14 April 1989.
46. "Exiting from the Poison"; Der Stern, 20 April 1989 (17).
47. Hardy, M. L. "Proceedings The Eighth Annual BCC Conference on Flame Retardants"; Business Communications Company, Inc.: Norwalk, CT, 1997.
48. Hoebel, J. F. "Proceedings The Eighth Annual BCC Conference on Flame Retardants"; Business Communications Company, Inc.: Norwalk, CT, 1997.
49. For ecolable websites see: Nordic Ecolabelling (WhiteSwan),http://www.sis.se/Miljo/Nordic/Nordic0.htm; TCO, http://www.tco info.com/tco95.htm; EuropeanEcolabelProgram, http://www.europa.eu.int/en/comm/dg11/ecolabel/index.html.
50. "Draft Regulation on the Avoidance, Reduction, and Utilization of Wastes From Used Electric and Electronic Equipment (Electronic Waste Regulation)"; German Federal Ministry for the Environment, Nature Conservation and Nuclear Safety: 1991; WA II 3-30 114-5.
51. "Regulation Regarding the Avoidance, Reduction and Recycling of the Waste/Used Electric and Electronic Equipment (Electronic Scrap Regulation)"; German Federal Ministry for the Environment, Protection of the Environment and Reactor Safety: Working Paper as of 15 October 1992; WA II 3-30 114/7.
52. Orgalime, Guidelines for the Working Group; "Priority Waste Stream on Electronic Waste"; 24 January 1994.
53. Jederlund, L. "Time to Do Away With Refuse", Current Sweden No. 417, September 1997, Svenska Institute, Stockholm, Sweden.
54. Appliance Manufacturer, 1994, 43, 55.
55. Kirby, R.; Wadehra, I. L. "1994 Fall Conference Proceedings"; Fire Retardant Chemicals Association: Lancaster, PA, 1994, 145-50.

56. Van Riel, H. C. H. A.; "1994 Fall Conference Proceedings"; Fire Retardant Chemicals Association: Lancaster, PA, 1994, 167-74.
57. Bopp, R. C. "1994 Fall Conference Proceedings"; Fire Retardant Chemicals Association: Lancaster, PA, 1994, addendum.
58. Christy, R.; Gavik, R. "1994 Fall Conference Proceedings"; Fire Retardant Chemicals Association: Lancaster, PA, 1994, 151-66.
59. Klason, C. "Recycling of Polymeric Materials"; Chalmers University of Technology, Sweden, April 1993.
60. Meyer, H.; Neupert, M.; Pump, W.; Willenberg, B. Kunststoff German Plastics, 1993, 83(4), 3-6.
61. Fact File, The European PVC Industry, Website http://www.pvc.org/factfile/euroindu.htm.
62. Tremblay, G. Plastics News, April 6, 1998, p. 6.
63. Petajan, J. H.; Voorhees, K. J.; Packham, S. C.; Baldwin, R. C.; Einhorn, I. N.; Grunnet, M. L.; Dinger, B. G.; Birky, M. M. Science, Feb. 28, 1975, 187(4178), 742-744.

EPOXY RESINS

J. L. MASSINGILL, JR. and R. S. BAUER[*]

Coatings Research Institute, Eastern Michigan University, Ypsilanti, MI 48197

Introduction

     The name epoxy resins has over the years become synonymous with performance; epoxy resins have established themselves as unique building blocks for high-performance coatings, adhesives, and reinforced plastics. Epoxy resins are a family of monomeric or oligomeric material that can be further reacted to form thermoset polymers possessing a high degree of chemical and solvent resistance, outstanding adhesion to a broad range of substrates, a low order of shrinkage on cure, impact resistance, flexibility, and good electrical properties.

[*]Deceased

Figure 1. U. S. Epoxy Resin Production

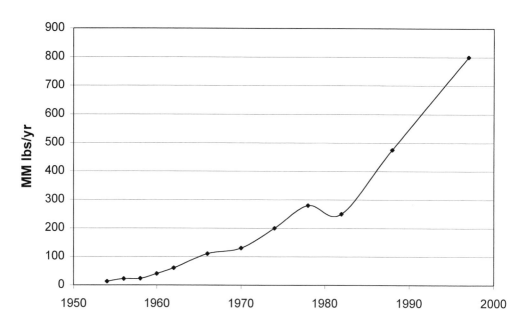

Figure 2. Global Epoxy Resin Markets 1997
1.5 BillionPounds

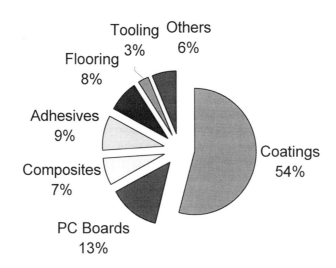

Introduced commercially in the United States in the late 1940s, epoxy resins had reached annual sales of about 13 million-lb by 1954, about 319 million-lb by 1981, and about 800 million-lb in 1997(1) (Figure 1).   Global production was estimated at 1.5 billion. The global markets in 1997 for epoxy resins were almost equally divided between protective coatings and structural end uses (Figure 2).

Resin Types and Structure

Epoxy resins are compounds or mixtures of compounds that are characterized by the presence of one or more epoxide or oxirane groups:

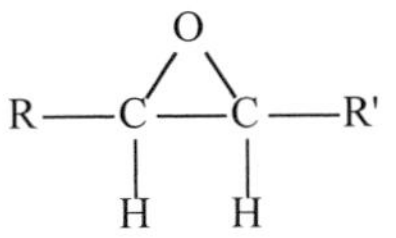

There are three major types of epoxy resins: cycloaliphatic epoxy resins (R and R' are part of a six-membered ring), epoxidized oils (R and R' are fragments of an unsaturated compound, such as oleic acid in soybean oil), and glycidated resins (R is hydrogen and R' can be a polyhydroxyphenol, a polybasic acid, or a polyol). The first two types of epoxy resins are obtained by the direct oxidation of the corresponding olefin with a peracid as illustrated by the following:

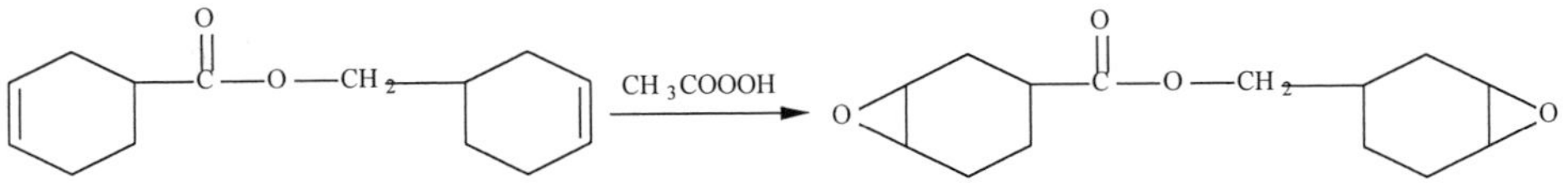

By far the most commercially significant of these resins, however, are the ones obtained by glycidation of bisphenol A with epichlorohydrin:

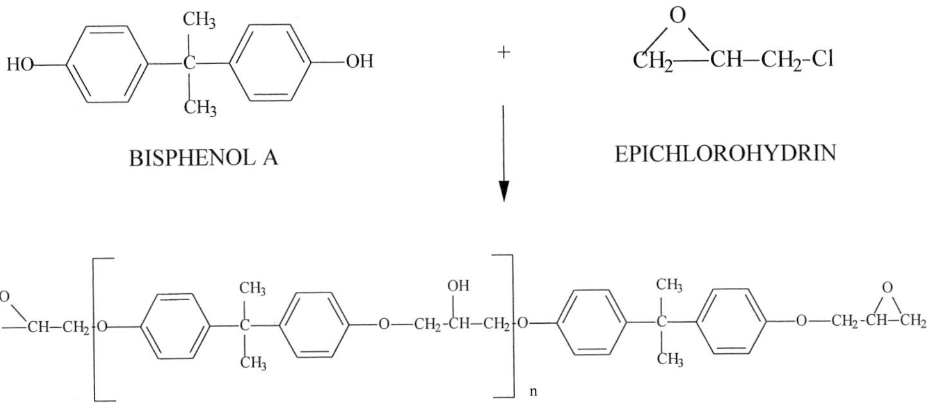

BISPHENOL A   +   EPICHLOROHYDRIN

Typical Bisphenol A Epoxy Resin

Table 1. Continuum of Epoxy Properties

% Epoxide	0.1	1.4	1.9	2.5	5	10	15	20	25
EEW	50,000	3000	2300	1720	860	430	287	215	172
M_n	25,000	6000	4600	3440	1720	860	574	430	344
N Value	50+	25		10.9	4.9	1.8	0.8	0.3	0
Softening Point, °C	150	140	130	100	80	40			
	9-type	8-type	7-type	4-type	2-type	1-type			

Phenoxy	Very Flexible Heat Converted Wire enamel	Flexible Moisture Resistant Heat Converted Can Coatings	Non-sintering Easily Powdered Good Melt Flow Heat Converted Powder Coatings Epoxy esters	High Solids Ambient cure............ Electrical Laminates Maintenance Coatings Marine Coatings CED	Liquid Highest XL Density General Purpose Coatings, Adhesives Laminating, Casting
Lacquer Thermoplastic					

Bisphenol Epoxy Resins

As can be seen from the structure given for a typical bisphenol A based epoxy resin, a spectrum of products is available. Commercial resins are generally mixtures of oligomers with the average value of n varying from essentially 0 to approximately 25. Available products range from low viscosity, low molecular weight epoxy resins all the way up to hard, tough, higher molecular weight phenoxy lacquer resins.

A list of the more common epoxy resins arranged in order of increasing molecular weight is shown in Table I. Although the structure drawn for epoxy resins depicts them as diepoxides, commercial epoxy resins are not 100% diepoxides. Other end groups can result from the manufacturing process, such as glycols derived from hydration of epoxide groups, unconverted chlorohydrin groups, and phenolic end groups from unreacted terminal bisphenol A molecules. The lower molecular weight resins may have a functionality of greater than 1.9 epoxides per molecule.

Reactions of Epoxides and Curing Mechanisms

Epoxy resins are reactive intermediates that, before they can be useful products, must be "cured" or cross-linked by polymerization into a three-dimensional infusible network with co-reactants (curing agents). Cross-linking of the resin can occur through the epoxide or hydroxyl groups, and proceeds basically by only two types of curing mechanisms: direct coupling of the resin molecules by a catalytic homopolymerization, or coupling through a reactive intermediate. Reactions used to cure low molecular weight epoxy resins occur with the epoxide ring:

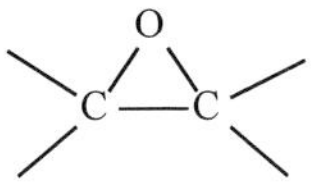

The capability of this ring to react by a number of paths and with a variety of reactants gives epoxy resins their great versatility. The chemistry of most curing agents currently used with epoxy resins is based on polyaddition reactions that result in coupling as well as cross-linking. The more widely used curing agents are compounds containing active hydrogen (polyamines, polyacids, polymercaptans, polyphenols, etc.) that react as shown in Reaction 1 to form the corresponding β-hydroxy -amine, ester, mercaptan, or β-phenyl ether.

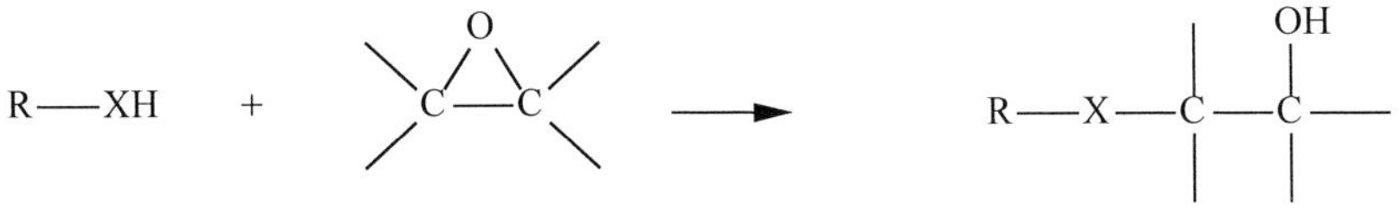

$$
R\!-\!XH \;+\; \text{(epoxide)} \;\longrightarrow\; R\!-\!X\!-\!C\!-\!C\!- \qquad (1)
$$

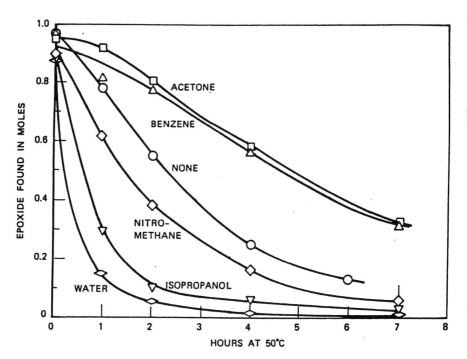

Figure 3.   Influence of reaction medium of amine-glycidyl ether reaction.

Epoxy resins and curing agents usually contain more than one reaction site per molecule, and the process of curing to form a three-dimensional network results from multiple reactions between epoxide molecules and curing agent. The specific reactions of the various reactants with epoxide groups have, in many cases, been studied in considerable detail and have been extensively reviewed elsewhere (2).

Addition Reactions of Amines

All amine functional curing agents, for example, aliphatic polyamines and their derivatives, modified aliphatic amines and aromatic amines, react with the epoxide ring by an addition reaction without formation of by-products. Shechter et al. (3) suggested that the amine addition reactions proceed in the following manner:

$$R-NH_2 \quad + \quad \underset{\displaystyle CH_2-CH-}{\overset{\displaystyle O}{\triangle}} \quad \longrightarrow \quad R-\underset{\displaystyle}{\overset{\displaystyle H}{N}}-CH_2-\underset{\displaystyle}{\overset{\displaystyle OH}{CH}}- \qquad (2)$$

$$R-\overset{H}{N}-CH_2-\overset{OH}{CH}- \quad + \quad \overset{O}{\triangle}_{CH_2-CH-} \quad \longrightarrow \quad \begin{array}{c} H\overset{}{C}-OH \\ | \\ CH_2 \quad OH \\ | \quad | \\ R-N-CH_2-CH- \end{array} \qquad (3)$$

$$-\overset{OH}{CH}- \quad + \quad \overset{O}{\triangle}_{CH_2-CH-} \quad \longrightarrow \quad -CH-O-CH_2-\overset{OH}{CH}- \qquad (4)$$

Subsequently, other workers including O'Neill and Cole (4) and Dannenberg (5) showed that Reactions 2 and 3 proceed to the exclusion of Reaction 4. The reactivity of a particular epoxide amine system depends on the influence of the steric and electronic factors associated with each of the reactants. It has been known for some time that hydroxyls play an important role in the epoxide-amine reaction. For example, Shechter et al. (3) studied the reaction of diethylamine with phenylglycidyl ether in concentrated solutions. They showed that acetone and benzene decreased the rate of reaction in a manner consistent with the dilution of the reactants, but that solvents such as 2-propanol, water, and nitromethane accelerated the reaction (Figure 3). They also found that addition of one mole of phenol to this reaction accelerated it more than addition of 2-propanol or water.

The "modest" acceleration of the amine-epoxide reaction by nitromethane was ascribed to the influence of the high dielectric constant of the solvent. The greater influence of hydroxyl-containing compounds in accelerating this reaction has been suggested to result from the formation of a ternary intermediate complex of the reactants with hydroxyl-containing material, such as that proposed by Smith (6) or Mika and Tanaka (7):

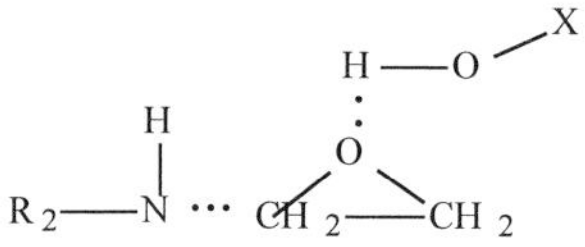

SMITH   (5)

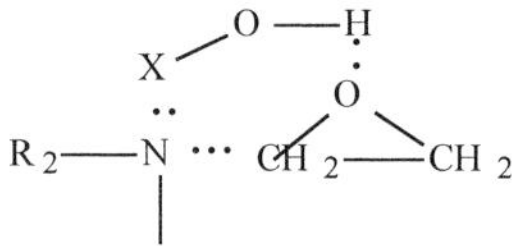

MIKA AND TANAKA   (6)

Addition Reactions of Polybasic Acids and Acid Anhydrides

Acid anhydrides are probably second in importance to amine-type curing agents; however, polybasic acids have little application as curing agents. However, esterification of epoxides with fatty acids to produce resins for surface coatings has great commercial significance. Shechter and Wynstra (8) proposed that the following reactions could occur with a carboxylic acid and an epoxide:

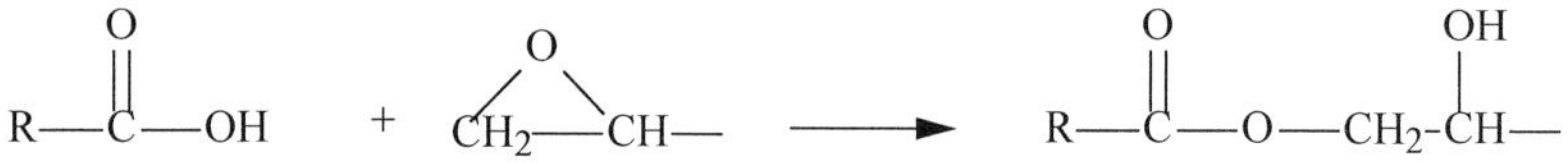

(7)

(8)

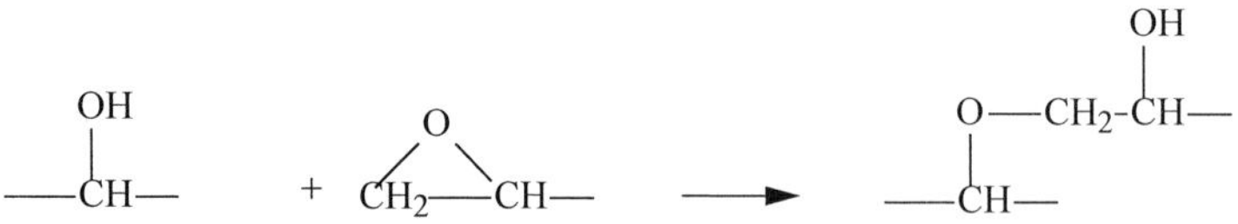

(9)

(10)

These workers showed that if water is removed during the reaction, Reaction 7-9 occurred in approximately the ratio 2:1:1 in an uncatalyzed system. A higher degree of selectivity for the hydroxy ester (Reaction 7) was observed to occur in the base-catalyzed reaction that was proposed to proceed as follows:

$$R-\overset{\overset{\displaystyle O}{\|}}{C}-OH \quad + \quad OH^- \quad \longrightarrow \quad R-\overset{\overset{\displaystyle O}{\|}}{C}-O^- \quad + \quad H_2O \quad (11)$$

$$R-\overset{\overset{\displaystyle O}{\|}}{C}-O^- \quad + \quad \overset{O}{\underset{CH_2-CH-}{\triangle}} \quad \longrightarrow \quad R-\overset{\overset{\displaystyle O}{\|}}{C}-O-CH_2-\overset{\overset{\displaystyle O^-}{|}}{CH}- \quad (12)$$

$$R-\overset{\overset{\displaystyle O}{\|}}{C}-O-CH_2\overset{\overset{\displaystyle O^-}{|}}{CH}- + R-\overset{\overset{\displaystyle O}{\|}}{C}-OH \longrightarrow R-\overset{\overset{\displaystyle O}{\|}}{C}-O-CH_2\overset{\overset{\displaystyle OH}{|}}{CH}- + R-\overset{\overset{\displaystyle O}{\|}}{C}-O^- \quad (13)$$

Although much more selective than the uncatalyzed reaction, the base-catalyzed reaction has some dependence on stoichiometry. At a ratio of epoxide to acid of 1:1, essentially all the product is the hydroxy ester. However, when an excess of epoxide groups is present, Reaction 7 proceeds until all the acid is consumed, after which the epoxide-hydroxyl reaction (Reaction 9) starts. This is illustrated in Figure 4.

The uncatalyzed reaction of acid anhydrides with epoxides is slow even at 200° C, however, with either acidic or basic catalysts the reaction proceeds readily with the formation of ester linkages. The reaction of acid anhydrides with conventional commercial epoxy resins is probably initiated by water or hydroxyl and carboxyl compounds present in the mixture. The following sequence is illustrative of initiation by a hydroxyl-containing material:

$$R-OH \quad + \quad \text{[ANHYDRIDE]} \quad \longrightarrow \quad \text{[ACID ESTER]} \quad (14)$$

HYDROXYL   ANHYDRIDE   ACID ESTER

$$\text{[ACID ESTER]} \quad + \quad \overset{O}{\underset{CH_2-CH-}{\triangle}} \quad \longrightarrow \quad \text{[product]} \quad (15)$$

ACID ESTER

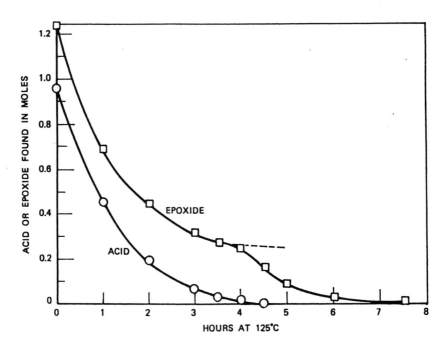

Figure 4.   Reaction of 1.25 mol of phenyl glycidyl ether with
             1.00 mol of caprylic acid plus 0.20% (by weight) of
             KOH catalyst.

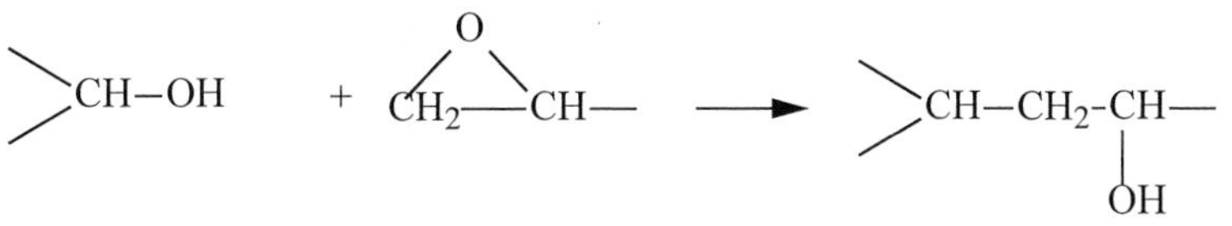

$$>CH-OH \quad + \quad CH_2—CH— \quad \longrightarrow \quad >CH-CH_2\text{-}CH— \qquad (16)$$
$$\qquad\qquad\qquad\qquad\qquad\qquad\qquad\qquad\qquad OH$$

Thus the reaction is essentially a two-step process involving the opening of the anhydride by reaction with the hydroxyl-containing material to give the half acid ester (Reaction 14) and the resulting carboxyl group reacting with an epoxide to form a hydroxy diester (Reaction 15). The hydroxyl compound formed from Reaction 15 can then react with another anhydride, and so on. Evidence has been presented (9), however, that indicates that in this type of catalysis consumption of epoxy groups is faster than appearance of diester groups because of Reaction 16. Base-catalyzed anhydride-epoxide reactions have been found to have greater selectivity toward diester formation. Shechter and Wynstra (8) showed that equal amounts of acetic anhydride and phenyl glycidyl ether with either potassium acetate or tertiary amines as a catalyst reacted very selectively. They proposed the following reaction mechanism for the acetate ion catalyzed reaction:

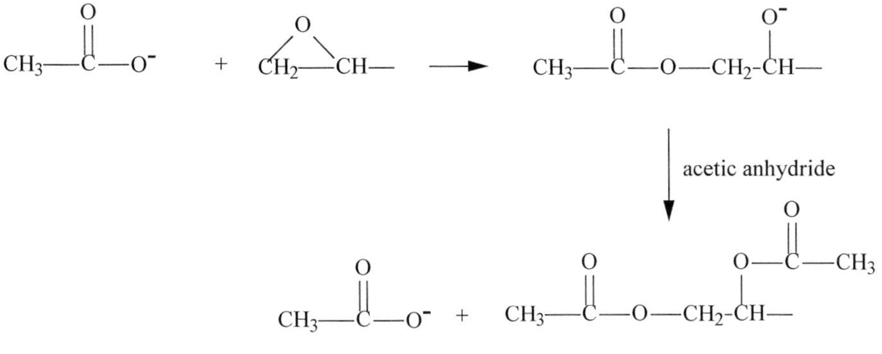

$$CH_3—\overset{O}{\overset{\|}{C}}—O^- \quad + \quad CH_2—CH— \quad \longrightarrow \quad CH_3—\overset{O}{\overset{\|}{C}}—O—CH_2\text{-}\overset{O^-}{\overset{|}{CH}}— \qquad (17)$$

$$\downarrow \text{acetic anhydride} \qquad (18)$$

$$CH_3—\overset{O}{\overset{\|}{C}}—O^- \quad + \quad CH_3—\overset{O}{\overset{\|}{C}}—O—CH_2\text{-}\overset{O—\overset{O}{\overset{\|}{C}}—CH_3}{\overset{|}{CH}}—$$

Catalysis by tertiary amines has been proposed by Fischer (10) to proceed by the base opening the anhydride ring to form an internal salt that then reacts with an epoxide group to yield an alkoxide ester as shown in Reactions 19-20. Thus, the direct reaction between epoxide and anhydride has been found to be very slow, and the anhydride ring must first be opened before reaction can occur. Ring opening can result from reaction with hydroxyl groups present in commercial epoxy resins; addition of basic catalysts, such as tertiary amines and carboxylate ions; or addition of Lewis acids (not discussed here). Ring opening by hydroxyl functionality ultimately results in both esterification and etherification, whereas base catalysis results predominantly in esterification.

$$(19)$$

$$(20)$$

$$(21)$$

The anhydride-epoxide reaction is complex because of the possibility of several reactions occurring simultaneously. Thus, appreciable etherification can result in undesirable amounts of unreacted anhydride and half-acid ester in the cured resin. On the basis of experimental results, Arnold (11) has suggested that for optimum properties anhydride to epoxide rations of 1:1 are needed for tertiary amine catalysts, and 0.85:1 for no catalyst.

Addition Reactions of Phenols and Mercaptans

As with polybasic carboxylic acids, phenols have not achieved significant importance as curing agents; however, the reaction of phenols with epoxides is technologically important. For example, the reaction of bisphenols with the diglycidyl ethers of the bisphenol is used commercially to prepare higher molecular weight epoxy resins (12).

Shechter and Wynstra (8) proposed two possible types of reactions between phenol and a glycidyl ether. One involves direct reaction of the phenol with the epoxide; the other involves direct reaction of the aliphatic hydroxyl, generated from the epoxide-phenol reaction, with another epoxide as shown in Reactions 22 and 23:

$$(22)$$

$$(23)$$

Using model systems, they found that without a catalyst no reaction occurred at 100° C. At 200° C epoxide disappeared at a much faster rate than phenol did (Figure 5); about 60% of the reaction was epoxide-phenol and the other 40% was epoxide-formed alcohol. Because alcohol was absent at the beginning of the reaction and only appeared when phenol reacted with epoxide, it was concluded that the phenol preferred to catalyze the epoxide-alcohol reaction rather than react itself.

The base-catalyzed reaction, however, proceeded readily at 100° C with 0.2 mole% of potassium hydroxide and exhibited a high degree of selectivity. As can be seen from Figure 6, disappearance of phenol and epoxide proceeded at the same rate throughout the course of the reaction. This phenomenon indicates that epoxide reacted with phenol to the essential exclusion of any epoxide-alcohol reaction. Shechter and Wynstra (8) proposed a mechanism in which the phenol first is ionized to phenoxide ion as shown in Reaction 24:

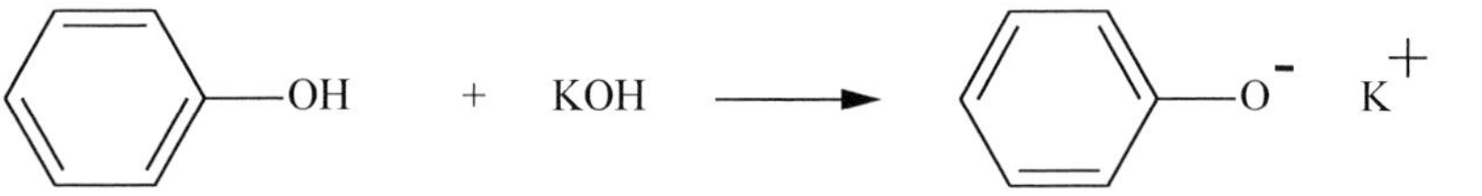

The phenoxide ion then attacks an epoxide as shown in Reaction 25:

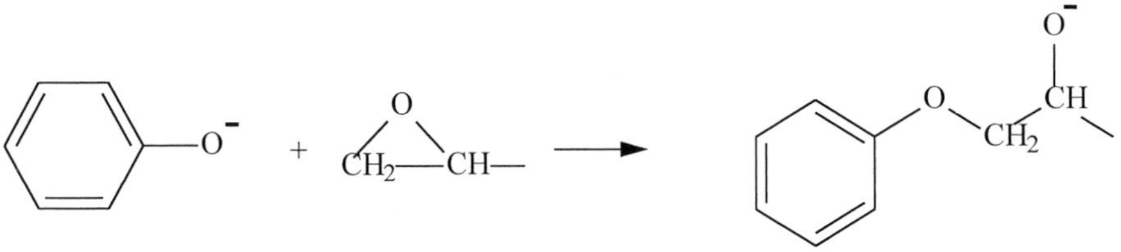

The highly basic alkoxide ion then immediately reacts with phenol to regenerate phenoxide to repeat the cycle (Reaction 26), and at the same time to exclude the possibility of side reactions taking place:

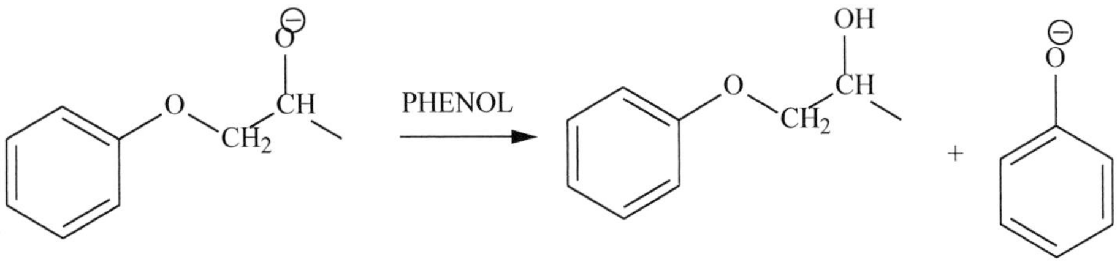

Shechter and Wynstra (8) also demonstrated that benzyldimethylamine was a somewhat more effective catalyst than potassium hydroxide, and the quaternary compound

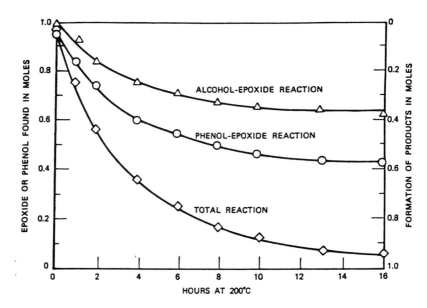

Figure 5.   Noncatalyzed reaction of equimolar amounts of phenol
             and phenyl glycidyl ether.

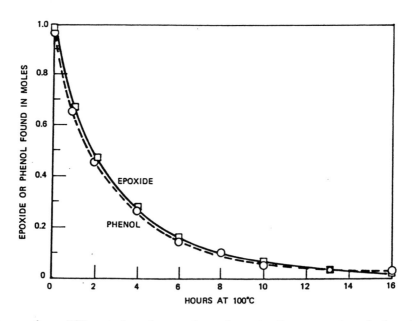

Figure 6.   KOH-catalyzed reaction of equimolar amounts of phenol
             and phenyl glycidyl ether plus 0.20% (by weight) of
             catalyst.

benzyltrimethylammonium hydroxide was even more powerful. Each reaction was highly selective.

Although the base-catalyzed reaction appears to be highly selective, more recent work (13, 14) has shown that the degree of selectivity is dependent on the catalyst, the addition of active hydrogen compound to the epoxide group, the temperature and other variables. In reactions of diglycidyl ethers with difunctional phenols to produce higher molecular weight homologs, branched products or gelation may result if the reaction is not highly selective. Alvey (13) has developed a method of determining the relative amounts of alcohol side reactions, and he has also rated the selectivity of a number of catalysts.

Analogous to the phenols, mercaptans react with epoxide groups to form hydroxyl sulfides as shown in Reaction 27:

$$R\text{---}SH \ + \ CH_2\overset{O}{\diagup\!\diagdown}CH\text{---} \quad\longrightarrow\quad R\text{---}S\text{---}CH_2\text{-}\overset{\displaystyle OH}{\underset{}{C}H}\text{---} \tag{27}$$

The epoxide-mercaptan reaction is highly selective, and there appears to be no concomitant epoxide-alcohol reaction. Thus, to obtain a cross-linked network with a diepoxide, the functionality of the mercaptan must be greater than two.

The epoxide-mercaptan reaction can be accelerated by amines which either react with the mercaptan to give a mercaptide ion (Reaction 28) that rapidly adds to the epoxide (Reaction 29), or by the amine first reacting with the epoxide (Reaction 31) to produce a reactive intermediate that then reacts with the mercaptan in a nucleophilic displacement reaction (Reaction 32).

$$R\text{---}SH \ + \ R_3N \quad\longrightarrow\quad R\text{---}S^{\ominus} \ + \ R_3\overset{\oplus}{N}H \tag{28}$$

$$R\text{---}S^{\ominus} \ + \ CH_2\overset{O}{\diagup\!\diagdown}CH\text{---} \quad\longrightarrow\quad R\text{---}S\text{---}CH_2\text{-}\overset{\displaystyle O^{\ominus}}{\underset{}{C}}H\text{---} \tag{29}$$

$$\downarrow \quad R_3\overset{\oplus}{N}H \tag{30}$$

$$R_3N \ + \ R\text{---}S\text{---}CH_2\text{-}\overset{\displaystyle OH}{\underset{}{C}}H\text{---}$$

$$R_3N \ + \ CH_2\overset{O}{\diagup\!\diagdown}CH\text{---} \quad\longrightarrow\quad R_3\overset{\oplus}{N}\text{---}CH_2\text{-}\overset{\displaystyle O^{\ominus}}{\underset{}{C}}H\text{---} \tag{31}$$

$$\downarrow \quad R\text{---}SH \tag{32}$$

$$R_3N \ + \ R\text{---}S\text{---}CH_2\text{-}\overset{\displaystyle OH}{\underset{}{C}}H\text{---}$$

Reactions of Alcohols

The uncatalyzed epoxide-alcohol reaction was shown by Shechter and Wynstra (8) to be "rather sluggish". A temperature of 200° C is necessary to realize a conveniently rapid rate. The reaction can be catalyzed by either acid or base to yield primary and secondary alcohols that will further react with the free epoxide to form polyethers (Reaction 33).

$$\text{(33)}$$

For example, in experiments with phenyl glycidyl ether and isopropyl alcohol (1:1 molar ratio) in which potassium hydroxide and benzyldimethylamine were used as catalysts, Shechter and Wynstra (8) found that after nearly complete consumption of the epoxide group approximately 80% of the charged alcohol was unreacted. This result indicates that the reaction was largely self-polymerization.

Acid catalysis of the epoxide-alcohol reaction is no more selective than base catalysis. The ratio of alcohols formed and the amount of polyether obtained vary with the type and amount of catalyst, epoxide-to-alcohol ration, solvent, and reaction temperature.

Although the epoxide-alcohol reaction is used industrially to prepare aliphatic glycidyl ethers, it is not an important curing reaction. Coatings based on the use of hydroxyl-functional phenol and amino-formaldehyde resins to cure the higher homologs of the diglycidyl ethers of bisphenol A were among the first to attain commercial significance. The curing chemistry of both the phenolic and amine-epoxy systems is based on the reactions of the cross-linking resin with the hydroxyl groups of the epoxy resin (Reaction 34), cross-linking resin with the epoxide group (Reaction 33), and the cross-linking resin with itself (19).

$$\text{(34)}$$

Catalytic Reactions

The reactions of the epoxide group involve addition reactions of epoxides with compounds having a labile hydrogen atom. Catalytic reactions are characterized by the reaction of the epoxide group with itself (homopolymerization). Although both Lewis-type bases and acids can catalyze homopolymerization by causing anionic and cationic propagation, respectively, the resultant structure is the same: a polyether. Catalytic polymerization of monoepoxides results in linear polymers, whereas diepoxides give a cross-linked network.

Anionic polymerization of epoxides can be induced by Lewis bases (usually tertiary amines) or by metal hydroxides. The amine-type catalysts are by far the most important type of catalyst for epoxide homopolymerization. The initiation of the polymerization of the epoxides has been proposed by Narracott (15) and Newey (16) to result from the attack by the tertiary amine on the epoxide (Reaction 35), with the resulting alkoxide amine being the propagating species (Reaction 36).

$$R_3N \ + \ \underset{CH_2 \ \text{---} \ CH \ \text{---}}{\overset{O}{\triangle}} \ \longrightarrow \ R_3\overset{\oplus}{N}\text{---}CH_2\text{-}\overset{\overset{\ominus}{O}}{\underset{|}{CH}}\text{---} \tag{35}$$

$$\tag{36}$$

$$R_3\overset{\oplus}{N}\text{---}CH_2\text{-}\underset{|}{CH}\text{---}O\text{---}CH_2\text{-}\overset{\overset{\ominus}{O}}{\underset{|}{CH}}\text{---}$$

Cationic polymerization of epoxides is initiated by Lewis acids, which are substances composed of atoms containing empty electron orbitals in their outer shells. Such atoms can form covalent bonds with atoms capable of donating pairs of electrons. Many inorganic halides are Lewis acids such as $AlCl_3$, $SbCl_5$, BF_3, $SnCl_4$, $TiCl_4$, and PF_5. In commercial practice the most important type of initiator for curing epoxides is BF_3, usually in the form of BF_3 complexes. Lewis acids initiate polymerization through the formation of carbonium ions, and Plesch (17) has proposed that a suitable co-initiator is necessary to produce these ions. The mechanism for the initiation of the homopolymerization of epoxy resins by BF_3 complexes has been proposed by Arnold (11) to proceed as follows:

$$BF_3 \ + \ H_2O \ \longrightarrow \ BF_3O\overset{\ominus}{H} \ + \ \overset{\oplus}{H} \tag{37}$$

$$BF_3 \ + \ R_2NH \ \longrightarrow \ BF_3\overset{\ominus}{NR_2} \ + \ \overset{\oplus}{H} \tag{38}$$

$$BF_3\overset{\ominus}{X}H \ \overset{\oplus}{H} \ + \ \underset{CH_2\text{---}CH\text{---}}{\overset{O}{\triangle}} \ \longrightarrow \ \underset{CH_2\text{---}CH\text{---}}{\overset{\overset{\oplus}{H} \ BF_3\overset{\ominus}{X}H}{\underset{\vdots}{\overset{O}{\triangle}}}} \tag{39}$$

$$\tag{40}$$

$$\underset{\overset{\oplus}{CH_2}\text{---}CH\text{---}O\text{---}(CH_2\text{---}CH)_n}{\overset{BF_3\overset{\ominus}{X}H}{\underset{\overset{|}{\ } \ \overset{|}{OH}}{}}} \longleftarrow \underset{CH_2\text{---}CH\text{---}}{\overset{OH \ BF_3\overset{\ominus}{X}H}{\underset{\ }{}}}$$

The polymerization chain is then propagated by the resulting carbonium ion that is stabilized by interaction with the anion produced in Reaction 40.

Photoinitiated catalytic cures have emerged as an important technology for applying solventless, low-temperature curing coatings. Initially photoinitiated or UV-initiated cure of epoxy resin systems employed free-radical polymerization of vinyl derivatives of epoxy resins because epoxy resins are not curable by typical free-radical chemistry. Therefore, the use of radical-generating photoinitiators is not applicable in directly effecting an epoxide cure. These free-radical cured systems are, generally, based on vinyl esters of diglycidyl ethers obtained as indicated in Reaction 41:

Acrylic acid

(41)

These systems have the disadvantage that they are sensitive to oxygen and require blanketing with nitrogen. Also, they have high viscosities and must be used with vinyl monomers as diluents, many of which present health hazards.

As early as 1965 Licari and Crepean (20) reported the photo-induced polymerization of epoxide resins by diazonium tetra-fluoroborates for use in the encapsulation of electronic components and the preparation of circuit boards. The use of these materials in coatings was pioneered by Schlesinger and Watt(21-23). When irradiated with UV light, these materials produce BF_3, fluoroaromatic compound, and nitrogen (Reaction 42):

$$Ar-N\overset{\oplus}{\equiv}N \ \overset{\ominus}{BF_4} \xrightarrow{h\nu} \quad BF_3 \quad + \quad N_2 \quad + \quad ArF$$

(42)

When BF_3 is produced in this fashion in an epoxy resin, it catalyzes the cationic polymerization of the resin as discussed earlier. Typically, a small amount of the diazonium compound is dissolved in the epoxy coating formulation and irradiated with UV light to form thin films (0.5 to 1 mil) deposited on metal, wood, or paper substrates. The high reactivity of the BF_3-type cure makes it possible to prepare hard solvent-resistant coatings in a few seconds exposure time under a standard 200-W/in. mercury vapor lamp.

The evolution of nitrogen on photolysis of the aryldiazonium salts limited the use of these systems to thin film applications such as container coatings and photoresists (23). Other efficient photoinitiators that do not produce highly volatile products have been disclosed (24-27). These systems are based on the photolysis of diaryliodonium and triarylsulfonium salts, Structures I and II, respectively.

$$\underset{\underset{Ar}{\overset{Ar}{|}}}{I^{\oplus}} \quad X^{\ominus} \qquad\qquad Ar\!-\!\underset{\underset{Ar}{\overset{Ar}{|}}}{\overset{\oplus}{S}} \quad X^{\ominus} \qquad\qquad X \;=\; BF_4,\ ASF_6,\ PF_6,\ SbCl_6,\ etc.$$

$$(I) \qquad\qquad\qquad\qquad (II)$$

These salts are thermally stable and upon irradiation they liberate strong Bronsted acids of the HX type (Reactions 43 and 44) that subsequently initiate cationic polymerization of the oxirane rings- free radicals are also generated and can initiate co-cure of vinyl groups:

$$Ar_2\overset{\oplus}{I}PF_6^{\ominus} \quad\longrightarrow\quad ArI \;+\; Ar\bullet + \; H^{\oplus}PF_6^{\ominus} \qquad\qquad (43)$$

$$Ar_2\overset{\oplus}{S}ASF_6^{\ominus} \quad\xrightarrow{UV}\quad Ar_2S \;+\; Ar\bullet \; + \; \overset{\oplus}{H}\; ASF_6^{\ominus} \qquad\qquad (44)$$

Unlike free-radical propagation, photoinitiated cationic polymerizations of epoxides are unaffected by oxygen and thus require no blanketing by an inert atmosphere. However, water and basic materials present in UV-curable epoxy formulations can inhibit cationic cures and should be excluded.

Curing Agents

The reactions just discussed describe the chemistry by which epoxy resins are converted into cross-linked polymeric structures by a variety of reactants. This cross-linking or "curing" is accomplished through the use of co-reactants or "curing agents". Curing agents are categorized into broad classes: active hydrogen compounds, which cure by polyaddition reactions, and ionic catalysts.

Most of the curing agents currently used with epoxy resins cure by polyaddition reactions that result in both the coupling as well as cross-linking of the epoxy resin molecules. Although these reactions are generally based on one active hydrogen in the curing agent per epoxide group, practical systems are not always based on this stoichiometry because of homopolymerization of the epoxide and other side reactions that cannot be avoided and that, in fact, are sometimes desired. In contrast to 1:1 stoichiometry generally required for active hydrogen-epoxide reactions, only catalytic amounts of Lewis acids or bases are required to cure an epoxy resin. The number of co-reactants developed over the years for epoxy resins is overwhelming. Selection of the co-reactant is almost as important as that of the base resin and is usually dependent on the performance requirements of the final product and the constraints dictated by their method of fabrication. Although the following is far from a complete list, the most common curing agents can be classified as follows:

1. <u>Aliphatic polyamines and derivatives</u>. These include materials such as ethylenediamine, diethylenetriamine, triethylenetetramine, tetraethylenepentamine, and several cycloaliphatic amines. These curing agents as a class offer low viscosities and ambient temperature cures. However, the unmodified amines present certain handling hazards because of their high basicity and relatively high vapor pressure. Less hazardous derivatives of aliphatic amines are obtained from the reaction products of higher molecular weight fatty acids with aliphatic amines. Besides having lower vapor pressure, these "reactive polyamide" and amidoamine systems will cure epoxy resins at room temperature to give tougher, more flexible products.

2. <u>Modified aliphatic polyamines</u>. These are room temperature curing agents formed by reacting excess aliphatic amines with epoxy-containing materials to increase the molecular weight of the amine to reduce its vapor pressure. The performance properties of amine adduct cured systems are not significantly different from those of aliphatic polyamines.

3. <u>Aromatic amines</u>. These include materials such as 4,4'-methylenedianiline, *m*-phenylenediamine, and 4,4'-diaminodiphenylsulfone. Aromatic amines are less reactive than aliphatic amines and usually require curing temperatures as high as 300° F.

4. <u>Acid anhydrides</u>. These are the second most commonly used curing agents after polyamines. In general, acid anhydrides require curing at elevated temperatures, but offer the advantages of longer pot lives and better electrical properties than aromatic amines. Illustrative of some of the more commonly used acid anhydrides are phthalic, trimellitic, hexahydrophthalic, and methylnadic anhydride.

5. <u>Carboxylic acids</u>. Formaldehyde-free and zero-VOC crosslinking agents for waterborne epoxy resins have been reported based on carboxylic acids. Acid functional oligomers/polymers can cure epoxy resins and react with backbone hydroxyl groups to form esters. Aromatic carboxylic acid resins based on trimellitic acid cure epoxy resins at 150 °C without additional catalyst.(42) Curing epoxies by transesterification has also been reported.(43)

6. <u>Lewis acid and base type curing agents</u>. Cures with Lewis acid and base type curing agents proceed by homopolymerization of the epoxy group that is initiated by both Lewis acids and bases. These curing agents can provide long room-temperature pot life with rapid cures at elevated temperatures, and thus produce products with good electrical and physical properties at relatively high temperatures (150-170 °C).

7. <u>Aminoplast and phenoplast resins</u>. This class represents a broad range of melamine-, phenol-, and urea-formaldehyde resins that cross-link by a combination of reactions through the hydroxyl group of the epoxy resin, self-condensation, and reaction through epoxide groups. These systems are cured at relatively high temperatures (325 to 400 °F) and yield final products with excellent chemical resistance. Also, a few miscellaneous types of curing agents, such as dimercaptans, dicyandiamide, dihydrazides, and guanamines, have found some limited industrial applications in electrical laminates and in powder coatings, these materials account for only a small volume of the total curing agent market.

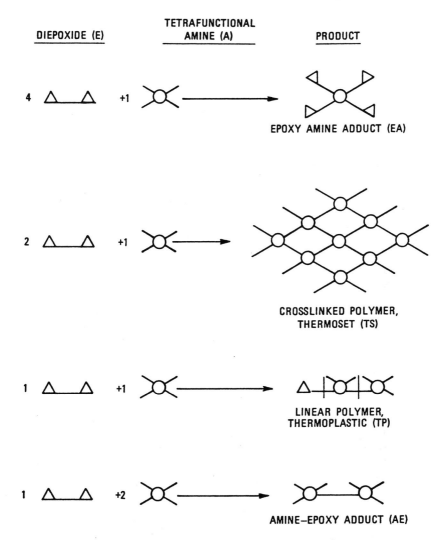

Figure 7.   Schematic representation of network formation.

8.   <u>Phenolic terminated epoxy resins</u>. Phenolic terminated aryl ether resins can be made by
      end capping bis A epoxy resins with bisphenol. Marx recommended these curing agents
      for powder coatings.(45)

<u>Curing of Epoxy Resin Compositions</u>

The properties that can ultimately be obtained from an epoxy resin system depend on the nature
of epoxy resin and curing agent as well as the degree of cross-linking that is obtained during
cure. The degree of cross-linking is a function of stoichiometry of the epoxy resin and curing
agent, and the extent of the reaction achieved during cure. To actually obtain a "cure," the
reaction of the curing agent with epoxy resin must result in a three-dimensional network. A
three-dimensional network is formed when one component has a functionality greater than two,
and the other component has a functionality of not less than two, as follows:

Primary and secondary aliphatic polyamines, their derivatives, and modified aliphatic
polyamines and aromatic amines react with and cure epoxy resins as indicated earlier. The
aliphatic systems usually give adequate cures at room temperature (7 days above 60° F);
however, under most conditions aromatic amines are less reactive and require curing temperature
of about 300° F to give optimum cured polymer properties. The importance of stoichiometry in
network formation can be illustrated with a difunctional epoxide and tetrafunctional curing
agent, which are represented schematically in Figure 7. As can be seen, a spectrum of products
is obtained in progressing from an excess of epoxide to an excess of a curing agent, such as a
tetrafunctional amine. At an excess of epoxide groups over reactive sites on the curing agent, an
epoxy-amine adduct is the predominate product. As the stoichiometry approaches one
equivalent of epoxide per equivalent of reactive amine sites, the molecular weight approaches
infinity, and a three-dimensional network polymer is obtained. As the ratio of curing agent to
epoxide is increased, the product approaches a linear polymer (thermoplastic), and finally with
an excess of curing agent, and amine-epoxide adduct is obtained.
       On the molecular level the cure of an epoxy resin system involves the reaction of the
epoxy groups (or hydroxyl groups in some cases) of the resin molecules with a curing agent to
form molecules of ever increasing size until and infinite network of cross-linked resin and curing
agent molecules is formed. As the chemical reactions proceed, the physical properties of the
curing resin change with time from a fluid to a solid. More specifically, as the cure proceeds, the
viscosity of the reacting system increases until gelation occurs, at which time the mass becomes
an insoluble rubber. Further chemical reaction eventually converts the rubbery gel into a glassy
solid (vitrification). Gelation correspond to incipient formation of an infinite network of cross-
linked polymer molecules, and vitrification involves a transformation from a liquid or a rubbery
state to a glassy state as a result of an increase in molecular weight. Vitrification can quench
further reaction.
       In an attempt to understand the cure phenomena Gillham (<u>28</u>) has developed the concept
of a state diagram. Such a time-temperature-transformation (TTT) diagram is given in Figure 8.
It is a plot of the times required to reach gelation and vitrification during isothermal cures as a
function of cure temperature, and it delineates the four distinct material states (liquid, gelled
rubber, ungelled glass, and gelled glass) that are encountered during cure. Also displayed in the
diagram are the three critical temperatures:

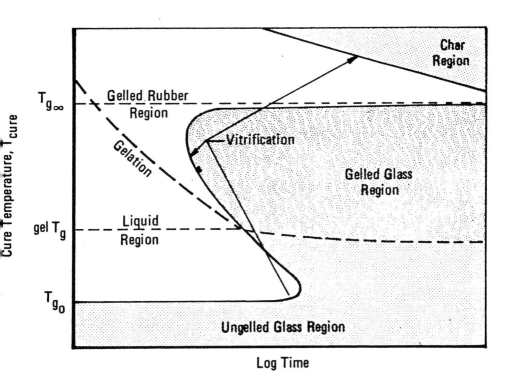

Figure 8.   Generalized TTT cure diagram.

1. $T_{g\infty}$, the maximum glass transition temperature of the fully cured system;
2.   Gel Tg, the isothermal temperature at which gelation and vitrification occur simultaneously; and
3. T_{go}, the glass transition temperature of the freshly mixed reactants.

Thus, during an isothermal cure at a temperature between gel Tg and $T_{g\infty}$, the resin will first cure and then vitrify. Once vitrification occurs, the curing reactions are usually quenched, which means the glass transition temperature (T_g) of the resin will equal the temperature of cure (such a material will not be fully cured). The resin, however, will not vitrify on isothermal cure if the cure temperature is above $T_{g\infty}$, the T_g of the fully cured resin. Above $T_{g\infty}$, the cure can proceed to completion, and the maximum T_g of the system is obtained. A more detailed discussion of the TTT cure diagram as it relates to epoxy resin curing can be found in the publications of Gillham (28, 29).

Applications of Epoxy Resins

Epoxy resins have found a broad range of application, mainly because the proper selection of resin, modifiers, and cross-linking agent allows the properties of the cured epoxy resin to be tailored to achieve specific performance characteristics. This versatility has been a major factor in the steady growth rate of epoxy resins over the years.

Besides this versatility feature, properly cured epoxy resins have other attributes:

1.   Excellent chemical resistance, particularly to alkaline environments.
2.   Outstanding adhesion to a variety of substrates.
3.   Very high tensile, compressive, and flexural strengths.
4.   Low shrinkage on cure.
5.   Excellent electrical insulation properties and retention thereof on aging or exposure to difficult environments.
6.   Remarkable resistance to corrosion.
7.   A high degree of resistance to physical abuse.
8.   Ability to cure over a wide range of temperatures.
9.   Superior fatigue strength.

Coatings

Even though epoxy resins are somewhat more expensive than many resinous materials used in coatings, epoxy resins have found commercial acceptance in a wide variety of high-performance protective and decorative surface coatings and are an important class of coating resins.(41) Epoxy resin based coatings have been established as premium coatings because of their excellent chemical and corrosion resistance and their outstanding adhesion compared to other coating materials. Epoxy coatings obtain their excellent properties through reaction with curing agents. The curing agents reacting with epoxide and hydroxyl functionality of the epoxy resins result in highly chemical- and solvent-resistant films because all the bonds are relatively stable carbon-carbon, carbon-oxygen (ether), and carbon-nitrogen (amine) linkages. Many of the more common epoxy resin coating systems and their end uses are summarized in Figure 9.

Epoxy resin coatings can be divided into two distinct types: Those that are cured at ambient temperature and those that are heat cured. The first type is cross-linked through the

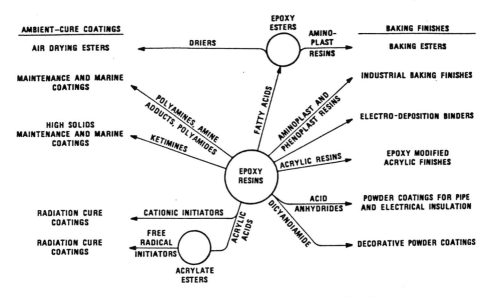

Figure 9.   Epoxy resins in surface coating applications.

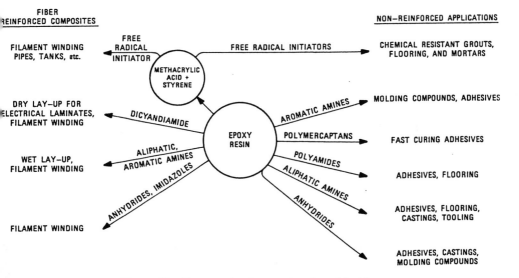

Figure 10.   Epoxy resins in structural applications.

oxirane ring by using polyamines, amine adducts, polyamides, polymercaptans, and catalytic cures. Heat-cured epoxy resin coatings are cross-linked through reaction of the hydroxyl groups, or in some cases a combination of the epoxy and hydroxyl functionality, by using anhydrides and polycarboxylic acids as well as melamine-, urea-, and phenol-formaldehyde resins.

Ambient Cured. Most people are familiar with the ambient cure or "two-package" type coating that is sold through retail stores. This coating was the first room-temperature curing coating to offer resistance properties previously obtainable only in baked industrial finishes. Coatings of this type are used extensively today as heavy duty industrial and marine maintenance coatings, tank linings, and float toppings; for farm and construction equipment, and aircraft primers and in do-it-yourself finishes for the homeowner. As the name implies, the "two-package" epoxy coatings are mixed just prior to use, and are characterized by a limited working life (pot life) after the curing agent is added. Commercial systems will have pot lives ranging from a few hours to a couple of days with typical working times ranging from approximately 8 to 12 h. By far the more common two-package coatings are based on epoxy resins having EEW values between 180 and 475 (see Table I) and cured with polyamides, amine adducts, and polyamides. The polyamine and amine adduct type cures provide better overall chemical resistance, but polyamides offer better film flexibility, water resistance, and are somewhat more forgiving of improper surface preparations. Environmental pressures and energy concerns have resulted in rapid changes in epoxy coating technology. Volatile organic compound (VOC) content of paints is being reduced by environmental laws. Metal primers using conventional solventborne epoxy formulations contain about 480 g/L VOC. New curing agents and epoxy resins are currently being developed for high solids, solventless, and waterborne coatings. Reactive diluents such as aliphatic mono- and polyglycidyl ethers, and epoxidized vegetable oils (44) have been used to reduce resin viscosity with a trade-off in reduced physical properties. Waterborne epoxy systems when first introduced had significant performance problems. Improvements to the technology have extended the utility of ambient cure two pack waterborne epoxy systems by improving performance while reducing VOCs by an order of magnitude.(32, 33)

Two other epoxy resin type coatings are classified as air dry, instead of ambient cure, because they do not involve a curing agent. The earliest of this type are the epoxy esters, which are simply higher molecular weight epoxy resins that have been esterified with unsaturated or drying type fatty acids, such as dehydrated castor acid. These esters are usually prepared from solid epoxy resins have EEW values in the range of 900 and can be considered as specialized polyols. As in conventional alkyd resin technology, these coatings are manufactured by esterifying the resin with a fatty acid at temperatures of 400-450 °F. Initially the fatty acid reacts with the oxirane ring at lower temperatures, and thereby forms a hydroxyl ester, and subsequently these hydroxyl groups and those already present in the resin backbone are esterified at the higher temperatures with the aid of esterification catalysts and azeotropic removal of the water formed. These epoxy ester vehicles are used in floor and gymnasium finishes, maintenance coatings, and metal decorating finishes, and air dry by oxidation of the unsaturated fatty acid, similar to the so-called "oil-based" paints.

Another class of air dry epoxy-derived finishes is based on very high molecular weight phenoxy resins which are linear polyhydroxy ethers (these ethers are essentially bisphenol A epoxy resins where n is approximately 30 or greater) having molecular weights greater than 50,000. Unlike the systems described earlier, these coatings may be cross-linked with small amounts of aminoplast resins for baking applications, as described later. Coatings of this type offer extreme flexibility with good adhesion and chemical resistance and are used for air dry pre-construction primers as well as baked container coatings and automotive primers. Phenoxy resins are also available as water dispersions.

Bake Cured. The higher molecular weight epoxy resins, that is, those with EEW values about 1750 or greater, generally are used in baking finishes. The concentration of oxirane groups is low, and cross-linking occurs principally through the hydroxyl functionality. Solid, high molecular weight epoxy resins co-reacted with phenoplast and aminoplast resins (phenol, melamine and urea-formaldehyde resins) through loss of water or alcohol to form ether linkages have been used for years for highly solvent- and chemical-resistant coatings. The phenolic converted systems are used as beverage and food can coatings, drum and tank linings, internal coatings for pipe, wire coatings, and impregnating varnishes. Although not as chemically resistant as the epoxy phenolic coatings, aminoplast resins are used in certain applications because of their better color and lower cure temperatures. Epoxy-aminoplast resin finishes are used widely as can linings; appliance primers; clear coatings for brass hardware, jewelry, and vacuum metallized plastics; foil coatings; and coatings for hospital and laboratory furniture. Phosphate esters of solid epoxy resins provide lower VOC water dispersible bake coatings with excellent properties.(30) Use of the epoxy phosphate esters in solvent borne coatings gave coatings with improved adhesion and flexibility.(31). In manufacturing processes, coatings can (1) be applied to flat metal sheets or coils before stamping or forming, (2) be applied to the fully formed article, or (3) be applied to the partially formed article that is then further deformed in a finishing operation. Since the first alternative is the simplest and most economical, there is constant pressure to develop more formable coatings. The subject of improving toughness of thermoset coatings for improved formability, post-formability, fabricability, impact resistance, and chip resistance was recently reviewed.(39)

Powder and UV Coatings. Epoxy powder coatings are used to essentially eliminate solvent emissions. Epoxy powder coatings were originally used for functional (protective) markets, but they are now used for decorative markets as well. Including the epoxy hybrids, epoxy powder coatings comprised over 50% of the thermosetting powder coating market by the early 1990s.(40) Solventless epoxy-acrylic esters and cycloaliphatic epoxy resins are applied as 100% solids and then cured by free radical or cationic photoinitiators with UV light or by heat.(34) Cycloaliphatic epoxy resins are one of the more expensive resins, however, epoxidized soybean oil was shown to be a low cost reactive diluent for cationic cured epoxy systems.(35) Techniques are also being developed to reduce baking temperatures of the heat-cured epoxy finishes. As a whole, however, all these coatings are still based on the chemistry already developed and described earlier. The resulting coatings still retain the properties for which epoxy resins have become known.

Structural Applications

Structural or non-coating type applications of epoxy resins are much more fragmented and more difficult to simplify than the coatings area. There are, basically, four major structural-type end uses for epoxy resin systems: bonding and adhesives; castings, molding powders, and tooling; flooring, paving, and aggregate; and reinforced plastics or composites (electrical laminates and filament winding). Again epoxy resins have found such broad application in the structural area because of their versatility (see Figure 10). They can be modified into low viscosity liquids for easy casting and impregnating, or converted to solid compositions for ease of laminating or molding. Depending upon choice of curing agent, they can be made to cure either slowly (hours) or very quickly (minutes) at room temperature to give a variety of properties ranging from soft, flexible materials to hard, tough, chemical-resistant products. High performance epoxy resin systems are inherently brittle and subject to catastrophic failure. Their use in engineering systems became possible by toughening the epoxy resin system with rubber dispersions.(36) Epoxidized crambe oil was shown to precipitate from epoxy systems to give a similar toughening effect.(37) The design of tough epoxy thermosets was recently reviewed.(38)

Bonding and Adhesives. Epoxy resin adhesives are most commonly used as two-component liquids or pastes and cure at room or elevated temperature. This type of adhesive is cured with a polyamide or polyamine, or in the quick setting type with an amine-catalyzed polymercaptan, and is available to the householder at hardware stores. Although the bulk of adhesives are of this type, a great deal of the more sophisticated epoxy adhesives are supplied as supported tape, that is, a glass fabric tape impregnated with adhesive, or as non-supported tape. These are "one-package" systems that contain a latent curing agent and epoxy resin. Latent curing agents such as dicyandiamide and boron trifluoride salts are used because they provide single-package stability and rapid cure at elevated temperatures. Adhesive systems of this type are used in the aircraft industry to replace many of the mechanical fasteners once used, and for automotive adhesives.

Casting, Molding Compounds, and Tooling. Epoxy resins maintain a dominant position in the electrical and electronic industry for casting resins, molding compounds, and potting resins. Their superior dielectric properties together with their low shrinkage upon cure and good adhesion make epoxy resins a natural for this end use. The low viscosity epoxy resins and reactive diluents mentioned above are used with bisphenol A resins where low viscosity would be advantageous.

Epoxy castings are made by pouring a resin, curing agent mixture into a mold, and curing to make the finished parts; subsequently, the mold is removed. In potting, the mold is retained as an integral part of the finished product. The epoxy systems are cast around parts that are in containers or housings. These applications use a liquid epoxy resin, usually cured at elevated temperatures, with aromatic amines or anhydrides that provide lower shrinkage, longer pot life, and lower exotherms on cure.

Molding compounds are based on solid epoxy resins cured with aromatic amines or anhydrides. They contain high filler loadings (as much as 50% by volume) and are manufactured by dispersing the components on a two-roll mill or in a heated kettle. The resulting mixture is cooled and then ground to the desired particle size. Currently, the major use of epoxy molding compounds is in the encapsulation of electrical components, the manufacture of pipe fittings, and certain aerospace applications.

Along with potting and encapsulation, one of the first two structural applications for epoxy resins was in tooling molds. For short production runs epoxy molds, drop hammers, stretch dies, jigs, and fixtures have replaced metal tools, especially in the aircraft industry and in automotive work. An epoxy tool can replace its metal counterpart on the basis of ease and speed of fabrication; however, a resin-based tool will not have the durability of metal in long-term production runs.

<u>Flooring, Paving, and Aggregates</u>. Two-package epoxy systems serve as binders for pothole patching; anti-skid surfaces; exposed aggregate; and industrial, seamless, and thin set terrazzo flooring. Typical formulations would be based on a liquid epoxy resin containing a reactive or nonreactive diluent, fillers, and special thickening agents; because a room-temperature cure is required, curing agents usually employed are amines or polyamides.

Epoxy terrazzo is being used in place of concrete flooring because it cures overnight and can be ground and polished the next day; its 1/4 to 3/8 in. thickness provides significant weight savings; it has superior chemical resistance; and it can be applied over many different surfaces. Industrial flooring provides anti-skid and wear surfaces and resistance to spills, such as in dairies and other food processing plants (18). A related application is the use of epoxy resin systems as binders for exposed aggregate patios, swimming pool decks, walks, and wall panels.

<u>Reinforced Plastics or Composites</u>. Including printed circuit boards for electronics, reinforced plastics represent the second largest market for epoxy resins (see Figure 2). The major markets are PC boards, laminating, and filament wound applications. Epoxy resin reinforced plastics exhibit low weight, good heat resistance, excellent corrosion resistance, and good mechanical and electrical properties (18).

Epoxy resin systems with fiber reinforcements are called "reinforced systems" or "composites". Composites are made by impregnating reinforcing fibers such as glass, synthetic polymer, or graphite fibers, by one of several processes with the desired epoxy resin system, and then curing in a heated mold or die. Epoxy composite systems are formulated with either liquid or solid resins with selection of the type of system dependent on the fabrication process, the cure temperature, and the final part application.

The several methods to convert resins and reinforcing fibers into composites are classified as either "wet lay-up" or "dry lay-up" methods. These processes are as follows:

1. "Wet lay-up" refers to a process in which liquid resin systems of low viscosity are used to impregnate the reinforcements, either before or after the reinforcements have been laid in place. The liquid resin penetrates the fibers and displaces the air. The distinctive feature of this method is that the composite object is shaped into its final configuration while the resin component of the lay-up is still liquid (uncured). Cure is effected after the lay-up is completely in place and conforms exactly to the mold.
2. "Dry lay-up" refers to a process in which the reinforcing material is preimpregnated with a resin solution. The solvent is then removed by heated air currents that also may partially cure the resin system. This removal produces a dry, resin-impregnated sheet (the "prepreg"). The prepreg stock is then cut and positioned in the desired configuration for cure. By application of heat under pressure, the resin softens and the curing agent is activated to complete the polymerization of the resin. Although either solid or liquid epoxy resins may be used for dry lay-up, use of low molecular weight solid resins is the most common. The pressure used in dry lay-ups gives a higher glass content and stronger composites than wet lay-up composites.

Filament winding processes use both wet and dry techniques. In this method, a continuous supply of roving or tape (formed from roving) is wound onto a suitable mandrel or mold. The glass fiber is either impregnated with the epoxy resin-curing agent system at the time of winding (wet winding), or has been preimpregnated and dried (as for dry lay-up laminating). This dry stock is wound onto the mandrel (dry winding). In either case, the mandrel may be stationary, but often it will revolve around one axis or more. The shapes of structures made by filament winding are not limited to surfaces of revolution, cylindrical shapes, and pipe; they may be round, oval, or even square.

Typically, a low molecular weight solid epoxy resin and latent curing agent, such as dicyandiamide dissolved in an appropriate solvent, are widely used in dry lay-up formulations for electrical laminates for computers, aerospace applications, and communications equipment. Wet lay-up systems are almost exclusively based on low-viscosity resins cured with aliphatic amines. These ambient cure resin systems are used primarily for manufacture of large chemical-resistant tanks, ducting, and scrubbers. Filament winding is used to produce lightweight chemical-resistant pipe that is generally produced from low molecular weight resin systems cured with anhydrides, aromatic amines, or imidazoles.

Although these applications will continue to use epoxy resins, another approach to reinforced thermosetting epoxy resin systems is the use of epoxy-based vinyl ester resins. these vinyl esters are derived from the reaction of a bisphenol A epoxy resin with methacrylic acid. This reaction results in a vinyl-terminated resin having an epoxy backbone. Usually the epoxy vinyl esters are used in solution in a vinyl monomer such as styrene as a co-reactant and viscosity reducer. Cross-linking is effected by free-radical generating initiators or high energy sources including UV- and electron-beam radiation. These resins provide the handling characteristics of polyesters resins, but with improvements in water, acid, base, solvent, and heat resistance over general purpose and isophthalic polyester resins. Their inherent chemical resistance and physical properties make the vinyl esters suitable for manufacture of equipment such as tanks, scrubbers, pipe, and fittings, or for tank and flue linings for application in a wide variety of chemical environments.

The scope of epoxy resin application is so broad that it has been possible to touch only upon a few major areas and types of epoxy resin systems. Epoxy resin as indicated in this article connotes not a single formulation but a wide range of compositions having properties specific to their chemical structure. Most of the important applications and typical curing agents used in the resin systems specific to an application are summarized in Figures 2, 9, and 10. The chemical versatility of the epoxy resins has been responsible for the diversified epoxy resin usage. It is this same versatility coupled with the ingenuity of chemists and engineers that ensures the continued growth of epoxy resins.

Literature Cited

1. Modern Plastics 1982, 59(1) and Resins Department, The Dow Chemical Co., 1999.
2. Tanaka, Y. and Mika, T. F. In "Epoxy Resins, Chemistry and Technology"; May C. A.; Tanaka, Y., Eds. Dekker: New York, 1973.
3. Shechter, L., et al., Ind. Eng. Chem. 1956, 48(1), 94.
4. O'Neill, L. A. and Cole, C. P., J. Appl. Chem. 1956, 6, 365.
5. Dannenberg, H., SPE Tran S. 1963, 3, 78.
6. Smith, I. T. Polymer 1961, 2, 95.
7. Tanaka, Y. and Mika, T. F. In "Epoxy Resins, Chemistry and Technology"; May, C. A.; Tanaka, Y., Eds. Dekker: New Your, 1973.
8. Shechter, L. and Wynstra, J. Ind. Eng. Chem. 1956, 48(1), 86.

9. Fisch, W. and Hofmann, W., J. Polym. Sci. 1954, 23, 497.
10. Fisher, R. F., J. Polym. Sci. 1960, 44, 155.
11. Arnold, R. J., Mod. Plastics 1964, 44, 149.
12. Somerville, G. F. and Parry, H. L., J. Paint Tech. 1970, 42(540).
13. Alvey, F. B., J Appl. Polym. Sci. 1969, 13, 1473.
14. Sow, P. N. and Weber, C. D., J. Appl. Polym. Sci. 1973, 17, 2415.
15. Narracott, E. S., Brit. Plastics 1953, 26, 120.
16. Newey, H. A., unpublished data.
17. Plesch, P. H. In "The Chemistry of Cationic Polymerization"; Plesch, P. H., Ed.; Pergamon: Oxford, 1963.
18. Somerville, G. R. and Jones, P. D. In: Applied Polymer Science"; Craver, J. K., Ed.; American Chemical Society, Advanced in Chemistry Series: Washington, D. C. 1975.
19. Nylen, P. and Sunderland, E, in "Modern Surface Coatings", Interscience: London 1965.
20. Licari, J. J. Crepean. P. C., U. S. Patent 3 205 157 (1965).
21. Schlesinger, S. I., U. S. Patent 3 708 296, 1973.
22. Schlesinger, S. I., Photo. Sci. Engr. 1974, 18(4), 387.
23. Watt, W. R. In "Epoxy Resin Chemistry"; Bauer, R. S. Ed.; ACS SYMPOSIUM SERIES 114, American Chemical Society: Washington, D. C., 1979.
24. Smith, G. H., Belg. Patent 828 841, 1975.
25. Crivello, J. V. and Lam, J. H. W., in "Epoxy Resin Chemistry"; Bauer, R. S., Ed.: ACS SYMPOSIUM SERIES 114, American Chemical Society: Washington, D.C. 1079.
26. Crivello, J. V. and Lam, J. H. W., J. Polym. Sci. 1976, 56, 383.
27. Crivello, J. V. and Lam, J. H. W., Macromolecules 1077, 10(6), 1307.
28. Gillham, J. K. In "Developments in Polymer Characterization-3"; Dawkins, J. V., Ed.; Applied Science: England, 1982.
29. Enns, J. B. and Gillham, J. K. J. Appl. Polym. Sci. 1983, 28, 2567.
30. Lucas, J. B., Industrial Finishing, 61(11), 37-40(1985).
31. Massingill, J. L. and Whiteside, R. C., J. Coating Technol., 65(824), 65-71(1993).
32. Galgoci, E. C., Shell Development Co. Brochure SC:2147-94 (1994).
33. Dubowik, D. A., and Ross, G. C., Paint & Coating Ind., 15(10), 60-72 (1999).
34. Eaton, R. F. and Lamb, K. T., Proceedings of the 23rd Waterborne, High-Solids, and Powder Coatings Symposium, pp252-264 (1996).
35. Raghavachar, R., et al., Proceedings of the 25th Waterborne, High-Solids, and Powder Coatings Symposium, pp.500-513 (1998) and RADTECH REPORT, 12(5), 36-40(1998).
36. Hung-Jue, S. et al., "Fracture Behavior of Rubber-Modified High-Performance Epoxies", in Polymer Toughening, Arends, C. B. ed., pp. 131-174. Marcel Dekker (1996).
37. Raghavachar, R., et al., "Rubber-Toughening Epoxy Thermosets with Epoxidized Crambe Oil", J. Am. Oil Chem. Soc. 76(4), 511-516 (1999).
38. Burton, B. L. and Bertram, J. L., "Design of Tough Epoxy resins", in Polymer Toughening, Arends, C. B., ed., pp. 339-379. Marcel Dekker (1996).
39. DuBois, R. A., et al., "Toughness in Thermoset Coatings", in ibid. pp. 381-409.
40. Misev, T. A., "Powder Coatings", p. 136. John Wiley & Sons (1991).
41. Wicks, Z. W., et al., "Organic Coatings", p. 208. Wiley-Interscience (1999).
42. Anderson, R. L. and Bohr, T., "Water SolubleAromatic Acids as Formaldehyde-Free Crosslinking Agents for Waterborne Epoxy Resins", Proceedings of the77th Ann. Mtg. Tech. Prog. of the FSCT, pp119-131 (1999).

43. Craun, G. P., "Epoxy Nucleophile Catalyzed Transesterification", J. Coat.Technol. 67(841), 23-30(1995).
44. Badou, I. in "Low VOC Coatings Demonstration Project Using Reactive Diluents Demonstration Project, EPA-600/R-98-043, pp. Ei-E17, ed. G. Roche.(1998).
45. Marx, E., "Preparing Hybrid Powder Coatings", Mod. Paint & Coat., Oct. 1999, pp. 41-44.

DEVELOPMENTS IN THE COMMERCIALIZATION OF WATER-SOLUBLE POLYMERS

J. EDWARD GLASS
North Dakota State University, Polymers and Coatings Department, Fargo, North Dakota 58105

Introduction
Synthetic Water-Soluble Polymers
Commodity Water-Soluble Polymers
Cationic Water-Soluble Polymers
Hydrophobe-Modification of Water-Soluble Polymers
Recent Products of Significance
References

Introduction

Those interested in a detailed discussion of water-soluble polymers are referred to two specific source[1,2] references. The following is a brief highlight of water-soluble polymer developments over the past 75 years. Water-soluble polymers did not just happen 75 years ago; they have been available for millennia in the form of polar carbohydrate polymers such as the carrageenans, alginates and other families, including the amylose and amylopectin component of starch. They exhibit a broad spectrum of properties arising from structural variations of essentially one type of monomer: the pyranose ring. The pyranosyl structure given in Figure 1 has all of the hydroxyls in the equatorial position and is designated glucopyranose or glucose, a molecule we all need in our metabolism sequence. The hydroxyls can be placed in other geometries to effect different reactivities, and in building a polymers from such rings, the linkages between rings can involve any of the hydroxyls.

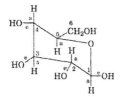

Figure 1. Structure of Glucopyranose Ring

From a recent industrial standpoint, the success of rayon and rayon acetate, as profitable fashion fabrics in the depression years undoubtedly placed some emphasis on exploring water-soluble derivatives of cellulose, the world's most abundant polymer. Cellulose (Figure 2) is an interconnection of the glucopyranose rings through the 1 and 4 hydroxyl positions; this leaves three hydroxyl per repeating glucopyranose ring. They form inter and intra hydrogen bonds in a planar projection of the chain due to the equatorial linkages

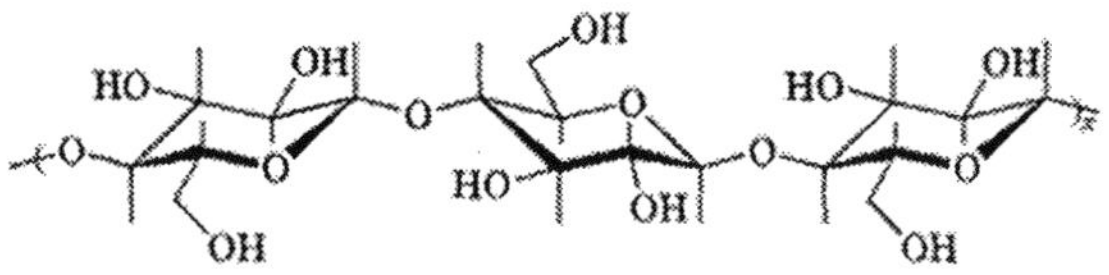

Figure 2. Structure of Cellulose

among the pyranose rings and the polymer is highly crystalline. This provides an complexity to the derivatization of cellulose that is addressed through variations of caustic, water and diluent phases. The first major efforts at commercial derivatizations of cellulose involved the addition of sodium hydroxide to form alkali cellulose (AC). In one area of derivatization the AC is reacted with methyl chloride or alpha-chloroacetic acid via a direct displacement reaction (Scheme I) to produce methyl cellulose and carboxymethyl cellulose, respectively.

$$RO^-Na^+ + R'X \longrightarrow ROR' + NaX$$

$$ROH + NaOH \rightleftharpoons RO^-Na^+ + H_2O$$

Scheme I Williamson Synthesis of Methyl and
           Carboxymethyl Cellulose

The most prominent of the early cellulose ether products was carboxymethyl cellulose (CMC) and the patent activity[3] from the thirties to the fifties in this area highlights the competitive nature of water-soluble polymer research. It is hard for the author to highlight any one contributor for the patent activities in this area were numerous. In another area of derivatization the AC is reacted with oxiranes (e.g., ethylene oxide or propylene oxide, Scheme II); this is a catalyzed addition and requires a much lower caustic to cellulose ratio than is used in scheme I. All four derivatives find numerous applications and there are other reactants that can be added to cellulose, including the mixed addition of the four adducts of commercial significance. In the commercial production of mixed ethers there are economic factors to consider that include the efficiency of adduct additions (ca. 40%), waste product disposal, and the method of product recovery and drying on a commercial scale. The products produced by Scheme I require heat and produce a mole of NaCl, a corrosive by-product., with each mole of adduct added. The products of scheme I are produced by a paste process and require corrosion resistant production units. The oxirane additions illustrated in Scheme II are exothermic, and with the explosive nature of the adducts, require a dispersion diluent in their synthesis.

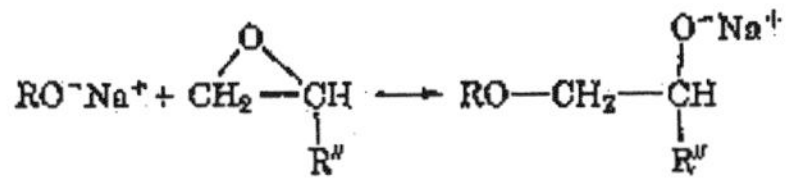

$$RO^-Na^+ + CH_2 \overset{O}{\underset{R''}{-}CH} \longrightarrow RO-CH_2 \overset{O^-Na^+}{\underset{R''}{-}CH}$$

Scheme II. Alkali Cellulose Reaction with Oxiranes

Cellulose as a consumer of ethylene oxide was proposed by A.E. Broderick[4] (Union Carbide) in 1937. He worked laboriously on developing a dispersion process (oxirane additions are exothermic and require a diluent for safety reasons). Unfortunately, a real commercial market was not realized until the early sixties, and Broderick retired in 1959. When large scale commercial acceptance came for hydroxy-ethyl cellulose [HEC] as a thickener in latex coatings, it was not anticipated, and the supply gap encouraged other producers into the field. Eugene Klug of the Hercules Corp. was quick to apply the caustic/water balancing strategies previously examined in CMC production to gain improved hydroxyethyl distributions to HEC prepared by a slurry process. These studies, complemented by more fundamental distribution studies across a broad spectrum of adducts[5,6,7] permitted the design of a process for the production of an improved bioresistant HEC[8]. This allowed removal of phenyl mercuric acetate from coatings formulations. With time other cellulose ethers were produced such as hydroxypropyl cellulose (a Eugene Klug, Hercules product designated Klucel). Similar controlled distribution of substituents studies were made by Sauvage (Dow Chemical) in developing methoxyl, hydroxypropyl cellulose derivatives. The addition of an epichlorohydrin derivative to hydroxyethyl cellulose by John Rutherford (POLYMER JR, Union Carbide) lead to a cationic cellulose water-soluble polymer that proved to be one of the most profitable water-soluble polymers of this century.

Cellulose ethers, in moderate molecular weights, exists as a random coil in aqueous environments, and as such, they have limited solution characteristics (e.g., poor viscosity retention with increasing temperature, and poor thermal and mechanical degradation behavior). The use of microorganisms has provided routes to structural variations that induce helix formation and thereby improve solution performance of carbohydrate polymers. The most successful commercial fermentation polymer is XCPS, synthesized by the bacteria, Xanthomonas campestris. The mainchain of XCPS is the same as cellulose, but XCPS has a three pyranosyl ring branched structure from the C-3 position (Figure 3) on every other repeating glucose ring. This arrangement promotes a helical conformation and a noted improvement in the solution properties mentioned above. With the proper ratio of nutrients and oxygen feed, a water-soluble polymer is produced and is accompanied by growth in the microorganism population. Both contribute to the viscosity of the medium and this limits the production process. Fermentation processes require more strenuous mixing and control conditions.

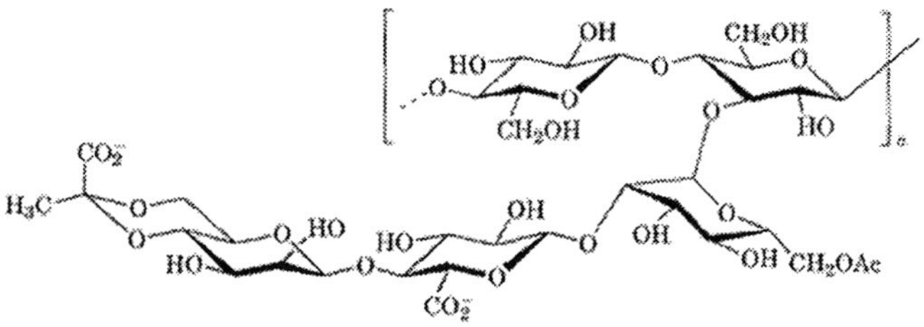

Figure 3. Structure of Xanthomonas Campestris PolySaccharide (XCPS).

Even though XCPS represented a marked improvement in solution properties and an excellent investment of tax payer funds at the USDA laboratories in Peoria, IL. , continued industrial fermentation synthesis studies have recently yielded more complex carbohydrate[9,10,11,12] structures , Wellan and Gellan, that provide additional improvements in solution performance properties. The structure of one of these products is given in (Figure 4).

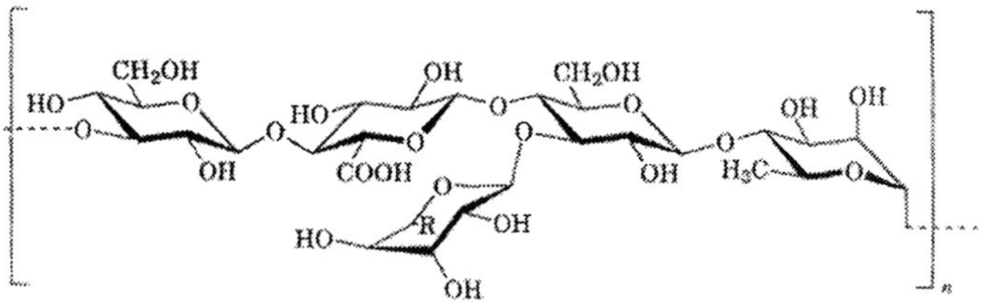

Figure 4. Primary structure of Wellan Gum

There has been growing activity in the biomodification of existing carbohydrate polymers, and while these types of studies may be too impractical to promote commercial activity in the near future, they are contributing to an understanding of structure/property relationships in aqueous media[13].

Synthetic Water-Soluble Polymers

Water solubility can be achieved through hydrophilic units in the backbone of a polymer such as - O- and -N- atoms that supply a lone pair of electrons for hydrogen bonding to water. Solubility in water is also achieved with hydrophilic side groups (e.g., -OH, -NH$_2$, -CO$_2^-$, -SO$_3^-$) These groups are found in carbohydrate polymers and in the synthetic polymers discussed below. In the carbohydrate polymers discussed above, their thickening efficiency was the parameter of importance. Among synthetic water-soluble polymers, poly(acryl-amide) partially hydrolyzed [HPAM], and poly(oxyethylene) [POE] are produced routinely in high molecular weights. These synthetic polymers at comparable molecular weights possess higher extensional viscosities than carbohydrate water-soluble polymers[14], and thereby are subject to greater mechanical degradation[15]. This difference is noted first on dissolution of the synthetic polymers, where attempts to dissolve high molecular weight water-soluble polymers resulted in mechanical degradation . This was addressed in HPAM synthesis by developing microemulsion polymerization techniques[16].

The interactions between water and polar groups contributes to their complex behavior in aqueous solutions. Comparison of four chemically similar synthetic W-SPs (Figure 5), poly(oxypropylene) [PPO, 5A], poly (vinyl ether) [PMVE, 5B], poly (vinyl alcohol) [PVALC,5C] and poly(oxyethylene) [POE, 5D] is given below.

-[(CH$_2$-CH)O]$_n$-   -[(CH$_2$-CH)]$_n$-
    CH$_3$ OCH$_3$
  A. PPO   B. PMVE

-[(CH$_2$CH]$_n$-   -[(CH$_2$-CH$_2$)O]$_n$-
  OH
  C. PVlc   D. POE

Figure 5. Structures of chemically similar synthetic W-S polymers

PPO with n>10 is not water soluble. The inclusion of the oxygen within the chain, and the shielding of its electron pair from interaction with water by the pendant methyl unit are responsible. Placing the oxygen pendant to the chain in PMVE produces a water-soluble polymer up to 50°C, even in high molecular weights.

Poly(vinyl alcohol) [PVAlc] is obtained by the hydrolysis of poly(vinyl acetate)[PVA]. The vinyl acetate monomer produces a very high energy radical during the chain growth propagation. A higher than normal head to head addition of propagating species occurs as well as grafting of some of the propagating species to polymer that is formed during an earlier stage of the polymerization. This leads to a more complex variation in structure than observed in PAM polymers (discussed in a section to follow). These differences with the differences that are realized with different methods of hydrolysis, can result in different PVAlc[17,18]. These factors, on hydrolysis, lead to PVAlc below 50,000 molecular weights (MW). Because of these complexities, more research has been expended on PVAlc than other W-S Ps. Descriptions of the industrial methods used in the synthesis of PVA polymer,[19,20] and the conversion of PVA to PVAlc have been discussed both from chemical and production aspects[21,22, 23]. PVAlc is isomeric with POE; however, PVAlc can enter into hydrogen bonds with both water and with the other hydroxyl units of the repeating polymer chain, forming both inter- and intra-hydrogen bonds. The extensive hydrogen bonding can lead to crystallinity, an occurrence that complicates its water solubility. Commercial PVAlc is essentially atactic; with most chain growth polymers, crystallinity is associated with stereo regularity, but the small size of the hydroxyl substituent group promotes crystallinity even in the atactic polymer. For this reason PVA is seldom hydrolyzed completely; it is manufactured retaining three ca. acetate levels: 25, 12 and 2 mole percents (these numbers of course represent averages). With higher acetate percentages the polymer is more surface active [important in its role as a suspending agent in poly(vinyl chloride) commodity resin production[24]] and it is readily soluble in water at ambient temperatures. With low acetate percentages, PVAlc is difficult to dissolve unless the water temperature is high, and on cooling the low acetate PVAlc may precipitate.

Poly(oxyethylene) [also referred to as polyethylene glycol (PEG) and poly(ethylene oxide}] is isomeric to PVAlc and has also been studied extensively. POE is truly unique in that it is soluble in almost any solvent. In water it exist in a nearly helical conformation. This occurs because of the unique interaction of water with the oxyethylene chain[25, 26, 27]. Anions are effective in disrupting the interaction between the oxyethylene units and water, particularly with increasing valence of the anion[28]. POEs and PEGs of various molecular weights have found numerous applications. The most recent interest has been in their use as anticoagulants through an effect on blood platelets[29,30]. There have also been extensive studies on and commercial use of ethylene oxide/propylene oxide block copolymers. The largest single commercial growth area for water-soluble polymer in the past decade has been their surfactant modification to produce "associative polymers". Before discussing the hydrophobically-modified, water-soluble polymers, a brief discussion of some of the commodity work horses of the water-soluble polymer area will be given.

Commodity Water-Soluble Polymers

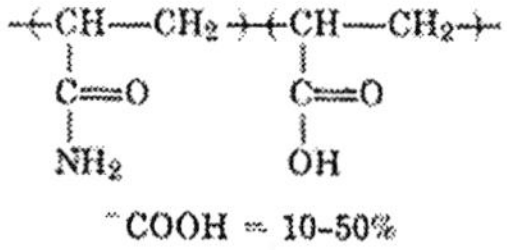

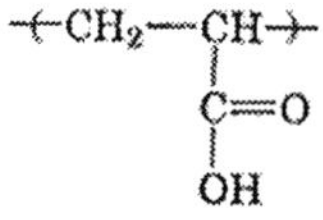

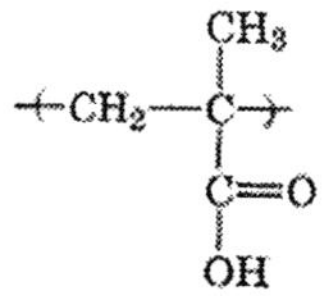

Figure 6. Commodity water-soluble polymers

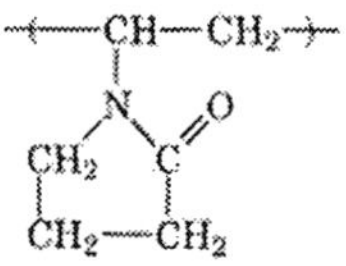

Hydrolyzed Poly(acrylamide) (HPAM)

HPAM can be prepared by a free radical process in which acrylamide (Figure 6A) is copolymerized with incremental amounts of acrylic acid or through homopolymerization of acrylamide followed by hydrolysis of some of the amide groups to carboxylate units. The carboxylated units, ionized, decrease adsorption on subterranean substrates, in proportion to the number of units, an important parameter in petroleum recovery processes. In waste treatment processes cationic acrylamide comonomer units are often used[31] to increase adsorption and thereby flocculation of solids in waste water. The favorable k_p, k_t and C_m characteristics of acrylamide facilitate the production of very high molecular weights[32]. The kinetic parameters are dependent on the pH of the aqueous media and salt effects. If HPAM is obtained by copolymerization with acrylic acid, there are strong salt effects, typical of ionogen polymerizations.

HPAM like POE discussed above exhibits an aging process in which the solution viscosity continues to change. This is most likely due to incomplete hydration of the dispersed polymer particle. Since the synthesis of HPAM can be conducted in water, the problem of dissolution in many applications is addressed by polymerizing the monomer in water in oil emulsions, where Aerosol OT, a branched anionic surfactant is the primary stabilizer. The thermal stability of poly(acrylamide) solutions are generally higher than most other water-soluble polymers; however, the high molecular weight species, like their POE equivalents, have very high extensional viscosities, favor-able for good drag reduction behavior, but this also reflects on their rapid degradation under mechanical deformation. The other negative aspect of PAM or HPAM aqueous solutions is hydrolysis of the amide group. This is most apparent in caustic flooding in secondary petroleum recovery applications. The hydrolysis of the amide to carboxylate groups is limited by the electrostatic repulsion of charged carboxylate groups, below 50 mole carboxylate units there is little hydrolysis[33] of the remaining amide units.

Poly(acrylic acid) [PAA] and Poly(methacrylic acid) [PMAA] .
Poly (acrylic acid) (Figure 6B) may be prepared by polymerization of the monomer with conventional free radical initiators using the monomer either undiluted[34] (with crosslinker for superabsorber applications) or in aqueous solution. Photochemical polymerization (UV sensitized by benzoin) of methylacrylate in ethanol solution at -78°C provides a syndiotactic form[35] that can be hydrolyzed to syndiotactic PAA. From academic studies, alkaline hydrolysis of the methyl ester requires a lower time than acid hydrolysis of the polymeric ester, and can lead to oxidative degradation of the polymer.[36]

Poly(methacrylic acid) [PMAA] is prepared only by the direct polymerization of the acid monomer; it is not readily obtained by the hydrolysis of methyl methacrylate. Free radical polymerization of the monomer is a commonly used route to prepare PMAA (Figure 6C) . Over the years this has been done for both PAA and PMAA in aqueous solution, using peroxy initiators such as hydrogen peroxide or persulfate ions. Hydrogen peroxide has the advantage over other initiators in that it does not leave organic or ionic impurities in the system. Redox systems also have been used, such as persulfate plus thiosulfate and ferrous ions plus hydrogen peroxide;[37,38] however, the latter system has the disadvantage that they lead to an appreciable iron content for the final polymer, which can lead to polymer degradation during applications[15]. Because of this the movement has been toward the use of water-soluble organic free radical initiators to avoid the transition metals present in application formulations. Both PAA and PMAA have multiple pK_as. This property has facilitated many application uses such as dispersing agents for pigments. Carboxylate groups have strong chelating tendencies with cations, particularly divalent ones. This may lead to precipitation of polyelectrolytes, such as PAA and PMAA, To take advantage of this while maintaining solubility, maleic anhydride is copolymerized with styrene. The copolymerization involves an electron-donor acceptor complex that produces an alternating copolymer[39]. The anhydride units are hydrolyzed to vicinal carboxylates which enhances the chelating ability over 1,3 carboxylate units, and the aromatic units are sulfonated to maintain solubility in high salinity environments. This type of product has found utility in boiler applications. Unfortunately, the degree of sulfonation of the aromatic unit has not been above 10 percent and this has limited its market potential.

Poly(Vinyl Pyrrolidone) [PVP] (Figure 6D) is another commercial polymer with significant usage. It was developed in World War II as a plasma substitute for blood[40]. This monomer polymerizes faster in 50% water than it does in bulk[41], an abnormality inconsistent with general polymerization kinetics. This may be due to a complex with water that activates the monomer; it may also be related to the impurities in the monomer (e.g., acetaldehyde, 1-methyl pyrrolidone , and 2-pyrroli-done) that are difficult to remove and that would be diluted and partitioned in a 50% aqueous media.

Cationic Water-Soluble Polymers

As noted above in the HPAM discussion cationic monomer are used to enhance adsorption on waste solids and facilitate flocculation[31]. There are several cationic that differ from the routes discussed above. One of the first used in water-treatment processes (Figure 7A) that is obtained by the cyclization of dimethyldiallyl-ammonium chloride in 60 - 70 weight percent aqueous solution.[42]. Another different cationic water soluble polymer, poly(dimethylamine-co-epichlorohydrin) (Figure 7B), prepared by the step-growth polymerization of dimethyl amine and epichlorohydrin also is used for adsorption on clays. This polymer stabilizes shale in drilling fluid formulations in petroleum applications. [43] In addition to these

Figure 7. Selected cationic water-soluble polymers.

A.   B.

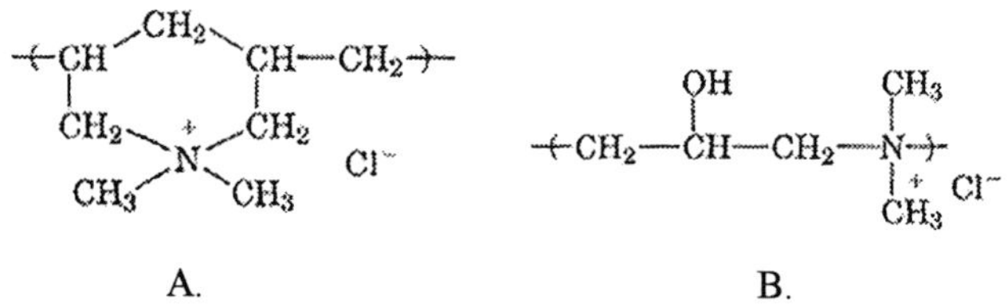

cationic polymers used in commodity applications, cationics find use in specialty products such as cosmetic applications. An example of this is the addition of an epichlorohydrin containing an amine group to hydroxyethyl cellulose (discussed earlier) At a molar substitution of 0.2 to 0.4 per repeating glucopyranosyl ring (Figure 1) the pendant amine provides substantivity to skin and hair[44] and has proven profitability comparable with Carbopol 940, an anionic hydrogel. The literature on the preparation and application of numerous synthetic cationic quaternary water-soluble polymers has been reviewed in detail.[45]

Hydrophobic Modification of Water-Soluble Polymers

The commercialization of a new water-soluble product takes time, as noted above in Broderick development of HEC. R. Evani (Dow) and others considered the surfactant modification of W-SPs for use in coatings and later in petroleum recovery processes. At the 50 year (Atlantic City ACS meeting in 1974) celebration of progress in Applied Polymer Science, Bob Lalk, now deceased, presented the Dow study of their surfactant-modified, water-soluble polymer, and the author (representing Union Carbide) gave a coatings rheology presentation. Ron Larson of Rohm and Haas asked numerous questions of both speakers. To my knowledge Larson's name did not appear on any of the patents issued to Rohm and Haas, but his interest reflected an interest by his organization, in defining a thickener for the coating industry. This product a Hydrophobically-modified, Ethoxylated Urethane (HEUR) is a well-accepted, commercial product. The general synthesis of this product by a step-growth process is given in Scheme 3.

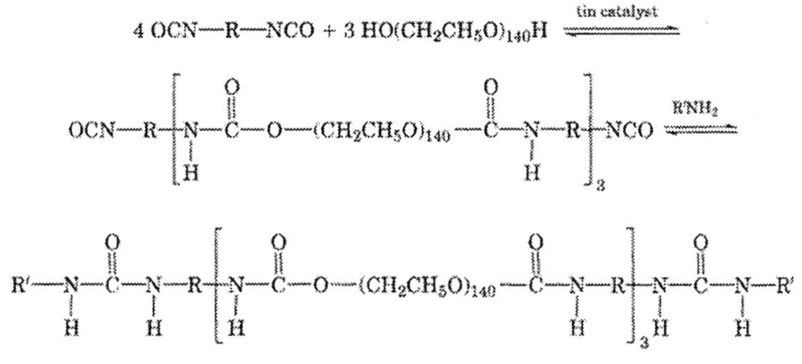

Scheme 3. Step-Growth Synthesis and Oversimplified Composite
          Structure of a Hydrophobe-Modified, POE W-SP.

This commercial product was followed by hydrophobically-modified, hydroxyethyl cellulose (HMHEC) and alkali-swellable emulsions (HASE). Significant scientific studies have been conducted in the area of hydrophobe modification of acrylamide. Much of this has been reviewed recently[46], and should have an impact on HASE technology.

Such products have met certain voids in a number of applications that include cosmetic, paper, architectural and Original Equipment Manufacturing coating areas, and have found unexpected application in the airplane deicers market. The driving force for the development of H-M, W-SPs in these areas is three fold in most applications areas:

1. The achievement of high viscosities at low shear rates without high molecular weights that were susceptible to mechanical degradation at high deformation rates. The regain of the high viscosities is lost when the application deformation rates are released if the high mole-cular weight polymer has been degraded. This is avoided with associative thickeners through the physical reuniting of the chains via hydrophobic associations.
2. Minimization of the elastic behavior of the fluid at high deformation rates that are present

when high molecular weight water-soluble polymers are used to obtain cost efficient viscosities at low shear rates

3. Providing colloidal stability to disperse phases in aqueous media, not achievable with traditional water-soluble polymers.

Associative polymers have been discussed in detail in the citations given in reference 2, including a recent symposium book based on a Boston symposium[47] held in 1998. Therefore, this class of water-soluble polymer will not be discussed in detail here.

Recent Products of Significance

As noted above the commercialization of new water-soluble polymers is often a slow process. In addition to the general production problems and appropriate market costs, new products face a variety of environ-mental controls today that add more constraints to market development. Two recently developed water-soluble polymers that have achieved limited market acceptance to date are poly(2-ethyl-2-oxazoline) [PEOX], prepared by the ring-opening polymerization of 2-ethyl-2-oxazoline (Dow) with a cationic initiator[48] (Figure 8), and poly(vinyl amine) and its copolymers with vinyl alcohol developed by Pinschmidt and coworkers[49] at Air Products.

Like POE, PEOX exhibits solubility in a variety of solvents in addition to water. It exhibits Newtonian rheology and is mechanically stable relative to other thermoplastics. It also forms miscible blends with a variety of other polymers. The water solubility and hot meltable characteristics promote adhesion in a number of applications.

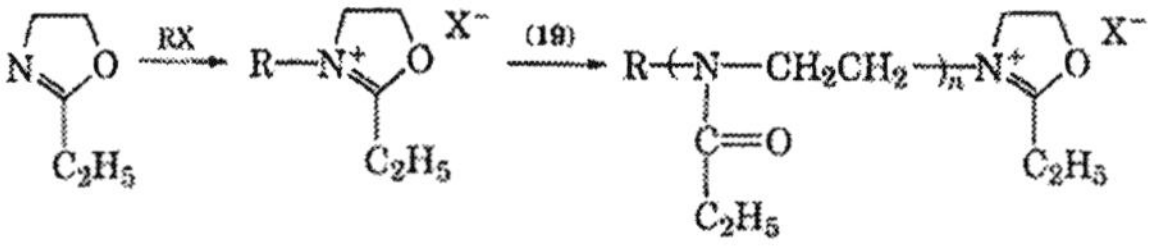

Figure 8. Synthesis of poly(2-ethyl-2-oxazoline)

Poly(vinyl amine) and poly(vinyl amine co-vinyl alcohol) have addressed the need for primary amines   and their selective reactivity. Prior efforts to synthesize poly(vinyl amine) have been limited because of the difficulty hydrolyzing the intermediate polymers. The vinyl amine polymer is prepared from N-Ethenyl-formamide, Figure 9, (prepared from the reaction of acetaldehyde and formamide) that is polymerized with a free radical initiator and then hydrolyzed.

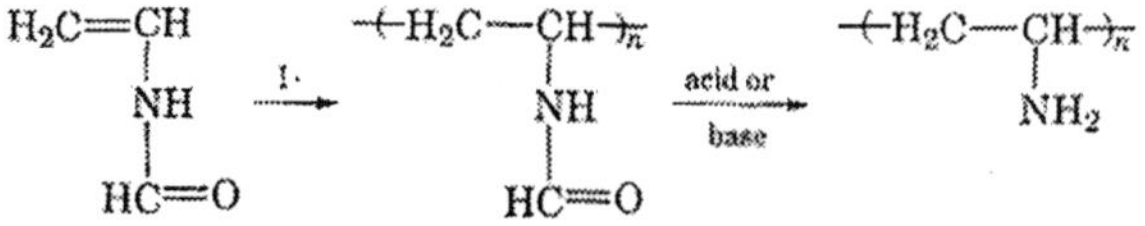

Figure 9. Synthesis of poly(vinyl amine)

The area of water-soluble polymers has grown at a steady rate over the past 50 years, and the polymers that have addressed particular application needs have been commercial successful. Some of the growth areas in the next millenium will be in the biological area. In more industrial applications manufacturing costs relative to established products will be a primary determinant in the success of any new product, but imaginative people will overcome this.

References

Molyneux, Philip, *Water-Soluble Synthetic Polymers: Properties and Behavior*, volumes I and II, CRC Press, Boca Raton, Florida, **1982**.

[2] A. Advances in Chemistry Series 213, *Water-Soluble Polymers: Beauty with Performance*; **1986**; B. Advances in Chemistry 223: *Polymers in Aqueous Media, Performance through Association*; **1989**; C. Advances in Chemistry Series 248: *Hydrophilic Polymers: Performance with Environmental Acceptance*, **1995**, all edited. by Glass, J.E., and published by the American Chemical Society, Washington, DC.

[3] Rafael Bury, M. S. Thesis, North Dakota State University, **1993**.

[4] A.E. Broderick, internal report issued within Union Carbide, August, **1937**.

[5] Klop, W.; Kooiman, P., Biochim. Biophys. Acta **1965**, 99, 102.

[6] Wade, C.P.; Roberts, E.J.; Rowland, S. P.; J.Polym.Sci., Part B; **1968**, 6, 829

[7] Ramnas, O.; Samuelson, O.; Sven. Papper stidn **1968**, 71, 829.

[8] Glass, J.E., Buettner, A.M., Lowther, R.W., Young, C. S., and Cosby, L.A., *Carbohydr. Res.*, **1980**, *84*, 245.

[9] Moorhouse, R., Industrial Polysaccharides: Genetic Engineering, Structure/Property Relations and Applications, edited by Yalpani, M., Elsevier Science Publishers B.V., Amsterdam, **1987**.

[10] Talashek, Todd A.; Bryant, David A.; Carbohydr. Res., **1987**, 303.

[11] Powell, J.W.; Parks, C.F.; and Seheult, J.M.; Soc. Petrol. Eng. Publ. No.5339, April, **1975**.

[12] Kang, K.S. Pettitt. D.J.; in Industrial Gums, Academic Press, Inc. **1993**.

[13] "Biopolymers as Advanced Materials",' Symposium conducted at the Anaheim, CA, Spring National ACS meeting , Preprints of presentations provided in Polym. Matl.:Science and Engin., Volume 72, **1985**.

[14] Soules, D. A., Fernando, R.H., Glass, J. E. Glass, "Application of and Complexities in Elongational Flow of Water-Soluble Polymers", Proc. ACS Div. Polym. Materials: Sci. & Engin., 57, 118-122(1987).

[15] Glass, J.E., Ahmed, A., Soules, D.A., Egland-Jongewaard, S.K. and Fernando,R.H., Soc. Petrol. Eng. Publ. No. 11691, **1983**.

[16] J.W. Vanderhoff, studies, originally at Dow Chemical and continuing studies at Lehigh University.

[17] Tubbs, R.K., Inskip, H.K., and Subramanian, P.M., in Properties and Applications of Polyvinyl Alcohol, S.C.I. Monograph No. 30, Finch, C.A., Ed., Society of Chemical Industry, London, **1968**, 88.

[18] Tubbs, R.K., J. Polym. Sci. Part A-1, 4, 623, **1966**.

19. Shohata, H., in reference 21, 18.
20. Noro, K., in Polyvinyl: Properties and Applications, Finch, C.A., Ed., John Wiley & Sons, London, **1973**, Chapters 3 and 4.
21. Hackel, E., in Properties and Applications of Polyvinyl Alcohol, S.C.I. Monograph No. 30, Finch, C.A., Ed., Society of Chemical Industry, London, **1968**, 18.
22. Pritchard, J.G., Poly(Vinyl Alcohol): Basic Properties and Uses, (Polymer Monographs, Vol.4, Gordon & Breach, New York, **1970**.
23. Finch, C.A., in Polyvinyl Alcohol: Properties and Application, Finch, C.A., Ed., John Wiley sons, London, **1973**, chap. 9.
24. Glass, J.E. and Fields, J.W., *J. Appl. Polym. Sci.*, **1972**, *16(9)*, 2269.
25. Bailey, F.E. and Koleske, J.V., Poly(Ethylene Oxide), Academic Press, New York,1976.
26. Koenig, J. L. and Angood, A.C., J. Polym.Sci.,1970, 8,Part A-2, **1787**.
27. Hey, M. J.; Ilett, S. M.; Mortimer, M. and Oates, G. ; J. Chem. Soc. Faraday Trans, **1990**, 86(14) 2673.
28. Lundberg, R. D.; Bailey, F.E.; Callard R.W., J. Polymer Sci., Part A-1, **1966,**1563.
29. Andrade, J.D.; Hlady, V. and Jeon, S.-I.; Chapter 3, reference 2C.
30. Li, J.; Carlsson, J., Huang, S.-C.; Caldwell, K. D.; Chapter 4 in reference 2C.
31. Rey, P.A. and Varsanik, Chapter 7 in reference 2A.
32. Beihoffer, Thomas W.; Lundberg, David J.; Glass, J. Edward; Chapter 8 in reference 2B.
33. Klein, J.and Heitzmann, R., Makromol. Chem., 1978, 179, **1895**.
34. Greenwald, H.L. and Luskin, L.S., in Handbook of Water-Soluble Gums and Resins, Davidson, R L., ed., McGraw-Hill, New York, **1980**, Chapter 17.
35. Monjol, P., C.R. ACAD. Sci. Ser. C, **1967**, 265, 1426.
36. Katchalsky, A. and Eisenberg, H.; J. Polym Sci., **1951**, 6, 145.
37. Oth, A. and Doty, P., J. Phys. Chem., **1952**, 56, 43.
38. Arnold, R. and Caplan, S. R.; Trans. Faraday Soc. **1955**, 51, 857.
39. Odian, G., Principles of Polymerization, 3 rd ed; John Wiley & Sons: New York, **1991**, Chap. 5.
40. Ravin, H.A., Seligman, A.M., and Fine, J., N. Engl. J. Med., 247, 921, **1952**.
41. Senogles, E.; Thomas, R., J.Polym. Sci. Sympos., **1975**, 49, 203.
42. Butler, G.; Zhang, N. Z.; in Water-Soluble Polymers, ed Shalaby, S.W.; McCormick, C.L.; Butler, G.B.; ACS Symposium Series **467**; chapter 2.
43. Beihoffer, T. W.; Dorrough, D. S.; Deem, C.AK. Schmidt, D.D.; Bray, R.P. ; Oil and Gas J.,**1992**, ,May 16 issue, 47.
44. Goddard, E.D.; Leung, P.S.; Langmuir, **1992**, 8,1499.
45. Hoover, M.F.; J. Macromol. Sci.-Chem., **1970**, A4, 1327.
46. Candau, F.; Selb, J,; *Adv. Colloid Interf.Sci.*, **1999**, *79(2-3)*149.
47. Associative Polymers in Aqueous Solutions, ACS Sympsoium 565, ed., Glass, J. Edward; American Chemical Society, August 2000.
48. Chiu, Thomas T.; Thill, Bruce P.; Fairchok, William J., reference 2A, Chapter 23.
49. Badesso, R.J.; Nordquist, A. F,; Pinschmidt, R.K.; Sagl, D.J.; reference 2C, Chapter 25.

Section 3

COATINGS

Section Editors: Kenneth N. Edwards and Hermenegildo Mislang

HISTORY OF COATINGS

KENNETH N. EDWARDS and HERMENEGILDO B. MISLANG

Dunn-Edwards Corporation
4885 East 52nd Place
Los Angeles, CA 90040

Introduction
Early Varnishes
Alkyd Resins
Emulsion Resins
Specialty Resins
Regulatory Influences
Technical Milestones

Introduction

To start, it may be valuable to have a few definitions to provide direction. They are as follows:

Paints, Varnishes, Lacquers and Stains - Coatings to protect an object or substrate from deterioration by its environment and/or to create an aesthetic effect.

Paints, or more exactly organic coatings are composed of vehicles, pigments, dryers, special effect additives, and solvents or dispersants which are formulated together to achieve a desired end product.

Vehicles - The vehicle can truly be called the heart of the coating, Thus, a history of coatings is in large part a history of the vehicles from which they were created. Whether the finish be the pure vehicle, as in varnish or in a clear lacquer, or whether it be pigmented, no coating can be better than the vehicle which binds it together. Basically, there are two types of vehicles (reactive and spirit varnishes); although for clarity's sake a third type (emulsion) is usually added and considered as a separate entity. It has become, in recent years, increasingly hard to place an organic finish in a specific category because of a trend by modern coating chemists to cross over set lines and combine resins with greatly differing properties to make coatings which are a combination of types.

Since before man emerged from the caves he was recording his dreams, his goals, his aspirations, and his gods through paintings on the walls; the paints probably created through the combination of bone marrow and/or egg with colored earths. Much of what we know about early man comes from such painted records.

<u>Early Varnishes</u>

The term varnish probably comes from the French vernis which in turn might come from the Latin vitrum which means possibly a glass. The other origin is possibly from mythology in the form of the story of Bernice whose hair was a glossy amber and went into the air to form the milky way. Varnish is called amber in German. The first written records of varnish appear about 350 BC when the writer Pliny told of applying a composition of his own. The Egyptians 1300 - 1400 BC had an oleo-resinous varnish which they applied, probably hot, and which has not cracked today. Even before this period, the Japanese had an Imperial Lacquer Maker. Lucca-resin dissolved in oil was a good varnish used in the early days.

A typical varnish of about the 7th and 8th century was as follows:

linseed oil	4 parts	sandarac	3 parts
turpentine resin	2 parts	cherry tree gum	2 parts
galbanum	2 parts	almond tree gum	2 parts
larch resin	3 parts	fir tree gum	2 parts
myrrh	3 parts	unknown substance	1 part
mastic	3 parts		

The gums were powdered and boiled with oil and then if the mix was not too thick it was strained. There was little oil and a great amount of resins.

A manuscript by the monk Theophilus from the Eleventh Century for a "Varnish Glutten" has survived (see R. R. Donnelly and Sons 1923 Glass Paint, Varnishes and Brushes - Their History Manufacture and Use. Lakeside Press for Pittsburgh Plate Glass Company). The entire formula and its manufacture is quoted in the original Applied Polymer Science Edition by Howard Gerhart and will not be repeated here. However, the actual formula is in the first sentence:

> "Put linseed oil into a small new pot and add, very finely
> powered, a gum which is called fornis, which has the
> appearance of the most lucid."

The "fornis" referred to by the monk was most probably sandarac gum. In processing this varnish it was boiled until about a third of the volume was lost. In the manuscript there is mention of cooling but the varnish was probably reheated and applied hot.

Oils with leads were used a bit later. By 1350 the varnishes were still very thick. In 1440, the formula of Theophilus was still used but now 1/4 ounce of alum and 1/2 ounce of incense were added to the oil when boiling and the mix allowed to boil for three or four hours more. The surface was burned for the length of time needed to say three Pater Noster's.

The first use of yellow amber as a primary resin came in the year 1575. That same century one chemist used:

1 lb. melted amber	He had no thermometer but held a piece of garlic a given
3 lb. linseed oil	distance from the oil and when the odor was enough to
1/2 oz.. turpentine	drive him from the room the mix was done.
1/2 oz. burnt alum	

A great amount of excellent research into varnishes of the middle ages was done by Joseph Michelman in his search for the secret of the great violin makers of the period 1550 to 1750 with regard to the finishes on their violins. (see Violin Varnish, Joseph Michelman 1946, Private publication for Joseph Michelman by W. B. Conkey Company, Hammond, Indiana, U.S.A.).

Without going into the many reasons for the nature of the finish, it is sufficient to observe that the pretreatment for the instruments was linseed oil either raw or sun-thickened and sun aged as were all of the subsequent coats also so cured.

The sophistication of the painters, varnishers and gilders of the period was interesting. The wooden substrate was apparently not dyed as we do today, but rather was coated with a transparent but tinted basecoat. In the case of violins, usually a yellow shade. The varnish for this usage were manufactured by the user and for the first time showed multistage production techniques. Using procedures common to the soap and dye industries of this period, apparently a potassium rosinate solution was made using potassium hydroxide and gum rosin. Then a second stage was prepared by precipitating aluminum or iron rosinates from the potassium rosinate solution. These rosinates were filtered, washed and dried. The varnishes were created by dissolving the dried rosinates in turpentine and raw or sun bodied linseed oil. Madder was well known to the dyeing industry of that time as well as the use of alum and zinc sulfate as mordants. Yellows and various shades of brown and red tones were thus available for decoration.

The resultant coatings became turpentine insoluble quite rapidly when exposed to the strong Italian sun but very slowly if not. In Northern Europe, cooked varnishes containing reinforcing resins would have been required in place of the raw linseed oil for satisfactory results.

Alberta was the first to use thinner after boiling. Watin in 1773 wrote an article on manufacture and prices. The first patent for varnish was issued in 1763.

In 1773 Watin in England described Varnish technology reciting formulations utilizing copals and amber, linseed, walnut, hempseed and poppyseed oils with turpentine as the solvent. His book was reprinted (with few and minor changes) fourteen times between 1786 and 1900.

With the advent of the lead pigments, the paint industry broke down into coatings formulated from elaborate blends of oils with resins from every part of the globe combined with numerous pigments on the one hand, and the (to us) traditional lead and oil mixture of white or red lead plus raw and or bodied linseed oil on the other hand. Semi-modern organic coatings (approximately the first 55 years of this century) were interesting combinations of science and art.

Many of the today's coatings systems depended upon research that was done in very early years, a lot of it in the 1800's, but had to wait for application to coatings until such time as other processes created a need for commercialization and thus the availability of the particular resin being considered. Styrene is a classical example of this.

Styrene monomer was first prepared in 1839 from a naturally occurring gum at which time its polymerization to a low molecular weight oligomer was also observed. It took nearly one hundred years to put the polymerization process of styrene on a scientific basis. By 1950 the U.S. production of styrene finally reached 50 million pounds per month.

In the comprehensive Mattiello series published from 1941 to 1946 there is no mention of styrene. Yet, only 25 years later styrene was either the workhorse or the modifier in many polymer coatings. In some respects it replaced rosin as the economic monomer of choice in the building block technique of resinous molecular structure. Inert, generally, coreactive, decently durable when attached to exceedingly durable monomers, its brittleness as a homo-polymer was controlled by plasticizing resins or chain modification; for instance, vinyl fluoride and vinylidene fluoride.

During the period surrounding World War II several significant pieces of research were adapted into coatings that have had significant influence on coatings in the later half of this century. These advances include alkyd resins, urethane and epoxy resins and emulsions. Because each of these have their own chapters, we will only treat them briefly in our discussion.

Alkyd Resins

The early research resulting in alkyd resins was done by a number of researchers led by Carothers[1] and Kienle[2] who provided a clearer understanding of the chemical reactions governing alcohol-acid polymeric resins from multi-functional monomers. Alkyd was a term coined by Kienle from alcohol and acid. In common usage alkyds refer to unsaturated polyesters where the unsaturation is in the side chain from attached natural oil fatty acids. The first alkyd resins in this country were the Glyptols developed by General Electric based upon the aforementioned research to develop better electrical insulation. By the early 1950's alkyds of all types were the predominate resins of the coatings industry.

By the mid seventies market and regulatory trends were working against alkyd resins. Rule 66 of Southern California based upon early smog chamber work and pushed by a desire to clean the Southern California atmosphere of its smog started to define the nature of solvents that could be used in coatings.

During the early eighties, coatings manufacturers throughout the regulated areas of the country developed and marketed higher solids and water reducible alkyd paints to meet the lower volatile organic contents required by the new regulations. These compliant alkyds were not readily accepted by the consumer markets because they exhibited much earlier degradation in color and gloss retention and poorer drying properties. The higher solids content also affected the proper application of these coatings. The increased "yellowing" of these compliant alkyds also kept them from being used in the decorative market heading towards very light pastel colors.

This trend has been amplified by EPA and ultimately prevented alkyds from being used in large areas of the surface coating markets. The great expansion of do-it-yourselfers who wished for easy cleanup created an explosion in the emulsion resin business and new generations of light fast, durable cross-linking resins with low solvent demand eroded what had been traditional alkyd markets.

Emulsion Resins

Probably the most important change of the last 40 years has been the introduction and growth of emulsions as a coatings resin source. The first scientific reference to surface tension and emulsification came in 1888 by Quiche[3]. It was not until the 1950's that commercial styrene-butadiene emulsions started getting serious attention.

Rohm and Haas Company had developed large scale production of acrylic and meth-acrylic polymers by the 1950's and by 1970 the solvent based versions were being replaced by emulsified offsets. Emulsions were initially homo-polymers with external plasticizers followed by co-polymers of large particle size. These were followed by increasingly smaller particle sizes for improved film forming and pigment binding properties. Recently, larger particle sized polymers have again been used to control penetration into a substrate. The addition of emulsifiable alkyds was also used to control chalking.

Emulsions did not form solid films therefore the coatings could pass water vapor without blistering. They were easy to apply, with moderately easy clean up, and had relatively low solvent content. They have become the number one product for architectural and industrial coatings. Architectural coatings manufacturers rely heavily on the use of emulsion resins in their products. Pure acrylic emulsions and vinyl acetate-acrylic copolymer emulsions have been most popular for use in housepaint products. In the early eighties, styrene-acrylic and other self-crosslinking emulsions offered the ability to improve the non-blocking characteristics of the latex non-flat enamels. In the late eighties, a major consideration of exterior coatings was their resistance to alkali attack and effloresence from masonry surfaces. This promulgated the use of alkali resistant emulsion based primers and waterborne epoxy modified emulsions for these surfaces. Since the early nineties, coatings manufacturers have introduced exterior coatings based on acrylic, vinyl-acrylic copolymers and terpolymers, ethylene-vinyl-acrylics and vinyl versatate based polymer emulsions providing improved resistance to alkali and the effects of alkaline attack.

Specialty Resins

There are a number of resins and techniques of application that are also significant in the last forty years. Catalytic epoxies which were developed simultaneously in Switzerland by Ciba and Jones Dabney in the United States have excellent chemical and food bacteria resistance and have become industrial and FDA standards for difficult areas. The development in the 1960's of aliphatic isocyanates by Bayer created a whole series of extremely color and gloss retentive coatings especially when cross-linked with acrylic polymers. Silicone resins provide durability and high temperature resistance. Teflons create low energy surfaces and prevent your eggs from sticking to the pan. Electro deposition of primer coatings allows the application of uniform primer, even in hard to reach corners.

Regulatory Influences

From 1970 thru today, the U.S. Environmental Protection Agency and other state regulatory agencies developed regulatory programs that greatly influenced the design and development of modern coatings products. Southern California's Air quality Management District was was foremost in leading these programs.

Because all of the subsequent regulatory rules including EPA's newly announced National Rule are based upon the Southern California Air Quality Management District's rules, especially Rule 1113, it is important to consider the background and effect of these rules on the coatings industry in California.

In 1975, a study by the Southern California Air Quality Management District assumed a substantial VOC inventory contributed by the coatings industry. These assumptions promulgated the development of new, more stringent regulations that would direct the reduction of solvents used in both architectural and industrial coatings. In 1977, it was Southern California Air Quality Management District's Rule 1113 that limited the volatile organic content of architectural and industrial coatings sold in areas under its jurisdiction and promulgated the development of higher solids solventborne raw materials and architectural coatings.

With Rule 1113 of Southern California in place, architectural alkyd coatings were formulated with higher oil content and at lower molecular weights to achieve lower viscosities because of the reduced solvent levels permissible. Another technical option was to develop water-reducible alkyds which had the ability to be carried in water because of amine-neutralized sites on the alkyd backbone which allowed miscibility in water.

The coatings products formulated with both higher solids and water reducible technology exhibited unacceptable performance in resistance to the effects of water, sunlight and UV radiation, erosive forces from the weather and physical contact. Blocking, adhesion and severe after-yellowing were also a common problem with these types of compliant coatings.

Resins and coatings initially developed and formulated had drawbacks in performance areas which were not acceptable to both coatings manufacturers and end users or consumers. Solventborne architectural coatings developed and manufactured to comply with the lower VOC's of Rule 1113 exhibited poor durability when compared to conventional products of the same resin kind.

Other coatings categories such as epoxies, polyurethanes, acrylics, and nitrocellulose were also forced to be formulated with lower solvent contents. The manufacturers of these coatings materials also tried to develop commercially acceptable high solids and water-reducible versions usable by the coatings manufacturers.

During the early eighties, waterborne and high solids epoxy and polyurethane coatings resins entered the marketplace. For regulated areas of the country, this allowed coatings manufacturers options to meet air quality regulations requiring the use of lower solvent content. Because of the lower performance levels of the new compliant coating systems, they did not achieve commercial acceptance by the marketplace. Many companies that produced these early lower VOC coatings encountered many field problems that produced unsatisfactory performance compared to their higher VOC counterparts. Some of the early lower VOC and zero VOC coatings exhibited early loss of gloss, poor color retention, poor corrosion resistance, poorer adhesion, slower drying properties, unwanted softness, poorer chemical and solvent resistance and difficulty in application. Most recent technology has improved some areas of performance, but in general, lower VOC and zero VOC coatings today do not possess the performance properties required to protect modern day surfaces and the exterior climates to which they are exposed.

Revisions of Rule 1113 have also impacted waterborne architectural coatings. Coatings based upon latex emulsions have also been forced towards zero VOC's by regulation. This reduction has been at the expense of coalescing solvents in the film. Major emulsion manufacturers developed specialized emulsions that did not require the use of coalescing solvents to form films. Versions of these emulsions offered usefulness in some interior paints, but were not practical for exterior purposes. Initial zero or low VOC paints provided the optical and aesthetic properties of conventional latex paints but lacked the performance properties needed to make them widely used.

Technical Milestones

When Howard L. Gerhart of PPG wrote a similar chapter to this one for the Fiftieth Anniversary issue[4] he included an excellent Table of Technical Milestones which is printed below and upgraded somewhat.

Table 1. Chronology of Technical Milestones in Coatings History

1804	Basic Carbonate White Lead, Dutch Process
1815	Varnish Production
1865	Coatings Patent Issued, U.S. 50068
1867	Ready-Mixed Paints
1908	Mie Theory
1910	Casein Powder Paints
1912	Patented Acrylic Resin

1915-1920   Phenolic Varnish Resins Conventional Spray Gun application

1921-1925   Low Viscosity Nitrocellulose Alkyd Resins

1921-1925   Anatase Titanium Dioxide Maleic-Rosin Gums

1926-1930   Oil Soluble Phenolics

1930-1935   Urea Formaldehyde and Alkyd blends, Chlorinated Rubber,   Molybdate
             Orange Pigments
             Kubelka-Munk Concepts of Opacity, Vinyl Chloride Copolymers
             Oil Based Emulsion Paints, Trichromatic Colorimeters, Phosphating Metal
             Treatments

1936-1940   Organotin Stabilizers, Phthalocyanine Pigments, Polyurethane Resins,
             Polytetrafluroethylene, Intumescent Coatings, Wash Primers
             Melamine Formaldehyde and, Alkyd Blends

1941-1945   Styrenated, Acrylated, Cyclopentadiene Oils
             Rutile Titanium Dioxide, Silicone Resins, U V Radiation Absorbers

1946-1950   Epoxy Resins, PVA Emulsion Paint

1961-1965   TwentyYearDurability, Coil Coatings, Thixotropic Emulsion Paints
             Electrodeposition Curtain Coating, Computer Color Control
             Electrostatic Powder Spray, Fluorocarbon Resins

1966-1970   Zinc Phosphate Pigments, Semigloss Latex Paints,
             Radiation Curable Coatings
             Non-Aqueous Dispersions, Rule 66 Reformulations

1970-1975   High Speed Dispersions, Aqueous Industrial Enamels
             Electron Beam Curing, Spray Application Efficiency
             Cationic Electrodeposition, Ultraviolet Curing, Pittmentized Paints

1976-1980   U V Curable Coatings, High Solids Coatings
             Silicone Latex Paints, Titanium Dioxide Extenders
             Water Reducible Resins, Water Soluble Resins

1981-1985   Vesiculated Beads, High Performance Pigments, Polyurea Resins
             High Solids Alkyd Paints, Glossy Latex Paints
             Non-Blocking Latex Paints, Low Odor Latex Paints

1986-1999   Core Shell Resins, Opaque Polymers, Waterborne Epoxy Coatings
             Waterborne Polyurethanes ,Zero VOC Paints

1986-1999   Elastomeric Latex Paints, Hydrophobic Latex Resins

Literature Cited

1.   Solomon, <u>The Chemistry of Organic Film Formers</u>, p. 75
2.   Solomon, <u>The Chemistry of Organic Film Formers</u>, p. 83
3.   Quiche, Wiediann 35, 589 (1888)
4.   Craver, J. Kenneth, and Roy W. Tess. <u>Applied Polymer Science</u>.

Additional Reading

1.   Bull, Sara, <u>Ole Bull, a Memoire</u> (1883)
2.   Heron-Allen, <u>Violin Making</u> (1885)
3.   Mach, <u>Monatsch</u>, 15, 633 (1894)
4.   Muller-Jacobs, <u>Journ. Soc. Chem. Ind</u>., 6, 138 (1887)
5.   Schweitzer, <u>Distillation of Resins</u> (1905)

HIGH PERFORMANCE INDUSTRIAL COATINGS

THOMAS M. SANTOSUSSO

Air Products and Chemicals, Inc.
7201 Hamilton Boulevard
Allentown, PA 18195-1501

Introduction
General Principles of Design of High Performance Coatings
Epoxy Binders
Polyurethanes
Melamine Resins
Fluoropolymers
Silicone Resins
Other High Performance Coating Resins
Conclusions

Introduction

Coating systems can be thought of as serving two major functions – protection and decoration. When the protective function clearly becomes the more important of the two, especially under service conditions of exposure to particularly aggressive agents of corrosion or chemical attack, or to extremes of temperature, radiative energy or wear, the appropriate coatings may be fairly labeled as "high performance." Such conditions are often found in coating applications such as aerospace, marine, and military equipment; power generation, waste treatment, and petroleum and chemical processing sites; and pipeline and tank facilities, including those involving secondary containment. The usual substrate for such applications is almost always a metal or concrete, although more recently protection of composite structures or ceramics has gained some importance. In this review, emphasis will be placed on the application and performance properties of the organic binder systems used in such coatings. Particular stress will be laid on field-applied coatings, as opposed to factory-applied or Original Equipment Manufacturer (OEM) coatings. Thus coatings which are capable of application and full property development under ambient conditions are treated most extensively here, though others will be discussed.

All the components of the coating systems used in high performance applications are important, from surface preparation and application methods to the components of the coating itself – pigment, binder, additives and carrier liquid. All these have evolved significantly over the years as the successful operation of our modern technological society has come to rely more and more on that "thin coat of paint." However, as mentioned above, in this review special (though not exclusive) attention will be paid to the polymeric binder materials, particularly those which have been developed in more recent times to meet those needs. For this reason, with the exception of the initial section which outlines the general considerations involved in designing a high performance coating, the review is organized by binder types. Although this approach inevitably leads to some overlap and possible ambiguity in organization, it does provide for easy accessibility and hopefully a good overview of the fields included. Each section includes a brief summary of the chemistry involved, a short history of the materials traditionally employed, a review of modern developments and finally some speculation as to the future of the technology.

General Principles of Design of High Performance Coatings

As alluded to above, a number of factors must be considered in the design of a high performance coating. These have been summarized in several excellent reviews.[1] The most significant design elements obviously are those associated with the overall properties of the intact coating film. The most obvious of these is the ability of the coating to act as a barrier to the external environment. The most important factor in that regard is the permeability of the film to agents which can cause degradation of the underlying substrate. These agents include: oxygen* ; solvents; acids, bases, oxidants and other reactive chemicals, including gases and vapors; and water vapor and aqueous solutions , particularly solutions of electrolytes. Just as important is the maintenance of these properties throughout the expected service life of the coating. This in turn depends upon the ability of the coating film to remain intact and adherent to the substrate in the face of attack by these agents while resisting changes due to weathering, temperature cycling, cleaning, abrasion and impact.

Film properties in turn are dependent on the properties and interactions of all the components present in the final cured coating. These include the binder resin; fillers and pigments, including color-carrying and opacifying pigments, inert and reinforcing fillers, and so-called active fillers which react with environmental agents to inhibit corrosion of the substrate; and the host of additives which can be present, such as pigment dispersants, wetting agents, flow and leveling agents, other surfactants, rheology modifiers, defoamers, deaerating agents, adhesion promoters, coupling agents, flash rust inhibitors, antioxidants, photostabilizers, plasticizers, catalysts, and so on. Although this review is centered on the properties of the organic binder, it must

---

* In metal coatings, the rate of oxygen transmission (as well as water vapor transmission) through most films is so high as to discount the simple concept of a barrier function as an effective means of corrosion control. However, with adequate wet adhesion and low ionic conductivity, the lower the permeability of oxygen and water, the better protection the film will provide. See W. Funke in *Surface Coatings*, A. D. Wilson, J. W. Nicholson and H. J. Prosser, eds., Elsevier, London, 1988, p. 107.

be remembered that all the other species present exert their effects, alone and in concert with the other components, on final film performance. The situation is further complicated by the fact that in most coatings the pigments and fillers represent a discontinuous phase which is mechanically coupled to the surrounding (usually) organic matrix to a greater or lesser degree, depending on the inherent binder-filler interaction and the presence or absence of pigment surface treatment and dispersing or coupling agents. This gives rise to the all important consideration of pigment volume concentration and its relation to critical pigment volume concentration, a topic which has been treated exhaustively.[2] The resulting paint film is of course really a complex composite. In fact, the mechanical properties of paint films as they relate to resistance to damage (of obvious interest in protective coatings) are probably best understood using methods first developed for structural composites, a relatively recent and promising approach.[3]

Even before considerations of final coating film properties, the details of film formation (the conversion of the free flowing liquid coating in the application/drying/curing stage to the final solid coating film) must be taken into account. The factors involved in the development of a dense, adherent, defect-free film during the coating application and drying process have been the subject of intensive study.[4] There are a number of often competing phenomena which operate during application and drying: the response of the coating liquid to the shear forces generated by the application method; the flow and leveling of the coating liquid into a conformal film; wetting of the substrate, with displacement of air, water, or impurities from the substrate by the coating liquid along with the release of gases trapped or generated during the coating process; in the application of solution or dispersions, the evaporation of the carrier liquid with concomitant generation of concentration and surface tension gradients; with dispersions, the coalescence of the individual dispersion particles into a coherent film; with reactive systems (as high performance coatings often are), the interdiffusion of the reacting species, especially as the rate of diffusion compares to the rate of reaction, the rate at which the system approaches final Tg, and with dispersions, the rate of coalescence; and finally, the so-called "densification" phenomenon[5], during which the last traces of volatile material evaporates, crystallinity or para-crystallinity develops, free volume decreases and the film assumes its final form. Obviously there are more occasions for forming film defects and for developing sub-optimum film performance then one cares to contemplate. Fortunately there is no lack of solutions provided by the judicious choice of coating components, especially the appropriate additives.[6]

Though it is beyond the scope of this review, the interaction between the coating and the substrate must also be taken into consideration. It is well known that surface preparation is a key factor in optimizing coating performance. The removal of low surface energy materials, electrolytes, and existing surface corrosion are important not only to promote continuous film formation and initial adhesion but to remove species which contribute to such in-use effects as penetration by water caused by osmotically-driven percolation (due to electrolytes); under-film corrosion (due to surface corrosion and other impurities); loss of adhesion (osmotic swelling or presence of a weak boundary layer due impurities or lack of initial wetting); and so on. Surface cleaning is accomplished in a variety of ways, both mechanical and chemical. These include particulate blasting, scarifying, degreasing, and application of a conversion coating. This latter method converts the surface of metals to a very thin insoluble inorganic layer which both resists further oxidation and acts to provide a clean, uniform surface suited for reception of the protective coating. Following along this line of thinking leads to the concept of the primer-topcoat system, where the primer is designed particularly for its protective action in preventing surface attack, either as a passive barrier film, or a chemically sacrificial or inhibitive one, or both; and the topcoat provides both protection for the primer and contributes to the aesthetic properties.

One perhaps obvious but typically overlooked practical element in coating performance is film thickness. All other things being equal (which admittedly they are usually not), a thick coating which remains both adherent and coherent during service will perform better than a thin one. (The qualifiers of adhesion and coherence are added since mismatches between modulus or coefficient of expansion between coating and substrate will tend to produce exaggerated effects– cracking or loss of adhesion during use – in thicker films.) The performance improvement is due to a number of factors, including the lower likelihood of film defects causing a discontinuity which exposes the substrate; the longer diffusion pathway for penetrants; and the greater film bulk which must be ablated before the substrate is exposed. In addition, while the protective nature of these coatings is being emphasized, the aesthetic considerations of gloss, color, resistance to dirt retention and overall appearance cannot be ignored. All in all, a tall order for a protective system that usually is not more than a small fraction of a millimeter thick.

Finally, a relatively new but far-reaching consideration has been added to the list of requirements for the development of any coating, but one with particular relevance for high performance coatings: that of control of environmental, health and safety effects.[7] From choice of raw materials, through manufacture, distribution, application and handling of waste streams to removal and replacement or recycling of the painted surface, "cradle-to-grave" lifetime impact analysis is being applied to all coating systems. This has a notable effect on the sophisticated compositions associated with high performance coatings, especially as these changes affect coating efficacy and durability.[8] In the U.S., the heaviest reformulation burden has arisen from the control of the level and type of volatile organic compounds (VOCs) and hazardous air pollutants (HAPS) imposed by federal and state regulations. This one change has had the most comprehensive effects of any that have occurred over the last twenty-five years. Inevitably, it is in light of this change that developments in the typical binder systems used in high performance coatings are reviewed below.

Epoxy Binders

Epoxy-based systems represent the prototypical binder for coatings used in demanding applications. They are among the earliest developed and most versatile of the high performance coatings and have been thoroughly reviewed.[9] They find their most use as primers and topcoats on structural steel and marine coatings; tank and drum linings; automotive and aerospace primers; concrete wall and floor paints; in flexibilized formulations as containment coatings; and as heat-cured formulations in coil and other OEM applications. The coatings typically exhibit low shrinkage, excellent adhesion, outstanding chemical and solvent resistance, good temperature resistance, and favorable curing characteristics with no emission of volatile by-products. These properties are derived from the combination of ring-opening crosslinking mechanism (little volume change during cure) and from the cyclic structures inherent in these coatings, with the resulting relatively high Tg and polar character. The weathering characteristics of the typical epoxies in thin film topcoats are not as desirable, especially on exposure to UV light, which manifests itself as discoloration and chalking. This response is associated with the aromatic structures which are derived from the epoxy-functional portion of the binder. Although improvements in external weathering of epoxy topcoats can be made, it is usually at the cost of limitations in other properties, and other more weatherable systems like the aliphatic polyurethanes are usually preferred for topcoat applications In the typical rigid epoxy, impact and abrasion resistance are also somewhat compromised, except where special efforts are made to modify the system, as in containment coatings.

Epoxy systems can be considered the paradigmatic two component coating. They consist of an epoxy-functional resin and a curative. The most important epoxy resins are those derived from the diphenol 2,2-bis(4-hydroxyphenyl)propane ("Bisphenol A") or other polyhydric phenols and epichlorohydrin;[10] depending on molecular weight and type, these are available as either liquid or solid materials. The versatility of epoxy coatings derives not only from the variations obtainable from the epoxy resins themselves but to an even greater extent from the type and number of curatives available.[11] These curatives can be generally classified into three types: multifunctional nucleophilic species containing active hydrogens (e.g., polyamines or polymercaptans) which undergo polyaddition reactions through opening of the epoxy ring; initiators which promote epoxy homopolymerization (e.g., boron trifluoride and its latent precursors); and hydroxy-reactive crosslinkers used with higher molecular weight epoxy resins having appreciable hydroxy functionality (e.g., polyisocyanates). The active hydrogen types are by far the most common, and their properties have considerable influence over the final coating characteristics. When the great variety of these curatives and the effects of modifications in stoichiometry are considered along with the diversity of diluents (reactive and non-reactive), catalysts, and other additives affecting the curing reaction, the various epoxy curing agents make for a profusion of possible coating types. Beyond that, the epoxy resins can be modified with resinous modifiers and curing agents to make hybridized materials, including polyester epoxies, epoxy-acrylics, epoxy-phenolics and various blends with coal tar or furfural resins, vinyl resins, fluorocarbons, or silicones.

Traditionally, epoxy coatings have been applied as solventborne systems. As with most other coatings, however, increasing importance has been given in recent years to low VOC and HAPS-free formulations, including high solids, 100% reactive and especially waterborne coatings[12]. In specialty applications, 100% reactive radiation-curable coatings are used; crosslinking of acrylic and polyester powder coatings with epoxy-functional crosslinkers could also be considered a solventless epoxy application. The trend toward lower and lower concentrations of organic volatiles in response to regulatory pressures also appears to have inspired an aesthetically-driven demand for low odor coatings, applicators and consumers having realized that such systems were possible; no doubt this trend will continue. In addition to concerns about air quality, issues of worker exposure to coating components and their possible acute and chronic toxicity, including mutagenicity and carcinogenicity, have become of increasing importance. This has led, for example, to severely restricted use of the formerly common curing agent methylene bis(4-aminobenzene) (methylenedianiline or MDA). Such restrictions have sparked a surge of development of amine curing agent alternatives which are not based on aromatic diamines, the most widely used of which are the cycloaliphatics.[13] In contrast, the safety of Bisphenol A epoxy resins in highly sensitive applications like food contact coatings[14] is so well established that, given the regulatory climate, very little activity in changing the fundamental chemistry of the resin backbones is currently being considered. Most recent epoxy resin research is concerned with reducing the viscosity of existing resins for high solids systems or modifying existing resins for waterbased applications. A promising area of future research involves the use of incompatible polymer blends of epoxy and thermoplastic resins to produce improved coatings, where the heterogeneity of the system can be predicted using phase diagram analysis.[15]

Polyurethanes

Polyurethane coatings are used in approximately the same quantities as epoxy coatings in the U.S., which testifies to their utility. While not exhibiting the same degree of thermal and chemical resistance as the epoxies, they often can be formulated to meet the requirements of high performance applications. In addition, they typically show outstanding abrasion resistance and, when based on aliphatic isocyanates, they demonstrate excellent weathering properties. This makes them very well suited as topcoats, particularly when maintenance of appearance is an important factor. The aliphatic versions are often used as topcoats in primer-topcoat systems; in many demanding applications, an epoxy primer and a polyurethane topcoat are considered the ideal combination. They find wide use in a variety of industrial maintenance applications and more recently, as automotive clearcoats. The various forms of polyurethane resins have been used in coatings since the discovery of the utility of isocyanates in polymer systems in the 1940s, and the topic has been reviewed at length.[16]

Polyurethanes are formed by the reaction of multifunctional isocyanates and so-called active hydrogen compounds. These latter are not unsimilar to the protic species used as epoxy crosslinkers and include amines, hydroxy compounds, mercaptans and under special conditions, carboxylic acids, ureas, and urethanes themselves.* Additionally, the isocyanates can undergo both self-reaction and reaction with the

---

* Strictly speaking, the term "urethane" refers only to the reaction product of a hydroxy compound and an isocyanate. The product of an isocyanate-amine reaction is a urea; an isocyanate-urea reaction, a biuret; an isocyanate-urethane reaction, an allophonate; and so on. Since all these reactions can occur simultaneously in the curing of the binder, the resulting coating tends to be called simply a "polyurethane". The separate reactions should be borne in mind, however, since the resulting polymer can have significantly different properties.

urethanes or ureas formed *in situ*, particularly under the influence of catalytic species, with the formation of polymer networks which take their structure from the resulting cyclic trimers (isocyanurates), dimers (uretdiones) and branched compounds (allophonates and biurets). The isocyanates can also react with water with the formation of an intermediate carbamic acid which decomposes at use temperature into carbon dioxide and an amine. The amine will then react in a very fast step with more isocyanate to form a polyurea. This is the basis of the so-called moisture curing polyurethane coatings (see below.)

The isocyanate-functional materials employed in coatings usually are limited in the amount of monomeric isocyanate they contain, since many isocyanates are irritating and can be respiratory sensitizers. For this reason, the monomeric isocyanates are typically not used as primary coatings components. Instead, a number of higher molecular weight adducts are employed. In many cases, the degree of polymerization of the self-reaction of multifunctional isocyanates can be controlled so that low molecular weight, isocyanate-functional isocyanurates and allophonates can be made; these serve as convenient and safe alternatives to the monomer. Alternatively, diisocyanates can be adducted to multifunctional hydroxy compounds, with molecular weight again controlled (through stoichiometry) to yield relatively low molecular weight isocyanate-functional adducts. A major distinguishing characteristic of polyurethane coatings is whether they are based on aromatic or aliphatic isocyanates. Aromatic isocyanates tend to give tougher coatings with better solvent and chemical resistance than the aliphatics, while the aliphatics exhibit better control of reactivity and are used exclusively in applications involving exposure to natural sunlight, where their color retention is superior to that of the aromatics which undergo severe photochemically-induced yellowing in exterior use.

The active hydrogen compounds used as co-reactants with the isocyanates in polyurethane coatings are often oligomeric. These include hydroxy functional polyesters, polyethers, and polyacrylates ("polyols") and telechetic mercaptans and amines. The choice of the particular polyol used has an important effect on overall coatings processing and final properties and leads to a classification of polyurethane coatings based on the polyol component, for example polyester-based polyurethanes, polyether-based polyurethanes, and so on. Lower molecular weight polyfunctional active hydrogen compounds may also be use in combination with the main polyol to control reactivity and to modify properties, particularly through control of crosslink density. In addition, catalysts, which are usually added to control isocyanate-active hydrogen reactivity, can also promote self-reaction of the isocyanate or reaction with in situ urethane or urea groups, thus giving a very different reaction profile and different final coating properties compared to the same system without the added catalyst.

Polyurethane coatings typically exhibit a unique combination of impact and abrasion resistance together with very good chemical resistance. This can be linked to the stability imparted by intermolecular hydrogen bonding and the self-reinforcing nature of the phase separated nature of many polyurethane networks.[17] In addition, the variety of chemical structures possible from variations in polyol backbone, crosslinkers, chain extenders, catalysts and isocyanates make polyurethanes among the most versatile of all coating binders in terms of final coatings properties. This versatility is mirrored by the several forms polyurethane coating systems can take. As with epoxies two component systems are the norm, the polyol and isocyanate components being packaged separately and not mixed until just before application. Traditionally, these have been solventborne systems, but recently two component waterborne systems have begun to appear.[18]

One component coatings can also be prepared using one of several formulation strategies. Moisture curable coatings take advantage of the isocyanate-water reaction mentioned above. An isocyanate-terminated oligomer ("prepolymer") is formed by reaction of a polyol with a stoichiometric excess of diisocyanate monomer, further formulated and then stored protected from adventitious water. When this is applied to a substrate, the water normally present in the atmosphere reacts with the isocyanate to chain extend and crosslink the coatings through formation of urea linkages. Alternatively, "blocked" isocyanates can be made by first reacting the isocyanate component of the coating with an active hydrogen compound that forms a thermally reversible urethane or urea bond (the urea derived from caprolactam is an example). The blocked isocyanate is stable in the presence of the polyol component but on heating undergoes irreversible reaction with the polyol and evaporation of the blocking agent. This approach can be used both in solvent systems and more recently in powder coatings.[19] As an alternative to reactive systems, fully reacted low solids solvent-based urethane lacquers have been used in the past, but these have been largely replaced by water-based polyurethane dispersions ("PUDs")[20], though solvent-borne urethane modified drying oil formulations are available.

Recent trends in polyurethane coatings, like epoxies, include environmental, health and safety issues, largely centered around the minimization of VOCs. In addition, there are concerns over the amount of monomeric isocyanates, especially in coating applications where respiratory exposure is an issue. The movement toward water-based systems to alleviate the former concern will probably continue. The latter topic has been addressed for some applications with the development of prepolymer systems which minimize the amount of monomer and while controlling the resin viscosity, even to the point of allowing the development of spray-applied one hundred percent reactive coatings.[21] Environmental issues aside, there has been a growing trend toward the development of polyurea coatings, formed from the reaction of multifunctional isocyanates and amines of moderated reactivity.[22] In some applications, these are beginning to rival epoxies and other highly crosslinked systems for corrosion resistance and durability.

Melamine Resins

Like epoxies, melamine resins offer a route to high Tg, highly crosslinked coatings with excellent overall properties. Unlike epoxies, and like the polyurethanes, they can be formulated into durable topcoats with good photochemical stability and weathering. They are also similar to polyurethanes in the sense that the melamine crosslinkers, which can be likened to the isocyanate crosslinkers of polyurethanes, will react with a variety of active hydrogen species, including many versions of hydroxy functional polyols; also like the isocyanates, they can undergo self-reaction. Additionally, their crosslinking chemistry is also very much influenced by catalytic species, especially strong acids. However, in contrast to epoxies and polyurethanes, almost all melamine coating systems require a bake cycle in order to achieve acceptable film properties. Another difference is that unlike most epoxy or polyurethane coatings, curing of the typical melamine results in the emission of volatile species (water or alcohols, with small amounts of formaldehyde) from the solid film, with possible implications for film quality. Despite these restrictions, melamine coatings have a long history of useful applications and have been the subject of numerous reviews.[23] They are used in a wide variety of OEM applications, including metal finishing and most importantly, automotive color coats.

Melamine crosslinkers are the most important class (for coatings) of the so-called amino resins, derived from the reaction of formaldehyde with amine-functional materials. The melamine types are prepared from the reaction of formaldehyde with melamine (2,4,6-triamino-1,3,5-triazine) to form N-methylolated species. These are then further reacted with alcohols, typically methyl or butyl alcohol, to form the corresponding methylol ethers. Depending on reaction conditions, the methylolation step can lead to various degrees of methylol substitution and to dimerization and oligomerization. Similarly, the type and degree of etherification can vary. This gives rise to a number of commercially available crosslinkers with a variety of functionalities and reaction characteristics. When considered along with the large number of polyol coreactants available, including polyesters, acrylics, alkyds, epoxies and urethane oligomers, it will be appreciated that the melamines offer a wide choice of coating options. The stoichiometry of reactive components is usually determined on an empirical basis, since the degree of polyol-crosslinker reaction vs. self-condensation of the crosslinker and the desired degree of total crosslinking is difficult to determine in advance. The melamines may be employed either as one component or two component systems. In one component coatings, package stability is extended by incorporation of monofunctional low molecular weight alcohols and the use of thermally labile blocked acid catalysts such as amine salts of sulfonic acids.

Typically, melamines have been employed as solventborne systems. The predominance of OEM applications with their engineering controls has mitigated concerns about solvent content to some degree, but VOC issues continue to be increasingly important in all coatings fields, and waterborne versions of the melamines are becoming more common. There is an issue centered around formaldehyde emissions during and after coating application, and alternatives to the traditional alkoxylated melamines have been proposed[24], those these have not been widely implemented. Other recent developments have been driven by the tendency of melamine-crosslinked coatings to undergo acid-catalyzed hydrolysis when subjected to acid etch conditions during weathering, such as are seen most noticeably with automotive coatings subjected to rain of low pH or to bird droppings. Improvements such as hybrid silane-melamine coatings have been developed and are used commercially[25]; whether they will offer a general, long term cost effective alternative remains to be seen.

Fluoropolymers

The chemical inertness, low surface energy, thermal stability, chemical and solvent resistance, resistance to both hydrophilic and oleophilic agents, and outstanding weathering characteristics of the fluorinated polymers are well known. This is due in part to the stability of the carbon-fluorine bond and to the very high molecular weight and crystalline nature of many of these polymers.[26] Indeed it is these factors that allow the fluoropolymers to be classified as high performance binders even though they are one of the few classes of linear, uncrosslinked polymers that fit this description. What epoxies, polyurethanes and melamines gain from crosslinking is replaced in the fluoropolymers by their typical fluorocarbon-like chemical and physical inertness.

Every since the discovery of poly(tetrafluoroethylene) (PTFE) in 1938, the practical utilization of the fluoropolymers has had to strike a balance between maintaining the excellent properties of the unmodified polymer and accepting the processing difficulties associated with them, or modifying the resin by backbone changes or blending but having to accept a diminution of the protective film properties. For example, PTFE is insoluble in almost all solvents and has a sintering temperature in excess of 400°C, and while it is almost completely inert, cannot be routinely used in most industrial maintenance applications. On the other hand, poly(vinylidene fluoride) (PVDF), with only half the fluorine substitution of PTFE, has a processing temperature on the order of 245°C and as an organosol coating is much more amenable to OEM metal finishing processes, particularly coil coating. These PVDF coatings exhibit outstanding weathering properties, though not nearly the heat or chemical resistance of PTFE, and are typically used on architectural fascia. Similar resins can be prepared by copolymerization of PVDF and hexafluoropropene.[27] Further compromises can be made by copolymerizing fluoroolefins, vinyl esters and other olefins to reduce polymer crystallinity. Although these polymers can be applied at relatively low temperature from solution or emulsion[28], their properties are not in the same class as PFTE. Additionally, functionalized fluoroolefin copolymers or telechetic fluoropolymer oligomers containing hydroxy groups can be incorporated into polyurethane coatings by crosslinking with isocyanates.[29] These should be considered as modified polyurethanes as opposed to fluoropolymer coatings. Similarly, other fluorinated polyols, siloxanes and epoxy resins can be incorporated into otherwise traditional coating systems, with a resultant improvement in properties.[30]

There have been a number of attempts to retain the desirable properties of fully fluorinated coatings resins while eliminating the difficulties associated with their lack of solubility and problems in film formation. Production of amorphous PTFE has been reported, for example.[31] Research into less highly fluorinated hydrocarbon polymers also continues.[32] Some of these can be formulated as low temperature-curing, water-based systems.[33] However, until lower cost, more generally applicable solutions are developed, the use of fluoropolymers in maintenance coatings will probably remain limited to current applications.

Silicone Resins

All of the resin systems discussed above have been predicated on the chemistry of carbon-based polymer backbones. Silicone binders of the type used in coatings, on the other hand, are highly branched polymers of polysiloxanes, the backbone repeat unit being an organo-substituted (usually methyl and phenyl) Si—O group[*]. Because of this "inorganic" character, they are noted for their high temperature resistance and resistance to chemical attack. Like the melamines, they usually require a bake cycle in order to achieve crosslinking, while like the fluoropolymers they are valued for their chemical inertness and weathering characteristics. Because of their highly crosslinked nature, they typically do not exhibit good impact resistance and have limited use over flexible substrates. Their use in coatings is of long standing and has been reviewed in a number of publications.[34] They find their most use in high temperature coatings such as those used on automotive mufflers and industrial heaters and chimneys.

Pure silicone coating resins are supplied as functional oligomers where a small percent of the organic substituent is replaced with an hydroxyl or alkoxyl group. On heating, these undergo condensation reactions with the formation of Si—O—Si bonds which form the basis of the crosslinking chemistry. The rate of crosslinking is dependent on the substituent groups and the presence of catalysts, usually organometallic compounds. The reactions tend to be relatively sluggish, and the polysiloxanes usually require relatively long bake cycles. In addition, a number of compromises must be made among the factors of desired degree of crosslinking, bake time and coating shelf life. Finally, the coatings have traditionally been solvent-based with relatively low solids content. Because of these limitations and the relatively high cost of the pure polysiloxanes, hybrid coating resins have been developed which are based on blends or copolymers of the polysiloxanes and carbon-based resins. These include polyesters, alkyds, acrylics and less importantly phenolics, epoxies and polyurethanes. Copolymerization can be accomplished via condensation of the hydroxyl groups on the base resin with the silanol of the reactive siloxane oligomer. The addition of the polysiloxane tends to markedly improve high temperature stability and, in the case of the modified alkyds, exterior durability. In addition, the hybrids usually require a less demanding baking cycle, and in some cases can be applied as ambient cure systems. They are typically used in coil coatings, cookware, tank coatings, marine applications and exterior architectural coatings.

The most important advances in silicone coating systems are centered around the development of resins with lower VOC content. The usual strategy being employed is modification of the resin so as to allow the formation of emulsion-based or water-reducible materials;[35] acrylic-silicone powder coating resins are also being developed.[36]

---

[*] Silicate binders based on a pure unsubstituted Si—O network from the hydrolysis of tetraalkoxy silanes or from condensation of alkali silicates will not be discussed here. They are important constituents of zinc rich primers and cementitious coatings but are truly inorganic in nature.

Other High Performance Coating Resins

The number of organic polymeric materials outside of traditional binder materials that could be developed to fit the definition of a high performance coating resin is probably without end. However only a few systems have been identified which have been successfully applied in actual commercial coatings. A short discussion of a selection of these is presented below.

Missing from the general discussion of high performance binders presented above is the topic of the chlorinated rubbers. In the past, these have been used extensively in marine, chemical processing, bridge and transportation applications. However, their lack of inherent thermal and photochemical stability, difficulties in production and availability only as relatively low solids solventborne systems make their use in current coatings problematic. Some attempt has been made to convert them to functional oligomers crosslinkable by isocyanates or melamines[37], but no general use of this technology appears to be emerging.

Similarly, alkyd resins are well known in the industrial coatings area, but their lack of external durability calls into question their unmodified use in high performance coatings, especially in the face of modern alternatives. They are attractive in that their curing mechanism, relying as it does on the autoxidation of the drying oils on which they are based, allows for a convenient one pack approach to an ambient-curing coating. Modification with silicones or isocyanates does allow for an upgrading of their properties, as mentioned above.

Styrene-unsaturated polyester resins can also be adapted for coatings purposes, although their main use is as the binder in fiberglass composites. They normally are applied as in-mold coatings (so-called "gel coats") where the oxygen inhibition of the typical peroxide-initiated cure is not a problem. For more typical coating applications, styrene-unsaturated polyesters alone are less suitable; their modification with allyl ether coreactants to reduce oxygen inhibition and preserve surface appearance is one alternative.[38]

They are a number of engineering thermoplastics which can adapted to use in coatings. These include polyphenylene sulfides[39] and polyimides[40]. Other similar polymers, which are not soluble in normal coating solvents, can be applied in a variety of thermal spray techniques.[41] The traditional engineering thermoplastics can also be modified to provide crosslinking for greater thermal stability.[42] The use of polyfluorophosphazenes and polycarbosiloxanes for aerospace applications has also been reviewed.[43]

Lastly, there has been a growing interest in the polyanilines, especially for corrosion control. This subject has been reviewed in several recent publications.[44]

Conclusions

It has been said that the best way to predict the future is to invent it. That certainly seems to be the approach being taken by scientists and technologists in the coatings arena, and especially in high performance coatings. Even a cursory review of the specialist literature reveals a host of exciting possibilities for future development.

Whether those approaches which will come to dominate will take the form of new resin systems, new crosslinking chemistries, hybrid coatings, new forms of coatings materials like powders or films, new application/curing methods like radiation or plasma – no one can say for sure at present. It is certain, however, that the next twenty-five years of development in high performance coatings will be at least as interesting as the last.

Literature Cited

[1] (a) C. G. Munger, "Resistant Coatings" in *Kirk Othmer Encyclopedia of Chemical Technology*, 3rd Edition, Vol. 6, John Wiley and Sons, pp. 455-481 (1979). (b) J. D. Kearne, ed., *Steel Structures Painting Manual*. Vol. I, *Good Painting Practices*, 2nd edition, 1983 and Vol. II, *Systems and Specifications*, 6th edition, 1991, Steel Structures Painting Council, Pittsburgh, PA.

[2] (a) G. P. Bierwagen and T. K. Hay, " The Reduced Pigment Volume Concentration as an Important Parameter in Interpreting and Predicting the Properties of Organic Coatings" , *Progress in Organic Coatings*, Vol. 3, 1975, p. 281-303. (b) G. P. Bierwagen, R. Fishman, T. Storsved and J. Johnson, "Recent Studies of Particle Packing in Organic Coatings, *Proceedings of the 24th International Conference in Organic Coatings*, Greek Society of the Paints Industry, 1998, p. 31-43.

[3] See for example (a) *Paint Research Association 2nd International Annual Research Colloquium*, 15/16 June, 1998, The Paint Research Association, Teddington, England. (b) M. E. Nichols, J. L. Gerlock, C. A. Smith and C. A. Darr, "The Effects of Weathering on the Mechanical Performance of Automotive Paint Systems", *Proceedings of the 24th International Conference in Organic Coatings*, Greek Society of the Paints Industry, 1998, p. 289-306.

[4] See for example, *"Film Formation in Waterborne Coatings"*, T. Provder, M. A. Winnik and M. W. Urban eds., ACS Symposium Series 648, American Chemical Society, 1996.

[5] L. C. E. Struik, *Physical Aging in Amorphous Polymers and Other Materials*, Elsesvier, Amsterdam, 1978.

[6] P. E. Pierce and C. K. Schoff, *Coating Film Defects*, Federation Series on Coatings Technology, Federation of Societies for Coatings Technology, Philadelphia, PA, 1988; *Handbook of Coatings Additives*, L. J. Calbo, ed., Marcel Dekker, New York, 1992.

[7] For an excellent treatment of the evolution of environmental regulations, see S. C. DeVito, "Present and Future Regulatory Trends of the United States Environmental Protection Agency" in *Proceedings of the 24th International Conference in Organic Coatings*, Greek Society of the Paints Industry, 1998, p. 105-116.

[8] J. W. Martin, S. C. Saunders, F. L. Floyd, and J. P. Wineburg, *"Methodologies for Predicting the Service Lives of Coating Systems"*, Federation of Societies for Coatings Technology, Blue Bell, PA 1966.

[9] (a) T. F. Mika and R. S. Bauer, *Epoxy Resins Chemistry and Technology*, 2nd Edition, C. A. May, ed., Marcel Dekker, New York, 1988. (b) *Waterborne and Solvent-Based Epoxies and Their End User Applications*, P. Oldring, ed., SITA

Technology Ltd., London, 1996. (c) H. Lee and K. Neville, *Handbook of Epoxy Resins*, McGraw Hill, New York, 1967 (reissued 1982).

[10] B. Ellis, ed., *Chemistry and Technology of Epoxy Resins*, Chapman & Hall, London, 1993.

[11] (a) J. B. Dickenson, "Curing Agents" in *Epoxy Resins Chemistry and Technology*, 3rd Edition, Marcel Dekker, New York, in press. (b) C. H. Hare, Paintindia, 46(12), 55-56, 58-60, 62-64 (1996). (c) C. H. Hare, Paintindia, 47(1), 51-52, 54-56 (1997).

[12] See for example, F. H. Walker and M. I. Cook, ACS Symposium Series 663, Technology for Waterborne Coatings, J. E. Glass, Editor, American Chemical Society, 1997.

[13] See for example, D. A. Dubowik, P. A. Lucas, A. K. Smith, (to Air Products and Chemicals) U.S. Pat. 5,280,091.

[14] Title 21 Code of Federal Regulations (C.F.R.) § 175.300.

[15] V. Verkholantsev, "The Use of Phase State Diagrams to Design Polymer/Polymer Heterophase Coatings", *Proceedings of the Twenty-Fourth International Waterborne, High-Solids & Powder Coatings Symposium*, February 5-7, 1997, Department of Polymer Science, The University of Southern Mississippi and the Southern Society for Coatings Technology, p.446-457.

[16] (a) G. Oertel, ed., *Polyurethane Handbook : Chemistry - Raw Materials - Processing - Application - Properties*, 2nd edition, Hanser Gardner, 1993. (b) M. J. Husbands, C. J. S. Standen and G. Hayward, "Polyurethanes" in *A Manual of Resins for Surface Coatings*, P. Oldring and G. Hayward, eds., Vol. 3, Chapter 9, SITA Ltd., London, 1987. (c) J. H. Saunders and K. C. Frisch, *Polyurethanes: Chemistry and Technology*, Interscience, New York, 1974. (d) R. T. Wojcik and A. T. Chen, "Urethane coatings for Metal Substrates", *Metal Finishing*, April, 1994, p. 22-27.

[17] In addition to Ref. 15, see R. Blokland, *Elasticity and Structure of Polyurethane Networks*, Gordon & Breach Science Publishers, 1969.

[18] (a) C. A. Hawkins, A. C. Sheppard and T. G. Wood, "Recent Advances in Aqueous Two-Component Systems for Heavy Duty Metal Protection", *Progress in Organic Coatings*, 32 (1997), p. 253-261. (b) C. R. Hegedus, A. G. Gilicinski and R. J. Haney, "Film Formation in Aqueous Two Component Polyurethane Coatings", *Journal of Coating Technology*, 68 (1996) p. 51. (c) W. O. Buckley, "Zero VOC Two Component Waterborne Polyurethane Coating Systems", *Modern Paint and Coatings*, Oct. 1996, p. 81-86.

[19] A. Wenning, J-V. Weiss and W. Grenda, "Polyisocyanates Today and Tomorrow", *European Coating Journal*, 4/98, p.244-249.

[20] I. Bechara, "Formulating with Polyurethane Dispersions", *European Coating Journal*, 4/98, p.236-243.

[21] S. L. Bassner, J. Kramer and T. M. Santosusso, "New Polyurethane Prepolymers for Ultra-Low VOC Plural Component Coatings" in *Proceedings: Low- and No-VOC Coating Technologies: 2nd Biennial International Conference*, U. S. Environmental Protection Agency, Office of Research and Development, October, 1998, Section 8, p. 70-82.

[22] (a) D. A. Wicks and P. E. Yeske, "Amine Chemistries for Isocyanate-Based Coatings", *Progress in Organic Coatings* 30 (1997), p. 265-270. (b) D. J. Primeaux II, "100% Solids Aliphatic Spray Polyurea Elastomer Systems" in *Polyurethanes World Congress 1991,* Society of the Plastics Industry, Inc., Polyurethane Division and European Isocyanate Producers Association, September, 1991, p. 473-477. (c) T. Santosusso, D. J. Finocchio and J. H. Frey, "Oligomeric Diamine-Based Polyureas", *Polyurethanes World Congress 1991,* Society of the Plastics Industry, Inc., Polyurethane Division and European Isocyanate Producers Association, September, 1991, p. 329-336

[23] (a) D. R. Bauer, "Melamine/Formaldehyde Crosslinkers: Characterization, Network Formation and Crosslink Degradation", *Progress in Organic Coatings,* 14, p. 193 ((1986). (b) W. J. Blank, "Reaction Mechanisms of Melamine Resins", *Journal of Coatings Technology,* 51, No. 656, p. 61 (1979).

[24] A. Essenfeld and K-J. Wu, "A New Formaldehyde-Free Etch Resistant Melamine Crosslinker", *Proceedings of the Twenty-Fourth International Waterborne, High-Solids & Powder Coatings Symposium,* February 5-7, 1997, Department of Polymer Science, The University of Southern Mississippi and the Southern Society for Coatings Technology, p.246-258.

[25] I. Hazen, "Low VOC - Super High Solids Clearcoats", *Proceedings of the 24th International Conference in Organic Coatings,* Greek Society of the Paints Industry, 1998, p. 125-143.

[26] (a) M. Howe-Grant, editor, "Fluorine Compounds, Organic (Polymers)" in *Kirk Othmer Encyclopedia of Chemical Technology,* 4th Edition, Vol. 11, John Wiley and Sons, 1994, p. 621-729. (b) W. W. Schmiegel, "Organic Fluoropolymers" in M. Hudlicky and A. E. Pavlath, *Chemistry of Organic Fluorine Compounds II, A Critical Review,* ACS Monograph 187, American Chemical Society, 1995, p. 1101-1118. (c) C. Tournut, P. Kappler and J. L. Perillon, "Copolymers of Vinylidene Fluoride in Coatings", *Surface Coatings International,* 1995 (3), p. 99-103.

[27] S. Gaboury, "High Performance Coatings: Novel VF2/HFP Copolymers", *European Coatings Journal,* June, 1997, p.624-626.

[28] S. Kuwamura, T. Hibi and T. Agawa, "Waterborne Fluorinated Polyolefins in Coatings", *Proceedings of the Twenty-Fourth International Waterborne, High-Solids & Powder Coatings Symposium,* February 5-7, 1997, Department of Polymer Science, The University of Southern Mississippi and the Southern Society for Coatings Technology, p.406-418.

[29] (a) S. Turri, M. Scicchitano, G. Simone and C. Tonelli, "Chemical Approaches toward the Definition of New High-Solid and High-Performance Fluorocoatings", *Progress in Organic Coatings* 32 (1997), p. 205-213. (b) V. Handforth, "Properties and Applications of Novel Fluoropolymer Resin", JOCCA, March 1993, p. 122.

[30] (a) J. F. Brady, Jr., "Properties Which Influence Marine Fouling Resistance in Polymers Containing Silicon and Fluorine", *Proceedings of the 24th International Conference in Organic Coatings,* Greek Society of the Paints Industry, 1998, p.59-66 and references therein. (b) R. F. Brady, Jr., "Formulation and Field

Performance of Fluorinated Polyurethane Coatings" in *Organic Coatings for Corrosion Control*, G. P. Bierwagen, ed., ACS Symposium Series 689, American Chemical Society, 1998, p. 282-291.

[31] P. R. Resnick, *Polymer Preprints*, 31, 1990, p. 312.

[32] Anonymous, "New Development in Fluorinated Hydrocarbons", *JOCCA*, April 1990, p.145.

[33] T. F. McCarthy, A. Thenappan, S. Murthy, K. Harris, R. Malec and D. Melick, "Poly(chlorotrifluoroethylene-vinylidene fluoride) Waterborne Coatings", *Proceedings of the Twenty-Fifth International Waterborne, High-Solids & Powder Coatings Symposium*, February 18-20, 1998, Department of Polymer Science, The University of Southern Mississippi and the Southern Society for Coatings Technology, p. 541-554.

[34] (a) W. A. Finzel, "Properties of High Temperature Silicone Coatings", *Journal of Protective Coatings and Linings*, August, 1887, p. 38. (b) H. L. Cahn, "Silicones" in *Technology of Paints, Varnishes and Lacquers*, Chapter 14, C. R. Martins, ed., Robert Krieger Publishing Company, Huntingdon, NY, 1974.

[35] (a) J. W. Adams, "VOC-Compliant Silicones: Recent Developments", *Proceedings of the 19th Annual Water-borne, Higher-Solids and Powder Coatings Symposium*, University of Southern Mississippi, February 24-26, 1993, p. 302-313. (b) M. E. Gage and D. J. Grulke, "An Environmentally Compliant, High Performance, Heat Resistant Coating", *26th International SAMPE Technical Conference*, 1994, p. 508-516.

[36] J. T. K. Woo, R. M. Marcinko, J. C. Reising and D. E. Miles, "Acrylic-Silicone Powder Coatings Resins", *Proceedings of the Twenty-Fifth International Waterborne, High-Solids & Powder Coatings Symposium*, February 18-20, 1998, Department of Polymer Science, The University of Southern Mississippi and the Southern Society for Coatings Technology, p. 231-242.

[37] S. F. Thames and Z. A. He, "Environmentally Compliant High Solids Chlorinated Rubber Coatings", *Proceedings of the 19th Annual Water-borne, Higher-Solids and Powder Coatings Symposium*, University of Southern Mississippi, February 24-26, 1993, p. 248.

[38] M. J. Dvorchak and B. H. Riberi, *Proceedings of the 16th International Conference on Organic Coating Science and Technology*, 1990, p.1.

[39] L. R. Kallenbach and M. R. Lindstrom, *American Chemical Society Polymer Preprints*, **28** (1987) No. 1, p. 63-64.

[40] E. Sacher and J. R. Susko, *Journal of Applied Polymer Science*, **26** (1981) No. 2, p. 679-686.

[41] See for example, R. H. Henne and S. Schitter, "Plasma Spraying of High Performance Thermoplastics", *Proceedings of the 8th National Thermal Spraying Conference*, 11-15 September, 1995, p. 527-531.

[42] F. W. Mercer, A. Easteal and M. Bruma, "Synthesis and Properties of New Alternating Poly(Aryl Ether) Copolymers Containing Cyano Groups", *Polymer*, Vol. 38, No. 3, 1997, p. 707-714.

[43] J. I. Kleinan, Z. A. Iskanderova, F. J. Perez and R. C. Tennyson, "Protective Coatings for LEO Environments in Spacecraft Applications", *Surface and Coatings Technology*, 76-77, 1995, p. 827-834.

[44] (a) T. P. McAndrew, S. A. Miller, A. G. Gilicinski and L. M. Robeson, "Polyaniline in Corrosion-Resistant Coatings" in *Organic Coatings for Corrosion Control*, G. P. Bierwagen, ed., ACS Symposium Series 689, American Chemical Society, 1998, p. 396-408. (b) S. Jasty and A. Epstein, "Corrosion Prevention Capability of Polyaniline (Emeraldine Base and Salt): An XPS Study", *Journal of Polymeric Materials Science and Engineering*, 72, 1995, p. 595-596.

BIBLIOGRAPHY

There are a number of publications which those interested in high performance coatings will find valuable. Some of these are listed below.

C. H. Hare, "Protective Coatings. Fundamentals of Chemistry and Composition", Technology Publishing Company, Pittsburgh, PA, 1994.

Z. Wicks, F. Jones and P. Pappas, "Organic Coatings Science and Technology", Wiley-Interscience, New York, Vol. I, 1992 and Vol. II, 1994.

S. Paul, "Surface Coatings Science and Technology", Wiley-Interscience, New York, 1985.

J. D. Kearne, ed., "Steel Structures Painting Manual. Vol. I, Good Painting Practices", 2nd edition, 1983 and Vol. II, "Systems and Specifications", 6th edition, 1991, Steel Structures Painting Council, Pittsburgh, PA.

Federation Series on Coatings Technology, Federation of Societies for Coatings Technology, Philadelphia, PA, 1986 to present.

B. Ellis, ed., "Chemistry and Technology of Epoxy Resins", Chapman & Hall, London, 1993.

G. Oertel, ed., "Polyurethane Handbook : Chemistry - Raw Materials - Processing - Application - Properties", 2nd edition, Hanser Gardner, 1993.

D. Scantlebury and M. Kendig, eds., "Proceedings of the Symposium on Advances in Corrosion Protection by Organic Coatings II, Proceedings Volume 95-13, The Electrochemical Society: New Jersey, 1995.

R. A. Dickie and F. L. Floyd, eds, "Polymeric Materials for Corrosion Control", ACS Symposium Series No. 322, ACS Books, Washington, DC, 1986.

C. G. Munger, "Corrosion Protection by Protective Coatings", National Association of Corrosion Engineers, Houston, TX, 1984.

National Research Council, "Coatings for High Temperature Structural Materials: Trends and Opportunities", National Academy Press, 1996.

J. H. Lindsay, ed., "Coatings and Coating Processes for Metals", ASM International, 1998.

J. Edwards, "Coating and Surface Treatment Systems for Metals: A Comprehensive Guide to Selection", National Association of Corrosion Engineers, Houston, TX, 1997.

Y. Saito, B. Onay and T. Maruyama, eds., "High Temperature Corrosion of Advanced Materials and Protective Coatings: Proceedings of the Workshop on High Temperature Corrosion of Advanced Materials", North-Holland, 1992.

L.H. Lee, ed., "Adhesives, Sealants and Coatings for Space and Harsh Environments", Polymer Science and Technology, Vol. 37, Plenum Publishing Corporation, 1988.

L. Pawlowski, "The Science and Engineering of Thermal Spray Coatings", John Wiley and Sons, New York, 1995.

INTERNET SITES

If there has been one change in the last twenty-five years, it has been the information technology supporting all of science and technology, including high performance coatings. The most rapid change has been the emergence of the Internet as a source of information. Some sites of interest are listed below.

Coatings Raw Material Suppliers

Aabor International - Produces organic & metallic pigments for color applications in printing inks, paints and coatings, plastic concentrates and dispersions. *http://aarbor.com/*
Air Products and Chemicals - Manufactures VOC-compliant resins, including waterborne emulsions and polyurethanes; epoxy curing agents; surfactants. *http://airproducts.com/*
Akzo Nobel - VOC-compliant waterborne coatings, high solids and powder coatings, adhesives, surfacing materials, printing inks, toners, and associated products and polymers. *http://www.akzonobel.com/*
Albright & Wilson - Operates internationally in three business areas: phosphates, phosphorus derivative, acrylics and surfactants. *http://www.albright-wilson.com/*
BASF - Resins, surfactants, and coatings. *http://www.basf.com/*

Bayer, Performance Products Division - Coatings and specialty products, including resins, curatives, organic dyes and pigments. http://www.bayerus.com/about/business/chemicals/products.html

Benjamin Moore - Resins and coatings. http://www.benjaminmoore.com/

Cargill - Resins. http://www.cargill.com/

Ciba Specialty Chemicals - Manufacture and marketing of innovative specialty chemicals including resins, and additives. http://www.cibasc.com/

Color Corporation of America - Pigments. http://www.ccofa.com/

Creanova Inc. - Resins, curatives and additives. http://www.creanovainc.com/

Degussa - Additives and pigments. http://www.degussa.com/

Derakane - A branch of Dow Chemical's Composite Group dedicated to epoxy vinyl ester resins. http://www.derakane.com/

Dupont - Major product areas include resins and pigments. http://www.dupont.com/

Elf Atochem North America, Inc. - Searchable site for resin information. http://www.elf-atochem.com/

Engelhard - Business groups include pigment and additives. http://www.engelhard.com/

Henkel Corporation - Organic specialty chemicals product groups including additives, curatives, and resins. http://www.henkel.com/

ICI Surfactants - Additives. http://www.surfactants.com/

Kentucky-Tennessee Clay - Produces ball clay, kaolin and feldspar. http://www.ceramics.com/kt/

Kronos - Pigments. http://www.nlink.com/kronos

Lubrizol Coating Additives - Additives for coatings, paints and inks. http://www.lubrizol.com/

Lyondell Petrochemical - Producer of high value-added specialty polymers, color concentrates and polymeric powders. http://www.lyondell.com/

Micro Powders - Technically advanced micronized wax additives. http://www.micropowders.com/

Millennium Chemicals Inc. - specialty chemicals. http://www.millenniumchem.com/

National Starch and Chemical Company - specialty synthetic polymers http://www.nationalstarch.com/

PPG Industries - Coatings and resins. http://www.ppg.com/

Reichhold Chemicals, Inc. - Online catalog of specialty polymer and adhesives listings. http://www.reichhold.com/

Rhodia Inc. - Resins, curatives, and coatings. http://www.rp.rpna.com/

Rohm and Haas - Resins and additives. http://www.rohmhaas.com/

Sun Chemical - Manufactures organic pigment products in a variety of physical forms. http://www.sunpigments.com/

Union Carbide - Producer of ethylene oxide and ethylene glycol (resins and additives). http://www.unioncarbide.com/

Valspar - Resins, colorants, and coatings. http://www.valspar.com/

Vianova Resins - Produces acrylic, epoxy, polyurethane, polyester and alkyd resins. http://www.vianova-resins.com/

W. W. Grainger - Additives - online catalog. http://www.grainger.com/

Wacker-Chemie - Additives - polymers and silicones.
http://www.wacker.de/english/0.htm

Organizations and Consortia

ASM International - A society for materials engineers who distribute technical information through electronic media, publications, conferences, training programs and chapter activities.
http://www.asm-intl.org/index.htm
ASTM - Developed and published over 10,000 technical standards, which are used by industries worldwide. *http://www.astm.org/*
Coating Alternatives Guide (CAGE) - Information about coating alternatives, namely solvent-borne, water reducible, powder, and other alternatives.
http://cage.rti.org/altern.htm
Corrosion, Protective Coatings and paint Resources on the Internet - Direct link listing of technical articles and papers; organizations and societies; publications; U.S. and foreign manufacturers. *http://www.execpc.com/~rustoleu/coatings.htm*
The Electrochemical Society - Society for electrochemical and solid state science and technology. Publications, books, technical meetings and awards
http://www.electrochem.org/
Federation of Societies for Coatings Technology (FSCT) - Trade organization with the monthly technical publication. *http://www.coatingstech.org/*
Industrial Paint & Powder - Online magazine, informational resource for producers of OEM paint and powder coatings and the finishers who apply them. Problem solver, supplier directory, listing of coming events and feature articles.
http://www.ippmagazine.com/
Inter-Society Color Council - Non-profit organization aiming to promote the practical application of coatings technology to the color problems arising in science and industry. *http://www.iscc.org/*
National Association of Corrosion Engineers (NACE) - Technical society dedicated to reducing the economic impact of corrosion, promoting public safety, and protecting the environment by advancing the knowledge of corrosion engineering and science. *http://www.nace.org/*
National Coil Coaters Association (NCCA) - Industrial trade organization representing the coil coating industry. *http://www.coilcoaters.org/*
National Metal Finishing Resource Center - A comprehensive environmental compliance, technical assistance and pollution prevention information source available for the metal finishing industry and technical assistance providers.
http://www.nmfrc.org/
National Paint and Coatings Association (NPCA) - Trade association representing the paint and coatings industry in the U.S. *http://www.paint.org/*
Organic Coatings Forum - A site on coatings information sponsored by Elsevier Publishing. *http://www.elsevier.com/locate/coatingsforum*

Paint & Coatings Industry Magazine - Online publication serving the manufacturer and formulator of paint, coatings and printing inks. *http://www.bnp.com/pci/*
Paint Coatings.Net - Related articles, directories, tradeshows, and product news. *http://www.paintcoatings.net/pcnmain.htm*
PaintExpo - Partnership between NPCA and PCI magazine created an on-line "trade show" concept. Provides information on companies (manufacturers, chemical suppliers and distributors), product information, industry news and calendar of key industry events.
http://www.paintexpo.com/
The Paint Research Association - Based in the UK, contains various links on research and technical information on paints, inks and adhesives.
http://www.pra.org.uk/index.htm
PaintWebs - With over 2,000 listings and hundreds of links for the paint & coatings industry; coating manufacturers, coating consultants, distributors, coating educators, equipment suppliers, related organizations, coating publications, and raw material suppliers.
http://www.jvhltd.com/paintwebs/default1.html
PowderCoating.COM - Powder coating resource for suppliers, shops, network forum or equipment. *http://www.powdercoating.com/*
Protective Coatings worldWIDE - An information resource for the protective coatings industry. Publications, news and information, and a coatings and equipment Directory. *http://www.protectivecoatings.com/*
Structural Steel Painting Council (SSPC) - Trade association; publications, conferences, and training. *http://www.sspc.org/*
Usenet Group for Coatings - Internet news. *news:sci.chem.coatings*

Universities

Eastern Michigan University Coatings Research Center - Research areas include: rheology and application, cross-linking, analysis and characterization, adhesion, corrosion, and design and modeling. *http://www.emich.edu/public/cot/crc.html*
Eindhoven University of Technology - Research programs include: Moisture curable sealants, Powder coatings, Waterborne paints, Organic/inorganic hybrid coatings, Self-stratifying coatings, Finishing in mold, Adhesion of coatings on metals, and Historical paints.
http://www.chem.tue.nl/coatings/
Massachusetts Institute of Technology, Department of Materials Science and Engineering - Current research activities include: Ceramics and Glasses; Composites and Joining; Device Materials, Thin Films; Economics of Materials; Environmental Interactions; History of Materials; Metals; Polymers and Biological Materials; and Theory and Modeling.
http://www-dmse.mit.edu/
North Dakota State University, Department of Polymers and Coatings - General research areas: corrosion protection of coatings, surface and interfacial

chemistry/spectroscopy in coatings, molecular level adhesion, scanning probe microscopy of coatings, cross-linking chemistry and cross-linked film properties, stabilization and rheology of dispersions and coatings, and new techniques to analyze VOC emission.

http://www.ndsu.nodak.edu/ndsu/nupoly/poly_coat/poly_coa.htm

University of Missouri-Rolla Coatings Institute - A variety of coatings research areas, including waterborne systems, pigments and additives, conductive coatings, rheology control agents, ultrasonics and microwaves in coatings, and transparent composites.

http://www.umr.edu/~coatings/

University of Southern Mississippi, Department of Polymer Science - Research areas include: Polymer Science and Chemistry, Polymer Synthesis and Characterization, Polymer Physical Chemistry Polymer Reaction Engineering, Polymer Characterization and Engineering and Heterogeneous Polymer Materials.

http://www.psrc.usm.edu/main.html

ADVANCES IN ALKYD RESINS

ROY C. WILLIAMS

Reichhold, Inc.
P.O. Box 13582
Research Triangle Park, NC 27707

Historical Background
Chemistry of Alkyd Resins
Manufacturing Technology
Chemical Modifications and Blends with Alkyd Resins
Alkyd Uses versus Higher Performance Polymers
Environmental Regulation and Compliant Alkyds
Future of Alkyd Resins

Historical Background

Upon the seventy-fifth anniversary of the Division of Polymeric Materials, it is staggering to contemplate the extent of changes that have occurred in the field of polyester science since the beginning of the twentieth century. At that time there were few commercially available synthetic polymers although chemical modification of polymers of biological origin such as cellulose, natural oils and resins, rubber, starch, collagen, polyamide, leathers, etc., were in considerable use in the materials industry. In the paint and varnish industry, which would consume the bulk of the future alkyd resin production, products were made from the oxidatively polymerizable linseed oil blended with hard natural resins, turpentine and/or or inorganic pigments such as white lead carbonate. These modifications gave opacity and/or film hardness. While slow drying, coatings of this type had been successfully used for hundreds of years. They were generally hand prepared and dispersed by the painter on location due to poor pigment wetting and the hard settling tendencies upon storage of these paints.

The rapid expansion of industrial civilization forced the paint art to move to ready made paints and less skilled applicators. In the first twenty-five years of the century, advances in several areas of industrial chemistry opened the way to the economical, synthetic material components, which make alkyd resins so useful. The petroleum and automotive industries brought low-cost, rapid evaporating and mostly odorless aliphatic or aromatic hydrocarbon solvents. Glycerin was increasingly available from the soap industry.   Perhaps the key development for alkyds was the practical process of the catalytic oxidation of naphthalene, available from the coke industry, into phthalic anhydride.   This aromatic difunctional acid imparted a higher molecular weight and glass transition temperature to the alkyd polymer than was available from the auto-oxidizable oils.   These paints and coatings were harder, faster drying and more durable.   The resulting higher viscosity in the paint was reduced by increasing the content in the new, more economical petroleum solvents.   Thus, an improvement over linseed oil was found for paints and varnishes.   While the glycerol phthalate polymer backbone reaction product showed some interest as a colorless, hard resin by itself, the plasticization obtained by the substitution of coconut, a non-oxidatively drying oil, for the linseed oil in the alkyd, gave a colorless and toughening co-resin for blending with nitrocellulose in lacquers, such as are still used on wood furniture today.

In the second quarter of this century, alkyd resins reached their greatest penetration into the coatings market, which has usually consumed 90% to 95% of the alkyds manufactured.   By 1950, the agricultural industry had made soybean oil available in large amounts and at low cost.   The Kraft paper process also made tall oil fatty acids and rosin acids available at lower cost.   The higher Tg and molecular weight of alkyds compensated for the slower drying aspects of these economical "oil" sources. These oils gave paints with lower initial color and less tendency to yellow on aging. Also, the development of alternatives to glycerin (such as pentaerythritol) were derived from formaldehyde and alcohols.   These polyols greatly improved the economics and performance of the alkyd polymers.   Maleic anhydride has also been used extensively as the dibasic acid base in alkyds, but it never grew to the extent of more than 10 percent of alkyd production and eventually fell out of favor.   Maleic's big use in resins would come to be seen in the unsaturated polyesters for reaction with styrene in reactive castings or molding liquids.

The third quarter of this century saw the decline of the alkyd and linseed oil domination of the architectural paint market.   Latex polymers resulting from the war effort was soon adapted to the manufacture of house paint.   Also, it should also be noted that as late as the 1960s, the advantages of linseed oil versus alkyds were still being debated in the market place and paint laboratories.   It is estimated that approximately 480 million pounds of oils and rosin derivatives were still being used as paint resins around 1980.   The use of these oils had also declined with the introduction of latex polymers.   Latex paints, based on SBR, acrylic and vinyl monomer, provide ease of application and clean up, light color, rapid dry and improved durability over typical alkyd paints.   These advantages were particularly important to the "do-it-yourself" market.   Latex paints ultimately reduced the role of the oils and alkyds to about 20% of the architectural paint market, an application that requires high gloss and water resistance.

Alkyd consumption, however, continued to increase in industrial baking enamels, specialty architectural paint and aerosol spray paints.   This compensated somewhat for the decline due to loss of the architectural market share.   In spite of considerable technical efforts, attempts to derive emulsified versions of the oils and alkyd polymers did not result in satisfactory paint products at that time.

This third quarter century also saw the rapid advancement of the high performance epoxy, acrylic and urethane polymers in the industrial and maintenance paint market. Their increase in market share outstripped the increase in alkyd consumption due to the overall growth in the economy.

The last quarter of the century brought proportionally diminishing use of alkyd resins for coatings in the USA and other industrialized countries. The total alkyd production, by sources such as the U.S. Tariff Commission appears to have peaked between 1960 and 1980 at around 600 to 700 million pounds/year. From 1980 to the present, with a 500 million pounds per year estimate for the year 2000, there has been a gradual decline in the USA. In contrast, the use of coatings polymers as a whole has increased from about 2,600 million to 4,100 million pounds in the same time period.

Environmental regulations, limiting the level of emission of volatile organic compounds (VOCs) from paints and coatings, have also had a negative impact on the consumption of solvent-borne alkyd paints. Further, environmental regulations have strongly encouraged R & D efforts in high solids and waterborne alkyds. In fact, recent decades have seen a significant growth in VOC compliant alkyds as well as higher performance polymers, such as epoxies, urethanes, acrylics, etc. Environmental regulations are still in a state of flux in the USA and have also spread to Europe and Japan. The future of alkyd resins may well follow the extent the regulations spread to the industry in the developing nations.

The Chemistry of Alkyd Resins

By today's perspective, the esterification chemistry of alkyd resins seems very simple. Polyester Example

$$n\ R(OH)_2 + n\text{-}1\ R_1(COOH)_2 \longrightarrow HO\text{-}(R\text{-}O\text{-}\overset{O}{\overset{||}{C}}\ R_1\text{-}\overset{O}{\overset{||}{C}}\text{-}O)_{n\text{-}1}\text{-}R\text{-}OH + 2(n\text{-}1)H_2O$$

It must be remembered that it was not until 1929 that Carothers[1] showed the relationship of polyester monomers reactive functionality to resultant polymer formation in condensation reactions. In 1927 Kienle[2,3] originally used the word "alkyd" to describe polymeric reaction products of polyhydric alcohols and polybasic acids. Then in 1930 he illustrated that the use of a trifunctional component can lead to gelation (infinite molecular weight of only a fraction of the polymeric mass) before the polyesterification reaction is complete. This assumes initial equimolar hydroxyl and carboxyl content. The result is in contrast to the more linear Gausian distribution of molecular weight as achieved in polyester fibers made with difunctional monomers. This is best illustrated by considering the chemical structure of these seven basic components that make up the vast majority of commercial alkyds. The natural glycerin based oils used are generally linseed, soya, coconut and tung but many others types are available.

Monofunctional

Benzoic Acid   Fatty Acids

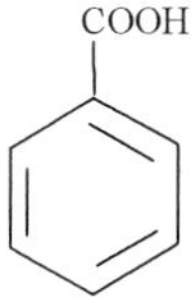

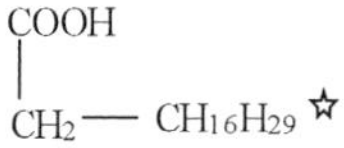

COOH

COOH
|
CH₂ —— CH₁₆H₂₉ ☆

---

* May also be H_{27}, H_{25} or H_{23} to denote varying levels of unsaturation.

Difunctional

OH — CH₂ — CH₂ — OH

Ethylene Glycol 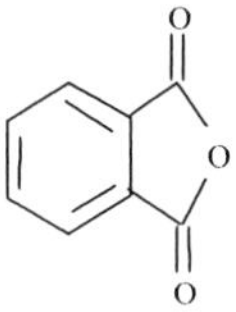   Phthalic Anhydride 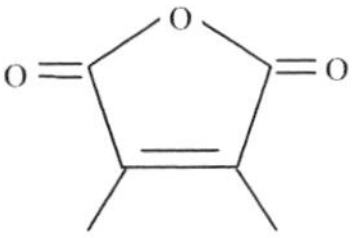   Maleic
                                                                Anhydride

Tri- and Tetrafunctional

Glycerine   Pentaerythritol

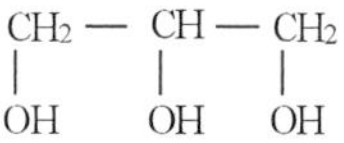

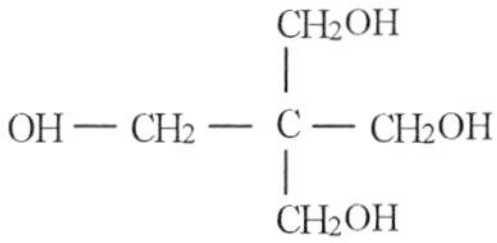

CH₂OH
|
CH₂ — CH — CH₂   OH — CH₂ — C — CH₂OH
| | | |
OH   OH   OH   CH₂OH

While the Tg, hardness and other coatings performance factors are in proportion to the weight percent of these components in the alkyd, the molecular weight (Mwt) and Mwt distribution are determined by the molar ratio of these components. The prediction of gel point is important because the best drying performance of alkyds is found near to the gel point. Carothers[4] advanced an equation to calculate the extent of the esterification reaction (P), which would occur before gelation ensued in the particular formulated alkyd's polymer mass. With the average functionality of the polymer listed as f and only stoichiometric (equal) equivalents of reactants, the equation stands as:

$$P = 2/f \qquad\qquad P^2 = 1/f-1$$

$$\text{Carothers} \qquad\qquad\qquad \text{Flory}$$

Flory[5] introduced probability statistics to the prediction of the gelation point and derived the above equation, which is much more accurate than the first but still lacking in accurate predictability. The considerably more complex equation below was prepared by Jonason.[6]

$$(P_A)gel = \left[\frac{\Sigma}{(1-\lambda)\{g_1(1-p-\theta)+ g_2\theta+ 2p-1\}} \right]^{1/2}$$

where

$\Sigma = \dfrac{\text{Total equivalent of hydroxyls}}{\text{Total equivalent of acids}}$ $\qquad\qquad$ $\lambda = \dfrac{\text{Equivalent of monofunctional acids}}{\text{Total equivalent of acids}}$

$g_1 = $ functionality of trihydric alcohol $\qquad$ $p = \dfrac{\text{Equivalent of hydroxyls in dihydric alcohol}}{\text{Total equivalent of hydroxyls}}$

$g_2 = $ functionality of tetrahydric alcohol $\qquad$ $\theta = $ Equivalent of hydroxyls in tetrahydric alcohol

$1-p-\theta = \dfrac{\text{Equivalent of hydroxyls in trihydric alcohol}}{\text{Total equivalent of hydroxyls}}$

Even with this equation, the variation from the kinetic theoretical gel point during the reactions varied from experimental by 2% to 9%. These differences are generally attributed to intramolecular condensations, unequal reactivity between differing carboxyl types and reversibility factors in the reactions. Others[7] found variations in results due to variations in reaction processing conditions and solvent content (intramolecular condensation from collapse of polymer chains in poor solvent medium). Thus the formulation of alkyd resins remained somewhat of an experimental art in spite of the desire to drive the reaction to as high and as narrow molecular weight distribution as possible for the best coatings dry rate performance properties.

Idealized Alkyd Polymer Structures Showing the Effects of the Tetrafunctional Polyols
in Increased Branching in the Polymer

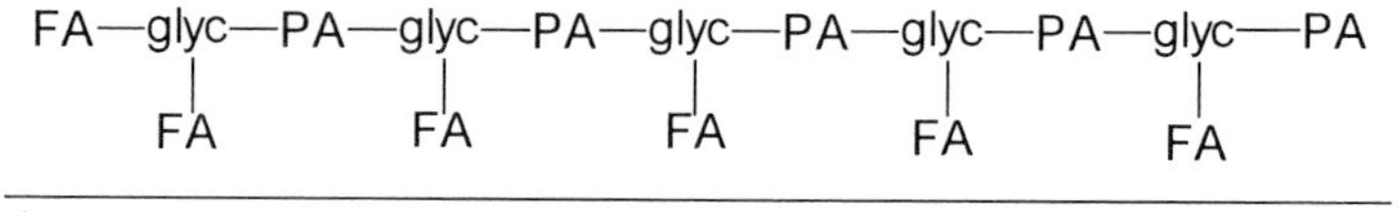

```
FA—glyc—PA—glyc—PA—glyc—PA—glyc—PA—glyc—PA
        |          |          |          |          |
        FA         FA         FA         FA         FA
```

Linear species

Branched with cyclic species

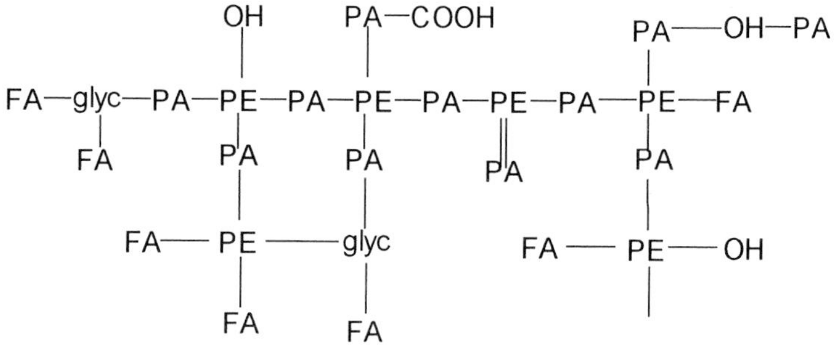

```
            OH       PA—COOH              PA——OH—PA
            |          |                   |
FA—glyc—PA—PE—PA—PE—PA—PE—PA——PE—FA
        |          |          |          |
        FA         PA         PA     PA        PA
                   |                        |
           FA——PE——glyc           FA——PE——OH
                   |      |                 |
                   FA     FA
```

FA = fatty acid esters   PE = pentaerythritol esters
PA = phthalic anhydride esters   Glyc = glycerin esters

A wider range of compositions is made possible by these multifunctional
components as opposed to difunctional components and their linear polymers. The
formulator can vary the content of the fatty acid component from the oil widely for a
variety of different coatings end users. A high fatty acid content ("long oil") e.g. 70%
gives a flexible alkyd compatible with odorless aliphatic solvents suitable for
architectural wood and metal painting. A low fatty acid content ("short oil"), having a
high aromatic acids content, gives a harder, faster drying alkyd soluble in xylene and
suitable for coating objects on an industrial assembly line. The multifunctional and
monofunctional reactants also allow considerable variation in end group residual
hydroxyl or carboxyl content at high molecular weight. This provides for crosslinking
sites with other reactive resins such as amino formaldehyde resins.

Another result of the multifunctional polyol content is a relatively wide molecular
weight distribution. On the surface, this at first sounds detrimental. However, this is
probably not the case as the high molecular weight species are particularly good for
pigment wetting and stability in terms of protecting the pigments surface and
counteracting the tendency toward flocculation and adverse pigment catalyzed
chemical reactions and drier metal absorption. The lower molecular weight species
helps keep the overall viscosity down and promotes substrate adhesion. The overall

versatility and compatibility of this with other polymers of the alkyds is thus helped by the wide polydispersity to make them the unique and important contribution to the coatings applications that they are.

Alkyd resin can be modified with vinyl monomer to improve dry speeds and resistance properties. Often maleic anhydride is included in the alkyd formulation to provide site for grafting of vinyl monomer. A wide variety of other acid or hydroxyl functional compounds are also used less frequently due to cost. Variation in the type of oil used varies the amount of diene or triene in the fatty acid component; thus, also the amount of film yellowing upon auto-oxidation. Taken all together the above features allow a wide spectrum of useful polymers that can be derived from the basic alkyd polymer type, accounting for the hundreds of different alkyd formulas sold today.

Manufacturing Technology

Alkyds can be made in almost any configuration of a stainless steel batch process reactor provided with a sufficient agitation device inside and means to heat the reactor to 210°C to 250°C remove water and then also cool it. A small amount (about 4% of the mass) of xylene as an azeotrope solvent is frequently used, along with a condenser to recycle solvent in order to decrease the reaction time. A larger reaction surface area to volume ratio speeds the equilibrium esterification reaction, which proceeds by removal of the water evolved. If higher solvent contents are needed, the reactor must be held under pressure. The alkyd reaction can also be conducted under "fusion" conditions without any solvent. The amount of inert oxygen free gas flow, which prevents yellowing discoloration in the alkyd, is usually increased with this process then and the entraining vapors "scrubbed" (falling water) to prevent escape of reactants and low Mwt polyester to the environment. Cycle times range from 8 to 24 hours.

Reactor size generally varies from 1,000 to 10,000 gallons and one can last as long as fifty years. Alkyd reactors generally are accompanied with a "thinning tank" (about twice the volume) in which the finished alkyd is mixed with solvent. This serves to cool the alkyd and gives the resultant liquid solution of alkyd a low enough viscosity for easy manufacture of coatings. The basic capital investment for alkyd reactors is not relatively high, but the infrastructure for compliance with safety and environmental regulations can discourage ready entry into this business in industrialized nations.

When oils are used, the esterification reaction is usually preceded by an ester interchange reaction (alcoholysis) with the other hydroxyl functional components to make the reaction mass compatible with the subsequently added dibasic acids. This reaction is usually conducted at about 210°C to 250°C in the presence of fatty acid calcium or lithium (0.15%) salt catalyst. The reactor takes about one hour. The resultant intermediate is predominantly a mixture of mono- and diglycerides. Following alcoholysis, the rest of the alkyd acids are added and the esterification then completed. With isophthalic acid (less compatible) as the diacid, a higher temperature 280°C acidolysis reaction (excess dibasic acids) is sometimes used. Various filtration devices are usually used to remove impurities from the alkyd while still hot before shipment to the user. The above alkyd polymers are usually sold at about 20 poise viscosity and between 40% and 75% non-volatile content in their respective solvents.

Chemical Modifications and Blends of Alkyd Resins

One of the advantages that alkyd resins have with their wide variety of structures and compatibility with other resins and solvents is that their coating performance can be enhanced by blending them or reacting them together with other polymer types. These blends or modifications then are able to maximize the alkyd's performance toward that particular coating property desired while still keeping the alkyd's basic ambient temperature, air curing property and ease of paint preparation properties. The following table lists resins blended or reacted into the alkyds.

Modifications of Alkyds

Modification	Primary Property Enhanced in Paints	Paint Uses	Approximate pounds Made in 1990, (MM)
Urethane Alkyd	Abrasion Resistance	Wood floors	68
Acrylic Alkyd	UV light resistance	Metal deco	22
Silicone Alkyd	Heat and high UV resistance	Exterior signs	10
Styrenated Alkyd	Fast dry rate, hardness	Lacquers	26
Vinyl Toluene Alkyd	Aliphatic solvent compatibility	Aerosol sprays	27
Phenolic Alkyd	Recoatability, water resistance	Primers, varnishes	20
Chlorinated Rubber Blend	Caustic resistance and fast dry	Traffic paints	25
Nitrocellulose Blend	High gloss and image	Wood furniture	14
		Total	212

Alkyd Uses versus Other Polymers

While alkyds have economical raw materials and the ability to further polymerize to crosslinked polymers by auto-oxidation, which enable them to capture a large portion of the coatings market's polymer use, they have significant weaknesses. One is that they are polyesters and as such are subject to acid or base catalyzed hydrolysis and depolymerization in the presence of strong acids or bases. Thus, they are subject to salty water corrosion over unprotected steel and saponification and decomposition on damp concrete. They are also subject to hydrolysis softening in contact with hot water in spite of their initial hydrophobicity.

Initially phenolic resins, particularly those made with butanol for compatibility, were developed for resin blends, and reactive types were made for crosslinking coatings by baking. Later, epoxy and polyvinyl chloride resins are solubilized by strong solvents (ketones, etc.) and those crosslinked with second component polyamines were developed for coatings to withstand the harsh chemical environments where alkyds would fail.

A second weakness of alkyds is their relatively low crosslink density due to the crosslinking mechanism between the 18 carbon fatty acid polymeric side chains. This subjects their films to swelling by strong solvents and permeability to organic molecules which leave stains. Further or tighter crosslinking with amino formaldehyde resins at approximately 150°C helps overcome this, but the rigid glycerol PE-phthalate alkyd polymer backbone still gives weakness in mechanical properties and relative brittleness in the coating. Urethane coatings resins were another development in the second half of the century which gave excellent mechanical properties and took coatings business from alkyds. Longer, linear chain, oil-free isophthalate polyesters were a development out of alkyds, and gave excellent flexibility and hardness for coil applied baking coatings, again replacing some alkyd consumption.

Another weakness in the alkyd polymer is its ultra-violet light resistance or exterior durability. The long sixteen carbon chain crosslinks also containing residual unsaturation are relatively readily degraded in sunlight in a method similar to polyethylene with discoloration and polymer chain scission. The use of coconut alkyds (short aliphatic sidechains) gave more durable coatings and were used in the automotive amino baking enamels for some time. These also gave way to acrylic polyol versions of these baking enamels because of the superior weathering resistance and long-term freedom from coatings discoloration and erosion.

Despite the drawbacks, alkyds maintained a substantial share of the coatings polymer business due to their good economics, pigment and substrate wetting and adhesion, and easy application properties using relatively innocuous solvent. Also, many industrially coated objects do not justify the cost of the high performance polymers. However, alkyds had to face another problem, which developed in the final quarter of this century.

Environmental Regulation and Compliant Alkyds

In the studies of the cause of smog creation in Los Angeles that led to "rule 66" regulation, xylene was identified as one of the principle ozone generating pollutants. The coatings industry was known to release substantial amounts of this from alkyds and other industrial coating polymers in use. Further studies indicated that the mild aliphatic solvents released into the air, given some time ("down wind effect") would degrade to equally detrimental organic species. Thus, the EPA took an initiative around 1974 to limit all solvent emissions from coatings according to what compromises in their performance could be tolerated; therefore, continued erosion of the alkyd business occurred.

With alkyds the end results were to force the molecular weight down to keep the coating viscosity in a suitable range for application. These solvent-borne systems were referred to as "high solids." An alternate method was to replace the majority of the solvent with a water cosolvent mixture with the alkyd in a dispersed particle form. Emulsion or latex forms of alkyds had been previously found undesirable in the 1950s. In either case, some detrimental effect to the alkyd's performance was seen. In high solids architectural coatings slower "dry" rates were seen and more objectionable color formation was also seen due to the higher unsaturation contents needed. This was for complete auto-oxidation to the final crosslink density required in the coating film. With water dispersion alkyd forms, molecular weight levels could be maintained, but hydrolytic stability in the paints and film water sensitivity due to residual hydrophilic alkyd polymer modifiers were performance obstacles. In spite of these factors, a substantial portion of the alkyd business (waterborne 20 million pounds with high solids 50 million pounds) had converted to the VOC compliant forms. These were primarily for industrial applications for metal and in some non-attainment areas, such as California, architectural alkyds (High Solids).

For preparing the water-dispersed alkyd forms, higher amounts of carboxyl groups were left unesterified in the polymer, and were converted into ammonia or amine salts in the waterborne paint. These salts "break" upon evaporation of the amine or ammonia and water. For such purposes trimellitic anhydride or dimethylol propionic acid were used in the alkyd to leave the residual carboxyls (see below).

Dimethylol Propionic Acid   Trimellitic Anhydride

Sometimes polyethylene oxide molecules are also esterified into the alkyd to give a non-ionic hydrophilic group.

In industrial amino-alkyd baking enamels, the "high solids" VOC compliance method forced conversion of a substantial amount of the alkyd business into low molecular weight polyester polyols in order to be able to obtain the lighter color and tighter crosslinking density required in these coatings films that were not available with the oil-based alkyds.

The Future of Alkyd Resins

In spite of the shift toward VOC-compliant alkyds occurring, it seems clear that substantial growth in the overall alkyd business is unlikely to occur in nations whose manufacturing specializes in high value-added goods. However, alkyd use in nations with developing manufacturing economies is likely to grow considerably due to the alkyd's low-cost and bio-renewable raw materials. However, the growing economies of these developing nations may come to consume large amounts of the lower cost alkyds where air pollution regulation will not be applied. Light industry may take advantage of the auto-oxidation cure mechanism to avoid energy intensity high temperature curing ovens. The additional factor of the hydrophobic alkyd films offering protection (particularly with zinc oxide pigment biocides) to the bio-renewable wood dwellings in humid tropical environments could be a strong factor in their growth. Soybean oil for alkyds, as a low-cost by-product of soya meal protein, may help keep alkyds at attractive prices if the growing world population outstrips the ability of the land masses to provide sufficient high protein foods. In regards to the chemical constituents of alkyds, not much change would be anticipated except perhaps in the sources of the biologically derived oils via genetic engineering and the increased need for the use of marginal lands for agricultural production. In the long run, it is this author's opinion that the alkyd production from bio-renewable resources will probably grow substantially on a worldwide basis due to increasing competition for scarce natural resources such as petroleum. Thus they should serve mankind well through the next century.

LITERATURE CITED

[1] W. H. Carothers, Trans. Faraday Soc. 3R, 43 (1936).
[2] R. H. Kienle and C. S. Ferguson, Ind. Eng. Chem. 21, 349 (1929).
[3] R. H. Kienle, Ind. Eng. Chem., 22, 590 (1930).
[4] W. H. Carothers, Collected Papers, High Polymers, Vol. 1, Interscience Publishers, Inc. N.Y. (1940).
[5] P. J. Flory, J. Am. Chem. Soc. 63, 3083, 3091, 3096 (1941); 69, 30 (1947); 74, 2718 (1952). Che. Rev. 39, 137 (1946).
[6] M. Jonason, J. Appl. Polymer Sci. 4, 129-140 (1950).
[7] S. L. Levy, Effect of Composition and Process Variables on Properties of Alkyd Resins, Ph.D. Dissertation, Case Institute of Technology, 1960.

WHITE PIGMENTS

JUERGEN H. BRAUN * and JOHN G. DICKINSON *

* Pigment Consultants, Inc.
614 Loveville Road, Bldg. E-1-H
Hockessin, Delaware 19707

**DuPont White Pigment & Mineral Products
Chestnut Run Plaza
Wilmington, Delaware 19805

Introduction

In 1985 and in this forum, Fred B. Steig[1] of N L Industries, reviewed the growth of "Opaque White Pigments in Coatings" from commercial art into a modern technology. Since then, white pigment technology continues to evolve from contributions of practitioners, theorists, engineers, scientists and businessmen. Within this orderly evolution there did occur a major development; a new manufacturing process causing profound changes in the white pigments industry but, because of proprietary concerns, the technical details were not revealed. The new process involved an engineering gamble against large odds, and progress was achieved through the cooperation of technically astute managers with innovative engineers and creative scientists.

White pigment technology was changing. In the 1930's titanium dioxide began to obsolete all other white pigments. The pigment was manufactured by a process based on the digestion of titanium ores in sulfuric acid, and in the late 1940's this "sulfate" process was growing into maturity. Meanwhile, the cold war prompted interests in titanium metal for aircraft and naval applications, the metal being made from titanium tetrachloride. Titanium dioxide pigment was a potential second outlet for the tetrachloride, and several manufacturers tried to develop processes to make pigment from it. Only two succeeded in their ventures – in the 1950's, DuPont and in the 1960's, American Potash, later Kerr-McGee. The chloride process, in turn, opened the door to decisive performance improvements for titanium dioxide pigments.

The last three or four decades have seen the decline of the "sulfate" process and the optimization of new "chloride" technology. The performance characteristics of titanium dioxide pigments were improved, and new grades were developed to suit the requirements of the plastics and paper industries and specific needs within coatings applications. This progress has come in part by opportunities inherent to chloride technology and in part by competitive pressures exerted by "chloride" into "sulfate" producers and vice versa.

Background

Chemical coatings are applied to surfaces to protect and to decorate. They are usually composed of two phases – continuous polymer and discontinuous pigment. The polymer component of the coating provides the protection; pigments supply aesthetics. Whether white, black or color, pigment particles hide the drabness and optical contrasts of the substrate.

Hiding by thin films is essential for virtually all architectural and industrial surfaces. This is a fairly straightforward issue with either dark or heavily colored coatings, simply because blacks or dark pigments effectively absorb all light impinging on the film. White or lightly tinted systems are another matter. The specific objective of a white pigment is to scatter light (as opposed to absorption) and this task is accomplished in two ways – refraction and diffraction. In the case of TiO_2, both parameters require extremely careful control of both composition and physical dimension. Titanium dioxide must be rutile phase – highest refractive index – highest refractive scattering. It must also have carefully controlled particle size since optical theory states that highest diffractive scattering occurs with particle dimensions that are approximately half that of the incident radiation to be scattered. This latter parameter, diffractive scattering implies a responsibility on the part of coatings, plastics, ink and paper manufacturers in that the pigment particle size furnished must be maintained in their finished products.

The Path of Progress

In white pigment technology theoretical insights developed slowly. At first, theory provided qualitative direction; now, chemical and optical theories are able to stake quantitative goals. With such help, optimization has occurred as an interplay of science with engineering technology.

Periodic reviews [2,3,4] and, more recently, a summary [5] have covered most aspects of titanium technology and pigment science. But, for proprietary reasons, not much has been published about manufacturing processes.

The Chloride Process

Most of the progress of the last several decades originated from development of an innovative new process for the manufacture of titanium dioxide pigment. This "chloride" process offers dramatic environmental, decisive quality and significant economic advantages.

Process development started in the 1940's and the first chloride plant began to produce in 1954. Other pigment producers and chemical manufacturers entered the technology and   currently 23 chloride process plants are operating in North America, Europe, Asia and Australia.

The chloride process can deal with a variety of titanium ores, rutile ore being preferred to take advantage of its low iron content. Because rutile does not easily dissolve in sulfuric acid, sulfate plants require ilmenite ores ($FeTiO_3$) or titanium slag (TiO_2, Fe_2O_3) whose high iron contents aggravate waste disposal problems. Raw material economics are affected decisively by local disposal options and by shifts in the price of titanium ores, some prompted by the ongoing conversion of the industry from "sulfate" to "chloride" and ilmenite to rutile ores.

The sulfate process's intermediate is an aqueous solution of titanyl sulfate that must be hydrolyzed to colloidal hydrous titania. The hydrolysis is seeded with rutile nuclei, and then calcined and crystal-grown into size optimized pigment particles. Eventually, vast quantities of iron sulfate and dilute sulfuric acid must be disposed of or recovered.

By contrast, the chloride process's chemical intermediate is anhydrous titanium tetrachloride, a high boiling liquid which is reacted with oxygen into a titanium dioxide particulate with exclusively rutile phase and pigmentary particle size. By-product chlorine is recycled and reused in the chlorination of the titanium ore.

In the early years chloride pigments outperformed their sulfate counterparts by wide margins. Currently the two manufacturing processes produce a range of products somewhat more closely matched.

The Chemical Reaction. Titanium dioxide can be made by an exothermic reaction, oxidation of titanium tetrachloride, carried out with a flame that generates a solid particulate from gaseous ingredients:

$$TiCl_{4\,(gas)} + O_{2\,(gas)}\ \text{->}\ TiO_2\ (rutile) + 2\ Cl_{2\,(gas)}$$

To sustain the flame, reactants must be preheated and the reaction proceeds quickly and almost completely. The chemistry is simple but engineering problems are legion; they include corrosion control, reactant mixing and flame stabilization. But the real challenges involve control of crystal phase, crystallite size and particle aggregation.

Flow Control and Mixing. The combustion generates a solid particulate that is susceptible to accumulation on whatever surfaces confine the flame and control mixing of the ingredient gas flows. Accumulating crusts, in turn, degrade intended flow patterns within the reactor, resulting in a loss of control over flow and mixing. Recirculation of particulate caused by turbulence near the flame front leads to particle growth beyond the intended size. Also, titanium dioxide crusts cannot be allowed to contaminate the pigment.

Crystal Phase. For outdoor durability as well as optimal hiding power in coatings and plastics applications, titanium dioxide pigment must be composed of rutile rather than anatase crystallites. Rutile has a higher refractive index than anatase, thus hides better.

From a hiding perspective, a few percent of anatase in rutile pigment would not really matter but even small concentrations of anatase in the rutile degrade the outdoor durability of paint films. Thus, crystal phase of the pigment must be controlled to better than 99% rutile.

Particle Size. The effectiveness with which white pigments hide is quite sensitive to particle size; primary rutile particles should be about $0.2\mu m$ in diameter. Particles below about $0.1\mu m$ or above about $0.5\mu m$ are wasted, the larger ones even detrimental. Thus, a narrow particle size distribution is essential. Size specifications were developed experimentally and confirmed theoretically.[6,7] Computers were essential;[8] without them, the calculations of particle size relationships to light scattering effectiveness could not have been accomplished.

For reasons involving color effects as well as hiding, optimal particle size of titanium dioxide pigment differs a bit with pigment volume concentration of the intended application. For uses at low pigment concentration, as in most plastics, crystallites are made to be a bit smaller than crystallites intended for coatings. Special pigment grades are made to cater to specific size requirements that exist even within coatings applications.

Aggregate Formation. The presence of aggregates of size-optimized primary particles, that is, sinter-bonded crystallites, causes several problems in the product:

- Sintered titanium dioxide crystallites, once formed, are extremely difficult to remove or grind to size.

- Aggregates decrease the hiding power of the pigment because they scatter light with the diminished effectiveness of larger-than-optimal particles.

- They can diminish gloss of paint films.

- They can increase the abrasiveness of pigment used in paper applications where aggregates dull slitting knives and in the delustering of textile fibers where they erode spinnerets.

The likelihood of aggregate formation must be diminished by the design of the burner, although one can grind burner discharge in a separate process step.

Engineering. The engineering challenges to process development were severe, where almost all problems had to be resolved at full manufacturing scale. The most serious problems involved the $TiCl_4/O_2$ burner where pigment particles must be generated in their final size and crystal phase. A pilot plant was not considered useful because this burner operates at high temperatures where volume-to-surface ratios are all but impossible to scale from pilot plant to production size facilities. Heat transfer problems around the burner are further complicated because, by its nature, pigmentary titanium dioxide is a superb thermal insulator.

The oxidation step, specifically its $TiCl_4$ burner, is the heart of chloride pigment technology. First, oxygen and titanium tetrachloride have to be heated separately to several hundred degrees centigrade without contaminating the hot gasses by corrosion products. Then in a flame front and within milliseconds, all the basic characteristics of the pigment must be fixed:

(1) The crystal phase of product must be regulated to produce rutile.
(2) The size of the pigment crystallites controlled, to serve a variety of product requirements.
(3) Aggregation of crystallites and the agglomeration of aggregates minimized to improve gloss performance and to reduce the costs of grinding.

Failures of crystallite size and phase control cannot be corrected because crystallites cannot be ground and pigmentary titanium dioxide cannot be phase converted.

Development of process engineering operations downstream from the oxidation step – wet-treatment, grinding and packaging – was continued in the traditions of conventional pigment technology. The products of the oxidation step, size optimized rutile crystallites, are "wet-treated," that is, subjected to a host of processes that are carried out in aqueous dispersion and that may include grinding operations. These operations enhance pigment performance in specific coatings, plastics and ink applications. Several of the wet-treatment processes evolved from sulfate pigment technology, other procedures were designed specifically for chloride pigments because the surface characteristics of chloride and sulfate pigments differ. Chloride pigment contains some alumina added for phase control of the crystallites, sulfate pigment some phosphate present for particle size control.

After wet-treatment, pigments are filtered, dried and ground in fluid-energy mills common to chloride and sulfate plants.

Product Considerations

Purity. Impurities in the pigment can degrade its color (brightness). Ordinary contaminants matter but far more detrimental are impurities that can substitute for Ti^{4+} ions within the lattice of the rutile crystal. In concentrations of only a few parts per million, such impurity cations – iron, chromium, nickel, vanadium and niobium among them – cause pronounced yellowing or graying of the pigment.

Since these interstitial impurities cannot be leached from the product, offensive elements must be removed from the process intermediates, titanium tetrachloride or titanyl sulfate solution. The purification is a task that is much easier accomplished in chloride than sulfate technology. Anhydrous $TiCl_4$ liquid can be purified readily by distillation and by chemical treatments whereas aqueous solutions of titanyl sulfate are quite difficult to treat effectively. Thus chloride pigments are whiter and brighter than their sulfate counterparts.

Durability. Titanium dioxide is a catalyst for the sunlight energized oxidation of organic polymers, and the semiconductor mechanisms involved are reasonably well understood.[9,10,11] At its surface, titanium dioxide transforms the energy of ultraviolet light into chemical energy. This chemical energy reacts with oxygen and water to generate two free radicals, hydroxyl and peroxyl:

$$\eta v[TiO_2] + O_2 + H_2O \nearrow OH^{\bullet} + HO_2^{\bullet}$$

The free radicals can, in turn, react with and destroy almost any organic molecule:

$$OH^{\bullet} + HO_2^{\bullet} + {-}CH_2{-} \nearrow CO_2 + H_2O$$

As a result, paint films pigmented with unprotected titanium dioxide are said to chalk, that is, turn into dust by prolonged outdoor exposure.[12,13] For anatase pigment, the effect is severe enough to all but preclude its outdoor use. Paint films pigmented with conventional rutile are less prone to degradation, but the chalking problems of titanium dioxide pigments were all but resolved by chemistry developed by Iler[14] and subsequent extensions. This chemistry made it possible to encapsulate certain inorganic particulates in shells of silica glass. Today, silica encapsulated rutile pigments perform exceedingly well in even the most demanding outdoor applications.

Dispersibility. By laws of physics, small particles stick to each other. These short-range attractions cause pigment crystals to be very sticky. They agglomerate into assemblies that are larger than intended, thus (1) become less hiding effective than individual pigment crystallites, (2) degrade gloss and (3) cause roughness of paint films. To break the agglomerates, pigment manufacturers grind their products in fluid energy mills. By a second, less intense grinding operation, pigment consumers disperse dry pigment into their appropriate media.

The pigment's tendency towards spontaneous agglomeration can be reduced. Just as a coating of feathers keeps wax balls from sticking to each other, so can a hydrous alumina coating reduce the stickiness of pigment particles.

Hydrous alumina coatings were advocated for titanium dioxide pigments for a variety of benefits, and predate the chloride process.[15] The explanation and optimization of their effectiveness required sophisticated techniques of modern electron microscopy.

Traditionally, paint grinding, that is, pigment dispersion into paint media, is performed by paint manufacturers. In 1970, DuPont commercialized aqueous slurry grades of titanium dioxide pigments. For waterborne paints and where the scale of paint manufacture justifies the installation of a slurry handling system, the burden of paint grinding can now be shifted to the pigment producer.

Gloss. Glossiness attracts attention. A high-gloss finish sells cars and objects intended to draw the eye of the observer. Obviously, pigment manufacturers would like to design the potential of high gloss performance into some of their products. However, highest levels of gloss can occur only on surfaces that are essentially amorphous in character, the surfaces of glasses, polished metals and clear liquids including clear paint films. Particulates in the paint film, pigments, can only degrade gloss.[16]

To be least detrimental to the gloss of its paint films, a pigment may not contain particles larger than necessary for hiding.[17] Ideally, the pigment should contain no particles that are larger than about 0.5μm, neither crystallites nor their assemblies into aggregates and agglomerates.

Also, the pigment must be capable of packing quite densely, that is, have the lowest possible "Oil Absorption," a measure of the packing density of pigment particles into liquid dispersion. This Oil Absorption requirement is difficult to meet for durable pigment grades. Only recently has a durable version of high-gloss pigment become available.

Pigment Grades. Titanium dioxide pigments are used in several industries and in many diverse applications and media. Since light scattering is their singular objective, all commercial products are appropriately particle size optimized. All rutile grades could serve the diverse needs of most industries and would do so at least moderately well. Optimal performance demands special grades designed for specific requirements: outdoor durability, high gloss, porous paint films, etc. In addition, customer convenience may be designed into the product: slurries that are intended for stir-in dispersion, dry-bulk products for use in plastics.

Alternatives to Titanium Dioxide

Classic Pigments. All the classic white pigments are now obsolete because they are neither as safe nor nearly as effective as titanium dioxide.[1] For two millennia, white leads – basic lead carbonate and sulfate - were the only white pigments that could deliver moderately durable whiteness and brightness into a drab world of grays and earth colors. Toxicity was recognized, but accepted. Eventually, white leads were displaced by zinc whites – zinc oxide, zinc sulfide and lithopone (an equimolar composite of zinc sulfide and barium sulfate.) Zinc whites are much less toxic than lead whites but still do not hide nearly as well as titanium dioxide.

A variety of composite pigments – intimate mixtures of titanium dioxide with calcium sulfate and with lithopone – were used extensively during a transition period from zinc whites to essentially pure titanium dioxide. They have effectively disappeared from the market.

Air Hiding. Air hiding, light scattering at the interfaces of particles with air rather than polymeric binder, occurs almost everywhere. Air hiding is the optical mechanism of fog, snow, textiles, paper, chalk marks and layers of dust. Air hiding contributes significantly to hiding by porous paint films.

In paint films, the advantage of air hiding is its low cost. Since porosity of the film is a prerequisite, coatings that involve air hiding have two principal disadvantages:

(1) They are brittle, thus lacking in mechanical strength;
(2) They do not protect because pores conduct contaminants into the coating and onto the substrate.

Nonetheless, modern porous paint films serve very well in applications where chemical and mechanical assaults are infrequent, for example on interior walls and ceilings.

Void Pigments. Paint films pigmented with microvoids,[7] "void pigments," are far less sensitive than porous films; their air voids are individually encapsulated by polymer:

·   Voids are sealed and, unlike pores, do not conduct contaminants into the film;
·   Capsules are spherical, thus mechanically strong.

Void hiding has also been demonstrated as an aqueous slurry of beads, each composed of polymer, voids and titanium dioxide. The voids enhanced the scattering effectiveness of the pigment.

Other Substances. To serve as a white pigment, a substance must satisfy stringent requirements. It must have an extreme refractive index, be colorless, chemically inert, stable, and nontoxic; it must be available as microscopic particulate. Only a few substances have refractive indices high enough for a pigment, most of them are compounds of titanium, zinc, and lead.

Titanium dioxide is uniquely qualified because it is nontoxic and rutile and anatase have the highest refractive indices of all colorless substances. Brookite, the third titanium dioxide phase has properties and stability characteristics that fall between rutile and anatase but promise no advantage over either. It can be manufactured in pigmentary particle size. Since an extreme refractive index is the essential feature of a white pigment, titanium dioxide has no direct competition.

Could Hyperbaric phases of titanium dioxide qualify? Yes, novel titanium dioxide phases made at extreme pressures and temperatures, are likely to combine higher densities with higher refractive indices. A pigment made from such a hypothetical titanium dioxide phase would be more hiding effective than rutile, but such pigment will not be available in the foreseeable future. Hyperbaric syntheses are incredibly costly and can make only milligram batches.

Outlook

Given the unique optical characteristics of rutile, titanium dioxide pigments are here to stay through the foreseeable future. No other substance or light scattering technology is on the horizon, let alone on the drawing board.

Titanium ores are plentiful. However, by-product disposal of pigment manufacture in a manner that is safe to the environment has become a worldwide issue. No longer is it acceptable to ocean dump by-product acids and acidic iron compounds. Deep well disposal is restricted severely, limiting deep wells to sites of exceptional geology. Minor but noxious contaminants in the ore have become major disposal problems. Pigment costs are now driven by waste problems and disposal costs are increasing quickly, much faster than selling prices. Progress in pigment development is likely to continue but, for good reasons, probably less rapidly than in the past:

For some time now, pigment grades have proliferated to meet specific requirements within the coatings, plastics, paper and even ink industries. Lately a consolidation of pigment grades is underway to offset inventory and distribution costs of grade proliferation.

Because of the urgency of disposal problems, the focus of pigment research and development has turned to waste disposal. The shift comes at the expense of long term research, application studies and product developments.

Meanwhile, the white pigments industry can take pride in the fact that the developments of the last fifty years have resulted in products that leave relatively modest room for improvement within the theoretical limits imposed by physics and chemistry.

Literature Cited

[1] Steig, F. B., Jr., "Opaque White Pigments in Coatings," ACS Symposium Series 285, Applied Polymer Science, 2d ed., edit. R. W. Tess and G. W. Poehlein, (94 references) 1985.

[2] Barksdale, J., "Titanium," The Ronald Press Company, New York, 2d ed., (691 pages) 1966.

[3] Patton, T. C., "Pigment Handbook," Vol. I, John Wiley & Sons, New York, pp. 1-108, 1973.

[4] Braun, J. H., A. Baidins, R. E. Marganski, "TiO_2 Pigment Technology – A Review," Prog. Organic Coatings, <u>20</u> [2], 105-138, (200 references) 1992.

[5]Braun, J. H., "White Pigments," Monograph in the Federation Series on Coatings Technology, Federation of Societies for Coatings Technology (USA), (43 pages) 1993.

[6]Ross, W. D., "Theoretical Computation of Light Scattering Power: Comparison between TiO_2 and Air Bubbles," J. Paint Technol., 43 [563], 50-66, (30 references) 1971.

[7]Ross, W.D., "Theoretical Computation of Light Scattering Power of TiO_2 and Microvoids," I&EC Product Research & Development, 13[3], 45-49, (12 references) 1974.

[8]Ross, W.D., "Kubelka-Munk Formulas Adapted for Better Computation," J. Paint Technol., 39 [511], 515-521, (8 references) 1967.

[9]Sullivan, W. F., "Weatherability of Titanium-Dioxide-Containing Paints," Progr. Org. Coatings, 1, 157-203 (238 references) 1972.

[10]Braun, J. H., "Titanium Dioxide's Contribution to the Durability of Paint Films," Progr. Org. Coatings, 15, 249-260, (9 references) (1987); "Titanium Dioxide's Contribution to the Durability of Paint Films – II. Prediction of Catalytic Activity." J. Coatings Technology, 62 [785], 37-42, (15 references) 1990.

[11]Diebold, M.P., "The Causes and Prevention of Titanium Dioxide Induced Photodegradation of Paints, Parts I and II," Surface Coatings International 1995 [6], 250-256, and 1995 [7], 294-299 (76 references) 1995.

[12]Kampf, G., W. Papenroth and R. Holm, "Degradation Processes in TiO_2 Pigmented Paint Films on Exposure to Weathering," J. Paint Technol., 46 [598], 56-63, (10 references) (1974).

[13]Voltz, H. G., G. Kampf, H. G. Fitzky, "Surface Reactions on Titanium Dioxide Pigments in Paint Films during Weathering," Prog. Org. Coatings, 2, 223-235, (41 references) (1973/4).

[14]Iler, R. A., "The Chemistry of Silica," John Wiley & Sons, New York, (866 pages) 1979.

[15]Farup, P., US Patent 1,368,392, "Titanium dioxide pigment coated with hydrous alumina," (1921)

[16]Braun, J. H., "Gloss of Paint Films and the Mechanism of Pigment Involvement," J. Coatings Technology, 63 [799], 43, (10 references) 1991.

[17]Braun, J. H. and D. P. Fields, "Gloss of Paint Films: II. Effect of Pigment Size," J. Coatings Technology, 66 [828], (9 references) 1994.

COLORED ORGANIC PIGMENTS

PETER A. LEWIS

Sun Chemical Corporation
Spring Grove Avenue
Cincinnati, OH 45232

Introduction
International Nomenclature – The Colour Index System
Classification of Organic Pigments by Color
Organic Red Pigments
Organic Blue Pigments
Organic Yellow Pigments
Organic Orange Pigments
Organic Green Pigments

Introduction

Significant progress in the chemistry surrounding the manufacture of organic pigments has been made within the past 75 years that has resulted in the availability of a wide variation of physical forms and new pigment types offering improved properties from novel color chemistry.

The year 1856 is often considered a milestone in the chronology of synthetic color chemistry. It was in this year that an event took place that many people thought impossible. A color that did not occur in nature was synthesized under laboratory conditions by the Englishman, William Henry Perkin. This 18 year old, working at home in a laboratory in his basement, was trying to oxidize aniline to produce quinine. He combined aniline, containing toluidine as an impurity, with potassium dichromate and sulfuric acid. The mauve colored dyestuff that resulted must have come as a considerable surprise to this young chemist.

Perkin's achievement marked a period of discovery that has been referred to as the "Dyestuffs Era," since the age between Perkin's synthesis and the start of the 20th century was a time for both the invention and commercialization of numerous dyestuffs, all based upon coal tar. Organic pigments were often produced as spin offs from these early developments, the major focus being the practical synthesis of dyestuffs. In fact the first color to be synthesized specifically as pigment in its own right was most likely Pigment Red 3, Toluidine Red, a color still in use today throughout the coatings industry. Notable inventions of the Dyestuffs Era are chronicled in *Table 1*, where organic pigments are listed alongside their date of discovery.

Before entering into any discussion relating to pigments it is first necessary to clearly define what is meant by a pigment as opposed to a dyestuff since in many earlier texts on color the terms "pigment" and "dyestuff" are used almost interchangeably.

A definition of a pigment has been proposed by the Color Pigments Manufacturers Association (CPMA). The following definition was developed in response to a request from the Toxic Substances Interagency Testing Committee, specifically to enable differentiation between a dyestuff and a pigment:

> *"Pigments are colored, black, white or fluorescent particulate organic and inorganic solids which usually are INSOLUBLE in, and essentially physically and chemically UNAFFECTED by, the vehicle or substrate in which they are incorporated. They alter appearance by selective absorption and/or by scattering of light.*
>
> *Pigments are usually DISPERSED in vehicles or substrates for application, as for instance in inks, paints, plastics or other polymeric materials. Pigments RETAIN a crystal or particulate structure throughout the coloration process.*
>
> *As a result of the physical and chemical characteristics of pigments, pigments and dyes differ in their application; when a dye is applied, it penetrates the substrate in a SOLUBLE form after which it may or may not become insoluble. When a pigment is used to color or opacify a substrate, the finely divided INSOLUBLE solid remains throughout the coloration process."*

Table 1 – Significant Years in Organic Pigment Synthesis

Year	Pigment Name	Year	Pigment Name
1856	Lab synthesis of Mauveine	1911	Dianisidine Blue
1872	Phloxine	1921	Pigment Green B
1876	Persian Orange	1924	Perinone Orange
1884	Stet Yellow	1925	Basic Dye Complexes
1885	Para Red	1931	Permanent Red 2B
1886	Alkali Blue	1935	Phthalocyanine Blue
1896	Peacock Blue (Erioglaucine)	1938	Phthalocyanine Green
1899	Lithol Red	1947	Nickel Azo Yellow
1901	Indanthrone Blue	1949	Red Lake C homologue
1902	Pigment Scarlet	1954	Azo Condensates
1903	Red Lake C	1955	Quinacridones
1903	Lithol Rubine	1956	Perylenes
1905	Toluidine Red	1956	Salicyloyl Yellow
1907	BON Maroon	1960	Benzimidazolones
1909	Hansa Yellows	1964	Isoindolinones
1909	Pyrazolones	1964	BON Red homologue
1910	BON Red	1973	Azomethines
1911	Diarylide Yellows	1974	Quinophthalones
1911	Diarylide Oranges	1986	Pyrrolo-pyrrole Red

International Nomenclature - The Color Index System.

Throughout this chapter use is made of the system as published as a joint undertaking by the Society of Dyers and Colourists (SDC) in the United Kingdom and the Association of Textile Chemists and Colorists (AATCC) in the United States. This system is known as the "Colour Index," and as such is a recognized Trade Mark.

The Colour Index, first published in 1926 and thoroughly revised in 1997, provides valuable information on all classes of colored materials, irrespective of the compounds chemical composition. The revised edition, now a single volume entitled "Pigments and Soluble Dyes," provides a valuable source of reference for all associated with the manufacture and use of both inorganic and organic colors. The Colour Index (C.I.) identifies each pigment by giving the compound a unique "Colour Index Name" (C.I. name) and a "Colour Index Number" (C.I. number). As such the identification of a pigment by mention of its C.I. name and number unequivocally identifies the chemical composition of the pigment.

The accepted convention is to abbreviate the Colour Index name for a pigment as:

PB = Pigment Blue	PBk = Pigment Black	PBr = Pigment Brown
PG = Pigment Green	PM = Pigment Metal	PO = Pigment Orange
PV = Pigment Violet	PR = Pigment Red	PW = Pigment White
PY = Pigment Yellow		

The colored pigments covered within the pages of this chapter are all organic in nature and as such contain a characteristic grouping or arrangement of atoms known as a "chromophore" which imparts color to the molecule. Additionally each molecule is likely to feature a number of modifying groups called "auxochromes," or auxiliary chromophores. These groupings alter the primary hue of the pigment in a more subtle way such as shifting a red to a yellower shade or a blue to a redder shade while still maintaining the primary hue of red or blue rather than pushing the hue over to an orange or a violet. An example of an auxiliary chromophore being used to cause a distinct shade shift in the molecule can be seen as bromine is introduced into the phthalocyanine green molecule. As each atom of chlorine is replaced by bromine the shade of the pigment moves gradually from that of a blue shade green to that of a very yellow shade green.

Possibly the most frequently encountered chromophore is the azo chromophore, -N=N-. Pigments based upon this chromophore include the naphthol reds, monoarylide and diarylide yellows, benzimidazolone orange and yellows, pyrazolones and azo condensation pigments. Of equal importance is the phthalocyanine structure based upon the compound tetrabenzotetra-azaporphin that forms the basis of copper phthalocyanine blues and greens. Pigments are also met that are based upon heterocyclic structures such as trans-linear quinacridone and carbazole dioxazine violet.

Classification of Organic Pigments by Color.

A convenient method of subdividing organic pigments is to list them by color rather than by chemistry. Most often the reader searching for information will find the data more easily when presented with a classification based upon color since this is most often the principle requirement of the search.

Organic Reds.

Metallized Azo Reds. Many of the common reds used fall within the chemical category of azo pigments since they contain the azo chromophore -N=N-, and as such are termed "azo reds". A further subdivision is possible into acid, monoazo metallized pigments such as Calcium Lithol (PR 49:2) and non-metallized azo reds such as Toluidine Red (PR 3). Typically, each of the acid, monoazo metallized pigments contain an anionic grouping, sulfonic ($-SO_3H$) or carboxylic acid (-COOH) which will ionize and react with a metal cation such as calcium, manganese, strontium or barium to form an insoluble metallized azo pigment. Conversely non-metallized azo reds do not contain an anionic group in their structure and therefore are unable to complex with a metal cation.

All azo reds are produced by a similar reaction sequence involving diazotization and coupling. A typical reaction sequence is represented in *Figure 1*, illustrating the synthesis of Barium Lithol from Tobias acid (2-naphthylamine-1-sulfonic acid) and 2-naphthol.

The initial reaction sequence is described as "diazotization" and involves reacting the primary aromatic amine, 2-naphthylamine-1-sulfonic acid in the example shown with nitrous acid, formed "in situ" by reacting sodium nitrite with hydrochloric acid, at low temperatures to yield a diazonium salt. Invariably the diazonium salt formed is inherently unstable and needs to be maintained at reduced temperatures, often below 5 degrees C, to avoid or minimize any decomposition. Typically, a diazotization reaction will be carried out at temperatures around zero degrees C. In the second half of the reaction sequence the diazonium salt is introduced to the second half of the pigment molecule, in this case 2-naphthol, called the "coupler," to yield the colored pigment. This "coupling" reaction takes place rapidly in the cold to yield the pigment. The sodium salt is formed initially as a result of the presence of free sodium ions in the coupling mixture due to sodium nitrite being used in the diazotization reaction.

The molecule is next metallized in order to confer improved properties on the pigment. In the example shown in *Figure 1* the pigment is produced as the barium salt by the addition of a stoichiometric amount of barium chloride to the suspension of the sodium based pigment. At this stage any refining treatments, such as opacification by boiling to promote particle growth, are given to the aqueous pigment suspension. The suspension is then filtered and washed to remove any residual inorganic salts or soluble by-products that may have been derived from the process. *Figure 2* illustrates the structures of a series of metallized azo reds that are of considerable commercial importance within the world of colored organic pigments. Each of these structures features a molecule based on the coupling of a naphthalene ring structure to a benzenoid structure.

Figure 1. Diazotization and coupling sequence to produce Pigment Red 49:1

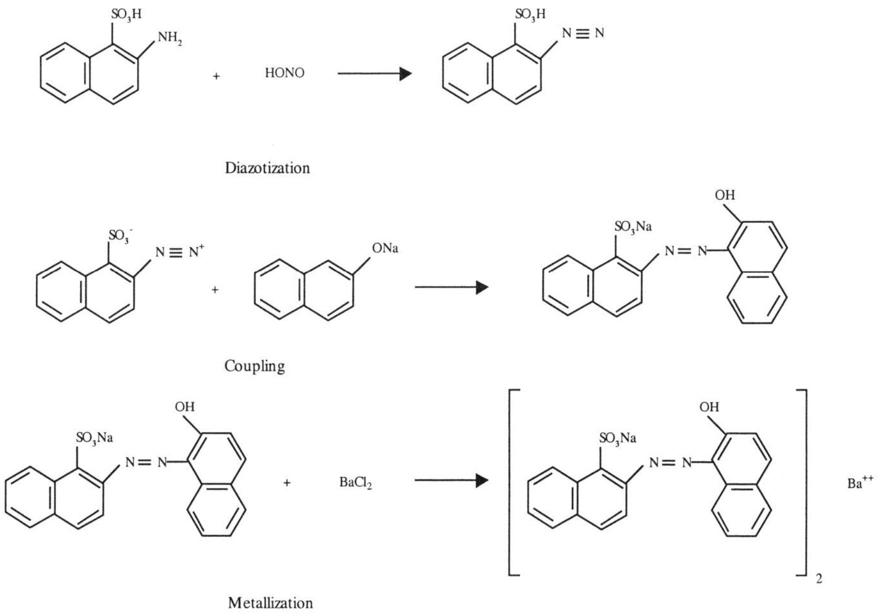

Diazotization

Coupling

Metallization

Figure 2. Structure of typical metallized azo reds.

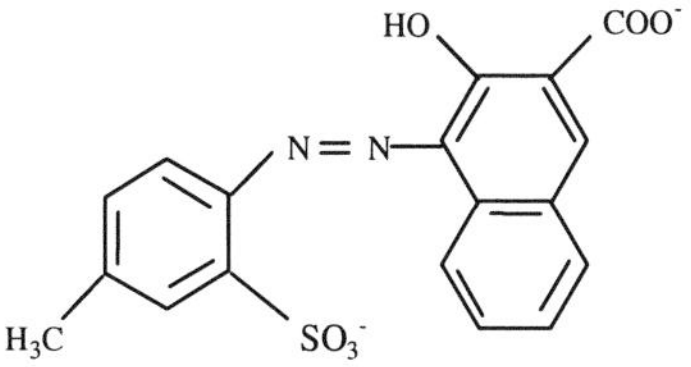

Lithol Rubine PR 57

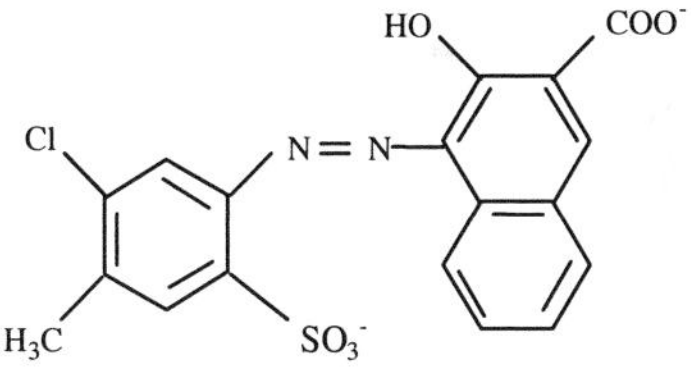

Red 2B PR 48

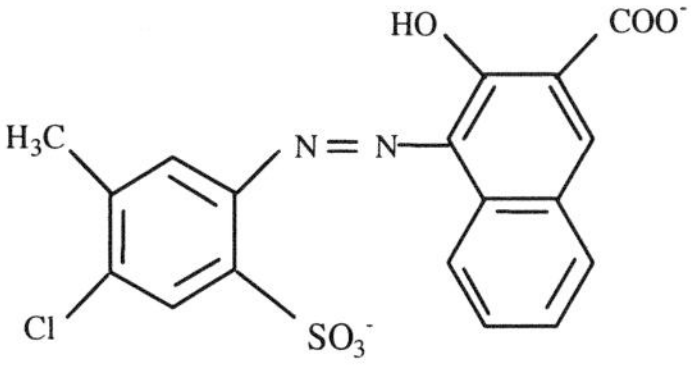

BON Red PR 52

The barium and calcium Lithol Reds, PR 49:1 and 49:2, were discovered in 1899. This pigment's major use is in the printing ink industry, the barium salt finding considerable use within the publication gravure industry where it is the red of choice for the four-color printing process. Neither the barium nor the calcium salt of the pigment can be recommended for use in outdoor situations. Additionally, neither can be used for applications in which high acid or alkali resistance is a requisite since such metallized red pigments will hydrolyse under such conditions to give a weaker yellower shade product.

The pigments are bright reds with high tint strengths and good dispersion properties, the barium salt is lighter and yellower in shade than the calcium salt, a medium red.

Permanent Red 2B, barium Red 2B, PR 48:1, calcium Red 2B, PR 48:2 and manganese Red 2B, PR 48:4 were all introduced in the 1920's. These "Red 2B" pigments are azo reds prepared from coupling diazotized 1-amino-3-chloro-4-methyl benzene sulfonic acid (2B acid) onto 3-hydroxy-2-naphthoic acid (BON) as can be seen from *Figure 2*.

Major outlets for the barium and calcium pigments are found in the coloration of plastics and for baked industrial enamels that are not required to feature appreciable outdoor fade resistance. The barium salt is characterized by a clean, yellow hue as compared to the bluer calcium salt. The barium salt has a poorer fade resistance when exposed to sunlight and weaker tinting strength but a slightly better bake stability as compared to the calcium salt.

The Manganese Red 2B has sufficiently improved fade resistance to allow it to be used in implement finishes and aerosol spray touch-up paints. The co-ordination of the manganese ion appears to enhance the lightfastness of the chromophore as compared to the equivalent calcium and barium salts.

Rubine Red, calcium Lithol Rubine, PR 57:1, is made by coupling 3-hydroxy-2-naphthoic acid (BON) onto diazotized 2-amino-5-methyl benzene sulfonic acid (4B acid), calcium Lithol Rubine is a blue shade red that was initially synthesized and made commercially available in 1903. In common with most of the other metallized azo reds this pigment has found widespread use in the printing ink industry. Lithol Rubine is the process "magenta" of the four-color printing process, both offset and packaging. Called Rubine 4B outside the United States, Lithol Rubine is a clean, blue shade red with high tint strength typical of such azo reds

BON Reds, calcium BON Red, PR 52:1, and manganese BON Red, PR 52:2 are manufactured by coupling diazotized 1-amino-4-chloro-3-methyl benzene sulfonic acid onto 3-hydroxy-2-naphthoic acid (BON). Such reds first entered the industrial scene in 1910, about the same time many similar pigments were being introduced into the marketplace from the European research laboratories. Credit for the discovery of BON Red is given to O. Ernst and H. Eichweide.

Both the calcium and manganese salts are characterized by outstanding cleanliness, brightness and color purity, the manganese salt, offers a very blue shade with improved fade resistance as compared to the calcium salt. As such the manganese salt is more suitable for outdoor applications.

BON Maroon, Pigment Red 63:1, is made by reacting the diazonium salt of 2-naphthylamine-1-sulfonic acid with 3-hydroxy-2-naphthoic acid (BON). As can be seen by comparing this structure to that of the Lithol Reds, the two pigments differ only in that the coupling component of BON Maroon contains the carboxylic acid group (-COOH). First synthesized in 1906 by Ernst Gulbransson of Farbwerken Meister, Lucius and Bruning, the manganese salt is the only one that enjoys real commercial significance rather than the calcium or barium variations. Its fade resistance is such that the pigment can be used in bright, high chroma red colors for implement and bicycle finishes. Over 40 years ago, when specifications were not as demanding, BON Maroon actually found application in automotive finishes. Overall, however, the importance of this pigment is declining and its use is diminishing within the United States.

Discovered in 1902, Pigment Scarlet, PR 60:1, is one of the few laked pigments still in production within our modern pigment industry. The pigment is a precipitate of Mordant Red 9 as the barium salt onto alumina hydrate. Anthranilic acid, 2-aminobenzoic acid, is diazotized in aqueous hydrochloric acid and then coupled onto an alkaline solution of 2-naphthol-3,6-disulfonic acid, R-salt, to form a bright red dyestuff, known as Mordant Red 9. This dyestuff is then isolated by precipitation with excess sodium chloride and laked in aluminum hydrate by co-precipitation with barium chloride under controlled pH conditions.

Pigment Scarlet is a clean, blue shade red that is used for the coloration of polymeric materials. Its acceptable heat resistance allows the pigment to withstand temperatures to a maximum of 500 degrees F for short periods. The marginal economics associated with the use of this pigment is all that restricts its more widespread acceptance within the US plastics industry. The pigment is suitable as a colorant for rubber and finds some use in vinyl plastics and cellulosics.

Pigment Red 53:1, Red Lake C, discovered in 1902 by K. Shirmacker, is a yellow shade red produced by the diazotization of C-amine, 2-amino-5-chloro para toluene sulfonic acid, followed by the subsequent coupling of the resultant diazo onto 2-naphthol. The coupling is carried out under carefully controlled pH and temperature conditions to initially form the sodium salt of the pigment. The insoluble barium salt is then formed by the replacement of the sodium cation by barium at elevated temperatures. By far the greatest market for this pigment, especially the rosin treated grade, is in the printing ink market where Red Lake C is known as a "warm red," due to its distinctive yellow shade. Within the plastics industry Red Lake C offers an economical azo red that finds use in LDPE and polyurethane where the end use allows such a pigment with relatively poor durability to be used as the system colorant. The moderate light stability and poor exterior durability in conjunction with a borderline, heat stability are the major factors restricting the more widespread use of this pigment.

Non-Metallized Azo Reds. As implied by their classification, the non-metallized azo reds do not contain a precipitating metal cation and, as such, offer increased stability to hydrolysis in highly acidic or alkaline environments. Synthesis of this sub-division of pigments follows the previously described classical method of diazotization of a primary aromatic amine followed by coupling the resultant diazonium salt. No anionic groups capable of accepting a metal cation are present in the molecule thus metallization is not a factor in their synthesis. Typical non-metallized reds are Toluidine Red (PR 3) and the wide range of Naphthol Reds as represented by Pigment Reds 17, 22, 23, 170, and 188.

Toluidine Red, Pigment Red 3, shown in *Figure 3*, is chemically the reaction product from coupling the diazonium salt of 2-nitro-4-toluidine onto 2-naphthol. The pigment was first synthesized in 1905 and has been a commercial product ever since. Various shades of Toluidine Red, described as "extra light, light, medium, dark and extra dark," are commercially available as are grades offering "haze resistance" and "easy dispersing" (ED) properties. Almost the entire U.S. production of Toluidine Red is consumed by the coatings industry. The pigment provides a bright, economical red of acceptable fade resistance properties, when used in full shade, bright reds, coupled with a high degree of color intensity and good hiding power. The pigment, however, is not fast to white overstriping since it will bleed through turning the white to pink. Toluidine Red is used in such coatings as farm implements, lawn and garden equipment, and bulletin paints where a bright, economical red of moderate lightfastness is required.

Figure 3. Structure of Pigment Red 3, Toluidine Red

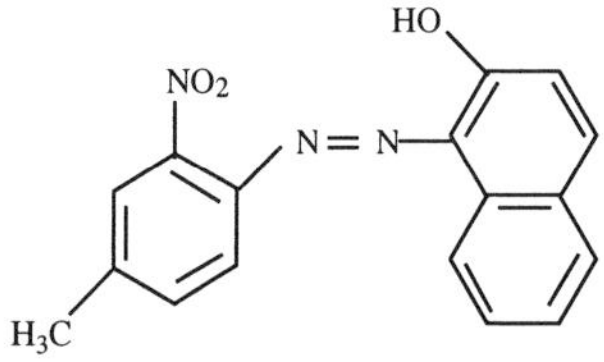

Para Red, PR 1, Chlorinated Para Red, PR 4, and Parachlor Red, PR 6 are shown in *Figure 4*, each of these pigments is based on the coupling of a primary amine to 2-naphthol. The position of the -chloro (-Cl) or -nitro (-NO$_2$) auxochromes on the molecule controls the shade of the pigment. Chlorinated Para Red differs from Parachlor Red only in the position of the -nitro and -chloro groups on the benzene nucleus. As such these pigments are isomers.

Figure 4. Structure of Para Reds, Pigment Reds 1, 4 and 6

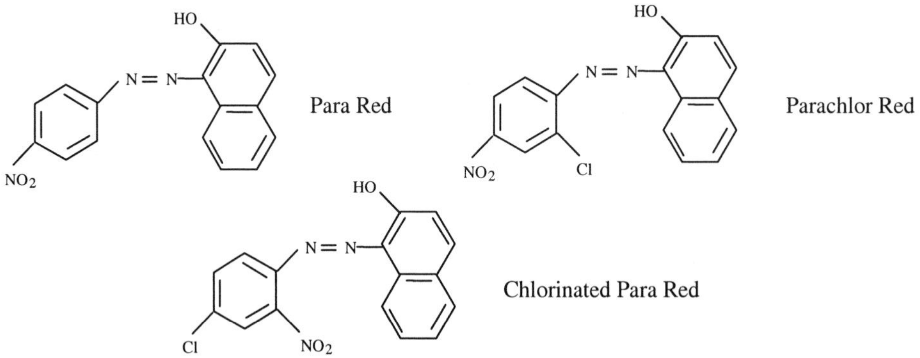

In 1880, Thomas and Robert Holliday of Read Holliday Ltd., of Huddersfield, Yorkshire, England, developed the technique of producing water insoluble azo dyes directly within the fibers of the cotton. They impregnated cotton with an alkaline solution of 2-naphthol and treated the cotton with a diazo solution such that particles of an azo pigment were formed within the cotton. In 1895, the process was modified by Meister, Lucius, and Bruning who used diazotized 4-nitroaniline to react with the 2-naphthol treated cotton to give Para Red, Pigment Red 1. This was the first organic red pigment synthesized and recognized as such by the Bradford Society of Dyers and Colourists.

Chlorinated Para Red was discovered by W. Hertberg and O. Spengler in 1907 and Parachlor Red by C. Schraube and E. Schleicher in 1906. Use of these three pigments has declined rapidly in the last 15 years due to the ever increasing and exacting demands placed upon colored finishes by the marketplace.

Naphthol Reds. Production of Naphthol Reds, chemically defined as monoazos of 2-hydroxy naphthoic acid N-arylamides without anionic salt forming groups, has been increasing on a worldwide basis to fulfill the needs of each of the three principle color consuming industries, printing inks, plastics and paints, for an economical, medium performance red. Their individual properties are dependent upon the specific composition of the pigment in addition to the conditioning steps used in their manufacture.

2-Hydroxy-3-naphthoic acid was synthesized in 1892 by the German chemist Schoepf. In 1911, the coupling of aniline and toluidine diazonium salts with 2-hydroxy-3-naphthoic acid by Laska, Winter, and Zitscher, working at the time with Naphtholchemie Offenbach, paved the way for today's "naphthol pigments."

Manufacture of the Naphthol Reds typically follows the diazotization and coupling sequences as described previously. The primary aromatic amine is reacted with acidic sodium nitrite to yield the corresponding diazonium salt. This salt is immediately coupled to the naphthoic acid anilide coupling component at a pH controlled between 5 and 7. The structure that is the backbone of all the modern Naphthol Reds is given as Figure 5.

The individual properties of the Naphthol Reds are dependent upon the specific composition of the pigment as well as the conditioning steps that the pigment is subjected to during the manufacturing process. As a class, they are a group of pigments that exhibit good tinctorial properties combined with moderate fastness to heat, light and solvents. Generally, it is observed that the naphthol based pigments containing an additional amide or sulfonamide group in the anilide ring show improved resistance to solvents and heat. Pigments such as Pigment Reds 170, 187, and 188 are examples of such structures. Unlike the metallized azo reds, these pigments are extremely acid, alkali, and soap resistant.

In terms of performance and economics, the Naphthol Reds lay between Toluidine Red at the lower end of the scale and Perylene and Quinacridone Reds at the higher end. Those Naphthol Reds that are considered of commercial significance may be briefly covered as:

Pigment Red 22, a light, yellow shade Naphthol used in printing inks and air drying alkyd and aqueous paints that can be satisfied with this pigments marginal light resistance characteristics.

Pigment Red 112, a newer Naphthol Red that possesses a very clean, yellow hue and that finds use in both industrial and architectural paints. The tendencies of this pigment to bloom or crock when used at high concentrations and its poor resistance to overstriping have limited its more widespread use.

Pigment Red 170, a medium performance, moderately priced organic red that has shown increasing commercial significance through all segments of the color using marketplace. There are actually two crystal modifications of this pigment that exhibit different hues. The first is a bluer, more transparent shade while the second is yellow shade that is more opaque with a lower tinting strength. The increased hiding power offered by the opaque grade results from the change in crystal modification rather than an increase in mean particle size of the pigment. The opaque grade finds widespread use in lead free formulations, in coatings for farm implements , and it has even been tried in automotive systems as an extender to the more expensive quinacridone and pyrrolo-pyrrole reds. Its use in coatings formulations in combinations with iron oxides has proved to be a practical solution to improving the light resistance, hiding, and economics of this pigment.

Pigment Red 188, a yellow, clean shade red with acceptable durability at all depths of shade. This pigment finds use in higher quality industrial paints and in colorant dispensing systems used to produce "in-store" customized architectural paints.

Figure 5. Generic Structure of Naphthol Red Pigments

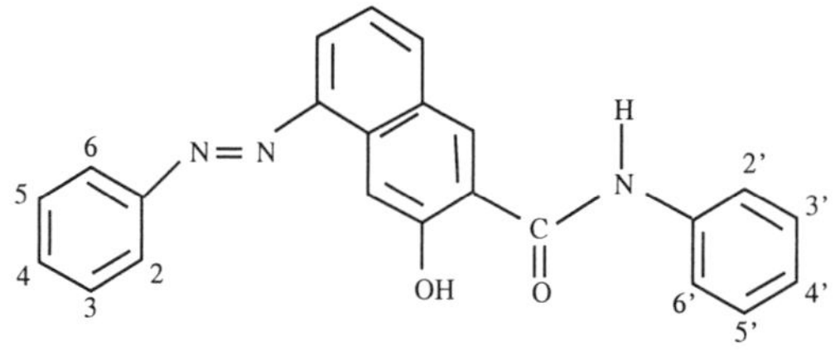

Generic Naphthol Structure

High Performance Reds. The two previous sections dealt with what are referred to as the low to medium performance reds. Although these pigments offer acceptable economics, they will never find a place in the coloration of a high performance coating such as that required to meet the exacting standards of today's automotive coatings. Pigments used in such formulations are required to show satisfactory durability to outdoor exposure in such states as Florida and Arizona for as little as two and as long as five years before finally being approved for use on an automobile body finish. Similar requirements are placed upon those pigments chosen for use in the majority of outdoor applications as are met in the plastics, ink and paint industries.

The high performance reds covered in this section fall into four basic classes: quinacridone reds and violets, vat dyestuff based reds such as perylenes, reds derived from benzimidazolone diazonium salts and disazo condensation reds.

DuPont first commercialized quinacridone pigmentss in 1958 by releasing a yellow shade red (Red Y), a blue shade red (Red B), and a violet (Violet R). The two reds were both gamma crystals, whereas the violet was a beta crystal modification. Of particular note about these early introductions was the Red Y, a pigment with outstanding exterior durability. Introduced in 1958, it is only now, in the late 90's, that the popularity of the pigment has tapered off as it is being replaced by color formulations containing the newer di-pyrrolo-pyrrole red, Pigment Red 254.

Quinacridone, *Figure 6*, is classically heterocyclic in structure since the molecule comprises a system of five fused alternate 4-pyridone and benzene rings, as such the ring atoms are dissimilar, being a combination of carbon and nitrogen rather than only carbon as we have seen in the previous pigments discussed. Addition of differing auxochromic groups such as methyl (-CH3) and chlorine (-Cl) gives Pigment Red 122 and Pigment Red 202 respectively, both described as magentas.

Figure 6. Trans-Linear Quinacridone, Pigment Violet 19

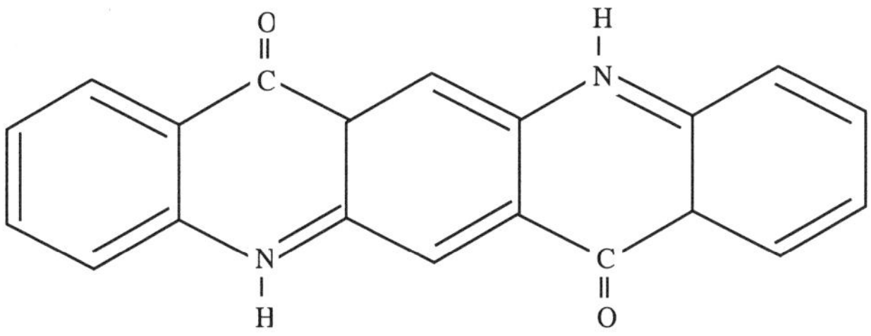

The theory behind the superior durability of pigments with the quinacridone structure invokes the proposal that considerable intermolecular hydrogen bonding occurs between molecules through the carbonyl (=C=O) and imino (=N-H) ring atoms, in addition, the strong dipolar nature of the pyridone rings results in strong intermolecular dipolar association through the crystal lattice.

One of the favored commercial routes in the synthesis of quinacridone (PV 19) involves the cyclization of 2,5- diarylaminoterephthalic acid. Subsequent conditioning leads to the product with the desired crystal morphology. Use of 2,5-dianilinoterephthalic acid at the ring closure stage yields the unsubstituted trans linear-quinacridone. Use of 2,5-ditoluidinoterephthalic acid yields the 2,9-dimethylquinacridone, Pigment Red 122, existing in only one crystal form.

As a group of high performance pigments the quinacridones find their primary outlets in plastics and the paint and coatings industries finding widespread use in automotive, industrial and exterior finishes. A minor use is in the preparation of quality furniture stains and finishes. The pigments combine excellent tinting strength with outstanding durability, solvent resistance, light, heat and chemical resistance.

Vat Reds. Vat Red pigments, based upon anthraquinone, include such structures as Anthraquinone Red (PR 177), Perinone Red (PR 194), Brominated Pyranthrone Red (PR 216), and Pyranthrone Red (PR 226), for examples see *Figure 7*.

Figure 7. Typical Vat Red pigments, Pigment Red 177 and Red 226

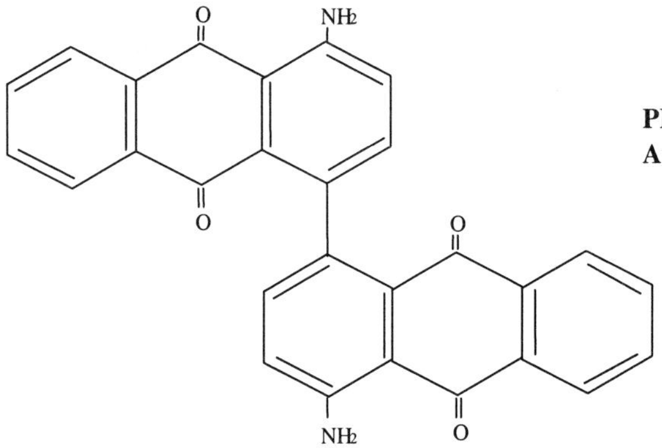

PR 177
Anthraquinone Red

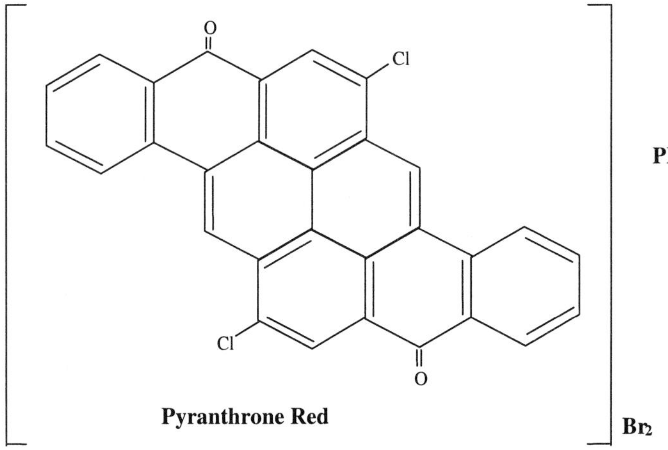

PR 226

Pyranthrone Red

Br₂

These vat pigments are related to the widely used vat dyestuffs as used originally in the dyeing of cotton. Although most the dyestuffs were developed around the turn of the century, it was not until the 1950's that they became commercially available in a pigmentary form and from the several hundred vat dyes known, only around 25 have been converted for use as pigments. Only their unfavorable economics limits the more widespread use of these pigments.

As with the quinacridones covered earlier, the product derived from the vat pigment process is obtained in a crude form. It is only rendered pigmentary after a subsequent conditioning stage in which the large crude crystals and aggregates are broken down into smaller crystals having a more uniform particle size distribution. Currently the most widely practiced conditioning process is that described as acid pasting. In this process the crude is dissolved in concentrated sulfuric acid, often requiring as much as 10 times the weight of crude to obtain a solution, the pigmentary particles are then obtained by drowning out this acid solution under carefully controlled conditions, at a controlled rate into a massive excess of water. The pigment thus obtained is isolated by filtration, washing and drying to obtain the dry color. Typical Vat reds as an example of the type are:

Dibromanthrone Red and Brominated Pyranthrone Red, Pigment Reds 216 and 168 respectively, both yellow shade reds with excellent weathering properties. Outdoor durability studies show that the pigments exhibit no tendency to fade or darken on prolonged exposure in either full or reduced shades. The pigments are ideally suited for the automotive industry.

Anthraquinone Red, Pigment Red 177; is a medium shade red with excellent all round durability characteristics. Again the color finds major use in the automotive industry for the production of clean, bright red coatings.

Perylene Reds. The perylene reds provide pure, transparent shades and novel styling effects when used in metallic aluminum and mica finishes. When formulated or styled in conjunction with other organic red and orange pigments, very pure brilliant red hues may be obtained. Perylene based pigments possess high color strength, good thermal stability, excellent light and weather resistance, and, with the possible exception of Pigment Red 224, excellent chemical resistance. Perylenes are the only class of vat pigments that were specifically developed for the pigment marketplace rather than as dyestuffs. Almost all of the perylenes have a structure as shown by the generic formula given as *Figure 8*, that is, they are based upon N,N'-substituted perylene-3,4,9,10-tetracarboxylic di-imide. An exception of note is Pigment Red 224, derived from the perylene tetracarboxylic dianhydride. Of interest is the fact that a relatively small change in the structure of the perylene side chain can result in markedly different colors being formed. The dimethyl perylene, Pigment Red 179, is a yellow shade red whereas the diethyl derivative is a black of unassigned Colour Index Name.

Figure 8. Generic Structure of Perylene molecule

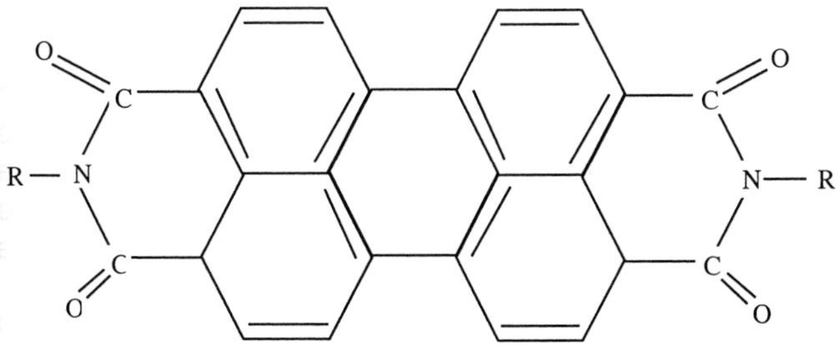

Generic Perylene Structure
R = methyl for PR 179

Perylenes were established in the marketplace in the 1950's with Hoechst introducing Pigment Red 149 (1957), Harmon introducing Pigment Red 190 (1959), and, BASF introducing Pigment Red 179 (1959) followed in 1966 by their introduction of Pigment Red 178.

The manufacture proceeds via a series of complex sequences. Acenaphthene is first oxidized to 1,8-naphthalic anhydride followed by ammonation to yield the naphthalimide. The naphthalimide is next condensed in a fused caustic medium to yield the perylene-3,4,9,10-tetracarboxylic acid di-imide which can then either be conditioned to convert the crude into pigment and produce Pigment Violet 29 or methylated to give Pigment Red 179. The di-imide may be hydrolyzed to produce the dianhydride, Pigment Red 224.

As with many of the pigments already highlighted, the perylenes have to be conditioned to obtain the necessary pigmentary form of the compound. The conditioning can involve an acid swelling or an attrition technique to achieve optimum particle size distribution. Conditioning can be a critical feature in obtaining the opaque, high hiding forms or conversely the transparent forms that are commercially available.

Benzimidazolone-Based Reds. The benzimidazolone pigments include such pigments as Pigment Reds 171, 175, 176, 185 and 208. Such pigments are azo reds that contain the benzimidazolone structure as part of their coupling component. Each is derived from the NAPHTOL® AS molecule, as illustrated in *Figure 9*. It is this structure which contributes significantly to the high molecular weight of the pigment and greatly influences the pigments durability and performance properties.

Figure 9. Typical structure of Benzimidazolone Reds, Pigment 171 and Red 175

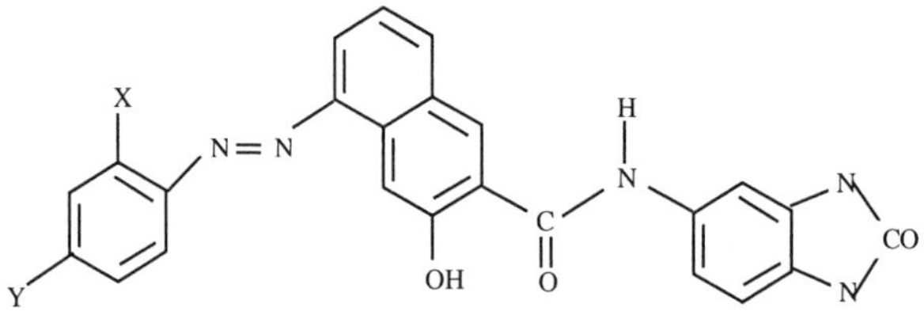

Benzimidazolone Reds Structure
X = Methoxy , Y = Nitro for PR 171
X = acetyl   , Y = H for PR 175

The first patent describing a benzimidazolone compound was applied for in 1960 by Hoechst and was to protect dyestuffs prepared from 2'-hydroxy-3'-naphthoic acid derivatives of 5-amino-benzimidazolone. Later applications by Hoechst discussed benzimidazolone pigments with aminobenzene as the diazo component.

Although benzimidazolone reds are used primarily in the coloring of plastics because of their outstanding thermal stability, some uses are found within the coatings industry. They show excellent resistance to light together with good outdoor durability. Such properties, coupled with outstanding heat stability, makes them the class of choice for many industrial finishes such as coil coatings, powder coatings, camouflage paints, automotive refinish and farm implements where cheaper, less stable pigments would be inadequate.

Pigment Red 175 is a highly transparent red with good lightfastness. This pigment finds application in automotive base coat/clear coat systems but it is not sufficiently durable for single, mono coat systems. Pigment Red 171 is also a transparent pigment, but with a maroon shade that finds application in high quality industrial finishes. Pigment Reds 176, 185, and 208 find considerable use in quality printing ink applications.

All the benzimidazolone pigments are prepared using the diazotization and coupling techniques described earlier. Specific auxiliaries such as dispersing agents are added during the coupling reaction sequence to custom make each individual type for it's designed end use application. Generally, after the coupling process, the pigment is treated to obtain a uniform, crystal growth and particle size distribution. Normally this is achieved by heating the aqueous suspension at a predetermined rate prior to filtering the suspension to isolate the pigment. Treatment temperatures can vary from 100 to 150 degrees C and often the reaction is carried out under pressure in order to achieve the higher temperature ranges.

Disazo Condensation Reds. The observation that pigment fastness increases and solubility decreases with increasing molecular weight has resulted in research to produce a range of pigments known as disazo condensation pigments with hues from red through yellow to violet These pigments feature such properties as high tint strength and superior resistance to heat and solvents.

Disazo condensation pigments have been available commercially in Europe since 1957 and in the United States since 1960. Disazo condensation reds have found considerable use as replacement pigments for inorganic, lead containing pigments. Their outstanding performance properties have resulted in their use in high quality industrial finishes and for the coloration of polymer systems that have to contend with severe exposure situations. This class of pigments also finds use in the coloration of PVC and polyolefins as a result of their fastness to migration and high heat stability. The low particle size of such pigments as C. I. Pigment Red 144 has also resulted in the use of this pigment for coloring synthetic fibers. Recently the disazo condensation reds have found new outlets in the plastics industry as formulators have been replacing cadmium containing inorganic reds with heavy metal free organic colorants.

Pigment Red 242, *Figure 10,* invented by Sandoz, now part of Clariant, is a bright yellow shade disazo condensation pigment with excellent fastness properties and is characteristic of this pigment class. This novel pigment continues to find increased use in high quality industrial finishes and as a lead replacement pigment for those colors that, for environmental considerations, must be formulated free of lead containing inorganic pigments. Also invented by Sandoz is Pigment Red 214, *Figure 11,* yet another example of a disazo condensation red with properties similar to Pigment Red 242.

Figure 10. Pigment Red 242

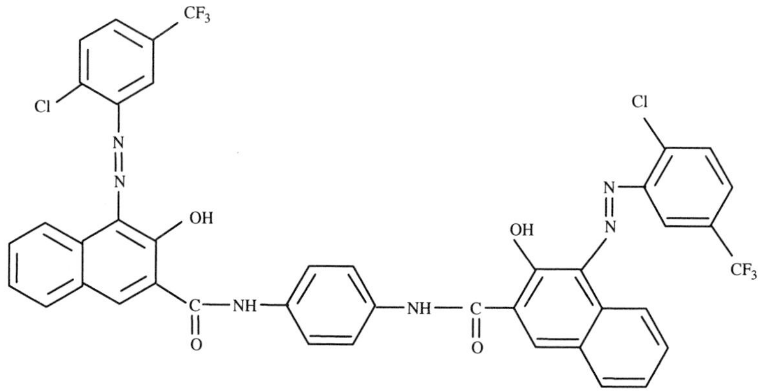

Figure 11. Pigment Red 214

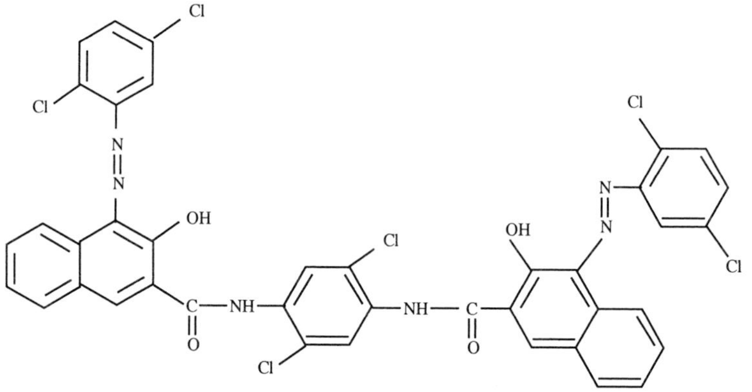

The synthesis of each of the disazo condensation pigments generally proceeds via a similar sequence. The azo components are initially coupled to yield monoazo dyestuff carboxylic acids, which are then converted to acid chlorides before final conversion to the disazo by condensation with the arylide component to yield the pigment in question.

Thioindigoid Reds. The thioindigoid chromophore serves as a nucleus for a wide range of red to violet pigments including such as Pigment Reds 86, 87, 88, 181, and 198, (*Figure 12.*) These pigments are all noted for their brightness of shade and generally good performance properties resulting in their use in the coatings industry with Pigment Red 88 once being the largest volume used followed by Pigment Red 198. Pigment Red 88 is widely used in automotive finishes but the bleed resistance of Pigment Red 198 limits its use. The commercially availability of these pigments has suffered in recent years with many products having been withdrawn from the marketplace by the almost exclusive supplier, Bayer.

Figure 12. Thioindigoid Red Pigments.. Pigment Red 88, 181 and 198

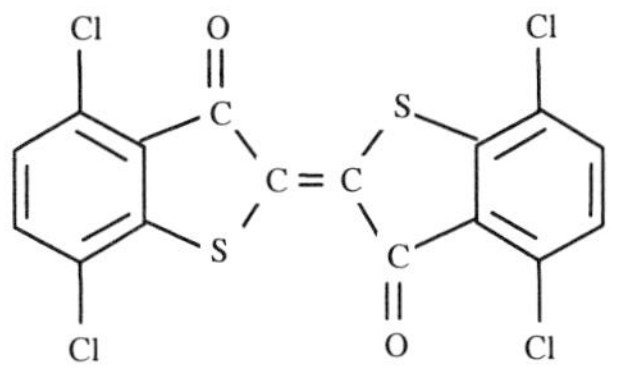

PR 88 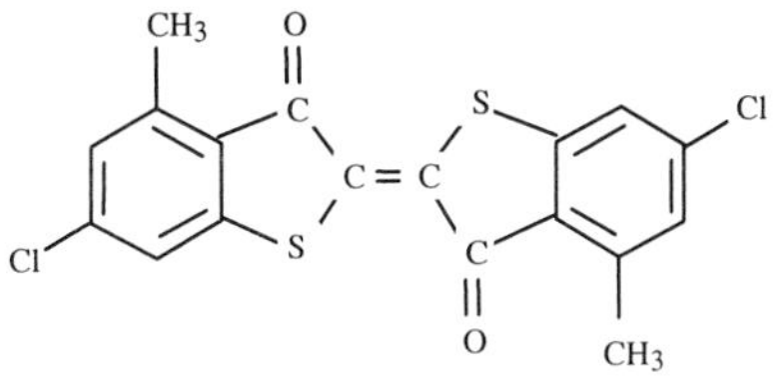   PR 181

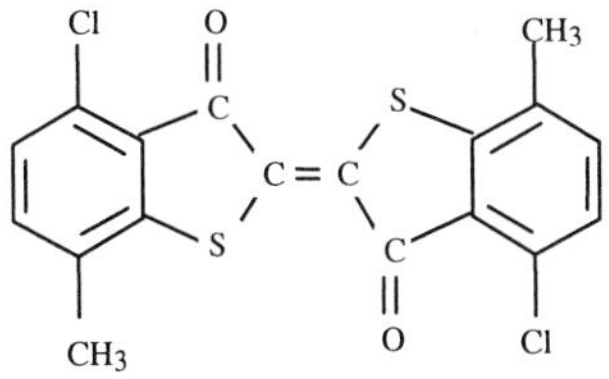

PR 198

Novel High Performance Reds. In more recent years several novel organic reds have been commercialized by such companies as BASF, CIBA and Sandoz targeted directly at the requirements of the coatings marketplace. Invented by Sandoz, Pigment Red 257 is a nickel complex pigment with a red-violet masstone and a magenta undertone that exhibits performance properties similar to those of quinacridone. The pigment is particularly useful in the formulation of durable, high quality industrial and automotive coatings. The pigment also exhibits excellent flow or rheological characteristics, enabling formulation of coatings using high pigment loading during dispersion to produce the mill base. Such good flow properties are required for the newer "high solids/low solvent" systems which are becoming the norm as a result of the industries drive to reduce the volume of organic solvents liberated from a paint as it dries to form a protective film.

High performance reds as Pigment Reds 251 and 252 are both based on the pyrazoloquinazolone structure. Both are monoazo compounds derived from pyrazolo(5,1-b)quinazolones as the coupling component and substituted anilines or polycyclic amines as diazo component. Each pigment exhibits excellent brightness of hue at full shades and good gloss retention. These features, combined with good light and weather resistance will result in these two novel pigments finding increased use in industrial and automotive coatings.

Ciba Specialty Chemicals has only recently patented a series of novel reds based upon the 1,4-diketo pyrrolo(3,4-c)pyrrole (DPP) structure, *Figure 13*. An essential feature common to these pigments, which have been synthesized in hues from scarlet to a very blue shade red, is the two fused five-membered ketopyrrole ring structure. Reds based upon this structure are currently finding widespread use in both automotive and industrial systems to provide brilliant, saturated red shades of outstanding durability. First reported in 1974, as a product formed in low yield from the reaction of benzonitrile with ethyl bromoacetate and zinc, DPP pigments are now made in a one step procedure from the treatment of a succinic acid diester with an aromatic cyanide in a strongly basic medium.

Figure 13. Di-pyrrolo-pyrrole based pigment, Pigment Red 254

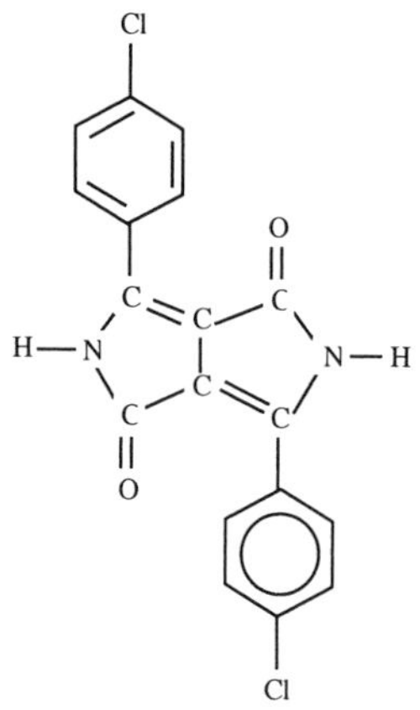

The first pigment to be marketed, Irgazin DPP Red BO, was a bright red with excellent color intensity that found use alongside quinacridones and perylenes in automotive formulations. Additionally the pigments thermal stability up to $500^{\circ}C$ is sure to meet the exacting demands of the powder coatings and plastics marketplace. The cleanliness and excellent hiding properties of this new pigment has resulted in the reduction in use of the opaque quinacridone reds (PV 19) formerly used in styling many bright reds. The excellent durability properties provided by a pigment of such small molecular weight are explained by the significant intermolecular hydrogen bonding shown to exist throughout the crystal lattice.

Organic Blues

Copper Phthalocyanine Blue. The most important and most widely used blue throughout all applications of the color consuming market is copper phthalocyanine blue, Pigment Blue 15, *Figure 14.*

Figure 14. Copper Phthalocyanine Blue, Pigment Blue 15

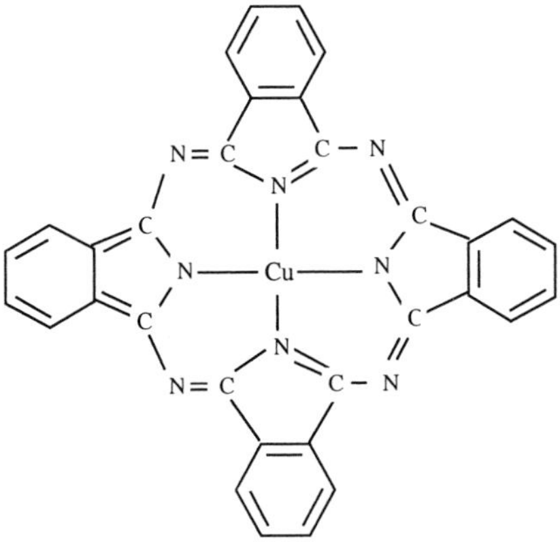

Recognition of phthalocyanine compounds as pigments of outstanding potential took place in 1928 when chemists working for the Scottish Dye Works (I.C.I.) observed a green impurity in phthalimide prepared from the reaction between molten phthalic anhydride and ammonia that took place in a chipped enameled, cast iron vessel. The iron phthalocyanine thus formed was isolated and its properties investigated. Significant research work resulted in the substitution of iron by copper to produce the bright blue pigment of outstanding performance characteristics known as copper phthalocyanine blue. This pigment has steadily increased in importance to become a product with worldwide significance in all markets and is the pigment of choice when a clean, economical green shade or red shade blue is required.

In the early 1930's, extensive industrial development of this pigment was entered into by ICI, by the German I.G. Farben cartel and, to a lesser extent, by the DuPont organization. The commercialization of copper phthalocyanine advanced rapidly but, in spite of all this activity, it is surprising that the beta crystal structure was not reported until 1968. The only metal derivative of significant commercial use is that of copper, derivatives of other metals having been shown by research to have less desirable shade or durability characteristics. Copper phthalocyanine is the most commercially significant organic blue pigment in today's marketplace. Metal free phthalocyanine, Pigment Blue 16, once found an outlet as a green shade blue but its inferior heat stability and its poorer chemical resistance, coupled with unfavorable economics, has resulted in a rapid decline in its consumption for all but very special applications. Metal free phthalocyanine blue (Pigment Blue 16) is normally manufactured via the phthalonitrile and sodium salt of phthalocyanine. Acid pasting is used to condition the crude and give the pigment.

Phthalocyanines are planar molecules, comprising four isoindole units connected by four nitrogen (-N) atoms to form a ring structure with the central atom being copper, in a planar arrangement as shown in *Figure 14*. Manufacture is comparatively easy despite the superficial complexity of the phthalocyanine molecule. Reaction of a phthalic acid derivative at temperatures approximating $190^\circ C$ with a source of nitrogen such as urea and a metal or metal salt is usually all that is required to produce the appropriate metal phthalocyanine. Molybdates, vanadates, and certain compounds of titanium have been found to be useful catalysts for this condensation reaction.

A typical process, known as the Wyler-Riley process, involves heating phthalic anhydride, urea, cuprous chloride, and a catalytic quantity of ammonium molybdate in a high boiling solvent such as kerosene or nitrobenzene. Urea acts as the source of nitrogen in the process, the carbonyl group being lost in the processing as carbon dioxide. The process results in the formation of copper phthalocyanine in a crude, non-pigmentary form. The product has thus to be "finished" or conditioned to give the pigment grade of choice. Typically, crude phthalocyanine blue is characterized by a crystal size of the order of 20 microns, a purity in excess of 92%, a surface area of the order of 1 - 2 square meters per gram, and poor tint strength.

At the end of the 1970's, it was discovered that the use of trichlorobenzene as a solvent for the production of copper phthalocyanine resulted in a crude that contained significant quantities of polychlorinated biphenyls (PCB's), principally hexachlorobiphenyl. Modern synthesis procedures have resulted in only minimal amounts of PCB's being formed during the reaction. Copper phthalocyanine is readily available that contains levels of PCB's well below the 25 ppm required by the Environmental Protection Agency.

Crude conditioning may be carried out using one of several alternative processes:

Salt Attrition - The crude is ground in a ball mill or double arm, sigma blade mixer in the presence of a large excess of salt and an organic solvent such as diethylene glycol or xylene to give the green shade, beta phthalocyanine blue.

Solvent Free Salt Attrition - The crude is ground in a ball mill or double arm, sigma blade mixer in the presence of a large excess of salt, sodium chloride, but in the absence of a solvent to give the red shade, alpha phthalocyanine blue.

Acid Pasting - The crude is dissolved in a large excess of concentrated sulfuric acid then re-precipitated by controlled drown out of the resultant acid solution into iced water to obtain the red shade, crystallizing grade of alpha phthalocyanine blue.

Acid Swelling - The crude is treated with a smaller volume of less concentrated sulfuric acid than is used in the acid pasting technique. Again, the acid swelled paste is added in a controlled manner to iced water to precipitate the copper phthalocyanine as the alpha, red shade product.

Acid Kneading - A process that is now of only minor significance relative to the other processes. Acid kneading involves intensive milling of the crude with urea and sulfuric acid to produce the less common and more expensive grade of pigment known as Epsilon Blue (PB 15:6).

Crystal forms of Phthalocyanine Blue. Copper phthalocyanine exhibits classical polymorphism, the most important crystal forms commercially available being the alpha and the beta forms. The alpha crystal is described as Pigment Blue 15, 15:1 and 15:2 and is a clean, bright red shade blue pigment. The beta crystal is described as Pigment Blue 15:3 and 15:4 and is a clean, bright green or peacock shade. The copper atom at the center of each beta crystal is coordinated to nitrogen (-N) atoms in adjacent molecules to form a distorted octahedron. No such octahedral coordination is possible in the alpha form, a fact which may contribute to the lower stability of this crystal form.

The beta form is the most stable crystal form and readily resists recrystallization. The alpha form is the metastable form, which readily converts to the more stable, green shade, beta crystal. As such the crystal requires special, proprietary treatments to produce a red shade product that is stable to both crystallization and flocculation and that will not revert to the more stable, green shade, beta crystal. Conversion from the alpha to the beta form is usually accompanied by an increase in crystal size with a subsequent loss of strength and shift to a greener hue. Use of any of the unstabilized grades in strong solvents or in systems that experience heat during dispersion or application will result in a shift in shade to the greener side and a loss of strength as recrystallization takes place within the unstabilized crystal.

Copper phthalocyanine approximates the ideal pigment, finding use throughout the pigment consuming marketplace, worldwide. It is a pigment that offers brightness, cleanliness, strength and economy with all round excellent performance as a color. Certain highly purified grades of copper phthalocyanine, purified to remove any free copper, are used in dissolvable sutures and soft contact lenses.

<u>Miscellaneous Blues.</u>   Indanthrone Blue. Pigment Blue 60, belongs to the class of pigments described as "vat pigments". As a blue this is a very red shade, non-bronzing, flocculation resistant, pigment with outstanding performance properties. Due to the pigment's expense relative to copper phthalocyanine, economic considerations are a limitation to its more widespread use. This pigment is used in colors that require only small amounts of an intense red shade blue as a shading pigment at low levels. Relative to copper phthalocyanine blue, Indanthrone Blue is considerably redder in hue and markedly superior in both bronze resistance and flocculation resistance.

Carbazole Violet, Pigment Violet 23, is a complex polynuclear pigment that is an intense red shade blue of high tint strength. The pigment possesses excellent fade resistance properties and only its relatively high cost and hard nature limit its more widespread use. Carbazole Violet is used as a shading component in high performance coatings, at very low levels as a "blueing" agent in white paints to produce a "brilliant" white, in high quality printing inks and in all types of plastic polymers.

<u>Organic Yellows</u>

Reference to Table 2 will illustrate that yellow pigments can be sub-divided into four broad classifications based upon their chemical constitution. These classifications consist of: monoarylide yellows, diarylide yellows, benzimidazolone yellows, and heterocyclic yellows.

Table 2 - Organic Yellow Pigments

Colour Index Name	Chemical Type	Colour Index Name	Chemical Type
Yellow 1	Monoarylide	Yellow 113	Diarylide
Yellow 3	Monoarylide	Yellow 114	Diarylide
Yellow 12	Diarylide	Yellow 116	Monoarylide
Yellow 13	Diarylide	Yellow 117	Azomethine
Yellow 14	Diarylide	Yellow 120	Benzimidazolone
Yellow 16	Diarylide	Yellow 126	Diarylide
Yellow 17	Diarylide	Yellow 127	Diarylide
Yellow 55	Diarylide	Yellow 129	Azomethine
Yellow 60	Heterocyclic	Yellow 138	Quinophthalone
Yellow 65	Monoarylide	Yellow 139	Isoindoline
Yellow 73	Monoarylide	Yellow 150	Diarylide
Yellow 74	Monoarylide	Yellow 151	Benzimidazolone
Yellow 75	Monoarylide	Yellow 152	Diarylide
Yellow 81	Diarylide	Yellow 153	Nickel dioxime
Yellow 83	Diarylide	Yellow 154	Benzimidazolone
Yellow 97	Monoarylide	Yellow 155	Bisazo
Yellow 98	Monoarylide	Yellow 156	Benzimidazolone
Yellow 101	Methine	Yellow 173	Isoindolinone
Yellow 106	Diarylide	Yellow 175	Benzimidazolone
Yellow 109	Isoindolinone	Yellow 182	Triazine
Yellow 110	Isoindolinone	Yellow 183	Monoazo

Monoarylide Yellows. Monoarylide yellows are all azo pigments; their manufacture is based upon the diazotization and coupling sequence as discussed earlier when dealing with azo reds. The structures of the major monoarylide yellows are represented in *Figure 15.* Of note is the fact that in each case the coupling takes place on the side chain of the coupling component rather than onto the aromatic ring as was the case with the azo reds.

Figure 15. Monoarylide Yellows

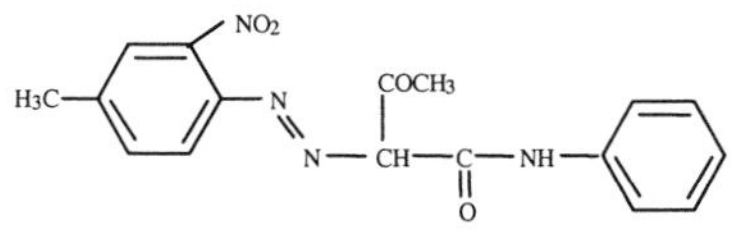

PY 1

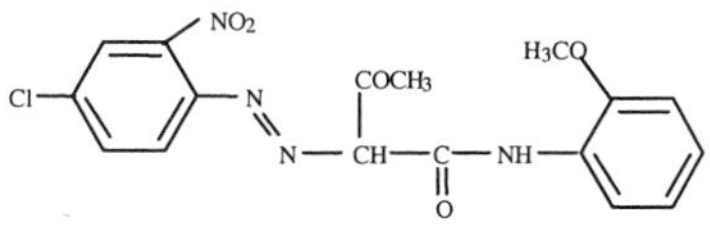

PY 73

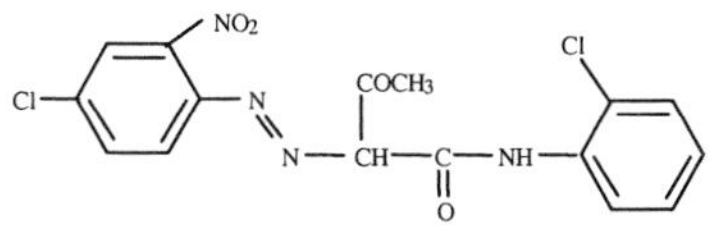

PY 3

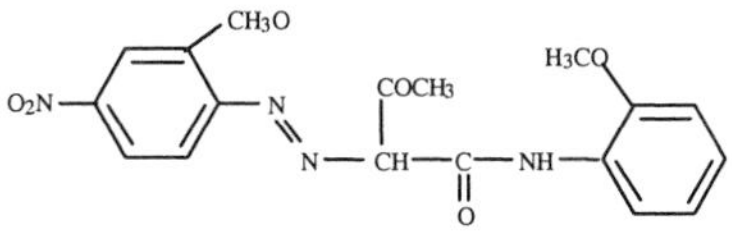

PY 74

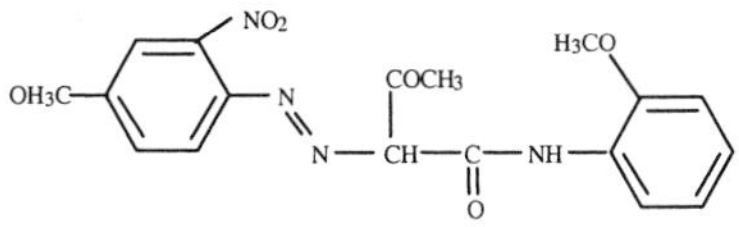

PY 65

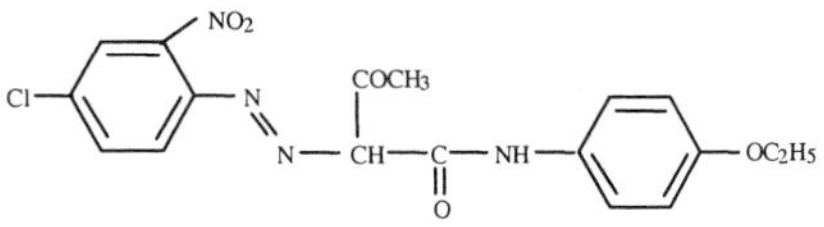

PY 75

In the early part of this century, pigment researchers switched their efforts from the synthesis of organic reds and blues to that of working to extend the range of yellows . Their work concentrated on replacing 2-naphthol, so common in the azo red coupling reaction, with acetoacetarylides. As a result, in 1909, the structure of the prototype of the monoarylides, Pigment Yellow 1, Hansa Yellow G, was made public and the pigment became commercially available.

Pigment Yellow 1, often referred to as Hansa Yellow G for historical reasons, is manufactured by coupling diazotized 2-nitro-4-methyl aniline onto acetoacetanilide. The product is a bright yellow pigment that has a major use in trade sales, emulsion and masonry paints. Its major disadvantages are its poor bleed resistance in most popular solvents, poor light resistance in shades tinted with white and markedly inferior bake resistance as a result of its tendency to sublime.

Pigment Yellow 3, Hansa Yellow 10G, has a heritage based upon the trade name under which both Pigment Yellows 1 and 3 were introduced by Hoechst when first marketed in 1910 and 1928, respectively. This pigment is very green shade yellow made by coupling the diazo of 4-chloro-3-nitro aniline onto 2-chloro acetanilide. Greener in shade than PY 1, this pigment is used in the same types of applications and suffers from the same deficiencies as PY 1 with the exception that Pigment Yellow 3 is suitable for use in exterior applications at high tint levels.

Pigment Yellow 65 is a newer monoarylide pigment that is finding increased use throughout the United States where inorganic, lead containing pigments have to be replaced and in road traffic marking paints that have been specified as being free of medium chrome yellow. Produced by coupling diazo 2-nitro-methoxy aniline onto 2-

acetacetanisidine the pigment offers a redder shade than the previous two yellows discussed, Pigment Yellow 65 is used in trade sales, latex, and masonry paints. The solvent bleed resistance and baking stability are little improved over PY 1 and PY 3.

Pigment Yellow 73 is manufactured by coupling diazotized 4-chlor-2-nitro aniline onto 2-acetoacetanisidine. The pigment is close in shade to that of PY 1 and again finds use in trade sales, latex and masonry paints. It is not, however, considered durable enough for exterior applications.

Pigment Yellow 74 is obtained by coupling diazo 4-nitro-2-anisidine onto 2-acetoacetanisidide, Pigment Yellow 74 offers the user a pigment suitable for outdoor applications that is considerably stronger and somewhat greener than PY 1. Major outlets, as with all the monoarylide yellows, are in latex, trade sales and masonry paints. Recently a new application has emerged for Pigment Yellow 74 in the production of offset printing inks where a yellow is required that offers superior performance characteristics over those of the conventional diarylide yellows more commonly used in offset printing.

Pigment Yellow 75 is a pigment produced by the coupling of 4-chloro-2-nitroaniline onto acetoaceto-phenetidide. A red shade yellow that has only recently found considerable application in the coatings industry as a replacement for lead containing medium chrome yellow (PY 34) as used in yellow road traffic marking paints. One of the few monoarylide yellows that has been found acceptable from the point of view of economy and durability, being able to withstand twelve to twenty four months exposure on a 100,000 vehicle a day highway when applied at the thickness levels typical of road traffic markings.

Pigment Yellow 97, first described in U.S. Patent 2,644,814, was introduced to the marketplace in the late 1950's alongside Pigment Yellow 98. This is a pigment that shows the advantages associated with smaller particle size, less agglomeration and an improved particle size distribution when compared to the other monoarylide yellow pigments. Pigment Yellow 97 has gained significant use in trade sales paint applications primarily because this pigment was available when demands for lead-free formulations were increasing. Additionally the pigment finds use in high quality wallpaper and decorative paints. This yellow is derived from the coupling of diazo 4-amino-2,5-dimethoxybenzene sulfoanilide to 4-chloro-2,5-dimethoxy acetoacetanilide.

Pigment Yellow 98, again another relatively new pigment, this molecule is produced from the coupling of diazo 4-chloro-2-nitro aniline onto 4-chloro-2-methyl acetoacetanilide. Similar in shade to Pigment Yellow 3, but considerably stronger and more heat stable, this pigment has, to date, only met with limited commercial success in the pigment consuming market.

Metallized Azo Yellows. Pigment Yellow 62 is a typical metallized azo yellow that finds use in coloration of rubber, LDPE, PVC and polypropylene fibers. The pigment is a calcium based azo yellow produced from the diazotization and coupling of 4-amino-3-nitro benzene sulfonic acid onto acetoacet-o-toluidide followed by conversion of the sodium salt to that of calcium. The use of this pigment within the plastics industry has increased over the past several years as a replacement for diarylide yellows, especially under those circumstances that may lead to thermal degradation of the colorant. The pigment is a green shade yellow with good economy and moderate resistance properties to light, acids and alkali.

Diarylide Yellows. The structures of these commercially important organic yellows are shown in *Figure 16*. Table 3 presents a summary of the properties of this class of yellows. Basically, as can be seen from an inspection of the structures, each of this class of yellows has a backbone structure based upon 3,3'-dichlorobenzidine with modifications to the shade and properties being achieved by variation of the coupling component used in the diazotization reaction. Properties that are common features of this group of yellows are low cost, reasonable heat stability, and moderate chemical resistance. The major worldwide market for the diarylide yellows is the printing ink industry. Diarylide yellows are approximately twice as strong as the monoarylide yellows; additionally, they offer improved bleed resistance and stability to heat. Nevertheless, none of the diarylide yellows have durability properties that would allow for their use in exterior situations and as such should never be considered for outdoor applications. Minor applications in the area of toy enamels and pencil coatings are found for the diarylide yellows, especially if a lead free formulation is specified.

Figure 16. Diarylide Yellows

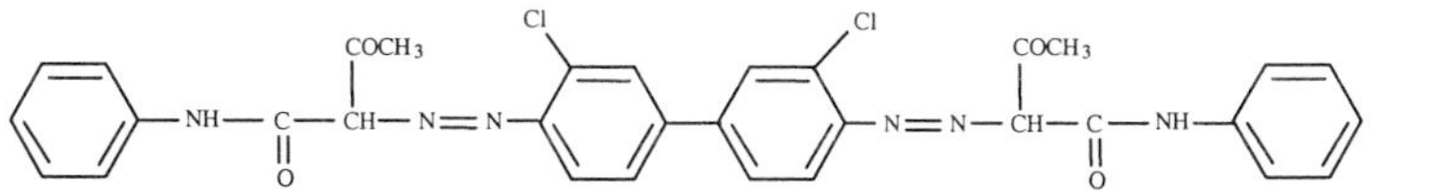

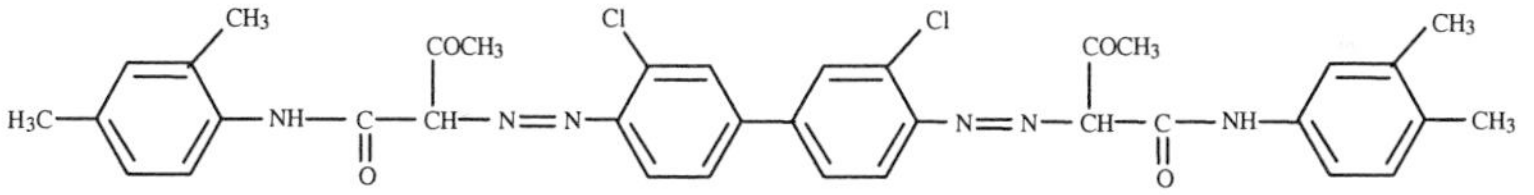

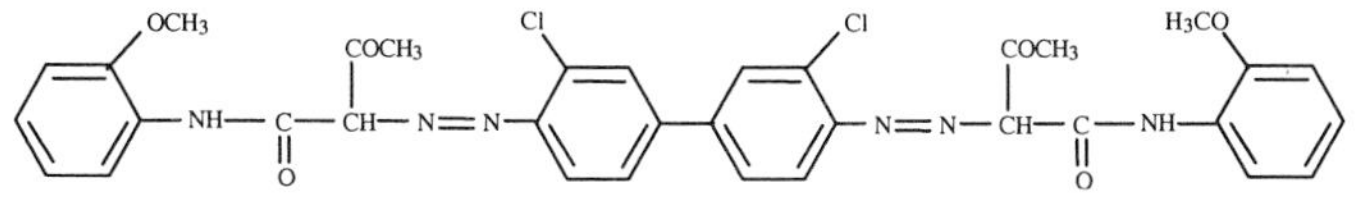

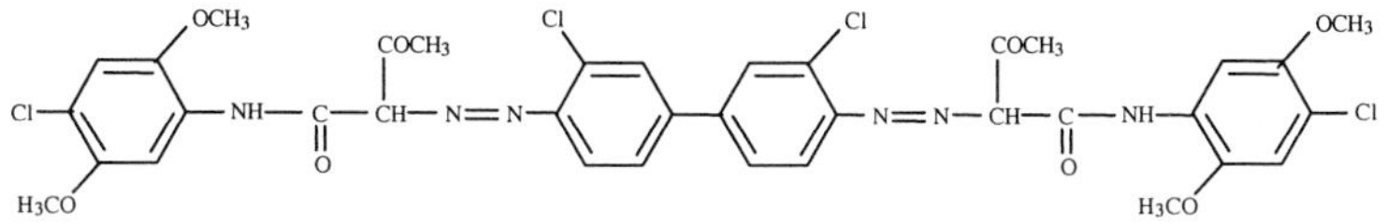

Benzimidazolone Yellows. *Figure 17* illustrates the structure of the organic yellows that fall into the classification of benzimidazolone yellows. As with the reds discussed earlier these yellows take their name from the fact that each features 5-acetoacetylaminobenzimidazolone within its structure. Additionally each is an azo pigment with an acetoacetylarylamide nucleus.

Figure 17. Benzimidazole based yellows and orange

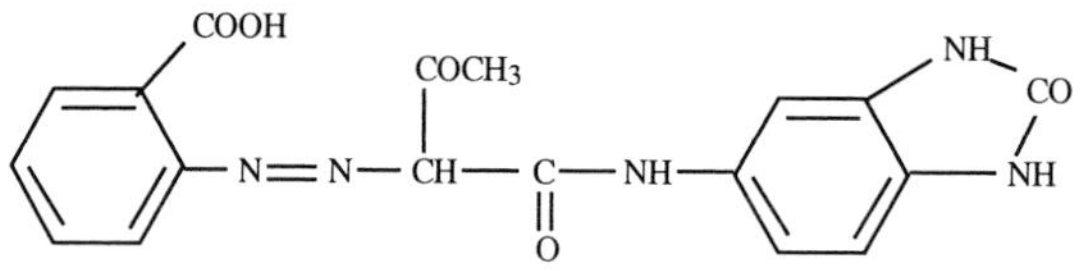

Benzimidazolone Yellows and Orange

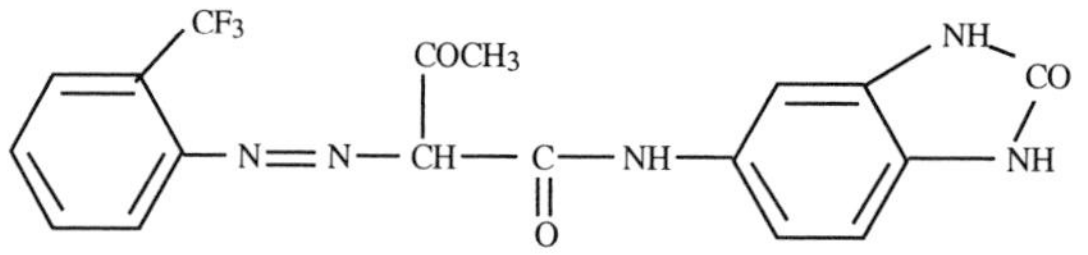

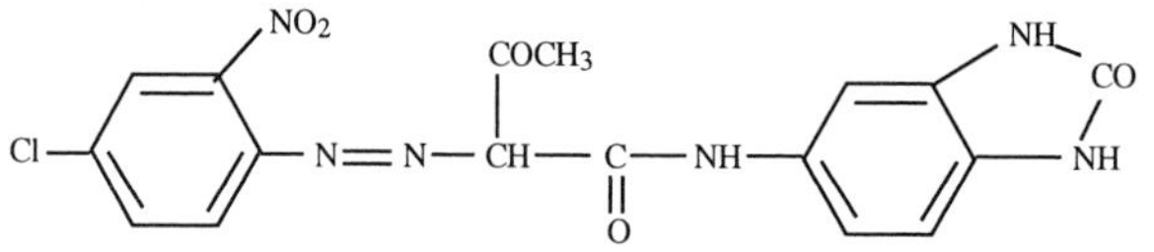

The exceptional resistance to heat and light in addition to the excellent weather resistance of this class of pigments, may be attributed to the presence of the benzimidazolone group in the structure. The benzimidazolone structure was first described in patents covering the synthesis of dyestuffs in 1964. However, it was not until 1969 that these pigments were offered to the marketplace.

Used initially for the coloring of plastics, these pigments are now finding increased use in the paint industry where their excellent durability, heat stability and light resistance are required as when formulating high quality industrial finishes

Each of the commercial benzimidazolone yellows is a monoazo pigment and is produced in the manner previously described, by diazotization and coupling of the appropriate primary aromatic amine and reaction of the resultant diazo with the coupling component in a controlled manner at a reduced temperature. The product resulting from the coupling reaction is, without exception, further conditioned to achieve a pigment of uniform particle size distribution that has a marked degree of opacity.

Heterocyclic Yellows. This class of yellows contains an assortment of pigments that all contain a heterocyclic molecule within their structure as shown by the two examples presented in Figure 18.

Figure 18. Heterocyclic Yellows, Quinophthalone Yellow and Isoindoline Yellow

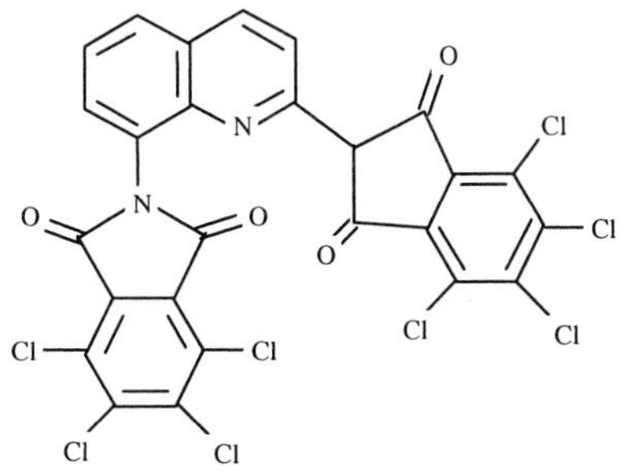

PY 138 Quinophthalone 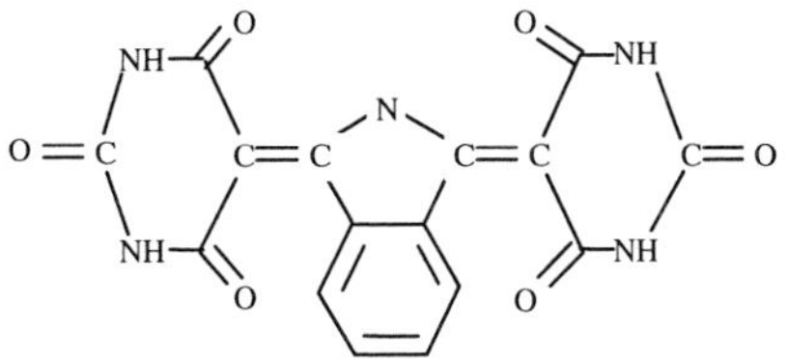   PY 139 Isoindoline

In spite of their apparent complexity, these newer high performance yellows continue to be introduced to satisfy the exacting demands of the marketplace. Pigments such as Quinophthalone Yellow (PY 138) and Isoindoline Yellow (PY 139) are typical examples of such complex, novel chromophores introduced as recently as 1974 and 1979 respectively. All of these yellows are equally complex structures that offer the user additional high performance pigments that find application where the end use can justify the economics of purchasing such high performance products. Table 4 summarizes the properties of the heterocyclic yellows currently commercially available.

Table 3, Summary of Diarylide Yellow Properties

Colour Index Name	Common Name	Properties
Pigment Yellow 106	Yellow GGR	Green shade. Major use in packaging inks.
Pigment Yellow 113	Yellow H10GL	Very green shade. More transparent than PY 12 and offering better heat & solvent resistance.
Pigment Yellow 114	Yellow G3R	Red shade. Improved solvent & lightfastness over PY 12. Major use in oil based inks.
Pigment Yellow 12	AAA Yellow	Poor light resistance. Major use in printing inks.
Pigment Yellow 13	MX Yellow	Redder than PY 12. Improved heat stability and solvent resistance. Major use in printing inks.
Pigment Yellow 14	OT Yellow	Green shade. Major use in packaging inks. Poor fade resistance.
Pigment Yellow 152	Yellow YR	Very red, opaque product. Poor light resistance. Some use in interior paints as lead chrome replacement.
Pigment Yellow 16	Yellow NCG	Bright green shade. Improved heat & solvent fastness. Used in full shade and deep tints for paints.
Pigment Yellow 17	OA Yellow	Green shade. Major use in packaging inks. Poor fade resistance.
Pigment Yellow 55	PT Yellow	Red shade. Poor light resistance. Isomer of PY 14.
Pigment Yellow 81	Yellow H10G	Bright green shade. Same shade as PY 3 but much stronger.
Pigment Yellow 83	Yellow HR	Very red shade. Improved transparency, heat stability and lightfastness over PY 12. Some use in interior paints. Major use is packaging

Table 4 Summary of the properties of the Heterocyclic Yellows

Colour Index Name	Common Name	Properties
Pigment Yellow 60	Arylide yellow	Very red shade. Moderate light and solvent resistance. Good acid and alkali stability.
Pigment Yellow 101	Methine yellow	Highly transparent and exceptionally brilliant. Used in Industrial paints. Only moderate solvent and alkali resistance.
Pigment Yellow 109	Tetrachloroiso-indolinone	Green shade. Excellent brightness, strength and durability. Automotive OEM and refinish uses.
Pigment Yellow 117	Azomethine Copper complex	Excellent chemical, light, and heat resistance. Used in industrial paints and finishes.
Pigment Yellow 129	Azomethine yellow	Very green shade. Excellent chemical, light, and heat resistance. Used in industrial and specialty paints.
Pigment Yellow 138	Quinophthalone	Green shade. Clean hue and excellent stability. Used for high quality industrial and automotive finishes.
Pigment Yellow 139	Isoindoline	Red shade. Similar in masstone to medium chrome (PY 34). Excellent light and solvent resistance. Used for industrial and automotive coatings.
Pigment Yellow 150	Pyrimidine Yellow	Very green shade. Good heat and light resistance. Used in industrial paints.
Pigment Yellow 153	Nickel dioxine	Red shade. Excellent overall properties. Used in specialty paints and baking enamels. Poor acid resistance.
Pigment Yellow 155	Azo Condensation Yellow	Green shade. Excellent overall properties in full shade. Used in industrial and specialty paints.
Pigment Yellow 173	Isoindolone Yellow	Very green shade. Excellent durability. Used in industrial and specialty finishes.
Pigment Yellow 182	Triazinyl Yellow	Medium shade. Excellent resistance at masstone level. Used in industrial finishes.

Organic Oranges

Table 5 lists those orange pigments that have commercial significance in the pigment marketplace of the 1990's. Once again, a broad subdivision can be made based upon the chemical features of the molecule as: (1) Azo based, exclusive of the benzimidazolone structures; (2) Benzimidazolone based and (3) Miscellaneous.

Table 5 Significant Organic Orange Pigments

Colour Index Name	Chemical Type	Colour Index Name	Chemical Type
Pigment Orange 13	Bisazo	Pigment Orange 49	Quinacridone
Pigment Orange 16	Bisazo	Pigment Orange 5	Azo
Pigment Orange 2	Azo	Pigment Orange 51	Pyranthrone
Pigment Orange 31	Bisazo condensation	Pigment Orange 52	Pyranthrone
Pigment Orange 34	Bisazo	Pigment Orange 60	Benzimidazolone (Azo)
Pigment Orange 36	Benzimidazolone (Azo)	Pigment Orange 61	Tetrachloroisoindolinone
Pigment Orange 38	Azo	Pigment Orange 62	Benzimidazolone (Azo)
Pigment Orange 43	Perinone	Pigment Orange 64	Heterocyclic hydroxy
Pigment Orange 46	Azo	Pigment Orange 67	Pyrazoloquinazolone
Pigment Orange 48	Quinacridone		

Azo Based Organic Oranges. The structures of two of the orange pigments that can be placed into the "azo" category are shown in *Figure 19*. These structures will appear familiar in that Pigment Orange 5 is similar to the Non-Metallized Azo Reds while Pigment Orange 16 is similar to the Diarylide Yellows. These superficial comparisons serve to illustrate the fact that an orange is merely a "yellow shade red" that has been afforded the hue designation "orange" by those responsible for compiling the "Colour Index". These oranges all have the azo chromophore (-N=N-) featured within the molecule.

Figure 19. Miscellaneous azo oranges, Pigment Orange 5 and 16

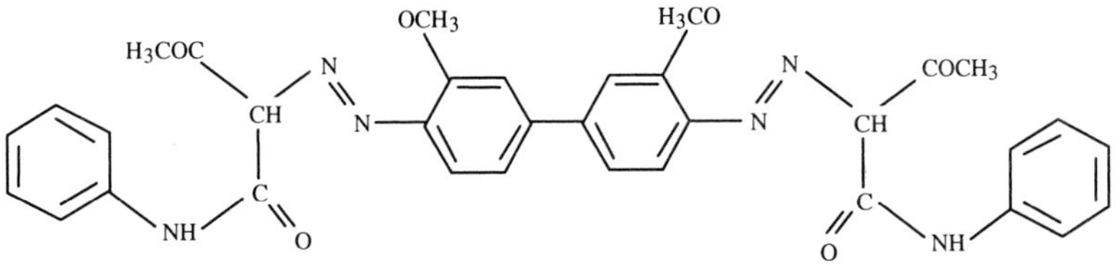

PO 16 Dianisidine Orange

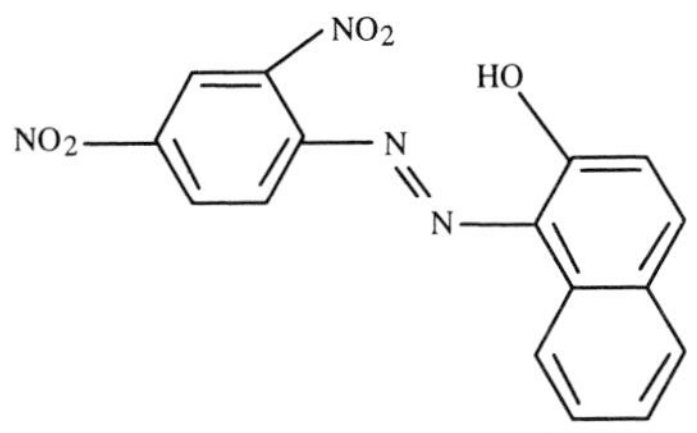

PO 5 DNA Orange

Pigment Orange 2, Orthonitraniline Orange, discovered by E. Bamberger and F. Miemberg in 1895. The pigment is prepared by the classical diazotization and coupling technique and is the product of coupling diazo orthonitro aniline onto 2-naphthol. Its major outlet is in printing inks and its use is not recommended in paints due to the pigments tendency to bleed in solvents and inadequate light stability.

Pigment Orange 5, Dinitroaniline Orange, was prepared in 1909 as outlined in the German Patent, 217,266 and the U.S. Patent, 912,138. Discovery of the pigment in 1907 is credited to R. Lauch. Once again, the pigment is manufactured by a diazotization and coupling sequence where diazotized dinitroaniline is coupled onto 2-naphthol. Dinitroaniline Orange does not suffer the poor solvent resistance shown by Orthonitroaniline Orange. The pigment offers good light resistance in full tone and moderate solvent resistance. As such Pigment Orange 5 finds widespread use in water-based latex paints and air dry industrial finishes and in packaging inks, especially water-based inks. Its poor heat stability rules out its use in high bake enamels.

Pigment Orange 13, Pyrazolone Orange, was discovered in 1910 and has enjoyed consistent popularity since its introduction to the marketplace. The pigment is synthesized by coupling tetrazotized 3,3'-dichlorobenzidine onto 3-methyl-1-phenyl-pyrazol-5-one. The pigment is a bright, clean yellow shade product that may be used for interior coatings, particularly as a replacement for lead based orange pigments.

Pigment Orange 16, Dianisidine Orange, is a diarylide orange produced by coupling tetrazotized 3,3'-dimethoxybenzidine onto acetoacetanilide. The pigment finds a use in baking enamels since its heat resistance is superior to other orange pigments with similar economics.

Pigment Orange 34, Tolyl Orange, is a diarylide pigment manufactured by coupling tetrazo 3,3'-dichlorobenzidine onto 3-methyl-1-(4' methylphenyl)-pyrazol-5-one. The pigment is a bright, reddish orange that offers fade resistance and good alkali stability but poor solvent resistance As such the pigment is used in specialty printing ink applications, particularly where a lead free formulation is specified.

Pigment Orange 38, Naphthol Orange, is manufactured by coupling diazo 3-amino-4-chloro-benzamide onto 4'-acetamido-3-hydroxy-2-naphthanilide. The pigment is a bright red shade orange that exhibits excellent stability when used in alkaline and acid media. Paint applications extend to baking enamels, latex and masonry paints.

Pigment Orange 46, Clarion® Red, is a metallized azo pigment originally marketed by American Cyanamid under the Clarion® trademark. The pigment is a metallized azo orange that is manufactured by coupling diazotized 2-amino-5-chloro-4-ethyl benzene sulfonic acid onto 2-naphthol followed by reacting this product with barium to yield the barium salt of the pigment. As such, the pigment is the ethyl homologue of Red Lake C (PR 53:1). This pigment finds its major outlet in the printing ink marketplace, particularly solvent packaging inks.

Benzimidazolone Derived Oranges. The benzimidazolone derived oranges contain the azo chromophore and are based upon 5-acetoacetylamino-benzimidazolone as coupling component. Pigment Orange 36 is the product of coupling diazotized 4-chloro-2-nitroaniline to the benzimidazolone. Pigment Orange 60 is the product from the coupling of 4-nitro aniline to the benzimidazolone.

Pigment Orange 36, Benzimidazolone Orange, is a bright red shade orange of high tint strength. In its opacified form this pigment offers excellent resistance to both heat and solvents and a hue similar to the lead-containing pigment, Molybdate Orange (PR 104). As such, over the last twenty years, Pigment Orange 36 has found increasing use in automotive and high quality industrial formulations which must be lead free and which were formerly made used the lead based pigment, Molybdate Orange.

Pigment Orange 60 is a transparent, yellow shade orange that also exhibits excellent heat and solvent resistance with an exterior durability that allows the pigment to be used in automotive and high performance industrial finishes.

Pigment Orange 62 is the newest of the benzimidazolone oranges, again a yellow shade pigment that has shares the fade resistance properties of PO 36 and PO 60, but which suffers from inferior solvent resistance and exhibits slight bleed in alkaline systems. Currently the pigment is being used in oil-based, offset inks and artists colors.

Miscellaneous Oranges. *Figure 20* illustrates two structures of those oranges that fall into a "miscellaneous" category. Table 6 summarizes the properties of this class of pigments, each of which is finding increased use within the paint and plastics industry.

Figure 20. Miscellaneous oranges, Perinone and Pyranthone, Pigment Orange 43 and 51

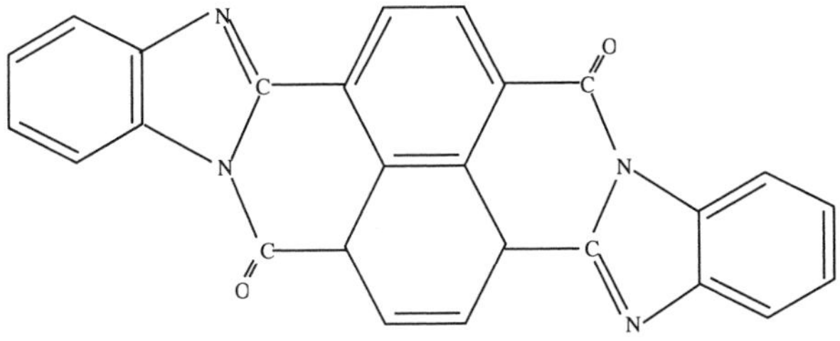

PO 43 Perinone

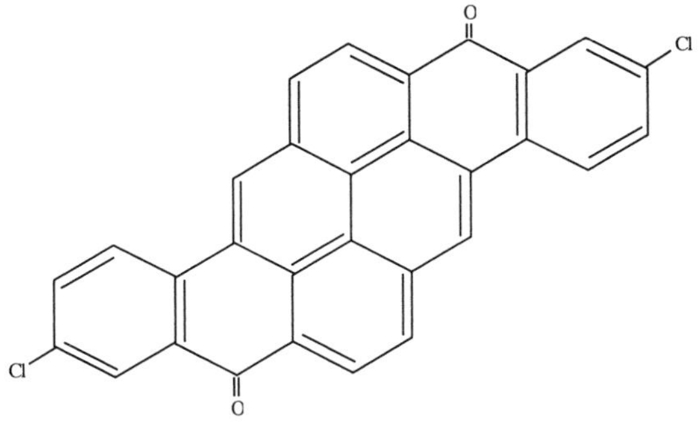

PO 51 Pyranthrone

able 6 Summary of the Properties of the Miscellaneous Oranges

Colour Index Name	Common Name	Properties
igment Orange 43	Perinone	Red shade. Strong, clean vat pigment with excellent stability. Used in metallized finishes and high quality paints. Shows slight solvent bleed.
gment Orange 49	Quinacridone Deep Gold	Red shade. Dull masstone. Excellent durability. Used in metallic finishes.
igment Orange 51	Pyranthrone Orange	Medium shade. Excellent solvent, light and heat resistance. Dull in tint. Exhibits slight solvent bleed. Used in air dry and bake enamels.
gment Orange 52	Pyranthrone Orange	Red shade. Vat pigment with excellent solvent, light and heat stability. Dull in tints. Slight solvent bleed. Used in air dry and bake enamels.
igment Orange 61	Tetrachloroisoindolinone	Medium shade. Exhibits some solvent bleed. Used in metallic auto finishes.
gment Orange 67	Pyrazoloquinazolone	Yellow shade. Excellent brilliance in full shade. Good gloss retention. Very good weather and light fastness in full shade. Used in industrial and automotive coatings.

Organic Greens

Copper Phthalocyanine Green. The major green pigment used as a self-shade throughout all pigment consuming industries is based upon halogenated copper phthalocyanine and, as such, is termed phthalocyanine green. The Colour Index names are Pigment Green 7 and Pigment Green 36.

Pigment Green 7, the blue shade green, is based upon chlorinated copper phthalocyanine with a chlorine content that varies from between 13 - 15 atoms per molecule of phthalocyanine. The higher the chlorine content of the molecule the yellower the hue of the resultant pigment.

Pigment Green 36 is a yellow shade green based upon a structure that involves progressive replacement of chlorine on the phthalocyanine structure with bromine. The composition of Pigment Green 36 varies with respect to the total halogen content, chlorine plus bromine, and in the ratio of chlorine to bromine. The shade is varied from a "bluer yellow shade" to a "greener yellow shade" by increasing the proportion of bromine in the molecule. The most highly brominated Pigment Green 36 has an extremely yellow shade and contains approximately 2.5 atoms of chlorine per molecule. *Figure 21* illustrates the currently accepted structures of the phthalocyanine greens. In practice, no single pigment consists of a specific molecular species. Rather, each pigment is a complex mixture of closely related isomeric compounds. As a result, phthalocyanine greens of similar shade with close technological properties can show considerable variation in chemical composition.

Figure 21. Copper Phthalocyanine Green, Pigment Green 7 and 36

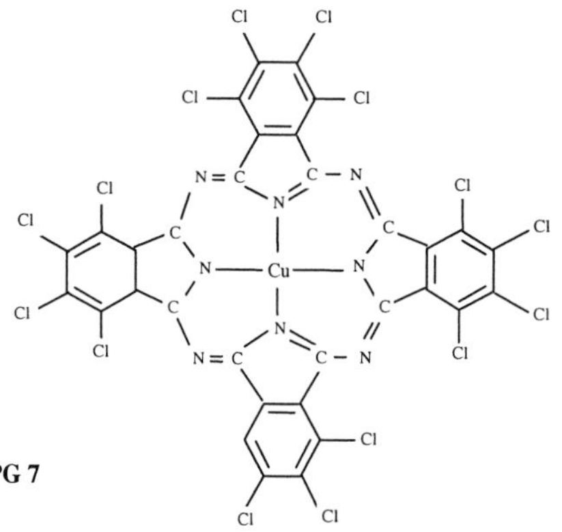

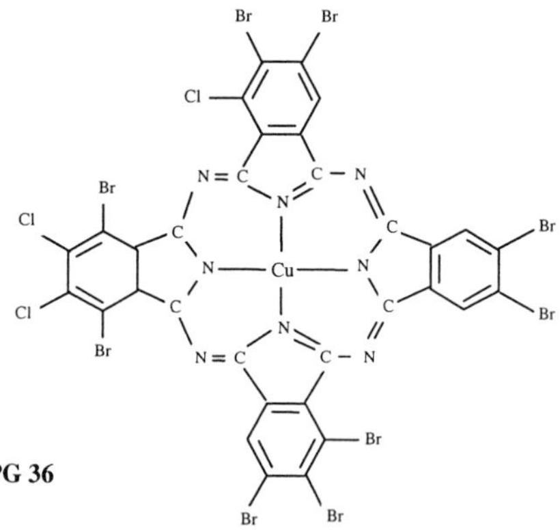

The blue shade green, Pigment Green 7, was developed in 1938, some ten years after the development of copper phthalocyanine blue. It was not until the late 1950's that the yellow shade chloro/bromo copper phthalocyanine green appeared in the marketplace.

It is estimated that approximately 50% of the worldwide production of phthalocyanine green is consumed by the paint and coatings industry due to the pigments excellent overall durability combined with outstanding economics of use.

Phthalocyanine greens are manufactured via a three-step process. The first step involves the manufacture of crude phthalocyanine blue. This crude is then halogenated to give the crude copper phthalocyanine green that is subsequently conditioned in the third stage to give the pigmentary product. Halogenation takes place in a molten aluminum chloride/sodium chloride eutectic mixture or in a solvent as severe as sulphur dioxide, thionyl chloride, or chlorosulfonic acid to give the crude green. Halogenation with chlorine gives the bluer shade, Pigment Green 7. Halogenation with bromine followed by the requisite amount of chlorine gives the yellower shade, Pigment Green 36. If the halogenation is carried out in a melt containing aluminum chloride, care must be exercised to avoid replacement of the copper with aluminum; otherwise, the durability of the resultant pigment will suffer markedly. Finally, the crude green is conditioned by either an attrition process or a solvent process to give the finished pigment.

Miscellaneous Greens. Table 7 gives a brief summary of the properties of other commercially available organic green pigments. Pigment Greens 1,2 and 4 are based upon triphenyl methane dyestuffs that are combined with a complex acid, usually phospho tungsto molybdic acid (PTMA), to give them some insolubility and allow their use as pigments. Their resistance properties are limited and, as such, the major outlets for these PTMA pigments are the printing ink industry and textile markets.

Pigment Green 8, the bisulfite complex of 1-nitroso-2-naphthol reacted with ferrous sulfate and then with sodium hydroxide is one of the oldest chelate type pigments that maintains some minor commercial significance as a colorant for cement. Exhibiting excellent alkali stability this pigment suffers from poor acid resistance.

Pigment Green 10, Nickel Azo Yellow or Green Gold, was developed by two researchers, Woodward and Kvalnes, working in the DuPont laboratories in 1945. The pigment is significant in being the most fade resistant azo pigment currently in commercial production. This excellent light resistance extends throughout the pigments use range from deep tones to pale tints.

Table 7 Summary of the Miscellaneous Green Properties.

Colour Index Name	Common Name	Properties
Pigment Green 1	Brilliant Green (TPM/PTMA)	Brilliant, blue shade. Poor alkali and soap resistance, solvent bleed and lightfastness. May be used in interior finishes.
Pigment Green 2	Permanent Green (TPM/PTMA)	Blend of PG 1 and PY 18. Bright yellow shade. Poor fastness overall.
Pigment Green 4	Malachite Green (TPM/PTMA)	Bright, blue shade. Poor fastness properties overall.
Pigment Green 8	Pigment Green B (Nitroso)	Yellow shade. Dull hue. Poor overall fastness. May be used in interior emulsions.
Pigment Green 10	Green Gold (Nickel Azo Complex)	Yellow shade. Loses metal in strong acid or alkali. Good lightfastness. Moderate solvent fastness. Used in automotive and exterior paints.

Conclusion

Worldwide pigment manufacturing is a mature process industry. This should come as no surprise when one considers the reason that pigments are used to color the articles and objects by which we are surrounded is that they offer a stable color with little or no tendency to change as they age or are subjected to prevailing weather. Many pigments, such as phthalocyanine blue, have been used for decades with every satisfaction by the end user. When such pigments work so well as colorants there is a great reluctance to change, and indeed in many cases there is little requirement for change. If "it isn't broke, why try to fix it" may well describe many users attitude to pigments as colorants.

With respect to future trends within the pigment manufacturing industry, the author sees research continuing to concentrate on providing the color consuming industries with pigments that provide "value in use" by being more easy to disperse, more durable, more economical, faster wetting, more compatible with the common vehicle/solvent systems, and that are environmentally acceptable. Further the pressures on our environment are not likely to decrease as more and more pigments, both organic and inorganic, are withdrawn from use due to concerns over their effect on the environment. While the use of lead containing pigments has been on the decline for several years perhaps this next decade will see their total replacement and cessation of manufacture.

SOLVENTS IN TODAY'S COATINGS

RON L. STOUT

Eastman Chemical Company
123 Lincoln Street
Kingsport, TN   37662

Introduction
Solvents used in the Coatings Industry
Solvent Properties
Solvent Selection in Today's Coatings
The Future

Introduction

The Clean Air Act (CAA) has been the driving force for change in the coatings industry for over twenty-five years. It addresses an assortment of air quality issues that include tropospheric (ground-level) ozone formation, and hazardous air pollution emissions. These two regulatory programs have had a significant impact on the use of solvents in the coatings industry.

Controlling VOC emissions is an important step in the control of ground-level ozone since VOCs are ozone precursors. The Environmental Protection Agency (EPA) defines a VOC as any volatile compound of carbon, excluding carbon monoxide, carbon dioxide, carbonic acid, metallic carbides or carbonates, and ammonium carbonate, which participates in atmospheric photochemical reactivity.

Ozone is formed when VOCs react in the atmosphere with nitrogen oxides under the influence of heat and sunlight. Ozone is a major component of smog and is considered a health and environmental hazard.

Most solvents used in the coatings industry are considered volatile organic compounds. However, coatings account for only about 15% of the man-made VOC emissions in this country. Other significant sources include vehicles, petroleum production, storage and distribution, and industrial fuel combustion. Some compounds are exempt as VOCs by the EPA because they have negligible photochemical reactivity. Acetone and methyl acetate are two common solvents that have been granted VOC exemption.

The objective of the National Emission Standards for Hazardous Air Pollutants (NESHAP) regulations is to minimize cancer and long-term health risk by restricting the emissions of hazardous air pollutants (HAPs). The EPA has established a list of 188 compounds as HAPs. This list is controversial because some of the compounds are considered to be very hazardous while others are not. It includes many of the most widely used solvents in the coatings industry, such as MEK, MIBK, toluene, xylene, and ethylene glycol ethers and their acetates. The EPA must develop Maximum Achievable Control Technology (MACT) standards for many industrial source categories to reduce emissions of these compounds. They will be applied nationwide to sources that have the potential to emit, on an annual basis, 10 tons/year of any HAPs or 25 tons/year of all HAPs. Furthermore, the CAA provides EPA the authority to add or remove compounds from the list. Under those provisions, the Chemical Manufacturers Association (CMA) has requested the removal of ethylene glycol monobutyl ether (EB), MEK, MIBK, and methanol from the list. However, at the time of this writing, the EPA has not acted on these petitions.

To comply with VOC and HAP regulations, solvent users may choose from several strategies such as capture/recycle, incineration, change to another coatings technology that emits less solvent, or reformulate the solvent system to reduce VOCs and/or HAPs. The greatest reduction in emissions might be achieved by using a combination of strategies.

The major coating technologies used to comply with current VOC and HAP regulations are high-solids, waterborne, and solvent-free coatings such as powder and radiation cure. In some applications, such as wood coatings, low-solids lacquers and enamels are still used.

Solvents used in the Coatings Industry

Hydrocarbons. The most widely used solvents in the coatings industry are derived from petroleum. The most common types are aliphatics and aromatics. Most aliphatic solvents are mixtures of paraffinic, cycloaliphatic, and aromatic hydrocarbons. The proportions of these three components will significantly affect solvency and other properties. Aromatics have the highest solvent power, cycloaliphatics intermediate, and paraffins the lowest. Aliphatic solvents function as diluents in most coatings. A diluent is not a solvent for the resin but is included in the solvent blend primarily to reduce cost.

Aromatic hydrocarbon solvents, mainly toluene and xylene, are stronger solvents than aliphatics, and will, therefore, dissolve some resins that aliphatic hydrocarbons won't. Toluene and xylene are on the HAPs list. Cycloaliphatics with similar evaporation rates are being promoted as potential replacements for these products. Common hydrocarbon solvents and some of their physical properties are shown in Table 1.

Oxygenated Solvents. Oxygenated solvents are an important group of solvents because they have a wide range of solvencies and volatilities. Their solvencies diminish with increasing molecular weight and with increased branching of the molecule. Their evaporation rates also decrease with increasing molecular weight but increase with branching of the molecule. Esters, generally, are more polar than hydrocarbons. Many resins that are not soluble in hydrocarbon solvents will dissolve in oxygenated solvents, because they have a wider range of solubility parameters and higher hydrogen bonding values. Oxygenated solvents are rarely used alone. They are normally blended with other types of solvents to obtain the proper solvency, overall evaporation rate, and cost. There are four major types of oxygenated solvents widely used in the coatings industry. They are ketones, esters, alcohols, and glycol ethers. The general structure for each of these families is shown in Table 8. The most widely used oxygenated solvents are discussed below.

Ester solvents used in coatings are mostly acetates. Acetate esters have moderate costs and pleasant, fruity odors. Their solvencies are less than ketones, in most cases, and their densities higher. They are less polar than ketones, alcohols, and glycol ethers, and consequently, have higher electrical resistances. The lower molecule weight esters are partially soluble in water. Common ester solvents and some of their physical properties are shown in Table 3.

Monohydric alcohols have a single hydroxyl group attached to a carbon atom. Alcohols are divided into three groups, primary, secondary, and tertiary, depending on their hydrocarbon segment configuration and the position of the hydroxyl group in the molecule. As a group, the alcohols have mild odors except for maybe the butanols. The lower alcohols are completely soluble in water. They also have low surface tensions, densities, and freezing points. Most form minimum boiling-point azeotropes with water. They also exhibit high hydrogen bonding which makes them important as solvents for many polymers. Some of the more common alcohols and some of their properties are shown in Table 4.

Glycol monoether solvents are synthesized by reacting ethylene oxide or propylene oxide with an alcohol. These solvents contain both an ether and an alcohol group as shown in Table 8. The ethylene glycol ethers are normally referred to as E-series glycol ethers and the propylene glycol ethers as P-series glycol ethers. The di and triglycol monoethers are formed as coproducts when the ethers further react with the remaining oxide and are separated by distillation. Glycol ether solvents typically have slow evaporation rates, high flash and freezing points, high densities, and low surface tensions. Almost all the E-series glycol ethers are completely soluble in water. The low molecular weight P-series glycol ethers are also completely soluble in water, but their higher homologs are only partially soluble in water.

Ethylene glycol monomethyl ether (EM) and ethylene glycol monoethyl ether (EE) have been removed from most applications because of concerns about their toxicity. However, extensive studies have found that the other ethylene glycol ethers do not exhibit the same toxicity as these two glycol ethers. However, these differences in toxicity are frequently not recognized. As a result, all the E-series glycol ethers and esters are on the HAPs list. Common E and P series glycol ethers and some of their physical properties are listed in Table 5.

The ethylene glycol and propylene glycol ether esters are produced by esterification of the hydroxyl group on a glycol ether with acetic acid. Sometimes the diglycol ethers are the basis of the molecule. The diglycol ethers are similar in solvency but slower evaporating than the derivatives of the simple glycols. Ethylene glycol monoethyl ether acetate (EE acetate), like EM and EE, has been removed from most

applications because of toxicity concerns. In most applications, it was replaced with PM
acetate. Other potential replacements are Ethyl 3-Ethoxypropionate (EEP), and Exxate
® 600 and 700 solvents. The common glycol ether acetates and some of their physical
properties are listed in Table 5.

The estimated consumption of hydrocarbon and oxygenated solvents in coatings
in 1997 is shown in Figure 1.

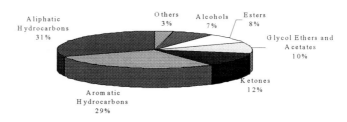

Figure 1 - U.S. Solvent Consumption by Type for Coatings in 1997
(Source: Skeist, Inc., 1998)

Solvent Properties

Solubility. The solvent system of a solventborne coating has two major functions. The
first is to dissolve the resin and provide a clear, homogeneous solution. The second is
to provide the proper coating viscosity.

Solubility parameters are often used to predict resin solubility in various
solvents. The solubility parameter is a measure of the total attractive forces between
molecules of a substance. These forces are of three major types: dispersion, polarity,
and hydrogen bonding. Dispersion (London) forces are the result of temporary dipoles
formed as continuous moving electrons within the molecule give rise to an unbalanced
electrical field. Polar forces represent a combination of dipole-dipole (Keesom) and
dipole-induced dipole (Debye) interactions. Hydrogen bonding is the attraction between
polar molecules in which hydrogen is covalently bonded to electronegative elements. In
order for a substance to dissolve in a solvent, the forces of attraction in the substance
should be of the same order of magnitude as the forces of attraction in the solvent.
Thus, dissolution will most often take place when the solubility parameters of the solute
and solvent are similar. The solubility parameter, as defined by Hildebrand[1], is the
square root of the cohesive energy density, and is the total molecular attraction per unit
volume.

The mathematical expression is:

$$\delta = (E/V)^{1/2}$$

δ = solubility parameter
E = cohesive energy/mole
V = molar volume
Units = (calories/cm^3)$^{1/2}$ or (MPa $^{1/2}$ cm ^{3})

The cohesive energy is equal to the energy of vaporization. It can be measured experimentally from the heat of vaporization if the substance is volatile, where $\Delta E = \Delta H - RT$ and ΔH = heat of vaporization.

The solubility parameter of a solvent is commonly determined at 25° C, but it varies with temperature. Each solvent has a unique solubility parameter. On the other hand, a resin will have a range of values. The width of the range will vary widely for different types of resins. Resins are not volatile, and their heats of vaporization and solubility parameters cannot be measured directly. Hence, resin solubility parameters are commonly estimated from experimental solubility data.

Hansen proposed that the total solubility parameter of a solvent be divided into the contributions from the three major molecular forces which was discussed earlier. He expressed the total solubility parameter as the square root of the sum of the squares of these interactions as follows:

$$\delta_t = (\delta_d^2 + \delta_p^2 + \delta_h^2)^{1/2}$$

where

δ_t = Total Solubility Parameter
δ_d = Dispersion forces
δ_p = Polar forces
δ_h = Hydrogen bonding forces

He determined values for these components empirically on the basis of experimental observations. Hansen solubility parameters for many common solvents are shown in Table 6.

Hansen represented solvents as points on a three-dimensional plot where the three parameters are axes. Solubility data determined experimentally on a given polymer in a spectrum of solvents permitted an estimation of the volume of solubility for that polymer in three-dimensional space. For most polymers, this volume of solubility is spherical if the scale of the dispersion axis relative to the other axes is doubled. Solvents within the polymer solubility sphere usually dissolve the polymer and those outside generally do not. So, a polymer's solubility parameter can be expressed as a center coordinate and a radius, which Hansen called the radius of interaction. Hansen solubility parameters for several polymers are shown in Table 7.

Solvency. A coating must have the proper viscosity for the selected application method. As the resin dissolves in the solvent, the solution viscosity increases—slowly at first then very rapidly. How fast this transition occurs depends on two factors: (1) the solubility of the resin, and (2) the activity of the solvent. These factors, plus the viscosity of the solvent and resin, are important properties that influence the final viscosity of the coating.

A major influence on viscosity is the effect the solvent can have on the hydrodynamic volume of the resin molecules. Resin solution viscosity is generally regarded as arising from molecular entanglements between resin molecules. In the presence of a good solvent, the molecular interactions are strong and the resin molecules are fully extended into the solvent which produces high viscosity. However in the presence of a poor solvent, the resin molecules stay coiled and the molecular interactions are weak, which produces low viscosity.

In dilute solutions, solution viscosity is proportional to the neat viscosity of the solvent. For example, isobutyl alcohol, has a neat viscosity more than five times that of VM&P naphtha, and solutions in these solvents will have viscosities that are proportional to the ratio of their neat viscosities.

The influence of the solvent's neat viscosity and solvency can be separated by calculating the specific viscosity[2]. The specific viscosity

$$\eta_{sp} = \frac{\eta_{solution} - \eta_{solvent}}{\eta_{solvent}}$$

is the ratio of the viscosity increase occurring from dissolution to the viscosity of the solvent. Therefore, specific viscosity is dimensionless and indicates the thickening power of the resin. The effect of solvent viscosity is suppressed by treating the viscosity increase relatively.

Two solvents can provide identical viscosity reduction curves for a resin if their neat viscosities and solvencies are identical, which is unlikely. Most oxygenated solvents have higher neat viscosities than hydrocarbons, but their solvencies are usually greater at a given molecular weight.

Neat Solvent Evaporation. Since ambient conditions vary widely, evaporation rates are measured and reported on a comparative basis. Measurements of relative evaporation rates are made with an evaporometer. A weighed sample is evaporated under carefully controlled conditions of temperature, humidity, and air flow rate. A plot of time versus weight percent evaporated is obtained. It is routine to make comparisons of evaporation rates at 90% evaporation so that high boiling impurities won't substantially effect the evaporation rate. The evaporation time of a solvent is compared to that of a reference material such as n-butyl acetate, which is commonly set to 1 or 100. Humidity affects the evaporation of water miscible solvents. However, humidity has no significant effect on the evaporation of water immiscible solvents.

Molecular interaction affects the evaporation of components from a mixture especially if one component is hydrogen bonded and another is not.[3] For ideal ystems, which occur only when the blend components are very similar, the vapor pressure and resulting evaporation rate are governed by Raoult's Law:

$$P_1 = P_1°X_1$$

Where P_1 = Partial pressure of a component in the mixture
        $P_1°$ = Vapor pressure of pure component
        X_1 = Molar concentration of liquid component

This simplest form of Raoult's Law says that the partial pressure of a component n a mixture will be proportional to its molar concentration and to its vapor pressure in he pure state. For non-ideal systems, it is necessary to include in the equation an activity coefficient:

$$P_1 = \gamma_1 P_1 ° X_1$$

Where γ_1 = Activity coefficient

The activity coefficient for a given component is the ratio of its actual partial pressure to its ideal partial pressure calculated from Raoult's Law.
Today most of the major solvent producers have computer programs that use a formula similar to the one above to calculate the evaporation rates of solvent blends. These programs typically have the physical properties of most commercially available solvents stored in files. Using this data, they estimate properties of the solvent blend such as solubility parameters, surface tension, neat viscosity, density, flash point, evaporation rate, evaporation profile (solvent composition versus weight percent evaporated), and cost. Some of these programs can even estimate the effect of emperature, humidity, and resin on solvent evaporation, and resin solubility as solvent composition changes during the evaporation process.

Solvent Evaporation from Solution. Solvent evaporation is significantly affected by the addition of a resin solute. Initially, evaporation is essentially equal to that of the neat solvents and is controlled by their volatilities. At some point in the process, solvent evaporation slows suddenly and diffusion becomes the limiting factor in this evaporative process. Resin plasticization increases solvent diffusion because solvent molecules will diffuse much faster through resins that are above their glass transition emperatures than through ones below. Small amounts of solvent are sometimes etained in the film for long periods of time. Their retention time depends on the types of resins and solvents selected.
During this evaporative process, the function of the solvent is to control flow characteristics as the film forms. If solvent evaporation is too fast, the film will not level or wet the substrate which can cause film defects such as poor gloss and adhesion. If solvent evaporation is too slow, the film will sag and perhaps become too thin. If the solvent's activity decreases significantly during evaporation, precipitation of the resin

can occur resulting in poor film formation. Thus, the evaporation rate of the solven
system can have a major impact on coating quality.

Solvent Selection in Today's Coatings

Lacquers. Lacquers are thermoplastic coatings that form films by solvent evaporation
In these coatings, a blend of solvents is typically used to achieve the proper solvency
evaporation rate, and cost. A good example is a thinner blend for a nitrocellulose resin
which contains three types of solvents: active, latent, and diluent. Ketones or esters are
active solvents for this resin. An alcohol is a latent solvent, which simply means that i
is not an active solvent, but in combination with an active solvent, will produce a
synergistic effect. The combination of the two will be a better solvent than either one
alone. Hydrocarbons are diluents and are included to reduce cost. A balanced blend o
high, medium, and low boilers is selected from the three types of solvents. The exac
combination depends on many factors such as application technique (spray or brush)
substrate to be coated, desired drying time, cost, and VOC restrictions.

High-solids Coatings. High-solids coatings are formulated with low molecular weigh
resins that have low polydispersity to obtain application viscosity at high solids. These
resins contain functional groups that are crosslinked to produce high performance
coatings. The viscosity of these coatings is highly dependent on temperature. Thus
many are often heated to reduce their viscosity to the appropriate range for application

According to Wicks, Jones, and Pappas[4], resins that contain hydroxyl o
carboxyl functional groups tend to associate with each other. This can cause
substantial increases in viscosity. To minimize these interactions, they sugges
choosing polar solvents with single hydrogen bonding acceptor sites such as ketones
ethers, or esters. Furthermore, they suggest that the effect the solvent has on the
hydrodynamic volume of the resin molecules can also influence viscosity. In addition
Sprinkle[5] reported that ketones are more effective in reducing the viscosity of high
solids resins for a given weight of solvent. He attributed these results to their highe
solvencies and lower densities. If high-solid coatings are applied with electrostatic
spray equipment, esters or hydrocarbon solvents are sometimes added to adjus
electrical resistance. In some high-solids coatings (polyol/melamine), alcohols are
commonly added to improve viscosity stability. In urethane crosslinked high-solids
coatings, urethane-grade solvents are utilized because low alcohol, water, and acic
content are important in these coatings.

Latex Coatings. Latexes are dispersions of high molecular weight polymers in water.
They are divided into two major categories by their glass transition temperatures (Tg),
low and high. The major end uses for low Tg latexes are interior and exterior
architectural house paints. These latex coatings don't form films at moderate to low
temperatures. Hence, solvents, known as a coalescing aids, are added to aid film
formation. They plasticize (soften) the polymer particles, allowing them to fuse together.
After film formation, the coalescing aids slowly diffuse to the film surface and

evaporate. Coalescing aids should have low solubility in water if the latex paint will be applied to porous substrates such as dry wall or cinder blocks. Coalescing aids are usually very slow evaporating solvents such as glycol ethers, glycol ether esters, and ester-alcohols. In addition to reducing the minimum film formation temperature of the latex coating, coalescing aids improve weatherability, scrubability, cleanability, touch-up ability, and color development. The most widely used coalescing aid for low Tg latex coatings is 2,2,4-trimethyl-1,3-pentanediol monoisobutyrate, commonly known as Texanol® ester-alcohol. It has very low solubility in water and excellent hydrolytic stability, allowing it to be used with a wide variety of latexes including those with high pH. Other solvents that are used as coalescing aids for low Tg latex coatings include: propylene glycol monophenyl ether (PPh), diethylene glycol monobutyl ether (DB), dipropylene glycol n-butyl ether (DPnB), ethylene glycol 2-ethylhexyl ether (EEH), and ethylene glycol monophenyl ether (EPh). Glycol ether esters such as DB acetate are also used as coalescing aids, but their usefulness is limited because of poor hydrolytic stability.

The major end uses for high Tg latexes are OEM and industrial maintenance coatings. High Tg latex coatings generally require higher levels of coalescing aids than their lower Tg counterparts for satisfactory film formation. Faster evaporating coalescing aids are often used with the high Tg latex coatings because faster drying and hardness development are required.

The type and level of coalescing aid are determined predominately by drying conditions (temperature and humidity), but polymer hardness and coating stability are also factors. Ethylene and propylene glycol ethers are typically used as coalescing aids for these types of latexes. Of these glycol ethers, ethylene glycol monobutyl ether (EB) is the most widely used. If the relative humidity is high (70% to 90%) or heat is applied to accelerate drying, the addition of higher-boiling coalescing aids such as Texanol® ester-alcohol may be necessary to ensure satisfactory film formation.

Water-reducible Coatings. Water-reducible coatings are used mainly as industrial coatings for cans and metal substrates. They are typically thermosetting coatings made from polymers with acid or amine functional groups on their backbones. These polymers become water dispersible when the pendant groups are neutralized with volatile amines or acids. Cross-linkers are then added and the coating is diluted with water. Ethylene and propylene glycol ethers are widely used as cosolvents for these water-reducible coatings. The most commonly used one is EB, but others are also used. Alcohols are sometimes included as part of the cosolvent mainly to improve coating stability, but they also reduce the surface tension of water and accelerate the evaporation of water by forming low-boiling azeotropes with water. The level and composition of the cosolvent are adjusted to control solution viscosity behavior, improve flow-out and leveling, control evaporation characteristics of the solvent/water system, keep the film open during drying, and control sagging and running.

The Future

The EPA is phasing out the current National Ambient Air Quality Standard for ozone (1-hour standard, 0.12 ppm limit) and replacing it with an 8-hour standard with a limit of 0.08 ppm. The old standard will not be revoked in a given area until that area has met the old standard for three consecutive years. The new standard will create more non-attainment areas and will continue to apply pressure on the coating industry to reduce VOCs from all major sources. Long-term, this change may reduce the size of the U. S. solvent market.

The EPA is also developing MACT standards for many source categories to reduce the emissions of HAPs. A number of surface coating processes will be affected such as can, coil, large appliances, etc. These regulations will most likely reduce the quantity of HAPs solvents used in the coating industry.

The California Low Emissions and Reactivity Program, commonly know as the CLEAR program, will provide a voluntary alternative method to comply with VOCs limits for some commercial and consumer products. This method allows the option of meeting photochemical reactive-based CLEAR limits rather than the mass based VOC limits.

If this compliance method spreads from California to other states, the solvents that are more photochemically reactive will be replaced with solvents that are less photochemically reactive. If this occurs, it could drastically change the makeup of the U. S. solvent market.

APPENDIX OF TABLES

Table 1—Aliphatic and Aromatic Hydrocarbon Solvents and Some of Their Properties

Solvent	API[a] Gravity	Specific Gravity, 60°F	lbs/gal	Distillation Range, °F	Flash Point TCC, °F[b]	Mixed Aniline Point, °F	Kauri-Butanol Value
Aliphatic Hydrocarbons							
Hexane	78	0.675	5.62	152-157	<0	150	30
Rubber solvent	71	0.699	5.82	175-225	<0	142	34
Lacquer diluent	57	0.751	6.25	200-220	20	120	38
VM&P naphtha	56	0.755	6.28	250-300	60	140	34
Stoddard solvent	53	0.767	6.39	310-370	100	148	35
Mineral spirits	52	0.771	6.42	310-400	100	150	36
140°F solvent	49	0.784	6.53	360-415	140	157	30
Deodorized kerosene	43	0.811	6.75	350-510	140	160	30
Aromatic Hydrocarbons							
Toluene	30.8	0.872	7.26	230-232	45	49	105
Xylene	31	0.871	7.26	281-284	80	51	98
Medium-flash aromatic naphtha	30.7	0.872	7.26	311-344	110	56	91
High-flash aromatic naphtha	25.4	0.902	7.51	362-410	150	60	95

(a)API = American Petroleum Institute. API gravity is an arbitrary, standard scale for petroleum products.
(b) TCC = Tag Closed Cup.

Table 2— Ketones and Some of Their Properties

Solvent	Specific Gravity, 20°C/20°C	lbs/gal 20°C	Boiling Point, °C	Flash Point TCC °F	Solubility at 25°C % by weight		Evaporation Rate (n-BuOAc=1)
					In Water	Water In	
Acetone	0.792	6.6	57	-4	Complete	Complete	5.7
Methyl ethyl ketone	0.802	6.67	80	16	27.1	12.5	3.8
Methyl n-propyl ketone	0.807	6.74	103	46	3.1	4.2	2.3
Methyl isobutyl ketone	0.802	6.67	116	60	2	1	1.6
Methyl isoamyl ketone	0.813	6.76	145	96	0.5	1.2	0.5
Methyl n-amyl ketone	0.818	6.8	150	102	0.5	1.3	0.4
Diisobutyl ketone	0.811	6.76	169	120	0.05	0.7	0.2
Cyclohexanone	0.948	7.89	156	111	2.3	8	0.3
Diacetone alcohol	0.94	7.82	158	126	Complete	Complete	0.12
Isophorone	0.922	7.67	215	179	1.2	4.3	0.03

Table 3—Esters and Some of Their Properties

Solvent	Specific Gravity 20°C/20°C	lbs/gal 20°C	Boiling Point, °C	Flash Point TCC °F	Solubility at 25°C % by weight		Evaporation Rate (n-BuOAc=1)
					In Water	Water In	
Methyl acetate	0.94	7.7	58	9	22	7.3	5.3
Ethyl acetate (99%)	0.901	7.51	77	24	7.4	3.3	4.1
Ethyl acetate (85%-88%)	0.884	7.36	75	27	7.4	3.1	4.2
Isopropyl acetate	0.873	7.26	88	35	2.9	1.8	3.0
-Propyl acetate	0.889	7.39	101	55	2.3	2.6	2.3
Isobutyl acetate	0.87	7.25	116	69	0.7	1.6	1.4
-Butyl acetate	0.883	7.34	127	100	0.7	1.2	1.0
-Butyl propionate	0.876	7.3	145	100	0.2	<0.02	0.45
-Amyl acetate (95%)	0.876	7.29	146	106	0.2	0.9	0.4
Isobutyl Isobutyrate	0.855	7.13	148	104	<0.1	<0.2	0.4
Exxate 600	0.874	7.3	170	134	0.02	0.7	0.17
Ethyl 3-ethoxypropionate	0.95	7.91	168	136[a]	2.9	2.2	0.12
-Ethylhexyl acetate	0.873	7.27	202	160	0.03	0.6	0.04
Exxate 800	0.875	7.3	201	160	0.02	0.35	0.03
Ethylene Glycol Diacetate	1.107	9.22	190	191	16.4	7.6	0.02
Dibasic Esters	1.092	9.09	214	212	5.3	3.1	0.007
Exxate 1000	0.871	7.26	235	212	<0.1	0.2	0.006
Texanol ester-alcohol	0.95	7.9	257	248[b]	<0.1	0.9	0.002
Exxate 1300	0.88	7.3	260	261	0	0.2	0.001

Setaflash Point
Cleveland Open Cup

Table 4—Alcohols and Some of Their Properties

Solvent	Specific Gravity, 20°C/20°C	lbs/gal 20°C	Boiling Point, °C	Flash Point TCC °F	Solubility at 25°C % by weight		Evaporation Rate (n-BuOAc=
					In Water	Water In	
Methanol	0.792	6.6	65	50	Complete	Complete	3.5
Ethanol, anhydrous	0.79	6.58	78	70	Complete	1.7	
Ethanol, 95%	0.812	6.76	78	75	Complete	Complete	1.9
n-Propanol	0.804	6.71	97	74	Complete	1.0	
Isopropanol, 99%	0.786	6.54	82	55	Complete	1.7	
Isobutanol	0.803	6.68	108	85	9.5	14.3	0.6
n-Butanol	0.811	6.75	118	97	7.9	20.8	0.5
Secondary butanol	0.81	6.73	100	72	20.6	30.7	0.9
Tertiary butanol	0.788	6.56	82	52	Complete	1.0	
Methyl isobutyl carbinol	0.808	6.73	132	103	1.6	6.3	0.3
Cyclohexanol	0.962	8.01	161	156	0.1	11.8	0.05
2-Ethylhexanol	0.833	6.94	185	164	0.1	2.6	0.01

Table 5—Glycol Ethers and Their Acetates, and Some of Their Properties

Solvent	Specific Gravity, 20°C/20°C	lbs/gal 20°C	Boiling Range, °C	Flash Point TCC °F	Solubility at 25°C % by weight		Evaporation Rate (n-BuAc=1
					In Water	Water In	
Ethylene Glycol							
Monoethyl Ether EE	0.931	7.75	134-136	110	Complete	Complete	0.3
Monoethyl ether acetate EE Acetate	0.973	8.11	150-160	130	23.8	6.5	0.2
Monopropyl ether EP	0.913	7.59	150-154	120	Complete	Complete	0.2
Monobutyl ether EB	0.902	7.51	169-173	143	Complete	Complete	0.09
Monobutyl ether acetate EB Acetate	0.941	7.84	186-194	160	1.1	1.6	0.03
Propylene Glycol							
Monomethyl ether PM	0.923	7.69	117-125	91	Complete	Complete	0.7
Monomethyl ether acetate PM Acetate	0.97	8.06	140-150	114	20	5.9	0.4
Monopropyl ether PnP	0.885 [b]	7.38 [b]	150 [c]	119	Complete	Complete	0.21
Monobutyl ether PnB	0.884 [b]	7.37 [b]	170 [c]	138	6.4	16	0.08
Monophenyl ether PPh	1.063 [b]	8.80 [b]	243 [c]	240	1.1	7	0.002

Diethylene Glycol

Monomethyl ether DM	1.021	8.5	192-196	191	Complete	Complete	0.02
Monoethyl ether DE	0.99	8.25	198-204	195	Complete	Complete	0.02
Monoethyl ether acetate DE Acetate	1.012	8.42	214-221	225[a]	Complete	Complete	0.008
Monobutyl ether DB	0.955	7.94	230-235	232 [a]	Complete	Complete	0.004
Monobutyl ether acetate DB Acetate	0.98	8.16	235-250	240 [a]	6.5	3.7	0.002

Dipropylene Glycol

Monomethyl ether DPM	0.950 b	7.91 [b]	184 [c]	167 [d]	Complete	Complete	0.02
Monopropyl ether DPnP	0.922 b	7.70 [b]	212 [c]	190 [d]	19	20.5	0.015
Monobutyl ether DPnB	0.906 b	7.55 [b]	229 [c]	212 [d]	5	12.5	0.01

[a]Cleveland Open Cup

[b] 25 °C

[c] Boiling Point

[d] Setaflash Point

Table 6—Hansen Solubility Parameters of Various Solvents (MPa$^{1/2}$ cm^3)

Solvents	δD	δP	δH	δT
Acetone	15.5	10.4	7.0	20.0
Aromatic 100	17.8	0.6	1.4	17.8
Aromatic 150	17.8	0.6	1.4	17.8
n-Butyl acetate	15.8	3.7	6.3	17.4
n-Butyl alcohol	16.0	5.7	15.8	23.1
sec-Butyl alcohol	15.8	5.7	14.5	22.1
γ-Butyrolactone	19.0	16.6	7.4	26.3
Cyclohexane	16.8	0	0.2	16.8
Cyclohexanol	17.4	4.1	13.5	22.5
Cyclohexanone	17.8	6.3	5.1	19.6
Diacetone alcohol	15.8	8.2	10.8	20.9
Dibasic esters (DBE)	16.2	4.7	8.4	18.8
Diethylene glycol n-butyl ether	16.0	7.0	0.6	20.5
Diethylene glycol butyl ether acetate	16.0	4.1	8.2	18.4
Diethylene glycol n-ethyl ether	16.2	9.2	12.3	21.9
Diethylene glycol methyl ether	16.2	7.8	12.7	21.9
Diethylene glycol n-propyl ether	16.0	7.2	11.3	20.9
Diethylene glycol ethyl ether acetate	16.8	6.4	15.8	19.2
Diethyl ketone	15.8	7.6	4.7	18.1
Diisobutyl ketone	16.0	3.7	4.1	16.9
Dipropylene glycol n-butyl ether	15.4	5.6	9.0	18.7
Dipropylene glycol methyl ether	15.5	5.7	11.2	20.0
Ethyl acetate	15.8	5.3	7.2	18.1
Ethyl alcohol	15.8	8.8	19.4	26.6
Ethyl benzene	17.8	0.6	1.4	17.8
Ethyl 3-ethoxypropionate	16.2	9.2	9.4	18.6
Ethylene glycol diacetate	16.4	10.5	12.9	19.4
Ethylene glycol n-butyl ether	16.0	5.1	12.3	20.9
Ethylene glycol butyl ether acetate	15.3	4.5	8.8	18.2
Ethylene glycol 2-ethylhexyl ether	16.0	4.1	5.1	17.2
Ethylene glycol n-propyl ether	16.0	6.2	13.3	22.7
2-Ethyl hexanol	16.0	3.3	11.8	20.3
2-Ethylhexyl acetate	15.7	2.9	5.1	16.8
n-Heptane	15.3	0	0	15.3
n-Hexane	14.9	0	0	14.9

Solvents	δD	δP	δH	δT
Isobutyl acetate	15.1	3.7	6.3	16.8
Isobutyl alcohol	15.1	5.7	16.0	22.7
Isobutyl isobutyrate	15.1	2.9	5.9	16.5
Isophorone	16.6	8.2	7.4	19.9
Isopropyl acetate	15.6	3.3	8.8	17.6
Isopropyl alcohol	15.8	6.1	16.4	23.5
Methyl acetate	15.5	7.2	7.6	18.7
Methyl alcohol	15.1	12.3	22.3	29.6
Methyl n-amyl ketone	16.2	5.7	4.1	17.6
Methyl ethyl ketone	16.0	9.0	5.1	19.0
Methyl isoamyl ketone	16.0	5.7	4.1	17.4
Methyl isobutyl carbinol	15.4	3.3	12.3	19.8
Methyl isobutyl ketone	15.3	6.1	4.1	17.0
Methyl n-propyl ketone	15.8	9.0	7.6	18.2
N-Methyl 2-pyrrolidone	18.0	12.3	7.2	22.9
Methylene chloride	18.2	6.3	6.1	20.3
Nitroethane	16.0	15.5	4.5	22.7
Nitromethane	15.8	18.8	5.1	25.1
2-Nitropropane	16.2	12.1	4.1	20.6
n-Propyl acetate	16.4	6.6	8.2	17.6
n-Propyl alcohol	16.0	6.8	17.4	24.5
Propylene glycol n-butyl ether	15.6	5.8	11.2	18.4
Propylene glycol t-butyl ether	15.4	6.8	9.3	19.6
Propylene glycol phenyl ether	18.7	5.7	11.3	21.5
Propylene glycol n-propyl ether	15.5	6.3	12.4	19.4
Propylene glycol methyl ether	15.6	6.3	11.6	20.5
Propylene glycol methyl ether acetate	16.1	6.1	6.6	19.2
Texanol ester-alcohol	15.1	6.1	9.8	19.0
Toluene	18.0	1.4	2.0	18.2
Xylene (mixed isomers)	17.6	1.0	3.1	17.8
VM&P naphtha	15.1	0	0.2	15.1

Table 7— Hansen Solubility Parameters of Polymers

Polymers	Tradename	Supplier	δD	δP	δH	Radiu
Acrylic (PEAM)	Elacite 2042	Du Pont	17.6	9.7	4.0	10.6
Acrylic (PMMA)	Perspex	ICI	18.6	10.5	7.5	8.6
Acrylic	Paraloid P400	Rohm&Haas	19.2	9.6	9.3	12.2
Alkyd, long oil (66% Oil Length)	Plexal P65	Polyplex	20.4	3.4	4.6	13.7
Alkyd, short oil (34% Oil Length)	Plexal C34	Polyplex	18.5	9.2	4.9	10.6
Nitrocellulose	H-23, 1/2 sec.	Aqualon	15.4	14.7	8.8	11.5
Epoxy	Epon 828	Shell	21.3	14.2	6.1	17.7
Epoxy	Epon 1001	Shell	20.0	10.3	10.1	10.0
Epoxy	Epon 1004	Shell	17.4	10.5	9.0	7.9
Hexamethoxymethyl melamine	Cymel 300	Cytec	20.4	8.5	10.6	14.7
Isocyanate, phenol blocked	Suprasec F5100	ICI	20.2	13.2	13.1	11.7
Phenolic	S. Beckacite 1000	Reichhold	23.3	6.6	8.3	19.8
Polyamide	Versamid 930	Henkel	17.4	-1.9	14.9	9.6
Polyester	Desmophen 850	Bayer	21.5	14.9	12.3	16.7
Polyvinyl butyral	Butvar B-76	Monsanto	18.6	4.4	13.0	10.6
Urea-formaldehyde	Plastopal H	BASF	20.8	8.3	12.7	12.7
Vinyl chloride	Vipla KR	Montecatini	18.2	7.5	8.3	3.5
Vinyl chloride / vinyl acetate	UCAR VYHH	Union Carbide	17.4	10.2	5.9	7.8

Table 8—General Chemical Structures for Various Types of Solvents

Hydrocarbons Groups		Oxygenated Groups	
Methyl	$- CH_3$		$\begin{matrix} & O \\ & \parallel \\ R - & C - R^a \end{matrix}$
Ethyl	$- CH_2\,CH_3$	Ketone	
Propyl	$- CH_2\,CH_2\,CH_3$	Alcohol	$R - OH$
Isopropyl	$- CH - CH_3$ $\quad\;\; \mid$ $\quad\;\; CH_3$	Glycol ether	$HO - R - O\,R$
Butyl	$- CH_2\,CH_2\,CH_2\,CH_3$	Ester (Acetate)	$\begin{matrix} O & & O \\ \parallel & & \parallel \\ CH_3 C - O - & & R \end{matrix}$
Isobutyl	$- CH_2\,CH\,CH_3$ $\qquad\;\; \mid$ $\qquad\;\; CH_3$	Ester (Propionate)	$\begin{matrix} & & O \\ & & \parallel \\ CH_3\,CH_2 - & C - & O - R \end{matrix}$
Amyl	$- CH_2\,CH_2\,CH_2\,CH_2\,CH_3$		

[a] R = Hydrocarbon (alkyl) group

LITERATURE CITED

(1) Hildebrand, J. H., "The Solubility of Non-electrolytes," Third Edition, Reinhold Publishing Corp., New York, 1949.
(2) Lambourne, R., "Paint and Surface Coatings: Theory and Practice," Ellis Horwood, Limited, Chichester, West Sussex, England (1993).
(3) Ellis, W.H. and Goff, P.L., "Precise Control of Solvent Blend Composition During Evaporation," JOURNAL OF PAINT TECHNOLOGY, 44, No. 564, 79 (1972).
(4) Wicks W. Z., Jones F. N., and Pappas S. P., "Organic Coatings, Science and Technology, Volume 1, Chapter XV, 1992.
(5) Sprinkle, G. P., "Selecting Solvents for High Solids Coatings," Modern Paint and Coatings, April, 44, (1983).
(6) Hansen, C. M., Private communication, 1987

RHEOLOGY AND COATING FLOWS[a]

RICHARD R. ELEY

ICI Paints North America
16651 Sprague Road
Strongsville, OH 44136

Introduction
Relevance of Rheology to Coatings
The Generalized Non-Newtonian Flow Curve
Mechanisms of Shear Thinning
Shear Thickening Fluids
Plastic (Yield) Behavior
Viscoelasticity and Industrial Processes
Creep Methods
Controlled Stress and Controlled Rate
Rheology of Coating Application and Film formation
Computer Modeling of Coating Flows
Conclusions

Introduction

Heraclitus (1) proclaimed "παντα ρει" (everything flows), and Deborah accurately prophesied that even "mountains flow before the Lord" (2), suggesting the pervasive nature of flow phenomena in natural processes. Not only do apparently solid mountains flow, resulting in the synclines and anticlines of geologic strata, but so also must the fluids vital to the functioning of a living body, and the viscoelastic properties of blood, mucus, synovial (joint) fluid[b] and the vitreous humor of the eye[c] are important for their proper function. Among man-made materials, flow behavior, often complex, can be a crucial element of commercial success. Paints and industrial coatings, creams and lotions, inks, adhesives, ceramic slips, solder pastes, foods, medicines, *etc.*, are examples of the range of materials whose commercial viability depends on having the "right" rheology. In turn, the required rheological properties must be defined with due regard to the flow or stress conditions which prevail during processing and application.

---

[a] Portions of text and figures are excerpted from "Principles and Methods of Rheology in Coatings", by R. R. Eley, in "The Encyclopedia of Analytical Chemistry: Techniques and Applications", copyright 2000 John Wiley & Sons Ltd. Reproduced with permission.
[b] an aqueous sol of polymerized hyaluronic acid and collagen, having excellent lubricating properties
[c] an aqueous hyaluronic acid-collagen gel

Rheology by definition is the study of the deformation and flow behavior of materials. An operational definition would be "the study of the response of certain materials to the stresses imposed on them". [d]   Rheology seeks to understand the relationship between applied force, or **stress**, and the resulting deformation, particularly for materials showing non-simple responses. **Applied rheology** endeavors to connect fundamental properties and real processes. This article will very briefly highlight some key topics and newer approaches to the rheological analysis of coatings with an eye to understanding coating flows, and other pertinent issues.

Relevance of Rheology to Coatings

To a degree matched by few other materials, rheology determines success for coatings. Though all other properties be acceptable, a coating will usually not meet with success if the rheology is not. Experienced formulators say that more than half the cost of new product development is consumed in "getting the rheology right". Moreover, apparently "minor" changes in a raw material or process can cause significant and unexpected variability in product rheology, a problem which will obviously require urgent resolution. For all these reasons, rheological analysis is a vital and cost-effective tool for the coatings industry.

In the times when the majority of paints and industrial coatings were solvent-borne, the array of solvents from which to choose was large, and provided great latitude in formulating for performance. Control of rheology, pigment dispersion stability, substrate wetting, open time, and film formation were relatively straightforward to achieve by solvent selection and blending. The large-scale move toward environmentally compliant coatings (waterborne, higher solids, reduced- or zero-VOC) has in general resulted in more complex rheology, while reducing the number of formulating options and at the same time generating a host of performance/application problems. Reduced-VOC aqueous coatings have consequent higher surface tension and stronger Marangoni effects, which make it more difficult to coat substrates of lower quality (lower or less uniform in surface energy), or substrates having sharp edges or small radii of curvature (holes or sharp corners). Difficult pigment wetting and stable foam are other consequences of reduced solvent content.

The purview of this article includes adhesives, sealants, and inks, as well as coatings. These products share a common task: they must be applied to substrates and function as a thin film. This process of application and film formation obviously requires not only a large total deformation, but also a high degree of control of flow, to achieve success. Flow cannot be controlled unless it can be properly measured. The objective for the applied rheologist, therefore, is to develop methods of rheological characterization that:
   1. yield accurate data for complex fluids, and
   2. are related specifically to the critical processes that paints must undergo.

---

[d] For definitions of basic rheological terms, see references (3) and (4).

To meet the latter objective requires characterization methods that cover a wide range of stresses and time scales. Modern rheological instrumentation is helping to achieve that goal.

However, the link between coating performance and rheology is still far from complete, for certain reasons. Among these reasons are (1) the sheer complexity of coatings processes, which complicates the understanding of the role of rheology in process outcomes, and (2) the inability to strongly link measured fundamental properties to real-world performance. One answer to the dilemma is computer simulation of coating processes, which utilizes the fundamental rheological data as the required input. A sophisticated model should take account not only of coating rheological and physicochemical properties but also of process details, complexities of substrate geometry, ambient environmental factors, and changes in properties with evaporation and temperature.

While computer modeling is perhaps the ideal approach, it is not readily accessible. Many coating problems can be understood and solved from shear viscosity and viscoelastic data alone, provided the experiments performed are well designed and the results properly interpreted. The following sections will elaborate on these points, with emphasis on the understanding and interpretation of steady-shear flow curve data.

The Generalized Non-Newtonian Flow Curve

Commercial fluid products comprise a wide variety of materials, with a wide range of consistencies. In general, polymer solutions and melts (above M_c), emulsions, colloidal dispersions, and other suspensions of particulate solids at useful concentrations will be *non-Newtonian*. For non-Newtonian materials, the viscosity is no longer a material constant, but a *material function*— in this case, a "function" of the shear rate (or shear stress). For non-Newtonian fluids, a viscosity measured at a single shear rate is not an adequate representation of the rheology of the system. For this reason, and because of other more fundamental shortcomings (3,4) the rugged but simple "single-point" viscometers commonly used in the industrial laboratory are generally not well suited for the characterization of non-Newtonian fluids. "Research-quality" rheometers measure some rheological property, or material function, such as the viscosity as a function of shear rate or shear stress. Normally, a curve will be produced representing the functional dependence of the measured property. A plot of viscosity against shear rate or shear stress (normally log-log) may be termed a "flow curve", and may be generated using equilibrium (as in Figure 1) or non-equilibrium (Figures 5 and 6) flow measurement methods (4).

The *Generalized Equilibrium Flow Curve* (Figure 1) represents the general features of the shear rate dependence of viscosity for non-Newtonian fluids, after Hoffman (5), Krieger (6), and Choi (7). "Equilibrium flow" means that any time-dependent or relaxation effects have been experimentally removed. This figure consists of:

1. a low shear rate Newtonian regime, Region I
2. a shear-thinning regime, Region II ("power law" regime)
3. a high-shear Newtonian regime, Region III
4. a possible shear-thickening regime, Region IV.

The chief limitation of the "power law" non-Newtonian shear viscosity models (Ostwald-DeWaele, Casson, Herschel-Bulkley) is that they are valid only in Region II of Figure 1. They cannot account for the upper or lower Newtonian regions and, in fact, predict infinite viscosity at zero shear rate and zero viscosity at infinite shear rate, both unrealistic limiting behaviors. Important coatings processes (various creeping flows and high-shear application processes) occur in regions I and III, outside the "power law" regime. More elaborate models are required in order to describe non-Newtonian behavior beyond the power law region.

Modern rheometers have made the low-shear Newtonian plateau viscosity η_0 more experimentally accessible, and models incorporating Region I are quite useful. Perhaps the two most successful of several proposed models encompassing Regions I, II, and III inclusively are the *Cross* (8) and *Carreau* (9) models. Hieber *et al.* (10) have written a general form of which the Cross and Carreau models are special cases (modified here to include Region III):

$$\eta = \eta_\infty + \frac{\eta_0 - \eta_\infty}{\left(1 + [\beta\dot{\gamma}]^a\right)^{(1-n)/a}} \tag{1}$$

where η_0 is the first (lower) Newtonian plateau viscosity and η_∞ the second (upper) Newtonian viscosity. The parameter n has the identical meaning as in the simple power law model, *i.e.*, it is the slope of the power law region in a log-log plot of shear stress *vs.* shear rate. The constant a determines the curvature of the transition region between the first Newtonian regime and the power law regime. The value of a can be a measure of the breadth of the molecular weight distribution of a polymer (10) or the particle-size distribution of a colloidal dispersion. Setting $a=1-n$ in the above expression yields the Cross equation, while $a=2$ gives the Carreau-B model. The constant β, with dimension of time, is a time constant characteristic of the mechanisms available to the system to accommodate to a shear field. Thus, β may be related to the diffusional or rotational relaxation time of the flow units (*e.g.*, colloidal particles or aggregates) or to the time for rupture of particle flocs or aggregates under shear. The characteristic shear rate of the transition from the initial Newtonian plateau (Region I) to the shear-thinning regime (Region II) in Figure 1 is determined by the value of β (11):

$$\dot{\gamma}_{tr} = \frac{1}{\beta} \tag{2}$$

For stable colloidal systems, β is the time constant for Brownian diffusion. The Stokes-Einstein equation leads to the value of β for a spherical-particle dispersion:

$$\beta = \frac{6\pi\eta R^3}{k_B T} \tag{3}$$

where η is the viscosity of the continuous phase, R is the particle radius, k_B the Boltzmann constant, and T absolute temperature. [e] Here, β is the time to diffuse a distance equal to a particle radius (12).

Mechanisms of shear thinning

As implied in the foregoing, there is a competition between thermal disorder and hydrodynamic order for colloidal systems under shear. At low shear rates, Brownian diffusion wins out and the viscosity does not change for small increases of shear rate (point **a** to point **b** in Figure 1). Since the structure is no less random anywhere in Region I than at zero shear rate, the viscosity equals η_0, the zero-shear value. For colloidal systems, shear thinning involves a transition from Brownian to hydrodynamic dominance of the microstructure. (This event corresponds to point **c** in Figure 1.) The dimensional shear rate of transition is $1/\beta$ (Equation (2)). Equations (2) and (3) suggest ways of controlling the onset of shear thinning for a dispersion. Any change that increases the value of β (*e.g.,* increasing the effective particle size, the continuous-phase viscosity, or lowering the temperature) will reduce the onset of shear thinning to lower shear rates. Decreasing β extends Newtonian behavior to higher shear rates. Although Equation (3) is strictly valid only for very dilute dispersions, it still provides qualitative guidelines for manipulating the rheology of dispersions. For concentrated dispersions, η can be taken as the viscosity of the dispersion (13,14) (with the qualification that it is a function of shear stress).

A broad particle size distribution will possess a range of β values, a corresponding range of $\dot{\gamma}_{tr}$ values, and consequently the transition to shear thinning will be spread out over a wide range of shear rates. Thus, the effect of polydispersity in the flow unit size distribution will be to make more gradual the transition from Region I to Region II. The mean shear rate of transition corresponds to a mean diffusion relaxation time, while the value of a in Equation (1) will vary according to the *distribution* of relaxation times. Thus, a colloidal dispersion having a wide particle size distribution will exhibit a broad, gradual transition from region I to region II. A similar effect will be seen in a polymer melt of broad molecular weight distribution.

The principal mechanisms for shear thinning in dispersions are:

1. For unstable (flocculating) systems, shear thinning results from the breakup of floc structures, releasing interstitially-trapped liquid phase, reducing the effective volume fraction, hence lowering the viscosity. Once flocs have broken down, shear thinning can proceed further by mechanism (2).

2. For stable (non-flocculating) colloidal systems, shear thinning results from the progressive domination of microstructure-randomizing Brownian motion by

---

[e] It should be said that Equation (3) is quantitatively correct only for very dilute dispersions. However, this expression is qualitatively useful for rationalizing the behavior of more concentrated dispersions that display Brownian motion.

hydrodynamic shear forces. This induces ordering of particles in "strings" (possibly ordered particle layers in concentrated dispersions), which is a more efficiently-flowing microstructure. Non-colloidal (non-Brownian) systems also shear thin by a hydrodynamic ordering mechanism.

The viscosity is a measure of the "energy cost" to flow. Shear thinning, therefore, implies a microstructural change that allows more efficient flow, consequently with less energy dissipation. The mechanism of this involves a shear-induced *increase* in order, or anisotropy, within the system. Thermal (Brownian) motion tends to keep systems disordered (of random order). Shear forces work against this, tending to impose order. For particulate dispersions, this means that individual particles tend to align with the shear field, in "string-of-pearls" fashion. (See Figure 2.) This alignment tends to reduce the degree of frictional interaction between individual particles and the surrounding fluid, making flow more "energy efficient", hence reducing the viscosity. For the case of polymers in solution, polymer random coils tend to stretch in the direction of shear and become ellipsoidal in shape, with the major axes aligned at a 45° angle to the flow direction. The result is a steadily-decreasing viscosity with increasing shear rate as the degree of order increases. This describes Region II in Figure 1. Ultimately, at high shear rates, the maximum amount of shear ordering possible is attained and the viscosity again becomes independent of shear rate (Region III—upper Newtonian plateau, η_∞, Figure 1). Particle dynamics simulations confirm this description in a general way, although some of the details are still in question (15). The general flow curve involves an initial constant-disorder regime (first Newtonian), followed by a disorder to order transition (to shear thinning regime), succeeded by a metastable constant-order regime (second Newtonian), and finally an order-disorder transition (shear thickening, Region IV). The onset of shear thickening occurs at progressively lower shear rates as dispersion concentration increases (16), with the consequence that one or more of the other regimes may be obliterated.

Shear-Thickening Fluids

It was described above how shear thinning results from a shear-induced increase in order of a system, which allows the fluid elements to flow with reduced expenditure of energy. Conversely, shear thickening can result from a decrease in order of a system. A disordered system flows less efficiently, hence is more viscous. An example of this is the catastrophic increase in viscosity observed by Hoffman in concentrated PVC dispersions (5,17), postulated to result from the "buckling" or collapse of ordered, layered arrays of particles.

A common type of shear thickening behavior is *dilatancy*. The term is sometimes applied generically to shear thickening, but dilatant behavior is, strictly, time-independent shear thickening accompanied by an increase in volume of the fluid (implying a decrease in packing efficiency or order). This specific behavior has been referred to as *volumetric dilatancy* and usually occurs in relatively concentrated disperse systems. In such a dilatant system, the disperse-phase particles are minimally wetted by the liquid continuous phase. At rest, the particles of the disperse phase are in a random close-packed structure, for which the interstitial volume is relatively minimal. If

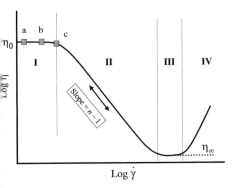

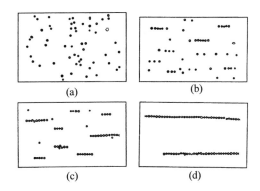

Figure 1. Generalized Non-Newtonian Equilibrium Flow curve. (Adapted from Ref. 7)

Figure 2. Glass microspheres in polymer solution between glass plates: (a) unsheared, particles randomly distributed; (b)-(d) increasingly vigorous side-to-side shear. (J. Michel, et al., Rheol. Acta 1977, 16, 317, by permission.)

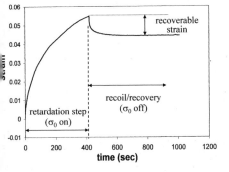

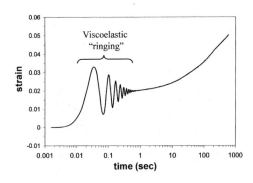

Figure 3. Creep experiment showing retarded compliance followed by recoil after removal of stress σ_0.

Figure 4. Showing "viscoelastic ringing", the damped strain oscillation after sudden imposition of stress σ_0. (also after stress removal)

the dispersion in this state is deformed slowly, adequate time exists for the meage
liquid phase to flow sufficiently to maintain the dispersed phase in a wetted state, an
the system is able to nearly maintain its close-packed structure. Faster deformatio
causes a liquid-starved condition because the interstitial volume increases when th
system is deformed or made to flow. There is no longer enough liquid to lubricate th
system. The particles are, therefore, incompletely wetted, may come into direct contac
and further deformation would ultimately create microscopic voids leading to fracture c
the material. The surface of a dilatant material may appear dry when stressed, due t
the withdrawal of surface liquid into the increased interstitial volume. This is seen whe
walking on wet sand on the beach. The resistance to deformation of the material ca
increase sharply due to these effects.

The mechanism for shear thickening described above clearly applies to tru
dilatant behavior such as can be observed in wet sand, as mentioned, and in cornstarc
dispersions, for example. For such systems, dilatancy appears when the system at res
or slowly flowing is disturbed. Shear thickening occurring in flowing systems wit
increasing shear rate has been termed *rheological dilatancy*, and the mechanism her
is probably somewhat different. Hoffman's postulated collapse of particle-layere
structures has been challenged by more recent neutron-scattering experiments as we
as by computer particle dynamics simulations. These studies indicate that shea
thickening is the result of particle cluster formation, occurring when hydrodynami
forces overcome stabilizing forces, allowing the formation of shear-induced aggregate
(15).

Plastic (Yield) Behavior

A *plastic* material has the properties of an elastic solid until a critical stress is applied
whereupon it suddenly "yields" and becomes fluid. This critical stress is the *yield stress*
the minimum stress necessary to initiate flow. Ideal plastic behavior includes th
following characteristics:
1. Hookean behavior below the yield stress, *i.e.* an ideal elastic solid; th
   deformation is linear with stress
2. The yield value is a material constant.

Metals typically exhibit ideal plastic behavior. Ironically, what we commonly cal
"plastics" (*i.e.* polymeric materials) do not. For most solid polymeric materials
deformation below the yield point is a combination of elastic strain and viscous flow
hence the measured yield stress is not a material constant but a material *function* an
will depend on the deformation history of the material. In a similar way, structured fluid
such as paints can exhibit "yield-like" behavior, apparently changing from solid-like t
readily-flowing fluid when a critical stress is exceeded. However, the measure
"apparent yield stress" is not single-valued but will vary depending on experimenta
conditions. This is because the interparticle forces ("secondary bonds") presume
responsible are of a range of types with a corresponding range of relaxation times (18)
Therefore, as stress is applied the material undergoes creep accompanied by viscou
relaxation, such that some of the "secondary bonds" rupture before overall "yield
occurs. As a consequence, the measured yield value will depend on the rate at which

the stress is increased up to the point where flow occurs: the faster the rate of stress increase, the higher the measured yield value, and *vice versa.*

Barnes and Walters (19) challenged the yield concept with the assertion that most materials with an apparent yield stress will be found, in reality, to have a high but finite viscosity if measured at sufficiently low stresses. Barnes recently published a comprehensive examination of the question (20). The idea is that few, if any, structured fluids possess true yield behavior. While, in principle, this view is likely correct (remember, even mountains flow), it is also true that, in practical terms, "yield" behavior can have important consequences for the processing, stability, and end use of materials. There seems to be, from practical experience if not from clear experimental evidence, some sort of rather sudden viscosification (flow discontinuity?) that can occur at low stresses, the magnitude of which is apparently related to the number and strength of interparticle attractive forces (18,21). The yield stress is therefore an "engineering reality" (22) that must be taken into account when formulating paints or dispersions. A yield stress may be desirable or undesirable depending on the process in question. Materials with an apparent yield stress will often exhibit thixotropy and viscoelasticity as well.

Yield Stress Measurement

A still rather novel technique for yield stress measurement employs a special vaned rotor (23,24). The vaned rotor consists of rectangular blades or vanes fixed to a rotating shaft. Advantages of this geometry are: 1) there is little disturbance of the sample when this type of probe is inserted, and 2) when the rotor turns, the material moves as a solid "cylinder". Thus, the yield surface is within the material itself, avoiding problems of wall slip. Barnes and Carnali (25) showed that the vaned geometry can be used in place of the bob-and-cup geometry for flow curve measurement of shear-thinning dense suspensions, as a remedy for problems of slip.

With the vaned rotor on a "controlled-strain-rate" type rheometer, the yield stress is measured by monitoring the torque as the geometry is rotated at a constant low rate. The torque will go through a maximum, corresponding to concerted yielding along the "virtual" cylindrical surface defined by the vane radii. The yield stress is related to the maximum torque T_m according to (24):

$$\sigma_0 = \frac{T_m}{K} \tag{4}$$

where K is the vane geometric constant given by:

$$K = \frac{\pi D^3}{2}\left(\frac{H}{D} + \frac{1}{3}\right) \tag{5}$$

D is the rotor diameter and H the vane height.

Using a controlled-stress rheometer, measurement of the yield stress using the vaned rotor is straightforward. The torque is ramped slowly and the stress at which a finite displacement is detected is reported as the yield stress.

Viscoelasticity and Industrial Processes

Coating materials and other structured fluids can display a mixture of elastic an
viscous character, in varying degrees. Hence, the term *viscoelastic* is applied to suc
materials. Viscous flow superimposed on elastic strain results in the "relaxation", c
gradual disappearance, of stress within the strained object. This is manifested as a
imperfect or "fading" stress memory, which is characterized by a *stress relaxation tim*
constant λ. In a viscoelastic fluid, elementary flow units interact together in som
fashion such that this structural relaxation time is measurably long. This may be due t
a flocculated or percolating particulate structure, polymer-particle interactions, or t
polymeric chain entanglement, for example.

The role of viscoelasticity in an industrial process depends on the rate (o
frequency) of deformation. Put another way, the response of a viscoelastic materia
depends on how long a stress is applied ($1/\omega$), relative to the time required for an
elastic "extra" stress to decay. This suggests taking a ratio of the stress relaxation tim
constant λ to the time (duration) of the process stress, t. This ratio is a define
rheological term known as the *Deborah Number, De*:

$$De = \frac{\lambda}{t}$$
(6

It is named for the Biblical prophetess Deborah, who prophesied that the "mountain
flow before the Lord" (2). This is an accurate statement, long before being verified b
the science of geology, of the fact that, on God's time scale, rock formations can be
observed to undergo permanent deformation, or flow. In other words, if the duration o
the applied stress t greatly exceeds the relaxation time λ ($De \ll 1$) the material wil
respond as a viscous fluid (because elastic stress has time to decay). Conversely, i
$De \gg 1$ a material behaves as if an elastic solid. When $De \approx 1$ the material will behave
viscoelastically (*i.e.*, stress relaxation will take place on the time scale of the process)
Thus, the Deborah number quantifies the proportion of elastic to viscous control of a
process. This is one reason why a determination of the viscoelastic properties of paints
and coatings is important. An equivalent way of writing the Deborah number is

$$De = \lambda\dot{\gamma}$$

where $\dot{\gamma}$ is the shear rate of a coating process. As an example of how the role o
elastic "extra stress" can be determined for a given process, let's assume a stress
relaxation time constant of about 10 s for a paint. Thus, for a pigment particle settling a
$\dot{\gamma} \approx 10^{-5} s^{-1}$, $De \ll 1$ and the paint would offer no elastic resistance to the settling
pigment (the particle will inevitably settle out). For leveling, on the other hand
($\dot{\gamma} \approx 10 s^{-1}$), $De \approx 1$, and elastic stress is likely to retard the leveling process.

Additional consequences of viscoelasticity are due to the stabilization o
otherwise unstable liquid structures by the elastic "extra stress". Thus, liquid fibers and
"webs" which would ordinarily collapse due to surface tension are stabilized, leading to
for example, rollcoat spatter (26) (or "misting") and ribbing, and also inhibiting
atomization of sprayed materials. Once again, the Deborah number gauges the

mportance of viscoelasticity for a particular process. Of course, both the magnitude
and lifetime of the elastic stress will be important, for together they will govern the
degree of stabilization.

Viscoelastic dispersions (27) and associative polymer solutions (28) can show
Maxwellian behavior with a single relaxation time. In general, however, real materials
may not exhibit simple exponential stress decay (*i.e.*, a single relaxation time), but
rather may possess a spectrum of relaxation times. However, the mechanical response
can be dominated by the longest relaxation time.

Many processes coatings undergo involve large strains, outside the range of
linear viscoelastic behavior. There is no general theoretical basis for treatment of nonlinear viscoelasticity at present. However, there are promising experimental methods,
based on an enhanced *creep* analysis, for characterizing nonlinear viscoelastic
behavior, even at high frequencies (29). It may be possible to build experimental
correlations between nonlinear viscoelasticity and coatings performance.

Creep methods

The viscoelastic character of a material can derive from various types of molecular
and/or particle interactions, representing a range of force constants, length scales, and
dissipative characteristics. Correspondingly, viscoelastic deformation can occur over a
range of strains and stresses. Another consequence is that viscoelastic materials will
display a range of relaxation times, referred to as a *relaxation spectrum*. An oscillatory
shear or *dynamic* method for determining linear viscoelastic behavior has a practical
lower limit of frequency ω, due to the time required to complete an oscillation at very low
frequencies. Recalling that the Deborah number $De = \lambda/t = \omega\lambda \approx 1$ for viscoelastic
response, it is clear that in order to measure very long relaxation times λ (characteristic
of polymers) very low frequencies are necessitated — probably below the practical limit
(we do not want to wait hours to complete enough oscillations to analyze). An answer
to this is the *creep method*, in which a constant stress σ_0 is applied to a material, which
then undergoes strain (Figure 3). Creep conveniently probes long relaxation time
scales, but also short- and intermediate-time behavior, as well.

Viscoelastic materials are not only sensitive to the rate of strain but also to the
total strain. Thus, there is a critical strain beyond which the material response is
nonlinear (non-Hookean), where the "springs" are "breaking" (*i.e.* the structures or
interactions responsible for the elastic character are breaking down). Such a system
displays nonlinear viscoelasticity, for which a general theoretical model is lacking. The
equations describing linear viscoelastic behavior are invalid in the nonlinear regime,
thus oscillatory methods are normally restricted to the linear regime. However, few real-
world processes for either coatings or other commercial fluids actually take place within
the linear regime. Many processes are high-strain or even infinite-strain, and therefore
the nonlinear viscoelastic properties will govern the material behavior. Creep offers a
way of characterizing materials in the nonlinear, as well as the linear, regime.

In a creep experiment a torque is suddenly applied to the specimen (or suddenly
removed, in the recovery step). The sudden acceleration, together with the
measurement system's inertia, causes a "strain overshoot", and for viscoelastic
materials this can result in "viscoelastic ringing", where the material undergoes a

damped oscillation as does a bowl of jello that is "bumped" (Figure 4). The oscillation can be rather short-lived, and may not be apparent unless the time axis is logarithmic Normand and Ravey (30) showed that this coupling between the instrumental inerti. and material elasticity can be analyzed according to standard viscoelastic models They also found that this modified creep analysis can both replace standard force. oscillation methods and extend the accessible frequency beyond the typical range c such methods. Baravian and Quemada (29) have shown in addition that the metho. can be used to characterize nonlinear viscoelasticity and to separate the effects c thixotropy from those of viscoelastic relaxation.

Controlled Stress and Controlled Rate

One has a choice of working principles in modern rotational rheometers: eithe "controlled strain" (or controlled strain rate) or "controlled stress". The difference is i whether the torque or the angular displacement is the controlled variable. In controlled-strain instrument the angular displacement (or perhaps angular velocity) i the independent (controlled) variable and the viscous drag-torque the dependen (measured) variable. "Controlled-stress" instruments in actuality control the torque, an measure the resulting angular displacement. Instruments of the controlled-strain type include, for example, the TA Instruments Weissenberg Rheogoniometer, Haake CV Rheometrics ARES, and Bohlin CVO. Instruments of the controlled-stress type include the TA Instruments AR-1000, Rheometrics SR5, Bohlin CSR, Haake RS, and Physic. MC rheometers.

Each of the two instrument types has characteristic advantages and limitations The choice depends on the material under test and the intended experiments. In term: of performance, controlled-stress (CS) instruments can measure much lower angula velocities than can controlled-strain-rate (CR) instruments, but CS instruments tend t be more limited at the high angular velocity and oscillation frequency end. CS is bette suited to measure long relaxation times, an advantage of its typically stable torque capability and high angular resolution (31).

CR instruments impose a shear-rate sweep while measuring the drag-torque response of materials. Structured fluids tend to be "shear-sensitive" (more precisely strain-sensitive). Consequently, as strain increases exponentially under a linear strain rate sweep protocol, fluid structure tends to collapse rapidly, with the result tha relatively few data points are obtained to provide information on structure. In contrast CS devices use linear (and logarithmic) rates of stress increase, a test mode which allows materials to "obey their own rules" of stress-strain response. CS instruments are especially useful for characterizing structured fluids and granular dispersions such a: paint, printing ink, adhesives, ceramic slips, coal slurries, cement, pigment and coloran dispersions, drilling muds, medicines, foodstuffs, personal care products and cosmetics solder pastes, etc. This is particularly true where materials exhibit apparent yiel behavior. CS instruments can, in principle, directly measure the stress at the onset o yield, avoiding errors associated with extrapolation or curve-fitting methods.

Taking coating flows as a case in point, it is important to realize that the prope variable for correlating coating rheology to real-world coating processes is the shea stress, not the shear rate. First of all, coating flows are the outcome of the sum o orces acting on a fluid coating layer. That is, the rheological response to those forces determines the resulting coating flow. Therefore, coating flows are not "driven" at a characteristic shear rate, but rather the observed shear rate is the resultant of the stress driving the process and the corresponding viscosity <u>at that stress</u>. Using stress as the controlled or independent variable is, in this sense, the more "natural" way to characterize coatings.

Rheology of Coating Application and Film Formation

In describing the way in which the coating rheology controls application and leveling, as well as undesirable flows leading to film defects (*e.g.*, sagging, cratering, crawling, edge withdrawal), there is an important point to be made. These flows are driven by specific shear stresses, which can be calculated from the forces acting (*e.g.*, gravity and surface tension) and the geometry of the film (<u>3,4</u>). However, these processes can occur over a wide range of shear rates, depending on the coating's viscosity at the acting shear stress. The shear <u>stress</u> acting on a coating layer (for a given process) is <u>independent</u> of the rheology. In contrast, the shear <u>rate</u> will be <u>dependent</u> on paint rheology. For this reason it is preferable to represent flow data as viscosity *vs.* <u>shear stress</u> plots, as opposed to viscosity vs. <u>shear rate</u>, which is the more common practice. The shear rate is a <u>dependent</u> variable, for real processes. The appropriate <u>independent</u> variable to use to differentiate the performance of paints according to their rheology is the shear stress. Not to do so is wrong in principle and will result in incorrect comparisons of paints with respect to their relative rates of, *e.g.*, sagging, leveling, pigment settling, or ease of application. This issue is key to using flow curve data correctly to understand coating performance.

Plotting with shear stress as the independent variable, as in Figure 5, allows straightforward and correct comparison of paints A and B <u>at the specific shear stress for a particular process</u>. The process stresses illustrated in the Figure are for (i) gravity-driven sagging of a 3-mil wet film, (ii) surface tension-driven leveling of surface roughness, and (iii) application by brush or roller. The point is that, no matter what the rheology of a paint, the shear stress acting on a coating layer for a given process is the same (for a given geometry, density, surface tension). The shear rates for these processes, however, are <u>not</u> the same, as illustrated in Figure 6. Figure 6 shows the same two paints as in Figure 5, this time with the viscosity plotted as a function of shear rate. Note that process shear rates are shifted to the <u>left</u> for the higher viscosity paint.

Sagging is driven by gravitational shear stress σ_g whose magnitude depends solely on the wet film thickness and density:

$$\sigma_g = \rho g h \cos\theta \qquad (8)$$

$\cos\theta = 1$ for a vertical substrate). Predictions of sagging from comparison of viscosities measured at an arbitrary shear rate will be misleading because paints sag at different shear rates, depending on their rheology. The proper way to predict relative sagging tendency is to first select the governing viscosity from a plot of viscosity *vs.* shear stress, rather than shear rate. The sagging shear stress calculated from Equation (8) determines the viscosity controlling sagging (η_{sag}) from the flow curve. Figure 5 compares two paints in this manner, with gravitational shear stress levels corresponding

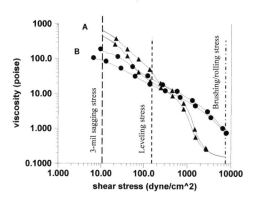

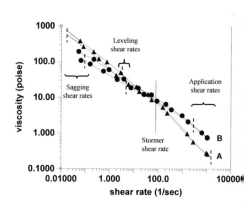

Figure 5. Viscosity vs. shear stress plot for two paints A (▲) and B (•). Vertical dashed lines indicate (i) gravitational shear stress driving sagging for 3-wet-mil paint layer; (ii) surface-tension-driven leveling stress; (iii) brushing/rolling application.

Figure 6. Viscosity vs. shear rate plot for paints A (▲) and B (•). Shear rates for saggi leveling, application are different for A and B, shifted to left for higher-viscosi Paint. Vertical dashed lines indicate sh rates for each paint and process. Vert solid line is the approximate shear rate the Stormer™ Viscometer.

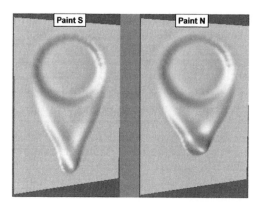

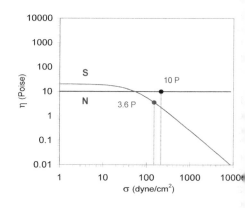

Figure 7. Sagging and dripmark formation for two paints, S (shear thinning) and N (Newtonian). Substrate simulates a "nailhead depression".

Figure 8. Viscosity-shear stress curves for paints S and N. Vertical lines indicate the geo etry-dependent gravitational shear stre driving the flow and corresponding viscosities governing the flow.

o 3 mils wet film thickness indicated. Figure 6 shows flow curves for the same two paints plotted as a function of shear rate, with the calculated sagging shear rates at 3 mils indicated ($\dot{\gamma}_{sag} = \sigma_{sag} / \eta_{sag}$).

The solid vertical line in Figure 6 approximates the shear rate of the Stormer™ viscometer. The viscosities of the two paints at the Stormer™ shear rate are about the same (Paint B actually slightly higher). At an arbitrary shear rate of 1 s^{-1}, paint A is about 50% higher, whereas at the <u>actual</u> sagging shear stress or shear rates paint A is 300% higher in viscosity. Clearly, comparison of viscosities measured at arbitrary shear rates can lead to incorrect predictions of relative sagging behavior, particularly when flow curves cross over. Comparison of coating rheology as a function of shear stress both simplifies the process and is more correct in principle.

Computer Modeling of Coating Flows

Computational Fluid Dynamics (CFD) represents perhaps the ideal fusion of process and rheology. The fundamental properties so painstakingly measured in the laboratory equal the required input for the numerical engine. Coating flows are three-dimensional and intrinsically nonlinear, as is the rheology of most systems, making it exceedingly difficult to predict outcomes using simple analytical equations. A nonlinear computer model containing the correct physics makes possible a realistic description of coating processes at unprecedented levels of detail, and is a potentially powerful learning and optimization tool for the coating or process engineer. (See, for example, references 32,33,34,35.) It is expected that increasing use will be made of flow modeling in the new century, as commercial codes improve and powerful computing and graphics platforms are readily available at reasonable cost. Several shear-thinning models are available with present codes, but viscoelastic models are generally lacking.

Figure 7 depicts a 3-D simulation of gravitational drainage flow and dripmark formation for an initially uniform liquid layer covering a shallow circular depression in an otherwise flat, vertical surface (36). The intent was to demonstrate the perturbing effect of a "nailhead depression" on the flow of a paint layer. Flow for a shear-thinning paint "S" and a Newtonian paint "N" is depicted. Figure 8 displays The viscosity-shear stress flow curves for "S" and "N", showing that, at low shear, paint "S" is about double (20 Poise) the viscosity of the Newtonian paint, having a constant viscosity of 10 Poise. At a characteristic stress, the viscosity of paint "S" begins to fall, until at the specific gravitational stresses driving the drainage (vertical dashed lines) paint "S" is actually lower in viscosity than is paint "N", and the longer dripmark for "S" than for "N" is the result. Had we compared the viscosities of "S" and "N" at an arbitrary shear rate of, say, 1 s^{-1}, we would have predicted the <u>reverse</u> result, because at 1 s^{-1} paint "S" is about twice the viscosity of "N", which can be computed from the flow curves shown. Furthermore, while a simple sagging velocity may be calculated without benefit of CFD, only the 3-D numerical simulation can show the developing width and shape characteristics of the dripmarks, <i>i.e.</i> the volumetric flow and appearance of the defect.

Conclusions

New tools and methods for coatings rheology characterization and for modeling of coating flows, with superior accuracy, sophistication, and power, make for exciting prospects both for improved coatings performance and better control of coatings processes in the 21st century. These developments are particularly timely, in view of the challenges facing the coatings industry as new technological solutions are being sought in response to continued demands for further reductions in organic solvent content.

1. Greek philosopher Heraclitus of Samos, 540-475 B.C.
2. The Holy Bible, Judges 5:5, translation of Marcus Reiner.
3. Eley, R. R., "Rheology and Viscometry", Chap. 33, ASTM Paint and Coatings Testing Manual, 14th ed., ASTM, Philadelphia, 1995, 333-368.
4. Eley, R. R., "Principles and Methods of Rheology in Coatings", in "Encyclopedia of Analytical Chemistry: Instrumentation and Applications", J. Wiley & Sons, R. A. Meyers, Ed., in press.
5. Hoffman, R. L., Trans. Soc. Rheol. 1972, 16, 99.
6. Krieger, I. M., "Rheology of Polymer Colloids", in "Polymer Colloids", Buscall, R., Corner, T., and Stageman, J. F., Eds., Elsevier Applied Science, New York 1986.
7. Choi, G. N., "Rheology of Sterically-Stabilized Model Colloidal Dispersions", Ph.D. Thesis, Case Western Reserve University, Cleveland, OH, 1982.
8. Cross, M. M., "Rheology of non-Newtonian Fluids: A New Flow Equation for Pseudoplastic Systems", J. Coll. Sci. 1965, 20, 417-437.
9. Carreau, P. J., "Rheological Equations from Molecular Network Theories", Trans. Soc. Rheol. 1972, 16, 99-127.
10. Hieber, C. A., and Chiang, H. H., "Shear-Rate-Dependence Modeling of Polymer Melt Viscosity", Polym. Eng. Sci. 1992, 32(14), 931.
11. Jenekhe, S. A., Polym. Eng. Sci. 1983, 23(15), 830.
12. Goodwin, J. W. and Ottewill, R. H., "Properties of Concentrated Colloidal Dispersions", J. Chem. Soc. Farad. Trans. 1991, 87(3), 357-369.
13. Krieger, I. M. and Dougherty, T. J., "A Mechanism for Non-Newtonian Flow in Suspensions of Rigid Spheres", Trans. Soc. Rheol. 1972, 3, 137.
14. Toussaint, A., "Choice of Rheological Model for Steady Flow", Prog. Org. Coat. 1992, 21, 255.
15. Silbert, L. E., Melrose, J. R., and Ball, R. C., "The Rheology and Microstructure of Concentrated, Aggregated Colloids", J. Rheol. 1999, 43(3), 673-700.
16. Cheng, D. C.-H., "Viscosity-Concentration Equations and Flow Curves for Suspensions", Chem. Ind., 17 May 1980, p. 403.
17. Hoffman, R. L., "Explanations for the Cause of Shear Thickening in Concentrated Colloidal Suspensions", J. Rheol. 1998, 42(1), 111-123.
18. Lang, E. R. and Rha, C., "Rheology", Vol. 2 "Fluids", Astarita, G., Marucci, G., and Nicolais, L., Eds., Plenum Press, New York 1980, p. 659.

19. Barnes, H. A., and Walters, K. "The Yield Stress Myth?", Rheol. Acta 1985, 24, 323.
20. Barnes, H. A., "The Yield Stress—a Review or 'παντα ρει'—Everything Flows?", J. Non-Newtonian Fluid Mech. 1999, 81, 133-178.
21. Zhou, Z., Solomon, M. J., Scales, P. J., and Boger, D. V., "The Yield Stress of Concentrated Flocculated Suspensions of Size-Distributed Particles", J. Rheol. 1999, 43(3), 651.
22. Astarita, G., J. Rheol. 1990, 34 (2), 275.
23. Nguyen, Q. D. and Boger, D. V., Rheol. Acta 1985, 24, 427.
24. Nguyen, Q. D., and Boger, D. V., "Yield Stress Measurement for Concentrated Suspensions", J. Rheol. 1983, 27(4), 321.
25. Barnes, H. A., and Carnali, J. O., J. Rheol. 1990, 34(6), 841.
26. Glass, J. E., J. Coat. Technol. 1978, 50(641), 56.
27. Strivens, T. A., "The Viscoelastic Properties of Concentrated Suspensions", Coll. Polym. Sci. 1983, 261, 74.
28. Annable, T., Buscall, R., Ettelaie, R., and Whittlestone, D., J. Rheol. 1993, 37(4), 695.
29. Baravian, C. and Quemada, D., "Using Instrumental Inertia in Controlled Stress Rheometry", Rheol. Acta 1998, 37, 223-233.
30. Normand, V. and Ravey, J.-C., "Dynamic Study of Gelatin Gels by Creep Measurements", Rheol. Acta 1997, 36, 610-617.
31. Macosko, C. W., "Rheology: Principles, Measurements and Applications", VCH Publishers, Inc., New York,1994, p. 352.
32. Eres, M. H., Weidner, D. E., and Schwartz, L. W., Langmuir 15(5), 1859-1871 (1999).
33. Schwartz, L. W., Weidner, D. E., and Eley, R. R., Langmuir 11, 3690-3693 (1995).
34. Weidner, D. E., Schwartz, L. W., and Eley, R. R., J. Coll. Interf. Sci. 179, 66-75 (1996).
35. Schwartz, L. W. and Eley, R. R., J. Coll. Interf. Sci. 202, 173-188 (1998).
36. Eley, R. R., Petrash, S., Schwartz, L. W., and Roy, L. V., unpublished results.

WATER-BORN COATINGS

PETER T. ELLIOTT AND J. EDWARD GLASS

Polymers and Coatings Department, North Dakota State University, Fargo North Dakota 58105

Introduction

The driving force for the development of water-borne coatings has been environmental (i.e., the lowering of volatile organic components in the traditional alkyd paints and the apparent limitations in achieving these goals with high solids coatings). This chapter begins by reviewing the synthetic aspects related to the development of water-borne coatings. Water based dispersion stability is then discussed in terms of the influence of surface stabilizers. The prior art on particle coalescence and film formation of latex particles is then reviewed and in the latter part of the chapter, the particle coalescence of step-growth oligomer dispersions is discussed.

Synthetic Aspects

Synthetic polymer coatings originally were applied from organic solvents. Solution viscosities, at a given concentration, are proportional to the polymer's molecular weight. Film properties, such as tensile strength and toughness also are proportional to the polymer's molecular weight. Therefore, to obtain the low viscosities needed for the application of a coating, the concentration of the polymeric resin was necessarily low, and the organic solvent emitted to the atmosphere (VOC, Volatile Organic Component) was high. As society changed and environmental concerns were emphasized, coating technology has advanced in several directions. Two areas considered VOC free are

UV cure (1) and powder (2) coatings. UV curing of solvent-free formulations is restricted, with only a few exceptions, to flat surfaces, with film thickness and pigmentation limitations. Powder coatings are subject to Faraday effects and are limited in the substrates they can be applied to, the materials that can be used synthetically, and appearance. Another direction has been in the transformation of solvent-borne coatings to water-borne coatings. This has proven to be a viable method for the reduction of VOCS.

Thermoplastic Water-borne Coatings (chain growth).

The effort to produce synthetic rubber during World War II led to the development of an emulsion polymerization process, and with it, water-borne coatings as they are designed in the architectural and paper coatings area. Such an emulsion polymerization recipe employs a surfactant, a monomer with low solubility in water, and a redox free radical initiator in the aqueous phase.

When a monomer with low solubility in the aqueous phase (e.g., styrene and butadiene) is employed, the emulsion polymerization occurs by a micellar (heterogeneous) process. After the initiator has reacted with the monomer in the aqueous phase, the monomer propagates to an oligomeric form ' and then enters a micelle, formed by the aggregation of excess of surfactant used in the polymerization recipe. The chain growth polymerization (chapter 3 in reference 3) of the olefinic monomer then occurs in the micelle. When the chain growth polymerization of olefinic monomers occurs in bulk or solvent, the free-radical concentration of the propagating species should be kept below 10^{-8} molar to minimize termination reactions. Polymerization in a micellar environment isolates the propagating species and pen-nits higher concentrations of propagating radicals. Thus, higher rates of polymerization (R_p) with simultaneous higher molecular weights (X_n, degree of polymerization) are realized, as described by the relationships (chapter 4 in reference 3) in equations 1 and 2.

$$R_p = 10^3 N n k_p [M]/N_A \qquad (1)$$

where: N = steady state concentration of micelles plus particles
           n = average number of radicals per micelle plus particle
           k_p = propagation rate constant
           $[M]$ = monomer concentration
           N_A = Avogadro number

$$X_n = r_p/r_i = N k_p [M]/R_I \qquad (2)$$

where: X_n = degree of polymerization
           r_p = rate of growth of a polymer chain
           r_i = rate at which primary radicals enter the polymer particle
           N = steady state concentration of micelles plus particles
           K_p = propagation rate constant

[M] = monomer concentration,
R_i = rate of initiation

and

$$N = k(R_i/\mu)^{2/5}(a_sS)^{3/5} \qquad (3)$$

where: k = constant
          μ = rate of volume increase of a polymer particle
          a_s = interfacial surface area occupied by a surfactant molecule
          S = total concentration of surfactant in the system

Dilution of these resins with t-butanol, the solvent used in synthesizing the low molecular weight W-R acrylic, results in the expected viscosity decrease with dilution; however, with water dilution of the t-butanol acrylic resin solution, abnormal viscosity

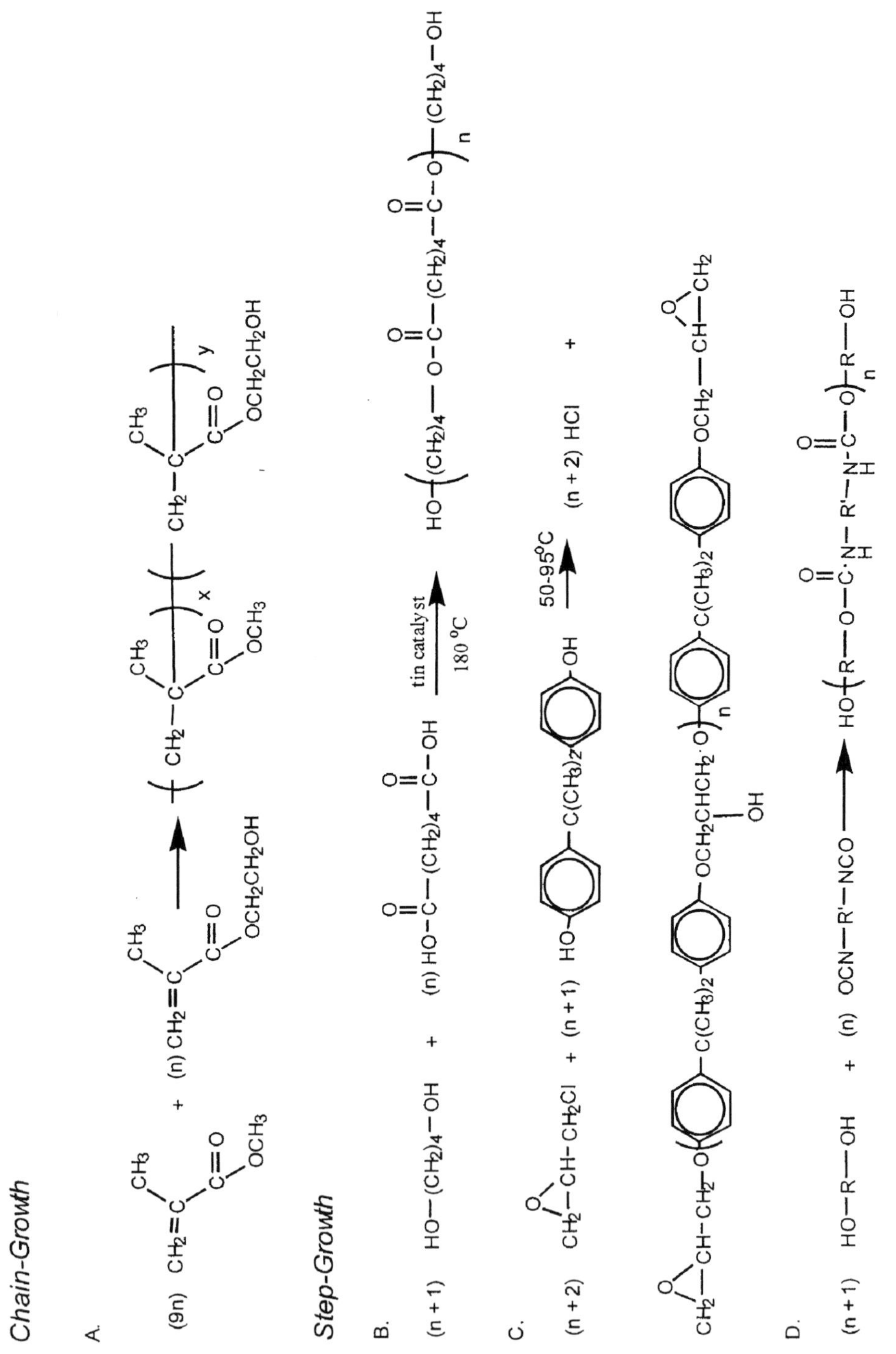

Chain-Growth

A.

$$(9n)\ CH_2=C\begin{smallmatrix}CH_3\\\\C=O\\OCH_3\end{smallmatrix} \ +\ (n)\ CH_2=C\begin{smallmatrix}CH_3\\\\C=O\\OCH_2CH_2OH\end{smallmatrix} \longrightarrow$$

$$\left[CH_2-C\begin{smallmatrix}CH_3\\\\C=O\\OCH_3\end{smallmatrix}\right]_x \left[CH_2-C\begin{smallmatrix}CH_3\\\\C=O\\OCH_2CH_2OH\end{smallmatrix}\right]_y$$

Step-Growth

B.

$$(n+1)\ HO-(CH_2)_4-OH\ +\ (n)\ HO-\overset{O}{\overset{\|}{C}}-(CH_2)_4-\overset{O}{\overset{\|}{C}}-OH \xrightarrow[180\,^{\circ}C]{tin\ catalyst}$$

$$HO\left[(CH_2)_4-O-\overset{O}{\overset{\|}{C}}-(CH_2)_4-\overset{O}{\overset{\|}{C}}-O\right]_n(CH_2)_4-OH$$

C.

$$(n+2)\ CH_2\overset{O}{-}CH-CH_2Cl\ +\ (n+1)\ HO\text{-}\bigcirc\text{-}C(CH_3)_2\text{-}\bigcirc\text{-}OH \xrightarrow{50\text{-}95\,^{\circ}C}$$

$$CH_2\overset{O}{-}CH-CH_2\text{-}O\text{-}\bigcirc\text{-}C(CH_3)_2\text{-}\bigcirc\text{-}O\left[CH_2CHCH_2\text{-}O\text{-}\bigcirc\text{-}C(CH_3)_2\text{-}\bigcirc\text{-}O\right]_n CH_2-CH\overset{O}{-}CH_2$$
$$OH$$

$$+\ (n+2)\ HCl$$

D.

$$(n+1)\ HO-R-OH\ +\ (n)\ OCN-R'-NCO \longrightarrow HO\left[R-O-\overset{O}{\overset{\|}{C}}-\overset{H}{\overset{|}{N}}-R'-\overset{H}{\overset{|}{N}}-\overset{O}{\overset{\|}{C}}-O\right]_n R-OH$$

Scheme 1. Preparation of Oligomers in the 3,000 to 5,000 Molecular Weight Range behavior is observed. Initially the viscosity decreases due to dilution; however, with continued dilution the viscosity increases. The addition of water to the t-butanol W-R acrylic solution causes a phase inversion and it is in this area that a viscosity abnormality occurs. The magnitude of the viscosity abnormality at a intermediate dilution level is dependent on the molecular weight of the W-R acrylic, on the monomer acid level in the resin (10,11), and on the extent of neutralization. The viscosity abnormality disappears as the acid content of the resin is increased. As the acid percentages in the resin is increased, fewer of the acid monomer units are isolated and trapped in the aggregated resin's interior, where with their salvation by water and reaction with the solubilized base, they help swell the particle. Interestingly, it has been observed in studies at NDSU that approximately 25% of the acid monomer is trapped within the acrylic disperse phase in both methacrylate latices and in water-reducible acrylics.

The surface acid units in W-R acrylics are not desirable crosslinking sites with multifunctional isocyanates and melamine resins. The reaction of isocyanates with carboxylic acids leads to carbon dioxide liberation and possible film defects. A reaction with melamine leads to ester linkages and possible hydrolysis during the life of the film. Reaction of the latter with the hydroxyl of the HEMA units leads to stable ether crosslinks that provide excellent chemical resistance, flexibility, and hardness. The goal, then, is to minimize the number of carboxylate groups, as far as particle stability in the aqueous phase will permit, and to maximize the number of hydroxyl groups of the HEMA or HEA units for crosslinking sites. There is a second problem to be addressed in these resins. The melamine crosslinking reactions are inhibited by the amine used to neutralize the carboxylate groups. This can lead to poor surface gloss when the monomers in the surface, where an amine of low volatility may evaporate, can polymerize without polymerization occurring throughout the film where amine is still present. The resulting immobilized surface film wrinkles forming an irregular surface when the subsurface, later free of amine by evaporation, polymerizes, and shrinks.

There also has been considerable discussion about the desirability of uniform crosslink density networks in OEM films. This is difficult to obtain in a free radical, chain-growth oligomerization. The inclusion of the hydroxy containing monomer in a given oligomeric chain is dependent on the monomer's relative reactivity ratio and concentration (chapter 5 *in reference 3*). Thus, in a low molecular weight chain, a HEMA or HEA unit may not be incorporated in many methacrylate oligomers and many chains may have more than two hydroxy containing HEMA or HEA units. In addition, there is a very low probability that hydroxyl units would be positioned at the terminal positions of the methacrylate chains, considered important to the fon-nation of a uniform network structure on curing with multifunctional crosslinkers. To this latter goal, researchers developed an anionic initiator (*12*), a ketene silyl acetal, that affected the synthesis of a monodisperse molecular weight O-MMA resin, typical of an anionic initiator, but also with the ability to place hydroxyl group at both ends of the chain. Cost is the primary parameter in coating formulations, and this innovative approach has not achieved a dominant commercial position.

B)Crosslinkable Acrylic Latices. Besides water reducible acrylic thermosetting systems, significant attention has been given to the development of crosslinkable, thermoset acrylic latices. To circumvent the problems associated with the reaction of isocyanates and melamines with carboxylic acid moieties, other technologies have been developed to react with the carboxylic acid groups on the polymer backbone of water based thermosetting acrylics including crosslinking with zinc acetate, aziridines (*13*), carbodiimides (*14*), cycloaliphatic epoxies, oxiranes (glycidyl methacrylate) (*15*), and oxazoline (*16*).   This area has received ample attention in recent years and understandably so since the presence of free carboxylic acid groups increases the water sensitivity of the final film.   One problem, though, is that functional groups designed to react with carboxylic groups generally also are reactive with water. This results in only marginally stability in one pack water based acrylic thermosetting systems.   To avoid this problem, new technology has again been developed that incorporates unsaturation into the acrylic latex particle. Although alkyds seem ideal, with their ambient cure only in the presence of oxygen, there are inherent disadvantages. Therefore significant research has gone into developing crosslinking chemistries that mimic the cure of alkyds without the disadvantages. One method has been to incorporate polymerizable methacrylate groups into the latex particle, for example, by copolymerizing with allyl methacrylate (*17,18*), reacting glycidyl methacrylate with an amine-containing (*19*) or carboxylic-containing (*20,21*) waterborne particle, or reacting acetoacetoxyethyl methacrylate with an amine-containing particle (*22*), that will cure by oxidation.   Another approach has been to convert the acetoacetoxy groups, which are susceptible to hydrolysis, to the more stable enamines that will also cure oxidatively in the presence of a fatty acid (*23,24,25*). Combinations of ammonia and a diamine have been shown to give improved heat stability (*26*). These types of systems offer the advantages of being thermosetting acrylic-based latices that cure at ambient conditions with adequate heat stability for commercialization.

Thermoset Water-borne Coatings (step-growth)

The synthesis of step-growth (S-G) thermosets provides an alternative approach that offers two advantages to the above methacrylate synthesis for water-reducible acrylics: 1. the functional groups are placed in terminal positions prior to crosslinking reactions with trifunctional additives, and 2. the partial segregation of hydrophilic and hydrophobic segments of the polymer chain result in association of the "hard" hydrophobic segments and improved physical properties in the applied film. Examples of their preparation are given in **Scheme 1** for the preparation of polyesters (**1B**), noted for their low costs and flexibility; for epoxies (**1C**), noted for their solvent resistance and adhesion; and for urethanes (**1D**), noted for their abrasion resistance and flexibility.

The reactive groups from these syntheses are generally hydroxyl units that are crosslinked with multifunctional compounds such as melamine or the isocyanate groups of an isocyanurate compound. Examples of such reactions are illustrated in **Scheme 2** between a polyester with terminal hydroxyls and a melamine resin (**2A**), between a methyl methacrylate chain containing multiple hydroxyl units and the isocyanate units

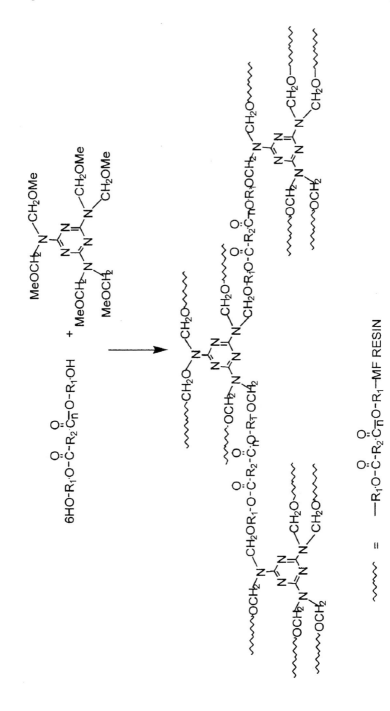

Scheme 2A. Crosslinking between a Hydroxy-terminated Polyester and a Melamine Resin.

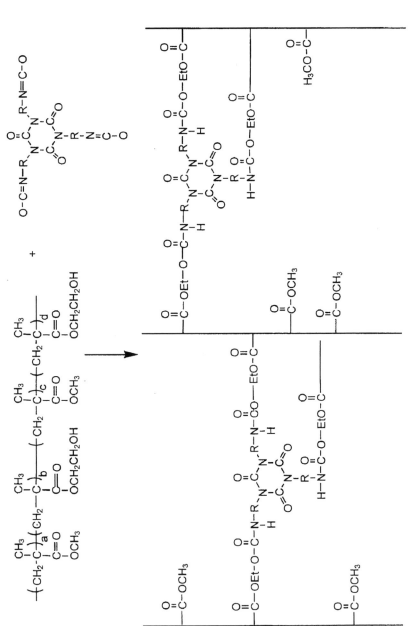

Scheme 2B. Crosslinking between a Methyl Methacrylate (with hydroxyl units) and an Isocyanurate

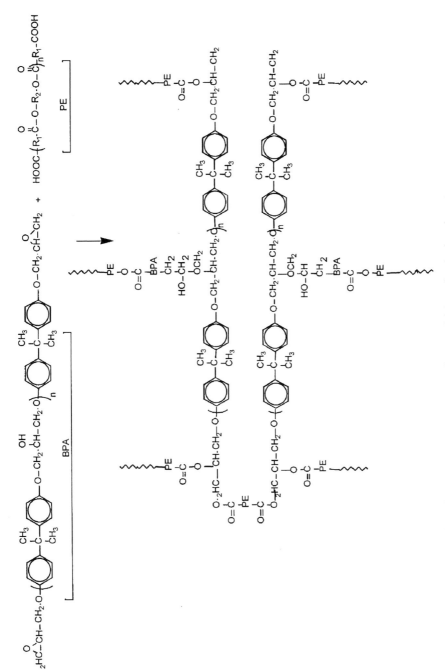

Scheme 2C. Crosslinking between an epoxy and an acid terminated polyester.

a. **Reaction with Alcohols**: most widely used in coatings

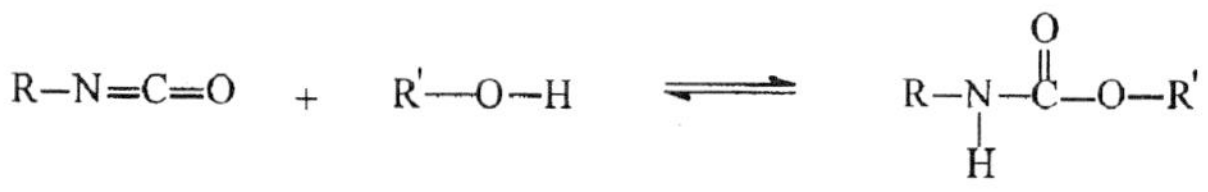

$$R-N=C=O \quad + \quad R'-O-H \quad \rightleftharpoons \quad R-N-\overset{\overset{\displaystyle O}{\|}}{C}-O-R'$$
$$\underset{\displaystyle H}{|}$$

urethane

b. **Reaction with Amines**: fast reaction limits use

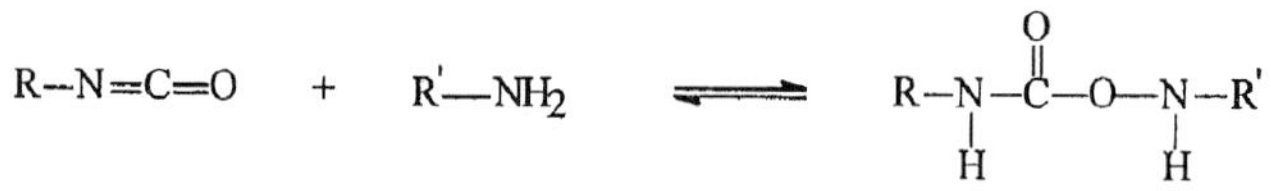

$$R-N=C=O \quad + \quad R'-NH_2 \quad \rightleftharpoons \quad R-N-\overset{\overset{\displaystyle O}{\|}}{C}-O-N-R'$$
$$\underset{\displaystyle H}{|} \qquad \underset{\displaystyle H}{|}$$

urea

c. **Reaction with Water**: water should be removed from polyols

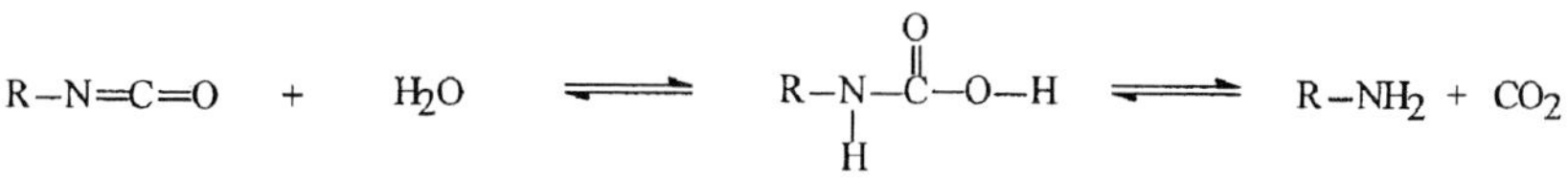

$$R-N=C=O \quad + \quad H_2O \quad \rightleftharpoons \quad R-N-\overset{\overset{\displaystyle O}{\|}}{C}-O-H \quad \rightleftharpoons \quad R-NH_2 + CO_2$$
$$\underset{\displaystyle H}{|}$$

d. **Reaction with Carboxylic Acids**: Polyols (especially polyesters) must have low acid content

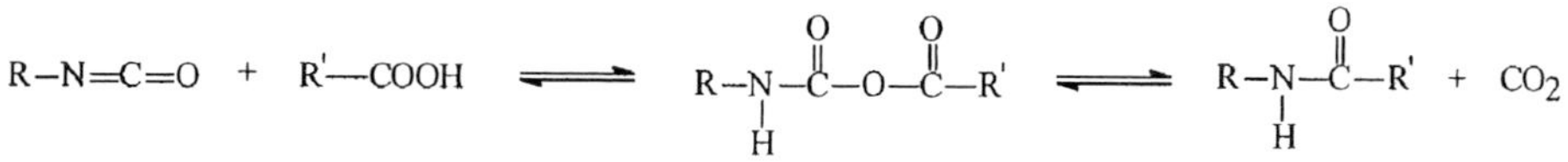

$$R-N=C=O \; + \; R'-COOH \; \rightleftharpoons \; R-N-\overset{\overset{\displaystyle O}{\|}}{C}-O-\overset{\overset{\displaystyle O}{\|}}{C}-R' \; \rightleftharpoons \; R-N-\overset{\overset{\displaystyle O}{\|}}{C}-R' \; + \; CO_2$$
$$\underset{\displaystyle H}{|} \qquad\qquad\qquad \underset{\displaystyle H}{|}$$

amide

e. **Reaction with urethanes**

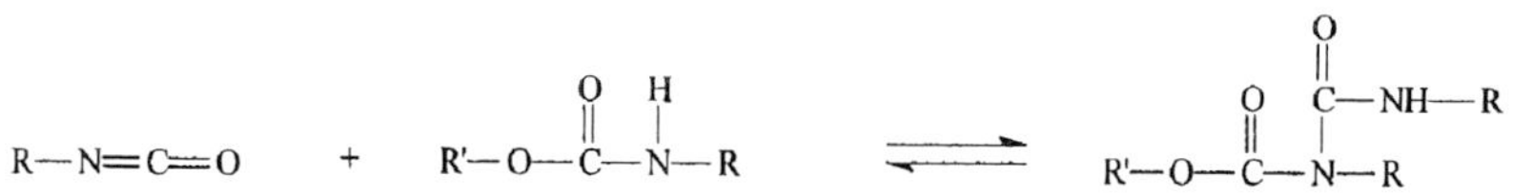

$$R-N=C=O \quad + \quad R'-O-\overset{\overset{\displaystyle O}{\|}}{C}-\overset{\overset{\displaystyle H}{|}}{N}-R \quad \rightleftharpoons \quad R'-O-\overset{\overset{\displaystyle O}{\|}}{C}-\overset{\overset{\displaystyle C-NH-R}{|}}{N}-R$$

allophanate

f. **Reaction with Urea**

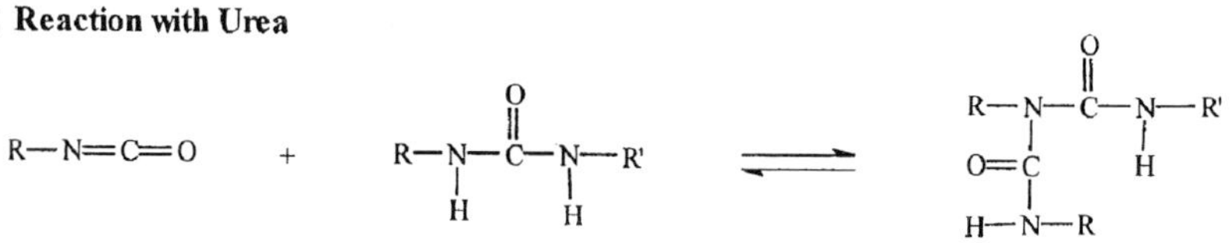

$$R-N=C=O \quad + \quad R-\overset{\overset{\displaystyle O}{\|}}{\underset{\underset{\displaystyle H}{|}}{N}}-C-\overset{\overset{}{\underset{\underset{\displaystyle H}{|}}{N}}}{-}R' \quad \rightleftharpoons \quad$$

$$R-\underset{\underset{\displaystyle O=C}{|}}{N}-\overset{\overset{\displaystyle O}{\|}}{C}-\underset{\underset{\displaystyle H}{|}}{N}-R'$$
$$\underset{\displaystyle H-N-R}{}$$

biuret

Scheme 3. Reactions of isocyanate with (a) alcohol (b) amine (c) water (d) carboxylate (e) urethane (f) urea of an isocyanurate (**2B**), and between an epoxy and an acid terminated polyester (**2C**) or with arnine fimctionalities (not illustrated). The scheme highlights only the reaction of the functional groups, in the coating these functional groups will be present at each end of the polymer chain and small molecules containing three or greater of the reactive functional groups will also be present. The most notable disadvantages in S-G syntheses are the broad molecular weight distributions and the inability to obtain high molecular weights until high conversions are reached. Lower molecular weights can be volatile and are toxic, especially isocyanates. The distribution is not a primary concern if the polymers are taken to high conversion and molecular weight. The most unique aspect of step-growth polymers is the latitude in designing hardness with flexibility into the resin.

The thermoset systems discussed more often in the open literature are waterborne urethanes. Polyurethanes are formed by the reaction of diisocyanates with diols (*27*, **Scheme 1D**), but in processing and application, many more isocyanate reactions can be involved (**Scheme 3**). There are several methodologies for producing an aqueous disperse polyurethane phase (PUDs)(*28*), which can generically represent the step-growth synthesis of OEM type resins. First, a medium molecular weight isocyanate terminated pre-polymer is formed by reaction of a diol with a stoichiometric excess of diisocyanate (*28*, **Scheme 4**). The second step is to chain extend this prepolymer, where problems in viscosity build up and the fast rate of reaction of the isocyanate units with the amine groups are encountered, and methodologies then vary with the process used. They include a solvent process using acetone, a ketimine/ ketazine process, a pre-polymer mixing process (**Scheme 4**), and a hot melt process. Although the acetone solvent process (*28*) yields reproducible PU dispersions, it is limited to those that are soluble in acetone, and consequently, the resulting urethane films are not very solvent resistant. The latter processes employ technologies that reduce or eliminate the use of organic solvents.

The Ketimine/Ketazine process does not require distilling off solvent or high shear equipment. The molecular weight builds in water so a viscosity build up is avoided and the solvent resistance of the final film can approach that cast from solvent; however, the process is sensitive, and subject to product variations.

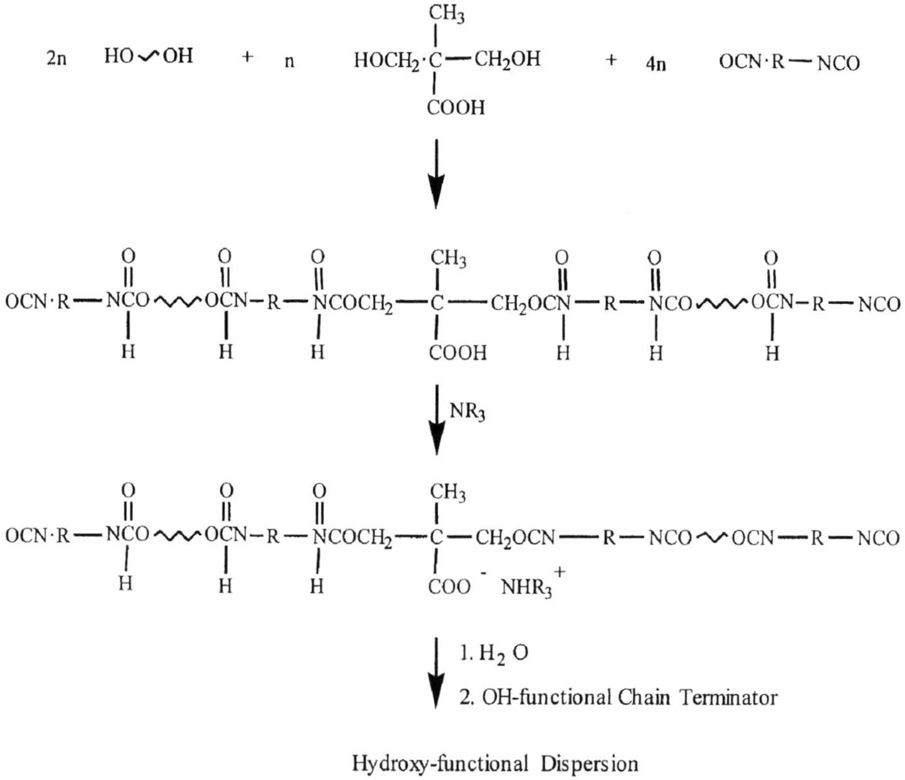

Scheme 4. Prepolymer Process: Preparation of hydroxy-functional dispersion.

Disperse Phase Surface Stabilizers

A. _In Chain-Growth Synthesis._ The stabilities of latices in their early development were related to the electrostatic repulsions of particles due to catalyst fragments and adsorbed surfactants on their surfaces. Commercial materials demand more stability to electrolyte concentration, mechanical deformation and temperature variations than these entities can provide. Therefore, early in their development, the stabilities of latices were increased by oligomeric surface stabilizers, or by chemically grafted water-soluble polymer segments. These surface segments, illustrated in **Figure 1**, provide electrosteric and steric stabilization.

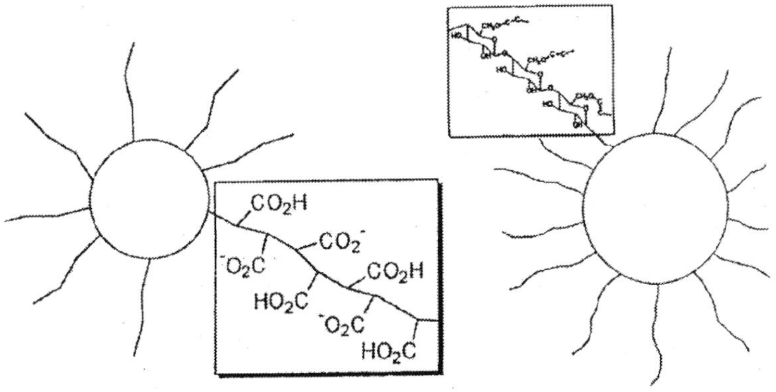

Figure 1. Electrosteric and Steric Stabilized Latices

The most universal approach was the inclusion of methacrylic or acrylic acid in the latex synthesis recipe, generally in the second stage of a starved, semi-continuous process. The acids polymerize in sequence runs of acrylic or methacrylic acid, not possible in the step-growth synthesis of polyester and polyurethane disperse phases, and position near or on the surface of the latex. When neutralized, they expand to provide electrosteric stabilization of the particle. The parameters that influence the positioning of the acid surface stabilizers near the water/latex interface are the hydrophilicity of the copolymer and the latex's glass transition temperature (29). Thus, for a methyl methacrylate copolymer latex having equivalent glass transition temperatures, the surface acid content and its extent of swelling will be greater in a copolymer composition having a more hydrophilic acrylate comonomer (where more is required of ethyl acetate than of butyl acrylate to achieve a given glass transition temperature). With the more hydrophobic acrylate comonomer, more of the 2 wt.% methacrylic acid charged is retained within the latex particle. The importance of the hydrophilicity of the latex composition also is evident in styrene and methyl methacrylate latices (not illustrated), both containing 50% ethyl acrylate. The ethacrylic acid segments are buried within the more hydrophobic styrene latex synthesized by a conventional batch process. In other compositions having a constant hydrophobicity but varying glass transition temperature, the importance of the glass transition temperature also is evident. Such stabilizers have been important in small particle (100 nm) latices used in formulations containing a variety of pigments, because of their high pigment binding power. The theoretical treatment of the parameters influencing the stability of this type of latex particle, and for those discussed in the next paragraph containing only nonionic grafted segments, is given in reference *30*.

While oligomeric surface acid stabilizers are used in vinyl acetate latices in the paper coatings area, where water sensitivity is not a factor in formulations containing high clay contents, and in methacrylate and styrene based latices in architectural latex coatings, vinyl acetate latices in low pigment volume concentration architectural coatings are water sensitive and surface acids are not the method of stabilization. An alternative approach with the more hydrophilic vinyl ester monomers is to chemically graft segments of water-soluble polymers to the growing polymer chain. Generally, cellulose ethers are used in the synthesis of vinyl acetate latices for architectural coatings; vinyl acetate/vinyl alcohol copolymers are grafted to latices used in adhesive applications. Grafting of the nonionic stabilizers to the latex surface also provides a means of increasing viscosities at high shear rates (10^4 s^{-1}) for vinyl acetate latices, and there was a desire to impart this to large particle size acrylic latices. The general technical understanding of the sixties impeded this development on a wide commercial scale. It was known that the vinyl acetate radical was high in energy and therefore responsible for the grafting of poly(vinyl acetate) to the cellulose ether. Efforts by most to graft acrylics or styrene to cellulose ethers were unsuccessful and this was generally considered to be related to the lower energy of these propagating radicals. In fact, the lowest energy radicals in these systems were those generated at the carbon acetal linkage of the cellulose derivative or next to the carbonyl group of the acetate segments of the vinyl alcohol copolymer (**Figure 2**). The primary deterrent to the commercial production of methacrylic latices was the relatively uninterrupted propagation of the methacrylate radical without transfer reactions. The high energy vinyl acetate radical transferred with everything in the system except water, and there was therefore, sufficient water-soluble polymer chemically grafted to ensure adequate stability of the final vinyl acetate latex. Once the error in understanding of the mechanism was understood (*31*), water-soluble chain transfer agents, such as triethanolarnine, were added to the polymerization recipe and methyl methacrylate latices, with sufficient cellulose ether grafts for stability, were successfully prepared.

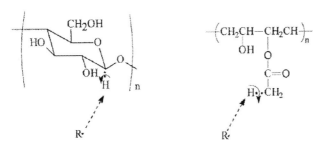

Figure 2. Low Energy Radicals Generated on Cellulose Derivative and Vinyle Alcohol Copolymer.

B. In Step-Growth Dispersions. Like traditional latices, "water-reducible, step-growth polymers" require anionic and/or nonionic surface stabilizers to maintain dispersion stability at reasonable volume solids in a coating formulation. This is not difficult to accomplish in the chain-growth polymerization of water-reducible acrylics but the achievement of stability of step-growth dispersions is more difficult. It is accomplished in step-growth polymers by the inclusion of a carboxylate (**Figure 3A**), sulfonate (**3B**), or nonionic oxyethylene (**3C**) group, placed within the chain by a diol or diamine structure. These disperse phases are not as stable as the acrylic disperse phases, for the surface stabilizers are isolated as one charge, and there is no steric contribution with these isolated anionic groups. They do not occur in sequence runs as observed in chain-growth acrylic polymers, and therefore, the particles are more susceptible to flocculation by mechanical and thermal forces and by electrolyte concentrations. There is also a particle size distribution difference among these different disperse phases, with the WR acrylic being much smaller than those obtained from an emulsion polymerization.

When an acrylic latex is thickened with a nonadsorbing polymer such as hydroxyethyl cellulose, the latex will flocculate by a depletion flocculation mechanism. This is usually not observed visually. With a step-growth disperse phase, separation is observed within hours (*32*). This does not happen when the thickener is a Hydrophobically-modified, Ethoxylated URethane (HEUR) polymer. HEUR polymers have been discussed in detail with respect to their synthesis, solution behavior and structural influences on architectural coatings rheology (*ref 33, chapters 10, 17, and 24, respectively*). The adsorption of an HEUR thickener on a latex is determined by its chemical composition (that defines the energy of the surface), the concentration of surface acids and the amount of surfactant present (*34*). The molecular weight, concentration and effective hydrophobe terminal size on the HEUR are also important. The relative concentration and size of the surfactant and of the effective hydrophobe terminal size of the HEUR are the key variables in the adsorption process. There is no particle dispersions in PUDs has resulted in some synergism. Some of the high T_g component is thought to be contained within the film of the low T_g component, acting somewhat like a composite for reinforcement. Scanning electron microscopy (SEM) is a fairly effective way of studying film formation, and thus the extent of particle coalescence. However, with the small particle size of the PUDS, atomic force microscopy (AFM), specifically tapping mode atomic force microscopy (TMAFM), has been found *(54)* to be a better alternative. The AFM studies have also proven that blending of soft (low T_g) and hard (high T_g) particle dispersions results in enhanced film formation over the hard particle dispersions by themselves, but the film formed is not as complete as observed with a disperse phase at its MFFT.

The addition of the PUD to an emulsion enhanced the physical properties of the acrylic polymer. Morphology studies indicated that there were distinct urethane and acrylic regions. In another study *(55)*, the acrylic monomers were added to a PUD and polymerization was initiated. The polymerization took place inside the PUD particles, creating a urethane stabilized acrylic particle. The overall MFFT and T_g can be controlled with the right balance between the urethane and the acrylic.

C. Thermoset Step-Growth. Two-Component Polyurethanes (2K). The 2K PUDs crosslink after the particles have coalesced to form a hard, continuous film. Diffusion of the polymer molecules, to ensure complete reaction of all functional groups, is very important in 2K PUDs. If formulated correctly (a motherhood statement borne of proprietary needs, but often used to hide questionable experimental results), 2K PUDs can have enhanced film properties, such as solvent resistance, over more conventional high molecular weight, water-borne PUDs due to their crosslinked nature. The formulation of 2K water-borne polyurethanes consists of combining a multi-functional polyisocyanate component, modified to be water dispersible, with a hydroxy-functional polyurethane dispersion. Typical hydroxyfunctional dispersions are polyurethane dispersions based on polyesters, polycaprolactones, acrylics, and poly(tetramethylene oxide). A common multifunctional isocyanate is a modified tritner of hexamethylene diisocyanate (HDI), an isocyanurate. The two components are mixed together, with a catalyst, prior to application. The reactivities of the functional groups dictate the length of the induction time (the time between mixing and application) in 2K waterborne polyurethanes. In spray coatings, formulations with induction times of less then a minute can be used, but require mixing of the two components at the spray head. During longer mixing periods the particles begin to coalesce (depending on T_g, % solids, various additives, etc.), and the isocyanate-functional groups react with both the hydroxyl-functional groups and water. Once the two reactive components in the mixed dispersion coalesce, interdiffusion can occur, thereby increasing the desired isocyanate-hydroxyl reaction *(56)*. After application, water and co-solvent evaporate, and film formation begins as described in the previous section. Coalescence and isocyanate reactions may continue, but ideally to a lesser extent. Problems can result during this stage if the crosslinking reactions and molecular weight build-up are too free surfactant used in the synthesis of a step-growth disperse phase, and therefore no impedance to adsorption and stabilization. The greater instability of step-growth dispersions can be improved by steric stabilization through adsorption of HEUR associative thickeners.

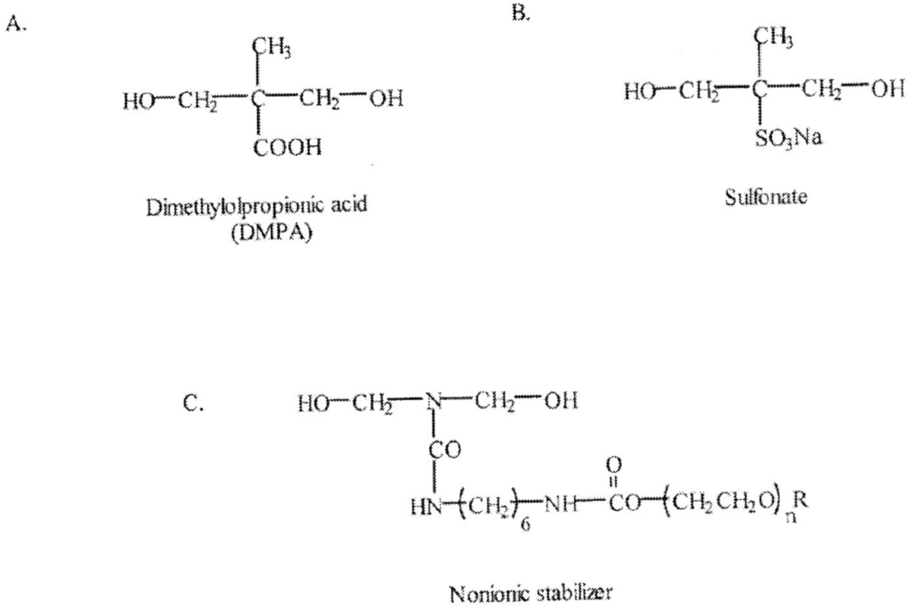

A.

$$HO-CH_2-\underset{\underset{COOH}{|}}{\overset{\overset{CH_3}{|}}{C}}-CH_2-OH$$

Dimethylolpropionic acid
(DMPA)

B.

$$HO-CH_2-\underset{\underset{SO_3Na}{|}}{\overset{\overset{CH_3}{|}}{C}}-CH_2-OH$$

Sulfonate

C.

$$HO-CH_2-\underset{\underset{HN-(CH_2)_6-NH-\overset{\overset{O}{\|}}{C}O-(CH_2CH_2O)_n R}{\overset{|}{CO}}}{\overset{|}{N}}-CH_2-OH$$

Nonionic stabilizer

Figure 3. Surface Stabilizers. (a) Carboxylate (b) sulfonate (c) nonionic
oxyethylene groups

   The disperse phases discussed above must coalesce to form a continuous film, and
to do so with sufficiently high glass transition temperature latices (or one that is
crosslinking) to provide reasonable film durability after application, requires a
coalescing aid. This contributes to the VOC of aqueous disperse phase coatings and
is discussed in this section.

COALESCENCE

A) Chain Growth Thermoplastics. *Latex Coalescence.* There have been several
descriptions of the latex particle coalescence process during the past half century. It is
generally accepted that the film-forming process occurs in three stages (Figure 4): (I)
evaporation of the water until the particles reach close-packing; (II) formation of particle
contacts and deformation of the latex particles as the particle volume fraction goes
above that of a close-packed structure; a polyhedral-foam-like structure is developed
with interparticle bilayers and Plateau borders; (III) gradual coalescence by
interdiffusion of polymer molecules between latex particles.

Aqueous latex dispersion

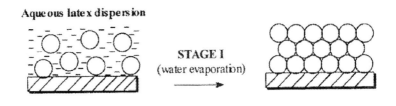

STAGE I
(water evaporation)

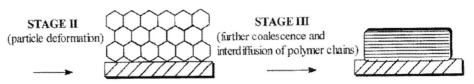

STAGE II
(particle deformation)

STAGE III
(further coalescence and
interdiffusion of polymer chains)

Figure 4. Three stages in the general film formation process of aqueous dispersions.

These stages are distinct if the stabilizing surfactant layer is effective, if there is no tendency to spontaneously form particle contacts or aggregates (as there is in depletion flocculation), and if the particles are not too hydrophilic.

During stage I, the water phase starts to evaporate and the concentration of the latex phase increases. As the volume fraction, ϕ, increases above that of close-packed spheres, capillary stresses develop which force the particles into contact. This type of stress can be thought of as arising from surface tension at the particle/water/vapor contact lines, at the top surface of the film. In effect, this tension pulls the film down (Figure 5), in the direction of the substrate, and leads to the "compression" of the latex particles. Latex films are typically prepared with thicknesses on the order of 50 μm. This is much larger than the particle diameter, and the film will comprise many layers of particles. Sheetz first proposed (*35*) that a thin layer of coalesced particles closes the surface of the drying latex. The remaining water evaporates after diffusion through this polymer layer, and the packing of particles is compressed like a piston.

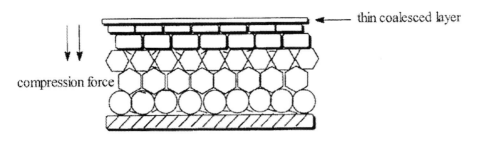

thin coalesced layer

compression force

Figure 5. "Compression" of latex particles due to surface tension forces

Up to this point, the particles have been stabilized by surface acid (electrostatic forces), surfactant molecules, and/or grafted water-soluble polymer fragments (steric forces). As the water evaporates and the particles pack closer together, the stabilizing forces are overcome and coalescence begins (stage II). At this point, descriptions of the primary forces driving coalescence have varied over the past four decades. In the early fifties, Bradford and coworkers modeled (36,37) film formation as a Frenkel viscous flow of contracting polymer spheres under polymer/air and polymer/water interfacial tensions. Brown complemented (38) this with arguments that water also contributed to the deformation process through capillary compression of the polymer assemblage by the serum/air surface tension. As the water evaporates, all of the interfacial tensions drive the particles to coalesce despite the viscoelastic resistance of the polymer to flow. To address the coalescence of large particles, Vanderhoff and his colleagues proposed (39) that as the water continues to evaporate, the forces due to the water-air surface tension pushed the particles together until the stabilizing layers were ruptured, resulting in polymer-polymer contact. Forces arising from the polymer/water interfacial tension then increased the pressure on the particles. It has also been proposed (40), that interfacial tension forces act along with the capillary force to cause film coalescence. Attempts have been made to verify the theories proposed. For example, the coalescence of core-shell latices (41) (core: St/BuA, shell: MAA/St/BuA) were studied, and the rate of coalescence decreased when the particle-water interfacial tension decreased.

Brown also observed that film formation occurs concurrent with the evaporation of water and is complete with total water evaporation. He observed that porous, incompletely coalesced films could be maintained by keeping the temperature lower than a certain critical value (below the glass transition temperature, T_g, of the resin) during the water evaporation stage, and that some voids would not coalesce when warmed to a temperature where coalescence would have occurred in the presence of water. His observation that liquid water was central to the deformation process led several industrial laboratories, in the past decade, to use 2-hydroxyethyl methacrylate and similar comonomers in latex synthesis. In addition, the ability of coalescing aids to decrease the rate of evaporation of water has been suggested to facilitate coalescence. Similar results have also been noted (42, 43) with hydrophilic compositions where it has been observed that higher humidity, which both Slows water evaporation and plasticizes the particles, will lower the MFFT. The minimum temperature at which film formation occurs is know as the Minimum Film Formation Temperature (MFFT); it generally occurs a few degrees below the T_g of the resin. If the MFFT is too far below the application temperature, a film will be formed with poor properties (e.g., mechanical, abrasion, solvent resistance, dirt pickup, etc.).

It has also been demonstrated that the hydrophilicity of the monomer is important to the morphology. For example, lower amounts of coalescing aid are needed for film formation with a methyl methacrylate/methyl acrylate (MMA/MA) porous copolymer than for a styrene/2-ethylhexyl acrylate (44) copolymer latex with a nonporous topography. Surprisingly, there had been a lack of studies on comparative compositional influences until this investigation and the one on hydroplasticization by water (43); both demonstrated the greater difficulty in film formation of a styrene/acrylate latex relative to a methacrylate/acrylate latex.

The interface between particles dissipates due to interdiffusion of macromolecular chains between particles (Stage III), driven by interfacial and surface tension forces. Film properties such as mechanical strength and chemical resistance begin to develop (*40*). A polyhedral-foam type structure is formed with the water contained in a network of bilayers, and Plateau borders. If the surfactant is effective, this is likely to occur only if the surfactant bilayer has been dissipated by diffusion of the surfactant into the particles or into the Plateau borders and interparticle interstices. This will occur at very low water contents, when the water of hydration of the surfactant is eliminated (*45*). The latter comment highlights a fact not generally considered: the colloid stabilizers that promote stability of the initial colloid particle (**Figure 1**). It would be harder for these types of stabilizers to disappear, and this has not been addressed in the literature. Based on the prior art, with only surfactant stabilization, it was concluded that the plateau boarders are so narrow and the water/vapor surface tension is so high that vapor penetration into the film is effectively impossible, and liquid latex particles with a surfactant layer will always form a film - independent of the particle size.

The importance of particle size on coalescence has generally been abandoned since the mid sixties; however, there have been recent commercial studies in support of the impact of the median size of latices on the minimum filming temperature (*43,46*). The drawback is the large amount of surfactant required to synthesize truly small particle sizes. The minimum filming temperature of vinyl acetate copolymer latices was shown to increase as the average particle radius increased (*47*). This has also been shown in the work of Eckersley and Rudin (*40*). In the latter study of methacrylate latices, the influence of the median particle size was small, but the particle size study stopped at 148 nm, and the particles contained methacrylic acid surface stabilizers. This effect is still under investigation due to questions raised, such as particle size distribution, concentration of surface stabilizer (grafted polymers), and type of stabilizers.

Controlling the modulus of the polymer particles has been the most practiced method of effecting the MFFT. Through the use of different monomer combinations, the T_g of the particle can be adjusted above a desired MFFT. Coalescing aids are then added to the dispersion to lower the T_g of the particle still further by plasticization and perhaps by retarding the rate of water evaporation. This creates a dispersion with the advantage of a low MFFT while maintaining a relatively high T_g after evaporation of the coalescing agent and glycol ethers. The most important properties (*44*) of a coalescing aid are its non-polar solubility parameter (that reflects the relative insolubility in water) and molar volume (that reflects its ease of diffusion through the film forming matrix). The hydrophilicity of the monomers used in the dispersion synthesis can also contribute to lowering the MFFT. These hydrophilic monomers will tend to be at the particle surface where water will plasticize the particle. For example, at the same T_g a methyl methacrylate/ethyl acrylate latex will form a film much more readily than a styrene/2-ethylhexyl acrylate copolymer latex (*44*). These studies are consistent with the ability of oligomeric acid stabilizers to reach the surface of related copolymer latices. It also has been reported that for a MMA/acrylate copolymer emulsion, the

MFFT displayed a greater increase with increasing MMA when the co-monomer was less hydrophilic (*48*).

Surfactant Effects. Early studies of nonylphenol surfactants by Bradford and Vanderhoff observed (*49,50*) that as the length of the ethylene oxide segment decreased, film integration with a styrene/butadiene copolymer increased. Other studies observed (*51*) that surfactants markedly reduced the MFFT and T_g of vinyl acrylic latices. In a recent study, a nonionic surfactant was observed (*52*) to plasticize an acrylic latex. There also have been studies of optimum surfactant concentrations. Post addition of surfactant to a surfactant-free PBMA latex observed (*41*) that the best film surface corresponded with a surfactant concentration providing full surface coverage of the latex, but the influence of protective colloids, used commercially, have not been reported to date.

B) Step-Growth Thermoplastics. *Aermoplastic Polyurethane Dispersions*. The "water-reducible" oligomers prepared by step-growth polymerizations must also undergo coalescence to form a continuous film. As noted earlier, detailed studies in the open literature are essentially restricted to aqueous polyurethane dispersions (PUDs). PUDs have similar film formation characteristics to that of conventional latices, but with significant differences that are unique to PUDS. The two most common types of PUDs used in water-borne coatings are the high molecular weight aqueous dispersion and the low molecular weight two component (2K) polyisocyanate/polyurethane systems.

The film formation process for PUDs is a combination of thermoplastic coalescence (as with latices) and in the 2K systems, thermoset crosslinking. It is in the areas of the stabilizers and the diffusion process were the PUDs differ most from the conventional latices. PUDs employ both anionic and nonionic stabilizers (**Figure 3**) in their synthesis. The nonionic stabilizers provide stability to salinity gradients but not to high temperatures. They are less frequently used because relatively high concentrations are needed for stable dispersions. The amount of dimethylolpropionic acid (DMPA, **Figure 3**), the most commonly used ionic stabilizer, determines the median particle size of the PUD phase and its stability. The acid functionalities are not contiguous, and the dispersions are more sensitive to pH changes than conventional latices. The contiguous acid groups in the chain-growth acrylic dispersions create areas with different ionization potentials; PUDs do not benefit from this neighboring group effect, and thus display a narrower ionization range (*53*). The water inside the dispersion (**Figure 6**) creates a more open particle, increasing the mobility of the polymer chains throughout the particle (*53*). The high molecular weight PUDs form films primarily by thermoplastic coalescence.

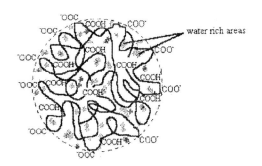

Figure 6. Representation of water swollen PUD particle

The particle size distribution of PUDs is generally broader than conventional latices. Since PUDs are dispersed mechanically, the emulsification process contributes to broader size distribution due to areas of non-uniform mixing. The particle size of the final dispersion is typically between 25-100 nm. The distribution may also be related to the statistical nature of the chain extension steps. The MFFT and polymer diffusion are directly related to the degree of swelling which is suggested to ultimately effect the final film quality. The parameters influencing the degree of particle swelling by water are hydrophilicity of the backbone, molecular weight, DMPA concentration, and level of co solvent (usually n-methylpyrrolidinone). Even though the particles are swollen by water prior to film formation, this appears to have little effect on the water sensitivity of the final film. The water sensitivity is determined, apparently, only by the hydrophilicity of the polymer backbone.

Blends and Hybrids. The use of blending techniques (acrylic latices with polyurethane dispersions) as well as hybrid systems has been investigated as a route to obtain systems with a lower MFFT while maintaining a higher, final T_g, similar to the use of coalescing aids in latices. If this can be achieved, lower VOC's would be realized. Recent studies have examined the blending of high and low T_g PUDs (*54*). This process also is employed in conventional latices as an attempt to combine good film formation and film hardness without the need for coalescing aids. The blending of low T_g and high T_g rapid. Cosolvent and CO_2 (from reaction of water with isocyanate) can become trapped in the film causing a decrease in the diffusion of reactive groups.

D) Thermoset Chain Growth. *Crosslinkable Latices.* An alternative to the use of coalescing aids and blending of acrylics has been the incorporation of crosslinkable functional groups in the polymer backbone. This approach is focused on attempting to effectively separate the MFFT of the latex from the "T_g" of the final film. The addition of either an acetoacetate functional monomer or an autooxidizable component, such as a drying oil, into the latex (*57,58*) were previously discussed above as ways to crosslink latices. The acetoacetate functional groups are converted to enamines when the latex is neutralized with ammonia or a primary amine. An excess of neutralizer is added to ensure enamine formation. The enamine functionalities have the advantage that they do not begin to crosslink significantly until coalescence is almost complete. As the enamine groups crosslink, the modulus (discussed in the literature incorrectly as the T_g) of the film increases. Therefore, a latex can be formulated to a low MFFT, yet have a hard final film. An autooxidizable component works in the same general manner, with crosslinking occurring among the unsaturation sites. The acetoacetate/enamine latices are designed to cure at ambient temperatures and with the aid of sunlight. Thus, the addition of photoinitiators enhances the curing process. The drying oil incorporated latices cure faster with the addition of a metal drier, such as cobalt. Both can be formulated to be stable, one package systems. In addition, the coalescing agent could be functionalized, such as dicyclopentenyloxyethyl methacrylate (*40,41*) and used as a crosslinker to obtain the hard film properties needed in industrial coatings.

CONCLUSIONS

This chapter has focused on past developments in the area of water-borne coatings. Synthetic aspects were discussed related to producing water-borne coatings as thermoplastic (emulsion polymerization) systems and as thermoset (chain growth and step growth) systems. Advantages and disadvantages were given. For application in commodity architectural and paper coatings areas, oligomeric surface structures are required. This is achieved by copolymerizing sequence runs of oligomeric acid or by grafting water-soluble polymers to the binder during the synthesis of the emulsion polymer. The procedures utilized to achieve this goal for different monomers are described in this chapter. Finally the issues related to the coalescence and film formation of water-borne coatings were discussed. The phenomenon of particle coalescence has been an area of study for some forty years. The recent studies described in this chapter have fine-tuned our understanding, but translation to any commercial improvements is unlikely. In the more commodity latex markets the greatest probability of success may be through the well studied core-shell approach with hydrophilic comonomers in the outer shell. The disperse phase must have colloidal stability and the influence of the type of stabilizers used commercially has not been reported in the open literature. Hydrophobically-modified ethoxylated urethane thickeners (*33*) have been observed to stabilize aqueous polyurethane dispersions (*32*), and their use in this mode is now generally one of the standard practices in formulating. HEURs also lower the MFFT of the formulation (*44*). The use of crosslinkable functional groups within each particle with the low MFFT systems is also promising.

REFERENCES

[1] Hoyle, C.E.; Kinstle, J.F., Eds., *Radiation Curing of Polymeric Materials,- ACS* Symposium Series 417, American Chemical Society: Washington, DC, 1990.

[2] Jilek, J.H., *Powder Coatings,-* Federation of Societies for Coatings Technology: Blue Bell, PA, 1991.

[3] Odian, G., Eds., *Principles of Polymerization,-* 3rd Edition, John Wiley & Sons, Inc.: New York, 1991, Chapters 3 and 4.

[4] Fitch, R.M.; Tsai, C.H. in *Polymer Colloids;* Fitch, R.M., Ed.; Plenum press; 197 1; pp.73 and pp. 103.

[5] Hansen, F.K., in *Polymer Latexes.- Preparation, Characterization, and Application,* ACS Symposium Series 492, Daniels, E.S.; Sudol, E.D.; El-Aasser, M.S.; Eds., American Chemical Society, Washington, DC, 1992; Chapter 2.

[6] Yeliseyeva, V.I., in *Emulsion Polymerization, Pirma,* I., Ed., Academic Press, New York, 1982; Chapter 7.

[7] Dunn, A.S., in *Emulsion Polymerization, Pirma,* I., Ed., Academic Press, New York, 1982; Chapter 6, and in *Polymer Latexes: Preparation, Characterization, and Application,* ACS Symposium Series 492, Daniels, E.S.; Sudol, E.D.; ElAasser, M.S.; Eds., American Chemical Society, Washington, DC, 1992; Chapter 4.

[8] Gardon, J.L., *"A Perspective on Resins in Aqueous Coatings, "* American Chemical Society Symposium Series No. 663: TechnoloQy for Waterborne Coatings, Glass, J.E. ed., American Chemical Society, Washington, D.C., 1997, Ch. 2.

[9] Durant, Y.G. and Sundberg, D.C., *"Progress in Predicting Latex-Particle Morphology and Projections for the Future, "* American Chemical Society Symposium Series No. 663: Technology for Waterborne Coatings, Glass, J.E. ed., American Chemical Society, Washington, D.C., 1997, Ch. 3.

[10] Richards, B.M.; Masters Thesis, North Dakota State University, 1977.

[11] Brandenburger, L.B., Ph.D. Thesis, North Dakota State University, 1977.

[12] Webster, O.W., Sogah, D.Y., *in Comprehensive Polymer Science,* Eastman, G.C.; Ledwith, A.; Russo, S.; Sigwalt, P., Eds., Pergamon Press, NY, 1989; Chapter I 0.

[13] Polyfunctional Aziridines, Cordova Chemical Company, Commercial Literature (1983).

[14] Taylor, J.W. and Bassett, D.R. American Chemical Society Symposium Series No. 663: Technology for Waterborne Coatina§, Glass, J.E. ed., American Chemical Society, Washington, D.C., 1997, Ch.8.

[15] Grawe, J.R. and Bufkin, B.G., J. Coat. Tech., **1978**, 50(647), 65.

[16] Nippon Shokubia Literature, 1994.

[17] Tillson, H.C. US Patent 3,219,610, **1965**.

[18] Taylor, J.W.;Collins, M.J.; and Clark, M.D. US Patent 5,53 9,073, **1996**.

[19] McGinniss, V.D. and Seidewand, J.R. US Patent 4,107,013, **1978**.

[20] Wolfersberger, M.H.; Schinder, F.J.; Beekely, R.S.; Novak, R.W. US Patent 5,306,744, **1994**.

[21] Mylonakis, S.G. US Patent 4,244,850, **1981**.

[22] Pears, D.A. and Overbeek, G.C. European Patent Application 442,653 A2, **1991**.

[23] Bors, D.A. Eur. Patent Appl. 492,847 A2, **1991**.

[24] Bors, D.A.; Lavoie, A.C.; Emmons, W.D. US Patent 5,484,849, **1996**.

[25] Del Rector, F.; Blount, W.W.; and Leonard, D.R., "Applications for the Acetoacetyl Functionality in Thermoset Coatings," *Proc. Of the Waterborne and Higher Solids Coatings Symposium,* New Orleans, **1988**.

[26] Esser, R.J.; Devona, J.E.; Setzke, D.E.; and Wagemans, L., *Eur. Coat. J,* **1998**, 10, 732.

[27] Dieterich, D., *Progress in Organic Coatings,* **1981**, *9,* 281.

[28] Rosthauser, J.W.; Nachtkamp, K., "Water-borne Polyurethanes," K.C. Frisch; Klempner, D.; Eds., *Advances in Urethane Technology,* **1987**, 10, 121.

[29] Hoy, K.L., Journal of Coatings Technology, **1979**, 51 (651), 27

[30] Goodwin, J.W. and Hughes, R.W., *"Particle Interactions and Dispersion Rheology, "* American Chemical Society Symposium Series No. 663: Technology for Waterborne Coatings, Glass, J.E. ed., American Chemical Society, Washington, D.C., 1997, Ch. 6.

[31] *Craig, D.H., in Water-Soluble Polymers.- Beauty with Performance,* Advances in Chemistry Series 213, Glass, J.E., Ed., American Chemical Society, Washington, DC, 1986; Chapter 18. Several additional studies were reported in ACS-PMSE national meeting reprints after this publication and the study was later republished in *J Coatings Techn.*

[32] Kaczmarski, J.P.; Fernando, R.H.; Glass, J.E.; *Journal of Coatings Technology,* **1993**, *65 (818),* 39.

[33] Wetzel, W.H.; Chen, Mao; Tarng, M.R.; Kaczmarski, J.P.; Lundberg, D.J.; Ma. Zeying; Alahapperuna, K. and Glass, J.E.; in *Hydrophillic Polymers.- Performance with Environmental Acceptability,* Advances in Chemistry Series 248, Glass, J.E., Ed., American Chemical Society, Washington, DC, 1996. Chapters 10, 17, 24.

[34] Ma, Z.; Chen, M. and Glass, J.E.; *Colloids and Surfaces,* **1996**, *112,* 163.

[35] Sheetz, D.P., *J Appl. Polym. Sci.,* 1965, 9, 3759.

[36] Dillion, R.E.; Matheson, L. A.; Bradford, E.B., *J Colloid Sci.,* 1951, *6,* 109.

[37] Henson, W.A.; Tabor, D.A.; Bradford, E.B., *Ind. Eng. Chem.,* 1953, *45, 73* 5.

[38] Brown, G.L. *J Polym. Sci.,* **1956**, *22,* 423.

[39] Vanderhoff, J.W.; Tarkowski, H.L.; Jenkins, M.C.; and Bradford, E.B., J *Macromol. Chem.,* **1996**, 361.

[40] Eckersley, S.T. and Rudin, A., *J Coat. Tech.,* **1990**, *62(780),* 89.

[41] Dobler, F.; Pith, J.; Lambla, M.; and Holl, Y., L., J *Colloid Interface Sci.,* **1992**, *152, 1.*

[42] Juhue, D. and Lang, J., *Langmuir ,* **1993**, 9, 792.

[43] Sperry, P.R.; Synder, B.S.; O'Dowd, M.L.; and Lesko, P.M., *Langmuir,* **1994**, 10, 2619.

[44] Alahapperuna, K. and Glass, J. E., *J Coat Tech.,* **1991**, *63(799),* 69.

[45] Crowley, T.L.; Sanderson, A.R.; Morrison, J.D.; Barry, M.D.; Morton-Jones, A.J.; and Rennie, A.R., *Langmuir,* **1992**, *8,* 21 10.

[46] Shah, P.K.; Blam, A.F.; and Yang, P.Y., B.F. Goodrich, Canadian Patent, (US) 07/333,376, 4/5/1989. And references therein.

[47] Nyugen, B., PhD Thesis, Univ. Waterloo (1986).

[48] Tongyu, C.; Yongshen, X.; Yuncheng, S.; Fu, L.; Xing, L.; and Yuhong, H., J *AppL Polym. Sci.,* **1990**, *41,* 1965.

[49] Bradford, E.B. and Vanderhoff, J.W., *J Macromol. Chem.,* **1966**, 1, 335.

[50] Bradford, E.B. and Vanderhoff, J.W., *J Macromol. Sci.-Phy.*, **1972**, *B6(4)*, 671.

[51] Vijayendran, B.R.; Bone, T.; and Sawyer, L.C., J *Disp. Sci. Tech.*, **1982**, 3(1), 8.

[52] Eckersley, S.T. and Rudin, A., *J Appl. Polym. Sci.*, 1993, 48,1369.

[53] Satguru, R.; McMahon, J.; Padget, J.C.; and Coogan, R.G., J *Coat. Tech.*, **1994**, *66(830)*, 47.

[54] Rynders, R.M.; Hegedus, C.R.; and Gilicinski, A.G., J *Coat. Tech.*, **1995**, *67(845)*, *59*.

[55] Jansse, P.L., *JOCCA,* **1989**, 72, 478.

[56] Jacobs, P.B. and Yu, P.C., J *Coat. Tech.*, **1993**, *65(822), 45.*

[57] Bors, D. A.; Lavoie, A.C.; and Emmons, W.D., Rhom and Haas, European Patent, Publication No. 0 492 847 A2, Application No. (913112926), 4/12/91.

[58] Bors, D.A.; Warminster, W.; Emmons, D.; and Edwards, S.S., Rhom and Haas, United States Patent, Patent No. 5,296,530, 3/22/1994.

Section 4

NEW MATERIALS
Section Editors: Elsa Reichmanis and Michael Jaffe

THE FUTURE OF APPLIED POLYMER SCIENCE

MICHAEL JAFFE

Department of Chemistry, Chemical Engineering and Environmental Science
New Jersey Institute of Technology, Rutgers University,111 Lock Street, Newark, NJ
07103

Contributors

Eric Baer, Department of Macromolecular Science and Center for Applied Polymer
Research, Case Western Reserve University, Cleveland, OH 44106

Ray A. Dickie, Ford Research Laboratories, Dearborn, MI 48121

Claus D. Eisenbach, Institute of Applied Macromolecular Chemisty, Universitat
Stuttgart, D-70569, Germany

Jean M. J. Frechet, Department of Chemistry, University of California, Berkeley, CA
94720-1460

J. P. Kennedy, Institute of Polymer Science, the University of Akron, Akron, OH
44325-3909

J. L. Koenig, Department of Macromolecular Science, Case Western Reserve
University, Cleveland, OH 44106-7202

Joachim Kohn, Department of Chemistry, Rutgers University, Piscatawy, NJ 08854-
8087

David A. Tirrell, Division of Chemistry and Chemical Engineering, California Institute
of Technology, Pasadena, CA 91125

Ian M. Ward, IRC in Polymer Science and Technology, University of Leeds, LS2
9JT, England

C. Grant Willson, Departments of Chemistry and Chemical Engineering, The University of Texas, Austin, TX 78712

Alan H. Windle, Department of Materials Science & Metallurgy, Cambridge University, CB2 3QZ, England

Introduction

The molecular long chain hypothesis of Professor H. Staudinger, for which he was awarded the Nobel Prize in 1953, marks the birth of modern polymer science [1]. The strategic importance of polymers became evident during the Second World War through the invention of artificial rubber, e.g., and led to the establishment of highly productive applied polymer research groups in industry (for example, at the duPont Company) and academe (for example at the University of Akron). The explosive commercial success of plastics, fibers and films in the decades after the war fueled a renaissance in polymer research that peaked in the 1960's and 1970's with the commercialization of a series of commodity polymers ranging from polyolefins to polyesters at volumes in the billions of pounds. The last decades of the twentieth century saw the polymer industry realign their priorities and dismantle much of the research infrastructure that had been assembled. Thus, for many companies much of their current participation in basic research lies in funding cooperative research with Universities. Concurrently, polymer science, long viewed in academe as plebian technology unworthy of academic consideration, became a fully recognized research focus in many traditional research universities, with exciting new discoveries ranging from controlled, self-assembling molecular architectures to polymer based electronic and photonic devices. These discoveries are adding new dimensions to the potential utility of macromolecules.

Many areas of polymer technical development, with high potential commercial impact, are limited not by component or system concepts but by the availability of reproducible, cost-effective materials to reduce the existing concepts to practice. In addition, the utility of many modern materials is limited by a lack of scalable fabrication technology and unattractive finished part costs.   In polymeric materials there are the additional problems of detailed specification of the molecular chain backbone, specification of the overall part morphology, and, in composite materials, controlling the interface between hard and soft segments. These issues are especially relevant for very high performance polymers, i.e., continuous fiber reinforced composites, materials designed for use under conditions of environmental extremes, and parts that need to function reliably for extended periods of time. Understanding of the inter-relationships of processing, properties and structure is the cornerstone of modern material science, often shown diagrammatically as a triangle (Figure 1), i.e.,

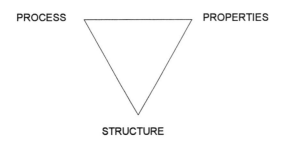

PROCESS   PROPERTIES

STRUCTURE

Figure 1

It is the recognition that performance is uniquely related to both the chemical and physical polymer structure, and that the structure of a polymer is a reflection of the totality of its processing history, which allows process and product development based on scientific principle rather than purely empirical methodologies.   The leg connecting process to structure represents process engineering (structure formation kinetics), while the structure-property leg represents materials chemistry and physics, hence the study of process-structure-property relationships is, by nature, highly interdisciplinary.   Once the process-structure-property database is established, a single material may be processed to satisfy a variety of market needs. An example of this is the production of molecularly identical, morphologically distinct poly (ethyleneterephthalate) fibers for applications as diverse as pliant textile yarns and stiff reinforcing cords for tires.

The use of a given polymer is ultimately limited by its intrinsic properties, which, in turn, are a reflection of fundamental chemical structure and molecular packing on a hierarchy of size scales. Incorporation of molecular structure variables into product and process development leads to an expansion of the process-structure-property space to include polymer synthesis (Figure 2).

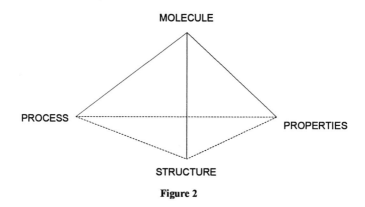

Figure 2

It is recognized that choice of a specific polymer chemistry for a given application is a non-trivial aspect of the total materials problem and cannot be addressed in isolation. Inherent in polymer choice is the ability to translate material properties into finished part performance through cost-effective part and component manufacturing. Historically, the polymer community has focused on either material science (structure-property relations) or materials engineering (material production and shaping), ignoring the inherent lack of utility of those polymers, even those polymers with extraordinary properties, for which no cost-effective fabrication technology exists. For modern polymers to significantly impact emerging systems technologies, the tasks of polymer development, part fabrication and part system design must become integrated and concurrent.

Biological systems represent a successful strategy for the design of polymers, the fabrication of parts and components, and the integration of parts into systems that meet complex performance criteria in a variety of environments. These evolutionary engineered realities have emerged slowly and represent a successful materials identification and fabrication strategy. Examples of biological "design rules" include:

- IDENTIFICATION OF MER SEQUENCES WITH DESIRED PERFORMANCE CHARACTERISTICS THROUGH EVOLUTIONARY ENGINEERING

- CREATION OF A BROAD RANGE OF STRUCTURAL AND FUNCTIONAL POLYMERS (PROTEINS) THROUGH CONTROLLED SEQUENCE ASSEMBLY (ENZYMATIC CATALYSIS) OF A FIXED RAW MATERIAL SUPPLY (20 AMINO ACIDS)

- INTEGRATED, *IN-SITU* CHEMICAL PRODUCTION (CELLS) OF MATERIALS, PARTS AND SYSTEMS WITH NANOSCALE CONTROL

- REPETITIVE USE OF PROVEN STRATEGIES AS EXEMPLIFIED BY HIERARCHICAL STRUCTURE

- EXTREME CONTROL OF THE PROCESSING ENVIRONMENT, ALLOWING STEADY STATE OPERATION FAR FROM EQUILIBRIUM

The possibility of "extracting" the principles of "evolutionary strategy" into modern engineering and scientific practice is an emerging focus of applied polymer science. Combinatorial synthesis may be viewed as computer enabled, real time evolutionary engineering. The broad range of performance characteristics now possible in polyolefins of controlled backbone architecture is simplistically analogous to control of protein function through control of primary structure. The rapid acceleration of computing power is allowing the simulation and zeroth order comparison of competitive materials concepts in cyberspace, allowing better prioritization of expensive experimental resources. Overall improvement in experimental techniques allows more accurate interpretation and generalization of experimental results. The labeling of the polymer industry as mature, hence generally unworthy of research or financial investment, is rejected. Directions and examples will be given of where the authors see opportunities for future industrial, as well as intellectual success. Emphasis is given to the emerging integration of polymer science, chemistry, engineering and biology as the important paradigm for applied polymer science in the twenty-first century.

The purpose of this introductory chapter in the Year 2000 edition of *Applied Polymer Science* is to speculate on where modern polymer science might bring the polymer industry in the twenty-first century. A group of leading polymer researchers were asked to contribute their insights, by specific example, by highlighting recent developments or by prediction of future directions. Much of what is summarized here was presented in the 1999 Symposium, "The Future of Polymer Science" as part of the Division of Polymeric Materials: Science and Engineering of the American Chemical Society celebration of its 75ᵗʰ anniversary.

Directions of Applied Polymer Science in the 21ˢᵗ Century

Synthesis : Jean M.J. Fréchet, University of California

The evolution of polymer synthesis over the past two decades has taken some predictable directions with more emphasis being placed on living polymerizations and the development of powerful and ever more capable catalytic systems. In the area of living polymerizations, the early work of Smets [2], Eniholopyan [3], Otsu [4], Rizzardo [5] and others, led to major developments in nitroxide or metal catalyzed "living" radical polymerizations. Clearly these methods of "living" polymerization will continue to attract much attention as they provide access to relatively well defined polymer and copolymer structures.

More versatility in the selection of monomer and monomer combinations will be required, and the use of systems involving only trace amounts of metal catalyst must be emphasized. A better understanding of the key thermodynamic and kinetic features of these polymerizations must also be acquired while systems more practical for use on an industrial scale are developed.

The unexpected development of divergently grown dendritic polymers [6], first reported by Tomalia in 1984 and Newkome in 1985, that led to a better understanding of the relationship between architecture, functionality, and properties, will continue with additional and significant synthetic efforts. The introduction of the convergent method of synthesis [7] by Hawker and Fréchet in 1989, enabled the construction of monodisperse synthetic polymer structures more precise than ever obtained previously. Today, dendrimers, hyperbranched polymers, dendrigrafts, and hybrid linear-dendritic structures are all being explored in a variety of contexts, from their use in advanced technologies and medicine to the preparation of novel polyolefins with unusual properties.

As the disciplines of chemistry, biology, and engineering come closer, the blending of targets and approaches has had a great effect on polymer synthesis. For example, biomimetic approaches and combinatorial chemistry will certainly claim a more prominent place in the polymer chemist's toolbox.

In the area of catalysis alone, combinatorial chemistry may be used for the development of more functional group tolerant systems, novel architecture, the combination of unusual or previously incompatible building blocks, or the development of materials with enhanced properties. The potential of combinatorial chemistry in the area of catalysts for olefln polymerization and copolymerization alone is tremendous. If olefin monomers, including branch points, are viewed as a fixed set of mers, single-site olefin catalysis has enabled the design of olefinic backbones meeting specific and enhanced performance requirements through control of mer sequence. This has significantly increased the value of olefinic polymers produced from traditional mers and illustrates the importance of controlled backbone architechture in polymer design. In areas of polymer chemistry other than catalysis, combinatorial approaches may be exploited with living systems for the creation of novel materials with controlled architectures, the development of polymers and copolymers with optimized properties, the development of new supports, probes and sensors and the matching of structure and applications.

Finally, polymer scientists must continue to learn from nature. The remarkable precision of some natural polymers, the functional efficiency of the biochemical machinery used for their production and their exquisite combination of structure and function must serve as models, not only for new synthetic targets, but also for new synthetic approaches. The synthetic approaches used by nature also constitute good models as well as challenges for their clever adaptation to unnatural systems. It is likely that approaches making use of encoded information for the development of synthetic unnatural polymers will receive much attention in the decade to come.

Modern polymer synthesis must be concerned not only with the cost-effective production of product but also with the environmental impact of the reaction scheme. The work described below, by J. P. Kennedy et al. of the University of Akron, is an example of modern polymer synthesis leading to improved product in an environmentally friendly reaction scheme.

Commerically acceptable production of high molecular weight polyisobutylene polymers has long been an elusive target. Serious limitations of conventional carbocationic polymerization are 1) the need for cooling to low (sometimes, cryogenic) temperatures and 2) the use of environmentally objectionable chlorinated solvents (i.e., methyl chloride, methylene chloride). Efforts to produce high molecular weight isobutylene polymers and copolymers at reasonable (close to ambient) temperatures in the absence of chlorinated solvents have remained largely unsuccessful [8], [9]. The objectives of this research were to find conditions, including initiating systems, which would lead to high molecular weight isobutylene polymers and copolymers at high rates at low temperatures in neat monomer. The solution to this ambitious aim came by combining the techniques of A) isobutylene polymerizations initiated by high energy (γ-ray) radiation and B) olefin polymerizations induced by organometallic initiators (e.g., metallocenes) acting in conjunction with weakly coordinating counteranions (WCA). γ-Ray initiated isobutylene polymerization yields the highest molecular weight polyisobutylenes in bulk, albeit slowly [8], [9]. The propagating sites are thought to be unencumbered carbenium ions due to the absence of counteranions. Very high molecular weight products arise because neither termination nor counteranion-assisted chain transfer is operating. Metallocene-induced olefin polymerizations usually proceed in the presence of WCAs, such as $[B(C_6F_5)_4]^-$. In these systems, the critical molecular weight building events (coordination, insertion and propagation) occur at the cationic metal center in the kinetic absence of the WCA. In view of these facts, γ-ray and metallocene induced olefin polymerization are similar. In line with this analysis, novel initiating systems have been sought that would induce efficient cationic polymerization in the absence of counteranions. A number of relevant publications have appeared recently [10], [11],[12], [13], [14],[15], [16]. They summarize recent work aimed at the preparation of highest molecular weight PIB and butyl rubber by the use of the *in-situ* prepared $(CH_3)_3Si^+[B(C_6F_5)_4]^-$ initiating system in the -35 to $-8°C$ range in the essential absence of chlorinated solvents. Details of polymerization techniques have been published [17].

It was hypothesized that the hypothetical Bronsted acid $H^+[B(C_6F_5)_4]^-$ would be the needed initator to mediate cationic polymerizations in the essential absence of a counteranion. Since this acid cannot be isolated, its precursor, $(CH_3)_3Si^+[B(C_6F_5)_4]^-$ was prepared by the following route:

$$(CH_3)_3SiCl + Li[B(C_6F_5)_4] \rightarrow (CH_3)_3Si^+[B(C_6F_5)_4]^- + LiCl$$

The initiating species was prepared *in situ* in the charge. Since $Li[B(C_6F_5)_4]$ is insoluble in neat monomer, it had to be dissolved in a little toluene or chloromethane. Thus in reality, the systems are only close-to-neat. It was found that isobutylene polymerizations can be readily initiated by adding $Li[B(C_6F_5)_4]$, dissolved in toluene or chloromethane, to neat monomer containing chlorotrimethylsilane at various temperatures (even under reflux at -8°C). A series of kinetic experiments were carried out under various conditions in which the rates of polymerization and polymer molecular weights obtained were determined. Interestingly, both the rates and molecular weights were higher when the lithium salt was dissolved in toluene (as compared to chloromethane). The reason for this may be that solvation of the propagating cation by chloromethane retards propagation and accelerates chain transfer (i.e., reduces molecular weights relative to those obtained by the use of toluene). The effect of temperature on molecular weight was determined. Figure 3 shows the Mn's obtained by γ-rays initiated isobutylene polymerization (solid lines) and values obtained in this work (filled-in points). For comparison the figure also shows the log Mv (viscosity average molecular weights) versus I/T line obtained by the use of AlCl₃ in methyl chloride [11].

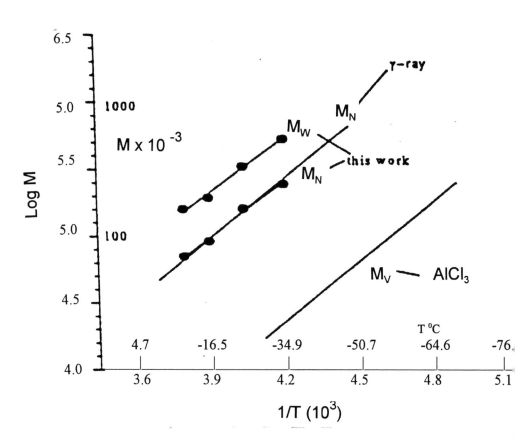

**Figure 3: Temperature dependence of PIB molecular weights with
different initiating systems in the –8 to –80 °C range**

It is concluded that $(CH_3)_3Si^+[B(C_6F_5)_4]^-$ prepared *in-situ* in close-to-neat isobutylene charges leads to highest molecular weight PIB at various temperatures (up to $-8°C$, reflux temperature). The presence of a proton trap inhibits polymerization, thus the initiating entity is most likely the hypothetical acid $H^+[B(C_6F_5)_4]^-$ arising from traces of moisture in the charge. The very high molecular weights are explained by proposing that counteranion-assisted (zero order) chain transfer does not occur in the presence of $[B(C_6F_5)_4]^-$ The high molecular weights are the consequence of a spontaneous (first order) transfer, a process less favorable than the zero order transfer.

Through the application of modern polymerization techniques and catalytic insight, an improved, environmentally friendly and commerical relevant PIB synthesis was demonstrated.

Molecular Driven Assembly, C.D. Eisenbach, Universität Stuttgart

Polymers have high potential as structural and functional materials. However, synthetic polymer systems are still far from emulating the elegant schemes of biological systems in self-organization phenomena and the formation of complex tertiary or quaternary structures with specific functions. In biopolymeric systems, supramolecular structure results from special sequences of amino acids along the polypeptide chain and/or periodic sequences of other building units along the chain. This results in highly organized structures which, in most cases, are stabilized by hydrogen bond formation or ionic interactions. A key problem in the creation of new polymeric structures with assembling principles analagous to biopolymeric systems is understanding how the overall morphology of both separated microphase and single phase systems is controlled. This understanding must include insights into the relationship of how the formation of supramolecular structures is defined by the conformation of the single chain, as controlled by the periodicity of repeat units or special building blocks along the chain, and the type and amount of interactions in a given macromolecular system. This is first illustrated in Figure 4 with the example of a segmented block copolymer consisting of a periodic sequence of two or more segments differing in their chemistry. In a hierarchical view [18], the simplest case of self-organization would be the formation of a phase separated system. The next higher level is represented by phase morphologies where the segments have adopted their thermodynamically preferred conformation and packing order. This is represented in Figure 4 by the domains formed from chain-extended crystallized segments, followed by double helices. It is obvious that the directionally controlled self-assembly of macromolecules has a high potential for novel structure formation in polymers.

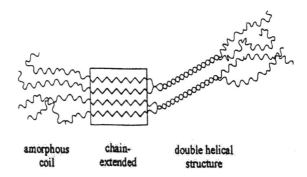

amorphous chain- double helical
coil extended structure

**Figure 4: Schematic of possible (supramolecular) strctures of macromolecules
composed of incompatible and/or self-recognizing segments.**

Another challenging example is illustrated in Figure 5 where the two limiting morphologies of a blend consisting of rigid rod and random coil macromolecules are depicted. It is well known that the miscibility of two component polymer systems decreases when the rigidity of one of the blend components increases. While this might be advantageous in the formation of microphase separated materials, it prevents the formation of a molecular dispersion of rodlike molecules in a coil molecule matrix, which is the ideal polymer-polymer composite (molecular composite) [19].

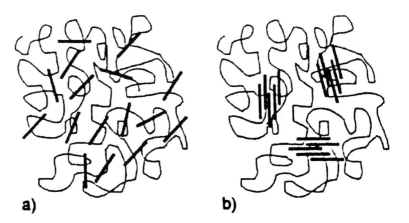

a)   **b)**

**Figure 5: Schematic of the morphology of a rigid-rod/flexible coil polymer blend: (a) molecularly
dispersed mixture and (b) microphase separated system with segregated rigid rods**

Block or segmented copolymers consisting of highly flexible polydisperse segments and molecularly uniform segments capable of either the formation of chain-extended and chain-folded crystallites or double helical metal complexes have been synthesized and the relationship of the suprastructure to the molecular architecture and sample history has been studied [20], [21]. It was demonstrated that the self-organization of macromolecular systems on the segmental level into phase-separated systems of nano to mesoscopic dimensions is determined by the constitution of the whole macromolecule. The incorporation of specific constitutional units into the polymer backbone allows the mimicking of conformations typically formed in biopolymers, such as a coil, chain-extended or chain-folded (β-sheet) crystallites, or even a double-helix [20], [22], [23].

The study of a homologous series of segmented polyurethane elastomers with molecularly uniform hard segments [24] has shown that the polyurethane segment crystallizes without chain folding (cf [25, 26]). It was found that chain folding is induced by replacing only a single tetramethylene unit by a more flexible unit such as hexamethylene or octamethylene (because this defect in the otherwise regular structure acts as a flexible joint) [27]The same morphology of chain-folded hard segment domains was also found for graft copolymers consisting of a poly(oxytetramethylene) backbone and bifurcated polyurethane grafts, molecularly uniform N-alkylpolyurethanes based on piperazine and butanediol bischloroformate. These were attached to the backbone to allow a symmetrical folding of the graft, as illustrated in Figure 6 [28]. Figure 6 also shows that the dynamic-mechanical properties of the linear segmented block copolymers and the graft copolymers are similar.

A-B-A triblock copolymers consisting of a polyether center block and oligo(bipyridine) outer blocks exhibit a chain extension reaction upon addition of Cu(I) ions, resulting in a block copolymer with alternating polyether segments and segments of tetra coordinated (oligo)bipyridine-Cu(I) complexes [29]. These ion complex segments segregate to form a supramolecular assembly, and the formed microphase separated system exhibits properties typical of a thermoplastic elastomer [23]. These studies are further extended through the encoding of a synthetic macromolecule with a special bipyridine sequence. The spontaneous self-assembly of bipyridine or bis(bipyridine) units is built into a polymer upon addition of copper(I) ions to give tetracoordinated copper(I) complexes. A double helical structure in the latter case, has been used in combination with the cooperativity of the ion complex formation to control selectivity and directionally controlled polymer-ion-complex formation. This is schematically illustrated in Figure 7 for a bipyridine/bis(bipyridine) model compound representing a simple encoding of a macromolecule. Complex formation leads to dimer formation, only as revealed by the MALDI spectrum; no other specimens, such as oligomers which could result from a series of anti-parallel complex formation events, could be detected. This directional control is retained up to a limiting spacer length between the bipyridine and the bis(bipyridine) moieties and is also observed for polymer-substituted bipyridine codes.

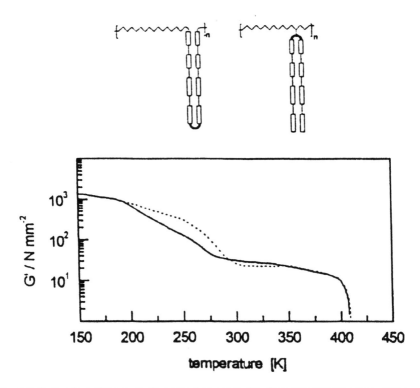

Figure 6: Comparison of the dynamic mechanical properties of a segmented polyetherurethane with monodisperse chain folded hard segment (solid line) and of polyether-g-polyurethane with bifurcated graft (dotted line); both systems have the same polyether/polyurethane volume fraction

Miscibility of otherwise immisicible rigid-rod and flexible random coil polymers can be achieved by acid-base interactions between the blend components. In such ionomer based blends, rod-like polymers can be dispersed in flexible polymer matrices. Examples of this approach are dispersions of rod-like molecules such as pyridyl substituted polydiacetylenes, sulfonic acid substituted poly(p-phenylene) or para or meta linked poly(phenylene/pyridyl-acetylene) in flexible matrix polymers carrying complementary ionic groups. Phase separation is prevented by the sufficently negative enthalpy of mixing, in agreement with theory [30]. The Young's modulus of these dispersions was increased by up to three orders of magnitude when compared to the pure matrix polymers [31]. In view of the impact of these findings, poly(arylene-ethinylene) copolymers with variable flexibilty as controlled by the ratio of para to meta linkages were synthesized and used as blend components. Figure 8a shows the stress strain curves of sulfonated polystyrene with the rigid p-linked polymer (PPPyPE) in comparison to the blend with the comparatively flexible polymers containing m-linkages. The modulus decreases significantly upon the incorporation of m-linkages in the reinforcer poly(phenylene/pyridyl-acetylene) (Figure 8b); the observed reinforcement factor as related to the mole ratio of the m-linked pyridyl units corresponds to the change of the Flory parameter for the rod-coil transition.

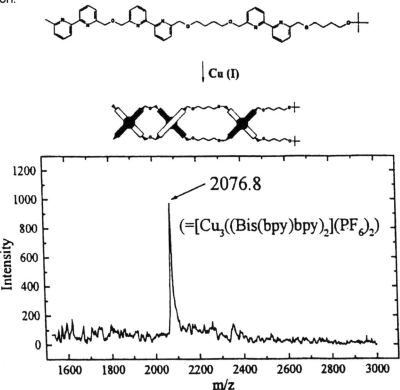

Figure 7: MALDI spectrum of the ion complex formed upon addition of Cu(I) to the bipyridine-bis(bipyridine) compound

Through the tailoring of the sequence distribution and the concentration and nature of functional groups incorporated into a polymer backbone a variety of self-assembled and phase separated morphologies can be achieved. A scheme has been defined to achieve the long elusive reinforcement of a flexible-coil polymer by rigid rod polymer, with the expected increase in mechanical properties noted. Further exploration of these phemnomena will form the basis for the design of novel and smart polymer systems in the twenty-first century.

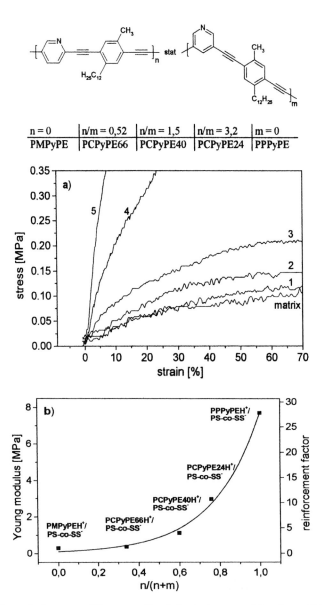

Figure 8(a)-Top: Stress-strain curves of the pure sulfonated polystyrene matrix polymer PS-co-SSH (11 mole-% sulfonation) and of blends (stoichiometry of acid/base groups) with poly(arylene-ethinylene) of different flexibility (n/(n+m)); curve 1: PMPyPEH⁺/PS-co-SS; curve 2: PCPyPE66H⁺/PS-co-SS; curve 3: PCPyPE40H⁺/PS-co-SS; curve 4: PCPyPE24H⁺/PS-co-SS; curve 5: PPPyPEH⁺/PS-co-SS. Figure 8(b)-Bottom: Dependency of the Young modulus of blends of sulfonated polystyrene/poly(arylene-ethinylene) on the flexibility (n/(n+m)) of the poly(arylene-ethinylene)

Mesophases, M. Jaffe, New Jersey Institute of Technology

The recognition of the importance of mesogenicity in the processing of linear polymers has been evolving since the recognition of the nematic character of stiff macromolecules in the late 1970's; see for example [32], [33], [34]. From initial concepts of routes to high strength and modulus in polymeric fibers, to current models of spider silk formation, mesogenic states play an increasingly important role in elucidating the key steps in the production of oriented polymer microstructures. For convenience, mesogenic polymers may be arbitrarily placed into four classes. These are:

- PERMANENT MESOGENS
- TRANSIENT MESOGENS
- ACCESSIBLE MESOGENS
- ASSEMBLED MESOGENS

Permanently mesogenic polymers have stiff molecular backbones with aspect ratios that satisfy the Flory [35] criteria for spontaneous nematogenic phase formation. All the polymers commonly referred to as "liquid crystalline polymers" or "LCPs", i.e. aramids, thermotropic polyesters, "ordered polymers", are members of this class. These polymers are characterized by high local order in the quiescent state, low viscosity in uniaxial flow fields and can be easily transformed into globally oriented uniaxial structures through extrusion.

To be processible these polymers must be in the mesogenic state at temperatures below their decomposition temperatures. This is achieved either through formation of lyotropic solutions in strong solvents or, in the case of the aromatic polyesters, the incorporation of comonomers into the p-phenylene ester backbone that decreases the crystalline melting point while preserving backbone linearity [33], [34].

In the solid state, fibers of these polymers typically exhibit high molecular orientation (f > 0.95), tensile moduli close to theory, tensile strengths of 3-4 GPa, and poor compressive and shear properties. Of emerging importance are the excellent gas barrier properties of LCPs, as well as anisotropic conductivities. Properties are controlled by inherent chain properties, molecular orientation, stress transfer mechanisms and defect structure. These polymers are now established in the marketplace, and industrial research is focused on cost-reduction and improved compressive properties.

Of interest are new mesogens such as the recently identified large bent-rigid or banana-like structures [36]and elucidation of stress transfer mechanisms. The much improved compressive properties in LCP fibers recently demonstrated by AKZO scientists [37], [38] through a highly hydrogen bonded, linear benzimidazole backbone confirms the importance of strong interchain interactions on mechanical performance. Concepts of ravels in LCP molecular ensembles are also an important emerging concept for the control of processibility and mechanical properties [39].The mechanism of the large (up to 5X) increases in tensile strength and strain to failure observed with thermotropic polyester fibers annealed close to their melting temperature is a continuing area of research focus, with the objective of identifying molecular and/or processing routes that lead directly to the desired level of tensile properties. Long associated with polymer molecular weight increases through solid state polymerization, recent work by Ward [40] and others suggests improved structural perfection (leading to increased levels of chain interactions) and the reorganization of ravel structures may also play a significant role.

Transient mesogens are mesogenic states locally induced into flexible, random coil polymers through perturbation of molecular trajectories. This perturbation is caused through the application of external fields, usually mechanical, and the resulting transient mesogenic phase is local in nature. Phase formation will be initiated when local chain orientation perturbations satisfy the geometric requirements for mesogenic phase formation, and will occur within the entangled network at points of minimum chain entropy, such as entanglements. Fibrillar in nature, the transient mesophase may be hypothesized to be the nucleating phase key to the formation of row and shish-kabob like structures in oriented polymer crystallization. While experimental evidence is beginning to emerge to support this hypothesis, the low concentrations and short lifetimes of these structures makes rigorous identification difficult. An understanding of transient mesophases would be most important in the understanding and control of high speed (high stress) melt spinning and provide the conceptual basis for models of structure formation that can describe both

FIBER SPINNING
STRUCTURE FORMATION

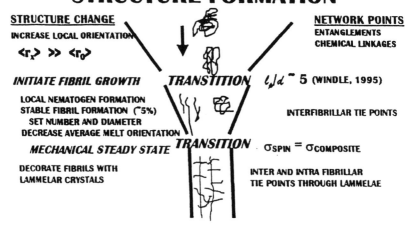

Figure 9: Melt Spinning Model

16

morphological observation and the nature of tie molecule populations. Consistent with these concepts, the melt spinning of conventional, flexible polymers under conditions where chain orientation precedes crystallization can be divided into three structure formation regimes:

- CHAIN ORIENTATION
- LOCAL, FIBRILLAR MESOPHASE FORMATION
- FOLDED CHAIN CRYSTAL DECORATION OF PERSISTENT FIBRILLAR MICROSTRUCTURE

A summary of the proposed model is illustrated in Figure 9.

Accessible mesogens are molecules that are thermotropic but have an accessible isotropic phase (clearing temperature below the decomposition temperature of the polymer). An example of such a polymer is the copolymer of p-hydroxybenzoic acid, isophthalic acid and hydroquinone, recently explored and abandoned as a commercial LCP. Such polymers allow processing from either the mesogenic state (with typical LCP processibility and performance) or quenching through the mesogenic state at a rate sufficient to preclude mesophase formation, leading to the formation of a metastable, isotropic glass or rubber [41]. In the metastable, isotropic state the polymer can be manipulated similar to conventional polymer without the restriction the inherently high local orientation of LCPs imposed on the creation of multiaxial or isotropic orientation distributions. As shown by the author and coworkers, annealing of the metastable, isotropic films leads to a global transformation of the microstructure from an isotropic to a nematogenic texture, with performance characteristics of the films reflecting the phase transformation [42]. An example of this is the two order of magnitude reduction in acetone vapor permeability observed in a film annealed at 200°C. Similar behavior is noted with other gaseous permeants, as illustrated in Figure 10 [43]. The transition from the isotropic to the nematogenic state in polymers is poorly understood, especially in the solid state. Kinetic studies suggest a mechanism more consistent with an aging process than a first order phase transition. Manipulation of the I→N transition is a fruitful area for future research.

Assembled mesogens are illustrated by the assembly of polymer coils or molecular moieties such as side-chains into geometries satisfying the criteria for mesophase formation. The most intriguing application of this concept is the suggestion that the lyotropic phase identified in the spinnerette of silk extruding spiders may represent a "string of beads" assembly of non-interpenetrating, proteinaceous coils [44]. If proven correct the assembly of polymeric molecules to form oriented, nematogenic-like arrays may be a completely new paradigm for the formation of oriented structures in synthetic polymers. Hydrogen bond assembly of side-chain polymers to form mesophases has also been demonstrated.

For all the molecular species described above, processibility implies a starting network capable of transmitting stresses across the polymer ensemble. This network is characterized by molecular chain trajectories, entanglement density (or equivalent), the distance between entanglements and the molecular motions controlling the network dynamics. Entanglements in conventional polymers have long been an area of intense study, but little attention has been focused on the tie-points between polymers in mesogenic states. Recent suggestions include persistent microcrystals as

HIQ40
GAS PERMEABILITY

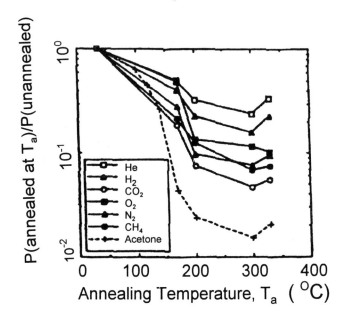

Figure 10: (Don Paul et. al.) Permeability of cast HIQ40 film as a function of annealing history

suggested by Wissburn [45] ravels as suggested by Collins [39] and the NPL concepts of Windle [44]. An understanding of the origins of the network and the control of network tie-points is fundamental to the understanding of mesogenic polymer processiblity and performance.

In conclusion, it is suggested that the polymer mesophase is omnipresent, and plays a critical role in most polymer processing. Mesogenicity may be local or global, and when mesogenic textures are quenched to the solid state they play a key role in determining both performance level and property balance. The mesophase may be permanent or transient and often serves as a template for morphology development.

Control and manipulation of polymer mesophases represents an attractive route to new products and processes, as well as a fruitful area for continuing research.

Molecular Modeling/Computer Simulation, A. Windle, Cambridge University

The rapid increase in computing power coupled with equally rapid decrease in computing costs has led to speculation of the computer replacing experimental science by accurate simulation of natural phenomena. While it is likely that computers will more and more lead us to invention and discovery, the methodologies allowing this to occur have not yet been demonstrated. It has been shown, however, that the simulation of isolated polymeric chains can give useful insight into probable polymer performance and is an effective first step in complex experimental design.

The 1990's saw rapid advances in the ability to model most probable polymer chain conformation [46] and to relate these conformations to simple properties such as Young's modulus [47]. Computing power also enables the increased use of theory; calculations that once took days can now be completed in minutes on desktop machines. It is reasonable to ask why one would model a phenomena where a credible theory exists. The answer lies in the flexibility of modeling, the ability to relate phenomena across orders of magnitude of size scale, the ability to access classes of non-equilibrium structures where theoretical description remains elusive and mostly, the versatility of models to describe real situations. The most desirable use of modeling today is in conjunction with theory and experiment as shown diagramatically below (Figure 11):

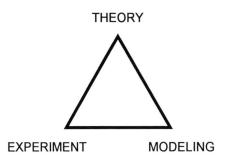

Figure 11

Figure 12 illustrates a scheme for the digital definition of materials, ranging from the subatomic through the prediction of macroscopic properties. Figure 13 shows the hierarchy of modeling size scales that must be accomplished if the properties of a polymer are to be usefully predicted from first principles.

DIGITAL DEFINITION OF MATERIALS

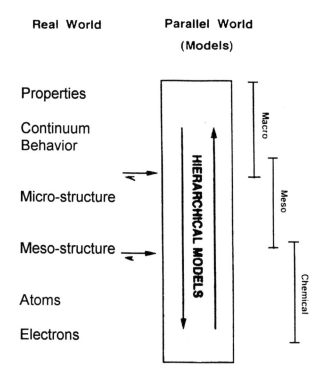

HIERARCHICAL MODELS   **Advantages:**

**Relates changes in atomic and molecular structure
right through to useful properties**

Figure 12

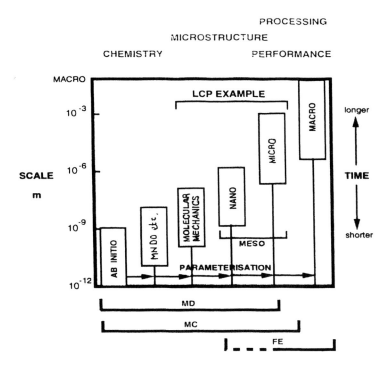

MODEL ING   HIERARCHY

Figure 13

Meso modeling is the generic term given to modeling at a sub macroscopic scale which does not include what might be termed 'chemical detail'. It is, perhaps, useful to subdivide 'meso' modeling into:

- Nano modeling - modeling on a scale small enough to enable a Monte Carlo approach to have meaning, i.e., the elements are of a size - of the order of a nanometer - where Brownian motion is significant. There is, however, no chemical detail which involves atoms as such. Examples of nano modeling are the modeling of polymer melts, where the molecule is represented by a chain, and models of the liquid crystalline state of the Lebwhol-Lasher type where each rod-like molecule is represented by its director only.
- Micro modeling, where the scale is larger, is distinguished by the fact that inter unit behavior is parameterised as materials properties, such as elasticity or viscosity. Examples of micro modeling are drawn from the relaxation of liquid crystalline microstructures and the influence of flow and boundaries on them.

Figure 14 shows how the modeling hierarchy described above relates to LCPs, from the molecular simulation of chains [48], [49] through a lattice model describing the liquid crystalline melt, complete with disclinations [50]

Model ing Hierarchy

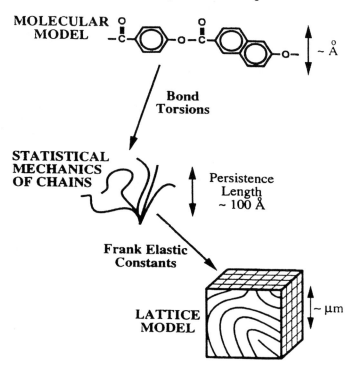

Figure 14

Figure 15 is a micrograph of a typical disclination observed in a nematic liquid crystalline melt while Figure 16 shows a lattice of directors that represents a similar defect. Please see [51], [52] for background details.

Figure 17 is a calculated representation of how such a defect may eradicate through escaping in the third dimension. Such results give useful insights into the production of more perfect structures and suggest directions for improved processing schemes. The theme of microstructural development must include both a description of the structural aspects of the nematic-isotropic transition, as well as the development and relaxation of topological features within the liquid crystalline phase. Meso modeling is able to draw on parameterization available from calculations made

Figure 15

Slice through Polymeric Case Model

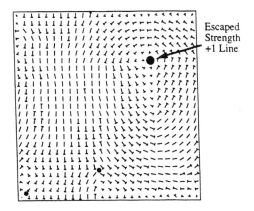

Escaped
Strength
+1 Line

Twist Escaped +1 Lines involve no splay

Expect them to be favoured in LCPs

Figure 16

Schematic of a Twist–Escaped Line

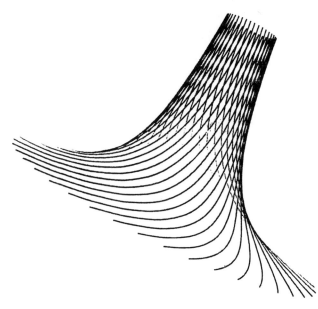

Figure 17

at the molecular level which provide rotation potentials as well as parameters such as persistence length. The interesting aspect of meso modeling at the micro level is that it maps seamlessly onto established finite element methods which are used routinely for macroscopic continua.

The twenty-first century will see rapid improvement in the modeling of materials and what are in 1999 considered to be advanced computing techniques will undoubtedly be routinely performed by students in the next several decades. Polymer modeling will transform from a technique that complements experiment to the methodology that will define experimental directions.

Biological Polymerization, D. Tirrell [53], California Institute of Technology

All synthetic polymerization reactions are statistical in nature, i.e., all lead to distributions in molecular weight and often to distributions in end-groups and mer sequencing. While this is not limiting in all applications, it restricts the ability to investigate and exploit the advantages of precise molecular specificity. With increasing technological interest in forming well defined architectures and surface chemistries for biological, and device applications [54], [55], [56] the desire to synthesize precisely defined macromolecular structures has increased. Biology represents an appropriate model for producing these polymers, in analogy with the precise chemical architectures (stereospecificity, mer sequencing, DP) associated with naturally occuring proteins and peptides. *In-vivo* protein synthesis may be viewed as a template directed polymerization in which messenger RNA directly encodes cellular DNA information. The ability to design and encode DNA sequences into cells using well established recombinant methods allows the use of *in-vivo* protein synthesis as a test-bed for assessing the utility and importance of polymer synthesis leading to exact and reproducible molecular structures. This methodology is shown diagramatically in Figure 18.

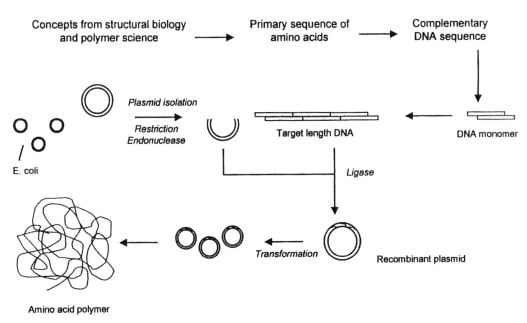

Figure 18: Schematic diagram of gene construction and protein synthesis

Recombinant DNA techniques have been employed to create artificial proteins that form well defined crystalline and liquid structures with attractive bulk, surface and biological properties [57], [58], [59]. Further exploration of the property limits of polymers with precisely controlled primary structure is enabled by the incorporation of non-naturally occurring amino acids into these artificial protein backbones. In this fashion, materials with unique surface and functional properties have been created [53]. Such results highlight the enormous potential of these methods to create unique polymers of commercial value in the future. Increased mechanistic understanding of biological template polymerization, i.e., enzymatic catalysis, could lead to the ability to create polymers of proteinaceous specificity in a variety of chemistries and herald a new renaissance in polymer research.

Biomaterials, J. Kohn, Rutgers University

Biomaterials are synthetic poymers designed for either *in-vivo* use or to come into intimate contact with with biological fluids, such as blood, in a living organism. Currently the majority of all implanted medical devices are permanently implanted prostheses, intended to remain within the body of the patient for extended periods of time. In the near future, advances in tissue engineering and tissue reconstruction will make it possible to replace some of these prostheses by temporary polymeric scaffolds that will support the regrowth of functional tissue. These new types of medical implants will most probably be based on carefully designed, new, resorbable polymers. The development of such biomaterials requires an interdisciplinary team of scientists and represents a complex challenge.

The design of tyrosine-derived polycarbonates illustrates some of the challenges in finding an optimum balance in basic material properties, processibility, degradation behavior, and biocompalibility. The synthesis of tyrosine-derived polycarbonates was published by Pulapura and Kohn in 1992 [60]. Methods for the *in-vivo* tissue compatibility testing in a subcutaneous rat model were reported by Silver et al. [61]. Detailed procedures for the evaluation of polymers as biomaterials are provided by Ertel [62]. A specialized set of experimental protocols for the evaluation of orthopedic biomaterials has been described by Ertel, Choueka, and James [63], [64], [65]; Hooper et al. describes the study of polymer degradation [66].

Tyrosine-derived polycarbonates are amorphous polymers, characterized by their relatively high strength and a stiffness which exceeds that of other degradable polymers, such as poly(ε- caprolactone) and poly(ortho ester)s, but is somewhat lower than the values reported for semicrystallinc materials such as poly(L-lactic acid) and poly(glycolic acid). These observations indicated early on that tyrosine-derived polycarbonates could be used in the design of low load-bearing orthopedic implants such as pins, rods, screws and plates. Tyrosine-derived polycarbonates have an alkyl ester pendant chain whose structure can be varied, providing a convenient means to fine-tune polymer properties (Figure19). The strongest and stiffest material among all tested polycarbonates was poly(DTE carbonate), the polymer having an alkyl ester pendant chain.

Small bone fixation devices usually need to retain significant strength for 6 to 12 months to allow sufficient time for bone healing. Thus the mechanism and rate of degradation and the degree of strength retention for various tyrosine-derived polycarbonates was investigated. These polymers were found to degrade on the order of months to years [62]. Using a combination of surface analysis (ESCA, ATR-FTIR), gel permeation chromatography, and kinetic modeling, the mechanism of degradation was elucidated [62]. Among all tested polymers, poly(DTE carbonate) provided the optimum strength retention profile. The more hydrophobic polymers, such as poly(DTO carbonate) degraded slower, but, surprisingly, tended to lose their strength faster. Thus, poly(DTE carbonate) was identified as the most promising candidate material for the intended application.

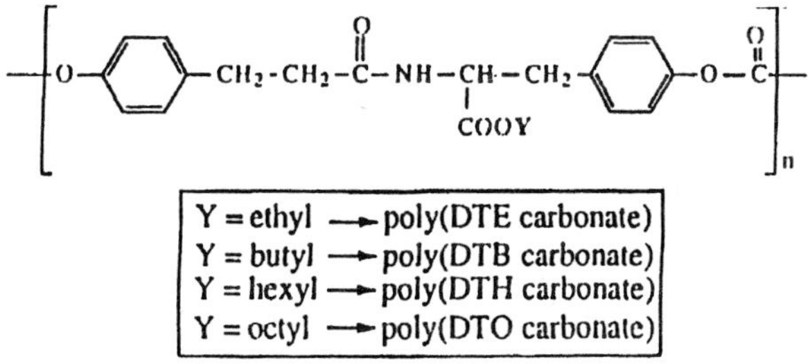

Y = ethyl ⟶ poly(DTE carbonate)
Y = butyl ⟶ poly(DTB carbonate)
Y = hexyl ⟶ poly(DTH carbonate)
Y = octyl ⟶ poly(DTO carbonate)

Figure 19: Chemical structure of tyrosine-derived polycarbonates monomer is an analog of tyrosine-dipeptide, consisting of desamino (hydroxyphenylpropionic acid) and L-tyrosine

Since tyrosine-derived polycarbonates are amorphous they are easier to process than semicrystalline materials. The wide gap between a relatively low glass transition temperature (below 100°C) and a relatively high decomposition temperature (above 250°C) makes the tyrosine-derived polycarbonates easy to extrude, compression mold, or injection mold. Pins, rods, and screws required for implantation into test animals could be obtained by extrusion and injection molding. Porous scaffolds were also prepared [67].

In order for tyrosine-derived polycarbonates to be utilized as medical devices, they must be able to withstand the conditions associated with sterilization. Currently used resorbable polymers such as poly(lactic acid) and polyy(glycolic acid) are sterilized by exposure to ethylene oxide. These materials degrade during sterilization by γ-irradiation. A recent study [68] showed that tyrosine-derived polycarbonates can be sterilized by γ-irradiation. This is a major advantage, considering the occupational hazard associated with the use of ethylene oxide.

The studies described so far provided a base for the selection of poly(DTE carbonate) as a lead candidate for prototype development. Appropriate devices have been fabricated, sterilized, and tested for biocompatibility in several animal models. In 1999 pins made of poly(glycolic acid) and polydioxanone are in clinical use in the USA., while poly(L-lactic acid) is being used in Europe. However, a number of publications linked the acidic nature of the degradation products and the release of crystalline particulate debris to a documented late inflammatory response [69]. Thus the release of acidic or toxic degradation products was a focus of studies and the biocompatibility of tyrosine-derived polycarbonates was evaluated in side-by-side comparisons with medical grade poly(L-lactic acid). An assay of biocompatibility in a subcutaneous rat model showed that the soft-tissue response to poly(DTH carbonate) is not significantly different from the responses elicited by medical grade poly(L-lactic acid) or polyethylene [61]. This initial study was followed by a detailed evaluation of the bone response to poly(DTH carbonate) in a side-by-side comparison with polydioxanone in a rabbit model [63] and by a comparative study of poly(DTE carbonate), poly(DTH carbonate), and medical grade poly(L-lactic acid) in a canine bone chamber model [64]. In all of these model systems, devices made of tyrosine-derived polycarbonates performed equal to or better than the currently used poly(glycolic acid), polydioxanone, or poly(L-lactic acid) based devices, as judged by the inflammatory response observed around the implant, the degree of bone apposition at the implant-bone interface, and the degree of long-term osteoconductivity.

In all biocompatibility assays, the performance of poly(DTE carbonate) was superior to that of all other tyrosine-based polycarbonates as well as the currently used polymers. The studies leading to the identification of poly(DTE carbonate) as a promising new biomaterial required the collaboration of interdisciplinary teams of chemists, material scientists, engineers, cell-biologists, and clinicians. Future work in this area will focus on decorating polymer surfaces with biorecognizable moieties that trigger specific protein adhesion and foster specific cell growth. This sophisticated mimicing of the minimum signals of the extra-cellular matrix (ECM) necessary to trigger a desired biological response will allow the regeneration of specific damaged tissues and organs. The polymer substrate needs to be tailorable in bioerosion rate, and mechanical performance; a lack of inflamatory *in-vivo* reponses is equally important for the scaffold or implant to be successful. The ability to successfully emulate the ECM will also trigger the identification of new diagnostic materials, sensor technologies and targeted drug delivery systems.

Polymer Characterization, J. Koenig, Case Western Reserve University

Today's polymer laboratories are striving to increase productivity, maintain quality, and simultaneously comply with an ever increasing number of regulatory mandates. They also are faced with other challenges: greater sample complexity, expanding numbers of samples and a decreasing number of skilled professionals.

The challenge is to develop an automated integrated spectroscopic machine capable of analyzing many samples without manual intervention in sample preparation. Spectroscopic chemical analysis can be divided into three main categories: 1) sample selection, separation and preparation for the specific spectroscopic system, 2) spectral detection and substance identification, and finally, 3) material quantification. Integrated procedures are sought where the entire process from sample preparation through chemical analysis to data handling and reporting can be realized in a **single automated step**.

Future instrumentation must involve **total integration and automation**. Such integrated and automated instrumentation can be termed a **virtual machine**. A virtual machine is one which receives the sample, ascertains the appropriate sampling technique, designs the most expedient measurements, performs the tasks in a prescribed way, carries out the needed data analysis and reports the results in a manner most easily assimilated by the user. All of the required components exist for integrating the analysis process into a virtual molecular spectroscopic machine.

Traditionally, the limitation has been the problem of converting the raw sample into a form suitable for spectroscopic analysis. This part of the analysis needs to be rethought. While it is true that better results will be obtained by having the sample as nearly ideal optically as possible; it will be at the expense of substantial loss of time and effort. The time is ripe to re-examine polymer sampling procedures in terms of the potential for time and labor savings. For many modern molecular spectroscopic methods, little sample preparation is required when reflection or emission or microscopic methods rather than the traditional transmission methods obtain the spectra. One should also consider fiber optic evanescent wave spectroscopy (FEWS) for sampling of liquids and streams where the optical fiber acts as an attenuated total reflection sensing element. Using newly developed optical fibers with optical losses of only 0.2 dB/m, one can obtain measurable multifrequency signals from a sample a few meters away.

Using computer-driven robotics for positioning and alignment of the sample (or the beam), automated sampling techniques can potentially be used. Robotic operations and transfer systems are common place and automatic sampling is available for most spectroscopic systems. Using robotics has several advantages including the obvious savings in human effort. Modern robotics cut costs dramatically while improving precision and cutting sample turnaround time in half. The average payout on automation is about 12 months. The decrease (improvement) in coefficient of the variation may be about a factor of five.

The design and evaluation of the spectroscopic measurements can be done by an **expert system**. An expert system is defined as a reasoning system that uses a knowledge base to capture and replicate the problem solving ability of human experts. It appears an expert system can be trained and developed based on a set of laboratory experiences which will allow the system to analyze the results of the spectroscopic measurements in terms of previously experienced sampling problems and alert the spectroscopist. An obvious extension is to have the expert system make the necessary corrections once the problem is recognized. Anyone who has used a computer for more than an hour has encountered the inevitable "error message." Seldom do such messages tell what is wrong, only that something is wrong. Why can't the message contain options suggesting possible sources of the error and allow the operator to select an appropriate "correction" which is then initiated by the computer. Or better yet, have the system do a self-diagnosis and correct the error itself. There is some risk here because a computer cannot read minds but, within a prescribed framework of experimental experiences, the expert system can recognize proper measurements relative to improper ones. Expert software systems coupled with the appropriate sensors can be used for establishing and modifying (when necessary) the measurement procedures. Networked tandem instrumentation allows several simultaneous determinations using independent sampling methods. Preprocessing of the spectral data can eliminate the nonspecific absorptions arising from optical inhomogeneities like particles and from the atmosphere or other contamination sources. Multivariant analysis systems and statistical testing methods can be used to automate the final data processing and calibration.

The virtual machine described above is one approach to solving some of the challenges of the current spectroscopic analysis laboratory. It can happen. It will happen.

Polymer Processing, I. M. Ward, University of Leeds

Polymer processing defines the structure and, hence, the performance of polymers in all possible end-uses. From the starting state of the dynamic molecular network that characterizes polymer melts and solutions to the detailed molecular chain topology that defines the interconnection of the final part morphology, processing plays the key role in defining shaped part performance. Part morpholgy is defined by crystal content, morphology and orientation, non-crystalline content and orientation and the all important interconnections between these morphological units. Property levels are determined by molecular chain orientation, and chain packing; crystallinity and strained chain conformation impact property retention in the neighborhood of the glass and melting transitions. The impact of polymer processing on performance is equal to that of polymer chemistry – the extruded and drawn polyethylene of trash bags and the gel-spun and ultradrawn 200+ Gpa modulus PE ballistic protection fibers are essentially identical chemically. Polymer processing in the future will center on the management of chain entanglements, structural defects, and process control.

Several aspects of molecular structure control are discussed in terms of their influence on solid phase processing of polymers: (1) the existence of a molecular network, (2) the initial morphology in crystalline polymers and (3) thermally activated processes. Key illustrations from recent research at Leeds University include the tensile drawing of polyesters, where the network defines a true stress-strain curve; tensile drawing, hydrostatic extrusion and die drawing of polyethylene where the effects of the network and initial morphology are controlled by molecular weight and thermal treatment, including annealing under high pressure.

It is very well recognized that the mechanical properties of fibers and films are greatly affected by the nature of the stretching processes employed to control the molecular orientation and, in the case of crystalline polymer, the nature and degree of crystallinity. For fibers and uniaxially drawn fibers, there is generally a tensile drawing stage and it is well-known that properties such as modulus which relate primarily to molecular orientation often show a very clear relationship to the draw ratio. A more useful parameter, which can take into account any preorientation of the sample which may have been introduced at the melt processing stage, is the network draw ratio. This implicitly assumes that tensile drawing involves the deformation of a molecular network, analogous to stretching a cross-linked rubber. For polymer processing this assumption leads to two useful ideas:

> • First, the physical properties of the drawn polymers, such as the Young's modulus, relate to the network draw ratio and, perhaps, also to other parameters such as crystallization or in the case of tensile strength, molecular weight.
> • Second, the flow stress required to model the mechanics of the drawing behaviour relates to the network draw ratio, because this represents accurately the degree of plastic strain, and also to the current strain rate and temperature defined by the processing conditions at each point in the processing route.

The network draw ratio can be determined by matching the strain hardening part of the true-stress true strain curves obtained from redrawing the oriented polymers obtained at different stages of orientation during melt spinning or melt spinning followed by hot drawing. An example of the procedure is illustrated by Figure 20 for polyester fibers and the corresponding relationship between Young's modulus and network draw ratio is shown in Figure 21, which is very straightforward compared to that obtained by plotting the relationship with the draw-ratio in the hot drawing stage. Such plots are a useful starting point for understanding the development of structure and properties in tensile drawing [70].

It is well known that the tensile drawing behavior of polyolefins is very much affected by the polymer molecular weight, so that in the case of polyethylene high draw ratios of 30 and more can only be easily obtained for comparatively low molecular weight polymers. This result has been attributed to the increasing number of molecular entanglements with increasing molecular weight. This conclusion receives support from the observation that the initial thermal treatment of the polymer is also important. For ease of processing it appears desirable to slow cool from the melt or more accurately to initiate the crystallization at a low degree of supercooling to encourage crystallization under conditions where a rather perfect chain-folded structure is produced rather than quenching to give a very entangled structure.

An even more dramatic illustration of the influence of initial morphology on the deformation behavior comes from recent studies at Leeds University on the deformation of polyethylenes which have been annealed at high pressures to produce structures with varying degrees of chain extension [71]. Figure 22 shows how the bending modulus of polyethylene hydrostatically extruded to a standard extrusion ratio of 7 changes with annealing pressure for annealing at 234°C. Examination of the morphology shows that as the pressure is reduced the pressure annealed polymer (before extrusion) changes in structure from chain folded to chain extended, with the gradual increase in lamellar thickness shown in the Figure 23.

The bending modulus of the extrudate also increases, but the greatest change occurs for the pressure change ca. 530 MPa, where the structure is transformed from chain folded to chain extended. It is notable that high moduli are then obtained for comparatively low extrusion ratios. This is due to a complete change in the route to obtaining high molecular orientation. In normal melt crystallized material the chains are reoriented by breakdown of the initial spherulitic structure and very high deformation ratios are required to produce high chain orientation. When the polyethylene is pressure crystallized, only comparatively low deformation ratios are required to align the chain extended crystals, which reorient in the pseudo-affine manner, retaining their individual structure. There is however, a very important similarity between the two routes of spherulitic and chain extended morphology. In both cases, deformation involves the stretching of a molecular network. For melt crystallized polyethylene, the spherulitic texture is completely destroyed during deformation and a new fibrillar structure forms, as well described by Peterlin and others. For the chain extended materials, the recent Leeds research has shown that it is essential to restrict the annealing process to intermediate pressures so that a compromise is achieved between increasing lamellar thickness and reducing the

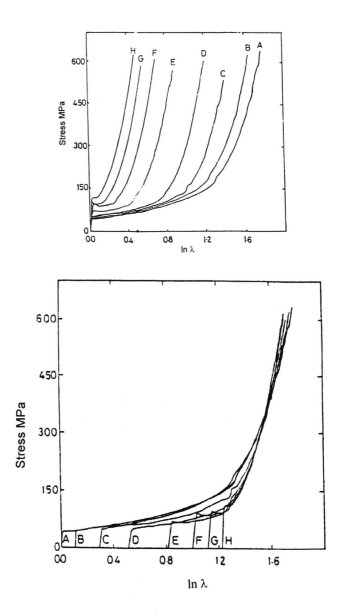

Figure 20(a): True stress-strain curves for spun yarns. Figure 20(b): Matching of true stress-strain curves for spun yearns.

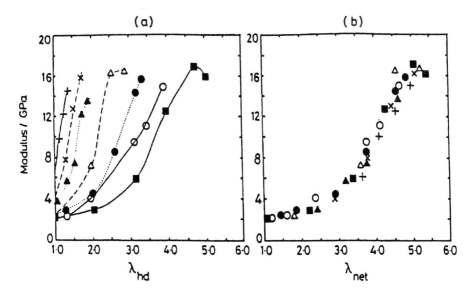

Figure 21: Initial modulus vs. actual draw ration λ_{hd} (a) and network draw ration λ_{net} (b) for pin-drawn yarns and their precursors

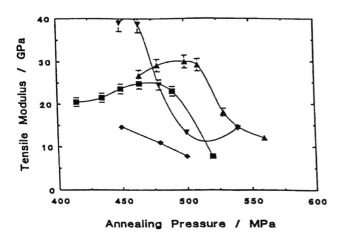

Figure 22: The extrusion pressure of annealed polyethylene (Grade A) as a function of extrusion rate. Samples annealed at 234°C for 1 h at the pressure indicated; +, 450; ▲, 480; ●, 515; ▼ 545; ♦, 580 MPa

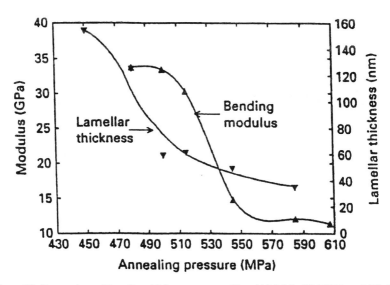

Figure 23: Comparison of lamellar thickness measured by: (▲) DSC, (●) TEM, and (■) GPC

coherence of the molecular network. The critical parameter appears to be the ratio of lamellar size to number average molecular chain length [72]. When this parameter extends about 0.5, the molecules do not complete two crystalline traverses and the network cannot sustain the deformation process, so that the polymer no longer holds together.

The application of polymer science principles plays an important role in the development of products with uniqely engineered molecular orientation in several major respects:

- Defining the theoretical limits of achievement in terms of properties, on the basis of structural considerations
- Stimulating new fabrication procedures
- Identifying new areas where property enhancements can be achieved.

These issues are described below by discussion of two areas of current development at Leeds University;

- Hot compaction to form a new class of polymer based composite materials
- Hydrostatic extrusion to form improved materials for bone repair

In the hot compaction process [70] fibers are heated under pressure to a temperature sufficient to melt a thin skin on the fiber surface. On cooling, this skin recrystallizes to form a matrix of a fiber/polymer composite. This is illustrated in Figure 24.

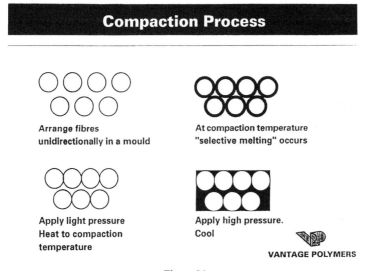

Figure 24

Ideally, only 10% of the initial fiber is melted, so that a very high proportion of the initial fiber remains intact, and there is a correspondingly high conversion of fiber properties into the composite properties. For academic studies, the fibers can be arranged in a parallel array in a matched metal mold to make a thick sheet for the determination of the elastic constants by an ultrasonic technique. The stiffness constants obtained in this way can then be compared with theoretical predictions [73]. A further parallel development is the use of a new analytical method which enables the elastic constants of the fibers in a unidirectional composite to be estimated from a knowledge of the elastic constants of the composite and the matrix resin. Following an analytical procedure established previously [71] for calculating close bounds for the elastic constants of unidirectional fiber composite, the fiber elastic constants can be calculated by an inverse method which optimizes the values of the elastic constants of the fibers to achieve the closest fit between theory and experiment. This method has been established for fibers where the elastic constants cannot be obtained by hot compaction [72] and was recently used to determine the elastic constants of the Akzo-Nobel M5 Fiber [74]. For practical applications of the hot compaction technology, where a good balance of stiffness and strength is required, it is more appropriate to use a woven fabric rather than to align fibers. Hot compacted woven fabrics in polypropylene have already found application in loudspeaker cones where the compacted sheets show an excellent combination of stiffness and damping characteristics. Another application for hot compacted polypropylene sheet now being tested is for interior panels in automobiles. The advantages of the hot compacted sheet include very high impact strength, especially at low temperatures, the ability to be postformed into a suitable shape and recyclability.

Hydrostatic extrusion was explored as a technique for producing oriented polymers in Japan and UK during the early seventies [75], [76], in parallel with ram extrusion by Porter and colleagues in the USA [77]. Because of the practical limitations of the process (batch process, slow production rates) it was not actively pursued until recently when it was shown to have possibilities for the production of a load bearing bone substitute material [40]This work stems from a joint research program between the IRC in Biomedical Materials at Queen Mary and Westfield College, London University and the IRC in Polymer Science & Technology. Professor Bonfield's group at QMC have shown that a composite material based on isotropic polyethylene (PE) reinforced with synthetic hydroxyapatite (HA) can provide a bone substitute material. This material has been commercialized under the name of HAPEX (Smith & Nephew, Richards, USA) and is finding increasing use in a range of minor load-bearing applications such as orbital floor reconstruction and otologic implants. There is therefore a genuine impetus to extend the use of this material to load-bearing applications. Hydrostatic extrusion of HA/PE provides a method of making a product with increased stiffness and strength. The hydrostatic extrusion of isotropic HA/PE, the hydrostatic extrusion of HA/PE fibers and the effects of high pressure annealing prior to HA/PE extrusion are described. It will be shown that there are several routes to the production of materials with stiffness and strengths in the range of cortical bone with the added advantage of high ductility.

Applications, E. Baer, Case Western Reserve University, R. Dickie, Ford Motor Company, M. Jaffe, New Jersey Institute of Technolgy, and C. Grant Willson, University of Texas

The issues and technologies involved in the development and commercialization of new structural polymers may be illustrated by discussions of liquid crystalline polymers, polymers with mechanical performance exceeding steel, and polymer matrices ("thermoplastic" polyimides) for continuous fiber reinforced composites for use at elevated temperatures above 200°C for thousands of hours in corrosive oxidative environments.

Liquid crystalline polymers are polymers which form fluid mesophases and, as such, spontaneously order in the melt or solution. From the initial identification of the aramids by the DuPont Company in the late '60s, thousands of mesophase forming polymer compositions have been identified and patented. When the mesophase order is nematic (non-correlated parallel packed rods) structures with very high molecular chain orientation can be easily produced through common polymer processing techniques such as fiber extrusion or injection molding (thermotropic polymers). These highly oriented structures possess extraordinary mechanical properties in the direction of the molecular chains (axial tensile modulus as high as 250 GPa), while exhibiting high levels of property anisotropy (transverse tensile modulus as low as 1 GPa). Also inherent in these polymers is low viscosity (a useful model is logs flowing down a river), allowing the filling of highly complex, thin walled molds. Challenges remaining to the utilization of liquid crystalline polymers are the control of molecular chain orientation during flow (hence, control of property anisotropy) and identifying "optimum" chain backbone chemistry.

The design of matrix polymers for high temperature composite usage has focused on polyimides where moieties between imide linkages control chain stiffness (hence Tg, processing temperature and use temperature) and chain end-capping, coupled with complete imidization retards high temperature induced chain extension or cross-linking, hence allowing "thermoplastic" part shaping. These materials tend to illustrate the classical fabrication (thermal) performance paradox, i.e., high performance is associated with poor fabrication and improvements in fabrication bring concurrent performance compromises. Polymer blending, reactive processing and novel part fabrication techniques offer approaches to solving the dilemma. Progress has been hampered by lack of backbone chain optimization, difficulty of producing sufficient quantities for manufacturing research and a lack of scaleable and cost-effective part manufacturing methods (as against part prototype fabrication). Further complicating successful materials development is small market size and high polymer cost.

What both of these examples have in common is the complexity of the enabling molecular architecture which strongly influences chain conformation, chain packing and phase formation, chain to chain interactions, and ultimately, the macroscopic properties of the final product. Recent progress in advanced computing (for example, molecular modeling and artificial intelligence) and high resolution analytical chemistry (for example, scanning tunneling microscopy and modern NMR) allow such material development programs to proceed in a much less empirical, less intuitive fashion than was possible even a half decade ago. Many of these techniques and the underlying concepts behind them originate in molecular biology. A major influence on polymeric materials development in the twenty-first century will come from the integration of the molecular biologists and material scientist's views of macromolecules.

Major trends in the evolution of organic coatings and adhesives have included the development and introduction of synthetic polymers development of analytical methodologies for bulk, surface and interface chemistry, and the emergence of a rich theoretical understanding of polymer structure, surface chemistry and adhesion.

Major trends of the future will include increasing reliance on analytical modeling and prediction, increased environmental consciousness in the design of materials and processes and an emphasis on cost reduction and sustainability. It is likely that coatings to present specific surface chemistries designed for biological compatibility, diagnostic/sensor activity or structural substrate protection will gain in importance. The vast opportunity represented by polymer blends, filled polymers and fiber reinforced composite materials has long been limited by a lack of sophistication in interface adhesion chemistry and interphase miocrostructure and it is likely that these deficiencies will be overcome in the next century.

Polymers will continue to play a defining and enabling role in the accelerating development of computing, communication and display devices. The integration of microelectronic devices into every aspect of our lives is sure to continue. The pace of this progress has followed a distinct function known as Moore's Law [78]. The semiconductor industry expends a great deal of effort in an attempt to predict the course of the technology evolution that is required to support continued growth of the microelectronics industry. This body publishes a "roadmap" that provides a detailed description of the projected advances in every aspect of microelectronic device technology as a function of time. It provides insights into the properties of the polymeric materials required to support each generation of devices and has proven valid for the decade of the 1990's. Polymeric inventions that have enabled these projections to be met include the micron to submicron and accelerated photoresists [79], [80] needed to provide the density of integrated circuitry that enables modern and future microelectronics, controlled dielectric constant materials for higher density IC substrates [81], and polymeric encapsulants [82] to protect circuitry from the environment and surrounding active components. Designer polyimides are the polymers of choice for IC applications [83]. Much of the required technologies were developed in the laboratories of the leading device manufacturers such as the IBM Company (see, for example the work of Willson [84], [85]), Bell Laboratories (now of Lucent Technologies – see, for example, the work of Reichmanis [86], [87], [88] often in cooperation with academic collaborators [89]). Scale-up, improvements and reduction to practice in these technologies historically took place at the major chemical companies. With the recent dimeunition of longer range research assets in the industrial sector there is some question as to how and where this very productive research and development area will be supported in the future.

Conducting polymers, discovered in the 1980's by MacDiarmid and others [90], [91] and the more recent development of polymeric semiconductors and light emitting diodes (see, for example the work of Heeger [92], [93], [94], Friend [95], [96], [97] opens the possibility of creating all polymeric devices and displays. The advantages of such materials lie in their processing flexibility and high figures of merit for certain device applications. All polymeric integrated circuits and displays could be produced via "roll to roll" processing, lowering costs and adding a new dimension to the rapidly accelerating computing and communications revolution in the twenty-first century.

Polymers also play a key roll in the area of photonics developments i.e., the manipulation of photons rather than electrons. Advantages of all optical switching systems are the very fast switching speeds associated with photon processes and the presumed robustness of all optical systems [98], [99]. Initiated though defense supported reseach in the early 1980's the field has been slowed by changing priories and very high materials costs. It is likely that all-optical devices will appear commercially early in this century, enabled, in part, by polymer developments.

Conclusions

The multidisciplinary field of polymer science and engineering will rapidly evolve into new opportunity areas early in the twenty-first century. Most notably these will include biomedical and pharmaceutical applications; microelectronic, photonic and display devices and systems; light weight composites, blends and composite materials for infrastructure revival; barrier membranes and packaging systems; and new coating and adhesives.   Vast opportunities for creating new polymers from inexpensive monomers exist due to the rapidly emerging "designed" heterogeneous catalysts and other novel synthetic tools. The vignettes presented here were chosen to show how novel innovations in the chemical and solid state structure of polymers will give the cost-effective performance spectrum needed to successfully penetrate these new areas of opportunity.

In addition to the engineering and chemical science issues highlighted above, several other factors will be pivotal to successful polymer developments in the future. These include:

- ENVIRONMENTAL IMPACT/PRODUCT RECYCLE/PRODUCT LIFECYCLE

  It is likely that in the future, product cost will include ultimate product disposal costs to the society. This will open the opportunity for material substitutions and materials developments not now cost-effective and reduce the attractiveness of materials of negative environmental impact

- PROTECTION OF INTELLECTUAL PROPERTY

  Introduction of new products, with concomitant manufacturing and marketing commitments is extremely expensive. If no secure mechanism for protection of innovation is in place, the advantages of "back engineering" new developments and introducing improved "second generation" products at reduced cost are obvious. This threat is exacerbated in small volume specialty markets where the continual dividing of a small pie ensures failure for all players. These problems can be avoided if an "optimum solution" rather than a "workable solution" to a given material need can be generated.

- POSITION OF THE BUSINESS IN THE VALUE ADDED CHAIN

  It is often the case with specialty materials that the availability of small volumes of a given material enables the success of a major system. For the material aspect of the overall business to be attractive, some mechanism must be devised to allow the material supplier to share profitably in this "value added", i.e., forward integration, joint venture, etc. In the absence of such a solution, there is no inducement for a material manufacturer to commercialize their product. High performance ceramics, especially ceramic fibers, usually derived from polymeric precursors, are an example of this dilemma.

- THE END OF THE COLD WAR

  Historically, the military has amortized the R&D costs associated with many advanced polymeric and composite materials, motivated by the performance criteria emphasis inherent in areas of national security.   With the slowing of military systems development and the shifting emphasis to commercial development, new means to amortize these costs will need to be identified.

The polymeric materials that will emerge in the twenty-first century will be based on backbones of controlled complexity, with a broad variety of products being produced from a smaller raw material base. The emphasis on materials development which characterized the R&D activities of the past several decades will be superceded by an emphasis on part "manufacturability" and an integrated "concurrent engineering" approach to new product development. Finally, ecological and product liability issues, and the impact of these issues on "true" product costs, will become an integral consideration of the polymeric materials development process.

References
1.   Staudinger, M., Die Hochmolekularen Organischen Berbindungen, 1932.
2.   Smets, G. and A.E. Woodward, J. Polym. Sci., 1954: p. 126.
3.   Enikolopyan, N.S., et al., J. Polym. Sci., 1981: p. 879.
4.   Otsu, T., M. Yoshida, and T. Tazaki, Makromol. Chem., Rapid Commun. 3, 1982: p. 133.
5.   Moad, G., E. Rizzardo, and D.H. Solomon, Macromolecules, 1982. **15**: p. 909.
6.   Tomalia, D.A., et al., Polym. J., 1985. **17**: p. 117.
7.   Frechet, J.M.J., et al. Molecular design of functional polymers. in IUPAC Int. Symp. 1989. Seoul, Korea.
8.   Kennedy, J.P., in Cationic Polymerization of Olefins. 1975, John Wiley: New York.
9.   Kennedy, J.P. and E. Marechal, Carbocationic Polymerization. 1982, New York: Wiley-Interscience.
10.   Wang, Q., et al., Organometallics, 1996. **15**: p. 693.
11.   Barsam, F. and M.C. Baird, J. Chem. Soc. Chem. Commun., 1995: p. 1065.
12.   Quyoum, R., et al., J. Amer. Chem. Soc., 1994. **116**: p. 6435.
13.   Shaffer, T.D., . 1994: USA.
14.   Schaffer, T.D., ACS Symposium Series, 1997. **665**: p. 96.
15.   Shaffer, T.D. and J.R. Ashbaugh, J. Polym. Sci., Part A: Polym. Chem., 1997. **35**: p. 329.
16.   Carr, A.G., D.M. Dawson, and M. Bochman, Macromolecules, 1998. **31**: p. 2035.
17.   Kennedy, J.P. and R. Smith, J. Polym. Sci. Polym. Chem Ed., 1980. **18**: p. 1523.
18.   Eisenbach, C.D., et al., in Macromolecular Engineering, M.K.M.e. al., Editor. 1995, Plenum Press: New York. p. 207.
19.   Kardos, J.L. and J. Raisoni, J. Polym. Eng. Sci., 1975. **15**: p. 183.
20.   Eisenbach, C.D. and E. Stadler, Macromol. Chem. Phys., 1975. **196**: p. 1981.
21.   Eisenbach, C.D., et al., Macromol. Symp., 1995. **98**: p. 565.
22.   Heinemann, T. and C.D. Eisenbach, Macromolecules, 1995. **28**: p. 2133.
23.   Eisenbach, C.D., et al., Colloid Polym. Sci., 1998. **276**: p. 780.

24. Eisenbach, C.D. and H. Nefzger, in *Contemporary Topics in Polymer Science*, W.M. Culbertson, Editor. 1990, Plenum Press: New York. p. 339.
25. Eisenbach, C.D., A. Ribbe, and C. Gunter, Macromol. Rapid Commun., 1994. **15**: p. 395.
26. Eisenbach, C.D., A. Goldel, and A. Ribbe, Kautsch. Gummi Kunst., 1996. **46**: p. 406.
27. Eisenbach, C.D., H. Hayen, and H. Nefzger, Macromol. Rapid Commun., 1989. **10**: p. 463.
28. Eisenbach, C.D. and T. Heinemann, Macromol. Rapid Commun., 1995. **196**: p. 2669.
29. Eisenbach, C.D. and U.S. Schubert, Macromolecules, 1993. **26**: p. 7372.
30. Eisenbach, C.D., *et al.*, Macromolecules, In Press.
31. Eisenbach, C.D., *et al.*, Polym. Prepr. (Amer. Chem. Soc., Div. Polym. Chem.), 1998. **3911**: p. 715.
32. Kwolek, S.L., . 1972, DuPont: US.
33. Calundunn, G. and M. Jaffe. *Robert A. Welch Conferences on Chemical Research XXVI*. 1983. Houston, TX: R.A. Welch Foundation.
34. Sawyer, L.C. and M. Jaffe, *Structure-Property Relationships in Liquid Crystalline Polymers*, in *High Performance Polymers*, E. Baer and A. Moet, Editors. 1991, Hanser: New York.
35. Flory, P.F. *Proc. R. Soc. London. Ser., Part A*. 1956. London.
36. Dingemans, T. and E. Samulski, J. Liq. Cryst., 2000. **27**(1): p. 131.
37. Ward, I.M., B. Brew, and P.J. Hine, Compos. Sci. Tech., 1999. **59**(7): p. 1109.
38. Northholt, M.G., *et al.*, Polymer, 1998. **39**(24): p. 5999.
39. Collins, G., J.D. Menczel, and S.K. Saw, J. Therm. Anal., 1997. **49**(1): p. 201.
40. Ward, I.M., W. Bonfield, and N.H. Ladizesky, Polymer International, 1997. **43**: p. 333.
41. Mukhija, S. and M. Jaffe, J. Appl. Polym. Sci., 1994. **53**: p. 609.
42. Cantrell, G.R., *et al.*, *The Influence of Thermal Annealing on Organic Vapor Sorption and Transport om a Nematogenic Copolyester*, in *Liquid Crystalline Polymers*, E. Cartagna, Editor. 1993, Pergamon.
43. Park, J.Y., *et al.*, J. Polym. Sci., Part B., 1996. **34**: p. 1741.
44. Windle, A.H., *et al.*, Dep. Metall. Mater. Sci., 1985. **79**: p. 55.
45. Wissbrun, K. and A.C. Griffin, J. Polym. Sci., 1982. **20**(10): p. 1835.
46. Hine, P.J., *et al.*, J. Mater. Sci., 1993. **28**: p. 316.
47. Hine, P.J. and I.M. Ward, J. Mater. Sci., 1996. **31**: p. 371.
48. Ward, I.M. and A.P. Wilczynski, J. Mater. Sci., 1993. **28**: p. 1973.
49. Wilczynski, A.P., I.M. Ward, and P.J. Hine, J. Mater. Sci., 1995. **30**: p. 5879.
50. Brew, B., P.J. Hine, and I.M. Ward, 1999.
51. Nakamura, K., K. Imada, and M. Takayanagi, J. Polym. Mater., 1972. **2**: p. 71.
52. Gibson, A.G., *et al.*, J. Mater. Sci., 1974. **9**: p. 1193.
53. Klick, K.J. and D.A. Tirrell, *Biosynthesis Routes to Novel Macromolecular Materials*, in *Material Science and Technology*, R.W. Cahn, P. Haasen, and E.J. Kramer, Editors, Wiley-VCH: New York.
54. Hubbe, J.A., Biotechnology, 1995. **15**: p. 565.

55. Langer, R., Chem. Eng. Sci., 1995. **50**: p. 4409.
56. Rataer, B.D., J. Biomatl. Res., 1993. **24**: p. 884.
57. McGrath, K.P. and e. al., J. Am. Chem. Soc., 1992. **110**: p. 121.
58. Ferraic and Cappello, in *Protein Based Materials*, M.e. al., Editor. 1997, Backhauser: Boston. p. 31.
59. Mellifield, R.B., Pure Appl. Chem., 1978. **80**: p. 645.
60. Pulapura, S. and J. Kohn, Biopolymers, 1992. **32**: p. 411.
61. Silver, F.H., *et al.*, J. Long-Term Effects Med. Implants, 1992. **1(4)**: p. 329.
62. Ertel, S.I. and J. Kohn, J. Biomed. Mater. Res., 1994. **28**: p. 919.
63. Ertel, S.I., *et al.*, J. Biomed. Mater. Res., 1995. **29(11)**: p. 1337.
64. Choueka, J., *et al.*, J. Biomed. Mater. Res., 1996. **31**: p. 35.
65. James, K.S., M.C. Zimmerman, and J. Kohn, in *Methods in Molecular Medicine: Tissue Engineering Methods and Protocols*, J.R.M.a.M.L. Yarmush, Editor. 1998, The Humana Press: Totowa, NJ. p. 121.
66. Hooper, K.A., N.D. Macon, and J. Kohn, J. Biomed. Mater. Res., 1998. **41(3)**: p. 443.
67. James, K. and J. Kohn, MRS Bulletin, 1996. **21(11)**: p. 22.
68. Hooper, K.A., J.D. Cox, and J. Kohn, J. Appl. Polym. Sci., 1997. **63(11)**: p. 1499.
69. Bostman, O.M., J. Bone Joint. Surg., 1991. **73(1)**: p. 148.
70. Long, S.D. and I.M. Ward, J. Appl. Polym. Sci., 1991. **42**: p. 1911.
71. Shahin, M.M., *et al.*, J. Mater. Sci., 1996. **31**: p. 5541.
72. Maxwell, A.S., *et al.*, J. Mater. Sci., 1997. **32**: p. 567.
73. Motashar, F.A., *et al.*, Polymer Eng. Sci., 1993. **33**: p. 1288.
74. Maxwell, A.S., A.P. Unwin, and I.M. Ward, Polymer, 1996. **15**: p. 3283.
75. Kakadjian, S., G. Craggs, and I.M. Ward, Proc. Inst. Mech. Eng., 1996. **210**: p. 65.
76. Sweeney, J. and I.M. Ward, Polymer, 1995. **36**: p. 299.
77. Capiati, N.J., *et al.*, J. Mater. Sci., 1997. **13**: p. 334.
78. England, J.S. and R.W. England, IEEE Int. Reliab. Phys. Symp. Proc., 1998. **36**: p. 1.
79. Merrem, H.J., R. Dammel, and G. Pawlowski, J. Inf. Rec., 1996: p. 481.
80. Willson, C.G., S.A. MacDonald, and J.M.J. Frechet, Acc. Chem. Res., 1994. **27**(6): p. 151.
81. Kohno, M., J. Photopolym. Sci. Technol., 1999. **12**(2): p. 189.
82. Gallagher, C., B. Shearer, and G. Matijasevic. *High-Temp. Electron. Mater. Devices. Sens. Conf.* 1998: Institute of Electrical & Electronics Engineers.
83. Meador, M.A., Annu. Rev. Mater. Sci., 1998. **28**: p. 599.
84. Willson, C.G., *et al.* . in *Advances in Resist Technology and Processing XVI*. 1999.
85. Willson, G., *et al. Step and Flash Imprint Lithography: A New Approach to High-Resolution Patterning.* in *Emerging Lithographic Technologies II*. 1999.
86. Reichmanis, E., *et al.*, Polym. Int., 1999. **48**(10): p. 1053.

87. Reichmanis, E., O. Nalamasu, and F.M. Houlihan, Acc. Chem. Res., 1999. **32**(8): p. 659.
88. Reichmanis, E., O. Nalamasu, and F.M. Houlihan, Polym. Mater. Sci. Eng., 1999. **80**: p. 301.
89. Willson, C.G., *et al.*, Polym. Eng. Sci., 1992. **32**(20): p. 1471.
90. MacDiarmid, A.G., S. Etemad, and A.J. Heeger, Annu. Rev. Phys. Chem., 1982. **33**(443).
91. Wessling, B., Synth. Met., 1997. **85**(1-3): p. 1313.
92. Heeger, A.J., *et al.*, Synth. Met., 1997. **85**(1-3): p. 1229.
93. Heeger, A.J. and N.S. Sariciftci, Synth. Met., 1995. **70**(1-3): p. 1349.
94. Heeger, A.J. and N.S. Sariciftci, Int. J. Mod. Phys., 1994. **8**(3): p. 237.
95. Friend, R.H., *et al.*, Nature (London), 1999. **397**: p. 121.
96. Friend, R.H., *et al. Mater. Res. Soc. Symp. Proc.* in *Electrical, Optical and Magnetic Properties of Organic Solid-State Materials IV.* 1998.
97. Friend, R.H., *et al.*, Solid-State Electron, 1996. **40**(1-8): p. 477.
98. Kaino, T., *et al.*, ACS Symp. Ser., 1997. **672**: p. 30.
99. Marder, S.R., *et al.*, Nature (London), 1997. **388**: p. 845.

POLYMER MATERIALS FOR MICROELECTRONICS IMAGING APPLICATIONS

E. REICHMANIS, O. NALAMASU

Bell Laboratories, Lucent Technologies
600 Mountain Avenue, Murray Hill, New Jersey

Introduction and Historical Perspective
Resist Design Requirements
Solution Developed Resist Chemistry
Negative Resist Chemistry
Single Component Crosslinking Resists
Image Reversal Chemistry
Positive Resist Chemistry
Main Chain Scission Resists
Dissolution Inhibition Resists
Chemically Amplified Resists
Positive Resists for Still Shorter Wavelengths
Dry-Developed Resist Chemistry
Non-Solvent Based Materials and Processes
Conclusion
Literature Cited

Introduction and Historical Perspective

The invention of the point contact transistor in 1947 heralded the dawn of the microelectronics era (1). The unprecedented advancements we have observed have had impact on every aspect of our lives and have been enabled through materials chemistry in general, and more specifically, organic and polymer chemistry. Over time, the business need to build devices with an increasingly larger number of circuit elements has led to increased device complexity and functionality concomitant with dramatically decreased minimum feature size (2). It was not that long ago that device dimensions of 5 to 6 μm were considered state-of-the-art. Today, advanced circuitry fabricated with minimum features of less than 0.18 μm are commonplace.

A modern integrated circuit is a complex three-dimensional structure of alternating, patterned layers of conductors, dielectrics and semiconductor films. The circuit is fabricated on an ultrahigh purity wafer substrate of a semiconducting material such as silicon. The performance of the device is to a large degree, governed by the size of the individual elements. As a general rule, the smaller the elements, the higher the device performance will be. The structure is produced by a series of steps used to precisely pattern each layer. The patterns are formed by lithographic processes that consist of two steps: i) delineation of the patterns in a radiation sensitive thin-polymer film called the resist, and ii) transfer of that pattern using an appropriate etching technique. A schematic representation of the lithographic process is shown in Figure 1 (2). Materials that undergo reactions that increase their solubility in a given solvent (developer) are called positive-tone resists while those that decrease their solubility are known as negative acting materials.

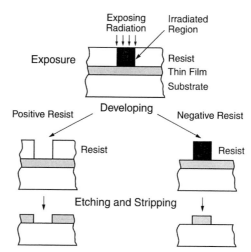

Figure 1: Schematic representation of the lithographic process.

An overwhelming preponderance of today's devices are fabricated via "conventional photolithography" employing 350-450 nm light. Incremental improvements in tool design and performance with concomitant refinements in what could be called the "industry standard", novolac/diazonaphthoquinone, resist materials chemistry and processing have allowed the continued use of this technology to produce ever smaller features. The cost of introducing a new technology, which includes the cost associated with the development and implementation of new hardware and resist materials, is a strong driving force pushing photolithography to its absolute resolution limit and extending its commercial viability.

The focus of this review concerns the design and selection of polymer materials that are useful as radiation sensitive resist films for advanced lithographic technologies such as short-wavelength (>250 nm) photolithography, scanning or projection electron-beam, x-ray or EUV, or ion-beam lithography (2,3). Conventional photoresists do not possess appropriate sensitivity and absorption characteristics to be practical, necessitating research investment in polymers that are designed to meet the specific requirements of each lithographic technology and device process. Although these requirements vary according to the radiation source and device process, the following are ubiquitous: sensitivity, contrast, etching resistance, shelf life, and purity (4). These properties can be achieved by careful manipulation of the polymer structure and molecular properties (5,6).

In developing new chemistry, it is also necessary to be reminded of the significant trade-off between optimum process performance and the chemical design of new resists. The ultimate goal of any lithographic technology is to be able to produce the smallest possible features with the lowest cost per level and widest process latitude. The best solution will invariably require compromises, and an understanding of materials and process issues is essential to select the correct compromise.

Resist Design Requirements

Resist chemical design must accommodate the specific requirements of a given lithographic technology. Materials issues that must be considered in designing resists with the appropriate properties are given below. The polymer resins must (1) exhibit solubility in solvents that allow the coating of uniform, defect free, thin films; (2) be sufficiently thermally stable to withstand the temperatures and conditions used with standard device processes; (3) exhibit no flow during pattern transfer of the resist image into the device substrate; (4) possess a reactive functionality that will facilitate pattern differentiation after irradiation; and (5) for photoexposure, have absorption characteristics that will permit uniform imaging through the thickness of a resist film. In general, thermally stable (>150°C), high glass transition temperature (T_g>90°C) materials with low absorption at the wavelength of interest are desired. If other additives are to be employed to effect the desired reaction, similar criteria apply. Specifically, they must be nonvolatile, be stable up to at least 175°C, possess a reactive functionality that will allow a change in solubility after irradiation, and have low absorbance. Table I provides a list of key lithographic properties and how each relates to the molecular characteristics of the material. The sections that follow outline many of the chemistries that have been applied to the design of resist materials for microlithography. The reader is referred to additional major overviews of the field (2,4,7-11).

Table 1: Key resist lithographic properties related to molecular characteristics.

LITHOGRAPHIC PARAMETER	MOLECULAR PROPERTIES
Absorption	Appropriate level of olefinic and aromatic groups to allow uniform light absorption
Etching Resistance	High structural carbon and/or aromatic content (low-oxygen content)
Aqueous Base Solubility	Presence of base solubilizing groups (OH, COOH, NH, etc.)
Adhesion	Materials polarity
Aspect Ratio	Surface tension effects, mechanical strength
Photospeed	Quantum yield, G value for scission for crosslinking, catalytic chain length
Post-exposure delay (PED) and Substrate Sensitivity	Catalytic chain length, polarity
Outgassing	Radiation/photochemical reaction products, additive chemistries
Low-Metal Ion Content	Synthesis/scale-up methodology

Solution Developed Resist Chemistry

Negative-Resist Chemistry

The radiation induced chemical processes that effect reduced solubility of the exposed areas of a resist leading to a negative image are cross-linking between polymer chains, image reversal, and use of a radiation generated species as a catalyst for cross-linking reactions in the resist matrix.

The first resists used during the early stages of the semiconductor industry (1957-1970) when the minimum size of circuit features exceeded 10 μm, were based on cyclized poly(cis-1,4 isoprene)/aromatic azide crosslinking chemistry (11). In one example, 2,6-bis(4-azidobenzal)-4-methyl cyclohexanone was the azide component in the Kodak resist, KTFR, which effectively initiated crosslinking of the matrix resin when exposed to near UV light. The reactions associated with such chemistry are outlined below.

$$R-N_3 \xrightarrow{h\nu} R-N\text{:} + N_2$$
$$\text{(azide)} \qquad\qquad \text{(nitrene)}$$

$$R-N\text{:} + R-N\text{:} \longrightarrow R-N=N-R$$

$$R-N\text{:} + H-\overset{|}{\underset{|}{C}}- \longrightarrow R-NH-\overset{|}{\underset{|}{C}}-$$

$$R-N\text{:} + H-\overset{|}{\underset{|}{C}}- \longrightarrow R-NH\cdot + \cdot\overset{|}{\underset{|}{C}}-$$

$$R-N\text{:} + |\underset{}{\overset{}{C}} \longrightarrow R-N \overset{}{\underset{}{\triangleleft}}$$

This highly sensitive two-component resist has limited resolution capability due to solvent induced swelling followed by stress relaxation of the developed resist images. Use of the arylazides was extended to the deep-UV by Iwayanagi, et al (12-14). Sub-micron imaging is obtained upon exposure and aqueous development of the poly(hydroxystyrene)/arylazide resists, but due to the high optical density of the material at 250 nm undercut resist profiles were obtained.

Single Component Crosslinking Resists. Concurrent with the rapid development of electron-beam lithographic tools for both optical mask making and direct write applications, was the commercialization of single-component negative resists. The electron-beam exposure requirements of these materials were compatible with the dose outputs (~1 μC/cm² at 10 KeV) of raster scan machines developed by AT&T Bell Laboratories (15). These resists typically contained epoxy, vinyl and halogen functionalities (16,17). Examples are presented in Figure 2.

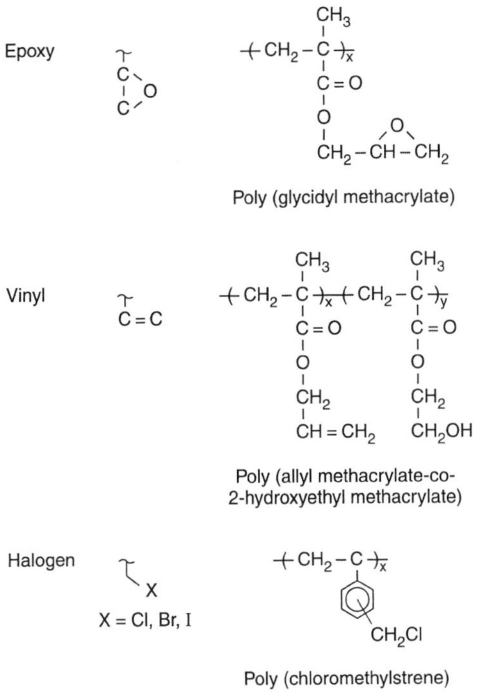

Figure 2: Representative examples of single-component negative resists.

The crosslinking reaction which predominates in epoxy (18-20), vinyl and allyl (21) containing polymers is one in which exposure produces radical or cationic species which can react with the same group on either the same (intramolecular crosslink) or neighboring (intermolecular crosslink) polymer chain. This process continues via a chain reaction leading to the formation of an insoluble polymer network. Though this reaction sequence affords high resist sensitivity, the propagation of radiation generated reactive species (dark reaction) continues in the vacuum environment of an electron-beam exposure tool. The consequence is that those features which were exposed first will have dimensions that are different from those exposed last (22), and this feature size difference can exceed the maximum allowable variation specified for a particular device level.

In sharp contrast, styrene based negative acting resists crosslink by a radiation induced reaction which involves radicals that recombine and do not propagate. The mechanism is based on experimental evidence obtained by Tabata, et al (23). Since the reaction sequence does not involve a post-exposure curing reaction, and since the addition of styrene to the resist improves the dry etching characteristics of the polymers, these materials are often favored over aliphatic based resists. Specific examples of halogenated styrene based resists are chlorinated (24) or chloromethylated (25) polystryrene, and poly(chloromethylstyrene) (26,27). In addition to being sensitive electron-beam resists, deep-UV (28) and x-ray applications were also explored (29). In one case, the halogenated styrene, chlorostyrene was copolymerized with glycidyl methacrylate to afford a very sensitive e-beam resist that exhibits little of the curing phenomenon typically observed with epoxy, crosslinking reactions (30-32).

A mechanism to improve the dry-etching characteristics of crosslinking single component resists has been reported by Hatzakis, et al. (33), who showed that polysiloxane polymers such as poly(vinylmethyl siloxane) readily provide e-beam sensitivity in the 1-2 $\mu C/cm^2$ range. These materials have a high silicon content (>30 w %), and as such have found use in bilevel (see reference 11 for a definition of bilevel) lithographic processes. Since the first reports by Hatzakis, several groups have designed silicon-containing polymers for microelectronic applications (34-36).

For the resists discussed in this section, lithographic performance is hindered by the extent to which the materials swell in organic developing solvents. This phenomenon is shown in Figure 3a which depicts a typical swollen and bridged feature in a styrene based negative resist. The type of pattern distortion shown in this figure ultimately limits the resolution capability of these resists. The extent of swelling observed by solvent developed materials can be minimized by proper choice of developer as proposed by Novembre and co-workers (32) who used a method based upon the Hansen 3-dimensional solubility parameter model. Developers found to be thermodynamically poor, and kinetically good, have afforded resist materials such as poly(glycidyl methacrylate-co-3-chlorostyrene), the ability to resolve very high resolution patterns. An example of the vastly improved resolution capability achieveable via this methodology is shown in Figure 3b. The approach described by Novembre simply yet elegantly facilitates selection of an optimal developer without the tedious trial and error approach that was commonly used.

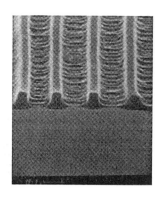   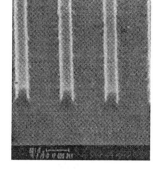

(a)   (b)

Figure 3:   SEM images of a styrene based negative resist exposed at a similar dose and developed (a) in a traditional 2-component solvent vs. (b) an optimizated single component system.

Image Reversal Chemistry. Through creative chemistry and resist processing, schemes have been developed that produce negative tone images in positive photoresist. One embodiment of these "image reversal" processes requires addition of small amounts of base additives such as imidazole to diazoquinone-novolac resists (37-40). The doped resist is exposed through a mask, based after exposure, flood exposed, and finally developed in aqueous base to generate high quality negative-tone images. A representative example of the chemistry and processes associated with this system is shown in Figure 4.

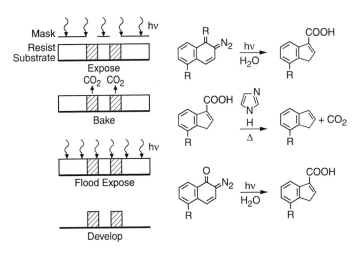

Figure 4: An example of image reversal chemistry utilizing diazonaphthoquinone chemistry with imidazole as a base additive.

Positive Resist Chemistry.
Resists that exhibit enhanced solubility after exposure to radiation are termed positive resists. The mechanism of positive resist action in most of these materials involves either chain scission or a polarity change. Positive photoresists that operate on the polarity change principle have been widely used for the fabrication of VLSI devices because of their high resolution and excellent dry etching resistance. Ordinarily the chain scission mechanism is only operable at photon energies below 300 nm where the energy is sufficient to break main chain bonds. Examples of polymer chemistries associated with positive resists are described below.

Main Chain Scission Resists. Two classes of polymers that undergo radiation induced main chain scission and have been investigated for resist applications include methacrylate and olefin-sulfone based resins. The mechanism of poly(methyl methylacrylate) (PMMA) main chain cleavage has been extensively studied and is described elsewhere (41). The methacrylate homopolymer has been shown to function as a positive acting material when exposed to electron-beam (42), x-ray (43), and deep-

UV radiation (44,45). Its primary attractive characteristics are that it only minimally swells during the wet development process, and that it exhibits extremely high resolution. The ability to delineate submicron images is, however, offset by low sensitivity and relatively poor dry etching resistance. Electron-beam and deep-UV exposure doses in excess of 50 $\mu C/cm^2$ at 10 KeV (42) and 1 J/cm^2 (44,45), respectively, have been required for patterning purposes.

Attempts have been made to improve the sensitivity of PMMA via increasing the efficiency of the main chain scission process. Examples include introduction of electronegative substituents such as Cl and CN at the backbone α-carbon (46-49), incorporation of bulky substituents into the polymer structure (50), which increases steric hindrance and concomitantly the probability for backbone bond scission, incorporation of fluorine into the ester group of PMMA (51-53) and generation of inter- and intra-molecular anhydride linkages (54-56). Enhancement of methacrylate polymer sensitivity to deep-UV radiation has been accomplished by copolymerizing MMA with other methacrylate based monomers having improved absorption characteristics in 220-260 nm range (58,59). Additionally, silicon has been incorporated into methacrylate polymers to afford positive bilevel resist systems (60). Materials based upon poly(methyl isopropenyl ketone)(PMIPK) are also known to undergo radiation-induced chain scission (61,62). The main deficiencies of most of these systems are the need to use organic solvent based image development processes, and inadequate etching resistance.

The second major class of chain scission resists are based on poly(olefin sulfone) chemistry. These resists are copolymers of an alkene or vinylaryl compound with sulfur dioxide. Interest in these materials as resists arose from published results by Brown and O'Donnell who reported a G(s) value of ~10 for poly(1-butene sulfone) (63). The proposed radiation induced degradation of the poly(olefin sulfones) involves cleavage of the main chain carbon-sulfur bond (64). This selective cleavage results from the relatively weak carbon-sulfur bond energy (60 kcal/mole) compared to that for a carbon-carbon bond (80 kcal/mole). Several poly(olefin sulfones) have been investigated including silicon containing analogs. Notably, the copolymer containing 1-butene (PBS) is a highly sensitive (<1 $\mu C/cm^2$ at 10 Kv) electron-beam resist (65) and continues as the predominant positive resist used in the fabrication of chromium photomasks (66).

$$\text{PBS:} \quad -\!\!\left(\!CH_2 - \underset{\underset{CH_2CH_3}{|}}{\overset{\overset{H}{|}}{C}} - SO_2\!\right)\!\!\!-_x$$

Dissolution-Inhibition Resists. Photolithography represents the workhorse technology for device manufacture and has traditionally used an Hg or Hg-Xe discharge lamp as the radiation source. The output of such lamps spans the range of 250 to 450 nm and the emission wavelengths at 436, 405 and 365 nm are typically used as the exposing radiation. Resist systems which have been developed to respond favorably to this energy spectrum are often called "conventional photoresists", and are typically comprised of two components; an aqueous alkali soluble resin, and a photosensitive dissolution inhibitor. The alkaline soluble resin is typically a novolac; prepared via condensation polymerization of a substituted phenol and formaldehyde (8). These resins and their modifications are formulated to exhibit low absorbance in the near and mid-UV region, are glassy amphorous materials at room temperature and can be dissolved in a variety of organic solvents useful for spin coating applications. Additionally, for oxygen reactive ion etching applications, silicon may be readily incorporated into the polymer (67,68). The second component of conventional photoresists is a hydrophobic, substituted diazonaphthoquinone (DNQ) dissolution inhibitor. Addition of this component to the novolac renders the polymer matrix insoluble in aqueous base developers. Upon irradiation, the diazonaphthoquinone undergoes a Wolff rearrangement followed by hydrolysis to generate the base-soluble indene carboxylic acid. The photogenerated acid accelerates the development of the exposed regions of the film in aqueous alkaline solution. The remaining, nonexposed regions are unaffected and do not swell in the developer. This chemistry affords high resolution, and as a consequence of the aromatic nature of the resin, good dry-etching resistance for pattern transfer processes. Figure 5 depicts the chemistry and processing sequence associated with conventional positive photoresists.

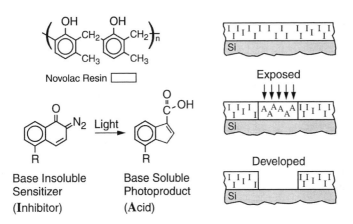

Figure 5: The chemistry associated with conventional novolac/ diazonaphthoquinone resists.

Use of conventional photoresist formulations in lithographic technologies utilizing <300 nm light is hampered by the high opacity of novolac resins, and the high absorbance of the DNQ and its photoproducts. Several approaches to designing deep-UV resists that overcome these problems have been reported. Initial efforts concentrated on the design of alternative dissolution inhibitors. Examples include 5-diazo-Meldrum's Acid derivatives (69) and 2-nitro benzyl carboxylate derivatives (70). In the latter case, exposure of the ester results in rearrangement and subsequent degradation to yield a carboxylic acid and a nitrosobenzaldehyde. Optimum results were obtained for ester derivatives of large-molecular organic acids such as cholic acid.

These esters are nonvolatile, and allow conversion of a relatively large volume fraction of resist from alkali insoluble to an alkali soluble state.

5-diazo-Meldrum's Acid   2-Nitrobenzyl carboxylate

Chemically Amplified Resists.   Aptly described as the first revolution in resist design that ultimately led to the adoption of short wavelength (248 nm) lithography as the technology of choice for advanced device fabrication was the announcement of what has been termed the "chemically amplified" resist mechanism.   The pioneering work relating to the development of chemically amplified resists based on deprotection mechanisms was carried out by Ito, Willson and Frechet (71). These initial studies dealt with the catalytic deprotection of poly((tert-butoxycarbonyloxy)styrene) (TBS) in which the thermally stable, acid-labile tert-butoxycarbonyl group is used to mask the hydroxyl functionality of poly(vinylphenol).   As shown in Figure 6, irradiation of TBS films containing   small   amounts   of   onium   salt,   such   as   diphenyliodonium hexafluoroantimonate with UV light, liberates an acid species that upon subsequent baking catalyzes cleavage of the protecting group to generate poly(p-hydroxystyrene). While this reaction will take place very slowly at room temperature, it is much faster at 100°C, requiring only a few seconds to reach completion. In the absence of an acidic species, the protected polymer undergoes no degradation during prolonged heating at that temperature.   Loss of the tert-butoxycarbonyl group results in a large polarity change in the exposed areas of the film. Whereas the substituted phenol polymer is a nonpolar material soluble in nonpolar lipophilic solvents, poly(vinylphenol) is soluble in polar organic solvents and aqueous base. This change in polarity allows formation of either positive or negative tone images, depending upon the developer.

Since the initial reports regarding chemically amplified resist mechanisms in 1980, numerous research groups have expanded on this revolutionary concept. Both alternate protective groups and parent polymers have been utilized (72). Generally, thermally stable, acid-labile substituents are desirable as protective groups for aqueous-base-soluble parent polymers. Some typical examples of protective groups that have been employed include tert-butyl ethers and esters, tetrahydropyranyl ethers, α-α-dimethylbenzyl esters, and ketals and acetals. Alternate polymer backbones include poly(hydroxystyrene), poly(vinylbenzoic acid), and poly(methacrylic acid). Additionally, high glass transition temperature (T_g) polymers based on N-blocked maleimide/styrene resins and substituted styrene-sulfone copolymers have been explored.

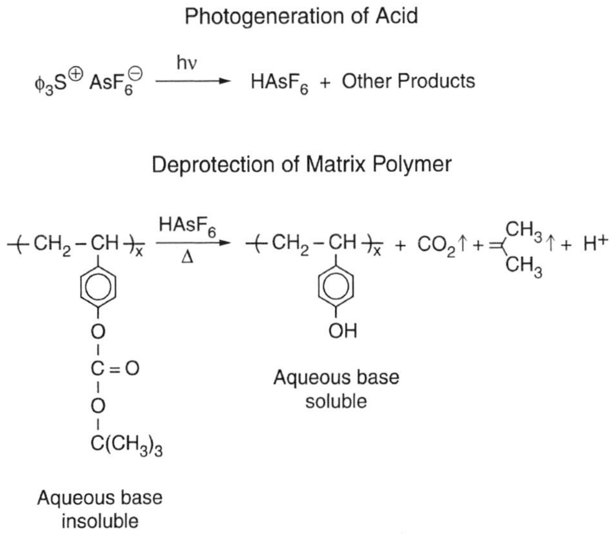

Photogeneration of Acid

$$\phi_3S^{\oplus} \; AsF_6^{\ominus} \xrightarrow{hv} HAsF_6 \; + \; Other \; Products$$

Deprotection of Matrix Polymer

Aqueous base
insoluble

Aqueous base
soluble

Figure 6:   Chemistry associated with the TBS based chemically amplified resist.

<u>Positive Resists for Still Shorter Wavelengths</u>. The drive towards still smaller features, fueled by a desire to maintain the availability of optical lithography in the device production environment, led to efforts aimed at developing a production worthy 193 nm lithographic technology (73). Within the materials arena, much of the research leading up to the design, development and implementation of 248 nm lithographic materials, laid the foundation for efforts relating to still shorter wavelength sensitive materials. The first experiments demonstrating the feasibility of UV light as an imaging source for lithography occurred at Bell Laboratories in 1975 (74). Bowden and Chandross demonstrated the concept using poly(butene-1-sulfone) which upon exposure to 185 nm light, exhibited a sensitivity $5mJ/cm^2$.

Later, the initial focus for designing 193 nm resists centered on derivatized acrylate and methacrylate copolymers that function via a chemical amplification mechanism (75).

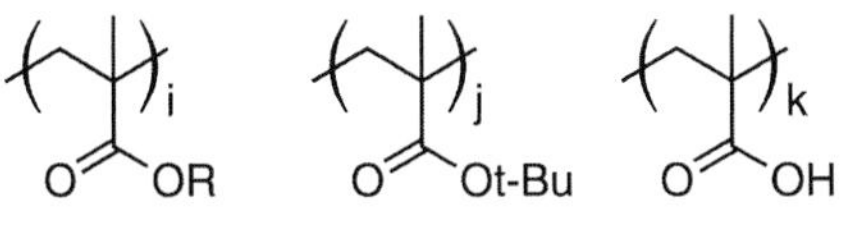

R = adamantyl, isobornyl, norbornyl, etc.

By and large, these polymers are effectively transparent at 193 nm and exhibit excellent resolution, but lack plasma-etching resistance and other requisite materials properties for lithographic performance. Attachment of alicyclic substituents onto the polymer backbone was the first aproach utilized to enhance etching performance of these materials. Examples of substituents that have been employed include methnyl (76), adamantyl (77), isobornyl (78), and tricyclodecyl (79).

More recently, efforts have focussed on polymers where the cyclo-olefin is incorporated into the polymer backbone. Cyclo-olefin-maleic anhydride alternating copolymers are one alternative to methacrylate-based matrix resins (80). Compelling features of these copolymers include a) facile synthesis via standard radical polymerization, b) a potentially large pool of cycloolefin feed stocks, and c) a generic structural motif that incorporates alicyclic structures directly into the polymer backbone and provides a latent water-solubilizing group that may also be useful for further structural elaboration. A large number of cycloolefins are known to copolymerize with maleic anhydride. As a rule, they yield high-T_g copolymers with a 1:1 alternating structure.

Aqueous base solubility can be induced in this alternating copolymer via incorporation of acrylic acid. Free radical polymerization of the cyloolefin and maleic anhydride in the presence of acrylic acid and/or its derivatives provides a controllable method for synthesizing aqueous base soluble resins (81). Such materials are readily soluble in standard organic solvents used to spin coat resist films, and additionally are soluble in aqueous base media such as 0.262 N tetramethyl ammonium hydroxide, the developer of choice for the electronics industry. Notably, thin films of these polymers cast onto quartz substrates display excellent transparency at both 248 and 193 nm, with the absorbance per micrometer of typical poly(norbornene-co-maleic anhydride-co-acrylic acid) materials being approximately 0.2 AU/µm.

These norbornene based matrix resins can be used in a variety of resist approaches (Figure 7). The strategies that have been examined include the use of a protected polymer in conjunction with a photoacid generator (PAG) in a "two-component" chemically amplified resist process; a three-component system using the parent acidic terpolymer, a dissolution inhibitor (DI), and a PAG; and a hybrid approach that uses both a DI and a partially protected polymer matrix with a PAG (81). High resolution imaging is readily achieved in such a system as shown in Figure 8.

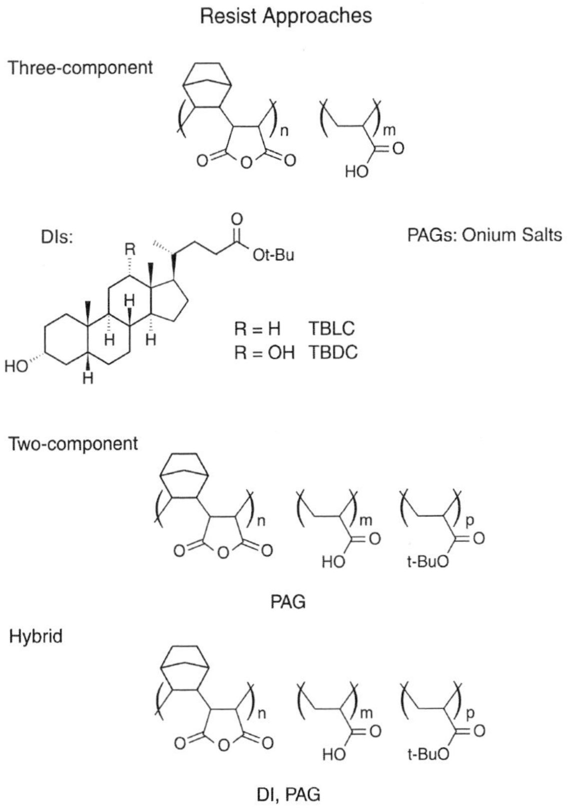

Resist Approaches
Three-component
n
m
O
O
O
HO
O
DIs:
O
R
Ot-Bu
PAGs: Onium Salts
H
H
H
R = H TBLC
R = OH TBDC
HO
H
Two-component
n
m
p
O
O
O
HO
O
t-BuO
O
PAG
Hybrid
n
m
p
O
O
O
HO
O
t-BuO
O
DI, PAG

Figure 7: Resist approaches based upon norbornene-maleic anhydride matrix resin chemistry.

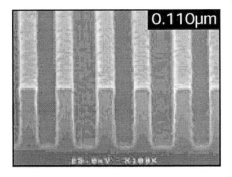

0.110μm

Figure 8: An example of the resolution capability of anhydride resist upon 193 nm exposure and aqueous base development.

Dry-Developed Resist Chemistry

The increasing complexity and miniaturization of integrated circuit technology is pushing conventional single-layer resist processes to their limit. The demand for improved resolution requires imaging features with increasingly higher aspect ratios and smaller linewidth variations over steep substrate topography. A number of schemes have been proposed to address this issue, namely; the use of polymeric planarizing layers, anti reflection coatings, and contrast enhancement materials. Since this topic has been the subject of several recent articles, this section will concentrate on the development of newer dry-developed resist chemistries, many of which are associated with top-surface imaging mechanisms rather than delve into the more traditional, multilevel approaches (82-84).

Non-Solvent Based Materials and Processes. Simplification of both tri- and bilevel resist processes can be achieved by incorporating all of the desirable features of these multilevel technologies into a single layer of resist. In one case, the advantages of multilayer processing can be extended to a single resist layer by selective incorporation of a vinyl organometallic monomer into the exposed regions of a resist film (85).

Incorporation of inorganic species into the resist can also be accomplished through gas-phase functionalization processes (86-89). For this process to work effectively, it is desirable to have an organic polymer that both contains a reactive functionality and is sufficiently absorbent to limit deposition of the irradiation dose to the topmost part of the resist film, allowing higher resolution imaging. One example involves the use of poly(t-butyl-p-vinylbenzoate) (90). When formulated with an onium salt, this material has an optical density of >>1 at 248 nm. Exposure and postexposure bake generates poly(p-vinylbenzoic acid) in the near surface regions of the resist, which may then be functionalized with a variety of reagents. After irradiation, treatment with a reactive inorganic or organometallic reagent such as $SiCl_4$, $TiCl_4$, or hexamtethyldisilazane (HMDS) results in a reaction between the exposed regions of the matrix and incorporation of silicone into these regions. When such films are subjected to an RIE environment, an etch barrier is created in the exposed areas, and a negative image is generated. Similar results were obtained with poly(t-butoxycarbonyloxystyrene) as the matrix polymer (88). A schematic representation of this process is shown in Figure 9.

The chemistry associated with a similar process using conventional positive photoresist is depicted in Figure 10 (91). The high absorption of a DNQ modified/novolac resist formulation at 248 nm limits exposure to the near surface. Heating the film after exposure effectively limits exposure to the near surface. Heating the film after exposure effectively cross-links only the nonirradiated regions of the film allowing selective silylation in the exposed regions. Patterning is subsequently accomplished via O_2 RIE processes. A modification of this process to generate positive tone patterns has been reported by Pierrat, et al. (92).

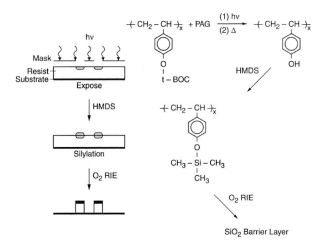

Figure 9: A schematic representation of the chemistry associated with silylation processing of the TBS based chemically amplified resist.

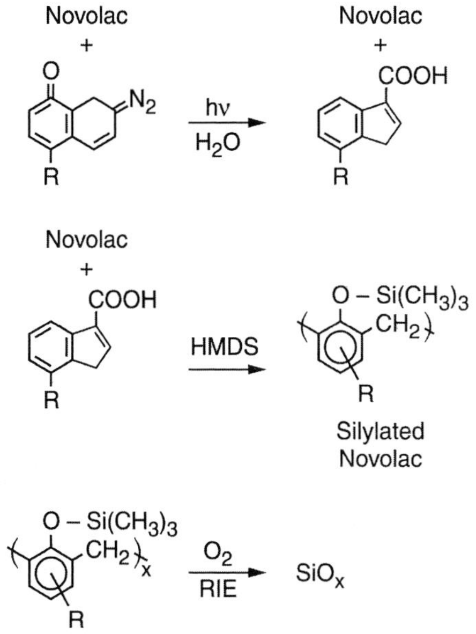

Figure 10: The chemistry associated with silylation processing of conventional positive photoresist.

In another embodiment of these top layer imaging processes, research at Siemens Corporation outlined a process referred to as Chemical Amplification of Resist Lines (CARL) (93). The thin top imaging CARL layer (170 nm) consists of an anhydride containing polymer spin coating over a thick aromatic planarizing resist layer. Following exposure and development of the imaging layer in aqueous base, the "amplification" step of silylation is accomplished using a siloxane which reacts with the remaining anhydride groups to create an etching resistant mask for pattern transfer into the thick planarizing layer.

Liquid-phase silylation, after either exposure or development has also been explored as an alternative process (94). The bifunctional silylating agent hexamethylcyclotrisilazane (HMCTS), acts as a crosslinking agent when incorporated into a standard diazonaphthoquinone-novolac resist and has been reported as a means to incorporate silicon into a patterned resist.

The dry deposition of silicon-containing, photodefinable materials has been investigated by several groups with a goal of developing environmentally sound processes to substitute for steps which traditionally create large volumes of waste volatile organic compounds in both the spin-on deposition and also in the post-exposure development steps. Several materials have shown promise as high performance dry developable photoresists. Thin films deposited from a low power plasma discharge of several liquid precursors, hexamethyldisilazane, hexamethyldisilane, tetramethylsilane and trimethylsilyldimethylamine reportedly resulted in polymers containing Si-C-Si moieties (95). These materials, which show no appreciable absorption above ~225 nm were studied as 193 nm UV resists. Negative tone development was accomplished using ordinary organic solvents or reactive ion etching in HBr or Cl_2. An all-dry positive tone development process was also realized at relatively high 193 nm exposure fluences (~15 mJ/cm^2/pulse) in single-pulse exposures. Self-developed sub-micron features, obtained in the positive tone, presumably arose from a process of volatilization of the organosilicon thin films.

The materials deposited from methylsilane in low power RF discharge produces a dual tone resist known as plasma polymerized methylsilane (PPMS) (Figure 11) (96). This material formed through the dehydrogenative polymerization of a monosubstituted alkylsilane has been characterized as an organosilicon hydride network with extensive Si-Si bonding. The plasma chemistry of methylsilane (CH_3SiH_3) carried out at relatively high radio-frequency power and temperatures tended toward the formation of amorphous SiC, not the photoreactive polymer (97).

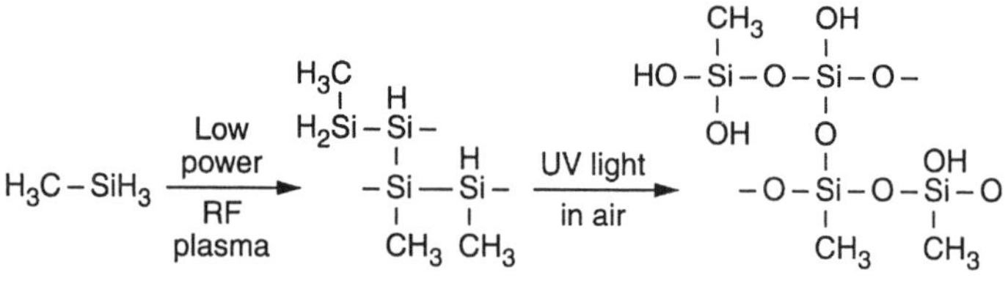

Figure 11: Plasma polymerization of methylsilane and its subsequent photo-oxidation to PPMSO.

The photoxidation of PPMS results in a networked siloxane material denoted as PPMSO, which is characterized as a glass-like Si-O-Si. Dry development in halogen based plasmas results in the selective etching of the unexposed regions. Photo-oxidation of branched and linear polysilanes containing Me, Et, Ph and cyclohexyl moieties appear to involve degradation of the polymer to form both siloxane and silanol groups. Total oxygen incorporation is higher for branched polymer films because of the higher density of Si-Si bonds in those polymers (98). The negative tone mode relies on the ability of halogen based plasmas to selectively react with unoxidized polymeric methylsilane while only slowly reacting with the exposed and oxidized polysiloxane type materials. Selectivities of about 4:1 were obtained using a Cl_2 plasma, and as high as 10:1 using Cl_2/HBr plasma chemistries. Positive tone imaging is accomplished with hydrofluoric acid (HF) chemistries that react preferentially with the exposed and oxidized glass-like polysiloxane material (99). The process sequence for this chemistry is shown in Figure 12, while its resolution capability is demonstrated by Figure 13.

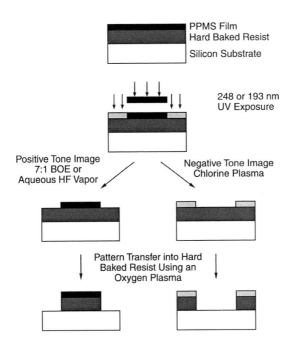

Figure 12:   Process sequence for negative and positive tone development of PPMS.

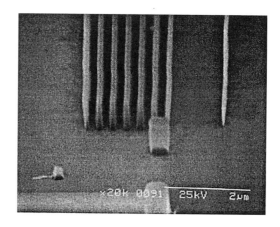

Figure 13:   SEM images depicting 0.15 μm features obtained in PPMS upon pattern transfer into an underlying polymerizing layer.

Conclusion

The unabated progress in design and integration of VLSI devices continues to demand increasingly smaller and more precise device features. At the present time, almost all commercial devices are made by photolithography utilizing UV radiation in the wavelength range of 365-436 nm. However, within the next 5-8 years, new lithographic strategies will be required to meet resolution needs that will likely extend well below 100 nm. Technologies under development include electron-beam, ion-beam, X-ray, and short-wavelength lithographies. Each of these alternative technologies will require new polymeric resist materials and processes. This chapter has attempted to discuss the varied chemistries that are available for the design of resist materials. The future of microlithography is bright and contains many challenges in the areas of resist research and associated processing. There is no doubt that within the decade, many new materials will be commonplace within the manufacturing environment.

Literature Cited

1.   Bardeen, J.; Brattain, W. H. "The Transistor, a Semi-Conductor Triode", Phys. Rev., 1948, 74(2), 230.
2.   Thompson, L. F.; Willson, C. G.; Bowden, M. J. In "Introduction to Microlithography", ACS Professional Reference Book, ACS, Washington, DC, 1994.
3.   Takigawa, T. J. Photopolym. Sci. Technol., 1992, 5(1), 1.
4.   Thompson, L. F.; Bowden, M. J. In "Introduction to Microlithography", ACS Symposium Series, Vol. 219, Thompson, L. F.; Willson, C. G.; Bowden, M. J., Eds., American Chemical Society, Washington, DC, 1983, 162.

5. a) Reichmanis, E.; Thompson, L. F. <u>Chem. Rev.</u>, 1989, <u>89</u>, 1273. b) Reichmanis, E.; Neenan, T. X.; In "Chemistry of Advanced Materials: An Overview", Interrante, L. V.; Hampden-Smith, M. J., Eds., Wiley-VCH Inc., 1998, 99.

6. Reichmanis, E.; Thompson L. F. In "Polymers in Microlithography: Materials and Processes", <u>ACS Symposium Series, Vol. 412</u>, Reichmanis, E.; MacDonald, S. A.; Iwayanagi, T., Eds., American Chemical Society, Washington, DC, 1989, 1.

7. "Electronic and Photonic Applications of Polymers", ACS Advances in Chemistry Series 218, Bowden, M. J.; Turner, S. R., Eds., ACS, Washington, DC, 1988.

8. Willson, C. G., In "Introduction to Microlithography", <u>ACS Symposium Series 219</u>, Thompson, L. F.; Willson, C. G.; Bowden, M. J., Eds., ACS, Washington, DC, 1983, 88.

9. Moreau, W. M., In "Semiconductor Lithography, Principles, Practices and Materials", Plenum, NY, 1988.

10. Reichmanis, E.; MacDonald, S. A.; Iwayanagi, T. "Polymers in Microlithography", <u>ACS Symposium Series 412</u>, ACS, Washington, DC, 1989.

11. Reiser, A., In "Photoreactive Polymers, The Science and Technology of Resists", John Wiley and Sons Inc., New York, NY, 1989, 22.

12. Iwayanagi, T.; Kohashi, T.; Nonogaki, S.; Matsuzawa, T.; Douta, K.; Yanazawa, H, <u>IEEE Trans. Elec. Dev.</u>, Ed., 1981 <u>28(11)</u>, 1306.

13. Nonogaki, S.; Hashimoto, H.; Iwanagi, T.; Shiraishi, H. <u>Proc. SPIE</u>, 1985, <u>539</u>, 189.

14. Hashimoto, H.; Iwayanagi, T.; Shiraishi, H.; Nonogaki, S., "Proc. Reg. Tech. Conf. on Photopolymers", Mid-Hudson Section, SPE, Ellenville, NY, 1985, 11.

15. Heriott, D. R.; Collier, R. J.; Alles, D. S.; Stafford, J. W. <u>IEEE Trans. Electronic Devices</u>, 1974, <u>ED-22</u>, 385.

16. Thompson, L. F.; Kerwin, R. E. <u>Annual Review of Materials Science</u>, Huggins, R. A.; Bube, R. H.; Roberts, R. W., Eds., 1976, <u>6</u>, 267.

17. Tagawa, S., In "Polymer for High Technology: Electronics and Photonics", <u>ACS Symposium Series 346</u>, Thompson, L. F.; Willson, C. G.; Frechet, J. M. J., Eds., ACS, Washington, DC, 1987, 37.

18. Hirai, T.; Hatano, Y.; Nonogaki, S. <u>J. Electrochem. Soc.</u>, 1971, <u>118(4)</u>, 669.

19. Feit, E. D.; Thompson, L. F.; Heidenreich, R. D. <u>ACS Div. of Org. Coat. and Plast. Chem. Preprint</u>, 1973, 383.

20. Taniguchi, Y.; Hatano, Y.; Shiraishi, H.; Horigome, S.; Nonogaki, S.; Naraoka, K. <u>Japan J. Appl. Phys.</u>, 1979, <u>28</u>, 1143.

21. Tan, Z. C.; Petropoulos, C. C.; Rauner, F. J. <u>J. Vac. Sci. Technol.</u>, 1981, <u>19(4)</u>, 1348.

22. Novembre, A. E.; Bowden, M. J., <u>Polym. Eng. Sci.</u>, 1983, <u>23</u>, 977.

23. Tabata, Y.; Tagawa, S.; Washio, M. "Materials for Microlithography", Thompson, L. F.; Willson, C. G.; Frechet, J. M. J., Eds., <u>ACS Symposium Series 266</u>, ACS, Washington, DC, 1984, 161.

24. Hartney, M. A.; Tarascon, R. G.; Novembre, A. E., <u>J. Vac. Sci. Technol.</u>, 1985, <u>B3</u>, 360.

25. Imamura, S. <u>J. Electrochem. Soc.</u>, 1979, <u>126(9)</u>, 1268.

26. Choong, H. S.; Kahn, F. J. <u>J. Vac. Sci. Technol.</u>, 1981, <u>19(4)</u>, 1121.

27. Feit, E. D.; Thompson, L. F.; Wilkins, C. W., Jr.; Wurtz, M. E.; Doerries, E. M.; Stillwagon, L. E. <u>J. Vac. Sci. Technol.</u>, 1979, <u>16(6)</u>, 1987.

28. Imamura, S.; Sugawara, S. J. Appl. Phys., 1982, 21, 776.
29. Yoshioka, N.; Suzuki, Y.; Yamazaki, T., Proc. SPIE, 1985, 537, 51.
30. Thompson, L. F.; Doerries, E. M. J. Electrochem. Soc., 1978, 126(10), 1699.
31. Thompson, L. F.; Yau, L.; Doerries, E. M. J. Electrochem. Soc., 1979, 126(10) 1703.
32. Novembre, A. E.; Masakowski, L. M.; Hartney, M. A. Poly. Eng. Sci., 1986, 26(6) 1158.
33. Hatzakis, M.; Paraszczak, J.; Shaw, J. M. "Microcircuit Eng. 81", Oosenburg, A. Ed., Swiss Fed. Inst. Technol., Lausanne, 1981, 386.
34. MacDonald, S. A.; Steinman, F.; Ito, H.; Lee, W-Y.; Willson, C. G., Preprints, ACS Division, Polymer Materials Science and Eng., 1983, 50, 104.
35. Suzuki, M.; Saigo, K.; Gokan, H.; Ohnishi, Y. J. Electrochem. Soc., 1983, 30, 1962.
36. Novembre, A. E.; Jurek, M. J.; Kornblit, A.; Reichmanis, E. Polym. Eng. Sci., 1989 23, 920.
37. Moritz, H. IEEE Trans. Electron Devices, 1985, ED-32, 672.
38. Takahashi, Y.; Shinozaki, F.; Ikeda, T. Jpn. Kokai Tokyo Koho, 1980, 88, 8032.
39. MacDonald, S. A.; Ito, H.; Willson, C. G. Microelectron. Eng., 1983, 1, 269.
40. Alling, E.; Stauffer, C. Proc. SPIE, 1985, 539, 194.
41. Ranby, B.; Rabek, J. F. In "Photodegradation, Photoxidation and Photostabilizatior of Polymers", John Wiley & Sons, New York, NY, 1975, 156.
42. Hatzakis, M. J. Electrochem. Soc., 1969, 116, 1033.
43. Moreau, W. M.; Schmidt, R. R., 138th Electrochemical Society Meeting, Extendec Abstracts, 1970, 459.
44. Mimura, Y.; Ohkubo, T.; Takanichi, T.; Sekikawa, K. Japan Appl. Phys., 1978, 17 541.
45. Lin, B. J. Vac Sci. Technol., 1975, 12, 1317.
46. Helbert, J. N.; Chen, C. Y.; Pittman, C. U., Jr.; Hagnauer, G. L. Macromolecules, 1978, 11, 1104.
47. Lai, J. H.; Helbert, J. N.; Cook, C. F., Jr.; Pittman, C. U., Jr. J. Vac. Sci. Technol., 1979, 16(6), 1992.
48. Helbert, J. J.; Wagner, B. E.; Caplan, J. P.; Poindexter, E. H. J. App. Poly. Sci., 1975, 19, 1201.
49. Chen, C-Y.; Pittman, C. U., Jr.; Helbert, J. N. J. Poly. Sci. Poly. Chem. Ed., 1980, 18, 169.
50. Moreau, W. M. Proc. SPIE, 1982, 333, 2.
51. Kakuchi, M.; Sugawara, S.; Murase, K.; Matsuyama, K. J. Electrochem. Soc., 1977, 224, 1648.
52. Tada, T. J. Electrochem. Soc., 1979, 126, 1829.
53. Tada, T. J. Electrochem. Soc., 1983, 130, 912.
54. Roberts, E. D. ACS Div. Org. Coat. and Plastics Chem. Preprints, 1977, 37(2), 36.
55. Moreau, W.; Merrit, D.; Moyer, W.; Hatzakis, M.; Johnson, D.; Pederson, L. J. Vac Sci. Technol., 1979, 16(6), 1989.
56. Namastse, Y. M. N.; Obendorf, S. K.; Anderson, C. C.; Krasicky, P. D.; Rodriquez, F.; Tiberio, R. J. Vac. Sci. Technol. B, 1983, 1(4), 1160.
57. Wilkins, C. W., Jr.; Reichmanis, E.; Chandross, E. A. J. Electrochem. Soc., 1980, 127(11), 2510.

58. a) Reichmanis, Wilkins, C. W., Jr.; E.; Chandross, E. A. J. Electrochem. Soc., 1980, 127(11), 2514. b) Reichmanis, E.; Wilkins, C. W., Jr., In "Polymer Materials for Electronics Applications", ACS Symposium Series 194, Feit, E. D.; Wilkins, C. W., Jr.; Eds., ACS, Washington, DC, 1982, 29.

59. Hartless, R. L; Chandross, E. A. J. Vac. Sci. Technol., 1981, 19, 1333.

60. a) Reichmanis, E.; Smolinsky, G.; Wilkins, C. W., Jr., Solid State Technology, 1984, 28(8), 130. b) Reichmanis, E.; Smolinsky, G. Proc. SPIE, 1984, 469, 38. c) Reichmanis, E.; Smolinsky, G. Electrochem. Soc., 1985, 132, 1178.

61. Tsuda, M.; Oikawa, S.; Nakamura, Y.; Nagata, H.; Yokota, A.; Nakane, H.; Tsumori, T.; Nakane, Y. Photogr. Sci. Eng., 1979, 23, 1290.

62. MacDonald, S. A.; Ito, H.; Willson, C. G.; Moore, J. W.; Charapetian, H. M.; Guillett, J. E., In "Materials for Microlithography", Thompson, L. F.; Willson, C. G.; Frechet, J. M. J., Eds., ACS Symposium Series 266, ACS, Washington, DC, 1984, 179.

63. a) Brown, J. R.; O'Donnell, J. H. Macromolecules, 1970, 3, 265. b) Brown, J. R.; O'Donnell, J. H. Macromolecules, 1972, B, 109.

64. Bowmer, T. N.; O'Donnell, J. H. Radiation Phys. Chem, 1981, 17, 177.

65. Bowden, M. J.; Thompson, L. F.; Ballantyne, J. P. J. Vac. Sci. Technol., 1975, 126(6), 1294.

66. Bowden, M. J.; Thompson, L. F. Solid State Technology, 1979, 22, 72.

67. Tarascon, R. G.; Shugard, A.; Reichmanis, E. Proc. SPIE, 1986, 631, 40.

68. Saotome, Y.; Gokan, H.; Saigo, K.; Suzuki, M.; Ohnishi, Y. J. Electrochem. Soc., 1985, 132, 909.

69. a) Grant, B. D.; Clecak, N. J.; Twieg, R. J.; Willson, C. G. IEEE Trans. Electron Dev., 1981, ED-25(11), 1300. b) Willson, C. G.; Miller, R. D.; McKean, D. R. Proc. SPIE, 1987, 771, 2.

70. Reichmanis, E.; Wilkins, C. W., Jr.; Chandross, E. A. J. Vac. Sci. Technol., 1981, 19(4), 1338.

71. a) Willson, C. G.; Ito, H.; Frechet, J. M. J., In "New UV Resists with Negative or Positive Tone", Digest of Technical Papers – 1982 Symposium on VLSI Technology, Kanagawa, Japan, 1982, 86. b) Willson, C. G.; Ito, H.; Frechet, J. M. J.; Tessier, T. G.; Houlihan, F. M. J. Electrochem. Soc., 1986, 133, 181.

72. Willson, C. G.; Bowden, M. J., In "Electronic and Photonic Applications of Polymers", Bowden, M. J.; Turner, S. R., Eds., ACS Advances in Chemistry Series 218, 1988, 75. b) Iwayanagi, T.; Ueno, T.; Nonogaki, S.; Ito, H.; Willson, C. G., "Materials and Processes for Deep-UV Lithography", ibid, 109. c) Reichmanis, E.; Houlihan, F. M.; Nalamasu, O.; Neenan, T. X. Chem. Mater., 1991, 3, 394.

73. Nakano, K.; Maeda, K.; Iwasa, S.; Ohfuji, T. Proc. SPIE, 1995, 2438, 433. b) Allen, R. D.; Wan, I. Y.; Wallraff, G. M.; DiPietro, R. A.; Hofer, D. C.; Kunz, R. R. J. Photopolym. Sci. Technol., 1995, 8, 623. c) Nalamasu, O.; Wallow, T. I.; Houlihan, F. M.; Reichmanis, E.; Timko, A. G.; Dabbagh, G.; Cirelli, R. A.; Hutton, R. S.; Novembre, A. E. Future Fab International, 1997, 1(2), 159.

74. a) Bowden, M. J.; Chandross, E. A.; J. Electrochem. Soc., 1975, 122, 1370. b) Feldman, M.; White, D. L.; Chandross, E. A.; Bowden, M. J.; Appelbaum, J., Proceedings, Kodak Microelectronics Seminar, Eastman Kodak, Rochester, NY, 1975, 40.

75. Kunz, R. R.; Allen, R. D.; Hinsberg, W. D.; Wallraff, G. M. Proc. SPIE, 1993, 1925, 167.
76. Shida, N.; Ushirogouchi, T.; Asakawa, K.; Nakase, M. J. Photopolymer Sci. Technol., 1996, 9, 457.
77. Takahashi, M.; Takechi, S. Proc. SPIE, 1995, 2438, 422. b) Takechi, S.; Takahashi, M.; Kotachi, K.; Nozaki, K.; Yano, E.; Hanyu, I. J. Photopolymer Sci. Technol., 1996, 9, 475.
78. Allen, R. D.; Wallraff, G. M.; DiPietro, R. A.; Hofer, D. C.; Kunz, R. R. Proc. SPIE, 1995, 2438, 474.
79. Mathew, J. P.; Reinmuth, A.; Melia, J.; Swords, N.; Risse, W. Macromolecules, 1996, 29, 2744. b) Safir, A. L.; Novak, B. M. Macromolecules, 1995, 28, 5396.
80. Houlihan, F. M.; Wallow, T. I.; Nalamasu, O.; Reichmanis, E. Macromolecules, 1997, 30, 6517. b) Wallow, T. I.; Houlihan, F. M.; Nalamasu, O.; Chandross, E. A.; Neenan, T. X.; Reichmanis, E. Proc. SPIE, 1996, 2724, 355.
81. Potter, G. H.; Zutty, N. L., U.S. Patent 3,280,080, 1996.
82. Reichmanis, E.; Novembre, A. E.; Tarascon, R. G.; Shugard, A.; Thompson, L. F., In "Silicon Based Polymer Science, A Comprehensive Resource", ACS Advances in Chemistry Series 244, Ziegler, J. M.; Fearon, F. W., Eds., American Chemical Society, Washington, DC, 1990, 265.
83. a) Havas, J. Electrochem. Soc. Ext. Abstr., 1976, 2, 743. b) Moran, J. M.; Maydan, D. J. Vac. Sci. Technol. A, 1979, 16, 1620. c) Ray, G. W.; Peng. S.; Burriesci, D.; O'Toole, M. M.; Lui, E. G. J. Electrochem. Soc., 1982, 129, 2152. d) Ting, C. H.; Liauw, K. L. J. Vac. Sci. Technol. B, 1983, 1, 1225.
84. Reichmanis, E.; Smolinsky, G.; Wilkins, C. W., Jr. Solid State Technol., 1985, 28(8), 130.
85. a) Taylor, G. N.; Wolf, T. M. J. Electrochem.Soc., 1980, 127, 2665. b) Taylor, G. N. Solid State Technol., 1980, 23(5), 73.
86. Stillwagon, L. E.; Silverman, P. J.; Taylor, G. N., Proceedings of the Regional Technical Conference on Photopolymers, Mid-Hudson Section, SPE, Ellenville, NY, 1985, 87.
87. Nalamasu, O.; Baiocchi, F. A.; Taylor, G. N., In "Polymers in Microlithography", ACS Symposium Series 412, Reichmanis, E.; MacDonald, S. A.; Iwayanagi, T., Eds., American Chemical Society, Washington, DC, 1989, 189.
88. MacDonald, S. A.; Schlosser, H.; Ito, H.; Clecak, N. J.; Willson, C. G. Chem. Mater., 1991, 3, 435.
89. Hartney, M. A.; Schaver, D. C.; Shepherd, M. I; Melngailis, J.; Medvedev, V.; Robinson, W. P. J. Vac. Sci. Technol. B9, 1991, 3432.
90. Ito, H. J. Photopolym. Sci. Technol., 1992, 5(1), 123.
91. a) Coopmans, F.; Roland, B. Proc. SPIE 631, 1986, 34. b) Roland, B.; Lombaerts, R.; Jakus, C.; Coopmans, F. Proc. SPIE 777, 1987, 69. c) Garza, C. M. Proc. SPIE 920, 1988, 233. d) Roland, B.; Vandendriessche, J.; Lombaerts, R.; Denturck, B.; Jakus, C. Proc. SPIE 920, 1988, 120.
92. Pierrat, C.; Tedesco, S.; Vinet, F.; Lerme, M.; Dal'Zotto, B. J. Vac. Sci. Technol. B7, 1989, 1782.
93. Sezi, R.; Sebald, M.; Leuschner, R.; Ahne, H.; Birkle, S.; Borndorfer, H. Proc. SPIE 1990, 1262, 84.

94. a) Shaw, J. M.; Hatzakis, M.; Babich, E. D.; Parasczak, J. R.; Witman, D. F.; Stewart, K. J.; J. Vac. Sci. Technol. B7, 1989, 1209.   b) La Tulipe, D. C.; Pomerene, A. T. S.; Simons, J. P.; Seeger, D. S. Microelectron. Eng. 17, 1992, 265. c) Seeger, D. E.; La Tulipe, D. C., Jr.; Kunz, R. R.; Garza, C. M.; Hanratty, M. A. IBM J. Res. Develop. 41, 1997, 105.

95. Horn, M. W.; Pang, S. W.; Rothschild, M. J. Vac. Sci. Technol. B, 1991, 8, 1493.

96. Weidman, T. W.; Joshi, A. J. Appl. Phys. Lett., 1992, 62, 372.

97. Delplancke, M. P.; Powers, J. M.; Vandentop, G. J.; Somorjai, G. A. Thin Solid Films, 1991, 202, 289.

98. Sartoratto, P. P. C.; Davanzo, C. U.; Yoshida, I. V. P. European Polymer Journal, 1997, 33, 81.

99. Dabbagh, G.; Hutton, R. S.; Cirelli, R. A.; Reichmanis, E.; Novembre, A. E., Nalamasu, O. Proc. SPIE, 1988, 3333, 394.

POLYMERS FOR ELECTRONIC PACKAGING IN THE 21ST CENTURY

C.P. Wong and Rao Tummala

School of Materials Science and Engineering and Packaging Research Center
Georgia Institute of Technology, Atlanta, GA 30332-0245

Introduction
Recent Advances in Polymeric Materials for Electronic Packaging
On-chip Passivation Encapsulants
Silicones (Polyorganosiloxanes)
Room Temperature Vulcanized (RTV) Silicones
Heat Curable Hydrosilation Silicones (Elastomers and Gels)
Epoxies
Polyimides
Silicone-Polyimide (New Modified Polyimides)
Benzocyclobutenes
Sycar (a Silicone-Carbon Hybride) Polymers
Bis-maleimide Triazine (BT) Polymers
Polycyclicolefins
High Performance No Flow Underfills for Low-cost Flip Chip Applications
The Next Generation of High Performance Polymers for Electronic Packaging
Simple Layer Integrated Module (SLIM)/System on Package (SOP) Base Substrate
Simple Layer Integrated Module (SLIM)/System on Package (SOP) Interlayer Polymer Dielectrics
Simple Layer Integrated Module (SLIM)/System on Package (SOP) Integrated Organic Compatible Passives
Simple Layer Integrated Module (SLIM)/System on Package (SOP) Flipchip and Underfill Materials
Heat Transfer Materials
Conclusion
References

Introduction

The invention of the transistor at Bell Labs in 1947 revolutionized the electronic industry. Today, this electronic industry is the most important industry that acts as a driving engine for science, technology, manufacturing and the overall economy. It currently accounts for over $993 billion world-wide and is expected to reach $2 trillion within the next decade (see Table 1). Polymers play a critical role and an integral part in the development of this semiconductor electronic technology.   Electronic packaging encompasses four key functions: (i) power, (ii) signal distribution, (iii) thermal management and (iv) protection.   It includes everything from the discrete integrated circuit (IC) to the electronic system: small systems like in cellular phones, PC's; large systems like mainframe computers and telephone electronic switching systems.   It is one of the most important electronic technologies. Polymers are the "heart and soul" of the next generation of electronic packaging which includes high density printed wiring boards, ultra-low dielectric constant dielectrics, integral passives (capacitors, resistors, inductors), flip-chip solder joints and alternative conductor adhesives, underfills, electrically and thermally conductive heat-transfer materials (1-3).

Figure 1 summarizes the packaging evolution during the past three decades demonstrating the ultimate measure of packaging, namely density of transistors on motherboard/density as transistors on IC. This is also called silicon efficiency, defined as the area of silicon divided by the area of the board. In this table, we can realize the silicon efficiency of the under 2% in 1970's with dual-in-linear packages (DIPs), about 10% in the 1980's with quard-flat-packages (QFP), about 20% with Ball-Grid-Array (BGA) in the 1990's, leading to 40% with both multichip modules (MCMs) and chip size/scale packaging (CSP) that are flip-chip bound to the ICs (see Figure 2). However the future electronic packaging system needs are dramatically changed in the computer, telecommunications, automotive, and consumer electronics industries.   Good representations of these are the four products such as auto navigation, digital wireless communications, high band width networks, and personal computers. All these require paradigm shifts in system technologies and only a system-level approach to these technologies involving design, fabrication, assembly, and test of both semiconductors and packages, in making up the above products, can address the market needs for cost, performance, size, and reliability. In computers, for example, the cost per unit of computing has come down from $4 million/MIP (millions instructions per second) with mainframes to less than $5/MIP with personal computers. Any future systems in this category must continue or exceed this trend.   The next generation of electronic packaging will involve many integrations to achieve high performance and low-cost systems. The Single Layer Integrated Module (SLIM) or System on Package (SOP) is an example (see Figure 3).   Polymeric materials are key to the success of this development.

Table 1(a): Worldwide Electronics Market

	1997		
	U.S.	World	World 2000
GDP	$7.5T	$33T	$39.3T
Electronics	$380B	$993B	$1.3T
Semiconductors	$73B	$152B	$300B
Equipment	$19B	$38B	$74B
Materials	$2B	$22B	$36B

Table 1(b): Worldwide Packaging Market

Worldwide Packaging Market

PWB	$ 30.0B
Flex Circuits	$ 3.2B
Assembly Equipment	$ 3.3B
Materials	$ 9.9B
Connectors	$ 23.4B
Opto Packaging	$ 10.0B
RF Packaging	$ 1.2B
Passive Components	$ 25.0B
Thermal	$ 3.0B
Total	$109.0B

Packaging Evolution

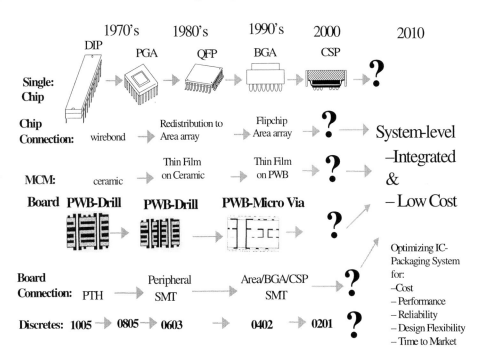

Fig. 1: Packaging Evolution

SLIM: Generation of Microelectronic Systems Packaging

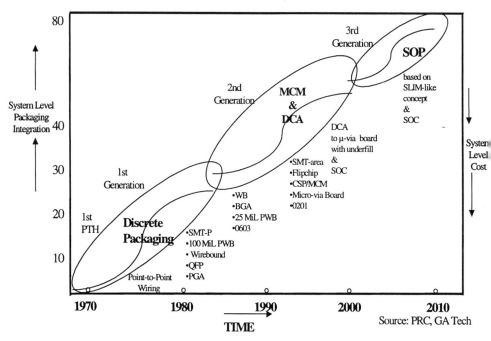

Fig. 2: Limits of Current Packaging Technology

Today's Vs. System on a Package (SOP)

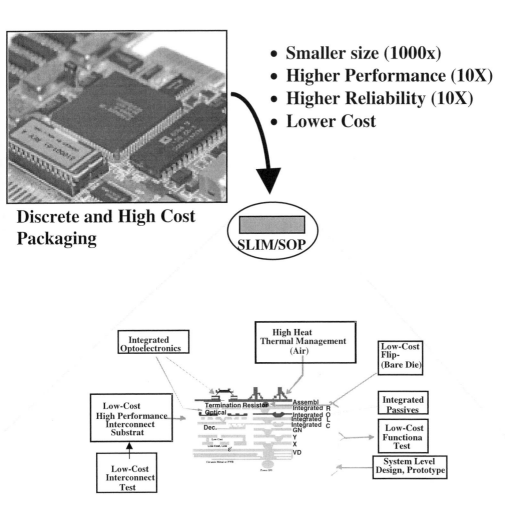

- **Smaller size (1000x)**
- **Higher Performance (10X)**
- **Higher Reliability (10X)**
- **Lower Cost**

Discrete and High Cost Packaging

SLIM/SOP

SLIM: Integrated and Low Cost

Fig. 3

This paper reviews some of the material challenges, particularly of polymers that must be addressed to enhance next generation of packaging. Since packaging is defined as interconnecting, powering, cooling and protecting semiconductor ICs to system needs, this paper starts out with some polymeric materials that we are using today in electronic packaging as well as some new materials for the next generation of electronic packaging.

<u>Recent Advances in Polymeric Materials for Electronic Packaging</u>

Most of the advances in polymeric material for electronic packaging are high glass transition temperature (Tg), photo definable materials such as polyimides, silicone-polyimides, silicones, benzocyclobutenes, epoxies, cyclic-polyolefins and liquid crystal type polymers. Each of these materials will be briefly reviewed as follows.

There are numerous organic polymeric materials that are used as electronic packaging materials. These materials are typically used as on-chip as well as off-chip packaging.

<u>On-chip Passivation Encapsulants</u>

Passivation materials are deposited on devices while they are still in the wafer form. This is usually done at the completion of the IC fabrication process. These type of materials are mainly used for the mechanical protection of the IC devices during dicing of the wafer (singulation process). Furthermore, the passivation layer also serves as corrosion protection of the device. Inorganic polymers such as silicon dioxide, silicon nitride, and silicon-oxynitride are usually used by the semiconductor industry. Although, silicon dioxide and silicon nitride are both excellent moisture barriers, silicon dioxide is still permeable to mobile ions such as sodium, particularly under bias conditions. The use of "getter" such as phosphorus doped silicon dioxide or P glass by either chemical vapor deposition (CVD), plasma and spin on types are readily available. However, silicon nitride eliminates these mobile ion diffusion problems. Recently, organic polymers such as polyimides, benzocyclobutenes, cyclic polyolefins, silicone-polyimides particularly, the photo-definable derivatives of these classes, are increasingly being used as passivating materials. The material usually deposits with a thickness of 0.25-2 μm, with multilevel coating for better reliability protection of the ICs. Bond pad areas of the IC are etched out for further interconnections. Because the passivation layers are not 100% pinhole-free or crack-free, corrosion of devices still occurs. As such, a second layer of the high performance organic encapsulant is still needed for their protection. This layer can also act as a buffer coating for a large IC, acting as a stress relief coating. Furthermore, molding compounds are widely used to package 90% of the worldwide ICs. Nevertheless, we will address both inorganic and organic materials for this type of application.

There are numerous organic polymeric materials that are used as electronic encapsulants. These materials are divided into three categories: (1) thermosetting polymers, (2) thermoplastics, (3) elastomers. Thermoplastic polymers are materials which when subjected to heat will flow and solidify upon cooling without crosslinking. These thermoplastic processes are reversible and the polymers become suitable engineering plastic materials. Examples include polyvinyl chloride, polystyrene, polyethylene, fluorocarbon polymers, acrylics, parylene (Union Carbide's poly-para-xylylene) and preimidized silicone-modified polyimides.

Thermosetting materials are crosslinking polymers which cannot be reversed to the original polymer after curing. Silicones, polyimides, epoxies, silicone-modified polyimides, silicone-epoxies, polyesters, butadiene-styrenes, alkyd resin, allyl esters, silicon-carbons (Sycars®) by Hercules and polycyclic-olefins by BF Goodrich are examples of electronic thermosetting encapsulants. Elastomers are thermosetting materials that have high elongation or elasticity. These types of materials consist of a long linear flexible molecular chain which is joined by internal covalent chemical crosslinking. Silicone rubbers, silicone gels, natural rubbers, and polyurethanes are examples. However, for IC technology applications, only a few of the materials in the above three groups which can be made ultrapure, such as epoxies, silicones, polyurethanes, polyimides, silicone-polyimides, benzocyclobutenes, parylenes, cyclicolefins, silicon-carbons and benzocyclobutenes have been shown to be acceptable IC and electronic encapsulants. In addition, the recently developed high performance liquid crystal materials (high performance engineering plastic materials) are also potential organic polymeric materials for electronic applications. We will discuss some of these in detail.

Silicones (Polyorganosiloxanes)

Silicone, with a repeating unit of alternating silicon-oxygen (Si-O) siloxane backbone, has some unique chemistry. The major types are discussed as follows:

Room Temperature Vulcanized (RTV) Silicones

RTV silicone is a typical condensation cure system material. The moisture-initiated, catalyst (such as organotitanate, tin dibutyldilaurate..etc.)-assisted process generates water or alcohol by-products which can cause outgassing and voids. However, by careful control of the curing process, one can achieve a very reliable encapsulant. Since the silicone has a low surface tension, it tends to creep and run over the encapsulated IC circuits. To better control the rheological properties of the material, thioxtropic agents (such as fumed silica) are usually added to the formulation. The thioxtropic agent provides a yield stress, increases the suitable G' (storage modulus), G" (loss modulus) and n* (dynamic viscosity) of the encapsulant. Filler-resin and filler-filler interactions are important in obtaining a well-balanced and well-controlled encapsulant. This rheological controlled material tends to flow evenly in circuit edges, covers all the underchip area, and prevents wicking and run-over of the circuits which is a critical parameter in coating production. In addition, pigments such as carbon-black and titanium dioxide are usually added as opacifiers to protect light sensitive devices. Organic solvents such as xylenes and Freons® are incorporated into the formulation to reduce the encapsulant viscosity. (4-8)

AT&T (now Lucent Technologies) has been using a RTV silicone system in electronic encapsulation. This RTV silicone has been used to protect their bipolar, MOS, hybrid IC for over 30 years. The ability of the RTV silicone to form chemical bonds with the coated substrate is one of the key reasons the material achieves excellent electrical performance. The reactive alkoxy functional groups of the silicone react with the surface hydroxyl groups to form a stable inert silicon-oxygen-substrate bond. In addition, this chemical reaction consumes the substrate surface hydrophilic hydroxyl groups which would hydrogen (H)-bond with diffused moisture. When sufficient diffused moisture is H-bonded with the surface hydroxyl groups, surface conduction could take place. Furthermore, diffused moisture could form a continuous path of thin water layers. Under such circumstances, and the presence of contaminant mobile ions and an applied electrical bias, corrosion of the IC metallization will result. However, when all surface hydroxyl groups are reacted with the silicone alkoxy groups, even though moisture continues to diffuse through the silicone matrix, no continuous water path is formed that could result in surface conduction and/or "H-bonding" of the diffused moisture. The diffused moisture will only diffuse in and out of the siloxane matrix polymer in an equilibrium fashion which does not cause electrical corrosion of the encapsulated devices. This is probably one of the reasons why RTV silicone is capable of achieving superior performance in temperature-humidity-bias (THB) accelerated electrical testing.

Heat Curable Hydrosilation of Silicones (Elastomers and Gels)

Heat curable hydrosilation silicone (either elastomer or gel – deliberately under crosslink system with ultra-low modulus) has become an attractive device encapsulant. Its curing time is much shorter than the RTV-type silicone. Heat curable silicones also tend to have slightly better stability at elevated temperatures than the conventional RTV silicone. With its jelly-like (very low modulus) intrinsic softness, silicone gel is a very attractive encapsulant in wire-bonded large chip size IC devices. The two-part heat curable system which consists of the vinyl and hydride reactive functional groups, and the platinum catalyst hydrosilation addition cure system provides a fast cure system without any by-product. To formulate a low modulus silicone gel, a vinyl-terminated polydimethylsiloxane with a moderate low viscosity range from 200 to a few thousand centipoise (cps), and a low viscosity (range from a few cps to ~100 cps) di- or multi-functional hydride-terminated polydimethylsiloxane are used in the formulation. The low viscosity hydride resin usually blends in with the higher viscosity vinyl resin to achieve an easier mixing ratio of part A (only vinyl portion) and part B (hydride plus some vinyl portion for ease of mixing). The key to formulate a low modulus silicone is the deliberate undercrosslinking of the silicone system. A few ppm platinum catalyst such as chloroplatinic acid or organoplatinum, is used in this system. This catalyst is usually incorporated in the Part A vinyl portion of the resin. However, a highly deactivated platinum catalyst system (by premixing a chelating compound such as 2-methyl-3-butyn-2-ol to coordinate the reactive platinum catalyst) is used to formulate a one-component system. This one-component silicone gel system provides less mixing, problem-free production material. This solventless type of heat curable silicone gel will have increased use in electronic applications. (4-8)

Epoxies

Epoxies are one of the most utilized polymeric materials in electronics. Their unique chemical and physical properties such as excellent chemical and corrosion resistant, electrical and physical properties, excellent adhesion, thermal insulation, low shrinkage, and reasonable material cost have made epoxy resins very attractive in electronic applications. The commercial preparation of epoxies are based on bisphenol A and F, which upon reaction with epichlorohydrin produces diglycidyl ethers. The repetition number, n, varies from zero (liquid) to approximately 30 (hard solid). The reactants' ratio (bisphenol A and F versus epichlorohydrin) determines the final viscosity of the epoxies. In addition to the bisphenol A resins, the Novolac resins with multifunctional groups which lead to higher cross-link density and better thermal and chemical resistance have gained increasing acceptance in electronic applications. Typical epoxy curing agents are amines, anhydrides, dicyanodiamides, imidazoles, triphenyl-phosphine, melamine/formaldehydes, urea/formaldehydes, phenol/formaldehydes, and catalytic curing agents. Anhydrides and amines are two of the most frequently used curing agents. (9)

Selecting the proper curing agents is dependent on application techniques, curing conditions, pot-life required and the desired physical properties. Besides affecting viscosity and reactivities of the epoxy formulations, curing agents determine the degree of cross-linking and the formation of chemical bonds in the cured epoxy system. The reactivity of some anhydrides with epoxies is slow; therefore an accelerator, usually a tertiary amine, is used to assist the cure. "Novolacs" and "Resole" are two major commonly used phenol/formaldehyde epoxies. A Novolac is a phenol/formaldehyde, acid-catalyzed epoxy polymer. The phenolic groups in the polymer are linked by a methylene bridge which provides highly cross-linked systems and a high temperature and excellent chemically resistant polymer. "Resole" is a base-catalyzed phenol-formaldehyde epoxy polymer. In most phenolic resins, the phenolic group is converted into an ether to give improved base resistance. Phenolic resins are cured through the secondary hydroxyl group on the epoxy backbone. High temperature curing is required in this system and it provides excellent chemical-resistance.

Recently developed high purity epoxies have become very attractive encapsulants for electronics. These new types of resin contain greatly reduced amounts of chloride and other mobile ions, such as sodium and potassium, and have become widely used in device encapsulation and molding compounds. The incorporation of the well-controlled spherical silica particles with bimodal size distribution as filler in the epoxy systems has drastically reduced the thermal coefficient of expansion of these materials and makes them more compatible with the IC die attached substrate materials. The incorporation of a small amount of an elastomeric material (such as, silicone elastomeric domain particles) to the rigid epoxy has drastically reduced the elastic modulus, reduced the thermal stress, and increased the toughness of the epoxy material. This new type of low stress epoxy encapsulant has great potential application in molding large IC devices. In addition, the biphenyl epoxy resin has very low viscosity, which allows high filler loading and ultra high purity for epoxy molding compound applications. The 'glob-top' (a glob of polymeric resin that covers the entire IC device, including the wire bonding, and flip-chip underfills where the encapsulant couples the IC and organic substrate for enhancement of cycle fatigue life is able to meet all the requirements as a device encapsulant) type epoxy material which is applied as an IC chip encapsulant is becoming increasingly more acceptable as an encapsulant for higher reliability chip-on-board and flip-chip type electronic devices and systems. When the epoxy materials are properly formulated and applied, and their stress related issues such as reduced stress and reduced thermal co-efficient of expansion have been properly considered and resolved, they could become a very attractive high performance encapsulant. The continuous advancements in epoxy material development will have a great impact in device packaging.

Polyimides

Polyimide was first developed at DuPont in the 1950's. During the past couple of decades, there has been tremendous interest in this material for electronic applications. The superior thermal (stability up to 550°C), mechanical and electrical properties of polyimide have made its use possible in many high performance applications, from aerospace to microelectronics. In addition, polyimides show very low electrical leakage in surface or bulk. They form excellent interlayer dielectric insulators and also provide excellent step coverage which is very important in fabrication of the multilayer IC structures. They have excellent solvent resistance and ease of application. They can be easily either sprayed or spun-on and imaged by a conventional photolithography and etch process.

Most polyimides are aromatic diamine and dianhydride compositions. However, by changing the diamine and dianhydride substitutes, one derives a variety of high performance polyimides. Polyamic acids are precursors of the polyimides. Thermal cyclization of polyamic acid is a simple curing mechanism for this material. Siemens of Germany developed the first photodefinable polyimide material. However, Ciba Geigy, Du Pont, Hitachi, Amoco, etc. have new types of photodefinable polyimide which do not require a photoinitator. Both of these photodefinable materials are negative resist type polyimides. A positive type resist polyimide which reduces the processing step in IC fabrication has recently been reported by Sumitomo Bakelite, Du Pont and others. An interpenetration network (IPN) of two types of polyimides is used to achieve the positive-tone material. Hitachi has developed an ultra-low thermal coefficient of expansion (TCE) polyimide which has some potential in reducing the thermal stress of the silicon chip and the polyimide encapsulant. The rigid rod-like structure of the polyimide backbone structure is the key to preparing the low TCE polyimide. By simply blending a high and low TCE polyimide, one is able to achieve a desirable TCE encapsulant which matches the TCE of the substrate, and reduces the thermal stress problem in encapsulated device temperature cycling testing. However, the affinity for moisture absorption due to the carbonyl polar groups of the polyimide, a high temperature cure, and high cost of the polyimide are drawbacks that prevent its use in low cost consumer electronic application. Preimidized polyimides which cure by evaporation of solvent may reduce the drawback of high temperature cure of the material. Advances in polyimide syntheses have reduced the material's moisture absorption and dielectric constant by the incorporation of siloxane segments into the polyimide backbone (see next section on silicone-polyimide). However, the affinity of the polyimide chemical structure to moisture is still a concern in its use in electronics. Nevertheless, polyimides are widely used as IC encapsulants, interlayer dielectrics, ion implant masks, and alpha particle getter applications. (11-12)

Silicone-Polyimide (New Modified Polyimides)

Combining the low modulus of the siloxane, and the high thermal stability of polyimide, the siloxane-polyimide (SPI) copolymers were first developed at General Electric. SPI copolymers have become very attractive IC device encapsulants. Silicone-polyimides are fully imidized co-polymers and are soluble in low boiling solvents such as diglyme which reduces the high processing temperature and eliminates the outgassing of water during the normal polyimide imidization (cure) process. The high processing temperature and outgassing of water are main drawbacks of the polyimides. Besides, the SPI has good adhesion to many materials, and eliminates the need for an adhesion promoter. Polycondensation and polyaddition processes are used to synthesize these materials. Thermoplastic and thermosetting SPI materials can be obtained by these processes. In addition, photo-and thermo-curable SPIs are obtaining by incorporating a photo-reactive functional groups into these types of materials. One can control the imide and siloxane blocks within the copolymer matrix to tailor the SPI properties. Since most of these are preimidized thermoplastic materials, their shelf-life is very stable. These materials have potential as IC device encapsulants, interlayer dielectrics, passivation in microelectronic applications. (13)

Benzocyclobutenes

High performance benzocyclobutene (BCB) polymers were recently developed by Dow Chemical Company. The crosslinking process is carried out by the thermal rearrangement of the dicyclobutyl monomer to form the reactive intermediate orthoquinodimethane, which can polymerize with the unsaturated functional group. Since it is based on the thermal rearrangement process, BCB requires no catalyst and there are no by-products during the curing process. The properties of BCB can be modified by the substituted group in its structure. Dimethysiloxane, substituted groups, are usually used as copolymer moieties in the BCB to enhance its adhesion and reduces its modulus. As such, BCB monomers are normally "B-stage" (partially thermally crosslinked) to enhance their viscosity stability. The commercially available BCB is usually in a "B-stage" diluted with xylene for spin-coating application. BCB has excellent physical, chemical and electrical properties which has found its use in microelectronic applications similar to polyimides. With its low dielectric constant (2.7), low moisture absorption (<1%) and good adhesion properties. BCB is currently widely used as an IC passivating encapsulant and interlayer dielectric for the current multichip modules and wafer level redistribution of flip-chip high I/O device applications. More recently, higher thermal oxidative stability with added antioxidant additives and photo - sensitive BCBs are also available. (14)

Sycar (a Silcon- Carbon Hybride) Polymers

Recently, Hercules (at Wilmington, DE) has developed a new class of silicon- carbon hybride material. This class of material consists of a backbone of the siloxane(-Si-O-Si-O-) structure and cross-links to the silicon siloxane backbones with hydrocarbons which provide excellent mechanical and solvent resistance properties, yet mantain its silicone-like excellent electrical properties. Blob-top, molding compounds, and high performance PWB are being made by this class of material. However, their actual use is not yet common, but it has potential for electronic applications. (15)

Bis-maleimide Triazine (BT) Polymers

A new resin used to prepare a high temperature Printed Wiring Board (PWB) vs the conventional FR(Fire retardant)-4 PWB is called bis-maleimide triazine (BT). The triazine polymer is mainly produced by Mitsubishi Chemicals in Japan. Trimerization of the monomers forms a high temperature, high Tg (>230C), triazine (Cyanurate ring) resin that is generally mixed with an epoxy resin to form a high performance PWB substrate for Ball Grid Array (BGA) and Multichip Module-Laminate (MCM-L) substrate applications. The mixture of BT/epoxy has good electrical and good thermal-mechanical properties. Epoxy is blended into the BT resin to provide good toughness. BT/epoxies have a regular Tg of 180-190°C. The process is compatible with PWB board manufacturing process and PWB wire-bonding processes. Furthermore, the BT/epoxies have a long history of good resistance to ionic conductive growth and popcorn testing results.

Polycyclicolefins

BF Goodrich has recently developed a new class of cyclicolefins based on the principle of polynorbornene chemistry. A transition metal catalyst is used to provide tightly controlled polymerization of the monomers to afford saturated polymers with excellent Tg (>350C), low dielectric constant (2.45), low moisture (<0.1), and low thermal coefficient expansion (50ppm). All the drawbacks of polyimide may be allieviated by this new material. Furthermore, the material has isotropic physical properties, i.e., the same properties in x, y, z directions, which are lacking in other high performance materials such as polyimides, BCB, etc. The incorporation of siloxane coupling agents provides excellent adhesion to copper, gold, silver, silicon and other oxides. The polymer can also be reactive ion etched (RIE) using a mixed gas plasma of oxygen and CHF_3 at 450 mtorr and 300 watt RF power for 2 minutes with a 12μm thick film. This material may be synthesized by fewer steps resulting in a low-cost, high-performance material for MCM applications. BF Goodrich is marketing these materials under the Tradename of Avatrel. (16)

High Performance No Flow Underfills for Low-Cost Flip-Chip Applications

Underfill encapsulants are polymeric materials used to reduce the shear stress of the solder joints between the chip and the organic substrate which generates from thermal mismatch. The encapsulant does not only provide dramatic fatigue life enhancement with minimal impact on the manufacturing process flow, but also extends its use to a variety of organic and inorganic substrate materials resulting in ten to hundredfold improvement in fatigue life compared to an unencapsulated package. Therefore, the underfill encapsulation has been key to the development of flip-chip DCA technology (17, 18).

No-flow underfill encapsulants are one type of encapsulant that is now becoming increasingly attractive due to process simplification that saves time and assembly cost. To develop no-flow underfill materials suitable for no-flow underfilling processing of flip-chip solder joint interconnects, several latent catalysts have been studied for such formulations. These metal chelate catalysts are reacted with epoxy resins (cycloaliphatic type epoxy), crosslinkers (anhydride) or hardeners, and other additives, such as adhesion promoters, silica fillers, self-fluxing agents, and surfactants, to form low-cost high performance underfills. (19)

However, current underfill encapsulants are dispensed on one or two edges of the assembled flip-chip package. This allows capillary action to draw the underfill into the gap between the chip and substrate of the assembled package to complete the encapsulation process as shown in Fig. 4. This process has two disadvantages: (1) The processes of flip-chip fluxing, reflowing solder bumping, deflux cleaning, and the processes of underfilling and curing the encapsulant are separated steps, which result in lower production efficiency. (2) The underfilling process takes a long time to complete the flow of the liquid underfill. Furthermore, larger ICs require even longer underfill flow times due to fine-pitch, lower bump height and long cure time. The No-flow underfilling process was invented to dispense the underfill materials on the substrate or the semiconductor devices first, then perform the solder bump reflow and underfill encapsulant curing simultaneously as shown in Fig. 5 (19). Therefore, the no-flow underfilling process not only eliminates the strict limits on the viscosity of underfill materials, process temperature and package size, but also improves production efficiency. Till now, however, the no-flow underfilling process has not been widely used in production. The reason mainly lies in the lack of successful no-flow underfill and reworkable materials. (19,20)

A successful no-flow underfill material should meet primary requirements described below. (1) Little curing reaction happens at the temperature below solder bump reflow temperature (~210-230°C). (2) Rapid curing reaction takes place after maximum solder bump reflow temperature. (3) Good adhesion of the underfill material to the chip, substrate, and solder joints. (4) Lower shrinkage of the material during curing, lower TCE, and reasonable modulus to minimize the thermal stress resulting from the curing process and finally, having the self-fluxing ability to passivate the substrate oxide conductor lines prior to solder reflow.

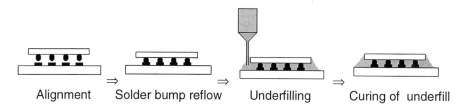

Alignment   Solder bump reflow   Underfilling   Curing of underfill

Fig. 4: The traditional underfilling process

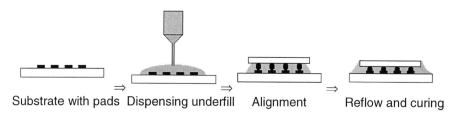

Substrate with pads Dispensing underfill   Alignment   Reflow and curing

Fig 5: The no-flow underfilling process

The Next Generation of High Performance Polymers for Electronic Packaging

The next generation of polymers are those that meet the needs of the next generation packaging structure, for the single chip integrated module or SLIM, for single-level integrated module as both package and board, or the so-called system on package (SOP). The cross-section of this package, as illustrated in Figure 3, requires major challenges.

The cross-section of this package, as illustrated in Figure 3, requires major challenges in materials at every level and function – base substrate, interlayer dielectric, interlayer conductor, integrated passives: capacitors, resistors and inductors, solder and alternative to solders such as conductive adhesives, underfills and thermal management materials. The following begins to address some of these, but only in organic packaging.

SLIM or SOP Base Substrate

Given the need for low-cost, which requires large-area processing among others, printed wiring board materials such as FR-4 are the leading candidates. FR-4 in particular, however, suffers from a number of deficiencies: low modulus, very high TCE in relation to silicon, surface smoothness and warpage, too low a temperature of stability. A number of materials are being proposed that include Teflon®, liquid-crystalline polymers with or without fillers and a variety of polymer-inorganic composites.

SLIM or SOP Interlayer Polymer Dielectrics

A dielectric and conductor are the two essential elements of an interconnection structure which is a wiring network embedded in a dielectric matrix. Polymers are an important class of dielectric because of the low dielectric constant and low dielectric loss associated with most polymers in addition to their low processing temperature. These are the two key electrical requirements in dielectrics. Among others are mechanical, thermal, chemical, environmental and process properties. In addition, cost should always be kept in mind.

A variety of dielectric polymers, such as in Table 2, have been developed and used as interlayer dielectrics on ICs, and for thin-film and PWB wiring substrates. Polyolefin is a new dielectric polymer candidate developed by B. F. Goodrich which exhibits many superior properties. Not listed here is polytetrafluroethene (PTFE) which has the best electrical performance and lowest dielectric constant (~ 2.0) among known polymers. However, its application is plagued by poor process properties. From an electrical designer's viewpoint, faster signal speed, higher wiring density and smaller feature sizes are all driving toward lower dielectric constant and loss. On the other hand, materials developers are facing a multi-faceted challenge. Many requirements need to be taken into consideration in order to develop a workable and manufacturable dielectric polymer, some of which may compromise electrical performance. More complex issues are involved when cost and compatibility with manufacturing infrastructure come into play. Nanostructured air-voids within a polymer matrix that reduce interlayer dielectric properties are an attractive material and process for ultra-low dielectric constant materials for the next generation of high frequency electronic applications.

	Dielectric Constant	Dielectric Loss (%)	T_g, °C	Water Absorption (%)	CTE, ppm/K	Metalization	Photo-sensitity	Modulus (Gpa)
Polyimide	2.9-3.7	0.2-0.5	260-550	0.5-3.0	37-50	sputter, plating	yes	2.4-3.4
Polyolefin	2.4-2.6	0.07	>350	0.1	50	sputter	no	1.4
BCB	2.7-2.8	0.08	>350	0.08-0.12	52	sputter	yes	2.0-2.2
Polyphenyl-quinoxaline	2.7	0.05	365	0.9	55	sputter	no	2.0
Cyanate Ester	2.7-3.1	0.1-0.7	>250	0.5-2.5	54-71	plating	no	2.7-3.5
Epoxy	3.2-4.5	0.7-2.5	120-180	~1.0	60-77	plating	yes	3.0-4.7

SLIM/SOP Integrated Organic Compatible Passives

Most passives in use today are discretes that are fabricated with high temperature ceramics using powder and thick film technologies. The SCIM (single chip integrated module) and SLIM or SOP packages described above however require a maximum processing temperature of about 200 °C to be PWB-compatible.

Some of the materials being explored for capacitors, inductors and resistors include those processed by thin film processes and thick film processes. Thin film materials include Ta_2O_3, BST, titantes, silicides, and nitrides and thick fims include polymer-ceramic composites.

SLIM/SOP Flipchip and Interconnect Materials

High lead solders have been the industry standard for flipchip bonding of VLSI chips for over two decades. They are being supplemented with lower temperature eutectic solders because of the need for PWB compatibility and conductive adhesives. There are certain intrinsic advantages of solders such as self-alignment, high electrical conductivity, excellent oxidation resistance, low-cost and good mechanical properties. The challenge for the newer materials therefore is how to meet or exceed these properties. Furthermore, environmental friendly lead-free alloy and conductive adhesives will play a critical role in the next decades for electronic interconnects.

Heat Transfer Materials

The SLIM/SOP package proposed is an all organic package. As such, the package is a poor thermal conductor and heat-transfer, therefore, must take place from the back of silicon IC. Direct conduction by means of high thermal conductivity heat sink materials such as Al, Cu, AlN, diamond, Cu-W/Mo, Cu-invar Cu and others is one way to accomplish this task. Microjet cooling using both single and multi-phase liquids is being practiced. There are a number of other ways as well that can effectively remove high heat flux modules, as well as the use of polymer and high thermal conductivity ceramic composite. (22)

Conclusions

The next generation of polymeric materials for electronic packaging will require high performance, such as high Tg, low loss, low dielectric constant for high frequency (in excess of 100 GHz for wireless and portable products). Nanostructure air gap ultra-low dielectrics will be developed. Furthermore, photo-definable with positive-tone materials for ease of process are needed for IC dielectric as well as sequential build-up high density PWB technology applications. High dielectric polymer composites (in excess of 100-200) with nanostructure inorganic filler will be needed for the capacitor application of building imbedded passives. Conductive polymers with high current density, high toughness and low conductivity fatigue will be needed for ultra fine-pitch replacements for solder joint interconnects. The electronics consumer will depend on all these materials which will be multi-functional, high performance, yet low cost. It is a challenge that all chemists, materials scientists, and chemical engineers will face in the 21ˢᵗ century. (23)

REFERENCES

1. C. P. Wong, Ed., "Polymers for Electronic and Photonic Applications," Academic Press (1993).
2. R. R. Tummala, E. Ramaszewski and A. Koperstein, "Handbook on Electronic Packaging," Chapman Hall (1997).
3. J. Lau, C. P. Wong, J. Prince, W. Nakayama, "Electronic Packaging: Design, Materials, Processes and Reliability," McGraw Hill (1998).
4. W. Noll, "Chemistry and Technology of Silicones," Academic Press, New York (1968).
5. J. Yilgör, J. E. McGrath, "Polysiloxane Containing Copolymers: A Survey of Recent Developments," in Advances in Polymers Science, D. Olivé, Ed., Vol. 86, Springer-Verlag, Berlin.
6. K. Otsuka, Y. Shirai, and K. Okutani, "A New Silicon Gel Sealing Mechanism for High Reliability Encapsulants," IEEE Trans. Comp. Hybrids, Manuf. Tech., CHMT-7, p. 249 (1984).
7. C.P. Wong, "Effect of RTV Silicone Cure in Device Packagings," Chapter 43 in Polymers for High Technology and Photonics (M. J. Bowden and S. R. Turner, eds.), Symposium Series, Vol. 346, p. 511, American Chemical Society, Washington, DC (1987).
8. C. P. Wong, J. M. Segelken, and J. W. Balde, "Understanding the Use of Silicone Gel for Non-hermetic Packaging of ICS," IEEE Trans. on Components, Hybrids and Manufacturing Technology, 4, p. 419 (1989).
9. C. A. May and Y. Tanaka, "Epoxy Resins," Marcel Dekker, New York (1979).
10. L. Manzione, "Plastic Packaging of Microelectronic Devices," Van Nostrand Reinhold, New York (1990).
11. K. L. Mittal, Ed., "Polyimides: Synthesis, Characterization and Applications," Vols. 1 & 2, Plenum Press, New York (1984).
12. S. Numata, K. Fujisaki, d. Makino, and N. Kinjo, "Chemical Structures and Properties of Low Thermal Expansion Polyimides," Proceedings of Second International Conference on Polyimides, Ellenville, New York, p. 492 (1995).
13. C. J. Lee, "Polyimide-siloxanes," The First International Society for Advanced Materials and Process Engineers, Vol. 1, p. 576 (1987).
14. T. Tessier, G. Ademon, and I. Turlik, "Polymer Dielectric Options for Thin Packaging Applications," IEEE Proc. of 39ᵗʰ Electronic Components Conference, p. 127 (1989).
15. J. K. Bard, R. L. Brady, J. M. Schwark, "Processing and Properties of Silicon-Carbon Liquid Encapsulants," IEEE 43ʳᵈ Electronic Components and Technology Conference, p. 742 (1993).

16.   R. A. Shick, B. L. Goodall, L. H. McIntosh, s. Sayaraman, P. A. Kohl, s. A. Bidstrup-Allen, N. R. Grove, "New Olefinic Interlevel Dielectric Materials for Multichip Modules," IEEE Proceedings on Multichip Conference, p. 182 (1996).

17.   D. Zoba, M. E. Edwards, "Review of Underfill Encapsulant Development and Performance of Flip-chip Development," Proceedings of 1995 ISHM (1995).

18.   C.P. Wong, M.B. Vincent, S. Shi, "Fast Flow Underfill Encapsulant: Flow Rate and Coefficient of Thermal Expansion," Advances in Electronic Packaging, Vol. 1, p. 301 (1997).

19.   C. P. Wong, S. Shi, G. Jefferson, "Novel No Flow Underfills for Flip Chip Applications," IEEE Trans. on CPMT, Vol. 21, No. 2, p. 450-458 (1998).

20.   L. Wang, C. P. Wong, "Novel Reworkable Underfills for Flip Chip on Board Applications," IEEE Trans. on CPMT, Part B, Vol. 22, No. 1, P. 46-53 (1999).

21.   J. Gonzalez, W. Black, "Study of Droplet Sprays Prior to Impact on Heated Horizontal Surface," Trans. ASME, Journal of Heat Transfer, Vol. 119, No. 2, p. 279-287 (1997).

22.   R. Bollampally, C. P. Wong, "High Thermal Conductive Encapsulants," IEEE Trans. on CPMT, Part B, Vol. 22, No. 1, p. 54-59 (1999).

23.   R. Tummala, C. P. Wong, "Materials in Next Generation of Packaging," IEEE Proceedings from the 3rd Int. Symp. on Adv. Packaging Materials Processes, Properties and Interfaces, p. 1-3 (1977).

ORGANOMETALLIC AND METAL-CONTAINING ORGANIC POLYMERS-AN INTRODUCTION

JOHN E. SHEATS, CHARLES E. CARRAHER, JR., CHARLES U. PITTMAN, JR., AND M. ZELDIN

Department of Chemistry, Rider University, Lawrenceville, NJ 08648

Department of Chemistry and Biochemistry, Florida Atlantic University, Boca Raton, FL, and Florida Center for Environmental Studies, Palm Beach Gardens, FL 33410

Department of Chemistry, Mississippi State University, P. O. Drawer CH, Mississippi State, MS 39762

Department of Chemistry, Hobart and William Smith College, Geneva, NY 14456-2842

Preface
Introduction
Silanes, Germanes, and Stannanes
Poly(silsequioxanes)
Living Polymerizations to Make Polyphosphates
Metallocene Polymers
Metallocene ROMP
Condensation Polymers
General Applications
Electrical
Analytical Reagents
Radiation
Biological Agents
Metal Deposition
Conclusions
Literature Cited

Preface

Classical polymers are composed of only about 10% of the available elements most of which are considered metals. There are a number of important non-metal inorganic polymers as well as a number of important metal-containing inorganic polymers that will not be covered here. Here we will focus on only synthetic organometallic and metal-containing polymers.

Introduction

As we approach the new millennium, we look back over recorded history and forward to the future. The development of human civilization has often been characterized by the materials used for tools, structures, and weapons. Thus, we have the Stone Age, the Bronze Age, the Iron Age, the Aluminum Age, and most recently, the Age of Plastics. Polymeric materials have in the last 50 years replaced wood, metal, and glass in many applications because of the ease of fabrication, lower density, higher strength, greater flexibility and resistance to breakage and the ability to achieve an attractive surface finish without painting or plating. Such diverse areas as clothing, children's toys, furniture, and automotive parts are now largely fabricated from polymers.

The need for new materials with lower density, higher strength, ability to withstand extremes of temperature and electrical conductivity ranging from semiconductors to superconductivity has stimulated research in polymer science. Indeed, both the National Research Council and the National Science Foundation have highlighted the need for high performance materials. A recent article in *C & E News* (1) stated, "One of the main thrusts in materials science research is the quest for organic or polymeric materials that could replace the inorganic materials that are the mainstay of today's electronic and photonic technologies..... for such applications as thin film transistors, holographic storage media, and optical fibers."

The polymers chosen for these applications are often organometallic or metal-containing organic polymers. These materials combine the low density and great variety of functional groups and structural variation of organic materials with the ease of achieving multiple oxidation states and electrical conductivity and the high temperature stability characteristic of inorganic compounds.

In order to keep abreast of this rapidly expanding field, PMSE, sometimes jointly with the Polymer and Inorganic Divisions, has sponsored symposia in 1971 (2), 1977 (3), 1979 (4), 1985 (5), 1989 (6), and 1994.(7) Several of the authors of this paper have participated in all of these symposia. Other groups have also sponsored symposia and written review articles and reviews.(8-10) The most comprehensive of these is a series of annual reviews by Ian Manners.(10,11,12) A new journal, *The Journal of Inorganic and Organometallic Polymers*, is now completing its eighth year. Growth in the general area of metal-containing polymers has been explosive and quite diversified. Thus, this paper attempts to review briefly developments in only five areas that are illustrative of the growth of the areas in general – Silanes, Germanes and Stannanes, Poly(silsequioxanes), Polyphosphazenes, Metallocene Polymers, and Condensation Polymers and to discuss potential applications of these materials.

Silanes, Germanes, and Stannanes

Carbon possesses a unique ability to form strong single, double, and triple bonds to itself. Organic chemistry is therefore filled with rings, chains and sheets, and three-dimensional structures to which a variety of functional groups can be appended. Because of the high C-C bond energy (80 kcal), once carbon chains are formed, they do not cleave readily. Early attempts to form similar structures with Si, Ge and Sn were largely unsuccessful because the low bond energies (20-30 kcal) led to facile cleavage and rearrangement. It was only in the 1980's that soluble, well characterized long silylene chains were produced. (13-15) The field has grown rapidly since then.

The great interest in silane, germane, and stannane polymers arises because they show σ-delocalization (16) and σ-π delocalization when they are conjugated with arenes or acetylenes. This property is not surprising since elemental Si possesses a covalent network structure similar to diamond but shows considerable electrical conductivity and elemental Ge and Sn exhibit metallic bonding. Indeed, stannane polymers have even been called "molecular metals." (17) The conductivity is greatly enhanced when holes are created by doping, (18) by illumination, (19) or by an electric field. Recent developments in the synthesis of these materials and their properties will therefore be summarized. Formation of silyl chains appears to be a more complex process than carbon chains, but substantial progress has been made in understanding the mechanisms involved. (20-24) More understanding is still needed before commercially useful material can be prepared. Four methods of preparing polysilylenes have been developed:

1. Reductive Coupling

   $$n\,R_2SiCl_2 + 2n\,Na \text{ or other metal} \longrightarrow 2n\,NaCl + (R_2Si)_n$$

2. Dehydrogenative Coupling

   $$n\,R_2SiH_2 + \text{metal} \longrightarrow n\,H_2 + (R_2Si)_n$$

3. Polymerization of Masked Disilenes

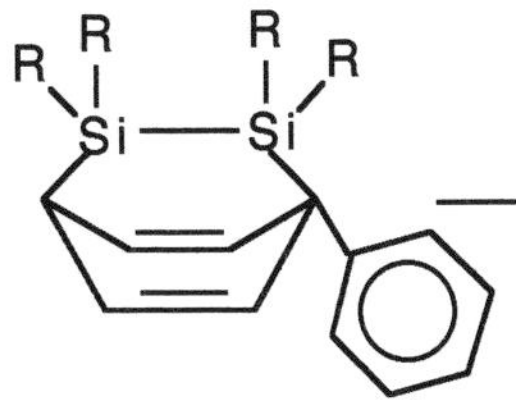

$$\longrightarrow -(R_2SiSiR_2)_n + n\,C_6H_5\text{-}C_6H_5$$

4. Anionic Ring-Opening Polymerization

$$\begin{array}{c} \underset{|}{C_6H_5}\ \underset{|}{H} \\ H-\underset{|}{Si}-\underset{|}{Si}-C_6H_5 \\ H-\underset{|}{Si}-\underset{|}{Si}-C_6H_5 \\ \underset{|}{C_6H_5}\ \underset{|}{H} \end{array} \quad A^- \quad \longrightarrow \quad \left(\begin{array}{c} C_6H_5 \\ | \\ -Si- \\ | \\ H \end{array}\right)_n$$

Reductive coupling initially employed Na or other alkali metals. R groups therefore could not contain any reactive groups. The reaction is heterogeneous, occurring at the surface of tiny metal globules and appears to follow a chain mechanism rather than a step mechanism.

Researchers appear to have gained considerable understanding of the process and how to control it, but much more development is needed before commercially useful material can be prepared reproducibly.

Coupling of dichlorodisilanes with dilithium salts of 1,2-diethynyl disilanes (25) or dilithio polythiophenes (26,27) produce alternating copolymers, which exhibit σ-π conjugation, earlier observed by Ishikawa et al. (28,29) for (–SiR$_2$C$_6$H$_4$SiR$_2$–)$_n$. The polymers also undergo solid state transitions to form liquid crystalline mesophases. (25)

$$LiC \equiv C - SiR_2SiR_2C \equiv CLi + ClSiR_2'SiR_2'Cl \longrightarrow$$
$$\{(SiR_2)_2\ C \equiv C\text{-}\ (SiR_2')_2\ C \equiv C\}_n$$

$$Li-\left(\begin{array}{c} S \\ \\ \end{array}\right)_n-Li\ +\ Cl(Me_2Si)_nCl \longrightarrow$$

n = 1, 2 or 4

$$-\left(\left(\begin{array}{c} S \\ \\ \end{array}\right)_n-(SiMe_2)_n-\right)_m$$

Water soluble silane polymers may be prepared by coupling silane monomers with short polyethylene glycol side chains. (30)

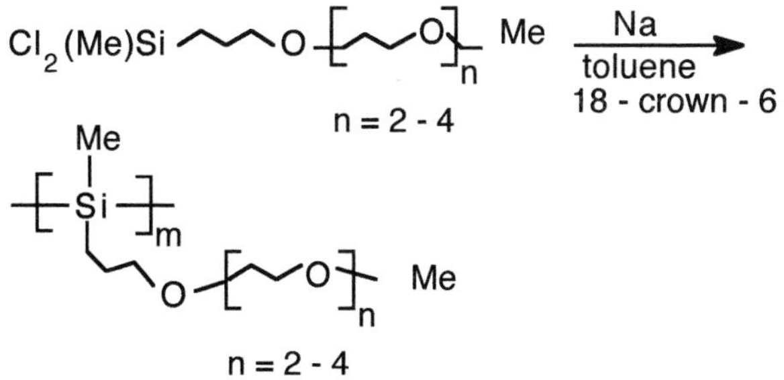

Dehydrogenative coupling of dialkyl and monoalkyl silanes (Method 2) in the presence of a transition metal catalyst has provided an alternate route to polysilylenes. (31-39) Titanocene and zirconocene (31-39) dialkyls have served as effective catalysts for the coupling, but hafnocene catalysts are, surprisingly, inactive. (39) Organolanthanide catalysts such as $(C_5Me_5)_2$ LaR (R = H, $CH(SiMe_3)_2$) are also effective. (39) Functionalized silane polymers may be prepared by this procedure. If the side chains contain amine functions which may be quarternized or carboxyl functions which may be deprotonated, the polymers can be water soluble.

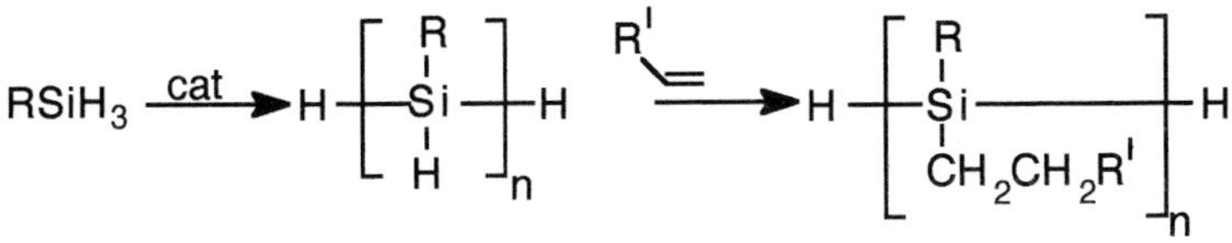

Network (41-44) and dendrimeric (45-47) silane polymers have also been prepared. These materials possess σ-conjugation in three dimensions and therefore possess greatly enhanced conductivity. They have been employed as negative-type photoresists, since exposure to light or air leads to insoluble materials. Synthetic routes to these polymers are shown below. (41-45)

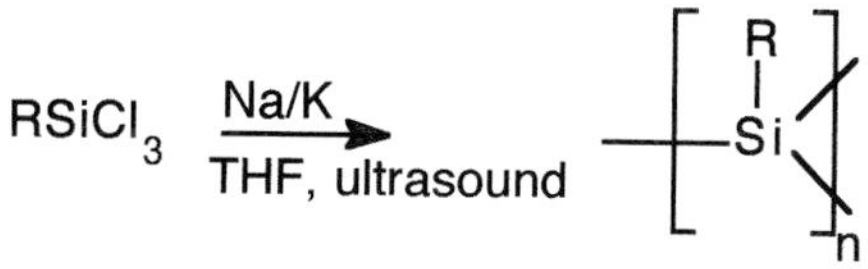

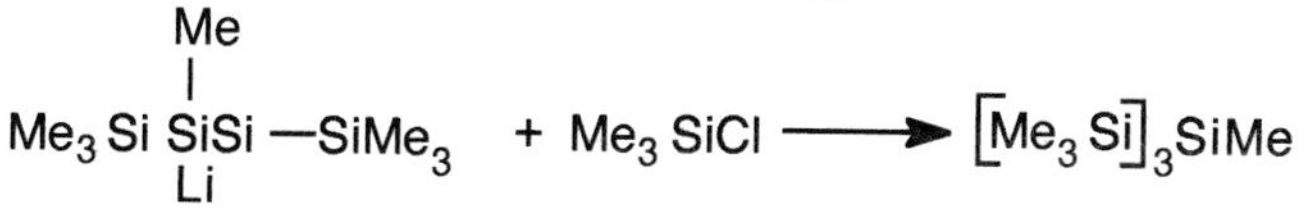

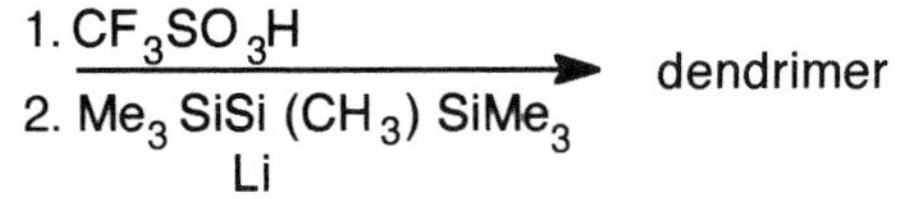
1. CF3SO3H
dendrimer
2. Me3 SiSi (CH3) SiMe3
Li

Polygermanes and polystannanes have been prepared by routes similar to the polystannanes. (48-51) These materials show enhanced σ conductivity relative to the silanes and many have useful applications as light emitting diodes. (52)

Polycarbosilanes have been prepared by a number of workers, usually as precursors to silicon carbide. The most exciting new development in this field is the preparation of poly(silaethylene) and several functional derivatives. (53-55)

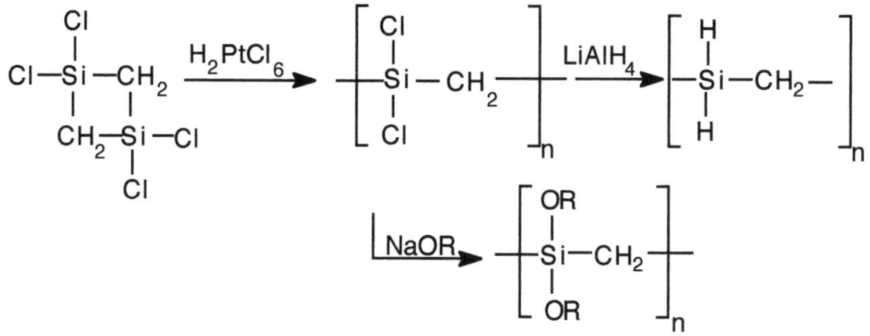
H2PtCl6
LiAlH4
NaOR

Polysilaethylene has an extremely low Tg (-135 to -140°C) and melts at room temperature Tm = 25°. The ease of substitution of the chloroderivative by alkoxide and possibly amines and other functionalities opens the door to a wide variety of materials and derivative chemistry similar to that known for the polyphosphazenes.

Fluorinated polysilaethylenes (56,57) have also been prepared. Poly(vinylidine fluoride), $(CF_2CH_2)_n$ is unique among existing commercial materials in its ability to form thin, tough, flexible films with very high piezo- and pyroelectric properties. These properties arise because the chains can allign in a zig-zag conformation such that all of the CF_2 dipoles point in the same direction. Thus, alligned films may be obtained by poling at high temperatures and cooling. An all-trans β-crystalline form is obtained. Since Si-F bonds are more polar than C-F bonds, the Si analog will have a greater dipole moment and may possess enhanced piezo- and pyroelectric properties. The synthesis of this polymer is shown below.

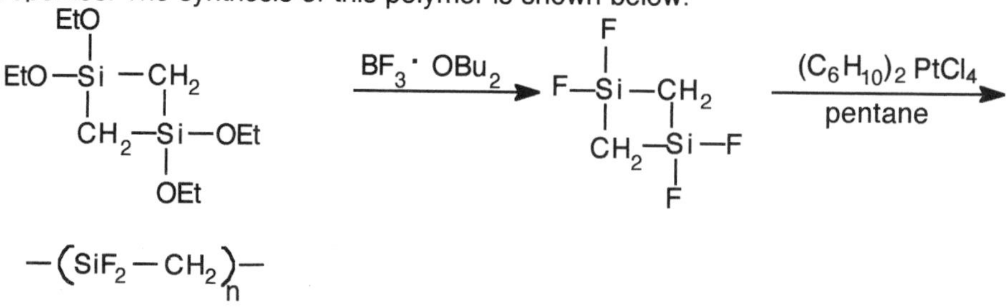
BF3 · OBu2
(C6H10)2 PtCl4
pentane

Powder X-ray diffraction measurements were consistent with a semicrystalline material with the desired zig zag conformation. Further studies of this material are now underway to determine the piezo- and pyelectric properties.

Although substitution of CH_2 groups in a silane chain should disrupt σ-conjugation, the disilane methylene polymer shown below still exhibits σ-conjugation and possesses properties intermediate between polyethylene and silane polymers. (58)

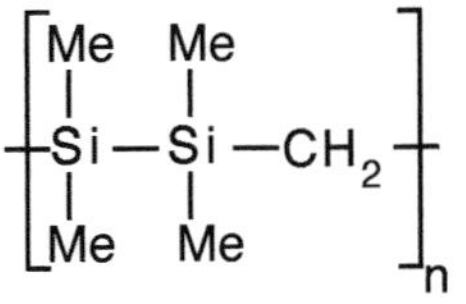

Poly(silsequioxanes

In recent years, the functionalization of silsequioxanes and their polymerization to a variety of interesting polymers containing the three-dimensional silsequioxane core has been extensively investigated. (59,60) Several groups have functionalized the T_8H cage to obtain eight armed "octopus" monomers with vinyl, allyl, hydrosilated, and epoxy groups. (61-63) Recently several octopus monomers have been prepared. Four examples are shown below.

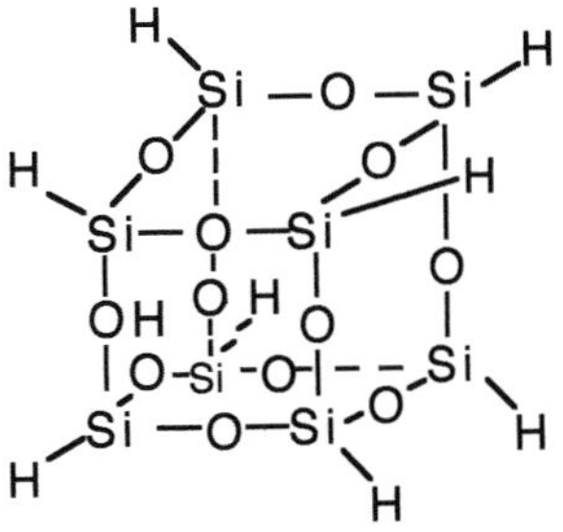

T_8H

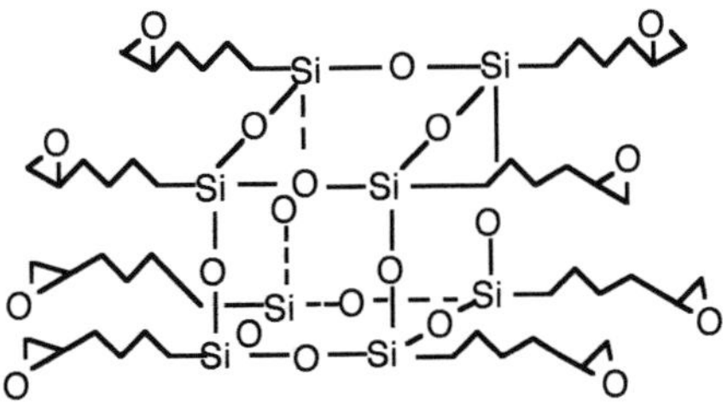

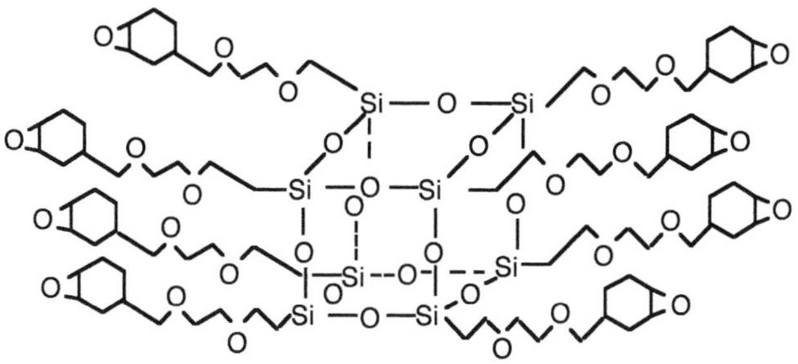

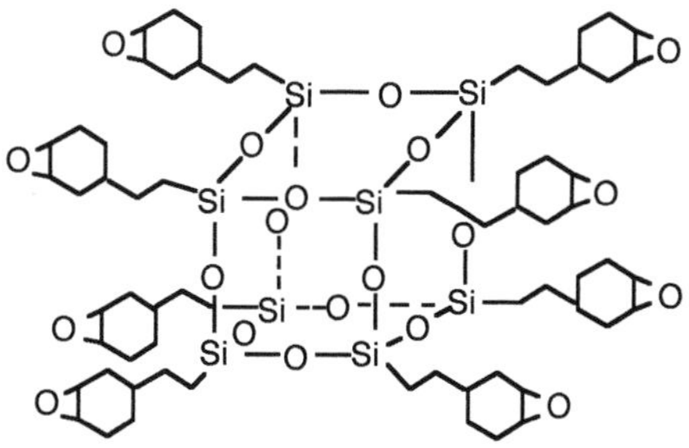

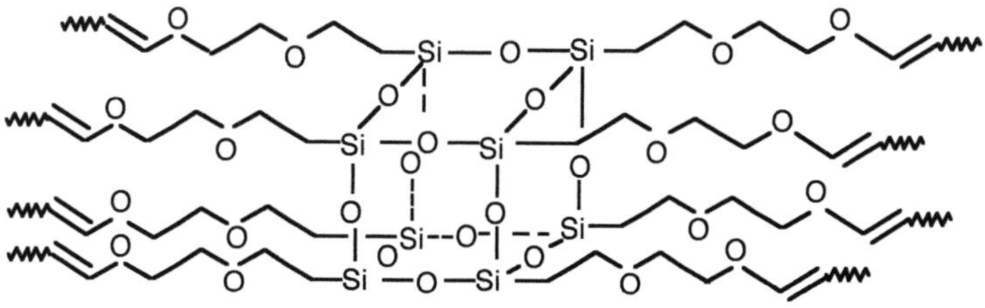

These vinylether and epoxy monomers undergo ready photoinitiated cationic polymerization. Limiting functional group conversions were around 78-90% which is quite high for octafunctional monomers where conversion is typically limited by high crosslink density.

While attempting to meet the demand for a new generation of light, high performance, thermally stable polymeric materials, the U.S. Air Force has pursued the development of the polyhedral oligomeric silsesquioxanes (POSS). (65-66) The POSS-reagets, like those shown above, have large three-dimensional cages (approximately 15Å diameter and 1000 amu) with a hybrid inorganic-organic nature. An overview of the types of POSS monomers available for bulk scale synthesis is represented below as the POSS-chemical tree.

POSS-Chemical Tree

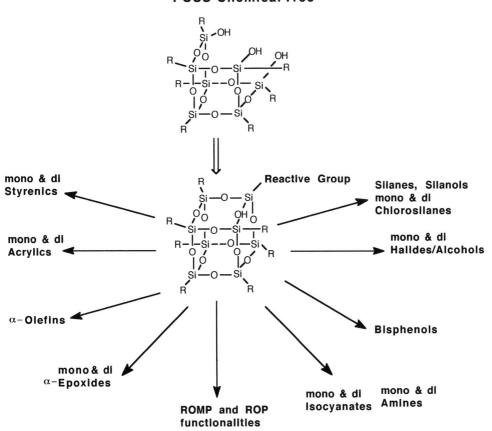

POSS structures have been used in glassy, semicrystalline, liquid crystalline and elastomeric types of polymers. The USAF has demonstrated difunctional POSS reagents for use in thermoplastics, sol-gel, and conventional thermoset systems. Several general property trends have become apparent with POSS polymers such as (1) flammability reduction, (66) lower thermal conductivity, (67) increased gas permeability, (68) increased Tg, (69) and improved heat distorsion and melt strengths. (70) The POSS cages themselves are not as susceptible to thermal motion (71) as are polymer segments so POSS polymers should be more oxidatively stable and retain their mechanical properties better at high temperatures.

Living Polymerizations to Make Polyphosphates

Recent development in "living" cationic ambient temperature syntheses of polyphosphazenes have permitted the first highly controlled architectures to be generated for this classic inorganic polymer backbone. Polyphosphazines are a well developed versatile class of inorganic/organic hybrid macromolecules prepared by thermal ring-opening polymerization of the cyclic trimer $(N=PCl_2)_3$ followed by macromolecular substitution of nucleophiles for chlorine. Over 700 such polymers have been prepared in the past 30 years. (72)

Broader development of polyphosphazenes has suffered from the lack of synthetic methods to prepare well defined materials (eg. controlled molecular weights and polydispersities, dendritic materials, block copolymers, and specific end groups.) Now that void is being bridged. Recent studies have shown the room temperature cationic "living" polymerization of phosphoranimines can be catalyzed with PCl_5. (73-75)

Use of the cationic species $[Cl_3P=N-PCl_3]^+$ $[PCl_6]^-$ as an efficient initiator for the polymerization of $Cl_3P=NSiMe_3$ has led to the development of multiple initiating site molecules which act as the core of star branched polyphosphazenes.

$$\bar{N}-(CH_2CH_2-\underset{\underset{R}{|}}{\overset{\overset{H}{|}}{N}}-\underset{\underset{R}{|}}{\overset{\overset{R}{|}}{P}}=NSiMe_3)_3 \xrightarrow{PCl_5} \begin{array}{c} \text{TRIBRANCHED} \\ \text{CORE} \\ \text{INITIATOR} \end{array} \xrightarrow[\text{2) NaOCH}_2CF_3]{\text{1) Cl}_3P=NSiMe_3}$$

$$\bar{N}-\left[CH_2CH_2\underset{\underset{R}{|}}{\overset{\overset{H}{|}}{N}}-\underset{\underset{R}{|}}{\overset{\overset{R}{|}}{P}}(N=\underset{\underset{OCH_2CF_3}{|}}{\overset{\overset{OCH_2CF_3}{|}}{P}})_n\right]_3$$

Star
polyphosphazine

Star branched polymers of narrow polydispersities (1.02-1.09) were achieved. Block poly(phosphazene-organic) structures with controlled architectures have now been made using commercially available polymers (such as polyethyleneoxide) with reactive end group amines. (76) These were converted to macroinitiators with $[R_3P=N-PC_3]^+[PCl_6]^-$ end groups which initiated the "living" polymerization of $Cl_3P=NSiMe_3$. (77)

$$CH_3O-(CH_2CH_2O)_n CH_2CH_2\underset{\underset{R}{|}}{\overset{\overset{H}{|}}{N}}-\underset{\underset{R}{|}}{\overset{\overset{R}{|}}{P}}=\overset{+}{N}PCl_3 \ PCl_6^- \xrightarrow[CH_2Cl_2]{Cl_3P=NSiMe_3}$$

$$CH_3O-(CH_2CH_2O)_n CH_2CH_2\underset{\underset{R}{|}}{\overset{\overset{H}{|}}{N}}-\underset{\underset{R}{|}}{\overset{\overset{R}{|}}{P}}(N=\underset{\underset{Cl}{|}}{\overset{\overset{Cl}{|}}{P}})_m \xrightarrow{NaOCH_2CF_3}$$

$$CH_3O-(CH_2CH_2O)_n CH_2CH_2\underset{\underset{R}{|}}{\overset{\overset{H}{|}}{N}}-\underset{\underset{R}{|}}{\overset{\overset{R}{|}}{P}}(N=\underset{\underset{OCH_2CF_3}{|}}{\overset{\overset{OCH_2CF_3}{|}}{P}})_m$$

The Manners (78) group has continued its research on polythionyl phosphazenes and has produced a wide variety of aryloxy, alkoxy and amino derivatives. The S-Cl bond is more difficult to cleave than the P-Cl bond and is not readily substituted by aryloxy groups. Complete substitution can be obtained by use of more basic alkoxy and fluoroalkoxy groups, but at the expense of chain degradation. Surprisingly, both S-Cl and P-Cl bonds are substituted readily by amines. The resulting methylamino polymers have a higher Tg ($\approx$ -46°C) than the corresponding polyphosphazenes (-63°C) but much lower than poly(oxothiazenes) (55-65°).

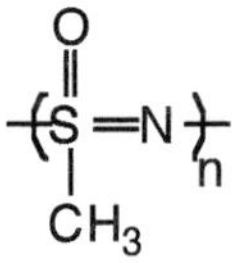

$$\left(\!\!\begin{array}{c} O \\ \| \\ S=N \\ | \\ CH_3 \end{array}\!\!\right)_n$$

Substitution of F for Cl in polythionyl phosphazenes increases the resistance toward substitution, but also lowers the glass transition by 10-15°C.

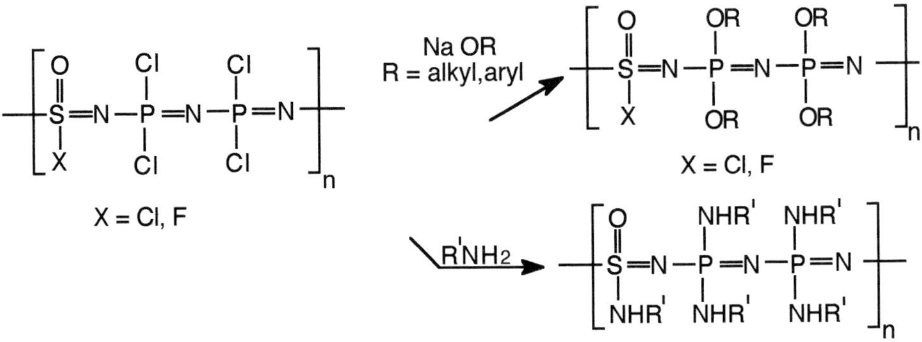

$$\left[\begin{array}{ccc} O & Cl & Cl \\ \| & | & | \\ S=N-P=N-P=N \\ | & | & | \\ X & Cl & Cl \end{array}\right]_n \xrightarrow[R = alkyl,aryl]{Na\ OR} \left[\begin{array}{ccc} O & OR & OR \\ \| & | & | \\ S=N-P=N-P=N \\ | & | & | \\ X & OR & OR \end{array}\right]_n$$

X = Cl, F

X = Cl, F

$$\xrightarrow{R'NH_2} \left[\begin{array}{ccc} O & NHR' & NHR' \\ \| & | & | \\ S=N-P=N-P=N \\ | & | & | \\ NHR' & NHR' & NHR' \end{array}\right]_n$$

    Thin films of amino-substituted polythionyl phosphazenes show an unusual permeability to molecular oxygen. (79) When a phosphorescent dye which can be quenched by oxygen such as $[Ru(L)_3]^{2+}$ (L = 4,7-diphenyl-1,10 - phenanthroline) is dispersed into the polymer film, the resulting material can be used as a visually interpretable oxygen sensor. Since oxygen is a constituent of the atmosphere in wind tunnels, a thin film coating an airplane wing or propellor can be used to detect variations in pressure across a wide surface area. Quenching of phosphoresence is a rapid process, so the sensor can respond rapidly to sudden changes in pressure. A color photograph provides a permanent record of the response under the experimental conditions employed and can be used for stress analyses.

    Phosphazene trimers containing B, As, and Sb have recently been prepared. (78) Attempts to polymerize these materials and to study their properties are currently under way.

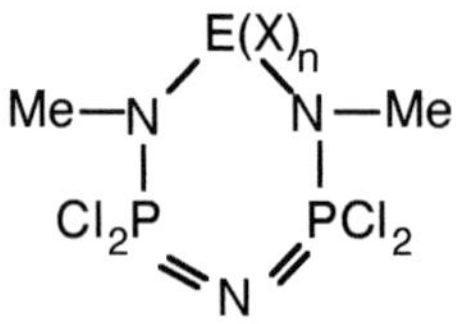

$$\begin{array}{c} E(X)_n \\ \diagup \quad \diagdown \\ Me-N \qquad N-Me \\ | \qquad\qquad | \\ Cl_2P \qquad\quad PCl_2 \\ \diagdown\ \ \diagup \\ N \end{array}$$

$EX_n = BCl_2$, AsF_4,

SbF_4 and SbF_3Cl

Metallocene Polymers

The most important recent advance in metallocene polymers has been the development of ring-opening polymerization of silicon-bridged [1]-ferrocenophanes. This is shown in equation 1. The unsymmetrical bridged metallocene **3** undergoes ROP at 150°C to produce the amorphous, regioirregular polymer **4** with three different Si environments.

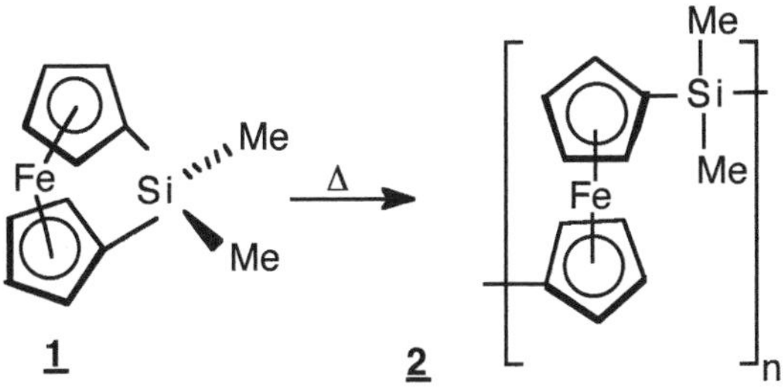

(1)

(2)

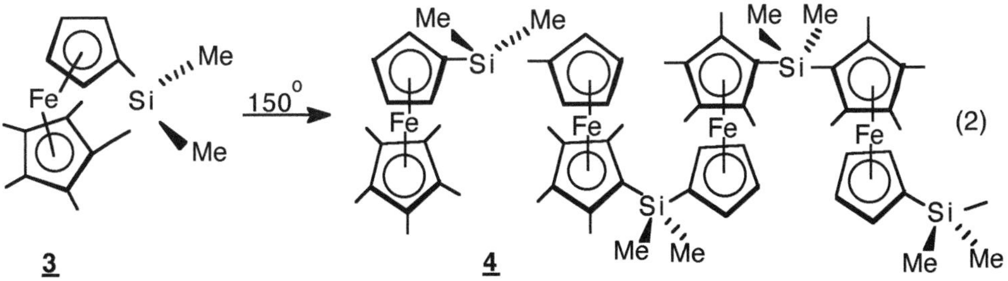

Manners et al. (81) have discovered the platinum-catalyzed ROP of bridged metallocene **3** which occurs at 25°C in toluene **3** produces regioregular **5** (equation 3) which is insoluble in organic solvents due to its high crystallinity (WAX scatterning) which also leads to a high Tg (285°C). Cleavage of the silicon to ring carbon bond during ROP only occurred at the unmethylated cyclopentadienyl ring. In the PtCl$_2$-catalyzed ROP of **1**, the molecular weight could be controlled by the addition of Et$_3$SiH. As the ratio of 1/Et$_3$SiH was increased, the molecular weight of **2** increased.

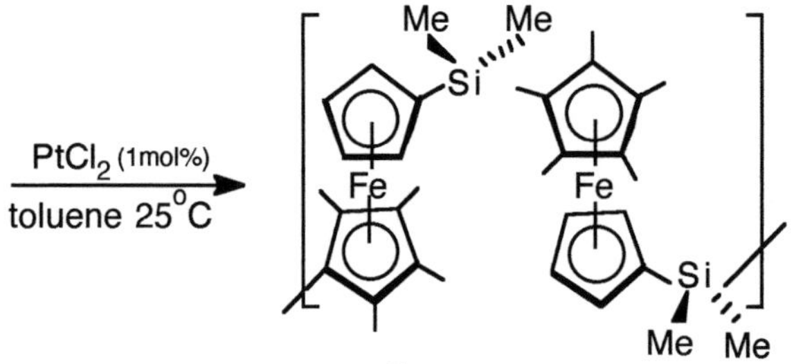

(3)

Both graft, star and block architectures of ferrocene-containing polymers have now been made. Thus, siloxane copolymer 6 when reacted with 1 in the presence of PtCo in toluene gave comb polymer 7 with metallocene side chains. (equation 4) Star polymer 9 formed when 1 was reacted with small amounts of cyclic siloxane 8. (equation 5)

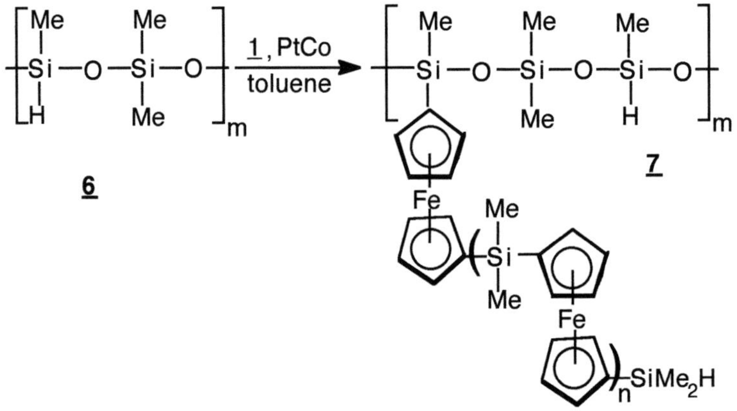

(4)

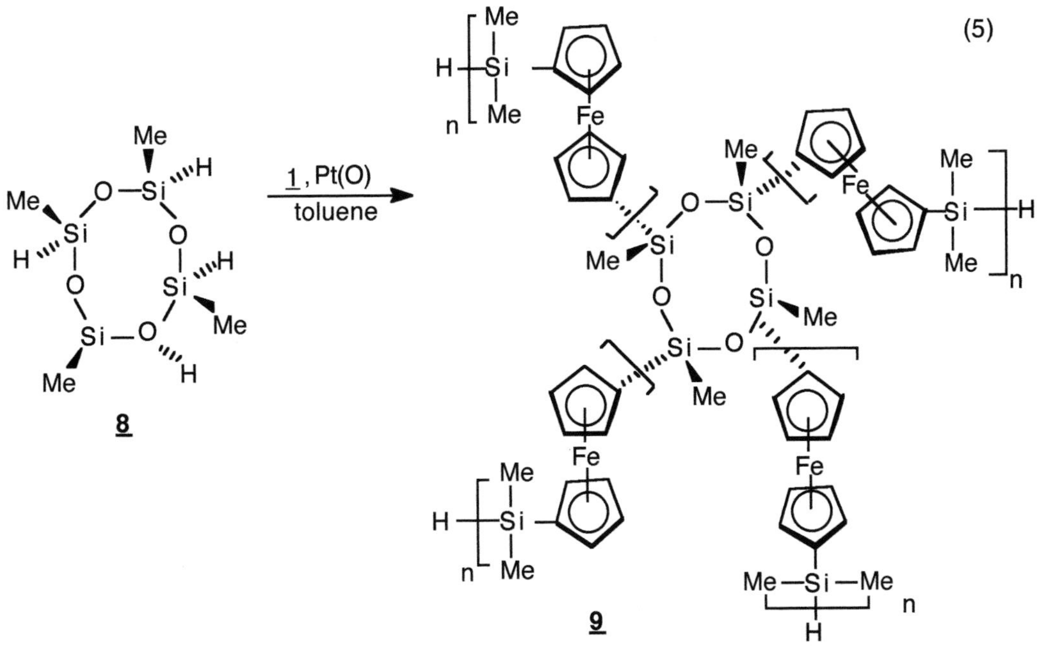

(5)

9

Random copolymers also have been made. In the example shown in equation 5, monomer **3** was copolymerized with **10** to give polymer **11** which contains regioregular ferrocenylsilane segments. Again, selective SiCp^H cleavage occurred in the presence of the PtCl$_2$ catalyst.

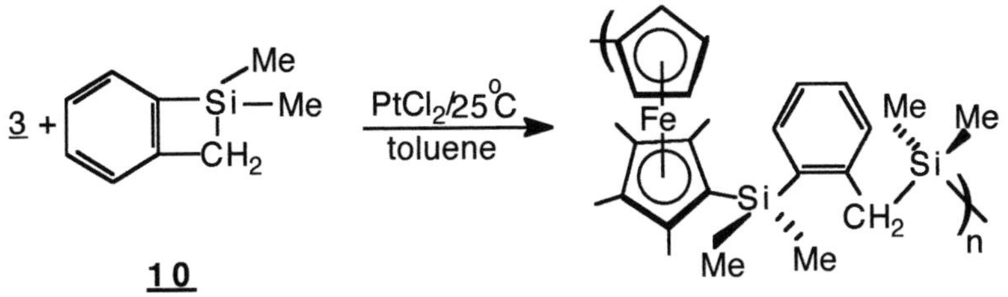

(6)

10

11

A synthetic route to alkoxy and amino-substituted ferrocenyl silane polymers (**7**) has now been discovered. (82) The way is now more open to prepare a rich array of derivatives by methods previously employed to prepare substituted polyphosphazenes.

$$(7)$$

Polymerization of silyl-bridged ferrocenes inside a porous silica matrix produces silica-bound monomers which undergo pyrolysis at 600° to form ferromagnetic iron particles imbedded in the silica matrix. (83) By adjusting the pore size and configuration and the monomer concentration, it may be possible to prepare wires or networks of conductive or ferromagnetic materials. If similar reactions are possible in other inorganic materials, a wide variety of ferromagnetic ceramics may be prepared.

$$(8)$$

Metallocene ROMP

Ring-opening metathesis polymerization (ROMP) has now been applied to the successful polymerization of 1,1'-[(1-t-butyl)1,3-butadienylene] ferrocene **12**, to produce the conjugated polymer **13** containing iron in the chain backbone. (84) Conjugated polymers that have transition metals in the backbone are a promising class of materials. (85)   Polymer **13** was prepared as a readily soluble high molecular weight system where the molecular weights were readily controlled by varying the monomer to catalyst ratio. Weight averaged molecular weights of 25,000, 152,000, and 309,000 with polydispersities of 1.57, 1.65, and 2.34 respectively, were obtained at ratios of 20:1, 100:1, and 200:1.

Polymer **13** exhibited a UV absorption at 320nm ($e=1.2 \times 10^4$) and high thermal stability with initial degradation occurring at about $300^{\circ}C$.

Condensation Polymers

Carraher et al. (86-88) have investigated condensation polymers for the past 30 years, employing a vast array of metals which can act as Lewis acids and Lewis bases. These have been prepared by interfacial polymerization with the Lewis acid in the organic phase and the Lewis base in the aqueous phase.

$$ML_2X_2 + BRB \longrightarrow -[ML_n-B-R-B]_m-$$

$n = 2, 3, 4$ $B = NH_2, NR_2, O^-, S^-$
$M = Ge, Sn, Pb, As, Sb, Bi, UO_2^{2+}$

Most of these materials are insoluble, but their lack of solubility combined with ease of degradation can sometimes be an advantage, i.e. controlled release of drugs (88,89) or growth hormones. (90)

Two of the more soluble polymers are shown below. (91,92)

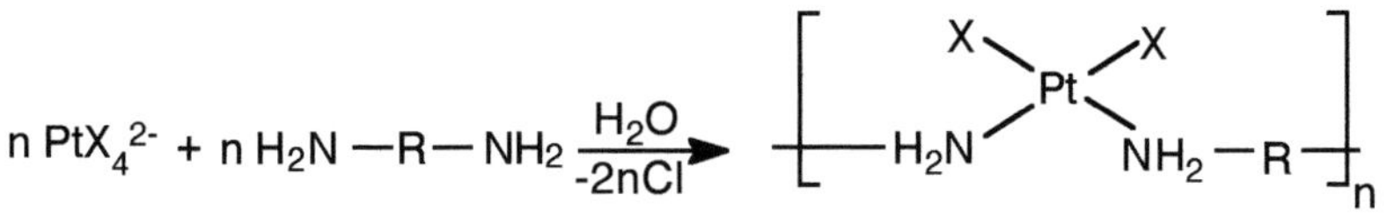

$$n\ PtX_4^{2-} + n\ H_2N-R-NH_2 \xrightarrow[-2nCl]{H_2O} \left[\begin{array}{c} X\diagdown_{Pt}\diagup X \\ H_2N \diagup \ \diagdown NH_2-R \end{array} \right]_n$$

$$n\ HS-R-SH + n\ \underline{cis}\text{-Ru(bpy)}_2\ Cl_2 \longrightarrow$$

$$\left(S-R-S-\underline{cis}\text{-Ru(bpy)}_2 \right)_n n\ HCl$$

$$R = n\text{-}C_8H_{16},\ (CH_2)_2\ O(CH_2)_2$$

In both cases, solubility is enhanced by the <u>cis</u>-coordination at the metal sites, which places kinks in the polymer chain, and by the flexibility of the aliphatic linkages, R.

Polymers with pendant organometallic species can be prepared by reacting polyvinyl alcohol with a monofunctional organometallic species such as R_3SnCl. (93) Crosslinked materials are obtained if a difunctional species such as R_2SnCl_2 is used. These materials often exhibit low solubility since Sn can expand its coordination sphere to 5 or 6 by crosslinking with oxygens in other parts of the chain.

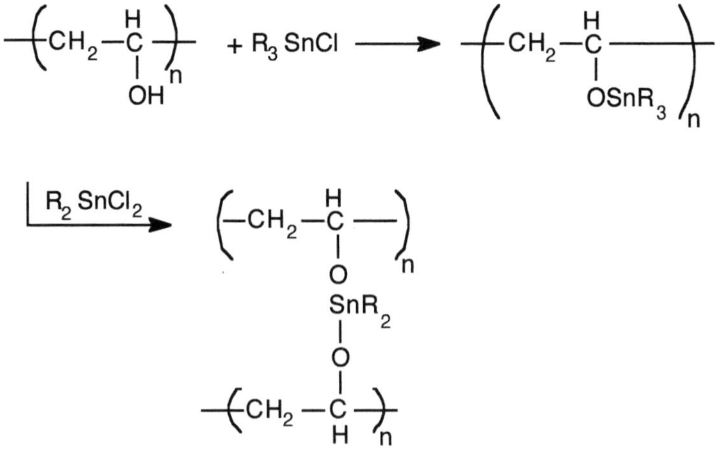

$$\left(CH_2-\underset{OH}{\overset{H}{C}} \right)_n + R_3\ SnCl \longrightarrow \left(CH_2-\underset{OSnR_3}{\overset{H}{C}} \right)_n$$

$$\Bigg\downarrow R_2\ SnCl_2$$

$$\left(-CH_2-\underset{\underset{SnR_2}{\overset{O}{|}}}{\overset{H}{\underset{|}{C}}} -- \right)_n$$

$$\overset{O}{\underset{|}{}}$$

$$\left(CH_2-\underset{H}{\overset{C}{}} \right)_n$$

"Shishkabob" structures have been formed from phthalocyanines, porphyrines and related macrocyclic structures. Some of these materials have been dissolved in acids and wet-spun into fibers. (94,95)

Other samples have been doped with iodine or other oxidants to levels approaching 50% of the metal centers. The electrical conductivity can be as high as 0.151 ohms cm, which approaches that of metals. (98,99)

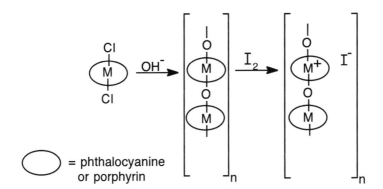

Applications

A number of unusual materials with varied applications have been described in the previous sections. Many others, with appropriate literature references are summarized in the table below. Brief summaries of some of the more developed areas are also given.

Table 1. Potential and real applications of metal-containing polymeric materials

Area	Comments	General References
Electrical	Both doped and non-doped systems	48-51, 94, 96-105
Analytical Reagents		106-108
Coloring Agents		109-111
Solar Energy Conversion		112-113
Drugs		116-126
Controlled Release Agents		127-130
Thickening Agents		131
Whiskers		132
Metal Deposition	Allows deposition at atomic level	133-134
Ceramic Precursors		134-140
Superconductivity		141
Photoresists		41-47
Light-emitting Diodes		52
Piezo and Pyroelectric Materials		56-57
Oxygen Sensors		79
Ferromagnetic Ceramics		33

Electrical

The design and synthesis of conducting and semiconducting metal-containing polymers is a recent research focus. Many conducting polymers have poor physical properties and may be insoluble or exhibit only limited solubilities. This is true of many of the metal-containing polymers formed by Carraher and co-workers which have bulk conductivities ranging to 0.01 s/cm and they exhibit high dielectric constants and dissipation factors. (96,97)   Cofacial polymers constructed from metal phthalocyanies and exhibiting a shishkabob structure have electrical, optical, and magnetic properties that are dependent on the ring-ring interplanar spacing, the nature of the off-axis counterions, and the band filling. (98)   The linear chains are electrically conducting along the stack axis. (98,99) Conductivities of the order of 0.1 s/cm have been reported. (99)

A number of applications are developing in the area of functionalized electrodes from electrocatalysis, photoelectrocatalysis, photovoltaic cells, and specialized electrode and sensor development. (100-105) For instance, a [1.1]-ferrocenophane-containing polystyrene, when applied as a film to the surface of a p-type semiconductor, gives a photoelectrolysis device from which hydrogen gas is generated from acid at 430mV more positive than for hydrogen gas evolution from platinum surfaces under the same conditions.

Poly[tris(5,5'bis(3-acrylyl-10-propoxy)carbonyl]-2,2'-bipyridine)ruthenium] is one of a series of polymerized electrochromic compounds that go through a range of different colors, from orange to cherry red. (105) Such polymeric complexes may be useful in generating multicolor displays in electronic display panels. (105) An extensive review of metal-organic photoconductivity by Jones et al. has recently been published. We will not attempt to summarize it here. (106)

Analytical Reagents

The formation of organometallic polymers can be utilized as a means to concentrate, isolate, purify, and separate metal-containing moieties. For instance, the natural water-soluble form of uranium is the uranyl ion. The uranyl ion has been effectively isolated using a variety of difunctional chelating agents such as terephthalic acid, polymers such as sodium polyacrylate, and a wide variety of carboxylate and sulfonate resins. (107,108)

Preformed poly(thiosemicarbazide) is highly effective in complexing the cupric ion from waste effluents from base mills. (105)

Radiation

A number of metal-containing polymers have been synthesized in which the Lewis base is derived from a dye. (110-112) These polydyes can be readily impregnated into cloth fabrics, paints, paper, and plastics. For instance, titanocene dichlchloride is a known "UV"-sink, acting to "attract" radiation. Many of these polydyes are designed to use the dye portion to attract the radiation, with the radiation subsequently dispersed through the metal or dye moiety. Possible uses include trace additives for identification, permanent coloring agents and in biomedical applications as stains and antibacterial agents. Some of these polydyes have been shown to exhibit potentially useful properties towards laser radiation. Depending on the wavelength of laser radiation, small amounts of selected polydyes can act to concentrate the radiation allowing "holes" to be cut in materials containing the polydye material. This might allow construction of electronic parts under milder conditions allowing greater detailing of individual parts. Conversely, the polydye material may act to disperse the radiation, allowing the impregnated material greater stability towards the radiation.

More recently, a number of ruthenium-containing polymers have been made which capture solar radiation and then rerelease it at such a wavelength which can break H-O bonds thus allowing the conversion of solar energy into storable energy. (113,114)

Organometallic polymers have been used as pre-heat shields for targets in inertial-confinement of nuclear fusion. (115)

Biological Agents

Many metal-containing polymers have been synthesized for use as biomedical agents - mostly as drugs.

One group of highly effective antibacterial agents have been synthesized based on the selective release of both a metabolite and the toxic agent through enzymatic and/or physical hydrolysis. (116-120) While many of these drugs contain organotin, other metals have been shown to be effective in some situations. Some of these drugs are active against all tested strains of staph MRSA. (119) Others exhibit selective inhibition against yeast-causing agents. (120)

A large group of polymeric derivatives of the widely used anticancer drug dichlorodiaminoplatinum II, cisplatin have been prepared. Coordination of potassium tetrachloroplatinate by poly[bis(methylamine)phosphazene] gives the structurally similar product that exhibits tumor inhibitory activity against mouse P388 lymphocytic leukemia and in the Ehrlich Ascites tumor regression test. (121) A wide variety of similar products have been synthesized from dinitrogen-containing compounds. (122-126) Some of these compounds have shown repressed replication of poliovirus I and L RNA virus while others show activity towards a wide variety of cell lines including the HeLa, WISH and L929 cell lines at concentrations well below those where the cisplatin itself exhibits activity. Further, many of these drugs are much less toxic towards test animals in comparison to the cisplatin. Some have acted as anti-virus agents in the treatment of EMC-D Virus-Induced diabetes. (124-126)

More recently, metal-containing polymers incorporating plant growth hormones have been used effectively to increase the growth rate, decrease decay, increase rate of seed germination, etc. for a number of plants (127-130). Again, the metal-containing moiety can act as an inert component or act to enhance a particular property.

Metal Deposition

The presence of metal atoms on a molecular basis is important in a wide variety of areas from catalysis to ceramics. Pittman and co-workers (133) have shown that metal atoms are deposited on an atomic level using a variety of polymers including linear and crosslinked chloromethylated polystyrenes containing metal carbonyls. Carraher and co-workers have used Mg and Zr-containing ionomer-like polymers in the production of partially stabilized zirconium ceramics. (134) Preparation of ferromagnetic ceramics has been discussed earlier in this paper.

Conclusions

The authors of this paper are all approaching or have exceeded the age of 60 and will probably not be active professionally in 2024 when the 100th Anniversary Issue is published, although we hope to be. It has been truly exciting to follow this field for the last 25 years. The young scientists who continue to explore and pioneer metalorganic polymers will find excitement on many fronts. Unexpected, unanticipated events will lead this field in multiple directions much like a complex fractal pattern that is continually expanding. The future should bring levels of achievement beyond our current imagination.

Literature Cited

1. Dagani, R. Chemistry and Engineering News, Nov. 30, 1998, pp 24-28.
2. The proceedings of the First Symposium on Inorganic Polymers are included in A.C.S. Org. Coastings & Plastics Chem. 1971, 31(2).
3. Carraher, C.E., Jr.; Sheats, J.E.; Pittman, C.U., Jr.; "Organometallic Polymers" 1978, Academic Press, New York.
4. Carraher, C.E., Jr.; Sheats, J.E.; Pittman, C.U., Jr.; "Advances in Organometallic and Inorganic Polymer Science" 1982, Marcel Dekker, New York.
5. Sheats, J.E.; Carraher, C.E., Jr.; Pittman, C.U., Jr.; "Metal Containing Polymeric Species" 1985, Plenum Press, New York.
6. Sheats, J.E.; Carraher, C.E., Jr.; Pittman, C.U., Jr.; Zeldin, M.; Currell, B.; "Inorganic and Metal-Containing Polymeric Materials" 1990, Plenum Press, New York.
7. Pittman, C.U., Jr. et al.; "Metal-Containing Polymeric Materials" 1996, Plenum Press, New York.
8. Zeldin, M.; Wynne, K.J.; Allcock, H.R.; Inorganic & Organometallic Polymers 1988, A.C.S. Symposium Series #360, Washington, D.C.

9.   Tsuchida, E.; "Macromolecular Complexes - Dynamic Interactions and Electronic Processes" 1991, VCH Publishers, New York.

10. "Macromolecule-Metal Complexes III," Selected Papers from Seminar of July 23-28, 1989, J. Macromol. Sci. Chem. 1990, A27(9-11), 1109-1446.

11. Manners, I.; Ann. Rep. Prog. Chem., Sect. A. Inorg. Chem. 1991, 88, 77-92; 1992, 89, 93-105; 1993, 90, 103-118; 1994, 91, 131-151; 1995, 92, 127-146; 1996, 129-146.

12. Manners, I.; Angew. Chem. Int. Ed. Engl. 1996, 35, 1602-1621.

13. West, R; David, L.D.; Djurovich, P.I.; Stearley, K.S.; Srinivasan, K.S.V.; Yu, H.; J. Am. Chem. Soc. 1981, 103, 7352.

14. Wesson, J.P.; Williams, T.C.; J. Polym. Sci., Polym. Chem. Ed. 1979, 17, 2833.

15. Trujillo, R.E.; J. Organometal. Chem. 1980, 198, C27.

16. Savin, A.; Jepsen, O.; Flad, J.; Andersen, O.K.; Preuss, H.; von Schnering, H.G.; Angew. Chem. Int. Ed. Engl. 1992, 31, 187.

17. Adams, S.; Drager, J.; Angew Chem. Int. Ed. Engl. 1987, 26, 1255.

18. Mark, J.E.; Allcock, H.R.; West, R.; Inorganic Polymers 1992, Prentice Hall, New York.

19. Suzuki, H.; Meyer, H.; Simmerer, J.; Yang, J.; Naarer, D.; Adv. Mater 1993, 5, 743.

20. Matyjaszewski, K. J.; Inorg. Organometal. Polym. 1991, 1, 463-85.

21. West, R.; Menescal, R.; Asuke, T.; Eveland, J.; J. Inorg. Organometal. Polym., 1992, 2, 29-45.

22. Hengge, E.F.; J. Inorg. Organometal. Polym. 1993, 3, 287-303.

23. Price, G.J.; J. Chem. Soc., Chem. Commun. 1992, 1209.

24. Ziegler, J.M.; McLaughlin, L.I.; Perry, R.J.; J. Inorg. Organometal. Polym. 1991, 1, 531.

25. West, R.; Hayase, S.; Iwahara, T.; J. Inorg. Organometal. Polym. 1991, 1, 545.

26. Wildeman, J.; Herrema, J.K.; Hadziioannou, G.; Schomaker, E.; J. Inorg. Organometal. Polyn. 1991, 1, 567.

27. Herrema, J.K.; van Hutten, P.F.; Gill, R.E.; Wildeman, J.; Wieringa, R.H.; Hadziioannou, G.; Macromolecules 1995, 28, 8102.

28. Shizuka, H.; Sato, Y.; Ueki, Y.; Ishikawa, M.; Kumada, M;. J. Chem. Soc. Faraday Trans., 1, 1984, 80, 341.

29. Shizuka, H.; Obuchi, H.; Ishikawa, M.; Kumada, M.; J. Chem. Soc. Faraday Trans., 1, 1984, 80, 341.

30. Van Walree, C.A.; Cleiji, T.J.; Zwikker, J.W.; Jenneskens, L.W.; Macromolecules, 28, 8696 (1995).

31. Aitken, C.; Harrod, J.F.; Samuel, E. J.; Organomet. Chem. 1985, 279, C11.

32. Aitken, C.; Harrod, J.F.; Samuel, E. J.; J. Am. Chem. Soc. 1986, 108, 4059.

33. Harrod, J.F.; A.C.S. Symposium Ser. 1988, 360, 89.

34. Woo, H.G.; Tilley, T.D.; J. Am. Chem. Soc. 1989, 111, 3757.

35. Tilley, T.D.; Acc. Chem. Res. 1993, 26, 22.

36. Banovetz, J.P.; Stein, K.M.; Waymouth, R.M.; <u>Organometallics</u> 1991, <u>10</u>, 3430.

37. Corey, J.Y.; Zhu, X.H.; Bedard, T.C.; Lange, L.D.; <u>Organometallics</u> 1991, <u>10</u>, 924.

38. Li, H.; Gauvin, F.; Harrod, J.F.; <u>Organometallics</u> 1993, <u>12</u>, 575.

39. Forsyth, C.M.; Nolan, S.P.; Marks, T.J.; <u>Organometallics</u> 1991, <u>10</u>, 2543.

40. Hsaiao, Y. and Waymouth, R.M.; <u>J. Am. Chem. Soc.</u> 1994, <u>116</u>, 9779.

41. Bianconi, P.; Weidman, T.W.; <u>J. Am. Chem. Soc.</u> 1988, <u>110</u>, 2342.

42. Bianconi, P.; Schilling, F.C.; Weidman, T.W.; <u>Macromolecules</u> 1989, <u>22</u>, 1697.

43. Szymanski, W.J.; Visscher, G.T.; Bianconi, P.; <u>Ultrasonics</u> 1990, <u>28</u>, 310.

44. Szymanski, W.J.; Visscher, G.T.; Bianconi, P.; <u>Macromolecules</u> 1993, <u>26</u>, 869.

45. Sekignchi, A.; Naugo, M.; Kabuto, C.; Sakurai, H.; <u>J. Am. Chem. Soc.</u> 1995, <u>117</u>, 4195.

46. Suzuki, H.; Kimata, Y.; Satoh, S.; Kuriyama, A.; Chem. Lett. 1995, 293.

47. Lambert, J.B; Pfing, J.L; Stern, C.L.; <u>Angew. Chem. Int. Ed. Engl.</u> 1995, <u>34</u>, 98.

48. Imori, T and Tilley, T.D.; <u>J. Chem. Soc. Chem. Commun.</u> 1993, 1607.

49. Devylder, N.; Hill, M.; Molloy, K.C; Price, C.J.; <u>Chem. Commun.</u> 1996, 711.

50. Lu, V. and Tilley, T.D.; <u>Macromolecules</u> 1996, <u>29</u>, 5763.

51. Reichi, J.A.; Popoff, C.M.; Gallagher, L.A.; Remsen, E.E.; Berry, D.H.; <u>J. Am. Chem. Soc.</u> 1996, <u>118</u>, 9430.

52. Fujino, M.; Hisaki, T.; Matsumoto, N.; <u>Macromolecules</u> 1995, <u>28</u>, 5017.

53. Interrante, L.V.; Wu, H.J.; Apple, T.; Shen, Q.; Ziemann, B.; Narsavage, D.M.; Smith, K.; <u>J. Amer. Chem. Soc.</u> 1994, <u>116</u>, 12025.

54. Wu, H.J. and Interrante, L.V.<u>; Macromolecules</u> 1992, <u>25</u>, 1840.

55. Rushkin, I.L. and Interrante, L.V.; <u>Macromolecules</u> 1995, <u>28</u>, 5160.

56. Lienhard, M.; Rushkin, I.; Verdecia, G.; Wiegand, C.; Apple, T.; Interrante, L.V.; <u>J. Am. Chem. Soc.</u> 1997, <u>119</u>, 12020.

57. Lienhard, M.; Weigand, C.; Apple, T.; Farmer, B.; Interrante, L.V.; <u>Polymer Preprints</u> 1998, <u>39(2)</u>, 808.

58. Isaka, H.; Teramae, H.; Fujiki, M.; Matsumoto, N.; <u>Macromolecules</u> 1995, <u>28</u>, 4733.

59. Feher, F.J. and Budzichowski, T.A.; <u>Organometallic Chem.</u> 1989, <u>379</u>, 33-40.

60. Dittman, U.; Hendan, B.J.; and Florke, U.; J. <u>Organometallic Chem.</u> 1995, <u>489</u>, 185-194.

61. Bassindale, A.B. and Gentle, T.E.; <u>J. Mater. Chem.</u> 1993, <u>3(12)</u>, 1319-1325.

62. Sellinger, A. and Laine, R.M.; <u>Polymer Reprints</u> 1994, 665-666.

63. Crivello, J.V.; Fan, M.; Bi, J.; <u>J. Applied Polym. Sci Part A: Polym Chem. Ed.</u> 1992, <u>44</u>, 9-16.

64. Crivello, J.V. and Malik, R.; <u>Polymer Preprints</u> 1998, <u>39(1)</u>, 486-487.

65. Lichtenhan, J.D.; "Silsesquioxane-based Polymers," Salamon, J.C.; Polymeric Materials Encyclopedia 1996, CRC Press, Vol. 10., p. 7768-7777.
66. Gilman, J.W.; Kashiwagi, T.; and Lichtenhan, J.P.; SAMPE 1997, 33, 40.
67. Gilman, J.W. et al.; J. Appl. Polym. Sci. 1996, 60, 591-596.
68. Lichtenhan, J.D. et al.; Mat. Res. Soc. Symp. Proc. 1996, Vol. 435, p. 3.
69. Haddad, T.S. et al.; Proc. Soc. of Plast. Eng. 54th ANTEC 1997, 1814.
70. Romo-Uribe, A.; Mather, P.T.; Haddad, T.S.; Lichtenhan, J.D.; Polym. Sci. Polym. Phys. (has appeared 1998); Mather, P.T. et al.; Mat Res. Proc. 1996, Vol. 425, pp 137-142.
71. Jones, P.F. et al.; Proc. 24th N. Am. Thermal Anal. Soc. 1995, pp 40.
72. Allcock, H.R.; Adv. Mater. 1994, 6, 106.
73. Allcock, H.R.; Nelson, J.M.; Reeves, S.D.; Honeyman, C.H.; Manners, I.; Macromolecules 1996, 29, 7740.
74. Nelson, J.M. and Allcock, H.R.; Macromolecules 1997, 30, 1854.
75. Allcock, H.R.; Reeves, S.D.; Nelson, J.M.; Crane, C.A.; Macromolecules 1997, 30, 22B.
76. Nelson, J.M. and Allcock, H.R.; Polymer Preprints 1998, 39(2), 631-632.
77. Nelson, J.M.; Primrose, A.P.; Hartle, T.J.; Allcock, H.R.; and Manners, I.; Macromolecules 1998, 31, 947.
78. Gates, D.P. and Manners, I.; J. Chem. Soc., Dalton Trans., 1997, 2525-32.
79. Pang, Z.; Gu, X.; Yektu, A.; Masoumi, Zo; Coll, J.B.; Winnik, M.A.; Manners, I.; Adv. Mater 1996, 8, 768-71.
80. Manners, I. et al.; J. Amer. Chem. Soc., 1995, 117, 7265 and Macromolecules, 1996, 29, 1894.
81. Manners, I. et al.; J. Amer. Chem. Soc., 1998, 120, 8348-8356.
82. Nguyen, P; Lough, A.J.; Manners, I.; Macromole. Rapid. Commun. 1997, 18, 953-9.
83. MacLachlan, M.J.; Aroca, P.; Coombs, N.; Manners, I; Ozin, G.A.; Adv. Mater 1998, 10, 144-9.
84. Heo, R.W. and Lee, T.R.; Polymer Preprints, 1998, 39(1), 169-170 also J. Am. Chem. Soc., 1998.
85. Allcock, H.R.; Science, 1992, 255, 1106.
86. Carraher, C.E. Jr.; Ref. 2, pp 79-86.
87. Carraher, C.E., Jr.; Louda, J.W.; Sterling, D.; Rivalta, A.; Zhang, Q.; Baker, E. PMSE 1994, 71, 386.
88. Carraher, C.E., Jr.; J. Chem. Ed., 1981, 58(11), 921.
89. Mbonyana, C.W.; Neuse, E.W.; Perlwitz, A.G.; Appl. Organomet. Chem. 1993, 7, 279.
90. Carraher, C.E., Jr.; Stewart, H.H.; Soldani, W.J., II; DeLaTorre, J.; Pandya, B.; Reckleben, L.; Hibiscus, F.; PMSE 1984, 71, 783.

91. Carraher, C.E., Jr.; Scott, W.J.; Schroeder, J.A.; <u>J. Macromol. Sci. Chem.</u> 1981, A15, 625.
92. Carraher, C.E., Jr.; Zhang, Q.; Párkányi, C.; <u>PMSE</u> 1994, <u>71</u>, 505.
93. Carraher, C.E., Jr.; "Polymer Chemistry: An Introduction," Dekker, NY, 1996.
94. Dirk, C.; Inabe, T.; Schoch, K.; Marks, T.J.; <u>J. Amer. Chem. Soc.</u>, 1983, <u>105</u>, 1539.
95. Takahashi, S.; Sonogashira, K.; "Polymeric Materials Encyclopedia," (Salamone J., Ed.), Vol. 6, 4804, CRC, Boca Raton, FL., 1996.
96. Carraher, C.E., Jr.; Nwufoh, V.; Taylor, J.R.; <u>PMSE</u> 1989, <u>60</u>, 685.
97. Carraher, C.E., Jr.; Manek, T.; Linville, R.; Taylor, J.R.; Torre, L.; Venable, W.; <u>Org. Coatings Plastic Chem.</u>, 1981, <u>44</u>, 753.
98. Peterson, J.; Scharamm, C.; Stojakovic, D.; Hoffman, B.; Marks, T.J.; <u>J. Amer. Chem. Soc.</u>, 1977, <u>99</u>, 286.
99. Diel, B.; Inabe, T.; Jaggi, H.; Lyding, J.; Schneider, O.; Hanack, M.; Kannewurf, C.; Marks, T.J.; Schwartz, L.; <u>J. Amer. Chem. Soc.</u>, 1984, <u>106</u>, 3207.
100. Kaneko, M.; Yamada, A.; <u>Adv. Polym. Sci.</u>, 1983, 55.
101. Wrightson, M.S.; "Interfacial Photoprocesses: Energy Conservation and Synthesis," American Chemical Society, Washington, DC, 1980.
102. Muller-Westerhoff, U.; Nazzal, A.; US Patent 4,379,740 (1983).
103. Anson, F.; <u>J. Amer. Chem. Soc.</u>, 1984, <u>106</u>, 59.
104. Savinova, E.; Kokorin, A.; Shepelin, A.; Pashis, A.; Zhdan, P.; Parmon, V.; <u>J. Mol. Catal.</u>, 1985, <u>32</u>, 149.
105. Spaulding, B.J.; <u>Chemical Week</u>, 1986, Oct. 8, 29; Elliott, C.M.; Abstracts, 193 ACS National Meeting, Denver, CO, April 5-10, 1987, INOR 232.
106. Jiang, B.; Yang, S.; Bailey, S.; Hermans, L.H.; Niver, R.; Bolcar, M.; Jones, W.E., Jr.; Choord. Chem., Rev., 1998, 171, 365. and Jones, W.E., Jr.; Hermans, L.H.; Jiang, B.; in Molecular and <u>Supramolecular Photochemistry: Vol.</u> <u>2</u>, V. Ramamurthy, K.S. Schanze, Eds., Marcel Dekker: New York, 1999 in press.
107. Carraher, C.E., Jr.; Tsuji, S.; DiNunzio, J.; Feld, W.; <u>PMSE</u>, 1986, <u>55</u>, 875.
108. Carraher, C.E., Jr.; Schroeder, J.; <u>Polymer P.</u>, 1975, <u>16</u>, 659; and Polymer Letters, 1975, 13, 215.
109. Donaruma, G.; <u>Polymer P.</u>, 1975, <u>22</u>, 1.
110. Carraher, C.E., Jr.; Taylor, A.; Medina, F.; Linville, R.; Randolph, E.; Kloss, J.; Stevison, D.; <u>Polymer P.</u>, 1993, <u>34</u>, 166.
111. Carraher, C.E., Jr.; Kloss, J.; <u>PMSE</u>, 1993, <u>68</u>, 253.
112. Carraher, C.E., Jr.;Foster, V.; Linvolle, R.; Stevenson, D.; Ventatachalam, R.S.; "Adhesives, Sealants, and Coatings for Space and Harsh Environments," Chpt. 19, 239, Plenum, NY, 1988.
113. Carraher, C.E., Jr.; Taylor-Murphy, A.; <u>PMSE</u>, 1997, <u>76</u>, 409.

114. Carraher, C.E., Jr.; Zhang, Q.; "Metal-Containing Polymeric Materials," 109, 1996, Plenum, NY.

115. Sheats, J.E.; Hessel, F.; Tsarouhas, L.; Podejko, K.; Porter, T.; Kool, L; Nolan, R.; "New and Unusual Monomers and Polymers," (Culbertson, B.; Pittman, C.U.; Eds.), 83, 1983, Plenum, NY.

116. Carraher, C.E., Jr.; Li, F.; Sigman-Louda, D.; Butler, C.; Ross, J.R.; PMSE, 1997, 77, 499.

117. Carraher, C.E., Jr.; Saurino, V.; Butler, C.; Sterling, D.; PMSE, 1995, 72, 192.

118. Carraher, C.E., Jr.; Butler, C.; Naoshima, Y.; Sterling, D.; Saurino, V.; "Industrial Biotechnical Polymers," Chpt. 8, 1955, Technomic Pub, Lancaster, PA.

119. Carraher, C.E., Jr.; Butler, C.; PMSE, in press.

120. Carraher, C.E., Jr.; Sterling, D.; Butler, C.; Ridgway, T.; "Biotechnical Polymers," Chpt. 13, 1995, Technomic Pub, Lancaster, PA.

121. Allcock, H.; "Organometallic Polymers," Chpt. 28, 1978, Academic Press, NY.

122. Carraher, C.E., Jr.; Williams, M.; "Cosmetic and Pharmaceutical Applications of Polymers," 269, 1992, Plenum, NY.

123. Carraher, C.E., Jr.; Sigmann, D.; Brenner, D.; Colvin, A.; Polner, B.; Strother, R.; "Inorganic and Organometallic Polymers," 335, 1991, Plenum, NY.

124. Carraher, C.E., Jr.; Trombley, M.; Bigley, N.; Giron, D.; PMSE, 1987, 57, 177.

125. Carraher, C.E., Jr.; Trombley, M.; Bigley, N.; Giron, D.; "Applied Biocative Polymeric Systems," Chpt. 15, 223, 1988, Plenum, NY.

126. Carraher, C.E., Jr.; Lopez, I.; Giron, D.; "Advances in Biomedical Polymers," 311, 1987, Plenum, NY.

127. Carraher, C.E., Jr.; Nagata, M.; Stewart, H.; Miao, S.; Carraher, S.; Gaonkar, A.; Highland, C.; Li, F.; PMSE, 1998, 79, 52.

128. Carraher, C.E., Jr.; Gaonkar, A.; Stewart, H.; Miao, S.; Mitchell, D.; Barosy, C.; Colbert, M.; Duffield, R.; Polymer P., 1997, 38(2), 572.

129. Carraher, C.E., Jr.; Stewart, H.; Soldani, W.; Dela Torre, J.; Pandya, B.; Reckleben, L.; "Metal-Containing Polymeric Materials," 93, 1996, Plenum, NY.

130. Stewart, H.; Carraher, C.E., Jr.; Soldani, W.; Reckleben, L.; "Inorganic and Organometallic Polymers," 267, 1991, Plenum, NY.

131. Allcock, H.; Allen, R.; O'Brien, J.; J. Amer. Chem. Soc., 1977, 97, 3914.

132. Carraher, C.E., Jr.; Chemtech, 1972, 744.

133. Pittman, C.U.; Felix, R.; J. Organometal. Chem., 1974, 72, 389 and 399.

134. Carraher, C.E., Jr.; Xu, X.; PMSE, 1996, 75, 182 and "Modification of Polymers," 85, 1997, Plenum, NY.

135. Yajima, S.; Hayashi, H.; Omori, M.; Chem. Letters, 1975; 931 and 1209.

136. West, R.; Am. Ceram. Soc. Bull. 1983, 62, 825.

137. Seyferth, D.; Wiseman, G.; J. Amer. Ceram. Soc., 1984, 67, 132 and US Patent 4,482,669; 1984.

138. Laine, R.; Blum, Y.; Chow, A.; Hamlin, R.; Schwartz, K.; Rowecliffe, D.; <u>Polymer P.</u>, 1987, <u>28(1)</u> 393.
139. Arai, M.; Sakurada, S.; Isoda, T.; Tomizawa, T.; <u>Polymer P.</u>, 1987, <u>28(1)</u>, 407.
140. Narula, C.; Paine, R.; Schaeffer, R.; <u>Polymer P.</u>; 1987, <u>28(1)</u>, 454.
141. Carraher, C.E., Jr.; Zhuang, H.; Medina, F.; Baird, D.; Pennisi, R.; Landreth, B.; Nounou, F.; <u>PMSE</u>, 1990, <u>62</u>. 633.

Section 5

SPECTROSCOPIC AND PHYSICAL CHARACTERIZATION
Section Editor: Clara D. Craver

FOURIER TRANSFORM INFRARED SPECTROSCOPY OF POLYMERS

VASILIS G. GREGORIOU
Chemical Research Division, Polaroid Corporation, Waltham, MA 02451

Introduction
Instrumentation for Infrared Spectroscopy
      Fourier Transform Infrared Instruments
      Dispersive Instruments
Sampling Techniques and Applications
      Transmission Spectroscopy
      Reflection Spectroscopy
      Photoacoustic Spectroscopy
      Infrared Microspectroscopy and Imaging
      Hyphenated Techniques
Infrared Characterization of Polymers and Polymeric Surfaces
      Identification of Composite and Morphology
      Phase Behavior of Polymer Blends
      Curing Studies
      Diffusion Studies
      Oxidation Studies
      Degradation of Polymers
      Isotopic Exchange
      Orientation Measurements
Dynamic Infrared Spectroscopy of Polymers
Two-dimensional Correlation Spectroscopy
Quantitative Analysis of Polymer Blends and Composites

Introduction

The field of using infrared spectroscopy for the characterization of polymeric materials has experienced tremendous growth in recent years, primarily due to the fact that a variety of sampling techniques and experimentation are now available.[1] Thousands of papers are published each year involving the use of infrared spectroscopy and polymer characterization. Due to the volume of this work, it is impossible to cover all aspects of the field in a monograph like this. Therefore, the focus of this chapter will primarily be on the background of these newer techniques along with some selected applications from the literature. These applications demonstrate the ability of infrared spectroscopy to provide answers to questions of both qualitative and quantitative polymer characterizations. More emphasis will be placed on the relatively recent research applications, (e.g. step-scan applications), whereas the more 'traditional" methods will be covered in less detail. Instead, the reader is directed to references that cover these techniques in considerable length. In addition, instruments based on the Fourier transform infrared (FT-IR) method have dominated the field of infrared spectroscopy, so this chapter will address many of the principles of interferometry as well.

Theory of Infrared Spectroscopy

Interaction of electromagnetic radiation with molecular vibrations gives rise to absorption bands throughout most of the infrared region of the spectrum. In the study of molecular vibrations, the classic model of the molecule, where the atoms are depicted as points with a finite mass, is widely used.[2] This model predicts that after subtracting the translational and rotational degrees of freedom, 3N-6 internal degrees of freedom for a nonlinear molecule and 3N-5 internal degrees of freedom for a linear molecule are left, where N is the number of atoms in the molecule. The above number of internal degrees of freedom of motion is then the same number of independent normal modes of vibration. In each normal mode of vibration all the atoms in the molecule vibrate with the same frequency and all the atoms pass through their equilibrium positions simultaneously. *In order for infrared absorption to occur, a molecular vibration must cause a change in the dipole moment of the molecule.* The absorption frequency is the molecular vibration frequency, whereas the absorption intensity depends on how effectively the infrared photon energy can be transferred to the molecule, which depends on the amount of dipole moment change and the symmetry of the vibration.

In general, it can be stated that the rich amount of information associated with structure, orientation state, and conformational order that can be extracted from infrared spectroscopy makes the technique very attractive as an analytical tool.[3] Infrared spectroscopy is generally more sensitive than NMR, while offering a similar level of sub-molecular (functional group) distinction. Furthermore, the use of polarized light in the study of aligned molecules provides information about the orientations of the individual transition dipole moments.[4] From the orientation of specific transition moments the average overall orientation of molecules can be deduced. While such information is valuable for the molecular understanding of the composition and structure of such materials, a still higher level of utility of infrared spectroscopy is achieved when spectra can be measured as a function of time.

Symmetry Considerations

The symmetry of a molecule plays a very important role in the determination of its infrared spectrum. It turns out that not every normal vibration is "active" and absorption of the incident infrared light takes place only under certain conditions. Arguments that have their origin in quantum mechanics dictate the rules for optical activity. The rules that govern the process are called *selection rules*. For instance, a vibration that is symmetric with regards to the center of symmetry of the molecule is forbidden in the infrared. Utilization of group theory, which is intimately related to the symmetry of the molecule, allows the prediction of active normal modes in the infrared for any given molecule.

Group frequencies

For many normal modes the vibrational motion is highly localized in particular parts of the molecule. This gives rise to the concept of *fundamental group frequencies*. The notion of these characteristic group frequencies is frequently used in assignments of the infrared spectra of unknown compounds. It has been experimentally observed that particular groups of atoms produce infrared bands in a characteristic region of the electromagnetic spectrum. When the high frequency region is considered, it is assumed that the remainder of the molecule does not move during the period of one vibration. In the case of a low frequency vibration, the molecule is considered at the average position of its high frequency vibration.

In addition, there are a number of absorption bands in an infrared spectrum that are characteristic of the individual compound and serve as a fingerprint in distinguishing this compound from a similar compound. A graphical representation of characteristic group frequencies can be found in Table 1.

Table 1
Characteristic group frequencies in the infrared

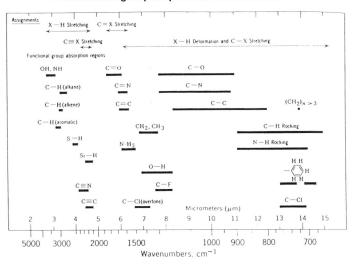

Vibrations of Polyatomic Molecules

In polymers the infrared spectrum of a macromolecule is often times remarkably simple, considering the large number of atoms involved.[5] This is the result of the fact that many of the normal vibrations have almost the same frequency and therefore appear in the spectrum as one absorption band. Additionally, the application of the selection rules prevents many of the vibrations from causing absorptions. Again, symmetry plays a very important role in the appearance of infrared spectra of macromolecules. Therefore, not only the chemical composition of the sample can be deduced, but also information about the distribution of individual units can be obtained.[6]

Instrumentation for Infrared Spectroscopy

Fourier Transform Infrared Instruments

Most commercial interferometers are based on the original Michelson design of 1891.[7] Interferometers record intensity as a function of optical path difference and the produced interferogram is related to the frequency of the incoming radiation by a Fourier transformation. The principle of a Michelson interferometer is illustrated in Figure 1. The device consists of two flat mirrors, one fixed and one free to move, and a beamsplitter. The radiation from the infrared source strikes the beamsplitter at 45°. The characteristic property of the beamsplitter is that it transmits and reflects equal parts of the radiation. One classic type of a beamsplitter, useful in the mid infrared spectral region, consists of a thin layer of germanium (refractive index, $n = 4.01$) on an infrared transparent substrate (e.g., KBr). The transmitted and reflected beams strike the above described mirrors, and are reflected back to the beamsplitter, where, again, equal parts are transmitted and reflected. As a consequence, interference occurs at the beamsplitter where the radiation from the two mirrors combine. As shown in Figure 1 when the two mirrors are equidistant from the beamsplitter, constructive interference occurs for the beam going to the detector for all wavelengths. In this case, the path length of the two beams in the interferometer are equal and their path difference, called the retardation (δ), is zero.

The plot of detector response as a function of retardation produces a pattern of light intensity versus retardation referred to commonly as the interferogram. The interferogram of a monochromatic source is a cosine function. The following Equation 1 describes the relationship

$$I(\delta) = B(\nu) \cos(2\pi\nu\delta) \tag{1}$$

where ν is the wavenumber in cm^{-1} and δ is the optical path difference, or retardation. The Fourier transform of the above expression is a peak at the frequency of the monochromatic radiation. I is the intensity in the output beam as a function of retardation (δ), whereas B is the intensity as a function of radiation frequency (ν). In contrast, the interferogram of a polychromatic source can be considered as the sum of all cosine waves that are produced from monochromatic sources. The polychromatic interferogram has a strong maximum intensity at the zero retardation point where all the cosine components are in phase as can be seen in Figure 2. This point is also known as the *centerburst* point. The expression for the intensity of the interferogram of a polychromatic source as a function of retardation is described by Equation 2:

$$I(\delta) = \int_{-\infty}^{+\infty} [[B(v)[1 + \cos(2\pi v \delta)]] / 2] dv \qquad (2)$$

Thus, in Fourier transform interferometry the data are "encoded" by the interference produced by the retardation and then "decoded" by the Fourier transform to yield the desired intensity signal as a function of frequency (or wavelength).

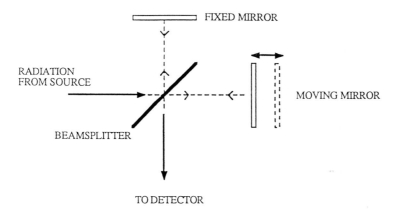

Figure 1: Block diagram of a Michelson interferometer.

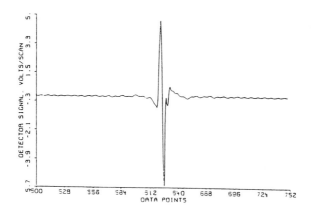

Figure 2: Typical interferogram showing the centerburst region.

Advantages of Interferometry

Two kinds of multichannel advantages exist in Fourier transform interferometry, compared with a dispersive instrument in which only a very narrow band of frequencies is observed at a time. The first, which is the biggest practical advantage of Fourier transform spectroscopy, is the simultaneous detection of the whole spectrum at once; it is called the *Fellgett* or *multiplex* advantage.[8] Even though a factor of <u>ca</u> two in signal strength is lost because half of the beam is reflected back to the source, the multichannel advantage is nevertheless 10^4 or higher. That is, theoretically an interferometer can achieve comparable signal-to-noise to a dispersive monochromator 10^4 faster.

In addition, the so-called *Jacquinot* or *"etendue"* advantage exists. This advantage is associated with the increase in source throughput.[9] During dispersive detection the throughput is severely limited by the area of the entrance slit. Even though the interferometer has an entrance aperture of its own, its throughput advantage ranges from 10 to 250 over the infrared frequency range. This was the reason that FT-IR spectra of astronomical sources, where very weak astronomical emission sources are present, were produced even before the Fast Fourier Transform (FFT) was invented.[10]

A third practical advantage of interferometry is the so-called *Connes* or *registration* advantage. The Connes advantage stems from the ability of interferometry using a monochromatic source (e.g. a helium-neon (HeNe) laser in today's spectrometers) to accurately and precisely index the retardation, resulting in a superior determination of the retardation sampling position. For example, if the above mentioned HeNe laser ($\lambda = 632.8$ nm) is used, zero crossings in the visible interferogram occur at intervals of 632.8/2 nm = 0.3164 μm. Because the Nyquist theorem demands at least two sampling points per cycle, the highest infrared frequency that would satisfy the Nyquist criterion is 15,804 cm^{-1}. For mid-IR use, sampling at every other zero-crossing (1 λ_{HeNe} intervals) produces a maximum Nyquist frequency of 7902 cm^{-1}. The Connes advantage allows tremendous reproducibility of interferogram sampling and data storage. This results in full realization of signal-to-noise problems from repeated scans.[11]

Apodization

The amplitude of the side lobes, which appear adjacent to absorption bands in the Fourier transform of an interferogram, can be drastically reduced if a mathematical manipulation is performed. This treatment is called *apodization* (from the Greek word απoδoς, without feet). This mathematical treatment is necessary because the Fourier transformation is performed over finite limits, even though the theoretical expression for the interferogram's intensity involves infinite limits. Therefore, when the interferogram is truncated, this sudden cut-off results in the appearance of oscillations around the sharp spectral features (absorption bands) in the transform.

When the interferogram is multiplied by the apodization function, the transform is essentially free of side lobes.   Two of the most popular and effective apodization functions are the *triangular* and the *Happ-Genzel* functions.   The amplitude of the first side lobe using triangular apodization is larger than that of the Happ-Genzel function, but the opposite is observed for the subsequent lobes.   In general, Happ-Genzel apodization is quite similar to triangular apodization and for most situations they give comparable results.[12] Overall, it can be stated that the biggest drawback of apodization is the worsening of the spectral resolution, since the contributions of the extremes of the interferogram wings are reduced.   Therefore, a trade off exists between the reduction in spectral distortion and the worsening of resolution.

Resolution

Resolution is defined as the minimum distinguishable spectral interval. The maximum retardation determines the resolution of the scan.   The maximum optical resolution achievable by a particular FT-IR spectrometer is given by $(\Delta max)^{-1}$ cm^{-1}, where Δmax is the maximum optical path difference attainable by the interferometer. Dynamic infrared methods have the potential to increase spectral resolution beyond the above limit, due to the existence of the possibility of different responses of the components of highly overlapped bands.

Phase Correction

The non-ideality of the beamsplitter in a real interferometer results in the introduction of sine components to an interferogram that in principle should consist only of cosine components.   Equation 3 shows the modified relation for the intensity of the interferogram

$$I(\delta) = \int\limits_{-\infty}^{+\infty} \left[B(v) \left[1 + \cos\left(2\pi v \delta + \Phi_{BS}(v)\right) \right] \right] / 2 \, dv \qquad (3)$$

where $\Phi_{BS}(v)$ is the wavelength dependent phase shift introduced by the beamsplitter.

Phase correction is the mathematical procedure to remove the sine components from the interferogram.   The Fourier transform of a complete double-sided interferogram provides the correct power spectrum, without any phase correction, since the ambiguity does not affect the magnitude.   However, when a single-sided interferogram is computed, some knowledge of the phase is required in order to compute the true spectrum.[13] Two of the most popular phase correction routines used in single-sided interferograms are the *Mertz* algorithm and the *Forman* algorithm. In the Mertz routine, the largest data point in the interferogram is assigned as the zero retardation point and the amplitude spectrum is calculated with respect to this point. A short double-sided interferogram is measured and its corresponding phase array is used to phase correct the entire single-sided spectrum. The Forman correction is essentially equivalent to the Mertz routine but it is performed in the retardation space.[14] Modifications to the Mertz phase correction have appeared in the literature and were originally applied to vibrational circular dichroism (VCD) spectra.[15] The result of these modifications is that the phase spectrum does not change sign if a quadrant boundary is crossed. As an alternative, a "stored" phase array can be used to produce proper phase correction for the transformed interferograms. This phase array is calculated from a double-sided reference interferogram. The procedure relies on the fact that the beamsplitter phase does not change from scan to scan.

Fast Fourier Transform

The breakthrough in the application of interferometry to spectroscopy came with the discovery by Cooley and Tukey of the fast Fourier transform (FFT) algorithm.[16] The Fourier transform of a 2048 point vector requires $(2048)^2$, or 4.2 million multiplications. The FFT algorithm reduces this amount to (2048) x log(2048), for a total of 24233 multiplications. Obviously, the great advantage of FFT can be appreciated as the number of data points gets larger and larger. Today's personal computers can calculate an array similar to the one described above in a fraction of one second.

Continuous-scan FT-IR

Most commercially available FT-IR spectrometers utilize the continuous-scan mode of operation, where the moving mirror is scanning at constant velocity. This type of scanning works very well for routine measurements. In the continuous-scan mode of interferometry the laser fringe counter is used to sense the accuracy of the scanning velocity. If a deviation is sensed, correction signals are generated that assure the proper operation (constant velocity). The consequence of this mode of operation is that each infrared wavelength (λ), is modulated at its own particular Fourier frequency, given by the following Equation 4:

$$f(\lambda) = 2v/\lambda \qquad (4)$$

where v is the mirror velocity. Continuous-scan FT-IR is the technique of choice when static spectral properties are determined. Co-addition of successive scans increases the signal-to-noise ratio (S/N) by a factor proportional to $\sqrt{t}$, where t is the time that the signal is averaged at each collection point.

Step-Scan FT-IR

In step-scan FT-IR data are collected while the retardation is held constant or is oscillated about a fixed value. Therefore, in order to apply the technique to mid-infrared and shorter wavelength measurements, a method for controlling the retardation and of implementing a special sampling rate comparable to that achieved in modern continuous-scan instruments is required. In recent years several different control methods have been reported. However, all of these rely basically on the use of the HeNe laser fringe pattern to generate the control signal and to determine the step size.[17] The biggest advantage of the step-scan mode is the separation of the time of the experiment from the time of the data collection.

Two types of experiments are possible with step-scan interferometry. One type is the *time-domain* or time resolved experiments where data are collected as a function of time at each mirror position. Sorting of the data results in interferograms that contain spectral responses at different times. The event under study must be a repeatable process in order for the experiment to work.

The other type of experiment capable with step-scan is the so-called *frequency domain* or synchronous modulation experiments. In these experiments, there are two ways to modulate the intensity of the infrared radiation in order to generate step-scan interferograms. One way is to use amplitude modulation (AM) which can be achieved by means of a chopper. When a chopper is used for intensity modulation, a lock-in amplifier is used to detect the signal before digitization occurs. The technique has the drawback that the signal is riding on top of a large DC offset, which has to be subtracted before any meaningful data can be obtained. This can be done by either calculating the average value of the interferogram and subtracting it from each sample point before the Fourier transform takes place or by setting the lock-in amplifier offset to zero, far from the interferogram. Even though the latter technique eliminates the problem of reduced dynamic range, the technique is still susceptible to DC drift.

Another way to achieve modulation of the radiation is by phase modulation (PM). Phase modulation is achieved in some step-scan instruments by a low amplitude oscillation of the moving mirror along the light path, but any other way of producing phase-difference modulation is acceptable. PM results are superior to AM results by at least a factor of 2 in S/N, when the experiment is detector-noise limited. This improvement stems from the fact that the PM interferogram is essentially the first derivative of the AM interferogram, therefore source intensity fluctuations and other variations of the beam intensity will cancel out.[18] Another parameter associated with PM modulation experiments is the so-called "phase modulation characteristic".[19] This refers to the connection between the amplitude of the phase modulation and the wavelength region of maximum modulation efficiency. For the mid-IR region, a PM modulation amplitude of 2 λ_{HeNe} (zero-to-peak) is appropriate (maximum modulation at 2300 cm^{-1}).

In contrast to the continuous-scan method, the advantages of step-scan operation include the ability, as mentioned above, to apply virtually any modulation frequency to the infrared radiation and to carry out multiple modulation experiments. Since the frequency of modulation is not a function of any retardation velocity (e.g., mirror scan speed), there is no dependence on radiation wavelength. In addition, the use of lock-in amplifier detection or digital signal processors (DSP), provides a high degree of noise rejection, analogous to the Fourier filtering effective in the continuous scan mode. Another advantage of lock-in amplifier or DSP detection is the easy retrieval of the signal phase. This is possible due to the fact that the beamsplitter (instrumental) phase is identical for the in-phase and quadrature (90° out of phase) components of the signal. These components are easily obtained as outputs of a two-phase lock-in amplifier. As a result, not only the magnitude M, but also the phase Φ can be easily obtained by following Equations 5 to 8:

$$M = (I^2 + Q^2)^{1/2} \tag{5}$$
$$\Phi = \arctan(Q/I) \tag{6}$$
$$I = M \cos\phi \tag{7}$$
$$Q = M \sin\phi \tag{8}$$

The signal-to-noise (S/N) ratio is increased by staying longer at each data collection point. The two modes of operation, step-scan and continuous-scan, should produce identical results under conditions of detector-limited noise. The time resolution of the step-scan technique is limited only by the rise-time of the detector, by the electronics, (especially by the A/D converter), and by the signal strength. Therefore, it is capable of measuring various relaxation processes that occur in the sub-microsecond regime and are closely associated with molecular-scale phenomena.

Dispersive instruments

Dispersive instruments are used only in selected applications nowadays, due to the fact the interferometric instruments offer all the above mentioned advantages for most of the applications. However, there are places where dispersive instrumentation is still used when the response at one wavelength or a short range of wavelengths is sought.[20]
The basic components of a dispersive spectrometer are the same as in a Fourier transform instrument with the exception of the interferometer. Any differences are within the elements and are the result of the different ways that the source radiation is detected. For instance, sensitive thermocouple detectors are common place in dispersive instruments whereas they are not appropriate for rapid-scanning instruments.[21] In a dispersive instrument a monochromator is used in the place of the interferometer. Before 1950, the monochromator was a rock salt prism for use in the mid-IR region of the electromagnetic spectrum and later was replaced by a diffraction grating. Older monograms provide excellent background information on the operation and maintenance of dispersive infrared instruments and the reader is instructed to consult them.[22, 23]

Sampling Techniques and Applications

A variety of new sampling techniques are available nowadays to the analyst who wants to use infrared spectroscopy for polymer characterization. The differences between these techniques depend on the angle of incidence of the infrared light as well as on the change in the refractive index as light propagates from one medium to the next. Figure 3 shows the propagation direction of light in some sampling techniques.

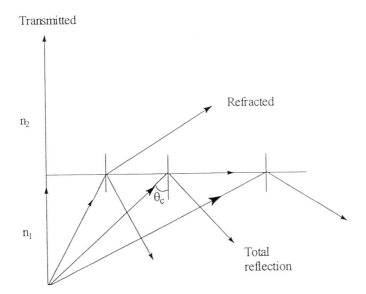

Figure 3: The propagation direction of IR light in different sampling methods. [Ref. 28]

<u>Transmission Spectroscopy</u>

Transmission spectroscopy involves the transmission of infrared radiation through the polymeric specimen. It is the oldest method used to obtain the infrared spectrum of a chemical compound and it is usually used first when polymers are also studied.[24] Polymers are usually analyzed in the form of a freestanding film by either solvent casting or melt pressing. The later requires an appropriate hydraulic press equipped with heated plates capable of reaching the softening point of the polymer. It is always the physical properties of the polymer that dictate the appropriate sample preparation technique. For example, cross-linked elastomeric materials are prepared either by microtoming or by cold grinding and pressed into a KBr pellet or made into a Nujol mull.[25,26]

One potential problem associated with transmission is the presence of interference fringes that can distort the appearance of the original spectrum. They have their origin in the multiple reflection of the light from the front and back surfaces of the sample. The interference can be removed by roughening the surface of the film, causing the reflection to become diffuse. Another way to address the problem is to remove the spike in the interferogram that transforms into the sine wave.[27] However, the presence of these interference fringes is not always undesirable, since the thickness of the sample can be determined from the period of the fringe.

Finally, the effect of "stray light" is not very important in FT-IR spectroscopy because only light that has passed through the interferometer (modulated light) is detected. Any problems associated with stray light are more apparent at the wavelengths where strong absorptions occur.

Reflection spectroscopy

Internal reflection

Internal reflection or better known as attenuated total reflectance (ATR) is a widely used technique in the characterization of polymeric materials, second only to transmission. The theory and practice of the technique has been extensively covered in other publications and only a brief overview will be presented here.[28,29,30] ATR is a contact sampling technique that involves an infrared transparent crystal that also has a high refractive index. The general idea behind the technique is that total internal reflection takes place when the angle of incidence θ is greater than the critical angle θ_c. As a consequence, the generation of a standing (evanescent) wave takes place. This wave extends in the rarer medium in accordance to the predictions of the Maxwell's equations. The electric field of this wave decays exponentially with distance from the interface. If the rarer medium absorbs at the wavelength(s) of the radiation, then loss of energy will occur. The depth of penetration is a very important parameter in ATR FT-IR spectroscopy and is defined as the distance necessary for the evanescent wave to reach 1/e of its initial value at the interface. The depth of penetration depends on the refractive index, the angle of incidence, and the wavelength of light. Since ATR is a contact technique, good quality spectra are obtained only when the contact of the polymer and the internally reflecting element is good. Figure 4 shows the optical path for an ATR experiment and the generated evanescent wave.[31]

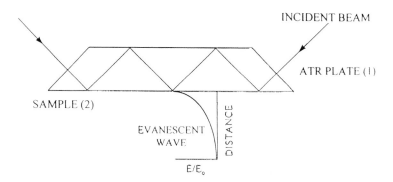

Figure 4: Optical path for an ATR experiment. [Ref. 31]

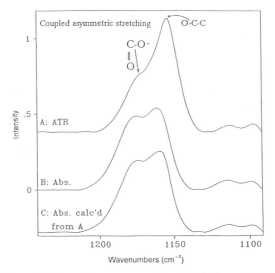

Figure 5: ATR and absorption spectra of poly(ethyl acrylate). [Ref. 28]

Caution must be exercised when comparing transmission spectra and ATR spectra. The later exhibit frequency shifts and intensity changes that can be corrected if the optical constants are known. Specifically, Kramers-Kronig transformations can provide the correct absorption spectrum from an ATR spectrum.[32] Figure 5 shows that application of this correction procedure to the spectra of poly(ethyl acrylate) in the C-O stretching region.

External reflection

Reflectance infrared spectroscopy has been primarily used as a characterization tool to study the molecular orientation, average conformation order, and chemical identity of thin films (<10 nm) on metallic surfaces.[33,34,35] The sensitivity of the technique comes from the fact that the intensity of the p-polarized component of the reflected infrared radiation (where the electric field is perpendicular to the plane of the metal surface) is enhanced at a metal surface at a high angle of incidence. This high angle of incidence is referred to as the "grazing angle". The sensitivity of the technique is such that the infrared spectra of a single monolayer have been reported in the literature.[36]

A recent extension of the above approach is the technique of phase modulated infrared reflection absorption spectroscopy (PM-IRRAS).[37,38] In this experiment, a photoelastic modulator (PEM) that induces a high frequency polarization modulation of the infrared beam between polarization states is used along with a lock-in amplifier that demodulates the spectral information. There are several advantages to the dynamic experiment when it is compared to a traditional static measurement. The most important of them is the sensitivity enhancement attainable in a dynamic measurement that reduces experimental times considerably. Another advantage is the elimination of

V.G. Gregoriou

atmospheric interference that allows the unambiguous interpretation of the spectral features. Figure 6 shows the experimental arrangement for PM-IRRAS spectroscopy.

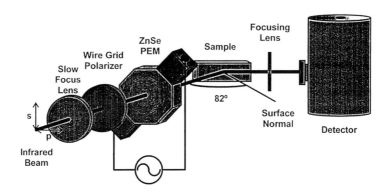

Figure 6: Experimental arrangement for external reflection infrared spectroscopy.

A recent application of infrared reflection absorption spectroscopy was in the elucidation of the effect of NaCl concentration in the morphology and orientation behavior of polymeric thin films of alternating bilayers of sulfonatedpolystyrene / poly(diallyldimethylammonium chloride) (SPS/PDAC) fabricated by ionic multilayer assembly.[39] It was found that the integrity of the underlying self-assembled monolayer depends on the ionic content of the initial solutions. Specifically, at higher salt concentrations (>1M) the loss of the lateral hydrogen bonding was observed, indicative of the disrupted nature of the alkanethiol packing. In addition, the presence of water was detected in the electrolyte samples, an indication of a high level of hydration in these films. Finally, an increase in the intensity of bands proportional to the salt content was attributed to a corresponding increase in the film thickness due to charge-shielding along the polymer backbone.[40] Figure 7 shows the PM-IRRAS spectra in the presence and absence of the electrolyte.

Another area where the techniques were applied was in polymer-metal interface reactions.[41] Polyimide-metal systems were studied that are relevant to electronic packaging. In addition, poly(phenylene vinylene) (PPV) derivatives used in electroluminescent devices were also studied. Emphasis was placed on the detection and interpretation of interfacial reactions that influence performance characteristics such as adhesion, stability, and light emission of polymer-metal structures. In particular, in-situ FT-IR reflection absorption spectroscopy was used to explore the formation of polyimide-on-metal interfaces, hydrolytic stability of such interfaces, and reactivity of PPV systems when exposed to ultraviolet radiation and oxygen.

Differences between polymer-on-metal and metal-on-polymer interfaces were discussed. Models of the IR optical processes in the thin film composites were used to distinguish between chemical and optical effects.

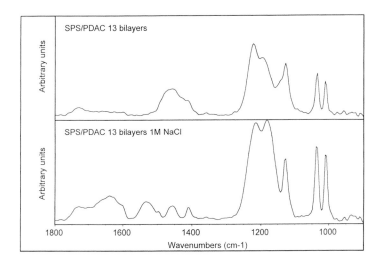

Figure 7: PM-IRRAS spectra of the SPS/PDAC system in the presence and absence of 1 M NaCl [Ref. 39].

Diffuse reflection

Diffuse reflection spectroscopy also known as DRIFTS (diffuse reflectance in Fourier transform spectroscopy) is associated with reflected or scattered radiation from internal surfaces at a finite depth in the material.[42,43] The practical utilization of the technique in the infrared spectral region became possible with the availability of Fourier transform instruments that provided the advantages mentioned in the instrumental section and produced high S/N spectra, primarily due to the high energy throughput that they offer. A variety of polymer samples have been studied and reported using this technique that affords the possibility of minimum sample preparation in many cases.

For qualitative analysis no sample preparation is necessary. There is a difficulty in reproducibility, since many factors contribute to reflection such as particle size, shape of the particles and packing density. KBr is used most of the times to dilute powdered samples. When we are dealing with quantitation, accurate weight and perfect mixing of the alkali halide is required.

The response of the technique is related to the concentration of the analyte via the Kubelka-Munk expression[44]. This expression is valid only when the amount of Fresnel reflectance is small. Fresnel reflectance is the reflection of the incident radiation from the surfaces of the particles and is the biggest contributor to the nonlinearity associated with DRIFTS spectroscopy. There are also two other types of reflected radiation produced on a DRIFTS experiment. The so called diffuse Fresnel reflectance that contains no information about the analyte, whereas the third and final class of radiation, the so called diffuse Kubelka-Munk reflectance contains all the information regarding the analyte and obeys the Kubelka-Munk relationship. Figure 8 shows these interactions in a schematic representation.[45]

One example where DRIFTS was implemented was in the surface characterization of nanostructured GaN.[46] The use of GaN/polymer nanocomposites in optoelectronics requires a perfect control of the GaN dispersion in the polymer matrix. This cannot be achieved without a good knowledge of the first layer of the GaN nanoparticles. That study dealt with the characterization of the surface chemical species of a GaN nanosized powder by diffuse reflectance IR Fourier transform spectrometry. In another report by the same group, the interaction of the GaN first layer with acetic acid was investigated as a preliminary step for the deagglomeration study of the nanoparticles in polymer matrixes.[47]

Another investigation was in the area of coupling agents for fiber reinforced plastics. Specifically, silica power was used to imitate glass fibers and this powder was subsequently treated with propyltrimethyloxysilane, which is a coupling agent used to increase reinforcement effectiveness. The amount of coating was monitored by following the disappearance of the free –OH groups on the glass surface and the simultaneous growth of $-CH_2$ and $-C=O$ bands due to the presence of the coupling agent[48]

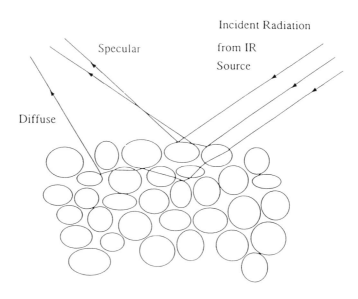

Figure 8: DRIFTS interactions [Ref. 45].

Photoacoustic Spectroscopy

Photoacoustic spectroscopy (PAS) is now commonly used in the analysis of various polymeric materials. It offers minimal or no sample preparation, the ability to look at opaque and scattering samples, and the capability to perform depth profiling experiments. In PAS the transformation of an optical event to an acoustic one takes place. Initially, modulated light is absorbed by the sample located in a sealed cell. The non-radiative decay of this absorbed light produces a modulated transfer of heat to the surface of the sample. This modulated thermal gradient produces pressure waves in the gas inside the cell that can be detected by the attached microphone. This microphone signal, when plotted as a function of wavelength, will give a spectrum proportional to the absorption spectrum of the sample. One main advantage of PAS is the ability to get information about the depth in the sample of the absorption. The amount of the sample contributing to the PA signal is proportional to the thermal diffusion depth. This thermal diffusion depth μ, is inversely proportional to the modulation frequency f. For example, for a typical organic polymer, a high modulation frequency will probe a thin layer (f = 1 KHz; $\mu \sim$ 4 μm,) while at a low modulation frequency, a much thicker layer is probed (f = 100 Hz; $\mu \sim$ 16 μm).

   Figure 9 shows a depiction of the PA signal generation as it relates to depth profiling. The model sample illustrated in this figure has a thermally thin surface layer (thickness << μ) on a bulk substrate. After the light has been absorbed, the heat has to diffuse from the point of absorption to the surface of the sample to be detected. Since this thermal diffusion is a slow process relative to the light absorption and nonradiative decay, an absorption in the bulk will have a phase lag between the time of absorption and the thermal signal. However, a surface absorption should not have a phase lag since the heat doesn't have far to travel to generate the detected pressure change in the transfer gas.[49]

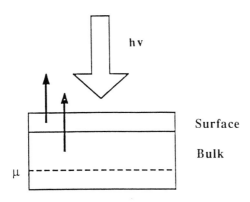

Figure 9: Photoacoustic signal generation.

In step-scan FT-IR spectroscopy, the same modulation frequency can be applied to the entire spectral range.[50] This is particularly useful for photoacoustic (PA) detection since, in the absence of saturation, the PA response at all wavelengths in step-scan FT-IR PA spectroscopy will correspond to the same depth in the sample, whereas in a typical continuous-scan FT-IR PA spectrum each wavelength corresponds to a different sampling depth. A step-scan FT-IR PA spectral depth profile can be obtained by changing the modulation frequency for different scans. Alternatively, depth information can be obtained from the phase of the PA signal. The step-scan mode of data collection permits easier access to the PA phase than does the continuous-scan mode. The combined use of modulation-frequency variation and analysis of the phase allows a wider range of depths to be probed. This type of spectrally resolved depth information is particularly useful when applied to polymeric materials with special surface properties. This includes not only characterization of surface-coated and laminar materials, but also studies of weathering, aging, curing, and the diffusion of species into or out of a polymer matrix.

One method for depth profiling uses the phase of the PA signal to distinguish between depths of different absorptions.[51] Theoretically, this can be done two different ways, by using the phase spectrum or by looking at different components of the signal by detecting at different phases with respect to the infrared light modulation.[52] It has been shown theoretically that the signal from a weakly absorbing thermally thin layer should be separated by $45°$ from the signal of the weakly absorbing substrate.[53]

In order to eliminate instrumental phase contributions, a reference phase must be established. This is done by placing a complete absorber into the PA cell and rotating the phase of the lock-in until all of the signal is in one channel of the lock-in, (e.g. in the quadrature channel, Q). By replacing the reference with the real sample and collecting the in-phase (I) and quadrature (Q) spectra, the bulk signal $(B/\sqrt{2})$ should be in one channel (I), while information proportional to the surface signal $(S+B/\sqrt{2})$ should be in the other channel (Q). As a consequence, the surface signal can be found by rotating the original interferograms by $45°$ and obtaining $S/\sqrt{2}$. Figure 10 illustrates the relationship between the various components of the step-scan photoacoustic signal.

An example of the sub-micron resolution capability of the technique was demonstrated for a coated system having a sub-micron layer as the top layer.[54] Specifically, step-scan FT-IR photoacoustic data were presented that showed the ability of successful isolation of the infrared signature of the top layer from the infrared spectrum of the bulk material. The system under study was a micron thick multi-layered structure of a mixture of acrylic polymers coated on a poly(ethylene terephthalate) (PET) base. The base is several tenths of microns thick, substantially thicker than that of the coated layer.

Figure 11 shows the results of the evolution of the photoacoustic signal as a function of the degree of rotation for the experiment performed with 750 Hz phase modulation. The evolution of the overtone of the carbonyl band at 3428 cm^{-1} that belongs to the PET substrate can be clearly followed as we move to phase delays that correspond to deeper depths. In addition, it is also easy to see the appearance of the C-H aromatic stretching modes above 3000 cm^{-1}, as the phase rotation angle increases. These bands were due to the PET base and, as expected, their intensity increased as the overall signal was dominated by the contribution of the polyester substrate.

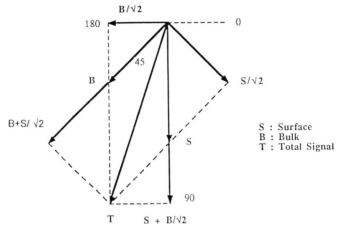

Figure 10: Photoacoustic phase relationships for a weakly absorbing thermally thin layer.

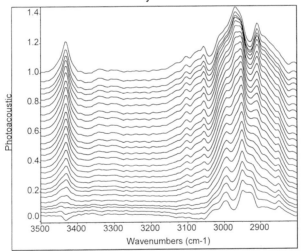

Figure 11: Evolution of the photoacoustic signal as a function of degrees of phase rotation. [Ref. 54]

Infrared Microspectroscopy and Imaging

Since the introduction by Bruker of the first FT-IR microscope interface in 1983, the infrared microscope has become on of the most valuable tools in polymer characterization, especially in industrial laboratories.[55,56] The development of an infrared microscope had already taken place in 1949, with the introduction of the first commercially available microscope by Perkin Elmer in 1953.[57,58] It is obvious that there are many advantages to using a microscope for the chemical characterization of small samples. The microscope is particularly useful in the analysis of contaminants, inclusions, small fibers, etc. In addition, chemical mapping of an area can take place using a computerized microscope stage that allows the detection of the spatial distribution of the species of interest.[59] One example of the use of FT-IR spectral mapping was in the study of polymers in human tissue associated with silicone breast implants.[60] Materials such as polyesters, polyurethanes, and silicone were identified by FT-IR spectroscopy, and their relative locations and distributions within tissue specimens were characterized by three-dimensional (3D) spectral maps.

Lately, a new imaging capability promises to revolutionize the field of mid-infrared imaging. The instrumentation for this technique is built around an InSb multichannel, focal-plane array (FPA) detector and a step-scan interferometer.[61] The technique is under constant development and rapid advances have been reported.[62] It is capable of generating high fidelity and high spectral resolution mid-infrared spectroscopic images. The spatial resolution of the images approaches the diffraction limit for mid-infrared wavelengths, while the spectral resolution is determined by the interferometer and can be 4 cm^{-1} or higher. In another development, a slightly different approach towards infrared spectroscopic imaging microscopy was introduced in which instrumentation was designed around the InSb FPA and a variable-bandpass dielectric filter.[63] The system can be used for either macroscopic or microscopic applications, and high fidelity, chemically specific images can be acquired in a short period of time. With the dielectric filter used in this assembly, continuous tuning was provided for the 4000-2320 cm^{-1} spectral region with spectral resolution of 35 to 18 cm^{-1} at the extremes of this wavelength interval.

Another exciting new development in the field of infrared micro-spectroscopy was the utilization of synchrotron radiation as a broadband infrared source. A synchrotron source uses accelerated electrons to produce high intensity radiation. The one thousand times improvement in source brightness over the typical globar infrared sources makes this approach particularly useful for microscopy experiments. Therefore, it was not surprising when a FTIR microspectrometer was first interfaced with the National Synchrotron Light Source that the performance at the diffraction limit had increased 40-100 times.[64] The synchrotron source transformed the infrared microscope into a true infrared microprobe, providing high-quality infrared spectra for probe diameters at the diffraction limit. Initial studies have shown that the technique can be used in a variety of material characterization applications, including polymers. The only drawback is that the synchrotron source does not easily fit into a FT-IR instrument!

Hyphenated techniques

The combination of different analytical techniques to tackle a problem results in the so-called hyphenated techniques. Obviously, the desired result is the coupling of two or more techniques without sacrificing the performance of them individually. In recent years, coupled or hyphenated techniques have become popular, and have been successfully applied to the solution of many analytical problems. HPLC-MS, tandem MS-MS, GC-FTIR, TLC-FID, GC-MS, etc. are some of the common names used for these double nature techniques. An overview of the most widely used combinations in polymer analysis today will be discussed here.

TGA/FT-IR

The coupling of thermogravimetric analysis (TGA) and Fourier transform IR spectroscopy (FT-IR) is a good practical example of such an instrumental approach for solving specific analytical problems. This hyphenated technique, (TGA/FT-IR), provides a quantitative assessment of the thermal process via the thermogram, and an identification of the decomposition products from the infrared signatures of the evolved gases. The gases are transferred from the TGA instrument by means of a heated transfer line to avoid the possibility of condensation. With such a combination, the sample is introduced into the TGA instrument without any form of chemical or physical modification. The application of the sequential infrared analysis adds a new dimension to thermogravimetric analysis by adding specificity, which it otherwise lacks. An alternate way of looking at the combination is to consider the TGA instrument as a sample handling or sample treatment front-end to the FTIR spectrometer, where one can make full use of the interpretive and diagnostic characteristics of infrared spectral analyses.[65]

There are many examples in the literature where the combination of TGA-FTIR was used successfully in polymer materials. One such example involved the characterization of weathered sealants.[66] In that study, TGA/FT-IR results were compared for silicone and polyurethane unweathered and weathered sealants (6000 h exposure time). The results indicated that the TGA/FT-IR combination is useful in the determination of the degradation changes occurring in the sealants due to weathering. In another study, the characterization of amine-activated epoxies as a function of cure took place using TGA/FT-IR.[67] It is known that the physical and chemical properties of cured diglycidyl ether of bisphenol A (DGEBA) are greatly affected by the initial and final cure temperatures and cure schedule. These properties are also affected by the deviation from the stoichiometric ratio of curing agent used. Analysis of a previously cured epoxy for these parameters has usually involved large samples and a larger amount of time. In that work, a cured epoxy was studied as it decomposed. During the TGA/FT-IR experiment, evolution profiles for specific gases were obtained, as well as the normal TGA weight loss profiles. Using this information, both the cure schedule and epoxy/activator cure ratios could be established from the analysis of the cured polymer. The particular material studied, a DGEBA polymer cured using a primary cycloaliphatic diamine, showed a curing mechanism similar to that obtained using an aromatic diamine. However, the decomposition behavior of the studied sample was more reminiscent of an epoxy cured by using an aliphatic diamine system. This work demonstrated that a cured polymer could therefore be characterized in terms of both thermal history and activator-resin ratio in a single TGA/FT-IR experiment.

LC/FT-IR

The combination of liquid chromatography with infrared spectroscopy has found extensive use in polymer characterization. One of the most popular systems currently in use is the so-called LC-Transform® device.[68] It consists of two modules, a sample collection module and a scanning module accessory. The fist module collects the eluant from the column as a thin trail around the perimeter of the sample collection disk. The solvent is removed by using a nozzle with a heated nebulizer. At the same time, the sample is deposited on the collection disk.[69] The collection disk is made of germanium and it is placed on a rotating stage inside the scanning module. This module consists of a scanning controller and of the necessary optics that couple the infrared light back to the FT-IR spectrometer. Figure 12 shows such an instrument.

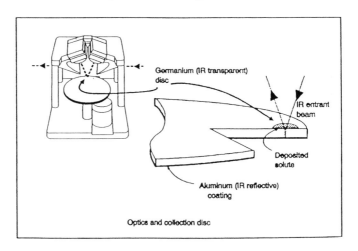

Germanium (IR transparent) disc

IR entrant beam

Deposited solute

Aluminum (IR reflective) coating

Optics and collection disc

Figure 12: LC/FT-IR technique.

Various examples in the literature show the power of the combination of liquid chromatography with infrared spectroscopy. Useful solutions to problems of determining polymer composition as a function of molecular weight for a range of polymers have been demonstrated by the technique. One study focused on the use of a solvent-evaporative interface in conjunction with a GPC-viscometer chromatograph and a FTIR spectrometer in order to provide functional-group information as a function of molecular weight.[70] The application of the GPC-viscometer/solvent evaporative interface/FTIR system to a variety of polymer and coatings systems as a tool for product problem solving and elucidation was presented. In addition, examples of the use of the solvent evaporative interface to elucidate compositional heterogeneity of copolymers will be illustrated. The potential use of the solvent evaporative interface in GPC/LC cross fractionation studies for very fine elucidation of polymer compositional heterogeneity was also explored.

This hyphenated technique has also been used in the characterization of asphalt binders based on chemical and physical properties.[71] The chemical composition and physical properties of unmodified and polymer modified asphalts were studied using a variety of techniques including GPC\FT-IR. Two viscosity-based asphalt grades and two polymers (styrene-butadiene-styrene and styrene-ethylene-butylene-styrene) were used to modify asphalt. The combination of GPC and FTIR was an excellent approach for fingerprinting and quality control of polymers and asphalt binders. In addition, the rheological properties of asphalt binders were good characteristics for determining the optimum polymer concentrations for effective modification.

A coupled GPC/FT-IR system was also developed to measure short chain branching as a function of molecular weight in polyethylenes and ethylene copolymers in relatively short time scales. Careful selection of the IR detector, use of a low volume flow-through cell with a large optical path length, and selecting GPC conditions to maximize the polymer concentration in the cell has enabled the characterization of polymers with very low average comonomer concentrations. A method for calibration of the infrared detector was presented and results for a series of polyethylenes of known average co-monomer content, VLDPE, LLDPE, MDPE, and broad molecular weight polyethylenes were also presented to illustrate the capability of the system. The quality of the data from the GPC/FTIR can be assessed with results on the same polymers obtained using other fractionation techniques. It was found that reliable results could be obtained above MW of approximately 10,000. However, at low molecular weights where chain end corrections become large for infrared measurements, values were confirmed with measurements obtained using NMR spectroscopy.

Infrared Characterization of Polymers and Polymeric Surfaces

Infrared spectroscopy in addition to Raman spectroscopy has been used for a long time in both the qualitative and quantitative determination of polymeric substances.[72] A few selected recent applications will be presented that illustrate the use of infrared spectroscopy as a practical characterization tool.

Identification of composition and morphology[73]

Subtraction studies

Spectral subtraction usually takes place when the isolation of a particular spectral feature in a polymer blend is desired. One of the most common problems associated with spectral subtraction is the interaction of the components of the system that results in frequency shifts and intensity changes. In addition, non-linearities in the measurements of the absorption bands will result in the generation of spectral artifacts.

However, a lot of times the analyst is not interested in the perfect spectrum since we often times are dealing with impurities, additives, plasticizers and fillers. Therefore, the technique of spectral subtraction in the infrared has been very useful in polymer characterization, especially lately when FT-IR instruments dominate the scene. An example of spectral subtraction is given in Figure 13 where the difference spectrum between annealed and quenched isotactic polypropylene is shown.[74]

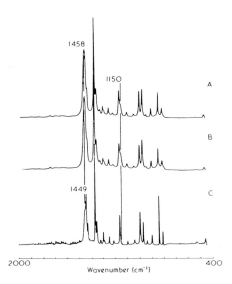

Figure 13: Difference spectrum (C) between annealed isotactic polypropylene (A) and quenched isotactic polypropylene (B). [Ref. 74].

Deconvolution of overlapping bands

This technique aids in resolution enhancement. It is based on the removal of the broadening function that is convoluted with the intrinsic line shape of the absorption band. Care must be taken to not overconvolute the absorption bands. Figure shows the result of the application of Fourier self-deconvolution to the infrared spectrum of urea clathrate type of polyvinyl chloride[75]

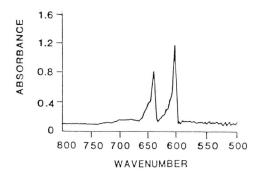

Figure 14: Fourier self-deconvolution infrared spectrum of urea clathrate type poly(vinyl chloride) [Ref. 75]

Another way to increase spectral resolution is through the use of derivative spectroscopy. Resolution enhancement of 2.7 is achieved using a second derivative and goes up to 3.8 for the fourth derivative. Care needs to be taken when such procedures are performed due to the deterioration in the signal-to-noise ratio and the possible appearance of spectral artifacts.[76]

<u>Phase behavior of polymer blends[77]</u>

FT-IR spectroscopy was used to test theoretical predictions about the miscibility of poly(4-vinyl phenol)(PVPh) and styrene-co-methyl acrylate (STMA) copolymers.[78] Figure 15 shows the carbonyl stretching region for a series of PVPh blends with different STMA copolymers, in addition to the spectrum of pure amorphous STMA. When there is substantial mixing at the molecular level, an extra band appears at 1715 cm^{-1} which is assigned to the hydrogen-bonded carbonyl groups. The favorable comparison between the experimentally determined and theoretically calculated fraction of hydrogen-bonded carbonyl groups as a function of blend composition is depicted in Figure 16.

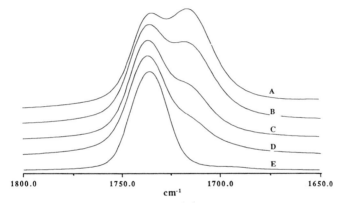

Figure 15: FT-IR spectra at 150°C of PVPh blends with STMA (91%wt) [ref. 78].

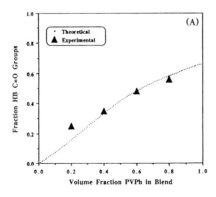

Figure 16: Theoretical versus experimental fraction of H-bonded carbonyl groups for PVPh blends with STMA (91%wt) at 150°C [ref. 78].

(PVPh) blends was examined by differential scanning calorimetry (DSC) and FT-IR spectroscopy.[79] PLLA/PVPh blends are partially miscible as characterized by shifts in the glass transition temperatures of the two component polymers. The Tg of the PLLA-rich phase increases with increasing PVPh content, while that of the PVPh-rich phase decreases with increasing PLLA content. The apparent melting temperature of PLLA is significantly depressed with increasing PVPh content. Weak hydrogen-bonding interaction exists between the carbonyl groups of PLLA and the hydroxyl groups of PVPh as evidenced in the FT-IR spectra. In the amorphous state, the hydrogen-bonded hydroxyl band of PVPh shifts to a higher frequency upon blending with PLLA, while the carbonyl stretching band of PLLA shifts to a lower frequency with the additon of PVPh. In addition, a new carbonyl stretching band attributed to the hydrogen-bonded carbonyl groups of PLLA is found at 1700 cm^{-1}. The crystallinity of PLLA in the blends displays a marked decrease with increasing PVPh content. When the PVPh content is above 40 wt%, crystallization of PLLA does not occur under isothermal conditions. Moreover, the cold crystallization process of the PLLA component is also affected significantly by the addition of PVPh.

The phase behavior of poly(methyl methacrylate) (PMMA) / poly(4-vinylphenol) (PVPh) blends was again investigated using a variety of techniques, including FT-IR spectroscopy.[80] Only one glass transition was observed by DSC in both pure states and blends. However, two loss tangent maxima were observed in DMTA relaxation spectra indicative of the existence of two phases in the blends. Thermal treatments influenced both the number of hydrogen bonds and the proportion of both phases in the blends: quick quenching leads to a lower concentration of hydrogen bonds and to an increase in the quantity of the minor phase.

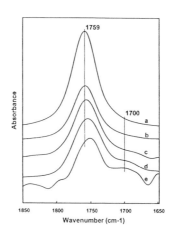

Figure 17: Carbonyl stretching region for PLLA/PVPh blends at 185 °C: a)PLLA, b)80/20 c)60/40 d) 40/60 e)20/80 [Ref. 79]

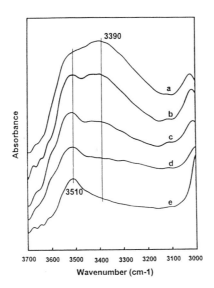

Figure 18: Hydroxyl stretching region for PLLA/PVPh blends at 185 °C: a)PVPh,
b)20/80 c)40/60 d) 60/40 e)80/20 [Ref. 79]

This minor phase gives rise to a tan δ peak appearing at low temperatures. It
has been assigned to domains composed of isotactic-like PMMA segments and PVPh
(*phase-i*). The Tg determined by the major loss tangent maximum exhibits positive
deviation from linearity and is likely due to the domain of syndiotactic-like PMMA
sequence with PVPh (*phase-s*). Using DMTA and X-ray scattering measurements, it
was established that less than 2% *phase-i* with average domain size 30 Angstroms is
dispersed in a *phase-s* matrix. This type of approach could be used to characterize
subtle differences in the behavior of optically active hydrogen-bond-forming polymer
blends related to tacticity.

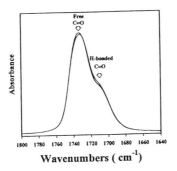

Figure 19: FT-IR spectra of PVPh-PMMA blends containing 30% wt PVPh; slowly
cooled and rapidly quenched samples. [Ref. 80].

Curing studies

Infrared spectroscopy is a very convenient way to monitor curing reactions in polymers. For instance, a polyfunctional aryl-acetylenic monomer, 1,2,4-tris(phenyl-ethynyl) benzene, thermally polymerizes by a free-radical mechanism to a highly crosslinked structure of interest as a precursor matrix for carbon/carbon composites. The polymerization reaction was characterized by both FT-IR and DSC.[81] The disappearance of the acetylenic stretching band at 2212 cm^{-1} was used successfully to monitor the cure reaction. The cure reaction follows first-order kinetics with an overall activation energy of 135 kJ mol^{-1}. Dynamic DSC analysis was carried out to establish that the activation energy of the cure reaction is 137 kJ mol^{-1}.

Another study where FT-IR was employed in reaction monitoring was in the area of surface modifications of polymers. Specifically, the surface modification of polymers by graft polymerization is a relatively new method for covalent immobilization of macromolecular chains on a polymer substrate.[82] Surface graft polymerization of various monomers (ionic and nonionic) onto fabrics, fibers, and films by use of simultaneous UV irradiation has been achieved. For this surface modification, a polymer substrate to be grafted is immersed in a monomer solution containing a small amount of $NaIO_4$ or riboflavin, followed by exposure to UV radiation without degassing. Principal factors affecting the surface photo-graft polymerization include the concentrations of $NaIO_4$ and monomer and the UV irradiation time. Both the characterization of grafted surfaces and the interfacial properties of the grafted surface/water were studied using ATR FT-IR spectroscopy.

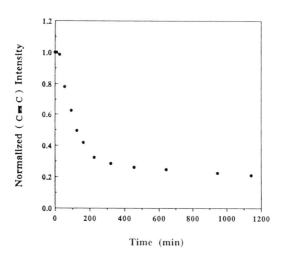

Figure 20: Plot of normalized intensity of the acetylenic absorbance band as a function of time (240 °C) [Ref. 81]

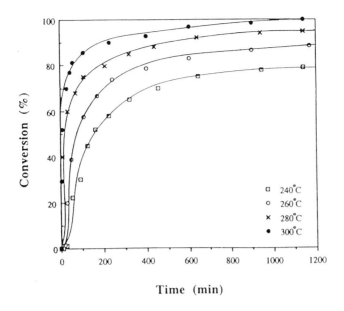

Figure 21: Conversion versus time plot for the isothermal temperatures of 240 °C, 260 °C, 280 °C and 300 °C [Ref. 81]

Diffusion studies

Monitoring the diffusion of small molecules in polymers is a very important task nowadays in polymer chemistry. Therefore, a variety of techniques have been implemented over the years, among them infrared spectroscopy.[83,84]

In one study, a cell was described for characterizing the diffusion of small molecules through thin polymer films using attenuated total reflectance (ATR) Fourier-transform IR spectroscopy. The cell was designed to be used with precast, commercially extruded, polymer films removing the need to cast the film directly onto the ATR crystal and allowing the transport properties of the film to be assessed.[85] Pressurized gas was used to maintain contact between the polymer film and the ATR crystal during the diffusion experiments. Data were presented to demonstrate the use of this cell for measuring the diffusion of a gas (CO_2), two liquids (amyl acetate and limonene), and simultaneous diffusion of individual components from a liquid mixture (50/50 amyl acetate/ limonene) through thin polymer films. In the last case, it was found that pure penetrant transport rates do not necessarily apply to diffusion of the same penetrants from a mixture. Diffusion coefficients obtained from the ATR cell compare favorably with values obtained gravimetrically for the same polymers and penetrants.

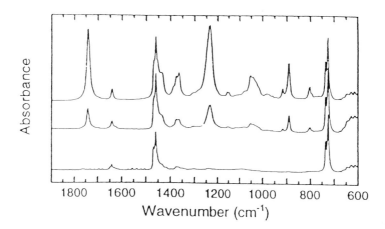

Figure 22: ATR FT-IR spectra of the diffusion of a 50/50(v/v) mixture of amyl acetate/limonene into LDPE. [Ref. 85]

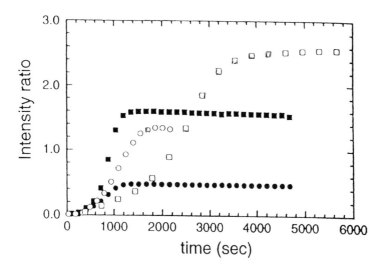

Figure 23: Absorption of limonene (circles) and amyl acetate (squares) into LDPE as neat liquids (open symbols) and as components of a 50/50 (v/v) mixture (filled symbols) using ATR FT-IR spectroscopy [Ref. 85]

Another study investigated the diffusion of a bisphenol A epoxy resin and a diamine curing agent into polysulfone (PSU) using ATR FT-IR.[86] The diffusion of epoxy or amine through the film was monitored in real time by measuring the changing intensity of selected characteristic absorbance bands in each of the polymers. The diffusion coefficients for the epoxy and amine were determined independently using the appropriate solution to the diffusion equation and the experimental data. Diffusivities were evaluated at several temperatures, and an Arrhenius relationship was observed. Results show that the diffusivity of amine is an order of magnitude greater than the diffusivity of epoxy at each temperature tested. Furthermore, the epoxy diffusion appears to be Fickian, while the diffusion of amine appears to be non-Fickian.

A recent application utilized mid-infrared imaging in the study of the diffusion of liquid crystals into polymers.[87] In that study, the focal plane array detection allowed the acquisition of high spatial resolution images in a very short period of time. In particular, the study focused on the diffusion of a low molecular weight liquid crystal, 4-pentyl-4'-cyanobiphenyl (5CB), into poly(butyl methacrylate) (PBMA). This system is interesting because of the implications of phase separation and aging in polymer-dispersed liquid crystal display devices. Figure 24 shows the infrared spectra of a) the liquid crystal and b) the polymer, whereas Figure 25 shows the normalized concentration of the liquid crystal as a function of distance from the polymer-liquid crystal interface. The –CN absorption band at 2225 cm^{-1} was used as a marker for the liquid crystal moiety. The solid line in Figure 25 is the theoretical Fickian diffusion profile that corresponds to a diffusion coefficient of 2.5 x 10^{-10} cm^{2}s^{-1}.

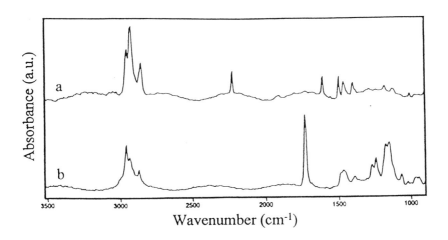

Figure 24: FT-IR spectra of a) liquid crystal and b) polymeric matrix. [Ref. 87]

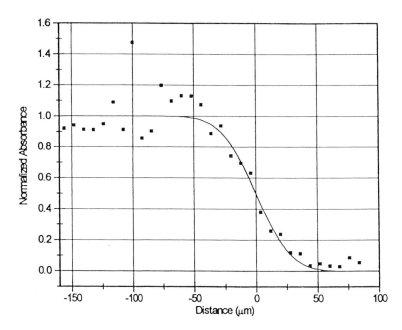

Figure 25: Normalized concentration of -CN absorption band of the liquid crystal across the polymer-LC interface and the ideal Fickian profile (solid line). [Ref. 87]

Oxidation Studies

The appearance of carbon-oxygen bonds is sought after in most oxidation studies characterized by infrared spectroscopy. In one example depicted in Figure 26, the oxidation of polytetrahydrofuran at 150 °C can be monitored as a function of time by following the appearance of the very intense carbonyl band at 1737 cm^{-1}.[88]

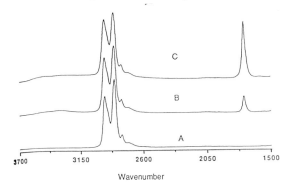

Figure 26: FT-IR spectra of polytetrahydrofuran in the carbonyl stretching region at 150 °C a) 0, b) 1.5 and c) 3.5 hours. [Ref. 88]

In addition, the effects of processing conditions, sterilization treatment, aging time, and post-sterilization aging environment on the oxidation behavior of ultrahigh molecular weight polyethylene (UHMWPE) were examined. Oxidation was again monitored by observing changes in the carbonyl peak and was found to be relatively insensitive to processing conditions but strongly influenced by sterilization treatments and aging parameters.[89]

Oxygen uptake by UHMWPE increased as a result of γ- or electron beam irradiation and continued to rise during subsequent aging at a rate influenced by the aging environment. A hydrogen peroxide ambient caused more severe oxidation than either air or hyaluronic acid. Control (unsterilized) sample and those sterilized in ethylene oxide were resistant to oxidation under all conditions except hydrogen peroxide aging.

Degradation of Polymers

TGA/FT-IR has been used to study the mechanism of degradation of a polymer in the presence or absence of an additive. In one particular study, the effect of flame retarders to the thermal degradation characteristics of poly(methyl methacrylate) (PMMA) was reported.[90] Poly(methyl methacrylate) alone degrades via different pathways, depending on the preparation process.[91] In the presence of additives, many different mechanisms have been proposed depending on the nature of the additive. For instance, there is fair evidence that the presence of radicals destroys polymer stability.[92]

In the case of Nafion-H and PMMA, TGA/FT-IR studies brought to light significant differences between the individual components and the blend. Inspection of the TGA curves shows that both PMMA and Nafion-H completely volatilize at 500 °C, whereas the blend has about 10% residue at 600 °C. Significant interactions between the Nafion-H and PMMA were found and a possible degradation mechanism was proposed based on the profiles of the evolved gases.

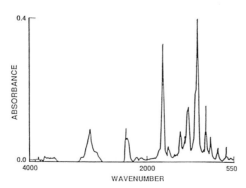

Figure 27: Infrared spectrum of the gases evolved from the blend of Nafion-H and PMMA at 367°C. [Ref. 90]

In another study, the process of PVC decomposition has been studied by FT-IR spectroscopy under dynamic conditions.[93] The infrared spectra have been directly obtained from PVC films pyrolyzed "in situ" in a specially designed IR cell. A qualitative description of the changes observed in the spectra was presented in that work. Below 270 °C, the only phenomena observed were shifts in the absorption bands involving chlorine atoms. Above that temperature strong modifications of the spectra were observed as a consequence of the changes produced in the structure of the polymer. Characteristic absorption bands of the polymer disappeared, whereas several bands corresponding to different vibration modes of C=C bonds appeared.

Deuteration

Isotopic exchange and direct deuteration in particular, are common techniques in infrared spectroscopic studies directed at elucidation of polymeric structure.[94] Frequency shifts upon isotopic substitution are expected due to the mass dependence of the vibrational frequency in the simple harmonic oscillator model. These frequency shifts can be predicted with reasonable accuracy from the changes in the atomic masses alone, since the force constants will remain essentially unchanged.

Krimm's rule[95] is an approximation rule that applies to hydrogen-deuterium substitution

$$v_k/v_{k'} = [1-(\Sigma\Delta T_i/\rho T)]^{-1/2} \tag{9}$$

where v_k and $v_{k'}$ are the zero order frequencies of the kth vibrations for -H and -D groups respectively, T is the total kinetic energy, $\Sigma\Delta T_i$ is the change in kinetic energy upon isotope exchange, and ρ is the ratio of isotopic to normal mass.

Table 2
Predicted versus actual frequencies for deuterated PE

Wavenumber Ratio (CH$_2$/CD$_2$)		Vibrational assignments
Observed	Predicted[Ref. 50]	
1.342	1.349	Asymmetric stretching
1.372	1.379	Symmetric bending
1.341	1.349	Bending
1.384	1.379	Rocking

For the case of polyethylene, Table 2 shows the predicted versus the actual frequency for deuterated polyethylene (PE). The FT-IR spectrum of low density / perdeuterated high density (LDPE/d*-HDPE) is shown in Figure 28. These values compare very well with these calculated by the application of Krimm's rule.

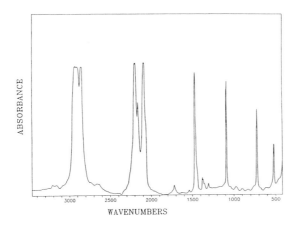

Figure 28: Infrared spectrum of low density / perdeuterated high density polyethylene (LDPE/d*-HDPE)

Orientation Measurements

Orientation in polymers can be measured by a variety of techniques, such as X-ray diffraction, NMR, birefringence, polarized fluorescence, Raman depolarization, sonic techniques and infrared dichroism.[96] The latter technique is one of the most frequently applied tools for the characterization of anisotropy in polymers. Most polymer systems are subjected to the application of stress during manufacturing. The stress is either applied in one direction (uniaxial stretching), or it can be applied along two perpendicular directions (biaxial stretching). The elucidation of the molecular mechanisms that take place during elongation is of great importance to the polymer industry.[97,98,99]

Classic rheo-optical studies emphasize the direct relationship between the perturbation and the spectral response.[100] In addition, dynamic FT-IR spectroscopy provides the element of time in the individual spectral responses of different parts of the molecules with respect to the external perturbation. Infrared has been proven capable of providing meaningful information not only for heterogeneous systems (e.g. micro-phase separated copolymers and semicrystalline polymers), but also for homogeneous systems (e.g. compatible polymer blends) as well.

Infrared Dichroism

In general, maximum absorption takes place when the electric vector is parallel to the transition moment of the specific normal mode (functional group) and no absorption will take place when the electric vector is perpendicular to the transition moment.[101] The absorption of each mode is proportional to the square of the dot product of the electric E and transition moment M vectors, according to the equation

$$I = (EM)^2 \cos^2 \theta \qquad (10$$

where k is a proportionality constant and θ is the angle between the two vectors.

In the case of a polymer macromolecule, the finally observed absorbance A is the sum of the intensity contributions from all the structural units of the polymer (n)

$$A_n = k' \int (EM)^2 \, dn \qquad (11)$$

The effect of the anisotropic distribution of the transition moments with respect to the direction of the electric vector E of the polarized radiation is characterized by the dichroic ratio R

$$R = A_{||}/A_{\perp} \qquad (12)$$

where $A_{||}$ and $A_{\perp}$ are the absorbances measured with radiation polarized parallel and perpendicular to the stretching direction. The value of R can range from zero (where there is no absorption in the parallel direction) to infinity (no absorption in the perpendicular direction). For random orientation, R = 1. If R is greater than 1 the band is called a parallel band; if R is smaller than 1 it is called a perpendicular band.[102] Another useful parameter is the dichroic difference $\Delta A = A_{||} - A_{\perp}$. It is used less often because its value is influenced by the thickness of the sample. In either case, several kinds of information can be acquired from the knowledge of these parameters, namely, the elucidation of the molecular geometry by the determination of the transition moment directions of particular functional groups with respect to the molecular axis and unambiguous assignment of various modes to specific symmetry types of the normal modes.

An important parameter for every absorption band is the so called *structural absorbance* A_0

$$A_0 = (A_x + A_y + A_z)/3 \qquad (13)$$

which represents the absorbance of the band without the contributions due to the orientation of the polymer. For uniaxially oriented sample, produced by stretching in one direction, the structural absorbance becomes

$$A_0 = (A_{||} + 2A_{\perp})/3 \qquad (14)$$

In practice, the orientation is never perfect and this effect can be simulated by supposing that, on average, all the molecular chains are displaced by the same angle from the preferred orientation (stretching axis). Figure 29 shows a schematic representation of the distribution of the molecular chains and their corresponding dipoles with respect to the draw axis. The Herman orientation function F is expressed by the following Equation 15

$$F = (3<\cos^2\theta> -1)/2 \qquad (15)$$

where θ is the angle between the draw direction and the local molecular axis chain. This orientation function can be related to experimentally measured quantities, such as the dichroic ratio R of the absorption band according to the following Equation 16

$$F = (R - 1)(R_O + 2)/(R_O -1)(R + 2) \qquad (16)$$

where $R_O = 2 \cot^2\alpha$ is the dichroic ratio for perfect uniaxial order. As α varies from 0 to $\pi/2$, R_O varies from infinity to zero. No dichroism is observed at the so-called magic angle (for $\alpha = 54°44'$, R_O becomes unity). [103, 104]

Studies point out the fact that the above described model is not able to adequately explain the observed behavior of many real systems. Recently, violations to the model have been noticed in both static systems and systems that underwent dynamic deformations .[105,106] Theoretical treatments that address the limitations of the above model are under way. [107]

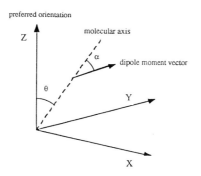

Figure 29: Molecular alignment with respect to the preferred direction.

Dynamic Infrared Spectroscopy of Polymers

Dynamic infrared spectroscopy is defined as the use of infrared spectroscopy to monitor a time-dependent process.[108]   The study of the dynamics of the vibrational excitation/de-excitation process itself is outside the scope of this chapter. However, since the time scale of this process is of the order of 10^{-13} sec or less, changes in the IR spectrum can be used to monitor the dynamics of slower processes, within the practical limits of the speed of the detector and electronics and the strength of the signal.

As mentioned before in the step-scan section, dynamic spectroscopy can be divided into experiments which use the impulse-response technique ("*time-resolved*" spectroscopy) and those which use synchronous modulation techniques ("*phase-resolved*" spectroscopy). In the first case the dynamic response to a perturbation is monitored as an explicit function of time; in the second case the phase and magnitude of the response with respect to that of the perturbation are measured. It has been previously stated that these two types of experiments are actually closely related (by the Fourier transform) and may be considered as limiting cases of a general modulation experiment which uses n frequencies.[109] For the impulse-response experiment $n \to \infty$, while for the synchronously modulated experiment, $n = 1$.

Until recently, dynamic infrared spectroscopy has been restricted to the study of either relatively slow processes or limited wavelength ranges. Dispersive spectrometers with point by point data collection or very slow scanning offer access to broad wavelength range but require very long data collection times to achieve this and are seriously limited by low throughput. However, for synchronous modulation experiments, over limited wavelength ranges, quite good results for time-resolution in the μs range have been achieved in reasonable times by use of dispersive IR.[110] Tunable laser radiation using either a gas phase laser (such as CO) or a diode laser is another possible approach to dynamic IR spectroscopy. The high intensity per spectral bandwidth of the laser make such techniques excellent from the point of view of both spectral and temporal resolution, but the requirement to make measurements essentially point by point (i.e., without any multiplex advantage) and the limited tuning ranges available, restrict the general utility of such techniques.

Two-Dimensional Correlation Spectroscopy

Two-dimensional correlation analysis in dynamic infrared spectroscopy (2D IR) is a relatively new concept.[111] 2D IR spectra are generated as the product of a pair-wise correlation between the time-dependent fluctuations of IR signals that occur during dynamic IR experiments. 2D IR was originally developed by I. Noda for analysis of dynamic rheo-optical data. However, as pointed out by Noda,[112] this type of analysis can be used on a wide variety of dynamic synchronous modulation measurements, either correlating spectroscopic responses to the same stimulus or two different stimuli.

The initial 2D IR studies were carried out by use of dispersive instrumentation. Although this allows for very high quality data to be obtained, correlations over wide spectral ranges can be extremely time consuming. The ability of the FT-IR spectrometer to sample wide wavelength regions simultaneously, while maintaining reasonable experimental collection times, is a distinct advantage over the dispersive technique in these instances.[113]

In 2D IR (or 2D FT-IR), the cross correlation function $\chi(t)$ for a pair of time-dependent variations of IR signals measured at two different wavenumbers $\tilde{A}(\nu_1, t)$ and $\tilde{A}(\nu_2, t)$ is defined as

$$\chi(\tau) = \lim_{T \to \infty} \frac{1}{T} \int_{-T/2}^{T/2} \bar{A}(v_1,t)\tilde{A}(v_2,t+\tau)dt \qquad (17)$$

where τ is the correlation time and T is the correlation period.

When a sinusoidal external perturbation, $\varepsilon(t)$, with a fixed frequency ω, is applied to the system, the dynamic variation of the IR signal becomes

$$\bar{A}(v,t) = A'(v) \sin \omega t + A''(v) \cos \omega t \qquad (18)$$

By substituting Eq. 18 into Eq. 17, the following expression for $\chi(\tau)$ results:

$$\chi(\tau) = \Phi(v_1,v_2) \cos \omega\tau + \Psi(v_1,v_2) \sin \omega\tau \qquad (19)$$

The terms $\Phi(v_1,v_2)$ and $\Psi(v_1,v_2)$ are referred to as the *synchronous* and *asynchronous* 2D IR correlation intensities, respectively. They are related to the dynamic fluctuations of the IR absorbance in the following way:

$$\Phi(v_1,v_2) = [A'(v_1) A'(v_2) + A''(v_1)A''(v_2)]/2 \qquad (20)$$

and

$$\Psi(v_1,v_2) = [A''(v_1) A'(v_2) - A'(v_1)A''(v_2)]/2 \qquad (21)$$

Plots of Φ as a function of v_1 and v_2 resemble those of 2D NMR spectra in many ways. However, in the 2D IR (or FT-IR) spectra it is important to remember that the dynamic processes correlated are those of transition dipole reorientation, in contrast to the nuclear excited state relaxation processes in 2D NMR. Although other responses such as frequency shifts and bandwidth changes are also possible, the use of polarized light in the proposed experiments makes them particularly sensitive to reorientation of functional groups with respect to the mechanical perturbation.

Diagonal peaks (*autopeaks*) in the synchronous correlation 2D IR spectrum Φ indicate which transition dipoles, and thus functional groups, have an orientational response to the perturbation. The sign of the autopeaks is always positive. The off-diagonal peaks (*cross peaks*) indicate the degree to which dipoles respond in phase or simultaneously with each other and, from their sign, the relative reorientation of these dipoles. In contrast to the autopeaks, the signs of cross peaks can be either negative or positive. The existence of a positive cross peak is an indication that the two corresponding dipole moments reorient parallel to each other. On the other hand, when the sign of the synchronous cross peak is negative, this indicates mutually perpendicular reorientations. It is interesting to note that the synchronous correlation map is symmetric with respect to the diagonal. Therefore, inspection of the lower half of the correlation map, for example, is enough to reveal all the information that can be extracted from a synchronous correlation map.

In contrast, different kinds of information can be obtained by the inspection of the asynchronous correlation map. The function Ψ is a measure of the degree of independence between the reorientation behavior of the corresponding dipole moments. Furthermore, the asynchronous 2D correlation spectrum has no diagonal peaks and produces cross peaks only to the extent that two transition dipoles reorient out of phase with each other. The signs of the asynchronous cross peaks give the relative rates of response of the two contributing dipoles. The asynchronous correlation map is antisymmetric with respect to the diagonal, thus making the upper half of the map the mirror image of the lower part. The details of interpretation of frequency correlation maps result from the properties of the corresponding functions, Φ and Ψ. Therefore, in addition to spectral resolution enhancement due to the incorporation of the second dimension, 2D IR spectra can provide information about the relative reorientation of transition dipole moments and the relative rates of inter- and intra-molecular conformational relaxations. One property of the synchronous correlation map is that it is symmetric with respect to the diagonal line. The sign of the autopeaks is always positive, as mentioned above. The asynchronous spectrum is antisymmetric with respect to the diagonal line. The big advantage of the 2D IR technique is the fact that the deconvolution of highly overlapped absorption bands is based on physical arguments instead of mathematical data manipulation techniques (e.g. curve fitting analysis, Fourier self-deconvolution etc.). Overall, it has to be kept in mind that the original dynamic spectra provide all the information found in the 2D spectra. Therefore, even though the two dimensional correlation highlights important features of the dynamic data, it has to emphasized that the S/N of the original data controls the quality and appearance of the 2D data.

Selected Examples

Dynamic infrared spectroscopy has been used successfully in the study of the molecular and sub-molecular (functional group) origins of the macroscopic rheological properties of organic polymers. In particular, the responses of polymer films to modulated mechanical fields have been examined. [114] The emphasis is on the use of step-scan FTIR spectroscopy for these measurements.

The application of step-scan interferometry to two dimensional FT-IR spectroscopy was first described in 1991[115]. The 2D FT-IR spectra for a composite film of isotactic polypropylene and poly(γ-benzyl-L-glutamate) subjected to a small-amplitude sinusoidal strain were presented.

The 2D FT-IR spectra clearly differentiate bands arising from the polyolefin and polypeptide. Overlapped bands were deconvoluted into individual components on the 2D spectral plane due to their different dynamic behavior. The applicability of step-scan 2D FT-IR to a variety of dynamic experiments was also discussed in this early paper.

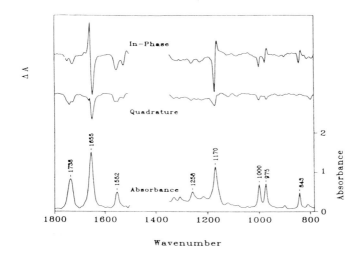

Figure 30: In-phase and quadrature dynamic FT-IR spectra of a poly(γ-benzyl-L-glutamate)/isotactic polypropylene composite sample. The absorbance spectrum is also included for comparison purposes. [Ref. 115].

The appearance of cross peaks in the correlation map between the poly(γ-benzyl-L-glutamate) bands and the isotactic polypropylene bands is a strong indication that these components are oriented at a different rate.

In another example, the dynamic FT-IR spectra of uniaxially drawn poly(ethylene terephthalate) (PET) under a sinusoidal strain were examined.[116] A very intense dynamic band at 973 cm⁻¹ assigned to the trans C-O stretching mode indicated stress-induced high mobility around the C-O bond in the ethylene glycol units. It was supposed that the bisignate skeletal bands observed in the dynamic spectra originated from the stress induced frequency shift. Two-dimensional correlation analyses of the dynamic spectra were also carried out and revealed that the phenyl ring 18a band at 1018 cm⁻¹ and the phenyl ring 19b band at 1410 cm⁻¹ were composed of three and two independent components, respectively. The correlation peaks between the phenyl ring and -CH₂ vibrational modes showed that orientation of the methylene group in the ethylene glycol unit, induced by mechanical stretching, is faster than that of the phenyl ring in the terephthalate unit.

It has also been advocated that the two dimensional plots give useful information but they are sometimes difficult to interpret.[117] The same information is present in both the two dimensional plots and in a phase and magnitude representation of the same data. The proposition was that the phase angle should be included to make easier the visualization of FT-IR experiments.

750

V.G. Gregoriou

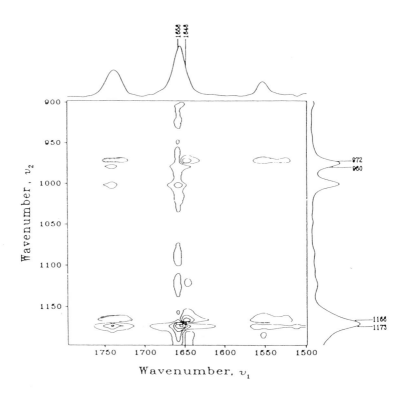

Figure 31: Asynchronous 2D FT-IR spectrum of the above composite sample, comparing poly(γ-benzyl-L-glutamate) and isotactic polypropylene responses. [Ref. 115].

Finally, a dynamic experiment will be presented where the changes take place in a matter minutes rather than ms. Specifically, a time resolved FT-IR experiment is described where the structural ordering of the physical gelation of syndiotactic polystyrene (SPS) dispersed in chloroform was studied.[118] It was found that the gelation process depends remarkably on factors such as the molecular weight of SPS (Mw), the polymer concentration (C), and temperature. It was found that the TTGG-type conformational ordering of the SPS molecules dispersed in the gel increased in time as measured by time-resolved FT-IR. The parameters describing gel-network structure, such as the radius of gyration of stars, the correlation length ξ', and the mass-fractal dimension D' inside the star were evaluated as a function of gelation time through a nonlinear least squares fitting. ξ' and D' exhibit a divergence-like abrupt change at a particular gelation time. At the same time the Q and the conformational order started to increase, indicating that the sol-gel transformation occurred at this point. The solution to gel transformation was delayed with lowering the molecular weight (Mw) or the concentration (C) or with raising the temperature.

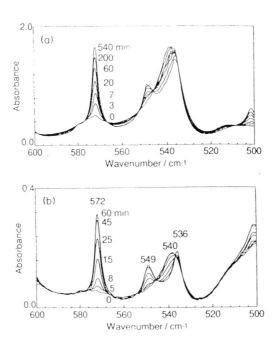

Figure 32: Time-dependent FT-IR spectra of the gelation of SPS in chloroform
a) c = 8.05 g/dL measured at 10 °C, b) c = 1.09 g/dL measured at -5 °C. [Ref. 118].

Quantitative Analysis of Polymer Blends and Composites

Infrared spectroscopy has been used extensively in the determination of concentration profiles of individual components of multicomponent systems. Lambert-Beer's law governs the relationship between the concentration, c, of an absorbing component and its absorbance, A,

$$A = abc \qquad (22)$$

where a is the absorptivity constant and b is the pathlength. Deviations from the Lambert-Beer law can arise from either the spectrometer or the sample itself. They have to be taken into account in the calibration in order to not affect the quantitation accuracy.[119]

In the case of multi component analysis the classical least squares method is applied when all the components and their spectra are known. This regression technique is also known as the *K*-matrix method. [120]   The procedure involves a calibration step, where the estimation of the pure component spectra for a known set of mixtures takes place and an analysis step where the estimation of the unknowns is performed. In matrix notation, the calibration equations can be written as

$$A = KC + E \qquad\qquad (23)$$

where **A** is the *nxm* matrix that has *n* infrared wavelengths and *m* number of calibration spectra. **K** is the *nxi* matrix of the *i* pure component spectra, **C** is the *ixm* matrix of the known component concentrations and **E** is the *nxm* matrix of error in the measured absorption values. The least-square method estimates **K** as

$$\underline{K} = AC'(CC')^{-1} \qquad\qquad (24)$$

where $\underline{K}$ is the estimate of **K** and **C'** is the transposed **C** matrix. $\underline{K}$ is used in the analysis step to calculate the concentration of the unknown sample

$$C = (\underline{K}\underline{K}')^{-1}\underline{K}'A \qquad\qquad (25)$$

where **A** is the unknown sample spectra.

If the inverse of Beer's law is used to relate A and *c*, then the inverse least squares method is used. This method is also referred to as the *P*-matrix method.[121] The advantage of this method is that the quantitation can be performed even if only one component of the calibration mixture is known. The disadvantage is that only a small number of wavelengths are used due to the fact that the inverted matrix has dimensions equal to the number of wavelengths. The selection of wavelengths is also a very important step, since colinearity and overfitting can occur.   Methods have been developed that help to avoid these problems.[122]

Other methods have also been developed. In the partial least-squares method, the calibration and prediction analyses are performed one component at a time.[123,124] Non-linear responses, component interactions and baseline shifts can be modeled to a degree using partial least-squares. Finally, cross-correlation, factor analysis, and rank annihilation methods have also been developed and used with success in multivariate quantitation analyses.[125,126,127]

One example of multicomponent analysis of polymeric materials involved the use of the so-called indicator function in the determination of the compatibility of blends that are made up with different weight fractions.[128] The indicator function is a parameter that shows if the contribution comes form the real eigenvalues or not.

In determining compatibility on a polymer blend, one would expect to find an incompatible blend to be phase separated and its spectrum to be the summation of the spectra of the individual homopolymers. Figure 33 shows the plots of the indicator function versus the number of components for compatible blends of polyoxyphenylene-polystyrene (PPO-PS) and the incompatible blend of polyoxyphenylene-poly(parachlorostyrene) (PPO-P$_4$ClS). Since mixing at the molecular level occurs for the compatible blend, an additional component is found.

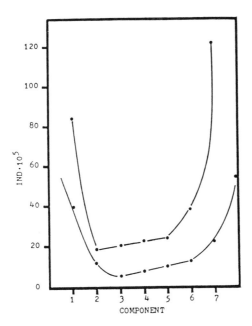

Figure 33: Plots of the indicator function versus the number of components. a) PPO-PS, b) PPO-P$_4$CIS. [ref. 128]

Acknowledgment

I would like to acknowledge the assistance of Sheila E. Rodman in the editing of this manuscript. I would also like to thank the management of Polaroid Corporation for their support.

Literature Cited

1   Griffiths, P.R.; de Haseth, J. "Fourier Transform Infrared Spectroscopy"; J. Wiley and Sons: New York, 1986.
2   Colthup, N.B; Daly, L.H.; Wiberley, S.E. "Introduction to Infrared and   Raman Spectroscopy"; Academic Press: 3rd Edition: Boston, 1990.
3   Brundle, C.R.; Evans Jr. C.A.; Wilson, S. "Encyclopedia of Materials Characterization"; Butterworth-Heinemann, Boston, 1992.
4   Painter, P.C.; Coleman, M.M.; Koenig, J.L. "The Theory of Vibrational Spectroscopy and Its Application to Polymeric Materials" Wiley: New York, 1982.
5   Garton, A. "Infrared Spectroscopy of Polymer Blends, Composites and Surfaces"; Hanser Publishers; Munich, 1992.
6   Schrader, B. "Infrared and Raman Spectroscopy"; VCH Publishers Inc.; New York, 1995.
7   Michelson, A.A. Phil. Mag., Ser. 5, 1891, 31, 256.

8   Fellgett, P. J. Phys. (Radium) 1958, 19, 187.
9   Jacquinot, P. J. Opt. Soc. Am. 1954, 44, 761.
10   Connes, J.; Connes, P. J. Opt. Soc. Am. 1966, 56, 896.
11   Connes, J. "Optical Instruments and Techniques"; Oriel Press; Paris, 1970.
12   Marsall, A. G.; Verdun, F. R. "Fourier Transforms in NMR, Optical, and Mass Spectrometry"; Elsevier Science; Amsterdam, The Netherlands, 1990.
13   Mertz, L. Infrared Phys. 1967, 7, 17.
14   Forman, M.L.; Steel, W.H.; Vanasse, G.A. J. Opt. Soc. Am. 1966, 56, 59.
15   McCoy, C.A.; de Haseth, J.A. Appl. Spectrosc. 1988, 42, 336.
16   Cooley, J.W.; Tukey, J.W. Math. Comput, 1965, 19, 297.
17   Palmer, R.A. Spectroscopy 1993, 8(2), 26.
18   Chamberlain, J. Infrared Phys. 1971, 11, 25.
19   Chamberlain, J.; Gebbie, H.A. Infrared Phys. 1971, 11, 57.
20   Noda, I.; Dowrey, A.E.; Marcott, C. Appl. Spectrosc. 1988, 42, 203.
21   White, R. "Chromatography/Fourier Transform Infrared Spectroscopy and   its Applications"; Marcel Dekker; New York, 1990.
22   Rao, C.N.R. "Chemical Applications of Infrared Spectroscopy"; Academic Press; New York, 1963.
23   Smith, A.L. "Applied Infrared Spectroscopy, Fundamentals, Techniques, and Analytical Problem-Solving";Wiley-Interscience; New York, 1979.
24   Coleman, P.B. "Practical Sampling Techniques for Infrared Analysis"; CRC Press; 1993.
25   ASTM method E 168-92.
26   Henniker, J.C. "Infrared Spectroscopy of Industrial Polymers"; Academic Press; 1967.
27   Moffatt, D.J.; Cameron, D.G. Appl. Spectrosc. 1983, 37, 586.
28   Urban, M.W. "Attenuated Total Reflectance Spectroscopy of Polymers-Theory and Practice"; Polymer Surfaces and Interfaces Series; American Chemical Society; Washington, DC, 1996.
29   Harrick, N.J. "Internal reflection Spectroscopy"; Wiley-Interscience; New York 1967.
30   ASTM method E 573-96.
31   Ishida, H. Rubber Chem. Technol. 1987, 60, 498.
32   Bertie, J.E.; Eysel, H.H. Appl. Spectrosc. 1985, 39, 392.
33   Greenler, R.G. J. Phys. Chem. 1966, 44, 310.
34   Greenler, R.G.; Rahn, R.R.; Schwartz, J.P. J. Catalysis 1971, 23, 42.
35   Young, J.T.; Boerio, F.J.; Zhang, Z.; Beck, T.L. Langmuir 1996, 12, 1219.
36   Rabolt, J.F.; Burns, F.C.; Schlotter, N.E.; Swalen, J.D. J. Chem. Phys. 1983, 78, 946.
37   Buffeteau, T.; Desbat, B.; Turlet, J.M. Appl. Spectrosc. 1991, 45, 380.
38   Huehnerfuss, H.; Neumann, V.; Stine, K.J. Langmuir 1996, 12, 2561.
39   Gregoriou, V.G.; Hapanowicz, R.; Clark, S.; Hammond, P.T Appl. Spectrosc. 1997, 51, 470.

40   Gregoriou, V.G.; Hapanowicz, R.; Clark, S.L.; Hammond, P.T. <u>Amer. Inst. Phys.</u> 1998, <u>430</u>, 551.

41   Cumpston, B.H.; Lu, J.P.; Willis, B.G; Jensen, K.F. <u>Mater. Res. Soc. Symp. Proc.</u>, 1995, <u>385</u> (Polymer/Inorganic Interfaces 2);103.

42   Fuller, M.P.; Griffiths, P.R. <u>Anal. Chem</u>. 1978, <u>50</u>, 1906.

43   Fuller, M.P.; Griffiths, P.R. <u>Appl. Spectrosc</u>. 1980, <u>34</u>, 533.

44   Kubelka, P.; Munk, F. <u>J. Opt. Soc. Am</u>. 1948, <u>38</u>, 448.

45   Mitchell, M.B. in "Structure-Property Relations in Polymers: Spectroscopy and Performance"; Advances in Chemical Series No. 236; Chapter 13; M.W. Urban and C.D. Craver Editors; American Chemical Society; 1993.

46   Gonsalves, K.E.; Carlson, G; Baraton, M-I. <u>Mater. Res. Soc. Symp. Proc.</u> 1998, <u>501,</u> 191.

47   Baraton, M-I.; Carlson, G. K.; Gonsalves, E. <u>Fr. Mater. Sci. Eng., B</u> 1997, <u>B50(1-3),</u> 42.

48   Siesler, H.W. <u>Mikrochim. Acta</u> 1988, <u>I,</u> 319.

49   Dittmar, R.M.; Chao, J.L.; Palmer, R.A. <u>Appl. Spectrosc.</u> 1991, <u>45</u>, 1104.

50   Palmer, R.A.; Dittmar, R.M. <u>Thin Solid Films</u>, 1993, <u>223(1),</u> 31.

51   Palmer, R.A. <u>Spectrosc</u>. 1993, <u>8(2),</u> 26.

52   Palmer, R.A.; Jiang, E.Y. <u>Anal. Chem</u>. 1997, <u>69</u>, 1931.

53   Mongeau, B.; Rousset,G; Bertrand, L. <u>Can. J. Phys</u>. 1986, <u>64</u>, 1056.

54   Gregoriou, V.G.; Hapanowicz, R. <u>Spectroscopy</u>. 1997, <u>12(5)</u>, 37.

55   Koenig, J.L. "Microspectroscopic Imaging of Polymers"; ACS Professional Series; 1998.

56   Katon, J.E. <u>Vib. Spectrosc</u>. 1989, <u>7</u>, 201.

57   Barer, R.; Cole, A.R.; Thompson, H.W. <u>Nature (London)</u> 1949, <u>163,</u> 198.

58   Coates, V.J.; Offner, A.; Siegler, Jr.; E.J. <u>J. Opt. Soc</u>. 1953, <u>43,</u> 984.

59   Ward, K.J. <u>Proc. SPIE Int. Soc. Opt. Eng.</u> 1989, <u>212</u>, 1145.

60   Ali, S.R.; Johnson, F.B.; Luke, J.L.; Kalasinsky, V. F. <u>Cell. Mol. Biol.</u> 1998, <u>44(1),</u> 75.

61   Lewis, E.N.; Treado, P.J.; Reeder, R.C.; Story, G.M.; Dowrey, A.E.; Marcott, C.; Levin, I.W. <u>Anal. Chem</u>. 1995, <u>67</u>, 3377.

62   Kidder, L.H.; Levin, I.W.; Lewis, E.N. <u>AIP Conf. Proc.</u> 1998, <u>430,</u> 148.

63   Lewis, E.N.; Levin, I.W. <u>Appl. Spectrosc</u>. 1995, <u>49(5)</u>, 672.

64   Reffner, J.A.; Martoglio, P.A.; Williams, G.P. <u>Rev. Sci. Instrum.</u>1995, <u>66</u>, 1298.

65   Materazzi, S. <u>Appl. Spectrosc. Rev</u>. 1997, <u>32(4),</u> 385.

66   Paroli, R.M.; Delgado, A.H. <u>Can. Polym. Mater. Sci. Eng</u>. 1993, <u>69</u>, 139.

67   Johnson, J.D.; Compton, D.A.; Cass, R.S.; Canale, P.L. <u>Thermochim. Acta,</u> 1993, <u>230</u>, 293.

68   Willis, J.N.; Wheller, L. <u>Polym. Mater. Sci. Eng</u>. 1993, <u>69</u>, 120.

69   Liu, M.X.; Dwyer, J.L. <u>Appl. Spectrosc</u>. 1995, <u>50</u>, 349.

70   Provder, T.; Whited, M.; Huddleston, D.; Kuo C-Y. <u>Prog. Org. Coat</u>. 1997, <u>32</u> 155.

71   Wei, J.B.; Shull, J.C.; Lee, Y.J.; Hawley, M.C. <u>Int. J. Polym. Anal. Charact.</u> 1996, <u>3(1),</u> 33.

72   Craver, C.D; Provder, T. "Polymer Characterization-Physical Property, Spectroscopic, and Chromatographic Methods"; in Advances in Chemical Series No. 227; American Chemical Society, 1990.

73   Koenig, J.L. "Spectroscopy of Polymers" American Chemical Society, Washington, DC., 1991.

74   Painter, P.C.; Watzek, M.; Koenig, J.L. Polymer 1977 ,18, 1169.

75   Compton, D.A.C.; Maddams, W.F. Appl. Spectrosc. 1986, 40, 239.

76   Gans, P.; Gill, J.B. Anal. Chem. 1980, 52, 351.

77   "Structure-Property Relations in Polymers: Spectroscopy and Performance", in Advances in Chemical Series No. 236; Urban, M.W.; Craver, C.D. Editors; American Chemical Society; 1993.

78   Coleman, M.; Zhang, H.; Xu, Y.; Painter, P.C. in "Structure-Property Relations in Polymers: Spectroscopy and Performance" in Advances in Chemical Series No. 236; Chapter 7; Urban, M.W.; Craver, C.D. Editors; American Chemical Society; 1993.

79   Zhang, L.; Goh, S.H.; Lee, S.Y. Polymer 1998, 39(20), 4841.

80   Li, D.; Brisson, J. Macromolecules 1996, 29(3), 868.

81   Sastri, S. B; Armistead, J.P.; Keller, T.M. Polymer 1995, 36(7), 1449.

82   Uchida, E.; Ikada, Y. Curr. Trends Polym. Sci. 1996, 1, 135.

83   Jabbari, E.; Peppas, N.A. Macromolecules 1993, 26, 2175.

84   High, M.S.; Painter, P.C.; Coleman, M.M. Macromolecules 1992, 25, 797.

85   Balik, C.M.; Simendinger, III, W.H. Polymer 1998, 39(20), 4723.

86   Immordino, K.M.; McKnight, S.; Gillespie, Jr, J.W. J. Adhes. 1998, 65, 115.

87   Snively, C.M.; Koenig, J.L. Macromolecules 1998, 31, 3753.

88   Painter, P.C.; Coleman, M.M. "Fundamentals of Polymer Science"; Technomic Inc.; Lancaster, 171, 1997.

89   Goldman, M.; Lee, M.; Gronsky, R.; Pruitt, L.J. Biomed. Mater. Res. 1997, 37(1), 43.

90   Wilkie, C.A.; Mittleman, M.L. in "Structure-Property Relations in Polymers: Spectroscopy and Performance"; in Advances in Chemical Series; No. 236, Chapter 28, Urban, M.W.; Craver, C.D. Editors; American Chemical Society; 1993.

91   Kashiwagi, T.; Inaba, A.; Brown, J.E.; Hatada, K.; Kitayama, T.; Masuda, E. Macromolecules 1986, 19, 2160.

92   ImcNeil, C. Dev. Polym. Degrad. 1977, 1, 171.

93   Beltran, M.; Marcilla, A. Eur. Polym. J. 1997, 33(7), 1135.

94   Krimm, S. J. Chem. Phys. 1960, 32, 1780.

95   Krimm, S. Adv. Polym. Sci. 1960, 2, 51.

96   Zbinden, R. "Infrared Spectroscopy of High Polymers"; Academic Press; New York, 1964.

97   Monnerie, L. Faraday Symp. Chem. Soc. 1983, 18, 57.

98   Jasse, B.; Koenig, J.L. J. Macromol. Sci. - Rev. Macromol. Chem. 1979, C17, 61.

99   Siesler, H.W. Appl. Spectrosc. 1990, 44, 550, 1990.

100   Fina, L.J.; Koeing, J. L. J. Polym. Sci. Part-B. 1986, 24, 2509.

101   Siesler, H.W.; Holland-Moritz, K. "Infrared and Raman Spectroscopy of Polymers"; Marcel Dekker; New York; 243-292, 1980.

102   Siesler, H.W. Macromol. Chem. 1989, 190, 2653.

103   Frasier, R.D.B. J. Chem. Phys. 1953, 21, 1511.

104   Stein, R.S. J. Appl. Phys. 1961, 32, 1280.

105   Cael, J.J.; Dalzell, W.; Trapani, G.; Gregoriou, V.G. Pittsburgh Conference in Analytical Chemistry and Applied Spectroscopy, Abst.#1204; March 1997; Atlanta, GA.

106   Noda, I.; Dowrey, A.E.; Marcott, C. Polym. Prep. 1984, 5(2), 167.

107   Cael, J.J.; Dalzell, W; Trapani, G.; Gregoriou, V.G. 24ᵗʰ FACSS Meeting, October 1997; Providence, RI.

108   Urban, M.W.; Provder, T. "Multidimensional Spectroscopy of Polymers"; in ACS Symposium Series No. 598; American Chemical Society; Washington, DC, 1995.

109   Manning, C.J.; Chao, J.L.; Palmer, R.A. Rev. Sci. Instrum. 1991, 62, 1219.

110   Noda, I.; Dowrey, A.E.; Marcott, C. J. Polym. Sci. Polym. Lett. Ed. 1983, 21, 99.

111   Noda, I; Dowrey, A.E.; Marcott, C. Microchim. Acta [Wien] 1988, 1, 101.

112   Noda, I. J. Am. Chem. Soc. 1989, 111, 8116; Noda, I. Appl. Spectrosc. 1990, 44, 550.

113   Gregoriou, V.G.; Marcott, C.; Noda, I.; Dowery, A.; Palmer, R.A. J. Polym. Sci.-B Polym. Phys. 1993, 31, 1769.

114   Palmer, R. A.; Gregoriou, V.G.; Fuji, A.; Jiang, E.Y.; Plunkett, S.E.; Connors, L.; Boccara, S.; Chao, J.L. ACS Symp. Ser. 598 (Multidimensional Spectroscopy of Polymers); 1995, 99.

115   Palmer, R.A.; Manning, C.J.; Chao, J.L.; Noda, I.; Dowrey, A.E.; Marcott, C. Appl. Spectrosc. 1991, 45, 12.

116   Sonoyama, M.; Shoda, K.; Katagiri, G.; Ishida, H. Appl. Spectrosc. 1996, 50, 377.

117   Budevska, B.O.; Manning, C.J.; Griffiths, P.R. Appl. Spectrosc. 1994, 48, 1556.

118   Kobayashi, M.; Yoshioka, T.; Imai, M.; Itoh, Y. Macromolecules 1995, 28, 7376.

119   ASTM method E 168-92.

120   Haaland, D.M.; Eastelning,R.G. Appl. Spectrosc. 1980, 34, 539.

121   Marris, M.A.; Brown, C.W.; Lavery, D.S. Anal. Chem. 1984, 55, 1694.

122   Honigs, D.E.; Freelin, J.M.; Hiefte, G.M.; Hirschfeld, T.B. Appl. Spectrosc. 1983, 37, 491.

123   Fuller, M.P.; Ritter, G.L.; Draper, C.S. Appl. Spectrosc. 1988, 42, 228.

124   Long, J.R.; Gregoriou, V.G.; Gemperline, P.J. Anal. Chem. 1990, 62, 1791.

125   Tyson, L.L.; Vickers, T.J.; Mann, C.K. Appl. Spectrosc. 1984, 38, 697.

126   Malinowski, E.R.; Howery, D.G. "Facor Analysis in Chemistry"; John Wiley and Sons Inc.; New York, 1980.

127   Ho, D.-N.; Gristian, G.D.; Davidson, E.R. Anal. Chem. 1980, 50, 108.

128   Koenig, J.L.; Tovar, M.J.M. Appl. Spectrosc. 1981, 35, 543.

RAMAN SPECTROSCOPY OF POLYMERS

RICHARD A. LARSEN

Richard Larsen Consulting, 6 South St. Suite 10, Danbury, CT 06810

Introduction
Theory
Raman Instrumentation
Polymer Identification
Structural Analysis
Quantitative Applications
Polymerization Reactions
Polymer Crystallinity Studies
Raman Imaging Experiments
Non-linear Raman Techniques

Introduction

As with infrared spectroscopy, Raman spectroscopy provides the ability to study the interactions of the vibrational and rotational energies of atoms or groups of atoms within molecules. In the simplest sense, the complete Raman spectrum is an emitted pattern of frequency displacements about a central excitation wavelength, usually a laser. The excitation energy 'diffraction' pattern is a result of molecular vibrations which cause a change in the induced dipole moment, or polarizability, of the molecule. Laser sources and the excitation wavelengths for Raman spectroscopy may range from the ultraviolet (UV) to the near-infrared (NIR) and the resultant spectral pattern is interpreted similarly to the infrared spectrum of the molecule (1-3).

The concept of the Raman effect was theorized as early as 1878 by Lomel (4) while Smekal (5) is given credit for the prediction of the physical process in 1923. Experimental work begun in that year by C.V. Raman (6) resulted in a Nobel prize awarded in 1930. First characterized as a 'weak fluorescence' by Sir Raman, the phenomenon was named for the scientist associated with the majority of the early experimental work (6, 7) and is known today as the Raman effect or Raman scattering.

Infrared and Raman spectroscopy are complementary in the structural determination of molecules as the data arise from two different physical effects in the molecule. Infrared spectra reflect vibrational motions that produce a change in the permanent dipole moment of the molecule. Raman spectra result from vibrational motions which cause a change in a source-induced molecular dipole moment. Infrared spectroscopy is a measurement of a simple absorption process. Raman spectroscopy is a measurement of the re-emission of the incident source energy. The majority of the peaks in both spectra generally correspond to the differences between adjacent vibrational energy levels (Figure 1) but are a result of different physical processes.

A limitation to infrared spectroscopy is the requirement for a finite thickness of the sample to control the absorption intensity of the molecular vibrations. Additionally, a majority of the infrared wavelengths are obscured by glass or quartz windows and cells, thus, hygroscopic infrared transmitting materials (KBr, NaCl, BaF_2, CaF_2, etc) must be used for the majority of sample handling applications. By contrast, Raman has no restriction with respect to sample size, shape or thickness and glass or quartz are ideal sample containment materials for Raman samples. Furthermore, infrared spectra exhibit extremely strong absorptions for water, making the analysis of aqueous solutions a challenge for infrared spectroscopy; conversely, water has very weak Raman bands and thus is an ideal solvent in Raman spectroscopy. (8-12)

Raman spectroscopy has finally earned a place as a common laboratory technique in the last few years due to reductions in the cost of laser sources and detector systems. This chapter outlines the theory of Raman spectroscopy and discusses the latest instrumentation developments and techniques that have enabled the explosive growth in the research, analytical and QC applications of Raman spectroscopy for evaluation of polymer chemistry. This discussion is not meant to be inclusive but rather to provide an overview of the field and offer some predictions as to the future use of Raman spectroscopy for polymer analysis.

Theory

Infrared and Raman spectroscopy are powerful qualitative and quantitative tools but primary differences in the theory and the informational content of each technique provide Raman spectroscopy with some particular advantages in the instrumental analysis of polymers.

As more complete treatments of the theoretical basis for Raman spectroscopy appear elsewhere, (1, 8-10, 12-14), this text covers only the essentials. The following also requires discussion of the infrared absorption process due to the complementary nature of infrared and Raman spectra and the interactions of both techniques with molecular vibrational energy levels.

When a molecule is exposed to an incident beam of light, energy can be absorbed by that molecule to enable a transition among electronic, vibrational or rotational energy levels. Absorption of the source energy can occur during a direct absorption process, such as infrared or ultraviolet-visible spectroscopies, where the energy of the source just happens to coincide with a specific energy level separation, or, during a light scattering process, e.g., fluorescence and Raman spectroscopies, where the energy required for absorption and the resulting energy level transition is drawn from, usually, a very intense source of high energy. This is one of the primary differences between infrared and Raman spectroscopy.

The absorption of infrared light and the interaction of that absorption with the vibrational modes of a molecule is the basis of an infrared spectrum. Infrared absorptions demand that the molecular vibrations produce a change in the permanent dipole moment of the molecule as the absorption process occurs. Transitions occur primarily among the ground and first excited vibrational state but overtone transitions among higher vibrational energy levels can and do occur in the mid-infrared spectral range of 4000 – 400 cm^{-1} (2.5 – 25 microns). Raman spectra reflect similar transitions among the vibrational energy levels of a molecule, but the energy exchange occurs in a completely different manner and usually only reflect transitions between the ground and first excited vibrational level.

Scattering of a source without a direct absorption of energy is classified as either inelastic or elastic scattering. Elastic scattering results in no interaction of the source radiation with the energy levels of a molecule; there is an excitation of the molecule with absorption of energy from the source, but the absorbed energy is immediately re-emitted and the molecule returns to its initial energy level. (Figure 1) The re-emitted energy is identical in wavelength to the incident source and the phenomenon is generally referred to as Rayleigh scattering.

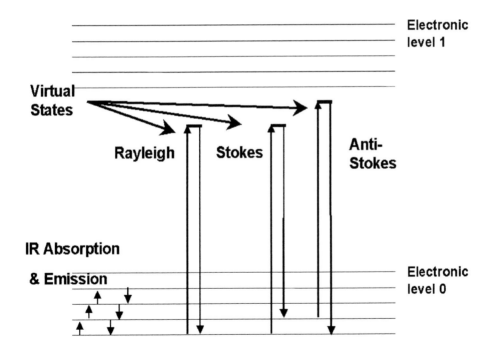

Figure 1. Vibrational energy level diagram demonstrating the transition mechanisms for infrared absorption and emission in addition to Rayleigh and Raman scattering.

Inelastic scattering is a direct interaction of the source radiation with the energy levels of a molecule. The molecule is raised to a 'virtual' level, absorption of energy from the source occurs and as the molecule changes energy levels during relaxation, light is re-emitted at a wavelength other than the incident source, the energy differences corresponding to the electronic, vibrational or rotational energy levels of a molecule. Specifically, when inelastic scattering is a result of interaction with the vibrational or rotational motions of a molecule, the Raman effect occurs.

The Raman effect relies upon the interaction of the incident beam with the vibrational modes of the molecules while the vibrational motion produces a change in the induced dipole moment of the molecule or atomic bond. In effect, how well the vibrational motion 'squeezes' or 'stretches' the electron cloud into another shape determines the intensity of the Raman peak corresponding to the vibrational motion. The term 'polarizability' is generally meant to express the capacity of the vibrational motion to change the electron cloud. If a molecular motion is especially polarizable and greatly alters the induced dipole moment, the Raman peak will be very intense.

More specifically, when a diatomic molecule is exposed to an intense monochromatic source, the electric field of the incident light induces a dipole moment in the molecule which oscillates in concert with the oscillation of the electric field of the source. The induced dipole moment of a molecule is given by

$$P = \alpha E \tag{1}$$

where E is the electric field of the source, and α the polarizability of the molecule.

Substitution for the energy term and incorporating the oscillation of the source electric field allows that

$$E = E_0 \cos 2\pi v_0 t. \tag{2}$$

The molecular dipole moment can then be described as

$$P = \alpha E_0 \cos 2\pi v_0 t \tag{3}$$

where E_0 is the incident electric field, v_0 the frequency of light and t is time.

The polarizability, α, is influenced by the position of the nuclei in the molecule during vibrational motions and thus is dependent upon the interaction of vibrational modes that alter the induced dipole moment of the molecule. Expressing the position of the nuclei of a diatomic molecule as a normal coordinate series, the series is expanded to give the approximation:

$$\alpha = \alpha_0 + (\partial \alpha / \partial x)x + \dots \tag{4}$$

The normal mode of vibration is time-dependent with a frequency of v_1. This time-dependence is then expressed as

$$x_1 = x^0_1 \cos 2\pi v_1 t \tag{5}$$

Where x^0_1 is the equilibrium position of the internuclear distance.

Substitution of these terms into equation 3, use of a trigonometric identity for the product of cosines and rearrangement gives the source scattering as

$$P = (\alpha_0 E_0 \cos 2\pi v_1 t) + (\tfrac{1}{2})E_0 x^0_1 (\partial\alpha/\partial x)[\cos 2\pi(v_0 + v_1)t + \cos 2\pi(v_0 - v_1)t] \quad (6)$$

This equation provides three different terms for, and outlines the three possible results of, interaction with a scattering diatomic molecule. The α_0 term represents the light that does not interact with the molecule and the created dipole. The incident light is re-emitted with no change in frequency (wavelength) and is the Rayleigh (elastic) scattering. (Figure 1)

The term $(\partial\alpha/\partial x)$ expresses the change in polarizability as a function of internuclear distance of the diatomic nuclei. When $(\partial\alpha/\partial x)$ is non-zero, the remaining two terms give rise to two scattering bands displaced from the incident light by the vibrational energy, v_1, of the molecular motion which perturbs the induced dipole. These two Raman bands are a result of inelastic scattering, the vibrational energy of the molecular motion being added or subtracted from the wavelength of the incident beam. In turn, the polarizability and subsequent Raman activity of a vibrational mode are a direct result of the interaction of the molecular motion and the strength of that interaction on the source-induced dipole moment of the molecule.

If a vibrational mode produces a change in the polarizability of a molecule in an excited vibrational state, energy is released as the molecule transitions to a lower energy level during the scattering process, and the sample emits a photon of light of a frequency greater than that of the incident excitation wavelength (anti-Stokes). Conversely, molecules that transition from the ground vibrational state to the first excited energy level absorb energy from the incident beam during the scattering process, transition to the higher energy level and re-emit a photon of a frequency less than that of the excitation source (Stokes). The remaining terms of equation 6, $(\tfrac{1}{2})E_0 x^0_1 (\partial\alpha/\partial x)(\cos 2\pi(v_0 + v_1)t)$ and $(\tfrac{1}{2})E_0 x^0_1 (\partial\alpha/\partial x)(\cos 2\pi(v_0 - v_1)t)$, represent the anti-Stokes and the Stokes frequencies, respectively. (Figure 1)

As the ground vibrational state has the greater population of molecules at room temperature, the Stokes lines provide the most intense Raman scattering. (Figure 2) The Stokes portion of the entire Raman spectrum is usually reset such that frequency values correspond to the notation used for infrared spectra, with the laser wavelength set as zero and the Stokes lines plotted from 0 to 4000 cm^{-1} to correspond to infrared spectra. (Figure 3)

Initially, interpretation of Raman spectra is somewhat confusing to those more familiar with infrared spectroscopy as the two techniques may be (and often are) mutually exclusive. Frequently, infrared spectra demonstrate weak absorptions for molecular vibrations that result in intense Raman peaks and vice versa. A closer examination of the polarizability of the molecular bond can provide the key to the interpretation of Raman spectra because the polarizability, $(\partial\alpha/\partial x)$, of the molecular vibration determines a.) the selectivity of the Raman effect and b.) the intensity of the Raman band resulting from the molecular vibration.

The strongest absorption bands in an infrared spectrum result from those bonds or molecules that have a very strong, highly oriented dipole moment. When a specific stretching or bending motion occurs that significantly strengthens or weakens the oriented dipole, an infrared absorption of a specific wavelength corresponding to the difference between two adjacent vibrational energy levels is feasible. With the proper instrumentation and sample preparation, an infrared spectrum can be recorded that shows the molecular motions that produce the described change in the highly oriented dipole. At room temperature, the molecule vibrates constantly and the absorption process that occurs naturally is amplified and recorded by the instrumentation. Strong absorptions occur for vibrations of highly oriented dipoles, e.g., O–H, N–H or C=O. Little to no infrared absorption is noted for those vibrational motions that minimally alter the dipole moment (e.g., the C–C stretch of a straight-chain alkane) or vibrations that symmetrically distort the oriented dipole (the symmetric S–S stretching or symmetric O=C=O stretch of CO_2). Naturally, diatomic molecules such as N_2 or O_2 have no infrared absorption spectrum.

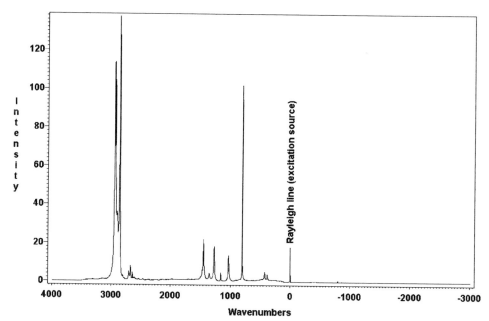

Figure 2. NIR FT-Raman spectrum of cyclohexane prior to final frequency scale correction. Note the drastic intensity difference between the Stokes and anti-Stokes Raman peaks. (The instrument filter system is a contributing factor.)

The most intense peaks in the Raman spectrum are a result of the molecular bonds or molecules that have a fairly diffuse dipole moment. With the lack of a strong orientation for an existing dipole moment, the source energy can easily create and begin oscillating an induced dipole moment in concert with the oscillation frequency of the source radiation. When a molecular vibration creates a significant distortion of the electron cloud disrupting the response of the induced dipole moment to the source, the conditions are prime for an interaction with the source energy.

Most of the time, the molecule absorbs energy from the source, is excited to a virtual level and immediately returns to the initial vibrational level, releasing a photon of the exact same wavelength as the laser, whatever frequency that happens to be. The result is Rayleigh scattering.

Occasionally (1×10^{-8}), the molecule will be excited to the virtual level and upon relaxation, the energy of a vibrational transition will either be added to (anti-Stokes), or subtracted from (Stokes) the energy of the source, resulting in Raman peaks. (Figure 2.) Although the vibrational movement occurs naturally, the observational environment must be created by subjecting the molecule to the dipole inducing oscillation of an intense, single-wavelength source, only then is the Raman spectrum observed. Raman peaks of fairly high intensity occur for vibrations of molecular bonds such as C–C, C–H, C=C and S–S. Highly oriented dipoles such as O–H, –P=O, C=O and the C–X (F, Cl, Br) bonds result in Raman peaks of moderately-weak to weak intensity. (Table 1)

Table 1. Comparison of Group Frequencies in Infrared and Raman Spectra

Structure	Frequency (cm^{-1})	Comments
OH	3300	Strong in IR; weak in Raman
NH, aliphatic	3300	Weak in IR; stronger in Raman
SH	2600	Weak in IR; strong in Raman
R-C≡C-R, R's equal	2200	Forbidden in IR; Strong in Raman
-C≡N	2200	Variable in IR; stronger in Raman
-C=O	1700	Strong in IR; medium in Raman
-C=C-	1640	Medium to absent in IR; strong in Raman
-P=O	1270	Strong in IR; weak in Raman
C-S	800-570	Weak in IR; strong in Raman
S-S	550-500	Weak in IR; strong in Raman

Reprinted with permission from Craver, C. D. in "Spectroscopic Methods in Research and Analysis of Coatings and Plastics" in Applied Polymer Science; Chapter 29; Tess, R. W., Poehlein, G. W. Editors; American Chemical Society, Washington, D. C.; 1985

As a general rule, infrared spectra highlight the functional groups of a molecule (OH, NH, CO, CH, CX, etc.) while Raman spectra offer greater information about the skeletal movements of the molecule. Sloane (15) gives examples of the group frequencies for which Raman offers advantages over infrared spectroscopy while Dollish, et. al. (16) have compiled extensive data of Raman group frequencies. Colthup et. al. (2, 3) treat infrared and Raman bands simultaneously. Table I outlines general comparisons of infrared and Raman spectral data. It should be recognized that the wavenumber values are listed in round numbers and that band positions and peak intensities may vary widely.

With molecules of high symmetry, infrared and Raman spectra are mutually exclusive and both methods are required to completely describe the vibrational motions of these molecules. (1, 2, 8, 11, 13, 16) Mutual exclusivity results in infrared and Raman spectra that contain peaks that are not found in the other spectrum. As an example, the infrared spectrum of cyclohexane has an absorption peak at 903 cm^{-1} for the asymmetric ring-stretching mode while the Raman spectrum has no peak in this area. Conversely, the Raman spectrum demonstrates the symmetric ring breathing mode at 802 cm^{-1} and no such vibrational motion is seen in the infrared spectrum. (Figure 3) The highly symmetric polyethylene structure also gives rise to mutually exclusive peaks in the infrared and Raman spectrum. In this case, the Raman spectrum shows the C–C stretching and C–C–C bending modes, these vibrational motions causing absorption peaks that are generally weak or absent in the infrared spectrum. (Figure 4) Additional examples of the complementary nature of infrared and Raman spectra include benzoic acid and dimethylaminopropyl methacrylamide (DMAPMA). (Figures 5 and 6)

Usually, however, the two techniques can be used interchangeably to determine qualitative and quantitative values for a wide range of analytical problems. Raman spectroscopy can be used to analyze most organic compounds and a wide variety of organometallic and inorganic complexes. Like infrared spectroscopy, Raman can be used to answer questions such as:

- "What is the sample?"
- "What is the concentration?"
- "Is this product identical to others?"
- "What happens if the physical or chemical environment of the molecule is changed?"

There are specific advantages to Raman spectroscopy with respect to sample handling and these advantages make Raman an attractive resource for the polymer laboratory:

- Raman is essentially a re-emission of the exciting source; samples of practically any size or shape may be examined as long as they can be exposed to the source and a detector can capture the Raman scattering.
- Very small (< 2 μm) samples can be examined with Raman microscopy systems, although the high intensity of the laser source may degrade the sample.
- Quartz fiber optics may be used for sampling.
- Aqueous solutions may be analyzed with little to no interference by water bands.

- Raman selection rules can be less restrictive than those for infrared spectroscopy, thus more information can be obtained for certain molecules.
- Glass or quartz cells and windows can be used to contain samples and/or used to construct sampling devices.
- Raman spectra can be collected in as little as 5 minutes or less.
- There is no rotational broadening of Raman peaks, Raman bands are generally very sharp and narrow, easily demonstrating spectral shifts due to molecular substitution.
- There is no interference from atmospheric H_2O and CO_2.
- Raman spectra do not require collection of an instrument 'background'.

With all of these advantages, Raman spectroscopy has not been a universal technique for chemical analysis mainly because of the fluorescence exhibited by a majority of samples. Often, the fluorescent component need only be present in low quantities to produce a strong signal, obscuring all or part of the Raman spectrum. To avoid fluorescence, using a different laser wavelength can sometimes reduce or even entirely diminish the fluorescence problems, but there are some compounds and sampling applications that simply cannot be examined using Raman spectroscopy.

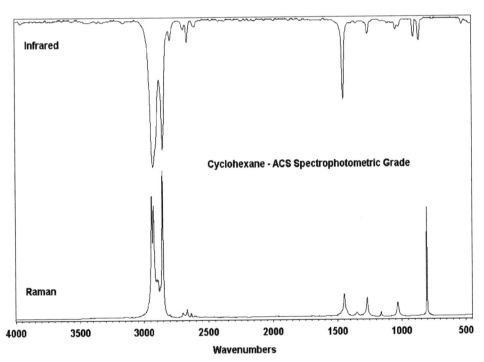

Figure 3. Infrared and Raman spectra of cyclohexane. Note the mutual exclusion of the asymmetric ring stretching mode at ~900 cm^{-1} in the Infrared and the ring breathing mode at ~800 cm^{-1} in the Raman.

Instrumentation Principles

The essential elements of the Raman experiment have remained much the same. (Figure 7) A monochromatic source of high intensity is focused onto the sample, the scattered radiation of various energies directed through some type of wavelength dispersion system and the intensities of the discrete wavelengths evaluated by a photon counting detector system. The instrument electronics are then used to display (or plot) a chart of scattering intensity versus wavenumbers from either side of the source wavelength. (Figure 2)

A primary objective for Raman instrumentation is the separation and/or extinction of the Rayleigh scattering signal from the Raman peaks. Secondly, a specified array of frequencies must be examined for detection of the relative intensities of the Raman scattering peaks. Since the Raman scattering is so weak compared to the Rayleigh signal ($\sim 10^{-6}$ to 10^{-9} of the Rayleigh), detectors used for Raman instruments must be extremely sensitive and are usually shielded in some manner from the high intensity of the Rayleigh scattering. Signal-to-noise limitations exist for all instrument designs and numerous instrument systems exist for the stimulation and detection of Raman scattering.

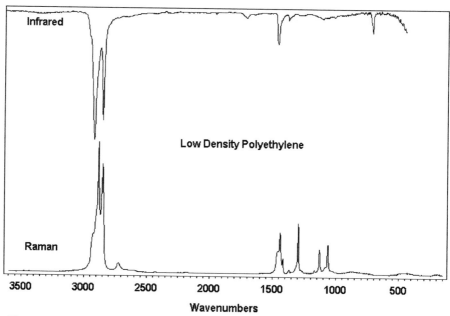

Figure 4. Infrared and Raman spectra of low density polyethylene. Note the mutual exclusion of some of the C-H stretching and bending modes in addition to the absence of C-C stretching modes in the infrared spectrum.

Infrared and Raman spectra courtesy of DigiLab, Sadtler Division.

The Laser Source

The development of the laser and its use as a Raman source solved a significant problem for Raman spectroscopy. A stable, coherent, monochromatic source of high intensity, the laser is a substantial improvement over the Hg arc source used prior to that time. The continued development of lasers – diode, gas, flash-lamp and dye – and the growing number of available excitation energies offers the ability to 'tune' the Raman excitation source to suit the particular analysis as required. If fluorescence is encountered when analyzing a specific chemical system, the laser is either changed or a new wavelength is dialed in and the experiment may proceed without interference from the fluorescence envelope. (Figure 8) In some cases, fluorescence cannot be avoided and the Raman spectrum cannot be collected.

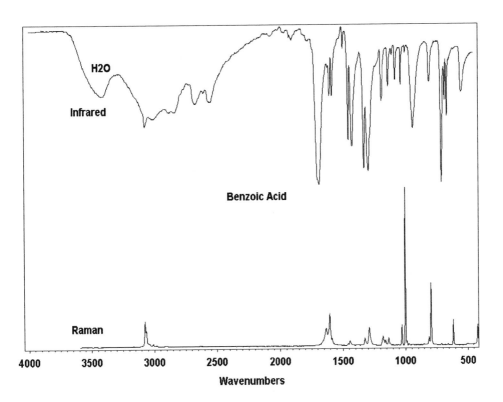

Figure 5. Infrared and Raman spectrum of benzoic acid. Note the intense infrared bands corresponding to the C=O and O–H functional groups and the corresponding lack of bands for the Raman spectrum. The strongest bands in the Raman spectrum correspond to the C–C skeletal modes.

Infrared and Raman spectra courtesy of DigiLab, Sadtler Division

Frequency (Wavelength) Dispersion Mechanisms

After collection of the Rayleigh and Raman scattering, some type of discrimination mechanism is used to resolve the separate wavelengths and allow detection of the intensity of the discrete Raman emissions. Prisms were used as a wavelength dispersion device prior to the development of grating systems; currently, the use of holographic replicates are the standard choice for Raman monochromator systems. Conventional Raman monochromator systems are comprised of a single, double or even triple set of grating(s), the additional gratings enhancing the stray-light rejection and resolving power of the Raman instrument. The monochromator system can be used to focus single frequencies onto a PMT or, a group of emission wavelengths can be directed onto a detector array. Numerous designs exist for monochromators and the grating systems used within. (2, 8-10, 14)

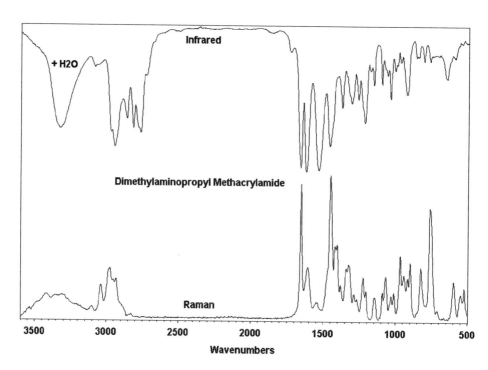

Figure 6. Infrared and Raman spectrum of dimethylaminopropyl methacrylamide (DMAPMA). The complementary nature of both techniques allows observation of all the vibrational modes of this compound in addition to providing a 'fingerprint' identification capability using either spectroscopic method.

Infrared and Raman spectra courtesy of DigiLab, Sadtler Division.

Instruments have also been developed that utilize acousto-optic tunable filter (AOTF) or liquid crystal tunable filter (LCTF) technology to provide a 'bandpass' for a selected range of wavelengths. These tunable filter systems allow a specified range of frequencies to reach the detector element(s) and reject those outside the transmission envelope, including the intense Rayleigh scattering of the laser source. The filter is 'tuned' to transmit the Raman bands of interest and additional filters can extend the range of laser excitation wavelengths. (17, 18) These types of filter 'monochromator' systems are extremely useful for quality control applications and are currently available in at least two commercial instruments.

Detector Devices

The detection of Raman scattering has evolved from the use of photographic plates to highly sensitive photo-multiplier tubes (PMT) or the latest in multi-channel charge-coupled devices (CCD) and intensified diode array detection systems.

The PMT is used as a 'single channel' detector; illuminated with a series of discrete wavelengths (ideally), or set of wavelengths, as the monochromator is 'scanned'. Resolution of the Raman spectrum is achieved by the wavelength discrimination of the grating(s) in the monochromator and the monochromator slit mechanism which defines the broadness of the allowed bandpass. The detector output is then digitized, recorded and plotted versus the scanned wavenumbers.

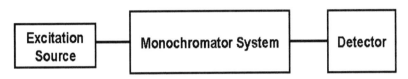

Figure 7. Schematic diagram of a Raman instrument.

With the exception of NIR FT-Raman, multi-channel detection is achieved with the use of diode array detectors or CCD's. In addition, a multi-channel detector uses a simpler monochromator system, usually only one or two grating sets. The Raman scattering is dispersed by the monochromator device across the span of detector elements (channels) and the signal intensity is integrated over a specified time. (19, 20) To reduce the intensity of the Rayleigh scattering, rejection filters (notch-filters) are sometimes used as an essential part of the Raman instrument. Notch filters 'reject' an extremely narrow set of frequencies corresponding to the laser (and Rayleigh) wavelength. Ideally, the rejection filter is specific to a certain laser frequency and will not transmit any energy at that frequency, providing a total extinction of the Rayleigh scattering. (21)

The frequency dispersion system, comprised of either grating(s), or an AOTF, LCTF or notch-filter and grating system, is used to disperse the Raman scattering across the elements of the diode array or the multi-channel CCD. Resolution for the multi-channel system is defined as a combination of the grating dispersion or the filter system bandpass in addition to the number of detector elements. Spectral resolution is sacrificed in multi-channel instruments (22), but the Raman scattering can be signal-averaged across the spectral range over a greater length of time as opposed to single channel detection systems. Higher resolution spectra are achieved by providing a greater dispersion of the scattered energy, then analyzing a multiple of smaller spectral ranges which are concatenated to reproduce the entire spectrum. (23) This gives a 'multi-channel' detection system, with the advantage of higher spectral resolution.

Avoiding Fluorescence

The application of Raman spectroscopy to the analysis of chemical problems is severely limited by the fluorescence of numerous samples, especially when using visible laser excitation (400 – 800 nm). Often, the fluorescence of these samples diminishes as the excitation wavelength approaches the NIR (1000 nm or, 1 μm). Shifting the excitation wavelength to the NIR can provide the ability to stimulate Raman scattering and decrease the possibility, or intensity, of the fluorescence signal. (Figure 8) Unfortunately, the ability to promote Raman scattering also decreases as a function of the excitation wavelength (~16 times less for 1.063 μm versus 514.5 nm excitation) as does the efficiency of PMT and diode array detector systems.

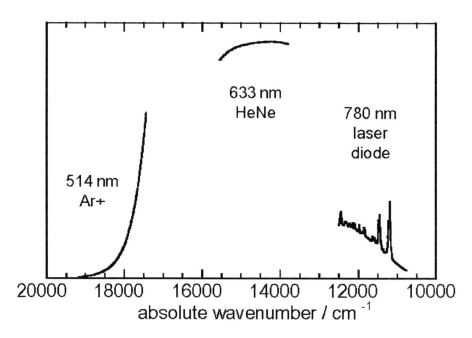

Figure 8. Raman spectrum of a diamond film demonstrating the presence of sample fluorescence with different visible excitation source frequencies. In this case, fluorescence cannot be totally avoided by changing the laser source.

Raman spectra courtesy of Renishaw.

NIR FT-Raman Instrumentation

Near-infrared Fourier Transform Raman spectroscopy (NIR FT-Raman) was demonstrated in a seminal article appearing in 1986, (24) the concept evolved to incorporate the reduction of fluorescence inherent to NIR excitation and the signal-averaging and throughput advantages of the optical interferometer. A Nd:YAG laser provides the NIR excitation, a quartz or other NIR beamsplitter is used in the interferometer and a NIR detector (usually, Indium Gallium Arsenide – InGaAs or Germanium - Ge) is used to detect and digitize the Raman interferogram. Dielectric notch filters are used to block, or reject, the Rayleigh line to prevent the intense (and noisy) scattering signal from being distributed evenly throughout the interferogram and resultant spectrum. Several FT-IR manufacturers offer the NIR FT-Raman capability as an 'accessory' for specific FT-IR models or even as a stand-alone instrument. The ability to collect the infrared and the Raman spectrum of the same material with the same instrument is extremely attractive and a distinct advantage over a stand-alone, visible excitation Raman instrument. However, fluorescence problems still exist for a number of chemical problems and darkly colored samples are often burned by the focused NIR laser beam.

Raman Microscope Systems

Raman microscopes are also manufactured for both dispersive and NIR FT-Raman systems. A natural extension of the Raman technique, the laser beam is focused to provide sample illumination on the μm scale and Raman scattering is collected with the instrument hardware to produce a spectrum of samples with a spatial resolution approaching 1-10 μm. Mapping of a sample surface is accomplished by examination of discrete sample sites and collection of multiple Raman spectra. Mapping of sample surfaces can provide structural data on a microscopic scale, similar to FT-IR microscopy mapping experiments.

Raman imaging experiments are used to produce a 'topographic' intensity map of the Raman response at a selected set of wavelength(s) versus the spatial position of the sampling area created as a result of a two-dimensional mapping of the sample. The variety of Raman microscope designs and microscopic methods provide the ability to routinely utilize Raman imaging and Raman microscopy for the investigation of polymer surface structure, or the defect and contaminant analysis of coatings, polymer films and laminates, etc.

There are numerous combinations of laser sources, monochromators and detectors used to construct a Raman instrument. The continued improvements of all technologies – laser, monochromator and detector – are reducing the complexity and cost of Raman instruments while broadening the application of Raman spectroscopy for analytical applications.

Applications for Polymer Analysis

There is a large amount of information that can be gained by using Raman spectroscopy for the analysis of polymers. Applications range from experimental confirmation of theoretical concepts to qualitative identification and/or quantitative analysis of co-polymers and polymer mixtures.   Kinetics studies of curing and polymerization can provide data to model these types of reactions in hopes of synthesizing new polymers with the desired properties. Raman spectroscopy can also provide the ability to examine the surface composition or conformational arrangement of polymer chains. Additionally, the simplicity of sample handling offers unique methods for the quality assurance/quality control (QA/QC) of both precursors and finished products.

Polymer Identification

A fundamental application for Raman spectroscopy has always been the analysis of the spectrum to elucidate the chemical structure of a molecule. Like infrared spectroscopy, the Raman spectrum can provide an exact identification of a specific chemical. Minor changes to the molecular structure usually result in a Raman spectrum that is clearly distinguishable from the spectrum of the original compound. Thus, Raman can be used to identify the presence of a specific chemical compound or mixture of compounds. Numerous texts containing collections of Raman spectra exist (3, 10, 16, 25) and the advent of NIR FT-Raman instruments has prompted the development of commercial spectral search libraries, a capability enjoyed by infrared spectroscopy for some time. More specifically, a recent study includes numerous FT-IR and NIR FT-Raman spectra of polymers. (26)

A unique application of the ability of Raman spectra to distinguish among chemical species is a study outlining the identification of several polymers as part of a post-consumer recycling program. (27)   The Raman spectra were collected for the various polymers and a repeatable identification using Raman spectra was demonstrated to be more reliable than a near-infrared method which was also evaluated in the paper. (Figure 9)

Structural Analysis

The complementary nature of infrared and Raman spectra allows the spectroscopist to realize all the vibrational motions of a specific molecule. With both infrared and Raman spectra in hand, it is possible to thoroughly analyze the vibrational motions of a molecule and assign these motions to the specific absorption and emission peaks present in the infrared and Raman spectra. (2, 3, 11, 28, 29) Although the influence of atomic or functional group substitution can drastically alter the spectrum of an original molecule, with the existing body of knowledge it is possible to interpret the spectral results of molecular substitution.   Numerous examples of the use of Raman spectroscopy to examine the structure of polymers are present in the literature and most of these include the vibrational assignment of the peaks in the Raman spectrum.

A very thorough discussion of the application of Raman spectroscopy for the analysis of the structure of crystalline polymers utilizes spectra collected for a series of linear polyethylenes. (30)   The structural assignment deduced as a result of the vibrational spectra allows the authors to refine a method for the calculation of the crystallinity for a majority of polymer systems.   The polyethylene series is fairly representative of most crystalline polymers and the proposed analytical model targets 3 specific areas of the Raman spectrum which demonstrate a quantifiable relationship to the crystallinity of the polymer. Another overview presented in the same text outlines the use of Raman spectra for the structural analysis of polydiacetylene polymers. (31) Raman spectra were collected at various excitation wavelengths ranging from the visible to the near-infrared for all of the polymers in the study. Raman peaks attributed to the polydiacetylene backbone were examined for the reaction products of diacetylene polymerization and related to the physical properties of the various polymers.

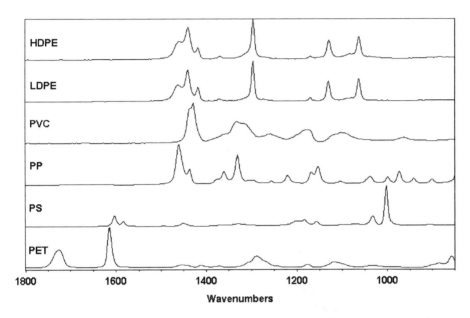

Figure 9. Raman spectra of common recyclable polymers. The unique spectral signature for each polymer allows a positive identification of each polymer with no sample preparation needed. The minor differences between the high and low-density polyethylenes is significant enough for differentiation.

Raman spectra courtesy of DigiLab, Sadtler Division.

A study of the vibrational data for a series of cyclic polysilanes including the infrared and Raman spectra and the corresponding vibrational assignment tables has been published. (32) The authors utilized many sources of information including the vibrational assignments for monosubstituted benzenes (33) in addition to comparisons amongst the spectra of the substituted analogs. The low-frequency Raman spectra of poly(methyl methacrylate) (PMMA) have been collected and the interpretation of the skeletal modes represented by these spectra has been discussed. (34) A similar treatment for the Raman spectra of amorphous poly(ethylene terephthalate) (PET) and additional interpretation of the spectrum of PMMA was presented. (35) A comparison of the low-frequency Raman spectra of amorphous and stretched PET films was discussed (36) and a complete characterization of the structure of propylene – ethylene copolymers was offered. (37) A complete normal coordinate analysis of the vibrational spectrum for polyethylene has been reported (38) in addition to structural studies of linear polyethylenes by factor analysis of the Raman spectra. (39)

Raman spectroscopy has also been used to study the interaction of other molecules with polymer systems or specific functional group activity. Raman spectra for the characterization of polarons, bipolarons and solitons in conducting polymers is reported (40) while the linear oligomers of poly(oxymethylene) with acetyloxy end groups were investigated with infrared and Raman spectroscopies. (41) The Raman spectra and associated structural studies of several liquid crystal polymers has been reported (42, 43) while hydrogen bonding in a series of polyamides has been investigated using Raman spectroscopy (44) in addition to an intensive examination of hydrogen bonding and resultant secondary structure for a series of commercial nylons. (45, 46)

The development of NIR FT-Raman spectroscopy has generated renewed interest into the application of Raman spectroscopy for the analysis of polymers. The reduction of fluorescence with the use of NIR laser excitation and the speed of FT-Raman data collection has prompted re-investigation of the Raman spectra of several polymer families. Examples include reports for the analysis of a series of nylons (47, 48), natural rubber (49), paints (50), a range of synthetic polymers (51), a selection of thin polymer films (52), and the characterization of the structure and properties of crystalline polymers (53).

Raman microspectroscopy has also been used to characterize polymers and these efforts have been extensively reviewed. (54, 55) Application of Raman microspectroscopy to the characterization of heterogeneous polymer blends has been reported (56) in addition to a study to differentiate the amorphous-oriented and crystalline components of PET fibers. (57) Raman spectra were used to characterize a series of γ-substituted propyl-trialkyloxysilanes (sol-gel precursors) while including the microspectroscopic analysis of a copolymer produced from two of the precursors. (58) In an attempt to clarify the vibrational assignment of the Raman spectra, the authors used polarization data to determine the vibrational motions that either retained or destroyed the existing plane(s) of symmetry of the molecule.

Quantitative Applications

A wide variety of analytical applications for Raman spectroscopy have been explored and there is great potential for the quantitative analysis of polymer systems. The relative peak intensity of Raman spectra can be directly related to concentration and this, in combination with sampling flexibility, has produced a number of application methods for the analysis of polymers and polymer chemistry. (8, 9, 13, 14, 25) Linearity of quantitative results is dependent upon the dynamic range of the detector system used in the Raman instrument, but generally, sampling conditions can be modified such that the intensities of the quantitative Raman peak(s) are within the linear range of the detector system. Multi-variate analysis techniques such as factor analysis or partial least-squares methods are especially useful because the spectral results of Raman spectroscopy are extremely reproducible. Examples of quantitative analysis systems are presented below.

The quantitation of polyethylene glycol is presented where very simple instrument components were used to accomplish the measurements. A fiber optic probe was used to transfer the laser excitation to a sample compartment and monochromator system equipped with a CCD used as the detector. (59) The authors used this simple instrument to analyze two sets of prepared standards ranging from 0-10% and 0-50%. With no changes to the experimental conditions, linear results were obtained for the 0-10% calibration curves, but some non-linearity was observed above 30% w/w concentration. Another quantitative analysis focused on the purity of ortho-xylene and the concentration of the common contaminants para- and meta-xylene. (60) The paper compares NIR and Raman analysis methods for the standards which were prepared to represent the bulk industrial purity of o-xylene. The Raman method proved to be much simpler to implement than the NIR and much easier to obtain reliable results. Another study demonstrated the quantitation of poly(dimethylsiloxanes) for turbid (highly scattering) samples (61) while NIR FT-Raman spectroscopy was used to develop a method for the quantitation of butadiene in styrene-butadiene copolymers. (62)

Polymerization Reactions

Raman spectroscopy can also be applied to the study of polymerization reactions and several examples are noted in the recent literature. The application of Raman spectroscopy to polymerization reactions is quite common and two excellent reviews of the technique are presented. (63, 64) The use of aqueous emulsions for polymerization reactions favors Raman analysis and the ability to study reactions in situ with fiber optic assemblies is particularly attractive.

An example of the monitoring of styrene concentration during a polymerization process using a fiber optic probe has been reported. (65) The aqueous emulsion was contained within a reaction vessel and the fiber optic probe was used to direct the laser excitation into the reactor with a subsequent transfer of the Raman scattering to a monochromator for dispersion onto an intensified diode array detector. Two studies detailing the explicit use of Raman spectroscopy and fiber optic probes for the monitoring of the emulsion polymerization of poly(vinyl acetate) are also described. (66, 67)

The emulsion polymerization of acrylate latexes has been well documented (68, 69) and the anionic polymerization of high-vinyl polybutadienes was reported (70) in addition to an FT-Raman analysis of the copolymerization of epoxy methacrylates. (71) The polymerization of styrene in a microemulsion system was studied (72) and the kinetics of the reaction process outlined. Another kinetics study investigated the photopolymerization of polyacrylamide. (73)

Curing processes are also of great interest and FT-Raman spectroscopy has been used to follow the curing of an epoxy resin. (74) The Raman spectroscopic results were favorably compared with a similar analysis utilizing NIR spectroscopy. Other references are available discussing the curing of divinyl ethers and epoxies (75-79) in addition to a unique study concerning the use of Raman spectroscopy to follow the progress of microwave curing. (80) A fiber optic probe was used to collect Raman spectra *in situ* of an amine-cured epoxide while the reaction occurred in a microwave oven. Raman spectroscopy with fiber optic probes for the *in situ* analysis of reaction systems has proven to be extremely useful and there are numerous applications for this type of sampling method.

Polymer Crystallinity Studies

The crystallinity of polymers can be directly related to the basic physical properties of the polymer and is of great interest as it can determine the flow characteristics during the thermo-mechanical processes used to produce manufactured goods from raw polymer material. The crystallinity of polymers can be directly related to the phase structure of the polymer system (81) and three distinct phase structures have been postulated:

- an ordered crystalline state
- a liquidlike, disordered region with an isotropic structure
- an interfacial region, involving chain units connecting the other two structures

Numerous analytical methods have been applied to understand the phase structure of polymers and while most of them contribute useful information, infrared and Raman spectroscopy have offered great sensitivity to the structural changes reflecting polymer crystallinity changes. In particular, Raman spectroscopy has been demonstrated to provide some of the most critical information as a result of the specificity for the polymer 'backbone' and the narrowness of the Raman peaks. The spectral changes reflecting the proportion of the three phase structures usually involve minor shifts in wavenumber and these shifts are often lost in the broadness of infrared absorption bands. Additionally, infrared spectroscopy often requires some method of sample preparation which may alter the crystalline structure of the polymer to be examined. Raman spectroscopy does not provide all of the data necessary to characterize the properties and structure of semi-crystalline polymers, but can be a very revealing probe of the degree of crystallization.

An excellent study (81) of numerous crystalline polymers was published as part of a volume on polymer characterization and utilized differential scanning calorimetry and Raman spectroscopy to examine the various polymer systems. The Raman spectra were used to determine the:

- phase structure internal modes
- longitudinal acoustical mode (LAM) representing the crystallite thickness distribution
- disordered longitudinal acoustic mode (D-LAM) as a measurement of the long-range conformational disorder.

Detailed explanations concerning the analysis of the Raman spectra and the modeling of the crystallinity can be found elsewhere (30, 81). Generally, polymer crystallinity can be related to the small wavenumber shifts of specific Raman peaks and the spectra are subsequently analyzed using multi-linear regression or principal component analysis methods.

A review of the application of Raman spectroscopy for the study of the phase structures of crystalline polyethylene was published (82) in addition to a more general review of Raman analysis of crystallinity. (83) The crystalline and amorphous phases of polycarbonate were studied with Raman spectroscopy (84) and a quantitative measurement of the crystallinity of poly(tetrafluoroethylene) was developed. (85) The use of vibrational spectroscopy for analysis of the crystallinity of poly(ethyl ether ketone) (PEEK) films was discussed (86) while the crystallization process in blends of PEEK and poly(ether imide) was studied using Raman spectroscopy. (87) A comparison of univariate and partial least-squares analysis of Raman spectra to model the crystallinity of PEEK polymers was also reported. (88)

The quantitative analysis of crystallinity and the effect of orientation on morphology were studied for high-density polyethylene (HDPE) (89) while morphology changes of high molecular weight polyethylene during melting and annealing processes were observed with Raman spectroscopy. (90) Factor analysis of Raman spectra produced a model for prediction of the crystallinity of linear polyethylene (91) and Raman spectroscopy was used for the on-line measurement of low-density polyethylene to provide guidance for the control of crystallinity during fabrication. (92)

A determination of the crystallinity of polylactide using NIR FT-Raman spectroscopy was published providing a comparison of univariate, multivariate and PLS methods of data analysis. The authors concluded that the PLS results provided the better model of the crystallinity of the polylactide compounds while outlining the advantages of the NIR FT-Raman experiment for polymer analysis. (93) NIR FT-Raman spectroscopy has also been used to study the cold crystallization of natural rubber and deproteinized natural rubber. (94)

Another natural application of Raman spectroscopy is the use of microspectroscopic systems for the analysis of polymer surfaces. A fiber-optic coupled Raman microprobe was used to provide a density mapping of poly(ethylene terephthalate) (PET) and PLS analysis was used to develop a crystallinity model based on the density measurements. (95)

Orientation and crystallinity measurements of uniaxially drawn PET fibers using polarized confocal Raman microspectroscopy was reported. (96) Confocal micro-spectroscopy allows a defined focal volume to be collected by the microscope optics (97, 98) and one can actually define the precise position of a sample to be examined. In the PET fiber study, confocal microspectroscopy was also used to collect Raman spectra throughout the depth of a film, providing a 'depth-profiling' of the PET polymer film. The depth-profiling advantages of Raman confocal microspectroscopy were also used to study polymer/polymer interfaces in laminates and the effect of changes in annealing temperatures. (99) Poly(acrylonitrile)/poly(acrylic acid) and poly(acrylo-nitrile)/poly(vinyl alcohol) laminates were annealed at specific temperatures and the Raman spectra of the laminate boundaries inspected for evidence of interdiffusion of the polymer systems.

Another study used a commercial Raman microscope system and an intermediate slit system to simulate confocal microscopy. (100) A defined area of approximately 1 μm in diameter was used to map the surface composition of several polymer blends including a rubber-toughened epoxy compound and a polyethylene-polypropylene copolymer. The spectral mapping results were then used to define a compositional map of the polymer blends.

Molecular structures can also reflect the results of stress and strain on the polymer system and measurements of the effects can be accomplished with Raman spectroscopy. The orientation and crystallinity of PET as a function of elongation has been reported (101) in addition to a study of the stretch-induced crystallization of PET films. (102) Rheo-optical analysis of PET films was performed using FT-IR and NIR FT-Raman spectroscopies. (103) The measurements were performed simultaneously and the infrared and Raman spectra demonstrated distinctive differences which are attributed to changes in the amorphous and crystalline phases of the polymer while being stretched.

Raman Imaging Experiments

Raman imaging consists of using a Raman microscope system and mapping the surface of a sample. The numerous spectra are analyzed to provide a correlation of spectral peak distribution to the existence of specific chemical compounds on the sample surface. Locating a specific peak that corresponds to the presence of a specific chemical bond of a molecule, the data can be analyzed and a two- or three-dimensional map of the intensity response of that peak is displayed by the instrument software. The intensity response represented by the spectral map is then used to define the presence or absence of a specific chemical on the sample surface. (104-106)

As examples, Raman imaging has been used to examine the surface of emulsion systems (107) and a series of glass-reinforced composites. (108) Another study examined the distribution of polyethylene and polystyrene in a copolymer using a Raman imaging system. (109) Principal component analysis (PCA) was used to evaluate the Raman spectra and the ability to distinguish among the discrete polymer molecules was demonstrated.

Non-linear Raman Techniques

Non-linear Raman spectroscopy such as surface-enhanced Raman spectroscopy (SERS), resonance Raman spectroscopy (RRS), coherent anti-Stokes Raman scattering (CARS) and stimulated Raman scattering (SRS) have limited application to polymer systems due to the particular requirements of these experiments. The non-linear experiments can provide enhancement of extremely weak Raman scattering, but the experimental design is complicated and the methods are not amenable to the majority of polymer samples. Non-linear Raman analysis methods can also help avoid the fluorescence problems associated with visible laser sources but the broader range of laser sources, the sensitivity of CCD detectors and the development of NIR FT-Raman instruments has reduced the interest in using non-linear Raman spectroscopies for the analysis of polymers. (9, 13, 14)

Conclusions

The development of NIR FT-Raman; laser sources; CCD and diode array detectors; and tunable filter systems have expanded the variety of Raman instrument systems. As a result, Raman spectroscopy has found a new home in many laboratories. Applications in all fields of chemical analysis have been reported and many more remain to be discovered.

The use of Raman spectroscopy to examine polymer systems can provide a wealth of information including:

- characterization of the vibrational activity and the determination of molecular structure
- an ability to identify and discriminate polymer compounds
- repeatable, simple methods for quantitative analysis
- the study of polymerization reactions
- the crystallinity of polymers

The sample handling advantages of Raman spectroscopy can outweigh even the most severe objection to the lack of universal applicability, especially when molecular structure data peculiar to Raman spectroscopy is desired. With an ability to examine samples *in situ*, Raman can reduce the sample handling effort to placement of a sample into the proper position for illumination by the laser beam. Raman spectroscopy has a very distinct advantage over every other instrumental method that requires sample preparation prior to analysis and offers a unique applicability for the evaluation of polymers and polymer chemistry.

Bibliography

1. Herzberg, G. "Infrared and Raman Spectra"; Van Nostrand Reinhold: New York, 1945.
2. Colthup, N., Daly, L., Wiberly, S. "Introduction to Infrared and Raman Spectroscopy"; 3rd Ed., Academic Press; New York; 1990.
3. Lin-Vien, D., Colthup, N., Fateley, W. G., Grasselli, J. G. "The Handbook of Infrared and Raman Characteristic Frequencies of Organic Molecules"; Academic Press; New York; 1991.
4. Lommel, E. Wiedemanns. Ann. Phys. 1878, 3, 251.
5. Smekal, A. Naturwiss 1923, 11, 873.
6. Raman, C. V., Krishnan, K. S. Nature 1928, 121, 501.
7. Venkataraman, G. "Journey into Light, Life and Science of C. V. Raman"; The Indian Academy of Sciences; Bangalore, India; 1978.
8. Brame, E. G., Grasselli, J. G. "Infrared and Raman Spectroscopy, Part A"; Marcel Decker; New York; 1976.
9. Grasselli, J. G., Snavely, M. K., Bulkin, B. J. "Chemical Applications of Raman Spectroscopy"; John Wiley and Sons; New York, 1981.

10. Freeman, S. K. "Applications of Laser Raman Spectroscopy"; Wiley-Interscience; New York; 1974.
11. Fateley, W. G., Dollish, F. R., McDevitt, N. T., Bentley, F. F. "Infrared and Raman Selection Rules for Molecules and Lattice Vibrations: The Correlation Method"; Wiley; New York, 1972.
12. Gardiner, D. J., Graves, P. R., Editors "Practical Raman Spectroscopy"; Springer-Verlag, Heidelberg, Germany; 1989.
13. Koenig, J. L. "Spectroscopy of Polymers"; American Chemical Society; Washington, D. C.; 1991.
14. Grasselli, J. G., Bulkin, B. J. "Analytical Raman Spectroscopy"; John Wiley and Sons; New York, NY; 1991.
15. Boerio, F. J., Koenig, J. L., Sloane, H., McGraw, G. E. In "Polymer Characterization: Interdisciplinary Approaches" Craver, C. D., Editor, Plenum, New York, 1971, Chapters 1-3.
16. Dollish, F. R., Fateley, W. G., Bentley, F. F. "Characteristic Raman Frequencies of Organic Compounds"; Wiley; New York; 1975.
17. Lewis, E. N., Treado, P. J., Levin, I. W. Appl. Spectrosc. 1993, 47, 539.
18. Morris, H. R., Hoyt, C. C., Treado, P. J. Appl. Spectrosc. 1994, 48, 857.
19. Harnly, J. M., Fields, R. E. Appl. Spectrosc. 1997, 51, 334A
20. Kim, M., Owen, H., Carey, P. R. Appl. Spectrosc. 1993, 47, 1780.
21. Everall, N. Appl;. Spectrosc. 1992, 46, 746.
22. Deckert, V., Kiefer, W. Appl. Spectrosc. 1992, 46, 322.
23. Panitz, J. C., Zimmermann, F., Fischer, F., Hafner, W., Wokaun, A. Appl. Spectrosc. 1994, 48, 454.
24. Hirschfeld, T., Chase, D. B. Appl. Spectrosc. 1986, 40, 133.
25. Chalmers, J., Everall, N. in "Polymer Characterisation" Chapter 3; Hunt, B. J., James, M. I. Editors; Blackie; Glasgow, U.K.; 1993.
26. Kuptsov, A. H., Zhizhin, G. N. "Handbook of Fourier Transform Raman and Infrared Spectra of Polymers"; Elsevier; Oxford, U.K.; 1998.
27. Allen, V., Kalivas, J. H., Rodriguez, R. G. Appl. Spectrosc. 1999, 53, 672.
28. Durig, J. R., Larsen, R. A. J. Mol. Struct. 1989, 238, 195.
29. Li, Y. S., Larsen, R. A., Cox, F. O., J. Raman Spectrosc. 1989, 20, 1.
30. Mandelkern, L., Alamo, R. G. "Structure Property Relations in Polymers: Spectroscopy and Performance", Advances in Chemistry Series No. 236; Chapter 5: M. W. Urban and C. D. Craver Editors; American Chemical Society, 1993.
31. Hankin, S. H. W., Sandman, D. J. "Structure Property Relations in Polymers: Spectroscopy and Performance", Advances in Chemistry Series No. 236; Chapter 8: M. W. Urban and C. D. Craver Editors; American Chemical Society, 1993.
32. Li, H., Butler, I. S., and Harrod, J. F. Appl. Spectrosc. 1993, 47, 1571.
33. Whiffen, D. H. J. Chem. Soc. 1965, 1350.
34. Papin, R., Rousseau, M., Durand, D., Nicolai, T. J. Raman Spectrosc. 1994, 25, 313.
35. Achibat, T., Boukenter, A., Duval, E., Lorentz, G., Etienne, S. J. Chem. Phys. 1991, 95, 2949.
36. Boukenter, A., Achibat, T., Duval, E., Lorentz, G., Beautemps, J. Polym. Commun. 1991, 32, 258.

37. Alamo, R., Mandelkern, L., Zachman, H., Stribeck, N. <u>Polym. Mater. Sci. Eng.</u> 1993, 69, 447.
38. Mohan, S., Prabakaran, A. R. <u>Asian J. Chem.</u> 1989, 1, 162.
39. Shen, C., Peacock, A. J., Alamo, R. G., Vickers, T. J., Mandelkern, L., Mann, C. K. <u>Appl. Spectrosc.</u> 1992, 46, 1226.
40. Furukawa, Y., Sakamoto, A., Ohta, H., Tasumi, M. <u>Synth. Met.</u> 1992, 49, 335.
41. Kobayashi, M., Matsumoto, Y., Ishida, A., Ute, K., Hatada, K. <u>Spectrochim. Acta, Part A</u> 1994, 50A, 1605.
42. Ellis, G., Lorente, J., Marco, C., Gomez, M. A., Fatou, J. G., Hendra, P. J. <u>Spectrochim. Acta, Part A</u> 1991, 47A, 1353.
43. Fontana, M. P., Antonioli, G., Jiang, Z., Angeloni, A. S. <u>Mol. Cryst. Liq. Cryst.</u> 1991, 207, 151.
44. Triggs, N., Valentini, J. <u>Isr. J. Chem.</u> 1994, 34, 89.
45. Brunsgaard, J., Christensen, D., Faurskov Nielsen, O. <u>Spectrochim Acta, Part A</u> 1993, 49A, 769.
46. Brunsgaard, J., Christensen, D., Faurskov Nielsen, O. J. <u>Chim. Phys. Phys.-Chim. Biol.</u> 1993, 90, 1797.
47. Hendra, P. J., Maddams, W. F., Royaud, I. A. M., Willis, H. A., Zichy, V. <u>Spectrochim. Acta, Part A</u> 1990, 46A, 747.
48. Maddams, W. F., Royaud, I. A. M. <u>Spectrochim. Acta, Part A</u> 1991, 47A, 1327.
49. Ellis, G., Hendra, P. J., Jones, C. H., Jackson, K. D. O., Loadman, M. J. R. <u>Kautsch Gummi Kundtst</u> 1990, 43, 118.
50. Ellis, G., Claybourne, M., Richards, S. E. <u>Spectrochim Acta, Part A</u> 1990, 46A, 227.
51. Agbenyega, J. K., Ellis, G., Hendra, P. J., Maddams, W. F., Passingham, C., Willis, H. A., Chalmers, J. <u>Spectrochim. Acta, Part A</u> 1990, 46A, 197.
52. Zimba, C. G., Turrell, S., Hallmark, V. M., Rabolt, J. F. J. <u>J. Phys. Chem.</u> 1990, 94, 939.
53. Mandelkern, L., Alamo, R. G. <u>Polym. Mater. Sci. Eng.</u> 1991, 64, 1.
54. Schaeberle, M., Karakatsanis, C., Lau, C. Treado, P. <u>Anal. Chem.</u> 1995, 67, 4316.
55. Meier,R., Kip, B. <u>Microbeam. Anal.</u> 1994, 3, 61.
56. Markwort, L., Kip, B. <u>J. Appl. Polym. Sci.</u> 1996, 61, 231.
57. Adar, F., Armellino, D., Noether, H. <u>Proc. SPIE-Int. Soc. Opt. Eng.</u> 1990, 1336, 182.
58. Posset, U., Lankers, M., Kiefer, W., Steins, H., <u>Appl. Spectrosc.</u> 1993, 47, 1600.
59. Melendez, Y., Schrum, K. F., Ben-Amotz, D. <u>Appl. Spectrosc.</u> 1997, 51, 1176.
60. Gresham, C. A., Gilmore, D. A., Denton, M. B. <u>Appl. Spectrosc.</u> 1999, 53, 1177.
61. Durkin, A., Ediger, M., Matchette, L., Petit, G. <u>Proc. SPIE-Int. Soc. Opt. Eng.</u> 1997, 2980, 217.
62. Lacoste, J., Delor, F., Pilichowski, J., Singh, R., Prasad, A., Silvaram, S. <u>J. Appl. Poly,. Sci.</u> 1996, 59, 953.
63. Hendra, P. <u>Vib. Spectrosc.</u>, 1993, <u>5</u>, 25.
64. Edwards, H., Johnson, A. Lewis, I. <u>J. Raman Spectrosc.</u> 1993, 24, 475.
65. Wang, C., Vickers, T. J., Schlenoff, J. B., Mann, C. K. <u>Appl. Spectrosc.</u> 1992, 46, 1729.

66. Dethomas, F., Hall, J., Grzbowski, D. Leaping Ahead Near Infrared Spectrosc. (Proc. Int. Conf. Near Infrared Spectrosc.) 6th, 1995, 364.
67. Sundell, T., Fagerholm, H., Crozier, H. Polymer 1996, 37, 3227.
68. Wang, C., Vickers, T. J., Mann, C. K. Appl. Spectrosc. 1993, 47, 928.
69. Claybourn, M., Massey, T., Highcock, J., Gogna, D. J. Raman Spectrosc. 1994, 25, 123.
70. Poshyachinda, S., Edwards, H., Johnson, A. Polymer 1991, 32, 338.
71. Sandner, B. Kammer. S., Wartewig, S. Polymer 1996, 37, 4705.
72. Feng, L., Ng, K. Colloids Surf. 1991, 53, 349.
73. Mailhot, G., Phillipart, J. L., Bolter, M. Polym. Commun. 1991, 32, 229.
74. Chike, K.E., Myrick, M. L., Lyon, R. E., Angel, S. M. Appl. Spectrosc. 1993, 47, 1631.
75. Nelson, E. W., Scranton, A. B. J. Raman Spectrosc. 1996, 27, 137.
76. Aust, J. F., Booksh, K. S., Stellman, C. M., Parnas, R. S., Myrick, M. L. Appl. Spectrosc. 1997, 51, 247.
77. Angel, S. M., Myrick, M. L. Proc. Annu. Meet. Adhes. Soc. 1996, 379.
78. Nelson, E. W., Scranton, A. B. IS&T's Annual Conference, The Society for Imaging Science and Technology Springfield, VA, 1996, 504.
79. Nelson, E., Scranton, A. Polym. Mater. Sci. Eng. 1995, 72, 413.
80. Stellman, C. M., Aust. J. F., Myrick, M. L. Appl. Spectrosc. 1995, 49, 392.
81. Mandelkern, L. in "Polymer Characterization: Physical Property, Spectroscopic and Chromatographic Methods", Advances in Chemistry Series No. 227; Chapter 22: C. D. Craver and T. Provder Editors; American Chemical Society, 1990.
82. Mandelkern, L., Alamo, R. Polym. Mater. Sci. Eng. 1996, 75, 8.
83. Stuart, B. Vib. Spectrosc. 1996, 10, 79.
84. Stuart, B. Polym. Bull. 1996, 36, 341.
85. Lehnert, R., Hendra, P., Everall, N. Polymer 1995, 36, 2473.
86. Damman, P., Fougnies, C., Moulin, J., Dosiere, M. Macromolecules 1994, 27, 1582.
87. Briscoe, B., Stuart, B., Rostami, S. Spectrochim. Acta, Part A 1993, 49A, 753.
88. Everall, N., Chalmers, J., Ferwerda, R., van der Maas, J., Hendra, P. J. Raman Spectrosc. 1994, 25, 43.
89. Rull, f., Prieto, A., Casado, J., Sobron, F., Edwards,. H. J. Raman Spectrosc. 1993, 4, 545.
90. Mutter, R., Stille, W., Strobl, G., J. Polym. Sci., Part B: Polym. Phys. 1993, 31, 99.
91. Shen, C., Peacock, A. J., Alamo, R. G., Vickers, T. J., Mandelkern, L., Mann, C. K. Appl. Spectrosc. 1992, 46, 1226.
92. Cakmak, M. Sehatkulu, F., Graves, M., Galay, J. Annu. Technol. Conf. – Soc. Plast. Eng. 1997, 55th, 1794.
93. Qin, D., Kean, R. T., Appl. Spectrosc. 1998, 52, 488.
94. Jones, C. H. Spectrochim. Acta, Part A 1991, 47(A), 1313.
95. Everall, N. J., Davis, K., Owen, H., Pelletier, M. J., Slater, J. Appl. Spectrosc. 1996, 50, 388.
96. Everall, N. J. Appl. Spectrosc. 1998, 52, 1498.
97. Tabaksblat, R., Meier, R. J. Kip, B. J. Appl. Spectrosc. 1992, 46, 60.

98. Williams, K. P. J., Pitt, G. D., Batchelder, D. N., Kip, B. J. Appl. Spectrosc. 1994, 48, 232.
99. Hajatdoost, S., Olsthoorn, M., Yarwood, J. <u>Appl. Spectrosc.</u> 1997, 51, 1784.
100. Garton, A., Batchelder, D. N., Cheng, C. <u>Appl. Spectrosc.</u> 1993, 47, 922.
101. Ellis, G., Roman, F., Marco, C., Gomez, M., Fatou, J. <u>Spectrochim Acta, Part A</u> 1995, 51A, 2139.
102. Boukenter, A., Archibat, T., Duval, E., Lorentz, G., Beautemps, J. <u>Polym. Commun.</u> 1991, 32, 258.
103. Hoffman, U., Pfeifer, F., Okretic, S., Volki, N., Zahedi, M., Siesler, H. W. <u>Appl. Spectrosc.</u> 1993, 47, 1531.
104. Schaeberle, M. D., Morris, H. R., Turner II, J. F., Treado, P. J. Anal. Chem. 1999, 71, 175A.
105. Schoonover, J. R., Weesner, F., Havrilla, G. J., Sparrow, M., Treado, P. J. Appl. Spectrosc., 1998, 52, 1505.
106. Morris, H. R., Hoyt, C. C., Miller, P., Treado, P. J. Appl. Spectrosc. 1996, 50, 805.
107. Andrew, J. J., Browne, M. A., Clark, I. E., Hancewicz, T. M., Millichope, A. J. <u>Appl. Spectrosc.</u> 1998, 52, 790.
108. Stellman, C. M., Booksh, K. S., Myrick, M. L. <u>Appl. Spectrosc.</u> 1996, 50, 552.
109. Drumm, C. A., Morris, M. D. Appl. Spectrosc. 1995, 49, 1330.

NMR CHARACTERIZATION OF POLYMERS

PETER A. MIRAU

Bell Laboratories, Lucent Technologies
600 Mountain Ave.
Murray Hill, NJ 07974

Introduction

NMR spectroscopy is an important analytical method that is extensively used to study the structure and properties of macromolecules. The first studies of polymers were reported by Alpert (1) only about a year after the discovery of nuclear resonance in bulk matter (2, 3). It was observed that natural rubber at room temperature gives a proton line width more like that of a mobile liquid than of a solid, but that the resonance broadens at temperatures approaching the glass transition temperature. This was recognized as being related to a change in chain dynamics. NMR methods developed rapidly after these initial observations, both for polymers in solution and in the solid state.

Solution NMR has emerged as one of the premier methods for polymer characterization because of its high resolution and sensitivity. It was observed in the early studies that the chemical shifts are sensitive to polymer microstructure, including polymer stereochemistry, regioisomerism and the presence of branches and defects. These observations led to an improved understanding of polymer microstructure and polymerization mechanisms. With the advent of higher magnetic fields and improved NMR methods and spectrometers, it has become possible to characterize even very low levels of defects in polymer chains. The NMR spectra are sensitive to the atomic level structure and many signals are observed in high resolution spectra. The chemical shift assignments in the early studies were established by comparing the spectra to those of model compounds. More recently the development of spectral editing methods and multidimensional NMR have made it possible to assign the spectra without resorting to model compounds. The assignments have been established not only for carbons and protons, but also for any silicon, nitrogen, phosphorus or fluorine atoms that may be present. The detailed microstructural characterization has led to a deeper understanding of polymer structure-property relationships.

The NMR analysis of polymers in the solid state emerged after the solution studies because solid state NMR spectra are more difficult to acquire and require more equipment and expertise. Solid state NMR has become such an important method that most modern spectrometers are capable of performing these studies. The interest in the solid state NMR of polymers is due in part to the fact that most polymers are used in the solid state, and in some cases the NMR properties can be directly related to the macroscopic properties. Solid state NMR provides information about the structure and dynamics of polymers over a range of length scales and time scales. Polymers have a restricted mobility in solids, and the chemical shifts can be directly related to the chain conformation. Solid state NMR is also an efficient way to monitor the reactivity of polymers, since the chemical changes give rise to large spectral changes. The relaxation times in solids depend not only the chain dynamics, but also on the morphology over a length scale of 20-200 Å. NMR has been extensively used to measure the length scale of mixing in blends and multiphase polymers, and the domain sizes in semicrystalline polymers. Solid state NMR methods have been greatly expanded with the introduction of multidimensional NMR (4). These studies have led to a molecular level understanding of the dynamics traditionally observed by dielectric and dynamic mechanical spectroscopy, and a better understanding of the relationship between polymer morphology and macroscopic properties.

Fundamental principles

The NMR phenomenon is possible because in addition to charge and mass, many isotopes possess spin, or angular momentum. Since a spinning charge generates a magnetic field, there is a magnetic moment associated with this angular momentum. According to a basic principle of quantum mechanics, the maximum experimentally observable component of the angular momentum of a nucleus possessing a spin (or of any particle or system having angular momentum) is a half- integral or integral multiple of h/2π, where h is Planck's constant. This maximum component is I, which is called the spin quantum number or simply "the spin." In general there are $2I + 1$ possible orientations or states of the nucleus. For spin ½ nuclei the possible magnetic quantum numbers are +½ and -½.

When placed in a magnetic field, nuclei having spin undergo precession about the field direction. The frequency of this so-called Larmor precession is designated as ω_0 in radians per second or υ_0 in Hertz (Hz), cycles per second ($\omega_0 = 2\pi\upsilon_0$). The nuclei can be made to flip over, i.e. reverse the direction of the spin axis, by applying a second magnetic field, designated as B_1, at right angles to B_0. The Larmor precession frequency or resonance frequency is given by

$$\omega_0 = \gamma B_0 \tag{1}$$

where γ is the magnetogyric ratio. The two quantities that determine the observation frequency for NMR signals are the magnetogyric ratio γ and the magnetic field strength B_0. Table 1 list some of the important nuclear properties of spins that are of interest to polymer chemists. The sensitivity depends both on the magnetogyric ratio and the natural abundance of the NMR active nuclei. Protons have the highest sensitivity because they have the highest magnetogyric ratio and natural abundance. At a field strength of 11.7 T (1 Tesla=10^5 Gauss) the NMR signals are observed at 500 MHz. Fluorine is the second most sensitive nuclei, but it is not a common element in polymers. Most polymers of interest contain carbon, and Table 1 shows that the sensitivity is very low compared to that of protons. However, the sensitivity of a modern NMR spectrometer is such that carbon spectra can be routinely observed. Nitrogen is also a common element in polymers but it is difficult to study because of its low magnetogyric ratio and natural abundance. ^{15}N NMR studies are possible, but only after isotopic labeling. The sensitivity of silicon and phosphorus is intermediate between that of protons and carbons, and provides a valuable probe of those polymers containing these elements.

The NMR spectrum is usually observed by applying a radio frequency (rf) pulse near the resonance frequency and observing a free induction decay (FID). The NMR spectrum, a plot of intensity vs. frequency, is obtained by Fourier transformation of the observed signal. The signal frequency (or chemical shift) is reported relative to some reference compound. The integrated signal intensity is proportional to the number of nuclei. This property is important because it allows us to use NMR as a quantitative tool.

The splitting of the energy levels in the presence of the magnetic field leads to a population difference between the upper and lower levels as determined by the Boltzman distribution. When the spin system is placed in a nonequilibrium position by

Table 1. Nuclear properties of interest in polymer science.

Isotope	Abundance (%)	Spin	Sensitivity[a]	Frequency (MHz)[b]
^1H	99.98	½	1.0	500.
^{19}F	100.0	½	0.83	470.2
^{29}Si	4.7	½	0.078	99.3
^{31}P	100.0	½	0.066	202.3
^{13}C	1.1	½	0.0159	125.6
^2H	0.015	1	0.00964	76.7
^{15}N	0.365	½	0.001	50.6

[a] The sensitivity relative to protons.
[b] The resonant frequency in a 11.7 T magnetic field.

the application of radio frequency pulses to interchange the populations of the upper and lower levels, the spin system relaxes toward equilibrium. This is termed *spin-lattice* or longitudinal relaxation and is designated by the symbol T_1. The relaxation times provide information about the molecular dynamics of polymers since they depend on the rate and amplitude of atomic fluctuations. The time scale of motion is related to the rotational correlation time τ_c. Chain dynamics are restricted in solids and the relaxation times can be very long.

A related property is the *spin-spin* or longitudinal relaxation time T_2 that is proportional to the inverse of the linewidth. As the chain motion is restricted, the linewidths become broader, limiting the resolution that can be obtained. Since the chain motion is a thermally activated process it is sometimes possible to acquire the spectrum at higher temperature to reduce the linewidth.

Solution NMR studies

Solution NMR is extremely useful for polymer characterization because the chemical shifts and relaxation times are sensitive to polymer microstructure. High resolution spectra were initially not expected since it is well known that the line widths depend on the third power of the diameter for a rigid sphere reorienting in solution (5). The means that sharp lines are expected for small molecules, but not for high molecular weight polymers. Fortunately, the relaxation in most polymers is not due to chain reorientation, but rather to more localized segmental motions such as *guache-trans* isomerization and librational motions. Since these fluctuations are usually in the fast motion limit ($\tau_c < 1$ ns), sharp lines are observed for high molecular weight polymers in solution.

High resolution spectra are observed because the linewidths are less than the chemical shift variations due to the polymer chain chemical structure and microstructure. The chemical shift resolution results from variations in nuclear shielding arising from variations in electron density that are affected by the presence of nearby electron withdrawing or donating groups. The chemical shift changes can be related to through-bond inductive effects and through-space interactions. The largest effects on the chemical shifts are the chemical type, and the resonances from methyl, methine, methylene and aromatic groups are often well separated in the NMR spectrum.

Carbon NMR has been extensively used for the solution characterization of polymers because it is one of the most common elements in polymers. Carbon has a low natural abundance (1% ^{13}C) but most polymers are sufficiently soluble that it is possible to record their NMR spectrum at high concentration ([C] > 0.05 M). The carbon spectra are often more highly resolved than the proton spectra because the carbon chemical shifts are spread over 200 ppm, rather than the 10 ppm commonly observed for protons. Nitrogen, silicon and fluorine also have large chemical shift ranges.

The NMR spectrum is also affected by the spin state of nearby atoms. For a carbon with a directly bonded proton, the local field it experiences will depend on whether the proton is aligned with or against the external magnetic field. Since there is an almost equal population of the proton spin states, the carbon resonance will be split into two signals. This phenomenon is known as scalar or J coupling. Scalar couplings provide important information about the number of attached protons and the identity of nearby groups. This information can be used to distinguish between a methine carbon with one attached proton that appears as a doublet and a methylene carbon with two attached protons that appears as a triplet. The number of lines is given by 2n+1 where n is the number of attached protons.

While the scalar couplings provide information about the number of attached protons, they also make the spectrum more complex. It is often desirable to suppress the coupling information by irradiating the protons during acquisition to obtain a spectrum with the highest resolution.

In a similar way the three bond couplings between pairs of protons can also give rise to splittings. However, since the protons are separated by three bonds, the proton-proton scalar couplings are smaller (2-15 Hz) than the one bond carbon-proton couplings (120-160 Hz). In addition, the proton-proton couplings contain conformational information since they depend on the torsional angle separating the pairs of protons (5, 6). These couplings are often valuable for establishing the resonance assignments using 2D NMR methods.

Solid state NMR

The NMR spectra of solids are fundamentally different from solutions because atomic motions are suppressed by neighboring chains. This means that the local interactions, such as the chemical shift anisotropy and dipolar couplings, are not averaged by molecular motion and some artificial means must be used to obtain a high resolution spectrum. It is difficult to record the NMR spectra of solids using the simple one-pulse experiment since the relaxation times for solids can be very long and it is necessary to wait for the spin system to return to equilibrium between pulses.

The chemical shift is actually *anisotropic* or directional, as it depends on the orientation of the molecule with respect to magnetic field direction. It is expressed as a tensor, a mathematical quantity having both direction and magnitude, and is composed of three principal components, σ_{ii}

$$\sigma = \lambda_{11}^2 \sigma_{11} + \lambda_{22}^2 \sigma_{22} + \lambda_{33}^2 \sigma_{33} \tag{2}$$

where λ_{ii} are the direction cosines of the principal axes of the screening constant with respect to the magnetic field. The principal axis systems may lie along the bond direction, but this is not necessarily so. The orientation of the axis system cannot be predicted *a priori*, and must be experimentally determined (7). By convention, the lowest field resonance is taken as σ_{33}. In solution, the chemical shift anisotropy is averaged by rapid molecular motion and the chemical shift is given by

$$\sigma = \frac{1}{3}(\sigma_{11} + \sigma_{22} + \sigma_{33}) \tag{3}$$

In molecules of any degree of complexity there will an anisotropy pattern for each carbon atom. These patterns typically overlap, producing a broad, unresolved spectrum. Under these circumstances it is desirable to sacrifice the anisotropy information in order to observe a high resolution spectrum. Under rapid sample rotation the orientations and chemical shifts become time dependent. The time average under rapid rotation is given by

$$\sigma = \frac{1}{2}\sin^2 \beta(\sigma_{11} + \sigma_{22} + \sigma_{33}) + \frac{1}{2}(3\cos^2 \beta - 1)$$

(4)

where β is the angle between the rotation axis and the magnetic field direction. When β is equal to the so-called magic angle (54.7°, or the body diagonal of a cube), $\sin^2\beta$ is 2/3 and the first term becomes equal to one-third of the trace of the tensor (i.e., the isotropic chemical shift) and the $(3\cos^2\beta\text{-}1)$ term is equal to zero. Thus, under magic angle rotation the chemical shift pattern collapses to the isotropic average, giving the high resolution spectrum.

The second factor leading to line broadening in solids is the dipolar couplings between nearby atoms. The local field experienced by a proton in the solid from a nearby proton is given by

$$\Delta E = 2\mu[B_0 \pm B_{loc}] = \left[B_0 \pm \frac{3}{2}\mu r^3 (3\cos^2 \theta - 1) \right]$$

(5)

where μ is the nuclear moment, r is the distance separating the nuclei and θ is the orientation of the vector connecting the spin pair relative to the magnetic field. This local field B_{loc} can be as large as 50 kHz, which is much greater than the proton chemical shift range. For an isolated pair of protons the spectrum would appear as a doublet. It is more common, however, for the proton to experience a number of interactions at different distances and orientations, resulting in a broad featureless line.

To observe high resolution spectra in solids it is necessary to remove the linebroadening from the dipolar couplings. In the carbon spectrum this is accomplished by high power proton irradiation to average the dipolar couplings. The strength of the proton rf field must be greater than the magnitude (50 kHz) of the dipolar couplings.

Because atomic motion is restricted, the spin-lattice relaxation can be very slow in solids, making it difficult to acquire a spectrum in a reasonable amount of time. We can overcome this limitation using the *cross polarization* pulse sequence shown in Figure 1. Cross polarization works by forcing the proton and carbon signals to precess at the same frequency in the rotating frame even though they do not have the same frequency in the laboratory frame. The means of doing this was demonstrated by Hartmann and Hahn in 1962 (8), when it was shown that energy transfer between nuclei with widely differing Larmor frequencies can be made to occur when

$$\gamma_C B_{1C} = \gamma_H B_{1H}$$

(6)

the so-called *Hartmann-Hahn condition* is satisfied. Since γ_H is four times γ_C, the Hartmann-Hahn match occurs when the strength of the applied carbon field B_{1C} is four times the strength of the applied proton field B_{1H}. When the proton and carbon rotating frame energy levels match, polarization is transferred from the abundant protons to the rare carbon-13 nuclei. Because polarization is being transferred from the protons to the

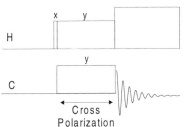

Figure 1. The pulse sequence diagram for cross polarization.

carbons, it is the shorter T_1 of the protons that determines the repetition rate for signal averaging.

The cross polarization pulse sequence begins with a 90_x pulse to the protons to tip them along the y' axis. The phase of the proton B_{1H} field is then shifted by 90° and the protons are *spin-locked* along the y' axis where they precess about the y' axis with a frequency $\omega_H = \gamma_H B_{1H}$. At the same time, the carbons are put into contact with the protons by turning on the carbon field B_{1C}. This causes the carbon magnetization to grow up in the direction of the spin-lock field.

The signal intensity during cross polarization represents a compromise between several processes, including the buildup of intensity from carbon-proton cross polarization and the decay of the proton and carbons magnetization in the spin locking fields. The intensity at cross polarization contact time t is given by

$$M = \frac{M_o}{T_{1C}} \cdot \frac{e^{-t/T_{1\rho}(H)} - e^{-t(1/T_{CH} + 1/T_{1\rho}(C))}}{1/T_{CH} + 1/T_{1\rho}(C) - 1/T_{1\rho}(H)} \tag{7}$$

where T_{CH} is the carbon-proton cross polarization time constant, and $T_{1\rho}(C)$ and $T_{1\rho}(H)$ are the carbon and proton rotating frame spin-lattice relaxation time constants. In most cases $T_{1\rho}(C) >> T_{CH}$ so Equation (7) simplifies to

$$M = \frac{M_o}{T_{1C}} \cdot \frac{e^{-t/T_{1\rho}(H)} - e^{-t/T_{CH}}}{1/T_{CH} - 1/T_{1\rho}(H)} \tag{8}$$

The time course of the magnetization is a rapid buildup in signal intensity from cross polarization followed by a slower decay due to $T_{1\rho}(H)$. The signal intensity depends on the relaxation rate constants that depend on the chain dynamics. It is often observed that the buildup and decay of magnetization is different for crystalline, amorphous and rubbery polymers, and cross polarization is one way to separate the signals from polymers with different chain dynamics.

Experimental methods

Solution NMR studies of polymers provide quantitative information about polymer microstructure. Sample preparation is relatively simple since most polymers are soluble in common organic solvents that can be purchased with deuterons in place of protons to facilitate the proton NMR analysis and to provide a frequency lock for the NMR spectrometer. The high solubility of polymers in organic solvents makes it possible to routinely obtain spectra for insensitive nuclei such as ^{13}C. In some cases the solutions of high polymers are very viscous and it is advantageous to obtain the spectra at a higher temperature.

Quantitative proton NMR spectra can be obtained by waiting five times T_1 between scans. This is usually not a limitation because the signal-to-noise ratio is high for protons and the relaxation times are on the order of a few seconds. However, this can be limitation in the carbon, silicon and nitrogen spectra where the relaxation times can be very long and many scans are required to obtain a spectrum with a high signal-to-noise ratio.

In addition, the signals for carbon, silicon and nitrogen atoms that are relaxed by nearby protons can be affected by the proton decoupling used to remove the scalar couplings. This is called the nuclear Overhauser effect (NOE) (9). Proton irradiation can change the signal intensities in a way that depends on the correlation time. Therefore it is possible to have a larger NOE for the more mobile side chain atoms relative to the more restricted main chain atoms. To avoid this possibility the spectra are acquired using gated decoupling, where the decoupler is only turned on during the acquisition period. The NOE leads to a large increase in signal intensity so the signal-to-noise ratio is higher in spectra gathered with decoupler irradiation during the entire pulse sequence.

The spectra in solids are often acquired with cross polarization. This has the advantage that they can be acquired more rapidly because it is necessary only to wait for proton relaxation between scans (which is usually on the order of 1-5 s). However, the signals in the cross polarization spectra depend on the chain dynamics (Equation 8), making the quantitative analysis difficult.

Chain motion is restricted in solids and the lines are broadened by the combination of chemical shift anisotropy and dipolar couplings. For many nuclei, such as carbon silicon and nitrogen, the chemical shift anisotropy is on the order of a few kHz. This is in the range that can be averaged by sample spinning at the magic angle ($\beta=54.7°$, Equation (4)). If the anisotropy is larger than the spinning rate (3-5 kHz) then the anisotropy pattern will be broken into a number of sharp sidebands. These sidebands can be identified by changing the spinning speed or removed from the spectrum using the TOSS sideband suppression pulse sequence (10). Rapid magic angle spinning (35 kHz) is now possible, but these experiments are most useful for proton and fluorine experiments since the rapidly spinning rotors have a small sample volume.

One of the fundamental limitations in polymer NMR studies is signal overlap resulting from the repeating sequence nature of polymer chains. In many cases the signal from the feature of interest may not be resolved from the main chain resonances. One way to overcome these limitations is to expand the frequency information into two or more frequency dimensions (nD NMR). There are many kinds of nD NMR experiments that can be categorized as (a) *correlated* or (b) *resolved* experiments (6, 11, 12). In correlated experiments the resonance frequency of one signal is related to those of its neighbors and molecular connectivities or distances between atoms can be determined. In resolved experiments the frequency axes show two different interactions.

In one kind of resolved experiment, for example, the carbon chemical shift may appear along one axis and the proton linewidth along the other (13). Extension of these same principles leads to 3D NMR experiments in which there are three independent frequency axes (14). This has the potential to improve the resolution in those 2D spectra where signal overlap is still a problem.

Polymer characterization in solution

Polymer characterization in solution has emerged as an important analytical method because many microstructural features can be observed in the high resolution spectra. The first step in the analysis of polymers is establishing the resonance assignments. This is a very important task and much effort has been directed towards developing new methods to make the assignments. The methods include comparison of the signal intensities with those expected for polymerization models (Bernoullian, Markov, etc.), comparison with model compounds, chemical shift calculations and multidimensional NMR.

Solution NMR has contributed to our understanding of polymerization mechanisms, and this in turn has helped with the resonance assignments in polymers. If the polymerization mechanism conforms to Bernoullian statistics, then the fraction of triads will be given by

$$(mm) = P_m^2 = (m)^2$$
$$(rm) = 2P_m(1-P_m) = 2(r)(m) \tag{9}$$
$$(rr) = (1-P_m)^2 = (r)^2$$

where P_m is the probability that the incoming monomer will add in a *meso* fashion to the end of the growing chain. Thus, the signal intensity for a given triad will depend on P_m. Table 2 compares the expected and observed intensities for poly(methyl methacrylate) polymerized via free radical polymerization (15). It is difficult to establish the peak assignments solely by intensity comparisons, but such comparisons can aid in the resonance assignments.

The comparison of the polymer spectra with the spectra from model compounds and chemical shift calculations are additional methods for making resonance assignments. This can be illustrated by considering 2-methyl hexane as a model compound for endgroups in polyethylene. If this compound is available we can measure the spectrum and compare the peak positions with those observed in the polymer to identify the possible endgroups. If the model compound is not available, we can compare the spectrum with the calculated values for the chemical shifts (16-19). A large number of organic compounds have been measured and empirical correlations have been established relating the chemical structure to the NMR spectrum. The carbon chemical shift is given by

$$\delta_C = B + \Sigma A_i n_i + \Sigma S_i \tag{10}$$

Table 2. Comparison the peak intensities for poly(methyl methacrylate) from a free radical polymerization with the intensities calculated from Bernoullian statistics.

	Stereochemistry	Observed	Bernoullinan Trial P_m=0.24
Dyad	*(m)*	0.24	
	(r)	0.76	
Triads	*(mm)*	0.04	0.06
	(mr)	0.36	0.36
	(rr)	0.60	0.58
Tetrads	*(mmm)*	~0	0.01
	(mmr)	0.07	0.09
	(rmr)	0.19	0.20
	(mrm)	0.04	0.04
	(mrr)	0.23	0.23
	(rrr)	0.43	0.44

where B is the chemical shift of methane (-2.3 ppm), n_l is the number of carbons at position I away from the carbon of interest, A_l is the additive shift due to carbon I and S_l is a term included to account for branching. The values for A_l and S_l are available in extensive tables. (19) The calculated chemical shift for the methine carbon in 2-methyl hexane is 39.1 ppm, which compares favorably with the observed value of 39.45 ppm.

The chemical shift assignments are also aided by the γ-*gauche* effect (20, 21). The γ-*gauche* effect is conformational in origin and arises from the proximity of a given carbon with its γ neighbor. The magnitude of the γ-*gauche* shift depends on the substituent and Table 3 lists the shifts expected for several types of substituents. Calculating the γ-*gauche* effect involves calculating the relative energies of the *gauche* and *trans* states, construction of a rotational isomeric state (RIS) model, calculation of the conformational probabilities and summing up all of the γ-*gauche* interactions (20, 21).

More recently 2D and 3D NMR have become important tools for establishing the resonance assignments (6, 14). These are powerful methods because they can be used to correlate both nearest neighbor and next nearest neighbor groups along the polymer chain. By correlating larger and larger chain segments, higher order sequences can be assigned without reference to model compounds.

There are many types of 2D experiments that use both through-space and through-bond correlations. Table 4 lists a few of the most useful experiments for establishing polymer assignments. COSY and TOCSY utilize three bond scalar

Table 3. The chemical shifts induced by the γ-gauche effect for several groups.

Group	$\Delta\delta_c$ (ppm)
-CH$_3$	-5.2
-OH	-7.2
-Cl	-6.8

Table 4. 2D and 3D NMR experiments of interest for the analysis of polymers.

Experiment[a]	Nuclei	Correlation
COSY	H-H	$^3J_{HH}$
TOCSY	H-H	$^3J_{HH}$ + relay
NOESY	H-H	r^{-6}
ROESY	H-H	r^{-6}
HMQC	C-H	$^1J_{CH}$
HMBC	C-H	$^2J_{CH}, ^3J_{CH}$
NOESY-HMQC	C-H, H-H	$^1J_{CH}, r^{-6}$
TOCSY-HMQC	C-H, H-H	$^1J_{CH}, ^3J_{HH}$ + relay
J-Resolved	C-H or H-H	$^1J_{CH}$ or $^3J_{HH}$

[a] The acronyms are for COrrelated SpectrsocospY (COSY), TOtal Correlation SpectrsocopY (TOCSY), Nuclear Overhauser Effect SpectrsocopY (NOESY), Rotating-frame Overhauser Enhancement SpectrsocpY (ROESY), Heteronuclear Multiple-Quantum Coherences spectrsocopy (HMQC) and Heteronuclear Multiple-Bond Correlation spectroscopy (HMBC).

couplings ($^3J_{HH}$) between nearby protons for magnetization exchange. In the 2D correlated spectrum of a vinyl polymer, for example, it may be possible to correlate the methine and methylene protons in several tetrads. The *mm* triad, for example, may be correlated with the *mmm* and *mmr* tetrads. In TOCSY and NOESY spectra the magnetization transfer is not restricted to directly coupled protons. This leads to not only the direct correlation observed in the COSY spectrum, but also more remote cross peaks such as those between the *mm* and *mr* methine peaks that are relayed via the *mmr* methylene protons.

Correlation between neighboring peaks can also be observed using NOESY or ROESY 2D NMR where the cross peak intensities are proportional to the inverse sixth power of the internuclear separation (9). These peaks can be either direct or relayed. These experiments are very useful for peak assignments in those cases where there is no through-bond coupling between the peaks of interest. In poly(styrene-co-methyl methacrylate), for example, NOESY cross peaks can be observed between the styrene aromatic protons and the methacrylate methoxy protons to establish the sequence assignments (22). The NOE depends on the correlation time, and there are some correlation times where the NOE is zero as it changes from positive to negative. In those cases the ROESY experiment can be used since magnetization exchange occurs during a spin-locking field and the rotating frame NOE has a different dependence on the chain dynamics (9).

The HMQC and HMBC experiments can be used to establish CH correlations, since the carbon chemical shifts appear along one axis and the proton chemical shifts along the other. If the carbon chemical shifts are known, then the proton chemical shifts can be directly assigned. The HMBC experiment correlations utilize the $^2J_{CH}$ and $^3J_{CH}$ couplings, making it possible to correlate the proton chemical shifts to carbons without directly bonded protons, such as those between carbonyl groups and methine protons in methacrylates (23, 24).

These experiments can be combined to give the 2D or 3D versions of the NOESY-HMQC or TOCSY-HMQC. In 3D NMR the spectral information is spread into three dimensions and a much greater resolution is possible (14).

The J-resolved 2D experiments are also useful for establishing the number of nearby protons and for measuring the coupling constants. The observed multiplet structure depends on the number of attached protons and methine groups appear as doublets while methylene carbons appear as triplets (25). For a more complete discussion of the technical details of nD NMR the readers are referred to a recent book (26).

Solution characterization of polymer microstructure

High resolution solution state NMR is an important analytical tool for the microstructural characterization of polymers. The NMR spectra are very sensitive to polymer stereochemistry, as well as the presence of defects and chain ends. Most polymers contain protons, and proton NMR is often used for polymer characterization. Carbon NMR has also been extensively used. Although carbon has a lower sensitivity, the range in chemical shifts is larger for carbons than for protons, so it is often possible to observe resolved resonances for the defect sites. If the polymer contains other nuclei, such as silicon, phosphorus or nitrogen, these nuclei can also be used for polymer characterization.

It is known from the early studies that the proton NMR spectra of polymers is sensitive to the polymer stereochemistry. Figure 2 shows the 500 MHz proton NMR spectra of a 10 wt% solution of poly(methyl methacrylate) plotted at two different gain levels (27). Figure 2(a) shows the upfield portion of the spectrum containing the resonances for the methylene and methyl protons. The methylene protons are nonequivalent in *m* centered stereosequences, and two peaks are observed at 2.32 and 1.61 ppm. The methyl protons appear as a single line, although a small peak is observed at higher field due to the methyl groups in *mmrm* sequences. The methoxyl peak is not shown, but appears at 3.4 ppm in the proton spectrum.

Although the poly(methyl methacrylate) in Figure 2(a) appears pure, plotting the spectrum with a 100-fold expansion in the vertical gain reveals the presence of many

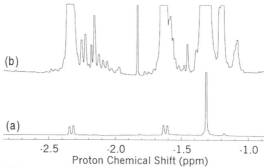

Figure 2. The 500 MHz proton NMR spectra of isotactic poly(methyl methacrylate) plotted at (a) 1x and (b) 100x vertical expansion.

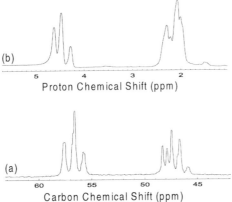

Figure 3. Comparison of the (a) 500 MHz proton and (b) 125 MHz carbon NMR spectra of poly(vinyl chloride).

stereosequences in the nominally pure material (27). These peaks can be assigned by comparison to model compounds and polymers, and by 2D NMR. This figure shows that very low levels of defects (< 1%) can be observed and assigned with high field proton NMR.

Carbon NMR is often used for microstructural characterization because carbon has a larger chemical shift range and the peaks are more sensitive to the local structure. This behavior is illustrated in Figure 3 which compares the carbon and proton spectra of poly(vinyl chloride) (28). In both spectra there are two groups of peaks that can be assigned to the methine and methylene signals. The rr, mr, and mm signals are well resolved in the methine region in both spectra. The resolution is not as good for the protons in the methylene region and no resolved peaks are observed due to overlap from the different stereosequences and the nonequivelent protons in m centered sequences.

2D NMR is an important tool for peak assignments because the assignments can often be established without recourse to model compounds or chemical shift calculations. The power of 2D NMR is illustrated in Figure 4 which shows the HMQC

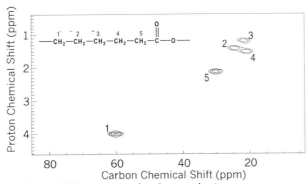

Figure 4. The 500 MHz HMQC spectra of polycaprolactone.

800

P.A. Mirau

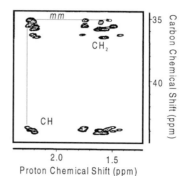

Figure 5. The HMQC-TOCSY spectrum of poly(acrylic acid). The solid line shows the correlation of the methine carbons and protons in the *mm* triad.

spectrum of polycaprolactone. HMQC is a method for correlating the carbon and proton chemical shifts (24). The data are similar to that observed using direct correlation of the carbons and protons, except the protons rather than the carbons are detected in HMQC experiments. This leads to a much higher sensitivity. The *m* centered sequences can be identified in the HMQC spectra of vinyl polymers because the carbons are correlated with two proton chemical shifts.

Higher order stereosequences can be identified with 2D NMR experiments that utilize relayed or longer distance correlations. This is illustrated in Figure 5 which shows the HMQC-TOCSY spectrum for poly(acrylic acid) (29). In addition to the direct correlations, we can now observe longer range correlations, such as between the *mm* triad and the *mmm* tetrad peaks. By following such connectivities it is possible to establish the resonance assignments for larger stereosequences.

The solution NMR spectrum is also sensitive to regioisomerism in polymers. This is illustrated in Figure 6 which shows the proton NMR spectra of regiorandom and regioregular poly(3-hexyl thiophene) (30). In this case the purity of regioregular polymers can be quickly established solely from the proton NMR spectra. The carbon chemical shifts are also sensitive to regioisomerism in polymers. In regioregular vinyl polymers, such as polypropylene, a methyene carbon will have two methine neighbors

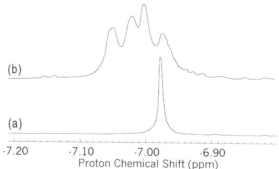

Figure 6. The proton NMR spectra of (a) regioregular and (b) regiorandom poly(3-hexyl thiophene).

Table 5. The chemical shifts for regioregular and regioisomers of polypropylene.

Carbon	Carbon Chemical Shift (ppm)		
	H-T	H-H	T-T
CH	28.5	37.0	—
CH_2	46.0	—	31.3
CH_3	20.5	15.0	—

and the methine will have two methylene neighbors. In a polymer with head-to-head defects there will be two neighboring methine carbons and the tail-to-tail defects will have neighboring methine carbons. Since the environments are very different for the defects and the regioregular chain, the peak will be well resolved from the main chain signals. Table 5 lists the chemical shifts for the polypropylene carbons in regioregular sequences and for the head-to-head and tail-to-tail defects (21). Note that large chemical shift changes are observed for all carbon types relative to the regioregular polymer. Polypropylene is a commercially important material and the defects have been extensively investigated using spectral editing methods and 2D NMR (31).

Another consequence of the head-to-head defects is that pairs of protons are near each other that are more distant in the regioregular polymer. These protons can be identified using either through-bond or through-space correlations in 2D NMR (32). It should be noted that nuclei other than protons are also used to investigate regioisomerism. The fluorine chemical shifts in poly(vinyl fluoride), for example, are very sensitive to regioisomerism (33).

The presence of other types of defects, such as branches and endgroups has also been studied by solution NMR. Polyethylene is an important polymers and the properties depend on both the number and type of branch defect. Figure 7 shows the carbon NMR spectrum a polyethylene sample produced by high pressure free radical polymerization at 270 °C (34). The main methylene peak is off scale in this plot to emphasize the number of defects that can be observed by carbon NMR. The

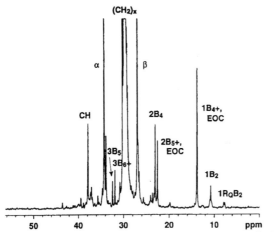

Figure 7. The carbon spectrum of free radical polymerized polyethylene.

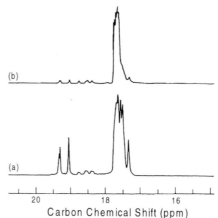

Figure 8. The carbon NMR spectra of poly(propylene oxide) of molecular weight (a)1000 and (b) 4000 Daltons.

nomenclature refers to the distance along the chain away from the branch and the position along the branch. As for the case of stereochemical isomerism, these peaks are assigned by comparison to model compounds and with chemical shift calculations.

The endgroups in polymers can be identified using the methods developed to identify stereochemical isomerism and regioisomerism. Endgroups can be readily identified because their intensities relative to the main chain vary with molecular weight. This is illustrated in Figure 8 which compares the carbon spectra of poly(propylene oxide) for molecular weights of 4000 and 1000 Da.

The NMR spectra are also sensitive to the polymer chain architecture, and very different spectra can be observed for polymers of the same composition in random, alternating and block copolymers. This is illustrated in Figure 9 which compares the 500 MHz proton NMR spectra for an alternating and random copolymer of styrene and methyl methacrylate (22). The lines are broadened in the random copolymer from the statistical distribution of copolymer sequence and stereochemistry.   Much better

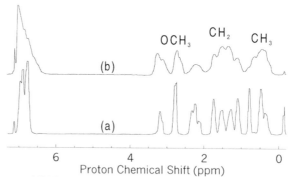

Figure 9. The proton NMR specta of (a) alternating and (b) random styrene-methyl methacrylate copolymers.

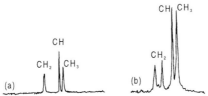

Figure 10. The solid state carbon NMR spectra of (a) isotactic and (b) syndiotactic polypropylene.

resolution is observed for the alternating copolymer. The assignments can be established via the through-space dipolar interactions measured by 2D NOESY NMR (35).

Polymer chain conformation in the solid state

The solid state NMR of polymers is a topic of great interest because it is often possible to relate the NMR properties to the macroscopic properties. The lines are broader in solids than in solutions, but high resolution spectra can be obtained with the combination of magic angle spinning and high power decoupling. Using the combination of chemical shifts, lineshapes and relaxation times it is possible to study the structure and dynamics of polymers over a range of length scales and time scales.

The chemical shifts for polymer solids depend on the same factors as for solution, including the carbon type and the nearest neighbors. In addition, the carbon chemical shifts depend on the chain conformation through the γ-gauche effect and on chain packing. This is illustrated in the carbon spectra of isotactic and syndiotactic polypropylene shown in Figure 10 (36). Both polymers are crystalline, but the isotactic polymer adopts a ...*gtgtgt*... 3_1 helical conformation while the syndiotactic polymer forms a 2_1 helix with a ...*ggttggtt*... conformation. In the syndiotactic polymer half of the methylene groups lie along the interior of the helix and are in a *gauche* arrangement with their γ neighbor, and half the methylene groups lie on the exterior of the helix and are *trans* to their γ neighbor (21). In isotactic polypropylene, the methylene groups are *trans* to one γ neighbor and *gauche* to another. A single resonance is observed for the methylene groups in the isotacitic polymer while two resonances separated by 8.7 ppm are observed for the syndiotactic material. The difference in chemical shift for the methylene carbons in syndiotactic polypropylene is approximately as large two γ-gauche effects (21). The methylene resonance for isotactic polypropylene appears midway between the two peaks in syndiotactic polypropylene, as expected for a methylene group that has one γ-gauche interaction. Chain packing can also influence the spectra. In the α and β forms of isotactic polypropylene, for example, the chains have the same all-*trans* conformation, so the chemical shift differences between the forms must arise from chain packing effects (37). The interchain packing distances are 5.28 ad 6.14 Å for the α and β forms.

There is often a large difference in the chemical shifts between the crystalline and amorphous phases. This is illustrated in Figure 11 which shows the carbon spectrum of polyethyene obtained with cross polarization and magic angle spinning.

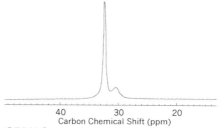

40   30   20
Carbon Chemical Shift (ppm)

Figure 11. The carbon CPMAS spectrum of polyethylene.

The chains exist in an all *trans* conformation in the crystals, giving rise to the sharp resonance an 33.6 ppm. The broad peak at 31 ppm is due to polyethylene in the amorphous phase. The broad line is a consequence of conformational disorder in the amorphous phase. This gives a broad line since the chemical shifts are not rapidly averaged as they are in solution. The spectra obtained with cross polarization are not quantitative, so the relative intensities cannot be used to estimate the crystallinity.

The chemical shifts in crystalline polymers are sensitive to the chain conformation and can be used to monitor solid-solid phase transitions. This is illustrated in Figure 12 which shows the ^{29}Si NMR spectra for poly(di-*n*-hexyl silane) at two temperatures (38). Poly(di-*n*-hexyl silane) exists in two crystalline forms that can interconverted with temperature. Form I has an all *trans* conformation and ordered sidechains, while Form II is disordered. The low temperature Form I has a peak at −22 ppm, while the high temperature Form II is observed at −25 ppm. The midpoint of the Form I to Form II conversion is 42 °C. The solid-solid phase transitions have been studied for a number of polymers, including 1,4-*trans*-butadiene (39), poly(butylene terephthalate) (40), polyacetylenes (41) and polyphosphazines (42).

Solid state NMR has also emerged as an important tool to study the reactivity in solid polymers, since chemical changes can give rise to large changes in the spectra. This is illustrated in Figure 13 which shows the curing of poly[(phenylsilylene)ethynylene-1,3-phenyleneethylnylene] (43). This polymer contains reactive Si-H bonds and acetylic groups that can react during curing. The spectra show

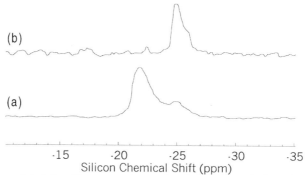

(b)

(a)

-15   -20   -25   -30   -35
Silicon Chemical Shift (ppm)

Figure 12. The solid state silicon spectrum of poly(di-n-hexyl silane) at (a) 25 and (b) 44 °C.

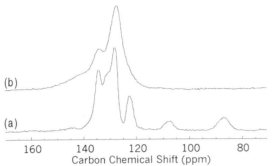

Figure 13. The carbon spectrum of poly[(phenylsilylene)ethynylene-1,3-phenyleneethylnylene] cured at (a) 150 and (b) 400 °C.

that the acetylinic carbons at 87.1 and 107.6 ppm are converted to aromatic groups, presumably from a Diels-Alder reaction. The curing can also be monitored by ^{29}Si NMR.

NMR also provides important information about the structure and dynamics of elastomers. The spectra are easier to acquire because the dipolar couplings and the chemical shift anisotropy are partially averaged by molecular motion (44). Also, these spectra can often be acquired without cross polarization and the resolution often approaches that observed for polymers in solution. In favorable cases it is possible to narrow the lines by magic angle spinning alone (44, 45).

The morphology of polymers in the solid state

Solid state NMR can provide not only information about the chain conformation of polymers, but also about the morphology and organization over longer length scales through proton spin diffusion. Proton spin diffusion is the process by which a proton can exchange magnetization with its neighbors. If one section of the sample is selectively excited by some means, then it is possible to monitor magnetization exchange to other sections of the sample. The rate of magnetization exchange depends on the spin diffusion coefficient, the density of protons, the chain dynamics and the length scale separating the domains (4).

The experiments for measuring polymer morphology typically have an excitation period, a spin diffusion time and a detection period. It is during the excitation period parts of the sample are selectively excited. This is most often accomplished by applying a series of pulses that saturates the signals from one domain. In crystalline polymers, for example, there is a large difference in the spin-spin relaxation rates for the crystalline and amorphous phases. After a 90° proton pulse, the magnetization from the crystalline domains quickly decays from spin-spin relaxation. If we then apply a 90° with opposite phase, we restore the magnetization from the more slowly relaxing component back along the z axis, and we have created a difference in polarization between the phases. After a spin diffusion waiting period we can detect the signal to determine how much magnetization has transferred from the mobile to the rigid phase.

Information about the domain structure is obtained by solving the diffusion equation

$$\dot{m}(r,t) = D\nabla^2 m(r,t) \tag{11}$$

where D is the diffusion coefficient and m(r,t) is the local magnetization density. Solutions to the equation are expressed in terms of the response function R(t) that is measured from the signal intensities as

$$R(t) = 1 - \frac{M(t) - M(\infty)}{M(0) - M(\infty)} \tag{12}$$

where M is the intensity of the signal from the more rigid (crystalline) phase. These equations can be solved analytically for the simple cases of diffusion in one, two or three dimensions, corresponding to lamellar, cylindrical or cubic morphologies. These approximations are often too simple and more complex models with interfaces are required. The recovery curves are often compared with numerical simulations.

Figure 14 shows a recovery plot for semicrystalline polyethylene (46). The data are plotted vs. $\sqrt{t}$ along with calculated recovery curves assuming a one-, two- or three-dimensional morphology. The best fits were to the 2D and 3D models. Using a diffusion coefficient was 8.2×10^{-12} cm^2/s, the domain sizes were calculated to be 90 and 144 Å for the two models. Such experiments give a good estimate of the domain sizes but the results must be considered approximate because of the assumed spin diffusion coefficient and the fact that the shape of the recovery curve are not as sensitive to the morphology as would be desired. In fact, many models can fit the data and the simplest one must be chosen.

More recently the structure of polyethylene samples has been studied using the dipolar filter pulse sequence (47). This pulse sequence works by applying multipulse decoupling to the sample during the excitation period (48). The multipulse decoupling is adjusted so that it works inefficiently for large dipolar couplings, and the result is that the

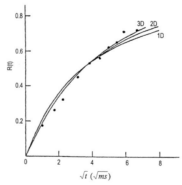

Figure 14. The spin diffusion recovery curve for polyethylene with fits to the 1D, 2D and 3D models.

signals from the rigid phase are saturated. The signals can then be detected in the carbon spectrum following cross polarization. The data from this polyethylene study were best fit to a model with three phases, a crystalline phase, an amorphous phase and an interface. Samples with crystallite thicknesses of 9 and 40 nm were studied, and in both cases the interface had a width of 2.2 ± 0.5 nm (47).

The length scale of mixing in polymers can also be measured via the proton T_1 and T_1 relaxation times. Chains that differ in their mobility often have large differences in their relaxation times. If the domain sizes are less than the lengthscale of spin diffusion, then magnetization exchange provides an efficient relaxation mechanism. The observed relaxation rate is then given by a weighted average of the values for the two chains

$$k = k_a \frac{N_a \phi_a}{N_a + N_b} + k_b \frac{N_b \phi_b}{N_a + N_b} \qquad (13)$$

where k is the relaxation rate ($1/T_1$ or $1/T_{1\rho}$), N is the total number of protons, and ϕ is the mole fraction for the chains a and b. The approximate length scale L of spin diffusion is given by

$$L = \sqrt{\frac{6D}{k}} \qquad (14)$$

where D is the spin diffusion coefficient. Assuming that the spin diffusion coefficient is on the order of 10^{-12} cm^2/s, the length scale of spin diffusion for a polymer with a T_1 relaxation time of 0.5 second is ca. 170 Å. The $T_{1\rho}$ relaxation times are typically much shorter (ca. 5 ms) so there is less time for spin diffusion before relaxation. The polymers must be mixed on a length scale of less than 17 Å to give an averaged $T_{1\rho}$ relaxation time. This behavior is illustrated in Figure 15 for blends poly(styrene-co-acrylonitrile) and poly(methyl methacrylate) (49). Within experimental error the measured T_1 relaxation rates are a weighted average of the mole factions, showing that

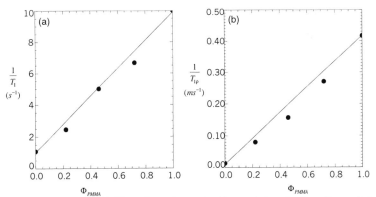

Figure 15. The (a) T_1 and (b) $T_{1\rho}$ relaxation rates for poly(methyl methacrylate)/poly(styrene-co-acrylonitrile) blends as a function of composition.

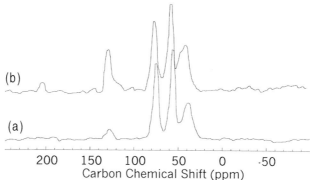

Figure 16. The cross polarization spectrum of (a) the mechanical mixture and (b) the miscible blend of polystyrene-d$_8$ and poly(vinyl methyl ether).

the polymers are mixed on a longer length scale. There is significant deviation from this behavior in the T$_{1\rho}$ relaxation, indicating that the polymers are not well mixed on a shorter length scale. In cases where the chains are partially mixed, multiexponential relaxation behavior is observed (50). This behavior has been observed in mixtures of poly(2,6-dimethyl phenylene oxide) (51) and poly(4-methyl styrene) (52).

    If the polymers are mixed on a molecular level, then other methods can be used to estimate the length scale of mixing. It is well known, for example, that miscible blends of polystyrene and poly(vinyl methyl ether) can be cast from toluene (53). If we make a mixture of deuterated polystyrene and protonated poly(vinyl methyl ether), then we can measure the length scale of mixing by cross polarization. Since there are no protons on the deuterated polystyrene, it cannot be directly cross polarized. However, if the polystyrene chains are in close proximity to the poly(vinyl methyl ether) chains, then the poly(vinyl methyl ether) protons can cross polarize the polystyrene carbons. This behavior is illustrated in Figure 16 which compares the cross polarization behavior for the miscible blend and a mechanical mixture (54). It can be seen that the polystyrene aromatic carbons have a large intensity for cross polarization times longer than 2 ms, indicating that there is intimate mixing of the chains.

<u>Polymer chain dynamics</u>

The chain dynamics are of considerable interest because they can often be directly related to the mechanical properties (55). The motions in polymers occur over a very wide time scale ranging from picoseconds to tens of seconds. While no single NMR method is able to measure the dynamics over this range, many different NMR methods can be used in combination to probe the dynamics over a range of time scales. Figure 17 summarizes the range in correlation times that various relaxation measurements, lineshape measurements and exchange experiments are sensitive to. The fastest motions can be probed with the T$_1$, T$_2$ and NOE relaxation rate measurements. Several lineshape and relaxation methods can be used to probe chain motion in the intermediate frequency regime. Solid state 2D exchange experiments that correlate the chemical shifts before and after some exchange process can be used to measure slow exchange. The upper limit on the timescale is determined by the T$_1$ relaxation time.

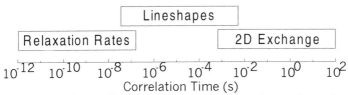

Figure 17. The time scales and methods for measuring polymer dynamics by NMR.

The chain dynamics of polymers in solution are often investigated via the T_1 and T_2 relaxation times and the nuclear Overhauser effect. The relaxation is caused by fluctuating magnetic fields from atomic motion. These motions can cause relaxation only to the extent that they have a component at the resonant frequency. The distribution of motional frequencies is given by the spectral densities $J(\omega)$

$$J(\omega) = \frac{1}{2}\int_{-\infty}^{\infty} G(\tau)e^{-i\omega\tau}d\tau \tag{15}$$

where $G(\tau)$ is the autocorrelation function of an internuclear vector connecting the nuclei. To use this equation it is necessary to adopt a model for $G(\tau)$. The models range form the most general to very specific models incorporating chemical features such as librations, *gauche-trans* isomerizations and longer range segmental motions. The simplest model is for a rigid sphere immersed in a viscous continuum where the loss of memory of the previous state is an exponential and is given by

$$G(\tau) = e^{-\tau/\tau_c} \tag{16}$$

A short value of τ_c corresponds to the motion of a small molecule or a flexible polymer chain, while a long value corresponds to a large molecule or a stiff chain. The relaxation times give information about the chain dynamics because they depend in different ways on the spectral densities. The dipole-dipole contribution to the spin-lattice relaxation of a carbon being relaxed by nearby protons is given by

$$\frac{1}{T_1^{DD}} = \frac{n}{10}\left(\frac{\mu_0}{4\pi}\right)^2 \frac{\gamma_I^2\gamma_S^2\hbar}{r^{-6}}\{J(\omega_H - \omega_C) + 3J(\omega_C) + 6J(\omega_H + \omega_C)\} \tag{17}$$

where n is the number of attached protons, μ_0 is the vacuum magnetic permeability, γ_H and γ_C are the proton and carbon magnetogyric ratios, ω_H and ω_C are the proton and carbon resonance frequencies and r is the internuclear distance. The spin-spin relaxation and the nuclear Overhauser enhancement depend in different ways on the spectral densities. The general strategy for measuring the chain dynamics is to measure as many parameters as possible at several magnetic field strengths and temperatures and find a model for $G(\tau)$ that fits all of the data.

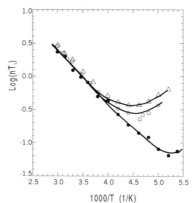

Figure 18. The dependence of the spin-lattice relaxation rate for polyisoprene on inverse temperature at (●)25, (○)90 and (△)125 MHz.

Figure 18 shows a plot of the spin-lattice relaxation time for the methylene carbons of polyisoprene as a function of inverse temperature at several frequencies (56). It should be noted that the carbon relaxation goes through a minimum and is quite sensitive to temperature. The lines through the data were calculated for a biexponential correlation function. Other models also gave a good fit to the data as long as the models had two well-separated correlation times. The activation energy for the chain motion can also be obtained from this analysis. The activation energy for the methylene carbon in polyisoprene is 13.3 kJ/mol (56).

It is often difficult to fit the dynamics of all of the carbons in a polymer with the same parameters. Polymers in solution are quite mobile and the side chains may experience motion relative to the main chain atoms. In such cases the side chain atoms will have different correlation times and activation energies than the main chain atoms. One of the advantages of using NMR to characterize chain dynamics is that it provides an atomic level measure of the chain dynamics.

NMR characterization of chain dynamics in the solid state uses these same relaxation measurements as well as several others. Deuterium NMR is a powerful tool for the study of chain motion because the lineshape depends on molecular motion in the kHz frequency regime. Deuterium is a quadrupolar nuclei and the frequency for a particular deuteron depends on its orientation relative to the magnetic field and is given by

$$\omega = \omega_0 \pm \delta\left(3\cos^2\theta - 1 - \eta\sin^2\theta\cos 2\varphi\right) \tag{18}$$

where ω_0 is the resonance frequency, $\delta = 3e^2qQ/8\hbar$, η is the asymmetry parameter, and the orientation of the deuterium bond vector with respect to the magnetic field is given by θ and φ. For an unoriented sample there will be a distribution in orientations and the lineshape is described as a "Pake" pattern as shown in Figure 19(a). If there is molecular motion that is fast compared to the deuterium linewidth (120 kHz) then an averaged lineshape will be observed. NMR is a powerful tool to study the molecular dynamics of polymers because the lineshape depends on both the rate and geometry of

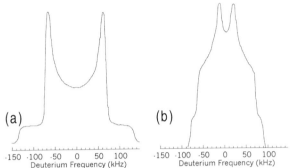

Figure 19. The simulated deuterium lineshape for (a) rigid deuterons and (b) deuterons undergoing 180° ring flips.

the molecular motion. This is illustrated in Figures 19(b) which shows the calculated lineshape for deuterons undergoing 180° flips, as in aromatic rings (57). As with the relaxation studies, it is desirable to make lineshape measurements over a range in temperature.

Figure 20 shows deuterium NMR spectra for the aromatic rings in polystyrene in a polymer where the rings have been specifically deuterated (57). The spectra were acquired with a short time between scans to enhance the signals from the rapidly relaxing aromatic rings. The intensity at the center of the spectra is due to rapidly flipping aromatic rings. The fraction of flipping rings and the correlation times for ring flips can be estimated from these data. These experiments can be combined with other measurements   relaxation and lineshape measurements to probe the chain dynamics on the kHz time scale (58).

2D exchange NMR can be used to probe the low frequency dynamics in polymers. The experiments are similar in principle to the NOESY experiments used to measure the structure of polymers in solution, except that the cross peaks are due to atoms than have changed position and frequency during the mixing time. This method has be used to study chain diffusion in polyethylene (59). Since resolved peaks are observed for the crystalline and amorphous domains, atoms that move between the

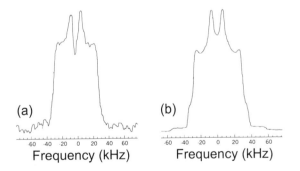

Figure 20. The (a) experimental and (b) simulated deuterium lineshape for ring-dueterated polystyrene.

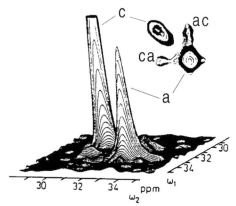

Figure 21. The 2D exchange spectrum for polyethylene showing exchange between the crystalline and amorphous phases. Both the stacked plot and a contour plots are shown.

crystalline and amorphous phase during the mixing time will show up as cross peaks in an exchange experiment. Figure 21 shows the 2D exchange spectrum for high density polyethylene at 90 °C obtained with a 1 s mixing time (59). Cross peaks (labeled *ac* and *ca*) are observed connecting the diagonal resonances for the crystalline and amorphous phases. These can be shown to arise from chain diffusion. The jump rates and activation energies (105 kJ/mol) agree well with the values previously reported for the α transition in polyethylene, and the NMR data show that the α transition is due to chain diffusion. The 2D exchange experiments have been used to assign the dielectric and dynamic mechanical relaxation peaks for a variety of polymers, including crystalline and amorphous polypropylene (60), poly(vinylidine fluoride) (61), poly(vinyl acetate) (62), poly(methyl methacrylate) (63) and poly(ethyl methacrylate) (64).

Summary

This brief review gives an overview of the applications of NMR for the determination of polymer structure and dynamics. NMR is a complex method that is being used for a wide variety of applications ranging from microstructural characterization to structure determination in solid polymers. On one level these NMR methods are relatively simple and can be performed on routine instruments. On another level there is a very active research effort to develop new methods to probe the structure and dynamics of polymers at an ever more detailed level. The new research efforts are aided by developments in magnets, probes, spectrometers and pulse sequences. This advance is driven in part by the desire to understand the molecular level structure and dynamics of polymers, and how these properties relate to the observed macroscopic properties.

References

1. Alpert, N. L. (1947) *Phys. Rev.,* **72,** 637-638.
2. Purcell, E. M., Torrey, H. C. and Pound, R. V. (1946) *Phys. Rev.,* **69,** 37-38.
3. Bloch, F., Hansen, W. W. and Packard, M. E. (1946) *Phys. Rev.,* **69,** 127.
4. Schmidt-Rohr, K. and Speiss, H. W. (1994) *Multidimensional Solid-State NMR and Polymers,* Academic Press, New York.
5. Bovey, F. (1988) *Nuclear Magnetic Resonance Spectroscopy,* Academic Press, Inc., New York.
6. Bovey, F. A. and Mirau, P. A. (1996) *NMR of Polymers,* Academic Press, New York.
7. Mehring, M. (1983) *High Resolution NMR in Solids,* Springer, Berlin.
8. Hartmann, S. R. and Hahn, E. L. (1962) *Phys. Rev.,* **128,** 2042-2053.
9. Neuhaus, D. and Williamson, M. (1989) *The Nuclear Overhauser Effect in Structural and Conformaitonal Analysis,* VCH Publishers Inc., New York.
10. Dixon, W. T. (1982) *J. Chem. Phys.,* **77,** 1800-1809.
11. Bax, A. (1982) *Two Dimensional Nuclear Magnetic Resonance in Liquids,* Delft University Press and D. Reidel Publishing Co., Dordrecht, Holland.
12. Bovey, F. and Mirau, P. (1988) *Acc. Chem. Res.,* **21,** 37-43.
13. Schmidt-Rohr, K., Clauss, J. and Spiess, H. (1992) *Macromolecules,* **25,** 3273-3277.
14. Saito, T. and Rinaldi, P. L. (1998) *Journal of magnetic resonance,* **132,** 41.
15. Frisch, H. L., Mallows, C. L., Heatley, F. and Bovey, F. A. (1968) *Macromolecules,* **1,** 533.
16. Grant, D. M. and Cheney, B. V. (1967) *J. Am. Chem. Soc.,* **39,** 5315.
17. Grant, D. M. and Paul, E. G. (1964) *J. Am. Chem. Soc.,* **86,** 2984 .
18. Cheng, H. N. and Bennett, M. A. (1987) *Makromol. Chem.,* **188,** 135.
19. Breitmaier, E. and Voelter, W. (1987) *Carbon-13 NMR Spectroscopy,* VCH, Weinheim.
20. Tonelli, A. E., Schilling, F. C., Starnes, W. H., Shepherd, L. and Plitz, I. M. (1979) *Macromolecules,* **12,** 78.
21. Tonelli, A. (1989) *NMR Spectroscopy and Polymer Microstructure: The Conformational Connection,* VCH Publishers, Inc., New York.
22. Heffner, S., Bovey, F., Verge, L., Mirau, P. and Tonneli, A. (1986) *Macromolecules,* **19,** 1628-1634.
23. Moad, G., Rizzardo, E., Solomon, D. H., Johns, S. R. and Willing, R. I. (1986) *Macromolecules,* **19,** 2494-2497.
24. Tokles, M., Keifer, P. A. and Rinaldi, P. L. (1995) *Macromolecules,* **28,** 3944-3952.
25. Bruch, M., Bovey, F., Cais, R. and Noggle, J. (1985) *Macromolecules,* **18,** 1253-1257.
26. Croasmun, W. R. and Carlson, M. K. (Eds.) (1994) *Two-Dimensional NMR. Applications for Chemists and Biochemists,* VCH Publishers, Inc., New York.
27. Schilling, F., Bovey, F., Bruch, M. and Kozlowski, S. (1985) *Macromolecules,* **18,** 1418.
28. Mirau, P. and Bovey, F. (1986) *Macromolecules,* **19,** 210-215.
29. Beshah, K. (1994) *Macromol. Symp.,* **86,** 35-46.
30. Chen, T.-A. and Rieke, R. D. (1992) *J. Amer. Chem. Soc.,* **114,** 10087-10088.

31. Asakura, T., Nakayama, N., Demura, M. and Asano, A. (1992) *Macromolecules,* **25,** 4876-4881.
32. Kharas, G. B., Mirau, P. A., Watson, K. and Harwood, H. J. (1992) *Polym. Int.,* **28,** 67-74.
33. Bruch, M., Bovey, F. and Cais, R. (1984) *Macromolecules,* **17,** 2547-2551.
34. McCord, E. F., Shaw, W. H. and Hutchinson, R. A. (1997) *Macromolecules,* **30,** 246-256.
35. Mirau, P., Bovey, F., Tonelli, A. and Heffner, S. (1987) *Macromolecules,* **20,** 1701-1707.
36. Bunn, A., Cudby, E., Harris, R., Packer, K. and Say, B. (1981) *Chem. Commun.,* **15,** 15.
37. Gomez, M. A., Tanaka, H. and Tonelli, A. E. (1987) *Polymer (British),* **28,** 2227.
38. Schilling, F., Bovey, F., Lovinger, A. and Zeigler, J. (1986) *Macromolecules,* **19,** 2660-2663.
39. Schilling, F. C., Gomez, M. A., Tonelli, A. E., Bovey, F. A. and Woodward, A. E. (1987) *Macromolecules,* **20,** 2954-2957.
40. Gomez, M. A., Cozine, M. H. and Tonelli, A. E. (1988) *Macromolecules,* **21,** 388.
41. Tanaka, H., Gomez, M. A., Tonelli, A. E. and Takur, M. (1989) *Macromolecules,* **22,** 1208-1215.
42. Tanaka, H., Gomez, M., Tonelli, A., Chichester-Hicks, S. V. and Haddon, R. C. (1989) *Macromolecules,* **22,** 1031.
43. Kuroki, S., Okita, K., Kakigano, T., Ishikawa, J. and Itoh, M. (1998) *Macromolecules,* **31,** 2804-2808.
44. English, A. D. and Debowski, C. (1984) *Macromolecules,* **17,** 446-449.
45. Mirau, P. A., Heffner, S. A. and Vathyam, S. (1998, in press) In *Advanced in Polymer Characterization: Methodologies and Applications*(Eds, Provder, T., Galuska, A., Mirau, P. A. and Urban, M.) American Chemical Society, Washington.
46. Cheung, T. and Gerstein, B. (1981) *J. Appl. Phys.,* **52,** 5517-5528.
47. Blumich, B., Hagemeyer, A., Schaefer, D., Schmidt-Rohr, K. and Spiess, H. W. (1990) *Adv. Mater.,* **2,** 72-81.
48. Egger, N., Schmidt-Rohr, K., Blumich, B., Domke, W. D. and Stapp, B. (1992) *J. Appl. Polym. Sci.,* **44,** 289.
49. McBrierty, V., Douglass, D. and Kwei, T. (1978) *Macromolecules,* **11,** 1265-1267.
50. Schaefer, J., Sefcik, M. D., Stejskal, E. O. and McKay, R. A. (1981) *Macromolecules,* **14,** 188-192.
51. Stejskal, E., Schaefer, J., Sefcik, M. and McKay, R. (1981) *Macromolecules,* **14,** 275-279.
52. Dickinson, L., Yang, H., Chu, C., Stein, R. and Chein, J. (1987) *Macromolecules,* **20,** 17571760.
53. Bank, M., Leffingwell, J. and Theis, C. (1971) *Macromolecules,* **4,** 43.
54. Gobbi, G., Silvestri, R., Thomas, R., Lyerla, J., Flemming, W. and Nishi, T. (1987) *J. Polym. Sci. Part C. Polym. Lett.,* **25,** 61-65.
55. Schaefer, J., Stejskal, E. O. and Buchdahl, R. (1977) *Macromolecules,* **10,** 384-405.
56. Gisser, D. J., Glowinkowski, S. and Ediger, M. D. (1991) *Macromolecules,* **24,** 4270-4277.

57. Spiess, H. (1983) *Coll. Polym. Sci.,* **261,** 193-209.
58. Schaefer, J., McKay, R. and Stejskal, E. (1983) *J. Magn. Reson.,* **52,** 123-129.
59. Schmidt-Rohr, K. and Spiess, H. (1991) *Macromolecules,* **24,** 5288-5293.
60. Schaefer, D., Spiess, H., Suter, U. and Flemming, W. (1990) *Macromolecules,* **23,** 3431-3439.
61. Hirschinger, J., Schaefer, D., Spiess, H. and Lovinger, A. (1991) *Macromolecules,* **24,** 2428-2433.
62. Schmidt-Rohr, K. and Spiess, H. W. (1991) *Phys. Rev. Lett.,* **66,** 3020-3023.
63. Schmidt-Rohr, K., Kulik, A. S., Beckham, H. W., Ohlemacher, A. O., Pawelzik, U., Boeffel, C. and Spiess, H. W. (1994) *Macromolecules,* **27,** 4733-4745.
64. Kulik, A. S., Beckham, H. W., Schmidt-Rohr, K., Radloff, D., Pawelzik, U., Boeffel, C. and Spiess, H. W. (1994) *Macromolecules,* **27,** 4746-4754.

MASS SPECTROMETRIC ANALYSIS OF POLYMERS

CHARLES L. WILKINS

Department of Chemistry and Biochemistry, University of Arkansas, Fayetteville, AR 72701

Introduction
Secondary Ion Mass Spectrometry
Laser Desorption Mass Spectrometry
Electrospray Ionization Mass Spectrometry
Conclusions

INTRODUCTION

An appealing approach to analysis of polymers is their direct analysis by mass spectrometry. Because mass spectrometry allows direct measurements relevant to end groups, cross-linking information, repeating units, and dispersity, it has many advantages lacked by non specific techniques. However, the need for mass spectrometric samples to be measured in the gaseous state has been a vexing problem and the single biggest hindrance to widespread use of mass spectrometry for polymer analysis. Some of the earliest attempts to deal with this problem were the pioneering efforts of Dole and coworkers about 30 years ago[1-3], when they reported investigations of electrospray ionization of polystyrenes with molecular weights in the 50,000 to 500,000 dalton range. Because mass analyzers of the time were not capable of dealing with such massive species, Dole used a Faraday cup detection scheme, together with retarding potential techniques, to estimate the kinetic energy of the ions produced and, indirectly, their masses. These experiments were the precursors of the widely used electrospray ionization sources of today. Other early approaches to mass spectrometry of intact polymers employed energetic ion or neutral beams with secondary ion mass analysis and a variety of types of analyzers. Depending upon the mass analyzer used, such methods provided either fingerprint information or more complete polymer distribution data. With the introduction of laser desorption techniques in the nineteen eighties, at about the same time that time-of-flight secondary ion mass spectrometry was developed, it became possible for the first time to obtain directly oligomer distributions and related information from mass spectral analysis of low molecular weight polymers. This chapter will briefly review these developments and consider the current state-of-the-art with respect to mass spectrometric polymer analysis.

SECONDARY ION MASS SPECTROMETRY (SIMS)

In its original applications to polymer analysis, secondary ion mass spectrometry (SIMS) was advocated as a direct method for obtaining fingerprint mass spectra of nonvolatile solids. In one of the earliest papers on this topic, Gardella and Hercules investigated a series of poly(alkyl methacrylates) where the molecular weights as estimated by light scattering ranged from 60,000 to 395,000 and the number average molecular weights from gel permeation or osmometry were in the 33,200 to 144,000 range[4]. In order to accomplish the analyses they reported, a combination of low primary ion current densities and low primary ion energies was used. These so-called "static SIMS" conditions allowed one to distinguish the various polymers examined and to identify common contaminants such as Na^+, K^+, and Al^+. However, because the standard quadrupole mass analyzers used with SIMS generally restricted the mass range to below m/z/ 500, no information regarding molecular weight distribution or end groups was available from these data.

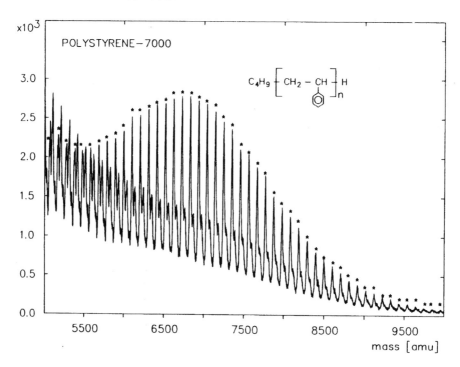

Figure 1. *Positive ion TOF-SIMS spectrum of polystyrene standard PS-7000, M_n =7390. The intense peaks are due to polystyrene oligomers cationized with Ag^+. Oligomer peaks are marked with an asterisk (*). Reprinted from Ref. 5 by permission of the copyright holder.*

Subsequently, high transmission time-of-flight mass analyzers with sensitive detection systems were adapted to SIMS. With this arrangement, it became possible to use even lower primary ion doses. As a consequence, fragmentation is minimized for TOF-SIMS and it is possible to detect polymer fragments containing a large number of repeat units. Thus, molecular weight distributions for polymers with number averages (M_n) up to 10,000 were reported using SIMS [5]. Furthermore, the distributions thus obtained compared favorably with those determined by gel permeation chromatography for the same polymers. Nevertheless, charging of polymer samples under SIMS is a problem and various experimental measures are required to compensate for this difficulty. Two methods which were explored included placing a conducting grid on the surface of the polymer and flooding the sample surface with thermal electrons[6].

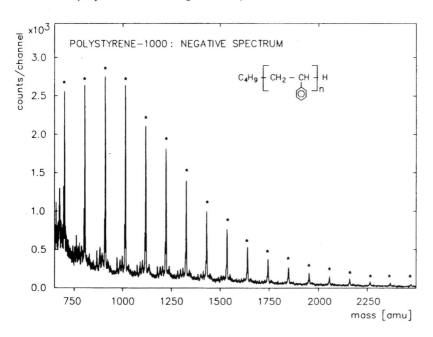

Figure 2. Negative-ion TOF-SIMS spectrum of polystyrene standard PS-1000, M_n= 930. The intense peaks are due to polystyrene oligomers anionized with F⁻. Oligomer peaks are marked with an asterisk (*). Reprinted from Ref. 5 by permission of the copyright holder.

One difficulty with TOF-SIMS for polymer characterization is that the mass resolution and detection efficiency of the instrument may bias molecular weight distributions toward low mass values. This can be corrected by appropriate calibration procedures. Calibration is necessary if high accuracy is required. Figures 1 and 2 are examples of typical positive and negative ion TOF-SIMS spectra[5].

LASER DESORPTION MASS SPECTROMETRY

Because a prerequisite for mass spectrometry is that the sample be volatilized for analysis, an attractive approach is to use a high power laser pulse of suitable wavelength to accomplish this. When the sample is placed upon a surface and a laser beam impinged upon it for desorption/volatilization, the process is generally called laser desorption. For practical applications, one must either use a mass analyzer responding on the time scale of the laser desorption event (several hundred microseconds or less), such as a time-of-flight mass spectrometer, or one must use an ion trapping mass analyzer, such as a Fourier transform or quadrupole ion trap mass spectrometer, to capture the desorbed ions and subsequently analyze them.

Direct Carbon Dioxide Laser Desorption - For polymer analysis by laser mass spectrometry, the earliest literature report employed a pulsed carbon dioxide laser and a Fourier transform mass spectrometer[7], using direct laser desorption. Figure 3 is a negative ion spectrum of Krytox 16140 which was included in that paper, demonstrating the capability of obtaining the oligomer distribution with a mass spectral envelope extending to m/z 7000. The oligomer ions detected correspond to the general composition $F-[CF(CF_3)CF_2O]_n^+$.

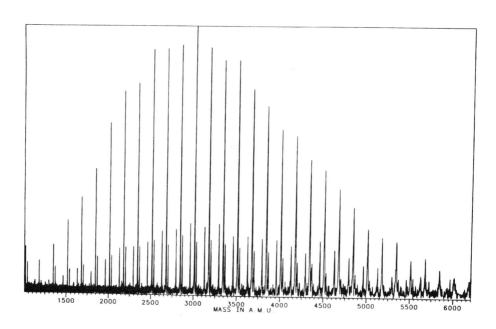

Figure 3. Negative-ion LD/FTMS spectrum of Krytox 16140. Reprinted from Ref. 7 by permission of the copyright holder.

A few years later, the same workers reported higher resolution carbon dioxide laser desorption spectra of poly(ethylene glycol) 8000 and some related polymers and included mass spectra having signals from oligomer ions with m/z up to 9,700[8]. The primary cause of this improved mass range and resolution was the substitution of a 7 Tesla magnet for the 3 Tesla magnet used in the earlier study. The results included in Figure 4 are close to the highest mass range achieved with direct CO_2 laser desorption FTMS of polymers. Furthermore, such results could only be obtained with

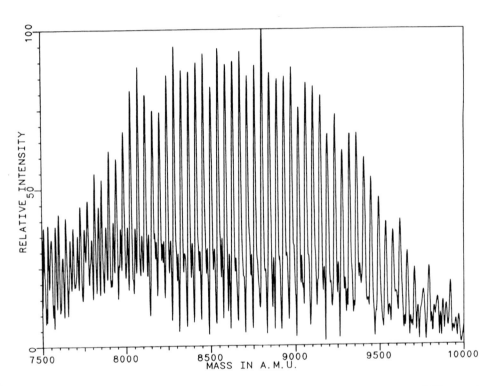

Figure 4. LD/FTMS spectrum of Poly(ethylene glycol) 8000. Average of 50 spectra acquired using a 7 Tesla magnet. Reprinted from Ref. 8 by permission of the copyright holder.

polar polymers, containing oxygen and nitrogen. For hydrocarbons, results were much less impressive. In virtually all cases, oligomer cationization plays an important role in the laser desorption ionization process. Sodium and potassium cationization are prevalent for the heteroatom-containing polar polymers. This may explain the difficulty in obtaining direct laser desorption mass spectra of hydrocarbon polymers, which have no convenient functional groups amenable to cationization.

However, more recently, a successful strategy for mass spectral analysis of hydrocarbon polymers has been developed. Essentially, the technique involves adding a suitable silver salt to the samples, prior to laser desorption mass spectrometry[9]. When this is done, silver-cationized oligomers can be detected. Weight (M_w) and number averages (M_n) obtained by integrating the mass spectral peaks show excellent correspondence to the values obtained by other methods. Figure 5 is an excellent example of the quality of mass spectrum that can be obtained with silver-cationized hydrocarbon polymers (a polybutadiene in this case) for samples which show only low mass fragment ions in the absence of the silver cationizing reagent.

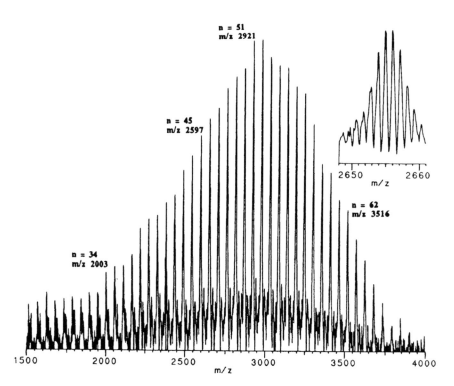

Figure 5. 7 Tesla LD/FTMS spectrum of Polybutadiene, GPC (M_w 2953). Inset shows the n=46 silver-attached oligomer molecular ion region with resolving power of 4773 at mass 2655. Reprinted from Ref. 9 by permission of the copyright holder.

As is evident from the spectra shown in Figures 3-5, direct carbon dioxide laser desorption can give a good deal of information about both polar and non polar polymer distributions with a minimum of sample preparation. However, based upon the extensive literature studies, it seems clear that this method is limited to polymers with average molecular weights below about 10,000 daltons. Subject to that constraint, direct carbon dioxide laser desorption is an excellent and reliable method for assessing polymer distributions.

<u>Matrix-assisted Laser Desorption/Ionization (MALDI)</u> - In the late nineteen eighties, two groups of researchers announced the development of a new approach to laser desorption mass spectrometry[10-12]. The method, which has become known as matrix-assisted laser desorption/ionization (MALDI) mass spectrometry, relies on the use of a low molecular weight absorbing material, chosen to absorb at the wavelength of the laser being used, in which the analyte is imbedded. These materials are typically co-crystallized on a surface and the sample desorbed using an ultraviolet laser. Common choices for the laser include nitrogen lasers (337 nm) or Nd-YAG lasers (usually 266 or 355 nm, although the fundamental has occasionally been used). In one of his earliest papers on MALDI, Tanaka reported spectra of both poly(propylene glycol) 4000 and poly(ethylene glycol) 20,000, resolving the oligomers in the first case, but only the envelope of the distribution for the higher molecular weight polymer[10]. Hillenkamp and coworkers also published one of the first papers demonstrating MALDI analysis of synthetic polymers, including poly(ethylene glycol), poly(methylmethacrylate), and poly(styrene)[13]. All of these early reports of MALDI involved use of time-of-flight mass analyzers, because of the need to obtain spectra during the limited time period ions are present following the desorption event. As a consequence, both then and now, those using MALDI-TOF instruments have had to accept lower resolution than might be possible with Fourier transform or

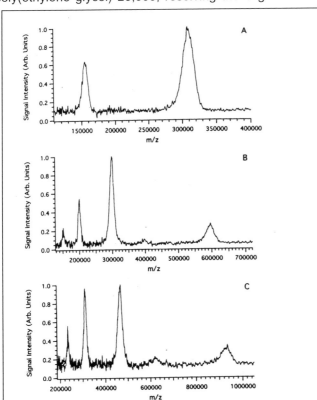

Figure 6. MALDI-TOF mass spectra of three poly(styrene) samples with nominal molecular weights of (A) 330,000, (B) 600,000, and (C) 900,000. Reprinted from Ref. 14 by permission of the copyright holder.

quadrupole ion trap mass analyzers. Nevertheless, mass range is exceptional, as is obvious from a recent paper which probably represents close to the practical mass limit for MALDI-TOF of poly(styrene) polymers[14]. That article reports MALDI-TOF analysis of a poly(styrene) standard with average molecular weight of 1.5 million daltons. As is seen in the three spectra contained in Figure 6, at such high masses resolving power is limited and the primary information supplied by the mass spectra is average molecular weight, rather than details of the oligomer distributions, dispersity, or end groups. In order to obtain these spectra, a silver cationizing procedure, similar to that mentioned earlier, was required. Also, notice that multiply charged distributions dominate the spectra.

Following the introduction of MALDI-TOF, numerous workers set to work to adapt the technique to Fourier transform mass spectrometry and that was accomplished by a gated trapping technique in 1992[15]. Subsequently, UV-MALDI-FTMS of poly(ethylene glycol) 6000 and poly(ethylene glycol) 10000 resolved oligomer distributions were demonstrated (Figure 7)[16]. Thus, it is apparent that there are some gains realized

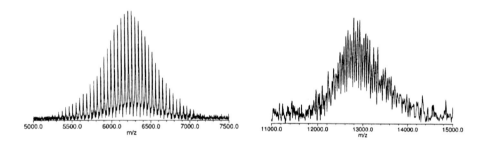

Figure 7. UV-MALDI-FTMS mass spectra for PEG 6000 (left) and PEG 10000 (right). Adapted from reference 16 by permission of the copyright holder.

by using the enhanced resolution of FTMS for MALDI of polymers with oligomer distributions extending to 15,000 daltons or so. Excellent mass measurement accuracy, routinely in the part-per-million range, can be obtained for mass spectra of low mass synthetic polymers by MALDI-FTMS[17]. Unfortunately, this high resolution/high accuracy performance for MALDI-FTMS has been limited to date to relatively low mass synthetic polymers. Nevertheless, it is an excellent tool for characterization of such materials.

ELECTROSPRAY IONIZATION (ESI) MASS SPECTROMETRY

For synthetic polymer analysis by mass spectrometry, the primary current alternative to MALDI is electrospray ionization(ESI). In this approach, a solution of analyte is dispersed in a bath gas by spraying it through a needle which is maintained at a few kilovolts, relative to the walls of the surrounding chamber and electrode. A consequence is that ions are eventually extracted and enter an appropriate mass analyzer through a skimmer. For macromolecules, these ions are very often multiply-charged, effectively extending the mass range, as a result of the fact that mass spectrometers usually detect m/z ratios. Fenn and coworkers published an excellent description of the process, its historical development, and a summary of early examples of electrospray mass spectra in an article about ten years ago[18]. As mentioned at the outset, electrospray derives from the much earlier work of Dole[2] and coworkers. In his 1989 paper, Fenn shows some interesting applications of ESI to analysis of a series of relatively low molecular weight poly(ethylene glycols) using a quadrupole mass analyzer[18]. Subsequently, Kallos and coworkers reported application of ESI for characterization of polyamidoamine starburst polymers[19]. It appears that the highest mass polymers measured thus far by ESI-MS are poly(ethylene glycols) with average molecular weights in the 4,500,000 to 5,000,000 dalton range[20, 21]. However, at the upper end of this mass range, the species observed must have between 2,200 and 4,000 charges per molecule, with a reported uncertainty of at least 400 charges[20, 21]. Thus, these results are of more theoretical than practical significance.

An interesting and analytically more useful result appeared in the report of Shi and coworkers of the analysis of glycidyl methacrylate and butyl methacrylate copolymers with masses up to 7000 daltons, using ESI with a 9.4 T Fourier transform

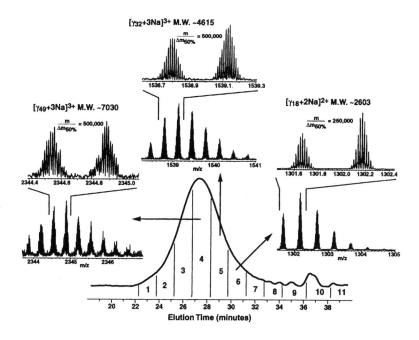

Figure 8. Gel permeation chromatographic separation of GMA/BMA copolymers (bottom) and high resolution FT-ICR mass spectra for several fractions. Adapted from reference 22 with permission of the copyright holder.

mass spectrometer[22].   In this study, mass resolving power of 500,000 was estimated for samples subjected to prior gel permeation chromatography fractionation.   This resolution is adequate to allow direct assignment of charge states, based upon the separations appearing in the isotopic multiplets.   Such a capability is essential for realistic ESI-MS analysis of complex mixtures such as the copolymer samples represented here.   Furthermore, with mass accuracy of 0.2 ppm, it is possible to answer very detailed questions regarding the nature of the various polymer sequences present. Figure 8 exemplifies the kind of analytical performance that is possible with high field FT-ICR analysis of ESI-generated ions.

CONCLUSIONS

There has been very substantial progress over the past twenty years in the capabilities of mass spectrometry for characterizing polymers.   These advances can primarily be attributed to the spectacular developments in ion source design which resulted in MALDI and ESI becoming routinely available with all types of mass spectrometers. Because it is now possible to carry out highly accurate and precise measurements of oliogmer distributions for masses up to ca. 15,000 daltons, mass spectrometry may now be the method of choice for detailed analysis of polymers in that mass range.   Much larger polymers, with masses as high as 5,000,000 daltons, have been successfully volatilized, but, at present, yield only rough estimates of molecular weights with existing methods.   Thus, practical mass spectrometry of such high mass polymers remains an unsolved problem at the time of this writing.

REFERENCES

1.   G. A. Clegg,M. Dole *Biopolymers* , 1971, ***10***, 821.
2.   M. Dole,L. L. Mack,R. L. Hines,R. C. Mobley,L. D. Ferguson,M. B. Alice *J. Chem. Phys.* , 1968, ***49***, 2240-2249.
3.   L. L. Mach,P. Kralik,A. Rheude,M. Dole *J. Chem. Phys.* , 1970, ***52***, 4977.
4.   J. A. Gardella,D. M. Hercules *Anal. Chem.* , 1980, ***52***, 226-232.
5.   I. V. Bletsos,D. M. Hercules,D. vanLeyen,B. Hagenhoff,E. Niehuis,A. Benninghoven *Anal. Chem.* , 1991, ***63***, 1953-1960.
6.   I. V. Bletsos,D. M. Hercules,J. H. Magill,D. VanLeyen,E. Niehuis,A. Benninghoven *Anal. Chem.* , 1988, ***60***, 938-944.
7.   C. L. Wilkins,D. A. Weil,C. L. C. Yang,C. F. Ijames *Anal. Chem.* , 1985, ***57***, 520-524.
8.   C. F. Ijames,C. L. Wilkins *J. Am. Chem. Soc.* , 1988, ***110***, 2687-2688.
9.   M. S. Kahr,C. L. Wilkins *J. Amer. Soc. Mass Spectrom.* , 1993, ***4***, 453-460.
10.   K. Tanaka,W. Hiroaki,Y. Ido,S. Akita,Y. Yoshida,T. Yoshida *Rapid Commun. Mass Spectrom.* , 1988, ***2***, 151-153.
11.   M. Karas,D. Bachman,U. Bahr,F. Hillenkamp *Int. J. Mass Spectrom. Ion Process.* , 1987, ***78***, 53-68.
12.   M. Karas,F. Hillenkamp *Anal Chem* , 1988, ***60***, 2299-2301.

13. U. Bahr,A. Deppe,M. Karas,F. Hillenkamp,U. Giessmann *Anal. Chem.* , 1992, **64**, 2866-2869.
14. D. C. Schriemer,L. Li *Anal. Chem.* , 1996, **68**, 2721-2725.
15. J. A. Castoro,C. Köster,C. L. Wilkins *Rapid Commun. Mass Spectrom.* , 1992, **6**, 239-241.
16. S. J. Pastor,J. A. Castoro,C. L. Wilkins *Anal. Chem.* , 1995, **67**, 379-384.
17. M. L. Easterling,T. H. Mize,I. J. Amster *Anal. Chem.* , 1999, **71**, 624-632.
18. J. B. Fenn,M. Mann,C. K. Meng,S. F. Wong,C. M. Whitehouse *Science* , 1989, **246**, 64-71.
19. J. Kallos,D. A. Tomalia,D. M. Hedstrand,S. Lewis,J. Zhou *Rapid Commun. Mass Spectrom.* , 1991, **5**, 383.
20. J. E. Bruce,X. Cheng,R. Bakhtiar,Q. Wu,S. A. Hofstadler,G. A. Anderson,R. D. Smith *J. Am. Chem. Soc.* , 1994, **116**, 7839-7847.
21. T. Nohmi,J. B. Fenn *J. Am. Chem. Soc.* , 1992, **114**, 3241-3246.
22. S. D. H. Shi,C. L. Hendrickson,A. G. Marshall,W. J. Simonsick,D. J. Aaserud *Anal. Chem.* , 1998, **70**, 3220-3226.

POLYMER CHARACTERIZATION BY FLUORESCENCE SPECTROSCOPY

JOHANNES W. HOFSTRAAT

University of Amsterdam, Institute of Molecular Chemistry, Molecular Photonics Group, Nieuwe Achtergracht 129, NL-1018 WS Amsterdam, The Netherlands

Outline

Introduction

Fluorescence methods are increasingly applied for the characterization of polymeric materials. In the present review a number of applications of fluorescence spectroscopy will be presented in an effort to illustrate the potential of this technique to study structure and properties of polymeric materials. In analytical chemistry fluorescence methods to date are widely applied to determine trace constituents in environmental and bioanalysis. Numerous examples are described in, for instance, the monographs edited by Lakowicz [1]. The technique can be applied to the determination of autofluorescent compounds or - since fluorescence by no means is a universal phenomenon - non-fluorescent compounds may be measured by making use of suitable fluorescent labels. Since fluorescence emission involves the measurement of light against a dark background, and is not based on difference measurements like most absorption techniques (e.g., IR spectroscopy, NMR spectroscopy, UV-VIS spectrophotometry), detection may be achieved on a very low level. Recently, even fluorescent measurements of single molecules have been reported [2]!

Similar fluorescence measurements as applied in chemical analysis and in bioanalysis can also be used to characterize polymeric materials. Here, two situations can be distinguished as well: the characterization of polymers containing fluorescent units, i.e., autofluorescent materials, and the characterization of polymers which do not fluoresce. The latter situation requires, as mentioned before, the use of fluorescent labels. In polymers the fluorescence technique has even more possible applications than in chemical analysis. In chemical analysis fluorescence spectroscopy is almost exclusively used for the quantification of low concentrations ($< 10^{-6}$ M) of analytes in homogeneous solutions. When higher concentrations of fluorophores are present, the quantitative application of the technique may be compromised as a result of intermolecular interactions or optical effects (e.g., inner filter effect). These processes have been described in detail in, for instance, the comprehensive book written by Lakowicz [3]. The intensity and spectral properties of fluorescence are very sensitive to interaction of the excited molecules with other excited molecules, with other molecules in the ground state and with solvent molecules. All these interactions limit the quantitative application of the technique, but on the other hand they provide opportunities to study other aspects of the system of interest. In polymers in particular, use can be made of these phenomena, as will be demonstrated in this paper.

The parameters which can be determined for a fluorescent molecule or system are: (1) the fluorescence intensity, (2) the fluorescence spectrum (excitation and emission), (3) the fluorescence lifetime and (4) the fluorescence polarization (to study ordered systems). By determination of the change in degree of polarization with time (or: fluorescence depolarization) also information is obtained on the mobility of fluorescent sidegroups or of fluorescent probes in polymers. In particular such "functional" determinations offer unique possibilities to characterize polymers.

All these aspects will be illustrated here. Use of fluorescence spectroscopy to characterize polymers with fluorescent sidegroups, i.e. intrinsically fluorescent materials, will be demonstrated. As examples, nonlinear optical active (NLO) polymers with fluorescent side groups and electroluminescent polymers, which are autofluorescent because they consist of highly conjugated main chains, will be discussed. Next, attention will be paid to polymeric materials which have been treated with a fluorescent label, i.e. extrinsically fluorescent materials. A special type of fluorescent probe which will be described in this paper is the so-called "Fluoroprobe" molecule (1-phenyl-4-(4-cyano-l-naphthylmethylene)piperidine), a molecule with a charge-transfer fluorescence transition, which is extremely sensitive to mobility and polarity changes in its immediate surroundings [4]. The Fluoroprobe molecule will be used to study polymerization reactions and properties of latices. Finally, the use of reactive fluorescent probes to determine concentration and distribution of functional groups on surfaces will be discussed. It will be shown that by application of flow cytometry even concentrations of fluorescently labeled functional groups on individual particles can be determined.

Experimental Aspects

Some experimental background is given to provide insight into the approaches which may be taken to obtain useful fluorescence spectra. Typical instrumentation and procedures for the acquisition of both steady-state and time-resolved fluorescence spectra are described. In addition, details are presented of aspects of preparation of samples for fluorescence measurements.

Instrumentation

Steady-state Fluorescence   Steady-state fluorescence measurements may be done using a conventional spectrofluorimeter, e.g. a Spex Fluorolog, an Edinburgh Analytical Instruments, or a PTI Alphascan spectrofluorimeter. The latter instrument, used in most experiments described in this paper, is equipped with an 0.25 m double monochromator for selection of the excitation light and a 0.2 m monochromator for separation of the emission light. The light is detected with a Hamamatsu R928 (250-800 nm) or Hamamatsu R2658 (500-1000 nm) photomultiplier tube. The instrument is equipped with full correction for wavelength dependence of the monochromators and detectors. Files are available to correct for the wavelength dependence for randomly polarized light and for S- and P-polarized light, according to the procedures as described in [5]. In particular for the reliable measurement of highly viscous polymer samples (in which the rotational movement of the fluorophoric groups occurs on a much longer time scale than the decay of the fluorescent state, so that polarized fluorescence is generated when the excitation light is anisotropic) the polarization state of excitation and emission light has to be controlled and correctly taken into account in the correction and evaluation of the experimental data. It is important to realize that the monochromators used to select the emission and excitation wavelengths are extremely anisotropic, so that even if no additional optics are applied to control the state of polarization of the excitation light, and hence of the emission light, is non-isotropic.

Time-resolved Fluorescence   Time-resolved fluorescence measurements can be done by two different approaches: phase modulation, in which the excitation light is modulated at a variable frequency and phase-resolved detection is applied, and pulse excitation methods, in which a short pulse of light is used to bring the fluorophores instantaneously into the excited state.

Phase-resolved fluorescence instruments are produced by ISS, Spex and SLM Instruments. In the experiments reported here the SLM Instruments 4850 multiharmonic Fourier transform (MHF) phase modulation fluorometer has been applied, using suitable interference filters for selection of the emission light and a 325-nm HeCd laser from Liconix for excitation [6]. For anisotropy measurements a T-format set-up has been used with the excitation polarizer set to pass vertically polarized light, whereas one of the emission polarizers is alternated between horizontal and vertical polarization, and the other is set to horizontal polarization as reference. Time-resolved detection has been done in an L-format using scattered light off a glycogen solution as reference, with the emission polarizer set to the magic angle to eliminate photoselection effects. The dynamic measurements have been done with a base modulation frequency of 5.0 MHz and using 30 frequencies in the range 5-150 MHz. The base cross-correlation frequency for detection has been set to 4.167 Hz. The fluorescence phase-angle differences and modulation amplitude ratios have been calculated as averages of 15 replicate experiments. Data analysis has been done with Globals Unlimited heterogeneity analysis (Version 3). The quality of the fit has been evaluated by examining the reduced χ-values and the distribution of the residuals across the modulation frequency range.

For short, pulse-induced, time-resolved fluorescence measurements not many commercially available instruments are available: from Edinburgh Instruments and from PTI. Mostly, however, home-built instrumentation is applied. The experiments described here have been done using a time-correlated single photon counting set-up [7]. Excitation is done with the ultrashort pulses of a synchronously pumped dye laser, operated with the stilbene 1 dye in the blue. Emission is detected under an angle of 90° with a Hamamatsu 1645U microchannel plate detector. At the moment the laser is fired, a time-to-amplitude converter starts a voltage ramp set to a time-base. Once the first signal from the photomultiplier tube is received the voltage ramp is stopped, the time span between the pulses converted to an analog pulse height, analyzed with an analog-to-digital converter and the data collected in a multichannel analyzer. In this way systematically the time-resolved profile is built-up. Wavelength selection is done using color filters. The measurement is calibrated using the short-living cyanine dye pinacyanol in methanol as lifetime standard. It is also used to correct for time-dependent signals resulting from reflections from the sample holder and from within the photomultiplier tube. Data analysis is done with the maximum entropy method (software from Maximum Entropy Data Consultants. Cambridge, UK).

Flow Cytometry   In many applications (e.g., in diagnostics) fluorescent polymer particles are applied. Fast single cell measurements have been done using a home-built flow cytometer. An argon-ion laser has been used for excitation. Detection of the fluorescence generated from the single particles has been effected in the orange part of the spectrum; in addition the perpendicular and forward light scatter and the fluorescence pulse duration (time-of-flight) have been measured. The latter measurements can be used as size estimates. The instrumentation is described in detail in [8]. Commercial flow cytometers are available from several sources, e.g. Becton Dickinson, Coulter or Partec, in general these instruments are optimized for the analysis of small particles (< 10-20 μm).

Materials

<u>Polymers</u> NLO polymers with fluorescent sidegroups were prepared as described in [9]. They were either measured in solution or as thin films, spincoated from a cyclopentanone solution onto a quartz substrate. The film thickness was about 1 μm. Details on the preparation of electroluminescent polymers can be found in [10]. The polystyrene latices were prepared as reported in [11]. A core-shell latex was prepared via subsequent seeded emulsion polymerization at 80 °C, with the polystyrene latex as seed, forming the core, and using styrene or glycidylmethacrylate to form the shell. The bulk radical polymerization of glycidylmethacrylate has been done in an oxygen-free environment using a carius tube and a freeze-pump-thaw approach to obtain an oxygen-free cuvette. Benzoylperoxide has been used as initiator and the polymerization has been carried out at 70 °C. Labeling of amino-groups on Spherosil porous silica particles (from Rhone Poulenc) has been done as described in [12]. Nylon samples have been prepared from caprolactam. following well-established procedures [13].

<u>Fluorescent Probes</u>   Fluoroprobe, or 1-phenyl-4-[(4-cyano-l-naphthyl)methylene]-piperidine, is a probe with strong charge-transfer (CT) fluorescence originating from a highly dipolar excited state. Preparation and photophysical characteristics of Fluoroprobe have been described in [14]. Excitation is done at 310 nm, emission is observed depending on the polarity and mobility of the solvent. Labeling of functional groups in polymers and on polymer surfaces has been done with Lucifer Yellow VS (vinyl sulphonate), reactive to amino-groups, and Lucifer Yellow CH, reactive to aldehyde groups. Lucifer Yellow is excited in the blue at 435 nm and emits green light (λ_{max} at 535 nm). In some experiments near-IR fluorescent labels have been used [15]. A cyanine dye with a hydrazide functionality was used for labeling of aldehydes; excitation is done at 665 nm, emission is observed at about 700 nm, after labeling. Another cyanine dye, with one ethylene bond more than the previous one, equipped with isothiocyanate functionality, has been applied to label amino-groups. The latter cyanine dye has been excited at 780 nm and emits at 800 nm.

<u>Preparation of Labeled Samples</u>   The latex was loaded with Fluoroprobe using 1,2-dichloroethane as carrier. The Fluoroprobe has been added to the aqueous latex as an emulsion in water by making use of Aerosol MA-80 surfactant (sodium dihexylsulfosuccinate, from American Cyanamid). Jeffamine T403, a multifunctional crosslinker containing primary amines, was labeled with modified Fluoroprobe, containing a maleimide functionality. The reaction occurred readily in ethanol; 1 % of the amino-functionality on Jeffamine was labeled. The Jeffamine was subsequently used in a curing reaction with a polyacrylate dispersion, equipped with acetoacetate groups. Aminogroups on functionalized silica spheres were labeled in 0.1 M sodiumhydrogencarbonate at room temperature during 2 h. In heterogeneous labeling reactions the excess label can be easily removed by application of a filtration procedure. The silica spheres have been functionalized with primary aminogroups, using γ-aminopropyltriethoxysilane [12]. Aldehyde functionalities on latices were labeled by reaction in water at 40 °C. A reaction time of 3 hours is sufficient for quantitative conversion of the functional groups. Excess label was removed by extensive washing with methanol. The methanol was checked for any residual probe by measuring the fluorescence. The nylons have been labeled in m-cresol. Reaction is complete after 4 h at 50 °C.

Results and Discussion

Two areas in which fluorescence spectroscopy may be applied to polymeric materials will be discerned in this paper: the first area comprises polymers which contain inherently fluorescent groups, i.e. polymers which are intrinsically fluorescent, the second area polymers which do *not* contain fluorescent groups. The latter group of polymers is made fluorescent by making use of a label, and is referred to as having extrinsic fluorescence. In the latter case first probes will be discussed with specific reactive groups, that give selective reactions to functional groups which are present in the polymeric material, and next a special type of fluorescent probes that show a distinctive fluorescence behavior depending on properties of the matrix in which they are applied.

Intrinsic Fluorescence

Many polymers contain fluorescent constituents, either as side groups attached to the polymeric backbone or as part of the backbone. Intrinsically fluorescent polymers are often encountered in biochemical sciences, since many biopolymers (e.g., DNA, proteins) contain fluorophores. Examples are the amino acids tryptophan and alanin [16]. At present more and more synthetic polymers are reported with fluorescent sidegroups. Important examples are polymers used for electro-optic applications, containing nonlinear optical (NLO) active [17] or electroluminescent moieties [18]. In most cases, however, electro-luminescent polymers are based on main-chain polymers consisting of fully conjugated units.

Fluorescence of Polymers with NLO-Side Chain Molecules
Most NLO polymers contain fluorescent charge-transfer molecules as sidegroups. Such materials show high NLO activities and at the same time are versatile and can be easily processed; as such they have important advantages over inorganic materials, in particular for the production of affordable optical communication components [19]. The most widely applied NLO active sidegroup is DANS, dimethylaminonitrostilbene, a molecule with charge transfer absorption. The difference in dipole moment between ground state (μ_g = 7 Debije) and excited state (μ_e = 25 Debije) is large. Hence, the fluorescence of this molecule is strongly dependent on the polarity of the solvent [20]. The maximum of the charge transfer absorption is at 437 nm (in chloroform, but not much different in solvents of different polarity); the maximum of the fluorescence may vary from about 460 nm in n-hexane to 718 nm in acotonitrile. From the dependence of the solvatochromism on the polarity of the solvent, the difference in dipole moment for the ground and excited state can be derived using the formulae given by Liptay [20]. Also, the NLO activity of the molecule can be estimated from these measurements. In fact, the solvatochromic method is the simplest way to derive the NLO activity of molecular systems and may even be applied to estimate the activity of polymers with NLO sidegroups. Since the environment plays an essential role in determining the photophysical properties of the NLO active molecules, such measurements are very informative. Alternative methods often rely on measurements in one solvent, which may not always be the most representative for the environment in which the NLO active material is applied [21,22].

An example of the application of the solvatochromic method on the DANS sidegroup is shown in Fig. 1.

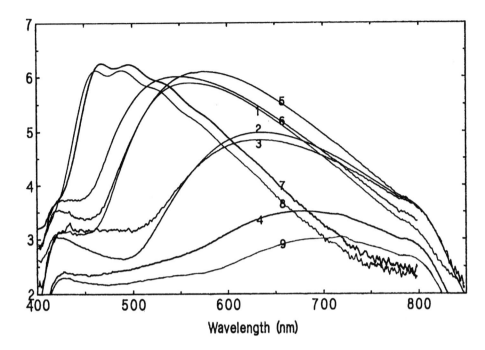

Figure 1. Fluorescence emission spectra obtained for solutions of dimethylamino-nitrostilbene (DANS) in solvents of different polarity. In all cases concentrations below 10^{-5} M, and absorptions less than 0.1 were applied. All spectra have been fully corrected for instrumental responses. The intensity values have been represented on a logarithmic scale, to compensate for the significant differences in fluorescence intensity obtained in the solvents used (1 = toluene, 2 = tetrahydrofurane, 3 = ethylacetate, 4 = chloroform, 5 = diethylether, 6 = dibutylether, 7 = cyclohexane, 8 = n-hexane, 9 = dichloromethane.

On the basis of the wavenumber shift measured in solvents of different polarity, the dipole moment of the excited state is calculated to be 17.8 D higher than that of the ground state. In the calculation only the emissions from the clear charge-transfer transitions (i.e., those not showing vibrational fine structure) have been used. The apolar solvents, cyclohexane and n-hexane show fluorescence with fine structure, which may indicate that in these solvents the energy of the charge transfer transition in this solvent is higher than that of one of the local (non-charge transfer) transitions of the molecule.

On the basis of the calculated $\Delta\mu_{g\text{-}e}$ = 17.8 D and using an estimated effective radius of 0.438 nm [20] an electro-optic coefficient β of 82.10^{-30} e.s.u. is calculated. With electric-field induced second harmonic generation (EFISH) measurements a β of 83.10^{-30} e.s.u. is determined for DANS in chloroform, which is indeed very close to the solvatochromic value [24].

Once the sidegroup is brought into a polymer the polarity of the environment is modified and, more importantly, interactions between sidegroups may occur. Both effects may influence the structure as well as the electro-optic and photophysical properties of the functional polymers. Interactions between the highly dipolar sidegroups may cause significant refractive index fluctuations in the material which may act as scattering centers. The occurrence of anti-parallel associations of sidegroups will counteract the effective electro-optic activity of the polymer, for which polar orientation of the sidegroups is required. Finally, energy transfer processes will influence the photochemical processes in such sidechain polymers, which may be submitted to visible radiation to define waveguiding patterns via photobleaching: under the influence of the irradiation the sidegroups are modified, so that the refractive index of the illuminated regions is lowered. The most direct way to study interactions between fluorescent sidegroups is via time-resolved spectroscopy.

Table 1 summarizes results obtained for sidechain copolymers containing different concentrations of DANS, by making use of dimethylaminobenzene (DAB) as comonomer. Fluorescence lifetimes of the polymers, with DANS concentrations varying from 100% to 2% (or at the most one DANS sidegroup per polymer chain, as the average molecular weight of the polymers was about 15,000), were determined in chloroform solution and spincoated onto quartz substrates. In both cases the decays were monoexponential. In solution, the DANS sidegroups in all polymers showed a lifetime of about 550 ps in chloroform solution, irrespective of the concentration of the sidegroups. Also, the emission maximum was the same for all polymers, and agreed with the maximum observed for the DANS molecule itself. However, the fluorescence lifetime of DANS in chloroform amounts to 850 ps, i.e. considerably longer than that of the sidegroups. The results indicate that the DANS sidegroups, though attached to the polymer backbone, in chloroform are fully solvated and do not show interactions. The only effect of the backbone is in providing somewhat more opportunity for radiationless processes (hence the slightly shorter lifetime of DANS sidechain molecules) and prolonging the rotational correlation time, most probably by increasing the local viscosity of the sidegroups. The rotational correlation time is determined by measuring the loss of polarization of originally photoselected molecules. Photoselection is brought about by using a polarized laser source for excitation. The rotational correlation time or fluorescence depolarization is mainly determined by the mobility of the sidegroups. For the dissolved DANS the rotational correlation time is shorter than 0.2 ns, indicative of very rapid reorientation of the molecule. The dissolved polymers show rotational correlation times of 0.6-0.7 ns, which are independent of the composition of the polymers.

Table 1. Fluorescence lifetimes and depolarization ratios obtained for DANS-side chain polymers.

	τ_f (ns)[a]	τ_{dep} (ns)[b]
1. Chloroform solutions		
100% DANS	0.55	0.6
25% DANS	0.56	0.6
2% DANS	0.54	0.7
DANS monomer	0.86	<0.2
2. Thin films		
100% DANS	0.38	
25% DANS	0.73	
10% DANS	1.1	
2% DANS	1.1	

a   Repeatability ± 0.03 ns in solution, ± 0.1 ns in film.
b   Repeatability ± 0.1 ns; not measurable in present set-up for film.

The situation in the polymer film is completely different. In the film the high local concentrations in the 100% DANS sidegroup polymer causes some energy transfer resulting in a shortened lifetime, of 0.38 ns, as compare to 1.1 ns in the 10% and 2% DANS-polymers, in which apparently no energy transfer interactions occur. The fluorescence intensity also is found to have decreased. The measurements on the polymer films are much more complicated from an experimental point of view, in particular due to the effects of scattered light and to the difficulty of finding a suitable reference sample. Presently, an improved measurement set-up is being constructed and more accurate data can be obtained. Also, rotational correlation times of the sidegroups in polymer films, which give information on the side chain dynamics, may then be obtained. First results, including a novel method to analyze the time-resolved data using stretched exponentials, are presented in [7]. It appears that the application of stretched exponential decay functions gives the best fit of fluorescence decays in amorphous (and hence heterogeneous) polymeric materials.

Electroluminescent Polymers As shown above the DANS-sidechain polymers give fluorescence spectra as expected: at high concentrations of the fluorophores in the polymer film fluorescence intensities are found to be reduced due to energy transfer phenomena. Relatively short fluorescence lifetimes and low quantum yields are observed, even though also in the 100% DANS polymer the fluorophores are separated by the transparent comonomer. Electroluminescent materials form a completely different class of polymers. In such materials the polymers consist of a chain of connected, conjugated, chromophores, which give luminescence by the grace of their joint electronic structure [18]. On the contrary, electroluminescent materials may show very high fluorescence quantum yields although the chromophores are adjacent. In fact, the fluorescence is produced by a number of connected, conjugated, monomeric units. The electroluminescent properties of the polymers can be studied most easily by making use of photoluminescent approaches: the electroluminescent emission spectrum closely resembles the photoluminescence spectrum [25] and the efficiency of the electroluminescence is directly related to that of the photoluminescence ($\Phi_{el} \sim 0.25\, \Phi_{pl}$ [26], due to the fact that the electroluminescent state, formed by recombination of charge, only has a 1:3 chance of ending in the excited singlet state). Unfortunately, in electroluminescence additional loss mechanisms are encountered, as compared to photoluminescence (e.g., due to interface phenomena, in particular near the − metal − electrode surfaces, due to traps that block the transport of charges towards each other in the thin polymer film, and due to unsuccessful recombination processes), so that the estimate based on photoluminescence sets an upper value to the attainable Φ_{el}.

To determine the electroluminescent properties, significant time has to be invested in the fabrication of a sample; the photoluminescence properties of the polymer, however, can be easily determined both in solution and for a thin film in the solid state. Fig. 2 shows the photoluminescence spectrum obtained for a poly(p-phenylene vinylene), PPV, type electroluminescent polymer. The quantum yield of photoluminescence was determined by making use of quininesulphate as reference, according to a well documented method, as depicted in the figure [27]. The method was calibrated using diphenylanthracene in degassed cyclohexane; the quantum yield of this compound was 0.84, close to the literature value of 0.83 [27]. It appeared that careful removal of oxygen was essential to obtain reproducible results for the measurement on the polymer as well. The Φ_f of the polymer was determined to be 0.67 in a solution saturated with argon. This value is indeed quite high and indicates that the polymer may be a suitable candidate for electroluminescent applications. Subsequent characterization requires the measurement of the photophysical properties of a film of the polymer. The polymer studied here has a relatively large Stokes' shift, and therefore negligible overlap of its absorption and emission spectrum. Reabsorption effects in the very thin layers applied in electroluminescent devices (typically 70-100 nm) are therefore relatively small. The *electroluminescence* efficiency is also determined by other factors, e.g. interface effects and loss of efficiency due to the only partial outcoupling of the light. The electroluminescence efficiency of conjugated polymer films is in the order of a few per cent.

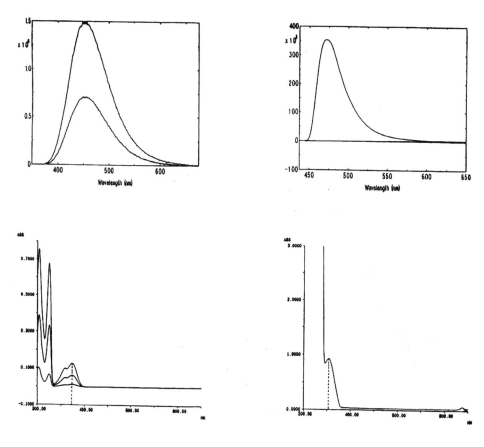

Figure 2. Quantum yield determination of electroluminescent polymer. Left figures show the emission spectrum (top) and absorption spectrum (bottom) of the reference quinine sulfate in 0.1 M sulfuric acid. Right figures show the emission (top) and absorption (bottom) spectra of the polymer. In the calculation differences in refractive index of the two solvents, and in the excitation intensity were corrected for.

Extrinsic fluorescence

Of course the applications of fluorescence methods would be limited if only intrinsically fluorescent materials could be studied. Fortunately, a wide range of fluorescent probes and labels is available which can be used to stain non-fluorescent polymers and study many different properties of these materials. The application of fluorescent labels in polymer chemistry to date still is limited, but from the (bio)chemical literature many interesting compounds can be derived [28]. In the present paper two types of application of fluorescent probes are discussed. First, a probe with solvent dependent charge-transfer fluorescence is used to study properties of polymers and polymerization reactions. This probe is generally added as such, and detects functional changes of the material it resides in. Next, the application of fluorescent probes to determine functional groups in materials and on material surfaces is described. In this application use is made of fluorescent probes with reactive groups which specifically react with the functional group of interest and yield a covalent bond.

Charge-Transfer Fluorescent Probes for Study of Polymer Properties In the section on intrinsically fluorescent polymers we discussed the use of solvent polarity effects on the fluorescence wavelength of the NLO-active DANS sidegroup as a means to estimate the difference in excited and ground state dipole moment and from that of the molecular β. The solvent polarity dependence of the fluorescence of intramolecular charge transfer molecules, however, can also be used as an extremely powerful probe for polymer characteristics. Studies reported before for one such probe, 1-phenyl-4-[(4-cyano-l-naphthyl)methylene]piperidine, further referred to as Fluoroprobe, have shown that the fluorescence wavelength and intensity of this molecule are sensitive to both solvent polarity and solvent viscosity [29]. The latter dependency can be understood if one realizes that the bathochromic wavelength shift in polar solvents is the result of the stabilization of the highly dipolar excited state by solvatation interactions. The extent of the bathochromic shift therefore is determined by the local polarity of the matrix and by its local viscosity (since the polar molecules have to reorient around the dipolar molecule to achieve maximum interaction). The fluorescence lifetime determines the time which is available for solvent reorientation. As will be shown the response of Fluoroprobe to (changes in) matrix polarity and viscosity may be used to obtain "inside" information on processes occurring during polymerization reactions etc.

As an example, observations will be discussed during interactions and subsequent reactions of polymer latices. Such suspensions of particles are used in the production of diagnostic tests and in waterborne coatings. Two phenomena are of interest in the study of the dispersed particles: the loading of the particles with solvents added to the aqueous dispersion and the subsequent reactions of the particles by seeded emulsion polymerization or by curing with a suitable multifunctional additive. The swelling of the latex particles is used to introduce dyes, which can be used for sensitive detection of selected constituents of bodily fluids via specific immunochemical interactions. Fig. 3a shows the shift of the fluorescence maximum of Fluoroprobe observed in a polystyrene latex swollen with increasing concentrations of dichloromethane. The final wavelength, reached for addition of 2.2 mol% of the organic solvent is exactly equal to the emission wavelength of the probe in the pure solvent, which indicates that Fluoroprobe is fully solvated by dichloromethane.

Swelling experiments in other solvents give similar results [11]. The wavelength of 445 nm in pure, unswollen, polystyrene latex is relatively short. The emission wavelength is partially shifted to short wavelength by the high viscosity of the matrix, as is indicated by the fact that the fluorescence of the Fluoroprobe in the latex is polarized: a degree of polarization of 0.17 is determined in polystyrene dispersion, as compared to a mere 0.015 in pure dichloromethane. Addition of toluene, with approximately the same polarity as polystyrene, causes rapid swelling and increased mobility of the dispersion, as is shown by a hypsochromic wavelength shift (see Fig. 3b). At the same time the polarization of the fluorescence decreases. Dynamic studies show that the fluorescence wavelength changes occur momentarily following the addition of the solvent. Alternative methods, like quasi-elastic light scatter and field-flow-fractionation, which rely on the measurement of increasing sizes of the latex particles upon swelling, only allow the observation of effects for high concentrations of added solvents. The in-situ fluorescent probe, however, unambiguously shows the presence of significant amounts of solvent inside the particles long before size changes become apparent [11].

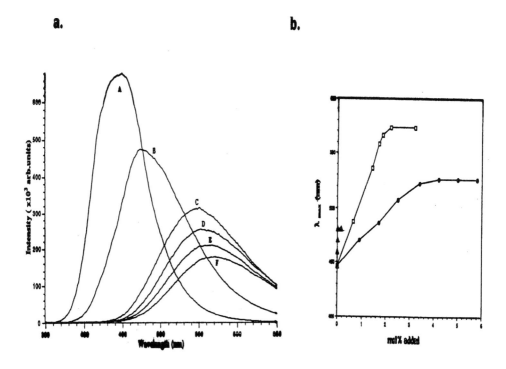

Figure 3. (a) Shift of the intensity of the fluorescence maximum of Fluoroprobe due to addition of dichloromethane. The curves A-F represent 0%, 0.7%, 1.4%, 1.7%, 2.0% and 2.2% of solvent added (w.r.t. amount of water in the dispersion). The spectrum obtained at 2.2% is the same as that in neat dichloromethane. (b) Shift of the fluorescence maximum of Fluoroprobe due to the addition of toluene (▲), dichloromethane (□), and diethylether (■) to the dispersion.

To obtain particles with immunochemically active groups first the polystyrene beads are supplied with a glycidylmethacrylate core. The thus formed latices are thought to consist of a polystyrene core surrounded by a glycidylmethacrylate shell, and hence are dubbed core-shell latices. The glycidylmethacrylate core can be functionalized to react with the immunochemical groups. The emulsion poymerization can be followed with the Fluoroprobe molecule as well, as becomes clear from Fig. 4. Directly after addition of glycidylmethacrylate monomer the fluorescence of the swollen latex shifts from 445 nm to 592 nm, due to the combined effects of greater mobility and higher polarity of the monomer. After the polymerization has been started, the viscosity increases and the fluorescence shifts to shorter wavelength. The final spectrum shows the presence of at least two species: one fluorescing at about 445 nm (shoulder, corresponds with the emission maximum of Fluoroprobe in "pure" polystyrene) and a pronounced transition at 416 nm, indicative of Fluoroprobe in a highly viscous environment. This environment cannot be the pure poly(glycidylmethacrylate) shell, since the fluorescence in this matrix is found at 512 nm (starting wavelength in glycidylmethacrylate solution was 645 nm). It is most likely that polystyrene forms a very tight interpenetrating network with glycidylmethacrylate with a very limited mobility. The fact that following the reaction the core-shell particles do not swell with, e.g. dichloromethane, supports this assumption. The presence of several fluorescent species in the latex is supported by fluorescent lifetime measurements: the lifetime analysis provides two distributions with reasonably long lifetimes, the relative contributions of these contributions shifting when the detected emission wavelength is shifted (Fig. 5). The Fluoroprobe molecule in neat solvents only shows a single exponential fluorescent lifetime distribution, as expected.

An exciting application of Fluoroprobe is in the study of curing reactions in coatings. A disadvantage of the application of fluorescent probes present as such is that one completely has to rely on the distribution of the probe over the different sample constituents. In order to get more control over the targeted regions a functionalized Fluoroprobe molecule has been synthesized. Details of the synthesis and the photophysical properties of the maleimido-Fluoroprobe will be presented elsewhere [30].

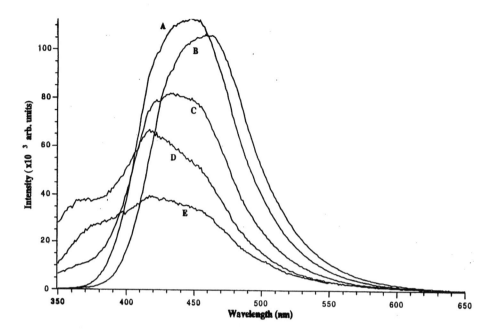

Figure 4. Shift of the fluorescence maximum of Fluoroprobe as a result of the polymerization of glycidylmethacrylate in the polystyrene latex. Curves A-E were taken at the start and after 20 min, 60 min, 180 min and 520 min of polymerization. The small shoulder at 380 nm is probably due to decomposition of the probe.

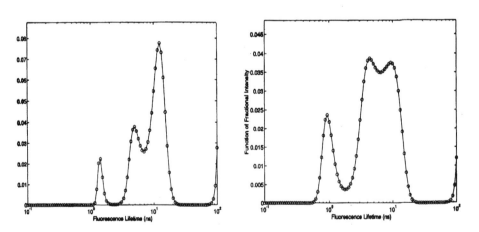

Figure 5. Fluorescence lifetimes determined for the core-shell latex obtained after 520 min of polymerization of glycidylmethacrylate, using detection wavelengths of 420 nm (left) and 470 nm (right). Two lifetime distributions, with average lifetimes of about 8 ns and 2 ns, respectively, are clearly discerned. The longer lifetime is predominantly observed when the shorter wavelength is monitored (i.e., in a more viscous environment). The very short component may be an artifact.

Here it is important to emphasize that the maleimido-Fluoroprobe is a fluorogenic compound, which exhibits charge-transfer fluorescence following reactions with amino- or thiol-groups. This probe molecule therefore has the important advantage that it is only observed after reaction with the target, which greatly enhances its specificity. With the maleimido-Fluoroprobe a coating system was investigated consisting of a polyacrylate latex with acetoacetate functional groups and Jeffamine T403 as crosslinker. Jeffamine contains trifunctional amino-groups and was labeled with a slight amount (1 mol%) of maleimido-Fluoroprobe. The labeled Jeffamine solution fluoresces at 576 nm. In the aqueous dispersion it also fluoresces, at about 572 nm, which is surprising since the fluorescence of the Fluoroprobe molecule itself is completely quenched in water and other protic solvents. Probably the less polar and nonprotic Jeffamine shield the Fluoroprobe molecules from the quenching effects induced by the water, maybe via the formation of micel-like structures. When the curing reaction takes place in solution the fluorescence maximum of the Jeffamine shifts to 511 nm (Fig. 6a). The subsequent loss of water from the coating dispersion when it is applied as a thin film can also be followed with the fluorescent probe. The film apparently loses the water quite fast: within 20-30 min the final fluorescent wavelength of about 495 nm is reached (Fig. 6b). The results obtained with the maleimido-Fluoroprobe agree very well with measurements done with alternative approaches, such as titration of functional groups used to monitor the reaction rate in solution and weighing to follow the evaporation of water from the film [31]. However, the fluorescence probe approach has as important advantages that it is extremely simple to apply and offers the possibility to visualize the curing and drying with microscopic techniques.

a.   **b.**

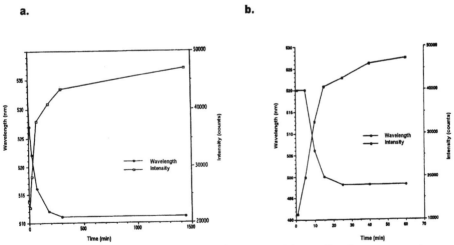

Figure 6. (a) Fluorescence wavelength shift (■) and intensity change (□), observed during the curing of a polyacrylate dispersion for maleimide-Fluoroprobe covalently bound to the Jeffamine crosslinker. (b) Fluorescence wavelength shift (□) and intensity change (■) observed during the curing and drying of a coating of poly-acrylate-Jeffamine (covalently labeled with the maleimide-probe) applied as thin film.

<u>Determination of Functional Groups</u> As last example the determination of functional groups will be discussed. Functional groups on surfaces are a major determinant of special properties of particles. A few examples are: (1) particles which are used in chromatographic systems, where functional groups may provide specific interactions with particular analytes, (2) latices which are used for diagnostic purposes, where particles are bound to antibodies to invoke immunochemical diagnosis, (3) functionalized latices applied in coatings with improved properties, such as better blocking or water resistance and (4) modified fiber surfaces with special adhesion properties. Functional groups on polymers determine properties like stability, entanglement, solubility, processability and possibilities for further reactions. Important information can be derived from the concentration and distribution of the functional groups. A number of analytical techniques are available to determine the concentration of functional groups. Most of these techniques have been developed for study of homogeneous systems, but can in some cases be applied to heterogeneous systems as well, provided that the concentration of the functional groups is sufficiently high. The most direct measurements are based on titrations, using a variety of detection principles (pH, conductivity, colorimetry, etc.) [32]. Alternatively, strongly colored (absorption) or fluorescent probes can be used, which form specific, covalent, bonds with the functional groups. In particular fluorescence spectroscopy has as advantage that it is sensitive and may be used to obtain images of labeled particles, so that even information on single particles may be obtained.

Lucifer Yellow-VS reacts readily with the amino-endgroups and has been applied to label functionalized silica particles. Emission spectra were obtained using 458 nm excitation, excitation spectra were recorded while monitoring the emission at 535 nm. The emission and excitation spectra thus obtained are shown in Figs. 6a and 6b. From the figures it is obvious that three concentration regions can be discerned for the fluorescence behaviour of LY on silica particles. In the first region, 0-40 mmol endgroup /kg, the fluorescence emission and excitation spectra have the same appearance and show a linear dependence with the concentration of endgroups. The emission spectrum peaks at 535 nm, the excitation maximum is at 430 nm. The third region starts beyond 76 mmol endgroup/kg, and shows strong evidence of intermolecular interactions: strong red shifts and a significant loss in fluorescence intensity. In this concentration range inner filter effects will play a role. In the second, intermediate, region the fluorescence intensity increases supraproportional and shows a blue shift (compared to the low concentration range of 0-40 mmol/kg). The emission maximum of the 76 mmol/kg sample is at 520 nm, i.e. 15 nm blue shifted. The most obvious explanation for this behaviour can be found in the mode of interaction of the LY groups with the silica surface [12]. The results underline that fluorescent labels can be used to quantitatively determine low concentrations of functional groups. At high concentrations interactions occur between the labels, which disturb the quantitative application, but offer possibilities to assess the distribution of the functional groups at the surface of particles or in latices. LY is highly polar and dissolves well in the aqueous medium applied in this study. The results obtained suggest that the LY molecules do not form clusters on the silica surface, but are homogeneously distributed.

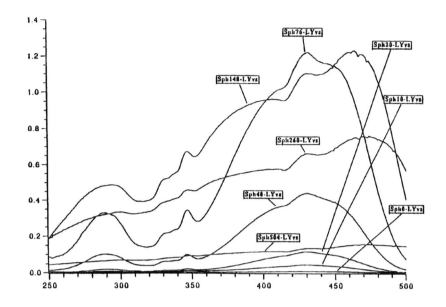

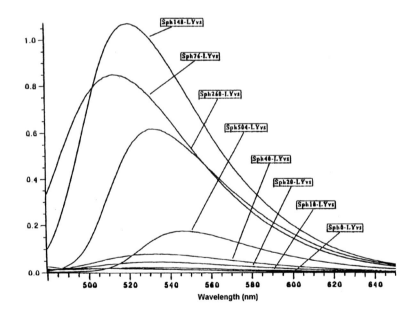

Figure 7. Fluorescence excitation (top) and emission (bottom) spectra observed for amino-functionalized silica spheres labeled with Lucifer Yellow VS (vinylsulfonate).

The functional group content of individual particles may be examined by application of flow cytometry. With this technique the fluorescence intensity of individual particles can be measured. Hence, the homogeneity of the functional group level on the particles can be examined. Fig. 8 shows typical bivariate plots obtained for flow cytometric analysis of the LY-labeled Spherosil beads. The measured intensities per particle are shown in the bivariate plots: every point in each of the figures represents a measurement. The total number of particles analyzed is 10,000, which takes approximately 2 min. A total of 4 parameters has been measured for every particle: the LY fluorescence (FBY), the length of the particle ("time-of-flight", TOF), forward light scatter (FLSB) and perpendicular light scatter (PLSB) for the blue laser. The figures also show (in the left-hand corner) data obtained for fluorescein containing standard beads. The standard beads are used for calibration of the measurements, i.e., to correct for possible variations in laser power or optical alignment. From the flow cytometric data it is obvious that the particles show a tremendously wide size distribution (typically 120±60 μm). The functional groups, however, are evenly distributed over the particles. The total concentration of functional groups per particle therefore is mainly determined by their size, as is illustrated by a strong decrease in the spread of the fluorescence intensity data after correction of the raw data with the size dependent variables [12].

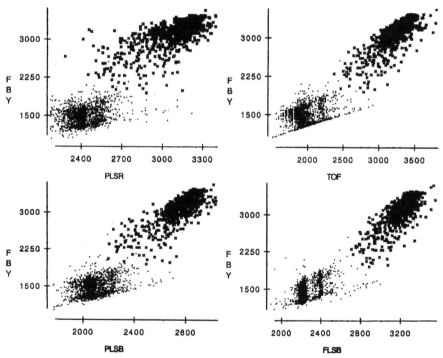

Figure 8. Bivariate plots obtained by flow cytometric analysis of Spherosil microspheres labeled with Lucifer Yellow VS. The plotted parameters are fluorescence blue-to-yellow (FBY), red and blue perpendicular light scatter (PLSR and PLSB, respectively), blue forward light scatter (FLSB), and time-of-flight (TOF). The intensities along the axes are logarithmic units.

Since quantitative data are available for an impressive number of particles use can be made of advanced statistical and computational approaches to derive information on individual particles and to extrapolate to bulk properties [33,34].

Moreover, the fluorescently labeled particles can also be directly visualized by application of microscopic techniques. With confocal laser scanning microscopy, for instance, 3-dimensional information can be obtained on the distribution of the functional groups in the particles [12].

A major limitation of the application of fluorescent techniques to study functional groups and other polymer properties is formed by background signals caused either by strong scatter of the particles or by strong absorption or, obviously, by background luminescence of the polymeric system. Most commercially available fluorescent labels are excited in the UV or blue and emit in the visible part of the spectrum. Just these relatively short wavelength ranges are subject to strong interferences. Recently, we have applied some newly developed specific labels which absorb and fluoresce in the near-infrared. The labels are based on cyanine-structures and are excited at about 650 nm (containing 5 conjugated C-atoms in the link) and 750 nm (containing 7 conjugated C-atoms) and fluoresce in the 670 nm and 790 nm region, respectively. Apart from having long wavelength emission, these labels also have very high extinction coefficients (typically 200,000 l mol^{-1} cm^{-1}), so that they can also be used for sensitive reflection and absorption measurements [15]. These labels can be applied successfully to study functional groups on scattering particles, as is demonstrated in Fig. 9a, for a core-shell latex with aldehyde-functionality in the shell, which has been labeled with a cyanine-hydrazide probe. The spectrum displays the near-infrared fluorescence of a latex containing only 0.044 mmol/kg of aldehyde groups, which have been quantitatively labeled. Recent results, obtained by making use of complexes containing lanthanide ions with emission far into the near-infrared (particularly Nd^{3+}, emitting at 880 nm, 1060 nm and 1330 nm, Yb^{3+}, emitting at 980 nm, Er^{3+}, emitting at 1530 nm), have been developed for sensitive diagnostic applications [35].

a.   **b.**

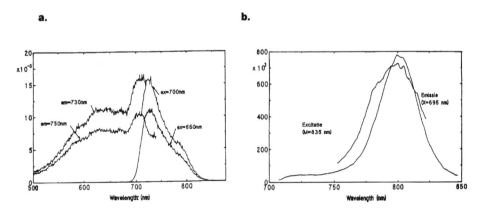

Figure 9. (a) Fluorescence emission and excitation spectra obtained for polystyrene core-shell latex, with aldehyde functionality, covalently labeled with a near-IR fluorescent fluorescent probe, based on a cyanine-hydrazide. (b) Fluorescence emission and excitation spectra obtained for polyamide covalently labeled with a near-IR emitting cyanine-isothiocyanate.

Fig. 9b shows the fluorescence spectrum obtained for a high molecular weight polyamide, the – low concentration – of amino-endgroups of which has been labeled with a cyanine-isothiocyanate label. Background fluorescence of the polymer, mainly occurring in the blue and green part of the spectrum, is absent. The application of fluorescent labels to study functional groups on polymers is not always straightforward. Most labels have been developed for biomedical applications. Particularly these labels are commercially available. The labels have been optimized to work well under physiological conditions, i.e. in aqueous solutions, pH of 7.5-8 and not too high temperatures. For latices such labels therefore can be used without many problems. For most polymers, however, only exotic solvents can be used. Polyamides only dissolve in solvents like formic acid, hexafluoroisopropanol, n-methylpyrrolidone or m-cresol. Only in the latter solvent the reaction with the isothiocyanate label can be performed. The acidic solvents render the amino-groups into their protonated forms, which precludes the labeling reaction, and the labels do not dissolve in n-methylpyrrolidone. An advantage of labeling polymers is that it is generally easier to remove excess label than in homogeneous systems. From dispersions or particles the excess label can be removed by simple filtration techniques. From labeled polymers the label can be separated by making use of washing procedures, using a solvent which dissolves the label and low molecular weight products and not the polymer. A simple precleaning procedure for homogeneously labeled systems is by making use of functionalized particles, which react with the excess label and are easily removed via filtration.

Conclusions

Fluorescent techniques are versatile and easy to use methods to characterize polymeric systems. Various methods of analysis may be applied, which make use of steady state fluorescence measurements, time-resolved measurements or polarized measurements, and can be used on materials in solution as well as applied in films. Of course, specific set-ups are required for optimal excitation and detection of the emitted light, depending on the kind of sample under consideration. A variety of properties of polymeric systems can be determined: structural aspects, in solution and in the solid state, interaction between fluorophores, dynamics of polymer sidechains and backbones (provided they contain fluorescent groups), specific properties of NLO-active and electroluminescent polymers etc.

By application of fluorescent labels even more areas of research come within reach. In-situ morphological characteristics of polymers as such and polymers in interaction with solvents or undergoing some sort of physical or chemical action may be studied. In addition, concentration and distribution of functional groups can be determined. The labeled systems can be imaged and characterized individually by application of microscopic or flow cytometric techniques. Application of recently developed near-infrared fluorescent labels allows the study of polymers which have interfering autofluorescence or strong absorption or occur as dispersed particles which produce strong scatter background signals. In particular the application of extrinsic fluorescence methods, using the wealth of available labels, offers many yet unexplored areas of research. On the other hand a lot of work remains to be done to find out how to use these labels, which almost exclusively have been developed for biomedical applications.

Acknowledgements

Mr. B.H.M. Hams and Dr. R.H.G. Brinkhuis (Akzo Nobel Central Research, Arnhem, The Netherlands) are acknowledged for preparation of the NLO- and electroluminescent polymers. Dr. A.J.W.G. Visser, Dr. E. Novikov and mr. A. van Hoek (Agricultural University Wageningen, The Netherlands) are thanked for the time-resolved fluorescence measurements, using time-correlated single photon counting. Prof L.B. McGown, of the Duke University, Durham NC. Dr. H.J. Verhey and Prof. Dr. J.W. Verhoeven (University of Amsterdam, The Netherlands) are acknowledged for their contributions to the use of charge-transfer fluorescent probes for the study of polymeric materials. Messrs. E.B. van der Tol and M.J. Nuijens (Akzo Nobel Central Research) are thanked for their assistance in the preparation of the labeled particles. Mr. W.J.M. van Zeijl (Public Works Department, The Hague, The Netherlands) is acknowledged for the flow cytometric measurements.

References

1.   J.R. Lakowicz (ed.), Topics in fluorescence spectroscopy, Vols. 1-3, Plenum Press, New York, 1991-1992.
2.   W.E. Moerner and L. Kador, Anal. Chem., 61 (1989) 1217A.
3.   J.R. Lakowicz, Principles of fluorescence spectroscopy, Wiley, New York, 1983.
4.   H.J. van Ramesdonk, M. Vos, J.W. Verhoeven, C.R. Möhlmann, N.A. Tissink and A.W. Meesen, Polymer, 28 (1987) 951.
5.   J.W. Hofstraat and M.J. Latuhihin, Appl. Spectrosc., 48 (1994) 436.
6.   J.W. Hofstraat, H.J. Verhey, J.W. Verhoeven, M.U. Kumke, G. Li, S.L. Hemmingsen and L.B. McGown, Polymer, 38 (1997) 2899.
7.   E.G. Novikov, A. van Hoek, A.J.W.G. Visser and J.W. Hofstraat, Opt. Commun., 166 (1999) 189.
8.   J.W. Hofstraat, W.J.M. van Zeijl, J.C.H. Peeters and L. Peperzak, Anal. Chim. Acta, 290 (1994) 135.
9.   C.T.J. Wreesmann and E.W.P. Erdhuisen, Eur. Pat. 88-200512 (1988).
10.   H.F.M. Schoo and R.J.C.E. Demandt, Philips J. Res., 51 (1998) 527.
11.   H.J. Verhey, B. Gebben, J.W. Hofstraat and J.W. Verhoeven, J. Polym. Sci., Part A: Polym. Chem. 33 (1995) 399.
12.   J.W. Hofstraat, G.D.B. van Houwelingen, E.B. van der Tol and W.J.M. van Zeijl, Anal. Chem., 66 (1994) 4408.
13.   G.D.B. van Houwelingen, J.G.M. Aalbers and A.J. de Hoog, Fres. Z. Anal. Chem., 300 (1980) 112.
14.   G.F. Mes, B. de Jong, H.J. van Ramesdonk, J.W. Verhoeven, J.M. Warman, M.P. de Haas and L.E.W. Horsman van den Dool, J. Am. Chem. Soc., 106 (1984) 6524.
15.   J.W. Hofstraat, G.D.B. van Houwelingen, M.J. Nuijens, C. Gooijer, N.H. Velthorst and G. Patonay, Polymer, 38 (1997) 4033.
16.   S.V. Konev, Fluorescence and phosphorescence of proteins and nucleic acids, Plenum Press, New York, 1967.
17.   P.N. Prasad and D.J. Williams, Introduction to nonlinear optics in molecular and polymeric materials, John Wiley, New York, 1990.

18. J.H. Burroughes, D.D.C. Bradley, A.R. Brown, R.N. Marks, K. Mackay, R.H. Friend, P.L. Burn and A.B. Holmes, Nature, 347 (1990) 539.
19. W.H.G. Horsthuis, H.M.M. Klein Koerkamp, J.L.P. Heideman, H.W. Mertens and B.H.M. Hams, Proc. SPIE, 2025 (1993) 516.
20. W. Liptay, in E.C. Lim (Ed.), Excited states, Vol. 1, Academic Press, New York, 1974. Chapter 4.
21. B.F. Levine and C.C. Bethea, J. Chem. Phys., 63 (1975) 466.
22. K. Clays and A. Persoons, Phys. Rev. Lett., 66 (1991) 2980.
23. R.M. Hermant, Thesis, University of Amsterdam, 1990.
24. L.T. Cheng, W. Tam, G.R. Meredith, G.L.J.A. Rikken and E.W. Meijer, Proc. SPIE, 1147 (1989) 6172.
25. N.F. Colaneri, D.D.C. Bradley, R.H. Friend, P.L. Burn, A.B. Holmes and C.W. Spangler, Phys. Rev. B, 42 (1990) 11670.
26. D.D.C. Bradley, A.R. Brown, P.L. Burn, R.H. Friend, A.B. Holmes and A. Kraft, Electronic properties of polymers, Springer series in solid state sciences, Vol. 107 (1992) 304.
27. J.N. Demas and G.A. Crosby, J. Phys. Chem., 75 (1971) 991.
28. W.R.C. Baeyens, D. de Keukeleire and K. Korkidis, (eds.), Luminescence techniques in chemical and biochemical analysis, Dekker, New York, 1991.
29. L.W. Jenneskens, H.J. Verhey, H.J. van Ramesdonk and J.W. Verhoeven, Macromolecules, 25 (1992) 6365.
30. H.J. Verhey, J.W. Hofstraat, C.H.W. Bekker and J.W. Verhoeven, New. J. Chem., 20 (1996) 7.
31. L.G.J. van der Ven, R.R. Lamping, P.J.A. Geurink and L. van Dalen, Proc. 22nd FATIPEC Conf., Budapest, 1994.
32. S. Siggla, Editor, Instrumental methods of organic functional group analysis, Wiley, New York, 1972.
33. J.W. Hofstraat, M.E.J. de Vreeze, W.J.M. van Zeiji, L. Peperzak, J.C.H. Peeters and H.W. Balfoort, J. Fluoresc., 1 (1991) 249.
34. J.R.M. Smits, H.W. Balfoort, L.W. Breedveld, J. Snoek, J.W. Hofstraat, M.W.J. Derksen and G. Kateman, Anal. Chim. Acta, 258 (1992) 11.
35. J.W. Hofstraat, M.P. Oude Wolbers, F.C.J.M. van Veggel, D.N. Reinhoudt, M.H.V. Werts and J.W. Verhoeven, J. Fluoresc., 8 (1998) 301.

PHYSICAL CHARACTERIZATION OF POLYMERIC MATERIALS

CHARLES E. CARRAHER, JR.

Department of Chemistry and Biochemistry, Florida Atlantic University, Boca Raton, FL 33431 and Florida Center for Environmental Studies, Palm Beach Gardens, FL 33431

Rheological and Physical Properties-Bulk
    General
    Rheology
    Testing
    Polymer Transitions
Surface Analysis
    General
    Auger Electron Spectroscopy and X-Ray Photoelectron Spectroscopy
    Scanning Probe Microscopy
    Secondary Ion Mass Spectroscopy
    Infrared Spectroscopy
    Raman Spectroscopy
Amorphous Region Determination
Size and Shape Determination
    General
    Light Scattering Photometry
    Chromatography
    Mass Spectroscopy
    Particle Size
    X-Ray Diffraction
    Complementary Molecular Shape Pairing

Rheological and Physical Properties-Bulk

General. Public and industrial acceptance of polymers is generally based on an assurance of quality based on a knowledge of successful long-term reliable tests. Research to achieve materials with desired and known properties is also based on a knowledge of long-term reliable tests. Many of these tests are standardized and appear as part of various national and international testing associations including the American Society for Testing and Materials, ASTM, International Standards Organization, ISO, American National Standards Institute, ANSI, Deutsch Normenausschuss, DNA, and the British Standards Institute, BSI. These tests are continually being developed and submitted to the appropriate committee within the testing society for verification through "round robin".

After sufficient verification they can be accepted by consensus as "standard tests".

Polymers are complex because of the varying microstructures, chain lengths and distributions, importance of previous treatment (history), presence of additives and modifiers, difference between interfacial surfaces and bulk structure and property, ... Commercial utilization of polymer-containing materials is dependent on particular polymer behavior under a range of possible conditions. Thus, a single characterization technique is not sufficient to adequately characterize polymers, but results from various techniques are needed to evaluate and characterize polymeric materials.

There is a move towards coupling instruments and in some instances forming instrument assemblies where three, four, and more instruments are interconnected. Such assemblies or instrumentation nodes will offer data that is more than the sum of the data output and associated characterization data of the stand-alone instruments. These instrumentation nodes will offer a decreased cost per individual analysis often simply because many of these analysis will become automated and the instruments will be run on an almost continuous basis.

But the cost of such instrument node sites will be high because of the
*increased number of expensive instruments within such an instrument node,
*increased need for expert maintenance, and
*increased need for highly skilled technicians, staff, and scientists associated with the instrument node.

Because of the high cost, most universities and smaller industrial and governmental sites will not have their own instrument node but rather will "plug-in" to other sites that, because of the electronic age, will not be limited by geographical location.

Such instrument nodes will contain both the instrument assemblies and numerous stand-along instruments that will deliver individual analyses.

Sample preparation can be done either by the originating scientist or at the node site. Results will be e-mailed or otherwise electrically communicated. Interruption can again be done either by the customer or at the instrument node.

Another trend will be the coupling of experimental results with modeling made possible by the increased scientific knowledge base and increased availability of high computing power. Such modeling may allow the validation of the model, elimination of areas of necessary testing, and point to areas in need of further testing. It will also help direct the ultimate synthesis of complex materials and combinations through creation of better structure-property relationships. Of caution, it will be important to connect the mathematical modeling and results to real molecular phenomena remembering that the model is only as good as it relates to reality. Scientific intuition will continue to play an essential role in this process.

Rheology. The introductory chapter of this volume describes briefly the topic of polymer testing including rheology. The reader is referred to it for a brief introduction and to references 1-5 for a fuller discussion.

While rheology is generally defined as the study of the deformation and flow of materials, for polymers it can be described as the study of the movement of polymers and polymer segments in solution or bulk. Movement or displacement of polymers and polymer segments can be of a liquid or solid type, that is flow or viscosity type and elastic type. Polymers are then described as being viscoelastic materials exhibiting either or both characteristics depending upon the operational conditions. For simple materials, below the glass transition temperature, Tg, polymers act as solids or elastic materials with the materials generally appearing as being brittle. Above the Tg the various segments are able to flow and, when given enough response time to an applied stress/strain, will act more flexible or viscous with the material being more pliable. Above the melting point, Tm, polymeric materials will flow acting as a viscous liquid.

A liquid can be defined as a material that continues to deform as long as the material is subjected to a stress/strain. Many materials will act as a liquid over only a specific range of temperatures and rate of applied stress/strain. Thus, even a plastic material that is above its Tg will behave as a solid when the rate of applied stress/strain is too fast to allow the various segments to move. For materials behaving as a liquid, applied shear stress produces a sliding of one molecular layer past another. The laminar flow in simple shear can be defined in terms of the force, f, required to move a plane of area A relative to another plane a distance, d, from the initial plane. This force is proportional to the area of the plane and inversely proportional to the distance. This is made into a direct mathematical relationship using a proportionality factor called the coefficient of shear viscosity or simply viscosity, η.

$$f = \eta \, (A/d) \tag{1}$$

Viscosity is then a measure of the resistance of a material to flow. It's inverse is called fluidicity. Viscosity is generally measured in the CGS unit of "Poise" which is dyne seconds per square centimeter. Another widely employed unit is the "Pascals" or "Pas" which is Newtons per square centimeter. The relationship is 10 Pois = 1 Pas.

Introductory texts contain the basic descriptions and relationships relating viscosity to molecular weight for dilute solutions. Briefly, the limiting viscosity number or intrinsic viscosity is related to molecular weight by the Mark-Houwink relationship

$$\eta = KM^a \tag{2}$$

Viscometry is still utilized as a relative molecular weight measure because of the accessability of a large number of constants that allow the viscosity to be related to molecular weight, ease of obtaining viscosity values, and the low cost of appropriate instrumentation.

Materials offer different flow relationships. For Newtonian materials the shear stress is directly proportional to shear rate. Most dilute solutions of polymers approximate Newtonian behavior. More concentrated solutions and melts of polymers can exhibit complex flow characteristics. Curves for some of these general classes of behaviors can be found in most introductory polymer texts.

Time-dependent behavior include irreversible changes such as degradation, mechanical instability, coagulation, and physical and chemical crosslinking. Reversible changes involve the breaking and reforming of networks and colloidal groupings.

Melt viscosity is important to a fabricator and manufacturer because many fabrications involve polymer and polymer mixtures that are "flowable" either with heat alone or with a combination of heat and pressure. Thus, the flow properties as a function of temperature, pressure, molecular weight and molecular weight distribution, and polymer structure and pre-history are important to know. While a number of methods have been applied to determining the melt viscosity of polymer melts the most employed techniques include the capillary extrusion methods (1), rotational methods (1), oscillatory (6,7) and controlled stress (8) methods.

For polymers with a narrow molecular weight distribution the limiting low or zero-shear viscosity of a polymer melt can be related to the weight average molecular weight by the Staudinger-like relationship

$$\eta = KM \tag{3}$$

for low molecular weight materials and for high molecular weight materials by the Mark-Houwink-like relationship

$$\eta = KM^{3.4} \tag{4}$$

with the transition between the two relationships occurring at a molecular weight called the critical chain length, Z. The critical chain length is important since below it the relationship between chain length and the amount of stress, and thus energy, is directly proportional.

For polymers with a broad molecular weight viscosity is related to some average between the weight average and z-average molecular weight.

As noted in the introductory chapter and references 2 and 9 the particular elements of a viscoelastic behavior can be described in terms of two elements and combinations of these elements. The Newtonian dashpot or frictionless piston corresponds to the viscous properties and is related to the movement and sliding of chains and chain segments past one another. The Hookean spring represents the elastic behavior and corresponds to bond flexing and other similar behavior. Comparing graphs made utilizing various combinations of these two elements with the actual stress/strain curves allows an approximate determination of the relative importance of bonding flexing and segmental slippage. Experimental curves are quite dependent on a number of factors including chain nature and kind, molecular weight and molecular weight distribution, conditions under which the experiment are carried out, and material prehistory.

Testing. A number of physical tests are important to the use of polymers (2,10). The results of such testing forms the basis of evaluating appropriate uses and applications. Such testing also allows quality control of the materials indicating that the particular polymeric materials will function as per specifications. Table 1 lists some of the common tests. Following is a brief review of some of these tests.

Table 1. Typical physical test categories for polymeric materials.

Chemical Properties and Resistance
Crystal/Amorphous Regions-Amount, Distribution, Size
Electrical
Molecular Weight and Distribution
Particle Size
Spectronic/Optical Characterization-Index of Refraction, Reflectance, Transmittance, Clarity
Transitions-Tg, Tm
Thermal-Conductivity, Degradation, Expansion, Heat Capacity, Heat Deflection, Softening-Tg, Melting-Tm, Flammability
Surface

Electrical. As polymers are employed as more than simply so-called "inert" materials surrounding and encasing electrical wires and devices the particular electrical properties become increasing important (2,11). Major electrical behavior characteristics include the dielectric properties of loss factor, dielectric loss, dielectric constant or specific inductive capacity, direct current (DC) conductivity, electric breakdown strength, and alternating current (AC) conductivity.

The dielectric behavior is often studied using charging or polarization currents. Since polarization currents depend on the applied voltage and the condenser it is customary to eliminate this dependence by dividing the charge Q by the voltage V giving the capacitance or capacity C

$$C = Q/V \tag{5}$$

and then using the dielectric constant which is simply the capacity of the condenser when the material is placed between its plates and the capacity of the condenser when it is empty. The terms dielectric constant and polarization constant are often used interchangeably in general discussions of the magnitude of the dielectric constant.

Dielectric polarization is the polarized condition in a material that results from application of a DC or AC current. The polarizability is the electric dipole moment per unit volume induced by the applied field or unit effective intensity. The molar polarizability is a measure of the polarizability per molar volume and is generally given per some repeat unit of a polymer.

Two types of charging currents and condenser charges are generally employed. These are rapidly or instantaneous forming polarizations and more slowly forming or absorptive polarizations. The total polarizability of the dielectric is the sum of contributions due to several types of displacement of charge produced in the polymer by the applied field. The relaxation time is the time required for polarization to form or disappear. The overall polarizability is a combination of electronic, atomic, dipole, and interfacial contributions. Only contributions by electronic polarizations are evident at high frequencies.

Today electrical testing can employ devices that introduce current under conditions that approximate use conditions or are used to evoke a special response allowing the experimenter to evaluate a particular property behavior. Along with the normal AC cycles, alternating series or applied voltages can be employed. Again, the more standard of these tests are contained with the ASTM manuals.

Thermal Analysis. Thermal analysis is the measure of a particular polymer property as a function of temperature that may include the measure as a function of heating rate, applied stress/strain, and atmosphere (2, 12-14). Depending upon the particular instrumentation employed many measurements are possible including phase changes, sample purity, crystal/amorphous ratio, specific heat, heat of reaction and transition, rate of melting/crystallization, reaction rate, activation energy, solvent retention, amount of cure, and sample identification.

Many of the newer techniques are so-called dynamic variations of already existing techniques. In both dynamic mechanical and thermal analysis, a sinusoidal effect is imposed on the sample. This sinusoidal effect may be stress/strain related in mechanical studies. Popular variations include changes in frequency, temperature, time, and magnitude of stress/strain.

In dynamic mechanical analysis, DMA, a sinusoidal applied stress or strain ais applied. Application of the sinusoidal signal to an ideal viscous liquid gives a response that is 90° out of phase while application of the signal to a perfectly elastic material gives a response that is proportional to and in phase with the applied signal. Since plastics are viscoelastic with both liquid and elastic properties, the responses of materials to application of sinusoidal applied stresses and strains is generally between the two. The degree of response that is out of phase is called the phase angle which in turn is a measure of the viscoelastic behavior of the material.

Many of the recent advances involve coupling of thermal techniques with other techniques. The recent commercialization of the thermal gravimetric analyzer and the mass spectrophotometer based on the earlier work of Carraher is illustrative of this (2).

Reading and co-workers have pioneered in the adaptation of nano-associated Atomic Force Microscopy, AFM, for thermal analyses (2). The tip of the AFM is replaced with a miniature resistive heater that is used to heat or measure temperature differences on a nano-scale. This allows differentiation between phases and (at times) individual polymer chains and segments (such as block and graft segments of copolymers), Tg and Tm and melting based on differences in thermal conductivity and diffusivity. This technique using the micro-thermal sensor, MTDSC, is called Calorimetric Analysis with Scanning Microscopy, CASM. The probe can also be used to measure certain mechanical properties performing a micro-Thermal Mechanical Analysis with Scanning Microscopy, MASH.

Phase Chances Because of the strong combined secondary forces, polymers do not form a true gaseous phase. Some undergo decomposition prior to melting including those with particularly strong secondary forces and some crosslinking. Even so, most linear polymers undergo at least two major phase changes with one associated with more local segmental movement, the glass transition temperature, and one related to wholesale entire chain mobility, the melting point or more properly "melting range". Many polymers can exhibit additional transitions associated with various rearrangements of the main chain, portions of the main chain, and groups attached to the main chain.

The particular temperature range for a given transition will vary depending on both the rate of temperature change and on the particular technique employed. In general, as the heating rate is decreased, the more time is allowed for the particular change to occur and the lower the temperature where the change is experimentally detected. As the temperature is lowered, a sample may undergo several transitions. By agreement, the first transition is called the alpha transition, the second the beta, etc. The glass transitions is most generally the alpha transition. These transitions are generally grouped together as second-order transitions whereas the melting range is called a first-order transition.

Transitions can be experimentally observed using any technique that allows a measure of a difference in some property as the material undergoes the transition. These measurements are necessarily carried out as a function of temperature. Table 2 contains a brief listing of some of the general categories used in determining polymer transitions.

Table 2. General categories of measurements that can be used to detect material transitions.

Mechanical Property Change (such as gas and vapor permeability)
Thermal Analysis (such as heat capacity)
Physical Property Change (such as volume and length/dilatometry; melt viscosity; refractive index, density, dielectric constant, penetration)
Spectral Analysis (such as broadening in NMR bands; X-ray; infrared and Raman spectroscopy)

As noted above, many polymer measurements are time as well as temperature dependent. Thus the viscoelastic behavior of materials is dependent on the time required for specific changes to occur. In general, the Boltzman time-temperature superposition approach can be used to predict behavior. Here master curves are used. An experimental curve is superimposed onto the set of master curves and behavior is then predicted under different conditions. While this approach is useful for some applications, it is sometimes used to predict long-term failure of materials. The curves are based on gradual changes and where the particular property is "gradually changing" this approach gives reasonable results. Polymer breakdown or failure is often catastrophic dependent on the "weak link within the chain" and for these situations, the use of such curves is only satisfactory prior to the catastrophic event and may not give accurate results even as the catastrophic event is approached.

Surface Analysis

General.   Surface is a general description describing the outermost atomic layers, including absorbed foreign atoms (2, 13). The surface makeup is very important to both the physical and biological properties of a material. Thus, biologically if incorporation into a structure is desired, then rugged or jagged surfaces are typically preferred, whereas when adherence is not wanted, smooth surfaces are preferred. In general, the nature of the surface is different from the bulk composition. The surface composition is dependent on the "forming" conditions. Thus, polymers with polar and non-polar segments will have a surface containing mainly polar components if the polymer is crystalized within air.

     Major problems addressed by surface analysis include the nature of the surface, impurities, wear, friction, corrosion, sites of deformation, catalysis structure, water content, effect and location of processing aids and additives, sites of environmental and chemical attack and chemical action and nature of interfaces.

     Surface characterization techniques can be divided into two broad categories-those that emphasize the outermost few layers and those that focus deeper into the surface.

Auger Electron Spectroscopy and X-Ray Photoelectron Spectroscopy.(2,13)   Auger Electron Spectroscopy, AES, and X-Ray Photoelectron Spectroscopy, XPS, are used to identify the elemental composition of the surface allowing identification of the amount and nature of the species present at the outermost surface (about 1 nm). In AES, incident electrons interact with the inner shell electrons of the sample. The vacancy created by an ejected inner shell electron is filled by an outer shell electron and a second outer shell electron is ejected leaving the atom in a doubly ionized state. The electrons ejected from the outer shells are the Auger electrons. Measure of the energy of these electrons allows the identification of the elements present in the outermost layers.

     XPS involves the polymer being bombarded by a beam of X-rays resulting in core electrons being ejected.   These core electrons are called X-ray photoelectrons.   The kinetic energies of these ejected electrons are characteristic of the particular atom and allows the determination of the nature and amount of surface atoms.

Scanning Probe Microscopy.   (2,13) Scanning probe microscopy, SPM, encompasses a group of surface-detection techniques that include atomic force microscopy, AFM, and scanning tunneling microscopy, STM, that allow the topographic profiling of surfaces. SPM techniques investigate only the outermost few atomic layers of the surface with nanometer resolutions and at times atomic level resolution.

     STM is typically used with electrically conductive materials applied to polymeric materials giving overlays consisting of conducting material layered over the surface of the sample.   STM experiments generally require very low pressures of less than 1×10^{-10} mbar. By comparison, AFM can be run under room conditions and does not require the use of electrically conductive material. In STM the metallic tip is held close (about 0.5 to 1 nm) to the surface. A voltage, applied between the tip and sample surface, drives a tunneling current. The conductive surface reconstructs the atomic positions via minimizing the surface free energy. This gives topographic superstructures with specific electronic states which are recorded as surface contours or images.

AFM can be run under room conditions. AFM can be performed in either of two forms-a contact mode and a non-contact mode. It does not require the use of electrically conductive material since (in the contact mode) the tip actually "touches" the surface rather than residing immediately above it as is the case in STM. In both the contact and the non-contact mode light is used as the sensing source rather than an applied voltage. In contact-AFM a cantilever with as sharp a point as possible is laid onto the sample surface with a small loading force in the range of 10^{-7} to 10^{-10} Newtons. Tips of differing size and shape are tailored made. Data is obtained optically by bouncing an incident laser beam onto the cantilever toward a quadrant detector on or into an interferometer. The AFM can work in two modes-either a contact mode or in a non-contact mode. In the non-contact mode the attractive force is important and the experiment must be carried out under low pressures similar to those employed in STM.

In the contact mode the tip acts as a low-load, high resolution profiler. Along with structure determination, the AFM can also be used to "move" atoms about allowing molecular architecture to occur at the atomic level. The AFM is also an important tool in the nanotechnology revolution.

Secondary Ion Mass Spectroscopy. (2,13) Secondary ion mass spectroscopy, SIMS, is a sensitive surface analysis tool. Here the mass analysis of negative and positive ions sputtered from the polymer surface through ion bombardment are analyzed. The sputtering ion beam is called the primary ion beam. This beam causes erosion of the polymer surface removing atomic and molecular ions. These newly created ions, composing what is called the secondary ion beam, are then analyzed as a function of mass and intensity. Depth of detection for SIMS is of the order of 20 to 50 Angstroms. Because it is the ions in the secondary ion beam that are detected, the mass spectra obtained from SIMS is different from those obtained using simple electron impact methods. The extent of particular ion fragments observed is dependent on a number of factors including the ionization efficiency of the particular atoms and molecules composing the polymer surface.

SIMS can detect species that are present on surfaces of the order of parts-per-million to parts-per-billion. Please see the chapter on mass spectrometry for a fuller discussion.

Infrared Spectroscopy. (2, 13-16) The infrared range spans the region bound by the red end of the visible region to the microwave region at the lower frequencies. Molecular interactions that involve vibrational modes correspond to this energy region. Infrared spectroscopy, IR, is one of the most common spectronic techniques used today to identify polymer structure. Briefly, when the frequency of incident radiation of a specific vibration is equal to the frequency of a specific molecular vibration the molecule absorbs the radiation. Today, most IR instruments are rapid scan where the spectra are Fourier transformed.

Following is a brief discussion of some of the more important techniques used in polymer analysis of surfaces. Please see the chapter on infrared spectroscopy for a more detailed coverage of the use of infrared spectroscopy.

Attenuated total reflectance, ATR-IR, allows the study of films, coatings, threads, powders, interfaces, and solutions (17). ATR occurs when radiation enters from a more-dense (that is a material with a higher refractive index) into a material that is less dense (that is with a lower refractive index). The fraction of the incident radiation reflected increases when the angle of incidence increases. The incident radiation is reflected at the interface when the angle of incidence is greater than the critical angle. The radiation penetrates a short depth into the interface before complete reflection occurs. This penetration is called the evanescent wave. The intensity is reduced by the sample at the frequencies where the sample absorbs.

Diffuse reflectance infrared Fourier transform spectroscopy, DRIFTS, is employed to obtain spectra of powders and rough polymeric surfaces such as textiles and paper. IR radiation is focused onto the surface of the sample in a cup resulting in both specular reflectance (which directly reflects off the surface having equal angles of incidence and reflectance) and diffuse reflectance (which penetrates into the sample subsequently scattering in all angles). Special mirrors allow the specular reflectance to be minimized.

Emission infrared spectroscopy is used for thin films and opaque polymers. The sample may be heated to increase the amount of energy emitted. The sample acts as the radiation source and the emitted radiation is recorded giving spectra similar to those of classical FTIR. In some cases, IR frequencies vary because of differences in the structures at different depths and interactions between surface and interior emissions.

Photoacoustic spectroscopy IR, PAS, is used for highly absorbing materials. Generally, modulated IR radiation is focused onto a sample contained in a cup inside a chamber containing an IR-transparent gas such as nitrogen or helium. The IR radiation absorbed by the sample is converted into heat inside the sample. The heat travels to the sample surface and then into the surrounding gas causing expansion of the boundary layer of gas next to the sample surface. The modulated IR radiation thus produces intermittent thermal expansion of the boundary layer creating pressure waves that are detected as photo-acoustic signals.

PAS spectra are similar to those obtained using ordinary FTIR except truncation of strong absorption bands because of photoacoustic signal saturation often occurs. PAS allows the structure to be studied at different sample depths because the slower the frequency of modulation, the deeper the penetration of IR radiation.

Specular reflectance IR involves a mirror-like reflection producing reflection measurements of a reflective material or a reflection-absorption spectrum of a film on a reflective surface. This technique is used to look at thin (from nanometers to micrometers thick) films.

Infrared microscopy allows the characterization of minute amounts of a material or trace contaminants or additives. Samples as small as 10 microns can be studied. The microscope, often using fiber-optics, allows IR radiation to be pin-pointed.

Today, there are many so-called hyphenated methods with IR acting to illustrate this. Hyphenated methods involving IR include GC-IR where the IR spectra are taken of materials as they are evolved through the column. Related to this is HPLC-IR., TG-IR, and MS-IR.

Raman Spectroscopy. (2,13-16,18) Raman spectroscopy is similar to IR spectroscopy in that it investigates polymer structure focusing on the vibrational modes. Whereas IR is a result of energy being absorbed by a molecule from the ground state to an excited state,

Raman spectroscopy is a scattering phenomenon where energy of photons is much larger than the vibrational transition energies. Most of these photons are scattered without change (so-called Raylleigh scattering). Some are scattered from molecular sites with less energy than they had before the interaction resulting in Raman-Stokes lines. Another small fraction of photons have energies that are now greater than they originally had leading to the formation of anti-Stokes lines. Only the Raman-Stokes photons are important in Raman spectroscopy. While many chemical sites on a polymer are both IR and Raman active, that is they give rise to bands, some are less active or even non-active because of the difference between groups that can absorb and those that scatter. These differences are generally described in terms of symmetry of vibration. Briefly, an IR absorption occurs only if there is a change in the dipole moment during the vibration where as a change in polarizability is required for Raman scattering to occur. A comparison of the two allows for additional structural characterization beyond that obtained from either of the techniques alone.

As in the case with IR spectrometers, there exists a wide variety of speciality techniques especially applicable to polymer analysis. Please refer to the chapter on Raman spectroscopy for a more detailed description.

In surface-enhanced Raman spectroscopy, SERS, samples are adsorbed onto microscopically roughened metal surfaces. The intensities and frequencies of scattered radiation originating from a sample that has been irradiated with a monochromatic source such as a laser make up the spectrum. SERS spectra are of molecules that are less than 5 nanometers from the surface.

Amorphous Region Determination
<u>Amorphous Region Determination</u>

There exists a variety of tools that allow various measures of the amorphous region (2,13, 25). Table 3 contains a listing of some of the more important techniques. In general, Rayleigh scattering can be described as elastic scattering where the scattered light is measured as a function of scattering angle. Brillouin scattering is in essence a Doppler effect that results in small shifts in the frequency of light. Raman scattering is an inelastic process where there is a shift in the wavelength due to chemical absorption or emission. Birefringence measures order in the axial, backbone direction. The birefringence of a sample can be defined as the difference between the refractive indices for light polarized in two directions 90 degrees apart. Thus, a polymer containing chains oriented in a preferential direction will exhibit a different refractive index along the direction of preferred chain alignment compared with that obtained at right angles to this. The degree of change gives information concerning the amount of disorder present in the material. Depolarized light scattering is a related technique where the intensity of scattered light is measured when the sample is irradiated by visible light. Techniques such as birefringence, depolarized light scattering, small-angle X-ray scattering, NMR relaxation, and Raman scattering give information related to the short-range (<2 nm) nature of the amorphous domains, while neutron scattering, electron diffraction, and electron microscopy yield information concerning the longer range nature of these regions.

Table 3.   Techniques employed to study the amorphous regions of polymers.

Bruillouin Scattering
Density
Depolarized Light Scattering
Electron Diffraction
Electron Microscopy
Magnetic and optical birefringence
NMR Relaxation
Raman Scattering
Rayleigh Scattering
Scanning Probe Microscopy-Atomic Force, Scanning Tunneling
Small and Wide-Angle X-Ray Scattering
Small-Angle Neutron Scattering

Application of many of the so-called nano-techniques such as ATM allows the determination of sites and abundance of order and disorder present at polymer surfaces and for broken samples, present within the bulk of the material.

Size and Shape Determination

General. It is the size of macromolecules that give them their unique and useful properties (2). Size allows polymers to act more as a group so that when one polymer chain moves, surrounding chains are affected by that movement. Size also allows polymers to be non-volatile since the secondary attractive forces are cumulative (for instance the London dispersion forces are about 2 kcal/mole of repeat units) and because of the shear size, the energy necessary to volatilize them is sufficient to degrade the polymer.

Generally, the larger the polymer, the higher is the molecular weight. Table 4 contains a listing of common molecular weight determining techniques.

Unlike smaller molecules that have a precise molecular weight, polymers samples are often composed of chains with varying lengths. Today there exists techniques that permit determination of polymer shape, size, and distribution of chain lengths. Here we will consider briefly some of techniques that allow these determinations. A fuller discussion of the types of molecular weight is given in most introductory texts and the accompanying chapter on particle size.

While there are a number of statistically described approaches arriving at differing molecular weight values owing to the different weight given to the various chain lengths three of these are most used by polymer chemists in describing the molecular weight and molecular weight distribution of polymers. These three statistical approaches also have a real-life correspondence to physical measurements that can be carried out.

Table 4. Common molecular weight determining techniques.

Method	Type of Average	Applicable Weight Range (Daltons)
Light Scattering	Weight-Average	to infinity
Coupled LS and Chromatography	All Averages & Distribution	to infinity
Vapor Phase Osmometry	Number-Average	to 40,000
Membrane Osmometry	Number-Average	ca 10^4 to 10^6
Electron & X-Ray Microscopy	All Averages & Distribution	100 to infinity
Isopiestic-Isothermal Distillation	Number-Average	to 20,000
Ebulliometry	Number-Average	to 40,000
Cryoscopy	Number-Average	to 50,000
Osmodialysis	Number-Average	500 to 25,000
End-Group analysis	Number-Average	to 20,000
Centrifugation		
Sedimentation Equilibrium	Z-Average	to infinity
Sedimentation Velocity	Gives real value for only monodisperse samples	
Archibald Modification	Weight & Z-Average	to infinity
Trautman's Method	Weight-Average	to infinity
SAXS	Weight-Average	
Chromatography	Calibrated, Distribution	to infinity
Viscometry	Calibrated	to infinity
Mass Spectroscopy, Matrix	All Averages & Distribution	to 10^6

The number-average molecular weight emphasizes the number of chains present and can be measured by any technique that "counts" these chains. Colligative properties are dependent on the number of particles present and are related then to the number-average molecular weight. Most thermodynamic-associated properties are related to the number of particles present and are thus closely associated with the number-average molecular weight.

The weight-average molecular weight emphasizes the size of the polymer chains and is determined from experiments where each chain makes a contribution to the measured result relative to its size, not number as in the case of the number-average molecular weight. Bulk properties associated with large deformations such as viscosity and toughness are particularly related to the weight-average molecular weight.

The Z-average molecular weight is the third moment or third power average. Some properties, such as melt elasticity, are more closely associated with the Z-average molecular weight.

For heterodisperse polymer samples Z-average> weight-average> number-average molecular weight with the ratio of weight-average/ number-average being a measure of the polydispersity of the polymer sample and given the name "polydispersity index". All so-called classical techniques require the polymer to be soluble. Matrix-assisted mass spectrometry and microscopy-related techniques do not require the polymer to be dissolved. Because many properties are dependent on both the particular average molecular weights, they are also dependent on the distribution of chain sizes. Thus it is important to know both the particular molecular weights and the distribution of these chain sizes.

For polydisperse polymer samples molecular weight values determined by light scattering photometry, colligative properties, mass spectrometry, microscopy approaches, and the appropriate ultracentrifugation data treatment give an "absolute molecular weight", while others such as chromatography and viscometry require calibration employing polymers of known molecular weight.

The basics of light scattering photometry, colligative approaches and other classical approaches are described in introductory polymer texts and will not be covered here. More recent approaches and modifications of some selected approaches will be briefly presented here.

Thermal field flow fractionation has been used to separate various chain lengths according to differences in the ordinary (Fick's) diffusion coefficient (13). When calibrated, this approach can be used to identify general molecular weight distributions. Diffusion coefficients can also be determined using photon correlation or quasielastic light scattering spectroscopy (13).

Light Scattering Photometry. (2,13,19) Light scattering photometry requires the measuring of scattered light from a polymer solution at different angles and then plotting the data. When the data is plotted as a double extrapolation to zero concentration and zero angle it forms what is called a Zimm plot. The extrapolation to zero angle corrects for finite particle size effects. The radius of gyration, related to polymer shape and size, can also be determined from this plot.

The second extrapolation to zero concentration corrects for concentration factors. The intercepts of both plots is equal to $1/M_w$.

The Zimm plot approach does not require knowing or having to assume a particular shape for the polymer in solution.

Related to the Zimm plot is the Debye plot. In the Zimm approach, different concentrations of the polymer solution are used. In the Debye one low concentration sample is used with $1/M_w$ plotted against $\sin^2(\theta/2)$, essentially one-half of the Zimm plot.

Low angle laser light scattering photometry, LALLS, and multiangle low angle laser light scattering photometry, MALS, takes advantage of the fact that at low or small angles, the form factor, P_ϕ, becomes one reducing the basic Rayleigh relation to

$$Hc/\tau = 1/M_w (1 + 2Bc + Cc^2 + ...) \tag{6}$$

or the analogous expression except where the constants are rearranged

$$Kc/R = 1/M_w (1 + 2Bc + Cc^2 + ...) \tag{7}$$

A number of automated systems exist with varying capabilities. Some internally carry out dilutions and refractive index measurements allowing molecular weight to be directly determined without additional sample treatment. The correct determination of dn/dc is very important since any error in its determination is magnified because it appears as the squared value in the expression relating light scattering and molecular weight.

These systems may also allow the determination of molecular conformation matching the radius and molecular weight to graphs showing the change in the root mean square radius of gyration, RMS, and molecular weight for different shaped molecules. The expression for the mean square radius of gyration is given as

$$r_g^2 = \Sigma r_i^2 m_i / \Sigma m_i \tag{8}$$

One of the most important advances in polymer molecular weight determination is the "coupling" of size exclusion chromatography, SEC, and light scattering photometry, specifically LALLS or MALS.

The LALLS or MALS detector measures tau, τ, related values, a differential refractive index, DRI, detector is used to measure concentration, and the SEC supplies samples containing "fractionated" polymer solutions allowing both molecular weight and molecular weight distribution, MWD, to be determined. Further, polymer shape can be determined. This combination represents the most powerful, based on ease of operation, variety of samples readily used, and cost, means to determine polymer size, shape, and MWD available today.

<u>Chromatography.</u> (2,13,20,21) As noted before, certain techniques such as colligative methods, light scattering photometry, special mass spectral techniques, and ultracentrifugation allow the calculation of specific or absolute molecular weights. Under certain conditions some of these also allow the calculation of the molecular weight distribution, MWD.

There are a wide variety of chromatography techniques including paper and column techniques. Chromatographic techniques involve passing a solution containing the to-be-tested sample through a medium that shows selective absorption for the different components in the solution. Ion exchange chromatography separates molecules on the basis of their electrical charge. Ion-exchange resins are either polyanions or polycations. For a polycation resin, those particles that are least attracted to the resin will flow more rapidly through the column and be emitted from the column first. This technique is most useful for polymers that contain charged moieties.

In high-performance liquid chromatography, HPLC, pressure is applied to the column that causes the solution to rapidly pass through the column allowing procedures to be completed in a fraction of the time in comparison to gravity-flow chromatography.

In affinity chromatography, the resin contains molecules that are especially selected to interact with the particular polymer(s) that is being studied. Thus, for a particular protein, the resin may be modified to contain a molecule that interacts with that protein type. The solution containing the mixture is passed through the column and the modified resin preferentially associates with the desired protein allowing it to be preferentially removed from the solution. Later, the protein is washed through the column by addition of a salt solution and collected for further evaluation.

When an electric field is applied to a solution, polymers containing a charge will move towards either the anode if they are a negatively charged species or to the cathode if the net charge is positive. This migration is called electrophoresis. The velocity at which molecules migrate is mainly dependent upon the electric field and change on the polymer driving the molecule towards one of the electrodes, and a frictional force dependent on the size and structure of the macromolecule that opposes the movement. In general, the more bulky and larger the polymer the greater is the resistance to movement and the greater the applied field and charge on the molecule the more rapid the movement. While electrophoresis can be conducted on solutions it is customary to use a supporting medium of a paper or gel. For a given system, it is possible to calibrate the rate of flow with the molecular weight and/or size of the molecule. Here, the flow characteristics of the calibration material must be similar to those of the unknown.

Gel Permeation Chromatography, GPC, is a form of chromatography that is based on separation by molecular size rather than chemical properties. GPC or Size exclusion chromatography, SEC, is widely used for molecular weight and molecular weight distribution, MWD, determination. In itself, SEC does not give an absolute molecular weight and must be calibrated against polymer samples whose molecular weight has been determined by a technique that does give an absolute molecular weight.

SEC is a high performance liquid chromatographic, HPLC, technique where the polymer chains are separated according to differences in hydrodynamic volume. This separation is made possible by use of special packing material in the column. The packing material is usually polymeric porous spheres often composed of polystyrene crosslinked by addition of varying amounts of divinylbenzene. Retention in the column is mainly governed by the partitioning (or exchanging) of polymer chains between the mobile (or eluant) phase flowing through the column and the stagnate liquid phase that is present in the interior of the packing material.

Through control of the amount of crosslinking, nature of the packing material and specific processing procedures, spheres of widely varying porosity are available. The motion in and out of the stationary phase is dependent on a number of factors including Brownian motion, chain size, and conformation. The latter two are related to the polymer chain's hydrodynamic volume- the real, excluded volume occupied by the polymer chain. Since smaller chains preferentially permeate the gel particles, the largest chains are eluted first. As noted above, the fractions are separated on the basis of size.

The resulting chromatogram is then a molecular size distribution. The relationship between molecular size and molecular weight is dependent on the conformation of the polymer in solution. As long as the polymer conformation remains constant, which is generally the case, molecular size increases with increase in molecular weight. The precise relationship between molecular size and molecular weight is conformation dependent. For random coils, molecular size as measured by the polymer's radius of gyration, R, and molecular weight, M, is R is proportional to M^b, where "b" is a constant dependent on the solvent, polymer concentration, and temperature. Such values are known and appear in the literature for many polymers allowing the ready conversion of molecular size data collected by SEC into molecular weight and MWD.

There is a wide variety of instrumentation ranging from simple manually operated devices to completely automated systems. Briefly, the polymer-containing solution and solvent alone are introduced into the system and pumped through separate columns at a specific rate. The differences in refractive index between the solvent itself and polymer solution are determined using a differential refractometer. This allows calculation of the amount of polymer present as the solution passes out of the column with the assumption that the full range of molecular weights have the same refractive index.

Conversion of retention volume for a given column to molecular weight can be accomplished using several approaches including peak position, universal calibration, broad standard and actual molecular weight determination by coupling the SEC to an instrument that gives absolute molecular weight.

In the peak position approach, well-characterized narrow fraction samples of known molecular weight are used to calibrate the column and retention times determined. A plot of log M versus retention is made and used for the determination of samples of unknown molecular weight. Unless properly treated, such molecular weights are subject to error. The best results are obtained when the structures of the samples used in the calibration and those of the test polymers are the same and the polymer is of uniform composition.

The universal calibration approach is based on the product of the limiting viscosity number, LVN, and molecular weight being proportional to the hydrodynamic volume. Benoit showed that for different polymers elution volume plotted againt the log LVN times Molecular Weight gave a common line. In one approach molecular weight is determined by constructing a "universal calibration line" through plotting the product of log LVN for polymer fractions with narrow MWDs as a function of the retention of these standard polymer samples for a given column. Molecular weight is then found from retention time of the polymer sample using the calibration line.

As noted above, probably the most accurate approach is to directly connect, couple, the SEC to a device, such as a light scattering photometer, that directly measures the molecular weight for each elution fraction. Here both molecular weight and MWD are accurately determined.

Mass Spectrophotometry. (2,13) Certain mass spectral, MS, procedures allow the determination of the molecular weight or molecular mass of oligomeric to polymeric materials (Table 5). A following chapter will deal in more depth with mass spectrometry, MS. Here we will concentrate on its use to determine molecular weight and molecular weight distribution.

In matrix-assisted laser desorption/ionization, MALDI, the polymer is dissolved, along with a "matrix chemical", and the solution deposited onto a sample probe. The solution is dried. MALDI depends on the sample having a strong UV absorption at the wavelength of the laser used. This helps minimize fragmentation since it is the matrix UV-absorbing material that absorbs most of the laser energy. Often employed UV-matrix materials are 2,5-dihydroxybenzoic acid, sinnapinic acid, picolinic acids, and alpha-cyano-4-hydroxy cinnamic acid. The high energy of the laser allows both the matrix material and the test sample to be volatilized. Such techniques are referred to as "soft" since the test sample is not subjected to (much) ionizing radiation and hence little fragmentation occurs.

Table 5. Mass spectral techniques employed in the determination of molecular masses of oligomeric and polymeric materials.

MS-Type	(Typical) Upper Molecular Weight Range (Daltons)
(Usual) Electron Impact, EI	to 2000
Electrospray Ionization, ESI	to 10^6
Fast Atom Bombardment, FAB	to 2000
Direct Laser Desorption, Direct LD	to 10^4
Matrix-Assisted Laser Desorption/Ionization, MALDI	to 10^6

Mass accuracy on the order of a few parts per million are obtained. Thus, chain content can be determined for copolymers and other chains with unlike repeat units. Polymer molecular weight distributions can also be determined using MALDI and related MS techniques.

Electrospray ionization mass spectrometry, ESI, is not a desorption technique but rather a spray ionization method. A solution containing the polymer is sprayed from a needle into an electrical field of several kilovolts. The ions are often multiply changed allowing the range to be increased over singly charged species allowing its range to be increased over FAB. It works better for polar polymers since they are more easily multiplied charged. For general work it does not appear to be as accurate as MALDI.

Recently, MS combinations have been available including the TG-MS combination developed by Carraher that allows the continuous characterization of evolved materials as a polymer undergoes controlled thermal degradation.

Particle Size. (2,13) Particle size is important in many polymer applications including coatings applications, creation of suspensions, and in a number of quality control procedures such as the determination of contaminates. Please see the chapter dealing with particle size determination for a fuller discussion.

There are many analytical techniques that allow the determination of particle size (Table 6). Before a technique is chosen, one needs to determine the relative particle size to be determined and the type of size information needed. Refractive index values are generally needed for size measurements based on light scattering. Densities are often needed for techniques based on acoustics and sedimentation. Simple microscopy generally allows the determination of the general range of size and shape. Most particle size determinations are dependent on particle shape with most techniques and related equations based on spherically shaped particles. Thus, the greater the deviation in particle shape from spherical , generally the greater the error. Particles that greatly differ in particle shape from spherical, such as needles and rods, are best analyzed using some form of microscopy. Properties of the particle in the suspending fluid are also important because some liquids may bring about changes in particle shape or encourage clustering of the particles. The suspending liquid should not dissolve or appreciably influence particle shape.

Table 6. Particle size and distribution determination methods.

Sedimentation	Capillary Hydrodynamic Chromatography
Microscopy	Light Obscuration
Fraunhofer Diffraction	Field Flow Fractionation
Phase Doppler Anemometry	Ultrasonic Spectroscopy

Light obscuration, LO, is one of the major techniques used to determine particle size. LO is based on the observation that particles whose refractive index is different from the suspending solvent scatter light. This scattering is the same as that employed in molecular weight determinations using light scattering photometry. In fact, light scattering photometry can also be used to determine particle size. Even so, LO instruments have been developed for the main purpose of determining particle size. Stirring is often required to maintain a somewhat homogeneous suspension. Wetting, often achieved by addition of a wetting agent, of the particles by the suspending liquid is also generally required. Dispersion of the particles is assisted by sonication for situations where a single particle size is needed. Sonication is not recommended if particle aggregation sizes are important.

Sedimentation techniques are also used to determine particle size distributions of the order of 0.1 to 50 micrometers. Capillary hydrodynamic chromatography, HDC, gives particle size distributions for particles of about 0.005 to 0.7 micrometers.

X-Ray Diffraction. (13) X-ray diffraction is a widely used tool for structural identification for almost all solids under the right conditions. X-ray diffractometers are generally either single-crystal or powder.

Single-crystal studies allow the absolute configurational determination of polymeric materials that have high degrees of crystallinity. Such determinations are costly with respect to time because of the complexity of polymeric materials.

Powder X-ray spectroscopy can employ smaller crystalline samples from 1 to several hundred nanometers. These crystallites have broadened peak profiles as a result of incomplete destructive interference at angles near the Bragg angle defined as

$$n\lambda = 2d \sin\theta \tag{9}$$

where n is the order of a reflection, λ the wavelength, d the distance between parallel lattice planes, and θ the angle between the incident beam and a lattice plane known as the Bragg angle. This broadening allows determination of crystallite size and size distribution. (Note that this is not particle size.)

X-ray analysis of proteins and nucleic acids is especially important as the absolute structure is needed for many advances in the field of medicine and biochemistry.

Complementary Molecular Shape Pairing. (22-25) Recent developments have produced a new method for determining the shape of some macromolecules. One of these involves the use of complementary molecular shape pairing with molecules, generally fixed on a column, of known shape. Libraries of known shapes are being produced that allow for more precise determinations of shape.

This approach is good for monodisperse molecules with rigid or fixed (by strong secondary bonding or primary bonding crosslinking) structures such as many of the biopolymers such as enzymes and nucleic acids.

<u>Literature Cited</u>

1.   Schoff, C. K., in "Encyclopedia of Polymer Science and Engineering", Vol. 14, Wiley, NY, 1986,p. 454.
2.   Wilkes, G., in "Encyclopedia of Polymer Science and Engineering", Vol. 14, Wiley, NY, 1986, p 542.
3.   Carraher, C. E.. "Polymer Chemistry", Dekker, NY, 2000.
4.   Bird, D. I., Curtiss, C. F., Armstrong, R. C., Hassager, O.; "Dynamics of Polymer Solutions", Wiley, NY, 1997.
5.   Rohn, C., "Analytical Polymer Rheology", Hanser Gardner, Cincinnati, OH, 1995.
6.   Maxwell, B.; Nguyen, <u>Polymer Engineering Science</u>, 1979, <u>19</u>, 1140.
7.   Maxwell, B, in "Polymerization Characterization, Advances in Chemistry Series No. 203", Craver, C. D., Ed., American Chemical Society, Washington, DC, 1983,p.149.
8.   Hanson, P.; Williams, M. C., <u>Polymer Engineering Science</u>, 1987, <u>27</u>, 586.
9.   Grosberg, A. (1998) "Theoretical and Mathematical Models in Polymer Research", Academic Press, Orlando, FL, 1998.
10.   Shah, V., ; "Handbook of Plastics Testing Technology, Wiley, 1998.
11.   Petty, M. C., Bryce, M., Bloor, D. , Eds.; "An Introduction to Molecular Electronics", Oxford University Press, Cary, NJ, 1995.
12.   Mathot, V.; "Calorimetry and Thermal Analysis of Polymers",:Hanser Gardner, Cincinnati, OH, 1994.
13.   Settle, F. (Ed) "Handbook of Instrumental Techniques for Analytical Chemists", Prentice Hall, Upper Saddle River, NJ, 1997.
14.   Turi, E., Ed. "Thermal Characterization of Polymeric Materials, 2nd Ed.", Academic Press, Orlando, FL., 1997
15.   Fawcett, A. H. ; "Polymer Spectroscopy", Wiley, NY, 1996.
16.   Koenig, J. "Spectroscopy of Polymers", Elsevier, NY, 1999.
17.   Urban, M.; "Attenuated Total Reflectance Spectroscopy of Polymers: Theory and

18. Craver, C. D.. (Ed); "Polymerization Characterization, Advances in Chemistry Series No. 203", American chemical Society, Washington, DC, 1983.
19. Brown, W.; "Light Scattering: Principles and Development", Verlag, NY, 1996.
20. Oliver, R.; "HPLC of Macromolecules: A Practical Approach, 2nd Ed.", Oxford University Press, Cary NY, 1998.
21. Wulff, G., Angew. Chem. Int. Engl.,1995, 34, 1812.
22. Schultz, P. G.; Lerner, R. A.; Science, 1995, 269, 1835.
23. Houghton, R. A. (Ed), Biopolymers, 1995, 37, issue 3.
24. Lowe, G.; Chemical Society Revs., 1995, 309.
25. Sperling, L. "Introduction to Physical Polymer Science", Wiley, NY, 1986.

General Reading

General
Mark, H.; Bikales, N.; Overberger, C.; Menges, G.; (Eds); "Encyclopedia of Polymer Science and engineering", Wiley, NY, 1988.
Salamone, J. C. ; "Concise Polymeric Materials Encyclopedia", CRC Press, Boca Raton, FL, 1998.
Sandler, S., Karo, W., Bonesteel, J., Pearce, E. M.; "Polymer Synthesis and Characterization", Academic Press, Orlando, FL, 1998.
Sperling, L. H., "Introduction to Physical Polymer Science", Wiley, NY, 1986.
Polymer Structure
Higgins, J., Benoit, H. C.; "Polymers and Neutron Scattering", Oxford University Press, Cary, NC, 1997
Tsujii, K. ; "Surface Activity", Academic Press, Orlando, FL, 1998.
Woodward, A.; "Understanding Polymer Morphology", Hanser Gardner, Cincinnati, OH, 1995.

Testing and Spectrometric Characterization
Adamson, A., Gast, A.; "Physical Chemistry of Surfaces, 6th Ed.", Wiley, NY, 1997.
Askadskii, A. A. ; "Physical Properties of Polymers: Prediction and Control", Gordon and Breach, NY, 1996
Bovey, F., Mirau, P. ; "NMR of Polymers", Academic Press, Orlando, FL, 1995.
Brandrup, J., Immergut, E. H., Grulke, E.,; "Polymer Handbook, 4th Edition", Wiley, NY, 1999.
Hilado, C.; "Flammability Handbook for Plastics", Technomic, Lancaster, PA, 1998.
Mark, J. E. (Ed.), ; "Polymer Data Handbook", Oxford University Press, NY, 1999.
Sibilia, J. P.; "A Guide to Materials Characterization and Chemical Analysis, 2nd Ed.", Wiley, NY, 1996.

Rheology and Physical Tests
Doi, M., See, H.; "Introduction to Polymer Physics", Oxford University Press, Cary, NC, 1996.
Pollack, T. C.; "Properties of Matter, 5th Ed.", McGraw-Hill, NY, 1995.
Strobl, G. R. ; "The Physics of Polymers: Concepts for Understanding Their Structure and Behavior, 2nd Ed.", Springer-Verlag, NY, 1997.
Van Dijk, M., Wakker, A.; "Concepts in Polymer Thermodynamics", Technomic, Lancaster,

PA, 1997.
White, M. A.; "Properties of Materials",Oxford University Press, Cary, NC, 1999.

CHALLENGES IN PARTICLE SIZE DISTRIBUTION MEASUREMENT PAST, PRESENT AND FOR THE 21ST CENTURY*

THEODORE PROVDER

Polymer and Coatings Consultants,
Olmsted Falls, OH 44138

INTRODUCTION
END USER INNOVATION-COMMERCIAL DEVELOPMENT, HDC, CHDF, FFF
REVITALIZATION OF OLDER INSTRUMENTS, SEDIMENTATION
EVOLUTION OF A RESEARCH INSTRUMENT INTO A ROUTINE USER-FRIENDLY INSTRUMENT, DLS, PCS, QELS
ATTEMPTS TO MEET EXTREMELY DIFFICULT TECHNICAL CHALLENGES
FOQELS, LASER LIGHT SCATTERING, ELECTROACOUSTIC, SPOS
CONCLUSIONS
LITERATURE CITED

Introduction

Over the last 15 years, new coatings technologies such as high solids, powder, waterborne, and radiation curable coatings have been developed to meet the challenges of: (a) governmental regulations in the areas of ecology (volatile organic compounds (VOC) emission); (b) long-term increasing costs of energy and petroleum based solvents; (c) more active public consumerism; and (d) the continual need for cost-effective high performance coatings in a highly competitive and global business environment. These new coatings technologies require the use of water as the major solvent with water soluble or high molecular weight latex polymers or the use of strategically designed low molecular weight polymers, oligomers and reactive additives which, when further reacted, produce high molecular weight and crosslinked polymers. This has led to a need for improved methods of materials characterization in diverse areas which include molecular weight distribution analysis, particle size distribution(PSD) assessment and characterization, rheology of coatings, film formation and cure process characterization, morphological surface and bulk characterization, and spectroscopic analysis, as well as a need for improved methods for modeling and predicting materials properties and processes.

*   Adapted   with permission from Elsevier Science Inc. from Provder,T <u>Progress in Organic Coatings</u>,1997, <u>32</u> , 143-153.

Concurrent with the major technological changes in the coatings industry was the significant increase in the rate of change in instrumentation technology. This change was driven by significant advances in electronics and computer and sensor technologies to produce computer-aided, more user-friendly, reliable and cost-effective instrumentation. The recent advances in instrumentation and computer technology are filling the need for improved polymer and coatings characterization methods in the context of the newer coatings technologies.

The newer coatings technologies have driven the need for improved methods of PSD assessment and characterization for each of the coatings technologies in terms of size ranges and component particulates as follows:

Waterborne coatings (0.01-50 μm)

- Latex
- Pigments (size and shape)
- Emulsions and dispersions

Powder coatings (0.1-100 μm)

- Resin-pigment composite particle: (a) dry powder; (b) wet concentrated dispersions

High solids

- Pigment grind particle size analysis (0.1-100 μm)

The particle size assessment and characterization needs resulting from the challenges presented by the new coatings technologies and their component materials leads to the following measurement requirements:

- Wide dynamic particle size range
- Improved resolution
- Measurements made in the concentrated dispersion regime
- Assessment of dispersion stability
- Description of structural and textural morphology

The renaissance in the field of PSD characterization and measurement, driven by the advances in electronics, computer and sensor technologies, can be characterized by at least four activities:

- End user innovation
- Revitalization of older instruments
- Evolution of research grade instrumentation into low cost, user-friendly instrumentation
- Attempts to meet extremely difficult technical challenges

The advances in PSD measurement methods will be discussed in the context of the four above-mentioned activities in terms of measurement principles and the strengths and weaknesses of these methods for characterizing PSDs.

End User Innovation - Commercial Development

The first activity in the renaissance of PSD characterization and measurement is the commercial development of instrumental methods originally developed in a few academic and industrial laboratories by individuals with high levels of skill and expertise. Innovations in instrumentation and methods usually are driven by the customers who are the technological and scientific leaders and prototype developers while the instrument vendors are the technological followers who have the engineering skills to do cost-effective commercial development. Examples of such methods are hydrodynamic chromatography (HDC), capillary hydrodynamic fractionation (CHDF), and field flow fractionation methods involving the use of sedimentation, flow, and thermal fields (e.g., SdFFF, FlFFF, and ThFFF).

Hydrodynamic chromatography (HDC)

HDC, first reported in 1976, was invented in an industrial laboratory by Hamish Small (1) to fulfill an analysis need of the Dow Chemical Company. It took ten years to be transferred from a laboratory method requiring a high degree of skill into a commercial instrument requiring a moderate degree of skill. The commercialization was carried out by Micromeritics Corporation. However, the commercialization failed in the marketplace because of technological limitations inherent in the method and because of competitive market pressures.

HDC instrumentation is comprised of a liquid chromatograph with an accurate and precise pumping system. Detection was accomplished with a sensitive fixed wavelength UV detector. Fractionation and separation of particles occurred in a column packed with uniform, non-porous beads. The separation of particles by size took place in the interstices between the beads and was primarily a function of the bead size and the ionic strength of the medium. The separation mechanism is shown in Fig. 1. The interstices between the beads can be treated as capillaries of varying sizes. Larger particles ride higher up on the parabolic flow profile and are eluted first while some smaller particles hug the walls of the capillary experiencing slower flow streamlines and exit later. Other contributing factors to the separation mechanism include electrical double layer effects and Van der Waals attraction. The fatal technical flaw in the methodology was the unpredicted occurrence and amount of particle deposition which took place in the packed columns. The overall features, benefits and limitations of this methodology are shown in Fig. 2.

Capillary hydrodynamic fractionation (CHDF)

The "Tubular Pinch Effect" mechanism operative in CHDF was first discovered by Segré and Silberberg (2) in 1962 and applied to particle size analysis in 1979 by Regnier and Ball (3). However, it did not develop into a viable commercial instrument until innovative experimental and theoretical work was carried out by Silebi and Dos Ramos (4,5). The commercial embodiment of CHDF has become available from MATEC Applied

Sciences over the last ten years. CHDF instrumentation is quite analogous to HDC in that a liquid chromatograph with precise, accurate and reproducible flow is required. In this instrument the separation does take place in capillaries of defined dimension using split-flow injection. The mechanism of this separation method is essentially the same as that postulated for HDC. However, now the column is an empty capillary so there is much less propensity for the particles to deposit in the column. The factors influencing the particle size fractionation are shown in Fig. 3 and the associated features, benefits and limitations of CHDF are shown in Fig. 4.

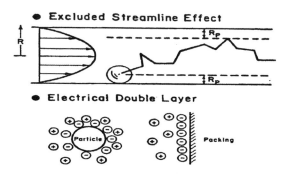

Fig 1. Schematic of hydrodynamic chromatography separation mechanism and operative forces.

FEATURES AND BENEFITS

- **Large Representative Sample Used (10^9 Particles)**
- **Samples Measured in Dispersion (No Particle Shape Change)**
- **Results are Independent of Particle Density**
- **Apparent PSD Directly Obtained**
- **Routine Methods for Plant Quality Control can be Developed (Equipment Easy to Set Up)**

LIMITATIONS

- **Particle Deposition in Packed Columns Probable**
- **Operating Conditions Somewhat Specific to Type of Samples Being Analyzed**
- **Calibration of Size Scale with Standards Required**
- **Detector Response Corrections Needed (Mie Scattering)**
- **Band Broadening Corrections Needed**
- **Complex Mathematics Required for Absolute PSD Analysis**

Fig. 2. Features, benefits and limitations of hydrodynamic chromatography for PSD analysis.

Capillary
- **Particle Radius/Tube Radius**
- **Tube Length**

Eluent
- **Liquid Viscosity (Temperature)**
- **Surfactant Type (Anionic, Ionic)**
- **Molecular Weight of Surfactant**
- **Ionic Strength**
- **pH**
- **Temperature**
- **Flow Rate - Accuracy – Constancy**

Detection
- **UV (Turbidity)**
- **Viscometer**

Data Analysis
- **Response Factor Correction**
- **Band Broadening Corrections**

Fig. 3. Operational factors and data analysis considerations influencing CHDF PSD analysis.

<u>Field flow fractionation (FFF)</u>
Field Flow Fractionation methods were invented by J. Calvin Giddings and co-workers and reported in 1967 (6-7). The technology transfer of FFF methods from an academic laboratory method to viable commercial instrument has taken a minimum of twenty years to occur. The DuPont Instrument Company commercialized a sedimentation FFF instrument (SdFFF) in 1986 based upon the pioneering work of J. Calvin Giddings and the subsequent research of Kirkland and Yau (8). Unfortunately, the DuPont Instrument Company's SdFFF instrument was a commercial failure and unavailable by 1991. However, a small entrepreneurial company known as FFFractionation, Inc. began making SdFFF instruments commercially available in 1988, followed by FlFFF (flow FFF) instruments in 1991 and more recently ThFFF(thermal FFF) instruments.

Field flow fractionation is a form of one-phase chromatography. The instrument is a liquid chromatographic system in which the separation takes place in a flat, narrow channel (typically one hundred to a few hundred micrometers in width). A parabolic flow profile is created in this narrow channel by the mobile phase. An external field is applied across the face of the channel such that the field extends over the channel's thin dimensions and is perpendicular to the flow field. Particles are driven toward the accumulation wall of the channel and form a diffuse cloud which has an exponential concentration distribution as a result of Brownian motion acting against the field. For particles less than 1-2 μm, smaller particles are displaced farther from the accumulation wall. When the flow field is turned on, the smaller particles which ride higher on the parabolic flow profiles elute first. The fields that have been used include sedimentation (SdFFF), thermal (ThFFF) consisting of a temperature gradient across the channel walls, cross flow (FlFFF), electric(FlFFF) and magnetic. Commercial instruments are available to perform SdFFF, ThFFF, and FlFFF.

A schematic of the instrumentation, separation mechanism and some channel configurations are shown in Fig. 5. When the particles are greater than 1 μm, effects of Brownian motion become negligible and the larger particles ride higher on the parabolic flow streamlines because the velocity dependent lift forces increase relative to the driving forces produced by the perpendicular external field. In the 1-100 μm size range, the larger particles elute first analogous to CHDF. This mechanism is termed the steric mode of FFF (StFFF). In the commercially available instrumentation, field programming options are available for extending the size separation range, reducing the analysis time and optimizing the resolution per unit of time. Examples of separation by SdFFF, FlFFF, ThFFF, and StFFF are shown in Fig. 6. The features, benefits and limitations of Field Flow Fractionation chromatography methods are shown in Fig. 7. For a review of FFF applied to particle characterization, the reader is referred to the many papers and review articles of Giddings and co-workers (9).

FEATURES AND BENEFITS

- **Large Representative Sample Used (10^9 Particles)**
- **Sample Measured in Dispersion (No Particle Shape Change)**
- **Apparent PSD Directly Obtained**
- **Routine Methods for Plant QC of Specific Products Easy to Develop**
- **High Resolution**
- **Very Little Sample Information is Required**
- **Results are Independent of Particle Density**
- **Analysis Time is Under 8 Minutes**
- **High Sample Throughput with Unattended Automatic Sample Injection**
- **Separation on Basis of Particle Surface Chemistry is Possible But Not Implemented**

LIMITATIONS

- **Possibility of Particle Deposition and Plugging of Capillary**
- **Operating Conditions Somewhat Specific to Type of Sample**
- **Calibration of Size Scale With Standards is Required**
- **Detector Response Corrections Needed (Mie Scattering)**
- **Band Broadening Corrections Needed**
- **Complex Mathematics Required for Absolute PSD Analysis**
- **Dynamic Size Range - 1.5 Decades for Particles Less Than 1.0μm**

Fig. 4. Features, benefits and limitations of CHDF for PSD characterization.

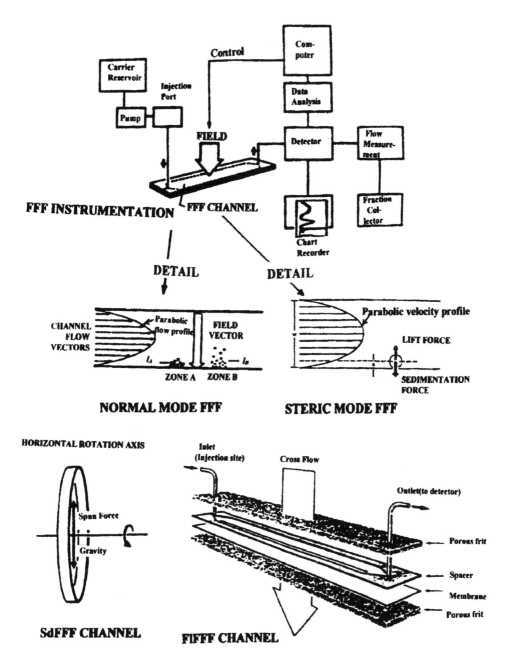

Fig 5. Schematic of FFF instrumentation, separation mechanisms and channel configurations for SdFFF, FIFFF.

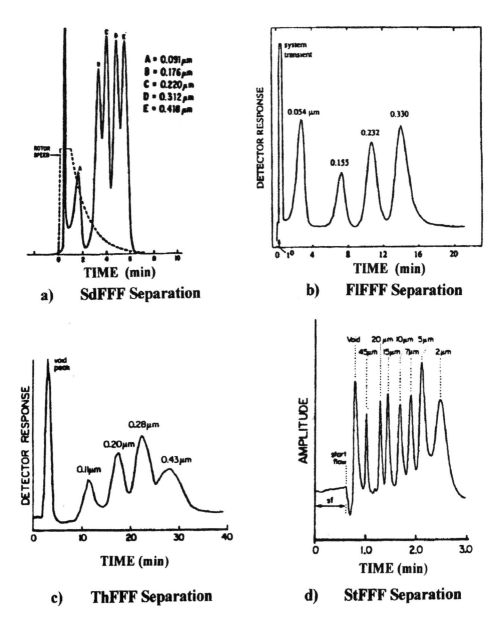

Fig. 6. FFF Separation of mixtures of monodisperse latex beads. a. Field programmed SdFFF with rpm = 10,000 at t = 0. b. FlFFF with polypropylene membrane at 42.9 ml/min. c. ThFFF in acetonitrile with ΔT = 170°C. d. StFFF at 38 ml/min and 1100 rpm.

FEATURES AND BENEFITS

- **Large Representative Sample Used (10' Particles)**
- **Samples Measured in Dispersion (No Particle Shape Change)**
- **Particle Density can be > or < Mobile Phase Density**
- **Apparent Particle Size Distribution Obtained Directly**
- **Retention (Particle Size) is Predictable**
- **Wide Dynamic Particle Size Range Achievable (.0l-1.μm for SdFFF, FlFFF, ThFFF, ~1-100μm for StFFF)**
- **PSD Calculations are Simple**

LIMITATIONS

- **FFF Channel Difficult to Construct with Precise Control Over Geometry**
- **Correction for Detector UV Response Needed (Mie-Scattering Correction)**
- **Upper Limit to Particle Size Range = 1 μm Unless Steric FFF Mode is Invoked**
- **Particles must be Spherical for Accurate Data Analysis**

Fig. 7. Features, benefits and limitations of FFF (Sd-, Fl-, Th- and St-FFF) for PSD characterization.

The commercialization of the instrumental techniques of HDC, CHDF, and FFF have attempted to address the market pull by many industries for PSD analysis methods, primarily in the 0.01-1.0 μm size range. The coatings industry material characterization needs, as a result of the growth of water borne coatings technology for improved particle size distribution analysis of latex, pigments and emulsions, is a component of the overall market pull for these techniques.

REVITALIZATION OF OLDER INSTRUMENTAL METHODS

Gravitational and Centrifugal Sedimentation
The second activity in the renaissance of PSD characterization and measurement is characterized by the revitalization of older instrumental methods such as gravitational and centrifugal sedimentation methods. Redesign, modernization with advanced electronics, and user-friendly computer-aided analysis have extended the instrument product life cycle. A good example is disc centrifuge photosedimentometry (DCP) (10,11). The basic technology has been around since the late 1950s and was embodied into a commercial instrument in the 1960s by the Joyce Loebel Company. During the 1970s, the Joyce Loebl Disc Centrifuge Photosedimentometer was the only commercially available instrument for obtaining PSD information by the DCP technique. This technique was particularly well suited for the particle size analysis requirements of the coatings industry for the analysis of latexes, pigments and emulsions (12). However, it had deficiencies that made it unsuitable as a plant quality control instrument. Over a

period of about ten years, industrial scientists at The Glidden Company developed an improved instrument (11), method of use (13,14) and user-friendly data analysis system. The Glidden Company's DCP technology was licensed to the Brookhaven Instruments Corporation in 1986 who made additional engineering improvements and successfully commercialized the enhanced and revitalized DCP technology. The current instrument can operate in both the line start and homogeneous start modes.

In addition, Brookhaven Instruments Corporation extended the technology by developing an X-ray detection system which facilitated the analysis of heavy small inorganic particles. The X-ray disc centrifuge photosedimentometer (X-DCP) became commercially available in 1991 and was based upon a prototype developed by Terry Allen at DuPont. The X-DCP can operate both in the gravitational and sedimentation modes.

The range of variations in sedimentation instrumentation mode, detection and experimental method is shown in Fig. 8. Also shown in Fig. 9 are Stokes' laws for gravitational and centrifugal sedimentation which govern the fractionation of particles by size in a gravitational or centrifugal force field. The line start mode (11,12) can produce very high resolution separations. An example of such a separation is shown in Fig. 10 for a mixture of nine Duke polystyrene latex standards covering a size range of 107 to 993 nm in approximately 100 nm increments. Baseline resolution was achieved in seven out of the nine standards. Another major advance in the method of operation is protecting the aqueous meniscus from evaporative cooling and disruption of the density gradient in the fluid by sealing the surface with a small amount of dodecane (usually 1 ml of n-dodecane, injected a few minutes after the density gradient has formed, in about 15-20 ml of an aqueous spin fluid). The sealing of the fluid surface inhibits evaporative cooling and thereby maintains a stable density gradient for several hours and extends the analysis size range. The homogeneous start mode is faster, covers a wider dynamic size separation range, but has less resolution than the line start mode for multi-modal separations. The features, benefits and limitations of sedimentation methods are shown in Fig. 11.

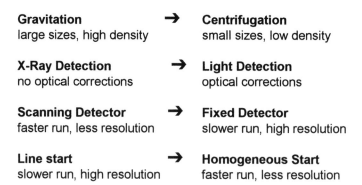

Gravitation → **Centrifugation**
large sizes, high density   small sizes, low density

X-Ray Detection → **Light Detection**
no optical corrections   optical corrections

Scanning Detector → **Fixed Detector**
faster run, less resolution   slower run, high resolution

Line start → **Homogeneous Start**
slower run, high resolution   faster run, less resolution

Fig. 8. Variations in sedimentation instrumentation and operational factors.

Gravitational Mode
(Rock in a Pond)

$$m_p \bullet\ d^2x/dt^2 = m_p\bullet g - m_f\bullet g - F\bullet dx/dt$$
$$\downarrow① \quad\uparrow② \quad\quad\uparrow③$$

$$\boxed{t=(18\bullet\eta\bullet x)/(g\bullet\Delta p\bullet D^2)}$$

Centrifugal Mode
(Rock in a Spinning Pond)

$$m_p\bullet\ d^2r/dt^2 = m_p\bullet\omega^2\bullet r - m_f\bullet\omega^2\bullet r - F\bullet dr/dt$$
$$\downarrow④ \quad\quad\uparrow② \quad\quad\uparrow③$$

→ $$\boxed{t=[18\bullet\eta\bullet\ln(r/s)]/(\omega^2\bullet\Delta p\bullet D^2)}$$

Where
① = Gravitational Force
② = Buoyancy Force
③ = Frictional Force
④ = Centrifugal Force
$F = 3\bullet\pi\bullet\eta\bullet D$, sphere
$\Delta p = p_p - p_f$
t = appearance time at detector
D = Particle Size Diameter
X = Sedimentation Distance to Detector Position

m_p = Mass of Particle
m_f = Mass of Fluid
p_p = Density of Particle
p_f = Density of Fluid
η = Viscosity of Fluid
g = Gravitational Constant
ω = Rotational Speed of Disc
r = Radius at Photodetector Position
s = Initial Radial Position of Particles determined by spin fluid volume

Fig. 9. Equations of motion and Stokes' law for gravitational and centrifugal sedimentation.

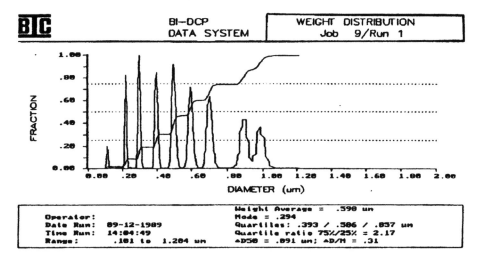

Fig. 10. Line start separation of a nine component mixture of polystyrene latex standards (107, 220, 298, 496, 597, 705, 895, 993 μm.)

An instrument designed for particle size analysis by sedimentation in a gravitational field was patented by Oliver, Hicken and Orr and commercially embodied, in 1969, by Micromeritics into an instrument known as the Sedigraph. This instrument produces a cumulative distribution of particle sizes and has been used heavily for the analysis of pigments such as titanium dioxide. Starting in 1988, the Sedigraph's design and function were revitalized by adding an automatic sample introduction system for unattended operation as well as improving the data analysis system.

FEATURES AND BENEFITS

- **Large, Representative Sample (10^9 Particles)**
- **Samples Measured in Dispersion (No Particle Shape Change)**
- **Scanning Detector, Fast Measurements Possible**
- **Fractionation Leads to High Resolution**
- **Wide Size Range, l0nm to 100μm**
- **Particle Size Predictable from Stokes' Laws**
- **PSD Calculations are Straightforward**

LIMITATIONS

- **Particle Size Limits Depend on Density, Viscosity**
- **Hydrodynamic Instabilities: Gradients**
- **Very Broad Distributions Require Long Measurement Times**

Fig. 11. Features, benefits and limitations of sedimentation methods for PSD characterization.

Evolution Of A Research Instrument Into A Routine User-Friendly Instrument

Dynamic Light Scattering (DLS, PCS, QELS)
The third activity is the evolution of a research instrument into a low-cost instrument that requires a minimum degree of skill to use. An excellent example of this process is the transformation of research-grade photon correlation spectrometers into low-cost, easy to use, limited-function instruments for routine analysis applications.

Dynamic light scattering (DLS) was first known as quasielastic light scattering (QELS) derived from the fact that when photons are scattered by mobile particles, the process is quasielastic. This gave rise to the acronym DLS, since QELS gave information on the dynamics of the scatterer. Since measurements are made with a digital correlator, the acronym PCS (photon correlation spectroscopy) is widely used. The first QELS measurements were made in 1964 by Cummins et al. (15). The first commercial instruments were available circa 1976 and only suitable for use by experts. The early measurements were concerned with obtaining translational diffusion coefficients of macromolecules and particles. The technique was used to gain information about particle size by relating the diffusion coefficient to particle size through the Stokes Einstein equation for spheres (16).

By the early 1970s, the fundamentals had been established. From the mid-1970s onward, improvements occurred in the technology for digital autocorrelators as well as in automation and integration of advances in microprocesses and laser technology. From the early 1980s onward, the PCS instrumentation was increasingly used for particle size analysis.

By 1985, low cost routine 90° instruments were commercially available for routine analyses. The PCS instrumentation was made more automatic and useful for the analysis of particle size and estimating PSD with the commercialization of an automatic sample dilution system invented by Nicoli and Elings(17).

The DLS method utilizes the fluctuations in light scattering intensity observed over a range of time intervals. The autocorrelation function of light scattering intensity at time t is compared to time zero, $C(\tau) = \langle I(t) \cdot I(0) \rangle$, for a range of Δt intervals. The autocorrelation function can be described by an exponential decay function of Δt. This process is schematically shown in Fig. 12. For a monodisperse sphere, the correlation function is represented by an exponential decay function as shown in Fig. 12 where D_T is the translational diffusion coefficient, q is the scattering wave vector. The particle size d is related to D_T through the Stokes-Einstein equation as shown in Fig. 12. It has been shown that particle sizes obtained on monodisperse spheres are accurate, precise, and highly reproducible. In addition, the PCS instrument is fast and has high throughput.

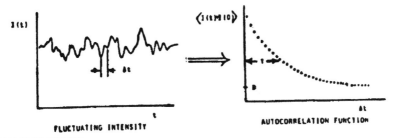

DLS THEORY

$$C(\tau) = B[1 + f^2 e^{-2\Gamma\tau}]$$
$$\Gamma = D_T q^2$$
$$q = (4\pi n/\lambda_o)\sin(\theta/2)$$
$$D_T = k_B T/3\pi\eta d$$

n = Refractive Index
λ_o = Wavelength of Light in Vacuum
θ = Scattering Angle
T = Temperature
η = Viscosity of the Medium
k_B = Boltzman Constant

Fig. 12. Simplified representation of scattering intensity I(t) and corresponding autocorrelation function $C(\tau)$.

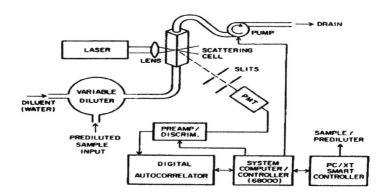

Fig. 13. Schematic diagram of a DLS particle sizing instrument, including autodilution (NICOMP Model 370).

A schematic of a PCS instrument with auto-dilution is shown in Fig. 13. This type of instrument also has been shown to have potential for online measurement of particle size for an emulsion polymerization reactor(18).

For polydisperse distributions, an average size and estimate of dispersity can be obtained from the method of cumulants. Inversion of correlograms to obtain PSDs for multimodal distributions has had limited success. If particle size modes are in a ratio of 3:1 or greater, then it should be possible to extract the peak modes from the data. This is a low resolution method compared to techniques such as CHDF, FFF, or Sedimentation in a centrifugal force field.

The features, benefits and limitations of DLS are summarized in Fig. 14.

FEATURES AND BENEFITS

- **Fast and Automatic**
- **Reproducibility**
- **No Calibration**
- **Large Size Range (3nm to - 3µm)**
- **Versatility**

LIMITATIONS

- **Low Resolution (~3:1 Modal Separation)**
- **Intensity-Weighted**
- **Not Counting Particles (Ensemble Method)**
- **Dust in H20**
- **Limited Shape Information**

Fig. 14. Features, benefits and limitations of DLS for PSD characterization.

Attempts to meet extremely difficult technical challenges

The fourth activity characterizing the renaissance in PSD characterization and measurement involves attempts to meet extremely difficult technological challenges as a result of advances in electronics and computing and sensor technologies. These technological challenges will be discussed below and can be cataloged as follows:

- Measurements in concentrated dispersions
- Wide dynamic particle size range measurement capability in a single instrument
- On-line and at line analyses
- Structural and textural morphology characterization of particle shape

Concentrated dispersion measurements

Fiber optic quasi-elastic light scattering(FOQELS)   The need to measure particles in concentrated dispersions has lead to the development of fiber optics QELS, given the acronym FOQELS by Brookhaven Instruments, for making measurements in concentrated dispersions up to 40% by weight. This commercial development is based on the work of Dhadwal et al. (19,20). Visible light from a laser diode is focused into the sample by a monomode fiber and the scattered light is collected by a second monomode fiber at an angle of 153°. The fluctuations in scattered light are analyzed by the photon correlation technique. As long as the dispersion has observable fluidity, translational diffusion coefficients can be measured and transformed into particle size information. The fiber optics technology enables remote measurement of processes. For small particles less than 100 nm, accurate measurements of size in concentrated dispersion can be obtained. As the particle size increases, deviations from absolute size are observed as the concentration increases due to multiple scattering effects. The precision, reproducibility, and size range of measurement associated with PCS measurements applies to the FOQELS measurements. FOQELS applications reported include particle size growth in emulsion polymerization, nucleation processes in metallic oxide manufacture and monitoring protein crystal growth (21).

Laser light diffraction   Over the last seven years, there has been intense activity to revitalize Fraunhofer light diffraction technology to measure concentrated dispersions over a wide dynamic particle size range (0.1-800 μm) by combining Fraunhofer diffraction technology with light scattering detectors to generate a hybrid instrument. There are at least ten instrument vendors involved in this marketplace. The market pull for this type of instrument has been the need to provide particle size distribution information above and below 1.0μ.m with a single instrument. Fraunhofer diffraction physics dates back to 1840. The advent of laser technology coupled with advances in computer technology has made this a viable commercial method of measuring particle size since 1972. The measurement is fast and can be used with dry powders and often the instrument of choice for powder coatings PSD measurements. Combining Fraunhofer diffraction with light scattering detection via Mie theory has enabled measurement to be made down to 0.06 μm, as reported by some vendors. The technique is a moderate to low resolution technique for extracting multimodal distribution information. Below 2μm one has to carefully check the validity of the specific instrument vendor's software to extract multimodal distributions.

Electroacoustic efforts   There have been at least three vendors (Matec Applied Sciences, Sympatec , and Malvern Instruments) who have produced instruments which take advantage of electroacoustic techniques in concentrated dispersion to measure both the particle size and the electrophoretic mobility distributions. These techniques are well suited for dense, inorganic materials such as titanium dioxide (22). Exploratory experimentation is underway to evaluate the technique for organic (low density) dispersions. The technique is limited with respect to bimodal or multimodal distribution analysis in the same way as is the DLS technique.

Other methods Dielectric Spectroscopy has some promise for particle size characterization in concentrated colloids, as reported by Sauer et al. (23). The authors showed that unique Cole-Cole plots could be obtained as a function of concentration and ionic strength for monodisperse latex particles.
    Rheological flow curve analysis using the Cross equation for concentrated dispersions shows correlations of one of the rheological parameters to particle size and distribution.
    Both dielectric spectroscopy and rheological flow curve analysis merit further study as possible methods for characterizing the PSD of concentrated dispersions.

Wide dynamic particle size range measurement capability
There has been a lot of activity in developing this capability manifested by the development of hybrid instruments to cover a broad particle size range, some of which has been previously mentioned.

- Fraunhofer Diffraction/Mie Scattering (0.06-8000 μm)
- High resolution gravitational/centrifugal sedimentation (0.05-100 μm)
- Combination of FFF modes (e.g., StFFF with FlFFF or SdFFF)
- Single-particle optical sensing/DLS (0.1-300 μ.m)

    The combination of single-particle optical sensing (SPOS) with DLS offers the possibility of analyzing a broad particle size range conservatively from 0.1-300 μm. The single particle optical sensing method is a photozone sensing method in which particles are counted by their ability to obscure light as they move through an orifice. This technique is automated through an auto-diluter. The sensing size range of SPOS is from about 1 μm to 300 μm. The SPOS technique is a high resolution technique with respect to extracting multimodal distributions and is extremely sensitive to low levels of large particles in the presence of small particles (contaminant analysis). A schematic of the autodilution SPOS apparatus is shown in Fig. 15 with a separation of a hexamodal mixture of monodisperse particles shown in Fig. 16 to demonstrate the resolution of the method. When coupled with DLS, this hybrid instrument can characterize the full PSD of particulate systems in which the main contributor to the PSD are small particles.

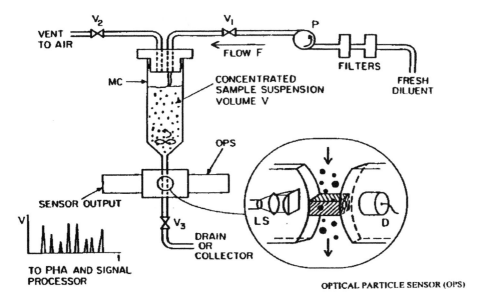

Fig. 15. Schematic diagram of the auto-dilution apparatus and optical particle sensor (light-blockage).

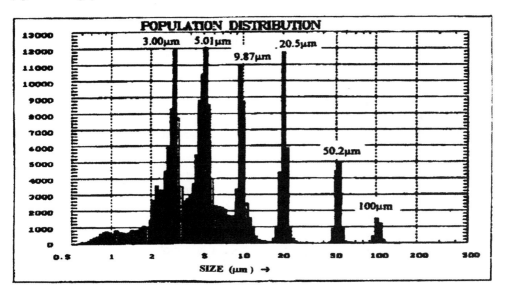

Fig. 16. Separation of a hexamodal mixture of monodisperse particle size standards by SPOS.

DLS provides characterization of the small particles while SPOS characterizes the small level of larger particles, barely discernible in the DLS size distribution. An example of this type of analysis is shown in Fig. 17.

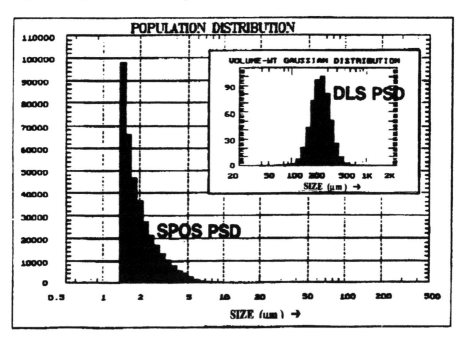

Fig. 17. Combined SPOS/DLS particle size distribution.

<u>On-line, at line analysis</u>
The main techniques available for on-line and at line analysis of particle size distributions have been discussed to some degree. These techniques are:

- Fiber optic QELS (FOQELS) for concentrated dispersions
- Automatic dilution PCS
- Automatic dilution turbidity analysis
- Laser sensor-backscatter measurements

    Angular scattering and absorbance measurements as a function of wavelength with photo diode array instruments and automatic dilution coupled with sophisticated mathematical analysis of PSDs is revitalizing the classical turbidity measurement and is the subject of ongoing research by Prof. Luis Garcia-Rubio at the University of South Florida at Tampa (24,25).
    Lasentech has developed a laser backscatter instrument to measure effective particle size (chord length across a particle) in concentrated dispersion from several microns to larger sizes and has found practical application for monitoring particle size changes in many processes.

Structural and textural morphology characterization of particle shape
The need for structural and textural morphology information has led to cost/performance improvements in automated image analyzers that now provide shape information using signature wave form characterization and fractal dimension characterization of shape and texture(26).

Conclusions

As we look forward to the 21st century and beyond, we can expect further advances in PSD characterization and measurement. It is anticipated that additional hybrid type instruments will be produced to cover a wide dynamic particle size range with improved resolution. The quest for PSD analysis methods for concentrated dispersions with good resolution over a wide dynamic range will continue and will be extremely difficult to achieve. However, advances will continue to be made in providing structural and textural information as a result of advances in computer technology. Another major technical challenge which will probably be met is the development of in-line or in-situ particle size monitoring technology for latex reactors.

In addition, we expect the instruments to be more user-friendly, easier to use, more automated and smarter with respect to programmed intelligence.

Acknowledgement

This paper is based partly on an American chemical Society (ACS) workshop on "Modern Methods of Particle Size Distribution: Assessment and Characterization" which originated as part of the ACS Division of Polymeric Materials: Science and Engineering technical programming. The author acknowledges the contributions of many of the workshop speakers over the last 15 years for their technical contributions to this evolving workshop and for numerous stimulating discussions of the subject. Specifically, the author wishes to acknowledge the following individuals: Dr. J. Gabriel Dos Ramos, Dr. David Fairhurst, Mr. Peter Faraday, Dr. J. Calvin Giddings, Mr. Kerry Hasapidas, Mr. Richard Karun, Dr. Brian Kaye, Dr. David Nicoli, Dr. Remi Trottier, Dr. Bruce Weiner, Dr. Kim (Ratanathanawongs)Williams, and Dr. Stewart Wood.

Literature Cited

1.   Small,.J. Colloid Interface Sci. 1976, 57, 337.
2.   Segré, G.; Silberberg, A. J. Fluid Mech. 1962,14,136.
3.   Regnier, E.; Ball, D., Pittsburgh Conference on Spectroscopy and Analytical Chemistry, 1977, paper No. 447.
4.   Dos Ramos, J. G.; Silebi, C. A. J. Colloid Interface Sci., 1989,133, 302.
5.   Silebi, C. A.; Dos Ramos, J. G. J. Colloid Interface Sci., 1989,130, 14.
6.   Thompson, G. H., Meyers, M. N; Giddings,J.C. Sep. Sci.1967, 2, 797.
7.   Giddings, J. C. J. Chem. Phys., 1968, 49, 81.
8.   Kirkland, J. J; Yau, W. W; Doerner, W. A.; Grant, J. W. Anal. Chem.,1980, 52 1944.

9.   Giddings, J. C., Ratanathanawongs, S. K.; Moon, M. H. KONA Powder Particle, 1991, 9, 200.
10.   Koehler, M E.; Zander,R.A.; Gill, T. T.; Provder, T.; Niemann T. F. ACS Symposium Series No. 332, T. Provder, Ed., 1987, 180.
11.   Koehler, M. E.; Provder,T.; Zander,R.A. US Patent 4,311,039 , 1982.
12.   Provder, T.; Holsworth, R. M. ACS Div. of Org. Coatings and Plastics Chemistry Preprints,1976, 36, 150.

13.   Holsworth, R. M.; Provder,T.; Stansbrey, J. J. ACS Symposium Series No. 332, T. Provder, Ed., 1987,p. 191.
14.   Holsworth, R. M.; Provder, T. US Patent 4,478,073, 1985.
15.   Cummins, H. Z.; Knable, N.; Yeh,Y. Phys. Rev. Let., 1964,12, 150.
16.   Foord, R.; et al. Nature, 1970, 227, 242.
17.   Nicoli, D. F.; Elings, V.B. US Patent 4,794,806, Jan. 3, 1989, "Automatic Dilution Systems".
18.   Nicoli, D .F.; Korti, T.; Gossen, P.;Nu, J.-S.; MacGregor J. F. "Particle Size Distribution II. Assessment and Characterization", ACS Symposium Series 472 T. Provder, Ed., 1991, 86.
19.   Dhadwal, H.; Ansari, R; Meyer, W. Rev. Sci-Instrum. 1991,62 (12), 2963.
20.   Dhadwal , H.;et al., Proc. SPIE, 1993,16,1884.
21.   Application Notes, Brookhaven Instruments, FOQELS.
22.   O'Brien R. W.; Rowlands, W. N.; Hunter, R. J. Proceedings of the NIST Workshop on "Electroacoustics for Characterization of Particulates in Suspensions", 1993.
23.   Sauer, B. B.; Stock, R. S.; Lim, K.-H.; Ray, W .H. J. Applied Polymer. Sci., 1990, 39, 2419.
24.   Sacato, P.; Lanza, F.; Suarez, H.; Garcia-Rubio, L. H. Proceedings of the ACS Division of Polymeric Materials: Science and Engineering, 1996, 75, 30.
25.   Bacon, C.; Garcia-Rubio, L. H. Proceedings of the ACS Division of Polymeric Materials: Science and Engineering 1996, 75, 32.
26.   Kaye, B. American Laboratory, April 1986,55.

THE THERMAL ANALYSIS OF POLYMERS

Edwin F Meyer III

ICI Paints Research Center
16651 Sprague Road
Strongsville OH 44136

Introduction
Differential Scanning Calorimetry
Dynamic Mechanical analysis
Thermomechanical Analysis
Thermogravimetric Analysis

Introduction

The manufacturers of polymeric products often demand that a particular product meet specifications that involve end-use properties such as: water resistance, chemical resistance, flexibility, permeability as well as the ability to absorb sound and vibrations -- all over a wide temperature range. Every one of these are specifications can be related back to fundamental chemical and physical properties of the polymer using thermal analysis. In general, thermal analysis is simply the characterization of the properties of a material as a function of temperature. This definition is frequently expanded to include isothermal experiments performed on conventional thermal analysis instruments. Standard techniques include: Differential Scanning Calorimetry (DSC), Thermogravimetric Analysis (TGA), Dynamic Mechanical Analysis (DMA), Dielectric Analysis (DEA) and Thermomechanical Analysis (TMA). Variations on these techniques include modulated temperature differential scanning calorimetry (MTDSC) and modulated thermogravimetric analysis (MTGA). In these two techniques the temperature program of the sample is modulated which provides superior results in virtually all cases.

Recent advances in thermal analysis include Micro-Thermal Calorimetry. This technique involves the marriage of modulated temperature spectroscopy and atomic force microscopy (AFM) by effectively placing a modulated temperature probe on the tip of an AFM probe.

Differential Scanning Calorimetry

History. A great step toward modern DSC was achieved in 1899 by William Chandler

Roberts-Austen, then Chemist to the Royal Mint, when he conceived the idea of measuring the temperature difference between the sample and a thermally inert material. Previously all attempts at determining the heating curve for a material -- at that time mostly clays and minerals -- involved only the measurement of the temperature of the material as a function of time. The advantage of the differential method is two-fold. First, a differential method is more precise and second, the differential method eliminates the need to use time as a dependent variable. That is, the differential method produces a DT vs T curve where T is the temperature of the material and DT is the temperature difference between the material and the thermally inert reference.

From this time until very recently the advances toward modern DSC were mainly instrumental. For example, better temperature measurement and control, control of the atmosphere within the cell and a greater symmetry between the sample and reference positions. Another great leap forward in differential scanning calorimetry came in 1992 when Dr. Mike Reading, then of ICI Paints, conceived the idea of superimposing a modulation on top of the linear sample temperature rate. Prior to this development the linear temperature rate performed two functions -- supply the temperature difference between the sample side and the reference side and scan the temperature regime of interest. The modulation of the temperature of the sample side can supply a useful temperature difference between the two cell positions allowing the experimentalist to independently adjust the rate at which the temperature regime of interest is scanned. Indeed the temperature scan rate or linear underlying temperature rate can be set to zero.

The technique of modulating around a fixed temperature and monitoring the output signal as a function of time has been called quasi-isothermal MTDSC. Of course, modulating the temperature of the sample side will produce a modulated temperature difference between the sample and reference. This modulated temperature difference acts as a continuous probe of the heat capacity of the sample -- irrespective of any kinetic events occurring within the sample. If the heat capacity of the sample increases, the amplitude of the thermal energy transfer between the furnace and the sample needed to produce the programmed temperature modulation will increase proportionally. This change in the amplitude of the thermal energy transfer will be manifest in the temperature difference between the sample and the reference.

DSC Fundamentals. The technique of differential scanning calorimetry is used to detect and quantify thermal events in a material. These thermal events include glass transitions, melts, crystallizations, stress relaxations, chemical reactions and volatilizations. The working part of the instrument is the cell, a schematic of which is shown in figure 1. The cell contains two positions -- sample and reference -- and is enclosed by a furnace block that, ostensibly, is thermally symmetric with respect to the sample side and reference side. Thermocouple junctions just underneath each position monitor both the temperature of the sample side, $T_{samp\ side}$, and the difference between the temperature of the sample side and the temperature of the reference side, $T_{ref\ side} - T_{samp\ side}$. If the cell is empty and thermally symmetric, the temperature of the reference side will always be the same as the temperature of the sample side irrespective of the manner in which the furnace block is heated or cooled.

To perform a typical DSC experiment a sample is placed in a small aluminum pan and the pan is then placed on the sample side. To thermally balance the pan containing the sample, an empty, identical "reference" pan is placed on the reference side. With this configuration any temperature difference between the reference side and the sample side can be attributed solely to the presence of the sample. Further, it can be shown, assuming Newton's Law of Cooling describes the thermal energy transfer, that this temperature difference is directly proportional to the rate of thermal energy transfer between the furnace and the sample.

Sample Preparation. A thermal analyst can divide all sample submissions into two groups -- commercial and research. Research samples are usually less problematical because their history is well known -- the samples have been prepared under laboratory conditions so their chemistry and thermal history is well known. Commercial samples offer a different set of challenges because the polymer may have been subjected to temperature extremes as well as harsh chemical environments.

For this reason, when characterizing commercial failure it is important to obtain a sample of the polymer that failed -- not a retain from the lab or the plant. For example, if the thermal analyst is to trouble-shoot the whitening of a protective polymeric film on the interior of garbanzo bean cans packed in Mexico, it is important that thermal analysis tests are performed on a sample of protective coating taken directly from a can that exhibited the problem. The reason for this is two-fold. First of all the polymer may have changed from the time it left the facility where it was manufactured and when it was applied and, second, virtually all the physical properties of the thermoset coating are a strong function of the bake conditions.

One method to obtain a coating that has been baked on a substrate is simply to carefully scrape off a small amount with a bladed instrument. Ideally, the sample mass for a DSC experiment is about 5 mg although lighter samples can be used. The sample is then placed in a cylindrical aluminum pan which, typically, is about 1 mm high and 5 mm in diameter. To optimize the thermal energy transfer, the sample should be spread out evenly in the pan. To ensure good thermal contact between the sample and the pan, an aluminum lid is placed on top of the sample and then "crimped" with the pan thus forming a aluminum sandwich with the sample. If the sample may contain volatiles it is good practice to make a hole in the pan lid before crimping. This will prevent a pressure increase inside the sample pan during heating.

Experimental Conditions. For both LTDSC and MTDSC the technician must choose a purge gas and a flow rate. The purge gas, which should be inert, will displace moisture and oxygen in the DSC cell and assist in the transfer of thermal energy among the furnace, the reference and the sample. Although helium, due to its high speed at a given temperature, provides a more efficient thermal energy transfer, nitrogen is conventionally used for virtually all standard applications. For specialized applications where fast heating rates or high modulation amplitudes are used, helium may be used. A typical nitrogen flow rate is 50 cc min⁻¹, but this can be adjusted based on the type of sample being run and the programmed temperature rate of the sample.

Another experimental parameter that is common to both LTDSC and MTDSC is the temperature range. For optimum results it is common practice to choose a starting temperature that is at least 50 °C before any transition and, of course, the final temperature should be sufficiently above any transitions of interest.

The lone experimental parameter that differentiates LTDSC and MTDSC is the temperature rate. In linear temperature DSC the operator selects only the linear temperature rate which is constant throughout the run. This temperature rate creates the steady state temperature difference between the temperature of the sample side and the temperature of the reference side -- in other words the output signal. So, a higher temperature rate provides a higher signal to noise ratio. Unfortunately, because the linear temperature rate is also the mechanism by which the temperature regime of interest is scanned, a high temperature rate tends to broaden the transitions and thus meld transitions that are close to each other. Often, when running a linear temperature DSC experiment the operator must compromise between sensitivity and precision.

In modulated temperature DSC the temperature difference between the sample side and the reference side -- the output signal -- can be generated solely by the modulation of the temperature leaving the operator free to choose a linear temperature rate that will scan the temperature regime of interest at a slower rate. Choosing the temperature protocol in an MTDSC experiment requires the assignment of three parameters: the underlying linear temperature rate, the amplitude of the temperature modulation and the period of the temperature modulation. The first consideration when assigning these parameters is whether the furnace is able to carry out the modulation. If the period is too small or the amplitude too high for a given linear temperature rate the actual temperature rate of the sample side will not follow the set protocol. That is, the sine wave will be distorted. A typical set of MTDSC temperature parameters may be an underlying heating rate of 2 °C/minute, a modulation amplitude of 1 °C and a modulation period of one minute. If a faster underlying temperature rate is desired the modulation amplitude may have to be lowered or the period of the modulation raised to ensure that the furnace can keep the sample side temperature on protocol. A simple technique to check if indeed the sample side temperature is a sine wave is to plot the modulated temperature of the sample side vs time.

Applications. In many instances the desired result from a DSC experiment is the T_g (or T_gs) of the polymer system. The T_g of a polymer is an important parameter for polymer engineers because the T_g gives an indication of the processing conditions needed in manufacturing. Due to the complexity of today's polymeric systems, the characterization of the T_g or T_gs of a particular coating system can be confounded by various thermal events. These include melts and crystallizations of additives, stress relaxations, volatilizations and all kinds of rearrangements of the polymer micromorphology with temperature -- especially near the T_g of one of the components. For this reason, Modulated Temperature DSC is the method of choice for thermal analysts. MTDSC can separate the heat capacity of the coating from kinetic thermal events and the T_g can then be obtained from the step transition in the heat capacity. A simple example that demonstrates this involves a pilot plant batch and a lab batch of a resin.

One of the problems frequently encountered by plant engineers is the scale-up of a reaction. Frequently, a resin made in a one gallon lab reactor will exhibit different

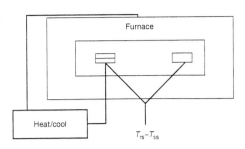

Figure 1 Simple diagram of the DSC sample cell. The output signal is the temperature difference between the sample side and the reference side.

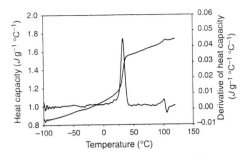

Figure 4 The heat capacity and its derivative for an 90/10 blend of VA/BA.

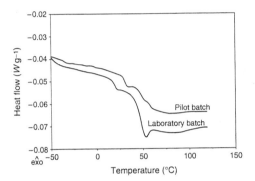

Figure 2 The total heat flow versus temperature curves for a pilot plant batch and a laboratory batch of a particular coating.

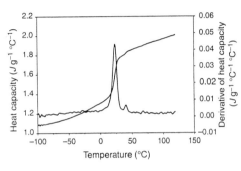

Figure 5 The heat capacity and its derivative for an 80/20 blend of VA/BA.

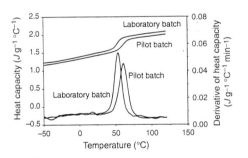

Figure 3 The heat capacity and its derivative versus temperature from the same experiment shown in Figure 2.

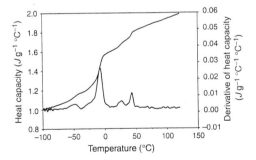

Figure 6 The heat capacity and its derivative for an 50/50 blend of VA/BA.

physical properties than one made in a 100 gallon pilot plant reactor or a 10,000 gallon plant reactor even though the "recipe" was the same. This is due to the fact that larger reactors often have different heat flow and mixing characteristics. When the product from two different reactors are significantly different, the thermal analyst is often called upon to characterize the two resins. Figure 2 shows the "total heat flow" vs temperature for two samples -- one manufactured in a one gallon lab reactor and the other manufactured in a 100 gallon pilot plant reactor. Here the assignment of T_g is problematical due to kinetic events near the T_g transition. The heat capacity, separated from the total heat flow, and its derivative, are shown in figure 3. Here the assignment of T_g is a trivial matter using the peak in the derivative curve. Clearly, the samples have different T_gs.

Another advantage MTDSC offers over LTDSC is sensitivity. In LTDSC the temperature scan rate through a region of interest must be kept high in order to produce a useful temperature difference between the sample and the reference sides. This precludes the accurate characterization of many minor thermal events. The first example that demonstrates the enhanced sensitivity of the MTDSC protocol involves the emulsion polymerization of vinyl acetate (VA) and butyl acrylate (BA) monomers with a ladder of concentrations. The ladder consists of five evenly spaced steps from 50/50 VA/BA to 90/10 VA/BA. The samples were prepared by drawing down the emulsion and allowing it to dry for three days.

Figure 4 shows a coplot of the heat capacity and its derivative with respect to temperature plotted vs temperature for the 90/10 concentration ratio. Here we have a sharp transition and an easily defined Tg. The 80/20 results, shown in figure 5, exhibits a major T_g from the co-polymer and a minor T_g from VA homopolymer (~40 °C). The VA homopolymer was from the pure VA seed used in manufacturing the emulsion. The results from the 50/50 sample are shown in figure 6. Here four transitions are evident that are identified as follows: the VA/BA copolymer (~-10 °C), the VA seed (~40 °C), a small amount of BA homopolymer at about -50 °C and a small transition at about 25 °C due to the region of interpenetration between the VA seed and the VA/BA co-polymer.

A second example that demonstrates the sensitivity of the MTDSC technique involves the homogenization of a latex with time. It is well known that a latex paint continues to "cure" for a couple of weeks after it is applied even though it is dry to the touch after a few hours. Coatings manufacturers can characterize this slow aging with MTDSC. Figure 7 shows the "total heat flow" traces for a latex sample at 4, 11 and 23 days after application. Due to the kinetic activity, the T_g is obscured. However, the derivative of the heat capacity traces, shown in figure 8, are free from the kinetic activity that confounds the total heat flow. From the gradual melding of the peaks with aging time we can imagine that the coating is becoming more homogeneous with time.

Another example that demonstrates the precision of the T_g determination using MTDSC involves the determination of the relative concentration of each monomer in a copolymer. An unknown sample was suspected to be a co-polymer of only two monomers. We'll call them monomer A and monomer B. The first experiment performed was to determine the T_g of the unknown. Next we tested three concentration ratios that were polymerized from the following monomer blend ratios: 100/0, 50/50 and 0/100. From a plot of the T_g vs concentration ratio using only these three points the estimated ratio of the

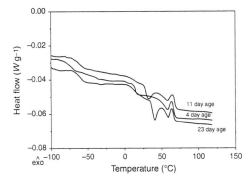

Figure 7 The total heat flow curves for a latex taken at three different aging times.

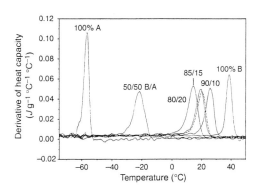

Figure 9 The derivative of the heat capacity curves for a series of experiments designed to determine the blend ratio of the unknown, which is shown as a dotted line in the plot.

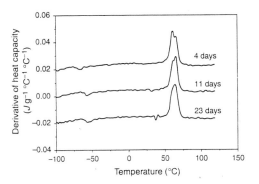

Figure 8 The derivative of the heat capacity curves for the runs shown in Figure 7.

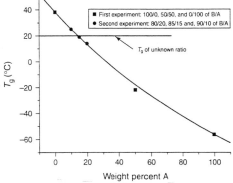

Figure 10 Determination of the polymer blend ratio with MTDSC. A plot of the T_g versus blend ratio of the controls; the horizontal line represents the T_g of the unknown.

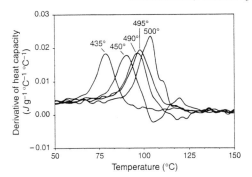

Figure 11 A co-plot of the derivative of the heat capacity for five control bakes of a particular coating.

unknown was about 84/16.   To check this, three polymers were prepared with concentration ratios of 80/20, 85/15 and 90/10. These results confirmed the original estimate of 84/16 as the concentration ratio of the unknown. A coplot of all the MTDSC results is shown in figure 9 and a plot of T_g vs concentration ratio is shown in figure 10. The curve in figure 10 is that calculated from the Fox equation and the horizontal line represents the T_g of the polymer with the unknown monomer blend ratio.

Whereas many manufacturing plants perform the entire process from synthesizing the polymers to manufacturing products from them, some facilities produce products from polymers purchased from a supplier. When the products fail to meet specifications, the supplier is often accused of supplying inferior or "out-of-spec" polymer and the supplier responds by accusing the manufacturer of poor quality control during the manufacturing process.   In these cases thermal analysis of the polymer can often assign the blame properly.   When a thermoset polymer system is used, the resulting physical properties depend strongly on the temperature program the polymer was subjected to during its cure. An underbake can result in a weak, soft and permeable product whereas an overbake can result in a brittle product that has a tendency to crack. The T_g of a coating can be a strong function of bake temperature and thus an indicator of the bake a coating received. The following example involves a corrosion problem with an interior food coating at a customer's plant. To investigate this complaint, bake extremes were prepared in the lab as controls and compared with the polymeric coating taken directly from a problem can. The temperature derivative of the heat capacity results are shown in figure 11. Clearly, the T_g is a function of bake and thus we can use the T_g of a coating that was baked at a customer's plant as a indicator of the level of bake it received.

The degree to which the T_g of a coating is a function of the bake depends upon the nature of the coating. The T_g of a thermoplastic coating increases with bake due to the loss of plasticizing agents, such as water, whereas the increase in the T_g of a thermoset coating during a bake is due mainly to an exothermic cross-linking reaction which transforms a low molecular weight liquid into an "infinite" network. The rate of the cure reaction will be limited by the reactivity of the components when its temperature is below its T_g and limited by diffusion when it has vitrified. However, since the T_g is changing due to crosslinking, a thermoset system may, if the conditions are right, vitrify, de-vitrify and then re-vitrify. That is, a chemical reaction causes the T_g to exceed the temperature of the coating and the coating vitrifies thus drastically altering the rate of reaction. Then the oven brings the coating above its current T_g where the reaction rate will again be governed by the reactivity ratio of the components. Finally, the cure proceeds to a point where the T_g again exceeds its temperature and the coating re-vitrifies. To optimize the efficiency of a thermoset cure it is helpful to know the T_g of the coating as a function of the degree of cure and the reaction rate of the components as a function of temperature. With such information the plant engineers can set the temperatures at various zones in the oven in a manner that minimizes the time the coating is in the oven and thus maximizes the efficiency of the oven.

To obtain these parameters requires access to the T_g as a function of time during the cure reaction. This makes it an ideal candidate for interrogation with MTDSC which can separate the heat capacity signal from the exotherm due to the cure reaction.

Indeed, modulated temperature DSC has been used to fully characterize the bake of a thermoset coating[1,2]. Note that this is virtually impossible with linear temperature DSC because the relatively small T_g transition is difficult to deconvolute from the larger exotherm.

Another application of DSC to coatings involves the detection of additives -- specifically waxes. These lubricants are added to industrial coatings for a variety of reasons. They provide slip during the metal-forming process in pre-coat post-form applications, they provide smooth transport through the trackwork that transports the work throughout the plant and they act as an interior release agent in food cans -- so the food slides out of the can easily. There are a tremendous variety of commercially available lubricants that range from natural carnauba wax to teflon. The characterization of a wax is accomplished not through its T_g but through its melt temperature. Because of this it is not necessary to run MTDSC -- a linear temperature rate can be used. One application where DSC can be useful is the characterization of the wax a competitor is using in their coatings. Another application involves the identification of a substance that has build up on the metal tooling at a customer's plant. Is it the wax? Is it the polymer? Or is it from some other source? Removing a few milligrams from the tool and running it in a DSC will, more likely than not, provide the answer.

The final example we will present involves a baseline shift which is difficult, if not impossible, to uncover using LTDSC. The total heat flow for a MTDSC run is shown in figure 12. It appears that there is a T_g at about 20 °C. However, the "event" at 20 °C had nothing to do with the polymer. For some reason the baseline shifted at about room temperature. That is, the symmetry of the cell changed at about room temperature. Regardless of the reason for this shift, a downward shift in baseline appears Tg-like when looking at the total heat flow. However, with MTDSC you will not be fooled because the amplitude of the modulation will not change during a baseline shift. Figure 13 shows the output signal from the cell vs the temperature of the sample side (averaged over a modulation) and its amplitude, which is simply the heat capacity. Here it is clear that the event at about room temperature is not a Tg, but rather a baseline shift because the amplitude of the output signal does not undergo a transition. The true T_g is at about 45 °C where the amplitude of the output signal increases significantly indicating that the sample's heat capacity has increased. That is, the amplitude of the thermal energy modulation of the furnace needed to produce a 1 °C modulation in the sample side temperature has increased.

Method Development. In modulated temperature DSC the temperature scan rate or linear underlying temperature rate can be set to zero. The technique of modulating around a fixed temperature and monitoring the output signal as a function of time has been called quasi-isothermal MTDSC. This technique can be used to elucidate the nature of crystallizations and glass transitions[3,4]. Typical experimental conditions for an isothermal step are an amplitude of a tenth of a Kelvin and periods ranging from 30 to 100 seconds. The entire quasi-isothermal experiment will consist of a series of isothermal steps across the transition of interest that are perhaps 0.1 to 0.2 Kelvin apart. Typically, each isothermal experiment will run for 20 minutes with data accumulation over the last ten minutes. In contrast with a continuous temperature scan which results in a continuous line representing heat capacity vs temperature, the results of a quasi-isothermal MTDSC experiment consist of a series of points each representing the heat capacity at (or around) a particular

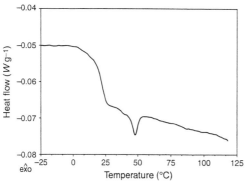

Figure 12 The total heat flow trace for a test coating.

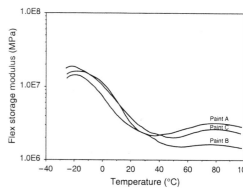

Figure 14 A coplot of the flexural storage modulus for te paints A, B, and C after drying at 30 °F.

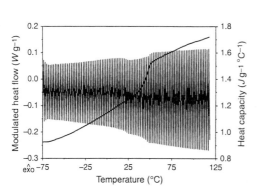

Figure 13 The modulated heat flow and heat capacity for a test coating from the same run shown in Figure 12.

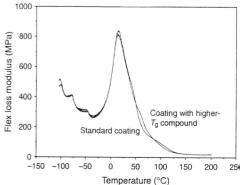

Figure 15 Coplot of the flexural loss modulus of a sta dard coating and an experimental coating with a higher-7 component.

temperature. The advantage of a series of quasi-isothermal steps is that a steady state is achieved by simply waiting and there is no need to deconvolute the effect of the underlying linear temperature rate from the effect of the modulation because the underlying linear temperature rate is zero.

Another novel development in MTDSC is the light-heating technique, developed by Saruyama[5]. This technique allows higher modulation frequencies to be accessed by employing a light beam directed on the sample which is attenuated by a polarizer rotating at a constant frequency. The intensity of the beam determines the modulation amplitude and the angular frequency of the rotating polarizer determines the period of the modulation. The underlying heating rate is supplied by a conventional heating source allowing access to high frequencies of modulation.

Still another, and perhaps the most technically impressive, development in this fertile field is the marriage of DSC and atomic force microscopy into a technique called Micro-Thermal Calorimetry. In this new technique the AFM tip is fitted with an ultra-miniature temperature probe that acts as both the heat source and the detector. Obvious applications include, characterization of micro-morphology and detecting and identifying contaminants. The accessible frequencies with this technique are much higher than with normal MTDSC simply because the mass whose temperature is being modulated is much smaller. Indeed, varying the frequency of the modulation is a means of depth profiling from the surface.

Dynamic Mechanical Analysis

History. The development of modern dynamic mechanical analysis began with the torsion pendulum. This instrument consists of a clamping mechanism that allows the sample to be deformed in the torsion mode (twisted). When the sample is released from the twisted state it recovers with a damped oscillation around its equilibrium position. The frequency of the oscillations can be related to the shear modulus of the sample and the damping of the oscillations (rate of decrease in amplitude) can be related to the loss modulus of the sample[6].

A torsion pendulum was used to study the mechanical properties of free films as early as 1966[7,8]. The development of the DMA has progressed significantly from the custom-made torsion pendula of the 1960s. Commercial instruments of the late 1990s have numerous deformation modes, precise temperature control and a wide range of deformation frequencies. Along with the sophisticated hardware comes software with the ability to analyze, fit and smooth the data.

DMA Fundamentals. In general, dynamic mechanical analysis (DMA) is a technique that is used to characterize the response of a material to a periodic deformation as a function of frequency, temperature or time. This deformation can consist of a constant oscillation amplitude at a fixed frequency or a free oscillation at the resonant frequency. The response includes both the elastic component, which is in phase with the spatial perturbation, and a viscous component, which is out of phase with the deformation.

To get quantitative results for the absolute value of the moduli from a DMA test the sample must be uniform and its geometry well known. If the sample is a solid polymer sheet, slabs of well known geometry can be produced and the results from the instrument will be quantitative. However, sometimes the polymer is a thin film or even a viscous liquid. In this case a supporting substrate may be used. The results obtained from such an experiment may be qualitative but in many cases the DMA testing involves a rank ordering of samples rather than a determination of the absolute moduli.

A couple of advantages the DMA technique has over other thermal analysis techniques is the ability to detect lower order transitions and the ability to simultaneously determine the viscous and elastic moduli of the material. The lower order transitions, which have been attributed to side chain motions of the polymer, are related to the ability of the polymer to respond to fast deformations without exhibiting catastrophic failure. A polymer may be subject to such deformations during a forming process in manufacturing, when exposed to vibrations, or when subject to a sharp impact. The DMA is a useful instrument to employ when developing, for example, sound absorbing and impact-resistant polymers as well as coatings for a flexible substrate.

Sample Preparation. If the polymer to be tested consists of a flat sheet of constant thinkness the sample preparation simply involves cutting an appropriate piece -- typically a centimeter by a few centimeters by a couple of millimeters. The more problematical samples are thin polymeric films and high viscosity liquids.

The solution usually employed to test such samples is to provide a support for the film. The two basic methods of supporting a coating are with a flat, solid substrate and with a mesh or absorbent fiber. An example of the former type is a coating applied to a metal panel and an example of the second type is a polymer-impregnated fiberglass mesh. With a flat substrate, we still have some information about the geometry of the film and we can get semi-quantitative results if the response from the substrate can be deconvoluted from the response from the coating.

A fiberglass or stainless steel mesh can be used to support viscous polymers. A mesh is useful when studying the cure or drying of a polymeric coating. Just prior to testing, a known amount of liquid coating, typically about 25 microliters, is spread evenly across the mesh. The surface tension of the coating is sufficient to keep the liquid from dripping off the braid. During the cure or drying of the film the elastic modulus of the coating will increase and this increase will be evident in the results.

Experimental Conditions. The difficulty in choosing the optimal experimental conditions has increased with the complexity of the commercially available instruments. First the operator has to choose the mode of deformation. The most common mode is the two point bend whereby one end of the sample is held fixed and the other end is oscillated sinusoidally. Other modes include the three point bend, shear, torsion, compression and tension. After the mode of deformation is chosen the operator must select the experimental conditions. These include the sample length, the amplitude of the deformation, the frequency or frequencies of the deformation and the temperature protocol. If the sample has a relatively low modulus, a shorter sample should be used and if the sample has a high modulus, a longer sample should be used. The amplitude of the deformation should be large enough to create a useful restoring force but not so large that it exceeds the linear response region. Consult the instrument manual for guidelines on setting experimental parameters for a given sample stiffness.

Ideally, a range of frequencies should be chosen. This will provide information regarding the relationship between both the elastic and loss moduli and the frequency of deformation. The temperature program may consist of a constant scan rate, a constant temperature or a combination of both. For example, the temperature program of a cure experiment may consist of a jump from room temperature to 400 °F, a 10 minute isotherm at 400 °F, a jump to -100 °F and finally a scan at 2 °C/minute from -100 °F to 400 °F.

Applications. Consumer coatings consist of paints, lacquers etc. that are applied by the consumer and dried at ambient temperature. Most industrial coatings are subjected to a bake which accelerates drying and, in the case of a thermoset, promotes a chemical reaction. DMA can be used either to characterize the final cured or dried film or to characterize the transition from a wet coating to a final film. The first example involves the ranking of exterior paints for low temperature film formation. That is, whether the films will dry completely at low temperatures. The temperature used in this study is a stretch target for exterior house paint application -- 30 °F.

There were three paints studied. The experimental protocol consisted of applying 25 microliters of coating to a fiberglass braid and holding its temperature at 30 °F for 700 minutes. Following the 700 minute isotherm, the temperature of the sample was scanned from -25 °C to 100 °C. The storage modulus during these temperature scans are co-plotted in figure 14. If the paint is completely dry, the modulus should not increase with temperature. Any increase in elastic modulus with temperature is evidence of an incomplete dry. From the results of the scan we can conclude that the paint B has the best low temperature film formation, paint C is second best and paint A has the most incomplete film formation at low temperature as its elastic modulus increased significantly during the temperature scan.

The second example involves the determination of whether a higher T_g additive would adversely affect the flexibility of an exterior house paint. An exterior house paint must be "soft" enough to expand and contract with the outside of the house as the exterior temperature changes. However, it also must be "hard" enough to resist dirt pick-up when the wind blows dust and dirt against the house. Often paint formulators combine hard and soft polymers in the formula in an attempt to produce a paint with both sufficient flexibility and low dirt pick-up. In this example, the formulators were considering replacing one of the components with a higher T_g polymer but were concerned that this would compromise the excellent flexibility of the product. To investigate this, the standard coating and the coating with the higher T_g component were dried on a fiberglass braid under identical conditions. A co-plot of the flexural loss modulus as determined during a temperature scan at a rate of 2 °C per minute is shown in figure 15. It can be seen that the low temperature loss modulus is virtually unaffected by the addition of the higher T_g component. Only at about 50 °C does the difference between the coatings become evident. Clearly, the higher T_g component can be used in the formula without sacrificing the ability of the coating to respond to the contraction and expansion of the house with temperature.

Method Development . Recent development of DMA has involved expanding the range of applied forces, increasing the range of allowable frequencies, increasing the temperature range over which experiments can be performed and increasing the number of modes of deformation. That is, much of the recent development has involved engineering rather than science. Current instruments have the ability to oscillate samples at 200 Hz using forces of up to 18 N with a temperature range from -150 °C to 600 °C. With modern instrumentation, all of this can be accomplished with a strain resolution in the range of 1 nanometer.

Along with these engineering developments comes an increase in the capability and speed of the software used both to control the instrument and analyze the data. Temperature programs can consist of rapid cools or heat-ups and they can contain numerous temperature excursions as well as isotherms. Software permits extensive co-plotting as well as advanced data handling and curve-fitting procedures.

Thermo-Mechanical Analysis

History. The precursor to the modern thermo-mechanical analyzer is simply a dilatometer that measures a sample's dimensions as a function of temperature. In this way the sample's linear or volume expansion coefficient can be determined. With the wide spread use of polymers in engineering applications, the need to monitor the sample's change in dimension as a function of time under a load at constant temperature (creep) became apparent. Further, the need to determine a material's response to either a tensile or compressive force as a function of temperature and time became necessary. These requirements were the driving force behind the development and commercialization of today's modern thermo-mechanical analyzers.

TMA Fundamentals. Conventionally, Thermo-Mechanical Analysis (TMA), involves the measurement of a linear dimension of a material as a function of temperature or time when the material is subject to a stress (either compressive or tensile). TMA is similar to DMA in that the instrument is monitoring the material's response to an applied force. However, the instrument-naming convention is such that instruments which perturb the sample with a force of constant magnitude are called TMAs and instruments that perturb the sample with a force that varies sinusoidally are called DMAs. A schematic of typical TMA sample cell set up to measure a sample under load is shown in figure 16. The sample under investigation rests on a stage and a probe is brought down on top of the sample with a pre-determined force. The measured parameter is the distance between the tip of the probe and the stage upon which the sample rests. If the sample expands, as it might when

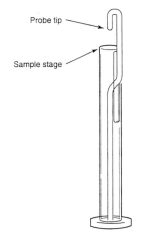

Figure 16 Diagram of a TMA cell.

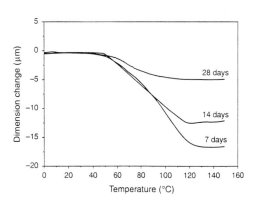

Figure 17 The dimension change versus temperature for a test coating at three drying times.

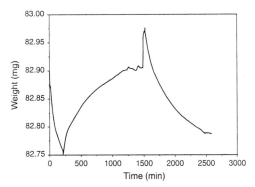

Figure 18 A plot of weight against time for a polymer in an effort to characterize its absorption of carbon dioxide.

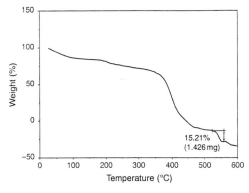

Figure 19 A plot of weight percent versus temperature, which shows a degradation between 530 °C and 550 °C.

heated, the measured distance increases and if the sample contracts or softens, the measured distance will decrease. A modern instrument comes with a variety of probe tips. Sharp probes with hemispherical or cylindrical tips are used to measure softening points and large flat probes are used to measure coefficients of linear expansion.

Sample Preparation. The only requirement to test a sample in a TMA is that it have two opposing flat surfaces, one to place on the stage and the other to contact the probe. Frequently, the only sample preparation required to analyze a polymer is to cut a piece that is small enough to fit on the stage (typically about 1 cm across). Also, when attempting to measure the softening point of a coating there is no reason to remove the coating from the substrate.

Experimental Conditions. The conditions chosen by the user are probe type, the force exerted on the sample by the probe, the temperature program. For a penetration experiment, the probe tip should be relatively sharp, for example, a hemisphere of small radius. The applied force should be great enough to eliminate the noise in the distance measurement but not so great as to cause premature damage to the coating. A typical value for the applied force when a hemispherical probe tip is used is 0.25 Newtons, which is the force of gravity on a mass of about 25 grams. Frequently, the temperature program consists of a simple linear scan through the temperature regime of interest. A typical scan rate is 5 °C per minute.

Applications. A polymeric coating may be dry to the touch after only a few hours but many coatings continue to develop physical properties for weeks after they are applied. To quantify this, a series of experiments were performed on a test coating after a series of aging times.
   The results are co-plotted in figure 17. Here we see that the probe starts to penetrate significantly into the coating at roughly 50 °C. At 7 days drying the penetration is deep: about 15 microns. After 14 days of drying the maximum penetration depth is less than at seven days and after 28 days of drying the penetration depth into the coating is less than 5 microns. From these results we can conclude that the coating is still building integrity after a month of drying. By adding specifically designed chemicals the "cure" of the coating can be accelerated and the effect quantified with TMA.

Method Development. A recent improvement in the software that controls the TMA instrument provides an option for the linear temperature scan through the regime of interest. This MaxRes[9] technique will lower temperature rate in direct proportion with the rate at which the distance between the probe tip and the sample stage is changing. That is, the temperature rate will be decreased during softening and increased when the sample dimension is constant. Thus, the temperature rate through the transitions is relatively small and the temperature rate through steady state regions is relatively large. This decreases the total time to run a series of experiments while increasing the sensitivity in the temperature regimes of importance.
   A significant advance in TMA is the capability to apply a dynamic load to the

sample. That is, a force that changes -- linearly or sinusoidally -- with time. Dynamic Load TMA (DLTMA) will provide the user with a tool to rapidly map the softening rate profile of a coating for any temperature and applied pressure.

Another recent advancement in TMA involves the monitoring of the temperature of the sample in a manner that allows the characterization of thermal events that occur within the sample. That is, when the temperature of the sample deviates from steady state, a thermal event is occurring within the sample. With such an instrument, simultaneous measurements of the displacement trace and the temperature difference trace can be plotted vs temperature. Thus, changes in the modulus of the sample can be related to thermal events within the sample.

Thermo Gravimetric Analysis

History. The concept that the mass of an object will change upon heating goes back at least 2000 years when Egyptians heated materials on a balance over a fire. The term thermo-balance was coined by Honda[10] in 1915 in his paper on the degradation of various oxides. The appearance of commercial thermo-balances occurred in 1963.

TGA Fundamentals. The TGA is designed to accurately measure the change in mass of a material as a function of temperature. The change in mass can be due to a number of phenomena: degradation, evaporation, oxidation, absorption, desorption, etc. Modern electrobalances used in quality TGAs are sensitive to a change in mass of 100 nanograms.

Sample Preparation. For most TGA experiments the sample preparation is very simple. All that is required is to isolate a representative piece of the material to be tested. Sample masses are typically tens of milligrams. Typically, the sample is placed in the center of a pan which is hanging from a microbalance. The pans are usually aluminum or platinum, the latter of which is required for high temperature applications.

Experimental Conditions. There are a number of factors which must be chosen to run a simple TGA experiment. First there is the temperature rate and the range over which the temperature is scanned. The linear scan rate should be small enough to accurately assign a temperature or temperature range to the phenomenon of interest. If the sample has a low surface-to-volume ratio (e.g., a large single piece), the temperature rate should be decreased when attempting to characterize a bulk phenomenon. On the other hand, if the sample has a high surface-to-volume ratio (e.g., a fine powder), higher temperature rates can be used. Often, an unknown sample is run first with a high temperature rate for the purpose of discovering the temperature (or temperatures) at which the mass changes. Once these are known the temperature can be scanned slowly through these regions. Of course, if the instrument is more than a simple TGA, there are other experimental parameters that must be chosen.

Applications. In the development of polymeric coatings, especially in coatings for food or beverage containers, it is important to know if the coating will absorb certain gases which may affect the flavor of the contents. In this example, a polymer is tested for selected absorbance of carbon dioxide. Over the first part of the experiment the polymer is exposed to high-purity nitrogen gas. After the mass of the sample has settled somewhat, the purge gas is switched to carbon dioxide. Following the apparent decrease in mass due to the buoyancy of the heavier carbon dioxide, there is a slow absorption of carbon dioxide into the polymer. Later in the experiment the purge gas is switched back to nitrogen. Here we can monitor the desorption of carbon dioxide from the polymer. The force on the balance is plotted vs time in figure 18. The jump in weight after the switch from carbon dioxide back to nitrogen is due to the greater density of the carbon dioxide. Various coatings and components thereof can be tested to qualitatively rank their relative permeability to carbon dioxide and the solubility of carbon dioxide within them.

A second example involves the detection of a contaminant in a coating. The contaminant is known to have a degradation temperature of about 530 °C -- which, fortunately, is higher than the rest of the coating. The contaminant was present at a concentration of only a few parts per million. At this concentration its presence would be virtually impossible to detect by TGA. However, the liquid was placed in a centrifuge which increased the level of contaminant in the remaining solids (if it was present) by a factor of roughly 1000. The concentrated sample was placed in the TGA and the temperature was scanned from room temperature to 600 °C at a rate of 10 °C per minute. The results are shown in figure 19. The presence of the contaminant is revealed by the mass loss between 500°C and 550°C. The amount present in the original sample can be approximated by dividing the concentration of contaminant in the sample tested by the factor by which the sample was concentrated with the help of the centrifuge.

Method Development. There are many possibilities for the combination of a TGA with other analytical instruments. By adding some thermocouples to a TGA, you can construct a simultaneous TGA/DTA. That is, a differential thermal analyzer as well as a TGA. With this set-up the user will be able to better characterize various weight losses. For example, even though both evaporation and degradation result in a loss of sample mass, an evaporation is endothermic whereas a degradation is exothermic. The difference in the character of the mass loss will be evident with the DTA.

By adding an FTIR or a mass spectrometer to the purge gas output from the sample cell, the gas coming off the sample can be analyzed for chemical composition. This is called evolved gas analysis (EVA).

Another advance in basic TGA is Controlled Rate TGA or CRTGA. With this method the temperature rate is controlled in such a way to make the mass loss constant. In regions where the mass loss is small the temperature rate will be high and in regions of degradation or evaporation the temperature rate will be slowed considerably. This has the effect of maximizing the sensitivity of the weight loss in regions of interest.

The final development in basic TGA is the addition of a modulated temperature signal to the linear temperature rate -- Modulated TGA. By subjecting the sample to a temperature modulation, the rate of weight loss is measured at a wide range of temperature rates rather than the single temperature rate used in ordinary TGA. This allows a more precise calculation of activation energies and pre-exponential factors.

Summary

A well-equipped thermal analysis laboratory has the potential to provide significant guidance to a coatings research and development team. However, to fully capitalize on the capability of the equipment, the users must possess the ability to skillfully prepare the samples, efficiently design an experiment and properly interpret the results.

LITERATURE CITED

1. Van Assche, G., Van Hemelrijck A., Rahier H. and Van Mele B., Thermochimica Acta 268 (1995) 121

2. Van Assche, G., Van Hemelrijck A., Rahier H. and Van Mele B., Thermochimica Acta, 286, (1996) 209

3. Ishikiriyama K., Pyda M., Zheng G., Forschner T., Grebowicz J and Wunderlich B., Heat Capacity of Poly-p-dioxanone", J. Macromol. Sci.-- Phys., B37(1), 27-44 (1998).

4. Wunderlich B., "The Heat Capacity of Polymers" Thermochimica Acta 300 43-65 (1997)

5. Saruyama Y., Thermochimica Acta, 283 (1996) 157

6. Nielsen L. E., Mechanical Properties of Polymers, Reinhold Publishing Corp., New York (1962)

7. Pierce P.E. and Holsworth R. M., "The Mechanical Properties of Coatings -- The Measurement of Torsion Modulus and Logarithmic Decrement of Free Films," Journal of Paint Technology, Vol. 38, No 496, pp. 263-268, May (1966)

8. Pierce P.E. and Holsworth R. M., "The Mechanical Properties and Performance of Wood Primers," Journal of Paint Technology, Vol. 38, No 501, pp. 263-268, May (1966)

9. Riesen B. and Schenker B., "MaxRes in TMA and TGA: A New Tool for Fast Maximum Resolution By Rate Adjustment," Proceedings of the 26th Conference of the North American Thermal Analysis Society, Pages 116-121 (1998)

10. Honda, K. (1915) Sci. Rep. Tohoku Univ. Ser. 1, 4, pages 97-105

DYNAMIC MECHANICAL PROPERTIES OF POLYMERS

LOREN W. HILL

Coatings Consultant, 9 Bellows Road, Wilbraham, MA 01095

Introduction
DMA Definitions and Methods
DMA Instruments and Procedures
Storage Modulus Plots and Their Interpretation
The Glass Transition and Tan Delta Plots
DMA Evidence for Heterogeneous Structure
Impact Resistance and Low Temperature Loss Peaks
Storage Modulus and Crosslink Density
Literature Cited

Introduction

Dynamic mechanical analysis (DMA) has been used in fundamental studies of relationships between polymer structure and polymer properties for many years (1-5). DMA has also been widely used for polymer characterization in industrial product research and development, especially in the field of coatings (6-9). Major changes over time have been automation of instruments with computer control of the run procedure and of treatment and handling of data (3-5). Many mergers and acquisitions have occurred among DMA instrument suppliers, over the last several years, but availability of automated and computer controlled instruments has continued to be a major reason for extensive use of DMA.

Recently (10) it has been stated that the relationships dynamic properties and polymer structure are pretty much already known from long term study. However, relationships between dynamic properties and practical performance properties are often still not known. For example, in the field of auto topcoats, structure changes needed to alter the dynamic properties are known, but we do not know how to shift the dynamic properties to improve car wash marring resistance. The emphasis is shifting to determination of relationships between dynamic properties and performance properties (10).

Compared to other physical test methods, such as stress-strain testing, DMA has the advantage of being easy to use in a temperature-scanning mode. The low amplitude of oscillatory deformation used in DMA tends to prevent fracture or yield behavior, and sample properties are not changed by the deformation. Therefore, data can be taken continuously on a single sample as temperature is scanned. In some papers, DMA is included under the general heading of thermal analysis, but the capability of DMA for understanding details of structure far exceeds that of other thermal analysis methods. DMA has been used for the following purposes (1-9):

Determination of dynamic properties, such as storage modulus, throughout the total range of polymer behavior from glassy to transition to rubbery plateau to (in some cases) viscous flow during a single temperature scan,

Elucidation of the viscoelastic nature of polymeric material and the (often unpopular) associated concept that a given polymer has many correct values of the glass transition temperature each associated with a particular rate of testing,

Determination of the glass transition temperature and of the broadness of the glass transition, which is related to sample homogeneity,

Identification of low temperature transitions (β or γ transitions) that indicate the occurrence of molecular motion of parts of the polymer that are smaller than the segments that begin relative motion at the glass transition,

Characterization of partially crystalline polymeric samples with determination of the melting point of crystalline domains and the T_g of amorphous domains during a single temperature scan,

Investigation of multiphase morphology in many types of mixed amorphous polymer systems such as blends, grafts, block copolymers, plasticized thermoplastic polymers, rubber toughened thermoplastic systems, interpenetrating polymer networks, network polymers containing phase-separated sol or phase-separated dangling ends,

Determination of the crosslink density of network polymers generally and for Industrial thermoset coatings in particular with (in some cases) identification of the gel point and subsequent build up of crosslink density during the curing process.

DMA Definitions and Methods

In dynamic testing, an oscillating strain is applied and the resulting oscillating stress is measured, or conversely an oscillating stress is applied and the resulting oscillating strain is measured. Definitions and mathematical treatments do not depend on which of these modes of operation is used. Relationships between strain, stress and time are sketched in Figure 1 for tensile DMA with application of strain and measurement of stress. The maximum applied strain is ε_o. The maximum resulting stress is σ_o. The oscillation is carried out as a sine wave as shown in Figure 1. The sample is held under sufficient tension so that it remains taut (not slack) even when the oscillating strain is at a minimum.

The sine waves for strain and stress have the same frequency, but for viscoelastic samples the waves are out of phase by an amount, δ, called the phase lag. Theoretically and experimentally, δ is zero for an ideal (Hookean) elastic solid. If an ideal (Newtonian) liquid could be tested in this way, δ would be 90°. For viscoelastic materials, δ lies between 0° and 90°, and the value of δ is a rather direct indication of viscoelastic character (1, 6).

Definitions of dynamic properties depend on the concept of resolving the stress wave of Figure 1 into two waves, one which is in phase with strain and one which is 90° out-of-phase with strain. The in phase resolved plot represents elastic response, and the 90° out-of-phase resolved plot represents viscous response. In terms of modulus, the separated responses result in the following definitions:

$$\text{Tensile Storage Modulus} = E' = (\sigma_o \cos \delta) / \varepsilon_o \qquad [1]$$

$$\text{Tensile Loss Modulus} = E'' = (\sigma_o \sin \delta) / \varepsilon_o \qquad [2]$$

$$\text{Loss Tangent} = E'' / E' = \text{Tan } \delta \qquad [3]$$

The term "storage" is associated with the elastic part of the response, E', because mechanical energy input to elastic materials is "stored" in the sense of being completely recoverable. The term "loss" is associated with the viscous part of the response, E'', because mechanical energy input to ideal liquids is totally lost through viscous heating. The ratio E''/E' is viscous response expressed relative to elastic response. This ratio reduces to $\sin \delta / \cos \delta$ which is Tan δ. Thus, the name "loss tangent" is appropriate.

Tensile deformation is convenient for free films that need no support. The maximum tensile strain during oscillation, ε_o, is kept very small, typically about 1×10^{-3} cm which corresponds to 0.05 % strain when the length of sample between the grips is 2.00 cm. In this way the sample neither breaks nor yields, and properties can be obtained continuously on a single sample as temperature is scanned. If the sample requires support, DMA is usually carried out in shear with the sample sandwiched between parallel plates. One plate is stationary with shear stress detectors attached. The other plate is mounted on a shaft that can rotate, but movement is limited to back and forth rotation through a very small angle to generate oscillatory shear strain (γ). For DMA in shear, the sine waves of Figure 1 are still appropriate, but maximum tensile strain (ε_o) would be replaced by maximum shear strain (γ_o). In this case, σ_o is maximum shear stress rather than maximum tensile stress. The DMA definitions for shear tests are similar to those for tensile tests:

$$\text{Shear Storage Modulus} = G' = (\sigma_o \cos \delta) / \gamma_o \qquad [4]$$

$$\text{Shear Loss Modulus} = G'' = (\sigma_o \sin \delta) / \gamma_o \qquad [5]$$

$$\text{Loss Tangent} = G'' / G' = \text{Tan } \delta \qquad [6]$$

In general, the relationship between E' and G' depends on Poisson's ratio (μ)

$$E' = 2G' (1 + \mu) \qquad [7]$$

For materials that do not undergo change in volume when exposed to moderate tensile elongation, $\mu = 1/2$. Inserting this μ value into eq. [7] results in E' = 3 G' (1, 6).

Tensile and shear modulus values can also be obtained in tests carried out with a linear increase in deformation with time (non-oscillatory). Symbols for modulus values from these stress/strain tests are E and G. There is a geometric mean relationship between primed and un-primed modulus values:

$$E = (E'^2 + E''^2)^{1/2} \qquad [8]$$

$$G = (G'^2 + G''^2)^{1/2} \qquad [9]$$

Usually stress/strain tests are carried out at or near room temperature, often in a constant temperature and humidity room. For polymeric solids, E'' is much smaller than E' except near the glass transition temperature. Therefore, unless T_g for the sample under study is near room temperature, the E'' term in eq. [8] adds very little to the E' term, and to a rather close approximation, E $\cong$ E'. By similar analysis, G $\cong$ G' by eq. [9] except for samples having T_g near room temperature.

<u>DMA Instruments and Procedures</u>

The availability of automated and computer controlled instruments greatly facilitates use of DMA for polymer characterization. In 1995 an ASTM chapter listed eighteen instruments and the supplier of each (6). Many changes have occurred recently with acquisitions and mergers among instrument suppliers as well as the usual replacement of older models by new and improved models. Instruments that determine dynamic properties in shear are often modifications of liquid rheometers, which can operate in a steady shear mode as well as in an oscillatory shear mode. These shear based instruments are often most suitable for samples that are mainly viscous liquids with just a limited amount of elastic or solid-like character. If a parallel plate sample holder is used for shear deformation of a material that is mostly solid with only a little liquid-like character, there is often insufficient adhesion of the sample to the plates. Slippage at the plates during shear deformation occurs, in some cases, invalidating results. Solid-like samples are often tested with oscillatory tensile deformation. Many studies involve free paint films (6), and thicker than normal paint films almost never cure in exactly the same way as films of normal paint film thickness (e.g. 25 μm $\cong$ 1.0 mil).

Instruments suitable for tensile DMA of thin films of include:

Rheometrics Solids Analyzer (RSA)™, Rheometrics Inc.;

Dynamic Mechanical Thermal Analyzer (DMTA)™, formerly Polymer Labs and recently acquired by Rheometrics;

Dynamic Mechanical Analyzer (DMA 2980)™, TA Instruments (now a subsidiary of Waters Corp.)

Dynamic Mechanical Analyzer (DMA 7)™, Perkin Elmer;

Rheovibron/Autovibron™ (DMA), Imass Inc.

Automated Rheovibron (DDV-II-C)™, Orientec Inc. (Japan)

Viscoanalyser™, Indikon/Metravib (Europe)

DMA has also been used to follow property development as it takes place during cure of thermoset coatings. With this approach an uncured sample is placed on a support (i.e. thin metal shim or metal spring) or impregnated into a supporting material (i.e. glass braid, cloth or paper support). The supported sample is then placed in a curing oven and dynamic properties are determined as a function of time. A Torsional Braid Analyzer (TBA)™ from Plastic Analysis Instruments has been used extensively to analyze the effects of vitrification (i.e. glass formation) prior to complete conversion during cure of thermoset epoxy resins (8, 9).

A common feature of all these instruments is oscillatory deformation as shown in Figure 1. Variable features include: free versus forced oscillation, frequency scan versus temperature scan versus either, sensitivity for thin film analysis, capability of transversing the entire range of property behavior (glassy to transition to rubbery) during a single temperature scan, breadth and rate of temperature scan, breadth and rate of frequency scan or range and number of frequency settings, versatility of sample holding devices, ruggedness versus flimsiness, amount of attention required once a run has been started, accuracy and versatility of the associated software for control during the run and data treatment and plotting after the run (6). In several cases, newer models permit determination of properties at several frequencies during a single temperature scan. Although it is necessary to use a rather low temperature scan rate in multiple frequency runs, the amount of data obtained is remarkable. Chances of acquiring an instrument that will actually function well in the intended experiments are greatly increased by asking suppliers about these variable features.

Carrying out a DMA run on an automated instrument is much easier than the grueling procedure required before computer controlled automation. Details depend on the particular instrument. Usually the associated software includes a "run" program which prompts the operator to input sample data (e.g. thickness) and settings for the run such as initial and final temperatures, frequency of oscillation, heating rate, etc.

After the input steps, there is usually a cooling period. The instrument takes over when the pre-set initial sub-ambient temperature is reached. Thereafter nothing is required of the operator until the run is finished. Usually initial and final temperatures are selected to span the glassy region, transition region and rubbery plateau region. For direct tensile DMA, the run program usually contains a tensioning sub-routine, which provides constant static tension sufficient to avoid slack in the sample in the glassy region and then decreasing static tension as the sample softens in the transition region. Modification of the tensioning sub-routine is often necessary. If tension is too high, films break in the glassy region or are pulled apart in the transition region. If tension is too low, slack results or the oscillatory stress falls below measurable values. Skilled operators soon develop several modified run programs with different tensioning parameters that are suitable for samples of various properties and dimensions (6).

Storage Modulus Plots and Their Interpretation

Storage modulus and loss tangent plots for a highly crosslinked coatings film are shown in Figure 2. The film was prepared by crosslinking a polyester polyol with an etherified melamine formaldehyde (MF) resin. A 0.4 x 3.5 cm strip of free film was mounted in the grips of an Autovibron™ instrument (Imass Inc,), and tensile DMA was carried out at an oscillating frequency of 11 cycles/second. The temperature scan rate was 2 °C/minute, and the computer controlled data acquisition system took data points at one minute intervals (11).

The full range of polymer network properties is exhibited in Figure 2: glassy behavior on the left, transition behavior in the center and rubbery plateau behavior at the right. In the glassy region the storage modulus, E', is about the same for all amorphous, unpigmented network polymers (approximately 2 to 4 x 10^{10} dynes/cm^2 which is equal to 2 to 4 x 10^9 Newtons/m^2). E' drops sharply in the transition region. For uncrosslinked, high molecular weight polymers, E' drops by more than three orders of magnitude. Crosslinking results in a smaller drop. In figure 2, the value of E'(min) is 1.47 x 10^8 dynes/cm^2, which is about two orders of magnitude below the glassy E' value. Cure studies of model coating formulations have shown that E'(min) is proportional to crosslink density (6, 12). Therefore, the value of E'(min) increases as extent of cure increases, and E'(min) can be used as a measure of the extent of cure. At higher temperatures in the rubbery plateau, E' rises for two reasons. A small increase in E' is attributed to the effect of temperature on rubbery retractive forces. A larger increase in E' results from additional curing during the temperature scan (6).

The Glass Transition and Tan Delta Plots

The Tan δ plot in Figure 2 shows a strong maximum in the transition region. The peak height, referred to as Tan δ (max), is also a sensitive measure of extent of cure. Tan δ (max) decreases as extent of cure increases because cure causes a decrease in liquid-like character of the sample. The point on the temperature axis where Tan δ reaches a maximum is often taken as the glass transition temperature, T_g as indicated in Figure 2. As extent of cure increases T_g also increases. A plot of E" versus temperature (not shown) also has a peak in the transition region. The temperature of the E" peak often lies about 15 °C below that of the Tan Delta peak, and some workers prefer to report this lower value as T_g. The lower value is in better agreement with T_g values obtained from differential scanning calorimetry (DSC), which is a widely used method for T_g determination.

The peak width at half height (PW in Figure 2) is an indication of network uniformity. MF resins mainly react with hydroxyl groups on the polyester co-reactant in a crosslinking reaction called co-condensation. However, there are several types of functional groups on MF resins, and they can react with each other in a process called self-condensation. Self-condensation results in very short chains between junction points in the network. Since some short chains along with mainly longer chains is a source of network heterogeneity, PW increases when the amount of self-condensation increases (11).

The position on the temperature scale of the Tan Delta peak depends on the frequency of oscillation used in the DMA test. Polymer Labs (now part of Rheometrics) DMTA™ instrument can acquire data at many frequencies during a single scan with a single sample as shown in Figure 3 (14). Five frequencies covering a little less than a 100 fold change were used. When multiple frequency runs are made, the scan rate is kept low, 1 °C/minute in this case. We have five values of T_g in Figure 3 ranging from 65 °C at 0.33 Hz to 78 °C at 30 Hz. If someone were to ask which is the real T_g value, the correct answer would be "all of them." Dependence of T_g on frequency is evidence of the viscoelastic character of polymeric material. A tenet of viscoelasticity is that properties depend on the rate of deformation in physical property testing. At higher rates, the sample is more solid-like whereas at lower rates the sample is more liquid–like. In Figure 3, higher frequency represents higher rate of deformation, and higher T_g represents more solid-like behavior.

Dependence of T_g on frequency is not a comfortable concept to many polymer chemists. It seems to make the value of T_g too uncertain. Perhaps the comfort level can be improved by noting that the low frequency values (obtained at 1.0 Hz or below) agree quite well with DSC values of T_g. A more enlightened attitude is that frequency dependence of T_g helps explain why we test polymer samples at several rates of deformation.

Figure 3 shows that peak width (PW) increases with increasing frequency of oscillation. If PW is being used as an indication of network homogeneity, results obtained at the same frequency should be compared.

DMA Evidence for Heterogeneous Structure

Non-homogeneity in structure can come in many degrees. A low degree of non-homogeneity, such as a broad distribution of the lengths of chains between junction points in a polymer network, is revealed by a moderate increase in peak width. A much higher degree of non-homogeneity occurs when there is phase separation in the sample so that well defined domains are present with sharp boundaries between the domains. DMA plots for this type of phase separation are shown in Figure 4 (2). The existence of two glass transitions is obvious. The high T_g is attributed to polystyrene and the low T_g is attributed to a styrene/butadiene copolymer (2). Plots are similar for commercial materials in which polybutadiene rubber is used to toughen polystyrene (high impact polystyrene, HIPS) or to toughen a styrene/acryrilonitrile copolymer (acryilonitrile/butadiene/styrene, ABS). The simplicity of Figure 4 is favored by the distinctness of the boundaries of the phase separated domains and the very large difference in T_g between the two domains. If the boundaries between domains are diffuse in the sense of a wide region of gradually changing chemical composition and if the domains have T_g values that do not differ much, then instead of observing two distinct transitions, a very broad poorly defined Tan Delta "peak" is observed, and it is difficult to assign any particular value from the plot as T_g. Films formed from some commercial aqueous polyurethane dispersions produce Tan Delta plots of this broad type (15).

Sperling and Corsaro (4) summarized interpretations of Tan Delta peaks in terms of domain sizes and phase mixing for interpenetrating polymer networks (IPNS). The concept of complete molecular level network interpenetration has been replaced by estimating domain sizes in the Angstrom range, which is called "microheterogeneous morphology." As the degree of interpenetration increases, the peaks move closer together, and the value of Tan Delta remains high between the peaks. It is reported (4) that development of a single broad transition spanning the transition range of both polymers can be achieved by extensive but incomplete mixing on the 100 Angstrom scale.

Other forms of non-homogeneity include co-existence of crystalline and amorphous domains in the same sample, and phase separated block copolymers. Storage modulus plots for both types are shown in Figure 5 (16). Curves A, B and C are schematics based on many examples whereas curves D and E are actual data for samples of the indicated composition. Curve C is for a partially crystalline sample that is not segmented. For C, the amorphous portion undergoes a glass transition at about -25 °C, but the drop in modulus is small because the crystalline regions are still solid. Finally at a temperature above 200 °C, the crystalline regions melt, and the modulus drops nearly straight down. The melting point, T_m, is indicated on the temperature axis. Curve A is for a linear amorphous polymer that has no crystalline regions. Once the glass transition starts the whole sample participates, and the modulus drops by over three orders of magnitude. Curve A shows a brief rubbery plateau due to chain entanglement from 0 °C to 50 °C, and then viscous flow begins. A low level of crosslinking prevents viscous flow as indicated by the curve labeled B.

Curves D and E of Figure 5 are for block copolymers of a type called polyurethane segmented elastomers. The hard segment is based on 4,4'-diphenylmethane diisocyanate (MDI) and butanediol (BD). The soft segment also contains MDI, but the polyol portion is a mixture of polytetramethylene adipate (PTMA) and butanediol. The PTMA/BD ratio is altered to give different MDI content at stoichiometric balance in films labeled D and E. For D, the PTMA/BD ratio is higher. This makes the soft segment content higher and reduces the percentage of MDI needed for a 1/1 isocyanate/hydroxyl ratio. The greater modulus drop for D (versus E) at T_g is attributed to its higher soft segment content. The hard segment for D melts around 100 °C to 125 °C as indicated by the second drop in modulus. For E, the soft segment content is lower which gives a smaller modulus drop at T_g. Sample D, with its higher MDI content, melts around 125 °C to 160 °C. The melting curves for the hard segments of D and E do not drop as sharply as the schematic curve for the non-segmented semi-crystalline polymer (Curve C). Curvature in the melting region is attributed (16) to some phase mixing at domain boundaries.

Impact Resistance and Low Temperature Loss Peaks

In the discussion of Tan Delta plots presented above, emphasis was placed on the peak associated with the main transition, which is called the glass transition. This peak is sometimes called the α peak. Many polymers show one or more additional peaks in loss modulus, E", or Tan Delta at lower temperatures. These are called β or γ peaks. These peaks are believed to represent a transition in which some part structure can begin to undergo motion. The part structure involved is smaller than the segments that begin relative motion at the glass transition. The occurrence of this motion may in some cases provide a means of responding to impact without bond fracture. This alternative means of response is believed to result in good impact resistance. Heijboer (17) studied a series of polymers that varied greatly in impact resistance and in the occurrence of low temperature loss peaks. Often but not always, polymers with low temperature loss peaks had good impact resistance. Heijboer concluded that some of the low temperature loss peaks represented side chain rather than main chain motion and that these side chain motions did not protect against chain rupture in impact tests. The issue is also complicated by the fact that the strengths of low temperature loss peaks vary greatly.

If polymers do not have intrinsic low temperature loss peaks, polymer chemists sometimes provide such peaks through rubber toughening. As shown earlier in Figure 4, a strong low temperature peak is provided by adding styrene/butadiene rubber to polystyrene.

Storage Modulus and Crosslink Density

In 1985, Flory (18) summarized years of work on the theory of rubber elasticity in the following words:

> "Rubber elasticity, identified as the capacity to sustain very large deformations followed by complete recovery, is exhibited exclusively by polymeric substances consisting predominantly of long molecular chains. Moreover, it is manifested under suitable conditions by virtually all polymers so constituted. The molecular theory of rubber elasticity rests on the premise, now fully validated by experiments, that alterations of the configurations of the chains comprising the network account for the elastic free energy and for the stress arising from deformation."

Introduction of crosslinks leads to a larger number of shorter chains. If more chains are trying to get back to their favored configuration (or conformation) the restoring force will be higher. The number-of-chains/restoring force relationship explains why crosslink density can be determined by measuring modulus in the rubbery plateau region.

Studies of thermoset coatings resulted in calculation of crosslink density from structure for networks made up of many very short chains (12).   Films of the short chain networks were analyzed by DMA (6,11,19). Values of E'(min) were used to calculate crosslink density with the most simple form of a rubber theory equation [10]:

$$\upsilon_e = E'(min)/ RT \qquad\qquad [10]$$

where υ_e = moles of elastically effective network chains
                per cubic centimeter of film,
       R = the gas constant (8.314×10^7 ergs/°K•mole in the cgs
                unit system or 8.314 Joules/°K•mole in the mks system),
       T = temperature (in degrees Kelvin) where E'(min) is taken.

In this treatment, crosslink density is expressed in terms of moles of elastaically effective network chains per cubic centimeter of film.   Agreement between υ_e calculated from structure and obtained by E'(min) determinations is excellent for polyols crosslinked by MF resins (19) and for many other short chain network structures of importance in thermoset coatings (6).

The success of eq. [10)] for short chain networks is surprising because such networks fall outside the realm of rubber elasticity as defined by Flory. These networks are not rubbery because they cannot sustain very large deformations followed by complete recovery. Small deformations, however, may be accommodated by changes in conformations of the short chains.   DMA provides modulus values with very small deformations of the sample. Alterations of the conformations of the short chains may account for the elastic free energy and for the stress arising from deformation. Larger deformations, which would cause changes in bond angles and breakage of bonds, are expected to invalidate the simple relationship of eq. [10].

Absence of entanglements is believed to contribute to success of the simple form of eq. [10] for short chain networks (12).   In rubber networks entanglements function as transient junction points (1). Rheological characterization has shown that linear polymers have a critical molecular weight for entanglement, usually about 10,000 to 20,000 g/mole depending on the polymer. The molecular weights of chains between junction points in short chain netwrorks are well below 10,000. Therefore, short chain networks are believed to have no entanglements.

The storage modulus plot of Figure 2 can be used along with eq. [10]. As noted previously the value of E'(min) is 1.47×10^8 dynes/cm^2 and the minimum occurs at 102 °C (375 °K). Substituting these values into eq. [10] along with R in cgs units gives υ_e = 1.24×10^{-3} moles/cm^3. In mks units the corresponding E'(min) value is 1.47×10^7 Pascal (or Newton/m^2).   Inserting the mks values of E'(min) and R gives υ_e = 1.24×10^3 moles/m^3. Note that the value crosslink density differs by a factor of 10^6 depending on use of the cgs versus mks system of units.   The factor corresponds to the conversion between cubic centimeters and cubic meters.

In summary, DMA has proven to be very useful for crosslink density determination especially in the field of thermoset coatings.

Literature Cited

1. Murayam, T. *Dynamic Mechanical Analysis of Polymeric Material,* Material Science Monographs 1; Elsevier: New York, 1978.
2. Cooper, S.L.; Estes, G.M., Editors, *Multiphase Polymers,* ACS Advances in Chemistry Series 176, American Chemical Society, Washington, DC, 1979.
3. Craver, C.D.; Provder, T., Editors, *Polymer Characterization: Physical Property, Spectroscopic, and Chromatographic Methods,* ACS Advances in Chemistry Series 227, American Chemical Society, Washington, DC, 1990.
4. Sperling, L. H.; Corsaro, R.D., Editors, *Sound and Vibration Damping in Polymers,* ACS Symposium Series 424, American Chemical Society, Washington, 1990.
5. Craver, C.D., Editor, *Polymer Characterization: Spectroscopic, Chrommatographic, and Physical Instrument* Methods, ACS Advances in Chemistry Series 203, American Chemical Society, Washington, 1983.
6. Hill, L.W. In *Paint and Coating Testing Manual;* Koleske, J.V., Ed.; 14th Edition; ASTM Manual Series: MNL 17; American Society for Testing Materials: Philadelphia, PA, 1995; pp 534-546.
7. Provder, T. *J. Coat. Technol.* 1989, *61,* No. 770, 33-50.
8. Enns,J.B.; Gillham, J.K., Chapter 2, *Torsional Braid Analysis: Time-Temperature-Transformation Cure Diagrams of Thermosetting Epoxy/Amine Systems,* in Reference 5, p. 27., 1983.
9. Gillham, J.K. *Poly. International Journal* 1997, *44,* 262-276.
10. Hill, L.W., *Mechanical Properties of Coatings,* Proceedings of the International Waterborne, High-Solids, and Powder coatings Symposium, March 1-3, 2000, New Orleans, LA, p.1.
11. Hill, L.W., *ACS Poly. Mater. Sci. Eng. Proceedings* 1997, *77,* 387-388.
12. Hill, L.W., *Prog. Organic Coat.* 1997, *31,* 235-242.
13. Hill, L.W.; Lee, S-B., *J. Coat. Technol.* 1999, *71,* No. 897, 127-133.
14. Wetton, R.E.; Croucher, T.G.; Furdson, J.W.M., Chapter 5, *The PL-Dynamic Mechanical Thermal Analyzer and Its Application to Study of Polymer Transitions,* in Reference 5, p.95, 1983.
15. Hill, L.W., *Dynamic Mechanical Analysis of Property Development During Film Formation,* a chapter in *Film Formation,* Provder, T., Editor, ACS Symposium Series, in preparation.
16. van Bogart, J.W.C.; Lilaonitkul; Cooper, S.L., Chapter 1, *Morphology and Properties of Segmented Copolymers,* in Reference 2, p.5, 1979.
17. Heijboer, J., *J. Appl. Poly. Sci.* 1968, *C16,* 3755.
18. Flory, P.J., *Polymer Journal* 1985, *17,* 1-12.
19. Hill, L.W.;Kozlowski, K., *J, Coat. Technol.* 1987, *59,* No. 751, 63-71.

DYNAMIC MECHANICAL ANALYSIS

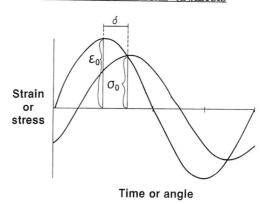

Time or angle

Figure 1. Sine wave plot of applied strain in DMA showing maximum strain, ε_0, and the plot for resulting stress showing maximum stress, σ_0. The lag in peak stress behind peak strain is called the phase lag, δ. (Reproduced from reference 19. Copyright 1987 Federation of Societies for Coatings Technology.)

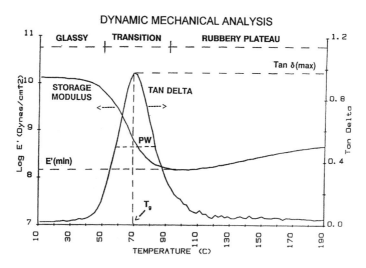

Figure 2. DMA plots of storage modulus, E', and loss tangent (Tan Delta) versus temperature for an unpigmented, highly crosslinked coating. (Reproduced from reference 11. Copyright 1997 Division of Polymeric Materials Science and Engineering, American Chemical Society.)

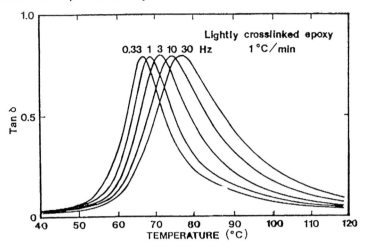

Figure 3. Tan Delta plots at five oscillation frequencies obtained in a single scan at 1 °C/minute on a Polymer Labs DMTA™ instrument. (Reproduced from reference 14. Copyright 1983, American Chemical Society.)

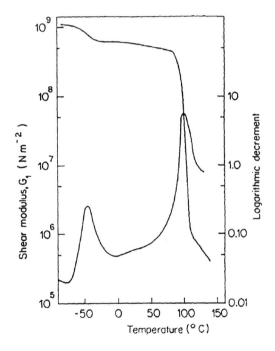

Figure 4. Shear storage modulus and logarithmic decrement plots for an immiscible polyblend of polystyrene and a styrene-butadiene rubber copolymer. (Reproduced from reference 2. Copyright 1979, American Chemical Society.)

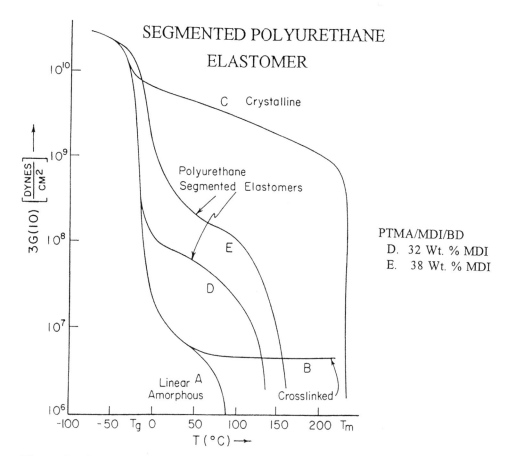

Figure 5. Storage modulus vs. temperature plots for: (A) linear amorphous polymer; (B) lightly crosslinked polymer; (C) semi-crystalline polymer; (D) PTMA/MDI/BD-segmented copolymer (32% MDI by wt.); (E) PTMA/MDI/BD-segmented copolymer (38% MDI by wt.). PTMA is polytetramethylene adipate, MDI is 4,4'-diphenylmethane diisocyanate, BD is butanediol. (Reproduced from reference 16. Copyright 1979, American Chemical Society.)

Section 6

POLYMERIZATION/POLYMERIZATION MECHANISM
Section Editor: Donald Schulz

FREE RADICAL POLYMERIZATION

KRZYSZTOF MATYJASZEWSKI and SCOTT G. GAYNOR

Center for Macromolecular Engineering, Department of Chemistry, Carnegie Mellon University, 4400 Fifth Avenue, Pittsburgh, PA 15213

Introduction

Chemistry of Conventional Free Radical Polymerization

Applications and Limitations of Conventional Radical Polymerization

Future Advances in Radical Polymerization

Chemistry of Controlled Radical Polymerization (CRP)

New Materials by Controlled/"Living" Radical Polymerization

Conclusions

Literature Cited

Introduction

Radical polymerization is the most important method for polymer synthesis.(1) Almost 50% of all synthetic polymers are made using radical processes. There are several reasons for the commercial success of radical polymerization, but probably the most important are a convenient temperature range (20 to 100 °C) and very minimal requirements for the purification of monomers, solvents, etc. which need only be deoxygenated. Radical polymerization is not affected by water and protic impurities and can be carried out in bulk, in solution, in aqueous suspension, emulsion, dispersion, etc. An important consideration is that a large range of monomers can be polymerized radically. Nearly all alkenes can be polymerized or copolymerized via a radical mechanism. This is due to the high tolerance of free radicals to many functional groups including acids, hydroxy, amino, etc. Moreover, numerous monomers can be easily copolymerized to yield a multitude of new materials. High polymer is formed at the early stages of the polymerization and does not require the long reaction times and high conversions typical for step-growth polymerization.

There are, however, some limitations for a radical polymerization, especially in comparison with ionic processes. Anionic, and more recently, cationic polymerization, enables the preparation of well-defined polymers and copolymers with predetermined molecular weights ($DP_n = \Delta[M]/[I]_o$) and low polydispersities ($M_w/M_n < 1.3$).(2-5) The main advantage of ionic reactions is the possibility of synthesizing segmented copolymers such as block and graft copolymers which microphase separate and can be used as thermoplastic elastomers, non-ionic surfactants, dispersants, lubricants, viscosity modifiers, blend compatibilizers, special additives, etc. These materials are prepared via so-called living polymerizations which, after relatively fast initiation, the chains only propagate and the contribution of chain breaking reactions such as transfer and termination is negligible.(6, 7) Unfortunately, ionic polymerization is limited to a handful of monomers; it requires very stringent drying, the exclusion of moisture from the reaction mixture and also very low temperatures ($< -70\ ^\circ C$).

On the other hand, until recently, it was not possible to prepare via radical methods, the well-defined polymers and copolymers available from ionic reactions. The main reason was the presence of unavoidable termination reactions between the growing radicals.   In a conventional radical process, radicals are slowly and continuously generated via thermal decomposition of suitable initiators such as peroxides or diazo compounds to minimize termination. This approach enables the synthesis of ill-defined high polymers but not block copolymers. In recent years, there has been a drive to make the novel well-defined materials available from ionic reactions via less demanding techniques.(6) Radical polymerization is an obvious choice due to the facile reaction conditions and also the large range of polymerizable monomers.(8)

In this chapter, we will cover the fundamentals of conventional radical polymerization and will also describe the most recent developments in the area of controlled radical polymerization, which enables the synthesis of many well-defined copolymers with novel architectures, compositions and functionalities.(9)

Chemistry of Conventional Free Radical Polymerization

Free radical polymerization has been known for more than sixty years. As far back as the 1950s, the basic theory and comprehension of radical polymerization was established.(10-12) It included the thorough understanding of the mechanism of the process, encompassing the chemistry and kinetics of the elementary reactions involved, with the determination of the corresponding absolute rate constants, the structure and concentrations of the growing species, as well as a correlation of the structure of the involved reagents and their reactivities.

Typically, nearly all compounds with C=C bonds can be either homopolymerized or copolymerized via a radical mechanism. They should fulfill two basic requirements: thermodynamic polymerizability and kinetic feasibility. The former indicates sufficiently negative free energy of polymerization and the latter sufficient reactivity of monomer, stability of the derived free radical and a low proportion of side reactions.

Most monosubstituted alkenes polymerize radically; some disubstituted alkenes either homopolymerize (methacrylates) or can be copolymerized (isobutene, maleic anhydride). Isobutene and α-olefins do not provide high molecular weight polymers radically because they propagate slowly and participate in transfer reactions. Copolymerization of these monomers with electron poor monomers is successful, however. Because radicals are tolerant to protons on heteroatoms, radical polymerization of unprotected hydroxyethyl (meth)acrylates, (meth)acrylic acid and some amino derivatives lead to high molecular weight polymers. Monomers can be polymerized in bulk or in aqueous suspension/emulsion but also in organic solvents. The choice of solvent is limited only by the propensity of the potential transfer reactions.

Initiators for radical polymerization include compounds which can readily be homolytically cleaved either in the presence of light or at elevated temperatures. Typical initiators are peroxides, diazo derivatives, redox systems, organometallics and photolabile compounds; they are usually used at concentrations between 1 and 0.01 mol%. Generally, the reaction temperature is correlated with the initiator structure in such a way that 50% of the initiator is typically decomposed within 10 hours. During that time more than 95% monomer conversion is often reached.

Under such conditions, polymers with molecular weights in the range of $M_n \approx$ 100,000 are generally formed. However, in the presence of transfer agents, it is possible to prepare polymers with much lower M_n. It is also possible to prepare polymers with $M_n > 10^6$, but this is often a difficult target due to transfer and termination between growing radicals. Polymers are generally linear, though transfer to polymer may provide branched structures; the best example of a (hyper)branched polymer is low density polyethylene prepared radically at T>200 °C. The lowest polydispersities in conventional radical polymerization are $M_w/M_n = 1.5$ for low conversions and termination by coupling (vide infra). However, in most systems polydispersities exceed $M_w/M_n > 2$, especially at higher conversion.

Radical polymerization consists of four basic elementary steps: initiation, propagation, termination and transfer. Initiation is usually composed of two processes: the generation of primary initiating radicals (In*) and the reaction of these radicals with monomer (Scheme 1).

Scheme 1

$$In\text{-}In \xrightarrow{k_d} 2\ In^*$$

$$In^* + M \xrightarrow{k_i} P_1{}^*$$

Typically, the former reaction is much slower than the latter and it is rate determining with typical values of $k_d \approx 10^{-5}\ s^{-1}$ and $k_i > 10^5\ M^{-1}s^{-1}$.

Propagation occurs by the repetitive addition of the growing radical to the double bond. It is considered to be chain length independent with typical values of $k_p \approx 10^{3\pm1}$ $M^{-1}s^{-1}$, Scheme 2.

Scheme 2

$$P_n{}^* + M \xrightarrow{k_p} P_{n+1}{}^*$$

Termination (Scheme 3) between two growing radicals can occur either by coupling (k_{tc}) or by disproportionation (k_{td}) with rate constants approaching the diffusion controlled limit, $k_t > 10^7\ M^{-1}s^{-1}$. Coupling is when two growing radicals simply combine to form a carbon-carbon bond. Disproportionation, however, is when one growing radical abstracts a β-H from another growing radical, leaving one chain end saturated and the other unsaturated. If termination by coupling dominates then polymers with higher molecular weights and lower polydispersities are formed.

Scheme 3

$$P_n^* + P_m^* \xrightarrow{\begin{array}{c} k_{tc} \\ \\ k_{td} \end{array}} \begin{array}{c} P_{n+m} \\ \\ P_n^= + P_m^H \end{array}$$

The fourth elementary reaction is transfer (Scheme 4). Transfer can occur to monomer or to polymer. In the latter case, the M_n value is not affected but polydispersities increase due to formation of branched and in some cases crosslinked polymers. Transfer can also occur to a transfer agent (TA). Typically, re-initiation ($k_{trTA'}$) is fast and transfer has no effect on kinetics but only on molecular weights. If re-initiation is slow, then some retardation/inhibition may occur (vide infra).

Scheme 4

$$P_n^* + M \xrightarrow{k_{trM}} P_n + P_1^*$$

$$P_n^* + P_x \xrightarrow{k_{trP}} P_n + P_x^*$$

$$P_n^* + TA \xrightarrow{k_{trTA}} P_n\text{-}A + T^*$$

$$T^* + M \xrightarrow{k_{trTA'}} P_1^*$$

Typically, kinetics are first order with respect to monomer and ½ order with respect to initiator (Eq. 1). The apparent rate constant is a function of the efficiency of initiation (f), the rate constant of initiation (k_d), propagation (k_p) and termination (k_t). Although the rate constant of initiator dissociation can be independently measured from GC, NMR, HPLC and other experiments, the kinetic measurements only provide the value of $k_p/(k_t)^{1/2}$ but not the absolute rate constants k_p and k_t. In a similar manner, molecular weights of the polymers also depend on the $k_p/(k_t)^{1/2}$ ratio. For termination occurring by coupling, the molecular weight is defined by Eq. 2.

$$R_p = k_p \, [M] \, (fk_d[I]_0/k_t)^{1/2} \tag{1}$$

$$DP_n = k_p \, [M] \, (fk_d[I]_0 k_t)^{-1/2} \tag{2}$$

Thus, the absolute values of the rate constants k_p and k_t can only be obtained under non-stationary conditions such as those employing rotating sector,(13) spatially intermittent polymerization(13) or pulsed laser photolysis techniques.(14) Another approach to obtain k_p values involves the direct measurements of the concentration of growing radicals ($[P^*] \approx 10^{-7}$ M) by ESR.(15)

Initiators and Initiation It may be instructive to follow a typical initiation pathway with benzoyl peroxide for the polymerization of styrene. Benzoyl peroxide decomposes to two benzoyloxy radicals with the rate constant $k_d \approx 10^{-5}$ s^{-1} at 70 °C and with the activation energy E_a= 140 kJ/mol. At 78 °C it has a half lifetime of 10 hours. The resulting radical can add to a monomer (for styrene there is a strong preference for tail addition), it can participate in aromatic substitution, in β–H abstraction, and it can also decompose by β-scission and form the reactive phenyl radical (Scheme 5).

Not all radicals initiate new chains. They can also terminate in the solvent cage before escaping and form inactive products, e.g., phenyl benzoate. Thus, radical initiation has a fractional efficiency (0<f<1). It usually decreases with monomer dilution and with conversion, due to lower [M] and higher viscosity. Peroxy compounds that are used as radical initiators include dialkyl peroxides, peresters, percarbonates, hydroperoxides and inorganic peroxides. Some relevant data on representative examples are listed in Table 1.

Scheme 5

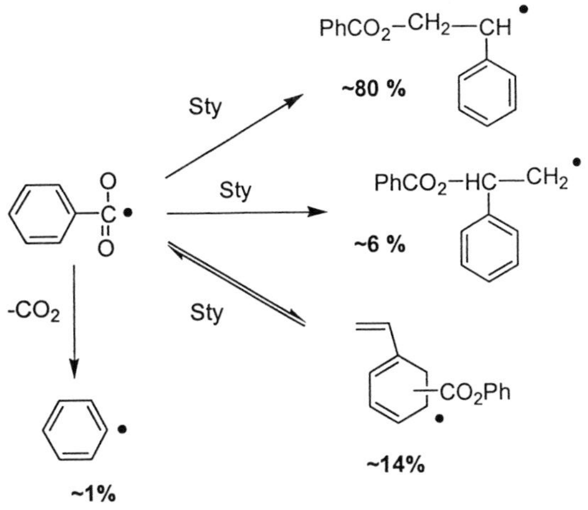

Table 1. Data for Selected Peroxides.[a]

Peroxide	Temp. Range (°C)	k_d (60 °C) $M^{-1}s^{-1}$	E_a kJ/mol	A (* 10^{15})	$\tau_{1/2}$=10h
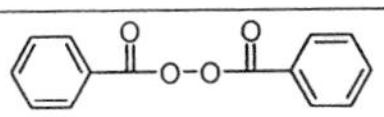	38-80	$1.5*10^{-6}$	139.0	9.34	78 °C
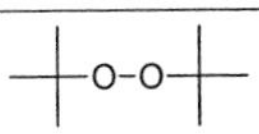	100-135	$2.5*10^{-9}$	152.7	2.16	125 °C
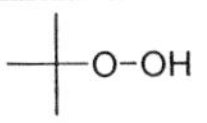	155-175	---	174.2	7.97	168 °C
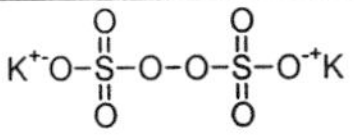	50-90	$4.4*10^{-6}$	148.0	709	69 °C

[a]Data taken from Ref. 1.

Alkoxy radicals have a much stronger affinity towards hydrogen than acyloxy radicals. For example, in the initiation of MMA polymerization with BPO, less than 1% of H-abstraction is observed vs. >99% addition, whereas initiation with di(*t*-butyl)peroxide provides 34% H-abstraction and 66% addition. The decomposition of peroxides can be significantly accelerated in the presence of electron rich compounds such as amines and can be also catalyzed by transition metals (Scheme 6).

Scheme 6

$$HO-OH \ + \ Fe(II) \longrightarrow HO^\bullet \ + \ HO^- \ + \ Fe(III)$$

$$HO-OH \ + \ Fe(III) \longrightarrow HOO^\bullet \ + \ H^+ \ + \ Fe(II)$$

Another class of useful radical initiators are diazo compounds. Azobisisobutyronitrile is the most representative and most widely used example. It decomposes with a rate constant of $k_d \approx 10^{-5}$ s^{-1} at 60 °C and has an activation energy E_a = 132 kJ/mol. At 65 °C it has a half lifetime of 10 hours, Table 2. The useful temperature range for diazenes depends very strongly on steric effects and ranges from room temperature to 200 °C. The decomposition of diazenes can be accelerated photochemically, presumably due to the isomerization of trans to the strained cis isomers. The efficiency of initiation is relatively modest with cumulative efficiency reaching 50% for high monomer conversions. The side products of AIBN initiation are ketenimine and the toxic tetramethyl succinonitrile.

Table 2. Data for Select Azo-Initiators[a]

Initiator	Solvent	Temp. (°C)	k_d (60 °C) $M^{-1}s^{-1}$	E_a kJ/mol	A*10^{15}	$\tau_{1/2}$=10h
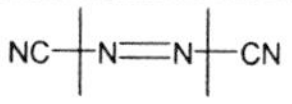 NC—N=N—CN	Benzene/ Toluene	37-105	9.7*10^{-6}	131.7	4.31	65 °C
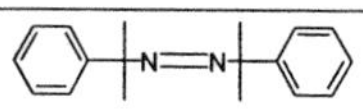 (ph)—N=N—(ph)	Toluene	40-70	170*10^{-6}	126.7	12.2	45 °C
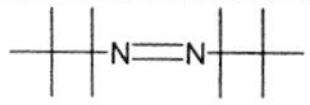 —N=N—	Diphenyl Ether	165-200	---	180.4	91.7	161 °C
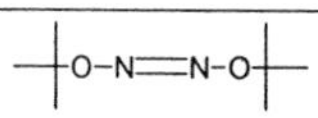 —O-N=N-O—	Isooctane	45-75	215*10^{-6}	119.5	1.17	42 °C

[a]Data taken from Ref. 1.

Other important classes of radical initiators include redox systems(Scheme 7).

Scheme 7

$$^-O_3S\text{-}O\text{-}O\text{-}SO_3^- + Fe^{2+} \longrightarrow Fe^{3+} + CCl_3 + SO^{4-}\text{•}$$

$$^-O_3S\text{-}O\text{-}O\text{-}SO_3^- + S_2O_3^{2-} \longrightarrow \text{•}S_2O_3^- + SO_4^{2-} + SO^{4-}\text{•}$$

$$HO\text{-}OH + Fe^{2+} \longrightarrow Fe^{3+} + HO^- + HO\text{•}$$

$$Cp_2Fe(II) + CCl_4 \longrightarrow Cp_2Fe(III)Cl + \text{•}CCl_3$$

Some monomers can self-initiate thermally. The best example is styrene, which participates in the Diels-Alder process forming the labile dihydronapthalene derivative, which further reacts with styrene to initiate polymer chains (Scheme 8).(16)   Other monomers do not readily undergo this Diels-Alder process, and consequently their rate of initiation is much slower than in the case of styrene.

Scheme 8

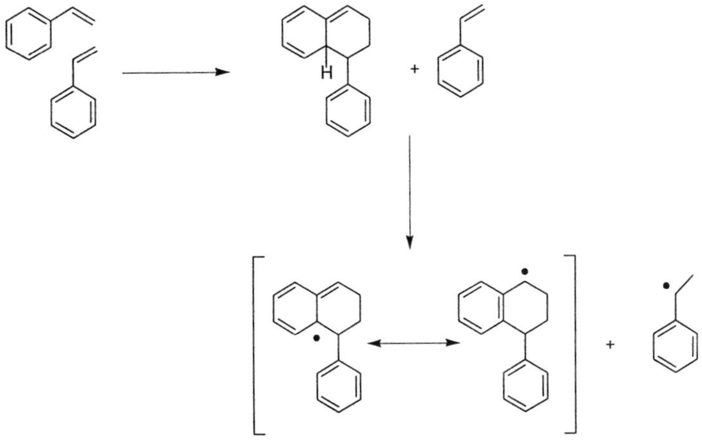

<u>Propagation</u>   Propagation is the most important reaction since it is responsible for the formation of high polymer. It occurs without interruption for more than 1000 times, meaning that the chemoselectivity of propagation exceeds 99.9%! Propagation is highly chemoselective and also relatively regioselective but has very poor stereoselectivity. The result is that atactic polymers are formed with a small preference for racemic addition for monosubstituted alkenes (P_r=0.5 to 0.6); disubstituted alkenes

like methyl methacrylate (MMA) favor syndiotactic arrangements due to steric effects (P_r=0.8).

Most polymers formed radically have a predominant head-to-tail connectivity. A small proportion of head-to-head units (1-2%) is formed for weakly stabilizing substituents such as in vinyl acetate or vinyl chloride. Even lower regioselectivity is observed for vinyl fluoride and trifluoroethylene (87%). Dienes polymerize to predominately form the 1,4-trans units. The proportion of 1,4-units in polybutadiene approaches 60%, 75% in polychloroprene, (Scheme 9) and 80% in polyisoprene. For the latter two polymers it can be further increased to 90% by polymerization at subambient temperatures.

Scheme 9

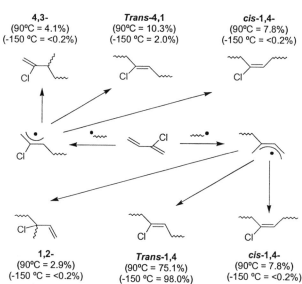

| 4,3-
(90°C = 4.1%)
(-150 °C = <0.2%) | Trans-4,1
(90°C = 10.3%)
(-150 °C = 2.0%) | cis-1,4-
(90°C = 7.8%)
(-150 °C = <0.2%) |

| 1,2-
(90°C = 2.9%)
(-150 °C = <0.2%) | Trans-1,4
(90°C = 75.1%)
(-150 °C = 98.0%) | cis-1,4-
(90°C = 7.8%)
(-150 °C = <0.2%) |

Precise information on the rate constants of propagation and activation parameters is extremely important for the proper control of commercial processes. The most recent data on rate constants of propagation are obtained from state of the art PLP measurements. The corresponding data are summarized in Table 3.(17)

Table 3. Propagation Rate Constants for Select Monomers Obtained by PLP Measurements[a]

Monomer	Sovlent	E_a	A	k_p (M^{-1}s^{-1}) 30 °C	50 °C	70 °C
Butyl Acrylate	Bulk	20	$2.5*10^7$	18 100	27 700	40 400
n-Butyl Methacrylate	Bulk	21.8	$2.47*10^6$	433	739	1 190
Butadiene	Chlorobenzene	35.7	$8.1*10^7$	56	138	295
Dodecyl Methacrylate	Bulk	21.7	$3.4*10^6$	622	1 060	1 700
Methacrylic Acid	Methanol	17.7	$6*10^5$	536	828	1 215
Methyl Methacrylate	Bulk	22.3	$2.7*10^6$	375	649	1 050
Styrene	Bulk	32.5	$4.3*10^7$	107	237	480
Vinyl Acetate	Bulk	20.4	$1.5*10^7$	4 570	7 540	11 700

[a]$A = L^{-1}mol^{-1}s^{-1}$, E_a in kJ mol^{-1}

A brief analysis of the data indicates that alkyl substituents in the corresponding (meth)acrylate esters have only minor effect on kinetics, but rates do increase with increasing ester group chain length. Disubstituted alkenes polymerize more slowly than monosubstitituted. There is also a general reciprocal correlation between the reactivity of monomers and the reactivity of the derived radicals (vide infra). The rate constants are more strongly affected by the radical's structure, meaning that less reactive monomers polymerize faster because they generate more reactive radicals, e.g., $k_{p(butadiene)} < k_{p(styrene)} < k_{p(MMA)} < k_{p(BA)}$.

<u>Termination</u> Termination between two free radicals can not be avoided. Usually, the rate of this reaction is limited by diffusion and it decreases with increasing conversion because of a progressive increase in the viscosity of the system. Termination rate coefficient (it is recommended to use this term rather than the termination rate constant

because it is *not* constant) depends on viscosity and also on the chain length and typically varies between $k_t = 10^7$ and 10^9 M^{-1}s^{-1}.

Termination can occur either by coupling or by disproportionation. For monosubstituted alkenes which generate secondary radicals, coupling is favored (>90%). However, due to steric reasons and the availability of β-H atoms for the disubstituted alkenes like MMA or α-methylstyrene, disproportionation becomes a dominant pathway due to the difficulty of formation, and instability of the hexasubstituted ethane derivatives at the junction points. Such termination generates weak points in the polymer chains and is responsible for the reduced thermal stability of the polymers. If the contribution of such structures is reduced or eliminated, e.g., by controlled radical polymerization, more stable polymers are formed.(18)

Transfer

<u>Transfer to Monomer</u> Transfer to most common monomers is not a very important reaction in radical polymerization. Therefore it is easy to prepare high molecular weight polymers with polymerization degrees DP>1000. The notable exception is vinyl chloride, which has a relatively high transfer to monomer constant, C_M = 10-25. However, longer chain lengths can be obtained when the polymerization is conducted at lower temperatures due to a high activation energy for transfer. Some allylic monomers efficiently participate in transfer due to their low propagation rate constants and the high lability of the allylic hydrogens. Thus, in the homopolymerization of allyl chloride, alcohol or esters only oligomers are formed, DP≈10.

The exact chemistry of transfer to monomer is not very clear. It may involve transfer of vinylic H-atom from monomer to radical or transfer of a β-H atom from the growing radical to monomer. The latter should be favored thermodynamically but the former can be kinetically preferred, as deduced from calculations for ethylene.(19)

<u>Transfer to Polymer</u> The by far more dominant form of transfer is transfer to polymer chains. Transfer to polymer can occur either *intra*molecularly or *inter*molecularly. The classic example of intramolecular transfer is found in the polymerization of ethylene. When ethylene is polymerized radically, the usual result is the formation of highly branched polyethylene, as a result of transfer to polymer. There are two factors which

favor this process, first is the high reaction temperatures (>200 °C) and the second is the high reactivity of the unsubstituted growing radicals. It can react with monomer (relatively slowly) or isomerize to more stable secondary or tertiary radicals by intramolecular H-transfer. The result is the formation of 10 to 20 branches per 1000 carbons. Ethyl and butyl branches dominate and are followed by pentyl and hexyl branches. The ethyl branches presumably do not originate from 1,3-H transfer but from two 1,5-H transfers followed by ethylene addition, since they are usually found in 2-ethylhexyl or 1,3-diethyl structures (as illustrated in Scheme 10).

Scheme 10

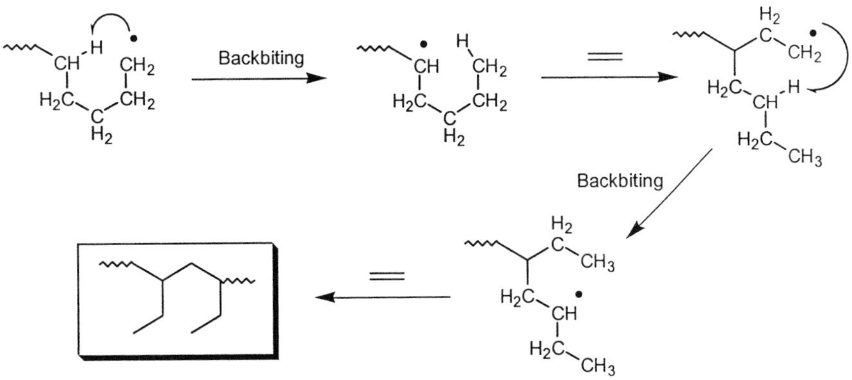

Reliable data on intermolecular transfer to polymer are not available but it is estimated that transfer coefficients are in the range of $C_P=10^{-4}$ to 10^{-3}. Thus, at low to moderate conversions, transfer should not significantly affect polymer structure. However, at higher conversions, and especially under monomer starved conditions in some emulsion systems, transfer to polymer should provide noticeable branching. For example, more than 1% of branched units in poly(butyl acrylate) and poly(vinyl acetate) have been recently detected by NMR under such conditions.(20, 21)

Transfer to Transfer Agents Transfer to a deliberately added transfer agent provides control of molecular weights and, in addition, end functionalities. The transfer process usually increases polydispersities, unless the transfer coefficient is very close to $C_T=1$. If $C_T>>1$, then the transfer agent is consumed very early, initially providing short chains and, then, subsequently, high polymer is formed. On the contrary, if $C_T << 1$, then high polymer is formed first, and only when monomer concentration is significantly reduced, does transfer becomes significant.

Transfer agents are relatively selective, which is related to radicals electrophilicity and nucleophilicity. For example, the transfer coefficient to acetone is twice larger for styrene (S) than for methyl acrylate (MA), but transfer to triethylamine is sixty times smaller for S than for MA.

The most typical mechanism of transfer includes atom or group transfer. The atom is usually hydrogen, which can be easily abstracted from molecules containing groups such as S-H, P-H, Sn-H, Si-H and in compounds with active β-H atoms next to carbonyl groups (acetone), phenyl groups (toluene, cumene) or vinyl groups (allyl derivatives) (Scheme 11). Polyhalocompounds such as CCl_4, CBr_4, etc. are also excellent transfer agents with even higher transfer coefficients.

Scheme 11

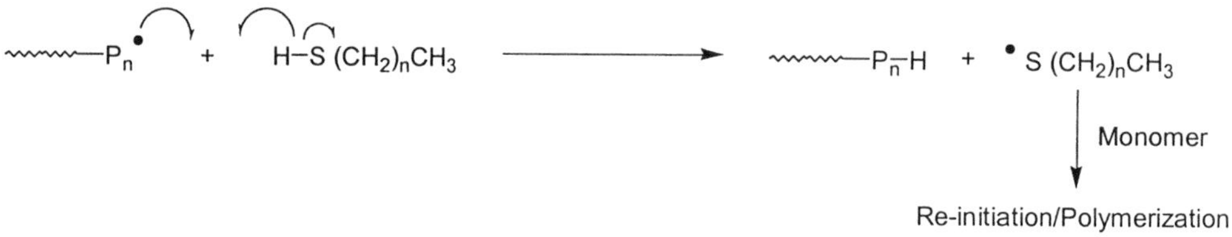

The group transfer process is most common for disulfides (Scheme 12):

Scheme 12

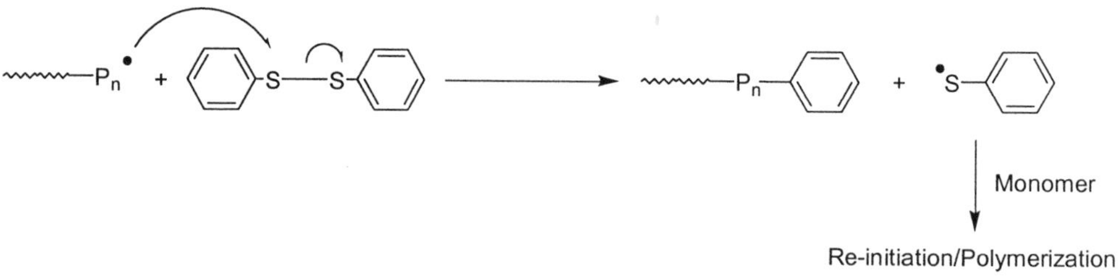

Another class of transfer agents employs addition fragmentation chemistry and has been used to incorporate useful functionalities as well as unsaturated structures for degenerative transfer processes(22) (Scheme 13):

Scheme 13

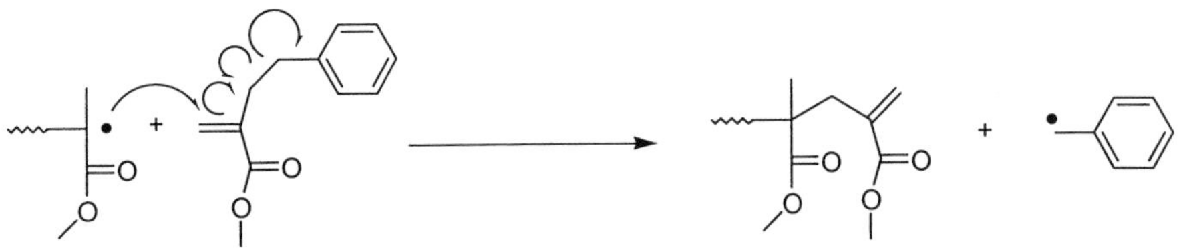

Finally, a relatively recent development is based on catalytic chain transfer in the presence of organometallic derivatives such as cobalt porphyrines and glyoximes (Scheme 14). The main advantage of this process is its extremely high efficiency ($C_T>10^3$), requiring very small concentrations of the transfer agent. However, it acts efficiently only for disubstituted alkenes such MMA and α-methylstyrene. For monosubstituted monomers such as acrylates and styrene which lack the α-methyl group, abstraction of the endo β-H atom is much slower and organometallic derivatives with a Co-C bond are formed reversibly. Nevertheless, use of substituted methacrylates such as acids, hydroxyethyl derivatives, etc. can lead to telechelic methacrylates and together with addition-fragmentation chemistry, provide for precise control of functionalities.(23, 24)

Scheme 14

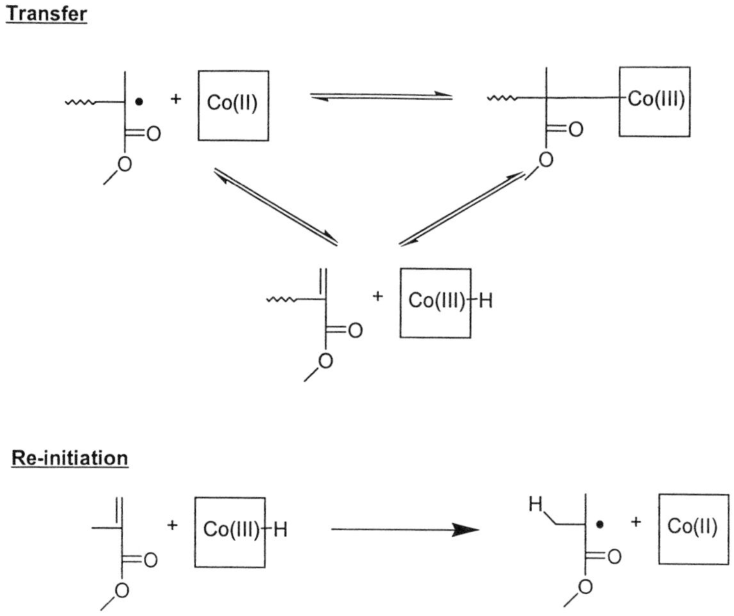

Transfer

Re-initiation

Other Reactions There are some other radical reactions not listed above. Some of them are usually not very important such as transfer to initiator but can still be used to make telechelic polymers. Some others, such as inhibition and retardation, can be used to control polymerization rates. The former process leads to the formation of relatively stable, unreactive radicals or employs trapping with persistent radicals such as nitroxides, galvinoxyl, diphenylpicrylhydrazyl, verdazyl, etc. Stable radicals can be formed either by abstraction of an atom or by an addition process. The most important inhibitor is oxygen which adds extremely fast to organic radicals and forms the relatively stable peroxy radicals which can further act as radical traps or abstractors. At higher temperatures, they decompose and can initiate polymerization (e.g., ethylene polymerization). Other compounds of this class include nitrones, nitro and nitroso compounds as well as quinones. Some of best inhibitors acting on the principle of atom abstraction are transition metal compounds such as $CuBr_2$ or $FeBr_3$. However, under some conditions (ligands, temperature) they can be converted to catalysts for controlled radical polymerization.(25)

Retarders behave similarly to inhibitors but usually with lower efficiency and sometimes reversibly. Thus, inhibitors stop polymerization until they are totally consumed, whereas retarders just slow it down.

<u>Copolymerization</u> Facile statistical copolymerization is one of the main advantages of free radical polymerization. The reactivities of the monomers, in contrast to ionic polymerization, for many systems are relatively similar. There is a tendency to alternation, especially for monomers with opposite polarities. Generally, electrophilic radicals (for example with -CN, -C(O)OR, Cl groups) prefer to react with electron rich monomers such as styrene, dienes or vinyl acetate, whereas nucleophilic radicals prefer to react with alkenes containing electron withdrawing substituents. For monomers with similar polarities, the reactivity ratios are close to an ideal copolymerization, $r_A = 1/r_B$, where $r_A = k_{AA}/k_{AB}$ and $r_B = k_{BB}/k_{BA}$ (Scheme 15). Thus, monomer reactivities are the same regardless the radical they react with. On the other hand, monomers with opposite polarities tend to alternate, $r_A \cdot r_B < 1$. For example, $r_S \approx r_{MMA} \approx 0.5$. An even more dramatic observation is for the copolymerization of styrene with maleic anhydride $r_S = 0.02$, $r_{MAH} < 0.001$, the latter value is additionally reduced for thermodynamic reasons do to poor homopolymerizability of the anhydride.

Scheme 15

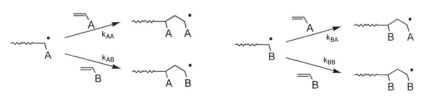

The difference in reactivity ratios can lead to some problems in homogeneity of the polymer composition. When the reactivity ratios are significantly different, nearly pure homopolymer of the more reactive monomer may be formed at the beginning of the polymerization. As the polymerization progress and the first monomer is consumed, then the second monomer will be polymerized. This can lead to the formation of a mixture of two homopolymers. Such an example would be in the copolymerization of styrene and vinyl acetate. Fortunately, the reactivity ratios for most monomers are relatively similar, allowing for the formation of copolymers. However, differences remain, and these must be considered when attempting to obtain uniform polymer chains in copolymerizations. This is generally overcome by attenuating the monomer feed throughout the polymerization by continuous addition of one or more monomers.

As mentioned before, monomer reactivities scale reciprocally with radical reactivities; more reactive monomers form more stable and less reactive radicals. There have been several attempts to predict the reactivities of monomers and especially the reactivity ratios between two comonomers. Perhaps the most important was the development of the Q-e scheme.(26) Q refers to the general monomer reactivity whereas e describes polar factors. More negative e values describe electron rich monomers (S=-0.8, VAc=-0.1), while positive e values were assigned to electron poor monomers (MMA=0.38, BA=1.06, AN=1.33). Values of Q and e for a new monomer can be estimated from the copolymerization with a monomer of known Q-e values (S was arbitrarily assigned as a standard with $Q=1.0$ and $e=-0.8$) and then compared to other monomers to predict reactivity ratios.

Further improvements of the Q-e scheme were made using a "pattern of reactivities" in which the rate constant of the known reaction, e.g., abstraction of H from toluene or cross-propagation with styrene) was correlated with the polar factor from the Hammett equation using two parameters typical for each monomer. In such a way it was possible to correctly predict many reactivity ratios and transfer coefficients.(27)

Applications and Limitations of Conventional Radical Polymerization

The commercial application of radical polymerization is very broad and it is impossible to go into significant detail here. However, some important aspects of radical polymerization, as applied commercially, will be covered. This includes the conditions under which radical polymerizations are conducted, some general uses for polymers prepared radically, and future developments in radical polymerization.

<u>Polymerization Processes</u> The four most common polymerization processes that are encountered in industrial processes are bulk (mass), solution, suspension, and emulsion polymerization. Bulk, or mass, polymerization is the polymerization of pure monomer without any added solvent or diluent. This type of process is desirable as the polymer can be directly obtained without contamination by solvent, or other impurities, which may require removal at the end of the polymerization. Unfortunately, since radical polymerizations are very exothermic, temperature control of the polymerization is difficult. Additionally, at higher conversions, autoacceleration (the Tromsdorff effect) becomes pronounced and can have adverse effects on the final properties of the polymer, i.e., M_n, M_w/M_n, etc. Further, as the monomer is consumed and converted to polymer, the reaction medium becomes quite viscous and can vitrify unless conducted at high temperatures. For these reasons, many bulk polymerizations are conducted to only low to moderate conversions. This allows for the monomer to act as a solvent providing better heat transfer/control, and to also lower the viscosity of the system. Such an approach however, requires that the remaining monomer be removed from the final polymer and recycled back to the reactor.

Solution polymerization is conducted in the presence of a solvent. The benefits of this type of polymerization is that the monomer can be completely consumed without obtaining extremely high viscosities as well as better temperature control of the reaction. If the solvent is required in the final product then the polymer/solvent mixture can be used directly, otherwise the solvent must be removed. Chain transfer to solvent can influence the final molecular weight of the polymer.

Suspension and emulsion polymerizations are similar in that they are generally conducted in a heterogeneous mixture of water and monomer and in the presence of a surfactant or stabilizer. In both cases, the mixture is stirred so that monomer droplets are formed. In suspension polymerizations, the initiator is usually oil-soluble, so it resides in the monomer droplets. The polymerization occurs in the droplets, which can be envisioned as mini-reactors where the bulk polymerization of the monomer occurs. The final particle size of the polymers is usually on the order of 10-500 μm in diameter.

Emulsion polymerizations are much more complex.(28) The most significant, practical difference between suspensions and emulsions is their particle size and stability. In emulsions, the final polymer particles are generally less than 1 µm in diameter. The particles in suspensions will settle if not stirred while emulsion systems will remain as stable colloids. Also, in emulsion polymerizations, the initiator is usually water-soluble. Both systems have their own benefits: suspension prepared polymers can often be isolated by simple filtration, dried and directly processed into its final form; emulsions of polymers can be directly applied, and upon evaporation of the water, result in coatings.

Specific Polymers

<u>Polyethylene</u> Polyethylene is usually prepared radically in a tubular reactor, under high pressures and at high temperatures. The prepared polymer is generally highly branched (*vide supra*) and is given the name low density polyethylene (LDPE). This polymer, T_g~-120 °C, T_m~100 °C, is flexible, resistant to solvents and water, has good flow properties and has good impact resistance. It is widely used as nearly 8 billion pounds were produced in the US in 1997.

The majority of LDPE is used in films such as plastic bags, food and beverage containers, and plastic wrapping for various containers (clothing, trash, etc.) LDPE is also extruded in a variety of applications, including toys, plastic containers and lids, and for coating electrical wires and cables. This is but a small listing of the applications that are used for LDPE. It should be noted that the other grades of polyethylene, i.e., HDPE, LLDPE, etc., are prepared using Ziegler-Natta type catalysts and not by radical polymerization.

<u>Polystyrene</u> Polystyrene is a desirable plastic as it is generally a rigid plastic, with good optical clarity (it is amorphous), and is resistant to acids and bases; it can be processed at moderate temperatures (T_g ~ 100 °C). It also is a good electrical insulator. One major use of styrene is as expandable styrene. This type of styrene is processed using a blowing agent, which results in the formation of a foam, commonly known as

styrofoam. Styrofoam has found many uses as an insulator, beverage containers, and in packaging. Over 6.5 billion pounds of polystyrene were produced in the US in 1997.

A drawback of polystyrene is that it is soluble in a variety of solvents (poor solvent resistance) and is relatively brittle. These problems are generally overcome by copolymerization of styrene with a variety of monomers or by addition of a variety of stabilizers. For example, copolymerization of styrene with acrylonitrile (< 40% AN) imparts good solvent resistance to the polymer, increases the tensile strength as well as the upper use temperature. Such copolymers find uses in appliances, furniture and electrical applications (battery casings). Nearly a hundred million pounds of SAN were produced in the US in 1997.

The flexibility of styrene is generally improved by copolymerization with dienes. Elastomeric copolymers of styrene and 1,3-butadiene or SBR rubbers. SBR rubbers generally contain low amounts of styrene (~25%) and have similar properties to natural rubber. These polymers are primarily used in the manufacturing of tires, where the SBR is vulcanized. Other automotive uses include belts and hoses, but SBR is also used in shoe soles, flooring materials and in electrical insulation. Higher contents of styrene in the copolymers are used in paints.

By combining the solvent resistance of SAN and flexibility of SBR, terpolymers of acrylonitrile, styrene and 1,3-butadiene have been prepared. Generally, styrene and acrylonitrile are copolymerized in the presence of a rubber (poly(1,3-butadiene), SBR, or poly(1,3-butadiene-*co*-acrylonitrile)). The copolymerizations are done as emulsions or suspensions; these copolymers are known as ABS polymers. ABS has excellent solvent and abrasion resistance and finds applications in houseware (luggage, hairdryers, furniture), construction (bathroom sinks/tubs, piping), electronic housings, automotive parts (light housings), and in recreation (hockey sticks, boat hulls). In 1997, 3 billion pounds of ABS were produced in the US.

Poly(vinyl chloride) PVC is generally produced by suspension polymerization of the gaseous monomer vinyl chloride. PVC and its copolymers are widely used with production topping 14 billion pounds in the US in 1997. Although PVC itself is very brittle and is relatively unstable to heat and light (with loss of HCl), these problems are overcome with the use of additives, i.e., plasticizers and stabilizers. By adding plasticizers (up to 40%), PVC can become very pliable and thus more useful. PVC is widely used in construction (drainage pipe, vinyl siding, window frames, gutters, flooring, etc.), packaging (bottles, boxes), wiring insulation, and in biomedical applications (gloves, tubing).

(Meth)Acrylates   Acrylic and methacrylic polymers are widely used in numerous applications. One of the benefits of the poly(acrylic esters) is the wide range of functional groups that can be incorporated into the ester group. For example, the polymers derived from ethylene glycol esters (2-hydroxyethyl (meth)acrylate) are water soluble or swellable. Additionally, long alkyl groups can be added which make the resulting polymer hydrophobic; glycidyl, or amine groups can be added to impart reactive functionality to the polymer.

Most notably, acrylics are used as comonomers in coatings (automotive, structural, and home), or paints. These acrylic paints are generally prepared as emulsions and the latexes can be applied directly to give good glossy coatings. They are also polymerized in solution for use as oil based paints. In 1997, the US paint industry sold $17.5 billion dollars worth of paint, with much of this stemming from acrylics. Acrylic copolymers can also be used as adhesives, oil additives, and caulks/sealants.

Homopolymers of methyl methacrylate are also widely used due to its optical clarity. PMMA prepared as sheets are commonly know as Plexiglas, and are used as safety glass to prevent breakage, vandalism and as barriers. They are also used in light fixtures, skylights, lenses, fiber optics, dentures, fillings and in contact lenses (MMA: hard contacts, cross-linked 2-hydroxyethyl methacrylate: soft contacts).

Sheets of PMMA are polymerized in stages to avoid autoacceleration and severe shrinkage. In the first stage partial polymerization of the bulk MMA leads to a viscous syrup which is subsequently poured into moulds where the polymerization is completed.

Others   There are a variety of other polymers prepared by radical polymerization, but in smaller quantities. Fluoropolymers are an example, with the polymer containing either a perfluorinated acrylic ester, or either a completely (tetrafluoroethylene) or partially (vinylidene fluoride) fluorinated double bond. In 1994, world sales of fluorinated polymers were valued at nearly $1.5 Billion, with half being in PTFE.

Other notable polymeric materials include polyacrylamides and poly(vinyl alcohol) (PVA). Both are widely used in aqueous solutions as thickeners, in biomedical applications, and in health and beauty products. PVA can not be prepared by direct polymerization of vinyl alcohol, but must be obtained by hydrolysis of poly(vinyl acetate) (PVAc). Generally, PVAc is not completely hydrolyzed, but is limited to 80-90% hydrolysis since complete hydrolysis results in a highly crystalline polymer that is less soluble in water.

Future Advances in Radical Polymerization

To date, many advances in radical polymerization have led to a startling array of materials and commercial products. These materials were originally developed by simple homopolymerization of vinyl monomers followed by copolymerization with two or more monomers. Copolymerization allowed for the tailoring of the bulk polymer's physical/mechanical properties. The polymers were then formulated with various additives and copolymer mixtures to prepare new, more advanced products. Developments in polymer processing techniques also made large strides and pushed the limits of the polymeric materials usefulness.

However, with all of the advances in formulation and processing, one is still left with the basic properties of the synthesized homo/copolymer. The next step in developing advanced polymeric materials is in designing the polymer with specific properties in mind. For example, block copolymers of hard and soft polymers can be used for elastomers, block copolymers of hydrophilic/hydrophobic monomers generate water swellable polymers, branched polymers have novel viscometric properties, and polymers with precise functional groups can be used in subsequent polymerizations, or to attach special molecules (drugs, sensors, energy converting molecular devices, etc.). To be able to develop these materials, however, requires precise control of the polymer chain's growth. Until only recently this level of control has been limited to a handful of polymerization systems, i.e., anionic, cationic polymerizations. Development of a controlled/"living" radical polymerization system has been the goal of many chemists for decades.

Chemistry of Controlled Radical Polymerization (CRP)

As concluded from the previous section, radical polymerization has many advantages. It can be used for the (co)polymerization of a very large range of monomers under undemanding conditions. This is why nearly 50% of all synthetic polymers are prepared using radical chemistry. They can form either hard plastics, fibers or elastomers, they can be of very high and also very low molecular weights, they can be used as thermoplastic but also as thermosetting materials, and they find uses in all ranges of applications including packaging, automotive industry (tires, coatings, bumpers, interior parts) cosmetics, pharmaceuticals, paints, agents for water treatment, printing (inks, binders), clothing, etc. The recent trend in chemistry is to exploit many soft or weak interactions which via molecular recognition and supramolecular chemistry can provide new organized structures. Such an approach can be also used in polymer chemistry provided that polymers with well-defined and uniform dimensions, topologies, compositions and functionalities will be used. Such (co)polymers self organize to develop many new morphologies and can further improve many properties of the materials listed above and also lead to entirely new applications. Thus, it is desirable to prepare well-defined polymers as in ionic living polymerizations, but by a radical mechanism. Attaining the control of a radical polymerization would open the door to the preparation of a virtually unlimited number of new polymeric materials.

There are several approaches to controlling radical polymerizations by suppressing the contribution of chain breaking reactions, together with assuring quantitative initiation. All of them employ dynamic equilibration between growing free radicals and some kind of dormant species. The equilibrium is established via activation (k_a) and deactivation (k_d) steps. We prefer to classify these reactions as controlled radical polymerizations (CRP) rather than as living radical polymerizations, due to the presence of unavoidable termination, which is intrinsically incompatible with the concept of living polymerizations.(29)

Currently three approaches appear to be most successful for controlling radical polymerization:

1. Thermal homolytic cleavage of a weak bond in covalent species which reversibly provides a growing radical and a less reactive radical (persistent or stable free radical) (Scheme 16). There are several examples of persistent radicals but it seems that the most successful are nitroxides,(30, 31) triazolinyl radicals,(32) some bulky organic radicals, e.g. trityl(33, 34) or photolabile C-S bonds(35) and organometallic species.(36)

Scheme 16

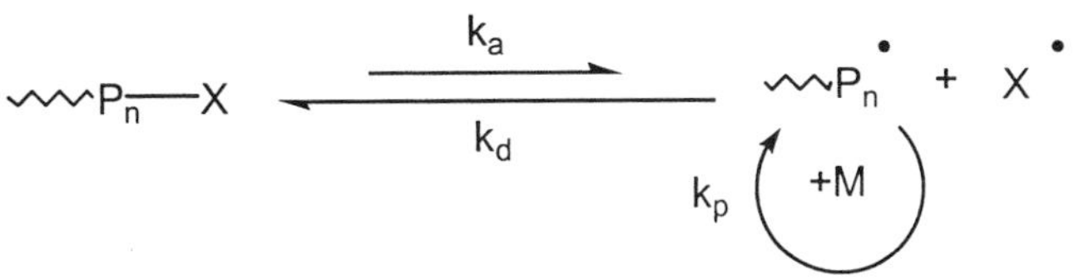

A subset of this process is the catalyzed, reversible cleavage of the covalent bond in the dormant species via a redox process (Scheme 17). As the key step in controlling the polymerization is atom (or group) transfer between growing chains and a catalyst, this process was named atom transfer radical polymerization (ATRP).(37, 38)

Scheme 17

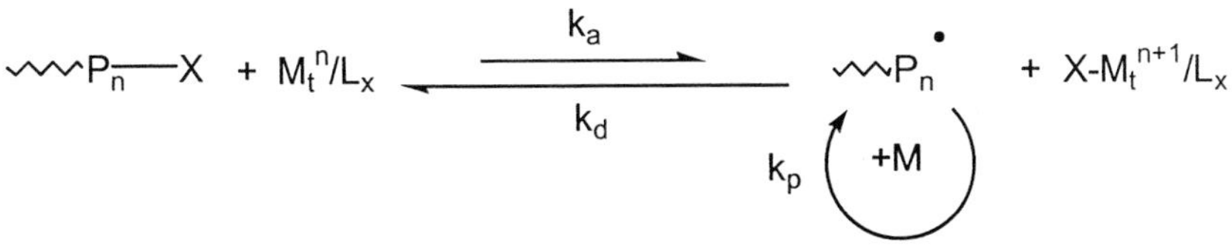

2. The second approach is based on a thermodynamically neutral exchange process between a growing radical, present at very low concentrations, and dormant

species, present at much higher concentrations (generally three to four orders of magnitude) (Scheme 18). This degenerative transfer process can employ alkyl iodides,(39) unsaturated methacrylate esters,(40) or thioesters.(41, 42) The latter two processes operate via addition-fragmentation chemistry.

Scheme 18

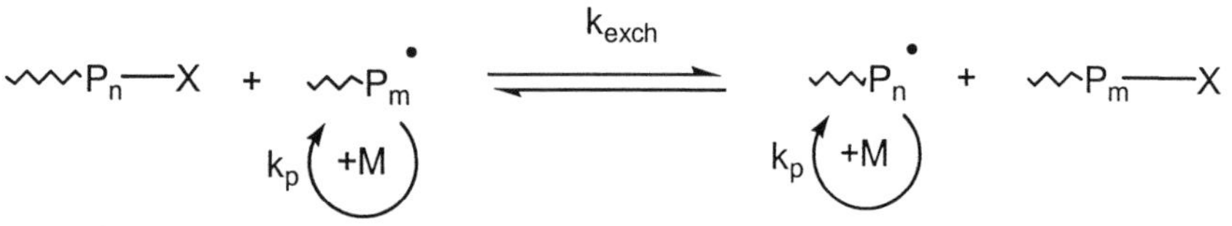

3. Finally, there is a third approach, which has not yet been as successful as the above systems. This process is the reversible formation of persistent radicals, which do not react with each other or monomer, by reaction of the growing radicals with a species containing an even number of electrons (Scheme 19). Here, the role of a reversible radical trap may be played by phosphites,(43) or some reactive, but non-polymerizable alkenes, such as tetrathiofulvalenes or stilbene.(44)

Scheme 19

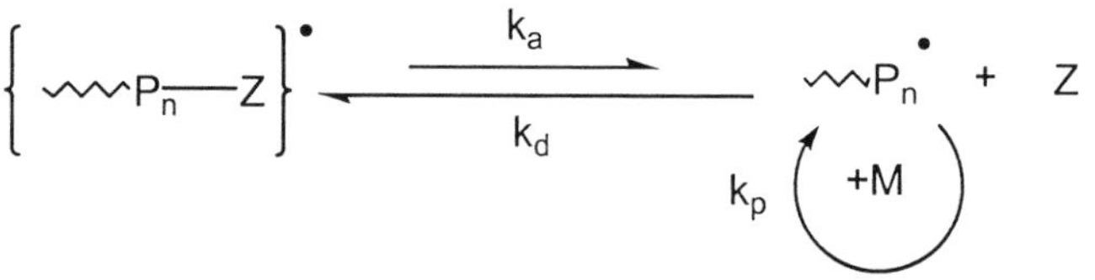

In the remaining pages of this chapter we will discuss the chemistry of the first two major approaches to controlled radical polymerization. Finally, some examples of new materials prepared by these techniques will be discussed.

Controlled Radical Polymerization Based on Reversible Thermal Cleavage of Weak Covalent Bonds   The homolytic cleavage of weak covalent bonds results in the formation of a growing radical and a counter radical which, in principle, should be only involved in the reversible capping of the growing chains. Such stable counter radicals should *not* react with themselves, with monomer to initiate growth of new chains or in other side reactions such as the abstraction of β-H atoms. These persistent radicals should be relatively stable, although some recent data indicate that their slow decomposition may help in maintaining appropriate polymerization rates.

There are several examples of persistent radicals used in controlled radical polymerization but perhaps the most successful are nitroxides, specifically TEMPO(31) (Scheme 20). Interesting results were also obtained with triazolinyl radicals,(32) bulky organic radicals, such as trityl and diarylmethyl,(34) and also with organometallic species, especially with paramagnetic high spin cobalt (II) compounds. However, often a particular trap acts efficiently only for one class of monomers. For example, Co(II) porphyrine derivatives are excellent for controlling the polymerization of acrylates(36) but poor for styrene, while for methacrylates they act as very efficient transfer reagents (catalytic chain transfer).(45) The nitroxide TEMPO is efficient only for the CRP of styrene and its copolymers, however, some new nitroxides are also successful for acrylates.(46, 47) Thus, the range of monomers is slowly expanding.

Scheme 20

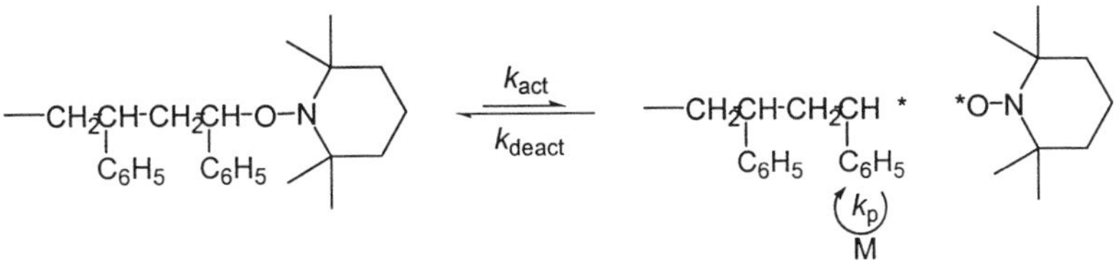

$$-CH_2CH\text{-}CH_2CH\text{-}O\text{-}N \quad \underset{k_{deact}}{\overset{k_{act}}{\rightleftharpoons}} \quad -CH_2CH\text{-}CH_2CH \cdot \quad \cdot O\text{-}N$$

with C_6H_5 substituents, and k_p, M.

Nitroxides were originally described in the patent literature as agents in the polymerization of (meth)acrylates.(30) However, only since the seminal paper by Georges using TEMPO in styrene polymerizations at elevated temperatures (>120 °C), have real advances in controlled radical polymerization been made.(31) The early results were most encouraging, as they employed very simple reaction conditions (bulk styrene, $[BPO]_o:[TEMPO]_o=1.3:1$ and simple heating) and gave the desired outcome ($DP_n=\Delta[Sty]/[TEMPO]_o$ in the range of $M_n=1,000$ to $50,000$ and with low polydispersities, $M_w/M_n<1.3$). The reactions were slow with rates being similar to the thermal polymerization of styrene. Under typical conditions, the majority of the chains are in the form of alkoxyamines which are covalent dormant species, while a very small fraction of radicals are continuously generated by thermal initiation and by the thermal cleavage of the alkoxyamines ($[P^*]\approx10^{-8}$ M).   Chains continuously terminate by coupling/ disproportionation and lead to an excess of TEMPO via the persistent radical effect ($[TEMPO]\approx10^{-5}$ M).   The chain ends can also slowly decompose and generate unsaturated structures and a hydroxylamine (Scheme 21).(48, 49) The latter can be reoxidized to TEMPO in the presence of traces of oxygen.

Scheme 21

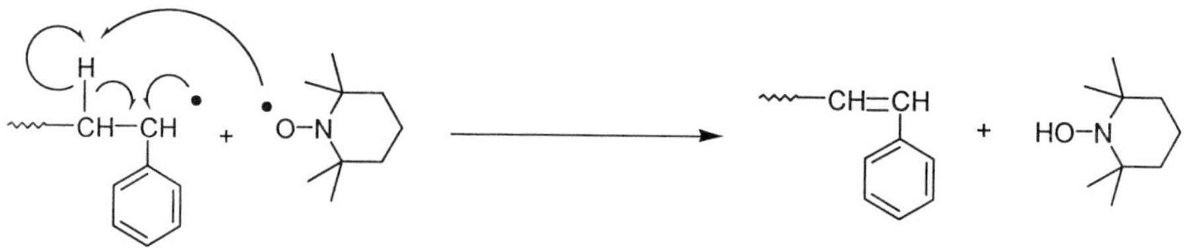

Control is relatively good but decreases with the progress of the reaction and molecular weights above M_n=20,000. Typically, above 80% of chains are in the form of dormant, potentially active species but this number drops with the chain length. However, under appropriate conditions it is possible to make chain extensions and block copolymers. Several improvements and alterations to the original system have been made. These include the use of different initiators such as AIBN instead of BPO, or even a simple pure thermal process(50, 51) as well as preformed alkoxyamines, so-called unimolecular initiators.(52)   Also di- and multifunctional initiators have been successfully used to make novel materials.   The rates of polymerization could be increased by using new nitroxides which are sterically more bulky, dissociate easier and provide a larger equilibrium constant. Examples include phosphoric and phosphonic acid derivatives.(46, 53)

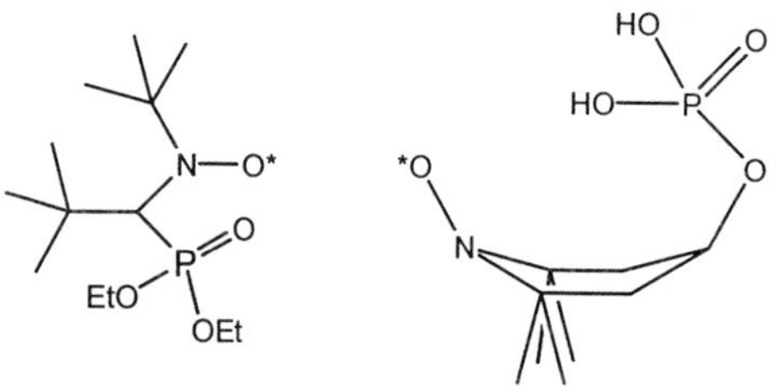

Rates in nitroxide mediated systems follow a simple law (Eq. 3) and depend on the concentration of radicals which are defined by the equilibrium constant (K_{eq}) and the concentration of dormant species [P-TEMPO] and TEMPO.

$$R_p = -d[M]/dt = k_p [M] [P^*] = k_p[M]K_{eq}[\text{P-TEMPO}]/[\text{TEMPO}] \qquad (3)$$
$$[P^*] = K_{eq} [\text{P-TEMPO}]/[\text{TEMPO}] \qquad (4)$$

However, the polymerizations are negligibly slow when the equilibrium constants are very small, as in the classic case of the TEMPO mediated polymerization of styrene, $K_{eq} \approx 10^{-11}$ M at 130 °C. In that case, the rate may be increased to an acceptable level by an additional flux of radicals either from thermal initiation of the monomer or by using conventional radical initiators which have appropriate lifetimes at the polymerization temperature, such as dicumyl peroxide.(48, 54, 55) In that case, the concentration of radicals is defined by the balance between rates of initiation and termination:

$$[P^*]=R_i/R_t \qquad (5)$$

A stationary concentration of TEMPO must self adjust and reduce to fulfill the equilibrium requirement and to obey both eqs (4) and (5).

Another approach to increase rates is to reduce the concentration of TEMPO by other reactions. The lower thermal stability of 4-oxoTEMPO results in its continuous decomposition, lower concentration and a shift of the equilibrium more towards growing radicals, and finally faster rates. The decomposition/dissociation may be also catalyzed intra- or intermolecularly by addition of acid derivatives and acetyl compounds (potentially acid generators).(56, 57) The principle of low thermal stability of persistent radicals was employed using triazolinyl radicals which decompose at elevated temperatures and spontaneously reduce their concentration.(32)

Probably the most important future direction of nitroxide chemistry will be the development of new compounds which will be applicable to a larger range of monomers and under milder reaction conditions. However, since these nitroxide compounds are at the end of each chain, these new compounds must be inexpensive, and introduce no adverse properties (color, poor thermal stability, etc.) to the final material.

Transition Metal Catalyzed Controlled Process – Atom Transfer Radical Polymerization

Atom transfer radical polymerization (ATRP) is based on the reversible transfer of halogen atoms between dormant alkyl halides (P_n-X) and transition metal catalysts (M_t^n/L) by redox chemistry. Alkyl halides are reduced to growing radicals and transition metals are oxidized via an inner sphere electron transfer process. In the most studied reaction, the role of the activator is played by a copper(I) species complexed by two bipyridine ligands and the role of deactivator by the corresponding copper (II) species. The scheme below shows such a system with the values of the rate constants of activation (k_a), deactivation (k_d), propagation (k_p) and termination (k_t) for a bulk styrene polymerization at 110 °C (Scheme 22).(58, 59) The rate coefficients of termination decrease significantly with the progress of reaction due to the increase in the chain length and viscosity of the system. In fact, the progressive reduction of k_t is one of the most important features of many controlled radical polymerizations.

Scheme 22

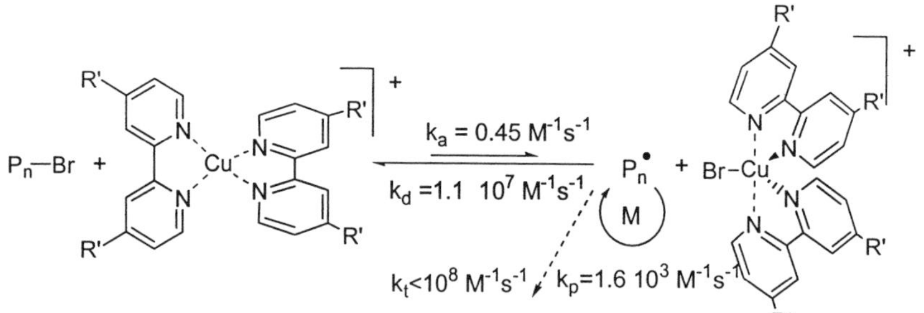

$$k_a = 0.45\ M^{-1}s^{-1}$$
$$k_d = 1.1\ 10^7\ M^{-1}s^{-1}$$
$$k_t < 10^8\ M^{-1}s^{-1} \quad k_p = 1.6\ 10^3\ M^{-1}s^{-1}$$

The main difference between TEMPO mediated systems and ATRP is that the latter can be used for a much larger range of monomers, including methacrylates and it is generally much faster. The rate of ATRP can be adjusted conveniently not only by the concentration of deactivator but also by the concentration of activator, since catalysis is at the very nature of ATRP.(58)

$$R_p = -d[M]/dt = k_p [M] [P^*] = = k_p [M] \{k_a [P\text{-}X] [Cu(I)]\}/ \{k_d [X\text{-}Cu(II)]\} \qquad (6)$$

Polydispersities in ATRP and in other controlled radical reactions depend on relative rates of propagation and deactivation.(60)

$$M_w/M_n = 1 + [(k_p [RX]_o)/(k_d [X\text{-}Cu(II)])] (2/p - 1) \qquad (7)$$

Thus, polydispersities decrease with conversion, p, with the rate constant of deactivation, k_d, and also with the concentration of deactivator, [X-Cu(II)]. However, they increase with the propagation rate constant, k_p, and with the concentration of initiator, $[RX]_o$. This means that more uniform polymers are obtained at higher conversions, when the concentration of deactivator in solution is high and the concentration of initiator is low. Also, more uniform polymers are formed when the deactivator is very reactive (e.g., copper(II) complexed by bipyridine or triamine rather than by water) and monomer propagates slowly (styrene rather than acrylate).

Chain breaking reactions do occur in these controlled radical systems.(61) Fortunately, at typical reaction temperatures, the contribution of transfer is relatively small. For example, in the polymerization of styrene, less than 10% of chains participate in transfer to monomer before reaching $M_n=100,000$. However, the contribution of transfer progressively increases with chain length, and therefore molecular weights must be limited by the appropriate ratio of monomer to initiator concentrations (for styrene $\Delta[M]/[I]_o < 1,000$).

Termination does occur in radical systems and currently can not be completely avoided. However, since termination is second order with respect to radical concentration and propagation is first order, the contribution of termination increases with radical concentration, and therefore also with the polymerization rate. Thus, most controlled radical polymerizations are designed to be slower than conventional systems. It is possible to generate relatively fast controlled radical polymerizations, but only for the most reactive monomers, such as acrylates, and/or for relatively short chains. For short chains, the absolute concentration of terminated chains is still high but their percentile in the total number of chains is small enough so as not to affect end functionalities and blocking efficiency. A typical proportion of terminated chains is between 1 and 10%, with a large fraction of those being very short chains that may not markedly affect the properties of the synthesized polymers and copolymers. It is possible to measure the evolution of concentration of terminated chains by following the copper(II) species by EPR.

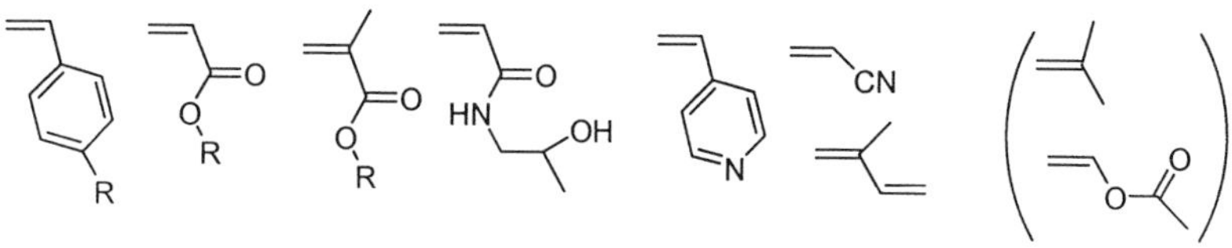

Figure 1. Monomers capable of (co)polymerization in ATRP.

The list of monomers polymerized successfully by ATRP is extensive and includes various substituted styrenes, acrylates, methacrylates, acrylamides, vinyl pyridine, acrylonitrile and dienes. In addition, several other monomers have been successfully copolymerized using ATRP and they include, for example, isobutene and vinyl acetate (Figure 1).

A big advantage of any radical process, and ATRP included, is its tolerance to many functional groups such as hydroxy, amino, amido, ether, ester, siloxy and others. All of them have been incorporated into (meth)acrylate monomers and successfully polymerized. The only exception is a carboxylic acid group which potentially complexes with the catalyst and disables ATRP and therefore it has to be protected.

Another advantage of ATRP is a multitude of available initiators. Nearly all compounds with halogen atoms activated by the presence of α-carbonyl, phenyl, vinyl or cyano groups have been used as efficient initiators. Also compounds with a weak halogen-heteroatom bond can be used, such as sulfonyl halides.(62) Very often they carry additional functionalities which can be incorporated at the chain end.

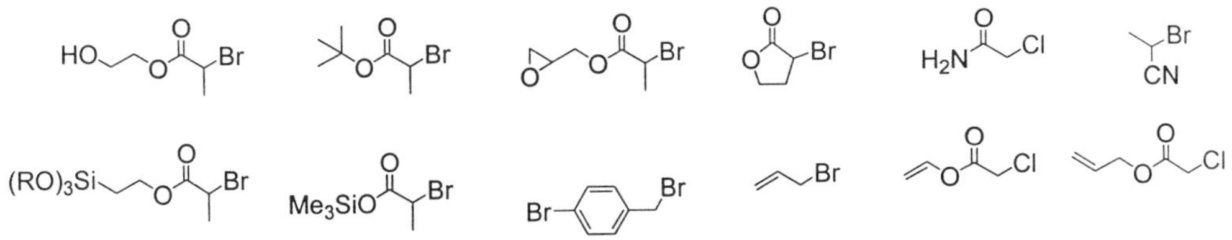

Figure 2. ATRP initiators.

In addition, there are many compounds with several active halogen atoms which have been used to initiate a bi- or multidirectional growth to form star-like polymers and copolymers.(63) Active halogens can be incorporated at the chain ends of polymers prepared by other techniques such as cationic, anionic, ring-opening metathesis and conventional radical processes to form macroinitiators. Such macroinitiators have been successfully used to chain extend via ATRP to form novel diblock, triblock and graft copolymers.(64-67)

The halogen atoms, at the chain ends, can be removed either by a reduction process or transformed to other useful functionalities as shown for styrene and acrylate systems (Scheme 23).(68)

Scheme 23

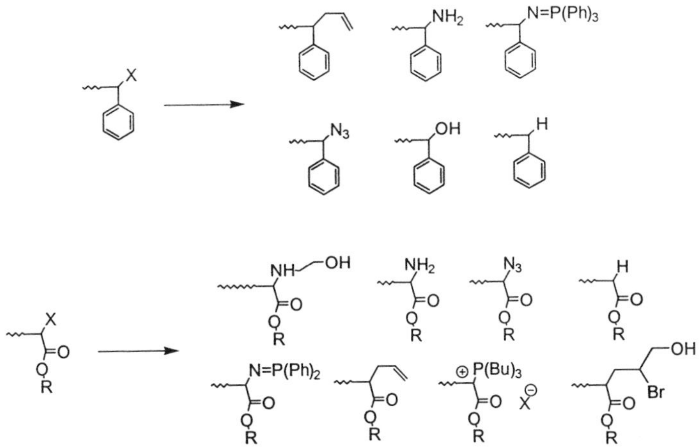

ATRP has been successfully carried out in bulk, in solution,(38, 69) as well as in aqueous emulsion(70) and suspension, and in other media (e.g., liquid or supercritical

CO_2).(71) The typical temperature range is from subambient temperature to 130 °C. Molecular weights range from 200 < M_n < 500,000 and polydispersities are low, 1.05 < M_w/M_n < 1.3, depending on the catalyst used, and also on the relative and absolute catalyst and initiator concentrations.

Copolymerization is facile and many statistical, gradient and block copolymers have been prepared.(18, 72) The reactivity ratios are nearly identical to conventional radical processes.(73)

The key feature of ATRP is a transition metal compound complexed by a suitable ligand. This ligand assures catalyst solubility and also adjusts its electronic and steric properties and should enhance the atom transfer chemistry by comparison with other reactions. Thus, it should allow for a dynamic atom transfer by the reversible expansion of the coordination sphere. As transition metals, Cu, Ru, Fe, Ni, Pd, Rh have been used so far.(37, 38, 69, 74-76) Ligands are usually mono or polydentate species such as ethers, amines, pyridines, phosphines and the corresponding polyethers, polyamines and polypyridines. The transition metal complex is very often a metal halide but pseudohalides, carboxylates and compounds with non-coordinating triflate and hexafluorophosphate anions have been also used successfully.(77)

Degenerative Transfer   Control by degenerative transfer (DT) carries perhaps the smallest alteration to a conventional free radical process. The principle of DT relies on a thermodynamically neutral (degenerative) transfer reaction. The key for control is the minimal energy barrier for that reaction.   Conventional free radical initiators are used, i.e., peroxides and diazenes, at temperatures typical for radical polymerization. At the same time, the polymerization is carried out in the presence of a compound with a labile group or atom which can be either reversibly abstracted or added-fragmented by the growing radical.   The simplest examples are reactions in the presence of alkyl iodides:(39)

Scheme 24

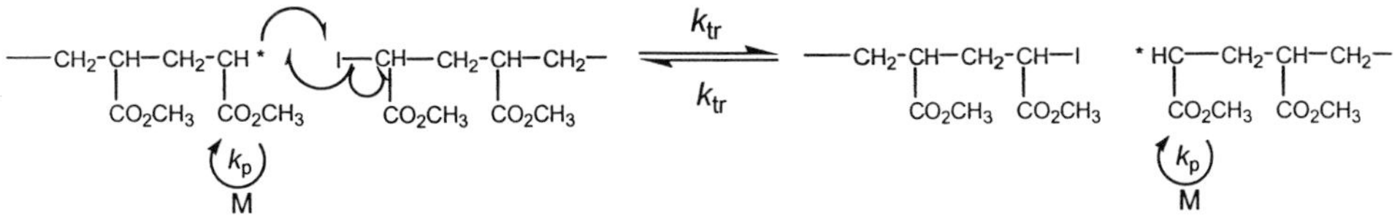

unsaturated methacrylate esters:(40)

Scheme 25

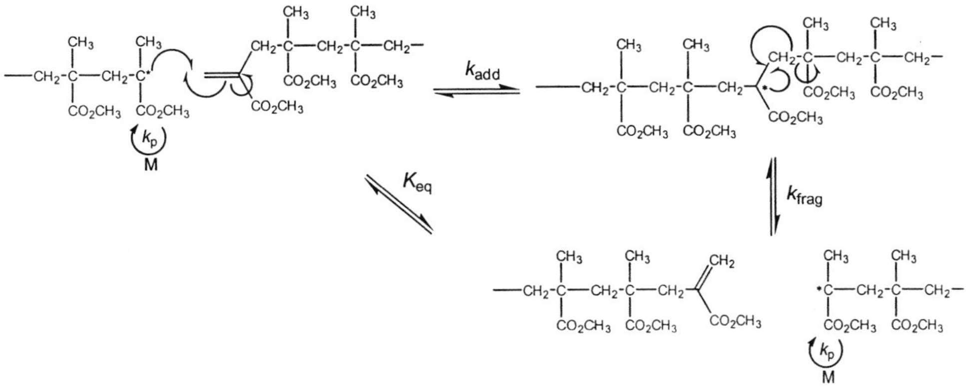

and dithioesters:(41)

Scheme 26

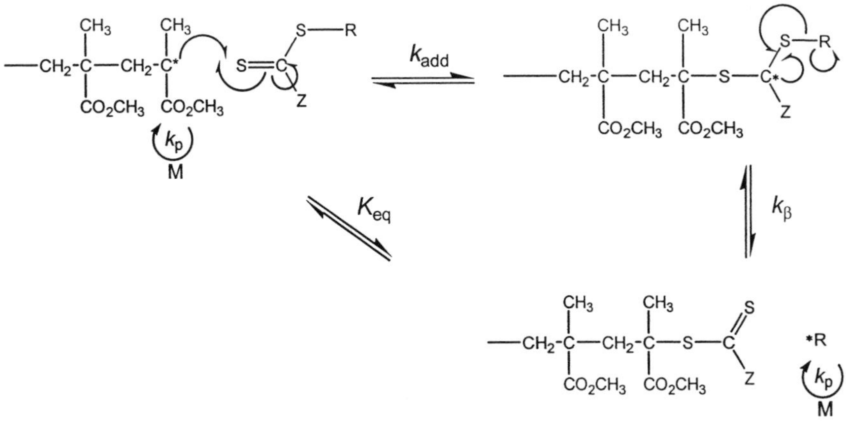

Polymerization rates in degenerative transfer are typically the same as in a conventional process however, molecular weights and polydispersities are much lower.(78) The degrees of polymerization are roughly defined by the ratio of the concentrations of converted monomer to the added transfer agent (more precisely a sum of concentrations of transfer agent and consumed initiator):

$$DP_n = \Delta[M]/([TA] + \Delta[I]) \tag{8}$$

Polydispersities do not depend on the concentration of transfer agent, since it defines both chain length and rate of deactivation:

$$M_w/M_n = 1 + (k_p/k_{tr}) (2/p - 1) \tag{9}$$

A key feature for degenerative transfer is the relative rate of transfer (k_{tr}) or of addition (k_{add}, usually fragmentation is faster than addition). Two factors determine the overall relative rate of degenerative transfer. One is the structure of the alkyl group in the initial transfer agent, the other one is that of transferable atom or group. It appears that for degenerative transfer, the only acceptable atom is iodine with the transfer coefficient in polymerization of styrene and acrylates being in the range of $k_{tr}/k_p \approx 2$ to 3. Degenerative transfer with bromine or chlorine was much too slow; the polymerizations behaved the same as without added transfer agent. Transfer coefficients for aryl chalcogenides are also relatively slow; rates for aryl sulfides correspond to that for chlorides, aryl selenides to bromides and potentially only tellurides could have sufficient transfer rates, similar to those for iodides.(79)

The other class of compounds are those with either C=C or C=S double bonds. Methacrylate derivatives transfer with rates similar to that of the propagation of methacrylates, and are successful only for the polymerization of methacrylates.(80) Due to steric effects the intermediate radical shown in Scheme 25 can not react directly with monomer but only fragmentate. Unfortunately, mono substituted alkenes such as styrenes and acrylates react with the intermediate radicals and give branched structures, i.e., there is inefficient fragmentation.

Among compounds with C=S double bonds, dithiocarbamates were originally used. This system was used by Otsu in the first studies of controlled radical polymerizations, and he termed them iniferters.(35) The main mode of action for these compounds was, however, a photochemical cleavage rather than bimolecular degenerative transfer (k_{tr}/k_p<0.1). Subsequent replacement of the electron donating group in dithiocarbamates (-NR$_2$) or xanthates (-OR) by an electron neutral (-Me, -Ph) or electron withdrawing (-CN) group increased enormously the relative rates of degenerative transfer to values of k_{tr}/k_p>100.(41) This new process, called reversible addition fragmentation transfer (RAFT) can be applied to the polymerization of many monomers including styrene, (meth)acrylates and vinyl benzoate as well as to the synthesis of new block copolymers. However, the efficiency of the block copolymer synthesis, as well as the consumption of the initial transfer agent depends strongly on the structure of the alkyl precursor. Thus, cumyl derivatives have been excellent transfer agents in RAFT but for example, isobutyrate derivatives were unsuccessful in polymerization of MMA. This example shows how important structure of the alkyl part may be.

Comparison of Various Methods of Controlling Radical Polymerization Currently, the three most efficient methods of controlling radical polymerization include nitroxides, ATRP and degenerative transfer. Each of these methods has advantages and also some limitations. The relative advantages and limitations of each method can be grouped into four categories. They include range of monomers, reaction conditions, activatable end groups and other additives such as catalysts, accelerators, etc.

Nitroxide mediated systems are best represented by TEMPO. TEMPO can be successfully applied only to styrene and copolymers due to its relatively small equilibrium constant. It yields mostly uncontrolled and unsaturated oligomers for acrylics and for methacrylates, requires the use of either special nitroxides with a higher equilibrium constant (phosphate derivatives) or those with a lower thermal stability (4-oxy TEMPO). Typical reactions are carried out in bulk and at high temperatures (>120 °C for TEMPO) because they are inherently slow. Polymerizations in solution, dispersion and emulsion have been also reported. As initiators, either a combination of conventional initiator and free nitroxide (1.3:1 ratio is apparently the best) or preformed alkoxyamines can be used. End groups in the dormant species are alkoxyamines and some unsaturated species formed by abstraction of β-H atoms or other inactive groups formed by side reactions, e.g. termination. Alkoxyamines are relatively expensive, the nitroxides are generally difficult to remove from the chain end, they are not commercially available and need to be synthesized. However, the process typically does not require a catalyst and is carried out at elevated temperatures. The polymerizations are usually slow, although some acceleration was reported in the presence of sugars, acyl compounds(81) and acids.(56)

ATRP has been used successfully for the largest range of monomers, although the polymerization of vinyl acetate and acrylic acids has not yet been successful. ATRP has been carried out in bulk, solution, dispersion and emulsion at temperatures ranging from −20 °C to 130 °C. Some tolerance to oxygen has been reported in the presence of zero-valent metals.(82) The catalyst must be available for the reaction and it should be sufficiently accessible (soluble) in the reaction medium. The catalyst is based on a transition metal which can regulate rate and polydispersities, can facilitate cross-propagation for the synthesis of difficult block copolymers, can scavenge some oxygen, but it should be removed or recycled from the final polymerization product. Perhaps the biggest advantage of ATRP is the inexpensive end group consisting of simple halogens. This is especially important for short chains due to the high proportion of the end groups. Additionally, there is a multitude of available initiators and macroinitiators for ATRP. Moreover, the halogen can be easily displaced with other useful functionalities using S_N2, S_N1, radical or other chemistries.

Degenerative transfer can be potentially used for any radically polymerizable monomer. However, reactions of vinyl esters are apparently more difficult and the RAFT of vinyl benzoate requires very high temperatures (T ~ 150 °C). It may be difficult to assure an efficient crosspropagation for some systems.(42) In principle, all classic radical systems can be converted to RAFT or to another degenerative transfer in the presence of efficient transfer reagents. The end groups are either alkyl iodides, methacrylates or thioesters. The latter are colored and can give some odor for low molar mass species and will require radical chemistry for removal and displacement.

Methacrylate oligomers are efficient only in the polymerization of methacrylates. No catalyst is needed for degenerative transfer but, in fact, the role of the catalyst is played by the radical initiator. It also means that the initiator may incorporate some undesired end groups and the amount of termination is governed by the amount of decomposed initiator. A potential disadvantage of degenerative transfer is that there is always a low molecular weight radical available for termination. By contrast, in the ATRP and TEMPO systems, at suffecent conversions only long chains exist, and, therefore, they terminate more slowly.

Thus, the main advantage of the nitroxide mediated system is the absence of any metal. TEMPO is applicable for styrene but new, more specialized nitroxides should be used for other monomers. ATRP may be especially well suited for low molar mass functional polymers due to the low cost of end groups and easier catalyst removal. It may be also very successful for the synthesis of "difficult" block copolymers and some special hybrids with end functionalities. However, it requires catalyst removal. Degenerative transfer and especially RAFT should be successful for the polymerization of many less reactive monomers and for the preparation of high polymers. Due to some limitations of the sulfur containing compounds the search for new efficient transferable groups should be continued.

New Materials by Controlled/"Living" Radical Polymerization

After all this discussion about radical polymerization and new methods to develop processes to obtain better control of the polymerization, the question remains: Why? Why should one use these exotic methods to polymerize vinyl monomers? The answer that comes first to mind is the replacement of anionic and cationic polymerization as the primary means of obtaining well-defined (co)polymers, by the radical polymerizations which are more tolerant of impurities, functional groups and are applicable to a wider range of monomers. However, this begs the question of why does one want well-defined (co)polymers.

Well-defined copolymers are generally recognized as polymers with molecular weights defined by $DP_n = \Delta[M]/[I]_o$, and with low polydispersities, say, $M_w/M_n < 1.3$ (an arbitrary figure). However, such homopolymers are of little interest commercially; in some instances, materials with broad molecular weight distributions are desired for various rheological reasons. What controlled/"living" polymerizations offer is the ability to prepare entirely new polymers with a myriad of compositions, architectures, and functionalities (Figure 3), with each polymer chain in the bulk having the same microstructure (composition, architecture and functionality), and not a random distribution of properties from chain to chain.

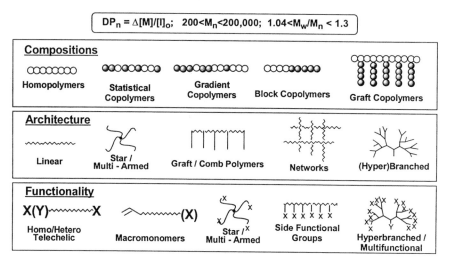

$$DP_n = \Delta[M]/[I]_o; \quad 200 < M_n < 200,000; \quad 1.04 < M_w/M_n < 1.3$$

Figure 3. Microstructures possible with controlled/"living" polymerizations.

<u>Compositions</u> When two or more monomers are combined and polymerized, statistical copolymers are formed where the relative compositions of the monomers in the polymer chain is a function of the reactivity ratios and the monomer feed. In conventional radical polymerization, high polymer is formed early in the reaction and then is irreversibly terminated. As one monomer is generally consumed faster than the other(s), there is a depletion of that monomer faster than the others from the monomer feed in the reactor. At higher conversions, that monomer will likely be present only in low amounts, while the other(s) will be present in higher amounts, which leads to polymers that contain lower (or none) amounts of the first monomer when compared to

the chains prepared early in the polymerization. This gradient of compositions from chain to chain can be overcome by continuously adding monomer(s) so that the monomer feed remains relatively stable throughout the polymerization.

In contrast, for controlled polymerizations, all chains grow at nearly the same rate, with little irreversible termination, although the relative rate of monomer consumption (based on the reactivity ratios) is nearly the same as in a conventional process. What is different is that the relative amount of monomer A vs. B in the polymer chains is not varied from chain to chain, but *along* the chains themselves. In these gradient copolymers, the composition gradually changes from one monomer to the other along the length of the chain. Such polymers have been prepared by both nitroxide based systems(83, 84) and by ATRP.(72, 85-87)

Instead of a gradual change in the composition, an abrupt transition from one monomer to another may occur as in segmented copolymers, i.e. block and graft copolymers. Block copolymers can be prepared in one of two manners: through the use of macroinitiators or by sequential addition of monomer. Macroinitiators can be prepared by a number of polymerization techniques, including controlled/"living" radical polymerization. In this case, a monomer is polymerized and the polymer is isolated then dissolved in a second monomer and used to initiate polymerization. In this manner, there is a very clean break between monomer units (blocks). Such a methodology has been used to prepare block copolymers as thermoplastic elastomers(88) and as amphiphilic copolymers.(88, 89) A second monomer can also be added at the end of the polymerization of the first monomer. This sequential addition of monomer may result in a slight taper of the transition from block A to block B if monomer A is not completely consumed. Novel materials may be developed by adjusting the length and degree of this taper. A special class of segmented copolymer are graft copolymers which can be prepared by use of a macroinitiator which contains multiple initiating sites *along* the polymer chain; initiation at these sites allows for the growth of polymer chains from the backbone.(88, 90, 91)

Architecture Another area where controlled/"living" radical polymerizations can make a significant contribution is in the development of polymers with unique architectures.

When an initiator site is incorporated into a monomer, branching of the polymer chain can be induced. When such functionalized monomers are homopolymerized, hyperbranched polymers are obtained.(92-97) When they are copolymerized with conventional monomers, polymers with random branching along the chain can be obtained.(90, 93) Polymerization of these monomers using techniques that do not consume the initiating sites for the controlled/"living" radical polymerization results in a polymer with initiating sites at every repeat unit. By using this polymer as a macroinitiator, very densely grafted polymer chains have been obtained. The macromolecules are very large (M_n=5,000,000 M_w/M_n=1.2) and the individual macromolecules have been resolved by atomic force microscopy with length in the range of 100 nm and width 10 nm (Fig. 4). Such macromolecules have been prepared by ATRP where the polymer has a methacrylic backbone with poly(butyl acrylate) grafts.(91)

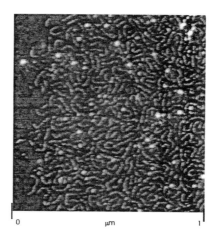

0   μm   1

Figure 4. AFM image of poly(butyl acrylate) brushes on mica.

<u>Functionality</u> Controlled/"living" radical polymerizations are important methods to produce polymers of lower molecular weight, but with high degrees of functionality.(68) Because of the precise control of the end groups in controlled radical polymerizations, this methodology is ideally suited to preparing telechelics. For example, poly(butyl

acrylate) with α, ω–hydroxyl groups can be used as a replacement for poly(ethylene glycol) in polyurethane synthesis.

A benefit of the relatively stable end groups of polymers prepared by controlled/"living" polymerizations, is that they can be isolated and stored with relative ease. Such is not the case for polymers prepared by ionic polymerizations; the active anion or cation will be quenched by advantageous moisture. This also allows one to modify polymers prepared by other methods so that they can become macroinitiators for controlled/"living" radical polymerization. Such "mechanism transformation" can be used to prepare a wide array of novel polymers; block copolymers of combinations of radically prepared polymers with those synthesized by step-growth polymerizations,(98, 99) ROMP,(100) cationic(101) and anionic polymerizations(102, 103) have been prepared.

Although this has been but a brief review of novel materials prepared using controlled radical polymerizations, one can easily see that, regardless of the type of controlled radical polymerization employed, these methodologies open the door to a wide range of novel polymers with unique properties. Only time will tell, but undoubtedly the question is not if such materials will find commercial uses, but one of when and how.

Conclusions

Radical polymerizations are widely used in industrial processes, accounting for the synthesis of nearly 50% of all polymeric materials. The widespread use of radical polymerization is due to its unique ability to easily and readily prepare high polymers from a variety of monomers, under relatively mild reaction conditions. Radical polymerizations are also simple to perform; they require only the absence of oxygen.

To extend the usefulness of radical polymerization, various systems have been developed to allow for the "control" of the polymerization such that termination and transfer processes can be avoided, or at least minimized. Towards this end, three main systems have shown the ability to solve this problem. These are the nitroxide mediated polymerization, atom transfer radical polymerization (ATRP) and radical addition-fragmentation transfer (RAFT). All three have their benefits and detractions, but each may be particularly suited for certain applications, i.e., high molecular weight polymers vs. low molecular weight oligomers, etc.

It has been demonstrated that novel compositions, architectures, and functionalized polymers can be readily prepared by using these methods. Although some functionality of the chains is lost due to unavoidable termination reactions, these materials may provide unique properties that will be good enough, or significant enough, to be used in new applications.

Literature Cited

1.   Moad, G.,Solomon, D. H. in "The Chemistry of Free Radical Polymerization"; Pergamon; Oxford, 1995.

2.   Szwarc, M. Nature, 1956, 176, 1168.

3.   Hsieh, H. L.,Quirk, R. P. in "Anionic Polymerization, Principles and Practical Applications"; Marcel Dekker, Inc.; New York, 1996.

4.   Kennedy, J. P.,Ivan, B. in "Designed Polymers by Carbocationic Macromolecular Engineering. Theory and Practice"; Hanser; Munich, 1992.

5.   Matyjaszewski, K. in "Cationic Polymerizations: Mechanisms, Synthesis and Applications"; Marcel Dekker; New York, 1996.

6.   Webster, O. Science, 1991, 251, 887.

7.   Matyjaszewski, K. J. Phys. Org. Chem., 1995, 8, 197.

8.   Greszta, D., Mardare, D.,Matyjaszewski, K. Macromolecules, 1994, 27, 638.

9.   "Controlled Radical Polymerization," Matyjaszewski, K., Ed.; ACS: Washington, D.C., 1998.

10.   Walling, C. in "Free Radicals in Solution"; Wiley; New York, 1957.

11.   Bamford, C. H. "Enc. Radical," Bamford, C. H., Ed.; Wiley: New York, 1988; Vol. 13.

12.   Bagdasarian, H. S. in "Theory of Radical Polymerization"; Izd. Akademii Nauk; Moscow, 1959.

13.   Fukuda, T., Ma, Y. D.,Inagaki, H. Macromolecules, 1985, 18, 17.

14.   Gilbert, R. G. Pure & Appl. Chem., 1992, 64, 1563.

15. Bresler, S. E., Kazbekov, E. N., Fomichev, V. N.,Shadrin, V. N. Makromol. Chem., 1972, 157, 167.

16. Mayo, F. R. Polym. Prepr. (Am. Chem. Soc., Div. Polym. Chem.), 1961, 2, 55.

17. van Herk, A. M. J. M. S.- Rev. Macromol. Chem. Phys., 1997, C37, 633-648.

18. Moineau, G., Minet, M., Dubois, P., Teyssié, P., Senninger, T.,Jérôme, R. Macromolecules, 1999, 32, 27.

19. Heuts, J. P. A., Pross, A.,Radom, L. J. Phys. Chem., 1996, 100, 17087.

20. Britton, D., Heatley, F.,Lovell, P. A. Macromolecules, 1998, 31, 2828.

21. Ahmad, N. M., Heatley, F.,Lovell, P. A. Macromolecules, 1998, 31, 2822.

22. Rizzardo, E., Chong, Y. K., Evans, R. A., Moad, G.,Thang, S. H. Macromol. Symp., 1996, 111, 1.

23. Haddleton, D. M., Topping, C., Kukulj, D.,Irvine, D. Polymer, 1998, 39, 3119.

24. Haddleton, D. M., Topping, C., Hastings, J. J.,Suddaby, K. G. Macromol. Chem. Phys., 1996, 197, 3027.

25. Xia, J.,Matyjaszewski, K. Macromolecules, 1997, 30, 7692.

26. Alfrey, T.,Price, C. C. J. Polym. Sci., 1947, 2, 101.

27. Jenkins, A. D. Europ. Pol. J., 1989, 25, 721.

28. Lovell, P. A.,El-Aasser, M. S. in "Emulsion Polymerization and Emulsion Polymers"; John Wiley and Sons; New York, 1997.

29. Matyjaszewski, K.,Müller, A. H. E. Polym. Prepr. (Am. Chem. Soc., Div. Polym. Chem.), 1997, 38(1), 6.

30. Solomon, D. H., Rizzardo, E.,Cacioli, P. USP 4,581,429, 1986.

31. Georges, M. K., Veregin, R. P. N., Kazmaier, P. M.,Hamer, G. K. Macromolecules, 1993, 26, 2987.

32. Steenbock, M., Klapper, M., Muellen, K., Bauer, C.,Hubrich, M. Macromolecules, 1998, 31, 5223.

33. Borsig, E., Lazar, M., Capla, M.,Florian, S. Angew. Makromol. Chem., 1969, 9, 89.

34. Braun, D. Macromol. Symp., 1996, 111, 63.

35. Otsu, T.,Yoshida, M. Makromol. Chem. Rapid Commun., 1982, 3, 127.

36. Wayland, B. B., Poszmik, G., Mukerjee, S. L.,Fryd, M. J. Am. Chem. Soc., 1994, 116, 7943.

37. Kato, M., Kamigaito, M., Sawamoto, M.,Higashimura, T. Macromolecules, 1995, 28, 1721.

38. Wang, J. S.,Matyjaszewski, K. J. Am. Chem. Soc., 1995, 117, 5614.

39. Matyjaszewski, K., Gaynor, S.,Wang, J. S. Macromolecules, 1995, 28, 2093.

40. Moad, C. L., Moad, G., Rizzardo, E.,Tang, S. H. Macromolecules, 1996, 29, 7717.

41. Chiefari, J., Chong, Y. K. B., Ercole, F., Krstina, J., Jeffery, J., Le, T. P. T., Mayadunne, R. T. A., Meijs, G. F., Moad, C. L., Moad, G., Rizzardo, E.,Thang, S. H. Macromolecules, 1998, 31, 5559.

42. Chong, B. Y. K., Le, T. P. T., Moad, G., Rizzardo, E.,Thang, S. H. Macromolecules, 1999, 32, 2071.

43. Greszta, D., D. Mardare,Matyjaszewski, K. Polym. Prepr. (Am. Chem. Soc., Div. Polym. Chem.), 1994, 35(1), 466.

44. Harwood, H. J., Christov, L., Guo, M., Holland, T. V., Huckstep, A. Y., Jones, D. H., Medsker, R. E., Rinaldi, P. L., Soito, T.,Tung, D. S. Macromol. Symp., 1996, 111, 25.

45. Davis, T. P., Kukulj, D., Haddleton, D. M.,Maloney, D. R. Trends Polym. Sci., 1995, 3(11), 365-73.

46. Benoit, D., Grimaldi, S., Finet, J. P., Tordo, P., Fontanille, M.,Gnanou, Y. Polym. Prepr. (Am. Chem. Soc., Div. Polym. Chem.), 1997, 38(1), 729.

47. Hawker, C. J., Benoit, D., Rivera, F., Chaplinski, V., Nilsen, A.,Braslau, R. Polym. Prepr. (Am. Chem. Soc., Div. Polym. Mater. Sci. Eng.), 1999, 80, 90.

48. Greszta, D.,Matyjaszewski, K. Macromolecules, 1996, 29, 7661.

49. Zhu, Y., Li, I. Q., Howell, B. A.,Priddy, D. B. ACS Symp. Ser, 1998, 685, 214.

50. Mardare, D.,Matyjaszewski, K. Polym. Prepr. (Am. Chem. Soc., Div. Polym. Chem.), 1994, 35(1), 778.

51. Devonport, W., Michalak, L., Malmstroem, E., Mate, M., Kurdi, B., Hawker, C. J., Barclay, G. G.,Sinta, R. Macromolecules, 1997, 30, 1929.

52. Hawker, C. J. J. Amer. Chem. Soc., 1994, 116, 11314.

53. Matyjaszewski, K., Gaynor, S., Greszta, D., Mardare, D.,Shigemoto, T. Macromol. Symp., 1995, 98, 73.

54. Fukuda, T., Terauchi, T., Goto, A., Ohno, K., Tsujii, Y., Miyamoto, T., Kobatake, S.,Yamada, B. Macromolecules, 1996, 29, 6393.

55. Greszta, D.,Matyjaszewski, K. J. Polym. Sci., Part A: Polym. Chem., 1997, 35, 1857-1861.

56. Georges, M. K., Veregin, R. P. N., Kazmaier, P. M., Hamer, G. K.,Saban, M. Macromolecules, 1994, 27, 7228.

57. Malmstrom, E., Miller, R. D.,Hawker, C. J. Tetrahedron, 1997, 53, 15225-15236.

58. Matyjaszewski, K., Patten, T.,Xia, J. J. Am. Chem. Soc., 1997, 119, 674.

59. Ohno, K., Goto, A., Fukuda, T., Xia, J.,Matyjaszewski, K. Macromolecules, 1998, 31, 2699.

60. Matyjaszewski, K. ACS Symp. Series, 1998, 685, 258.

61. Matyjaszewski, K., Davis, K., Patten, T. E.,Wei, M. Tetrahedron, 1997, 53, 15321.

62. Percec, V.,Barboiu, B. Macromolecules, 1995, 28, 7970.

63. Wang, J. S., Greszta, D.,Matyjaszewski, K. Polym. Mater. Sci. Eng., 1995, 73, 416.

64. Coca, S.,Matyjaszewski, K. Polym. Prepr. (Am. Chem. Soc., Div. Polym. Chem.), 1997, 38(1), 693.

65. Coca, S., Paik, H.,Matyjaszewski, K. Macromolecules, 1997, 30, 6513.

66. Gaynor, S. G.,Matyjaszewski, K. Macromolecules, 1997, 30, 4241.

67. Matyjaszewski, K. Macromol. Symp., 1998, 132, 85.

68. Matyjaszewski, K., Coessens, V., Nakagawa, Y., Xia, J., Qiu, J., Gaynor, S. G., Coca, S.,Jasieczek, C. ACS Symposium Series, 1998, 704, 16.

69. Matyjaszewski, K.,Wang, J.-S. Macromolecules, 1995, 28, 7901.

70. Gaynor, S. G., Qiu, J.,Matyjaszewski, K. Macromolecules, 1998, 31, 5951.

71. Xia, J., Johnson, T., Gaynor, S. G., Matyjaszewski, K.,DeSimone, J. M. Macromolecules, 1999, 32, 4802.

72. Arehart, S.,Matyjaszewski, K. Macromolecules, 1999, 32, 2221.

73. Matyjaszewski, K. Macromolecules, 1998, 31, 4710.

74. Matyjaszewski, K., Wei, M., Xia, J.,McDermott, N. E. Macromolecules, 1997, 30, 8161.

75. Granel, C., Dubois, P., Jerome, R.,Teyssie, P. Macromolecules, 1996, 29, 8576.

76. Ando, T., Kamigaito, M.,Sawamoto, M. Macromolecules, 1997, 30, 4507.

77. Woodworth, B. W., Metzner, Z.,Matyjaszewski, K. Macromolecules, 1998, 31, 7999.

78. Gaynor, S., Wang, J. S.,Matyjaszewski, K. Macromolecules, 1995, 28, 8051.

79. Curran, D. P. in "Comprehensive Organic Synthesis"; Pergamon; Oxford, 1991.

80. Krstina, J., Moad, G., Rizzardo, E., Winzor, C. L., Berge, C. T.,Fryd, M. Macromolecules, 1995, 28, 5381.

81. Malmström, E. E.,Hawker, C. J. Macromol. Chem. Phys., 1998, 199, 923.

82. Matyjaszewski, K., Coca, S., Gaynor, S. G., Wei, M.,Woodworth, B. E. Macromolecules, 1998, 31, 5967.

83. Fukuda, T., Terauchi, T., Goto, A., Tsujii, Y.,Miyamoto, T. Macromolecules, 1996, 29, 3050.

84. Hawker, C. J., Elce, E., Dao, J., Volksen, W., Russell, T. P.,Barclay, G. G. Macromolecules, 1996, 29, 2686.

85. Greszta, D.,Matyjaszewski, K. Polym. Prepr. (Am. Chem. Soc., Div. Polym. Chem.), 1996, 37(1), 569.

86. Greszta, D., Matyjaszewski, K.,Pakula, T. Polym. Prepr. (Am. Chem. Soc., Div. Polym. Chem.), 1997, 38(1), 709.

87. Greszta, D., Matyjaszewski, K.,Pakula, T. Polym. Prepr. (Am. Chem. Soc., Div. Polym. Chem.), 1997, 38, 709.

88. Gaynor, S. G.,Matyjaszewski, K. ACS Symposium Series, 1998, 685, 396.

89. Mühlebach, A., Gaynor, S. G.,Matyjaszewski, K. Macromolecules, 1998, 31, 6046.

90. Grubbs, R. B., Hawker, C. J., Dao, J.,Frechet, J. M. J. Angew. Chem. Int. Ed. Engl., 1997, 36, 270.

91. Beers, K. L., Gaynor, S. G., Matyjaszewski, K., Sheiko, S. S.,Moller, M. Macromolecules, 1998, 31, 9413.

92. Hawker, C. J., Frechet, J. M. J., Grubbs, R. B.,Dao, J. *J. Amer. Chem. Soc.*, 1995, 117, 10763.

93. Gaynor, S. G., Edelman, S. Z.,Matyjaszewski, K. *Macromolecules*, 1996, 29, 1079.

94. Gaynor, S. G., Kulfan, A., Podwika, M.,Matyjaszewski, K. *Macromolecules*, 1997, 30, 5192.

95. Gaynor, S. G., Matyjaszewski, K.,Muller, A. H. E. *Macromolecules*, 1997, 30, 7034.

96. Gaynor, S. G.,Matyjaszewski, K. *Macromolecules*, 1997, 30, 7042.

97. Matyjaszewski, K., Pyun, J.,Gaynor, S. G. *Macromol. Rapid Commun.*, 1998, 19, 665.

98. Gaynor, S. G.,Matyjaszewski, K. *Macromolecules*, 1997, 30, 4247.

99. Nakagawa, Y., Miller, P. J.,Matyjaszewski, K. *Polymer*, 1998, 39, 5163.

100. Coca, S., Paik, H.,Matyjaszewski, K. *Macromolecules*, 1997, 30, 6513.

101. Coca, S.,Matyjaszewski, K. *Polym. Prepr. (Am. Chem. Soc., Div. Polym. Chem.)*, 1997, 38(1), 693.

102. Kobatake, S., Harwood, H. J., Quirk, R. P.,Priddy, D. B. *Macromolecules*, 1998, 31, 3735.

103. Acar, M.,Matyjaszewski, K. *Macromol. Chem. Phys.*, 1999, 200, 1094.

STEP-GROWTH POLYMERIZATION

TIMOTHY E. LONG

Department of Chemistry, Virginia Polytechnic Institute and State University, Blacksburg, VA 24061-0212

S. RICHARD TURNER

Polymers Technology Group, Research Laboratories, Eastman Chemical Company, Kingsport, TN 37662

Introduction
Requirements for Successful Step-Growth Polymerization
Polyesters
Polycarbonates
Polyamides
Polyurethanes
Polyimides

Introduction

Step-growth polymers are produced by the reactions of functional groups of monomers in a stepwise progression from dimers, trimers, etc. to eventually form high polymer. When this polymerization process is accompanied by the elimination of small molecules during the reaction step, the process is called polycondensation. An excellent review of the history of step-growth polymers is contained in Morawetz's "Polymers, The Origins and Growth of a Science" (1). The first example of step-growth polymer dates back over 100 years with the reaction by Baeyer of resorcinol and formaldehyde (2). Years later Baekeland patented and commercialized these chemistries when they were done under pressure and with suitable fillers present (3).

The struggle of Staudinger to gain acceptance of macromolecules is well documented and perhaps is a reason why most of the early major discoveries and development of the science involved in step-growth polymers were made in industry. Carothers, armed with the concept of macromolecules, as championed by Staudinger, envisioned that large molecules could be synthesized by the direct addition of diols and diacids to form polyesters and diamines and diacids to form polyamides (4,5). Out of this pioneering work by Carothers (and Flory who joined Carothers at DuPont) came the basic principles of condensation polymerizations (6, 7) and the initial commercial linear synthetic step-growth polymer, Nylon 6,6.

It is interesting that polyesters were the original focus of Carothers' research, but were abandoned since the melting points of the resulting polyesters were too low to be useful. Whinfield (8) from ICI discovered that changing the aliphatic diacids, which were the basis of Carothers' studies, to an aromatic diacid (terephthalic acid) led to a high melting polymer (256 °C) when it was condensed with ethylene glycol. This discovery led to the development of poly (ethylene terephthalate) (PET) which, even today, is one of the fastest growing, if not the fastest growing, step-growth polymer.

Another of the basic step-growth polymer families, polyurethanes and polyureas, arose from basic work at Farbenfabriken Bayer by Bayer (9). These chemistries were investigated as alternatives to the polyamides of DuPont. Polycarbonates and polyimides are two other important step-growth polymers of considerable commercial importance that will be discussed in this chapter.

Today the commercial volume of basic step-growth polymers is approximately 85 billion pounds (1996) and is growing at a healthy rate (10). Progress continues to be made along many fronts with step-growth polymers. Unlike polyolefins and other vinyl polymers, where catalyst and process changes can lead to different types of monomer enchainments with concomitant mechanical and other property changes, significant variation in the properties of step-growth polymers requires the incorporation of new monomers into the polymer backbone. For the majority of the engineering resin applications, where most step-growth polymers find applications, the development and commercialization of new step-growth monomers and polymers are challenging tasks.

This chapter will review the basics of the chemistries that govern the build-up of molecular weight in step-growth polymerizations. This is followed by summary discussions on the major classes of step-growth polymers i.e., polyesters, polycarbonates, polyamides, polyurethanes, and polyimides. Other step-growth polymers such as poly(arylene ethers), polyketones, and polysulfones are also important commercially, but are not discussed in this chapter.

Requirements for Successful Step-Growth Polymerization

As discussed in the Introduction, Carothers (and Flory) derived several mathematical equations that facilitate our understanding of the experimental and manufacturing requirements for successful preparations of high molecular weight macromolecules (6,7). Equation 1 defines the number average degree of polymerization (X_n) in terms of the extent of reaction (p) in a difunctional step growth polymerization process, and it is quickly appreciated that very high levels of monomer conversion (p > 0.99) lead to higher number average degrees of polymerization and molecular weight.

Consequently, the ability to achieve high (greater than 99%) degrees of functional group conversion is essential for the preparation of high molecular weight polymers. In most instances, the thermal and mechanical properties of many thermoplastics increase linearly and eventually plateau with an increase in molecular weight. Molecular weight is known to influence many thermal and mechanical properties of commercial products including crystallization rate, strain-induced crystallization, impact properties, toughness and other mechanical tensile and flexural properties.

$$X_n = 1/1\text{-}p \qquad\qquad \text{Eq. 1}$$

where X_n = number average degree of polymerization
                p = the degree of functional group conversion

       The importance of achieving high degrees of functional group conversion and high molecular weight products points to several other related requirements for successful step-growth polymerization. Industrial and economic considerations require that the polymerizations proceed quickly in order to achieve high degrees of conversion at economically feasible rates. In addition, perfect reaction stoichiometries or potential processes that in-situ generate perfect stoichiometries are required. Consequently, careful monomer charging and ultra-pure monomer sources are required in step-growth polymerization processes. The presence of side reactions also will deleteriously effect the reaction stoichiometry resulting in the inability to prepare high molecular weight products. Although one could intuitively propose the use of longer polymerization times to achieve higher molecular weight products, the importance of relatively slow side reactions during the polymerization process becomes more important when using either longer polymerization times or higher polymerization temperatures. For example, the manufacture of PET is accompanied by the formation of diethylene glycol (DEG), vinyl esters, and acetaldehyde during the polymerization process. DEG is incorporated into the chain during the polymerization and its presence in the backbone has been shown to negatively effect the chemical resistance, crystallization rate, thermal stability, and ability for strain-induced crystallization in many PET product applications.

       As discussed earlier in the introduction, step-growth polymerizations proceed via the reaction of any two species in the reaction mixture, and step-wise reactions result in the formation of high molecular weight polymers. In many instances, step-growth polymerization results in the formation of a condensate, and earlier nomenclatures referred to step-growth polymerization processes as condensation polymerization processes. Although the current scope of suitable chemistries has eliminated the strict requirement for the formation of a condensation by-product, many commercial step-growth polymers including polyesters and polyamides do involve an equilibrium between the monomeric reactants and polymerization by-products. Consequently, the presence of this equilibrium requires that the synthetic process completely eliminate the polymerization by-product in order to achieve high molecular weight.

       A salient feature of step-growth polymerization processes is the gradual increase in molecular weight throughout the polymerization process. This observation is a direct result of the step-wise addition of reactive species (monomers, dimers, trimers, tetramers, etc.) to form high molecular weight. In contrast, high molecular products are obtained very quickly in chain polymerization processes and are typically accompanied by highly exothermic reactions. Consequently, the gradual increase in melt viscosity and the absence of highly exothermic reactions facilitates the use of bulk polymerization processes in step-growth polymerization. For example, polyesters and polyamides are manufactured commercially in the absence of solvents and the final products are directly usable without further isolation and purification. Although molecular weights gradually increase throughout most of a step-growth polymerization process, the final stages of the polymerization process involve the rapid increase in molecular weight. It is well known that the melt viscosity for thermoplastics increases with the 3.4 power of the weight average molecular weight. It is not surprising, therefore, that significant attention has been devoted in recent years to the development of novel agitation and reactor designs to facilitate the transport of the viscous melt and the transport of polymerization by-products in the latter stages of step growth polymerization processes (11).

The requirements for a successful step growth polymerization process are summarized as follows:
(1) High conversion reactions (greater than 99%)
(2) Absence of deleterious side reactions resulting in the loss of functionality
(3) Controlled functional group stoichiometry
(4) High monomer purities
(5) Efficient removal of any polymerization condensates
(6) Relatively fast polymerization rates

Although high molecular weight is often required for the maximization of thermal and mechanical properties, the synthesis of difunctional oligomers (less than 10,000 g/mole) is accomplished in step-growth polymerization using monofunctional reagents or a stoichiometric excess of a difunctional monomer. The difunctional oligomers are suitable starting materials for the preparation of crosslinked coatings, adhesives, or segmented block copolymers. Due to the reactive nature of many polymers prepared using step-growth polymerization processes, reactive endgroups and internal reactive functionalities offer the potential for subsequent deriviatization, chain extension, or depolymerization. For example, the depolymerization of polyesters is easily accomplished with the addition of a suitable difunctional acid or glycol (12). In addition, polyester and polycarbonate blends are easily compatibilized in a twin screw extruder to prepare optically clear coatings (13).

It should also be noted that commercial processes often require that the polymerization process minimize the use of solvents and the in-situ formation of salts. In most instances, the final polymeric products are directly charged to extrusion and molding operations without further expensive purification steps.

It is easy to understand, based on the above discussion, the serious restrictions that severely limit the number of suitable organic reactions that have been used for the successful preparation of high molecular weight products via step-growth polymerization. Although many synthetic organic reactions appear to be suitable for the preparation of macromolecules via a step-growth polymerization process, most organic reactions do not meet all the necessary requirements and have not been utilized in commercial products. A challenge remains to broaden the scope of suitable polymerization chemistries and processes leading to new families of high performance polymeric products.

<u>Polyesters</u>

<u>Commercial importance and applications.</u> As stated in the Introduction, polyesters were apparently the first targets of Carothers (4), but were not attractive for commercial applications because of low melting points. The insertion of aromatic groups into the backbone using terephthalic acid was the discovery that overcame this deficiency (8). From this beginning poly(ethylene terephthalate) (PET) has by far become the most important polyester with over 60 billion pounds produced in 1997. The first major application area for PET was in textile fibers and today about 70% of the PET produced is used in fibers. The fastest growing area is for application in food and beverage containers consuming around 20% of the PET produced with a growth rate of 10-12 %/year. The remaining 10 % are consumed as extruded film (photographic, magnetic tape, etc.) and in engineering plastics, primarily as glass fiber reinforced composites (14). The high growth rate in the container area is based on ease of processing, barrier, aesthetic properties of the containers, recycle ease (12), and cost. The reheat blow molding process, used to produce containers, utilizes the unique strain hardening characteristics of PET and permits the high speed manufacture of high volumes of containers. Copolyesters containing small amounts of 1,4-cyclohexanedimethanol (CHDM) or isophthalic acid (IPA) are often used instead of the homopolymer of PET in the bottle applications because these copolyesters process easier.

Other, specialty polyesters are increasingly becoming important as commercial step-growth polymers. The commercialization of 2,6-dimethylnaphthalene dicarboxylate (DMN) has made the commercialization of poly(ethylene naphthalate) (PEN) possible. This polyester has a Tg of 125 °C compared to PET with a Tg of 80 °C. In addition its barrier to oxygen in containers is approximately 5 times that of PET (15). PEN homopolymer, copolymers, and blends are currently being investigated in refillable containers, beer containers, and small volume containers requiring enhanced barrier properties. Future obstacles to the larger penetration of PEN and copolymers in the marketplace will include the necessity for less expensive monomer grades in commodity beverage container markets and the wider application of 2,6-naphthalene monomer in higher value, niche applications. In addition, higher processing temperatures required for PEN for higher melt viscosities often hamper manufacturing and molding operations.

Opportunities in packaging of oxygen sensitive foods have led to considerable interest in *bis*-hydroxyethyl resorcinol (HER), which has been shown to provide significantly improved barrier (16). Other specialty monomers that have been recently commercialized include 1,4-cyclohexanedicarboxylic acid (CHDA), which has found applications in tough coatings (17) and 1,3-propane diol (PDO). PDO has been available for many years, but recent developments in catalytic (18) and biochemical (19) processes have lowered the cost of PDO to permit the commercialization of poly(trimethylene terephthalate) (PTT) as a carpet fiber (20).

In addition to applications in containers and fibers, semi-crystalline polyesters, which are reinforced with glass fibers (GFR), find many uses in the automotive and electronics industry. Poly(butylene terephthalate) (PBT), which was originally commercialized in the mid 1970s as a replacement for phenolic resins in automotive applications, is used extensively in these kinds of applications. Some semi-crystalline PBT is used unfilled, but the majority is used in GFR composites. The unique combination of toughness, fast crystallization rates, resistance to creep, and ease of compounding in additives such as flame retardants has led PBT to be the polyester of choice for glass filled applications in the electronics, automotive, and fiber cable industries, despite having a lower melting point than PET. The global demand for PBT in 1997 exceeded 7 billion lbs. and the long-term average annual growth rate is predicted to be approximately 6.2% (21). GFR PTT has been shown to have some unique properties when compared to GFR PBT and PET and proposed to have considerable potential for applications in the composite area (22). The polyester of TPA and CHDM, poly(1,4-cyclohexyldimethylene terephthalate, (PCT), which has a melting point of 290 °C, has been found to be particularly useful, as a GFR composite, in electronic and automotive applications where higher temperatures are needed (23).

Amorphous polyesters are also important in polyester applications. PET can be extruded into amorphous sheet (aPET) and this clear, transparent sheet can be thermoformed into useful articles. Copolyesters that contain mixtures of EG and CHDM with TPA or IPA (glycol modified PET or PETG) have proven to be excellent clear sheeting materials for signs, etc. and excellent materials with wide processing windows for thermoformed articles (24).

Polyesters derived in part from unique rigid aromatic monomers including hydroquinone and biphenyl derivatives offer the potential for new families of high temperature, high performance polyester resins. Although liquid crystalline polyesters (LCPs) were discovered in the late 1960s (25), this family of engineering thermoplastics continues to receive intense academic and industrial attention. In addition to inherent flammability resistance, LCPs offer exceptional moldability due to the shear-induced alignment of the rigid polyester backbones. Therefore LCPs have become very important in the manufacture of small parts for the electronics and other industries. Ticona (formerly Hoechst-Celanese) has pioneered the commercialization of all-aromatic liquid crystalline polyesters based on 2-hydroxy-6-naphthaoic acid (HNA) and p-hydroxybenzoic acid (PHB) (26). Many other industrial competitors including Eastman Chemical and DuPont have recently entered this technical arena based on recent bullish projections for LCP market growth.

Synthesis. Most commercial polyesters are prepared by the direct esterification of diacids with diols or the transesterification of methyl esters and diols. Many other polyesterification reactions are known to fit the rigorous reaction requirements to give high molecular weight polymers (27). Acid chlorides, in particular, are used to prepare polyesters that are not stable in the melt phase. Other chemistries using silyl derivatives can also be employed (28). Equation 2 depicts the metal catalyzed polymerization of terephthalic acid and ethylene glycol. Typical manufacturing operations, for semicrystalline polyesters, involve the preparation of a low molecular weight oligomer in the melt phase followed by polymerization in the solid state (29). The solid state polymerization process offers two advantages. First, the melt viscosities for the high molecular weight products are avoided. Secondly, the solid state process occurs at lower temperatures than the melt phase leading to reduced levels of polymerization by-products such as acetaldehyde. In fact, the solid state process also provides a mechanism for the devolatization of by-products that are formed in the higher temperature, melt phase polymerization process. A variety of metal catalysts have been employed to obtain PET and other polyesters at commercially acceptable rates with minimization of side reactions (30).

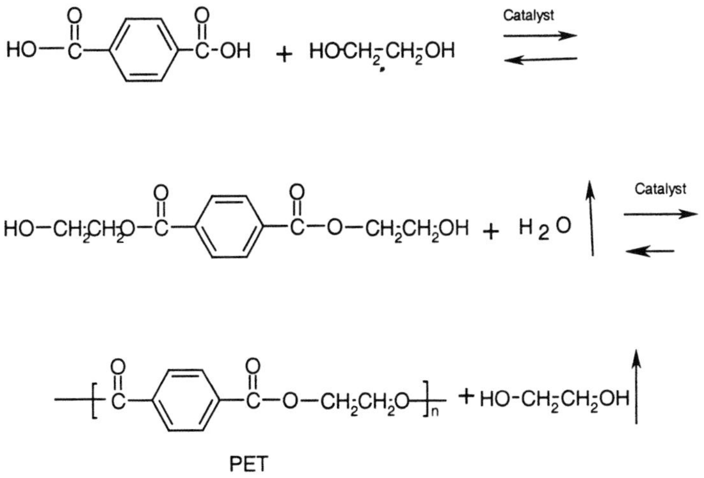

$$HO-\overset{O}{\underset{\|}{C}}-\underset{}{\bigcirc}-\overset{O}{\underset{\|}{C}}-OH \;+\; HOCH_2CH_2OH \;\underset{\longleftarrow}{\overset{Catalyst}{\longrightarrow}}$$

Eq. 2

$$HO-CH_2CH_2O-\overset{O}{\underset{\|}{C}}-\underset{}{\bigcirc}-\overset{O}{\underset{\|}{C}}-O-CH_2CH_2OH \;+\; H_2O \;\Big\uparrow\; \underset{\longleftarrow}{\overset{Catalyst}{\longrightarrow}}$$

$$\Big[\overset{O}{\underset{\|}{C}}-\underset{}{\bigcirc}-\overset{O}{\underset{\|}{C}}-O-CH_2CH_2O\Big]_n \;+\; HO-CH_2CH_2OH\Big\uparrow$$

PET

Ester formation based on the reaction between an aromatic carboxylic acid and an aromatic phenol is not suitable for polyester manufacture, and all-aromatic polyesters require the in-situ acetylation of the aromatic phenols. Condensation of the acetate and aromatic carboxylic acid readily occurs at 250 °C with the liberation of acetic acid, and the polymerization mechanism is often referred to as acidolysis. Although the polymerization mechanism appears straightforward, Hall and coworkers (31) have recently elucidated the very complex nature of this polymerization process. Equation 3 summarizes the polymerization chemistry to prepare all-aromatic polyesters via the acidolysis route. Industrial attention has focused on the development of suitable manufacturing processes that can handle the corrosive reaction environment and the high polymerization temperatures required in viscous melt phase. Significant attention has also been devoted to the preparation of liquid crystalline polyesters derived from aliphatic glycols and biphenyl dicarboxylic acids.

$$HO-\overset{O}{\underset{\|}{C}}-\langle C_6H_4\rangle-\overset{O}{\underset{\|}{C}}-OH \quad + \quad HO-\langle C_6H_4\rangle-OH \quad \xrightarrow{\;X\;}$$

Eq.

$$HO-\overset{O}{\underset{\|}{C}}-\langle C_6H_4\rangle-\overset{O}{\underset{\|}{C}}-OH \;+\; CH_3\overset{O}{\underset{\|}{C}}-O-\langle C_6H_4\rangle-O-\overset{O}{\underset{\|}{C}}-CH_3 \;\rightleftharpoons$$

$$-\overset{O}{\underset{\|}{C}}-\langle C_6H_4\rangle-\overset{O}{\underset{\|}{C}}-O-\langle C_6H_4\rangle-O- \quad + \quad CH_3\overset{O}{\underset{\|}{C}}{\sim}OH \;\uparrow$$

Polycarbonates

Commercial importance and applications. Polycarbonates had their industrial birth almost 2 decades later than the other major step-growth polymers, polyamides, polyesters, and polyurethanes with their commercialization based on the work of Fox and Schnell. (32,33)

Since the early 1960's polycarbonates have become an extremely important and fast growing clear amorphous thermoplastic for injection molding and extruded sheet products. The workhorse polycarbonate resin is based on bisphenol A (BPA PC) and has a unique set of properties, which include a very high glass transition temperature of 145 °C and excellent toughness. These properties along with the low color and excellent clarity of products produced from BPA PC have propelled consumption to approximately 2.6 billion lbs. in 1996 with a growth rate of 8-10% a year (34).

Conventional BPA PC has found wide acceptance as the plastic of choice for applications where a combination of optical properties (color, transparency), impact resistance, and resistance to thermal flow are important (35).

Specific applications include the replacement for glass in areas where impact resistance is important, e.g. in areas where the resistance to vandalism is needed. BPA PC finds many applications in transportation such as instrument panel covers in automobiles and other vehicles. Currently major research and development efforts are underway to develop scratch resistant coatings for polycarbonate to permit the replacement of glass in automotive window applications (36).

Polycarbonate resins are used widely as substrates for data recording. Compact disc (CD) technology is based on BPA PC as the substrate. Specially designed polycarbonates, based on bisphenols other than BPA, have been studied for the requirements for advanced optical data storage systems. Higher heat deflection temperatures as well as better flow and lower birefringence are all desired features for these applications (37).

Considerable research has been done using different bisphenols to change these important properties(38,39) but generally flow and impact suffer as the Tg is raised.

The enchainment of 4,4"-(3,3,5-trimethylcyclohexylidene)diphenol into the polycarbonate backbone has been found to raise the Tg from 150 °C to 239 °C without sacrificing the flow characteristics and impact properties of BPA PC (40).

Synthesis. There are several excellent reviews describing laboratory and industrial synthetic process for polycarbonates (41). Polycarbonates are prepared in solution or interfacially using bisphenol and phosgene in the presence of a base to react with the hydrochloric acid that is liberated. Currently the commercial process of choice is the interfacial method using the sodium salt of bisphenol A with phosgene with methylene chloride serving as the organic solvent (Eq. 4). Due to the toxicity of phosgene and methylene chloride and the problems with disposal of the large amounts of sodium chloride that are generated during production of polycarbonate, there has been a very large interest in developing melt phase processes (42, 43, 44). Catalyzed transesterification based on the reaction of bisphenol A with diphenyl carbonate is accomplished at temperatures that go as high as 320 °C under vacuum in order to obtain the high molecular weights needed for good mechanical properties. Various lithium salts and other additives are used as catalysts. Considerable research continues in this area to minimize the degradation reactions, color formation, etc. that accompany these high temperature melt phase processes.

Eq. 4

Polyamides

Commercial importance and applications. As stated earlier Nylon 6,6 was the first of the modern day thermoplastics with properties based on an intentionally designed molecular structure. Nylon nomenclature is based on the number of carbon atoms in the diacid and the diamine, e.g. Nylon 4,6 is the polyamide of tetramethylene diamine and adipic acid. Polyamides remain very important commercially today with a host of various backbones based on different diamines and diacids being available in the market place. Close to 6 billion lbs. of synthetic polyamides (this number includes Nylon 6, which is produced by a ring opening polymerization) were produced in 1998 (45). Nylons are used in a wide variety of applications. Originally nylons were used in fiber applications and this still constitutes a large market for nylons. Their high tensile strength and good dyeability make them superior in fiber performance. By varying the diamine and the diacid in aliphatic nylons a wide variety of polyamides with much different property profiles can be prepared. Systematic variation in melting points, glass transition temperatures, water uptakes, and mechanical properties has led to the ability to design these polyamides with very specific sets of properties for applications and has resulted in aliphatic nylons finding a wide range of applications (46).

Partially aromatic nylons are generally based on mixtures of terephthalic acid and isophthalic acid with various linear diamines (47). Very high melting points and excellent resistance to solvents and other chemicals are important characteristics of these kinds of nylons. This has led to their development as useful high temperature thermoplastics. Thus partially aromatic nylons find many applications where high temperature resistance and solvent resistance are required, e.g. automotive under the hood applications and electronic applications where resistance to high temperature processing is important. In many cases these applications are as a glass fiber reinforced (GFR) composite. Crystalline nylons often form superior GFR composites because of the ability to tailor glass surface chemistries to give strong interactions with the polar amide functionalities along the backbone. In addition these rigid structures have been found to be readily toughened with functionalized low Tg olefins, which are designed to interact or react with terminal amine groups in the polyamide (48).

The high level of hydrogen bonding in nylons often leads to excellent barrier properties. Nylons therefore find use as food packaging resins. A partially aromatic nylon (I) based on adipic acid and m-xylylene diamine is of considerable current interest because of its high barrier to oxygen and to carbon dioxide. This material has great potential as a high barrier layer when sandwiched between two poly(ethylene terephthalate) layers in containers used for beer and other foods that easily spoil when

exposed to oxygen (49).

I

Nylons that crystallize very slowly and thus can be used as transparent films are based on copolyamide structures containing branched diamines or mixtures of diamines and diacids. Such transparent structures also find use in the food packaging areas.

All aromatic nylons, or aramids, possess the highest tensile strengths of any man-made synthetic step-growth polymer and thus now enjoy a very high level of usage in the high performance fiber area. Due to the extremely high melting points of aramids (decomposition occurs in most cases before melting), all synthesis and processing of aramids is done via solution based processes. The first materials of this class were the all meta linked polymers, i.e. m-phenylenediamine and isophthalic acid (Nomex, DuPont). Fibers based on Nomex are important because of their fire and chemical resistance. The *para* linked analogues (*p*-phenylenediamine and terephthalic acid) have extremely high tensile moduli and have become fibers of considerable importance in high performance composites (Kevlar, DuPont and Twaron, Akzo) (50). (There are many good references on aramids, this one details the history of the development of the synthesis and properties of aramids.)

Synthesis. There are many excellent reviews that have been written on the synthesis of nylons (51). Most aliphatic and partially aromatic nylons are synthesized commercially by melt phase polycondensation processes. In cases where crystalline polyamides are prepared and melt phase viscosities limit desired molecular weight formation, solid state polymerization processes have been employed (29). For nylons, the formation of the "nylon salt" is very useful in purifying the monomers and obtaining the exact stoichiometry to be able to get a high degree of polymerization (Eq. 5). Unlike polyesters, the equilibrium in a polyamide condensation lies far to the right and thus the polymerizations can be charged stoichiometrically and initially run under pressure to react all the diamine and maintain the stoichiometry.

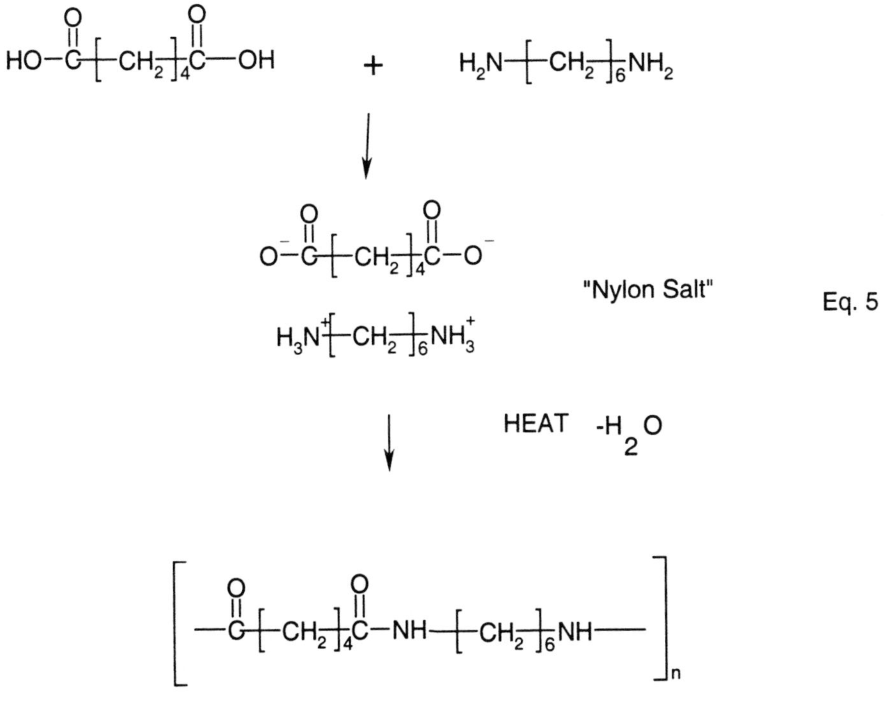

$$HO-\overset{O}{\overset{\|}{C}}-[CH_2]_4\overset{O}{\overset{\|}{C}}-OH \quad + \quad H_2N-[CH_2]_6NH_2$$

$$\downarrow$$

$$^-O-\overset{O}{\overset{\|}{C}}-[CH_2]_4\overset{O}{\overset{\|}{C}}-O^-$$

$$H_3\overset{+}{N}-[CH_2]_6\overset{+}{N}H_3$$

"Nylon Salt"

Eq. 5

$$\downarrow \quad \text{HEAT} \quad -H_2O$$

$$\left[-\overset{O}{\overset{\|}{C}}-[CH_2]_4\overset{O}{\overset{\|}{C}}-NH-[CH_2]_6NH- \right]_n$$

Nylon 6,6

Solution phase polymerization processes can be used to prepare step-growth polyamides and, for aramids, is the only way that these materials can be prepared since most examples decompose before they melt. There are several published routes to make polyamides in solution. The most utilized is based on starting with the diacid chloride and the diamine. It is necessary to use a base to take up the liberated acid in these reactions in order to keep from protonating the diamine. Interfacial techniques, with the acid chloride in the non polar solvent and the diamine in water with a base such as sodium hydroxide present, is a facile way to quickly obtain high molecular weight polyamides. Aramids are prepared directly from the acid chlorides and diamines in strong polar aprotic solvents such as N-methylpyrrolidone (NMP) or dimethylformamide (DMF). It is necessary to use salts such as lithium chloride or calcium chloride to assist in solublizing the aramid so that it does not precipitate and stop the polycondensation at low molecular weight (52).

A solution method that involves the direct polymerization of aromatic diacids and aromatic diamines, as well as of aromatic amino acids, has been studied extensively (53). When, for example, an aromatic diacid is reacted with an aromatic diamine in NMP which contains an aryl phosphite such as triphenyl phosphite and a base such as pyridine in the presence of a salt such as lithium chloride, high molecular weight aramids can be formed (54). These kinds of polymerizations have been found to be very sensitive to the conditions and the monomer/polymer structure and have not found commercial success.

A much different solution process involves the use of diaryl halides and diamines to prepare aramids via the Heck carbonylation reaction (55). This route to aramids was first disclosed (56) using dibromo aromatics. Relatively low molecular weight polymers were formed. Diiodo aromatics were found to react at a much higher rate and to give polymers of much higher ultimate molecular weight, presumable due to fewer side reactions (57) (Eq. 6). This process eliminates the use of corrosive and moisture sensitive acid chlorides, but requires the use of expensive palladium catalysts.

Polyamides continue to show steady growth in applications in the marketplace. Recently introduced ethylene-carbon monoxide alternating copolymers (58) have been targeted at many basic polyamide applications because of a combination of high melting point and good solvent resistance along with good barrier properties. Based on the inexpensive starting materials, these copolymers should have significant cost advantages over nylons if "world-scale" volume production is reached. Despite this and other threats, thermoplastic and amorphous polyamides will likely continue to be attractive commercial thermoplastic engineering resins in the future. The great flexibility in designing specific structures with specific properties from readily available building blocks by well-proven condensation techniques will assist in the continued growth of these materials.

$$\text{(aryl diiodide)} + \text{NH}_2\!\!-\!\!\text{Ar}\!\!-\!\!\text{O}\!\!-\!\!\text{Ar}\!\!-\!\!\text{NH}_2 \quad\xrightarrow[\text{DMAC, Base, CO}]{\text{PdCl}_2\text{L}_2} \quad \left[\text{polymer}\right]_n \qquad \text{Eq. 6}$$

Polyurethanes

<u>Commercial importance and applications.</u> O. Bayer in Germany originally investigated polyurethanes and polyureas as alternatives to DuPont's polyamides in the 1930's (59). The polyureas turned out to be very difficult to process and characterize, in contrast to polyurethanes, which could readily be molded and spun into fibers. In 1998 approximately 9 billion lbs. of polyurethanes (PURs) were produced for three major classes of applications in rigid and flexible foams, elastomers, and coatings (60).

The great flexibility in choosing the starting polyisocyanate and the polyol leads to the capability to design polyurethanes with a wide-range of properties. Most of the flexible foam is based on toluene diisocyanate (TDI) with various polyols. These foams are used primarily for cushioning applications, e.g. car seats, furniture cushions, bedding, etc. The technology for blowing these foams, flame retarding, stabilizing, etc. is very involved and key to the enormous commercial success. Rigid PUR foams generally are based on polymeric methylenediphenyl isocyanates (PMDI) and are used as insulation in transportation vehicles, appliances, etc. These foams are characterized by their dimensional stability, structural strength, and insulation performance (61). The polyols most widely used are generally based on polyether or polyester backbones.

Polyurethane elastomers are based on hard-soft segment type polymeric structures and can exist as cast elastomers or as thermoplastic elastomers. These elastomers generally possess good chemical and abrasion resistance and maintain their properties over wide temperature ranges. The hard segments that phase separate in the elastomer are primarily based on methylenediphenyl isocyanate (MDI).

The reversible nature of aromatic isocyanate and alcohol reactions has been exploited by the commercial development of engineering thermoplastic polymers that are designed to break down at melt temperatures to form an easily processable low viscosity material and then reform as the polymer solidifies (62). The low melt viscosity achieved by this reversible linkage allows the injection molding of smaller parts and decreases the cycle time of making parts. This basic concept has been extended to "thermally reversible" polyesters (63).

Synthesis. The synthesis of polyurethanes is an example of a step-growth polymerization that is not an actual condensation since no small molecules are eliminated during the reaction. This reaction is shown in Eq. 7. Urethane formation can be done in solution, in bulk, or interfacially (64). The reactions are very fast and can proceed far below room temperature at high rates. Catalysts have been developed to allow the reaction rates to be varied from seconds to hours. The inclusion of these catalysts is very important in many applications that involve the reactive processing of the monomers. Another way the reactivity of these systems is controlled is by blocking the isocyanate group with a group that comes off on heating to regenerate the reactive isocyanate group. This chemistry has found considerable application in coating technologies.

$$HO-R-OH \ + \ O{=}C{=}N-R'-N{=}C{=}O \longrightarrow \left[-R-O-\overset{\overset{O}{\|}}{C}-\overset{\overset{H}{|}}{N}-R'-\overset{\overset{H}{|}}{N}-\overset{\overset{O}{\|}}{C}-O- \right]_n$$

Eq. 7

Since isocyanates can react with many nucleophiles the control of purity of the monomers is very important to obtain the desired structures. The presence of water leads to amine formation and the elimination of carbon dioxide. This reaction can lead to urea links in the backbone and the released carbon dioxide can serve as a foaming agent. An excellent review has been published of the myriad of chemistries that are involved in polyurethane technologies (65).

Polyimides

Commercial importance and applications. Efforts to prepare aromatic polyimides via the imidization of a soluble poly(amic acid) precursor were initiated in 1956 in the DuPont laboratories and within only one month, A. L. Endrey prepared the first poly(amic acid) film (66). Research efforts at DuPont were fueled by the presence of parallel research efforts in novel diamines and high temperature polyamides. Polyimides have since received significant attention over the past three decades in the aerospace and electronics industries due to the exceptional thermal and chemical stability of the rigid heterocyclic backbone. Polyimides are recognized as one of the most thermally stable organic polymeric materials available today and research activities dealing with polyimide homo- and co-polymers continue for many high performance applications. Although DuPont's Kapton polyimides were initially developed in the late 1950s and commercialized in late 1965, this family of high temperature polymers continues to receive significant research attention and many review articles have been published that summarize their interesting structure-property relationships (67,68,69). Recent efforts have focused on the control of the dielectric constant and water sensitivity for polyimides in microelectronics applications and the improvement of solution and melt processability while maintaining thermal stability in composite applications (70). Other polyimide applications include gas separation membranes, photosensitive materials, electronic packaging, and Langmuir-Blodgett films. These very diverse applications are due to low relative permittivity and high breakdown voltages in combination with their chemical resistance, tough and flexible mechanical performance, and thermal stability exceeding 450 °C (71). Other Kapton applications have included aerospace wire and cable insulation, substrates for flexible printed circuits, electrically conducted films, and various applications requiring flame resistance.

Segmented block copolymers containing polyimides and poly(dimethyl siloxanes) have received intense attention during the past two decades for applications ranging from adhesives and composites to circuit boards and protective coatings (72,73). In addition, NASA has devoted significant attention to the utility of polyimides as new aerospace materials due to their intrinsic stability to both high-energy radiation and aggressive atomic oxygen. In general, polyimides derived from dianhydrides and diamines produce insoluble products after imidization and are generally processed by casting the poly(amic acid) intermediate onto a suitable substrate and subsequently heating to induce quantitative imidization. Recent advances include the preparation of thermoplastic polyimides and subsequent melt processing. General Electric has devoted significant attention to the commercial manufacture of Ultem polyether imide and it is often recognized as a premiere soluble and processible thermoplastic material.

Chemistry and Synthetic Methodologies. Condensation polyimides are classically prepared via the addition of an aromatic dianhydride to a diamine solution in the presence of a polar aprotic solvent such as NMP, DMAc and DMF at 15-75 °C to form a poly(amic acid) (74). The poly(amic acid) is either chemically or thermally converted to the corresponding polyimide via cyclodehydration. The general chemistry for this two-stage, step-growth polymerization process for the preparation of Kapton polyimide is depicted in Equation 8. It is important to note that the formation of the poly(amic acid) is an equilibrium reaction and attention must be given to ensure that the forward reaction is favored in order to obtain high molecular weight poly(amic acids). If the final polyimide is insoluble and infusible, the polymer is generally processed in the form of the poly(amic acid). Caution must be exercised when working with classical poly(amic acid) solutions due to their hydrolytic instability, and shelf life is limited unless properly stored at low temperatures. This is due to the presence of an equilibrium concentration of anhydride and their susceptibility to hydrolytic degradation in solution. On the other hand, poly(amic diesters) can be stored for indefinite periods of time without degradation due to the inability to form an intermediate carboxylate anion and have been utilized reproducibly in microelectronics applications (75). Earlier studies have shown that the formation of tri- and tetramethylesters of dianhydrides increased the likelihood of N-alkylation side reactions and a corresponding decrease in the molecular weight and mechanical properties of the final product. A similar side reaction has been observed in attempts to prepare polyamide esters using dimethyl esters of terephthalic acid (76).

Melt processible thermoplastic polyimides are prepared by the addition of flexible units, bulky side groups, or as mentioned earlier, flexible difunctional oligomers. Examples of these modifications include GE's Ultem (ether units), Amoco's Torlon (amide units), Hoechst-Celanese's P150 (sulfone units), and GE's Siltem (siloxane segments). Ultem polyimide is manufactured by General Electric and is an injection moldable thermoplastic poly(ether imide). This commercial product exhibits high mechanical properties including modulus and strength, excellent ductility, and high thermal stability. Although polyimides have been prepared using a myriad of other synthetic

Eq. 8

methodologies including aryl coupling with palladium catalysts, poly(amic silylesters), silylethers with activated halides, and trans-imidization, this chapter has focused on only two of the commercial polyimides in the marketplace today. Many specialty polyimides for composite applications have also evolved including LARC-TPI and CPI (Mitsui Toatsu), and Hoechst-Celanese's fluorinated polyimides, Due to the tremendous scope of polyimide research and applications, several excellent comprehensive texts have been devoted to this family of high temperature polymers.

Literature Cited

1. Morawetz, H. "Polymers, The Origin and Growth of a Science"; John Wiley and Sons, 1985.
2. Baeyer, A.v. Ber.1872, 5, 1094.
3. Baekeland, L. H. U.S. Patent 942,699, 1907.
4. Carothers, W. H.; Hill, J. W. J. Amer. Chem. Soc. 1932, 54, 1559.
5. Carothers, W. H. U.S. Patents 2,130,947 and 2.130, 948 (to DuPont), 1938.
6. Carothers, W. H. Trans. Farad. Soc. 1936, 32, 44.
7. Flory, P. J. J. Amer. Chem. Soc. 1936, 58, 1877.
8. Whinfield, J. R.; Dickinson, J. T. Br. Patent 578,079, 1946.
9. Bayer, O. Angew. Chem.1947, 59A, 257.
10. This number is an estimate obtained by adding together the 1996 worldwide consumption of the major step-growth resin discussed in this report. The consumption values were all obtained from SRI International reports.
11. Jones, E. B.; Burch, R. R., US Patent 5,602,199 (to DuPont), 1997.
12. Scheirs, J. "Polymer Recycling: Science, Technology and Applications," John Wiley and Sons Ltd. Baffins Lane, Chichester, West Sussex, 1998.
13. Barnum, R. S.; Barlow, J. W.; Paul, D. R. J. Appl. Polym. Sci. 1982, 27, 4065.
14. SRI Consulting, World Petrochemicals 1998, January.
15. Schiller, P. R.: Spec. Polyesters '95, Proc. 1995, 319.
16. Hashimoto, M.; Kaneshige, N. US Patent 5,115,047 (to Mitsui Petrochemical), 1992.
17. Johnson, L.K.; Sade, W. T. Proc. Water-Borne, Higher-solids, Powder Coat. Symp. 1991, 65.
18. Powell, J. B.; Slaugh, L. H.; Foschner, T. C.; Lin, J.; Thomason, T. B.; Weider, P. R.; Semmple, T. C.; Arhancet, J. P.; Fong, H. L.; Mullin, S. B.; Allen, K. D.; Eubanks, D. C.; Johnson, D. W. US Patent 5,77,182 (to Shell Oil Co.), 1998.
19. Gatenby, A. A.; Haynie, S. L.; Nagarajan, V.; Ramesh, V.; Nakamura, C. E.; Payne, M. S.; Picataggio, S. K.; Dias-Tores, M.; Hsu, A. K.; Lareau, R. D. PCT Int. Appl. WO 9821339 (to DuPont and Genencor), 1998.
20. Chuah, H. H.; Brown, H. S.; Dalton, P. A. International Fiber Journal, 1995, October, 50.
21. Van Berkel, R. W. M.; Van Hartingsveldt, E. A. A.; Van Der Sluijs, C. L. Plast. Eng. (N.Y.), Handbook of Thermoplastics, 1997, 465.
22. Dangayach, K; Chuah, H.; Gergen, W.; Dalton, P.; Smith, F. Antec97, 1997, 1010.
23. Martin, E. V.; Kibler, C. J.; Man-Made Fiber, Sci. Technol. 1968, 3, 83 and Minnick. L. A.; Seymour, R. W. PCT WO910502 (to Eastman Kodak Company), 1991.
24. Light, R. R.; Seymour, R. W. Polym. Eng. Sci. 1982, 22, 14.
25. Jackson, W. J., Jr.; Kuhfuss, H. F. J. Polym. Sci., Polym. Chem. Ed. 1976, 14 2043.
26. Yoon, H. N.; Charbonneau, L. F.; Calundann, G. W. Adv. Mater. 1992, 4,3, 206.
27. Pilati, F. *"Polyesters"* in "Comprehensive Polymer Science", Allen,G.; Bevington, J. C., Eds., Pergamon, Vol. 5, 1989; Chapter 17.
28. Kricheldorf, H. R. in "Silicon in Polymer Synthesis", Kricheldorf, H. R., ed.; Springer-Verlag: Berlin, Heidleberg, New York, 1996, Chapter 5.

29. Pilati, F. *"Solid-state Polymerization,"* in "Comprehensive Polymer Science", Allen, G.; Bevington, J. C. Eds., Pergamon, Vol. 5, 1989; Chapter 13.
30. Wilfong, R. E. J. Polym. Sci. 1961, 54, 385.
31. Han, X.; Williams, P. A.; Padias, A. B.; Hall, Jr., H. K.; Linstid, H. C.; Lee, C.; Sung, H. N. Macromolecules 1996, 29, 8313. Zimmerman, H.; Kim, N.T. Polym. Eng. Sci. 1980, 20, 680.
32. Christopher, W. F.; Fox, D. W. "Polycarbonates"; Reinhold, New York,1962.
33. Schnell, H. "Chemistry and Physics of Polycarbonates"; Interscience, New York, 1964.
34. Ring,K.-L.; Janshekar, H.; Takei, N "Chemical Economics Handbook"; September 1997, SRI International.
35. Sikdar, S. K. Chemtech 1987,112.
36. Katsamberis, D.; Browall, K.; Iacovangelo, C.; Neumann, M.; Morgner, H. Proc.-Int. Conf. Org. Coat.: Waterborne, High Solids, Powder Coat., 23rd 1997, 271.
37. Kampf, G.; Freitag, D.; Fendler, G.; Sommer, K. Polymers for Advanced Technologies 1992, 3,169.
38. Morbitzer, L.; Grigo, U. Angew. Makromol. Chem. 1988, 162, 87.
39. Stueben, K.C. J. Polym. Sci., Part A3 1965, 2309.
40. Freitag, D.; Westeppe, U. Makromol. Chem., Rapid Commun. 1991,12, 95 and Kampf, G.; Freitag, D.; Fengler, G. Kunstoffe 1992, 82, 5, 385 .
41. Clagett, D. C.; Shafer, S. J. *"Polycarbonates"* in "Comprehensive Polymer Sci.", Allen G.; Bevington, J. C., Eds. Pergamon, Vol. 5, 1989; Chapter 20, 345.
42. Brunelle, D. US Pat. 4321356, (to General Electric), 1982.
43. Kim, Y. Choi, K. Y. J. Appl. Polym. Sci. 1993, 49, 747.
44. Fischer, T.; Bachmann, R.; Hucks, U.; Rhiel, F.; Kuehling, S. Eur. Pat. Appl.96-104205, (to Bayer) 1996.
45. Davenport, R. E.; Feneleon, S.; Takei, N. "Chemical Economics Handbook"—SRI International, 1998.
46. Heym, M. Die Angew. Makrom. Chem. 1997, 244, 67.
47. Mewborn, Jr., J. W. Modern Plastics Mid-November 1996, B-54.
48. Oshinski, A. J.; Deskkula, H.; Paul, D. R. J. Appl. Polym. Sci. 1996, 61, 623.
49. Masahiro, H. Plast. Eng. 1988, 44,1, 27.
50. Morgan, P. W. Chemtech 1979, 316.
51. Gayman, R. J.; Sikkema, D. J. in *"Aliphatic Polyamides"* in "Comprehensive Polymer Science", Allen, G.; Bevington, J. C., Eds., Pergamon, Vol. 5, 1989; Chapter 21, 357.
52. Vollbracht, L. *"Aromatic Polyamides"* in "Comprehensive Polymer Science", Allen, G.; Bevington, J. C., Eds. Pergamon, Vol. 5, 1989; Chapter 22, 375.
53. Yamazaki, N.; Higashi, F.; Kawabata, J. J. Polym. Sci., Polym. Chem. Ed. 1974, 12, 2149.
54. Krigbaum, W. R.; Kotek, R.; Mihara, Y.; and Preston, J. J. Polym. Sci. Polym. Chem. Ed. 1985, 23, 1907.
55. Heck, R. F. "Palladium Reagents in Organic Synthesis", Academic Press, New York, 1985.
56. Yoneyama, M.; Kakimoto, M.; Imai, Y. Macromolecules 1988, 21, 1908.

57. Turner, S. R.; Perry, R. J.; Blevins, R. W. <u>Macromolecules</u>, 1992, <u>25</u>, 4819. Perry, R. J.; Turner, S. R.; Blevins, R. W. <u>Macromolecules</u>, 1993, <u>26</u>, 1509.
58. Ash, C. E. <u>Intern. J. Polymeric Mater</u>. 1995, <u>30</u>, 1.
59. Bayer, O. <u>Angew. Chem.</u>, 1947, <u>A59</u>, 275.
60. Chinn, H.; Jakobi, R.; Mori, S. in "Chemical Economics Handbook"—SRI International, 1996. Connolly, E.; Kalat, F. P.; Sakuma, Y. in"Chemical Economics Handbook"—SRI International, 1996.
61. Schmelzer, H. G.; Taylor, R. P. <u>Modern Plastics</u> November 1996, B-17.
62. Moses, P. J.; Chen, A. T.; Ehrlich, B. S. <u>ANTEC'89</u>, 1989, 860.
63. Markle, R. A.; Brusky, P. L.; Creameans, G. E. US Patent 5,097,010 (to Battelle Memorial Institute), 1992.
64. Frisch, K. C.; Klempner, D. *"Polyurethanes"* in "Comprehensive Polymer Science", Allen, G.; Bevington, J. C., Eds., Pergamon, Vol. 5, 1989; Chapter 24, 413.
65. Backus, J. K.; Blue, C. D.; Boyd, P. M.; Cama, F. J.; Chapman, J. H.; Eakin, J. L.; Harasin, S. J.; McAfee, E. R.; McCarty, C. G.; Nodelman,N. H.; Rieck, J. N.; Schmelzer, H. R.; Squiller, E. P. *"Polyurethanes"* in "Encyclopedia of Polymer Science and Engineering", Vol. 13, Mark, H. F.; Bikales, N. M.; Overberger, C. G.; Menges, G. Eds. Wiley-Interscience, New York, 1988.
66. Sroog, C. E. in "Polyimides: Fundamentals and Applications", Ghosh, M. K.; Mittal, K. L., eds., Marcel Dekker, New York, Basel, Hong Kong, 1996; Chapter 1 and Endrey, A. L. U.S. Patent 3,410,826 (to DuPont) 1969.
67. Mittal, K. L. , ed., "Polyimides: Synthesis, Characterization, and Applications", Plenum Press, New York, London, 1984 and Ghosh, M. K.; Mittal, K. L., eds., "Polyimides: Fundamentals and Applications", Marcel Dekker, New York, Basel, Hong Kong, 1996.
68. McGrath, J. E.; Dunson, D. L.; Mecham, S. J.; Hedrick, J. L. <u>Adv. Poly. Sci</u>. 1999, <u>140</u>, 62.
69. Wilson, D.; Stenzenberger, H. D.: Hergenrother, P. M. eds.,"Polyimides", Blackie, Glasgow, 1990.
70. Yamada, Y. <u>High Perform. Polym</u>. 1998, <u>10</u>,1, 69.
71. Takekoshi, T. *"Synthesis of Polyimides,"* in "Polyimides: Fundamentals and Applications" Ghosh, M. K.; Mittal, K. L. eds, Marcel Dekker, New York, Basel, Hong Kong, 1996; Chapter 7.
72. Arnold, C. A.; Summers, J.D.; McGrath, J.E. <u>Polym. Eng. Sci</u>. 1989, <u>29</u>(2) 1413.
73. Gilman, J. W.; Schlitzer, D.S.; Lichtenhan, J. D. <u>J. Appl. Polym. Sci</u>. 1996, <u>60</u>, 4, 591.
74. Sillion, B. *"Polyimides and Other Heteroaromatic Polymers*, "in "Comprehensive Polymer Science", Allen, G.; Bevington, J. C. Eds., Pergamon, Vol. 5, 1989; Chapter 30.
75. Labadie, J. W.; Hedrick, J. L. <u>Electron. Compon. Technol. Conf</u>. 1990, <u>40</u>[th] ,1, 706.
76. Turner, S. R. unpublished results.

IONIC POLYMERIZATION

RUDOLF FAUST* AND HELMUT SCHLAAD[‡]

* Department of Chemistry, University of Massachusetts Lowell
  One University Avenue, Lowell, MA 01854, USA
[‡] Max-Planck-Institute of Colloids and Interfaces, Colloid Department
  Am Muehlenberg, D-14476 Golm, Germany

Introduction
Anionic Polymerization
Cationic Polymerization
Combination of Anionic and Cationic Polymerization

Introduction

Historical Background

Although modern polymer science is considered to be founded by Staudinger in 1920 [1], cationic polymerization dates back to 1839, when the first cationic polymerization of styrene was achieved [2]. Anionic polymerization developed much later mainly by the pioneering work of Ziegler in 1928, who also reported first on terminationless anionic polymerization [3]. Ironically, this discovery was not appreciated until 28 years later when Szwarc introduced the concept of living polymerization and outlined its ramifications [4]. Although the discovery of living cationic ring opening polymerization followed shortly, cationic polymerization of vinyl monomers matured more slowly. It was a long standing view that living carbocationic polymerization is inherently unattainable due to the instability of the growing carbocationic centers. It is a matter of record that history proved this notion wrong: the first living cationic polymerization of isobutyl vinyl ether was reported in 1984 [5], and that of isobutylene in 1985 [6]. Since then, rapid advances have been made toward the synthesis of well-controlled materials in ionic polymerizations.

Definition of Living Polymerization

Living anionic and cationic polymerizations are chain polymerization processes that proceed in the absence of chain transfer to monomer and irreversible termination [3,7,8]. The irreversibility of termination is important since contrary to conventional living anionic polymerization, in living cationic polymerizations and contemporary anionic polymerization of (meth)acrylates the concentration of active species is often very small. Most of the chain ends are in a dormant, inactive form rapidly equilibrating with the active species. Although the term "living polymerization" was originally used to describe polymerizations in the absence of chain breaking reactions, a polymerization with reversible chain breaking reactions would show all the characteristic features of living polymerizations, i.e., a linear increase of M_n with conversion and linear first order plots. When initiation and exchange reactions are relatively fast compared to propagation, polymers with molecular weights controlled by the [monomer]/[initiator] ratio and narrow molecular weight distribution (MWD) are obtained ($M_w/M_n \sim 1.0$).

Diagnostic proof for the absence of chain transfer to monomer may be obtained by plotting M_n *vs.* conversion of polymerization. A linear M_n *vs.* conversion plot starting at the origin proves that the over-all concentration of polymer chains remains constant during the polymerization, i.e., chain transfer to monomer is absent. Propagation is usually first order in monomer; the first order plot is linear when initiation is rapid and the concentration of growing center remains constant, i.e., termination is absent.

Living polymerization is a most valuable technique in polymer synthesis. In addition to the controlled and uniform size of the polymers, living polymerizations provide the simplest and most convenient method for the preparation of block copolymers and functional polymers.

Anionic Polymerization

An anionic polymerization generally demands an inert atmosphere or vacuum techniques since the initiators and active chain ends are sensitive to moisture, oxygen, and carbon dioxide [9,10]. Whether the polymerization of a specific monomer then proceeds in a living and controlled manner depends substantially on the proper choice of initiator, counterion, solvent, temperature, and additives [11,12]. For example, the polymerization of a styrene monomer at 50 °C with lithium as the counterion is free of side reactions in cyclohexane, whereas transfer and termination reactions occur in toluene or tetrahydrofuran (THF) unless working at low temperatures. In the case of anionic polymerization of polar vinyl monomers or heterocyclic monomers, the presence of electrophilic functional groups may lead to side reactions of the anionic chain ends, like the intramolecular Claisen-type condensation of the ester enolate end group of a growing poly(meth)acrylate chain with the pen-penultimate ester carbonyl group ('back-biting'). In order to prevent termination during (meth)acrylate polymerization, the reaction conditions have to be optimized very rigorously and additives, which alter the kinetics of the reaction and the reactivity of the active sites, have to be employed. However, conditions for a living and controlled polymerization have been found for a large variety of monomers, among them styrenes, vinylpyridines, dienes, (meth)acrylates, oxiranes, lactones, and cyclic siloxanes – a recent overview is given in [11].

In the following two sections, the use of anionic polymerization of vinyl monomers in the synthesis of well-defined polymers and in commercial applications is described.

Synthesis of Well-Defined Polymers

Since the living anionic polymerization systems generate stable active chain ends, they can be used to produce a large number of functionalized polymers, block copolymers, and star, and graft (co)polymers. Some basic procedures are presented here, more detailed information is available in [11,13].

Functionalized Polymers

Chain end-functionalized and asymmetric telechelic polymers can be prepared by using lithiated initiators with protected functional groups and by end-quenching living polymer solutions with specific electrophiles. Symmetric telechelic polymers result from the coupling of living, functionalized polymer chains with e.g. dialkylsilyl halides or from the specific quenching of difunctionally initiated polymerizations (cf. section 2.1.2).

The success of the quenching reaction with a specific electrophile depends considerably on the nucleophilicity of the living chain ends and whether or not side reactions can be prevented. The use of functionalized initiators ensures a quantitative functionalization of the polymer, but the availability of such initiators is rather limited. In addition, most of the functional groups of interest are not stable in the presence of anionic sites and therefore require the use of suitable protecting groups [14]. The use of silyl halides and 1,1-diphenylethylenes (DPE) carrying functional groups provides a general and versatile method to obtain polymers with high functionalities. Since the addition of DPE derivatives to organolithium compounds leads again to a carbanionic species (Scheme 1), it is a living functionalization reaction which can be used to prepare functionalized initiators and in-chain or end functionalized polymers [15].

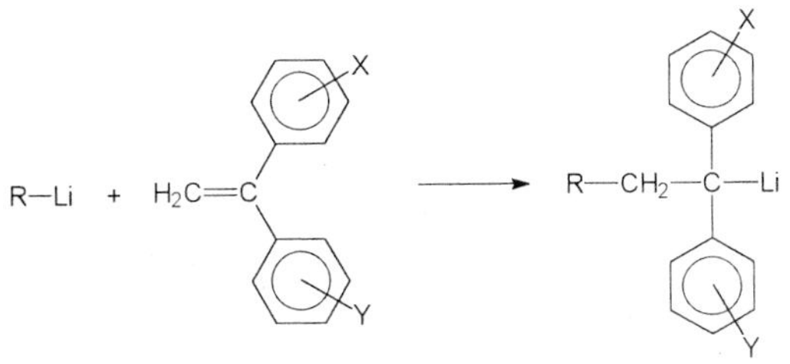

Scheme 1. Reaction of lithium organyls with DPE derivatives.

All the methodologies described above have been employed to synthesize a large number of functionalized and telechelic polymers with e.g. hydroxy [16–21], amino [22–29], and carboxy end groups [30,31] and macromonomers [32–38].

Block, Star, and Graft Copolymers

Linear block copolymers

One of the most important features of living polymerizations is the synthesis of linear di- and triblock copolymers by sequential monomer addition. For the successful design and synthesis of block copolymers, not only the proper reaction conditions (solvent, additives, and temperature) but also the mode of monomer addition is of importance. In general, the cross-over from one living chain end to an other occurs efficiently only when the newly formed active species exhibits equal or lower reactivity; the relative reactivities of the anionic sites can be estimated by their conjugate acid pK_a values [39]. For example, to prepare a diblock copolymer of styrene (pK_a of toluene: ~43) and methyl methacrylate (pK_a of ethyl acetate: ~30) [40], styrene has to be polymerized first and methyl methacrylate afterwards because an ester enolate chain end would not add styrene.

Basically, there are three synthetic routes to prepare linear triblock copolymers: (i) three-step sequential monomer addition, (ii) two-step sequential monomer addition and subsequent coupling, and (iii) difunctional initiation and two-step sequential monomer addition. 1,4-*Bis*(chloromethyl)benzene, Me_2SiCl_2, I_2 [41], or terephthaloyl chloride [42], for example, can be used as coupling agents, and sodium naphthalene [43] or the adduct of 2 moles *sec*-butyllithium (sBuLi) to 1,3-*bis*(1-phenylethenyl)benzene (*bis*-DPE) (Scheme 2) [44] as difunctional initiators.

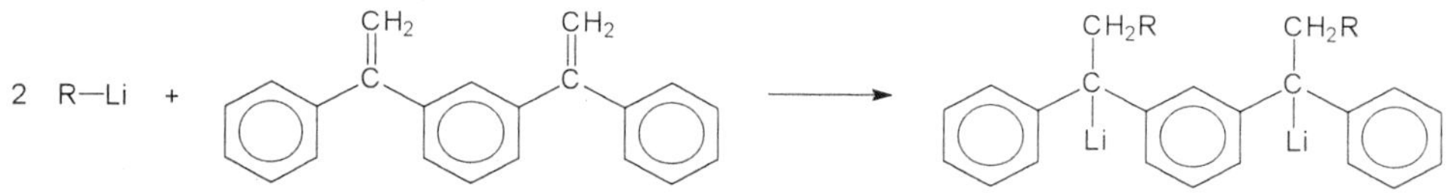

Scheme 2. Addition of lithium organyls to *bis*-DPE.

The ABC triblock copolymers PSt-*b*-PB-*b*-PMMA (St: styrene, B: butadiene, MMA: methyl methacrylate) [45] and PSt-*b*-PE-*b*-PεCL (E: ethylene, εCL: ε-capro-lactone) [46], for example, were prepared by a three-step sequential anionic poly-merization. The final products displayed narrow MWDs (M_w/M_n < 1.12) and were not significantly contaminated with residual diblock copolymer. ABA triblock copolymers such as PSt-*b*-PMMA-*b*-PSt [29] and PαMeSt-*b*-PDMS-*b*-PαMeSt (αMeSt: α-methyl-styrene, DMS: dimethylsiloxane) [47] can only be obtained by coupling the respective living diblock precursor polymers. The purity of these triblock copolymers was reported to be > 95%. For the synthesis of ABA triblock copolymers such as PEO-*b*-PSt-*b*-PEO (EO: ethylene oxide) [48] and PtBMA-*b*-PI-*b*-PtBMA (tBMA: *tert*-butyl methacrylate, I: isoprene) [49], difunctional initiation and two-step sequential monomer addition is necessary. PSt-*b*-PB-*b*-PSt instead can be prepared by any of the three methods [50]. A listing of other anionically prepared block copolymers is given in [51].

Star Copolymers

One synthetic route to prepare star-branched polymers is to link living polymers with multifunctional electrophiles ("arm first" method) like 1,2,4-*tris*(chloromethyl)benzene,

PCl$_3$ (f = 3), dimethyl phthalate, SiCl$_4$ (f = 4), hexachlorodisilane (f = 6), or tetra(methyl-dichlorosilylethane)silane (f = 8); for more linking agents, see [41]. Silyl halides in particular have been used extensively to obtain regular and heteroarm star polymers based on polystyrenes and polydienes [52–68]. The efficiency of a given silyl halide generally decreases with increasing steric bulkiness of the living polymer chain end in the order PB-Li > PI-Li > PSt-Li. Although these agents are more effective than the respective halogenated hydrocarbons, side reactions still may occur – this is summarized and discussed elsewhere [69].

p/m-Divinylbenzene has been used to link living PSt chain ends and, with lower efficiencies, polydienes [70–76]. The reaction involves three steps, namely the cross-over to divinylbenzene, the block copolymerization of divinylbenzene, and the linking of carbanionic chain ends with pendant vinyl groups. However, the functionality of divinylbenzene depends substantially on the reaction conditions and may vary between 3 and 56, i.e., the control over the exact number of arms in the star polymer is rather limited [77].

The living polymerization of divinylbenzene with BuLi in benzene [78] or potassium naphthalene in THF [79,80] yields a soluble microgel which then has been utilized as a polyfunctional initiator for styrene ("core first" method, Scheme 3). By linking polymeric organolithium compounds with divinylbenzene [81], heteroarm star copolymers such as *star*-PSt-*star*-PtBMA have been synthesized [73,74,79].

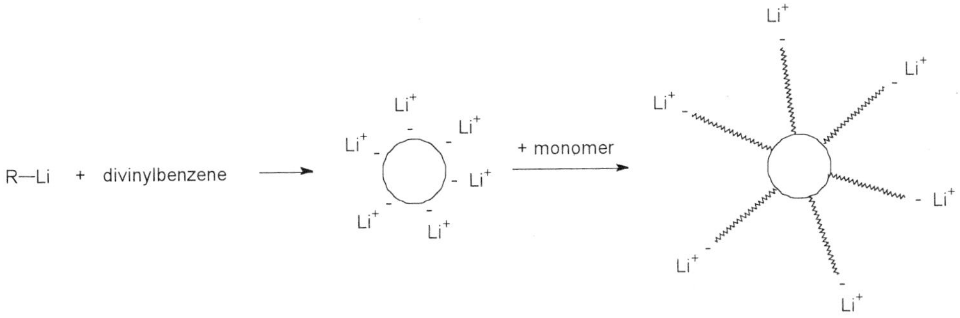

Scheme 3. Synthesis of star-shaped polymers via the "core first" method.

For the synthesis of well-defined hetero 3- and 4-armed star copolymers, the methodology of linking living polymers with DPE-functionalized macromonomers and *bis*-DPE compounds, respectively, followed by the polymerization of a second monomer has been utilized [82–85].

Graft Copolymers

Graft copolymers can be obtained by three different methods: (i) in-chain metalation of a polymer backbone and subsequent addition of the graft-forming monomer, ("grafting from"), (ii) linking of living anionic polymers with in-chain functional groups on a polymer backbone, ("grafting onto"), and (iii) copolymerization of a macromonomer with the backbone-forming monomer, ("grafting through").

The "grafting from" method involves the generation of anionic sites along a polymer backbone either by metalation of C–H or C–halogen bonds or by addition of organometallic compounds to reactive vinyl groups [10,86]. For example, PB and PI have been metalated with sBuLi in the presence of N,N,N',N'-tetramethylethylene-diamine (TMEDA) at room temperature and used as initiators for the "grafting from" of styrene [87–89]. However, this procedure suffers from side reactions, which result in chain degradation and PSt contamination of the graft copolymer [90], and the graft branches usually exhibit broad MWD [9].

For the "grafting onto" of a PSt-Li on a chloromethylated PSt, for example, side reactions like chain coupling and termination have been observed [91]. These problems of direct anionic coupling reactions can be circumvented when using the potassium salt of a carboxyl-terminated PSt in the presence of a crownether instead [92,93]. PBs functionalized with chlorosilane groups have been successfully coupled with PSt- or PαMeSt-Li [94]. This method provides graft copolymers with uniform grafts that are randomly distributed along the backbone.

Since macromonomers can be prepared by living polymerizations, "grafting through" would also lead to graft copolymers with well-defined side chains. However, only a few examples describing the syntheses of graft copolymers by an anionic "grafting through" [32,95,96] have been reported up to the present, mainly due to experimental difficulties of purifying the macromonomers.

Recent Developments in (Meth)acrylate Polymerization

There are many practical uses for block copolymers of methacrylates and acrylates, which is why they are of such great commercial interest. Up to now, these polymers could not be produced via a living and controlled anionic polymerization without considerable cost because it usually requires the use of low temperatures, ethereal solvents [12,97], and additives like LiCl [98], lithium *tert*-butoxide [99], diethyl zinc [100], tertiary amines [101], crown ethers, or cryptands [102]. The effects of these additives on the structure, reactivity, and association behavior of the living ester enolate chain ends as well as the kinetics of propagation and side reactions have been described extensively in the literature [9–12].

The major problems to be solved concern the low temperatures and the ethereal solvents. During the last two decades, some promising routes have been reported that make a living/controlled anionic polymerization of alkyl (meth)acrylates possible at ambient or elevated temperatures, some of them even for hydrocarbon solvents.

(1) The metal-free polymerization of methyl methacrylate, which employs large organic tetrabutylammonium [103], tetraphenylphosphonium [104], or phosphazene-based counterions (Chart 1) [105] in polar solvents, takes advantage of the fact that the ratio of propagation over termination rate increases with the size of the counterion [12]. The livingness of the polymerization may be preserved up to 60 °C, and the resulting polymers exhibit monomodal and narrow MWDs ($M_w/M_n < 1.2$). Similar results can be obtained for some primary acrylates when a tetrabutylammonium counterion is used in toluene at room temperature [106]. Polymers with M_ns exceeding 20,000 g/mol are, however, difficult to obtain, and the block copolymerization of methacrylates and acrylates also proceeds less than satisfactorily.

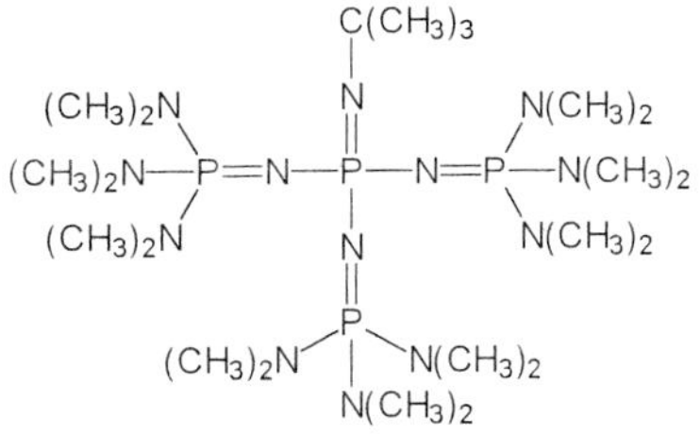

Chart 1

(2) The group-transfer polymerization (GTP) [107] of acrylic monomers may be considered as a metal-free anionic polymerization although its mechanism is still under discussion [108,109]. The polymerization is initiated by a silyl ketene acetal like 1-methoxy-1-(trimethylsilyloxy)-2-methyl-1-propene, MTS (Chart 2), activated by a nucleophilic or electrophilic catalyst. It proceeds in a living and controlled manner from −100 to +80 °C. The polymerization of methacrylates is preferably carried out in polar solvents with nucleophilic catalysts such as azides, cyanides, bifluorides, or oxyanions, while that of acrylates provides best results in non-polar solvents with electrophilic catalysts such as aluminum organyls, zinc or mercury halides [110]. GTP produces homo and block copolymers with low polydispersities but suffers from the same deficiencies as the metal-free polymerizations mentioned above.

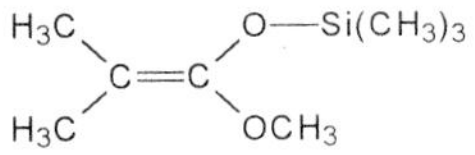

Chart 2

(3) Up to 0 °C, aluminum compounds like triethyl-, triisobutyl- or (2,6-di-*tert*-butyl-4-methylphenoxy)diisobutylaluminum can support a living and controlled polymerization of methacrylates initiated by lithium compounds in toluene [111,112] or toluene/ester mixed solvents [113]. The complex formation with the aluminum alkyl is assumed to substantially decrease the reactivity of the living chain end and with it the occurrence of side reactions. This reaction is best suited for the synthesis of homo and block copolymers with M_ns below 20,000 g/mol, as its half-life is in the range of several hours. Polymers with higher molecular weights can be obtained using tetra-alkylammonium halide-aluminum alkyl complexes as additives in toluene [96]. This procedure allows the polymerization of methacrylates in a living and controlled manner near room temperature, although primary acrylates still require the use of low temperature (< −70 °C).

(4) Lithium alkoxy alkoxides like lithium 2-(2-methoxyethoxy)ethoxide (Chart 3), which combine the properties of a common ion salt and that of a polyether or crown-ether, support a living anionic polymerization of methacrylates and acrylates in toluene at or above room temperature [114]. The extremely high rates of polymerization demand the use of a flow-tube reactor in order to achieve good control of the reaction [115]. Contrary to the metal-free systems, block copolymers of methacrylates and acrylates with high molecular weight and very narrow MWD can be synthesized by sequential monomer addition.

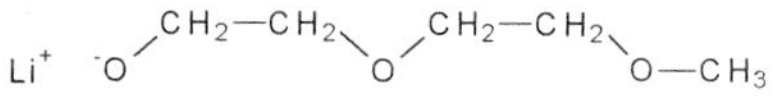

Li⁺ ⁻O—CH₂—CH₂—O—CH₂—CH₂—O—CH₃

Chart 3

Industrial-Scale Syntheses

Up to the present, commercial exploitation of anionic polymerization is mainly limited to the production of block copolymers of styrene, butadiene, and isoprene (see section 2.1.2). Depending on molecular weight, composition, architecture, and microstructure of the polymers, they are used as synthetic rubbers, thermoplastic elastomers, or impact-modifying resins – the properties and the industrial applications of styrene/diene co-polymers have been described in detail elsewhere [116]. The patent literature has been recently reviewed in [117].

The monomers are purified by washing with dilute NaOH (to remove inhibitors) and subsequent passing over molecular sieves. Cyclohexane, the preferred solvent is distilled and also passed over molecular sieves. The initiator used is typically a ~1% solution of nBuLi or sBuLi, but the difunctional initiator shown in Scheme 2 is also utilized. BuLi is usually added in slight excess to remove any residual impurities in the solvent and monomer. The polymerizations are carried out under nitrogen in either stirred batch, tubular, or continuous reactors. The molecular weight of the polymer is controlled by the rate of the initiator feed and the block size by the flow of the monomers. 1,4-*Bis*(chloromethyl)benzene and divinylbenzene are often used as coupling/linking agents to produce linear ABA triblock and star-branched copolymers, respectively.

Established industrial-scale anionic processing of polar monomers is scarce. A copolymer P(MMA-*co*-BMA)$_{40}$-*b*-PGMA$_4$ (BMA: butyl methacrylate, GMA: glycidyl meth-acrylate) is prepared by GTP in glyme at temperatures up to 80 °C. It is then reacted with 4-nitrobenzoic acid to give a dispersant for organic pigments. PEOs and PPOs (PO: propylene oxide), which are used for cosmetical and pharmaceutical applications and in the production of polyurethanes, are synthesized via anionic ring opening poly-merization initiated by KOH in an alcohol at 100 to 200 °C [117].

Cationic Polymerization

Cationic polymerization may be induced by a variety of chemical as well as physical methods, the most important being the cation donor (initiator)/Lewis acid (coinitiator) initiating system. Most propagating carbenium ions are very strong electrophiles, and react with monomer very rapidly. Reported propagation rate constants k_p are typically between 10^4 - 10^6 L·mol^{-1}·sec^{-1}, but recent results indicate that some k_ps, especially for monomers such as isobutylene (IB) and styrene which give rise to very reactive propagating cations, may be underestimated and propagation is diffusion controlled or close to it [118,119]. In contrast to anionic polymerization, where free ions are much more reactive than ion pairs, free and paired cations possess similar reactivity, largely independent of the nature of the counter anion [120].

It is apparent that, due to the extremely rapid propagation, if all chain ends were ionized and grew simultaneously, monomer would disappear at such a high rate that the polymerization could be considered adiabatic and uncontrollable. In living cationic polymerization, a dynamic equilibrium exists between a very small amount of active and a large pool of dormant species. Mechanistically, the equilibrium concentration of cations is provided by the reversible termination/reinitiation or activation/deactivation (Scheme 4).

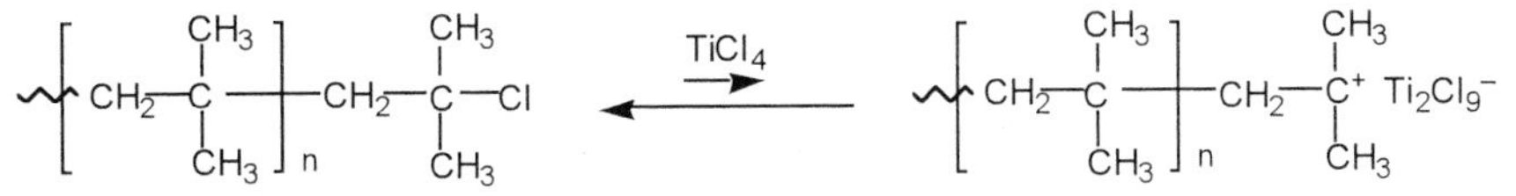

Scheme 4. Dormant-active chain end equilibrium in the living polymerization of isobutylene (dissociation of ion pairs not shown)

For a specific monomer, the rate of exchange as well as the position of the equilibrium and, to some extent, the zero order monomer transfer constants depend on the nature of the counter anion in addition to temperature and solvent polarity. Therefore, initiator/coinitiator systems that bring about living polymerization under a certain set of experimental conditions are largely determined by monomer reactivity.

Since the first report of living carbocationic polymerization in 1984 [5], the scope has been expanded to different vinyl ethers, N-vinylcarbazole, styrene, α-methylstyrene and their ring substituted derivatives, indene, isobutylene, etc. Weak Lewis acids such as I_2, or zinc halides may be necessary to effect living polymerization of the more reactive vinyl ethers, and $SnBr_4$, a moderate Lewis acid, efficiently mediates living polymerization of α-methylstyrene [121]. They are ineffective however to induce polymerization of the less reactive monomers such as isobutylene and styrene. Living cationic homo- and copolymerization of isobutylene, styrene, and styrene derivatives requires the use of much stronger Lewis acids such as BCl_3 and $TiCl_4$. It is therefore apparent that the activity of an initiating system and the "livingness" of the polymerization is affected by the nature and strength of the Lewis acid. Based on the $\nu(C=O_{stretching})$ difference determined by IR spectroscopy between 9-xanthone and its complex with Lewis acids, the following Lewis acidity scale was established for metal chlorides [122]: $ZnCl_2 < SnCl_4 < AlCl_3 < TiCl_4 < BCl_3 < SbCl_5$. The acidity scale may be, however, different when based on other properties of Lewis acids (e.g. heat of complex formation with amines). Moreover, the Lewis acidity is ambiguous for some Lewis acids, notably aluminum and titanium halides, that may form dimers and/or dimeric counter anions under polymerization conditions [123, 124]. Since the negative charge is dispersed more effectively in the dimeric gegenions, the dimeric counter anions are less nucleophilic than the monomeric equivalents. Consequently, the equilibrium in Scheme 4 is shifted to the right resulting in an increase in polymerization rate.

In order to obtain polymers with controlled molecular weight and narrow (Poisson-) MWD, the rate of initiation should be comparable to and preferentially higher than the rate of propagation. The selection of the initiator therefore is very important in living cationic polymerization that usually needs to be determined for each monomer. Initiators that have a structure similar to the dormant macromolecular end are usually effective. It is important to note that back strain, i.e., the release of steric strain upon ionization, may contribute significantly to the ease of ionization of the dormant macromolecular end, for instance PIB-Cl or PαMeSt-Cl. Due to the absence of back strain, *tert*-butyl chloride and cumyl chloride are inefficient initiators for the living polymerization of isobutylene and α-methylstyrene, respectively. In contrast, the corresponding dimeric chlorides (2-chloro-2,4,4-trimethylpentane and 2-chloro-2,4-diphenyl-4-methylpentane) are excellent initiators [125, 126]. Since adventitious water is always present as a proton source, most living cationic polymerizations are carried out in the presence of a proton trap, e.g., 2,6-di-*tert*-butylpyridine (DTBP), to prevent initiation and other side reactions by protic impurities.

Recently, it was reported that boron halides alone can initiate the living polymerization of isobutylene [127], styrene, and styrene derivatives [128, 129] in the presence of DTBP. Kinetic and mechanistic studies supported the proposed initiation mechanism via haloboration [130].

Based on the above considerations living cationic polymerization and the formation of well defined polymer with controlled M_n and narrow MWD ($M_w/M_n \sim 1.1$-1.2) have been reported with a large variety of initiating systems. A detailed review is available [125].

Functionalized Polymers

Chain end-functionalized and asymmetric telechelic polymers can be obtained in a manner similar to that described for anionic polymerization, i.e., by employing cationic initiators with protected functional groups and by end-quenching living polymers with specific nucleophiles. Symmetric telechelic polymers may also be obtained from the coupling of α-functional living polymer chains. While it is relatively easy to obtain end-functionalized polymers by living anionic polymerization, it is more difficult in cationic polymerization, especially with the less reactive monomers. The malonate anion [131], certain silyl enol ethers [132] and silyl ketene acetals [133] have been successfully used as functionalizing agents for vinyl ethers. Up to recently, however, only chloro- and allyl-functional PIB and chloro functional PSt have been reported. Allyl-functional PIBs have been obtained by the reaction of living PIB with allyltrimethylsilane [134], tetraallyltin or allyltributyltin [135]. In-situ functionalization using other nucleophiles, on the other hand, has been more difficult. This is due to the facts that the equilibrium position in Scheme 4 is strongly to the left and that the added nucleophiles react with the Lewis acid rapidly and shift the ionogenic equilibrium to the dormant chlorides. Although there are various methods available to modify the resulting chloro chain ends, they usually involve a number of steps and are rather cumbersome [136].

In-situ functionalization of the living ends by a variety of nucleophiles was only recently realized via capping with non(homo)polymerizable monomers such as DPE, substituted DPEs, or 2-alkyl furans followed by end-quenching [137]. The stable and fully ionized diarylcarbenium ion, obtained in the capping of PIBCl or PStCl with DPE, is readily amenable for chain-end functionalization by quenching with appropriate nucleophiles as shown in Scheme 5 [138]. Using this strategy, a variety of chain-end functional PIBs, including methoxy, amine, carbonyl, and ester end-groups, have been prepared. It is also notable that, when living PIB is capped with DPE, organotin compounds can also be used to introduce new functionalities such as –H, –N(CH₃)₂, and furan [139].

α,ω-Telechelic PIBs are readily available when a difunctional initiator is used. By the rational combination of haloboration-initiation and capping techniques, a series of α,ω-asymmetrically functionalized PIBs have been prepared [140-142]. Polymers prepared by haloboration-initiation invariably carry an alkylboron head group [127-129] which can easily be converted into a primary hydroxy [128] or a secondary amine group [141,142]. To functionalize the ω-living ends, the functionalization strategy shown in Scheme 5 is applicable and has been used [139] to incorporate methoxycarbonyl groups as ω-functionality.

Symmetric telechelic polymers can also be prepared by coupling of α-functional living polymer chains using any of the recently discovered coupling agents (see later).

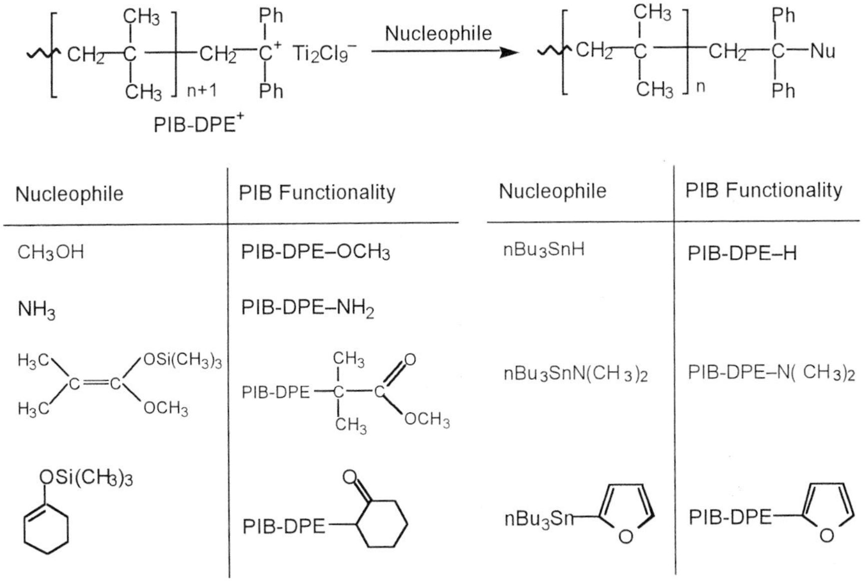

Nucleophile	PIB Functionality	Nucleophile	PIB Functionality
CH₃OH	PIB-DPE–OCH₃	nBu₃SnH	PIB-DPE–H
NH₃	PIB-DPE–NH₂		
(image)	(image)	nBu₃SnN(CH₃)₂	PIB-DPE–N(CH₃)₂
(image)	(image)	nBu₃Sn—(furan)	PIB-DPE—(furan)

Scheme 5. Synthesis of chain-end functionalized PIBs

Coupling Reaction of Living Cationic Polymers

The synthetic application of non(homo)polymerizable monomers was further extended to *bis*-DPE compounds, such as 2,2-bis[4-(1-phenylethenyl)phenyl]propane (BDPEP) and 2,2-bis[4-(1-tolylethenyl)phenyl]propane (BDTEP) (Chart 4), consummating the living coupling reaction of living PIB [143,144]. It was demonstrated that living PIB reacts quantitatively with BDPEP or BDTEP to yield stoichiometric amounts of *bis*(diarylalkylcarbenium) ions, as confirmed by the quantitative formation of diaryl-methoxy functionalities at the junction of the coupled PIB. Kinetic studies indicated the coupling reaction of living PIB by BDPEP is a consecutive reaction where the second addition is much faster than the first one. As a result, high coupling efficiency was also observed with excess BDPEP.

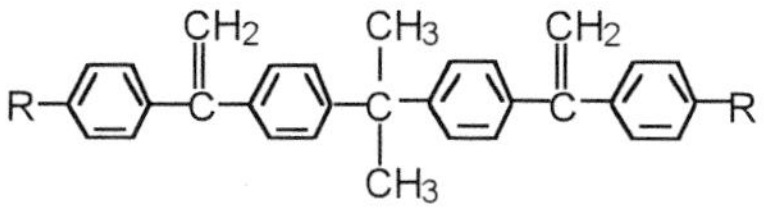

R = H (BDPEP), CH₃ (BDTEP

Since 2-alkylfurans add rapidly and quantitatively to living PIB yielding stable tertiary allylic cations, the coupling reaction of living PIB was also studied using *bis*-furanyl compounds [145]. Using 2,5-*bis*[1-furanyl)-1-methylethyl]-furan (BFPF) (Chart 5), coupling of living PIB was found to be rapid and quantitative in hexane/CH_3Cl (60/40 or 40/60, v/v) solvent mixtures at –80 °C in conjunction with $TiCl_4$, as well as in CH_3Cl at –40 °C with BCl_3 as Lewis acid. For instance, in-situ coupling of living PIB, prepared by haloboration-initiation using the BCl_3/CH_3Cl/–40 °C system, with BFPF yielded α,ω-telechelic PIB with alkylboron functionality. After oxidation, this telechelic PIB was converted to α,ω-hydroxyl PIB. The synthesis of α,ω-telechelic PIBs with a vinyl functionality was also achieved by the coupling reaction of living PIB, prepared using 3,3,5-trimethyl-5-chloro-1-hexene as an initiator in the presence of $TiCl_4$.

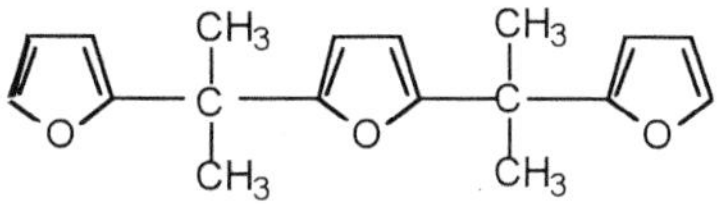

BFPF

Chart 5

Block, Star, and Graft copolymers

Linear Block Copolymers

Living cationic sequential block copolymerization is generally recognized as one of the simplest and most convenient methods to obtain well-controlled block copolymers with high structural integrity. However, the efficient synthesis of block copolymers by sequential monomer addition requires that the rate of a crossover reaction to a second monomer (R_{cr}) be faster than or at least equal to that of homopolymerization of a second monomer (R_p). This can be controlled by the judicious selection of poly-merization condition; mainly by selecting the appropriate order of monomer addition. For instance, since the crossover from living PIB to α-methylstyrene is not efficient, PIB-*b*-PαMeSt could not be prepared by a simple sequential block copolymerization. In contrast, the crossover from living PαMeSt to isobutylene is efficient and PαMeSt-*b*-PIB could be synthesized.

PIB based triblock copolymer thermoplastic elastomers (TPEs), predicted to have superior properties, have been the focus of recent interest. The synthesis of PSt-*b*-PIB-*b*-PSt TPE has been accomplished by many research groups [146-149]. The synthesis invariably involved sequential monomer addition using di- or trifunctional initiator in conjunction with $TiCl_4$ in moderately polar solvent mixture at low (–70 to –90

°C) temperatures. Structure-property relationships indicated that the tensile strength is controlled by the molecular weight of the PSt segment and is independent of the PIB middle block length. Phase separation starts when the M_n of the PSt segment reaches ~5,000, and it is complete when the M_n reaches ~15,000. Representative triblocks exhibited 23-25 MPa tensile strength [149], similar to that of commercially available styrenic TPEs obtained by anionic polymerization.

The synthesis of various PIB based TPEs containing PInd (Ind: indene) (T_g ~ 220 °C) [150,151], PpMeSt (pMeSt: *p*-methylstyrene) (T_g ~ 108 °C) [151-152], P(Ind-*co*-pMeSt) (T_g varied by composition) [153], PptBuSt (ptBuSt: *p-tert*-butylstyrene) (T_g ~ 144 °C) [152, 154] and P(tBuSt-*co*-Ind) [152], PpClSt (pClSt: *p*-chlorostyrene) [155] and PpFSt (pFSt: *p*-fluorostyrene) [156] outer glassy segment have also been reported by a procedure essentially identical to that used for PSt-*b*-PIB-*b*-PSt. All exhibited typical TPE properties. The preparation and properties of these TPEs have been reviewed recently. [157]

Sequential block copolymerization of isobutylene with more reactive monomers, such as *p*-methylstyrene [158,159], α-methylstyrene [160,161], isobutyl vinyl ether (IBVE) [162] or methyl vinyl ether (MeVE) [163], usually leads to a mixture of block copolymer and homo-PIB due to the unfavorable ratio of R_{cr}/R_p. As shown in Scheme 6, a general scheme was developed to increase the relative ratio of R_{cr}/R_p, especially when the second monomer is more reactive than the first. This process involves the capping reaction of living PIB with DPE or DTE, followed by tuning of the Lewis acidity to the reactivity of the second monomer. The purpose of the Lewis acidity tuning is to generate stronger nucleophilic counterions which ensure a high R_{cr}/R_p ratio as well as the living polymerization of a second monomer. This has been carried out using three different methods: (i) by the addition of titanium(IV) alkoxides (Ti(OR)$_4$), (ii) by the substitution of a strong Lewis acid with a weaker one, or (iii) by the addition of tetra-butylammonium chloride (nBu$_4$NCl).

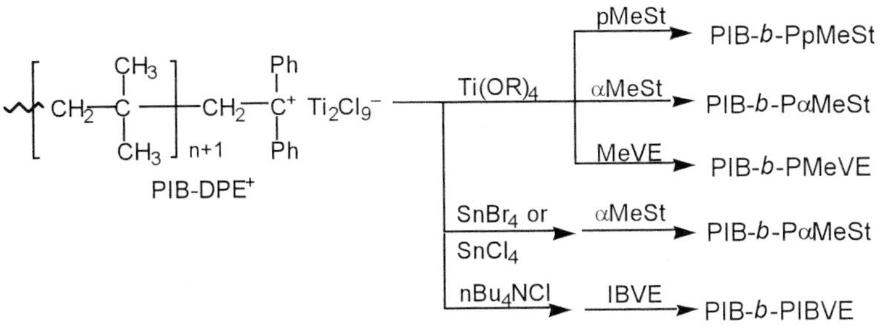

Scheme 6. Synthesis of block copolymers by sequential monomer addition

The first method has been successfully employed in the block copolymerization of isobutylene with α-methylstyrene, *p*-methylstyrene, or methyl vinyl ether [158–161,163]. The substitution of TiCl$_4$ with a weaker Lewis acid (SnBr$_4$ or SnCl$_4$) also proved to be an efficient strategy in the synthesis of PIB-*b*-PαMeSt diblock and PαMeSt-*b*-PIB-*b*-PαMeSt triblock copolymers [160,161]. When SnCl$_4$ was employed, it was necessary to keep [SnCl$_4$] equal to or below 0.5[chain end] to increase the relative ratio of R_{cr}/R_p. Mechanistic studies indicated that, when [SnCl$_4$] ~ 0.5[chain end], a double charged counterion, SnCl$_6^{2-}$, is involved during the crossover reaction which then is converting to a single charged counterion, SnCl$_5^-$, during the polymerization of α-methylstyrene [164]. The block copolymerization of isobutylene with isobutyl vinyl ether was also achieved by Lewis acidity tuning using nBu$_4$NCl [162]. The addition of nBu$_4$NCl reduces the concentration of free and uncomplexed TiCl$_4$ ([TiCl$_4$]$_{free}$), and mechanistic studies indicated that, when [TiCl$_4$]$_{free}$ < [chain end], the dimeric counterion, Ti$_2$Cl$_9^-$, is converted to a more nucleophilic monomeric TiCl$_5^-$ counterion.

Block copolymerization of isobutylene with methyl vinyl ether was also carried out using 2-methylfuran or 2-*tert*-butylfuran as a capping agent [165].

Star and Graft Copolymers

The synthesis of branched polymers by cationic polymerization of vinyl monomers has been reviewed recently [166]. Similarly to their use in anionic polymerization, difunctional monomers such as divinylbenzene have been found to be efficient in the synthesis of star (co)polymers having a cross-linked core from which homopolymer or block copolymer arms radiate outwards. However, so far only "core last" methods have been reported in cationic polymerization.

In contrast to anionic multifunctional initiators, well-defined soluble multifunctional cationic initiators are readily available. These multifunctional initiators with 3-8 initiating sites have been sucessfully applied for the synthesis of 3-8 arm star homo- and block copolymers of vinyl ethers, styrene, styrene derivatives, and isobutylene. By subsequent end-functionalization a variety of end-functionalized A$_n$ or (AB)$_n$ star-shaped structures can be obtained.

Well-defined star-branched polymers have also been obtained by utilizing multifunctional coupling agents with the nucleophilic functions well separated to avoid steric hindrance. For example, the synthesis of a PIB star with up to 8 arms was accomplished in two steps [167]. First, allyl terminated PIB was prepared by reacting living PIB with trimethylallylsilane. Linking was effected by hydrosilylation of the allyl-functional PIB with cyclosiloxanes carrying 6 or 8 Si–H groups.

Recently, a new concept in cationic polymerization, the concept of living coupling reaction, was introduced. According to the definition originally proposed for living anionic polymerization [82], a living coupling agent must react quantitatively with the living chain ends, the coupled product must retain the living centers stoichiometrically and must be able to reinitiate the second monomer rapidly. It was reported that living PIB reacts quantitatively with *bis*-DPEs where the two DPE moieties are separated by an electron-donating spacer group to yield stoichiometric amounts of *bis*(diarylalkylcarbenium) ions. Since the resulting diaryl cations have been successfully employed for the controlled initiation of styrenic monomers and vinyl ethers, the concept of the living coupling reaction was proposed as a general route for the synthesis of A$_2$B$_2$ star-block copolymers. As proof of this concept, an amphiphilic A$_2$B$_2$ star-block copolymer (A = PIB and B = PMeVE) has been prepared by the living coupling reaction of living PIB followed by the chain-ramification polymerization of methyl vinyl ether at the junction of the living coupled PIB as shown in Scheme 7 [144].

Architecture/property studies were also carried out by comparison of micellar properties of star-block copolymer and the linear diblock analogues. These studies suggested that block copolymers with star architecture exhibit a decreased tendency to association than their corresponding linear diblock copolymers.

Graft copolymers by cationic polymerization may be obtained by the "grafting from", and "grafting onto" methods and by (co)polymerization of macromonomers. For example, PIB with pendant functionalities could be prepared by copolymerization of isobutylene with a functional monomer such as bromomethylstyrene or chloromethyl-styrene in CH_2Cl_2 at −80 °C with BCl_3. In subsequent initiation of 2-methyl-2-oxazoline, water soluble amphiphilic graft compolymers have been obtained [168].

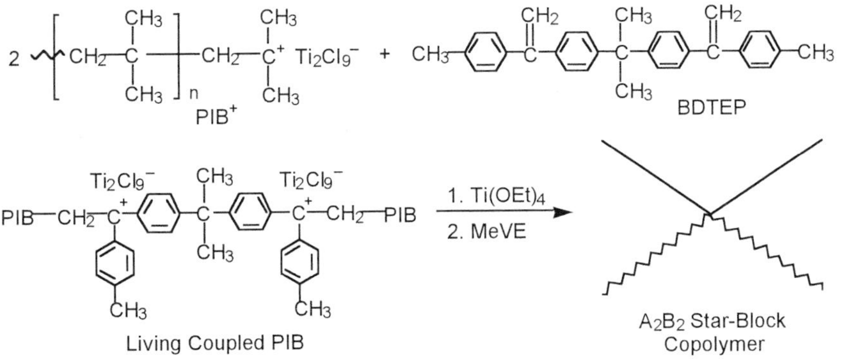

Scheme 7. Synthesis of A_2B_2 star-block copolymers via the living coupling reaction

Macromonomers have been synthesized by living cationic polymerization by three different techniques: by (i) using a functionalized initiator, (ii) employing a functionalized capping agent or by (iii) chain end modification. The simplest but most efficient technique is the first one, since it involves only one step. Using this technique PSt, PpMeSt, poly(vinyl ether), poly(β-pinene) and PIB macromonomers containing a methacrylate end group have been synthesized. A polymerizable group may also be incorporated by a reaction between a functionalized capping agent and the living end of the polymer. A detailed review can be found in [166].

Highly branched, so called "hyperbranched" macromolecules have recently attracted interest, due to their interesting properties, which closely resemble those of dendrimers. Vinyl monomers with pendant initiating moieties, for example 3-(1-chloroethyl)-ethenylbenzene, have been reported to give rise to hyperbranched polymers in a process termed "self-condensing vinyl polymerization" [169].

Industrial-Scale Syntheses

The major industrial products are butyl type elastomers, polybutenes and other hydrocarbon resins. Industrial manufacture and uses of these materials have been recently reviewed [170]. Commercialization of new technologies based on living cationic polymerization has just begun. Allyl-, and alkoxysilyl-telechelic curable PIB elastomers (Epion) are made by Kaneka Corporation (Japan), and PSt-*b*-PIB-*b*-PSt block copolymer TPE (TS-Polymer) is test marketed by Kuraray Ltd. (Japan).

Combination of Anionic and Cationic Polymerization

The combination of living cationic and anionic techniques provides a unique approach to polymers not available by a single method. Coupling of living PSt-Li and living poly(ethyl vinyl ether) for instance, has been reported to yield di- and tri- block copolymers with high efficiency [171]. Comb-like copolymers have been prepared by Schappacher and Deffieux by grafting PSt-Li onto poly(2-chloroethyl vinyl ether) [172]. Recently it was reported that DPE-capped PIB carrying methoxy or olefin functional groups, can be quantitatively metalated with alkali metals in THF at room temperature [173]. The stable macrocarbanions resulting from metalation with K/Na alloy have been used to initiate the living anionic polymerization of tBMA, yielding PIB-*b*-PtBMA block copolymers. Replacing K$^+$ by Li$^+$ with excess LiCl gives a PIB macroinitiator suitable for anionic polymerization of MMA, and has been employed to prepare PMMA-*b*-PIB-*b*-PMMA triblock copolymers [174].

Another facile route for the preparation of block copolymers unavailable by cationic or anionic only techniques is the coupling reaction of living cationic polymers with living anionic polymers. This approach was employed in the synthesis of e.g. PIB-*b*-PMMA in which the original PMMA was prepared by group transfer polymerization (GTP) [175]. The coupling reaction of two living homopolymers with antagonistic functions can be an expedient alternative to the site-transformation technique.

Literature Cited

1. Staudinger, H. Ber. dtsch. chem. Ges. 1920, 53, 1073.
2. Deville, M. Ann. Cim. France 1839, 75, 66.
3. Ziegler, K.: Bahr, K. Chem. Ber. 1928, 61, 253.
4. Szwarc, M. Nature 1956, 178, 778.
5. Miyamoto, M.: Sawamoto, M.: Higashimura, T. Macromolecules, 1984 17, 265.
6. Faust, R.: Kennedy, J.P., Abstract, ACS Meeting New York 1985.
7. Szwarc, M.: van Beylen, M., Ionic Polymerization and Living Polymers, Chapman and Hall, New York, 1993.
8. Matyjaszewski, K.: Müller, A.H.E. Polym. Prepr. 1997, 38, 6.
9. Morton, M.: Fetters, L.J. Rubber Chem. Tech. 1975, 48, 359.
10. Wakefield, B.J., The Chemistry of Organolithium Compounds, Pergamon Press, New York, 1974.
11. Hsieh, H.L.: Quirk, R.P., Anionic Polymerization: Principles and Applications, Dekker, New York, 1996.
12. Müller, A.H.E., in: Comprehensive Polymer Science, Vol. 3, Eastmond, G.C.: Ledwith, A.: Russo, S.: Sigwalt, P. , Eds., Pergamon Press, New York, 1989, p. 387.
13. Patil, A.O.: Schulz, D.N.: Novak, B.M., Eds., Functional Polymers: Modern Synthetic Methods and Novel Structures, ACS Symposium Series 704, American Chemical Society, Washington D.C., 1998.
14. Schulz, D.N.: Sanda, J.C.: Willoughby, B.G., in: Anionic Polymerization: Kinetics, Mechanisms, and Synthesis, McGrath, J.E., Ed., ACS Symposium Series 166, American Chemical Society, Washington D.C., 1981.
15. [11], p.278.
16. Schulz, D.N.: Halasa, A.F.: Oberster, A.E. J. Polym. Sci., Polym. Chem. 1974, 12, 153.
17. Anderson, B.C.: Andrews, G.D.: Arthur Jr., P.: Jacobsen, H.W.: Melby, L.R.: Playtis, A.J.: Sharky, W.H. Macromolecules 1981, 14, 1599.
18. Quirk, R.P.: Ma, J.-J. J. Polym. Sci., Polym. Chem. 1988, 26, 2031.
19. Quirk, R.P.: Zhu, L. Makromol. Chem. 1989, 190, 487.
20. Richards, D.H.: Eastmond, G.C.: Steward, M.J., in: Telechelic Polymers: Synthesis and Applications, Goethals, E.J., Ed., CRC Press, Boca Raton, Florida, 1989, p. 33.
21. Quirk, R.P.: Takizawa, T.: Lizzaraga, G.: Zhu, L.-F. J. Appl. Polym. Sci., Polym. Symp. 1992, 50, 23.
22. Schulz, D.N.: Halasa, A.F. J. Polym. Sci., Polym. Chem. 1977, 15, 2401.
23. Richards, D.H.: Service, D.M.: Steward, M.J. Brit. Polym. J. 1984, 16, 117.
24. Hirao, A.: Hattori, I.: Sasagawa, T.: Yamaguchi, K.: Nakahama, S.: Yamazaki, N. Makromol. Chem., Rapid. Commun. 1982, 3, 59.
25. Quirk, R.P.: Summers, G.J. Brit. Polym. J. 1990, 22, 249.
26. Quirk, R.P.: Chen, P.L. Macromolecules 1986, 19, 1291.
27. Eisenbach, C.D.: Schnecko, H.: Kern, W. Makromol. Chem. 1975, 176, 1587.
28. Steward, M.J.: Shephard, N.: Service, D.M. Brit. Polym. J. 1990, 22, 319.

29. Quirk, R.P.: Zhu, L. Brit. Polym. J. 1990, 23, 47.
30. Quirk, R.P.: Yin, J. J. Polym. Sci., Polym. Chem. 1992, 30, 2349.
31. Hirao, A.: Nagahama, H.: Ishizone, T.: Nakahama, S. Macromolecules 1993, 26, 2145.
32. Gnanou, Y.: Lutz, P. Makromol. Chem. 1989, 190, 577.
33. Asami, R.: Takaki, M.: Hanahata, H. Macromolecules 1983, 16, 628.
34. Asami, R.: Takaki, M. Makromol. Chem., Suppl. 1985, 12, 163.
35. Andrews, G.D.: Melby, L.R., in: New Monomers and Polymers, Culbertson, B.M.: Pittman, Jr., C.U., Eds., Plenum Press, 1984, p. 357.
36. Schulz, G.O.: Milkovich, R. J. Appl. Polym. Sci. 1982, 27, 4773.
37. Radke, W.: Müller, A.H.E. Makromol. Chem., Makromol. Symp. 1992, 54/55, 583.
38. Quirk, R.P.: Wang, Y. Polym. Int. 1993, 31, 51.
39. Cram, D.J., Fundamentals of Carbanion Chemistry, Academic Press, New York, 1965.
40. Bordwell, F.G.: Algrim, D.J. J. Am. Chem. Soc. 1988, 110, 2964.
41. Hsieh, H.L. Rubber Chem. Tech. 1976, 49, 1305.
42. Varshney, S.K.: Kesani, P.: Agarwal, N.: Zhang, J.X.: Rafailovich, M. Macromolecules 1999, 32, 235.
43. Szwarc, M.: Levy, M.: Milkovich, R. J. Am. Chem. Soc. 1956, 78, 2656.
44. Leitz, E.: Höcker, H. Makromol. Chem. 1993, 184, 1893.
45. Auschra, C.: Stadler, R. Polym. Bull. 1993, 30, 257.
46. Balsamo, V.: Müller, A.J.: von Gyldenfeldt, F.: Stadler, R. Macromol. Chem. Phys. 1998, 199, 1063.
47. Morton, M.: Kesten, Y.: Fetters, L.J. Appl. Polym. Symp. 1975, 26, 113.
48. Wilhelm, M.: Zhao, C.-L.: Wang, Y.: Xu, R.: Winnik, M.A.: Mura, J.-L.: Riess, G.: Croucher, M.D. Macromolecules 1991, 24, 1033.
49. long, T.E.: Broske, A.D.: Bradley, D.J.: McGrath, J.E. J. Polym. Sci., Polym. Chem. 1989, 27, 4001.
50. [11], p. 308.
51. [11], p. 322.
52. Quirk, R.P.: Kinning, D.J.: Fetters, L.J., in: Comprehensive Polymer Science, Allen, G.: Bevington, J.C., Eds., Pergamon Press, Vol. 7, 1989, p. 1.
53. Masuda, T.: Ohta, Y.: Onogi, S. Macromolecules 1971, 4, 763.
54. Fetters, L.J.: Morton, M. Macromolecules 1974, 7, 552.
55. Roovers, J.E.L.: Bywater, S. Macromolecules 1972, 5, 384.
56. Roovers, J.E.L.: Bywater, S. Macromolecules 1974, 7, 443.
57. Hadjichristidis, N.: Roovers, J.E.L. J. Polym. Sci., Polym. Phys. 1974, 12, 2521.
58. Hadjichristidis, N.: Guyot, A.: Fetters, L.J. Macromolecules 1978, 11, 668.
59. Hadjichristidis, N.: Fetters, L.J. Macromolecules 1980, 13, 191.
60. Toporowski, P.J.: Roovers, J. J. Polym. Sci., Polym. Chem. 1986, 24, 3009.
61. Roovers, J.: Hadjichristidis, N.: Fetters, L.J. Macromolecules 1983, 16, 214.
62. Nguyen, A.B.: Hadjichristidis, N.: Fetters, L.J. Macromolecules 1993, 26, 963.
63. Pennisi, R.W.: Fetters, L.J. Macromolecules 1988, 21, 1094.
64. Khasat, N.: Pennisi, R.W.: Hadjichristidis, N.: Fetters, L.J. Macromolecules 1988, 21, 1100.

65. Mays, J.W. Polym. Bull. 1990, 23, 247.
66. Iatrou, H.: Hadjichristidis, N. Macromolecules 1992, 25, 4649.
67. Iatrou, H.: Hadjichristidis, N. Macromolecules 1993, 26, 2479.
68. Wright, S.J.: Young, R.N.: Croucher, T.G. Polym. Int. 1994, 33, 121.
69. [11], p. 347.
70. Young, R.N.: Fetters, L.J. Macromolecules 1978, 11, 899.
71. Quack, G.: Fetters, L.J.: Hadjichristidis, N.: Young, R.N. Ind. Eng. Chem. Prod. Res. Dev. 1980, 19, 587.
72. Martin, M.K.: Ward, T.C.: McGrath, J.E., in: Anionic Polymerization: Kinetics, Mechanisms, and Synthesis, McGrath, J.E., Ed., ACS Symposium Series 166, American Chemical Society, Washington D.C., 1981, p. 558.
73. Tsitsilianis, C.: Chaumont, P.: Rempp, P. Makromol. Chem. 1990, 191, 2319.
74. Tsitsilianis, C.: Graff, S.: Rempp, P. Eur. Polym. J. 1991, 27, 243.
75. Masuda, : Ohta, Y.: Yamauchi, T.: Onogi, S. Polym. J. 1984, 16, 273.
76. Bi, L.-K.: Fetters, L.J. Macromolecules 1976, 9, 732.
77. [11], p. 335.
78. Eschwey, H.: Hallensleben, M.L.: Burchard, W. Makromol. Chem. 1973, 173, 235.
79. Lutz, P.: Rempp, P. Makromol. Chem. 1988, 189, 1051.
80. Tsitsilianis, C.: Lutz, P.: Graff, S.: Lamps, J.-L.: Rempp, P. Macromolecules 1991, 24, 5897.
81. Eschwey, H.: Burchard, W. Polymer 1975, 16, 180.
82. Quirk, R.P.: Hoover, F., in: Recent Advances in Anionic Polymerization, Hogen-Esch, T.E.: Smid, J., Eds., Elsevier, New York, 1987, p. 393.
83. Quirk, R.P.: Lee, B.: Schock, L.E. Makromol. Chem., Macromol. Symp. 1992, 53, 201.
84. Quirk, R.P.: Yoo, T. Polym. Bull. 1993, 31, 29.
85. Quirk, R.P.: Yoo, T.: Lee, B. J. Macromol. Sci. Pure Appl. Chem. 1994, A31, 911.
86. Brandsma, L., Preparative Polar Organometallic Chemistry, Springer Verlag, Heidelberg, Germany, 1990.
87. Falk, J.C.: Schlott, R.J.: Hoeg, D.F.: Pendleton, J.F. Rubber Chem. Tech. 1973, 46, 1044.
88. Hadjichristidis, N.: Roovers, J.E.L. J. Polym. Sci., Phys. 1978, 16, 851.
89. Tate, D.P.: Halasa, A.F.: Webb, F.J.: Koch, R.W.: Oberster, A.E. J. Polym. Sci, A-1 1971, 9, 139.
90. Halasa, A.F.: Mitchell, G.B.: Stayer, M.: Tate, D.P.: Oberster, A.E. J. Polym. Sci, Polym. Chem. 1976, 14, 497.
91. Takaki, M.: Asami, R.: Kuwata, Y. Macromolecules 1979, 12, 378.
92. Roovers, J. Polymer 1979, 17, 1107.
93. Roovers, J. Polymer 1979, 20, 843.
94. Cameron, G.G.: Qureshi, M.Y. Makromol. Chem., Rapid Commun. 1981, 2, 287.
95. Wofford, C.F.: Hsieh, H.L. J. Polym. Sci., A-1 1969, 7, 461.
96. Schlaad, H.: Schmitt, B.: Müller, A.H.E. Angew. Chem. 1998, 110, 1497; Angew. Chem. Int. Ed. 1998, 37, 1389.
97. [11], p. 641.

98. Varshney, S.K.: Hautekeer, J.P.: Fayt, R.: Jérôme, R.: Teyssié, P. Macromolecules 1990, 23, 2618.
99. Lochmann, L.: Müller, A.H.E. Makromol. Chem. 1990, 191, 1657.
100. Ozaki, H.: Hirao, A.: Nakahama, S. Macromol. Chem. Phys. 1995, 196, 2099.
101. Gia, H.-B.: McGrath, J.E., in: Recent Advances in Anionic Polymerization, Hogen-Esch, T.E.: Smid, J., Eds., Elsevier, New York, 1987, p. 173.
102. Wang, J.-S.: Jérôme, R.: Teyssié, P. Macromolecules 1994, 27, 4902.
103. Reetz, M.T.: Herzog, H.M.: Könen, W. Macromol. Rapid Commun. 1996, 17, 383.
104. Zagala, A. P.: Hogen-Esch, T. Macromolecules 1996, 29, 3038.
105. Pietzonka, T.: Seebach, D. Angew. Chem. 1993, 105, 741; Angew. Chem. Int. Ed. 1993, 32, 716.
106. Reetz, M.T.: Knauf, T.: Minet, U.: Bingel, C. Angew. Chem. 1988, 100, 1422; Angew. Chem. Int. Ed. 1988, 27, 1373.
107. Webster, O.W.: Hertler, W.R.: Sogah, D.Y.: Farnham, W.B.: Rajan-Babu, T.V. J. Am. Chem. Soc. 1983, 105, 5706.
108. Sogah, D.Y.: Farnham, W.B., in: Organosilicon and Bioorganosilicon Chemistry, Sakurai, H., Ed., Wiley, New York, 1986, p. 219.
109. Quirk, R.P.: Ren, J. Macromolecules 1992, 25, 6612.
110. Farnham, W.B.: Sogah, D.Y., US patent 4524196 (1985); Farnham, W.B.: Sogah, D.Y.: Middleton, W.J., European patent 184863 A2 (1986); Dicker, I.B.: Farnham, W.B.: Hertler, W.R.: Laganis, E.D.: Sogah, D.Y.: Del Pesco, T.W.: Fitzgerald, P.H., US patent 4588795 (1986), assigned to DuPont.
111. Ballard, D.G.H.: Bowles, R.J.: Haddleton, D.M.: Richards, S.N.: Sellens, R.J.: Twose, D.L. Macromolecules 1992, 25, 5907
112. Haddleton, D.M.: Muir, A.V.G.: O'Donnell, J.P.: Richards, S.N.: Twose, D.L. Macromol. Symp. 1995, 91, 91.
113. Schlaad, H.: Schmitt, B.: Müller, A.H.E.: Jüngling, S.: Weiss, H. Macromolecules 1998, 31, 573.
114. Wang, J.-S.: Jérôme, R.: Bayard, P.: Patin, M.: Teyssié, P. Macromolecules 1994, 27, 4635.
115. Maurer, A.: Marcarian, X.: Müller, A.H.E.: Navarro, C.: Vuillemin, B. Polym. Prepr. 1997, 38, 467.
116. [11], p. 395.
117. Haddleton, D.M.: Muir, A.V.G.: Richards, S.N., in: Macromolecular Design of Polymeric Materials, Hatada, K.: Kitayama, T.: Vogl, O., Eds., Dekker, New York, 1997, p. 123.
118. Roth, M.: Mayr, H. Macromolecules 1996, 29, 6104.
119. Faust, R., unpublished
120. Mayr, H., in: Cationic Polymerization. Mechanism, Synthesis, and Application, Matyjaszewski, K., Ed.; Marcel Dekker: New York, 1996, p. 51.
121. Higashimura, T.: Kamigaito, M.: Kato, M.: Hasebe, T.: Sawamoto, M. Macromolecules 1993, 26, 2670.
122. Cook, D. Canad. J. Chem. 1963, 41, 522.
123. Gyor, M.: Wang, H.-C.: Faust, R. J. Macromol. Sci. 1992, A29, 639.
124. Creaser, C.S.: Creighton, J.A. J. Chem. Soc. Dalton 1975, 402.

125. Matyjaszewski, K.: Sawamoto, M., in: Cationic Polymerization. Mechanism, Synthesis, and Application, Matyjaszewski, K., Ed.; Dekker, New York, 1996, p. 265.
126. Fodor, Zs.: Faust, R. J. Macromol. Sci., Pure Appl. Chem. 1998, A35, 375.
127. Balogh, L.: Faust, R.: Wang, L. Macromolecules 1994, 27, 3453.
128. Wang, L.: Svirkin, J.: Faust, R. PMSE Preprint 1995, 72, 173.
129. Koroskenyi, B.: Wang, L.: Faust, R. Macromolecules 1997, 30, 7667.
130. Balogh, L.: Fodor, Zs.: Kelen, T.: Faust, R. Macromolecules 1994, 27, 4648.
131. Shohi, H.: Sawamoto, M.: Higashimura, T. Macromolecules 1992, 25, 58.
132. Fukui, H.: Sawamoto, M.: Higashimura, T. Macromolecules 1993, 26, 7315.
133. Verma, A.: Nielsen, A.: McGrath, J.E.: Riffle, J. Polym. Bull. 1990, 23, 563.
134. Iván, B.: Kennedy, J.P. J. Polym. Sci., Polym. Chem. 1990, 28, 89.
135. Faust, R.: Iván, B., unpublished results.
136. For a review see Kennedy, J.P.: Iván, B., Designed Polymers by Carbocationic Macromolecular Engineering, Hanser Publishers, Munich, 1991
137. For a review see Bae, Y.C.: Hadjikyriacou, S.: Schlaad, H.: Faust, R., in: Ionic Polymerization and Related Processes, Puskas, J.E., Ed., Kluwer Academic Publishers; Dordrecht, Netherlands, 1999
138. Hadjikyriacou, S.: Fodor, Zs.: Faust, R. J. Macromol. Sci., Pure Appl. Chem. 1995, A32, 1137.
139. Koroskenyi, B.: Faust, R., ACS Symp Series 704, American Chemical Society, Washington D.C., 1998, p. 135.
141. Koroskenyi, B.: Faust, R. Polym. Prepr. 1998, 39, 492.
142. Koroskenyi, B.: Faust, R. J. Macromol. Sci., Pure Appl. Chem. 1999, A36, 471.
143. Bae, Y.C.: Fodor, Zs.: Faust, R. Macromolecules 1997, 30, 198.
144. Bae, Y.C.: Faust, R. Macromolecules 1998, 31, 2480.
145. Hadjikyriacou, S.: Faust, R., Method for Coupling Living Cationic Polymer, 1998, US Patent Application # 965,443
146. Kaszas, G.: Puskas, J.E.: Kennedy, J.P.: Hager, W.G. J. Polymer. Sci. 1991, A29, 427.
147. Everland, H.: Kops, J.: Nielsen, A.: Iván, B. Polym. Bull. 1993, 31, 159.
147. Storey, R.F.: Chisholm, B.J.: Choate, K.R. J. Macromol. Sci., Pure Appl. Chem. 1994, A31, 969.
149. Gyor, M.: Fodor, Zs.: Wang, H.-C.: Faust, R. Polym. Prepr. 1993, 34, 562; J. Macromol. Sci., Pure Appl. Chem. 1994, A31, 2053.
150. Kennedy, J.P.: Midha, S.: Tsunogae, Y. Macromolecules 1993, 26, 429.
151. Kaszas, G.: Puskas, J.E.: Kennedy, J.P.: Hager; W.G. J. Polym. Sci., Polym. Chem. 1992, 30, 41.
152. Tsunogae, Y.: Kennedy, J.P. Polym. Bull. 1992, 27, 631.
153. Tsunogae, Y.: Kennedy, J.P. J. Macromol. Sci., Pure Appl. Chem. 1993, A30, 269.
154. Kennedy, J.P.: Meguriya, N.: Keszler, B. Macromolecules 1991, 24, 6572.
155. Kennedy, J.P.: Kurian, J. J. Polym. Sci., Polym. Chem. 1990, 28, 3725.
156. Kurian, J., Ph.D. Thesis, The University of Akron, 1991.
157. Kennedy, J.P., in: Thermoplastic Elastomers, 2nd Ed., Holden, G.: Legge, N.R.: Quirk, R.P.: Schröder, H.E., Eds., Hanser Publishers, Munich, 1996, p. 365.
158. Fodor, Zs.: Faust, R. J. Macromol. Sci., Pure Appl. Chem. 1994, A31, 1983.

159. Fodor, Zs.: Faust, R. J. Macromol. Sci., Pure Appl. Chem. 1995, A32, 575.
160. Li, D.: Faust, R. Macromolecules 1995, 28, 1383.
161. Li, D.: Faust, R. Macromolecules 1995, 28, 4893.
162. Hadjikyriacou, S.: Faust, R. Macromolecules 1995, 28, 7893.
163. Hadjikyriacou, S.: Faust, R. Macromolecules 1996, 29, 5261.
164. Li, D.: Hadjikyriacou, S.: Faust, R. Macromolecules 1996, 29, 6061.
165. Hadjikyriacou, S.: Faust, R. Polym. Prepr. 1998, 39, 398.
166. Charleux, B.: Faust, R. Advances in Polymer Science 1998, 142, 1.
167. Omura, N.: Lubnin, A.V.: Kennedy, J.P., ACS Symposium Series 665, American Chemical Society, Washington D.C., 1997, p. 178.
168. Nuyken, O.: Sanchez, J.R.: Voit, B. Macromol. Rapid Commun. 1997, 18, 125.
169. Fréchet, J.M.J.: Henmi, M.: Gitsov, I.: Aoshima, S.: Leduc, M.R.: Grubbs, R.B. Science 1995, 269, 1080.
170. Vairon, J.-P.: Spassky, N., in: Cationic Polymerization. Mechanism, Synthesis, and Application, Matyjaszewski, K., Ed.; Dekker, New York, 1996, p. 683.
171. Creutz, S.: Vandooren, C.: Jérôme, R.: Teyssié, P. Polym. Bull. 1994, 33, 21.
172. Schappacher, M.: Deffieux, A. Macromol. Chem. Phys. 1997, 198, 3953.
173. Feldhusen, J.: Iván, B.: Müller, A.H.E. Macromolecules 1997, 30, 6989.
174. Feldhusen, J.: Iván, B.: Müller, A.H.E. Macromolecules 1998, 31, 4483.
175. Takács, A.: Faust, R. Macromolecules 1995, 28, 7266.

AN OVERVIEW OF TRANSITION METAL-MEDIATED POLYMERIZATIONS: CATALYSTS FOR THE 21ST CENTURY

LISA S. BOFFA

Exxon Research & Engineering Company
Route 22 East
Annandale, NJ 08801

Introduction
Olefin Addition (Coordination-Insertion) Polymerization
Olefin / Carbon Monoxide Copolymerization
Olefin Metathesis
Addition Polymerization of Polar Monomers
Step-Growth Addition / Elimination (Coupling) Polymerization
Beyond Addition and Condensation: Unique Organometallic Chain-Building Mechanisms

Introduction

The use of transition metal complexes in polymer synthesis is one of the most actively investigated areas in macromolecular science today. Over the last two decades, organometallic initiators have revolutionized strategies for controlled polymer syntheses, dramatically increased the available degree of control of polymer architectures and properties, and expanded the range of conditions under which commercial polymerizations can be carried out. In doing so, they have also greatly enhanced the technological applications of plastics and our fundamental chemical understanding of polymerization mechanisms.

Transition metal-mediated polymerizations may be defined as those having an organometallic active center covalently bound to the growing polymer chain. Depending on the nature of the organometallic fragment, the polymer-metal bond may be formally more covalent or ionic in character, but may be contrasted with anionic and cationic polymerizations in which the counterion is typically a dissociated entity. The useful feature of metal-mediated polymerization is the ability to influence the structure and / or kinetics of chain growth through the attached organometallic species. The metal's non-initiating or spectator ligands provide the steric and electronic means to do so. Organometallic polymerizations may be thought of as a logical refinement of anionic and cationic polymerizations in the sense that very active, "naked" sites have been replaced by more robust and selective propagating centers.

Many monomers that are susceptible to unwanted side reactions when polymerized ionically may in fact be polymerized in a "living" manner through the use of a transition metal initiator.

Organometallic initiators are also a refinement of traditional heterogeneous catalysts, which tend to give polymers with broad property distributions due to the many types of active sites on their surfaces. Since it is impossible to know the exact roles of all sites, heterogeneous catalysts are very difficult to understand and modify in a rational manner. In contrast, discrete transition metal initiators have only a single site at which polymerization takes place. This not only gives a more uniform polymer product, but also allows the relationships between catalyst structure and polymer properties to be understood and manipulated for improvement.

Organometallic polymerizations have a number of advantages over traditional radical and ionic techniques but are not without challenges. Transition metal initiators can be quite complex, requiring several synthetic steps, and many catalysts need expensive cocatalysts or activators. As a result of their structural fine tuning, organometallic initiators that function well for certain monomers may be completely inactive for others. Additionally, a tradeoff often exists between catalyst selectivity and activity, in that making a catalyst more well-behaved may also mean making it slower (however, recent achievements in olefin polymerization have demonstrated that this obstacle is not insurmountable).

Transition metal complexes that carry out addition polymerization act as initiators, inserting a monomer unit into a metal-organic bond to begin propagation. However, most of them also activate the monomer unit first through precoordination, and in this sense also act as catalysts. Due to the fascinating variety of organometallic reaction mechanisms, transition metal catalysts can carry out a number of other chain-building mechanisms besides addition polymerization. Metal-mediated organic couplings based on two-electron addition-elimination steps provide an organometallic analogue to condensation polymerization, and there are also a number of recently discovered metal-mediated polymerizations that have no analogues in classical ionic-based polymerization. These systems perhaps best illustrate the possibilities that make organometallic polymerization initiators worthy of the phrase "catalysts for the 21st century."

Olefin Addition (Coordination-Insertion) Polymerization

Metallocenes By far, the most significant example of transition metal-catalyzed polymerization is the use of metallocenes for polyolefin synthesis (1). Polyethylene and polypropylene are the world's largest volume thermoplastics, and taken together comprise an over 40 billion-dollar-a-year industry (2). At this time the fraction of polyolefins produced commercially by metallocenes is still very small (< 10%) compared to Ziegler-Natta and radical-based processes (3); however, this number is certain to increase and greatly underrepresents the role metallocenes have played in advancing the understanding and development of olefin polymerization.

Metallocenes are soluble, single-site analogues of the traditional Ziegler-Natta initiators ($TiCl_4$ / AlR_3 / support) used to produce high-density and low-density polyethylene, isotactic polypropylene, and EPDM rubbers. The active polymerization sites in Ziegler-Natta systems are heterogeneous, numerous in character, and generated *in situ*, and are thus very difficult to manipulate. Metallocene initiators arose from the early study of soluble Ziegler initiators comprising titanocene or zirconocene dichlorides (Cp_2TiCl_2, or Cp_2ZrCl_2; $Cp = C_5H_5$) with a trialkylaluminum activator, which converts the Group IV complex to a highly active cation upon addition. It was later found that if methylaluminoxane ("MAO," a complex oligomer with the general structure [-Al(CH_3)O-]) is used instead of a simple aluminum alkyl, the resultant polymerization systems exhibit amazingly high rates and activities (4). In fact, examples of metallocenes producing yields in excess of one ton polyethylene / g catalyst•hour and total efficiencies of 25 tons polyethylene / g catalyst have been noted (5). In the last decade, single-component metallocene initiators have been developed which do not require the use of MAO, which is structurally poorly understood. These systems comprise preformed Group IV metallocene cations combined with a discrete, weakly coordinating borane counterion (6). Although metallocenes, boranes, and MAO are all atmospherically sensitive, the enormous activities shown by these systems makes them industrially viable despite the need to exclude water and air from the polymerization process.

The term "metallocene" originally denoted a transition metal sandwich compound having two cyclopentadienyl (Cp) or substituted cyclopentadienyl ligands, but in practice refers to any catalyst having two ligands performing a similar steric and electronic function. Some representative metallocenes used for olefin polymerization are shown in Figure 1. These initiators illustrate the utility of a modifiable, covalently bound organometallic endgroup (7-12). By adding sterically bulky, electron-donating,

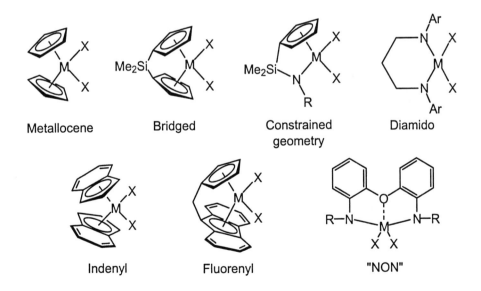

Metallocene   Bridged   Constrained   Diamido
                          geometry

Indenyl   Fluorenyl   "NON"

Figure 1. Generic types of metallocenes used in olefin polymerization (M = Ti, Zr, Hf; X = halide, alkyl).

or electron-withdrawing groups to the Cp groups (or Cp equivalents), rates of polymerization, catalyst activities, molecular weights, comonomer incorporation rates, and branching in the resultant polyolefins may be controlled to give monodisperse polymers or polymers with molecular weight distributions optimized for processing. Similarly, if the geometry of the catalyst is constrained by linking the Cp equivalents, catalysts with chiral or prochiral active sites for monomer insertion may be prepared that result in tactic polyolefins.

The active species in olefin polymerization is a cationic metallocene alkyl, generated by reaction of a neutral metallocene dialkyl with a borane Lewis acid, or similarly from a metallocene dichloride and a large excess of MAO, which also acts as an alkylating agent and scavenger for protic impurities (Eq. 1,2). As shown for

$$(1)$$

$$(2)$$

ethylene, initiation and propagation proceed through precoordination and insertion of the olefin into the alkyl group / polymer chain (Eq. 3). Chain transfer through β-hydride elimination produces metal hydrides and long-chain α-olefins (9). By modifying the

$$(3)$$

structure of the catalyst, these α-olefin macromers can be excluded from the polymerization process (giving linear high-density polyethylene, HDPE) or reincorporated as comonomers to give long-chain-branched polyethylene. Alternately, a variety of comonomers including propylene, 1-hexene, 1-octene, and 1,5-hexadiene can be directly added to the ethylene feed to produce branched (linear-low density, LLDPE) or cyclic polyolefins with a wide range of properties. Homopolymerization of cycloolefins such as norbornene and cyclopentene with metallocenes is also possible and produces polymers with extremely high melting points (above 380 °C) (13). These polymers are not amenable to processing, although copolymers of cycloolefins with ethylene show improved properties such as transparency, chemical resistance, and high service temperatures, and are used in optoelectronic applications (1).

Commodity metallocene polyethylene is currently being produced in the U.S. for film applications, originally by Exxon in 1991 (*Exact* HDPE and *Plastomer* plasticized polyethylene, made with the *Exxpol* process) and subsequently by Dow (*Affinity* and *Engage* LLDPE, made through *Insite*), Phillips, Mobil, and a large number of licenses, alliances, and joint ventures (e.g., Exxon / Union Carbide Univation) (14). Exxon and Fina are also selling polypropylene for fiber applications. A number of specialty products, not available through traditional catalysts, are in development. These include Hoechst / Mitsui's *Topas* ethylene / norbornene copolymers (3), syndiotactic polystyrene from Dow / Idemitsu (15), and syndiotactic polypropylene from Fina (16).

Late Transition Metal Initiators A major drawback of Group IV metallocene catalysts is that, despite their extremely high activities for olefins, most are very air- and moisture-sensitive. They are also highly intolerant to heteroatom-containing monomers, whose lone pairs preferentially coordinate to the Lewis acidic metal center in place of the olefin monomer. Electron-rich late transition metal organometallics are more tolerant in these regards, but traditional thinking has held that these complexes are too limited in activity to compete with metallocenes. However, some breakthroughs in the last several years have shown that with proper spectator ligands to control sterics at the active site, middle- and late-transition metal complexes are quite useful for the polymerization of olefins and are promising catalysts for the potential copolymerization of olefins and polar monomers (10).

In 1995, Brookhart and coworkers developed a class of cationic Pd(II) and Ni(II) initiators incorporating α-diimine spectator ligands with N-aryl groups (Figure 2a) (17, 18). The 2,6-aryl substituents on the diimine ligands are crucial for controlling the steric hindrance above and below the active sites of the square planar complexes and allow for the precise tailoring of branching, activity, and molecular weight. The polyethylene produced by some types of these initiators is unique in that it is much more highly branched even than LDPE. These initiators are also unusual in that they are able to copolymerize small amounts (< 12 mol %) of acrylic monomers, which are selectively incorporated at the ends of the branches (19). Very recently, Grubbs and coworkers reported *neutral* salicylaldiminato nickel(II) complexes that polymerize olefins (Figure 2b) (20, 21). These single-component initiators require only the presence of a phosphine sponge for activation, and as a consequence of their neutral and therefore more tolerant character, can incorporate functionalized norbornene

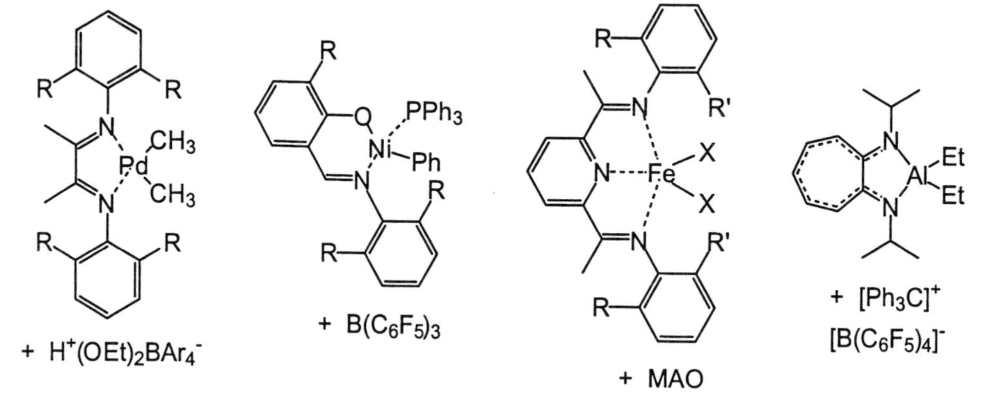

Figure 2. Late transition metal and main group catalysts used for olefin polymerization.

monomers and show polymerization activity in the presence of alcohols, ethers, and even water.

Midgroup transition metals and even main group elements are also being investigated for olefin polymerization. The groups of Brookhart and Gibson have developed extremely active iron- and cobalt-based catalysts incorporating bis(imino)pyridine ligands (Figure 2c) that function similarly to the nickel catalyst diimine ligands in Fig. 2a (22-24). These catalysts produce linear polyethylene and isotactic polypropylene, and exhibit activities comparable to Ziegler-Natta systems (turnover frequencies exceeding 10^7 monomer units per hour at 600 psi ethylene). Since iron is a relatively low-cost metal, these catalysts may have great industrial significance and are being examined for development by DuPont. Cationic aluminum catalysts with similar di- or trinitrogen ligands have also been reported by Gibson and Jordan (Figure 2d) (25, 26). Although these initiators show only low activities, they are a good demonstration of how far "life exists beyond metallocenes" with proper mechanistic understanding and catalyst design.

Other Monomers (Acetylene, Dienes, Isobutylene) In contrast to the huge body of work that has been carried out on transition metal-mediated olefin polymerization, the analogous polymerization of alkynes and dienes has not been exhaustively studied. Butadiene polymerization has been carried out in a "living" manner with [(η^3-allyl)Ni(trifluoroacetate)]$_2$ to give *trans*- or *cis*-1,4-microstructures under varying conditions (27-29). Transition metal-mediated routes into polyacetylenes and polyalkynes have been studied more frequently, particularly the polymerization of functional acetylenes that give dopable polymers showing enhanced stability and solubility. Most of the organometallic initiators used are similar to the classical initiators used for ring-opening metathesis polymerization (ROMP) and comprise midgroup or late-transition metal chlorides, carbonyls, or alkoxides which form metal carbenes *in situ* when combined with an alkylating cocatalyst (30-34). These carbenes are the true active species of the polymerization; it is not clear whether polymerization takes place through an addition or metathesis mechanism.

Very recently, organometallics have been studied as initiators for the pseudocarbocationic polymerization of isobutylene. Cationic metallocenes ([Cp'$_2$MMe]$^+$; M = Zr, Hf) (35, 36) and aluminocenes ([Cp'$_2$Al]$^+$) (Cp' = C$_5$H$_5$, C$_5$Me$_5$) (37, 38) have been used to prepare both high molecular weight homopoly(isobutylene) and isobutylene / isoprene copolymers. These systems utilize weakly coordinating borane counterions that play the same advantageous role as in metallocene polymerizations.

Olefin / Carbon Monoxide Copolymerization

The greater polar monomer tolerance of late transition metals makes it possible to copolymerize olefins with carbon monoxide, giving aliphatic polyketones. The copolymerization of ethylene and CO by radical means at elevated temperatures and pressures has been known for several decades, but the polymers formed have a random structure containing polyethylene sequences. However, perfectly alternating olefin / CO copolymers may be obtained through the use of organometallic nickel-and palladium-based catalysts (39-42). The alternating polyketone copolymers have a number of unusual properties due to their regular and functional structure, including very high crystallinities, melting points ($T_M \sim 260$ °C for ethylene-CO), high tensile strengths, photodegradabilty, and amenability to modification through the keto group. In addition, the low cost of ethylene and carbon monoxide makes these polymers commercially attractive, particularly since the palladium catalysts are not atmospherically sensitive.

Alternating olefin / CO copolymerization is optimally carried out with dicationic palladium(II) complexes incorporating a chelating dinitrogen or diphosphine ligand in the presence of methanol. These complexes can typically be prepared *in situ* from palladium(II) acetate, free ligand, and the Brønsted acid of a weakly coordinating anion. Methanol plays a key role in the initiation step of polymerization, assisting in the formation of monocationic palladium hydride and alkoxide initiating species. Polymerization (Eq. 4) involves the rapid and reversible insertion of coordinated CO into a palladium alkyl (or initial palladium alkoxide) bond to give a palladium acyl; carbon monoxide is selected over the olefin in this step due to its higher coordinating ability. Ethylene is then irreversibly inserted into the acyl. A subsequent second insertion of CO into the growing polymer chain does not occur for thermodynamic reasons (39).

Substituted olefins can also be copolymerized with carbon monoxide to give alternating terpolymers. Propylene, styrene, allenes (43), functional olefins (44), and diolefins (giving cyclocopolymers) all may be incorporated with varying regio- and stereoregularity, depending on the nature of the ligands used (39-42). Since straight poly(ethylene / CO) is an extremely insoluble material (soluble only in hexafluoro-

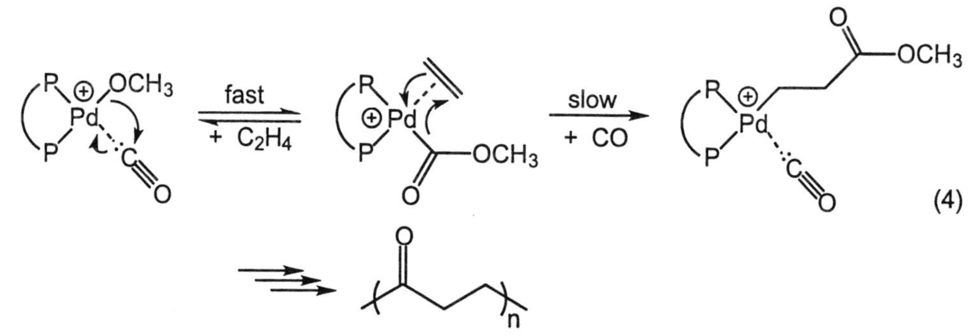

(4)

isopropyl alcohol), these less crystalline, lower melting materials are of much greater potential use for industrial applications.

Shell has developed a commercial process for the production of polyketones utilizing cationic palladium initiators. Although the catalysts used in olefin / CO copolymerization do not possess the extremely high turnover frequencies of metallocene systems, activities high enough to give acceptably low catalyst residues under relatively mild conditions (90 °C, 44 atm) are obtained, and the polyketones can be melt processed (39). Shell's thermoplastic *Carilon*, an ethylene / CO copolymer containing a small amount of propylene, was brought into production in 1996 and is projected to have automotive, industrial, and electrical applications. BP is currently developing *Ketonex*, a similar class of aliphatic polyketones (45).

Olefin Metathesis

Metathesis ("exchange") polymerization, unlike olefin addition polymerization, does not couple monomer units by sacrificing a double bond to make two single bonds. Rather, it builds a chain by clipping a double bond in half and attaching each side to one terminus of another double bond. In this manner, a polymer is built by "exchanging" the two olefin carbons of a monomer for others. The important feature is that the resultant polymer retains the unsaturation present in the original monomers.

Metathesis has been most widely carried out on olefins. The active species in the process is a metal carbene. Ring-opening metathesis polymerization, or ROMP, involves an addition chain building mechanism from cyclic monomers containing a single double bond. Acyclic diene metathesis (ADMET) polymerization uses a condensation mechanism to polymerize monomers containing two terminal olefinic groups.

Ring-Opening Metathesis Polymerization (ROMP) The general exchange reaction carried out in olefin metathesis may be represented as:

$$(R_1)HC{=}CH(R_1) \; + \; (R_2)HC{=}CH(R_2) \; \rightleftharpoons \; 2\,(R_1)HC{=}CH(R_2) \qquad (5)$$

If a cyclic olefin is substituted for the generic olefin and each side of the double bond is thought of as an R_1 or R_2 moiety, it is apparent that a polymeric chain can be built (46, 47):

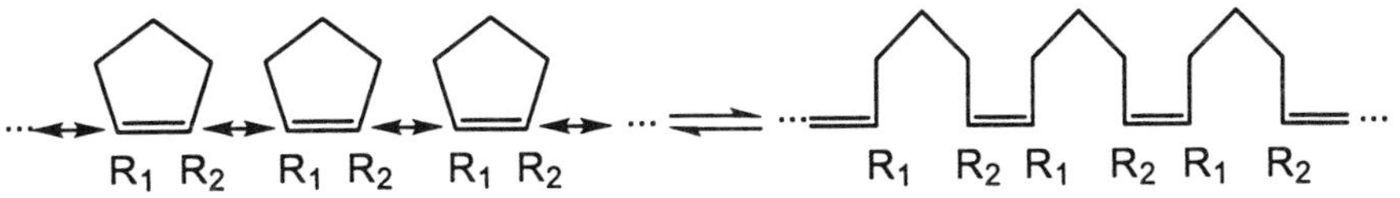

$$\tag{6}$$

Since metathesis is an equilibrium process, the regiochemistry of the double bonds formed is ultimately subject to thermodynamic control. Polymers with varying amounts of cis and trans linkages are formed depending on catalyst and monomer structure. The driving force for polymerization is the release of monomer ring strain; therefore cycloalkenes such as norbornene and cyclobutene are better ROMP substrates than less-strained monomers such as cyclohexene.

ROMP was originally noted as a side reaction to olefin addition polymerization during the early years of Ziegler-Natta catalysis. Modern ROMP initiators are typically tungsten, molybdenum, or ruthenium carbenes bearing spectator ligands specifically tailored to provide "living" steric and electronic active site control. The most important of these are the Grubbs (48) and Schrock (49) initiators. Catalysts may also be prepared *in situ* by combining metal chlorides (particularly tungsten) with aluminum alkyls to give metal alkyls that undergo elimination to carbenes (50). Although ROMP polymerizations in aqueous media have been achieved (51-53), most catalysts and processes are highly air- and water-sensitive. Recently, a number of functional group-tolerant polymerizations have been reported (54, 55).

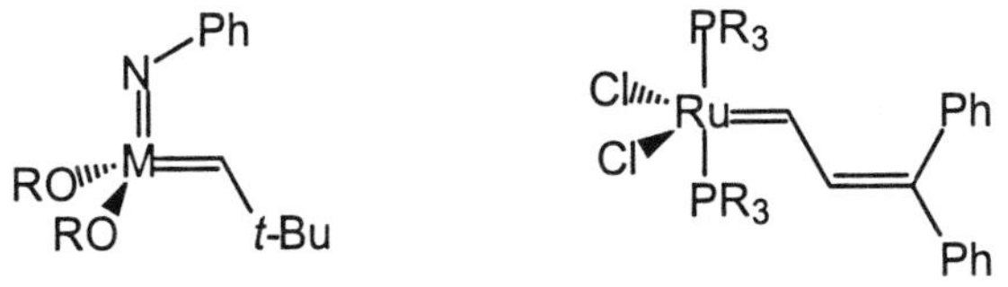

M = W, Mo
"Schrock"

"Grubbs"

The ROMP mechanism involves metallacyclobutane formation via a [2+2] cycloaddition reaction. The metallacyclobutane then decomposes through a retro [2+2] mechanism where bond cleavage occurs in the opposite direction from formation. The polymerization may be terminated upon completion with an aldehyde to give a metal oxo complex (eq. 7).

Since unsaturation is preserved when olefins are polymerized through ROMP, it is a useful technique for producing conjugated polymers. ROMP particularly serves as a versatile route for poly(acetylene) synthesis, either through the direct polymerization of cyclooctatetraene or via a route involving the ROMP synthesis of a

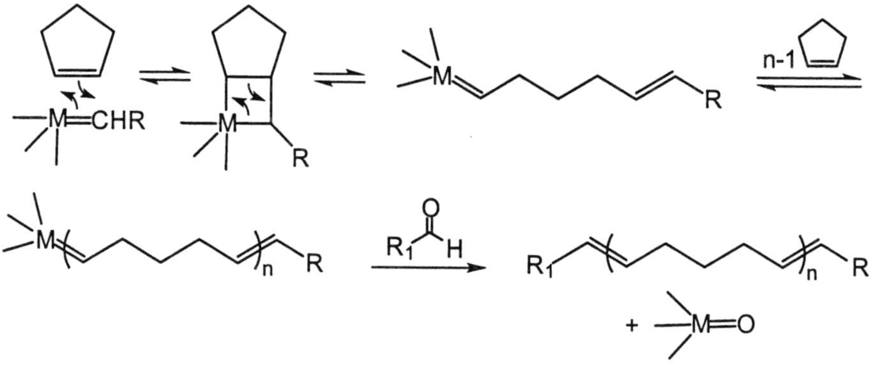

(7)

soluble precursor, poly(benzvalene) (56, 57). When the latter polymer is treated with $HgCl_2$, its backbone bicyclo[1.1.0]butane units are isomerized to butadiene groups, giving poly(acetylene).

ROMP methods are used in industry to prepare a number of specialty polymers generally known as *polyalkenamers* (58). *Trans*-poly(cyclopentene), or *Polypentenamer*, is a commercial elastomer, while copolymers of poly(dicyclopentadiene) (*Telene*, BF Goodrich, and *Meton*, PPD Hercules), which can be crosslinked through the second double bond, are RIM thermosets used for snowmobile bodies and auto parts. *Norsorex* (polynorbornene, CdF Chimie) is marketed as a specialty rubber, and *Vestenamer* (poly[1-octylene], Hüls) is used in elastomer blends with SBR and natural rubbers for gaskets and hoses.

Acyclic Diene Metathesis Polymerization (ADMET) ADMET, developed and extensively studied by Wagener et al. (59-61), is a metathesis polymerization which utilizes a condensation mechanism rather than the addition-type mechanism employed in ROMP. A straight chain monomer having two terminal olefin groups is reacted with a metal carbene, giving a metallacyclobutane product in a similar manner to ring-opening metathesis polymerization (Eq. 8). In this case, however, the decomposition

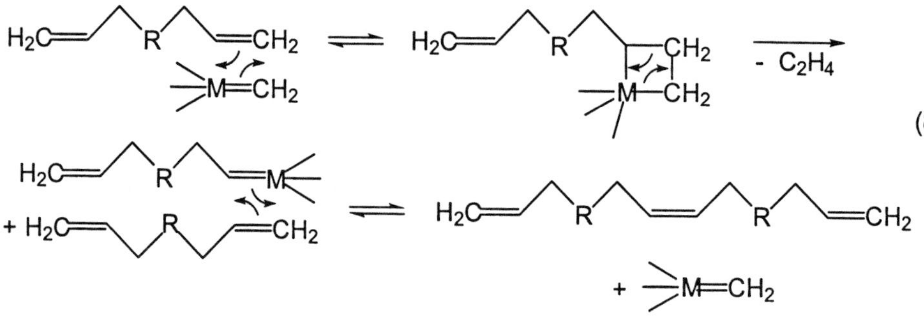

(8)

of the metallacycle produces ethylene in addition to a new carbene. A second monomer unit reacts with the product carbene and ethylene is again released to form a dimer. As the condensation process continues, removal of ethylene at reduced pressure pushes the ADMET system to high conversions and high polymer.

Equation 8 shows ADMET polymerization of a symmetrical monomer; however, asymmetrically substituted monomers can also be used. In this case, condensation may occur to give different regiochemistries depending on which "ends" of the monomers are joined together. As expected from an equilibrium polymerization, ADMET produces polymers with a high content of thermodynamically favored *trans* double bonds, typically 80:20 *trans:cis* (59).

ADMET polymerization may be carried out using the preformed Schrock and Grubbs catalysts, as well as classical tungsten-based, tin-activated initiators (62). With the proper catalyst choice it is tolerant to the presence of functionality at the monomer backbone center, as long as at least two CH_2 units are present between the olefin and central functionality (63). Monomers incorporating ether, alcohol, sulfur, silane, and stannane groups have been polymerized (64-66). When the backbone double bonds of the resultant polymers are quantitatively hydrogenated, unusual functional polyethylenes are obtained. Similarly, ADMET has also been used to prepare a number of interesting segmented copolymers from α,ω-olefin-substituted macromonomers (67, 68).

Another interesting use of ADMET stems from the ability to control the spacing between the incorporated R groups through the number of methylene units in the monomer. By choosing monomers with a variety of spacer lengths and a single methyl substituent on the central carbon, polymers with precisely spaced sidechains are formed (Eq. 9). These polymers can be hydrogenated to give polyethylenes with methyl branches on every *n*th carbon. Such "perfectly imperfect" polyethylenes serve as useful models for understanding the relationships between branching and thermal properties (69-72). Similar strategies using alcohol- and acetate-substituted monomers have provided property models for industrially important ethylene-vinyl alcohol and -vinyl acetate copolymers (69, 73).

(9)

Branch on every...	T_M
11th	11 °C
15th	39 °C
19th	57 °C
21st	62 °C
Linear	134 °C

Branch on every (2n+3) carbon

Metathesis of Alkynes Well-defined ROMP and ADMET have also been carried out on alkynes, although metal carbyne catalysts for well-controlled polymerizations are not as highly developed as those for alkene polymerization (74-76). Recently, ADIMET (acyclic diyne metathesis) has been used as an alternative to Pd-catalyzed couplings for the synthesis of high molecular weight poly(p-phenyleneethynylenes) (74, 77).

Addition Polymerization of Polar Monomers

Group Transfer Polymerization (GTP) of (Meth)acrylates Traditional group transfer polymerization (GTP) is a technique for the "living" polymerization of (meth)acrylates involving a silyl ketene acetal initiator (producing a silyl enolate active species) and a carbonyl-activating catalyst (typically a Lewis acid) as an independent second component (78, 79). Until recently it was believed that GTP was similar to the other organometallic coordination-addition polymerizations discussed here, in that the silyl endgroup remained firmly bound to the growing polymer chain during propagation. However, recent work has shown that it may be better described as an anionic polymerization stabilized by a reversibly bound silyl group (80, 81). However, when early transition metal complexes are used as initiators, "living" group transfer polymerizations of acrylic monomers are obtained which in fact do involve robust organometallic active sites, allowing for the control of polymer microstructure. Elegantly, the Lewis acid character of the organometallic initiators allows them to simultaneously serve as the second catalyst component of the GTP system.

Collins has developed a transition metal GTP technique for the polymerization of methyl methacrylate (MMA) which utilizes a neutral zirconocene enolate as an initiator and the conjugate zirconocene cation as a catalyst (82, 83). Each monomer addition step interconverts the two organometallic components (Eq. 10). The poly(methyl methacrylate) (PMMA) obtained is predominantly syndiotactic, although isotactic PMMA has been obtained by using chiral indenyl zirconocenes in combination with non-zirconocene Lewis acids (84).

(10)

Yasuda has developed a second GTP system based on lanthanocenes, in particular Cp*$_2$SmR complexes (Cp* = C$_5$Me$_5$) (85-88). In this case, the large and highly electropositive organosamarium center can serve simultaneously as both the initiator and catalyst, and a second Lewis acid equivalent is not needed (Eq. 11). The

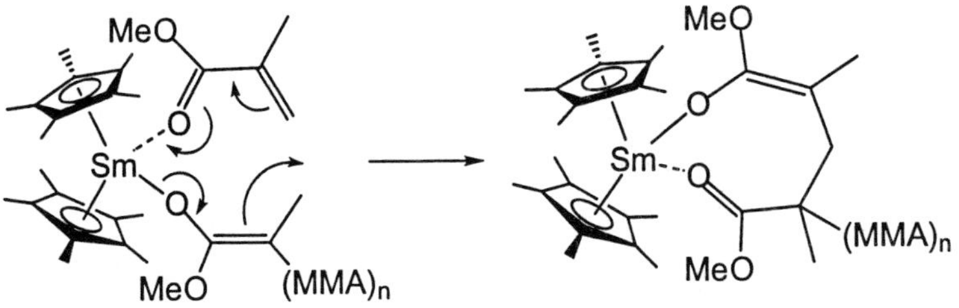

(11)

PMMA produced by samarocene GTP is monodisperse and highly syndiotactic (up to > 96% rr at lower temperatures) and the polymerizations are very rapid, although extremely sensitive to water and air. Acrylates and lactones are also polymerized by these catalysts, and since the metallocene-like Cp*$_2$SmR complexes are active for ethylene polymerization, block copolymers of ethylene and these polar monomers may be prepared (89).

<u>Ring Opening Polymerization</u> Lactones and lactides are polymerized in a ring-opening fashion by organometallic initiators. The mechanism of this process is a transesterification in which an organometallic alkoxide attacks the carbonyl group of a coordinated monomer (e.g. Eq. 12). These polymerizations are typically quite "living"

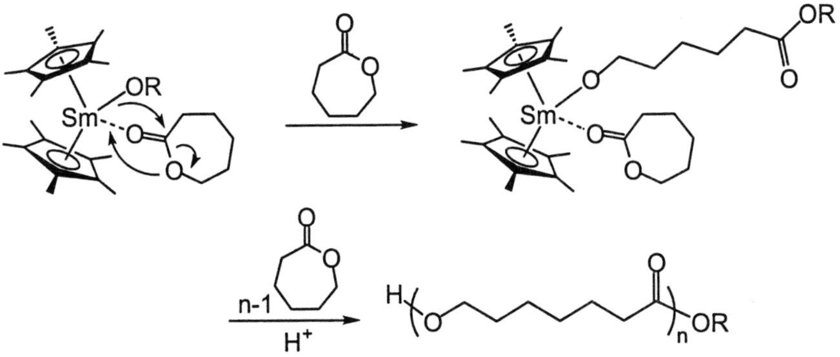

(12)

except that the polymer chain is susceptible to degradation by further transesterification after the concentration of free monomer has dropped. Polymerizations based on half-sandwich titanium alkoxides (90), Cp*$_2$LnOMe (91), aluminum porphyrins (92, 93), R$_2$AlOR (94, 95), and Ln(OR)$_3$ (96, 97) initiators have been reported (98). Cyclic carbonate homopolymerizations and copolymerizations are also possible (99).

Oxiranes are also polymerized in a ring-opening manner, most spectacularly by aluminum complexes with porphyrin or Schiff base ligands as initiators in conjunction with a second aluminum compound to act as a monomer-activating Lewis acid (92, 100). The steric protection provided by the bulky tetradentate ligands is so great that the polymerizations (termed "immortal" rather than "living") may be carried out in the presence of protic compounds such as water and even HCl. Aluminum porphyrin initiators also catalyze the alternating copolymerization of oxiranes with carbon dioxide (98, 101, 102).

Heterocumulenes Heterocumulenes and heterounsaturated monomers may be polymerized by mechanisms analogous to olefin addition polymerization. Novak and coworkers have developed a CpTiCl$_2$OR-based system for the "living" room temperature polymerization of isocyanates (Eq. 13) (103, 104) remarkable since well-

controlled anionic polymerizations are possible only at -78 °C (105). This example illustrates the mitigative powers of organometallic polymerization catalysts nicely: in the anionic process, the extremely reactive "naked" amidinate chain end backbites into the polymer backbone at a carbonyl group to form cyclic trimers, while the less reactive titanium amidate present in the organometallic process is sterically and electronically prevented from doing so. Carbodiimides are also polymerized by CpTiCl$_2$OR-type initiators to produce polyguanidines (Eq. 13) (106); both types of poly(heterocumulenes) exhibit interesting helical chain conformations and rigid rod behavior.

Isocyanides, which cannot be polymerized anionically, may be polymerized in a "living" manner with allyl nickel carboxylate initiators using a system also developed by Novak et al. (Eq. 14) (29, 107). The nickel initiators are active for butadiene polymerization as well, allowing for the synthesis of polyisocyanide-polybutadiene elastomers. Polyisocyanides also exhibit helical behavior when bulky sidechains are incorporated (108).

Step-Growth Addition / Elimination (Coupling) Polymerization

Late transition metal organometallic complexes commonly undergo catalytic two-electron redox cycles in which two separate organic components are first oxidatively added to the metal center, then reductively eliminated as a coupled fragment. These "addition-elimination" reactions are very useful for forming carbon-carbon bonds, and may be employed in polymerization by choosing fragments with reactive groups on both "ends" rather than only one. The resultant process provides a step-growth, rather than addition, method for the synthesis of all-carbon-backbone polymers.

The most common types of organometallic-catalyzed condensation polymerization are based on palladium and nickel organometallics that can cycle between oxidation states 0 and 2 (109). These processes are typically referred to as "cross-couplings" since the two organic moieties undergoing addition and elimination are not identical. Suzuki coupling polymerization involves the cross-coupling of aryl halides and aryl boronic acids to give polyarylenes with retention of regiochemistry (Eq. 15) (110, 111). Stille couplings (aryl halides and aryl tin reagents) (112, 113) and Heck couplings (aryl halides and alkenes) (114, 115) have also been employed similarly. These methods provide important routes into conjugated rigid-rod polymers, including poly(p-phenylene)s, poly(p-phenylene vinylene)s, and poly(p-phenylene ethynylene)s which are of importance for their engineering, conducting, LED, and NLO properties. Many unusual macromolecules, such as water-soluble rigid rod polyarylenes (116), graphite ribbons (117), and chiral biaryl-backbone polymers (118) have been synthesized through coupling routes.

Transition metal-catalyzed step-growth polymerizations involving other metals and monomers are also known. Meyer and coworkers have recently extended

(15)

palladium-catalyzed coupling methodology to aryl amines to produce linear *m*-polyaniline (Eq. 16) (119). Weber and coworkers have developed Heck-type coupling polymerizations with divinylsiloxanes or -silanes and acetophenones catalyzed by $RuH_2(CO)(PPh_3)_3$ (Eq. 17) (120, 121). In this regioselective process, coupling involves the activated *ortho*-protons of the acetophenone rather than an aryl halide, giving a net addition of the aryl C-H bond across the double bonds of the α,ω-diene. Similar methodology involving hydrosilation polymerization (adding a Si-H bond across a C=O group) produces polymeric silyl ethers (122, 123).

$$(16)$$

$$(17)$$

Beyond Addition and Condensation: Unique Organometallic Chain-Building Mechanisms

The vast majority of traditional polymerization processes may be classified as simple addition or condensation, double bond addition being the prevailing type of the former and bimolecular condensation of the latter. The addition / elimination polymerizations described in the last section illustrate how organometallic initiators can carry out these processes through mechanisms unavailable to ionic or radical initiators. However, organometallics can also catalyze polymerizations that fall into new mechanistic categories.

Shea and coworkers have developed a novel synthesis of "polyethylene" in which the macromolecular backbone is built from one CH_2 unit, rather than an olefin (CH_2CH_2) unit, at a time (124). The resultant polymer is actually *polymethylene*, which unlike polyethylene can contain an odd number of monomer-derived CH_2 units. The polymerization is carried out by a boron-catalyzed polyhomologation on a methylene ylide (Eq. 18) and is "living" in nature. Polymethylenes with an M_n of over 3000 have been prepared.

An example of a novel condensation organometallic polymerization mechanism is the cycloaddition copolymerization methodology developed by Tsuda (125-127). In this process, a cobaltocene or a nickel(0) complex (typically $Ni(COD)_2$-

$$\text{B-CH}_2\text{CH}_3 + \overset{\ominus}{\text{CH}_2}\overset{\oplus}{\text{-S(O)Me}_2} \longrightarrow \text{B} \overset{\ominus}{\underset{\text{CH}_2\text{-S(O)Me}_2}{\overset{\text{CH}_2\text{-CH}_3}{\Big\langle}}} \xrightarrow{\text{- S(O)Me}_2}$$

(18)

$$\text{B} \diagup \diagdown \text{CH}_3 \xrightarrow[\text{NaOH}]{\overset{\ominus \ \oplus}{\text{n-1 CH}_2\text{S(O)Me}_2}} \xrightarrow{\text{-OOH /}} \text{Et-}(\text{CH}_2)_n\text{-CH}_2\text{OH}$$

$(\text{PCy}_3)_2)$ is used to join two diynes and a heterocumulene or nitrile electrophile (Eq. 19). Poly(2-pyrone)s, poly(2-pyridone)s, and polypyridines are produced having various aliphatic, aromatic, ladder (128, 129), or ether (130) linkages. The regiochemistry of the polymerization is somewhat disordered, with four substitution patterns possible for the cyclized unit arising from the monomer alkyne substituents.

(plus other regioisomers) (19)

Literature Cited

1. Kaminsky, W. *J. Chem. Soc. Dalton Trans.* 1998, 1413-1418.
2. *Chemical & Engineering News,* June 29, 1998, p. 42-81.
3. Benedikt, G. M.; Goodall, B. L. "Metallocene-Catalyzed Polymers: Materials, Properties, Processing, & Markets"; Plastics Design Library: Norwich, NY, 1998; p. v-viii.
4. Sinn, H.; Kaminsky, W. *Adv. Organomet. Chem.* 1980, *18*, 99-149.
5. Feldman, D.; Barbalata, A. "Synthetic Polymers: Technology, Properties, Applications"; Chapman & Hall: London, 1996; Chapter 1.
6. Ewen, J.; Elder, M. J.; Jones, R. L.; Haspelaugh, L.; Atwood, J. L.; Bott, S. G.; Robinson, K. *Makromol. Chem., Macromol. Symp.* 1991, *48/49*, 253-295.
7. Bochmann, M. *J. Chem. Soc. Dalton Trans.* 1996, 255-170.

8.   Britzinger, H. H.; Fischer, D.; Mülhaupt, R.; Rieger, B.; Waymouth, R. M. Angew. Chem Int. Ed. Engl. 1995, 34, 1143-1170.
9.   Gupta, V. K.; Satish, S.; Bhardwaj, I. S. J. Macromol. Sci., Rev. Macromol. Chem. Phys. 1994, C34, 439-514.
10.   Britovsek, G. J. P.; Gibson, V. C.; Wass, D. F. Angew. Chem. Int. Ed. Engl. 1999, 38, 429-447.
11.   Soga, K.; Shiono, T. Prog. Polym. Sci. 1997, 22, 1503-1546.
12.   Jordan, R. F. Adv. Organomet. Chem. 1991, 32, 325-387.
13.   Kaminsky, W.; Bark, A.; Arndt, M. Makromol. Chem., Macromol. Symp. 1991, 47, 83-93.
14.   Chemical & Engineering News, July 6, 1998, p. 11-16.
15.   Ishihara, N.; Kuromoto, M.; Uoi, M. Macromolecules 1988, 21, 3356-3360.
16.   Ewen, J. J. Am. Chem. Soc. 1988, 110, 6255-6266.
17.   Johnson, L. K.; Killian, C. M.; Brookhart, M. J. Am. Chem. Soc. 1995, 117, 6141-6415.
18.   Johnson, L. K. et al. (DuPont). World Patent Application 96/23010, 1996.
19.   Johnson, L. K.; Mecking, S.; Brookhart, M. J. Am. Chem. Soc. 1996, 118, 267-268.
20.   Wang, C.; Friedrich, S.; Younkin, T. R.; Li, R. T.; Grubbs, R. H.; Bansleben, D. A.; Day, M. W. Organometallics 1998, 17, 3149-3151.
21.   Bansleben, D. A.; Friedrich, S. K.; Younkin, T. R.; Grubbs, R. H.; Wang, C.; Li, R. T. World Patent Application 98/42665, 1998.
22.   Small, B. L.; Brookhart, M.; Bennett, A. M. A. J. Am. Chem. Soc. 1998, 120, 4049-4050.
23.   Small, B. L.; Brookhart, M. Macromolecules 1999, 32, 2120-2130.
24.   Britovsek, G. J. P.; Gibson, V. C.; Kimberley, B. S.; Maddox, P. J.; McTavish, S. J.; Solan, G. A.; White, A. J. P.; Williams, D. J. Chem. Commun. 1998, 849-850.
25.   Ihara, E.; Young, V. G. J.; Jordan, R. F. J. Am. Chem. Soc. 1998, 120, 8277-8278.
26.   Bruce, M.; Gibson, V. C.; Redshaw, C.; Solan, G. A.; White, A. J. P.; Williams, D. J. Chem. Commun. 1998, 2523-2524.
27.   Hadjiandreou, P.; Julemont, M.; Teyssie, P. Macromolecules 1984, 17, 2455-2456.
28.   Fayt, R.; Hadjiandreou, P.; Teyssie, P. J. Polym. Sci., Polym. Chem. Ed. 1985, 23, 337-342.
29.   Deming, T. J.; Novak, B. M.; Ziller, J. W. J. Am. Chem. Soc. 1994, 116, 2366-2374 and references therein.
30.   Tang, B.-Z.; Kotera, N. Macromolecules 1989, 1989, 4388-4390.
31.   Nakayama, Y.; Mashima, K.; Nakamura, A. Macromolecules 1993, 26, 6267-6272.
32.   Masuda, T.; Yoshimura, T.; Tamura, K.; Higashimura, T. Macromolecules 1987, 20, 1734-1739.
33.   Yang, M.; Zheng, M.; Furlani, A.; Russo, M. V. J. Polym. Sci., Polym. Chem. Ed. 1994, 32, 2709-2713.
34.   Xu, K.; Peng, H.; Tang, B. Z. Polym. Mater. Sci. Eng. 1999, 80, 485-486.

35. Carr, A. G.; Dawson, D. M.; Bochmann, M. <u>Macromolecules</u> 1998, <u>31</u>, 2035-2040.
36. Shaffer, T. D.; Ashbaugh, J. R. <u>J. Polym. Sci., Polym. Chem. Ed.</u> 1997, <u>35</u>, 329-344.
37. Bochmann, M.; Dawson, D. M. <u>Angew. Chem. Int. Ed. Engl.</u> 1996, <u>35</u>, 2226-2228.
38. Burns, C. T.; Shapiro, C. J. Abstracts of Papers, 215th National Meeting of the American Chemical Society, Dallas, TX; American Chemical Society: Washington, DC, 1997; INOR 5.
39. Drent, E.; Budzelaar, H. M. <u>Chem. Rev.</u> 1996, <u>96</u>, 663-681.
40. Sommazzi, A.; Garbassi, F. <u>Prog. Polym. Sci.</u> 1997, <u>22</u>, 1547-1605.
41. Sen, A. <u>Acc. Chem. Res.</u> 1993, <u>26</u>, 303-310.
42. Amevor, E.; Bronco, S.; Consiglio, G.; DiBenedetto, S. <u>Makromol. Chem., Macromol. Symp.</u> 1995, <u>89</u>, 443-454.
43. Kacker, S.; Sen, A. <u>Macromolecules</u> 1997, <u>119</u>, 10028-10033.
44. Kacker, S.; Jiang, Z.; Sen, A. <u>Macromolecules</u> 1996, <u>29</u>, 5852-5858.
45. Bonner, J. G.; Powell, A. K. <u>Polym. Mater. Sci. Eng.</u> 1997, <u>76</u>, 108-109.
46. Schrock, R. R. <u>Acc. Chem. Res.</u> 1990, <u>23</u>, 158-165.
47. Amass, A. J. in "Comprehensive Polymer Science", Allen, G.; Bevington, J. C., Eds.; Pergamon: Oxford, 1989; Vol. 4, p. 109-134.
48. Gilliom, L. R.; Grubbs, R. H. <u>J. Am. Chem. Soc.</u> 1986, <u>108</u>, 733-742.
49. Schrock, R. R.; Murdzek, J. S.; Bazan, G. C.; Robbins, J.; DiMare, M.; O'Regan, M. <u>J. Am. Chem. Soc.</u> 1990, <u>112</u>, 3875-3886.
50. Calderon, N.; Ofstead, E. A.; Judy, W. A. <u>J. Polym. Sci., Polym. Chem. Ed.</u> 1967, <u>5</u>, 2209-2217.
51. Lynn, D. M.; Mohr, B.; Grubbs, R. H. <u>J. Am. Chem. Soc.</u> 1998, <u>120</u>, 1627-1628.
52. Lynn, D. M.; Kanaoka, S.; Grubbs, R. H. <u>J. Am. Chem. Soc.</u> 1996, <u>118</u>, 784-790.
53. Novak, B. M.; Grubbs, R. H. <u>J. Am. Chem. Soc.</u> 1988, <u>110</u>, 7542-7543.
54. Perrott, M. G.; Novak, B. M. <u>Macromolecules</u> 1996, <u>29</u>, 1817-1823 and references therein.
55. Maughon, B. R.; Grubbs, R. H. <u>Macromolecules</u> 1997, <u>30</u>, 3459-3469 and references therein.
56. Klavetter, F. L.; Grubbs, R. H. <u>J. Am. Chem. Soc.</u> 1988, <u>110</u>, 7807-7813.
57. Swager, T. M.; Dougherty, D. A.; Grubbs, R. H. <u>J. Am. Chem. Soc.</u> 1988, <u>110</u>, 2973-2974.
58. Ofstead, E. A. in "Encyclopedia of Polymer Science & Engineering", Mark, H. F.; Kroschwitz, J. I., Eds.; Wiley-Interscience: New York, 1988; Vol. 11, p. 287-214.
59. Wagener, K. B.; Boncella, J. M.; Nel, J. G. <u>Macromolecules</u> 1991, <u>24</u>, 2649-2657.
60. Tindall, D.; Pawlow, J. H.; Wagener, K. B. in "Trends in Organometallic Chemistry: Alkene Metathesis in Organic Synthesis", Fuerstner, A., Ed.; Springer-Verlag: Berlin, 1998; p. 183-198.
61. Davidson, T. A.; Wagener, K. B. in "Synthesis of Polymers", Schluter, A.-D., Ed.; "Materials Science and Technology: A Comprehensive Treatment", Cahn, R. W.; Haasen, P.; Kramer, E. J., Eds.; Wiley-VCH: New York, 1999; Vol. 19, Chapter 4.

62. Gómez, F. J.; Wagener, K. B. Macromol. Chem. Phys. 1998, 199, 1581-1587 and references therein.
63. Wagener, K. B.; Brzezinska, K.; Anderson, J. D.; Younkin, T. R.; Steppe, K.; DeBoer, W. Macromolecules 1997, 30, 7363-7369 and references therein.
64. Valenti, D. J.; Wagener, K. B. Polym. Prepr., Am. Chem. Soc. Div. Polym. Chem. 1996, 37 (2), 325-326 and references therein.
65. Wolfe, P. S.; Wagener, K. B. Macromol. Rapid Commun. 1998, 19, 305-308.
66. Wolfe, P. S.; Gómez, F. J.; Wagener, K. B. Macromolecules 1997, 30, 714-717.
67. Tindall, D.; Wagener, K. B. Polym. Prepr., Am. Chem. Soc. Div. Polym. Chem. 1998, 39 (2), 539.
68. Gómez, F. J.; Wagener, K. B. Polym. Prepr., Am. Chem. Soc. Div. Polym. Chem. 1998, 39 (2), 540-541.
69. Wagener, K. B.; Valenti, D.; Watson, M. Polym. Prepr., Am. Chem. Soc. Div. Polym. Chem. 1998, 39 (1), 719-720.
70. Wagener, K. B.; Valenti, D. Macromolecules 1997, 30, 6688-6690.
71. Wagener, K. B.; Valenti, D. J.; Hahn, S. F. Polym. Prepr., Am. Chem. Soc. Div. Polym. Chem. 1997, 38 (2), 394-395.
72. Wagener, K. B., University of Florida, personal communication, 1999.
73. Valenti, D. J.; Wagener, K. B. Macromolecules 1998, 31, 2764-2773.
74. Weiss, K.; Michel, A. M.; Auth, E.-M.; Bunz, U. H. F.; Mangel, T.; Müllen, K. Angew. Chem. Int. Ed. Engl. 1997, 36, 506-509.
75. Zhang, X.-I.; Bazan, G. C. Macromolecules 1994, 27, 4627-4628.
76. Krouse, S. A.; Schrock, R. R. Macromolecules 1989, 22, 2569-2576.
77. Kloppenburg, L.; Song, D.; Bunz, U. H. F. J. Am. Chem. Soc. 1998, 120, 7973-7974.
78. Webster, O. W.; Hertler, W. R.; Sogah, D. Y.; Farnham, W. B.; RajanBabu, T. V. J. Am. Chem. Soc. 1983, 105, 5706-5708.
79. Webster, O. W.; Sogah, D. Y. in "Comprehensive Polymer Science", Allen, G.; Bevington, J. C., Eds.; Pergamon Press: Oxford, 1989; Vol. 4, p. 163-169 and references therein.
80. Müller, A. H. E. Macromolecules 1994, 27, 1685-1690.
81. Quirk, R. P.; Bidinger, G. P. Polym. Bull. 1989, 27, 63-70.
82. Collins, S.; Ward, D. G. J. Am..Chem. Soc. 1992, 114, 5461-5462.
83. Li, Y.; Ward, D. G.; Reddy, S. S.; Collins, S. Macromolecules 1997, 30, 1875-1883.
84. Yano, T. et al. Polym. Prepr. Jpn. 1994, 43, 141.
85. Yasuda, H.; Yamamoto, H.; Yamashita, M.; Yokota, K.; Nakamura, A.; Miyake, S.; Kai, Y.; Kanehisa, N. Macromolecules 1993, 26, 7134-7143.
86. Ihara, E.; Morimoto, M.; Yasuda, H. Macromolecules 1995, 28, 7886-7892.
87. Boffa, L. S.; Novak, B. M. Macromolecules 1997, 30, 3494-3506.
88. Boffa, L. S.; Novak, B. M. Tetrahedron 1997, 53, 15367-15396.
89. Yasuda, H.; Furo, M.; Yamamoto, H.; Nakamura, A.; Miyake, S.; Kibino, N. Macromolecules 1992, 25, 5115-5116.
90. Okuda, J.; König, P.; Rushkin, I. L.; Kang, H.-C.; Massa, W. J. Organomet. Chem. 1995, 501, 37-39.

91. Yamashita, M.; Takemoto, Y.; Ihara, E.; Yasuda, H. Macromolecules 1996, 29, 1798-1806.
92. Aida, T.; Maekawa, Y.; Asano, S.; Inoue, S. Macromolecules 1988, 21, 1195-1202.
93. Endo, M.; Aida, T.; Inoue, S. Macromolecules 1987, 20, 2982-2988.
94. Dubois, P.; Ropson, N.; Jérôme, R.; Teyssié, P. Macromolecules 1996, 29, 1965-1975.
95. Duda, A.; Florjanczyk, Z.; Hofman, A.; Slomkowski, S.; Penczek, S. Macromolecules 1990, 23, 1640-1646.
96. McLain, S. J.; Drysdale, N. E. Polym. Prepr., Am. Chem. Soc. Div. Polym. Chem. 1992, 33 (1), 174-175.
97. Stevels, W. M.; Ankoné, M. J. K.; Dijkstra, P. J.; Feijen, J. Macromolecules 1996, 29, 3332-3333.
98. Kuran, W. Prog. Polym. Sci. 1998, 23, 919-992.
99. Evans, W. J.; Katsumata, H. Macromolecules 1994, 27, 4011-4013.
100. Sugimoto, H.; Kawamura, C.; Kuroki, M.; Aida, T.; Inoue, S. Macromolecules 1994, 27, 2013-2018.
101. Kuran, W.; Listos, T.; Abramczyk, M.; Dawidek, A. J. Macromol. Sci., Chem. 1998, A35, 427-437.
102. Aida, T.; Ishikawa, M.; Inoue, S. Macromolecules 1986, 19, 8-13.
103. Patten, T. E.; Novak, B. M. J. Am. Chem. Soc. 1991, 113, 5065-5066.
104. Patten, T. E.; Novak, B. M. J. Am. Chem. Soc. 1996, 118, 1906-1916.
105. Shashoua, V. E.; Sweeny, W.; Tietz, R. J. Am. Chem. Soc. 1960, 82, 866-873.
106. Goodwin, A.; Novak, B. M. Macromolecules 1994, 27, 5520-5522.
107. Deming, T. J.; Novak, B. M. Macromolecules 1991, 24, 5478-5480.
108. Novak, B. M.; Goodwin, A. A.; Schlitzer, D.; Patten, T. E.; Deming, T. J. Polym. Prepr., Am. Chem. Soc. Div. Polym. Chem. 1996, 37 (2), 446-447.
109. Percec, V.; Hill, D. H. ACS Symp. Ser. 1996, 624, 2-56.
110. Rehahn, M.; Schlüter, A.-D.; Wegner, G.; Feast, W. J. Polymer 1989, 30, 1060-1062.
111. Goodson, F. E.; Wallow, T. I.; Novak, B. M. Macromolecules 1998, 31, 2047-2056.
112. Yu, L.; Bao, Z.; Cai, R. Angew. Chem. Int. Ed. Engl. 1993, 32, 1345-1347.
113. Babudri, F.; Cicco, S. R.; Farinola, G. M.; Naso, F. Makromol. Rapid Commun. 1996, 17, 905-911.
114. Bao, Z.; Chen, Y.; Cai, R.; Yu, L. Macromolecules 1993, 26, 5281-5286.
115. Fu, D.-K.; Xu, B.; Swager, T. M. Tetrahedron 1997, 53, 15487-15494.
116. Wallow, T. I.; Novak, B. M. J. Am. Chem. Soc. 1991, 113, 7411-7412.
117. Goldfinger, M. B.; Swager, T. M. J. Am. Chem. Soc. 1994, 116, 7895-7896.
118. Hu, Q.-S.; Vitharana, D.; Liu, G.; Jain, V.; Pu, L. Macromolecules 1996, 29, 5075-5082.
119. Spetseris, N.; Ward, R. E.; Meyer, T. Y. Macromolecules 1998, 31, 3158-3161.
120. Guo, H.; Weber, W. P. Polym. Bull. 1994, 1994, 525-528.
121. Guo, H.; Tapsak, M. A.; Weber, W. P. Macromolecules 1995, 28, 4714-4718.
122. Gupta, S.; Londergan, T. M.; Paulasaari, J. K.; Sargent, J. R.; Weber, W. P. Polym. Mater. Sci. Eng. 1999, 80, 445-446.

123. Paulasaari, J. K.; Weber, W. P. Macromolecules 1998, 31, 7105-7107.
124. Shea, K. J.; Walker, J. W.; Zhu, H.; Paz, M.; Greaves, J. J. Am. Chem. Soc. 1997, 119, 9049-9050.
125. Tsuda, T.; Maruta, K.; Kitaike, Y. J. Am. Chem. Soc. 1992, 114, 1498-1499 and references therein.
126. Tsuda, T.; Hokazono, H. Macromolecules 1993, 26, 1796-1797 and references therein.
127. Tsuda, T.; Maehara, H. Macromolecules 1996, 29, 4544-4548 and references therein.
128. Tsuda, T.; Yasukawa, H.; Hokazono, H.; Kitaike, Y. Macromolecules 1995, 28, 1312-1315.
129. Tsuda, T.; Hokazono, H. Macromolecules 1993, 26, 5528-5529.
130. Tsuda, T.; Yasukawa, H.; Komori, K. Macromolecules 1995, 28, 1356-1359.

Author Index

Keyword index